f the World

D0161740

60°
Severnaya Zemlya
120°
180°
Arctic Ocean
Franz Josef Land
Novaya Zemlya
New Siberian Islands
Wrangel Island
nland
Russia
Estonia
Latvia
Lithuania
Belarus
60°
Aleutian Islands (USA)
Ukraine
Moldova
vak.
Romania
ugo.
Bulgaria
Kazakhstan
Mongolia
Kuril Islands
Georgia
Armenia
Azerbaijan
Uzbekistan
Kyrgyzstan
Turkmenistan
Tajikistan
reece
Turkey
N. Korea
S. Korea
Japan
North Pacific Ocean
Cyp.
Leb.
Israel
Syria
Iraq
Iran
Afghanistan
China
Jordan
Kuwait
Pakistan
Nepal
Bhu.
a
Egypt
Qatar
U.A.E.
Bang.
30°
Saudi Arabia
Oman
India
Myanmar (Burma)
Taiwan
Laos
Sudan
Eritrea
Yemen
Thailand
Philippines
ad
Djibouti
Vietnam
Andaman Islands (India)
Cambodia
Ethiopia
Sri Lanka
Brunei
Federated States of Micronesia
Marshall Islands
C.A.R.
Maldives
Guam (USA)
Somalia
Malaysia
Uganda
Singapore
Rwanda
Kenya
Kiribati
0°
Zaire
Burundi
Indonesia
Papua New Guinea
Tanzania
Seychelles
Solomon Islands
Malawi
ola
Zambia
Mozambique
Madagascar
Fiji
Zimbabwe
otswana
Mauritius
Indian Ocean
New Caledonia
Swaziland
Australia
South Africa
Lesotho
30°
New Zealand
Tasmania
Îles Crozet (France)
60°
ctica
60°
90°
120°
150°
180°

Projection

NATURAL RESOURCE CONSERVATION

NATURAL RESOURCE CONSERVATION

MANAGEMENT FOR A SUSTAINABLE FUTURE

Eighth Edition

DANIEL D. CHIRAS
University of Denver

JOHN P. REGANOLD
Washington State University

OLIVER S. OWEN
University of Wisconsin, Eau Claire

PRENTICE HALL Upper Saddle River, New Jersey 07458

Library of Congress Cataloging-in-Publication Data
Chiras, Daniel D.
Natural resource conservation : management for a sustainable future / Daniel D. Chiras, John P. Reganold, Oliver S. Owen.—8th ed.
p. cm.
Rev. ed. of: Natural resource conservation / Oliver S. Owen, Daniel D. Chiras, John P. Reganold. 7th ed. 1998.
Includes bibliographical references (p.).
ISBN 0-13-033398-0
1. Conservation of natural resources. 2. Natural resources—Management. 3. Sustainable development. 4. Environmental protection. 5. Conservation of natural resources—United States. 6. Natural resources—United States—Management. 7. Sustainable development—United States. 8. Environmental protection—United States. I. Reganold, John P. II. Owen, Oliver S., Natural resource conservation. III. Title.

S938 .O87 2002
333.7'2—dc21 2001019899

Executive Editor: Teresa K. Ryu
Editor in Chief, Biology: Sheri L. Snavely
Project Manager: Travis Moses-Westphal
Vice President of Production and Manufacturing: David W. Riccardi
Executive Managing Editor: Kathleen Schiaparelli
Marketing Manager: Jennifer Welchans
Manufacturing Buyer: Michael Bell
Director of Design: Carole Anson
Director of Creative Services: Paul Belfanti
Manufacturing Manager: Trudy Pisciotti
Art Director: Jonathan Boylan
Managing Editor, Audio/Visual Assets: Grace Hazeldine
Art Tech Support: Debra Lowenfish
Text/Cover Designer: Alamani Design
Cover Photograph: Clouds over Tufa Towers at Millennium Sunset, Mono Lake, California, photograph by Leping Zha
Photo Researcher: Linda Sykes
Editorial Assistant: Colleen Lee
Text Composition: WestWords
Index: Linda M. Stuart

© 2002, 1998, 1995, 1990, 1985, 1980, 1975, 1971 by Prentice-Hall, Inc.
Upper Saddle River, New Jersey 07458

All rights reserved. No part of this book may be reproduced, in any form or by any means, without permission in writing from the publisher.

Printed in the United States of America
10 9 8 7 6 5 4 3

ISBN 0-13-033398-0

Prentice-Hall International (UK) Limited, *London*
Prentice-Hall of Australia Pty. Limited, *Sydney*
Prentice-Hall Canada Inc., *Toronto*
Prentice-Hall Hispanoamericana, S.A., *Mexico*
Prentice-Hall of India Private Limited, *New Delhi*
Prentice-Hall of Japan, Inc., *Tokyo*
Pearson Education Asia Pte, Ltd.
Editora Prentice-Hall do Brasil, Ltda., *Rio de Janeiro*

Brief Contents

Contents

Preface

Natural Resource Conservation is written for the introductory resource conservation course. It is designed toprovide comprehensive coverage of a variety of local, regional, national, and global resource and environmental issues from population growth to wetlands to sustainable agriculture to global air pollution.

The first edition of this book was published in 1971, a year after the first Earth Day by our esteemed colleague, the late Oliver S. Owen. To many observers, Earth Day marked the formal beginning of the environmental movement in the United States. Since that time, impressive gains have been made in air and water pollution control, species protection, forest management, and rangeland management.

Despite this progress, many environmental problems still remain. Many others have grown worse. In 1970, for instance, the world population hovered around 3 billion. Today it has exceeded the 6 billion mark and is growing by approximately 84 million people a year. Hunger and starvation have become a way of life in many less developed nations. An estimated 12 million people die each year of starvation and disease worsened by hunger and malnutrition. Species extinction continues as well. Today an estimated 40 to 100 species become extinct every day. In the United States and abroad, soil erosion and rangeland deterioration continue.

Added to the list of growing problems are a whole host of new ones that have cropped up along the way. Groundwater pollution, ozone depletion, acid deposition, global warming, and growing mountains of urban trash top the list. Yet, along with the new problems are new and exciting solutions.

If we work together in solving these problems, then there is much hope. However, to address these problems in meaningful ways will require dramatic changes in the way we live our lives and conduct commerce. We need a way that is sustainable—a way of doing business and living on the planet that does not bankrupt the Earth. Most people call this *sustainable development.* Sustainable development is about creating a new relationship with the Earth. It is about creating a sustainable economy and a sustainable system of commerce. It is about creating sustainable communities and sustainable lifestyles. It requires new ways of managing resources using the best available scientific knowledge and understandings of complex systems and how they are maintained, even enhanced, over time. It will entail changes in virtually every aspect of our society, from farming to forest management to energy production.

We believe that establishing a sustainable relationship with the Earth will require us to use resources more frugally—using only what we need and using all resources much more efficiently than we do today. Creating a sustainable way of life will very likely mean a massive expansion of our recycling efforts, not just getting recyclables to markets, but encouraging manufacturers to use secondary materials for production and encouraging citizens to buy products made from recycled materials.

Creating a sustainable society will also very likely mean a shift to clean, economical renewable energy supplies, such as solar and wind energy. Another vital component of a sustainable society is restoration—replanting forests, grasslands, and wetlands—to ensure an adequate supply of resources for future generations as well as for the many species that share this planet with us.

Essential to the success of our efforts to create a sustainable society are efforts to slow down, even stop, world population growth. But that means stopping population growth in all nations, not just the poorer less developed nations. Population growth, in the rich nations, combined with our resource-intensive lifestyles, is contributing as much to the current global crisis as population growth in the less developed nations.

Curtailing population growth also entails efforts to better manage how we spread out on the land—that is, how and where our cities and towns expand. By adhering to judicious growth measures we can preserve farmland, forests, pastures, wildlands, and fisheries—all essential to our future and often crucial to the well-being of the countless species that share this planet with us.

In this book, we present the case for building a sustainable future based on conservation, recycling, renewable resources, restoration, and population control. We dub these the operating principles of a sustainable society. We believe that by putting these principles into practice in all sectors of our society, from agriculture to industry to transportation, we can build an enduring relationship with the planet.

The operating principles, however, must be complemented by a change in attitudes. No longer can we afford to regard the earth as an infinite source of materials for exclusive human use. Many of the Earth's resources, upon which human beings depend, are finite. The Earth offers a limited supply of resources. We ignore this imperative at our own risk.

We and many others believe that humans must adopt an attitude that seeks cooperation with, rather than domination of, nature. Our efforts to dominate and control nature are often in vain and sometimes backfire on us. Cooperation may be one of the keys to our long-term success. By cooperation, we mean fitting into nature's cycles—creating production systems, for instance, on farms that more closely correspond with nature's cycles.

Finally, we believe it is time to rethink our position in the ecosystem. Humans are not apart from nature but a part of it. Our lives and our economy are vitally dependent on the environment. The Earth is the source of all goods and services and the sink for all of our wastes. What we do to the environment we do to ourselves. The logical extension of this simple truth is that planet care is the ultimate form of self-care.

Despite the wonderful accomplishments of human society over many centuries, it is time to realize that humans are not the crowning achievement of nature, but rather members in a club comprised of all of Earth's living creatures. To achieve a sustainable relationship, many observers argue, it is time to recognize and respect the rights of other species to exist and thrive alongside humans. In this sense, natural resources may be viewed as the Earth's endowment to all species. Such a view may mean curbing our demands and finding new ways to live on the planet. In the long run, such changes will benefit all of us.

Focus on Principles, Problems, and Solutions

This book describes many important principles of ecology and resource management, concepts that will prove useful throughout your lifetime. It also outlines many of the local, regional, national and global environmental problems and offers a variety of solutions to these problems. Solutions take three basic forms: legislative (new laws and regulations), technological (applying existing or new and improved technologies), and methodological (changing how we do things). Applying these solutions is a responsibility we all have in common. It is not just the domain of government. Citizens, business people, and government officials all have an important role to play in solving the environmental crisis and in building a sustainable society.

On the personal level, what we do or what we fail to do can have a remarkable impact on the future. We encourage you to take active steps to find ways to reduce your impact.

Learning Aids

To help students learn key terms and concepts, we have included three learning aids: key words and phrases, chapter summaries of key concepts, and critical thinking and discussion questions. To help students deepen and broaden their knowledge, we have included Ethics in Resource Conservation boxes, a section on critical thinking, Case Studies, and numerous Suggested Readings.

Key Terms

At the end of each chapter is a list of key words and phrases. We recommend that students read this list before reading the chapter. After reading the chapter, take a few moments to define the terms and phrases.

Summary of Key Concepts

Each chapter in the book also contains a summary of important facts and concepts. These short summaries will help students review material before tests. Before reading the chapter, we think it is a good idea to read through the summary or study the major headings and subheadings to orient yourself.

Critical Thinking and Discussion Questions

Discussion questions at the end of each chapter also provide a way of focusing on important material and reviewing concepts and crucial facts. We have written many questions to encourage you to tie information together and to draw on personal experience. We have also included a number of questions that ask you to think critically about various issues.

Ethics in Resource Conservation

This book contains eight essays on ethics and resource management. These brief pieces present important ethical issues that confront resource managers and people like yourself on a daily basis. The ethics boxes were designed to encourage you to think about your own values and how they influence your views. They will help you understand others, too.

Critical Thinking

Critical thinking is a vital skill for all of us, but it is especially important in resource conservation and management. In Chapter 1, we present a number of critical thinking rules that will help you analyze the material we present.

Case Studies

The case studies delve into controversial issues or provide detailed information that may be of interest to students pursuing a career in natural resource management. In this edition, we have removed outdated case studies and replaced them with newer ones. We have also eliminated a few to keep the number more manageable.

Suggested Readings

The Suggested Readings section in each chapter lists articles and books that are worthwhile reading for students who want to learn more about the environment.

New to the Eighth Edition

Because this field changes rapidly, we have carefully updated the text with recent statistics, recent examples, and new photographs. In addition, we have expanded coverage of pressing issues such as global climate change, ozone depletion, acid deposition, species extinction, and wetlands protection. We've added material on carrying capacity, genetic engineering, genetically modified crops, brownfield development, environmental justice, alternative fuels, and alternative vehicles.

In this edition, we have added information on geographic information systems and remote sensing. Chapter 1, for instance, presents an overview of these resource management tools. GIS and Remote Sensing case studies re'searched and written by John Hayes at Salem State University and Dr. Chiras give examples of the application of these tools.

This edition contains expanded coverage of policy. New international treaties, new federal laws, and other policy tools are discussed in appropriate chapters.

This edition also greatly expands previous coverage of ecosystem management and watershed management. We have continued to look for ways to expand the critical thinking theme and have, as we have in previous editions, tried to maintain an objective approach, offering both sides of many issues. The reader will also find useful our new web page: http://www.prenhall.com/chiras.

Finally, we have made a special effort to expand the scope of this book to include more examples of environmental and resource issues and solutions from other countries. In short, we have attempted to "internationalize" this book. Many examples from Canada were added in this effort.

Acknowledgments

We thank the staff at Prentice Hall, especially our editor, Teresa Ryu, who has helped us improve the quality of this book in many ways. Her insights and enthusiasm have been most appreciated. Teresa has been a pleasure to work with throughout this project, for which we are eternally grateful. We also thank Teresa's assistant, Colleen Lee, for her help throughout the project. Many thanks to our photoresearcher, Linda Sykes, for her hard work and persistence in researching photographs, and to Martin Barr, who edited the manuscript. Many thanks to Kandis Elliot and James Jaeger, our artists. Patrick Burt of WestWords, handled the production of this title expeditiously and thoughtfully. It was a pleasure working with him.

Many thanks to Linda Klein for co-authoring Chapter 9 and to Linda Stuart for her diligence and thoroughness in updating the statistics and preparing the index. Many special thanks to Professor John Hayes for his excellent assistance with the newest feature of the book, the GIS and Remote Sensing case studies. John presented us with numerous excellent ideas for boxes, then researched and wrote first drafts, which we then massaged to be consistent with the writing style of the book.

Finally, we thank our families for their love and support during the writing and production of this book.

Reviewers

Donald F. Anthrop, *San Jose State University*

Thomas B. Begley, *Murray State University*

Ronald E. Beiswenger, *University of Wyoming*

Mikhail Blinnikov, *St. Cloud State University*

Michael Brody, *Montana State University*

Peter T. Bromley, *North Carolina State University*

Conrad S. Brumley, *Texas Tech University*

Neal E. Catt, *Vincennes University*

Thomas Daniels, *SUNY, Albany*

Ray DePalma, *William Rainey Harper College, Illinois*

Donald Friend, *Minnesota State University*

Eric Fritzell, *University of Missouri*

Ken Fulgham, *Humboldt State University*

Jerry D. Glover, *Washington State University*

Paul K. Grogger, *University of Colorado*

Jeanne Harrison, *Rockingham Community College*

John Hayes, *Salem State College*

Bill Kelly, *Bakersfield College*

William E. Kelso, *Louisiana State University*

Linda R. Klein, *LRK Communications*

John Lemberger, *University of Wisconsin, Oshkosh*

Jim Merchant, *University of Kansas*

Frederick A. Montague, Jr., *Purdue University*

Gary Nelson, *Des Moines Area Community College*

Wanna D. Pitts, *San Jose State University*

Jerry Reynolds, *University of Central Arkansas*

David W. Willis, *South Dakota State University*

Gary W. Witmer, *USDA Animal and Plant Health Inspection Service, Fort Collins, Colorado*

Richard J.Wright, *Valencia Community College, Florida*

Biographies

Dan Chiras earned his Ph.D. in reproductive physiology in 1976 from the University of Kansas Medical School. He is currently an adjunct professor at the University of Colorado in Denver and at the University of Denver, where he teaches courses on sustainable development and global environmental issues. He has published 17 books and over 200 articles in journals, magazines, newspapers, and encyclopedias. Dr. Chiras lectures about a variety of topics, including ways to build a sustainable society. His newest book is *The Natural House: A Complete Guide to Healthy, Energy-Efficient, Environmental Homes*. Besides his scientific and environmental pursuits, Dr. Chiras is a river runner, cross-country skier, bicyclist, organic gardener, and musician. He and his sons live in a nearly self-sufficient home in Evergreen, Colorado, overlooking the snowcapped Rocky Mountains.

John Reganold received his Ph.D. in soil science from the University of California at Davis in 1980. As a professor of soil science at Washington State University, he teaches courses on introductory soils, land use, and soil management and conducts research in land use and sustainable agriculture. He also advises undergraduate and graduate students in soil science and environmental science. His excellence in teaching and research has been recognized by several awards from Washington State University. Dr. Reganold's research focuses on the effects of alternative and conventional farming systems on soil and crop quality, farm profitability, environmental quality, and energy efficiency. In addition to his research, he enjoys spending time outdoors, swimming, cycling, and backpacking.

Chapter 1

Natural Resource Conservation and Management: Past, Present, and Future

The late Aldo Leopold once defined conservation as "a state of harmony between man and the land." For Leopold, conservation required equal portions of reflection and action. Leopold believed strongly that effective conservation depends primarily on a basic human respect for natural resources. He called such respect a land ethic. Each of us, he said, is individually responsible for maintaining "the health of the land." A healthy land has "the capacity for self-renewal." "Conservation," he concluded, "is our effort to understand and preserve that capacity." It is this concept of conservation that has guided and influenced the writing of this book over the past three decades.

1.1 A Crisis on Planet Earth?

Effective conservation and management of natural resources in the United States and other countries is becoming more and more urgent, for many reasons. First and foremost, the human population is growing by leaps and bounds. Eighty-four million new people are added to the planet each year. Second, along with this growth is an unprecedented growth in the human economy. As the world's population expands and our economic activity increases, human society is rapidly degrading the natural environment. The environment is the source of all the resources that fuel the economy and make our lives possible, and a sink for all of our wastes. In short, the Earth is vital to our well-being. The damage we create threatens our own future and the future of our children and the many species that share the planet with us.

Ironically, humankind prides itself on conquering outer space and on its many new technologies that make space exploration possible. Yet, after two centuries of technological progress, we still fail to adequately manage the space around us here on planet Earth. This failure has led to an environmental crisis that results from three interrelated problems: (1) a large and rapidly growing human population, (2) excessive resource consumption and depletion, and (3) local, regional, and global pollution.

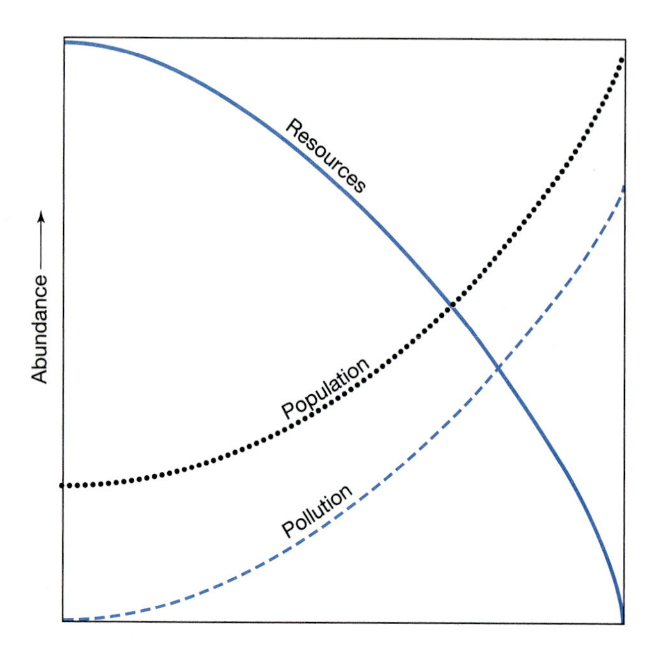

FIGURE 1.1 *Population, resources, and pollution. This simple diagram illustrates a very fundamental relationship between people, the resources they require, and environmental destruction and pollution.*

Before we examine each of these facets of the crisis, it is important to point out that even if there were no crisis, it would be important to manage natural resources and the environment better and to create a more positive relationship between people and the planet. The planet with the living organisms that inhabit it is, as just mentioned, the basis of all life. Our health and well-being are intimately tied to the planet's health. So, whether you believe there is an environmental crisis or not, this book will help you understand ways to live on the planet that are keenly important to a healthy human community.

Population Increase

The human population is growing rapidly. At the current rate of growth, global population will surge from more than 6 billion in the year 2000 to more than 8 billion by the year 2025. This cancerous growth of the human population clouds the future on planet Earth and is the main driving force behind the depletion of resources and the pollution of our planet. Why is population growth such an important force?

As a general rule, every increase in population results in an increase in the demand for food, water, clothing, shelter, and other goods and services. In meeting these needs, we draw on the Earth's natural resources, many of which are already in short supply or are declining in quality. Meeting our demands for these resources also increases environmental pollution (Figure 1.1). Evidence of this simple but powerful relationship is all around us. For example, the rapid increase in population in many less developed countries (LDCs) such as those in Africa result in rampant environmental damage. Population growth in LDCs as well as the wealthier more developed nations also causes problems, among them overcrowding and degraded landscapes (Figure 1.2). In fact, there's not an environmental problem that can't be linked to human population growth. Even many social problems such as drug abuse, mental illness, crime, and suicide are all thought to increase as a result of overcrowding.

The evidence is pretty clear: rising population is resulting in a decline in the overall standard of living in virtually all nations of the world. Unless population growth is halted within the very near future, even the most soundly conceived and effectively implemented conservation and environmental practices will be to no avail.

How fast is the human population growing? By this time tomorrow, nearly 225,000 people will join the global family; in a week, 1.6 million more will be here; and by next year, an additional 84 million will be making demands for food and other necessities. On Memorial Day, our nation honors the memory of

(A)

(B)

FIGURE 1.2 *Overpopulation and other factors like poverty cause a host of environmental problems from degraded landscapes (A) to crowded, squalid living conditions (B).*

those Americans who gave their lives for their country on the world's battlefields. The fatalities have indeed been numerous—57,000 in the Vietnam War alone. Yet, the rate of population growth is so high that all the battlefield deaths of soldiers the world over since the voyages of Christopher Columbus will have been replaced in about 6 months.

Excessive Resource Consumption and Depletion

All people need resources. The most noticeable demand for those resources comes from the world's industrialized or more developed nations (MDCs), which are consuming many natural resources (coal, oil, gas, copper, zinc, and cobalt, for example) at an accelerating pace. The United States ranks first in per capita consumption—that is, consumption per person. Although our nation has only 5 percent of the global population, it consumes 30 percent of the world's resources.

Americans are the most overfed, overhoused, overclothed, overmobilized, and overentertained people in the world. Our enormous consumption of cars, color television sets, dishwashers, air conditioners, golf carts, home computers, swimming pools, CD players, and video cassette recorders satisfies a wide range of longings, far beyond our basic needs. Through such excessive production and consumption, the United States and other highly industrialized nations such as Canada and Japan are accelerating the depletion of our planet's resources.

Resource demands are extraordinary in the heavily populated less developed nations, too. But in this instance, demands are due in large part to meeting basic needs of the people for food, shelter, and clothing. A large part of the demand is also due to the exportation of raw materials and goods for industrial nations. Whatever the cause, it is clear that the natural environments and resource bases of less developed nations such as India, China, and Bangladesh, for example, all suffer enormously under the strain of large and rapidly growing populations.

FIGURE 1.3 *The United States and other nations have polluted their lakes and streams with sewage, industrial wastes, radioactive materials, heat, detergents, agricultural fertilizers, and pesticides. Massive fish kills are often the result.*

FIGURE 1.4 *Like many cities, Los Angeles is often blanketed in a thick layer of pollution from cars, busses, trucks, motorcycles, lawnmowers, factories, powerplants, backyard grills, and other sources.*

Pollution

People produce pollution directly and through the extraction and use of resources (Figure 1.1). Rich and poor nations alike are responsible for widespread pollution problems. The United States, the world's most affluent nation, has also become its most effluent (Figures 1.3 and 1.4). Like other industrialized nations, we have degraded our environment with an enormous variety and volume of contaminants. We have polluted lakes, streams, oceans, and groundwater with sewage, industrial wastes, radioactive materials, heat, detergents, fertilizers, pesticides, and plastics. Millions of tons of sulfur dioxide and carbon dioxide are spewed into the air each year from the combustion of fossil fuels, such as coal and oil, and are causing serious environmental effects, not only in the United States but in other nations as well. Our dependence on nuclear power, as well as on nuclear arms, has led to the accumulation of large amounts of radioactive waste.

Pollution also abounds in less developed nations where sewage, animal waste, and sediment from farms and deforested lands contaminate the air, water, and land. Making matters worse for the environment, many less developed nations are industrializing. As industry expands and standards of living improve, the environment often becomes more polluted.

1.2 Differing Viewpoints: Are We on a Sustainable Course?

The Earth and its ecosystems are the life support system of the planet. Can the Earth and its ecosystems support the reasonably high standard of living many of us now enjoy through the year 2050? Can they support the rising level of affluence in less developed nations? Will they be able to support the human population by the year 2100? These important questions are almost impossible to answer with any degree of certainty. Why?

The reason for our inability to answer these simple questions is that are so many interacting variables that influence the issue. In 1972, a research group at the Massachusetts Institute of Technology set out to find an answer to such questions using computers. They published their results in a landmark book, entitled *The Limits to Growth.*

The researchers showed through computer analysis that the human population would exceed the planet's carrying capacity within a century if exponential growth continued. Figure 1.5 summarizes the findings. The computer program they devised shows that as the world population expands, resource supplies begin to fall. This is accompanied by a decline in the amount of food available on an individual basis (food per capita) and a decline in industrial output per capita. In time, the human population begins to decrease in number, largely as a result of starvation.

What would happen if resource supplies were much larger than the researchers estimated? To address this question, the team doubled their estimated available supply of nonrenewable resources—things like oil and minerals. What they found was that the human population would still overshoot the Earth's resource supplies, just a couple decades later. In another scenario, the authors assumed that world resources were unlimited. Under these conditions, population growth was still halted by rising levels of pollution.

The conclusions of the MIT study were unequivocal. Any way you look at it, if human civilization continues to increase in number, we will reach a perilous state where we exceed the planet's ability to support human life—the human carrying capacity of the planet.

All of this suggests that our current path cannot be sustained. Put another way, our society is unsustainable. Don't get us wrong: this doesn't mean we're doomed and that you should abandon hope. It means our course cannot be sustained. Steering onto a sustainable course will require efforts by individuals like you, business, and governments. That's largely what this book is about—outlining problems and proposing personal, governmental, and corporate solutions. But action must occur soon. Although there are areas of marked improvement, the level of environmental destruction continues at an unsustainable rate.

Moreover, there are signs that we've already overstepped the planet's carrying capacity (its ability to support life) in several vital areas. In fact, in 1992, three members of the original *Limits to Growth* team re-examined their findings and re-analyzed the state of the world in a book entitled *Beyond the Limits.* Their conclusion: that their earlier projections were wrong. Humans have already exceeded critical thresholds and are dangerously close to others. Put another way, the authors' previous projections had underestimated the hazards of continued population growth with its accompanying rise in resource demand and pollution.

Viewpoint of the Optimists

Not everyone agrees with these somber projections and the need for swift action. In fact, the *Limits to Growth* study has been severely criticized by many. Especially vocal are those who believe that technology can solve our resource and environmental problems. History, they like to point out, is full of examples showing how necessity fosters new inventions and cultural changes. We call such folks the optimists. They argue that the Western world is on the brink of another tech-

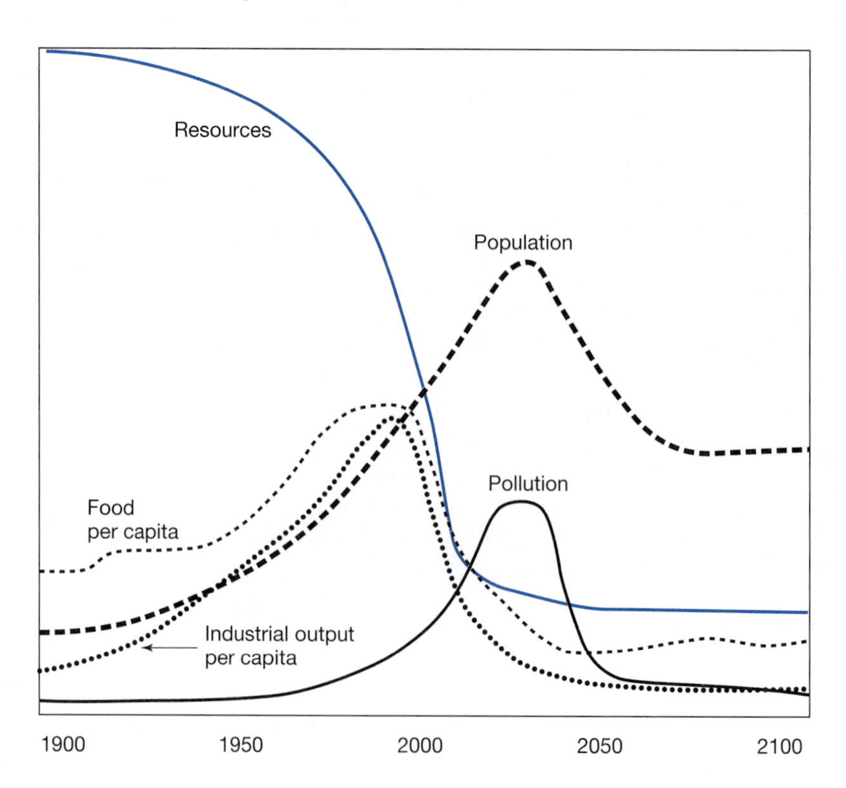

FIGURE 1.5 The Limits to Growth *study. Researchers used the computer to predict the fate of human society if current trends continue. This graph shows that if the population continues to grow, resources will decline dramatically. Pollution levels will increase. The combined effect is a decline in human population and considerable environmental damage.*

nological revolution that could save us from impending doom. After all, isn't the current crisis the greatest in human history? If small crises result in small innovations, then today's composite population-resources-pollution dilemma may be the necessary stimulus for the greatest technological breakthroughs of all time. The optimists confidently claim that whenever something goes awry, technology will provide a fix.

Athelstan Spilhaus of the University of Minnesota is a leading spokesperson for this school of thought. He has suggested that if enough cheap energy is available, all things can be accomplished, pollution will be controlled, food will be available for all, and clothing and shelter for the needy millions will be provided. Indeed, some nuclear power enthusiasts have predicted virtually unlimited energy supplies once the breeder and fusion reactors have been perfected, although neither technology seems to be doing very well for reasons made clear in Chapter 21. To increase food production, the optimists suggest a variety of schemes ranging from fish farming to synthesizing food in test tubes, from growing yeast and algal culture to irrigating deserts, from draining swamplands to using genetic engineering to produce miracle wheats and supercorn. Our options are endless, say optimists. They are limited only by our ingenuity. According to the optimists, we can always depend on human ingenuity and skill to pull another rabbit out of our technological hat.

Viewpoint of the Pessimists (or Realists?)

On the other side of the fence is a group we call the pessimists. Some think they're infinitely more in touch with the reality of the situation and prefer to call this group of doomsayers the realists.

Whatever they're called, they believe that technology will not—and cannot—solve all of our problems. For one, there isn't enough time to find technological fixes to problems that need solutions today.

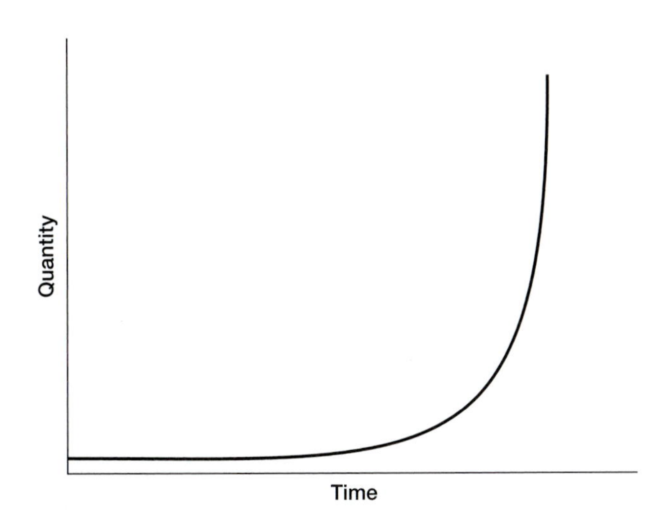

FIGURE 1.6 *Exponential growth curve. Exponential growth is a fixed percentage growth. Each increase is added to the base amount. Exponential growth is deceptive. The entity in question grows slowly over a long period; but once it reaches a certain level, even small percentage increases result in remarkable increases in the size.*

Why is time so crucial? The key to the answer is in the words *exponential growth.* Figure 1.6 shows a J-curve. It represents exponential growth—of a population, of resource depletion rates, of pollution. **Exponential growth** occurs when something—be it a population or a bank account, grows by a fixed percentage and the increase is added to the base amount. For example, an interest-bearing bank account is growing exponentially if the interest is added to the principle. Exponential growth is surprisingly deceptive. It begins slowly but over time the net increase gets larger and larger as the base amount increases. All of a sudden, the base amount gets so large that each yearly increase becomes enormous. The item being measured is said to have rounded the bend of the growth curve. Human population growth has clearly rounded the bend. Even though the global population is only growing at 1.4% per year, there are over 6 billion people. Thus, each year adds nearly 84 million more people.

Global population, resource consumption, and many forms of pollution are all growing exponentially. Even though many of the world's scientists, technologists, ecologists, sociologists, and economists are struggling to find solutions, the annual increase in human numbers and growth in economic activity are becoming overwhelming. In short, we and the problems we create, such as pollution, habitat destruction, and the loss of forests, are increasing faster than solutions.

Viewpoint of the Moderates

Which viewpoint, optimistic or pessimistic, is closer to the truth? Unfortunately, we cannot be sure. Perhaps, however, as in some instances, a moderate viewpoint is more correct. The moderates view our current resource-environmental posture with justifiable concern. Yet they feel that there is still sufficient time, if we start now, to shift from today's spendthrift society to a sustainable society—one that lives within Earth's limits. A **sustainable society** is defined here as one that meets its needs without preventing future generations and other species from meeting their needs. It lives without robbing Peter (future generations) to pay Paul (to meet our needs and desires). It's a simple goal, but given human nature, ethics of many people, economics, existing laws, and a host of other factors, it will be enormously complicated and difficult to achieve.

Fortunately, there's no shortage of writers, teachers, and even a few political leaders who are proposing, even experimenting with, strategies for building a sustainable society. In *Lessons From Nature: Learning to Live Sustainably on the Earth* and other writings, the senior author of this book has outlined six key ecological principles of sustainability based on studies of natural systems. They are: (1) conservation, (2) recycling, (3) renewable resource use, (4) restoration, (5) population control and management, and (6) adaptability. We call them the **biological principles of sustainability,** for they explain in a general way why natural systems sustain themselves. For example, natural biological systems or ecosystems tend to persist because organisms use what they need and use resources with efficiency. We call this the **conservation principle.** Although there are notable exceptions, for example, the grizzly that eats only the eggs of a freshly killed salmon then goes off to hunt again, most species are pretty frugal. (They don't have the technology we have to exploit resources, either). Another reason why natural systems tend to be sustainable is that life forms recycle wastes for reuse. There is no waste in nature. Virtually everything is used over and over again. Waste from one species is food for another. Natural systems tend to persist because life relies in large part on renewable resources. Natural systems also restore damage. Finally, organisms can adapt to changes.

We believe that these principles can be successfully applied to human societies and that they could help steer us on to a sustainable course. In this book, we attempt to show you how these and other principles can be used to create solutions in a wide range of arenas, from forest management to range management to waste management.

In the *Limits to Growth* and *Beyond the Limits* studies, the authors argue that we can achieve a sustainable society by stopping population growth and making drastic changes in other sectors of our society. However, lifestyles in this sustainable society would be much simpler than those of today—more like an average European lifestyle. But such a goal can be met only if serious changes are made, and soon. The new lifestyle would be ecologically sound rather than ecologically short-sighted, some say, suicidal.

Our task—to achieve a truly sustainable society—is difficult and challenging. It requires the dedicated, highly coordinated, and long-sustained efforts of many different types of people, from factory workers to business executives, from college students to farmers, from scientists to politicians, from food specialists to geographers. It requires imaginative and inspirational leadership from government leaders at all levels, from small-town mayors to presidents of the United States and leaders of other industrialized nations, from village tribal chiefs in Africa to benevolent despots in South America. Some of the required changes have already begun. Much of the ground work for change began many years ago, as you shall soon see.

1.3 A Brief History of the Resource Conservation, Environmental, and Sustainability Movements

Conservation in the Nineteenth Century

The 1700s and 1800s were restless times in the United States. The new land stretched for miles on end. It appeared to be a seemingly endless source of resources. Settlers spread westward, cutting trees to build homes and towns and to make room for farms. Huge forests were leveled and the prairies were converted to farmland. Swamps were drained and endless roads cut into the landscape. The prevalent attitude of the times was use it up and then move on.

In the 1800s, the need for conservation became evident, at least for a small number of visionaries. In the early 1800s, for example, George Washington and Thomas Jefferson employed effective methods to control soil erosion on their farms, notably contour farming—plowing and planting across the slope rather down the fall line of a hill to reduce soil erosion. In 1864 the diplomat-naturalist George Perkins Marsh wrote *Man and Nature,* a book that did much to draw attention to the fragile nature of our resources and how they can be abused by humans. It served as a catalyst for the fledgling conservation movement.

Another key figure in this period was John Muir. Muir was born in 1838 in Scotland and immigrated to the United States, settling first in Wisconsin at age 11. In 1867, he walked from Indiana to the Gulf of Mexico, recording his observations of plant and animal life in a book. He then moved to California and became enamored with its exquisite wildlands. Devoting his life to conservation, especially the preservation of forest land in the West, Muir lobbied hard for the establishment of Yosemite and Sequoia National Parks. Muir's love for wilderness which he expressed in his books and articles influenced many people, and continues to appeal to many readers.

In large part because of Muir, the United States Congress established three national parks in the 1800s: Yellowstone, the world's first, in 1872, followed by Yosemite and Sequoia in 1890. In 1891, Congress established 28 forest reserves, later to be designated as the nation's first national forests. In 1892 John Muir founded the Sierra Club, an organization that today is one of our country's most politically active conservation groups.

Conservation in the Twentieth Century

By far the most significant advances in natural resource conservation have been made in this century. They have occurred primarily in four waves. The first (1901–1909) developed under the dynamic and forceful leadership of President Theodore Roosevelt, the second (1930s) occurred during the presidency of Franklin D. Roosevelt, and the third (1970–1980) was given impetus by the Nixon, Ford, and Carter administrations. The newest wave is global in scope. In the United States, Vice President Al Gore and President Bill Clinton did much to promote a more sustainable economy and way of life.

The First Wave (1901–1909) In 1908, President Theodore Roosevelt convened the **White House Conference on Natural Resources,** a high-water mark for the cause of conservation (Figure 1.7). Several developments influenced Roosevelt's decision to call the conference. Among them were (1) the deep concern among scientists over the severe depletion of timber in the Great Lakes states; (2) the study of arid western lands by Major J. W. Powell in the 1870s, which stimulated great interest in growing vegetables and other crops in the deserts of the West with the aid of irrigation; (3) the 1907 report by the Inland Waterways Commission pointing out that the excessive use of water would inevitably have a negative impact on other resources, such as timber, soils, and wildlife; and (4) a growing apprehension that our nation's resources were being grossly mismanaged and that severe economic hardship would be the inevitable result.

Invited to the White House Conference were governors, congressional leaders, scientists, anglers, hunters, and resource

FIGURE 1.7 *President Theodore Roosevelt, outdoorsman, hunter, and ardent conservationist, at Yosemite National Park.*

FIGURE 1.8 *President Franklin Delano Roosevelt. During his administration (1933–1945) many of the United States' resource problems were dealt with in creative ways.*

experts from several foreign nations. As a result of the conference, a 50-member **National Conservation Commission** was formed, composed of scientists, legislators, and businessmen. Inspirational leadership was provided by Gifford Pinchot, a professional forester who profoundly influenced the way forests are now managed. Pinchot is credited with introducing scientific principles to forest management. He favored conservation and future growth—the use of forest resources but with measures taken to ensure regrowth so forests could provide a steady supply of resources to people. His personal friendship with President Theodore Roosevelt, himself an avid conservationist, was crucial in forging public policy.

The commission completed the United States' first comprehensive **Natural Resources Inventory.** The White House Conference also resulted indirectly in the formation of 41 state conservation departments, almost all of which are still operating today.

The Second Wave (1933–1941) Franklin D. Roosevelt is a good example of the right man in the right place at the right time (Figure 1.8). In 1934, Roosevelt established the National Resources Board, which completed the nation's second comprehensive Natural Resources Inventory. In its report, the board identified serious resource problems plaguing the country and described methods for solving them. Roosevelt also created an imaginative nationwide program to create jobs and simultaneously solved many natural resource problems affecting the nation. Much of the impetus for these programs came from the dust bowl era and the great depression of the late 1920s. The dust bowl, described in Chapter 7, resulted from extended drought (1927–1932) that caused crops to fail. In the ensuing years, massive amounts of soil were eroded from farmland laid barren by drought occurring in the Great Plains. Here are some examples of Roosevelt's programs:

1. The **Prairie States Forestry Project** was begun in 1934. Its goal was to establish shelter belts of trees and shrubs (thin bands of trees and shrubs to reduce wind erosion) on farmland along the one-hundredth meridian extending from the Canadian border of North Dakota south to Texas. This project did much to reduce soil erosion from wind.
2. The **Civilian Conservation Corps (CCC),** which was established in 1933 and functioned until 1949, was organized into 2,652 camps of 200 men each. Many were located in the national parks and forests. The forest workers constructed fire lanes, removed fire hazards, fought forest fires, controlled pests, and planted millions of trees. The park workers constructed bridges, improved roads, and built hiking trails. In addition, the CCC made lake and stream improvements and participated in flood-control projects.
3. In 1935, Roosevelt established the **Soil Conservation Service (SCS).** The time was ripe for such a program. The frequent occurrence of severe dust storms over the Dust Bowl of the Great Plains in the late 1920s bore testimony to the vulnerability of the nation's soils. The SCS conducted soil conservation demonstrations to show farmers the techniques and importance of erosion

control. (The Soil Conservation Service remains today, but is now known as the Natural Resource Conservation Service.)

4. The establishment of the **Tennessee Valley Authority (TVA)** in 1933 was a bold experiment, unique in conservation history, to integrate the use of the resources (water, soil, forests, and wildlife) of an entire river basin. Although highly controversial at the time, it has received international acclaim and has served as a model for similar projects in India and other nations.
5. The **North American Wildlife and Resources Conference** was convened by President Roosevelt in 1936. Attended by wildlife management specialists, hunters, anglers, and government officials, it set out to develop an inventory of the nation's wildlife resources and a statement on wildlife and other conservation problems, including policies by which those problems might be solved. This conference meets annually to this day.

The Third Wave (1960–1980) During the 1960s, the U.S. conservation and environmental movement really took off. Several highly influential books and essays were written during this time that sensitized the general public to the gravity of the nation's problems. Rachel Carson's *Silent Spring* (1962), a runaway best-seller, alerted the general public to the potentially harmful effects of pesticides such as DDT on both wildlife and humans. Noted ecologist Paul Ehrlich, of Stanford University, wrote *The Population Bomb,* which warned of the environmental degradation that would result if society did not control the worldwide surge in population. Garrett Hardin's classic essay, "The Tragedy of the Commons," proposed that any resource shared by many people would eventually be exploited and degraded.

In 1969 Senator Gaylord Nelson (D–Wis.) called for a nationwide environmental teach-in in an attempt to marshal the energies of the nation's college students "to halt the accelerating pollution and destruction of the environment." Co-organized by Denis Hayes, the event was called Earth Day. It continues to be celebrated each year on May 22 and offers an opportunity to examine new issues and renew one's commitment to environmental protection.

Congress also responded by enacting so many important laws to upgrade our resources and control pollution from 1970 to 1980 that this period has been called the decade of the environment. Much of the progress during this period occurred during the Nixon administration. A partial list of these acts is shown in Table 1.1.

One of the major advances in environmental protection during this period was the establishment of the **Environmental Protection Agency (EPA),** which was formed during the early 1970s from environmental branches of key federal agencies. Over the years, the EPA has grown to become a prime mover in environmental protection with enormous power and oversight over numerous issues. Although EPA's focus has primarily been that of a watchdog and regulatory agency, it is now moving in new directions to promote a more proactive approach, preventing problems and working with businesses rather than merely punishing them for lack of compliance.

TABLE 1.1 *Major Environmental Acts Passed During the Decade of the Environment (1970–1980)*

Category	Act
Air Quality	Clean Air Acts of 1970, 1977, 1990
Control of Noise	Noise Control Act of 1972
	Quiet Communities Act of 1978
Control of Toxic Substances	Toxic Substances Control Act of 1976
	Resource Conservation and Recovery Act of 1976
Control of Solid Wastes	Solid Waste Disposal Act of 1965
	Resources Recovery Act of 1970
Energy	National Energy Act of 1978
Land Use	National Coastal Zone Management Act of 1972
	Forest Reserves Management Acts of 1974, 1976
	Federal Land Policy Management Act of 1976
	National Forest Management Act of 1976
	Surface Mining Control and Reclamation Act of 1977
	Endangered American Wilderness Act of 1978
Water Quality	Federal Water Pollution Control Act of 1972
	Ocean Dumping Act of 1972
	Safe Drinking Water Act of 1974
	Toxic Substances Control Act of 1976
	Clean Water Act of 1977
Wildlife	Federal Insecticide, Fungicide, and Rodenticide Control Act of 1972
	Marine Protection, Research, and Sanctuaries Act of 1972
	Endangered Species Act of 1973

Although many great environmental laws were passed during this period, we must remember, of course, that the mere existence of an environmental law does not ensure environmental protection. A law is only as good as its enforcement. Enforcement requires money to support agencies. Even the most superbly orchestrated act must be enforced if its purpose is to be realized, and effective enforcement requires money for personnel and equipment.

The Fourth Wave (1980–Present): The Beginnings of a Sustainable Revolution? From an environmental standpoint, the 1980s represented some of the best of times and the worst of times. The 1980s, for example, was a period of intense resistance to environmental protection, especially in the United States. After suffering through a crippling period of global inflation in the late 1970s and early 1980s, many politicians and business leaders cast a skeptical eye on environmental protection, perceiving it as counterproductive to economic progress.

All the while, a growing number of observers, including key environmentalists like Fred Krupp, executive director of Environmental Defense, and Barry Commoner, a professor and author, were beginning to speak out on an important finding: Despite the billions of dollars being spent to protect the environment, many gains were being offset by economic growth or growth in the population. Some solutions, critics noted, only shifted problems from one medium to another. For example, pollution control devices were installed at many power plants to remove sulfur dioxide, which causes acid rain. Although effective in cleaning up the air, scrubbers simply convert sulfur dioxide into a solid waste that is typically disposed of in landfills, posing a potential threat to groundwater. Some solutions were incomplete. Catalytic converters, pollution control devices on cars, for instance, converted carbon monoxide, a poison to people, into carbon dioxide, a pollutant that traps heat in the atmosphere and may be contributing to global warming (Chapter 19). Catalytic converters burned off unburned hydrocarbons in the exhaust of vehicles, helping to reduce smog, but didn't remove nitrogen oxides, which contribute to acid rain and acid snow (Chapter 19).

During the 1980s and 1990s, it became increasingly clear that environmental protection was simply not working as well as it could—and it was costing us a fortune. Couldn't we find a way to protect the environment without exacting such a huge price tag? And couldn't we find ways to prevent problems in the first place?

During the 1980s and 1990s, many experts began to note that, judging from current trends, modern society was on a fundamentally unsustainable course. One of the first to call attention to this phenomenon was Lester Brown, an agricultural economist who founded the Worldwatch Institute in Washington, D.C. (Figure 1.9). Brown's book, *Building a Sustainable Future,* outlined the persistent erosion of the Earth's life support system and proposed a strategy for building a sustainable society. His organization has gone on to produce a series of books and papers on a wide range of environmental topics that are translated into many languages. With uncanny accuracy and vision, they outline problems and sustainable solutions, and have influenced thinking and policy worldwide. They've been especially influential in renewable energy and agriculture.

Pioneering work in energy during the 1980s and 1990s was performed by Amory Lovins, a physicist who co-founded the Rocky Mountain Institute. Lovins raised awareness of the environmental and economic importance of energy conservation and served to redirect the thinking of many power companies the world over. In the 1990s, he turned his attention to the development of super energy efficient cars, which we'll examine in Chapter 21.

FIGURE 1.9 *Lester Brown. A leader in the sustainability movement, Lester Brown and his staff at the Worldwatch Institute have probably done more to promote an understanding of sustainable development than anyone else in the world.*

Another highly influential force emerged in 1983 when the United Nations assembled a commission to address the issues and propose actions to promote a sustainable future. Called the **World Commission on Environment and Development,** it produced a monumental piece of work, *Our Common Future,* which stirred intense global interest in alternatives to the environmentally unsustainable development taking place worldwide.

This interest, in turn, was largely responsible for the 1992 **United Nations Conference on Environment and Development,** commonly called the **Earth Summit** (see Case Study 1.1). Attended by officials from nearly 180 nations, the Earth Summit was the largest international meeting on the environment in the history of human civilization. Participants produced agreements on a wide range of issues, including global climate change, biodiversity, and deforestation. One of the most impressive outcomes was an 800-page document called *Agenda 21.* It outlined over 4,000 action items for achieving sustainable development.

Since the Earth Summit, numerous cities, towns, states, and national governments have adopted sustainable development as an official policy. **Sustainable development** is defined here as a strategy to meet one's needs without preventing future generations and other species from meeting their needs. Sustainable development requires a long-term approach to resource management and human development. It requires us to think in terms of systems, too. Systems thinking, which we discuss in this book, requires an understanding of how human and natural systems operate. It requires us to understand our dependence on natural systems and how we affect them, positively

CASE STUDY 1.1 THE EARTH SUMMIT AND BEYOND

All the world's people belong to just one ecological system. This means that whatever environmental damage you create in your home town may eventually, either directly or indirectly, have a harmful effect on people living elsewhere on this planet, even though they may be half a world away. For example, suppose that you burned up 1,000 gallons of gasoline driving your family car last year. As you sped along the city streets, roads, and highways, the exhaust pipe on your car belched many billions of molecules of carbon dioxide into the air. These molecules then trap heat that otherwise would have escaped into outer space. The net result is a warming of the Earth's atmosphere, a phenomenon popularly known as the **greenhouse effect** (Chapter 19). And not only will the atmosphere over your town warm up to a very slight degree, but so eventually will the atmosphere around the entire planet Earth!

It is apparent, therefore, that future resource and environmental management cannot be accomplished effectively without an integrated global effort, rather than a piecemeal and localized approach. To this end the United Nations convened the first Conference on the Environment and Development (UNCED) in Rio de Janeiro, Brazil, in 1992. This conference, popularly referred to as the Earth Summit, was worldwide in scope. It was attended by official environmental delegates from nearly 180 nations. It also attracted many heads of state, including then President George Bush, more than 8,000 science reporters, thousands of environmentalists and concerned citizens, who attended a parallel conference put on by the nonprofit community. And what were the results?

Participants in the Earth Summit passed two international agreements, one on global warming and another on protecting nonhuman species (biodiversity). The climate treaty called on the nations of the world to hold greenhouse gas emissions such as carbon dioxide to 1990 levels by the year 2000. Although some nations have taken this seriously, most made little progress toward this goal. Since the Earth Summit, a new agreement on climate change has been drawn up. Known as the Kyoto Protocol and signed in 1997, it calls on industrial nations and former Eastern bloc countries to cut emissions of greenhouse gases by 5.2 percent below 1990 levels between 2008 and 2012.

The biodiversity agreement calls on nations to create inventories of their plants and animals and to develop strategies to protect them. Political wrangling over these documents was intense, and as a result, both agreements are considered to be rather weak by many critics, but they are a step in the right direction.

Participants of the Earth Summit also created a massive document called *Agenda 21.* This report outlines over 4,000 actions that can be taken to create a sustainable future. In addition, the participants agreed on a set of forest management and protection principles. It, too, was watered down by certain participating nations. Finally, the Rio conference produced a declaration of legal principles essential to creating global sustainability.

Even though the outcome of the Rio conference was much less than proponents of a strong global stance on sustainable development had hoped for, the fact that the event took place at all is testimony to the heightened environmental awareness and concern of the international community. It is hoped that the second UNCED will convene in the not too distant future. As Maurice F. Strong, Secretary General of the conference, so eloquently stated in his opening address to the delegates:

> The Earth Summit is not an end to itself, but a new beginning. The measures you agree on here will be but first steps on a new pathway to our common future. Thus, the results of this conference will ultimately depend on the credibility and effectiveness of its follow-up. The road beyond Rio will be a long and difficult one. It will also be a journey of renewed hope, of excitement, challenge and opportunity. As we move into the 21st century it will lead us to the dawning of a new world in which the hopes and aspirations of the world's children for a more secure and hospitable future can be fulfilled.

International conferences and treaties are only as good as the action they inspire. Fortunately, the Earth Summit has stimulated an incredible amount of action on the part of individuals, communities, cities, states, and nations. In the United States, the interest in sustainable development that emerged in communities has resulted in numerous strategies, for example, community plans that outline steps for sustainable development. Soon after being elected, President Clinton established a national commission, the President's Council on Sustainable Development, which brainstormed on policies and actions required to build a sustainable future. In Holland, government agencies and businesses have hammered out a plan to drastically cut pollution without hurting the national economy. Other countries were also spurred to take similar steps.

Despite these gains, there is much more to be done. We're a long way from building a sustainable society. The Worldwatch Institute tracks environmental trends in their annual publication, *Vital Signs.* Despite growing awareness, many key areas of the environment are deteriorating. We can solve these problems, say some observers, but we must redouble our efforts and find ways to strike at the roots of the problems to create lasting solutions.

and negatively. It is preventive in nature—seeking solutions that prevent problems in the first place. Numerous entities have undertaken sustainable development projects, many of which are outlined in this book.

Important gains have been made in Europe and in the less developed countries of the world. One of the most promising has been the gradual slowing of the growth in the human population in many parts of the world. Numerous wildlife preserves have been set up in the United States and in less developed countries thanks to efforts of The Nature Conservancy and other organizations.

Although progress toward sustainable development is moving slower than many would like, progress *is* occurring. Many examples cited in this book illustrate this fact. If progress continues, some observers believe that the 1980s and 1990s could mark the beginning of a Sustainable Revolution, not unlike the Agricultural and Industrial revolutions in its impact on the shape of human civilization. This change, already underway, seeks strategies that allow people to live well while protecting, even enhancing, our natural resources and environment so vital to our future and the future of all other life forms.

On a final note, the 1980s and 1990s seem like decades of great environmental progress. But, this is not the entire story. Although many people and government agencies and businesses rallied behind sustainable development, powerful antagonism against environmental protection has arisen. At the state level, there's been a great deal of back peddling, as businesses and powerful lobbying groups have attempted, and often succeeded, in weakening environmental laws. At the national level, anti-environmental efforts have been well funded and extensive. Legislators and environmentalists who have favored renewing many key pieces of environmental legislation like the Endangered Species Act have found themselves locked in long, difficult battles with those who see environmental protection as a hindrance to economic progress and personal property. Powerful public personalities like conservative commentator Rush Limbaugh , often with little scientific information to back arguments, swayed many people's opinions on key issues. This sentiment so far continues into the 21st century and may be with us for a long time.

If many businesses and lobbying groups were anti-environmental, most of the U.S. public was complacent during the 1900s. After disastrous years in the 1980s, witnessing the Valdez oil spill, record drought and fires, problems with garbage barges, and medical waste washing up on the shore's of the nation's eastern seaboard, which spurred considerable environmental interest, many people slipped into complacency. A thriving economy caused many to abandon their environmental views and adopt a more consumptive lifestyle. Many purchased large gas-guzzling vehicles and huge homes. Although recycling continued to rise, individual consumption climbed. Complacency continues today.

TABLE 1.2 *Classification of Natural Resources*

Renewable

Resources whose continued harvest or use depends on proper human planning and management. Improper use and/or management results in impairment or exhaustion, with resulting harmful social and economic effects.

1. Fertile soil. The fertility of soil can be renewed, but the process is expensive and takes time.
2. Products of the land. Resources grown in or dependent on the soil.
   a. Agricultural products. Vegetables, grains, fruits, and fibers.
   b. Forests. Source of timber and paper pulp. Valuable as a source of scenic beauty, as an agent in erosion control, as recreational areas, and as wildlife habitat.
   c. Rangeland. Sustains herds of cattle, sheep, and goats for the production of meat, leather, and wool.
   d. Wild animals. Provide aesthetic values, hunting sport, and food. Examples are deer, wolves, eagles, bluebirds, and fireflies.
3. Products of lakes, streams, and oceans. Example are black bass, lake trout, salmon, cod, mackerel, lobsters, oysters, and seaweed.
4. Ground water and surface water.
5. Ecosystems. Natural ecosystems provide many free ecological services. Wetlands, for instance, help reduce flooding, filter surface water, increase groundwater discharge, and provide habitat for animals.
6. Certain energy resources such as wind energy, solar energy, geothermal energy, tidal energy, and hydropower.

Nonrenewable

Amount of resource is finite. When destroyed or consumed, such as the burning of coal, the resource cannot be replaced.

1. Fossil fuels. Produced by processes that occurred millions of years ago. When consumed (burned), heat, water, and gases (carbon monoxide, carbon dioxide, and sulfur dioxide) are released. The gases may pose serious air pollution problems. These resources cannot be recycled.
2. Nonmetallic minerals. Phosphate rock, glass sand, and salt. Phosphate rock is of crucial importance as a source of fertilizer.
3. Metals. Gold, platinum, silver, cobalt, lead, iron, zinc, and copper. Without these, modern civilization would be impossible. Zinc is used in galvanized iron to protect it from rusting, tin is used in toothpaste tubes, and iron is used in cans, auto bodies, and bridges. These resources can be recycled.

1.4 Classification of Natural Resources

To begin to understand the challenge of building a sustainable society, we must first examine the resource base itself. Conservationists recognize two major kinds of resources: renewable and nonrenewable. Be sure to take a look at Table 1.2 for a more detailed explanation of renewable and nonrenewable resources.

Renewable resources are those resources that can be renewed by natural processes and include soils, rangelands, forests, fish, wildlife, air, and water. Although these resources can be regenerated, they can also be depleted as a result of overuse by humans. Overharvesting fish populations, for example, can result in their decline. By the same token, renewal can also be facilitated by humans. That is to say, human actions can enhance fish populations. Some resources may be renewed much more rapidly than others. For example, it may take 1,000 years for just 1 inch of fertile topsoil to form. A pine forest may develop from the seedling stage to maturity in 100 years. On the other hand, a herd of deer, under optimal conditions, could build up from 6 head to 1,000 head in only 10 years!

Several energy sources are also classified as renewable resource, for example, solar, wind energy, and hydropower. Although solar energy comes from a finite source, the sun, we still consider it a renewable resource because the sun's projected lifespan is huge, perhaps a couple billion years. We won't need to worry about running out of solar energy for a long time.

Wind and hydropower, both discussed in more detail in Chapter 22, are produced largely as a result of solar energy. Winds, for example, result from the differential heating of the Earth's surface. Hydropower, energy captured from flowing water, depends on evaporation and precipitation. Evaporation is powered by sunlight. Both wind and flowing water will continue so long as the sun survives. They're renewed daily by the sun.

Nonrenewable resources occur in a fixed amount. They cannot be renewed by natural processes at all, or not rapidly enough to be usable by current human society. They include fossil fuels (oil, coal, natural gas), nonmetallic minerals (phosphates, magnesium), and metallic minerals (copper, aluminum). Among nonrenewable resources, there are some major distinctions worth considering. One of the most important is that some nonrenewables are recyclable. Metals, for instance, are finite, but they can be recycled. We can use them over and over again. Fossil fuels, however, are not. They're burned to release energy and are lost forever.

In building a sustainable society, we must find strategies to better manage both renewable and nonrenewable resources. As you might suspect, they can be very different. We must also give serious consideration to the exhaustibility of a natural resource. Some, like solar energy, wind, and tidal energy are inexhaustible. As long as the sun shines and the moon stays in orbit around the Earth, these sources of energy will be available to us. Renewable resources are not a panacea. Some can be exhausted by overexploitation and poor management practices. Forests, fish, topsoil, ground water, and many other renewable resources we depend on, while capable of being renewed, can be depleted. Cut down a rain forest and let the soil wash away, and you may have destroyed a valuable renewable resource forever. Other resources such as coal and oil are nonrenewable (finite). There's only so much of each in the Earth's crust. Coal and oil are therefore clearly exhaustible. When they're burned, they're lost forever. But not all nonrenewable resources are like coal and oil. Minerals, for instance, are nonrenewable or finite, but they can be recycled. We treat them differently. Keep these distinctions in mind as you ponder the fate of the planet and strategies to steer our society onto a sustainable path. The next section begins to uncover the strategies that Americans and others the world over have employed in the past, and then outlines the rudiments of a sustainable management/living strategy, which provides important background for the rest of this book and your study.

1.5 Approaches to Natural Resource Management

As you read this book and talk to others about resource issues, you will find that opinions vary considerably about the value of natural resources and the importance of safeguarding them while we meet our needs. During the past two centuries, four resource management ideologies have emerged in the United States and elsewhere: (1) exploitation, (2) preservation, (3) the utilitarian approach, and (4) the ecological or sustainable approach. Understanding them will help you understand where we've been and where we're going as a society. They'll help you understand different perspectives, too.

Exploitation: A Human-Centered Approach

The exploitation approach suggests that a given resource should be used as intensively as possible to provide the greatest profit to the user. This philosophy prevailed early in the United States' history and continues today in some areas of the world, especially in the less developed nations that are beginning to industrialize.

In the early days, little if any concern was given to the adverse effects of exploitation such as soil erosion, water pollution, or wildlife depletion. For example, the slogan of the early loggers of the 1800s who exploited our primeval forests was: "Get in, log off the trees, and get out." When the trees were gone, the loggers moved westward and repeated the process. Why not? The nation's forests seemed inexhaustible (Figure 1.10).

The exploitive ideology is human-centered. It seeks to maximize human gain with little, if any, concern for nature and natural resources. It assumes an unlimited supply of resources and that nature is here to serve humans. It also assumes that nature is of no importance to us other than as a source of commodities to make our lives better.

FIGURE 1.10 *Loggers felling a huge Douglas fir tree in an Oregon forest about 1920—an example of "cut and get out." During the 1800s, many thousands of acres were cleared, then abandoned as timber companies moved on to virgin forests for more timber.*

Preservation: A Nature-Centered Approach

The preservation approach to resource management suggests that resources should be preserved, set aside, and protected. A forest, for example, should not be used as a source of timber. It should be preserved in its natural state as a wilderness. In the 1880s, the naturalist John Muir, who founded the Sierra Club, proposed that federal lands of unique beauty should be withdrawn from exploitation by timber, grazing, and mining interests and converted into national parks. As such, they would be preserved virtually unchanged for the enjoyment of future generations. Partly as a result of Muir's influence, Congress established Yosemite and Sequoia as national parks in 1890 and 1891. Those who promote wilderness protection seek a similar goal for millions of acres in the United States and abroad, often over the loud protests of development interests, especially mining and timber companies.

Preservation ideology is the antithesis of the exploitive approach. In fact, it probably emerged in reaction to exploitation and the observed effects of it. Preservation remains alive today in various forms. A national environmental movement, which calls itself Earth First!, proffers such a view. They promote dramatic actions, including tree spiking, to stop the continued destruction of natural systems (Figure 1.11). Although most environmental and conservation groups today support the protection of wilderness, their tactics are more mainstream. That is, they prefer to work within the system to achieve change. At

FIGURE 1.11 *Environmental protection can be achieved by many means. Most nonprofit organizations work within the system, lobbying legislators or influencing policymakers like those in the US Forest Service. Some take more direct action like protests or civil disobedience. Others pursue more direct confrontation and sabotage like spiking trees—driving nails in trees. Spikes catch chain saws and may injure loggers or may damage blades in saw mills. Because of this, spiked trees are clearly marked.*

the same time, they recognize the need for resources and promote wise stewardship. In short, their stand is for intelligent, sustainable use of resources and preservation of some areas. This is sometimes called a utilitarian approach.

Utilitarian Approach

The utilitarian approach to resource management began during the late 1800s and early 1900s, spearheaded by Gifford Pinchot, who was head of the U.S. Forest Service from 1898 to 1910, and Theodore Roosevelt. The key organizing principles behind the utilitarian approach is that of **sustained yield.** This concept suggests that renewable resources (soil, rangelands, forests, wildlife, fisheries, and so on) should be managed so that they will never be exhausted. Careful management will ensure their replentishment, so they can serve future generations ad infinitum. When a forest has been logged off, the site must be reseeded naturally or artificially so that a new forest can develop over time and provide timber for future generations. Similarly, when fish are harvested from a lake or stream, they must be replaced, either by natural reproduction or by stocking with hatchery-reared fish.

The utilitarian approach is fairly human centered, but a step in the right direction. The idea behind this approach is simple: Protect a nation's resources by harvesting them at rates that can be sustained over the long run, a task that has proved to be much more difficult than once thought. Scientific research has shown that systems are more complex than once thought and they're connected to other systems. To manage them sustainably requires many more considerations than ever imagined—and it requires better management of ourselves. A good understanding of the science of ecology is essential to the newest approach, the ecological or sustainable approach.

The Ecological or Sustainable Approach: Managing Natural Resources and Ourselves

Ecology is the study of the interrelationships between organisms and their environment. Modern conservation strategies often operate within an ecological framework. What that means, in part, is that they are based on ecological principles. Chapter 3 outlines key principles and describes ways they're currently applied. The resource management chapters in this book also focus on principles that are essential to sustainable management of the world's natural resources.

The sustainable approach to resource management is complex. It is an evolving art, changing as our knowledge expands. One key concept in this approach is that policies promoted to improve management are often designed to protect more than harvestable species. Efforts are made to protect entire ecosystems, those that renewable resources we're interested in harvesting depend on. Soil in a forest must be protected, for example, to ensure a steady supply of lumber from it. If we let it erode away, lumber production will fall off.

More than 100 years ago, the American diplomat-naturalist George Perkins Marsh observed how humans had abused agricultural lands in Europe and Asia. He further observed how this abuse resulted in soil erosion, dust storms, water pollution, and a decline in many nations' economic well-being. Upon returning to the United States, Marsh wrote *Man and Nature* (1864), mentioned earlier. In it he argued that humans cannot degrade one part of the environment without harming other parts. Marsh argued that although our natural environment is highly varied and infinitely complex, it is a dynamic and organic whole. Today's conservationists realize that Marsh was correct. Soil conservation measures on forested land, not only help to ensure a sustainable yield, they help protect neighboring streams, rivers, and lakes. Fish and other organisms that depend on these waterbodies are protected by sustainable forest practices.

The ecological approach therefore requires a systems approach. It calls on managers to think in terms of whole systems, not just isolated parts of them. When working on a problem at any one level (species, population, or ecosystem), managers must seek to understand and protect the connections between all levels. In the ecosystems approach, for example, efforts to protect a species of fish in a river or lake would encompass more than simply improving in-stream habitat. In fact, such measures may be meaningless if outside influences poison a stream. An ecosystems approach to stream protection, therefore, would include steps to protect the surrounding watershed, which is part of the ecological integrity of the water body. Stream protection begins many miles from stream banks and requires measures to eliminate careless road building, to control sources of industrial pollution, or to modify ecologically unsound farming practices. It may require measures to eliminate human development, which often has many adverse impacts on neighboring biological systems. Or, it may require ways to better manage our own activities—for example, find alternative ways of meeting our needs that don't alter neighboring ecosystems.

Another important aspect of ecosystem management, especially important in wildlife management, is the notion of protecting and managing entire ecosystems to protect species. Organisms, for example, depend on two or more ecosystems to survive. Protecting all of them may be essential to the survival of the organism. For example, grizzly bears in and around Yellowstone National Park depend on a 2-million-hectare (5-million-acre) area that includes the Park and surrounding areas. Protecting the grizzlies requires efforts to protect and manage these areas, a portion of which is privately owned.

Another key concept of ecosystem management is carrying capacity. **Carrying capacity** is defined as the ability of an ecosystem to support life indefinitely. It is determined by resource availability and the ability of the ecosystem to absorb wastes, known as source and sink functions, respectively. Each ecosystem has a carrying capacity for individual organisms. Altering one component in an ecosystem, we've learned, may help one species, but to the detriment of another. Once a common practice for popular species, like fish and huntable wildlife, single-species management is falling out of favor. "Protect the ecosystem and protect the species" is a view growing more and more popular in resource management agencies.

Maintaining carrying capacity and stability within an ecosystem requires steps to maintain its ecological integrity. This often means that efforts must be made to protect the fertility of soils, the purity of water supplies, the diversity of species, populations, and ecosystems in a given management area. This may require efforts to protect viable populations of native species and maintain natural disturbances such as periodic fire. This may require the removal of nonnative species and the reintroduction of native species.

The ecological approach to resource management means that resources such as soil, water, rangelands, wildlife, fisheries, and forests are used in ways that ensure their long-term health and vitality. In other words, there are ways of harvesting the Earth's generous supply sustainably and without causing long-term damage. Sustained yield of a desired species is not as important as sustaining the environment and its many interrelationships. This is no easy task.

The ecological approach to resource management requires a long-term view. It embraces the concept of multiple uses. A forest, for example, is not only a source of timber, as the utilitarians suggest, but has many other values as well, such as wildlife habitat, scenic beauty, and flood and erosion control. Forests must, therefore, be managed so that these other values can be realized. A forest ecologist, for example, would ask the following questions regarding the effects of the proposed timber harvest:

1. Will water runoff be accelerated? If so, will this result in flood damage in the valley towns? Will valley farmers have enough water for their crops and livestock?
2. Will the resultant soil erosion destroy trout spawning beds in the streams below the cut?
3. Will soil fertility at the site be reduced because of erosion and jeopardize the healthy development of seedlings on the logged-off site?

4. What will happen to deer, grouse, and other wildlife when the forest that provided food, cover, and breeding sites is removed?
5. Will the scar left after the stand has been logged off cause visual pollution for the thousands of motorists who travel along the nearby highways?
6. What effect will this timber harvest, and thousands like it throughout the world, have on the buildup of carbon dioxide in the global atmosphere and the progressive warming of the planet?

To summarize, in the ecological approach to conservation, resources (soil, rangelands, forests, wildlife, fisheries) should be used without adversely affecting the physical and biological environments, or should be used in ways that do not result in irreparable damage. Much of this book focuses on this difficult objective.

The ecological approach to resource management, now commonly referred to as the ecosystem approach or **ecosystems management,** embraces the notion of multiple use, as noted above. However, it also may require actions that restrict or limit human activity, sometimes drastically. In other words, it recognizes the need for preservation. As an example, to protect vital biological resources such as birds and wildlife, it is often necessary to establish core reserves in forested areas. A **core reserve** is a region left exclusively to plants and animals. It is a region that is off limits to human exploitation. Surrounding these reserves are **buffer zones,** areas of minimal human intrusion and impact. Limited tree harvesting may occur in such areas. Surrounding the buffer zones are regions where human activities are permitted.

Managing ecosystems requires good scientific information and vigilant monitoring of conditions so that if management strategies are not working, they can be adjusted. The use of such information to monitor conditions and make policy changes is called **adaptive management,** and is a keystone of good ecosystem management. Unfortunately, it is not yet widely used. In many cases, management agencies establish policies, then manage accordingly. Changes in our understanding of systems may not result in immediate changes in policy because of bureaucratic inertia.

As you will see in other chapters, ecosystem management requires the cooperation of various government agencies that control public land because, as noted earlier, a species's habitat often spans two or more jurisdictions. The involvement of private landholders will also be essential to protect ecosystems.

In a sense, an ecological approach helps us live within limits while sustaining human culture. It is a human-centered and nature-centered approach. It gives equal consideration to both. Ecosystems management is part of a larger sustainable development strategy for human survival and prosperity on a finite planet.

Creating a sustainable society requires better ways of managing resources, but it also requires better management of ourselves—our systems, our society. We'd be remiss if we were to suggest that all we need to do is find ways to better manage our resources. That's only half of the task. To create an enduring human presence, we must live our lives and conduct our business affairs in ways that do not deplete the Earth's resources or foul its air, water, and soil. Earlier in the chapter, we mentioned this task in conjunction with the Limits to Growth Study. Many chapters in this book point out the problems created by our own systems such as transportation and waste management. To create a sustainable society we must also address these matters. We believe that the biological principles of sustainability, outlined earlier, can be used to put human civilization back on a sustainable path. These principles include conservation, recycling, renewable resource use, restoration, and population control.

Because they are so important, we will repeat them here. In this book, the conservation principle refers to two basic notions: using only the resources we need (the frugality principle) and using resources efficiently (the efficiency principle). Recycling, of course, means ensuring that the materials we use are not discarded but rather placed back into manufacturing systems, thus virtually eliminating wastes. Building a sustainable society will also require efforts to use and protect renewable resources. Especially important, say proponents, is the use of renewable energy, which is a clean and safe form of energy. Restoration simply implies restoring damaged ecosystems to enhance their productive capacity and ecological functions, many of which (like oxygen production by plants) benefit us. It addresses the need to restore millions of acres the world over that have been exploited in the past and abandoned. Finally, population control means limiting human population growth and better managing how we spread out on the land.

In closing, these principles apply to the management of natural resources like farmland and forests, but they also apply to the management of human systems such as transportation, energy, waste management, and the like. This book primarily deals with natural resources, but we remind you early on to remember that the challenge of creating a sustainable future entails much more than simply managing farms and forests and rangeland and fisheries in ways that ensure their ecological integrity and long-term vigor. It requires changes in us. As future chapters will show, these principles hold great promise for redirecting many human systems and bringing our lives and our businesses back in line with the ecological limits of our planet.

1.6 Changing Realities: The Nemesis Effect

The need for sustainable resource management and other sustainable strategies is greater than ever in human history. With the continued growth of the human population and growth in the global economy, the need for a new way of conducting human affairs will only increase. This book will present ample evidence to support these claims. But there's another reason for striking out in new directions, and doing it quickly, that's only beginning to be discussed among scientists, conservationists, environmentalists, and others. Chris Bright of the Worldwatch Institute in Washington, D.C. calls it the **nemesis effect.** Put in simple terms, the nemesis effects are environmental outcomes that result from the interaction of several changes. For example, acid rain, global warming, and habitat loss all contribute to the loss of species. But they may also combine to drive species more rapidly into extinction.

"If there is comfort to be found in the prospect of environmental decline, it lies in the idea that the process is gradual and predictable," writes Bright. "But this way of thinking is sleepwalking." Many factors will combine forces to produce unanticipated results, impacts that occur more rapidly than predicted. "When one problem combines with another problem, the outcome may be not a double problem," writes ecologists Norman Myers, "but a super-problem." Although ecologists, scientists who study ecosystems and the relationships in them, have barely begun to identify potential super-problems, writes Bright, "in the planet's increasingly stressed natural systems, the possibility of rapid, unexpected change is pervasive and growing."

Keep this idea in mind as you observe environmental change in your area, read the newspaper, listen to the television, and study this book. It may become a key idea in the coming years as decades of environmental perturbation begin to manifest and interact. This possibility, combined with a host of others, also argues for the need to achieve root-level solutions—solutions that strike at the root of environmental problems.

1.7 New Tools for Resource Management: Geographic Information Systems and Remote Sensing

Solutions to environmental problems are many and varied. There are technological solutions, personal solutions, corporate solutions, and governmental solutions through policies and other measures. You will learn about many viable solutions in upcoming chapters. Two tools that are of increasing value, and are used in many areas of resource management, are geographic information systems and remote sensing. We'll discuss GIS first.

A **geographic information system (GIS)** is a computer system consisting of hardware and software that is used to assemble and store information, but not any old information. It stores geographically referenced information, that is, data identified according to its location. This includes a vast collection of information about the Earth including vegetation types, land disturbance, human settlement patterns, and surface water temperature. Geographic information systems do more than assemble and store information, however. They allow operators to display, manipulate, and analyze geographically referenced information. They're a valuable scientific tool and have many practical applications, which we'll be pointing out throughout the text.

GIS can use information from many different sources and can help us analyze trends, detect change, and formulate policy. For example, rainfall and stream flow data from your state can be used to determine which wetlands identified by aerial photos (images taken from airplanes) might run dry in times of drought. Federal and state air quality control agencies can use GIS to study the effects of air pollution on childhood asthma by comparing air quality data and hospital admission records.

GIS identifies locations based on coordinates such as latitude and longitude. But that's not all. Even ZIP codes or postal codes and highway mile markers can be used to identify locations. It doesn't matter. GIS can operate using any data as long as it can be located spatially, and the computer software is set up to handle the data. If the data isn't **georeferenced,** linked to specific locations, it must be before it can be used. All data used in a GIS program must be similarly georeferenced, so it can be easily compared and analyzed.

Data can be entered either from existing maps, aerial photos, or in digital form, for example, digital satellite images can be used to generate useful maps. Hydrologic data or air pollution data contained in tables, for instance, can also be used to generate maps as long as it is georeferenced. GIS can be used to understand and underscore spatial relationships between mapped variables, for example, air pollution and asthma attacks in children.

One of the most time consuming parts of GIS is called data capture, simply defined as entering georeferenced information into the system. Maps, for instance, can be digitized, that is, converted into a numerical data set or hand-traced with a computer mouse, to collect the coordinates of features. Geographers create an abstract picture of our world, essentially a map in two dimensions that can be viewed on screen or printed to produce a hard copy. Lines and points and areas on the map represent features of our world. Points on the map indicate the location of fire hydrants, telephone poles, buildings, and cell phone towers. Lines represent roads, power lines, streams, and trails. Regions, for example, regions of a certain soil type or vegetative type, are indicated by polygons. Three-dimensional representations are also commonly made. Watersheds, for example, may be drawn in three-dimension to show mountains and hills and stream gradient.

GIS is used by a number of government agencies at the federal level, including the U.S. Forest Service, Bureau of Land Management, U.S. Geological Survey, Environmental Protection Agency, and the National Center for Atmospheric Research. State and local government also use GIS as do private companies. A considerable amount of this information is also available to the general public on Web sites maintained by these organizations.

Although GIS may not seem like much more than fancy map making, "its power of spatial analysis extends far beyond using a computer to do cartography," says John Hayes, a geographer at Salem State College in Salem, Massachusetts. According to the USGS, "traditional maps are abstractions of the real world, a sampling of important elements portrayed on a sheet of paper with symbols to represent physical objects." People who use maps must interpret these symbols. Topographic maps, for example, show the shape of land surfaces with contour lines. The actual shape of the land can be seen only in the mind's eye. "Graphic display techniques in GIS's make relationships among map elements visible, heightening one's ability to extract and analyze information," according to the USGS. In addition, "a GIS makes it possible to link, or in-

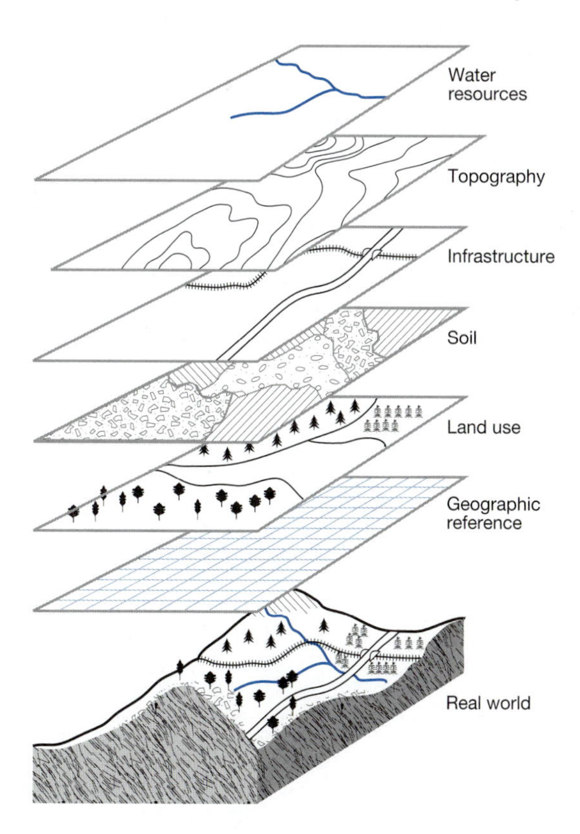

FIGURE 1.12 *Overlay analysis allows GIS users to combine a number of georeferenced data sources to produce a more complex and often more useful view of the world we live in. Each data set can be viewed as a thematic layer, each linked to a common georeferencing system.*

tegrate, information that is difficult to associate through any other means. Thus, a GIS can use combinations of mapped variables to build and analyze new variables." For example, "by using GIS technology and water company billing information, it is possible to simulate the discharge of (potentially harmful) materials in the septic systems in a neighborhood upstream from a wetland." Water bills indicate how much water is used at each address. This, in turn, is an approximate measure of the amount of waste that is discharged into each septic tank and leach field. Areas of heavy septic discharge can therefore be located using a GIS. These and other spatial combinations are often referred to as **overlay analyses,** or **overlay operations.** Figure 1.12 illustrates this concept.

The computer also makes it easier to integrate information from different sources. For example, property ownership might be indicated on city maps drawn with a different scale than county soils maps prepared by the U.S. Department of Agriculture. The software in a GIS can alter the scale of the property map, manipulating it so that it registers, or fits, with the county soil maps. More conventional maps can also be registered with maps generated from data acquired from satellites. The computer may perform other manipulations as well, for example, **projection conversions** (described next).

Projection refers to a complex mathematical technique to transfer information from the Earth's three-dimensional curved surface to a two-dimensional medium, such as paper or a computer screen. Different projections are used for different types of maps, depending on the needs of the cartographer (map maker) and the user. One of the more common projections is known as the Mercator projection. It is widely used in textbooks and wall maps of the world because it accurately represents the shapes of the continents. Unfortunately, it distorts their relative sizes (Figure 1.13).[1] In contrast, a planar projection gives an accurate view of size. "Since much of the information in a GIS comes from existing maps," says the USGS, "a GIS uses the processing power of the computer to transform digital information, gathered from sources with different projections to a common projection."

GIS also allows us to retrieve specific georeferenced information. "With a GIS, for example, you can point at a location, object, or area on the screen and retrieve recorded information about it from off-screen files. "Using scanned aerial photographs as a visual guide," notes one USGS report, "you can ask a GIS about the geology or hydrology of the area or even about how close a swamp is to the end of a street." GIS can help you locate activities, past and present, allowing you to make conclusions about potential impacts. Or, it can help you determine the suitability of future activities. If the analysis reveals the presence of an abandoned factory with contaminated soils, for instance, the land might be viewed as a potential future factory with a little clean up. It might not be a good site for housing development, however. In summary, GIS permits one to analyze conditions of adjacency (what is next to what), containment (what is enclosed by what), and proximity (how close something is to something else).

With appropriate software, GIS can simulate the route of movement of hazardous materials through surface or ground water. Retrieval functions and modeling of chemical movement allow one to draw conclusions more readily than might otherwise be possible from other data sets, as it puts a great deal of information at your fingertips.

GIS can also be used to map sensitivities of various ecosystems. A map that looks at topography, soil type, and other features in and around wetlands, for instance, could be used to assess the sensitivity of various wetlands to human activities. All of this can be printed on paper, providing a graphic view. This, in turn, can be used to predict possible outcomes of various development patterns and can profoundly influence land use decisions by state, local, even federal governmental agencies. In one case, wildlife biologists use transmitter collars and satellite receivers to track migration routes of polar bears over a two-year period. This data was then used to determine impacts of projected oil development on the bears.

GIS can also be used in emergency planning. For example, satellite images of earthquake zones are used to pinpoint the location of emergency services, roads, and a variety of hazards. This information can be used to determine evacuation routes. GIS may be used to determine areas that are most vulnerable.

[1]Mercator projection is not commonly used in GIS. The more popular maps are the Universal Transverse Mercator and State Plane Coordinate.

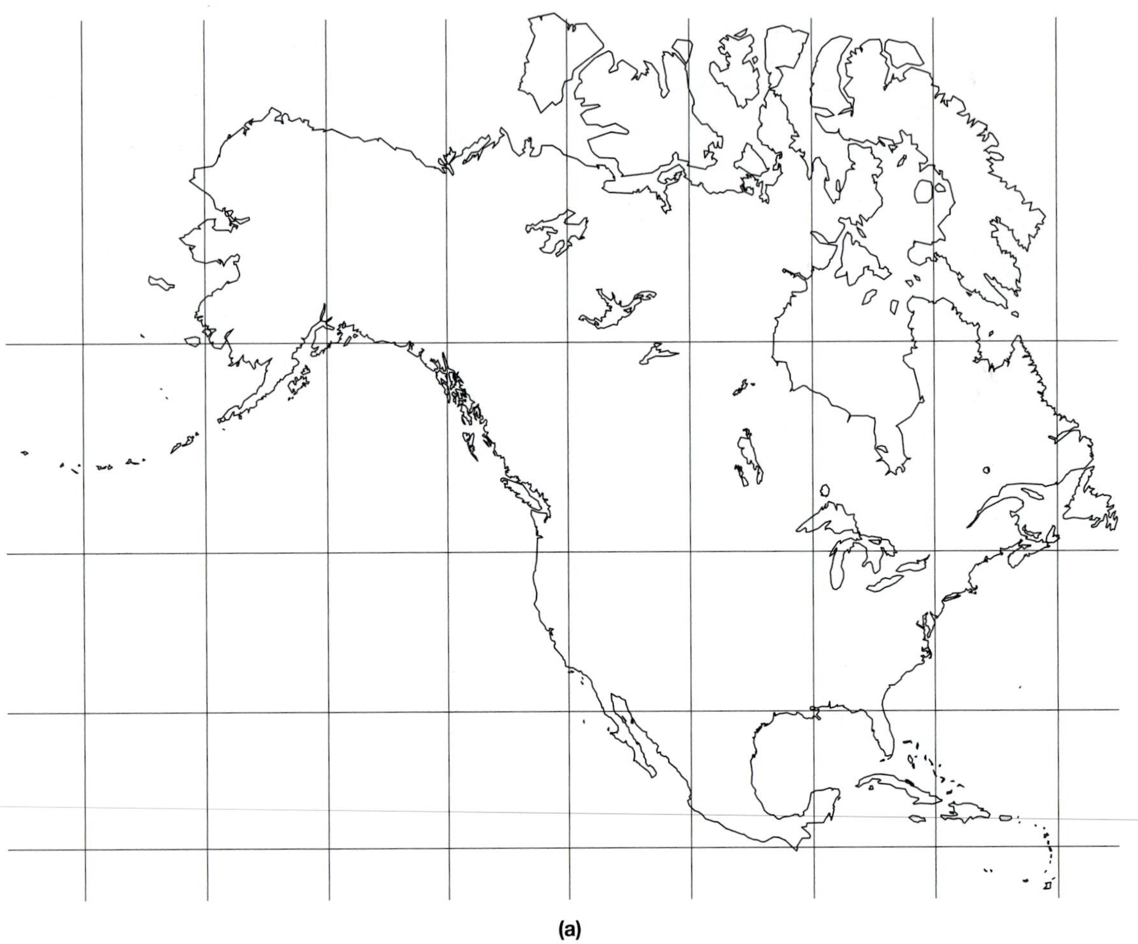

(a)
Map of North America, Mercator projection

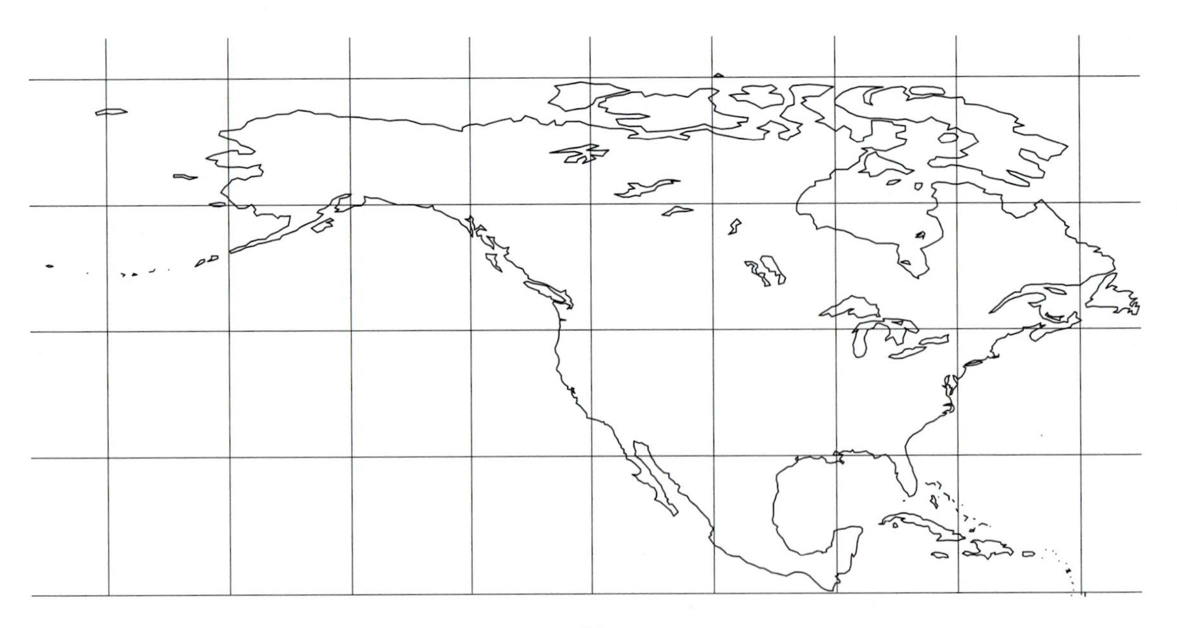

(b)
Map of North America, Planar projection

FIGURE 1.13 *A World Apart. Maps can be drawn in different ways to project the complex three-dimensional relationships on a two-dimensional surface (paper or a computer screen). (a) The mercator projection is commonly used. However, it distorts the size of land masses at the northern and southern latitudes. (b) A planar projection gives a more accurate view of the size of land masses.*

Now that we've briefly discussed GIS, it is time to turn our attention to a second tool that is being used to manage and protect natural resources, namely remote sensing. **Remote sensing** is a means of gathering information on conditions on Earth from remote (distant) locations. Aerial photographs are an example, as are digital images taken from satellite. Satellites, as you well know, are launched into outer space on rockets, then released into orbit. Circling around the Earth, satellites are typically powered with electricity generated from solar panels, known as PVs (photovoltaics), described in Chapter 22 of this book.

Aerial photos are usually created using film. To be useful in GIS, they must be digitized–scanned and converted to digital files. Satellites are equipped with various sensors, such as infrared scanners, that monitor the Earth. Information is digitized, then sent to Earth where it is collected and stored in massive computers. Digital data is processed into maps or other useful images. Today, GIS and remote sensing are used by various government agencies and laboratories.

With GIS, remote sensing, and a host of other tools, we head into the 21st century facing a challenge of enormous magnitude. As Professor Jerry Reynolds of the University of Central Arkansas likes to remind his Conservation and Land Use students, present generations face critical resource problems that have never been faced before. Together, we will have to develop new, perhaps more creative solutions to solve these problems in order to create a sustainable society—one that lives well without undermining the future.

Summary of Key Concepts

1. The global environmental crisis is the result of three major factors: (1) a large and rapidly growing human population, (2) depletion of natural resources, and (3) rising levels of pollution that threaten our climate and our planet's ecological health.
2. The global population will surge from more than 6 billion in the year 2001 to 8 billion by the year 2025. Rising population places increased demands on an already strained resource base.
3. Resource depletion occurs in both the more developed and less developed nations. The more developed nations, like the United States and Canada, whose citizens enjoy high standards of living compared to the rest of the world, use a much larger share of the world's resources than the less developed nations. The United States, for instance, has only 5 percent of the world's population but consumes 30 percent of the world's resources.
4. Resource depletion is a threat to economies, people, and the wealth of wild species that share this planet with us.
5. The acquisition and use of resources (for example, the combustion of fossil fuels) results in pollution that threatens ecosystems locally, regionally, even globally. These ecosystems are vital to the welfare of other species and are also vital to human well-being and the well-being of the global economy. The United States is degrading the environment more rapidly than any other nation on Earth.
6. People's views on solutions to the environmental and resource problems facing the world today vary widely. Some believe that technology can solve virtually any problem and are quite optimistic about the future. Others are less optimistic. They argue that because population, resource use and depletion, and pollution are increasing at an exponential rate, the chance that technological breakthroughs will solve them is rather remote.
7. Surely technology is a part of the solution, but it cannot be relied on in its entirety. Society must also make political and economic changes and people must take actions and learn to live more sustainably on the Earth.
8. Natural resources can be placed into two broad categories, renewable and nonrenewable. Renewable resources are represented by soil, rangelands, forests, fish, wildlife, air, and water. They also include energy resources such as wind, solar energy, hydropower, and geothermal energy. Nonrenewable resources include fossil fuels and metallic and nonmetallic minerals.
9. In a broad sense, conservation can be defined as "a state of harmony between man and the land."
10. The most significant conservation developments in the United States were made in this century in four waves: (1) under the leadership of Teddy Roosevelt (1901–1909), (2) during the presidency of Franklin D. Roosevelt in the 1930s, (3) under the impetus provided by the Nixon, Ford, and Carter administrations during the environmental decade of the 1970s, and (4) in the years following the United Nations Conference on Environment and Development.
11. Although the conservation movement progressed rather slowly during the nineteenth century, some advances were made. They included (a) the publication of *Man and Nature,* by George Perkins Marsh, (b) the establishment of 28 federal forest reserves, later designated as national forests, and (c) the founding of the Sierra Club by John Muir.
12. The 1960s and 1970s saw many changes, including the establishment of Earth Day and the passage of numerous environmental laws. So many environmental laws were passed by Congress from 1970 to 1980 that this period has been called the decade of the environment.
13. Development during the decade of the 1980s led to the realization that despite substantial progress in environmental protection, many gains were inadequate. Environmental conditions were not improving as was once hoped and in some cases were actually deteriorating. It became clear that most of the world's nations were on an unsustainable course.
14. This realization prompted a flurry of work on sustainable development, including the 1992 United Nations Conference on the Environment and Development held in Rio de Janeiro. Attended by nearly 180 nations, UNCED resulted in several treaties on key issues and an extensive action agenda for the world community.
15. Attitudes about the environment and the value of resources varies dramatically. Exploitation of resources, a human-centered approach, was commonplace early in this country's history and is commonplace elsewhere as well.

The opposing view, preservation, seeks to set aside nature for its own sake. It is a hands-off view. The utilitarian view is human- and nature-centered, promoting the wise use of resources to ensure lasting supplies. The ecological or sustainable approach is a more modern approach that seeks to manage resources and ourselves in ways to ensure that both survive well.

16. The ecological/sustainable approach to resource management requires a long-term systems view. Ecosystems are managed to protect multiple components. Protecting the stability and diversity of ecosystems are vital to this approach.
17. Creating a sustainable society will also require changes in human systems and may take many decades to complete, but positive steps are underway now to create a more enduring human presence. These involve changes in how we manage natural resources such as forests, farmfields, and fisheries, and also changes in how we design and manage our own systems such as transportation, water supply, energy, and the like.
18. Urgency may be required given the rapid growth of human population and the economy and the potential for combined problems to create a greater than anticipated effect, a phenomenon known as the nemesis effect.
19. Two tools that are proving to be helpful in solving local, regional, even global environmental problems and improving resource management at these levels are GIS (geographic information systems) and remote sensing. The computer-based systems allow researchers and others to acquire and depict vast amounts of information about the world we live in. This information is used to analyze spatial relationships and predict future outcomes of current and projected actions. GIS relies on data from many sources, but especially remote sensors aboard satellites.

Key Words and Phrases

Adaptive Management
Biological Principles of Sustainability
Buffer Zone
Carrying Capacity
Civilian Conservation Corps (CCC)
Conservation
Core Reserve
Decade of the Environment
Earth Summit
Ecological Approach to Conservation
Ecology
Ecosystem
Ecosystem Management
Ecosystems Approach
Environmental Protection Agency (EPA)
Exploitation Approach to Conservation
Exponential Growth
Geographic Information System (GIS)
Georeference
Leopold, Aldo
Limits to Growth, The
Man and Nature
Marsh, George Perkins
Muir, John
Nemesis Effect
Nonrenewable Resource
Overlay Analysis
Population Bomb, The
Population Growth
Prairie States Forestry Project
Preservation Approach to Conservation
Projection
Projection Analysis
Projection Conversion
Remote Sensing
Renewable Resource
Resource Consumption
Silent Spring
Soil Conservation Service (SCS)
Spendthrift Society
Sustainable Development
Sustainable Society
Tennessee Valley Authority (TVA)
"Tragedy of the Commons, The"
United Nations Conference on Environment and Development
Utilitarian Approach to Conservation
Waves of the Conservation Movement

Critical Thinking and Discussion Questions

1. Define conservation.
2. Discuss the major theme of the *Limits to Growth* study.
3. Many people think that humankind is in the midst of an environmental crisis. Why do they believe this? What evidence is there to support this assertion? Do you agree with this assertion?
4. In what ways are population growth, resource demand and depletion, and pollution linked? How can they be delinked, or can they?
5. Debate the statement: "Even if there were no environmental crisis, there would be good reason to manage our resources more sustainably."
6. Describe four major advances in the conservation movement that occurred during the administration of Franklin D. Roosevelt.
7. Name five major environmental acts that were passed during the decade of the environment (1970–1980).
8. Does the passage of an environmental law guarantee that the goals of the law will be met? Discuss.
9. Discuss the basic difference between renewable and nonrenewable resources. Name four resources of each type.
10. Name and describe the four basic approaches to conservation in the United States.
11. What is sustainable development? Describe the principles of sustainability and how they might be applied to various human systems such as agriculture.
12. Define the Nemesis Effect. Can you think of any examples?

Suggested Readings

Bowermaster, J. 1993. Is Carol Browner in over her head? *Audubon* 95(5): 58–63. Provides insights into the daunting responsibilities of the head of the federal Environmental Protection Agency.

Bright, C. 2000. Anticipating Environmental 'Surprise,' In *State of the World 2000,* eds. Linda Starke, W.W., pp. 22–38. Norton: New York, 2000. Extremely interesting and thorough discussion of the way environmental disturbances and alterations combine to produce unanticipated or more rapid deterioration than one might expect.

Brown, L. R., Renner, M. and Halweil, B. 1999. *Vital Signs 1999. The Environmental Trends that are Shaping Our Future.* W.W. Norton: New York. This annual publication tracks numerous important environmental indicators and is very valuable reading for those interested in seeing where we are headed.

Carson, R. 1962. *Silent Spring.* Boston: Houghton Mifflin. A classic. Alerted the public to the harmful consequences of the use of pesticides on humans and wildlife.

Chiras, D. D. 1990. *Beyond the Fray: Reshaping the American Environmental Response.* Boulder, Colo.: Johnson Books. A critique of the present-day environmental movement with suggestions for strengthening it.

Chiras, D. D. 1992. *Lessons From Nature: Learning to Live Sustainably on the Earth.* Washington, D.C.: Island Press. Describes a set of ecological principles of sustainability and their application.

Chiras, D. D. 1993. Toward a Sustainable Public Policy. In *Environmental Carcinogenesis and Ecotoxicology Reviews,* C11(1): 73–114.

Chiras, D. D. 1995. Principles of Sustainable Development: A New Paradigm for the Twenty-First Century. In *Environmental Carcinogenesis and Ecotoxicology Reviews,* C13(2): 143–178.

Cortner, H. J., and Moote, M. A. 1999. *The Politics of Ecosystem Management.* Washington, D.C.: Island Press. Examines the political challenges facing ecosystem management as it moves from theory to practice.

Ehrlich, P. R. 1968. *The Population Bomb.* New York: Ballantine Books. A classic. Alerted the public to the urgent need to control the population explosion so that massive starvation and environmental abuse are prevented.

Grumbine, R. E. 1994. What Is Ecosystem Management? *Conservation Biology* 8(1): 27–38. Overview of the history of ecosystem management and its underlying principles.

Hayes, Denis. 1999. *The Official Earth Day Guide to Repair the Planet.* Washington, D.C.: Island Press, United Nations Environment Programme. *Global Environment Outlook 2000.* New York(?): Earthscan. Region-by-region analysis of the state of the world's environment with recommendations for change.

Willers, B. 1999. *Unmanaged Landscapes: Voices for Untamed Nature.* Washington, D.C.: Island Press. An interesting book well worth reading. It examines the effects and implications of resource management and the need to keep some landscapes free from human interference.

Web Explorations

Online resources for this chapter are on the World Wide Web at: **http://www.prenhall.com/chiras** *(click on the Table of Contents link and then select Chapter 1).*

Chapter 2

Economics, Ethics, and Critical Thinking:

Tools for Creating a Sustainable Future

Resource managers often complain that they are amazed how often management decisions boil down to politics. What they found is that much of the science they learned in college to manage natural resources is meaningless against powerful political forces. In other words, although scientific findings may suggest a particular course of action, political considerations, which are typically based on economic or ethical concerns, prevail (Ethics in Resource Conservation Box 2.1). In practice, this means that decisions to cut old growth forests, to drill for oil in environmentally sensitive regions, or to harvest fish above a certain level are often made on economic grounds. Such an approach can be ecologically suicidal, not only for the ecosystem in question but, in the long run, for people themselves. Why?

The Earth's ecosystems and its web of life are the infra-infrastructure of our society. They provide us with a wealth of free services, and they are the foundation of our economy. History has shown time and again that societies that destroy their topsoil or forests end up destroying themselves. It may be hard to believe that a similar fate could befall modern society. Most of us think we're immune to such problems. Part of the reason for this is the globalization of the human economy. Sure, forests are being depleted worldwide, but there's not worry. We'll get the wood we need from some other location. The same thinking characterizes the view of many as they ponder other resources such as oil and oceanic fish. Eventually, though, the planet will be in ruins and there will be no place to turn. It would be far better to start using and managing resources more judiciously.

Unfortunately, the debate over resource management is often framed in terms of immediate survival. If loggers cannot cut the remaining old growth forests, their families will suffer. Local economies will collapse. People will be displaced—all to save a few owls or a few trees, they say. Even though such concerns are very real and very important, critics point out that long-term considerations must also be factored into decision making. Ecosystems management and sustainable development, introduced in the last chapter, require a long-term view. Balancing short-term and long-term needs creates enormous frustration and controversy.

ETHICS IN RESOURCE CONSERVATION 2.1

ETHICS VERSUS ECONOMICS?

You'll hear people say this many times in your life: One of the root causes of the current problems facing the environment is economics. The general line of reasoning is that the pursuit of economic wealth often results in rampant resource consumption, which is accompanied by massive environmental destruction. In addition, economic decision making often fails to take environmental considerations into account and thus often results in production systems that run roughshod over the Earth. Our singular dedication to economic growth and our failure to consider environmental concerns are, say many critics, wreaking havoc on the planet.

An astute historian, however, pointed out that it was not fair to blame the environmental crisis purely on economics. Ethics, he asserted, underpins all economic decisions.

Which is it, economics or ethics?

The answer is, both. Economics has a lot to do with the way the world operates, as this chapter points out. But our economic system is based on ethics. Put another way, economic decisions and the functioning of our economy are driven, in part, by ethics. In most instances, however, the underlying ethical assumptions are often forgotten or ignored, and all we see in operation are economic forces. What are some of the underlying ethical driving forces that influence economic decisions?

One of the predominant ethical positions that drives economic growth is the notion of unlimited abundance, a central tenet of the frontier ethics described in this chapter. When people view the Earth as an unlimited supply of resources, economics is given a kind of moral free rein. This, in turn, often leads to environmentally unsound and environmentally costly actions—overexploitation of a wide assortment of natural resources.

Another ethical underpinning of modern economics is the notion that humans are apart from nature and superior to other life forms. By viewing humans and nature as separate entities, we act irresponsibly, destroying ecosystems that are vital to our survival.

Another component of frontier ethics is that the key to success is domination and control. Huge levees built along the Mississippi River, for instance, point to a persistent desire to tame the river. Over the years, people have tamed other rivers with massive dams and levees. We have also subdued vast acreage of wilderness. But the domination of nature has led to serious ecological backlashes. One of the most disheartening was the channelization of the Kissimmee River in Florida.

Once a meandering river with ample wetlands that supported huge flocks of waterfowl and numerous fish species, the Kissimmee River was channelized by the Army Corps of Engineers in an effort to control flooding. The river was straightened and dammed so that its overall length was cut in half.

As the Corps put the finishing touches on its multi-million-dollar project, the impact began to be evident. Waterfowl disappeared as wetlands dried up. Fish catches plummeted. Pollution in Lake Ochechobee, into which the river drained, began to increase.

Ecologists realized that the Corps had made a colossal mistake. Human actions had deadened a once-teeming biological system that supported countless species of wildlife, controlled flooding, and absorbed pollutants rather well on its own.

After facing the evidence of this failure, officials have begun to restore the river, but at an enormous cost. Had humans left the river alone, millions of dollars would have been saved and a valuable habitat for countless species would have been protected.

Individual ethics also figures in individual economic decision making. For example, the primacy of the individual, a tenet firmly rooted in American culture, drives us to make many ecologically foolish decisions. Viewing our personal pleasure as the paramount concern in our lives, for instance, leads us to acquire wealth and material goods at levels that cannot be sustained. Rare is the individual who says, although I can afford a bigger house, I won't buy one, because the Earth cannot sustain such a lavish lifestyle.

In brief, individualism fuels consumption, and consumption is at the heart of our economic system as well as the environmental crisis. A more inclusive philosophy, in which personal decisions are made on the basis of environmental realities and the welfare of future generations, might temper consumer desires and could contribute to a more sustainable lifestyle.

Economic competition is yet another cultural norm that, while seemingly good for the economy, spells disaster for the environment. In U.S. society, competition is considered an unquestioned good. But cooperative ventures could save our society and business enormous resources. Consider an example.

In Colorado, hundreds of individual water districts have been formed to provide water to cities, towns, and farms. Each district has its own sources, and many have their own dams in the mountains. It is possible that had the communities joined forces, rather than competed for water, they could have provided the same amount of water with far fewer dams, and at a lower cost. A cooperative ethos might have promoted a more sustainable system of water supply.

These are a few of the negative ethical underpinnings of economics. However, positive ethical views could be instilled and could lead to environmentally sustainable economic behavior—for example, if humans are viewed as part of nature, environmental protection is recognized as a means of protecting ourselves and future generations.

We encourage you to take some time to consider your ethics, especially with regard to economics. What are the underlying rules that drive your decisions? Do they promote a sustainable future or detract from it?

This chapter examines economics, one of the central issues in debates over resource management. It presents important economic principles that affect both resource management and pollution control. It also describes a new and emerging economics that seeks solutions that make sense from a financial standpoint as well as from environmental and cultural perspectives. In other words, such an economics seeks to satisfy three bottom lines—people, nature, and economy. We refer to this new economics as **sustainable economics.** One point that will become clear is that sustainable economics takes a long-term view, with an eye on what's good for people *and* the environment.

This chapter also examines ethics, describing just what ethics is and why it is important. It examines different ethical systems and outlines an ecological ethic or, more appropriately, a sustainable ethic. Finally, in this chapter we present overviews of the scientific method and critical thinking. This discussion provides important background information that will help you analyze complex environmental and resource management issues.

2.1 Understanding Economics

Although economics is a complex science, a few basic principles provide the foundation of economic literacy. First and foremost, **economics** is a science that seeks to understand and explain the production, distribution, and consumption of goods and services. Historically, economics has been concerned with two major elements: inputs and outputs. **Inputs** include the commodities companies need to produce goods and services, including raw materials, labor, and energy. **Outputs** include goods and services. Although this may sound simple, the human economy is massive and complex. It includes a wide range of goods and services traded locally, regionally, even globally.

Economists recognize two basic types of economic systems: command and market. **Command economies,** like that of Cuba, China, and the former Soviet Union, rely on central control of the means of production. Not only do governments own and operate the factories, they make all the decisions regarding production. They determine *what* is produced and *how much* is made at any particular time.

Market economies, on the other hand, rely in large part on monetary signals to control the production and distribution of goods and services. Companies produce an assortment of goods and services, but *what* they produce and the *quantity* they produce each year are determined by demand from consumers and the potential for profit. Although there may be a high demand for a particular product or service, if a company cannot make a profit it will not waste its time.

Command and market economies are polar opposites. But in reality, most economies have elements of both. Market economies contain some elements of command economies and vice versa. The market economy of the United States, for instance, is highly influenced by public policy. Some laws, for instance, provide subsidies for the production of goods. Mining companies, for instance, are given special tax breaks as they deplete their mineral reserves. The economic incentives are intended to provide money for further mineral exploration and development. No matter whether you agree or disagree with the policy, it does represent governmental interference with the free market economy that alters the cost of producing goods. As another example, privately owned oil vessels carrying oil, owned by private corporations, that will be refined in the United States, are protected by the U.S. military as they travel the waters of the Persian Gulf, at considerable expense—around $50 billion per year. This, too, represents a means of government intrusion in the free market economy. If oil companies paid the cost of protecting their oil themselves, the cost of fuel oil, diesel, jet fuel, gasoline, and hundreds of other products would be much higher. Instead, taxpayers pay the bill. To be fair, not all policies promote such activities. There are some environmentally sound policies that provide subsidies that help foster a sustainable future. In Colorado, companies that install recycling equipment, for instance, are given a tax credit (a deduction from their annual tax bill) as a government incentive.

Economists also point out that there are **subsistence economies** or **hunter–gatherer economies.** For most of human history the human economy did not deal in dollars and did not require jet airplanes, massive ships, and trucks to transport goods and services. People living in small groups acquired or made what they needed to survive from the natural environment. They generally met their needs without bankrupting the Earth in the process or threatening their own future, although there are some notable exceptions to this rule.

Because populations were small and technological development was virtually nonexistent, subsistence economies were rather benign, environmentally speaking. But over time, trade picked up. Money began to be used, and the market economies of the world emerged. In remote corners of the world subsistence economies still exist, although many are now endangered by market economies.

One of the key principles of market economies is the law of supply and demand. It is a relatively simple idea that explains a great deal of economic activity at all levels. The **law of supply and demand** basically says that the price of a good or service is determined by the interaction of supply and demand. If demand exceeds supply, the price increases. If supply exceeds demand, the price falls. Therefore, the price is said to be **elastic,** which means that it can change with fluctuations in supply or demand. Examples of the law of supply and demand abound. A rare car, for instance, will fetch a higher price than a mass-produced vehicle simply because, in the case of the former, supply is limited.

Although economics is far more complicated than this discussion may suggest, this is enough information to start with. More basic economics will be introduced as we examine flaws in the economy as seen from an ecological perspective.

What's Wrong with the Economic System: An Ecological View

The economy is large, complex, and a major influence in our lives and in the future of the environment. It provides us with a wealth of goods and services from TV sets, to computers, to coffee, to toothpaste, to bed sheets. It makes our lives better, providing opportunities to work to earn money to pay for our homes, automobiles, food, clothing, and recreation. But economic activity does not come without a cost. In this section, we're going to be taking a fairly critical look at the economy with one purpose in mind: to consider ways it can be altered to create an enduring human presence.

In his book *Ecological Literacy,* biologist David Orr notes that command economies have failed (as have the political systems that supported them) in large part because they produced too little. Shortages created dissatisfaction, which contributed to the collapse of the communist system and its pervasive belief in command economies.

Today, market economies have become the main component of the global economy. Their persistence, though, does not mean they are perfect, or environmentally benign. Market economies and capitalism in general are flawed, according to various critics, because they are predicated on unlimited growth. Market economies have produced extraordinary wealth and unlimited expectations for material possessions. In a world of unlimited resources and pollution-free production, these outcomes might be sustainable, but in reality, our resources are not unlimited and few systems of production are pollution free. Economies slavishly dedicated to constantly increasing material production and consumption are bound to fail.

Many others have criticized market economies for other reasons. A brief examination of a few of these criticisms is helpful in forging a sustainable, ecologically sensitive economic system.

One of the principal complaints of ecologists is that business economists have a rather short time horizon for planning. For a business, a long-term plan might list goals and objectives to be met over a one-to-five-year period. Decisions are made within this time frame.

Ecologists, conservationists, and environmentalists, on the other hand, take a long-term view in decision making. To them, 5 years is a drop in the bucket of ecological time. For any decision to make sense from an ecological standpoint, it must make sense not just 5 years from now but 100, perhaps 200 years from now. They argue that short-term economic decisions often turn out to be ecologically disastrous. For example, it may make monetary sense for a rancher to overgraze his or her land to increase sales, but the ecological costs in the long term, not the least of which is the permanent destruction of once productive rangeland, are daunting. Resource managers find themselves in this bind over and over again.

The short-sightedness of the economy is also manifested in its almost single-minded dependence on the law of supply and demand to determine prices. Supply and demand economics reflects the immediate picture, that is, the immediate supply of a resource in relation to its demand. Very little consideration is given to long-term supplies. To understand the importance of this point, consider an example.

Oil prices are determined by supply and demand. Although oil supplies appeared to be adequate during the 1990s, they are ultimately finite. Peak of oil production could occur between 2005 and 2010. But the economy is blind to that reality. As a result of the interaction of immediate supply and demand, however, oil was priced low throughout the 1990s. This in turn encouraged consumption and waste, with many environmental consequences.

Rangeland, forests, fisheries, and other natural resources have been driven to the brink of ecological ruin the world over in large part because they were once abundant and their services and products were underpriced. Now that many resources are on the edge of ruin, we must spend billions to restore these systems. In fact, far more money will be needed to restore them than would have been required to manage them properly in the first place. So, supply and demand economics has not only resulted in a foolish and wasteful use of resources, with pollution and environmental destruction the undesirable outcome, it also ends up costing us more than is necessary in the long run! More careful management, sustainable management, might have yielded lower returns, but the lower returns would have been offset by a much longer productive period, perhaps even an infinite period of production.

Another criticism of capitalism as it is currently practiced is its narrow notions of the input and outputs of an economy. Critics point out that at least three essential factors are missing from the input/output analyses of mainstream economics: environmental impacts, social and cultural impacts, and pollution. In other words, when businesses calculate the costs of manufacturing their products, they typically fail to take into account environmental and social costs. Air pollution from a factory that makes DVD players, skateboards, BMX bikes, CDs, PlayStations, cars, and computers, for instance, is not factored into the cost of doing business and thus the cost of the product to the consumer (Figure 2.1). Instead, air pollution costs, including those incurred by damage to human health, buildings, and ecosystems, are ignored and essentially passed on to society at large. Such costs are generally referred to as **economic externalities.** Automobiles, for example, create huge economic externalities few of us are aware of. Cars pump out carbon monoxide that affects the health of the aged and infirm, causing an increase in chest pain in those with heart disease. They emit millions of tons of carbon dioxide, a gas that may be causing the Earth's temperature to increase, which, in turn, may be causing the sea level to rise, which floods islands and coastal regions. Global warming may also be upsetting weather patterns and creating much more violent storms, which cause billions of dollars worth of damage and cause the death of many people each year. Automobiles also produce sulfur dioxide and nitrogen oxides, two pollutants that are converted into acids in the atmosphere. These rain down on the land, causing widespread damage to lakes and the species that depend on them in certain areas of the world. All of these impacts have a cost, but the economic externalities are not paid by automobile users at all. Substituting busses and trains for cars, wherever practical, can greatly reduce the economic externalities associated with the transportation of people in automobiles because they move people more efficiently (with less fuel consumed per passenger mile).

Although efforts are being made to eliminate economic externalities in cars and a host of other sources like power plants through pollution controls and other measures, many outside costs are still unaccounted for and left unpaid.

Yet another shortcoming of traditional capitalism is its single-minded dedication to measuring success by the **gross national**

FIGURE 2.1 *The price of progress? Waste from an abandoned aluminum factory abutting nearby backyards of homes causes serious health problems in residents.*

product, (GNP). The GNP is the sum total of all goods and services produced by a nation's economy, including all government expenditure and business activities occurring in other countries. The GNP includes such varied items as washing machines, solar panels, textbooks, and ice cream cones. It also includes the cost of oil spill cleanups and the billions of dollars spent on cleaning up hazardous waste. It is therefore an aggregate measure of all goods and services. As a side note, the **gross domestic product (GDP),** is a subset of the GNP, as it includes all economic activity within the borders of a nation.

The problem with the GNP is that it fails to distinguish between good and bad economic activities. In other words, as the GNP is structured, a dollar spent on a piano lesson for a future Mozart is just as important as a dollar spent on a pack of cigarettes.

Because the GNP is an indiscriminate measure, it fails to track a nation's true progress. That is, it fails to tell us exactly what we are getting from our economy. To illustrate this point, let us compare two hypothetical nations. The first nation has a robust economy with a high output of pollution. Each year 50,000 people die from urban air pollution and thousands of others are poisoned by pesticides, hazardous wastes that have leaked into the water supply, and workplace chemicals. Vast sums of money are being spent on producing goods and services. Huge amounts are also being spent on emergency medical help, hospital bills, coffins, and cemetery plots. The net effect is that this economy boasts an exceptional GNP.

The second nation has a robust economy as well, but it is on a sustainable path. Its industries produce products without pollution. They use clean, renewable energy. As a result, the nation's air is clean and its water is pure. Pests are controlled on farms without pesticides. Few people die from industrial accidents. Although the per capita GNP, the economic output per person, may be lower than the first nation's, the second nation is far better off. The economy is serving people, not poisoning and killing them.

In summary, distinguishing the good from the bad helps one determine just how sustainable an economy is. Because the GNP fails to make this distinction, it is a false measure of success.

Studies that attempt to sort out good economics from bad show that this distinction is more than academic. Economist Herman Daly, for instance, developed a comprehensive **Index of Sustainable Economic Welfare (ISEW)** for the United States for the 1970s and 1980s. Although the procedure for calculating the ISEW is quite complex, it basically entails adding up all of the beneficial economic gains in a year and subtracting the negative economic output. The ISEW is therefore an approximation of nation's true economic welfare. This can be compared to the GNP to determine whether growth in GNP is actually improving our lives and the condition of the planet. When Daly compared the ISEW to the GNP of the United States, he made a startling discovery. In the 1970s, the GNP grew annually by 2 percent per capita. In the 1980s, the per capita GNP grew by 1.8 percent per year. The ISEW during the 1970s, however, climbed only 0.7 percent per capita (Figure 2.2A), indicating that only about one third of the annual growth in GNP improved the lives of Americans. Largely because of environmental degradation, the ISEW declined by 0.8 percent per year in the 1980s. During the early 1990s, the latest period for which data is available, the trend continued. What does this mean?

Frankly, although our economic output is increasing, and economists and politicians would have us believe that we are better off, our lot in life is actually declining. In other words, despite a growing economy, which most people think is good news, we are actually worse off.

The United States is not the only country to show a discrepancy between the GNP and ISEW. But it is one of the few countries for which the ISEW continues to plummet. In Germany, the ISEW is actually increasing faster than the GNP (Figure 2.2B).

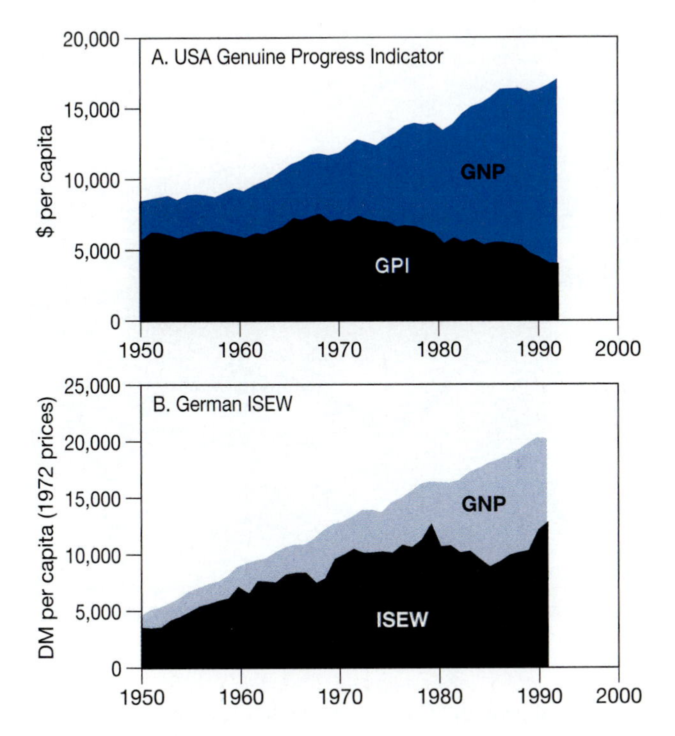

FIGURE 2.2 *Economic growth is good for us—or is it? (A) As this graph shows, while the GNP continues to climb, our overall welfare, as measured by the Index of Sustainable Economic Welfare, declines. (B) In Germany, the ISEW is actually increasing due to many progressive initiatives designed to reduce the impact of the German people on the environment.*

The GNP also fails to take into account natural capital. **Natural capital** is a nation's ecological wealth. It includes topsoil, forests, grasslands, open spaces, farmlands, fisheries, and wild species—resources of great aesthetic and economic importance. In many ways, natural capital is like money in the bank. Nations draw on this capital to fuel their industries and to satisfy human needs for goods and services. But most nations are dipping into the bank account, taking not the interest but the principal itself. Thus, a nation can be rapidly depleting its natural capital while posting a high GNP, and there is no way of knowing it. Herman Daly once wrote that "Most nations are treating the Earth as if it were a corporation in liquidation." In other words, most nations are selling off their natural assets such as their forests and farmland to make a quick buck. Although their GNPs may be impressive, their futures are not bright. Sooner or later, they will have no productive capacity left.

Myths About Economy and Environment

While we are on the subject of the economy and its faults, it is important to look at the ideas that support environmentally and socially unsustainable economic activities—that is, myths we have about our economy and the environment.

Economics Versus Environment: Rethinking an Old Debate Several economic myths pervade our society and prevent progress toward a sustainable future. One of those myths is that *environmental protection is bad for the economy.* Many examples show that environmental protection as it has been practiced can be quite costly. But failure to institute environmental safeguards can also result in extremely costly cleanups. An ounce of prevention is worth many pounds of cure.

Experience also shows that environmental protection and economic health need not be antagonistic forces. Careful attention to the design and operation of factories , for example, can prevent pollution and make costly pollution controls and cleanups unnecessary, a step that saves companies millions of dollars a year in the United States and many other countries. Using resources more efficiently, yet another sustainable strategy, has many benefits. In fact, two of the healthiest economies among the industrialized nations, those of Germany and Japan, are also the most energy-efficient ones.

All of this is to say, it is important to rethink the old ideas. Critical thinking rules warn us to avoid oversimplification, to look at the big picture, and to question conclusions. This is a case in point.

The Economic Pursuit of Quality Another common myth in our society is that *environmental protection is about quality of life and economics is about survival.* That is to say, environmental protection is a less important endeavor than manufacturing, which provides jobs and income. In even simpler language, environment is a luxury and economics is a necessity. Speaking about wildlife preservation, one local government official in Colorado noted, "Unfortunately, preservation of wildlife falls at the bottom of the list of priorities when it comes to staying ahead of the competition in creating a successful development effort. Wildlife and open space protection efforts are just too costly since they lack revenue-generating components that justify protecting them."

In his book, *The Economic Pursuit of Quality,* Thomas Michael Power, chairman of the Economics Department at the University of Montana, shows that virtually all economics in the industrial nations is about the pursuit of quality. What he means is that most of the money people spend on food, clothing, shelter, and other areas goes to improving the quality of our lives. Power shows that if we were only concerned about survival, our economic outlay would be much lower, perhaps as much as 70 to 80 percent lower than it is today. In other words, we can live (survive) on much less. Instead, we lavish on ourselves fashionable clothes, elaborate homes, and foods far beyond our survival needs. The conclusion of this line of reasoning is simple: For most of us, modern economics goes far beyond survival needs. It is about attaining aesthetic qualities.

Power also dispels the notion that environmental quality is far from frivolous from an economic standpoint. He points out that beautiful vistas, open space, clean air, and other environmental values have economic value. Homeowners, for instance, will spend many thousands of dollars to buy beachfront property or a home along a lake, or to live in a low-pollution setting. When they do, they are paying for quality (Figure 2.3). Many resort towns recognize this connection as well. People

FIGURE 2.3 *This scene from Atlantic City, New Jersey, shows the extent of human encroachment on natural biological systems.*

travel to visit nature resorts, not just to ski or kayak but to enjoy the untrammeled spaces and scenic vistas.

In summary, the environment and its protection is not a frivolous matter beyond the realm of economics. It is another quality we pursue in our lives and pay for.

The Myth of Economic Growth Yet another dominant myth in our society is that *economic growth is good, indeed essential.* Without a doubt, economic growth has made our lives better than those of our predecessors. But economic growth has not come without a cost, and some of its most significant costs are the excessive consumption and depletion of natural resources, the loss of countless species, the pollution of lakes and rivers, and the fouling of the air. From a social perspective, economic growth often fails to deliver on its promises of widespread prosperity. Economic growth, for instance, is touted as a means of eliminating poverty and raising the welfare of the middle class. It is promoted as a means of solving unemployment. But its success in these areas is questionable in recent years. Statistics show that economic growth often disproportionately benefits the wealthy and fails to deliver on other promises. In his book, *The Economic Pursuit of Quality,* Power cites studies of economic growth (job growth) and per capita income in many western states and finds, almost uniformly, that despite the creation of many new jobs, per capita income increased very little, if at all. From 1950 to 1977, for instance, the number of new jobs in Arizona increased 5.5 times the U.S. average, but the per capita income increased only 2 percent above the national average. In California, job growth was 2.3 times greater than the national average, but per capita income fell 13 percent compared to the national average.

If growth cannot deliver the promises its proponents make and singularly undermines our future, we must ask the question: Can we create an economic system that is not predicated on continual growth? Can we create an economy that exists in harmony with the environment and yet supplies us with the goods and services we need to survive and prosper? Before we embark on an explanation of an alternative, remember the "we" referred to in this paragraph includes well over 6 billion people.

2.2 Creating a Sustainable Economy

The preceding analysis of the shortcomings of capitalism is not meant to be an indictment of economists or those who participate in the economy, which is all of us. Rather, it is intended to set the stage for a discussion of ways to revamp the economy to create an environmentally sensitive system of commerce, or a sustainable economic system.

In general, a **sustainable economy** is one that produces goods and services in a manner that does not foreclose on future generations. A sustainable economy is one that is based in part on the principles introduced in the last chapter: frugality, efficient resource use, recycling, and use of renewable resources, especially energy. Creating a sustainable economy also hinges on our becoming better able to manage our own population growth and better able to balance our demands for goods and services and living space with our needs for clean air, water, food, and the like. These principles, when seriously applied to all of the basic systems that compose the economy—from industrial production to transportation to housing to energy and waste management—can dramatically reshape how we live on planet Earth.

Much of this book outlines the principles and practices of sustainable natural resource management. Other discussions suggest ways to alter human systems, such as industry and transportation, to make them more sustainable.

To set the stage for this discussion, we will begin with a general discussion of the economics of resource management and pollution control. These sections will help you better understand what a sustainable economy entails, as well as some of the barriers to its creation. We begin with a look at the economics of natural resource management.

Economics of Resource Management: Going Beyond Business as Usual

Economics plays a crucial role in the management of natural resources—for example, how a farmer manages his farmland or how an owner of a logging company manages forests. The future resource manager reading this book will become a practicing economist. Three guiding principles will affect your decisions: time preference, opportunity costs, and discount rates.

Time Preference Time preference is a fairly straightforward concept. What **time preference** refers to is one's preference for economic returns on an investment, expressed in time. An immediate time preference means that one prefers immediate returns.

Time preference is influenced by several factors, with need being one of the most pressing. A less developed nation that needs money to pay off its debt to a more developed nation, for instance, is likely to deplete its resources to do so. Little consideration is given to long-term production. Paying off the debt is paramount. Similarly, a farmer with pressing financial needs (perhaps a child in college) may ignore expenditures in soil erosion control and adopt strategies that result in less capital outlay and higher short-term returns. A logging company that has to pay back junk bonds that were offered to investors to finance a corporate buyout may recklessly cut its forests to create a high rate of return.

Unfortunately, as the previous examples show, immediate time preference often leads to environmentally, and sometimes economically, suicidal practices. To create a sustainable economy, economic incentives may be needed to compensate for lost income resulting from a shift in time preference—that is, a concern for the ecological integrity of forests and other natural resources that is such a key element of ecosystem management and sustainable development. For example, tax credits for farmers willing to shift to sustainable agricultural practices may alter the economic picture sufficiently so that they can afford to invest in environmentally sensible strategies.

The previous discussion, however, is not meant to imply that all ecologically sensible and sustainable practices require a long payback period or exceptionally high initial investments. On the contrary, many practices require only a modest investment and offer rapid and sometimes substantial economic return. As an example, environmentally sound pest control techniques can offer rapid payback. Thus, a sustainable approach may be economically more beneficial than the traditional approach. So, before one assumes that doing the right thing will cost too much, it is wise to look at the economics. It might be the other way around.

Opportunity Costs Another related factor is something economists call **opportunity cost,** which is the cost of lost opportunities resulting from certain policies and actions. Suppose, for instance, you are a farmer and you realized a $10,000 profit from last year's crops. You could invest that $10,000 in a mutual fund and earn 15 percent per year, perhaps more. You could also invest it in soil conservation measures that may increase your yield by 10 percent. A strictly economic view would suggest that you would be wiser to invest your money in mutual funds. Put another way, if you chose to invest in your farm, you would be losing an opportunity to reap slightly higher economic benefits.

We like to point out that opportunity costs should be looked at in other ways, too. Consider what happens when a resource is depleted—in other words, consider the cost of lost opportunities by embarking on the exploitive resource management strategy outlined in Chapter 1. The large-scale cutting of tropical rain forests, for instance, may make sense from an immediate economic standpoint, but this activity also creates a loss of opportunity for you and future generations. Why? Tropical rain forests have and will continue to provide many life-saving medicines which have been extracted from plants that grow in jungles. New discoveries could be worth billions of dollars and could save thousands of lives worth countless billions. Tourism also lost because of clear-cutting (removing all of the trees) could have been worth millions of dollars in revenue each year as well.

Tremendous future opportunities are lost when an ocean fishery is depleted, or when a farm field is destroyed by erosion caused by careless farming practices that reflected an immediate time preference and a narrow view of opportunity costs.

In summary, the term opportunity costs typically pertains to potential economic losses incurred when one pursues a particular course of action, usually a long-term action rather than an action that yields higher profit in the short term. Short-term exploitation, however, reduces long-term opportunities. Short-term exploitive action that may yield higher immediate profits rob us of the potential for long-term benefits, both ecological and economic. In other words, in our nearly singular focus on immediate financial opportunity costs, we may be bankrupting the Earth and our society. We are, in effect, assigning no value to resources that could sustain human life and other species indefinitely. Nor are we assigning a value to sustained economic benefits, so essential in our quest to build a sustainable economic system.

An understanding of the true meaning of opportunity costs could help business economists shift their time frame. In addition, it could help them venture beyond the narrow realm of economics to consider factors that are not traditionally seen as economic—for example, the aesthetic pleasure we derive from wildlands.

Discount Rates Economists use a tool called the discount rate to quantify returns on various economic strategies. Basically, the **discount rate** is the present economic value of different economic strategies—for instance, cutting down a tropical rain forest, or harvesting wood from it sustainably over 20 years.

Unfortunately, under this complicated process immediate income is the most important determinant. The World Bank, an international lending agency financed by many nations, currently invests about 20 billion dollars a year in the less developed countries. It uses a discount rate of 10 percent as its measure of economic rationality of the projects it funds. Thus, if a project yields 10 percent on investment, it is deemed acceptable. Lower returns from strategies that might be more ecologically sensible are viewed as unacceptable.

Using the notion of discount rate, it is more rational for a business person to liquidate a resource than to harvest at a rate that brings a lower rate of return. The strategy, then, is to make your money and run, investing it in another high-rate-of-return venture. This technique allows resource managers and others to compare the opportunity costs of various strategies.

Economics and ecology clash when arbitrary rates of return conflict with natural replenishment rates. For example, if a company must make 10 percent on its investment in a logging practice but the forest's sustainable yield is 4 percent, it won't be long before the forest loses out.

Creating a sustainable system of resource management requires rethinking this system of calculating the value of different opportunities. Education can provide economists and resource managers with a new view of time preference, as well as alternative (sustainable) strategies to achieve respectable economic returns without depleting ecological resources. Bearing in mind that planet care is the ultimate form of self-care, even a way of ensuring our species's survival, many critics argue that we must learn to assign value to long-term considerations and factor these into economic decisions. One means of doing this is to calculate the replacement or mitigation cost of various actions. Replacement cost is an estimate of the ultimate value of a resource. How much would it cost to replant a tropical rain forest and reestablish the complex ecosystem? How much would it cost to mitigate (offset or repair) the damage caused by soil erosion from careless farming practices?

By factoring these costs into the economic calculus used to determine discount rates and opportunity costs, resource managers get a more accurate picture of the economics of various strategies. In this light, many profitable ventures will seem foolish. Sustainable strategies will appear much more attractive under a more ecologically sane means of accounting.

Economics of Pollution Control: Finding New Ways

This book is concerned primarily with natural resource management, but it also delves into other important environmental issues, among them pollution control. Because pollution threatens the long-term prospects of humans and other species, we will next examine the economics of traditional pollution control measures and propose alternative methods with a much more favorable economic outlook.

Pollution control in today's world is heavily driven by economics. The goal of pollution control is to reduce emissions and ambient pollution in the most cost-effective ways. From a purely economic standpoint, then, investment in pollution control devices (capital costs) and their operation should be equal to the benefits received. Put more bluntly, a factory owner would not want to install a $30 million pollution control device to prevent $1 million worth of damage. In such instances, the cost of control would clearly exceed the benefits.

The relationship between the cost of pollution control and pollution reduction is shown in Figure 2.4. Part A shows the cost of damage caused by different levels of pollution. As illustrated, the higher the pollution level, the higher is the cost.

Figure 2.4B illustrates another important consideration, the cost of reducing pollution. This graph shows that the more money one invests, the more pollution is removed. It also shows that a relatively small initial investment results in a substantial reduction in pollution, but as one tries to reduce pollution emissions further, the cost per unit of pollution removed increases much faster. This is known as the **law of diminishing returns.**

Figure 2.4C overlays the two graphs from parts A and B. The intersection of the two lines is the point at which the cost of control and the economic benefits of pollution reductions are equal. This is the **break-even point.** If society desires even lower levels of pollution, it must pay more. However, because costs of removing pollution increase disproportionately as one approaches zero emissions, the economic costs can be quite high.

Although balancing costs and benefits may sound simple, it isn't. One of the chief problems with this process, known as **cost-benefit analysis,** lies in determining the full extent of damage, that is, determining *all* of the costs. To do so, we must study and quantify all of the impacts an activity has. Many hidden or less obvious effects might be missed. Another problem lies in determining the economic value of the damage, for instance, the economic cost incurred by polluting a stream or eliminating a species. How can we assign an economic value to a healthy fish population in a stream or a healthy forest? What is the value of your health?

One technique that helps solve the dilemma of underpricing damage is to calculate the costs of mitigation. Thus, while it may be impossible to assign an economic value to the survival of a species of salamander, we could calculate what it would cost to capture the remaining salamanders and raise them in captivity for eventual release in a cleaner environment. This is the mitigation cost.

As another example, a forest in a watershed that supplies drinking water to a city may help keep the water clean. If the forest were cut down, erosion would increase, and that would increase filtration costs. These are mitigation costs. Mitigation costs, however, only indicate part of the value of the forest. What about the loss of wildlife and recreational

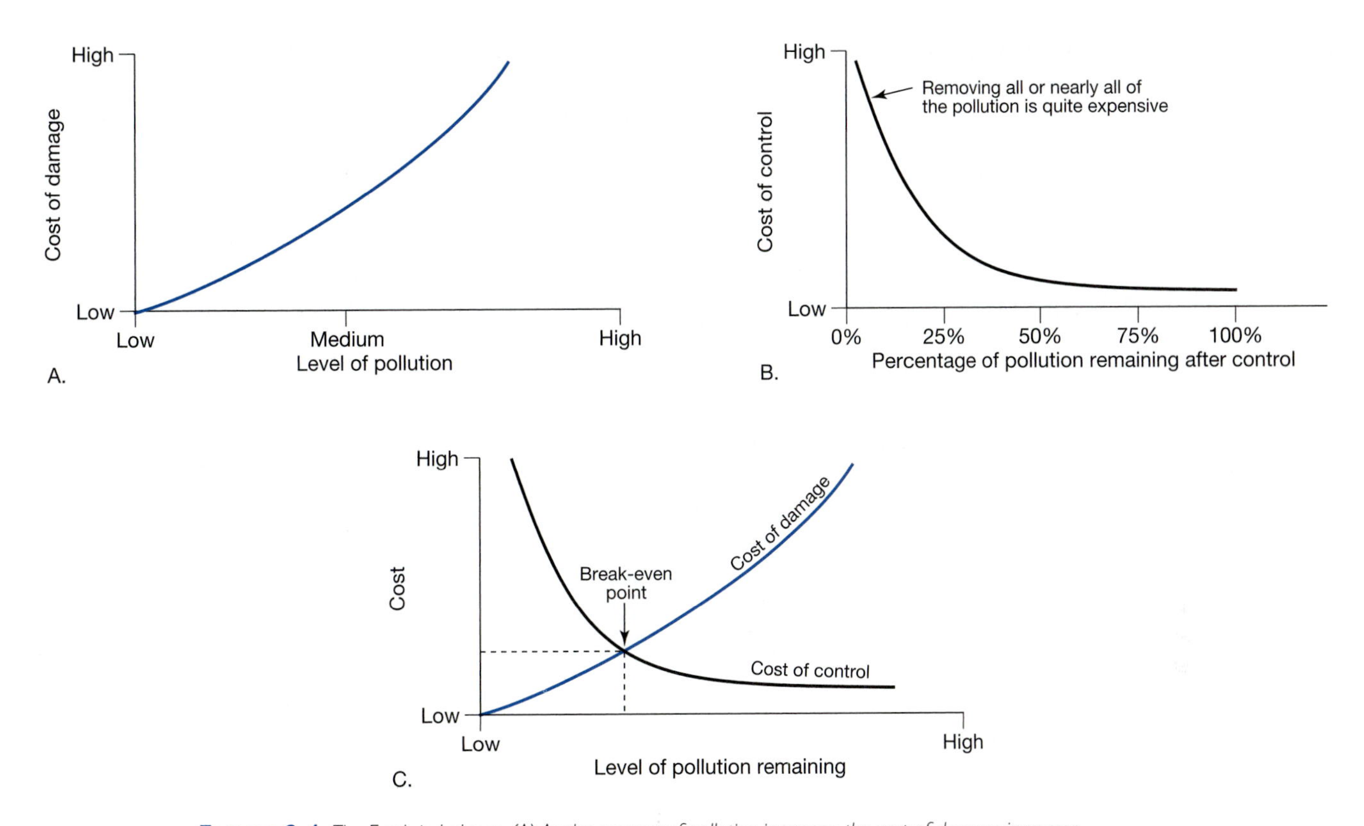

FIGURE 2.4 *The Earth in balance. (A) As the amount of pollution increases, the cost of damage increases. (B) Removing pollution also costs money. After a certain point, the cost of removing a unit of pollution climbs dramatically. (C) The goal of pollution control is to find a break-even point where cost equals benefits. Pollution prevention is an alternative measure. It results in far greater reductions at a lower price.*

activities and increased flooding caused by clear cutting the forest?

Because of these and other problems, more often than not, those involved in this economic analysis fail to include all of the damage and therefore underestimate the real costs to society and the environment. Investments in pollution control would underprotect us.

If the full costs and extent of human damage were known and could be assigned an economic value, the break-even point in graph C would very likely shift to the left. It would, therefore, be worth spending much more on pollution control. Until the science of predicting and quantifying costs of human actions like farming and forestry improve, however, most policy based on cost-benefit analysis will result in an investment that falls short of the true costs.

One fact that many companies are realizing is that there are many ways to achieve a desirable goal. In other words, there are many ways of reducing pollution emissions, and some of them yield substantially greater environmental benefits with minimal economic investment. They fall under the rubric of pollution prevention. **Pollution prevention** is a technique that differs fundamentally from pollution control, the most common strategy until recently. Pollution prevention seeks to eliminate the production of pollution altogether. **Pollution control** seeks to capture pollutants after the fact, when it's too late.

Pollution prevention is achieved in one of several ways. For example, in factories, slight adjustments in manufacturing processes can reduce pollution output enormously. This is an operational change. In some cases, nontoxic chemicals can be substituted in chemical manufacturing or other industrial processes for more toxic substances. This is a design change.

Pollution prevention works and can save enormous sums of money. In Yorktown, Virginia, for example, Amoco and the Environmental Protection Agency (an agency of the federal government that monitors and regulates many environmental pollutants) dropped their normally adversarial roles to find ways to reduce toxic emissions. They found that a $5 million investment in pollution prevention proposed by the oil company could cut emissions of the toxic chemical benzene five times more than a $35 million traditional pollution control device. In other words, at one-seventh the cost, Amoco could cut pollution emissions five times more than with the traditional strategy. Unfortunately, the law required the company to install the more expensive control technology. Fortunately, this example raised the ire of many people and has resulted in a dramatic shift in policy to avoid such unnecessary calamities in the future.

Similar examples illustrate the economic and environmental benefits of pollution prevention. They also illustrate the need for regulatory flexibility in meeting goals. If, for example, governmental agencies set strict standards for emissions, but companies are allowed to apply their ingenuity in meeting those standards, far greater gains may be made in reducing pollution than is possible with standard pollution control methods. The government of Holland, mentioned in the last chapter, is embarking on a daring experiment with their many industries. Together, Dutch companies and the government have set strict standards for reducing industrial pollution, but the companies have been left to their own devices to figure out how to do it in the most economical manner.

Additional Measures for Promoting Sustainable Economics

To create a sustainable economy requires the changes outlined in this book—pollution prevention, sustainable resource management, recycling, the use of renewable energy, restoration, and the like. Changes in ways of thinking about economics itself will also help promote an enduring human presence.

Alternative Measures of Progress At the top of the list are new measures of success. The Index of Sustainable Economic Welfare, described earlier, is one example. The ISEW provides a more comprehensive picture of the state of a nation than its GNP.

Several nations, including Germany and France, are developing alternatives to the GNP. The U.N. Statistical Commission drew up guidelines for nations interested in calculating alternatives. Although alternative measures of prosperity are not a cure-all, they are essential to keep direct corporate and public policy along more sustainable lines. At this writing, over 120 cities and towns and states in the United States, among them the state of Oregon; Jacksonville, Florida; and Seattle, Washington have developed alternative measures designed to track social, economic, and environmental trends, giving citizens and government officials a more realistic picture of their progress—or lack thereof (Figure 2.5). Similar programs are underway in London, Stockholm, Vienna, and Zurich. These indicators of sustainable development and quality of life are not aggregate measures, however. They are dozens of measurements of key conditions in various communities that allow people to track their progress toward or away from sustainability. Combined with other measures, they could help steer us back onto a sustainable course.

National Inventories of Natural Resources Another important movement underway that addresses weaknesses in the GNP are national inventories of natural resources. Designed

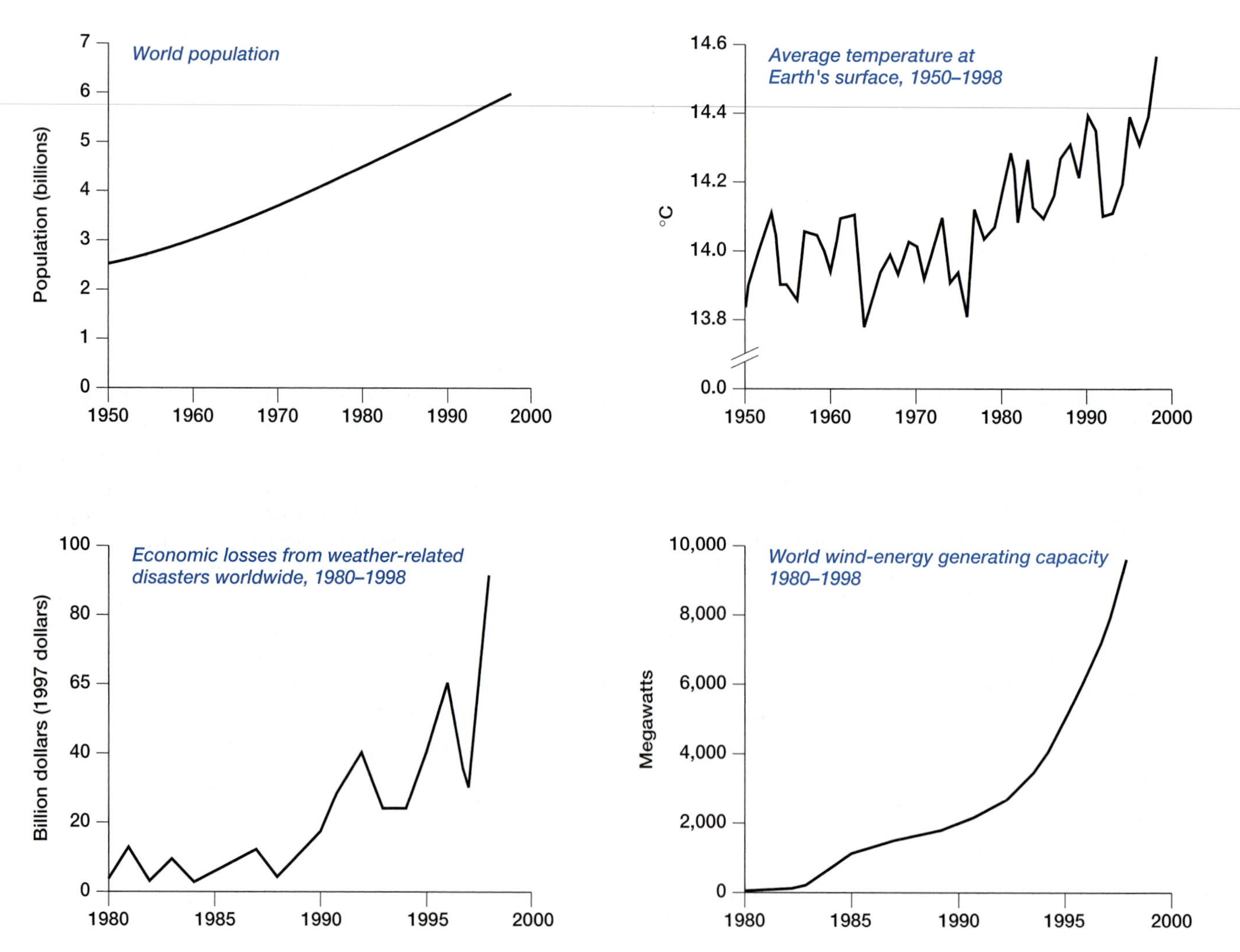

FIGURE 2.5 *Indicators of sustainable development. Many cities, states, and nations are starting to track key indicators of their social, economic, and environmental health to determine where they're going and to help them prioritize limited financial and human resources. Such exercises help nations plot their progress toward or away from sustainability.*

in large part to determine if economic growth is depleting natural capital accounts, natural resource inventories are being conducted in Australia, Canada, France, the Netherlands, and Norway. To create a truly sustainable global economy, such efforts are needed in virtually every nation on Earth.

Green Taxes Yet another strategy for sustainable economic development invokes user fees or green taxes. **User fees** or **green taxes** are taxes levied on raw materials. They are paid by producers, for example, mining companies. Green taxes artificially increase the price of a raw material. Although this may seem counterproductive from an economic standpoint, from an environmental viewpoint, they make perfect sense. One benefit is that green taxes may help promote more efficient use of resources. Reduced consumption reduces impacts and extends the lifespan of resources. Green taxes can also be levied on undesirable activities—for example, factories that generate pollution or autos that burn excessive amounts of gas and produce inordinate amounts of pollution.

Green taxes offer other benefits in addition to stimulating a more efficient use of resources. Revenues from green taxes, for example, can be used to develop alternatives or to offset the impacts of resource development (for example, mining) and pollution. Several western states, for instance, charge a severance tax (essentially a user tax) on coal. This tax reimburses local communities for the additional cost of infrastructure—roads, schools, and the like—required by the population growth that accompanies resource development and provides funds for alternative economic strategies when the coal seams run out.

Because the costs imposed on businesses are ultimately passed on to consumers, green taxes provide a useful market signal and help offset one of the principal weaknesses of supply-and-demand economics, namely, the underpricing of goods and services derived from finite resources.

Green taxes are not popular in the United States but are more common in Europe, where an estimated 50 green taxes are currently levied on a variety of products. Nonetheless, there are some examples. In 1990, for instance, the U.S. Congress passed sweeping amendments to the Clean Air Act. One provision of the new law was a substantial tax on chemicals used by industry that deplete the ozone layer. The **ozone layer** is a region of the upper atmosphere that filters out harmful ultraviolet radiation, which can cause severe burns and cancer. By imposing an incremental tax on chemicals that deplete the ozone layer, the U.S. government hoped to encourage many companies to search for alternative chemicals that performed as well without damaging the ozone layer. So far, many companies have been successful in finding alternatives that, besides being better for the environment, are cheaper to use.

Carefully designed economic disincentives could be implemented to discourage a wide range of activities, among them tropical deforestation, overgrazing, fossil fuel combustion, disposable products, and energy inefficiency.

Full-Cost Pricing As noted earlier, many goods and services reach consumers at considerable expense to the environment as well as to the communities we live in. That is, they come with unreconciled economic externalities. Put still another way, what we pay for products is not a reflection of their full cost—especially the environmental costs.

Green taxes, covered in the previous section, not only help adjust demand and discourage activities that are environmentally unsustainable, they can help us pay for external costs. In other words, they help us achieve **full-cost pricing,** defined here as a pricing scheme in which the price of a good or service is a reflection of all of the costs associated with its production. For example, a coal tax might be levied to pay for repairing damage to forests and lakes from acid deposition or to offset losses caused by global climate change caused by carbon dioxide emitted from coal-fired power plants.

Far better than applying taxes such as these after the fact are measures that seek to incorporate full-cost pricing at the outset, in the planning stage. Many states, for instance, have regulations that require utility companies to assess the costs of various options needed to produce electricity. When a utility company seeks permission to expand its electrical generating capacity, it must acquire approval from the Public Utilities Commission. But it must select the least costly approach, which frequently does not include environmental costs. To account for economic externalities, several states require that an additional 15 percent be added to the cost of coal and nuclear energy strategies. Because of this least-cost-planning provision and the adjustment to account for economic externalities, energy conservation and several renewable energy strategies often appear much more attractive.

Full-cost pricing is a means of examining the life-cycle costs of various technologies or strategies for meeting human needs. **Life-cycle costs** include all costs from the point of origin of a material to its point of disposal—"from cradle to grave." For example, life-cycle cost analysis forces business economists to examine long-term environmental damage and costs incurred by mining, manufacturing, transportation, and waste disposal, considerations rarely made in most economic decisions.

Full-cost pricing is particularly useful in cost-benefit analysis (described earlier). Cost benefit analysis is a procedure companies, government agencies, and legislative bodies use when analyzing a particular strategy or considering various options. As noted earlier, it requires an examination of all costs and all benefits. Historically, most economists have looked only at the cost of inputs and the economic benefits accrued by using inputs to create valuable goods and services. They've traditionally ignored or undervalued the costs of environmental damage, as noted earlier.

Economic Incentives Economic incentives are also useful in promoting a sustainable economy and sustainable resource management. **Economic incentives** are measures that encourage environmentally compatible goods and services. They do so by providing economic benefit. Most governments currently offer a variety of incentives to businesses. In some instances, governments provide outright grants—money to support the research and development of new products. The Japanese government, for instance, has spent millions of dollars to promote sustainable technologies like solar energy among businesses. Governments also offer various tax breaks to businesses that are developing or manufacturing environmentally desirable products. These savings on a company's annual tax bill can be used to encourage companies to develop new technologies or to offer environmentally beneficial goods and services. In other instances, governments enter into partnership with business. For example, a local government might donate land it owns to a company to set up a recycling or composting facility. This donation of land makes it economically feasible for the company to offer a valuable service.

Incentives can also be given to individuals. Federal and state tax credits of the 1980s for the installation of solar energy systems on homes in the United States, for example, cut the cost of these systems, giving people a huge financial incentive to go solar.

Tradable or Marketable Permits Another market approach to sustainable development is the **tradable** or **marketable permit** for companies that produce air and water pollution. What is a tradable permit and how is it different from permits currently issued to businesses?

Many governments currently issue permits to industries to release certain amounts of pollution into the air and water. This is the way government regulates pollution emissions. Marketable permits are similar: they are licenses granted by governments to businesses to release certain amounts of pollution. What makes marketable permits different from permits already issued by a government is that they can be bought and sold. So, if your company can reduce pollution below the level stipulated on your permit, you can sell the remaining pollution emission allowance to another company. What's the advantage of that? Consider a specific example.

In 1990, the United States Congress passed sweeping amendments to Clean Air Act. One of them established marketable permits for sulfur dioxide, a pollutant that is converted into sulfur acid in the atmosphere, a compound that causes all sorts of problems. The marketable permit provides companies an opportunity to buy and sell sulfur dioxide pollution emissions on the Chicago Board of Trade (sales began in 1994).

Under the previous system, companies were granted permits to release certain amounts of sulfur dioxide by state and federal agencies. Companies had no incentive to release less. The marketable permit, however, provides companies with an economic incentive (profit) to reduce pollution below the permitted level. Suppose you own a factory that releases 2,000 tons of sulfur dioxide into the atmosphere each year. To clean up the air in the city in which your factory is located, the government (usually a state regulatory body like a department of the environment or natural resources) issues you a permit to release no more than 1,000 tons per year. However, unlike the situation in the past, when the government also required you to install a pollution control device to reach this goal, you are now free to achieve this reduction any way you want. After some deliberation, one of your company's engineers finds a way to reduce the company's annual air emissions to 500 tons per year. Because your annual limit is 1,000 tons, you now have 500 tons of emissions "credit" that, under the marketable permit system, you can sell (market) at a profit to another company to help it meet its emission allowances. If, for example, another company has an emissions limit of 1,000 tons per year but is unable to lower emissions below 1,500 tons, it can buy your unused tons to meet its legal limit.

The benefit of the marketable permit system is that it allows flexibility in achieving pollution emission goals and at the same time creates an economic incentive to meet or exceed those goals. Not only has your company found a cheap way of reducing pollution, you can make a profit from selling the unused portion of the permit to another company, which would then be able to meet its goals thanks to the engineers in your company. Your company, in turn, profits by the sale of part of its permit to another company.

In the United States, only one pollutant is covered by the marketable permit: sulfur. When marketable permits first appeared on the scene, one enterprising environmental group bought up several hundred thousand tons of pollution emissions. The group did not own a factory: It simply wanted to ensure that those pollutants would be permanently removed from the environment. Another environmental group more recently offered members an opportunity to invest in permits to help rid the air of potential troublesome contaminants.

Removing Market Barriers and Subsidies Removing market barriers and hidden subsidies is another market-based solution for a sustainable economy. Surprisingly, significant market barriers exist to environmentally responsible business. Federal regulations, for instance, make it legal for freight haulers to charge more for scrap metal bound for a recycling plant than for raw ore. Because scrap metal can be fashioned into useful products with much less energy, pollution, and damage to the environment than raw ore, this legal loophole is a major obstacle to creating an economically viable system of recycling in the United States. It gives raw ore an economic advantage over recyclable scrap metal.

Subsidies are ways that governments support private enterprises. They include tax breaks and outright grants, that is, economic incentives. Although subsidies can be used to encourage environmentally useful goods and services, most are not. In fact, tens of billions of dollars in subsidies in one form or another are given to oil, coal, and nuclear power industries. Automobiles, for instance, are subsidized by public taxes that pay for roads, parking lots, traffic lights, police, and emergency protection on the order of $1,500 per year per car. This money comes not from gasoline taxes but rather from general revenues. Whether you own a car or not, you pay the subsidy.

Removing these and other subsidies to create a level economic playing field will make environmentally compatible industries more competitive with their less environmentally compatible cousins (Figure 2.6).

These are just a few of the ideas needed to build a sustainable economy. In many countries they have already been implemented. Existing efforts, however, are small compared to the task at hand.

FIGURE 2.6 *In Portland, Oregon, electric trains whisk passengers to and from the city with a fraction of the air pollution of the private passenger vehicle. Systems such as this are vital to building a sustainable system of transportation. They often can't compete with automobiles, but that may be due to the huge subsidies automobile use receives.*

2.3 Toward Sustainable Ethics

Decisions about the environment are not all economic. Many are made on the basis of what we think is right or wrong, that is, **ethics.** We learn ethics from our parents, friends, ministers, and teachers. We also learn ethics through experience, reading, and thinking. Even the study of science can teach us about Earth ethics—what is right or wrong from an ecological perspective.

Frontier Ethics

Ethics serves a practical purpose in our society—it helps preserve social order by influencing the way we treat one another. In this respect, ethics is utilitarian. That is, it serves us well.

Ethics can also influence how we treat the Earth. In many parts of the world, the modern ethics by which many people live their lives and by which businesses and governments operate is a rather dangerous proposition. In the United States and other industrial nations, for example, many people view the Earth as an unlimited (infinite) supply of resources exclusively for human use. If the Earth's resources are unlimited, so then are our options for tapping into those resources. Another key tenet of modern ethics in many places is that humans are apart from, rather than a part of, the natural world. We are somehow superior to it. Our needs are of tantamount importance. To those who hold this dangerous view, nature is merely a backdrop for human civilization and is therefore of little consequence to their lives, when in fact the future of humankind is intricately linked to the future of the planet. Another key tenet of frontier ethics is that the key to success is the domination and control of nature (Figure 2.7). To meet our needs, we must conquer nature, leaving hardly a trace of its former self. We reduce complex ecosystems to highly simplified systems that produce food, fiber, and other resources required to meet our needs. We level mountains to get at underlying minerals. We dam rivers and cut roadways through forests.

In this book, we refer to this ecologically preposterous system of ethics as **frontier ethics,** because it reflects the naive, optimistic, and careless attitudes of the American frontier people who had little, if any, notions of limits. Today, although most of our frontiers have vanished, frontier ethics remains, influencing countless decisions every day. Unspoken ethical tenets drive innumerable environmentally unsustainable decisions of government officials, resource managers, business people, and individuals. In fact, frontier ethics—especially the notion that the world possesses an unlimited supply of resources for human use—is a prime driving force of the national

FIGURE 2.7 *This surface mine is a symbol of human domination over the environment. Technological development has allowed us to alter the face of the Earth in dramatic ways.*

economies of many major economic powerhouses. On the surface, frontier ethics appear to be quite utilitarian, but they may prove to be counterproductive.

Sustainable Ethics

Developing a sustainable, worldwide economy will require a much greater awareness of our problems and a dramatic increase in our understanding of the importance of natural systems to our lives. It will also require a new system of ethics, a sustainable ethic.

Sustainable ethics holds that the Earth has a limited supply of resources. In other words, the Earth's resources are finite and should be managed carefully. Another key tenet of sustainable ethics is that the Earth's resources are not for the exclusive use of humans. Other species require a share of the Earth's wealth to survive and prosper. Some people claim that other species have a right to live and prosper, too. Sustainable ethics also holds that humans are a part of nature—we are dependent on natural systems in many ways. Another key element of this ethical system is that human success requires cooperation with nature, rather than domination and control. The core value of the new ethical proposition is that humans and nature are inextricably linked: we are dependent on natural systems, and our future depends on a healthy, well-functioning ecosystem. Conversely, nature is also dependent on us. Human society has reached a size and level of technological power so great that we literally hold the future of the planet in our hands.

Sustainable ethics is not a new notion by any means. In fact, it may have been the predominant approach throughout much of the several-million-year history of the human species. Judging from the practices of Native Americans and other indigenous cultures, sustainable ethics has been central to the survival of people. Only in the past few centuries have we drifted away from this more humble, Earth-centered view.

In the twentieth century, numerous writers have proposed a return to a sustainable ethic. One of the most influential was the late Aldo Leopold, whom we first mentioned in Chapter 1 (Figure 2.8). A wildlife ecologist by trade, Leopold proposed a land ethic in 1933, 70 years ago. Leopold's **land ethic** held that humans were a part of a larger community that included the soil, water, plants, and animals—in short, the land. Recognizing that humans are a part of nature, Leopold argued that the role of human beings would shift from "conqueror… to plain members and citizens of it."

Perhaps best known for his book, *A Sand County Almanac,* Aldo Leopold has become the ideological progenitor of modern conservationists and environmentalists. His ideas go beyond the well-meaning but human-centered ideology of Theodore Roosevelt and Gifford Pinchot, who sought to protect natural resources primarily for their value to humans.

For Leopold, conservation required equal amounts of reflection and action. In other words, he promoted responsibility and action. Leopold's suggestions for action reflect the needs of time when resource management was the primary concern and the phrase "global environmental problems" had not been invented. Today, the extent of the planet's environmental ills and the threat they pose to our future call for individual responsibility and unprecedented action at all levels of society. Sustainable ethics is therefore something of an extension of the land ethic. To summarize, sustainable ethics rests on four **directive principles,** described earlier: (1) the Earth's resources are limited, there is not always more; (2) humans are a part of nature; (3) the key to success is cooperation with nature; and (4) natural systems are essential to human welfare. But how does one put sustainable ethics into practice?

This book proposes five **operating principles** for putting sustainable ethics into action: (1) conservation, (2) recycling, (3) renewable resource use, (4) restoration, and (5) population control. These guidelines for action are not all that is needed; they are only some of the most important steps humans can take to create an environmentally sustainable society.

Together, the directive and operational principles could result in a measure of restraint that is necessary to shift away from the resource-intensive society that now threatens its own long-term well-being. Restraint means doing what is right not just for the individual, but for future generations and for other species. This concept is known as fairness to future generations or **intergenerational equity.**

One of the central tenets of sustainable development, intergenerational equity asserts that present generations hold the Earth in common with all generations, past and future. We are,

FIGURE 2.8 *Conservationist Aldo Leopold (shown here) has inspired many individuals over the past 50 years to rethink their role in the environment through his accessible writings, such as A Sand County Almanac, published in 1949. His books continue to sell well today.*

in short, part of a long line of planetary occupants with common rights and obligations. For example, present generations have the right to benefit from the Earth—to profit from its riches and enjoy its beauty. At the same time, we have an obligation to protect the Earth for future generations so that they may benefit as well. If you're having trouble grasping this concept, think of the Earth as an heirloom that is handed down from generation to generation. As with an heirloom, the rights and obligations of the Earth's custodians are attached to the gift.

In *Lessons From Nature: Learning to Live Sustainably on the Earth,* Dr. Chiras proposes a notion of **intragenerational equity**—fairness among members of the present generational cohort. The idea behind this concept is that because the Earth's human community shares the global commons—the air, water, and land—and because local actions often have global impacts, people have responsibilities to one another in the present. More specifically, we have a responsibility to act in ways that do not adversely affect other people, no matter how distant or unrelated they may be.

As an example, consider the use of fossil fuels by the United States and other industrialized countries. Burning fossil fuels may cause a rise in sea level through the release of carbon dioxide, which in turn causes global warming. Flooding may inundate the coastal farmlands of millions of people in less developed countries—people who have never burned a drop of oil or an ounce of coal. Do users of fossil fuels have an obligation to these people? Under the doctrine of intragenerational equity, the answer is yes.

One type of intragenerational equity that is of great concern to many is known as **environmental justice** or **environmental equity.** Environmental justice is defined as fair treatment for people of all races, cultures, and socioeconomic levels in the execution of environmental laws, regulations, and policies. It has also been defined as the pursuit of equal justice and equal protection under environmental laws and regulations without discrimination based on race, ethnicity, gender, and/or socioeconomic status. Now spurring an international movement, the notion of environmental justice grew out of concern that minority populations and/or low income populations (often one in the same) bear a disproportionate amount of adverse health and environmental effects.

Concern for environmental justice began in the early 1980s in Warren County, North Carolina when a predominantly African-American community was targeted as a proposed site for disposal of PCB contaminated soil from 14 other sites in the state. Protests emerged over the siting. Although protestors were not able to stop it, the fury sparked further inquiry. The U.S. General Accounting Office studied the issue in eight southern states and found that three out of every four hazardous waste landfills were located near predominantly minority communities. Since then, numerous studies have shown that hazardous waste facilities, polluting factories, power plants, incinerators, and other potentially harmful facilities are frequently sited in or near low-income minority communities. Race and economic status have emerged as primary determinants in siting decisions of businesses, as poor minority communities typically lack the financial resources and political savvy to mount opposition to such plans.

A study by the *National Law Journal* showed further evidence of environmental discrimination in the government. Their researchers found that it took the EPA 20% longer to nominate an abandoned hazardous waste site as a priority area in need of clean up if it was in a minority community than a white community. The researchers also found that industrial polluters in minority communities paid fines that were on average 54% lower than polluters in white communities.

Since the early 1980s, environmental justice has become a priority in the EPA. Today, in fact, the EPA has an Office of Environmental Equity that has initiated a number of environmental justice projects in the United States. In 1994, President Clinton issued an executive order requiring all federal agencies to make environmental justice part of their mission. Numerous grassroots organizations have also emerged to address the issue in the United States and in less developed nations as well, where the practice is also commonplace. The U.S. Congress and state legislators have also begun to grapple with the issue and two states, Arkansas and Louisiana, have passed environmental justice laws.

A variety of solutions have also been proposed, including pollution prevention and improved stakeholder participation in the public decision-making process—in other words, more participation by minorities in communities that would be affected by proposed projects. Improving access to information could help also (see Case Study 2.1). Yet another solution is improved enforcement and compliance assurance (making sure companies comply with laws and regulations).

Yet another ethical idea is ecological justice. **Ecological justice** holds that people have an obligation to other species, present and future. Put more forcefully, other species have a right to exist, and we must act in ways that ensure their survival.

Intergenerational equity, intragenerational equity, and ecological justice contradict many peoples' most basic ethical ideas. They fly in the face of the frontierist notions and principles upon which capitalism is based. For example, the idea that we must act in ways that protect the interests of present and future generations contradicts the prevalent ethical notion that the Earth is an infinite source of materials. It also contradicts the economic notion of competition, in which each person, present and future, must compete for his or her own survival and well-being. It calls for cooperation, not competition.

Creating a Global Sustainable Ethic

The task of creating a global sustainable ethic is among the most pressing of our time. It will require efforts by millions of educators from primary education through college worldwide. It will depend on the efforts of the personnel of museums and nature centers. Religious leaders can play a role. The entertainment industry and media (newspapers, magazines, and television) could participate. Creating a global sustainable ethic will also require the cooperation of parents, not just in the United States but in all nations, rich and poor.

Given the conflict in many parts of the world (the Middle East, for example) and the decaying conditions under which many people live, convincing people that they must live more sustainably will not be easy. Meeting immediate needs often takes precedence over urgent environmental concerns, but ironically, meeting immediate needs may worsen environmental conditions, creating a downward spiral.

CASE STUDY 2.1 GEOGRAPHIC INFORMATION SYSTEMS AND ECOLOGICAL JUSTICE

There's a river in Mississippi that pours into the Gulf of Mexico. It is known as the Escatawpa. Along its banks are two communities, one white and rich, the other composed of African Americans who are largely poor. In the late 1970s, the mostly white affluent community, known as Pascagoula, decided to build an incinerator to burn its trash. Many residents, however, opposed siting the facility in their town, and the contract eventually went to a chemical company located three miles from Moss Point, the nearby poor, African-American town.

Little controversy was stirred by the decision, until in 1991 the city council of Pascagoula voted to permit the burning of medical waste in the incinerator in addition to their trash. Residents of Moss Point became enraged by the odors from the incinerator. Concern erupted over the potential for contamination from mercury, cadmium, and dioxin released from the facility, too. Local doctors worried about further health problems in an area that had an already high incidence of respiratory disorders. They raised concern over potential long-term health effects.

The controversy along the banks of the Escatawpa River is a classic example of environmental injustice, or environmental racism. A wealthy, predominantly white community wants an incinerator but not in their backyard. The incinerator ends up in an area surrounded by people of color with no money for health care or legal representation.

Environmental racism is on the decline. Those who are active in the fight for justice are receiving assistance from a high-tech tool, GIS or geographic information systems, specifically a GIS known as LandView(TM) III. Discussed in Chapter 1, GIS is a computerized system of mapping that permits us to store huge amounts of information in map form to study relationships, make predictions, and better plan human development.

LandView III is described as community right-to-know software tool by the EPA, which sponsored its development. According to the EPA, LandView III places a "wealth of important environmental information at the fingertips of local decision-makers and the public." It provides database extracts from the EPA, Bureau of Census, the U.S. Geological Survey, the Nuclear Regulatory Commission, the Department of Transportation, and the Federal Emergency Management Agency. These databases are presented on maps that contain jurisdictional boundaries, detailed networks of roads, rivers, and railroads, census information, schools, hospitals, airports, landmark features, and much more.

LandView III is accessible to anyone with a modern computer and contains a tutorial so individuals can learn how to use the system. It can be used to learn about the demographics of an area—that is, the age, income, and ethnicity of an area—and to learn about potential health impacts from a wide variety of sources. The environmental information available to the user includes air pollution emissions, wastewater discharges, toxic release data, hazardous waste sites, nuclear sites, watershed assessment, and air quality monitoring sites. LandView III has the capacity to display multiple layers of information for decision makers and it is proving to be a useful tool in the effort to develop currently unused industrial sites, known as **Brownfields,** that have been contaminated by toxic substances. For example, LandView III can be used to locate potential brownfield sites in a city, county, or state and assess their potential impact. Sites slated from development are typically cleaned up, but such strategies must be communicated to nearest neighbors. LandView III enables brownfield developers to assess the demographics of the surrounding regions and tailor informational programs to allay fears and initiate public involvement.

LandView III is available on line and in CD-ROM from the Bureau of Census. For information on purchasing a copy through the Bureau of Census, see our Web Page.

Nevertheless, fostering sustainable ethics may be one of the most important steps we can take to build a sustainable future. The United Nations Conference on Environment and Development, described in Chapter 1, was a good start. Many nonprofit organizations, teachers groups, and business organizations have begun the process of global attitude change.

2.4 Critical Thinking and Sustainable Development

This book will help you understand environmental and resource management issues. During your reading, you will encounter many scientific facts and principles that are essential to understanding and solving problems. Understanding issues and solutions requires more than remembering scientific facts and having a grasp of policies and practices that promote sustainable development. To become a resource manager and a responsible citizen, it is essential that one learn how to analyze issues and solutions. This section outlines six "rules" for critical thinking.

Critical thinking means many things to many people. We define it as a means of analyzing information to distinguish between beliefs (what people believe is true) and knowledge (facts supported by scientific observations). Critical thinking is used to analyze the results of scientific research, environmental issues, and proposed solutions. It can be used to analyze what is said in newspaper articles, speeches, and classroom lectures. Critical thinking is the most ordered thinking. It permits us to look for weaknesses in reasoning. This section presents a few rules of critical thinking (Table 2.1).

Gather Sufficient Information

In analyzing issues and assertions or in solving problems, it is important to learn as much as you can about a subject. Hence, the first rule of critical thinking, and the one frequently broken, is to gather as much information as possible. Learn everything you can about a subject before you make decisions or criticize.

Part of the process of gathering information is to understand all terms and concepts. Many an advocate of a particular viewpoint has a shaky understanding of science and speaks

TABLE 2.1 *Critical Thinking Guidelines*

Gather sufficient information and define all terms
Question the methods by which facts and conclusions were reached
Question the conclusions of studies
Question the source of information; look for hidden biases
Tolerate uncertainty
Examine the big picture

more from emotion than from facts. This is true of environmentalists as well as their critics. The deeper you dig on many issues, the more you will understand and the better your solutions will be. Whatever you do, do not mistake ignorance on a topic for perspective.

Question the Methods

In separating fact from fiction, or belief from knowledge, it is often essential to question the methods by which the information came into being. Many people, for instance, draw sweeping conclusions from casual observations. They may have read an article in the newspaper about a company that closed down because of tighter pollution control regulations. From this observation, they conclude that environmental regulations are too costly to society. Regulations rob us of jobs and income.

A more accurate study of many companies shows that environmental regulations have very little negative impact on most companies. Those that close down are often those that are in danger anyway because they are not operating efficiently or their market is drying up. Environmental regulations are merely the straw that broke the camel's back. So, beware of individual case studies and anecdotal information. They may not be representative of the larger truth.

Rigorous scientific proof is often necessary to distinguish between knowledge and beliefs regarding many environmental issues—for example, the effects of pesticides on human health and the environment. Such proof comes in the form of carefully conducted experiments. Scrutinize all reports to determine if the experiments were adequately performed. Was the sample size large? If not, the results may be a fluke. Did the experimenters control for all variables? For example, did the researchers of a study that demonstrated that a particular pollutant caused a health effect eliminate the possibility that another factor, such as cigarette smoking, was responsible?

Careful experiments, whether they use laboratory rats or human beings, require two groups. The first is the **control group.** The second is the **experimental group.** In good experiments, the control group and the experimental group are as similar as possible. They differ only in one regard, the **experimental variable.** For example, suppose you wanted to determine if a particular pesticide was harmful to mice. You would conduct an experiment with two groups identical in age, sex, weight, diet, and so forth. The experimental group would receive daily injections of the chemical under study. The control group would receive injections as well, but the injections would not contain the pesticide under study. Thus, any observed differences could be attributed to the chemical.

To be valid, studies must also use an adequate number of subjects. Generally, ten experimental animals is enough, but the more the better. Studies must also be repeated by others to determine if the results are replicable.

Scientists perform experiments to test various ideas—or **hypotheses** (high-poth-eh-seas). For example, a scientist might hypothesize that a certain chemical in the water of a lake is responsible for structural defects in bird embryos. To test this hypothesis, she might perform an experiment using control and experimental groups. If the results of the experiment do not support the hypothesis, the scientist might change it. For example, she might assume that another chemical was to blame. She would then test this hypothesis.

Scientific hypotheses that have been supported through experimentation often tell us many facts about the world we live in. Hypotheses related to the same subject may lead to the formulation of a theory. A **theory** is a larger explanation of a phenomenon. For example, the many experiments on the atom have led to a theory that explains what an atom consists of and looks like. This is called atomic theory.

Theories are supported by many observations and, as a rule, cannot be refuted by a single experiment. But theories are not immutable. Some cherished theories have been modified or replaced as new evidence accumulated.

A great deal of information today also comes from computer modeling. The Limits to Growth Study discussed in Chapter 1 is a good example. Scientists use computers to simulate natural systems, such as forests, and to predict changes. Building computer models based on complex mathematics is useful. That is, it helps explain our world and helps us try to understand how human activities and natural phenomena can alter natural systems and humans systems. They're terrific tools for prediction. But they have their short comings, too. Computer models are only as good as the assumptions that are made when formulating them. If faulty assumptions are made, then outcomes and predictions will be faulty. Climate models used to predict the effects of increasing concentrations of pollutants in the atmosphere, for example, were primitive at first, but have grown in sophistication and precision, although they still are only mathematical models of an extremely complex system, the atmosphere and climate.

Question the Conclusions, Question the Source

Just because an experiment is correctly designed does not mean that the conclusions drawn from its results will necessarily be valid. Scientists are human; they make errors, and they have hidden biases that may taint their interpretation of the results.

Scientists may also fail to take into account contributing factors. This is especially true in studies of the effects of chemicals on people. Unlike commercially available laboratory rats and mice, people are a mixed lot. Our genetic makeup varies.

Our life histories vary, as do our exposures to various chemicals. Of course, it is not ethical to expose people to chemicals in controlled experiments. It is not possible to cage a hundred people, control their diet, and intentionally expose them to toxic chemicals.

Because of these and other constraints, studies of the effects of chemicals on humans are often difficult to perform. In most cases, scientists locate populations of individuals who have been inadvertently or accidentally exposed—say, at work—and then compare them with a similar population (a control group) that was not exposed. To draw conclusions, scientists must take into account a variety of variables—for example, possible exposure at home, or exposure to other factors that might have the same effect as the chemical under study. Such studies on people fall within the realm of **epidemiology** (ep-eh-deem-ee-ol-oh-gee).

Because epidemiological studies are so difficult, it often requires dozens to demonstrate a connection between an environmental pollutant and a health effect. To date, over 40 studies have shown that smoking causes lung cancer, and yet some people still dispute the results. This leads to another important point, namely, that people with an axe to grind—for example, representatives of the tobacco industry or their political allies—are particularly prone to interpreting results in ways that suit their needs.

The lesson here is to question the conclusions. Ask if the facts really support the conclusions, or if a hidden bias or a hidden agenda might be tainting the interpreter's viewpoint.

Tolerate Uncertainty

Critical thinkers must also tolerate uncertainty as they work through an issue. Formal science is a relatively recent phenomenon in human history, and scientific knowledge is quite skimpy in many areas. As our culture becomes more complex and our impacts grow, scientists are barely able to keep up with the many questions we need answered to make intelligent decisions. In other cases, for instance global climate, the interactions scientists must understand are so many and so complex that decades may be needed to come to an informed decision. Of course, by that time action may be too late.

The rule here is to be aware that uncertainty exists. Tidy answers are not always available. When necessary, we must make decisions based on incomplete information.

Understand the Big Picture

Too often political and scientific debates focus on small pieces of the puzzle while ignoring the big picture. For example, in the debate over nuclear energy, many people focus on the issue of reactor safety (Chapter 21). To address this issue, the nuclear industry has proposed smaller, supposedly safer nuclear power plants. Proponents of nuclear power argue that by installing these power plants, we can avoid serious and costly accidents.

Although that logic may be appealing, it misses an important point—one that is evident when you examine the big picture. The nuclear energy issue involves much more than reactor safety. It includes the serious and as yet unresolved issue of waste disposal: How do we safely dispose of waste that will remain radioactive for 10,000 years? And what about radiation exposure to miners and workers in nuclear fuel–processing plants? These and at least ten other issues are relevant to the debate and can only be appreciated if one examines the issue from the largest possible perspective.

Big-picture analysis also helps one make sense of other issues and come to important conclusions. For instance, many people are currently debating the rate of tropical deforestation. Some say that present estimates of 17 million hectares (about 45 million acres) lost each year are too high. But even if the actual rate of loss is half the estimated rate, it is still too high to be sustained. Stop and look at the big picture before you get lost quibbling over details.

Critical thinking requires us to think in larger terms, often in terms of entire systems. By breaking out of narrow perspectives and looking at how humans affect entire systems, we gain a broader perspective than can lead us to more comprehensive and potentially lasting solutions.

One area in desperate need of a systems approach is environmental protection. It is clear that we can no longer depend on narrow solutions that strike only at the symptoms of environmental problems. It is time to look at entire systems—for example, forest management, wildlife management, waste management, energy, transportation, housing, and the like—and find ways to revamp them in their entirety, not one piece at a time, which is usually inadequate. Ecosystems management, described in the previous chapter, requires systems thinking.

Economics and ethics are two systems where critical thinking will come in handy, and where systems reform is essential. The question is, do we have the political will to make the changes, or will we continue to muddle along, applying Band-Aids to problems that are often growing worse with every day.

Summary of Key Concepts

1. Resource management requires an understanding of economics and ethics and an ability to think critically.
2. Economics is a science that seeks to understand and explain the production, distribution, and consumption of goods and services.
3. Historically, economics has been concerned with inputs and outputs. Inputs include the commodities (raw materials, labor, and energy) companies need to produce goods and services. Outputs include goods and services.
4. One of the guiding principles of market economies is the law of supply and demand, which describes the manner in which the price of a good or service is determined by the interaction of supply and demand.
5. Despite its many successes, market economics has several fundamental weaknesses when viewed from an environmental perspective.
6. One of the principal complaints of ecologists is that business economists have a rather short time horizon for planning.

The short-sightedness of the market economy is evident in our almost single-minded dependence on the law of supply and demand. Pricing established by supply and demand typically reflects only the immediate supplies of resources in relation to their demand. Long-term supplies are essential to the future of humankind but are not factored into prices.

7. Oil, rangeland, forests, fisheries, and other resources have been depleted or driven to the brink of depletion, in large part because they were once abundant and have historically been undervalued.
8. Critics also point out that at least three essential factors are missing from the input/output analyses of mainstream economics: environmental damage, social and cultural impacts, and pollution. Thus, when businesses calculate the costs of manufacturing their products, they typically fail to take into account environmental and social costs—that is, economic externalities.
9. Yet another shortcoming of traditional capitalism is its single-minded dedication to measuring success by the gross national product, or GNP. The GNP is the sum total of all goods and services produced by an economy.
10. The problem with the GNP is that it is a crude measure that fails to distinguish between "good" and "bad" economic activities and thus fails to track a nation's true economic progress. Growth in the GNP may occur at the expense of society and the environment.
11. The GNP also fails to take into account natural capital, a nation's natural resource and ecological wealth. Nations draw on this capital to fuel their industries and to satisfy human needs for goods and services. But many nations are depleting their natural capital. Although they may boast high GNPs, their long-term prospects for growth are dim.
12. To live sustainably we need an environmentally sensitive system of commerce, a sustainable economic system.
13. In general, a sustainable economy is one that produces goods and services in a manner that does not foreclose on future generations. A sustainable economy uses resources efficiently, recycles materials, and eliminates waste. It also depends on sustainable management of natural resources and the restoration of renewable resources.
14. Most advocates of sustainability believe that a sustainable economy will very likely be based on a clean, renewable energy supply. Creating a sustainable economy also hinges on our becoming better able to manage our own population growth and better at balancing our demands for goods and services and living space with our needs for clean air, water, food, and aesthetic pleasure.
15. Three principles affect the management of resources: time preference, opportunity costs, and discount rates.
16. Time preference is the preference for economic returns on an investment, expressed in time. An immediate time preference means that the individual prefers immediate returns and is influenced by several factors, with need being one of the most pressing. An immediate time preference often leads to ecologically, and sometimes economically, suicidal practices.
17. To create a sustainable economy, economic incentives may be needed to compensate for lost income resulting from a shift in time preference caused by a concern for the ecological integrity of forests and other natural resources. However, not all environmentally sensible practices involve a long payback period or exceptionally high initial investments.
18. Opportunity cost is the economic cost of opportunities that are lost when one embarks on a particular path. Many people act in ways to ensure they maximize their economic opportunities, but such actions often lead to resource depletion, which results in lost opportunities of incredible proportion.
19. Economists use a tool called the discount rate to quantify the return on investment of different investment strategies. Unfortunately, immediate income is the most important determinant. Using the notions of discount rate, it is more rational for a business person to liquidate a resource than to harvest at a rate that brings a lower rate of return.
20. Education can provide economists and resource managers with a new view of time preference, as well as alternative (sustainable) strategies to provide acceptable economic returns without depleting natural resources.
21. One means of taking into account lost opportunity costs of various actions is to calculate replacement costs—that is, how much it would cost to replace all of the topsoil lost on a farm or to replant a tropical rain forest and reestablish the complex ecosystem.
22. The goal of pollution control is to reduce emissions and ambient pollution in the most cost-effective ways. From a purely economic standpoint, investment in pollution control devices (capital costs) and their daily operation should be equal to the benefits received. This is the break-even point.
23. Although balancing costs and benefits may sound simple, it is not. Two of the chief problems with this process are determining the full extent of damage and the economic value of damage.
24. One alternative to pollution control that brings far greater economic and environmental benefits is pollution prevention. Pollution control seeks to capture pollutants after the fact, when it's too late; pollution prevention seeks to eliminate the production of pollution altogether.
25. Several economic myths pervade our society and prevent progress toward a sustainable future. One of those myths is that environmental protection is bad for the economy. Another myth is that environmental protection is a luxury of lesser value than economic growth. A third myth is that environmental quality is noneconomic. A fourth myth is that economic growth is good, indeed essential. This chapter provided evidence to counter these myths.
26. To create a sustainable economy requires the changes outlined in this book—pollution prevention, sustainable resource management, recycling, the use of renewable energy, restoration, and the like. Changes in economics itself will also help promote an enduring human presence. Among such changes are alternative measures of progress, national inventories of natural resources, green taxes, and full-cost pricing.

27. Efforts are also needed to harness market forces through the use of economic disincentives, economic incentives, tradable and marketable permits, removal of market barriers, and removal of hidden subsidies.
28. Some noted economists argue that we simply need to alter economics a bit and come up with new technologies to steer society onto a sustainable course and allow us to continue to grow economically ad infinitum.
29. Others argue that changes in ethics are also required. The term ethics refers to what people view as right and wrong. Ethics serves a practical purpose in our society—it helps preserve social order by influencing the way we treat one another. Ethics can also influence how we treat the Earth.
30. In the United States and other industrial nations, many people subscribe to frontier ethics, in which the Earth is viewed as an unlimited (infinite) supply of resources for exclusive human use. It also views humans as apart from rather than a part of nature and sees humans as dominators who must control nature, to meet their needs.
31. These often unspoken tenets drive innumerable decisions of government officials, resource managers, business people, and individuals and are at the root of our present crisis of unsustainability as much as the many economic factors described earlier.
32. Steering onto a sustainable path will require a much greater awareness of our problems and will also require a new system of ethics, a sustainable ethic.
33. Sustainable ethics holds that the Earth has a limited supply of resources, and that not all of these resources are for humankind. It asserts that other species share the Earth's wealth and have a right to prosper as much as we do. It also holds that humans are a part of nature and that the key to success is through cooperation with nature, not domination. The core value of the new ethical proposition is that humans and nature are inextricably linked: we are dependent on natural systems, and our future depends on a healthy, well-functioning ecosystem.
34. One of the most influential thinkers on ethics was the late Aldo Leopold, a wildlife ecologist who proposed a land ethic in his writings 70 years ago. The land ethic held that humans were a part of a larger community that included the soil, water, plants, and animals—in short, the land. By recognizing that humans are a part of nature, Leopold argued that the role of human beings would shift from "conqueror… to plain members and citizens of it."
35. This book proposes five operational principles for putting sustainable ethics into action: (1) conservation, (2) recycling, (3) renewable resource use, (4) restoration, and (5) population control.
36. One of the ethical underpinnings of sustainable ethics is the notion of intergenerational equity. It asserts that the present generations hold the Earth in common with all generations, past and future, who share certain rights and obligations. Most important, we have the right to benefit from the Earth—to profit from its riches and enjoy its beauty. We also have an obligation to protect the Earth for future generations so that they may benefit as well.
37. Creating a sustainable ethic will require worldwide efforts on the part of citizens, teachers, government officials, business people, and others.
38. Critical thinking is also essential to building a sustainable future. We define critical thinking as a means of analyzing information to distinguish between beliefs (what people believe is true) from knowledge (facts supported by scientific observations).
39. Critical thinking is used to analyze the results of scientific research, environmental issues, and proposed solutions.
40. Table 2.1 summarizes six critical thinking rules.

Key Words and Phrases

Brownfields
Command Economies
Control Group
Cost-Benefit Analysis
Critical Thinking
Discount Rate
Ecological Justice
Economic Externalities
Economic Incentives
Economics
Environmental Equity
Environmental Justice
Epidemiology
Ethics
Experimental Group
Experimental Variable
Frontier Ethics
Full-Cost Pricing
Green Tax
Gross Domestic Product (GDP)
Gross National Product (GNP)
Hypothesis
Index of Sustainable Economic Welfare
Intergenerational Equity
Intragenerational Equity
Land Ethic
Law of Diminishing Return
Law of Supply and Demand
Life Cycle Cost
Marketable Permits
Natural Capital
Opportunity Cost
Pollution Control
Pollution Prevention
Subsistence Economy
Sustainable Ethics
Theory
Time Preference
Tradable Permits
User Fee

Critical Thinking and Discussion Questions

1. What does it mean to say that the Earth's ecosystems are the infra-infrastructure of our society?
2. Describe the law of supply and demand. What are its weaknesses from an environmental standpoint? How can these be corrected?
3. What is an economic externality? How can such costs be internalized? Do pollution control devices internalize externalities? How do pollution prevention techniques compare with pollution control techniques?
4. Critically analyze the following statement: "The gross national product is not a measure of our welfare, despite what economists say." Do you agree or disagree with this statement? Why?
5. Outline general measures required to build a sustainable economic system.

6. How do time preference, opportunity costs, and discount rates influence resource management and environmental decisions?
7. Not all opportunity costs relate to a loss of present income. Explain what this statement means.
8. Draw a graph of pollution costs and the cost of pollution control. What is the break-even point? Is the break-even point an accurate representation of the equalization of costs and benefits? Why or why not? If not, how can it be adjusted?
9. Critically analyze the assertion: "Environmental protection is bad for the economy."
10. What is meant when one says that environmental quality is an important economic factor?
11. How could natural resource inventories, green taxes, and full-cost pricing help correct flaws in the economic system?
12. What is meant by harnessing market forces to protect the environment? Give some specific examples.
13. Outline your beliefs regarding humans and the environment. Do they correspond more closely with those of frontier ethics or sustainable ethics?
14. Where have your values come from?
15. In what ways are ethics and economics complementary forces? In other words, how can these two be used together to foster a transition to a sustainable society?
16. Describe the tenets of sustainable ethics. Are these values utilitarian?
17. Define the following terms: intergenerational equity, intragenerational equity, social justice. Do you agree with these principles? How can they be put into action?
18. What is critical thinking? Describe each of the rules outlined in the chapter.
19. Define the following terms: hypothesis, experiment, and theory.

Suggested Readings

Atkisson, Alan. 1999. *Believing Cassandra: An Optimist Looks at a Pessimist's World.* White River Junction, VT: Chelsea Green. An entertaining and thoughtful book on sustainability.

Berry, W. 1987. *Home Economics.* Berkeley: North Point Press. Thoughtful collection of essays on wise stewardship.

Brown, L. R., C. Flavin, and S. Postel. 1999. *Saving the Planet: How to Shape an Environmentally Sustainable Economy.* New York: Norton. Contains considerable information on sustainable economics.

Chapman, A. R., R. L. Petersen, and B. Smith-Moran. 1999. *Consumption, Population, and Sustainability: Perspectives from Science and Religion.* Washington, D.C. Important reading on ethics.

Chiras, D. D. 1992. *Lessons From Nature: Learning to Live Sustainably on the Earth.* Washington, D. C.: Island Press. See Chapters 2 and 3 for a detailed discussion of sustainable ethics.

Chiras, D. D. 1992. Teaching critical thinking in the biology and environmental science classrooms. In *American Biology Teacher* 54(8): 464–468.

Chiras, D.D. (Ed.). 1995. *Voices for the Earth: Vital Ideas from America's Best Environmental Books.* Boulder, Colo.: Johnson Books. A collection of essays on sustainability with several important contributions from ecological economists.

Cobb, C., T. Halstead, and J. Rowe. 1995. If the GDP is up, why is America down? *The Atlantic Monthly* 276(4): 59–78.

Cole, K. C. 1985. Is there such a thing as scientific objectivity? *Discover* 6(9): 98–99. Insightful look at science and the scientific method.

DeVall, B. 1998. *Simple in Means, Rich in Ends: Practicing Deep Ecology.* Salt Lake City: Peregrine Smith. Important discussion of deep ecology and ways to put reverence for nature into action.

Ford, A. 1999. *Modeling the Environment: An Introduction to System Dynamics Modeling of Environmental Systems.* Washington, D.C.: Island Press. Introduction to system dynamics designed to help students learn principles of modeling and develop models themselves.

Goodstein, E. 1999. *The Trade-Off Myth: Fact and Fiction about Jobs and the Environment.* Washington, D.C.: Island Press. Examines a deeply held belief, notably that environmental protection threatens jobs.

Gowdy, J. 1998. *Limited Wants, Unlimited Means: A Reader on Hunter-Gatherer Economics and the Environment.* Washington, D.C.: Island Press. A detailed look at subsistence economics mentioned in the chapter.

Hanna, S., C. Folke, and M. Karl-Goran. 1996. *Rights to Nature: Ecological, Economic, Cultural, and Political Principles of Institutions for the Environment.* Washington, D.C.: Island Press. Discusses rights, rules, and responsibilities that guide and control human use of the environment.

Krishnan, R., J.M. Harris, and N. R. Goodwin, (Eds.). 1995. *A Survey of Ecological Economics.* Washington, D.C.: Island Press. A useful collection of writings on ecological economics.

Leopold, A. 1966. *A Sand County Almanac.* New York: Ballantine. Collection of essays on nature and conservation.

Leopold, A. 1999. *For the Health of the Land: Previously Unpublished Essays and Other Writings.* Edited by J. Baird Callicott and Eric T. Freyfogle. Washington, D.C.: Island Press. A collection of Leopold's writings that focuses on the notion of land health and ways to protect private land.

Meadows, D. 1991. *The Global Citizen.* Washington, D.C.: Island Press. Collection of short, insightful essays on a variety of environmental topics, especially values.

Nash, R. F. 1989. *The Rights of Nature: A History of Environmental Ethics.* Madison: University of Wisconsin Press. In-depth analysis of the history of environmental ethics.

Pinchot, G. 1999. *Breaking New Ground.* Washington, D.C.: Island Press. The autobiography of Gifford Pinchot. Essential reading for those interested in learning the basis of our present national forest policy.

Repetto, R. and W. B. Magrath. 1988. *Natural Resources Accounting*. Washington, D.C.: World Resources Institute. An investigation into the true economic value of our natural resources.

Rolston, H. 1987. *Environmental Ethics: Duties to and Values in the Natural World.* Philadelphia: Temple University Press. Explains the rights of other creatures.

Sarewitz, D., R. A. Pielke Jr., and R. Byerly Jr. 2000. *Prediction: Science, Decision Making, and the Future of Nature.* Washington, D.C.: Island Press. Examines predictive science and how it is used and can be better used in making policy.

Schumacher, E. F. 1977. *Small Is Beautiful: Economics as if People Mattered.* New York: Harper and Row. One of the best books ever written on the subject of a sustainable society and new ethical systems.

Web Explorations

Online resources for this chapter are on the World Wide Web at: **http://www.prenhall.com/chiras** *(click on the Table of Contents link and then select Chapter 2).*

Chapter 3

Lessons From Ecology

To understand environmental problems and their solutions, you must have a firm understanding of ecology. **Ecology** is a field of study that attempts to uncover important relationships in the living world—more specifically, relationships between organisms and their environment. Ecologists study how organisms interact with one another and how they interact with their environment. A good understanding of basic ecological concepts will help you to appreciate the problems facing conservationists, environmentalists, and resource managers as they grapple with complex resource management issues. It will also help you understand how problems in human systems—for example, transportation, housing and waste management—might be solved and how society might be nudged onto a sustainable path.

3.1 Levels of Organization

To begin to develop our understanding of ecology, it helps to begin by examining one of the outstanding characteristics of living organisms, their organization. In ascending order of complexity, the organizational levels of an organism are the atom, molecule, cell, tissue, organ, organ system, and organism. Although ecologists are concerned with each of these levels, their attention focuses primarily on levels of organization above that of the organism: the population, community, and ecological system or ecosystem (Figure 3.1). That is to say, they examine interactions occurring in populations, communities, and ecosystems (defined shortly). Let's consider each one.

Population

When the laypersons use the term **population,** they are invariably referring to the number of humans in a given locality. For example, they might be referring to the number of people in Detroit, Michigan or Reno, Nevada, or Stuart, Florida. Ecologists, however, use the term to include any organism, human or nonhuman. They may talk about (and study) the population of white-tailed deer in New York or the population in the entire country. Or

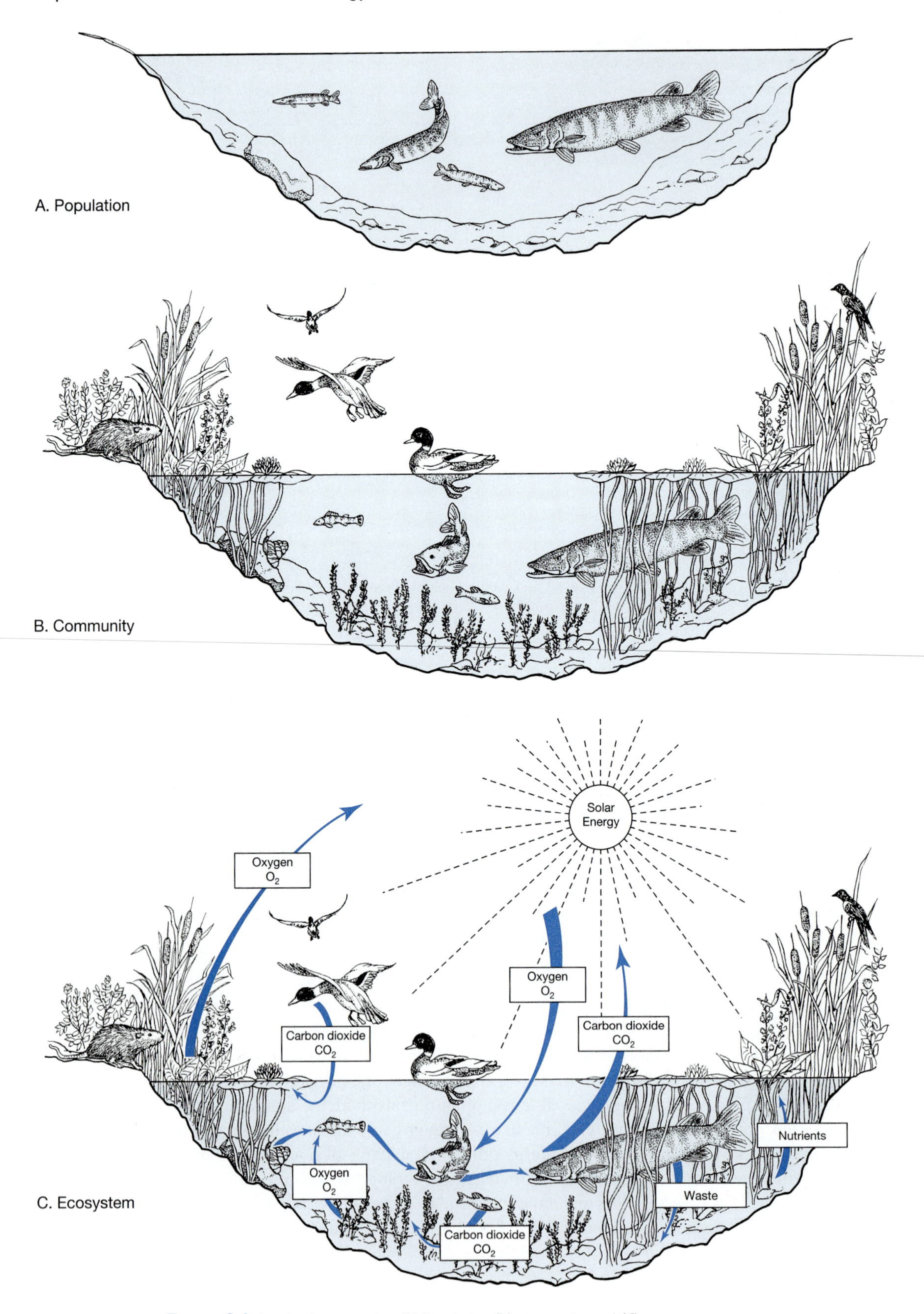

FIGURE 3.1 *Levels of organization. (A) Population, (B) community, and (C) ecosystem.*

they may refer to populations of white pine in a forest, pike in a river, or fleas on your pet dog or cat (Figure 3.1A). Some ecologists specialize in studying populations—for example, researching factors that influence their growth and decline. Wildlife ecologists are especially concerned about populations.

Community

Ecologists are also concerned about biological communities. A **biological community** is defined as all of the living organisms occupying a given locality. As you can see from Figure 3.1B, a biological community is comprised of two or more populations living in a certain location. Communities typically contain populations of plants, animals, and microorganisms that interact in many ways. These interactions can be quite crucial to the survival of the whole, and for this reason more and more resource management plans are being redrafted to protect the community of life in a region in its totality, rather than focusing narrowly on one or two populations. Ecosystem management, mentioned in Chapter 1, requires community protection.

Ecosystem

You can't protect a community of organisms while ignoring its physical environment. The two are interdependent. Together, organisms and their environment comprise **ecological systems** or **ecosystem,** for short (Figure 3.1C). Therefore all ecosystems are said to consist of two interactive components, the living (biotic) and nonliving (abiotic).

Like so many things, ecosystems are often delineated artificially by humans. We might talk about the ecosystem of a lake or the ecosystem of a meadow or forest. As we will see in later chapters, most measures applied by modern science-based resource managers are forms of ecosystem manipulation. A good example is the removal of snow cover from an icebound lake to prevent the winter kill of fish. This permits sunlight to penetrate the ice, making it available to aquatic plants for photosynthesis. The resultant increase in dissolved oxygen may prevent massive fish mortality.

In reality, all ecosystems are part of one large global ecosystem, the **biosphere** or **ecosphere.** The biosphere extends from the bottom of the oceans to the tops of the tallest mountains, although life is scarce at the extremes. Most life forms exist in a much narrower range where conditions are more conducive to survival.

In the living world, organisms may move from one ecosystem to another, whether immediately adjacent or thousands of miles distant. Many waterfowl, for example, spend their summers in Alaska or Northern Canada, but winter in the warmer waters of the southern United States. Chemical nutrients and energy, which we'll discuss shortly, also move from one ecosystem to another. This is accomplished by biological (animal migration), meteorological (dust storms and hurricanes), and geological (flowing rivers and volcanic eruptions) processes. Thus, topsoil containing nutrients from an Oklahoma wheat field may be washed by spring rains into a nearby stream, which flows into larger rivers that reach the ocean. Phosphorus originating in deep marine sediments may eventually be transferred to terrestrial ecosystems in bird droppings from seabirds that feed on fish, which in turn are nourished by crustaceans that eat algae, which absorb phosphorus from the ocean. Because of this, all ecosystems on Earth are linked and ecosystem disruption can have far-reaching effects.

3.2 Scientific Principles Relevant to Ecology

With this brief overview of the organization of life and the living world, we now turn out attention to a few scientific principles vital to our understanding of ecology. We begin with a discussion of matter and energy—two key components of the systems you will be studying in this book.

The Law of Conservation of Matter

Organisms and their physical environment are all composed of matter. **Matter** is defined as anything that occupies space and is perceptible to the senses. Ice cream, for instance, is a form of matter, as are water and lead.

Matter exists in two basic forms: organic and inorganic. Organic matter refers to substances composed primarily of carbon, hydrogen, and oxygen. Proteins in the cells of your body or the starch in rice are two forms of organic matter. Inorganic matter consists of mineral-based substances like salt or aluminum ore.

Matter can be altered in many ways, but one thing scientists have learned is that it cannot be created or destroyed. That is, you cannot magically create matter, nor can you destroy it. All you can do is change it from one form to another. This important law is known as the **law of conservation of matter.**

Restated, *the law of conservation of matter states that although matter can be changed from one form to another, it can neither be created nor destroyed by ordinary physical and chemical means.* As an example, the flesh of one organism can be consumed by another, at which time it becomes part of the second organism. The first organism hasn't been destroyed, it has been eaten and its components reassembled to make parts of a new organism. By the same token, the molecules in your body are not new creations, they're made from atoms and molecules that have been around since the Earth formed. They're just being recycled over and over again to support new generations of living things. Without this recycling process, life could not exist.

The law of conservation of matter has many important implications. First, it helps us understand the massive pollution problems facing the world. The United States, for example, is consuming many of its natural resources (matter) at a record-breaking pace to support a lifestyle enjoyed by few other nations on the Earth. But the matter does not disappear. Some of it is converted into useful products. When they lose their usefulness or fall out of fashion, it becomes waste—for example, the solid waste accumulating in landfills throughout the nation. The law of conservation of matter reminds us that waste remains, in one form or another, forever. Some ecologists remind us that when we throw something away, there really is no "away." Things we discard do not miraculously vanish. They're always somewhere—and they may come back to haunt us. For example, if a landfill is not properly designed, some of the waste may leach into the groundwater that we later drink. You can't even burn trash and think you've truly gotten rid of it. Incineration certainly reduces the volume of waste to a few ashes, but the combustion process generates large amounts of smoke and gases that linger in the atmosphere or rain down upon the land causing a host of health and environmental problems.

The Laws of Energy

The leap of a tiger, the beat of a heart, the scream of an eagle, the turn of a wheel, the dip of a canoe paddle—all of these seemingly diverse events have something in common: they require energy. What is energy?

Physicists define **energy** as the ability to do work or cause change. Lifting a mountain bike to climb a steep path may be your idea of having fun, but to a physicist, it's work.

Energy plays a vital role in the living world. It is used by organisms to make molecules and it provides motive force—that is, it allows organisms to move about in their environment.

Energy can be found in a highly concentrated state such as in foods or fuels such as gasoline. It can also be found in a more dispersed or disorganized states such as heat (also known as thermal energy). Energy in a concentrated state can perform a great deal of useful work and is considered to be of high quality. Energy in the dispersed condition, on the other hand, cannot perform as much work and therefore is considered to be of relatively low quality. Don't let these labels confuse you, both forms are essential to life on Earth.

Energy exists in many different forms. The two main types are potential energy and kinetic energy. **Potential energy** is stored energy. Coal, gasoline, jet fuel, and food contain lots of potential energy. The potential energy in fuels is released when it is burned. The potential energy of food is released when it is broken down in cells of the body. **Kinetic energy** (kin-et-ick) is the energy of motion. A moving car has kinetic energy. As does a ball flying through the air or a hammer pounding a nail.

Physicists, engineers, and ecologists apply other labels to different forms of energy. You will read about chemical energy, mechanical energy, thermal energy, nuclear energy, and so on as you work your way through this book.

First Law of Energy

Energy, like matter, is governed by several laws, known as the **laws of thermodynamics** or, simply, the **energy laws.** As you shall soon see, these laws profoundly influence the structure of the biosphere—and the living organisms themselves. The **first law of energy** states that *energy cannot be created or destroyed, it can be converted from one form to another.* Just like matter! Energy locked in the fuel molecules in gasoline, for instance, is released in the car's engine, but it is not destroyed. It is converted into heat and mechanical energy which propels the car along the highway.

Just so you really understand the concept of energy conversions, take a look at Figure 3.2. On the right side of the figure is a lamp, like the one you study under. As illustrated, the light given off by the lamp comes from a power plant (generator). The power plant burns coal. When coal is burned at a power plant, the heat (thermal energy) it produces boils water, creating steam that turns a turbine. The turbine is attached to an electrical device known as generator, so named because it generates electricity by simply rotating a magnet through a huge mass of wires. But where did coal get its energy?

The energy contained in the coal comes from sunlight, more specifically sunlight that shone on the Earth several hundred million years ago. This energy was captured by plants in a process known as **photosynthesis** (defined more fully later). During photosynthesis plants use sunlight, water, and carbon dioxide to produce organic molecules. The chemical bonds in these molecules contain energy that can be used by animals that feed on the plants. In ancient times, the plants growing around swamps that were not eaten were often buried in sediment. Heat and pressure from the accumulating sediment converted the plant matter to coal. Today, when this coal is burned by the electrical power industry, the sunlight energy locked in the organic molecules in the plant matter is released. That is, it is converted into heat energy, which is used to generate steam in power plants.

This example shows the true origin of much of the fuel that powers our society, the sun, but it also illustrates the conversion of energy from one form to another. In fact, in this sequence energy undergoes six conversions.

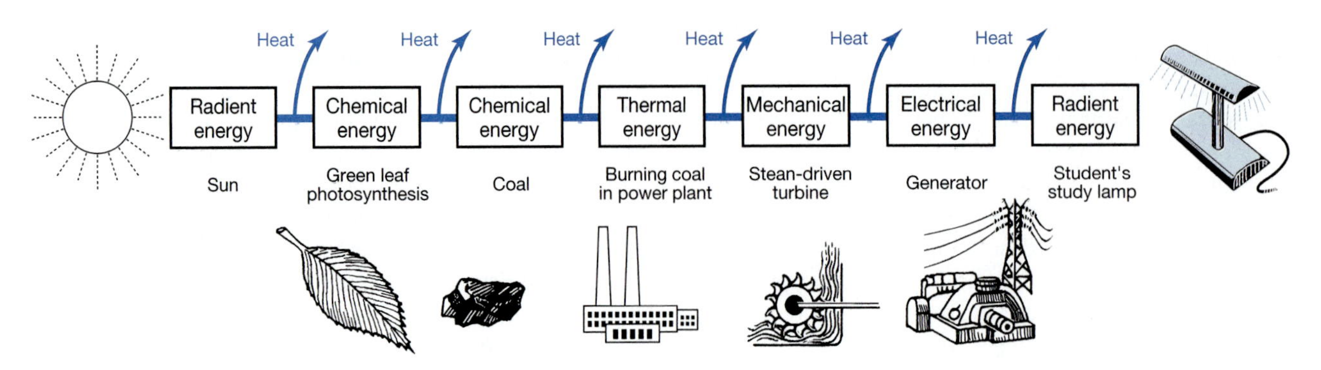

FIGURE 3.2 *Different forms of energy. Energy can be changed from one form to another. However, with each change a certain amount of heat is lost as heat.*

Take a moment to study Figure 3.2 again to see the changes. As you do, you may notice something else that occurs in this process. Note that heat is lost at each conversion. This brings us to the second law.

Second Law of Energy The **second law of energy** states: whenever energy is converted from one form to another, a certain amount is lost in the form of heat. It is as inevitable as death and taxes. Put another way, no conversion is 100 percent efficient. In fact, most conversions are closer to 20 to 40 percent efficient.

When we say that "heat is lost during an energy conversion," what we really mean is that heat is given off. Heat isn't lost, *per se.* It can even be captured to perform additional work. For example, heat given off by a boiler designed to provide hot water to heat an office building can be captured and used to boil water to make electricity. This process is known as cogeneration and is discussed in Chapter 22.

The important point in all of this is that during the successive energy conversions, heat is given off and the total amount of high quality energy tends to decrease, so overall there's less high quality energy available to do work. Let's test your new found knowledge. Which would provide more energy for cooking food: natural gas burned directly at the burner of a gas range or electricity made from natural gas burned at an electrical generating plant? If you answered natural gas, you're right, because there's one less conversion.

Before we leave the topic, let's examine another common device, the automobile. You may be surprised to learn that only about 25 percent of the high-quality chemical energy in the gasoline "consumed" by most car engines engine is converted into high-quality kinetic energy that propels you along the highway. About 75 percent of the energy in the gasoline is converted to low-quality heat energy—useless as power but nice to have for warming your car on a winter trip. Eventually this heat energy radiates from the car and cannot be used again. As a side note, some innovators are looking for ways to greatly increase the efficiency of automobiles so that instead of getting 20 to 30 miles per gallon, which is typical, new cars could get as much as 100 to 200 miles per gallon (more on this in Chapter 22).

The second law of energy also operates in living systems. For example, during photosynthesis, high-quality solar energy is converted to high-quality chemical energy by green plants. When you eat plants, whether beans or bananas, the chemical energy of the food is eventually converted by your body to the high-quality chemical energy stored in the cells of your body. Sooner or later, your body converts this energy into kinetic energy (energy of motion) that powers such life-sustaining activities as breathing, the beating of the heart, and the muscular contractions of such organs as the stomach, intestine, and throat. However, here again, with each conversion, from solar energy to chemical energy and from chemical energy to kinetic energy, a certain amount of energy is lost as heat. Of course, the heat is not wasted—it keeps our body temperature at about 37°C (98.6°F). Eventually, however, it radiates into the surrounding atmosphere and then into outer space. It isn't lost; but it is no longer available to perform work here on Earth.

Entropy This brings up another important concept, known as **entropy** (en-trow-pee). Scientists use the term entropy to refer to matter and energy. More specifically, they use it to refer to the degree of disorder in a system, known as entropy (Table 3.1). All systems move toward maximum disorder or maximum entropy. Think of your room as an example. It grows disordered slowly but surely over time.

Biological systems tend to move toward disorder as energy is continuously being lost from the bodies of organisms in the form of heat. If the energy in a particular system is largely in a dispersed state—that is, its in the form of heat—we say the system shows a high degree of disorder.

Biological systems also tend toward disorder with respect to matter. But how is it that living organisms retain their structure, their organization, their order which is so essential to the continuation of life? They do it because they receive a constant input of energy to keep making molecules needed for the structures of their bodies—to repair and maintain themselves. Animals receive high-quality energy from food they eat. The energy in the food they eat comes from plants that obtain their energy from sunlight.

Matter and Energy Laws: What Can We Learn from Them? The activities of all organisms, from beans to bananas, from hummingbirds to humans, are under the control of the basic laws of matter and energy we just described. These laws also control the activities of all ecosystems on Earth. In addition, they affect a whole series of environmental problems. These laws provide us with a key to understanding (1) the urgency of environmental problems facing all of us and (2) how those problems can be solved or brought under control.

To recap, the law of conservation of matter, tells us that there is no "away." What we throw away ends up somewhere. The carbon molecules in the fossil fuels we burn in our cars, jet planes, factories, and power plants end up in the atmosphere as billions of tons of carbon dioxide. Here, it has the power to alter climate, returning the planet to a much hotter era, similar to that when the dinosaurs roamed the Earth. Simply put, all that carbon that was in the atmosphere back then and was responsible for the warmer climate, is now being released by the combustion of fossil fuels and may well be causing the Earth to warm up again.

TABLE 3.1 *Concept of Entropy*

High Organization Low Entropy		Low Organization High Entropy
Gasoline	→	Movement of car
Electricity	→	Light and heat from study lamp
Waterfall	→	Water flowing downstream
Volcanic eruption	→	"Rain" of volcanic ash
Candy bar (sugar)	→	Body heat

The first law of thermodynamics tells us that energy is neither created nor destroyed, but the second law says that energy is degraded at each conversion. For a society that relies primarily on high-quality energy from finite supplies of fossil fuels, the laws of energy tell us that each time we burn fuel we're depleting our supply. In other words, we're using up valuable supplies that cannot be replenished. Energy cannot be recycled as it all eventually dissipates into space as heat. Sooner or later, we'll have to find new sources.

3.3 The Flow of Energy Through Ecosystems

The sun is the source of virtually all energy in the biosphere. It heats the Earth and causes winds that turn giant windmills or wind machines that generate electricity. It evaporates water, allowing for rainfall and stream flow. The energy of flowing water can be tapped by dams equipped with turbines and generators to produce electricity. More important for living organisms, sunlight is captured by plants and stored in the molecules they make. This energy is used by plants and a wide assortment of animals and even most microorganisms.

The sunshine that warms you on your way to class reached your skin only 8 minutes after leaving the sun's surface, roughly 155 million kilometers (93 million miles) away. The sun releases many forms of energy. As shown in Figure 3.3, the visible light (sunshine) we are familiar with forms only a small portion of the sun's total energy emission or radiant energy, known as the **electromagnetic spectrum.**

The types of energy represented in the spectrum range from low-energy radio waves to high-energy gamma waves. Each type of electromagnetic radiation has two vital characteristics: wavelength and energy level. Let's begin with wavelength.

Electromagnetic radiation from the sun is transmitted in the form of waves. They are like the waves in the ocean. The distance between two successive wave peaks is known as the **wavelength.** As shown in Figure 3.4, radio waves have a long wave length. The wavelength of visible light is intermediate. Gamma rays have a relatively short wavelength.

Energy levels also vary. Radio waves are low energy. Light is intermediate. Gamma rays are the highest energy radiation the sun produces.

Although the sun produces a wide range of electromagnetic radiation, visible light and heat (infrared radiation) are generally the most useful. Visible light is detected by animals (although some can perceive ultraviolet as well). Some colors of visible light are absorbed by plants during photosynthesis. Gamma rays, X-rays, and ultraviolet radiation, however, are all harmful. We'll learn more about them in later chapters.

Solar Energy Flow

As shown in Figure 3.4, not all of the sun's radiant energy makes it to the Earth's surface. In fact, 32 percent is reflected back into space by dust and clouds in the atmosphere and from the Earth's surface—either from land, water, soil, or vegetation. This reflectivity is known as the **albedo** (al-bee-doe).

As shown in Figure 3.4, 67 percent of the incoming solar energy is absorbed by the Earth's atmosphere, land, water, and vegetation. It is converted to heat. Although this heat will eventually dissipate into space, it performs some vital functions before it escapes. For example, heat from absorbed sunlight warms the Earth's surface (both land and water). It also causes evaporation of water from land and water. The water, of

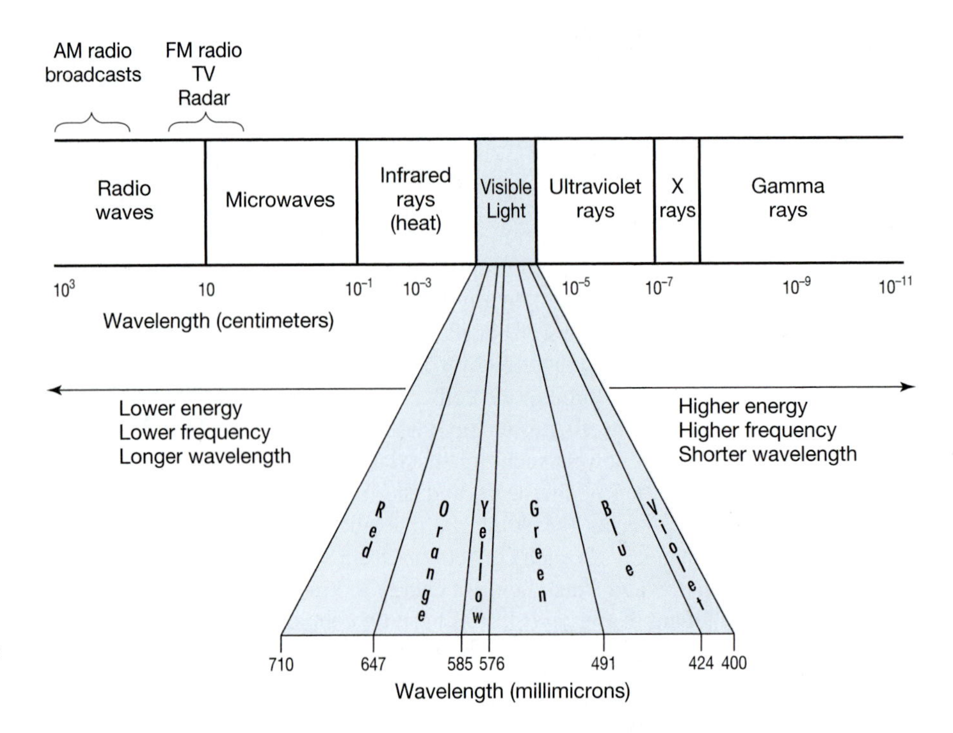

FIGURE 3.3 *The electromagnetic spectrum. The sun produces many different forms of radiation, some useful, some potentially dangerous.*

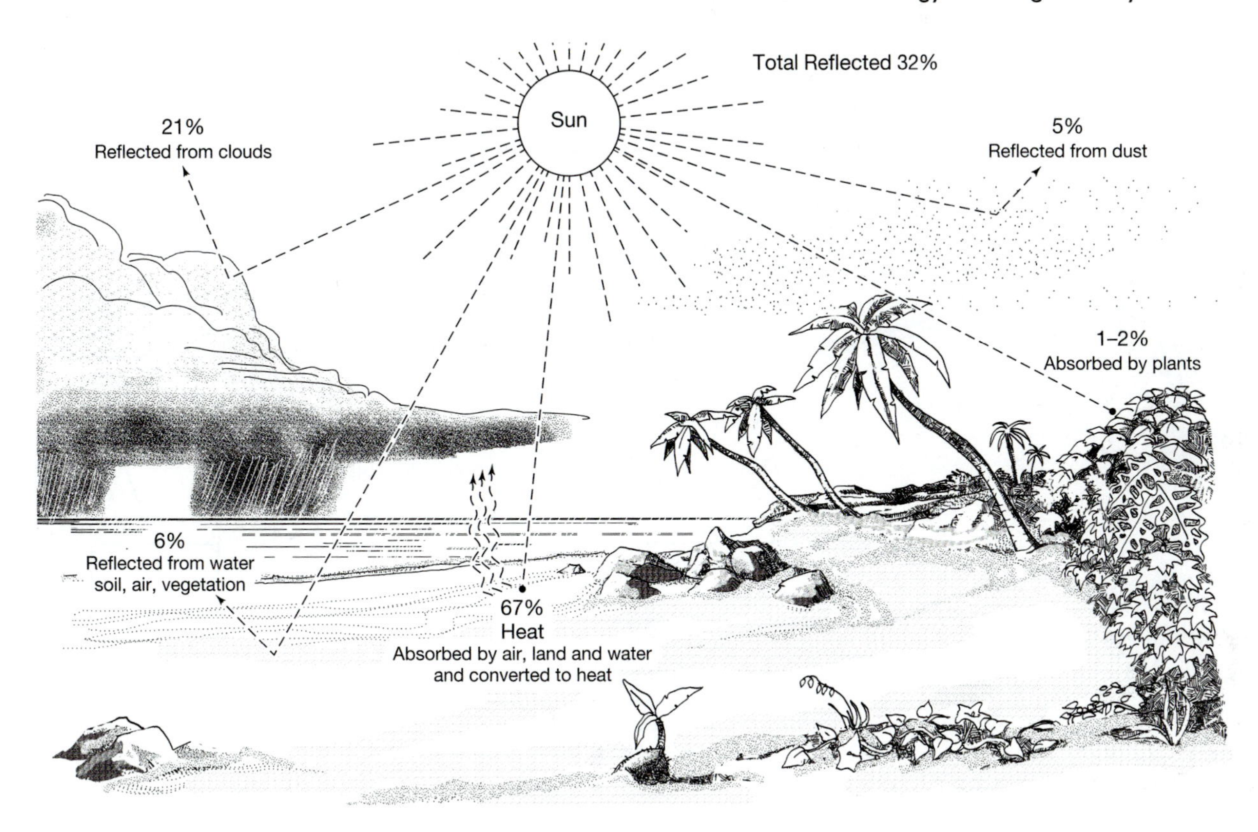

FIGURE 3.4 *The global flow of energy. Virtually all energy that strikes the Earth escapes as heat. Only a tiny portion of the sun's energy is absorbed by plants for photosynthesis.*

course, is later deposited as rain and snow. Heating of the Earth's surface also produces winds, waves, and weather. Interestingly, only a tiny fraction of the sunlight striking the Earth is used by photosynthetic organisms—somewhere around 1 to 2 percent, according to some estimates.

Solar energy flow and heat loss can be altered by human activities. Agricultural practices that result in a loss of vegetative cover, such as overgrazing, for example, can decrease albedo, causing the Earth's surface to heat up. Tropical deforestation has a similar effect. Pollutants in the atmosphere can also alter these energy flows. Over 100 years ago, scientists found that certain naturally occurring chemicals in the atmosphere, such as carbon dioxide molecules and water vapor, retard the release of heat into outer space. Like a giant blanket surrounding the Earth, these chemicals cause heat to be retained in the Earth's atmosphere, creating a climate suitable for life. In fact, without them the Earth would be much much colder and virtually uninhabitable.

In the last 100 years, however, humans have begun to burn massive quantities of fossil fuels. This, in turn, has resulted in a substantial increase in atmospheric carbon dioxide. There's a growing body of scientific evidence that suggests that the release of massive quantities of carbon dioxide (and other factors) may be causing the temperature of the Earth's atmosphere to increase, a phenomenon known as **global warming.** This, in turn, could have dramatic effects on ecosystems, economies, and human civilization as you shall see in Chapter 19. As hard as it is to believe, humans may be profoundly influencing solar energy flow and the climate of the entire planet.

Photosynthesis and Respiration

As noted earlier, all of the energy that powers the biosphere, from the growth of a cabbage to the beating of the human heart, can be traced back to its original source, the sun. Sunlight energy is captured by plants during photosynthesis. During photosynthesis, plants use solar energy to convert carbon dioxide and water into sugar, mostly glucose. With a few minor exceptions, this process can occur only in the presence of **chlorophyll,** a green pigment found in algae, some bacteria, and plants (Figure 3.5). Chlorophyll acts like a solar collector in the cell, gathering up sunlight energy. The general equation for photosynthesis is

$$\text{solar energy} + \underset{\text{(carbon dioxide)}}{6CO_2} + \underset{\text{(water)}}{6H_2O} \rightarrow \underset{\text{(sugar)}}{C_6H_{12}O_6} + \underset{\text{(oxygen)}}{6O_2}$$

Photosynthesis produces two important substances, sugar (glucose) and oxygen. Glucose is used as a source of energy by the plant itself and by animals feeding on plants and one another.

Some of the oxygen produced by photosynthesis is also used directly by the plant, but the rest passes from the leaf through microscopic pores into the atmosphere. Here the oxygen may be used by other organisms, from bacteria to humans. As you shall soon see, oxygen is used to help us breakdown sugars to produce energy.

An understanding of photosynthesis makes it clear why plants are so important to us and why it is important to protect

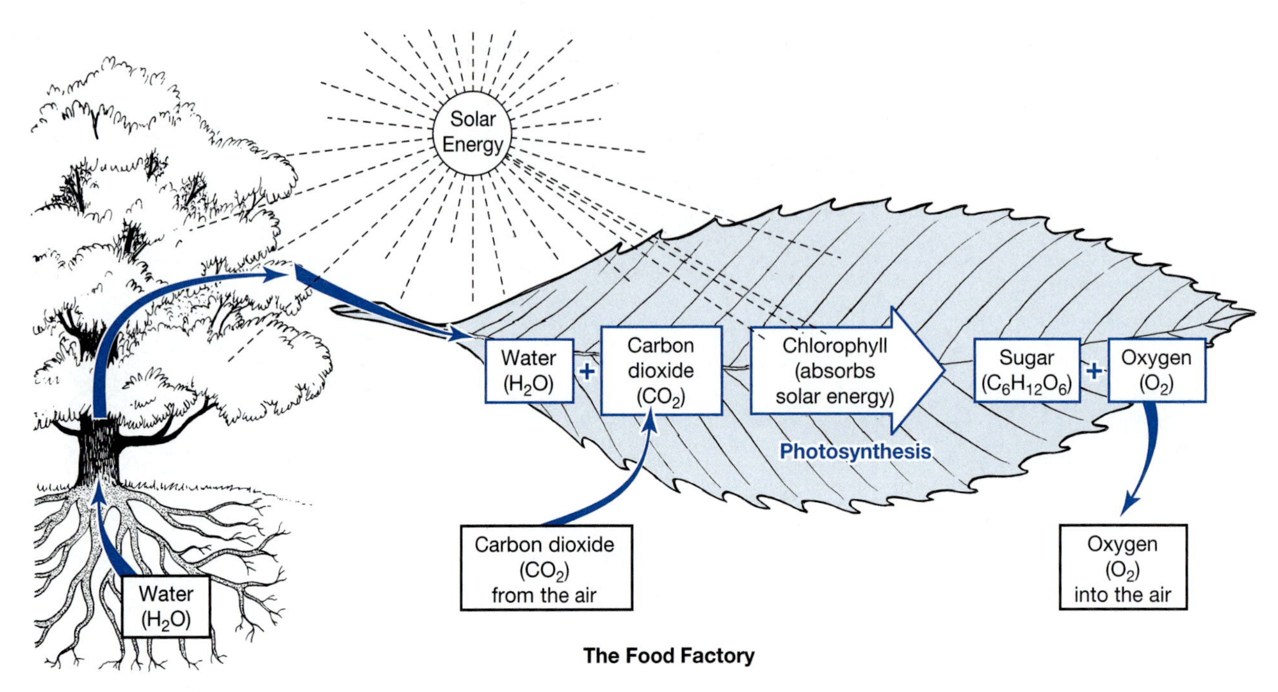

FIGURE 3.5 *The food factory. Leaves absorb carbon dioxide and sunlight. Water is taken up by the roots. The solar energy powers the process of photosynthesis by which water and carbon dioxide are used as raw materials in the production of sugar.*

the Earth's terrestrial and aquatic plant life. Unfortunately, many plants are in trouble. Large tracts of forest have been cleared to produce timber or make way for mines, farms, or human settlements, wiping out native vegetation and decreasing the planet's ability to produce oxygen. Some ecologists are concerned that the progressive contamination of the oceans with chemical pesticides and industrial wastes may sharply reduce the photosynthetic activity of marine (saltwater-inhabiting) algae and, over time, diminish the Earth's supply of atmospheric oxygen—a harmful consequence for all animals, including us humans.

The breakdown of sugars by plants and animals to produce energy is called **cellular respiration.** The general equation for respiration is:

$$\underset{\text{(Sugar)}}{C_6H_{12}O_6} + \underset{\text{(Oxygen)}}{6O_2} \rightarrow \underset{\text{(Carbon dioxide)}}{6CO_2} + \underset{\text{(Water)}}{6H_2O} + \text{energy}$$

If this formula appears familiar, it should. It is exactly the opposite of the equation for photosynthesis. In the living world, then, plants produce glucose and oxygen, which are inputs of cellular respiration. The products of cellular respiration, in turn, become the raw materials of photosynthesis, creating a nifty recycling of materials that is vital to the continuation of life on Earth.

Primary Production and Net Production

Each year, terrestrial and aquatic plants produce an estimated 243 million metric tons of organic matter. This is known as the biosphere's **primary production.** Organic matter serves as a source of organic building blocks and as a source of energy for other organisms. But not all of the organic material plants produce is available to other organisms. Plants themselves consume some of the material in cellular respiration occurring in their own cells. They use it for growth (for example, root growth and budding) and general cell maintenance (maintenance of cell walls, for instance). For this reason, ecologists distinguish between **gross primary production,** the total amount of biomass produced and **net primary production,** that which remains after plants use their share to provide food and energy for cellular processes. The equation NPP = GPP − R describes the phenomenon mathematically (R = cellular respiration). Remember, unlike green plants, animals are able to capture energy only by consuming other organisms.

Primary production varies considerably by ecosystem. The most productive are tropical rain forests. They produce on average about 2,200 grams of biomass per square meter per year. Deciduous forests like those of New England produce about half as much, about 1,200 grams per square meter per year. Evergreen forests produce about 800 grams per square meter per year, while deserts produce only about 90 grams per year. Farmland averages out at about 650 grams per square meter per year.

Food Chains and Food Webs

Now that you see where food in the biosphere comes from, let's see what happens to it, looking first at food chains. A **food chain** is a feeding sequence in ecosystems. It illustrates who eats who and, in so doing, shows the path through which energy and nutrients move in ecosystems. Ecologists need to know about them to understand the relationships between organisms in the ecosystems they study. Let's consider a food chain that might operate in a wet meadow:

grass →	grasshopper →	frog →	snake →	hawk
(producer)	(primary consumer)	(secondary consumer)	(tertiary consumer)	(quaternary consumer)
	herbivore	carnivores		

In this food chain, shown in the center of Figure 3.6, grass is classified as a **producer** because it produces organic food molecules that nourish all other organisms. The remaining organisms that either feed directly on grass or feed on other organisms higher up in the food chain are known as **consumers.** Consumers are "ranked" by their location in the food chain. The grasshopper in the food chain is known as a **primary consumer.** It is also classified as an **herbivore** because it is a plant eater. The flesh-eating frog, snake, and hawk are all classified as consumers, too, but they're referred to more specifically as secondary, tertiary, and quaternary consumers. Because they only eat other animals, they're also known as **carnivores.** Food chains like this one that begin with a photosynthetic organism—like a green plant or algae—are known as **grazer food chains.**

In grazer food chains, plants are the main source of food. The chemical energy in plant material (derived from the sun) and organic matter they generate during photosynthesis are available to all "members" of the food chain. The plant matter generated in an ecosystem contains the energy that either directly or indirectly powers the activities of all organisms on the Earth.

Another type of consumer, not represented in the food chain diagram, is the **detritus feeder** (deh-TRITE-us). These organisms obtain their energy and nutrients from waste materials and the dead bodies of plants and animals—that is, detritus. Detritus feeders in a forest include microscopic organisms such as bacteria and fungi (mushrooms and molds), as well as macroscopic organisms such as maggots and termites. As a result of the activities of these organisms, large, complex molecules in plant and animal remains and waste products are broken down (decomposed) into smaller molecules such as nitrates. These compounds, in turn, are released into soil and water and are absorbed by plants, ensuring the continuation of life. Detritus feeders, therefore, are said to assist in nutrient recycling. In some shallow lakes, detritus is represented by decaying vegetation. This material is eaten by crayfish and snails, which in turn are consumed by fish.

Any food chain that has its base in the dead remains of plants and animals is known as a **detritus food chain** (Figure 3.7). Together, detritus and grazer food chains account for the movement of all nutrients and energy (except heat) in the biosphere. In an oak woods, roughly 90 percent of the **biomass** (dry weight of living organisms) eventually dies and enters detritus food chains. The remaining biomass is channeled through the grazer

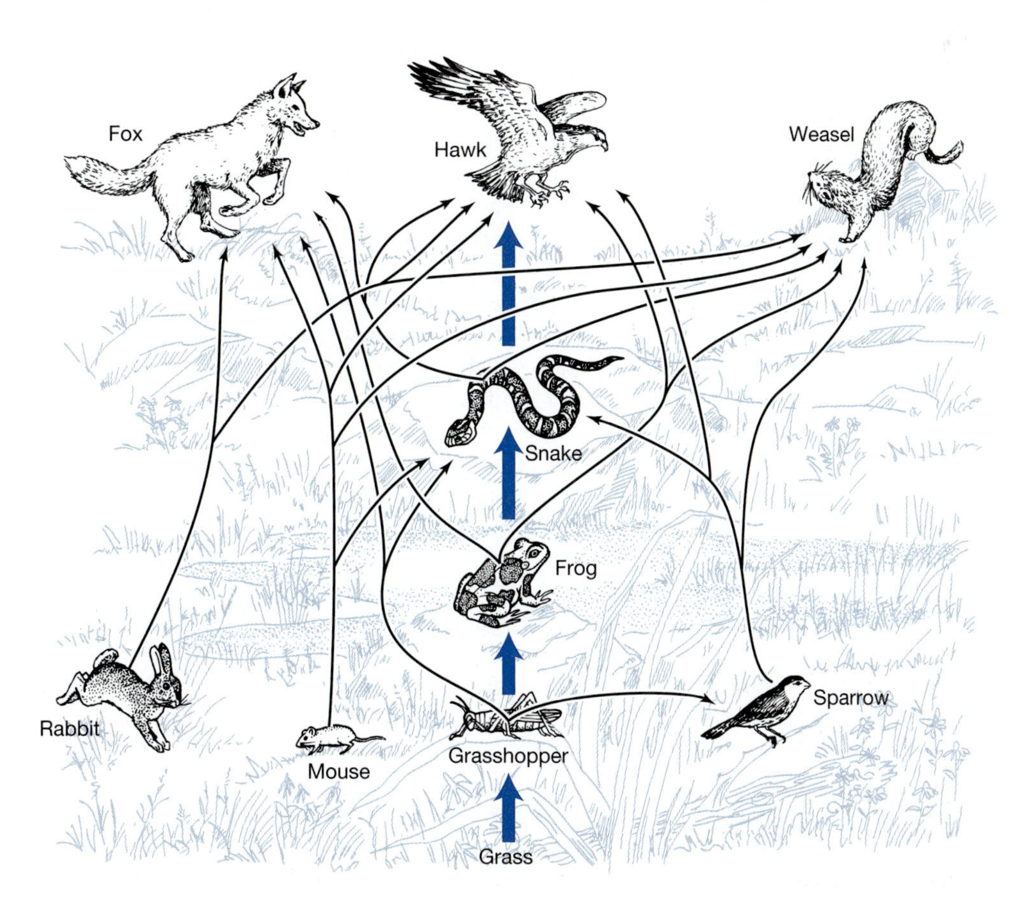

FIGURE 3.6 *Food webs and Food Chains. In the center of the drawing is a food chain. It is really a part of several other food chains, together forming a food web.*

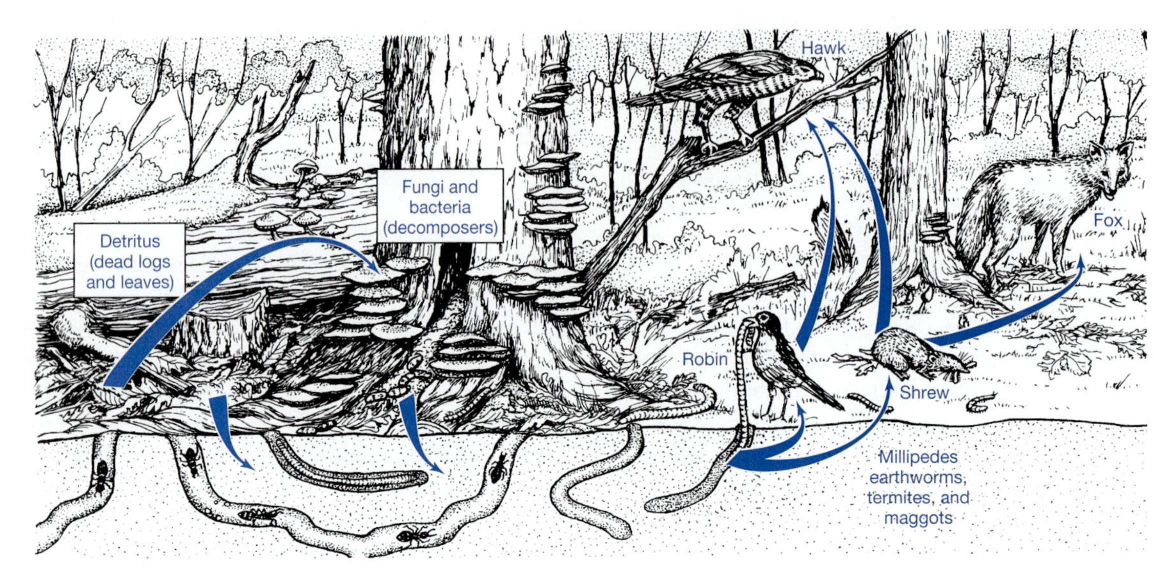

FIGURE 3.7 *Detritus food web of an oak woods. Highly simplified.*

food chains. In the open water of a lake, on the other hand, the primary food chains are grazer food chains. There the floating algae represent the principal producers. Roughly 90 percent of the algal producers are consumed by small crustaceans, which in turn are eaten by fish.

Food chains are easy to understand and help shed light on ecological relationships, but in the real world food chains virtually never exist as isolated entities. They're usually part of larger food webs, more complex feeding relationships. Consider the food chain corn → pig → human. You might assume by the diagram that a pig eats corn and, in turn, is eaten by humans. In reality, pigs eat much more. Pigs roaming the farm will eat rats, mice, insects, grubs, earthworms, baby chicks, grass, weeds, garbage, and even feces. Humans have a varied diet, too. The corn → pig → human food chain is really just part of a much more complex and interconnect food web.

We like to think of a food web as a collection of interconnected food chains. A food web provides a more accurate picture of the feeding relationships and the flow the energy and nutrients in ecosystems. It illustrates the fact that in ecosystems energy and nutrients travel through multiple pathways. The grass-grasshopper-frog-snake-hawk food chain of a wet meadow, discussed earlier, for example, in nature is part of a more complex food web (Figure 3.6). Even this food web shown in Figure 3.6 is still grossly oversimplified. The complete food web of a wet meadow includes hundreds of species.

As a general rule, the greater the number of channels through which energy can flow, the greater the stability of the food web and the ecosystem. Why? The reason is simple. The more strands in a food web, the stronger it is. The loss of one strand is not felt. Let us explain by example. In the wet meadow described earlier, foxes prey on rabbits, mice, grasshoppers, sparrows, frogs, and snakes. Suppose, however, that the rabbit population was reduced because of adverse weather during the breeding period or because wild dogs moved into the area. Under those conditions, the fox population would shift the predatory pressure it had exerted on rabbits to some (or all) of its alternative prey without suffering from nutritional hardship. Reduced predatory pressure on the rabbit population, in turn, might permit it to rebound quickly when breeding conditions were favorable. If the fox did not have other prey, however, rabbits might have been exterminated, or the foxes would have left the area or died.

As a general rule, the more organisms that are removed from a food web, the more vulnerable it is. Ecologists refer to the loss of species in a food web as **ecosystem simplification.** Meadows, for instance, contain many different species of plants, among them herbs, forbs, grasses, and wild flowers. Plowing the field and planting it in a crop of one species represents an extreme form of simplification that has many ecological ramifications. It not only destroys the plant life, it eliminates most, if not all, of the animals that depend on them. The new single-species ecosystem is also much more vulnerable to attack by insect or disease organisms. The Irish potato famine of the 1840s is an illuminating example of this phenomenon.

The potato was introduced to Ireland in the late sixteenth century. For many years, it was a staple of the Irish diet and yielded more calories per acre than any other crop. However, in 1845 a disease called the potato blight, caused by a fungus, began attacking Ireland's potato crop. With acre upon acre of field in potato, the fungus spread rapidly, causing massive destruction that lasted for the next five years. Because the Irish were highly dependent on only a few varieties of potato, which the fungus liked, and because few alternative foods were available, more than 1 million people died of starvation or disease. Another 1 million Irish left the country. The overall effect of the blight was a 25 percent decline in Ireland's population in just 5 years. It was all due to ecosystem simplification brought on by the planting of huge crops of a single food that allowed disease organisms to proliferate.

Trophic Levels and the Pyramids of Energy and Biomass

Before we leave the concept of food chains and food webs and move on to nutrient cycles, we must examine a couple additional concepts and terms. To do this, we begin with a food web in a lake ecosystem (Figure 3.8). As illustrated, each feeding level in that community is called a **trophic level** (trow-fic; trophic means to nourish). Thus, all the producers (algae, rooted plants, water lilies, cattails, and so on) form the first trophic level. All the plant eaters, known as primary consumers or herbivores, such as crustaceans and insects, form the second trophic level. The secondary consumers (carnivores), such as fish, feed on the primary consumers and form the next trophic level, and so on.

So far, our description of tropic levels has been greatly simplified for clarity. You must not get the idea that a given species of organism can belong to only one trophic level. For example, although some fish are either exclusively herbivorous or exclusively carnivorous, other fish, such as the carp, are omnivorous—they feed on both plants and animals. Therefore, carp are both primary and secondary consumers in the food web of a lake.

One thing that ecologists have discovered is that not all the energy and biomass (organic matter) in a given trophic level ends up in the next level. *In fact, as a general rule, only about 5 to 20 percent of the biomass and energy in one trophic level is transferred to the next. Ten percent is a typical transfer rate.* Thus, the primary consumers (herbivores) contain only 10 percent of the biomass and food energy of the producer (green plant) level. Similarly, the secondary consumers (carnivores) contain only ten percent of the biomass and food energy present in the primary consumer level. The energy contained in the various trophic levels of a food web, graphically represented, forms a pyramid, the **energy pyramid** (Figure 3.9). The biomass forms a **biomass pyramid** (Figure 3.10).

A good example is the corn-pig-human biomass pyramid. For example, 1,000 kilograms (2,203 pounds) of corn is needed to produce 100 kilograms (220 pounds) of pork and ham, which in turn can be converted into 10 kilograms (22 pounds) of human flesh (Figure 3.10).

Why does each consumer level contain only about 10 percent of the biomass and food energy of the lower level? At least four reasons account for the low efficiencies.

First, not all producers in a food web are consumed by primary consumers and not all primary consumers are eaten by secondary consumers. Plants, for example, deter herbivores with spines, thorns, thick protective bark, irritating secretions, or foul odors. Animals such as insects, trout, and antelope fly, swim, or run from the approaching predator at high speed. Some animals rely on protective coloration to escape detection by predators. This adaptation is well exemplified in certain species of moths whose colors make them almost invisible against the bark of a tree. Similarly, many ground-nesting birds, like pheasants and grouse, are extremely difficult to detect while they are incubating their eggs, even from a few feet away. If an organism escapes predators, it will eventually die from some other cause, such as disease or starvation. Its carcass will then be decomposed by bacteria and/or fungi and will enter the detritus food web.

The second reason is that not all the body of the food organism is digestible. For example, many herbivores, such as porcupines, beaver, deer, and rabbits, cannot digest the cellulose in plant cell walls. Therefore, many cellulose fibers are voided from the body as waste.

Third, not all material is accessible. As a rule, the roots of plants are not eaten by herbivores and thus retain much of the energy captured from the sun.

Fourth, the low efficiency in energy transfer can be partially explained by the fact that all organisms respire. You will recall that cellular respiration is the process by which organisms oxidize, or burn, energy-rich organic compounds to release the energy they need to power their life activities. Cellular respiration

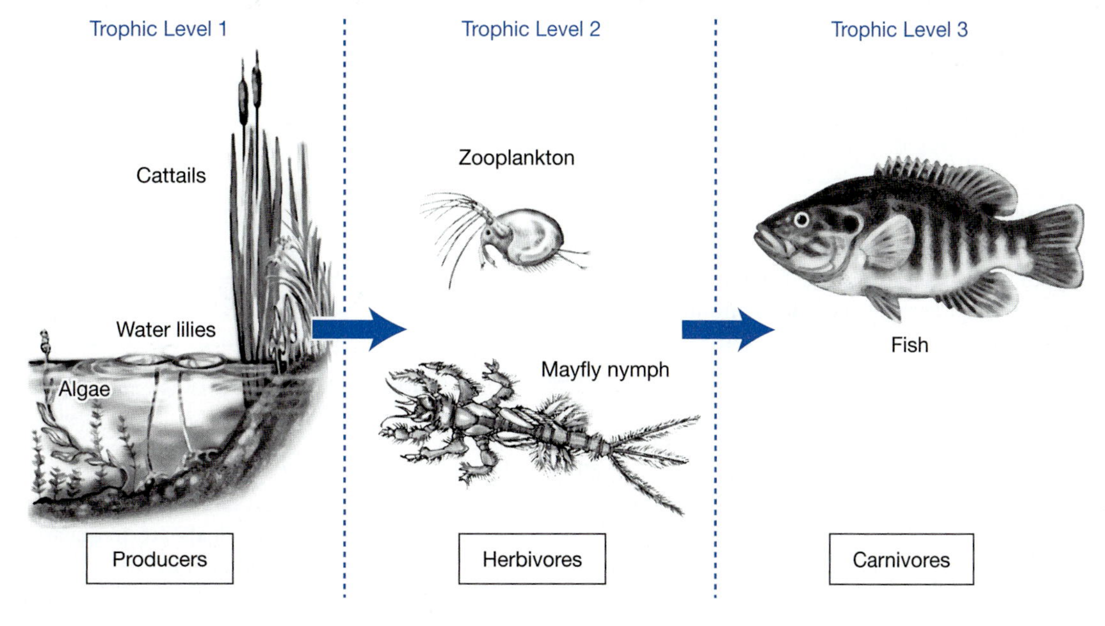

FIGURE 3.8 *Trophic levels of a lake ecosystem.*

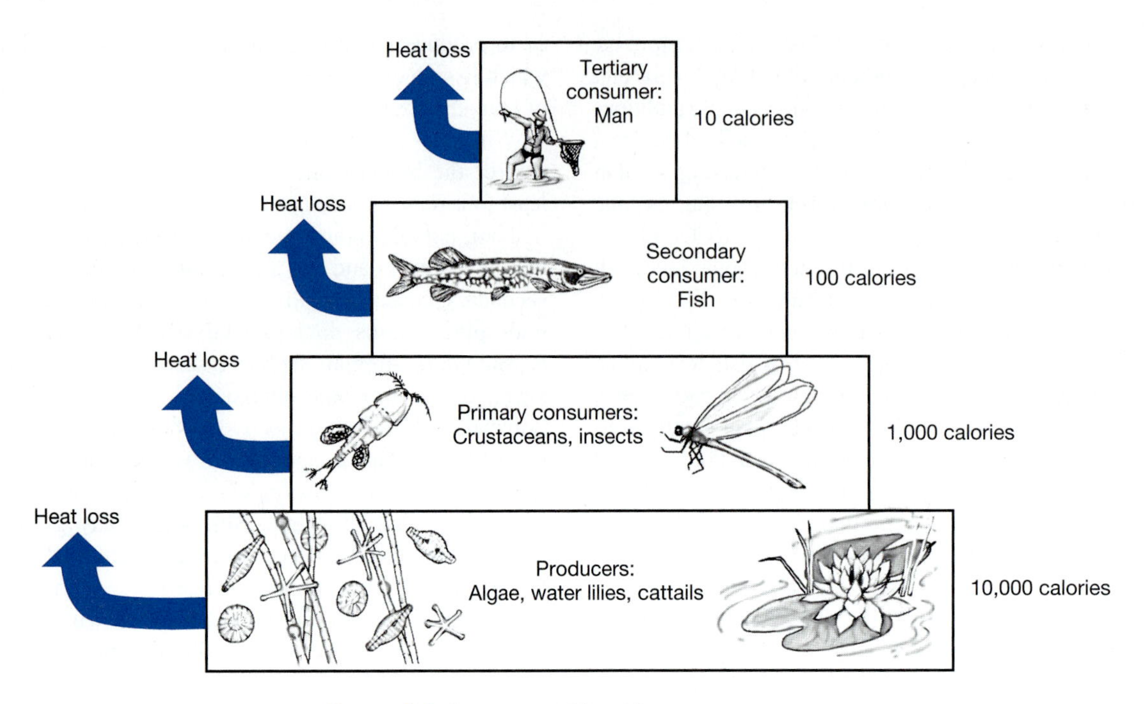

FIGURE 3.9 *Energy pyramid for a lake ecosystem.*

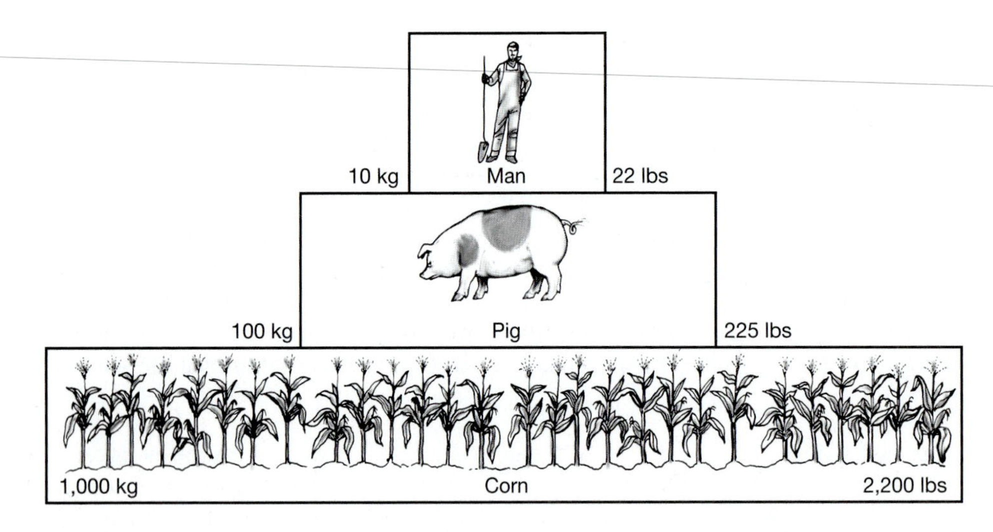

FIGURE 3.10 *Biomass pyramid. It takes 1,000 kilograms of corn to produce 100 kilograms of pork needed to produce 10 kilograms of human biomass.*

itself is a relatively inefficient process. For example, only about 40 percent of the energy in a given sugar molecule may actually be put to work in an organism. The remainder is largely converted to heat that escapes into the environment.

Biomass and energy pyramids may, in some instances, be upside down or inverted (Figure 3.11). This occurs in some aquatic ecosystems where the food webs are based on billions of microscopic algae. The algae reproduce very rapidly, gobbling up nutrients and making new organic matter rapidly. Their populations will double every few days. The consumers (crustaceans), however, feed on the algae almost as rapidly as they are produced. As a result, the algal (producer) trophic level has less biomass than the crustacean (consumer) trophic level (Figure 3.11).

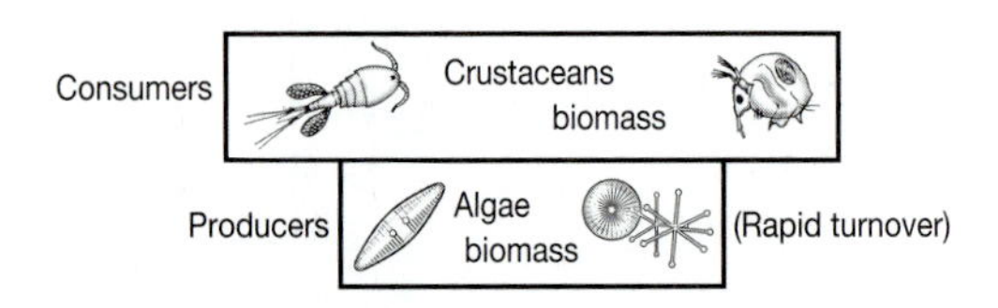

FIGURE 3.11 *Inverted biomass pyramid of a lake ecosystem.*

Pyramid of Numbers In typical food webs based on small green plants or algae and ending in predators like hawks and fish, the numbers of individuals are frequently greatest at the producer level (base), smaller at the herbivore level, and smallest at the carnivore level (top). When graphically represented this forms a **pyramid of numbers.** Figure 3.12A illustrates the

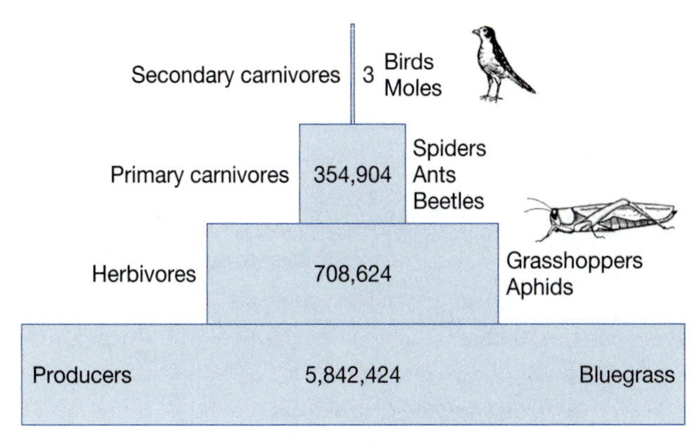

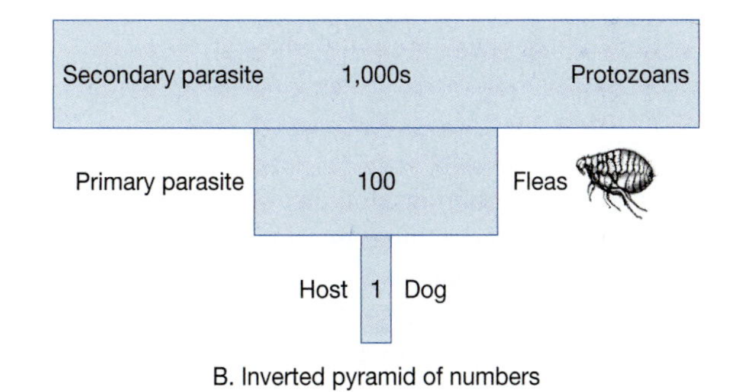

FIGURE 3.12 *Two kinds of number pyramids. The upper pyramid (A) is for a weedy field. The lower pyramid (B) is for a dog-parasite system, and is inverted.*

pyramid of numbers in the food web of 0.4 hectare (1 acre) of bluegrass in Michigan.

In food webs ending in parasites, the pyramid of numbers is upside down. This is evident in the dog-flea-protozoan food chain, where the dog is the food source of the fleas and fleas are the food source of the protozoans (Figure 3.12B). Interesting as they are, number pyramids do not provide accurate pictures of either biomass or energy content in the trophic levels, and therefore are of somewhat limited value.

Pyramid of Energy and Human Nutrition Over the years, ecologists have consistently found that most terrestrial food chains have only three or four trophic levels. Why? The reason is found in the second law of energy described earlier in the chapter. As you may recall, the second law says that during each energy conversion some energy is lost as heat. Therefore, the total amount of high quality energy decreases with each trophic level.

This simple phenomenon imposes a limit on the length of food chains in nature. For example, in the unusually long food chain clover-grasshopper-frog-snake-hawk, the hawk uses only 0.0001 percent of the solar energy captured by plants. There's simply not enough energy captured by producers to support many organisms past three or four trophic levels. The reasons cited for the 10 percent loss of energy and biomass also come into play.

Knowledge of these energy relationships enables us to understand the striking difference between Indian and American diets—the former based primarily on grains, the latter based on ample servings of meat. If, as an American, you live to be 70, you probably will have consumed 10,000 pounds of meat, 28,000 pounds of milk and cream, and thousands of pounds of grain, sugar, and speciality foods. However, a 70-year-old Indian probably would have eaten only 1 percent as much meat as you.

In overpopulated nations such as India and China, it is no accident that humans rely primarily on grains. In such instances, the food chain frequently has only two links: grain → people. Can you guess why?

As shown in Figure 3.13, herbivorous humans have replaced the herbivorous cow, sheep, or pig typically found in the American diet. The reason is that there is more energy available to the Asian further down on the food chain, closer to the producer base.

Americans still live in a land of milk and honey (as well as pork chops and tuna). However, some futurists believe that the time may well come, as a result of the mushrooming U.S. population (expected to reach 300 million by the year 2010), when we will be faced with the ecological ultimatum: Shorten our food chains or tighten our belts. Eating lower on the food chain uses fewer resources and is also healthier if done right.

Nutrient Cycles

Earlier in the chapter you learned that matter is neither created nor destroyed. In the biosphere, scientists have found that matter cycles over and over again in global nutrient cycles, the subject of this section. To understand the importance of nutrient cycles, suppose that Figure 3.14A represents the mass of planet Earth. Now suppose that we draw another figure to represent the total amount of living substance (protoplasm) that has ever existed since the first living organism evolved 3.5 billion years ago (Figure 3.14B). This sphere, of course, would include your own body, as well as those of Stone Age people, ancient tree ferns, and dinosaurs. How big do you think this ball of protoplasm would be? You might be surprised to know that it would be considerably larger than the Earth itself.

Of the 103 elements known to science, only about 35 contribute to the formation of protoplasm, living tissue of plants, animals, and microorganisms. Of these, carbon, hydrogen, oxygen, and nitrogen form about 96 percent of the human body. (You can use the letters COHN to remember them.) Elements such as sodium, calcium, potassium, magnesium, sulfur, cobalt, zinc, iron, iodine, and many others occur in smaller amounts. All the elements forming the bodies of living organisms were derived either from the top few feet of the Earth's crust (soil); from the rivers, lakes, oceans, and aquifers (groundwater); or from the atmosphere, the thin blanket of air that envelops the Earth. If this is true, then the only way we can explain a cumulative ball of protoplasm being larger than the planet Earth is to assume that the elements forming the protoplasm were used over and over again. It's recycled. An atom of nitrogen that was once part of a dinosaur's jawbone might eventually have formed part of a professor's brain and, at some time in the future, may form part of an apple your great-grandson will have for a snack.

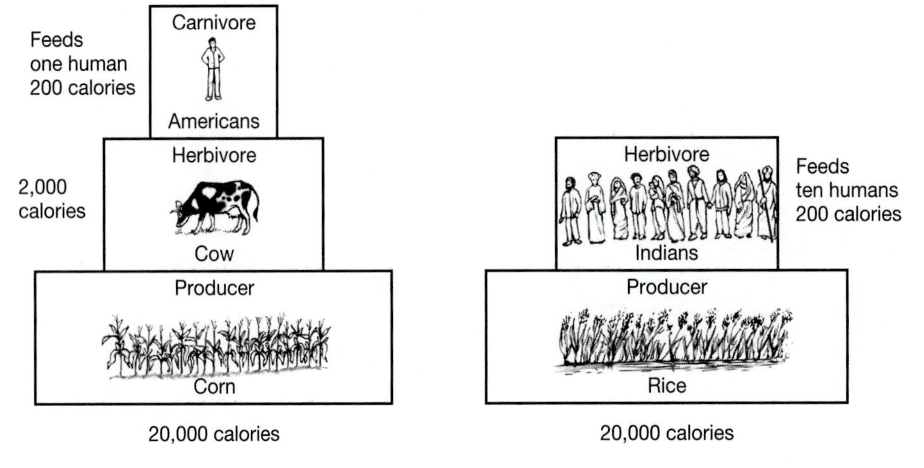

FIGURE 3.13 *The diagram above illustrates a theoretical situation in which Americans are exclusively carnivorous and the people of India are exclusively herbivorous. Note that the removal of one link in the food chain permits the Indians to get ten times as many food calories as the Americans receive from equal amounts of grain.*

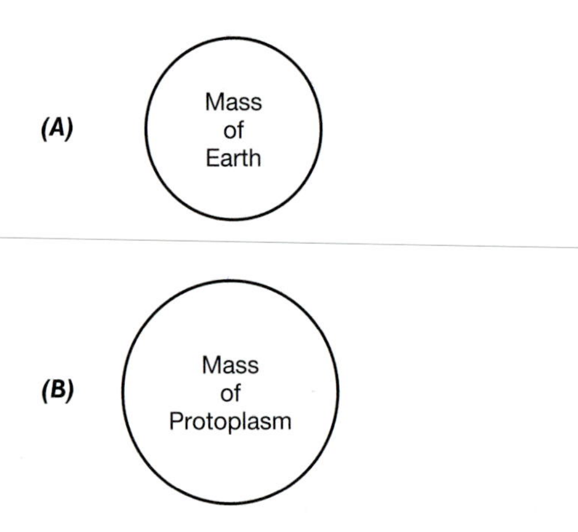

FIGURE 3.14 *There's only one explanation. (A) Mass of the Earth. (B) Mass of all organisms that have lived on the Earth since the beginning. How can the mass of living protoplasm exceed that of the Earth? There's only one possibility, matter is recycled.*

The circular flow of an element from the nonliving (abiotic) environment such as rocks, air, and water into the bodies of living organisms and then back into the nonliving environment once again is known as a **nutrient cycle.** The continuation of life depends on the function of these cycles, also known as **elemental** or **biogeochemical cycles.**

For eons, nutrient cycles were in equilibrium. Therefore, the same amount of an element moved into the various elemental reservoirs as moved out. However, in the last 200 years or so, humans have caused imbalances of some cycles. The result has been the buildup of large concentrations of certain elements, such as carbon, nitrogen, and phosphorus, in certain parts of the cycle where they can be harmful to humans and other organisms. Imbalances in the carbon cycle, you will soon see, could even be upsetting global climate in a major way. To understand cycles and our impact on them, we will examine three cycles, the nitrogen cycle, the carbon cycle, and the phosphorus cycle.

The Nitrogen Cycle Atomic nitrogen (N) is an essential component of many important compounds such as chlorophyll in plants and hemoglobin, insulin, and deoxyribonucleic acid (DNA, the heredity-determining molecule) in animals. Nitrogen in the bodies of plants and animals comes from atmospheric nitrogen. Nitrogen gas is extremely abundant. It is colorless, tasteless, and odorless gas and constitutes about 80 percent of the atmosphere. In the atmospheric air column above each 0.4 hectare (1 acre) of the Earth's surface there are approximately 31,000 metric tons of nitrogen. One would suppose, therefore, that securing adequate supplies of nitrogen would be relatively simple for living organisms. The problem is that nitrogen gas is chemically inactive. It does not combine readily with other elements, and it cannot be used by most organisms in this form.

How, then, can we and the great majority of living organisms make use of gaseous nitrogen to synthesize life-sustaining proteins? First, the nitrogen has to be converted to a usable form. We say that is has to be **fixed.** There are several mechanisms by which nitrogen is fixed in nature.

One type of nitrogen fixation is **atmospheric fixation.** Atmospheric fixation is a naturally occurring phenomenon caused by lightning or sunlight. Energy from these sources causes the nitrogen to combine with oxygen to form nitrate (NO_3). Globally, about 7.6 million metric tons of nitrate are formed annually in this way. This nitrate is then washed to Earth by rain and snow and is absorbed by the roots of growing plants.

A second type of nitrogen fixation is **biological fixation.** Biological fixation is much more important than atmospheric fixation, for about 54 million metric tons of nitrogen are fixed annually by this process—about seven times more than atmospheric fixation. Biological fixation is accomplished by microscopic organisms, primarily bacteria and cyanobacteria (also known as blue-green algae), which occur abundantly in soil and water.[1] During nitrogen fixation, nitrogen is first combined with hydrogen to form ammonia (NH_3) by certain bacteria. Plants can use this ammonia to produce amino acids.

[1]Cyanobacteria were first called blue-green algae. These tiny bacteria are capable of photosynthesis.

FIGURE 3.15 *Root nodules on soybean plants and other legumes incorporate atmospheric nitrogen. Bacteria living inside the nodules convert the nitrogen to forms the plant can use to make amino acids and other biologically important molecules.*

Nitrogen-fixing bacteria also live inside the root systems of many plants. At least 190 species of plants, including legumes (alfalfa, peas, beans, soybeans, and clover), some pines, and alders, contain these nitrogen-fixing bacteria (Figure 3.15). These plants fix nitrogen for themselves and the bacteria in their roots. They also produce excesses that are released into the soil and are therefore available for other plants as well. By growing legumes, a farmer may increase the nitrogen content (and hence the fertility) of his or her soil by 90 kilograms per hectare (80 pounds per acre) per year.

The fertility of aquatic ecosystems may also depend on nitrogen fixation. Certain mountain lakes, for instance, may depend on the nitrogen-fixing bacteria living in the roots of the alders along its shores.

A third type of nitrogen fixation is **industrial fixation,** a process in which nitrogen is combined with hydrogen to form ammonia. Later the ammonia is converted into ammonium salts that can be used as fertilizers. Such commercial production of fertilizer, which requires large amounts of energy (natural gas), has increased enormously since World War II.

Let us now, with the aid of Figure 3.16, follow nitrogen through a hypothetical cycle involving biological fixation by clover. The numbered steps that follow refer to the numbers on the figure.

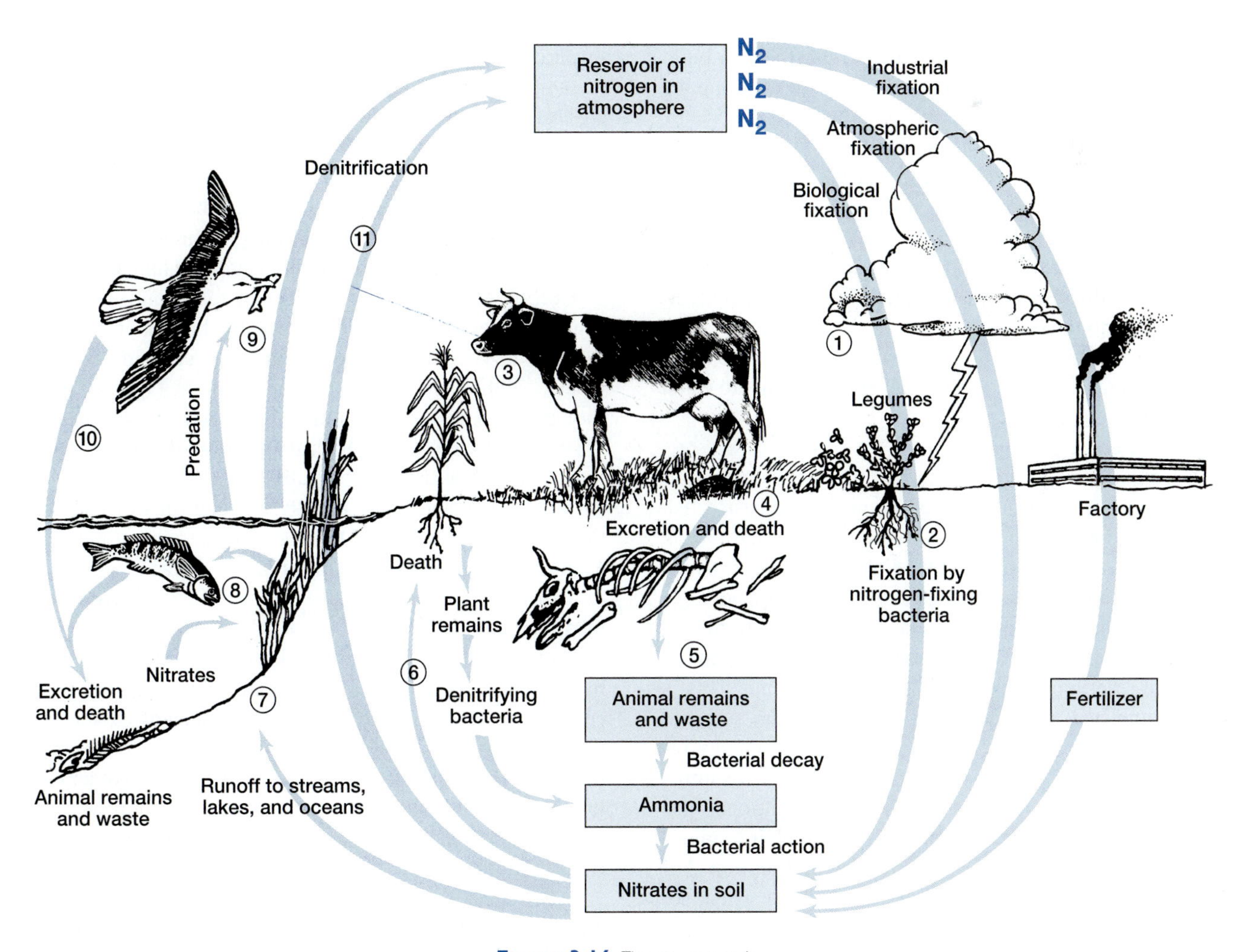

FIGURE 3.16 *The nitrogen cycle.*

1. Nitrogen (N_2) diffuses from the atmosphere into the air spaces of soil.
2. It enters small swellings on the roots of the clover plants. These are called **root nodules,** and are home to the nitrogen-fixing bacteria. The nitrogen-fixing bacteria combine the nitrogen with hydrogen to form ammonia and eventually incorporate the nitrogen into amino acids, the building blocks of proteins. The clover plant therefore builds up its own protein from the surplus ammonia not used by the bacteria.
3. A cow (or other consumer) feeding on the alfalfa digests the plant protein, using the amino acids to make its own proteins. The cow uses protein to build muscle, make milk, and produce enzymes. Some protein is broken down in the cells of the cow, releasing amino acids.
4. When the amino acids are broken down, nitrogen-containing urea is formed. It is excreted in the urine. The cow's feces are also rich in nitrogen, which comes from undigested protein and bacteria from the cow's digestive tract.
5. The urea and large, complex, nitrogen-containing protein molecules in the wastes (or carcass) are then eventually broken down (decomposed) by successive groups of soil bacteria into nitrates, a process called nitrification.
6. A plant, such as corn, wheat, or oak, can now absorb the soluble nitrates through its roots and use them to build up its own essential protein compounds, thus starting the cycle over again.

   Nitrogen can flow from one ecosystem to another. For example, it may flow from a terrestrial ecosystem to an aquatic ecosystem and back to a terrestrial ecosystem again, as shown in Figure 3.16. Thus:
7. The soluble nitrate salts formed by the decay of the carcass of a cow may be washed into a stream and eventually carried to the ocean.
8. The nitrates may be absorbed by marine algae. The algae may be consumed by crustaceans that, in turn, are eaten by fish.
9. Fish are consumed by gulls.
10. The gulls then fly back to their nesting colony on the California coast and feed some of the partially digested fish to their young. Or the adult bird may excrete some waste as it flies over the water and the California farmland, thus contributing slightly to the fertility of water or land.
11. The flow of nitrogen is circular. All nitrogen in plants, animals, soil, and water eventually re-enters the atmospheric reservoir from which it originally came. This is accomplished by denitrifying bacteria in the soil and water that break down nitrates. They use the energy released to sustain their own life processes. Gaseous nitrogen (N_2) is given off as a byproduct. The nitrogen gas then escapes into the atmosphere from which it originally came. This process is known as **denitrification.** Nitrogen is also released into the atmosphere whenever organic material, like trees, grasses, and animals, is consumed by fire.

Soil bacteria that fix nitrogen are essential for the continuation of life. Researchers in Florida have shown that some chlorinated hydrocarbon pesticides are detrimental to nitrogen-fixing soil bacteria. This suggests that toxic chemical pesticides should be used cautiously or perhaps even eliminated. Why? If populations of soil bacteria are greatly diminished, the nutrient cycle on which plants, and eventually animals and humans, depend could be severely disrupted.

Humans can disrupt the nitrogen cycle by poisoning the soil. We can also disrupt it by spreading excess fertilizer on the land—a process that overwhelms the cycle. When too much fertilizer is applied, it can run off into streams and lakes, causing serious water pollution. It can also seep into groundwater.

Another form of fixation occurs when fossil fuels are burned. In such instances, atmospheric nitrogen combines with oxygen (a reaction driven by the heat of combustion) to form nitrogen dioxide, which is later converted to nitrates and nitric acid. Nitrates may nourish terrestrial and aquatic plants, causing excess growth and if severe enough ecological disruption, a subject discussed in more detail in Chapter 10. Nitric acid can poison many life forms as you will see in Chapter 19.

For thousands of years before commercial fertilizer production and the modern automobile, the nitrogen cycle was in dynamic equilibrium, a steady state in which the amount of nitrogen leaving the atmosphere by nitrogen fixation was balanced by the amount of nitrogen entering the atmosphere as a result of denitrification. But the artificial fixation of increasing amounts of nitrogen has caused a dramatic imbalance. To create a sustainable future, we must find ways to regain the balance.

The Carbon Cycle Carbon is a key element in all organisms from bacteria to humans, where it forms the backbone of many molecules from DNA to protein to sugars. In fact, carbon atoms constitute 49 percent of the dry weight of the human body.

Carbon exists in several different reservoirs in the environment, including the atmosphere, the bodies of organisms, the ocean, ocean sediments, and as calcium carbonate (rocks, shells, and skeletons), as shown in Figure 3.17. Although the actual size of atmospheric and organismic reservoirs is relatively small, the rate of flow of carbon into and out of those reservoirs is relatively high—in other words, carbon is recycled fairly rapidly. Fifteen tons of carbon (in the form of carbon dioxide) occur in the air column above each hectare (2.5 acres) of the Earth's surface. One hectare of lush vegetation can remove 50 tons of carbon from the atmosphere annually.

To trace the flow of carbon through the carbon cycle, we begin with carbon dioxide in the atmosphere. Carbon dioxide enters tiny pores in the leaves of plants, such as clover. Inside the leaves, the carbon of the carbon dioxide molecules is combined with hydrogen from water to form sugar and other organic molecules, as illustrated in the equation of photosynthesis on page 51. As noted earlier, energy to drive this reaction comes from sunlight.

When the clover leaves are consumed by an animal, such as a deer, some of the carbon-containing organic compounds of the clover are digested and converted into deer protoplasm.

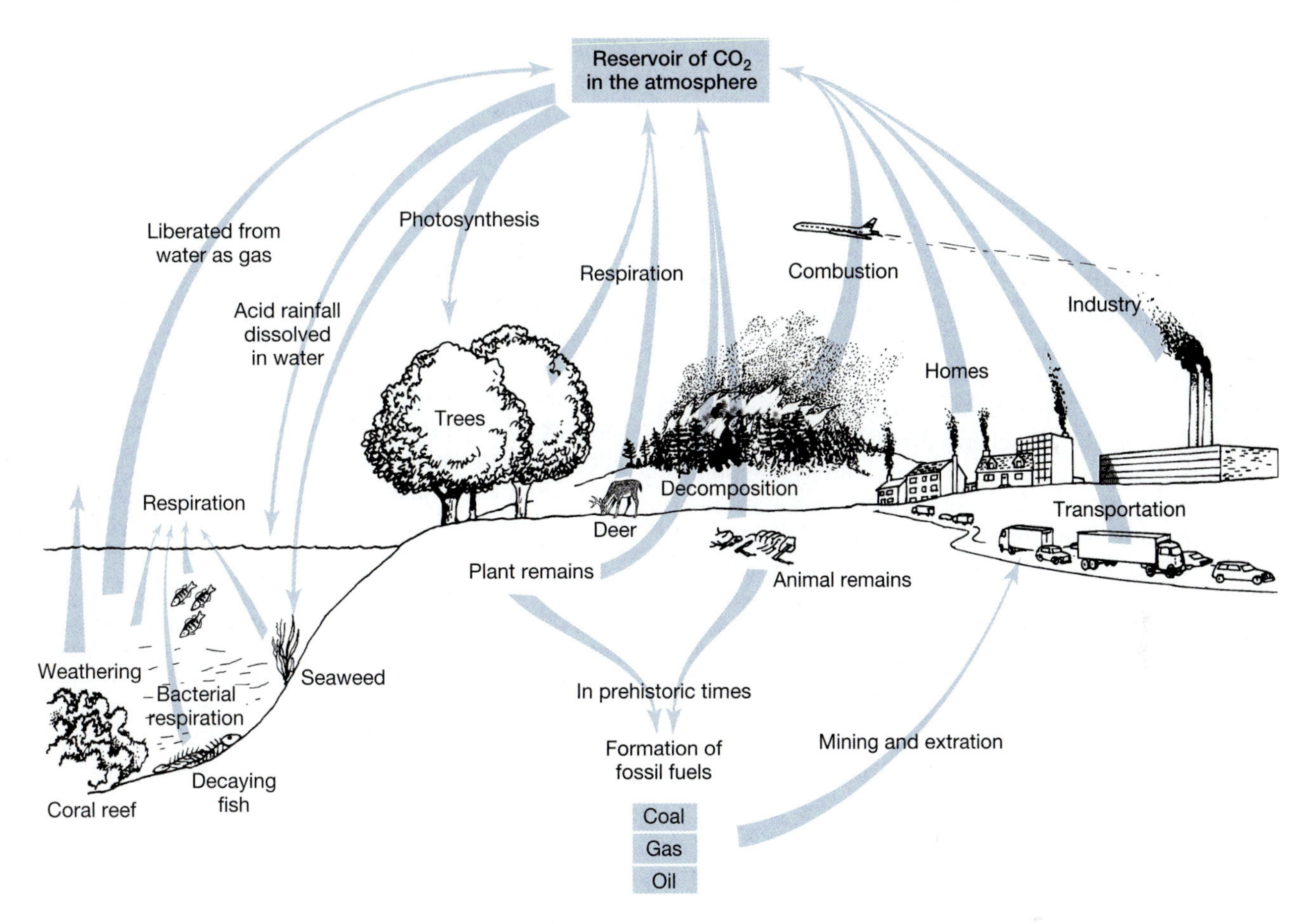

FIGURE 3.17 *The carbon cycle.*

The rest is converted to energy in cellular respiration. When humans or coyotes eat the deer meat, the meat is then transformed into human protoplasm.

In all of these organisms—clover, deer, and humans or coyotes—some carbon is released during cellular respiration as shown in the equation on page 52. In plants, carbon is released as carbon dioxide to the atmosphere via pores in the leaves; in animals, the carbon dioxide is exhaled from the lungs. Carbon also returns to the atmosphere when an organism (clover, deer, humans, and so on) dies and decomposes. In addition, it may re-enter via wastes (feces and urine) of animals which are broken down by the bacteria and fungi that occur abundantly in soil, air, and water. Such decomposition releases carbon into soil and air. It can then re-enter the biological world through plant photosynthesis.

About 250 to 300 million years ago, during the Carboniferous Period, giant tree ferns and other plants grew in what is today Pennsylvania, West Virginia, Ohio, Kentucky, Tennessee, Indiana, Illinois, Wyoming, New Mexico, Colorado, and South Dakota. Many of those plants were buried by sediment and therefore escaped decomposition. They were eventually converted into coal. In a somewhat similar fashion, both plant and microorganisms were converted to crude oil and natural gas. The carbon in these fossil fuels was removed from circulation. Two hundred years ago, however, humankind discovered the remarkable potential of fossil fuels. Society today depends almost entirely on these fuels. Humans have also removed enormous amounts of forests, which absorb and store carbon. As discussed in Chapter 19, our accelerated combustion of fossil fuels and the loss of forests has resulted in an increase of about 30 percent in the amount of carbon dioxide in the atmosphere from 1870 to 2000. As noted earlier in be chapter, this increase may result in climatic changes that could be harmful to humans and ecosystems.

The oceans, which cover 70 percent of the Earth's surface, also serve as a carbon reservoir. Carbon dioxide dissolved in the ocean may move through an algae-crustacean-fish food chain. The fish, in turn, may be eaten by such organisms as sharks, tuna, waterfowl, whales, or humans. Some of this carbon is returned to the ocean by the respiration of marine organisms and some is used by clams, oysters, scallops, and corals to build limestone shells and skeletons. Tremendous quantities of carbon are locked up in coral reefs off the coasts of California and Florida. The Great Barrier Reef off the Australian coast—a mass of limestone 56 meters (180 feet) wide and 2,100 kilometers (1,260 miles) long—is composed of billions of coral skeletons. Eventually, as a result of weathering processes that operate for millennia, small amounts of the carbon from coral reefs and the shells of clams and oysters are returned to the ocean waters.

Carbon dioxide is continuously being exchanged between the atmosphere and the ocean. When atmospheric carbon dioxide increases, more is dissolved in the ocean. Conversely, when atmospheric carbon dioxide decreases, more carbon dioxide is liberated from the ocean. By this mechanism, the carbon dioxide in the atmosphere was maintained at a fairly constant level for thousands of years—until the last century, that is.

The Phosphorus Cycle Approximately 1 percent of the human body is composed of phosphorus, an essential component of such compounds as the hereditary molecule DNA and the energy-rich molecule that powers virtually all organisms, adenosine triphosphate (pronounced ah-den-oh-seen try-phoss-fate and also known as ATP).

Like other nutrients, phosphorus is recycled in a giant, global nutrient cycle, the phosphorus cycle (Figure 3.18). To understand how this cycle works, we begin with its reservoir in phosphate rock above the soil (shown as erosion weathering in the diagram). Raindrops dissolve some of the phosphate in the rock and wash into the soil. There it represents a potential plant nutrient, just like nitrogen. Phosphorus atoms pass into the plants and then into the animals that feed on them. When plants and animals die, their bodies are decomposed by bacteria and fungi. Phosphorus is released to the soil as a result. Some of this phosphorus may be taken up by other plants and some of it may be washed into a stream after rainstorms. The stream may transport it to a lake or to the ocean. In these aquatic ecosystems, the phosphorus may be absorbed by algae and rooted plants. Plants may be eaten by fish. When aquatic organisms die and decompose, phosphorus-containing materials are released back into the water. Aquatic animals also excrete phosphorus-containing wastes into the water. Phosphorus in freshwater systems may eventually make their way to the oceans of the world. Phosphorus in seawater often settles to the ocean floor, forming part of the sediment. Over a period of millions of years, these sediments may eventually form phosphate rock. Ultimately, geological process such as upheaval, may expose the rock to the atmosphere. Then, by the action of weathering and erosion, some of the phosphorus in the rock may become part of the soil once again.

For eons, the amount of phosphorus moving from its phosphate rock reservoir into the bodies of living organisms was in equilibrium with the amount moving in the reverse direction. The total amount of phosphorus in circulation was relatively small. However, in the past few decades the intensive fertilization of agricultural lands with phosphorus-containing chemicals to feed a mushrooming human population, as well as the use of phosphorus-containing detergents, has caused an increase in phosphorus in our lakes and streams, discussed more fully in Chapter 10.

3.4 Principles of Ecology

The previous material has provided an overview of scientific principles and how energy and matter flow through ecosystems. We now turn our attention to specific ecological

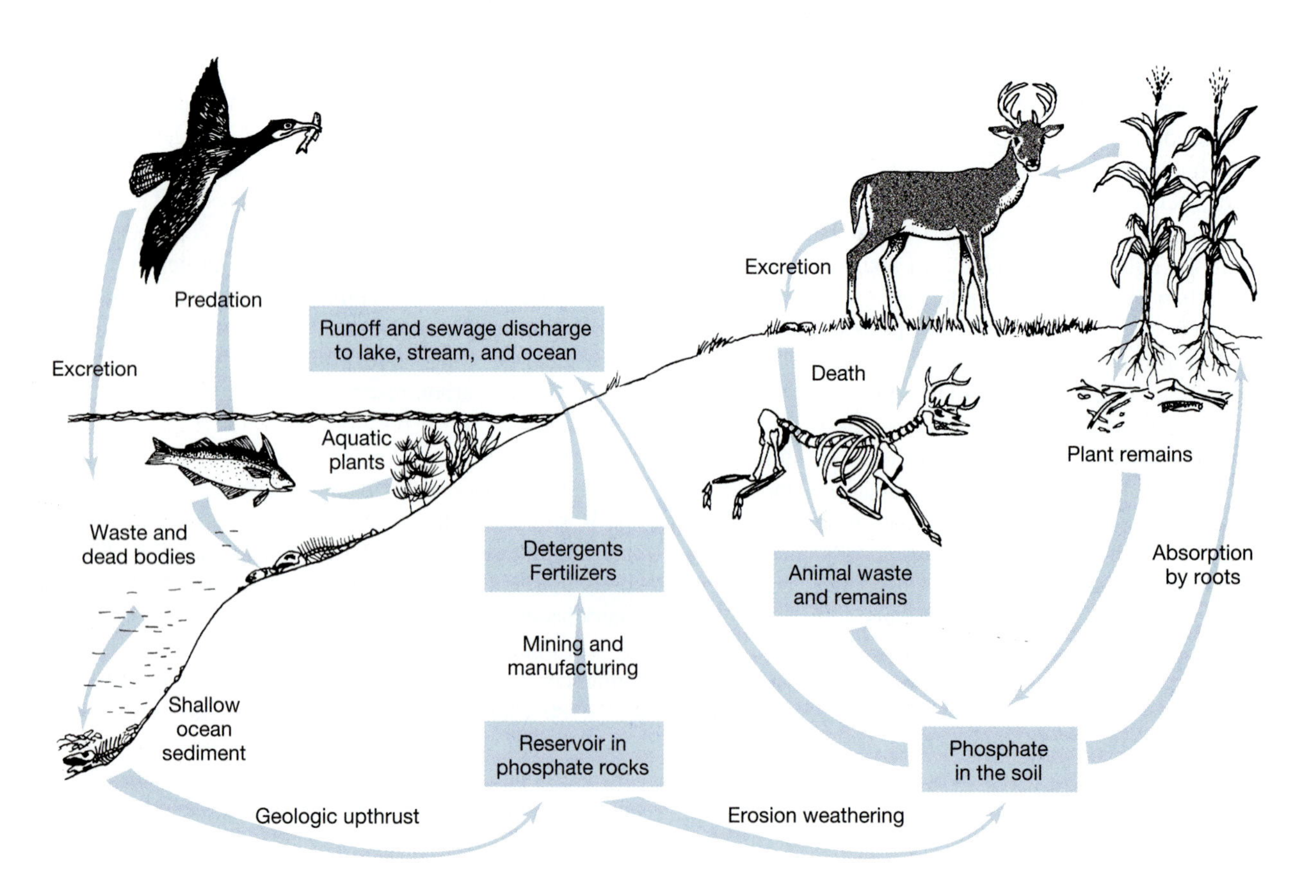

FIGURE 3.18 *The phosphorus cycle.*

principles that will help you deepen your understanding of ecosystems, a knowledge essential to creating a more sustainable way of life.

The Law of Tolerance

The survival of any organism depends on many essential factors in its physical environment, such as water, temperature, oxygen, and nutrients. The concentrations of these factors may vary greatly. For each factor, a given species has a **range of tolerance**—defined as a range of conditions it can tolerate. The organism is adversely affected when a factor approaches values beyond its tolerance limits. It can even die. Figure 3.19 illustrates the concept. As you can see, the center of the range is the optimal conditions for survival. Above and below the optimum range are the zones of physiological stress. Organisms survive, but are in conditions not entirely conducive to survival. Heading further outward on the range of tolerance is the zone of intolerance in which death occurs.

The tolerance of a given species for an environmental factor may vary with the age of the organism. Thus, newly hatched salmon are much more vulnerable to such water contaminants as heat, toxic metals, and pesticides than are adults. Tolerance may also vary with the genetic makeup of the organism. For example, when the insecticide DDT is first sprayed on a population of houseflies, almost all of them are destroyed. Those that survive do so because they are genetically resistant. They may produce resistant offspring. Higher doses of chemical pesticides such as DDT (now banned in the United States and many other countries) may kill many of these offspring. Once again, however, survivors genetically resistant to the higher doses of pesticide will produce a new generation of pesticide-resistant flies. Eventually, after many generations of flies have been sprayed with pesticide, a generation develops that is highly resistant to the chemical.

Habitat and Niche

Ecologists also frequently refer to an organism's habitat and its niche. The two are very different. The **habitat** of an organism is its "address"—the place where it lives. For example, the habitat of the Red-winged Blackbird is a cattail marsh; a cold-water stream is the habitat of a trout. The term **niche** is more difficult to define and has been defined by different people in different ways. Some ecologists consider the niche an organism's functional role—what it does, and its relationship to its food and its enemies. The niche is really much more and can only be appreciated by surveying a set of questions that illustrate many aspects of the niche of an organism:

1. What is the animal's range of tolerance to temperature, humidity, solar radiation, and wind velocity?
2. What type of food does it consume?
3. With which species (other than its own) does it compete for food and nesting or breeding sites?
4. Where does it produce its young? In a den? In a nest?
5. If it is a fish, where does it lay its eggs? On the stream bottom? On aquatic vegetation? Or do the eggs simply float on the water's surface?
6. If it is a nest builder, what types of material does it use in nest construction?
7. What parasites plague the animal?

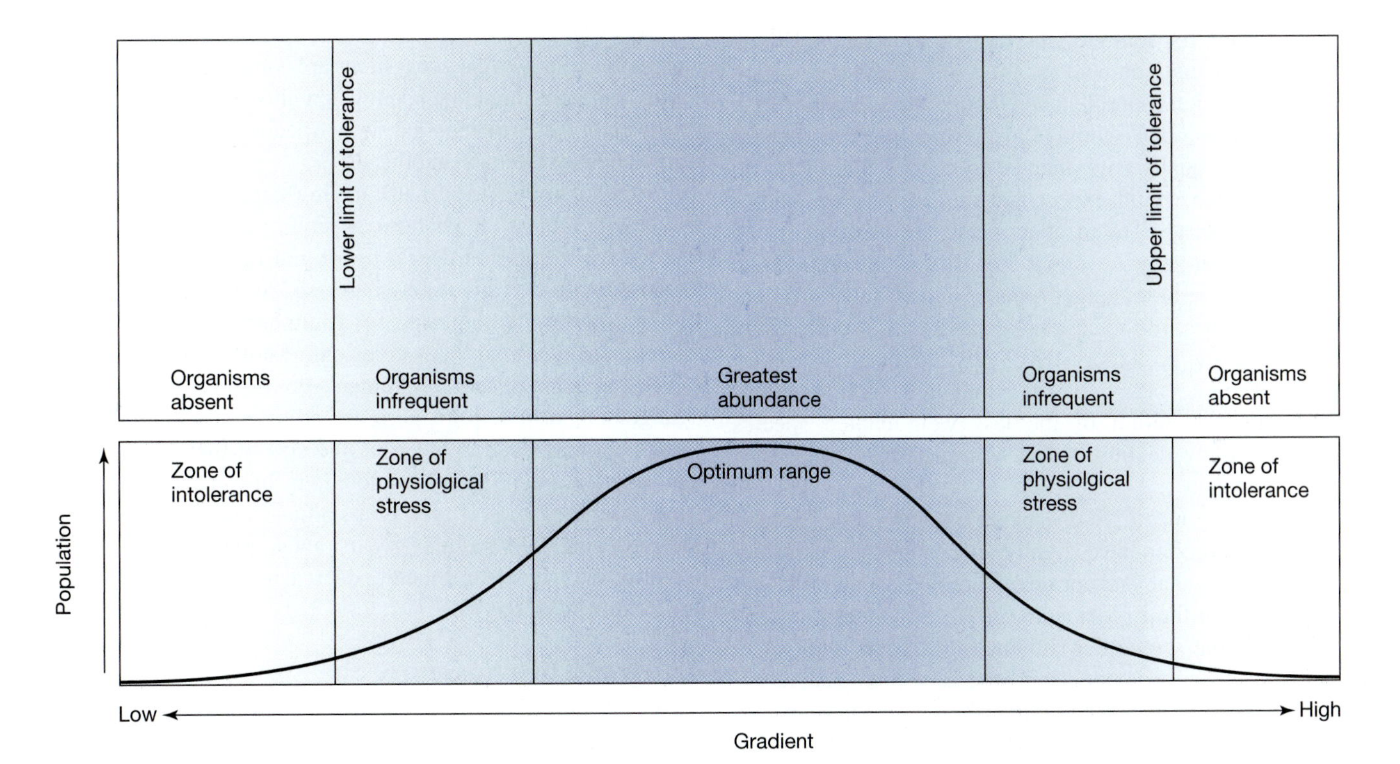

FIGURE 3.19 *Range of tolerance.*

8. To what type of predators is the animal vulnerable?
9. How does the animal protect itself from predators? By "freezing," by its coloration, by hiding under vegetative cover, by fighting, or by swimming, running, or flying away?
10. What type and volume of waste does the animal excrete?
11. How does this waste affect the surrounding plant and animal life?

Field ecologists spend a great deal of time studying organisms in an attempt to find answers to these and other questions. Understanding an organism's niche is not just a scientific exercise, it is a way of building our knowledge base so that humans can manage other species better and, increasingly, can fashion their own systems in ways that do not interfere with other species' niches. With these considerations in mind, let us now examine an important principle based on the concept of the niche.

The Competitive Exclusion Principle

Imagine that you had two species that occupied the same niche and lived in the same habitat. What would happen to them? Studies show that two species of plants or animals cannot occupy the same ecological niche indefinitely. Sooner or later, the population of one of the species will decline to zero. This phenomenon is called the **competitive exclusion principle.**

Competitive exclusion, the elimination of one of the species whose niche corresponds to the nice of another species, occurs for two reasons. First, competition between two species with identical niches would be intense. Plants would compete for sunlight, soil moisture, nutrients, and so on. Animals would compete for food, breeding sites, cover, and so on. Second, as a rule, even though the species occupy identical niches, one would expect that the two species would not be equally well adapted to occupy that niche. In other words, we would expect that one of the two species would be better able to meet its needs. For example, one species of plant may be more effective in absorbing soil moisture because of its longer roots or may be more efficient in photosynthesis. In the case of predators, one species may be able to pursue prey more effectively than another species because of its keener vision and longer legs. In such instances, the population of the less well-adapted species would eventually decline and be eliminated by the better-adapted organism.

Because of competitive exclusion, organisms that occupy the same habitat often have slightly different niches. Consider an example. The shag and cormorant are fish-eating birds that nest on cliffs on the British coast and feed in nearby waters. But studies have shown that although they occupy the same habitat, they actually do not occupy the same ecological niche. The shag, for example, feeds primarily on eels and herring-like fish that it finds in the upper waters. The cormorant preys on shrimp and flatfish that live on or near the ocean bottom. Moreover, the shag nests at lower levels on the cliffs than the cormorant.

In actuality, very few well-documented examples of the competitive exclusion principle have been reported. One such example involved competition between house mice and meadow mice that were coexisting in a field in California. These two species competed for the same space, food (weed seeds), and breeding sites. However, the meadow mouse was much more aggressive in this competition. As a result, the house mouse population declined sharply from about 825 per hectare (330 per acre) to zero in only 14 months.

Competitive exclusion is important when considering introducing a species into a new habitat—for example, introducing a sport fish into waters to which it is not native. The major concern is not just how well the new species will do, but how it will affect native species whose niches may be very similar or identical.

Carrying Capacity

Another important concept in ecology is carrying capacity. Briefly mentioned in Chapter 1, **carrying capacity** is defined as the number of organisms an ecosystem can support. When an ecologists speaks of the carrying capacity of an ecosystem for Red Foxes or White-tailed Deer, she's referring to how many of each species the system can provide for.

Carrying capacity is determined by a variety of factors, such as food supply, nesting sites, water supplies, climatic conditions, and waste assimilation. We think of it as being determined by source and sink functions—the term *source,* referring to the resources an organism needs to survive and reproduction, and the term *sink* referring to the ability of an ecosystem to get rid of waste. These factors can be manipulated by resource managers to increase or decrease numbers. But experience shows that ecosystems are complex interactive systems and care should be taken when altering such factors. Altering one factor to benefit a single species can have unforseen and adverse effects on another.

Bear in mind, too, that carrying capacity is a dynamic number. That is, it changes as conditions change. In dry years, the number of deer a forest and field can support decreases. In wet years, it may increase. Fortunately, in ecosystems numbers are adjusted by a number of mechanisms so organisms don't eat themselves out of house and home. Sustainable management of human ecosystems, for example, rangeland, also require periodic adjustment. In dry years, ranchers may need to cut back the size of their herd or find more grazing land to accommodate it. In wet years, they can increase herd size. If they don't adjust, they run the risk of overgrazing the land in dry years, causing damage that could last for many years, ultimately decreasing their overall carrying capacity of the land.

Population Growth

An understanding of ecosystems, communities, and populations requires an understanding of population growth, a topic discussed more fully in the next chapter. All organisms are governed by the same principles. Their populations grow when conditions are favorable. They decline when they're not.

Numerous **biotic factors** (biological) and **abiotic factors** (physical and chemical) stimulate growth causing a population to reach what ecologists call its **biotic potential**—the maximum reproductive rate. The ability to find food or to hide from predators, for example, are two biotic factors that favor survival and reproduction. They will cause a population to grow, just as favorable light or temperature, two abiotic factors (Figure 3.20).

In ecosystems, factors that promote growth tend to be balanced by biotic and abiotic factors that reduce growth (Figure 3.20). These are sometimes referred to as **environmental resistance.** The population size is determined by the interplay of environmental resistance and the biotic potential of a species (Figure 3.21). With these concepts in mind, we now turn our attention to another important concept, biological succession.

Biological Succession

Ecosystems are dynamic entities. Organisms grow and die. Populations increase and decrease. Conditions change over time in concert with the seasons. One form of change is called biological succession. **Biological succession** is the replacement of one community of organisms by another in an orderly and predictable manner. Two types of succession are recognized by ecologists, primary and secondary.

Primary Succession Succession that occurs in areas not previously occupied by organisms—for example, a newly formed volcanic island—is known as a **primary succession.** During primary succession, life becomes established on previously lifeless areas such as jagged outcrops of granite, lava-covered slopes, rubble left in the wake of a landslide, or even on the waste heaps of an open-pit mine.

To understand this process, we will trace a primary succession that might occur on a rocky surface of the eastern United States after the retreat of the glaciers (Figure 3.22). The first stage is the **pioneer community.** Organisms of this stage are adapted to withstand great extremes of temperature and moisture. A typical pioneer organism that might become established on a bare, windswept, rocky outcrop is the **lichen,** an odd little organism that clings to rocks and other lifeless substrates. Lichen consists of two organisms, one living inside the other. The main body is a fungus. Inside the fungus are many photosynthetic algae. The reproductive bodies of lichens—spores—are blown around in the wind and thus can be blown into the barren, rocky area. The spores settle on the rocks and begin to develop.

The lichens gradually form a crust on the rocks. Many are grayish green, although some lichens are bright orange, even powder blue. Lichens are remarkable organisms. Even if the rock is completely dry, spores can develop so long as adequate atmospheric moisture is available. Once established, lichens begin to modify the immediate environment around them (Figure 3.22.) Weak carbonic acid (H_2CO_3) produced by the lichen begins to dissolve the underlying rock. Lichens growing side by side trap particles of windblown sand, dust,

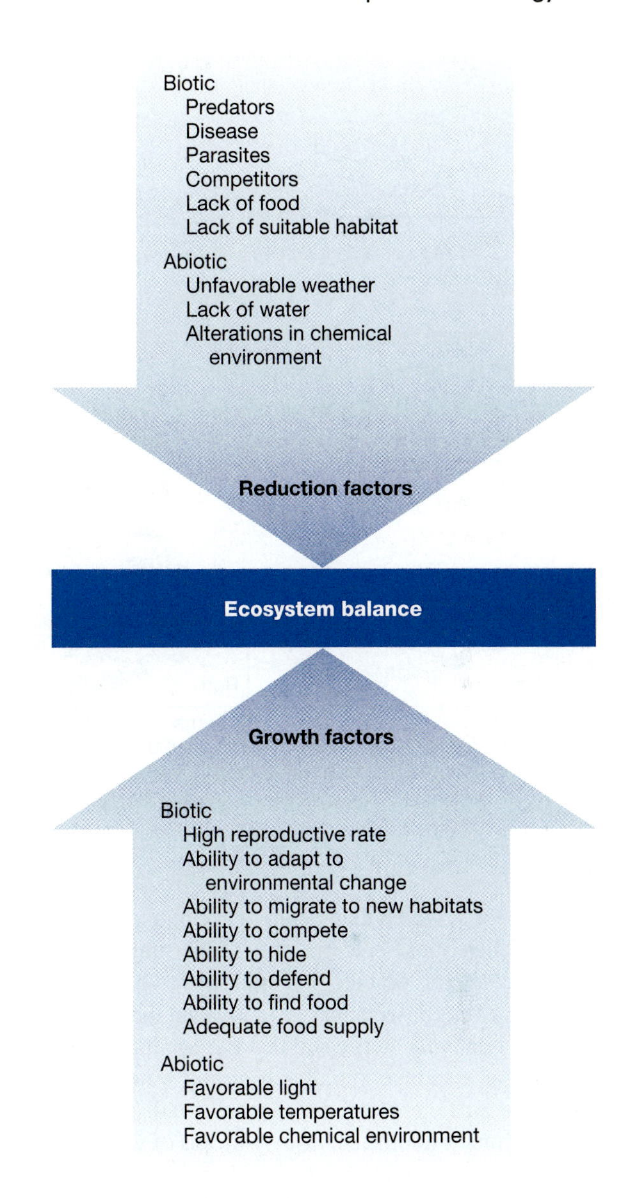

FIGURE 3.20 *Factors that affect populations. Numerous factors stimulate growth and cause reductions in the size of populations.*

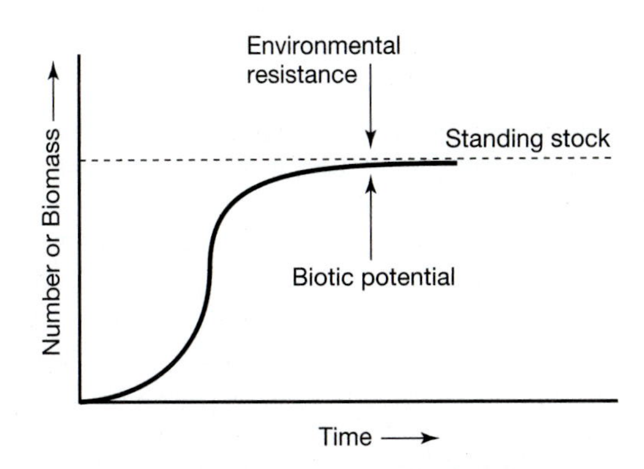

FIGURE 3.21 *Biotic potential and environmental resistance. Factors that cause populations to decline check growth or achievement of the full reproductive or biotic potential of an organism.*

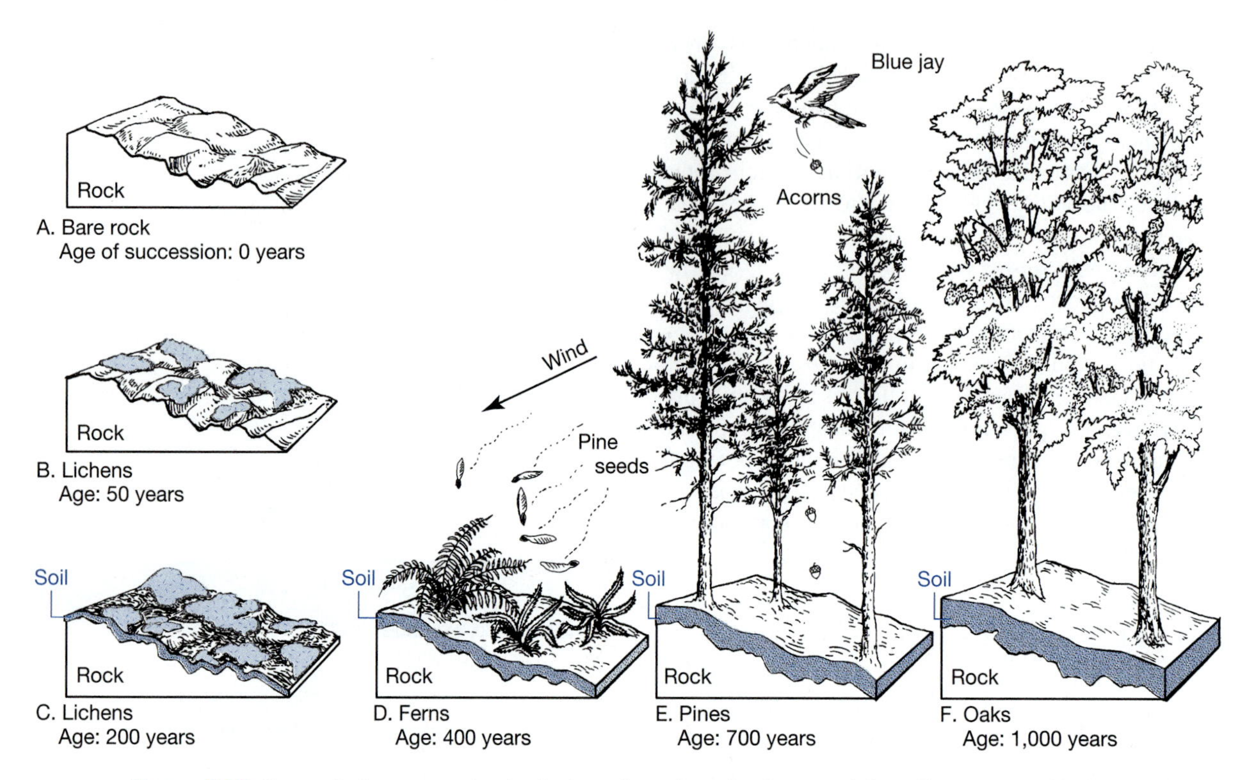

FIGURE 3.22 *Stages of primary succession that begins on bar rocks and ends in an oak forest. This process could take 1,000 years to complete.*

and organic debris, which then begin to accumulate. When an occasional lichen dies, bacteria and small fungi cause its decay. The resultant organic material and the excreta of minute lichen-eating insects that have invaded the microhabitat enrich the relatively sterile soil that has accumulated. This soil now acts as a sponge, rapidly absorbing water from dew or rain. Once sufficient soil has accumulated, mosses and ferns may become established, also by means of wind-distributed spores. Ferns eventually shade out the lichens and replace them in the succession. The soil becomes further enriched each fall with the decay of the ferns. Eventually, as the decades pass, windblown pine seeds from a hilltop pine forest may fall in the area. They may have been dropped by seed-eating birds, such as the pine siskin, as they flew overhead. The young sun-loving pine seedlings, in turn, compete with the ferns. Eventually the pines shade out the ferns and replace them in the succession.

With the passage of time, gray squirrels may enter the area to bury acorns brought in from a neighboring oak woods. Acorns may also be accidentally dropped by a wandering raccoon or by a blue jay during a flight over the young pines. The acorns germinate readily in the relatively fertile soil, which by now has been developing for centuries since the succession began. The young oak seedlings grow well in the shade under the pine canopy. The oaks (and other species such as hickory, red maple, and red gum) have long roots and therefore can make use of soil moisture that is unavailable to the shallow-rooted pines. As an occasional pine dies, its position in the forest community is filled by an oak. Over time, an oak forest, with its characteristic complement of plants and animals, will become established as the relatively stable terminal community, or **climax community,** of the succession. Thus, after hundreds of years, the bare, windswept rock outcrop is replaced by a mature oak forest.[2]

Secondary Succession Succession in an area that was previously occupied by organisms is called **secondary succession.** Much more common than primary succession, secondary succession occurs when a given ecosystem is partially destroyed by natural or human forces. The main causes are fires, volcanoes, hurricanes, deforestation, and agriculture. Chemical contamination caused by smelters or hazardous waste may also cause severe environmental damage that leads to secondary succession. Because the top soil is not generally destroyed, secondary succession usually requires less time to complete than primary succession. Secondary succession is occurring in the forests around Mount St. Helens in Washington. This volcano erupted in May 18, 1980, blowing down forests and spreading ash on much of the surrounding land (see Case Study 3.1).

With the help of Figure 3.23, let us examine the secondary succession that develops in abandoned cotton fields in the Piedmont region of Georgia. During the first year following abandonment, the land is populated by two pioneer species: crabgrass and horseweed. During the second year, they are replaced by aster, a type of wildflower. Tall grass, in turn, replaces the aster by the third year. Shrubs and young pines become established in about the fifth year and gradually shade

[2]Most ecologists use the terms *early, intermediate,* and *late stages* of succession to reflect the current thinking on the subject.

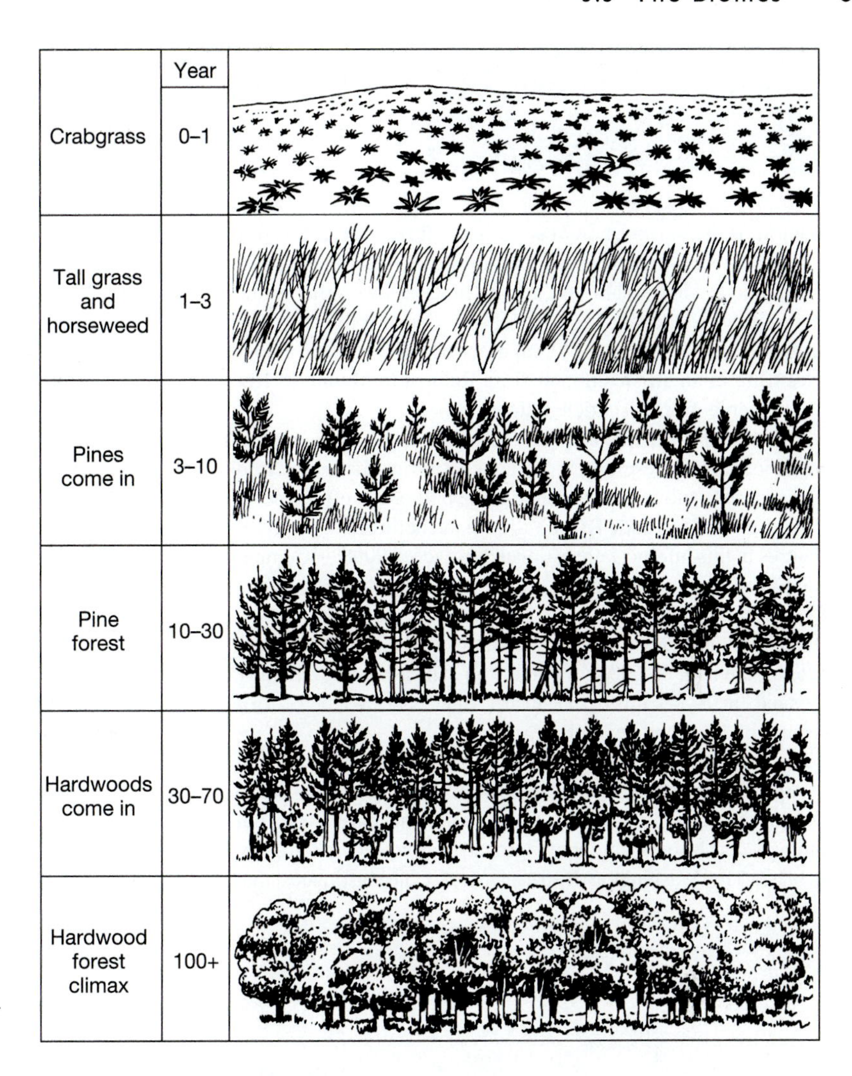

FIGURE 3.23 *Stages of a secondary succession that begins on abandoned farmland and ends, after only 70 years, in a hardwood forest.*

out the tall grass. The pine forest dominates the succession until it is about 50 years old. At this point, sun-loving pine seedlings can no longer survive in the dense shade on the forest floor. They are replaced by the shade-loving seedlings of oaks and hickories. With the passage of time, mature pines die, one by one, and their place is taken by young oaks and hickories. About 100 years after the succession began, all remaining pines are shaded out by the oaks and hickories. The latter species eventually form the climax stage of the succession. Note that this secondary succession required only 100 years to move from the pioneer stage to the climax because of the presence of soil. By contrast, a primary succession that starts on bare rock or sand may require 500 to 1,000 years before the climax stage is reached. The reason for this is that many of the components of an ecosystem, the soil in particular, were already present.

Although we have stressed the succession of plants because vegetational changes are more basic and conspicuous, it should be emphasized that animal species in the community change as well. This occurs because different animal prefer different types of vegetation—not only as food, but for protective cover and breeding sites. In the Georgia study, marked changes in breeding bird populations were observed in the various stages. Mammal species also vary during succession. In the example we've been discussing, cottontails and grass snakes of the grass stage were succeeded by such oak–hickory forest inhabitants as the opossum, raccoon, and squirrel. A summary of some of the major trends occurring in a succession is shown in Figure 3.24.

3.5 The Biomes

Anyone who has driven from New England to California is acutely aware of the marked changes in landscape that unfold, from the evergreen forests of Maine to the beech–maple woodlands of Ohio, from the windswept Kansas prairie to the hot Arizona desert. Each of those distinctive areas is part of a different life zone or biome. A **life zone** or **biome** is a large terrestrial community characterized by its climate and unique assemblage of plants and animals. The biomes of the world are shown in Figure 3.25 and summarized in Table 3.2. The role of climate in determining biome type is shown in Figure 3.26.

Tundra

The Siberian word *tundra* means "north of the timberline." The term is highly appropriate because this biome extends from the timberline in the south to the belt of perpetual ice and snow in the north. The flat to slightly rolling terrain of the tun-

CASE STUDY 3.1 LIFE RETURNS TO MOUNT ST. HELENS: A DRAMATIC EXAMPLE OF SUCCESSION

Mount St. Helens erupted on May 18, 1980 (Figure 1). It was one of the most spectacular volcanic eruptions witnessed on the planet. Mortality to both plant and animal life was awesome. Some ecosystems were literally knocked back to ground zero and had to start from scratch. Roger del Moral described the dramatic comeback of life in an article entitled "Life Returns to Mount St. Helens." Excerpts from his article follow.

At 8:32 A.M. superheated groundwater close to the magma flashed into steam, resulting in a lateral explosion that pulverized rocks and trees and sent a hurricane-force bolt of ash off the north face and across the Toutle River Valley to the north and west. Temperatures in this inferno were estimated to exceed 9008°F. Comparable to a 400-megaton blast, the explosion blew down trees in a 160° arc up to 14 miles north of the crater. As the summit of the mountain collapsed, two vertical columns of gas and steam were ejected more than 65,000 feet into the air. The ash was eventually deposited, in layers up to five inches thick, over 49 percent of Washington State and beyond.

The number of animals killed as a direct result of the explosion was high. Subterranean animals, such as pocket gophers, appear to have survived in many places even within the blast zone, but mammals and birds living above ground had no protection from the blast. The Washington Department of Game estimates that among the more important casualties were 5,200 elk, 6,000 Black-tailed deer, 200 black bears, 11,000 hares, 15 mountain lions, 300 bobcats, 27,000 grouse, and 1,400 coyotes. The eruption also severely damaged 26 lakes and killed some 11 million fish, including trout and young salmon.

FIGURE 1 *Mount St. Helens after its historic 1980 eruption. Barren land in the foreground will slowly revegetate and over decades return to its previous condition.*

Larger vertebrates may form a crucial link in the process of vegetation recolonization. Where heavy ash or mudflows dried to form a hard, uniform crust, there are few cracks to shelter germinating seeds. But large animals wandering in search of food or water make tracks that trap seeds.

Higher terrestrial life may be scarce in the blast zone but dead organic matter is abundant, and such a resource is never unexploited for long. Here, where many humans died, where entire ecosystems ceased to exist in a matter of seconds, Dave Hosford of Central Washington State College has found a mushroom growing from the ash, slowly decomposing organic matter found there and beginning a terrestrial succession.

Of all the regions on the mountain, the most severely affected was the blast zone immediately north of the crater, including Spirit Lake. Every type of volcanic behavior displayed by the mountain has assailed this terrain. All life was seemingly obliterated. Trees were pulverized and soil vaporized. A 100-meter wall of volcanic material lies over the surface.

A decade and a half later, the devastated ecosystems around Mt. Saint Helens appear to be on the mend. Studies of the recovery (ecological succession) show that the rate of recovery depends on the degree of disturbance. The greater the disturbance, the slower the recovery. In areas protected by snow cover or areas along streams, for example, vegetative recovery has occurred more readily. Another area of notable recovery is the steep terrain where erosion washed ash away and exposed soil and plants. But the region immediately north of the volcano, called the pumice plains, remains only sparsely vegetated.

During the first few years after the blast, recovery was accelerated by surviving plants. These plants spread rapidly and were soon joined by colonizers, plants whose seeds had been blown in on the wind or deposited in bird feces. Since 1990, the number and variety of both survivors and colonizers in the blast zone has increased. Trees have also begun to grow in previously forested areas. Elk and deer have returned, as have a number of bird species. About one-third of the rodent species survived; their populations are expanding as are insect populations.

Despite these successes, the first decade and a half of vegetative recovery must be viewed as the beginning of a long process of successional change that will take 200 to 500 years, depending on the amount of disturbance. Eventually, the now battered landscape will support a rich old-growth forest as it had prior to the fateful day in 1980.

dra, occupying about 10 percent of the Earth's land area, extends around the globe in the northern latitudes.

Because of its relative simplicity, the ecology of the tundra is better understood than that of other biomes. The principal limiting factors are the small amount of solar energy and the bitter winter cold. During June and July, the tundra on the edge of the Arctic Circle is a "land of the midnight sun." In January, however, this same region is a "land of daytime darkness," for the sun never rises. Annual precipitation is less than 10 inches, most of which falls in summer or autumn as rain.

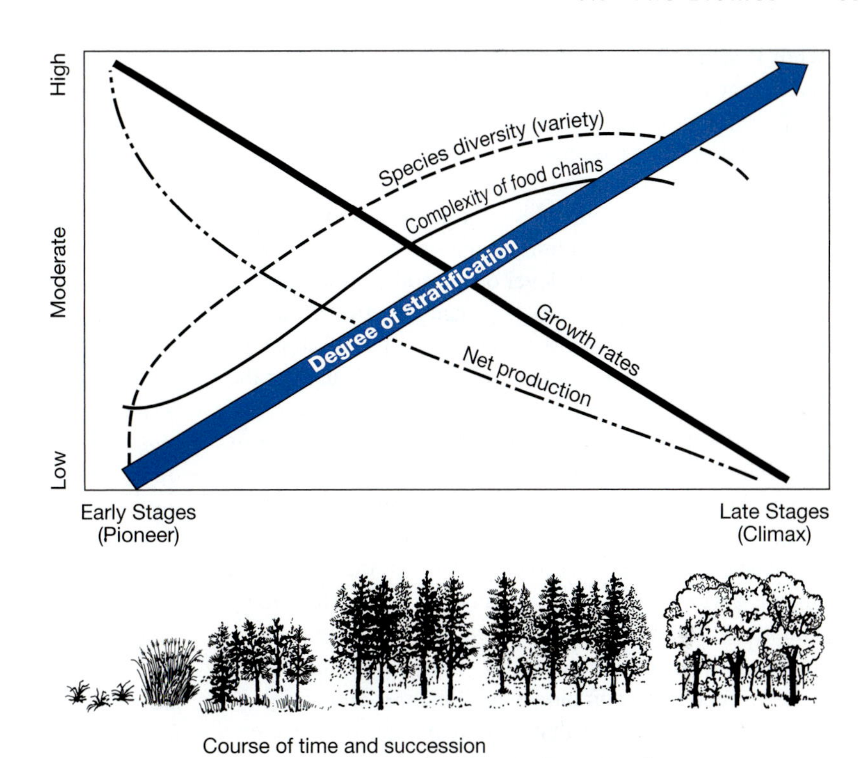

FIGURE 3.24 *Summary of some major trends occurring during a secondary ecological succession. The number of species, complexity of food chains, and degree of vegetational stratification (layering) increase as the succession proceeds. However, rates of growth and net production decrease.*

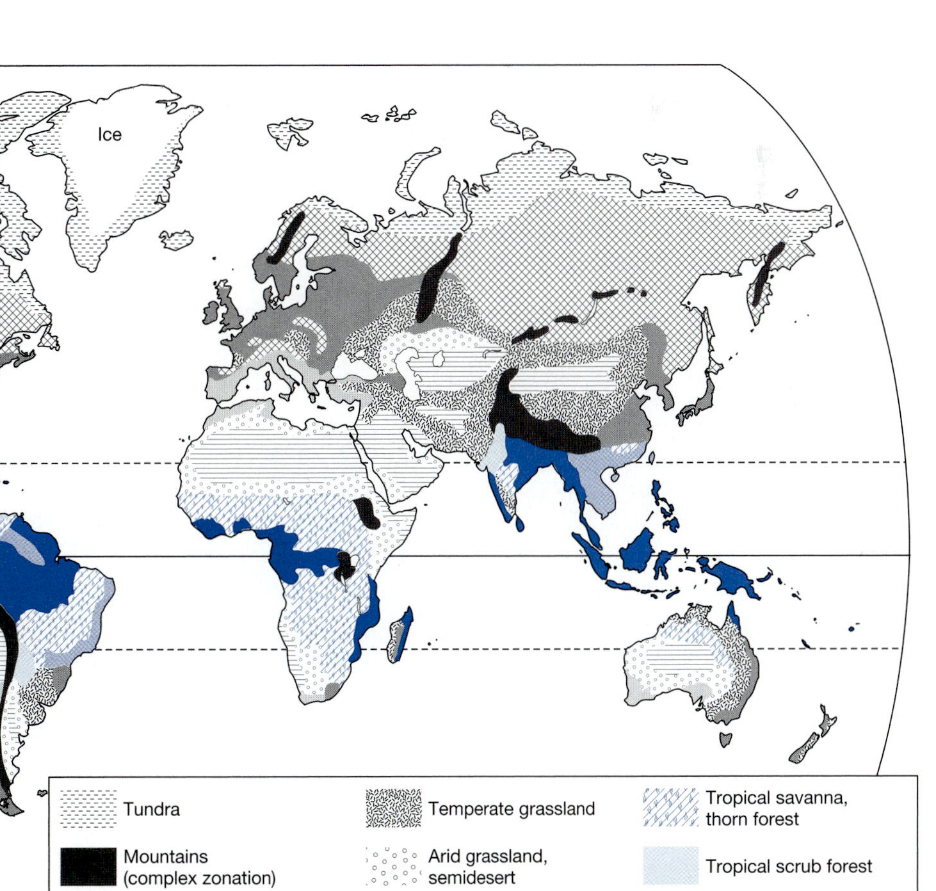

FIGURE 3.25 *The biomes.*

Snowfall is scant. Because of the cold weather and the short, six-week growing season, the vegetation in the northern tundra is sparse. As a result, this region is sometimes referred to as an Arctic desert. In fact, plant productivity in the tundra is just slightly higher than in the desert biome. The low temperature slows down the chemical and biological activities that are necessary for the formation of mature soils. Each year during late spring, the upper level of the soil begins to thaw, but the soil beneath it remains permanently frozen and is called **permafrost.** Permafrost occurs at a depth of about 15 to 45 centimeters (6 to 18 inches). In spring and summer the thawed-out ground turns into a quagmire. The meltwaters of late spring form thousands of tiny lakes because of poor drainage and low evaporation. Dwarf willows are characteristic producers. Although only a few meters high, some of these willows may be more than 100 years old. Representative consumer species include herbivores such as the lemming, ptarmigan, caribou, and musk ox (Figure 3.27). The snowy owl and arctic fox are characteristic carnivores in the area.

The tundra is a fragile ecosystem. One reason is the very slow rate of recovery from human disturbance. Thus, wagon-wheel tracks formed more than 120 years ago by early explorers and inhabitants are still plainly visible.

The tundra is currently under assault by U.S. oil companies' quest for more oil. In 1968 the richest oil deposit in the Western Hemisphere was discovered on Alaska's North Slope. A 1,262-kilometer (789-mile) pipeline was constructed to transport this oil to the port of Valdez on Alaska's southern coast. This caused the destruction of much vegetation and valuable wildlife habitat and resulted in excessive soil erosion. Today, oil companies have their sights set on a large segment of the Arctic National Wildlife Refuge and hope someday to develop it as they did the North Slope oil deposits.

In addition to oil, the tundra contains two thirds of the North American continent's known reserves of coal, and substantial amounts of zinc, lead, and copper. These resources could be exploited in the not too distant future. The damage from such development could be severe.

Northern Coniferous Forest

As shown on the global biome map (Figure 3.25), the **northern coniferous biome** or **taiga** forms an extensive east–west belt just south of the Arctic tundra in North America, Sweden, Finland, Russia, and Siberia. Characteristic physical features include an annual rainfall of 37 to 100 centimeters (15 to 40 inches), average temperatures of 26.6°C (20°F) in the winter to more than 21°C (70°F) in the summer, and a 150-day growing season. Dominant climax vegetation includes black spruce, white spruce, balsam fir, and tamarack. The evergreen trees are well adapted to survive; their branches are so flexible that they bend under the burden of a heavy snowfall without snapping. Lightning-triggered crown fires are fairly common because the dead, dry needles, which persist on the branches, are easily ignited. Over 50 species of insects are adapted to feed on the conifers. The populations of some species, such as spruce budworm, tussock moth, pine bark beetle, and pine sawfly, may suddenly erupt, causing considerable mortality of spruce, fir, and pine. White birch and quaking aspen are representative of the earlier successional stages. Typical animals include the moose, snowshoe hare, lynx, pine grosbeak, and red crossbill.

The taiga is currently the site of intensive tree cutting and mineral extraction, and, although it is a large biome, some individuals are concerned that the taiga faces a future of considerable exploitation that could have a devastating effect on ecosystems.

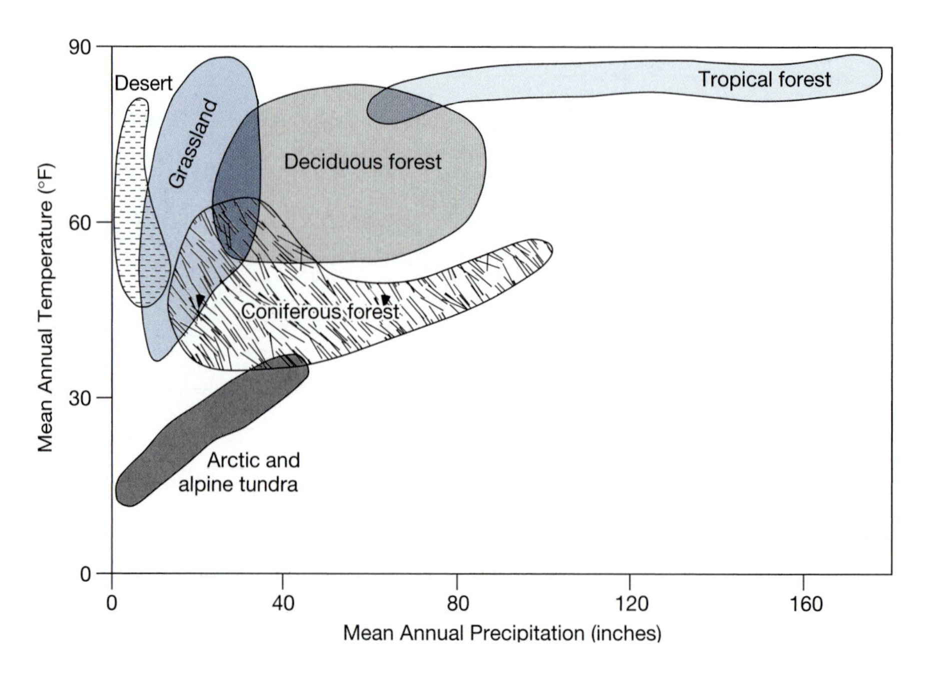

FIGURE 3.26 *Influence of temperature and rainfall in determining biome types.*

TABLE 3.2 *Biome Summary*

Biome	Temperature	Rainfall	Typical Plants	Typical Animals
Tundra	−57–16°C	10–50 cm	Lichenes, mosses, dwarf willows	Ptarmigan, snowy owl, lemming, caribou, musk ox, Artic fox
Coniferous forest	−54–21°C	35–200 cm	Black spurce, white spruce, balsam fir, white birch, aspen	Spruce budworm, tussock moth, moose, snowshoe hare, lynx
Decidous forest	−30–38°C	60–225 cm	Oak, hickory, beech, maple, black walnut, yellow poplar	White-tailed deer, gray squirrel, skunk, opossum, black bear
Grassland	40–60°C	30–200 cm	Little and big bluestem, grama grass, buffalo grass	Meadowlark, burrowing owl, pronghorned antelope, badger, jackrabbit, coyote
Desert	2–57°C	0–25 cm	Prickly pear cactus, saguaro cactus, creosote bush, mesquite, sagebrush	Diamond-backed rattlesnake, Gila monster, roadrunner, kangaroo rat, wild pig
Savannah	13–40°C	25–90 cm	Baobab tree, acacia tree, grasses	Zebra, giraffe, wildebeest, elephant, antelope
Tropical rain forest	18–35°C	125–1250 cm	Great diversity	Great diversity

FIGURE 3.27 *The ptarmigan, a characteristic tundra herbivore. It is consumed by snowy owls, Arctic foxes, and other predators.*

Deciduous Forest

The global biome map shows that the **deciduous** (broadleafed) **forest biome** covers eastern North America, all of Europe, and parts of China, Japan, and Australia. In the United States, the deciduous forest biome attains its greatest development east of the Mississippi River and south of the northern coniferous forest.

In precolonial times, the deciduous forest was virtually continuous and unbroken. Today, however, as a result of settlement, agricultural development, logging, mining, and highway construction, the biome has been reduced to a small fraction of its original area. Fingers of deciduous forest extend westward into the prairie country along major river valleys in response to increased levels of soil moisture. Because of abundant precipitation (at least 75 centimeters [30 inches]) and the fairly long growing season, plant productivity is the highest of any biome in North America except the temperate rain forest of the Pacific Northwest. Characteristic trees are oak, hickory, beech, maple, black walnut, black cherry, and yellow poplar. Representative consumers include the gray squirrel, skunk, black bear, and white-tailed deer.

Tropical Rain Forest

The **tropical rain forest biome** is located in the equatorial regions wherever there is more than 200 centimeters (80 inches) of rainfall annually. As shown in Figure 3.25, rain forests are located primarily in Central America, in South America along the Amazon and Orinoco rivers, in the Congo River basin of Africa, in Madagascar, and in Southeast Asia. The day–night temperature variations are greater than those of summer–winter. Heavy rains may fall daily throughout the rainy season. Because so much sunlight is screened out by the canopy, the forest floor is relatively dark and poorly vegetated.

Plant life is highly diverse in the rain forest. Two hundred species of trees may be found in 1 hectare (2.47 acres), compared with only ten species per hectare in the temperate deciduous forests of the United States. The forests are highly layered, or stratified, with the tree crowns occurring at three or even four different levels.

Animal life is abundant and highly diverse. At least 369 species of birds have been identified in just 2 hectares of Costa Rican rain forest—more than are found throughout Alaska.

Some insects are extremely large. In the Amazon rain forest, the wingspread of one species of moth is nearly 1 foot, and some spiders are large enough to feed on birds caught in their webs!

Unfortunately, humans are cutting the rain forest, leaving only small fragments in many countries. Tropical rain forests are cut down or destroyed to make room for human settlements, to acquire wood, to establish farms and pastureland, and to clear land for mining operations. Agricultural ventures usually result in failure, not only because the soil is very infertile, but also because it frequently contains iron compounds that bake brick hard under the tropical sun during the dry season. Deforestation has contributed to the atmospheric buildup of carbon dioxide and the warming of the planet (Chapter 19).

Tropical Savannah

A **savannah** (sah-van-ah) is a warm-climate grassland characterized by scattered trees. As shown in the global biome map (Figure 3.25), savannahs are located primarily in South America, Africa, India, and Australia. Rainfall averages 100 to 150 centimeters (40 to 60 inches) per year. Wet seasons alternate with dry seasons, and fires are common during the prolonged dry spells. Plants and animals that live in the savannah must therefore be drought and fire resistant. As a result, the diversity of species is not great.

The African savannah is characterized by the picturesque baobab and thorny acacia trees. This savannah supports the greatest variety and largest number of hoofed herbivores in the world, including the zebra, giraffe, wildebeest, and many species of antelope. In the struggle to produce more food, many people in Africa and India have converted extensive acreages of savannah into farms and livestock pasture. Those disturbances, along with illegal hunting, have caused a rapid decline in the herds of many species.

Grassland

The **grassland biome** is a huge expanse of land characterized, as its name states, by grasses. This land of no trees, except along rivers and where planted by humans around farms, cities, and towns, consists of flat to rolling hills. The main grasslands of the planet include the North American prairie of Canada and the United States, the pampas of South America, the steppes of Eurasia, and the veldts of Africa (Figure 3.25). Major grasslands occur in two regions in the United States: the Great Plains, a vast area extending from the eastern slopes of the Rockies to the Mississippi River, and the more moist portions of the Great Basin—a huge area that nestles between the Sierras to the west and the Rockies to the east. In north temperate latitudes, grasslands are the vegetational expression of an average annual precipitation that is excessive (over 25 centimeters [10 inches]) for the development of desert vegetation and inadequate (under 75 centimeters [30 inches]) for the development of forest. Winter blizzards and summer drought can be severe. Scientific studies suggest that over time, devastating fires periodically have burned the prairie. But the deep root systems of the plants of the grassland, which tap into groundwater supplies during periodic drought, also permit the grasses to regrow after fire.

The dominant vegetation includes the big bluestem, little bluestem, buffalo grass, and grama grass. The horned lark, meadowlark, and burrowing owl are characteristic birds in that grassland. Dominant mammals include the pronghorned antelope, badger, white-tailed jackrabbit, coyote, and prairie dog (Figure 3.28).

Today, only scattered remnants of the climax grassland biome remain. Humans have replaced the original wild grasses with cultivated "grasses" such as corn and wheat. In addition, humans have almost eradicated the bison and replaced it with domesticated herbivores such as cattle and sheep.

Desert

As shown in the global biome map, the world's largest deserts are located in the United States, Mexico, Chile, Africa (Sahara), Asia (Tibet and Gobi), and Australia. American deserts are located in the hotter, drier portions of the Great Basin and in parts of California, New Mexico, Arizona, Texas, Nevada, Idaho, Utah, and Oregon.

Deserts occur primarily leeward of prominent mountain ranges, such as the Sierra Nevada and the Rocky Mountains. The prevailing warm, humid air masses from the Pacific Ocean cool as they move up windward slopes and release their moisture as rain or snow. The region leeward of the mountains lies in the rain shadow, where precipitation is minimal. A desert community

FIGURE 3.28 *A prairie dog, resident of the grassland biome at the edge of its craterlike burrow entrance. The mound of earth not only serves as a lookout post but also prevents flash-flooding of its burrow.*

generally receives less than 25 centimeters (10 inches) of annual precipitation. Rainfall, moreover, may not be uniformly distributed, but may fall periodically in cloudbursts that cause flash floods and severe erosion. An extremely high evaporation rate aggravates the severe moisture problem. For example, the water that theoretically could evaporate from a given hectare in one year may be 30 times the actual amount received as precipitation.

Summer temperatures range from about 10°C (50°F) at night to about 48°C (120°F) during the day. Desert floor temperatures reach 62°C (145°F) in summer. Only organisms that have evolved specialized structural, physiological, and behavioral adaptations to extreme heat and dryness can survive in the desert. Characteristic producers are prickly pear cactus, saguaro cactus, creosote bush, and mesquite. Conspicuous among desert consumers in the United States are the rattlesnake, Gila monster, roadrunner, jackrabbit, kangaroo rat, and wild pig.

Altitudinal Biomes

By traveling several thousand miles from Texas northeastward to northern Canada, you pass through a series of different biomes—desert, grassland, deciduous forest, coniferous forest, and tundra. These biome changes reflect progressive changes in temperature and precipitation, which in turn result in dramatic differences in primary productivity (Figure 3.29). Such a series of biomes also

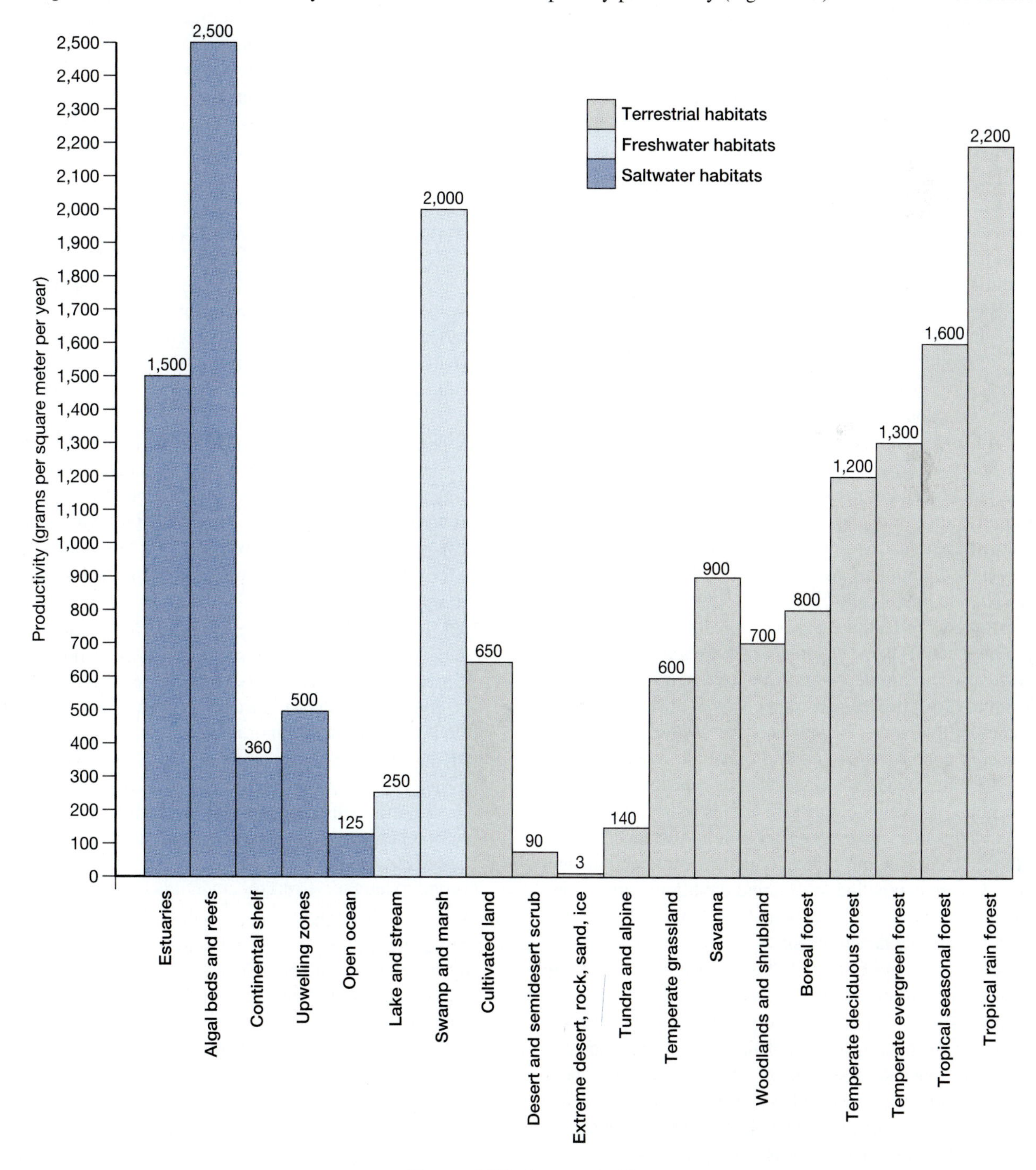

FIGURE 3.29 *Biome productivity.*

exists on the slopes of tall mountains, such as the 4,200 meter (14,000-foot) Colorado Rockies. Here again, the gradual changes in biome type are dictated by climatic factors, which in this case change progressively with altitude rather than latitude.

3.6 Ecology and Sustainability

Ecology is a science that examines relationships between organisms and their environment. This is another way of saying that ecology concerns itself with the way natural systems work. Few fields of study could be more important to resource management and sustainable development than ecology. We believe that a solid grounding in basic ecological principles can help resource managers better predict the impacts of various actions—for instance, the elimination of a predator from a given ecosystem. A basic understanding of ecology helps resource managers predict how human actions like building a dam or bulldozing a road through a forest will affect species on the site and in neighboring areas. A little knowledge can even help us avoid disastrous consequences. It may suggest alternative approaches or may cause us to rethink plans entirely.

Ultimately, the study of ecology could help human civilization steer onto a sustainable course. For example, an understanding of the laws of energy and the loss of biomass in food chains suggests that we could more easily feed the world's growing population if people ate more fruits, grains, and vegetables. An understanding of the nutrient cycles and of how they can be adversely altered through industrial and agricultural development could help us devise strategies that do not disrupt important cycles of life.

A knowledge of succession and the productivity of different successional communities has proved useful in managing resources for maximum yield. That knowledge, combined with an understanding of factors essential to the long-term health of ecosystems, could help us achieve optimum **sustainable yields**—that is, optimum productive rates that do not threaten the health of farms, forests, and other ecosystems.

The law of tolerance reminds us of the need to act in ways that preserve conditions essential to other life forms. By exceeding the limits, we become agents of destruction. The niche concept helps us better understand a species's full requirements and helps us manage ecosystems more sustainably.

With the understanding that an organism's interactions are complex, we are compelled to examine the big picture before we embark on any course of action. This systems view is essential to sustainable development.

The preceding chapters described a set of biological principles that would ensure the sustainability of natural systems. These principles are conservation, recycling, the use of renewable resources, restoration, and population control. In Chapter 1, we pointed out that natural systems tend to endure because most organisms use what they need and use resources efficiently. This is the conservation principle. Although there are exceptions, a grizzly that kills salmon and only eats its eggs, for instance, most organisms are by nature fairly frugal. In addition, in natural systems all wastes are recycled. Thus, the waste products of one organism become the food source of another. Consequently, all nutrients cycle continuously through natural systems.

Nature also sustains itself because it relies on renewable resources: the sun, soil, water, and plants. This renewable base ensures life's continuation. In addition, natural systems repair damage. They restore themselves. Finally, natural systems control numbers within the carrying capacity of the environment. Collectively, these principles seem to ensure the continuation of life.

But life is not static. Organisms change by evolution. In other words, organisms can adapt to meet changing conditions. This, too, is essential to life's continuation. Humans must also adapt to a changing world, but we do not have the luxury of time needed to adapt biologically. We must make social, economic, political, and even personal changes to ensure our long-term survival. By applying these rules to our own society, we can adapt to the changing conditions of a world threatened by our own actions.

Summary of Key Concepts

1. Ecology is the study of the interrelationships in the natural world. They examine relationships between organisms and between organisms and their environment. Ecologists focus their studies primarily on three levels of organization: the population, the community, and the ecological system or ecosystem.
2. A population is the total number of individuals of one species in a given area. A community includes all living organisms in a given area—including plants, animals, and microorganisms. An ecosystem is the biological community and the environment with which it interacts.
3. All of the ecosystems of the Earth form the biosphere or ecosphere.
4. All living things are comprised of matter. Matter, in turn, is anything that occupies space and is perceptible to the senses. Matter cannot be created nor destroyed, but it can be converted from one form to another. This is known as the law of conservation of matter.
5. Living things depend on energy. Energy exists in many forms. Energy obeys certain laws. The first law of energy states: Although energy cannot be created or destroyed, it can be converted from one form to another. The second law of energy states: Whenever one form of energy is converted to another, a certain amount is lost as heat.
6. Virtually all of the energy required by living organisms comes from photosynthesis. Photosynthesis is the process by which green plants convert the raw materials carbon dioxide and water into glucose and other organic molecules in the presence of sunlight. Sunlight energy is stored in the chemical bonds of the molecules produced by plants. Some of it used by the plants, much of the rest will be used by animals and microorganisms in various food chains.
7. A food chain is an sequence of organisms through which nutrients and energy move. In ecosystems, food chains form interconnected networks known as food webs.

8. The stability of a food web increases in proportion to its complexity. Simplifying a food web makes an ecosystem more vulnerable.
9. During photosynthesis, green plants convert energy from the sun to organic food energy that can be used by other organisms. Because of this process, green plants are called producers. Plant-eating animals, such as the rabbit, are called herbivores, or primary consumers. Animals like the wolf, which feed on other animals, are known as carnivores, or secondary consumers.
10. The total dry weight of living substance (protoplasm) in a given organism, population, or community is known as the biological mass, or biomass. Studies of ecosystems show that there is a progressive reduction in biomass at each succeeding feeding level, or trophic level, of a given food web. This is known as the pyramid of biomass.
11. Because biomass contains potential energy and because living organisms are only about 10 percent efficient in converting the energy of their food into the energy of their own biomass, there is a progressive reduction in the amount of food energy in successive trophic levels (feeding levels) of a given food web. This is known as the pyramid of energy. In other words, more energy is available to herbivores than to carnivores. One of the practical outcomes of this is that more food energy would be available to human beings if we would move down the food chain, closer to the producer base.
12. The circular flow of an element from the nonliving environment into the bodies of living organisms and then back into the nonliving environment is known as a nutrient cycle. For millions of years before the advent of human technology, the nutrient cycles were delicately balanced. However, in recent years our technology has caused many nutrient cycles to become unbalanced. In other words, we have greatly increased the amount of certain nutrients in its cycle. The result is the pollution of air, water, and land—often with serious consequences.
13. For each organism there exists a specific tolerance range for any essential environmental factor. This is known as the range of tolerance.
14. Organisms occupy a specific area, known as their habitat. Their niche is a more complex concept. It consists of many factors, including the foods they eat, their predators, its nesting requirements, and so on. The niche describes an organism's unique role in an ecosystem–its requirements and its relationships.
15. Ecologists have found that organisms cannot occupy the same niche, known as the competitive exclusion principle.
16. The replacement of one community by another under constant climatic conditions is known as biological succession. The initial stage of a succession is known as a pioneer community. The final stage of a succession is known as the climax community. This stage is stable. In theory it will persist indefinitely unless the climate changes or some geological process or human action destroys it.
17. Succession occurring in an area where living organisms were not previously present, such as bare rock, is known as a primary succession. However, a succession that occurs in an area that once supported life, such as a burned-over forest or an abandoned corn field, is known as a secondary succession.
18. A biome is the largest terrestrial community that is easily recognized by a biologist. It is the biological expression of the interaction of climate, soil, water, and organisms. Representative biomes discussed in this chapter are: tundra, northern coniferous forest, deciduous forest, tropical rain forest, tropical savannah, grassland, and desert. Each biome is characterized by a unique assemblage of plants and animals. On the slopes of tall mountains, such as the Rockies, a series of altitudinal biomes can be recognized.
19. An understanding of ecology helps us better understand human impacts on the environment and ways to better manage ecosystems. Ultimately, it can help us steer onto a sustainable course. By practicing ecological principles of sustainability we can create a human culture that respects and operates within the limits imposed by global ecosystems.

Key Words and Phrases

Abiotic Factors
Altitudinal Biomes
Atmospheric Fixation
Biogeochemical Cycle
Biological Fixation
Biological Succession
Biomass
Biome
Biosphere
Biotic Factors
Biotic Potential
Carnivore
Cellular Respiration
Chlorophyll
Climax Community
Community
Competitive Exclusion Principle
Coniferous Forest Biome
Consumer
Decomposer
Decomposer Food Chain
Denitrification
Desert
Detritus
Detritus Feeder
Detritus Food Chain
Dynamic Equilibrium
Ecological Succession
Ecology
Ecosphere
Ecosystem
Ecosystem Simplification
Electromagnetic Spectrum
Elemental Cycle
Energy
Entropy
Eutrophication
First Law of Energy
Food Chain
Food Web
Fossil Fuel
Grassland Biome
Grazer Food Chain
Gross Primary Production
Herbivore
Industrial Fixation
Kinetic Energy
Law of Conservation of Matter
Laws of Energy
Legume
Limiting Factor
Matter
Mechanical Energy
Net Primary Production
Niche
Nitrification
Nitrogen Cycle
Nitrogen Fixation
Nutrient Cycle
Omnivore
Permafrost
Phosphorus Cycle
Photosynthesis
Pioneer Community
Population
Potential Energy
Primary Consumer
Primary Production

Primary Succession
Producer
Pyramid of Biomass
Pyramid of Energy
Pyramid of Numbers
Radiant Energy
Range of Tolerance
Reservoir
Respiration
Root Nodules
Savannah
Second Law of Energy
Secondary Consumer
Secondary Succession
Stomata
Succession
Sustainable Yield
Taiga
Trophic Level
Tropical Rain Forest
Tundra
Wavelength

Critical Thinking and Discussion Questions

1. Define and give an example of each of the following levels of organization: community, population, ecosystem.
2. Define the biosphere. Why are most life forms restricted to a narrow band within the biosphere?
3. For many years, a lake has been operating as a balanced ecosystem. Suppose now that a disease kills off all the plants. What effect would this have on the animals in that ecosystem? Suppose that instead of killing off all the plants, a disease destroyed all the bacteria in the lake. What effect would this have on the lake ecosystem? Discuss your answer.
4. What is energy? What is mass?
5. State the first and second laws of energy. What do they mean to you? How do they affect you? How do they affect ecosystems?
6. Trace the radiant energy from your study lamp backward in time to its ultimate source in solar energy.
7. Define photosynthesis with respiration. How are they dependent on each other?
8. What is a food chain? What is a food web?
9. Distinguish between grazing food chains and detritus food chains. Give an example of each type. Which type is more easily observed? Which type of food chain is most important in a forest ecosystem? Which type is most important in a marine ecosystem?
10. How many different species of organisms can you identify that are living either in your dormitory or in your home? Are they producers, herbivores, carnivores, omnivores, or detritus feeders? Compare your results with those of other members of your class.
11. List all the food organisms you consumed, in part or in entirety, during your last three meals. Now construct a food web, using this list and yourself as a starting point. You may include other organisms not actually represented in your meals. Identify the trophic level of each species in the web.
12. The total biomass of the animals living in the English Channel is five times greater than that of the plants. Can you offer any explanation for this seeming contradiction of the biomass pyramid concept?
13. Suppose that you are in charge of a unique wildlife refuge that is completely isolated from other ecosystems. This refuge is composed of one kind of producer (grass), one kind of herbivore (rabbits), and one kind of carnivore (hawks). Assume that one rabbit weighs as much as one hawk. Assume further that the rabbits consume only grass and that the hawks feed only on rabbits. Suppose that you originally stocked the refuge with three rabbits and six hawks, in other words, a 2:1 ratio of hawks to rabbits. After about five years, would you expect that this same ratio would exist? Why or why not? What ratio would you expect?
14. As you learned in this chapter, elements are recycled through ecosystems but energy is not. Suppose, however, that energy was recycled and elements were not. What would be the effects on ecosystems? Discuss your answer.
15. Suppose that the sun stopped shining. What effect would this have on life on the Earth? Would life still be possible? Discuss your answer.
16. Suppose that 20,000 calories of solar energy is available to grass for photosynthesis. Suppose that grass converts 1 percent of the solar energy available to it. Suppose further that animals can store 10 percent of the food energy they consume in their own protoplasm. How many food calories would you get if you consumed the grass? If you lived on beef derived from grass-eating cattle?
17. What is a nutrient cycle? Why is it important to know about them?
18. Make labeled diagrams of the shortest nitrogen, carbon, and phosphorus cycles you can devise.
19. Describe three ways in which nitrogen can be fixed.
20. Trace a given carbon atom from the air that existed over Pennsylvania 250 to 300 million years ago to the fried egg you had for breakfast this morning.
21. In one sense, pollution can be defined as an imbalance of a nutrient cycle caused by humans. Explain. In this context, can you give one example each of land, air, and water pollution? In general, how do humans upset nutrient cycles?
22. Discuss the range of tolerances that you have for various physical conditions in the daily environment to which you are exposed.
23. Why is it important to understand the range of tolerance of species in an ecosystem?
24. What is the competitive Exclusion Principle?
25. Define the term biological succession.
26. Describe three sites at which a primary succession might start. Describe three sites at which a secondary succession might occur.
27. Construct a chart listing four basic differences between the pioneer and climax stages of an ecological succession.
28. Suppose that a square mile of oak woods in Ohio is removed and replaced with a huge slab of polished marble. One thousand years pass. Will the marble slab still be visible? Why or why not? Give a detailed explanation of the probable events that occurred.

29. What is a biome? Which biome do you live in? What are its major characteristics?
30. Describe the physical features of the tundra biome. Why is it sometimes called an Arctic desert? Why is it considered a fragile ecosystem? Would it be a good place to study succession? Why or why not? Explain.
31. Which of the biomes discussed in the chapter has the greatest diversity of animal life? Which features of the biome make this diversity possible?

Suggested Readings

Chiras, D. D. 1992. *Lessons From Nature: Learning to Live Sustainably on the Earth.* Washington, D. C.: Island Press. Discusses how to build a sustainable society.

Dodson, S.I., et al. 1998. *Ecology.* New York: Oxford University Press. Great reference for students who are eager to learn more about ecology.

Dodson, S. I, et al. (eds.). 1999. *Readings in Ecology.* New York: Oxford University Press. Collection of essays including classical studies and new, interesting thinking about ecology.

Ehrlich, P. R. 1986. The *Machinery of Nature.* New York: Simon and Schuster. Great general discussion of ecology.

Ehrlich, P. R., and J. Roughgarden. 1987. *The Science of Ecology.* New York: Macmillan. Higher-level coverage of ecology.

Morgan, S. 1995. *Ecology and Environment: The Cycles of Life.* New York: Oxford University Press. Very general, but very beautiful, introduction to basic principles of ecology.

Odum, E. P. 1993. *Ecology and Our Endangered Life-Support Systems.* Sunderland, Mass.: Sinauer Associates. Good introduction to ecology.

Smith, R. L. 1992. *Elements of Ecology,* 3rd ed. New York: Harper and Row. Advanced readings on ecology and environmental problems.

Southwick, C. H. 1995. *Global Ecology in Human Perspective.* New York: Oxford University Press. A very readable introduction to ecology.

Web Explorations

Online resources for this chapter are on the World Wide Web at: **http://www.prenhall.com/chiras** *(click on the Table of Contents link and then select Chapter 3).*

Chapter 4

The Human Population Challenge

Humans have unique abilities among the animal kingdom. One of the most distinguishing is our ability to reason. Most important is our uncanny ability to assess problems, devise solutions, and then put those solutions into operation. Unfortunately, at this critical period in human history, we have failed to fully apply our talents to solving one of the most pressing of all problems: our rapidly expanding global population. Countless experts and numerous studies strongly suggest that, if left uncontrolled, the population could skyrocket even further. This, in turn, could cause further widespread hardship to people and extensive damage to the life support systems of the planet.

How serious is the problem? Consider a few statistics. In the next ten seconds, the world population will expand by 27 people. That's nearly three additional people (2.7 to be more precise) every second. In a single day, the world population will expand by nearly a quarter of a million people (230,000 to be more precise). In a week, the human population grows by 1.6 million. In a year, the Earth and its ecosystems must support another 84 million people. Each of these individuals must be fed, and each one must be given shelter and clothing. Just to give each person in this *daily increase* one glass of milk a day would require 15,000 additional dairy cows. To give each person a loaf of bread would take more than 300 acres of wheat.

Fortunately, not all of the news about population is alarming. Although the global human population is increasing dramatically, since 1970 the rate of growth has slowly decreased. Most of the slowdown has occurred in the less developed countries (LDCs) of Asia and South America. Family planning programs have assisted greatly in this shift. The populations in some European countries have stopped growing altogether, a condition called **zero population growth (ZPG),** thanks to family planning programs and individual actions. Growth has slowed in Africa as well, but the news there is not good. The slow down has been largely due to increased mortality (death) resulting from starvation, civil war, and infectious disease, notably AIDS. Russia and the former Eastern bloc nations have also witnessed dramatic decreases in growth. In fact, the population of Russia has actually declined. The decline in population has been attributed in large part to the difficult transition to democracy and capitalism, which has left Russia in economic ruin.

Important as the decline in population growth is, it does not mean that the Earth's population will stop growing. In fact, the Population Reference Bureau, a nonprofit organization in Washington, D.C. that tracks population growth, predicts that the current population of about 6.1 billion in 2000 could reach 7.8 billion in the next 25 years. The reason for continued growth despite a decline in the rate of growth will be made clear soon.

4.1 Understanding Populations and Population Growth

Population is a serious matter. Before we look at the full impacts of the exploding human population and suggest some strategies to control it, however, we will discuss some concepts and measurements used to describe populations. These, in turn, will help you understand the importance of the problem and ways various solutions will help solve it.

Birth and Death Rates

Two of the most fundamental terms you must understand are birth rates and death rates. The **birth rate** of a country is simply the number of persons born per 1,000 individuals in the population in given year. The birth rate of the United States, for instance, is currently 11. Thus, for every 1,000 people in the population, there are 11 births.

The **death rate** is the number of persons per 1,000 individuals who die in a particular year. In the United States, the current death rate is seven.

The difference between the birth and death rate of a country is known as the **rate of natural increase** (or decrease). **Demographers,** scientists who study populations, use the following equation to determine natural increase:

Natural increase = growth rate − death rate

In the United States, the rate of natural increase is 11/1,000 − 7/1,000 which equals 4 per 1,000. That means, every year the population expands by four people for every 1,000 people currently living in the country.

This example shows that if the birth rate is higher than the death rate, the population will expand. But just the opposite may occur. The birth rate may be lower than the death rate. In such instances, the population will decrease.

Now, take a moment to study Figure 4.1. As you can see, the birth rate, minus the death rate equals the natural increase. But this figure shows that there's another factor that determines population growth rate of a country, and that's the **net migration**—that is, the number of people moving into or out of a country. Demographers use the following equation to determine the population growth in a country:

Growth rate = birth rate − death rate + net migration

In the United States, net migration is well over a million people a year. This causes further expansion of our population, which reached 280 million in the year 2000.

Now that you understand a little about population growth, let's use your new-found knowledge to explore the global human population. According to the Population Reference Bureau, the birth rate of the human population in 2000 was 22 per 1,000 in 2000. The death rate was 9 per 1,000. This results in a natural increase of about 13 people per 1,000.

The population growth rate of a nation can be calculated *as a percentage,* as follows:

percent annual growth rate
= birth rate − death rate × 100

Therefore, the percent annual growth rate for the world in 2000 is determined as follows:

22/1,000 − 9/1,000 = 13/1,000
13/1,000 × 100 = 1.3 percent[1]

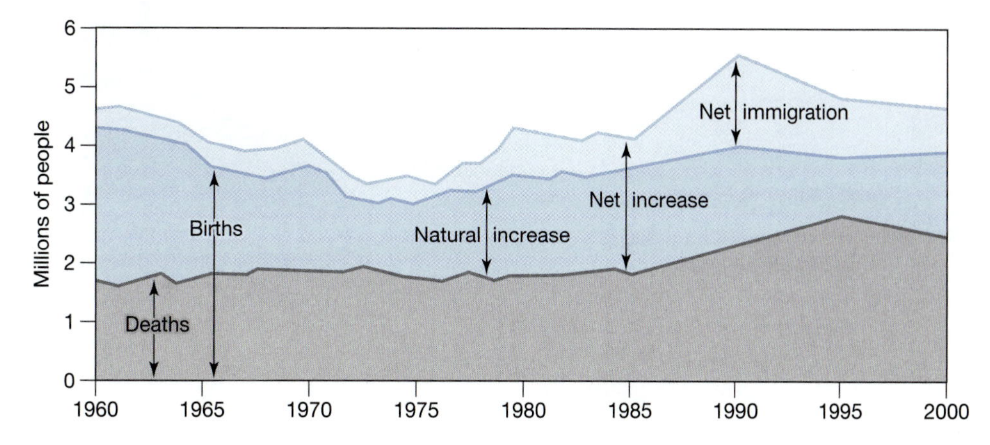

FIGURE 4.1 *Components of population change in the United States, 1960–1995. U.S. population growth is a product of birth rates, death rates, and net immigration.*

[1]Because of rounding errors in the growth rate and death rate, the world population growth rate is really closer to 1.4%.

Exponential Growth

Figure 4.2 shows the growth of the human population over the past millennium. As illustrated, the human population remained small for many thousands of years. Only in the last 200 years has it begun to climb—reaching 6.1 billion in the year 2000.

Demographers call this type of growth pattern exponential growth, a term you may have already heard. What exactly is exponential growth and why all of the fuss over it? **Exponential growth** occurs any time something grows by a fixed percentage but only if the annual growth (annual increase) is added to the base amount. A bank account growing at five percent a year, for example, is said to be growing exponentially if the interest is retained in the account so you earn interest on your interest, too. As shown in Figure 4.2, exponential growth results in a **J-shaped curve.**

Exponential growth is quite deceiving. As the graph shows, initial growth is quite slow, but there comes a time in the growth process when the population—or whatever else you're examining—experiences a very rapid upswing. Once the population rounds the bend of the J-curve. The rapid upswing seems to occur from nowhere, taking many people by surprise. Why does growth suddenly seem to accelerate or does it?

In exponential growth the rate of growth does not change to produce this mysterious effect. What happens is that after a long period of slow growth in absolute numbers, the base amount gradually reaches a size at which even small percentage increases result in very large increases in the absolute

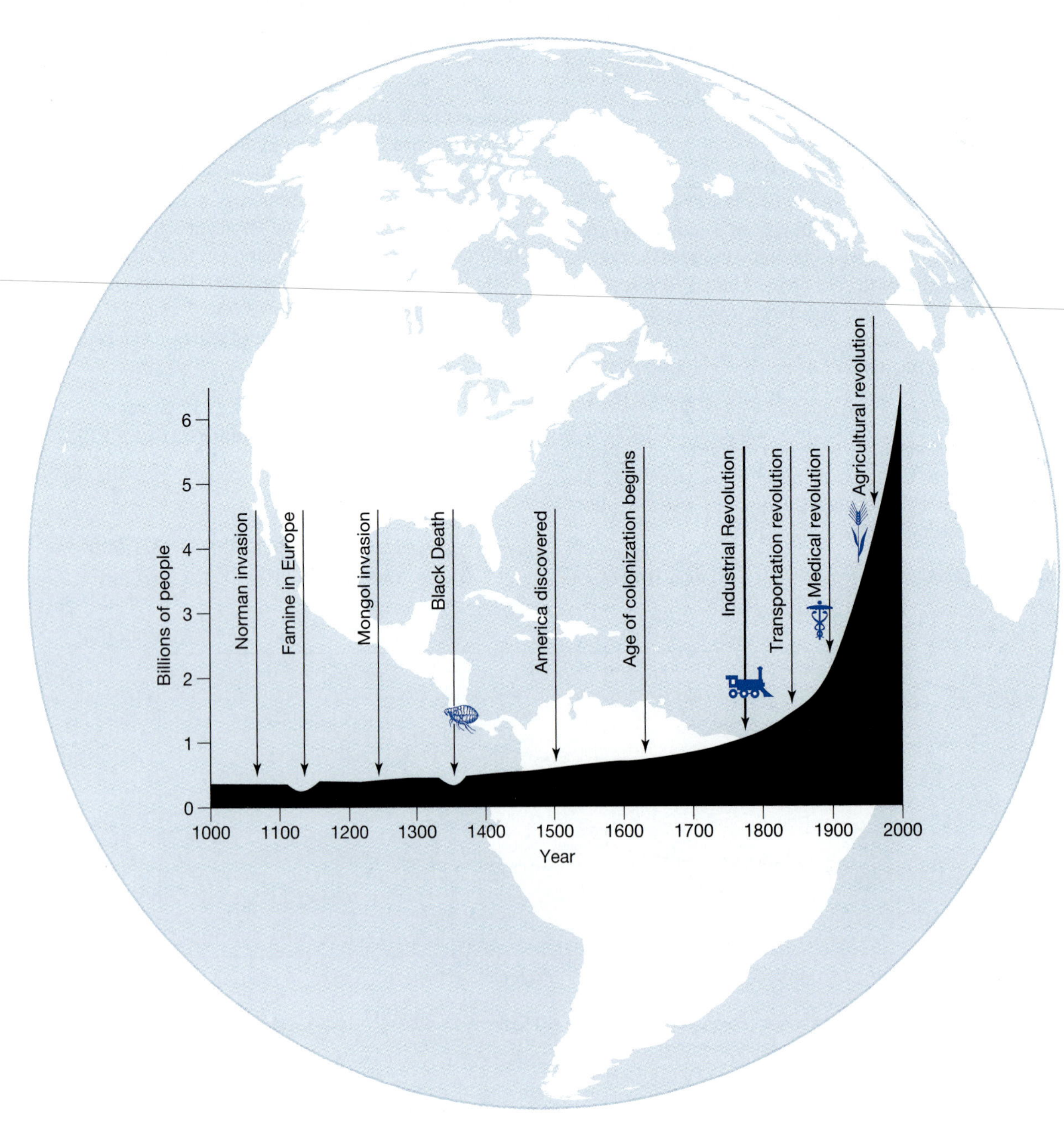

FIGURE 4.2 *Exponential growth of the human population. Growth has skyrocketed in the past 200 years as a result of improvements in sanitation, medicine, and agricultural production.*

number. Let us put it another way: even though the entity may be growing at a rate that seems quite insignificant, if the base amount has gotten large enough, a small percentage increase will result in a huge numerical increase. The human population, for example, is now 6.1 billion. Even though it is growing at a rate of 1.3% per year, which seems small, the net increase is huge—a whopping 84 million people a year!

Exponential growth is one of the most important concepts you will learn in your study of natural resource conservation. It describes growth of the human population and growth of many other aspects of the current environmental crisis as well such as resource depletion and pollution.

Doubling Time

Demographers also use another measure to help understand populations and predict their growth, the doubling time. **Doubling time** is the time it takes a given population to double in size. To determine the doubling time of a population you simply divide 70 (the demographic constant) by the percent annual growth rate. The result is the doubling time. For the global population (as of 1996) the doubling time is

$$70/1.3 = 50 \text{ years}$$

At its current rate of growth, the world population is projected to double in about 50 years.

Doubling times vary considerably from one nation to another. On average, the more developed nations are growing at a rate of about 0.1% per year, which yields a doubling time of about 580 years. The less developed nations are growing at a rate of about 1.7% per year, doubling in about 40 years. Table 4.1 lists some other important features of developed and less developed nations.

Why has Global Population Skyrocketed?

As noted earlier in the chapter, the human population has skyrocketed in the last 200 years increasing from about one billion to six billion. What has caused this phenomenon?

For years, human population was held in check by disease, starvation, wars, and other factors. Infections caused by viruses, bacteria, and protozoa killed many children early in life and took the lives of many adults, too. In Medieval times, for example, plagues killed over 25 million people in Asia and Europe. Smallpox spread like wildfire and killed one out of every four afflicted persons until Jenner developed his vaccine at the close of the eighteenth century. As a result, a child born in 1550 had a life expectancy of only 8.5 years. Disease continued to cause problems into the early 1900s. In fact, in 1900 tuberculosis was a prime killer, with pneumonia running a close second. As late as 1919, the influenza virus took a toll of 25 million people worldwide.

Although birth rates were high during this time, the human population remained small because of the extremely high death rates. But then came a series of advances in agriculture, medicine, and sanitation. The invention and perfection of the plow and other farm machinery, for example, made more food available, helping to stave off hunger and starvation. Later on, new drugs and vaccines combated viral and bacterial infections. Today, tuberculosis and pneumonia are largely controlled by antibiotics. Effective vaccines have been developed against smallpox, tetanus, diphtheria, whooping cough, influenza, and many other diseases. Better sanitation methods have had a positive effect, too. In particular, the treatment of human waste and the purification of drinking water have helped to reduce the incidence of infectious disease. Combined, these factors resulted in dramatic decrease in death rates. Largely because of these changes, the world death rate of 25 per 1,000 in 1935 fell to about nine per 1,000 today.

Chemical pesticides have also aided in the expansion of the human population. Early in the twentieth century, for example, the mosquito-borne disease malaria was either directly or indirectly responsible for 50 percent of all human mortality. After World War II, this disease was held in check by insecticides such as DDT. In Sri Lanka (formerly Ceylon), an intensive malaria-control campaign launched in 1946 resulted in a decline in the mortality rate from 22 to 13 per 1,000 in only six years. The use of DDT resulted in a dramatic increase in the population of nations of Europe and other wealthier nations.

Slowly but surely, humans conquered the high death rate, but the decrease in mortality was not accompanied by a decrease in birth rates—at least for 100 to 150 years. And, as we noted earlier, it is the difference between birth rates and death rates that determines the growth of a population.

Demographic Transition: Reestablishing the Balance The nations that benefited from advances in agriculture, medicine, and sanitation, eventually did bring birth rates down. Today, many of these nations continue to post low growth rates and about 50 of them have stable or even shrinking populations. What caused the transition?

Demographers have found that industrialization and rising wealth have had a profound effect on the birth rate of many na-

TABLE 4.1 *Comparison of More Developed and Less Developed Nations in 2000*

	More Developed	Less Developed (Including China)
Population	1.2 billion	4.9 billion
Average Birth Rate	11/1000	25/1000
Average Death Rate	10/1000	9/1000
Average Population Growth Rate	<0.1%	1.7%
Doubling Time	800 years	40 years
Infant Mortality Rate	8/1000	63/1000
Total Fertility Rate	1.5	3.2
Life Expectancy at Birth	79 years	66 years
GNP per capita	$19,500/yr	$1,260/yr

tions, resulting in a population change called a demographic transition. **Demographic transition** is defined here as a shift in both birth rates and death rates over time going from high birth and death rates to low birth and death rates, as shown in Figure 4.3. Virtually all of the industrialized European nations, as well as the United States, Canada, and Japan, have undergone the demographic transition.

The demographic transition consists of four stages, shown in Figure 4.3 which plots data from Finland as an example. In stage I (pre-industrial), large families prevailed because many children were required to work on farms. They also provided financial security for parents in their old age. The birth rates and death rates, however, were both high and the population growth rate was small.

In stage II of the demographic transition, death rates began to fall because of improvements in sanitation, food production, and disease control mentioned earlier. The discrepancy between birth rates and death rates results in rapid population growth.

In stage III, rising affluence began to drive the birth rates down. The reason for this shift is that as food production techniques improved, fewer farmers were needed to meet demand. Many farmers and their families abandoned their way of life and moved to urban areas to find work. In the city, however, children were no longer needed to help with farm chores. A large family became an economic drain. Couples voluntarily reduced the number of children they had. As a consequence, birth rates and death rates gradually came into balance and population growth slowed dramatically. This is stage IV of the transition.

The demographic transition began 100 to 150 years ago in Europe, the United States, and the rest of the industrial nations. In the less developed nations, however, the demographic transition began much later, after World War II as modern medicine, pesticides, and other advances already enjoyed by the Western world began to be introduced to these nations. But most LDCs never made it through the demographic transition. They remain stuck in stage II (transition) experiencing lower death rates thanks to advances in medicine and sanitation but fairly high birth rates (Figure 4.4). As you would predict, the result is a rapid increase in population among these people. So rapid is the growth that 90% of all growth occurs in the LDCs (Figure 4.5).

The hope among many people was that these countries would industrialize and that this, in turn, would lower the birth rates, as it has in many more developed nations. However, many experts assert that many LDCs such as Ghana, Chad, and India simply cannot depend on industrial development to carry them through the demographic transition and bring about a reduction in birth rates. Why? There are several reasons:

1. These countries often lack the large number of highly trained people (engineers, for example) on which industrial development depends.
2. They also lack the energy base represented by such resources as coal, oil, and natural gas that has been used to fuel industrial development.
3. They do not have enough time. Substantial industrial development cannot be accomplished in less than several decades, even with trained personnel and abundant energy resources. With the populations of many of these nations doubling every 30 to 40 years or less, time is running out!
4. Many countries also lack the financial resources needed to develop economically.

So, what will help these nations reduce their growth rates? Rising death rates, already being seen in some nations, could help bring the populations back in balance. As noted earlier,

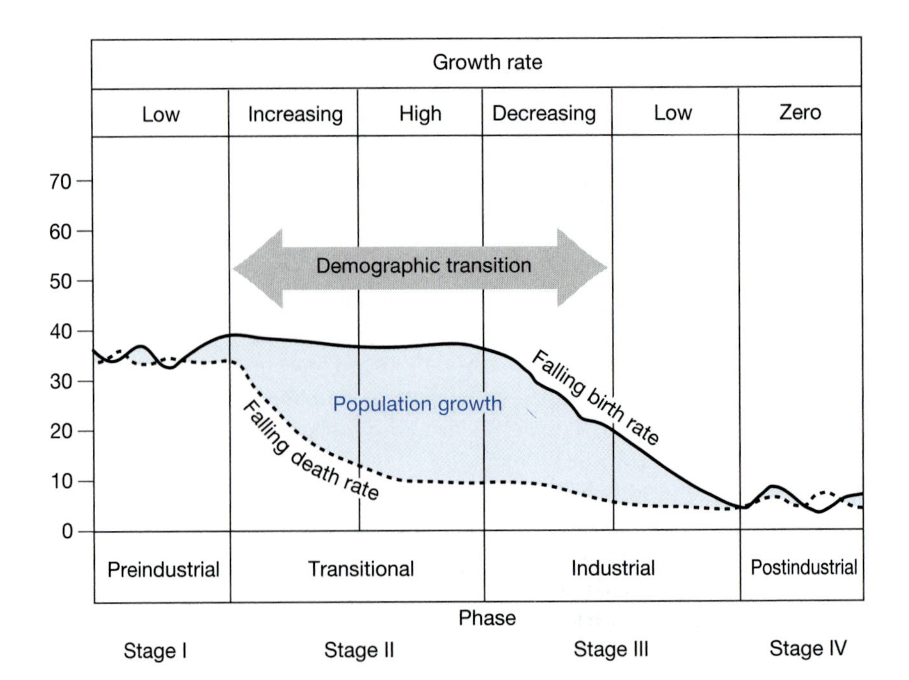

FIGURE 4.3 *Characteristic features of the demographic transition.*

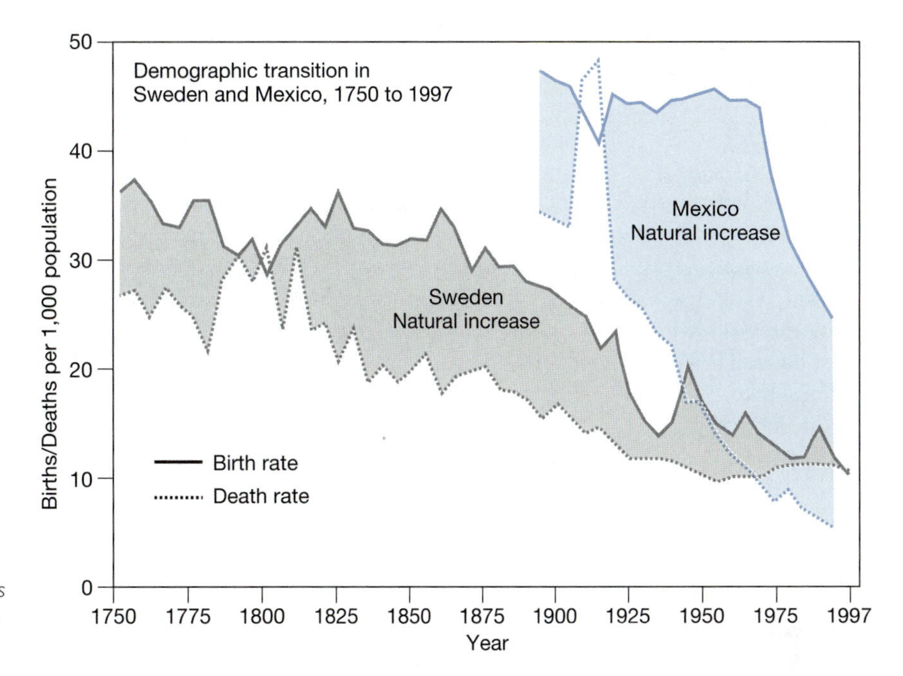

FIGURE 4.4 *Demographic transition in Sweden and Mexico. Note that Mexico like many other LDCs is stuck in Stage II of the transition and is experiencing rapid population growth.*

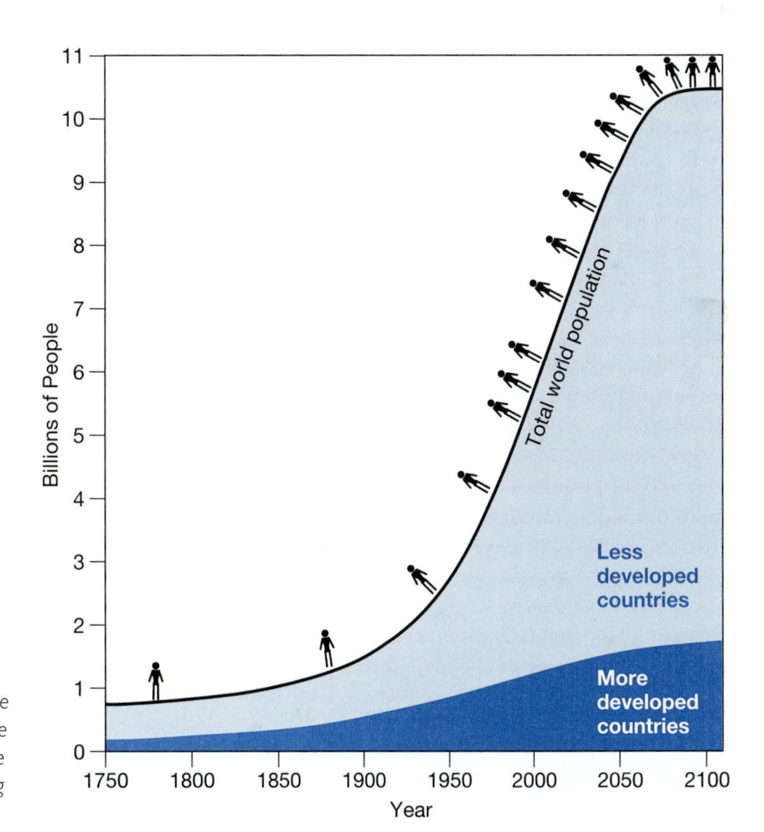

FIGURE 4.5 *Growth in the LDCs vs MDCs. Despite the fact that the world's annual population growth rate shows a decreasing trend, the number of people in the world is still increasing rapidly. Most growth is occurring in the less developed nations.*

many African nations are already experiencing an increase in death rates due to starvation, infectious disease, civil war, and AIDS. **Family planning programs,** which allow couples to determine the number and spacing of children they desire, could help stabilize population growth by bringing birth rates down. Sustainable economic development could also assist. Improvements in education and job opportunities for men and women alike are also vital. These options will be described in detail shortly. First, let's take a look at AIDS.

Will AIDS Correct the Imbalance? In the early 1980s, a new human disease known as **acquired immunodeficiency syndrome (AIDS)** began spreading rapidly through the African population and then throughout the rest of the world. AIDS is caused by a virus known as the **human immunodeficiency virus (HIV)** that is transmitted during sexual intercourse and anal intercourse, through the sharing of contaminated needles used for drug injections, through blood transfusions, and from an infected pregnant woman to her fetus. HIV attacks the immune system, greatly diminishing one's protection against infectious diseases such as tuberculosis and pneumonia. Most patients who contract the disease have little hope of surviving it. HIV is nearly 100% fatal, especially in less developed nations where medical care is limited and access to new drugs designed to combat it are too expensive to be available.

The spread of this disease is accelerating. Many consider it to be an epidemic. By 1999, 14 million people worldwide had died from the disease. Another 34 million are currently infected with the virus with over 6 million new cases a year. The hardest hit area is sub-Saharan Africa. Seven of every ten new infections take place here. Nine of every ten AIDS-related deaths occur in this region as well. In a dozen African nations, one of every ten adults is infected with AIDS. In Zimbabwe and Botswana, one of every four adults has the virus. It is in these nations that AIDS will have its greatest impact on population growth, some say, causing it to come to a halt.

Other areas of the world are also witnessing a rise in AIDS. Myanmar, Viet Nam, Cambodia, and India are experiencing a rapid increase. Eastern Europe has witnessed a dramatic increase as well. In the United States and Western Europe, AIDS-related deaths have declined largely due to new antiviral drugs that prolong the onset of the disease after initial infection. However, even here HIV infections are on the rise.

Not all of the news is bad. In about half of the less developed countries HIV has not spread into the general public. It is most common among prostitutes, their clients, and addicts. Several countries have mounted preventive campaigns to address AIDS and have seen substantial decreases in HIV infections.

Unfortunately, a vaccine for AIDS may not be developed for 20 years, and by that time AIDS may have killed 20 percent of the black population of Africa, and millions of people on other continents as well. Indeed, AIDS ultimately may kill many more people than the Black Death in the fourteenth century. This devastating disease may eventually be a factor in bringing our global population more in balance with the resource supply on which it depends.

Total Fertility Rate and Population Histograms

Before we examine the impacts of population growth, you must become acquainted with a couple more concepts and terms. One measure that is extremely helpful in understanding population growth and predicting the future size of the human family is the total fertility rate. The **total fertility rate (TFR)** is the number of children a woman is expected to produce in her life based on fertility rates in various age groups at the present time. It is reported as an average for all women. In the United States, for instance, the TFR in 2000 was 2.1. That is, women alive today are expected on average to give birth to two children during their lifetime.

The number of children required to replace a mother and father in the United States is currently 2.1. This is known as **replacement level fertility.** This means that each woman must have 2.1 children to replace herself and her husband or that ten women must have 21 children to replace themselves and their husbands. The extra child makes up for deaths that occur prior to reaching reproductive age. Since 1972, the TFR of American women has been below the replacement level fertility.

Another predictive tool that helps us understand where populations are going is the age structure. The **age structure** is the number of individuals occurring in each age class within the population. Males and females are generally enumerated separately. Plotted as a graph, this forms a **population histogram** (hiss-toe-gram).

By examining the profile of an age-structure diagram, it is possible to determine whether a given population is growing rapidly, growing slowly, remaining stable, or even declining (Figure 4.6).

When considering age structure, it is useful to divide the population into three major groups: pre-reproductive (ages 1 to 15), reproductive (ages 16 to 45), and post-reproductive (ages 46 to 85+). Current population growth, of course, depends on the number and fertility of the females in the age group 15 to 45. Future population growth, on the other hand, depends ultimately on females who are now in the age group 1 to 15. In other words, it is the size of this pre-reproductive group that tells the future of a population. At the present time, about 31 percent of the world's population is under 15 years of age. In the less developed countries, the average is 34 (In Kenya the figure is 50 percent!). In the more developed countries it is 19.

An age-structure diagram for a typical LDC such as Mexico has a very broad base and a narrow apex. This triangular shape is characteristic of a rapidly growing population. The doubling time for these populations is about 20 to 40 years. The histogram is said to be expansive (Figure 4.6A).

Countries such as the United States are characterized by slightly expansive histograms. They are shaped like those of other faster-growing nations, but their bases are not as wide and the sides of the triangles are steeper, indicating that numbers are increasing, but more slowly than fast-growing nations with doubling times ranging between 40 to 120 years.

Age-structure diagrams for slower growing, nearly stable more developed countries (MDC) such as Canada have narrower bases and steeper sides. The middle histogram in Figure 4.6 is an example. As you can see, the narrow base and steeper sides indicate that the size of the pre-reproductive and reproductive age groups is relatively stable. The histogram is said to be near stationary. The age-structure diagram of many European countries has a narrow base, which indicates an extremely slow growth rate, with doubling times of 121 to 3,000 years.

A declining population would show a consistent narrowing at its base. Sweden, shown in the right of Figure 4.6, is an example. It's histogram is said to be constrictive.

Population histograms of many countries, especially the less developed nations of the world, indicate the presence of a built-in mechanism for explosive population growth in the near future. From the standpoint of population control, this is an extremely distressing situation. After all, there are already more men and women in the reproductive age group on this planet than there have ever been—about 2 billion.

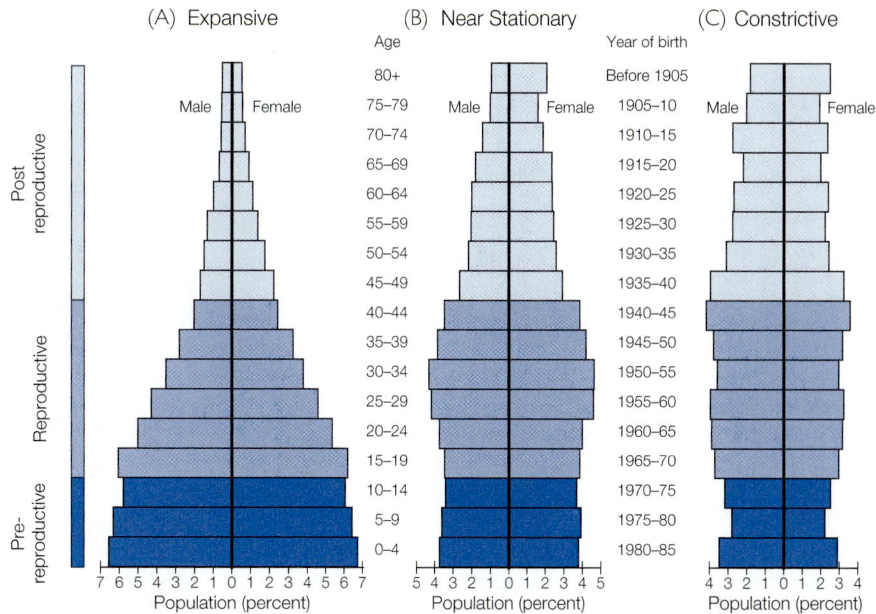

FIGURE 4.6 *Characteristic population age-structure profiles for (A) expansive, (B) near stationary, and (C) constrictive populations.*

4.2 The Impacts of Overpopulation

With these basic terms and concepts in mind, we now turn our attention to the impacts of population. We will begin by examining a phenomenon commonly referred to as overpopulation.

Overpopulation in the Less Developed Countries

When discussing environmental issues, many people assert that one of the key problems is overpopulation—too many people. More precisely, **overpopulation** refers to a condition in which the population size exceeds the carrying capacity of the environment. As noted in Chapter 3, carrying capacity refers to the ability of our environment to supply resources and assimilate waste. *Overpopulation, as it pertains to people, means too many people for the available resources. It also means too many people for the planet's waste assimilation/detoxification mechanisms.* What are the symptoms?

In the less developed nations, which are generally poor, rural, and agricultural, one of the most obvious symptoms of overpopulation is food shortage and its consequences—hunger, malnutrition, and disease. In 1798, Robert Malthus, a British economist, stated that populations tend to increase faster than food supplies. He asserted that the only way a population could come into balance with the available food supply would be through a massive die-off resulting from starvation, disease, war, or some other calamity. Although this concept may not be entirely true, population balance has been achieved without these calamities, overpopulation in the less developed nations is often referred to as **Malthusian overpopulation.** Malthusian overpopulation is described as the result of too many stomachs and not enough food (Figure 4.7). For example, at present growth rates, the populations of Kenya, Nigeria, Tanzania, and Uganda will double in 23 years or less. Such a population surge has resulted in widespread malnutrition and starvation.

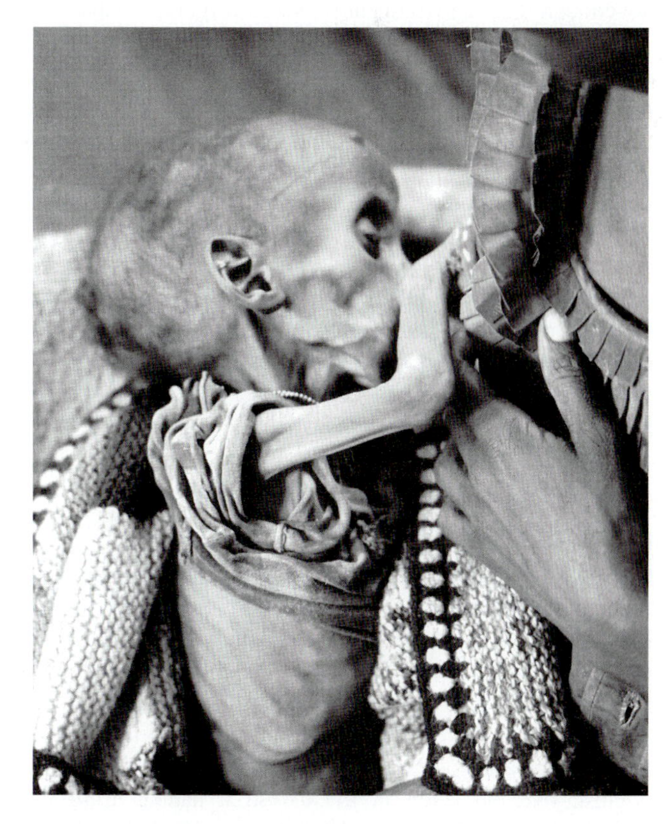

FIGURE 4.7 *An African mother with her skeletal child, now close to death from starvation due to Malthusian overpopulation (Beyah, Ethiopia, June 1991).*

The effects of Malthusian overpopulation are grimly described in the following extract by Lee Ranck, an executive of an international relief organization. The following is merely one of many similar tragedies that he observed while on an extended tour through several of the approximately 100 LDCs, where a majority of people live in appalling poverty and chronic malnutrition (Figure 4.8).

> Hunger is more than cold facts and awesome statistics. Hunger has a face. I know. I have looked into it. Hunger is a Bengali face—a little mother named Jobeda whom I found in the shade of a tattered lean-to in a refugee camp in Dacca. A small withered form lying close beside her whimpered and stirred. Instinctively she reached down to brush away the flies. Her hand carefully wiped the fevered face of her child. At six years, acute malnutrition had crippled his legs, left him dumb, and robbed him of his hearing. All that was left was the shallow, labored breathing of life itself, and that, too, would soon be gone. But death is no stranger to Jobeda. She has seen starvation take away her husband and five of her seven children. . . .

Today 33,000 people will die either directly or indirectly from starvation and malnutrition. Tomorrow 33,000 more. Although the causes of these deaths are multiple, certainly the basic cause is Malthusian overpopulation: too many human stomachs and not enough food, a problem discussed at length in Chapter 7.

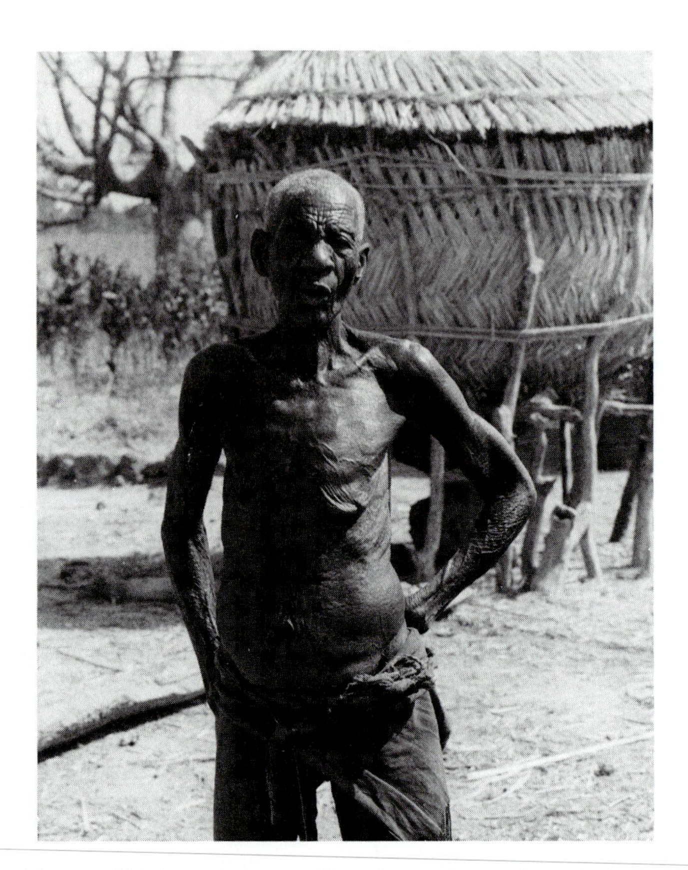

FIGURE 4.8 *An aged woman suffering from undernourishment in famine-plagued Africa.*

Overpopulation in the less developed nations is much more complex than the discussion above indicates. Although hunger and starvation are the most visible symptoms, large populations also take their toll on local rivers and streams, which human waste often makes unfit for drinking. Intense demand for wood for fuel often results in deforestation and erosion. Wildlife species may be exterminated in the rush to meet rising demand for food, fiber, and wood. Urban centers in the less developed nations are often crowded and highly polluted. For this reason, some educators refer to the complex of problems in the LDCs as **population-based resource degradation.** In a phrase, this would be described as the environmental consequences of too many people. Even with meager needs, enormous damage can result.

Overpopulation in the More Developed Nations

Overpopulation is also a fact of life in the more developed nations of the world such as Japan, Canada, England, Germany, Russia, and the United States. In these and other industrial nations, starvation and hunger occur, although much less frequently than in the less developed nations such as Ghana and Somalia. The symptoms of overpopulation in the more developed nations are, however, just as serious (Figure 4.9). To distinguish overpopulation in the industrial nations from that occurring in the less developed nations, some ecologists use the term **technological overpopulation** because of the technological dependence of such nations and the impacts of the use of many technologies like automobiles on the environment and the resource base. Others prefer the term **consumption-based resource degradation** to reflect the importance of our high rates of consumption. In more developed nations, then, the most important factors when considering environmental impact are the resources used and the pollution generated per person (per capita). As noted in previous chapters, the United States and other MDCs use more resources per person than citizens of the less developed nations—often 20 to 40 times more. *A 1-mile drive in your automobile to pick up a quart of milk or a six-pack of soda uses more energy than most people in the less developed nations use in an entire day to perform all of the tasks required to survive.* Because of the higher per capita consumption, a relatively small population may therefore cause considerably more damage. Instead of dying from starvation, people are killed by pollutants their technology has generated. Their environment is threatened by pollution and resource degradation resulting from technology and consumption. Because environmental problems are really based on both the level of technology and high rates of consumption, the term **technological-and-consumption-based resource degradation** might be best.

The effects of overpopulation in the more developed nations are visible all around us. Highway congestion may be one of the most obvious. Urban air pollution is also a visible reminder, as is the loss of open space, farmland, and wildlife at the fringes of major cities and in rural areas as new towns and resort areas expand. Look a little closer and you will find other, less visible signs of the problem. Species are rapidly going extinct. Rivers turn brown after rains as a result of sediment washed into them

FIGURE 4.9 *A sea-going tug pushes a barge carrying over 2,000 tons of garbage from Key West Florida. Excess trash production is a major problem in rich, industrialized nations.*

from nearby development. The rain has turned acid in many parts of the world, too, as a result of pollutants generated by power plants, factories, cars, trucks, busses, jets, and homes. Even the protective ozone layer above our heads has begun to thin as a result of the release of chemicals known as chlorofluorocarbons from refrigerators, air conditioners, and other sources. And finally, there's the rapid warming of the planet, known as global warming, resulting from the release of carbon dioxide from a wide range of human sources, including cars, homes, and factories, and from deforestation—the loss of carbon-dioxide assimilating plants.

Clearly, these and other impacts are a result of our resource-intensive way of life, made possible partly by our heavy dependence on environmentally unfriendly technologies such as coal-fired power plants and automobiles. Taken together, these impacts constitute an environmental horror story as serious as the starvation and malnutrition problem in the poorer nations of the world.

We'd be remiss if we were to leave the discussion at this point with the mistaken notion that population size and population growth in the MDCs are not major contributors to environmental degradation. Although many people struggle under this misconception, in reality, *we think that population growth is an important consideration in the MDCs—perhaps even more important than in the LDCs.* In other words, we can't blame all of our problems on consumption and technology. Consider some facts: as noted earlier, U.S. citizens consume, on average, 20 to 40 times the resources of an average resident of a less developed nation. They therefore have 20 to 40 times the impact on the environment. Thus, each new child has the environmental impact of 20 to 40 residents in the less developed world. Each birth averted in this country can make tremendous inroads into environmental problems. Numbers in MDCs do matter!

People who don't view population growth as a problem in the more developed nations often assert that we can solve our problems by reducing levels of consumption—for example, through efficiency, pollution prevention, and recycling. Although that may be true, many people believe that we must also reduce population growth to create a more sustainable society (Ethics in Resource Conservation Box 4.1). Such actions have a far greater impact on the overall quality of the environment than our most aggressive energy and resource conservation strategies.

Throughout the remainder of this book, you will see examples of these types of overpopulation in discussions of soil erosion, air and water pollution, wildlife extinction, the energy crisis, the toxic chemical problem, or global starvation. We urge you not to lose track of the fact that overpopulation is at the root of all environmental problems.

We also point out that population size is not the only issue. The rate of growth has profound implications, too. Rapid growth makes matters worse. Problems will escalate and solutions will be more difficult as the number of people increases. This is especially true in the less developed nations where economic resources are lacking and where corruption in government is prevalent. In a nutshell, then, the population problem can be summarized by six words, "Too many people, reproducing too quickly."

4.3 Population Growth in the More Developed Nations: A Closer Look

With these facts and concepts in mind, let us take a closer look at the population situation of the more developed nations. For the most part, the news from the more developed nations is encouraging (Table 4.1). For example, studies show that the TFR has steadily decreased in the more developed nations of Western

ETHICS IN RESOURCE CONSERVATION 4.1

IS REPRODUCTION A PERSONAL RIGHT?

A pamphlet on wildlife recently published by the Colorado Division of Wildlife presents engaging descriptions of the state's mammals. The back cover also discusses what has become a central concern among citizens in Colorado—the loss of wildlife habitat. It notes that Coloradans have changed the environment and points out that "What were once elk calving grounds are now shoppettes below ski areas; deer migration routes are cut by six-lane highways." It goes on to say that as many as "30,000 acres of traditional wildlife habitat are converted to human use in Colorado every year." It concludes by saying that because of this, "we must manage our wildlife resources more carefully than ever."

Leo Tolstoy once wrote something to the effect that everyone dreams of changing the world, but no one dreams of changing himself. The pamphlet displays this fundamental human tendency.

To build a sustainable society, most people agree, we have to do something about ourselves. We have to manage ourselves better. One key area in need of better management is population growth. Most experts agree that, in the very least, we must find ways to slow the rate of human population growth. Eventually, we may have to stop growth altogether. In fact, this was one of the recommendations of the President's Council on Sustainable Development in 1996. After we stabilize, we may even want to find ways to reduce population size humanely—for example, by having smaller families so that the birth rate is lower than the death rate.

Population control brings up ethical issues of incredible magnitude. The overriding question, of course, is this: Is it ethical to control population growth?

As a rule, people tend to line up in one of two camps—those who think it is not ethical to place limits on family size and those who think that it is.

Reasons for opposing limits are many. Some people, for example, object to population control primarily on religious grounds. That is, they object to it because of religious ethics. It is their belief that governments have no right to limit human numbers. The Catholic Church, for instance, opposes population control and all forms of family planning, such as contraceptives or birth control pills. The only form of family planning it deems acceptable is the rhythm method or natural method, which involves sexual abstinence around the time of ovulation to prevent pregnancy.

Some advocates of the viewpoint that population control is unethical base their arguments on personal freedom. The right to reproduce at will is, they say, a basic human freedom. No one should dictate another person's family size. Certain racial groups feel that limits are discriminatory in nature.

Those who argue that it is ethical to limit human population growth say that limitations on population growth will ensure a better quality of life for people who are alive today. In other words, limiting family size is for the greater good. The right to reproduce at will should be curtailed when it interferes with the welfare of society. Advocates of this viewpoint like to turn the tables and ask, Is it ethical not to limit population growth? In other words, Is it ethical to continue as we are, breeding rapidly and quickly overrunning the world and wreaking havoc on the environment? Unbridled population growth inevitably results in widespread environmental destruction and robs future generations of the opportunities many of us now take for granted. Advocates of such a view note that population growth without adequate resources and services could very well result in disaster. So, not only is it ethical to control numbers, it is prudent.

Advocates of limiting growth also note that measures to reduce population growth prevent unwanted pregnancies and end the cruel fate of those born into families that cannot support and nurture them. Moreover, limitations protect the welfare of other species that share this planet with us.

Advocates of population control and stabilization, such as ourselves, are quick to point out that they need not involve methods that are morally repugnant to people. As noted in the text, improving education, health care, job opportunities, and women's rights can play a huge role in slowing, even stopping, world population growth.

Take a few moments and summarize your views on the subject. What is your position, and why?

Europe, Canada, and the United States over the past two to three decades. At the current rate of growth it would take 580 years for the populations of the more developed nations (now about 1.2 billion) of the world to double. As noted earlier in the chapter, in some countries growth has slowed considerably, even stopped. Others are experiencing what demographers call negative growth. In other words, their populations are declining. Today, more than 50 developed nations are stable or declining. The populations of Austria, Spain, and Greece are stable, for example, while Germany, Romania, and Russia are all on the decline. Overall, Europe's growth rate is -0.1% per year.

But there are notable exceptions among the more developed nations. The U.S. population, for example, is growing naturally at a rate of about 0.6 percent per year—one of the fastest growth rates in the developed world. When immigration is included, the growth rate shoots up to 1.1 percent per year. In the sixth edition of this book, we cited projections by the U.S. Census Bureau which held that the US population would stabilize at about 300 million by the year 2050. However, because of a shift in trends, the U.S. Census Bureau now projects a population of 392 million by 2050—120 million more than live in the United States today. Zero population growth is no longer in sight.

What caused this change in predictions? Two things. A slight rise in the total fertility rate and an increase in immigration, both legal and illegal. About 1,000,000 people enter the

United States legally and illegally each year.[2] Only by maintaining the total fertility rate well below 2.1 and eliminating or drastically curtailing immigration will the United States achieve a stable population size.

4.4 Population Growth in the Less Developed Nations: A Closer Look

Population growth in the less developed nations is currently occurring seventeen times faster than in the developed nations. Dozens of nations in Central and South America, Africa, and Asia are caught in a kind of demographic trap in which reductions in death rate have occurred without equal reductions in birth rate. In most of these nations, natural resources are being overtaxed, and per capita food supplies and income are declining. Why are birth rates still high?

Large Family Size

In contrast to the MDCs, where the desired number of children per family ranges from zero to three, the LDCs of Asia, Africa, and Latin America, with a few exceptions such as China, prefer to have much larger families. For example, couples are having, on average, 4.7 children in Syria (Middle East), and 5.4 in Ghana and Sudan (Africa).

Large families result from several factors. One of the most important is the desire for them. In agricultural nations, children are still needed to help with household tasks and farm work, such as gathering wood and dried cow dung for fuel or bringing water from distant streams. Children are also needed to provide security to the parents in old age. In contrast, most MDCs such as the United States, England, and Sweden have well-developed social security programs to support their older citizens when their wage-earning years come to an end. However, in many of the LDCs such programs are lacking.

Yet another reason for large family size in the less developed nations has to do with the macho image desired by many males. The larger the family, the greater the status of the father. Many children are fathered out of wedlock as well. A man named Denja, from the province of Nyanza, in Kenya (Africa), boasts of being the father of 497 children!

Another reason for large family size in the LDCs is that birth control methods such as chemical contraceptives, sterilization, and abortion are strictly forbidden by the religion of the prospective parents. For example, in Mexico, Kenya, and the Philippines, where populations are soaring, the Catholic Church is firmly opposed to any artificial method of birth control, although many people ignore the Church's dictates. A similar stand has been taken by the Muslim fundamentalists in Pakistan, Egypt, and Iran.

[2]According to the Immigration and Naturalization Service, approximately 300,000 persons enter the United States illegally and 720,000 enter legally each year.

Lester Brown, president of the Worldwatch Institute, a highly respected environmental organization, views the continued population surge as a dire threat to the world's resources and the quality of human life. He strongly urges countries to adopt a small-family policy with the number of children averaging one to two per family. To achieve this goal, the governments of the less developed nations must launch effective mass education programs on birth control in cities and in rural areas. Although such efforts have begun in some nations, they must be greatly expanded and intensified.

Africa: A Continent in Danger

No continent has suffered more from the adverse impact of rapid population increase than Africa. At current growth rates, the African population will increase sharply from over 800 million in 2000 to nearly 1.3 billion in 2025—nearly doubling in 25 years! Such rapid growth will put enormous strains on resources. Millions of acres of land will be converted to farmland, squeezing out precious wildlife. Already crowded, filthy, and crime-ridden cities will become a nightmare for those who make their homes in them. Rivers and streams will become polluted with human waste.

Demographers believe that Africa's birth rate of 39 per 1000 will remain dangerously high well into the new millennium if current trends continue. Even if strong population curbs could be initiated immediately, the continent's population will continue to soar far into the next century because 45 percent of Africa's population is under the age of 15 and will soon be entering their reproductive age. The only mitigating factor, discussed earlier, is the death rate, which too is predicted to increase, primarily as a result of HIV.

4.5 Controlling the Growth of the World's Population

Controlling the growth of the world's population is an enormous task. This section discusses two major approaches: family planning through birth control and abortion, and a host of sustainable development strategies that have an equally powerful effect.

Birth Control

As we view (and feel) the crush of people on this planet, it may come as a surprise that humans have practiced various methods of birth control for centuries. **Birth control** refers to techniques and devices that prevent births by preventing conception—fertilization of the egg. Birth control was advocated on ancient Egyptian papyri dating back to 5,000 B.C. A great variety of techniques were tried. For example, a combination of wool fibers and alligator dung was used by women as a vaginal barrier to sperm. Men often used condoms fashioned from animal bladders. Foul-tasting concoctions made from bark fibers, weeds, and ground gallbladders were gulped down prior to intercourse in the erroneous belief that conception would be prevented.

According to the Population Reference Bureau, a Washington, D.C.-based nonprofit organization, 58% of married women worldwide use some form of birth control. Most are using modern methods. Of course, a much greater variety as well as more effective methods are available today than in ancient times. Many have been developed in the past 30 to 40 years. A major breakthrough in birth control was the **pill**—a chemical contraceptive taken orally that prevents sexually mature females from ovulating (releasing ova or eggs) each month. It is the most widely used method of birth control in the MDCs. The worldwide rate of use ranges from zero percent in Mauritania (Africa) to 13.5 percent in the United States and a high of 40 percent in Austria. The pill is popular, in part, because it is easy to use and highly effective. It does have side effects like weight gain, but efforts to reduce other more serious side effects have proven successful.

Another type of contraceptive is a plastic or nylon loop that is inserted into the uterus. Known as **intrauterine devices (IUDs),** they apparently prevent the newly formed embryo from embedding in the uterine lining. IUD manufacturing plants have been built in some of the LDCs so that these contraceptives can be readily available. The late Oliver Owen, who first published this book in 1971, had an acquaintance who wore IUDs as earrings to coffee parties in order to emphasize her position on the importance of birth control! The rate of IUD use among women of childbearing age ranges from zero percent in Mauritania (Africa) to 4.8 percent in the United States to 30 and 37 percent in China and Cuba, respectively. IUDs are effective, but do cause problems like uterine perforation. As a result, their use has been on the decline in many countries.

Sterilization is another major method of birth control. Basically, it involves cutting and tying off the sperm ducts in the male or the oviducts or fallopian tubes in the female (Figure 4.10). Its worldwide use among couples of childbearing age ranges from zero percent in the Ivory Coast (Africa) to 28 percent in the United States to a high of 35 percent in Canada. Sterilization has become more popular in the United States in recent years. It is the method of choice in India, El Salvador, and Korea. Sterilization is popular because it is highly effective and requires no effort. Once the surgery has been performed a person is relatively safe.

Another widely used method of contraception is the barrier method, typified by the condom. A **condom** is a thin rubber sheath worn over the penis during sexual intercourse. It prevents ejaculated sperm from entering the vagina. Condoms also help reduce the spread of sexually transmitted diseases such as genital herpes, syphilis, gonorrhea, claymydia, and AIDS.

Another barrier method is the diaphragm and vaginal sponge. A **diaphragm** is a rubber cap that fits over the cervix, the end of the uterus through which sperm must pass to get from the vagina into the uterus. Diaphragms are generally coated with a thin layer of a sperm-killing (spermicidal) jelly that greatly increases their effectiveness. **Vaginal sponges** are small, spongelike devices that fit over the end of the cervix. They are impregnated with a spermicidal chemical. Barrier methods are easy to use, but not as effective as sterilization and the pill.

Some couples practice birth control techniques that are behavioral in nature. The **rhythm method** involves timing sexual intercourse so that it occurs well before or after ovulation to prevent sperm and ova from uniting. Body temperature can be used to monitor a woman's time of ovulation. **Abstinence** is a technique often preached to teenagers and unmarried couples. Unfortunately, both techniques have a high failure rate.

Abortion

In case of contraceptive failure, a woman may terminate her pregnancy by an abortion rather than give birth to an unwanted child. An **abortion** is a premature expulsion of the fetus from the womb that results in fetal death. Many abortions occur naturally. These so-called spontaneous abortions are believed to be nature's way of eliminating defective fetuses—nearly 40 percent of the spontaneously aborted fetuses have

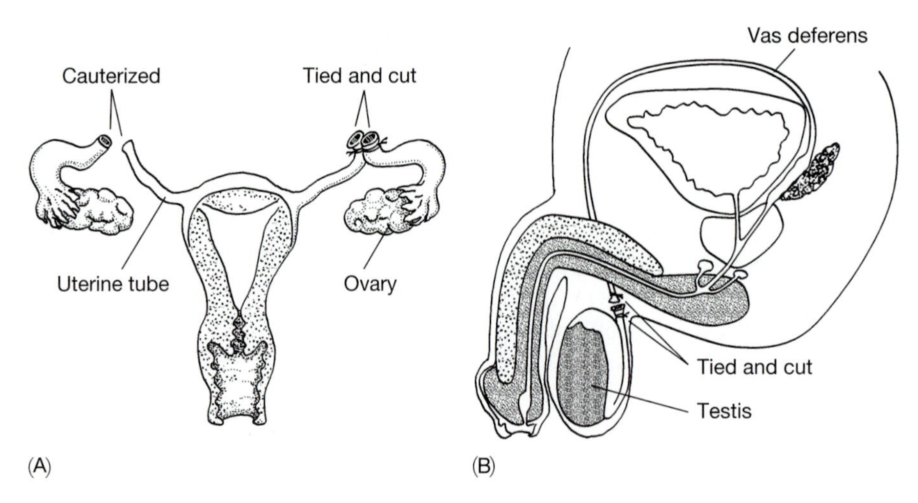

FIGURE 4.10 *Male and Female Sterilization. In females, the oviducts, which transport ova and sperm, are cut and tied. In males, the sperm-transporting ducts, the vas deferens, are cut and tied. Both operations can be reversed through microsurgery, although not with 100% success.*

some structural impairment or birth defect. Abortions resulting from human intervention result in the death of unborn fetuses.

The U.S. Supreme Court legalized abortions in 1972. They are legal in most other countries of the world as well. Readers know that there are strong ethical, moral, and religious arguments against abortion. Anti-abortion activists ask, "What right does a human who is living today have to take the life of another human, who, except for the violent act of abortion, would be alive tomorrow?" In the eyes of many eminent persons in the law, medicine, and religion, the willful act of abortion is murder, except perhaps in the case of incest or a life threat to the mother. Opponents argue that abortion "gives women a right to control their own bodies." And, it prevents unwanted births and suffering. In 1989, the Supreme Court upheld the states' right to restrict abortions funded out of state monies, which has touched off what is proving to be a long, drawn-out battle.

On a global basis, abortion is the third most widely used method of birth control after sterilization and the pill. In fact, were it not for the practice of abortion, the resources of this planet would have to sustain another 50 million people every year. Abortion is being used primarily by the women of Asia, Africa, and Latin America. Abortion rates are high in Latin America, even though the act is illegal. Unfortunately, many of the illegal abortions are self-induced with wires, coat hangers, or pointed sticks and are 75 times more dangerous to the mother than legal abortions performed by the medical profession.

Newer Methods of Birth Control

Each of the methods of birth control has some drawbacks. Some like the condom and diaphragm, are inconvenient and prone to failure. Others, like the pill and intrauterine device, while more convenient and more reliable, may pose a health risk to some women. The rhythm method can be unreliable. Because of this, scientists have continued to search for safe, effective, long-term birth control methods, especially methods suited to the rural poor of the less developed nations. Let's examine several new antifertility techniques that will probably be used with increasing frequency in the near future, not only in poor nations but in the rest of the world as well.

In 1988 a French pharmaceutical firm developed a pill known as RU-486. Although it is known as an abortion pill, RU-486 actually prevents an early embryo from implanting in the wall of the uterus. It works like an IUD, actually. RU-486 had been approved for use in several countries (France, Sweden, China, and Great Britain). At this writing (September 2000), it is being tested in Canada and was just approved for use in the United States with some restrictions.

That said, we should point out that there already is a "morning after pill" that is available in the United States. It's ordinary oral contraceptives used by millions of women. However, they're taken in high doses after unprotected sex and for a short period. Hospitals and doctors can provide one of at least four oral contraceptives and tell patients how to use them to achieve the desired effect. One of the most important use criteria is that the pills be taken within 72 hours after unprotected intercourse to block implantation.

Still another promising birth control method is DepoProvera, a hormone-containing solution that, when injected under the skin, prevents a pregnancy for up to 3 months. It has been used by women in at least 90 countries. Birth control advocates in the United States breathed a collective sigh of relief when the Food and Drug Administration (FDA) at long last approved it for use in 1992.

Norplant is yet another contraceptive that holds considerable promise. It consists of six matchstick-sized capsules that are inserted under the skin of a woman's upper arm (Figure 4.11). These capsules gradually release a hormone that prevents conception for at least 5 years. The insertion procedure can be completed in less than 10 minutes and can be performed by a specially trained nurse. Norplant has been approved for use by at least 57 nations, including Indonesia, Thailand, and China. In 1991 the FDA certified the use of Norplant in the United States. The city of Milwaukee provides free Norplant services to teenage mothers who wish to prevent another pregnancy.

Sustainable Development Is the Best Contraceptive

The rapid growth of the human population cannot be solved by just supplying contraceptives. Underlying factors—people's beliefs and their opportunities for education and health care—are also powerful forces that can be enlisted to control population growth worldwide.

Sustainable development is a strategy that seeks to better people's lives without undermining the environmental life support systems of the planet. Among other things, sustainable development seeks to increase the educational level of women and provide men and women with greater economic opportunities.

FIGURE 4.11 *Norplant—an effective method of birth control. The 1-inch capsules are inserted into a woman's arm. The chemicals that slowly exude from the capsules prevent the woman from conceiving for at least 5 years. This birth control method has been approved by the World Health Organization.*

Increasing educational levels has many practical benefits. It will, for instance, help a women read and understand instructions on her contraceptive package. It will also help her seek employment, which may delay childbearing and thus effectively reduce the number of children a woman has in her lifetime. Increasing economic opportunities gives men and women other options as well. Where women can be gainfully employed or operate their own businesses, birth rates decline.

Small scale economic development, designed to meet the needs of people in ways that do not harm the environment, can also help. They provide jobs and income. With rising income, say population experts, family size will fall. Several nonprofit organizations are helping in this regard. One program that seems to be helping provides small loans to women in rural countries in Asia and Latin America. Women use the loans to start small businesses to support their families and with great success! Women participating in such programs realize that more children makes it more difficult to improve their lot.

Changes in the status of women in a society also have a bearing on the problem. In many cultures, a women's value derives primarily from her ability to produce offspring for the husband. Cultural shifts, although difficult, may be a powerful force in slowing the growth of the human population.

Improvements in health care, another component of sustainable development, also yields population benefits. Increasing the accessibility and affordability of health care in poor rural villages, for example, reduces infant death and, over time, reduces the number of children a women must have to produce enough surviving offspring to take care of the mother and father in their old age. Health care clinics are also an avenue for learning about and receiving contraceptives. Because childbirth is a leading cause of maternal death in the less developed nations, birth control is now also being seen as a means of increasing a woman's long-term prospects.

These changes and several others can help LDCs currently stuck in stage II of the demographic transition move beyond this dangerous phase. As with so many environmental problems, the answer is not easy. Deep, systematic changes are needed to reduce the underlying impetus for large families—the social, economic, and cultural factors that create runaway population growth.

The importance of family planning and deeper changes, such as those discussed in this section, was expressed in a document produced by the United Nations' Conference on Population and Development that was convened in September 1992. More than 20,000 representatives of over 150 nations assembled in Cairo, Egypt, for the conference. The United States was represented by a 35-person delegation.

The major conclusions of the conference, contained in a 113-page Voluntary Action Plan, are as follows:

1. To check population growth, family planning services should be advertised and provided to all interested families.
2. Teenagers should be given essential information concerning the use of both chemical and mechanical contraceptives.
3. Women of all nations should have access to safe abortions. (Forty percent of the world's people now have such access.)
4. Sex education should begin within the family unit, the community, and the school at an early age in the life of the individual.
5. The status of women in all nations should be raised above that of "baby-making machines." Such empowerment of women will reduce their fertility rates.
6. The more developed nations must accept much of the blame for resource exploitation and the pollution of land, water, and air.
7. The less developed nations of Asia, Africa, and South America must acknowledge their shortcomings in population control.
8. The more developed nations will assist the less developed nations to utilize their resources in a sustainable manner so that they can improve the quality of life for their people.
9. There should be a more equitable distribution of the world's wealth, resources, and technology.

Population is truly one of the most pressing issues of our times. Bringing it under control is one of the greatest challenges facing us today.

4.6 Human Population and the Earth's Carrying Capacity

To create a sustainable society, we must learn to live within the Earth's carrying capacity. Unfortunately, no one knows the Earth's carrying capacity for *Homo sapiens.* Some experts believe that this planet could sustain at least 10 to 50 billion people. Others think the number is closer to 500 million. Still others argue that a lower number might be more accurate.

At the risk of a gross oversimplification, let us assume that our global population can follow two alternative paths, as shown in Figure 4.12. In Scenario A, our population continues to grow exponentially and overshoots the Earth's carrying capacity by a wide margin. This would eventually result in a die-off caused by starvation, disease, pollution, and resource depletion, which could reduce our population to the carrying capacity level. In such an instance, the carrying capacity might even be reduced as a result of widespread damage. In other words, the planet might not be able to support anywhere near as many people as it could if we had kept from overshooting its limits.

There's another point worth making, too. Carrying capacity is about more than food supplies. It is about the planet's ability to provide all resources and rid all of our wastes safely. Simply expanding food production may permit more people to be sustained over the short term. However, the long-term result would be a human-dominated ecosystem that would be severely out of balance for, in the process of producing more and more food that feeds more and more people, we would severely deplete our natural resources, generate massive amounts of air and water pollution. These changes would cause the extinction of many forms of wildlife.

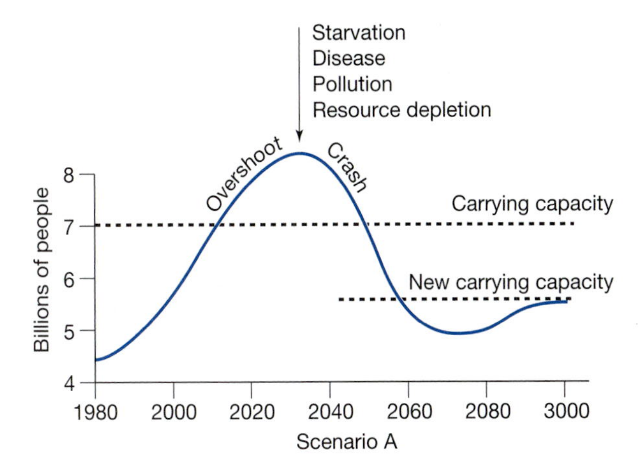

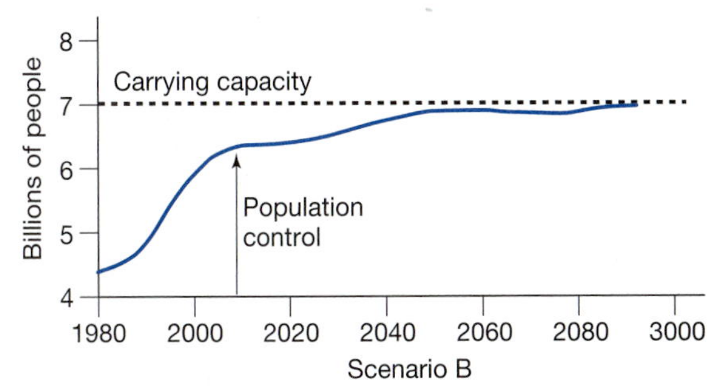

FIGURE 4.12 *Population growth and carrying capacity. Scenario A. Population overshoots carrying capacity. Eventually there is a massive die-off due to starvation, disease, pollution, and resource depletion. A new carrying capacity may be established because of massive damage caused by population growth. Scenario B. Before the population reaches carrying capacity, intensive efforts are made at population control (family planning, use of contraceptives, delayed marriage, and so on). As a result, the population gradually reaches carrying capacity and die-off is prevented.*

The preferred alternative, shown in Scenario B, is a gradual reduction in the rate of population growth, long before it reaches the carrying capacity, by means of the strategies discussed earlier in this chapter. Eventually our population would stabilize at the carrying capacity, at which point the human species and the environment with which it interacts would become part of an ecological system that could be sustainable far into the future.

Unfortunately, many scientists believe that human populations have already exceeded the carrying capacity in many regions. Excessive soil erosion, the conversion of vast acreages of grassland to desert, species extinction, loss of fish populations, and other regional problems are blatant signs of a people living beyond the ability of their bioregion to provide for them. Some scientists believe we may have exceeded global limits as well. Acid deposition, ozone depletion, and global warming, discussed in Chapter 19, are three of the most alarming signs of this transgression.

If these assertions are true, human society is clearly engaged in Scenario A. Our choice may be to voluntarily control numbers, reduce environmental damage through technological and behavioral change, and restore the Earth to ease our population and its impact to within tolerable limits, or to face a severe crash. The decision is ours, and time is limited. (See Case Study 4.1.)

Summary of Key Concepts

1. The World population was 6.1 billion in 2000, and is growing at a rate of about 1.3% per year. This results in the addition of 84 million people a year.
2. The birth rate is the number of persons born per 1,000 individuals in a given year. The death rate is the number of deaths per 1,000 individuals in a given year.
3. The rate of natural increase (or decrease) is the difference between the birth and death rates. The world's rate of natural increase at present is about 14 per 1,000.
4. The doubling time for the global population presently about 49 years.
5. Population increases exponentially. Exponential growth occurs when any entity increases by a fixed percentage, but only if the annual increase is added to the base amount. Exponential growth occurs very slowly at first, but once the item being measured rounds the bend growth in absolute numbers is rapid.
6. Human population remained small for many years as birth rates and death rates were in equilibrium. Decreases in death rates, without accompanying increases in birth rates, resulted in a massive increase in the human population in the last 200 years. Advances in agriculture, medicine (especially the development of vaccines and antibiotics), and sanitation all contributed to the present population explosion on Earth by reducing the death rate.
7. When nations become industrialized, they undergo a demographic transition. This is characterized by a reduction in both birth and death rates. Many of the LDCs are stuck in stage II of the demographic transition, which is characterized by a falling death rate combined with a high birth rate.
8. One sign of hope is that global population growth has slowed in recent years in many parts of the world. This trend is due to a number of factors, including better family planning, economic prosperity, and rising death rates, especially in Africa. Rising death rates result from starvation and disease. AIDS is taking its toll on many African nations. Unless a cure for AIDS is found, the disease is expected to destroy 20 percent of the African black population by 2010, according to some projections.
9. Total fertility rate is an estimate of the number of children women will have. It is based on current fertility trends in different age groups. Replacement level fertility, the number of children required to replace a couple, in developed countries is about 2.1.
10. The age structure of a population is the number of individuals occurring in each age class in a population. It is graphically represented in a population histogram, a useful tool for predicting trends.
11. The age-structure diagram of a rapidly growing population has a broad base, whereas the diagram of one that is growing very slowly has a narrow base. Populations that are declining have a narrowing base.
12. At the present time, about 31 percent of the world's population is under 15 years of age.

CASE STUDY 4.2 CHINA: ONE OF FAMILY PLANNING'S SUCCESS STORIES?

During China's great famine of 1958–1962, 30 million people starved to death. This event was a grim warning to its leaders that the theory advanced by Thomas Malthus is probably correct, although it may have had more to do with the Mao Tse Tung cultural revolution than anything else. However, it was not until the late 1970s that the nation's State Birth Planning Commission initiated the most comprehensive, the most rigidly enforced, and probably the most effective population control program in human history. The results have been impressive: the TFR has been reduced from 5.9 in the late 1960s to only 1.8 today, the crude birth rate has fallen from 36.9 to 16, and the annual population growth rate has declined from 2.6 to 1.0. Why was this program so successful?

Several features attribute to the success of China's program. One of the most important is the mass education program launched to foster public understanding of the effects on living standards if China's population was not curbed. Long-term projections were made public concerning how much food, water, energy, and other resources each person would have if China continued on its course into the demographic trap. The education program emphasized the benefits of postponing marriage to reduce the average family size. Another educational objective was to make the one-child family the norm among recently married couples. Couples who made the commitment to a one-child family received multiple benefits, including (a) cash payments, (b) free family-planning education, (c) larger old-age pensions, so that extra children would not be needed to provide security to the parents in their later years, (d) better housing and employment, and (e) free schooling for children. Family planning education was made readily available and publicized widely throughout all available media. Abortions, sterilizations, IUD insertions, and the dispensation of birth control pills were performed by local people who had been trained as nurses or paramedics.

China's government took some harsh steps to ensure small family size. For example, penalties were imposed on couples who had more than two children after entering the program. Among the penalties were (a) an increase in taxes, (b) compulsory sterilization for either the father or the mother, (c) the return to the state of all financial benefits awarded to couples who had agreed to curb family size, (d) intense peer pressure for women pregnant with their third child to have an abortion, and (e) reductions in food, employment, and educational benefits for parents and children.

China's family planning program gained momentum rapidly, especially in the urban areas. By 1982, for example, 70 percent of the couples in China's three largest cities—Shanghai, Bejjing, and Tientsin—with an aggregate population of more than 20 million, had agreed to have no more than one child. The goal of the Birth Planning Commission was to achieve zero population growth by 2000, with a population of 1.2 billion, followed by a decline to 800 million by 2100. Currently, however, China's population is 1.4 billion and is projected to rise to nearly 1.6 billion by 2025. Whether or not they will successfully reduce population size remains to be seen. At its current rate of growth, it will double in 73 years. With the one-child family policy no longer being strictly enforced, it is unlikely that China will see a substantial decline in population.

Although China's population control program has been eminently successful from a technical standpoint, many democratic nations, including the United States, have been deeply concerned about human rights violations, such as the Chinese government making sterilization mandatory after a woman has had her second child and rumors of forced abortions. The United States had funded population control programs in the LDCs for many years through such organizations as the World Bank, the Agency for International Development, and the United Nations. However, because of China's coercive sterilization and abortion policy, President Ronald Reagan terminated the United States' support of the United Nations Fund for Population Activity (UNFPA), which provided millions of dollars each year for family planning in less developed nations. This hurt not only China, but many other countries as well. Fortunately, President Clinton has reinstated support for UNFPA.

13. Malthusian overpopulation means too many people for the available food supply. It is characteristic of the LDCs of Asia, Africa, and South America. Although hunger and starvation are the major manifestations of overpopulation in such nations, environmental destruction is often significant. Some scholars prefer to speak of population-based resource degradation to be more inclusive.
14. Technological overpopulation refers to an overabundance of people, usually in industrialized countries, who, because of their use of advanced technology, have a harmful effect on the environment.
15. Technological overpopulation includes depletion of resources; pollution of air, land, and water; defilement of scenic beauty; wildlife extinction; and the release of chemicals that may be hazardous to human health. The terms consumption-based resource depletion and technological-and-consumption-based resource degradation are also used.
16. Population growth in the more developed nations of the world has slowed considerably in recent decades. In many European countries, population growth has either come close to stabilization, stabilized, or begun to decline.
17. Although growth in the developed nations of the world has slowed, some nations, such as the United States and Canada, continue to grow fairly rapidly. Population growth will continue in the United States for many years to come despite the fact that the total fertility rate has been at or below replacement level fertility for three decades. Two facts are responsible for continued growth: an increase in the number of women in the reproductive age group and immigration.
18. Population growth in the less developed nations has slowed, but still continues at a rapid pace. Growth is especially rapid in Africa.
19. Many women have access to and use modern birth control measures. Although many birth control options are avail-

able to reduce family size and slow the growth of the human population, changes in the status, educational opportunities, employment opportunities, economic well being, and health care of women and men are also needed.

20. The major features of China's population control program, which reduced TFR to replacement level (2.0) in 1994, are the following: (a) mass education on family planning, (b) delayed marriage, (c) multiple educational, health, and economic benefits for one-child families, (d) family planning services provided by specially trained local people, and (e) mandatory sterilizations or abortions for families having more than two children.
21. The human population may already have exceeded the Earth's carrying capacity. Continued overshoot could result in environmental damage that will ultimately lower the planet's capacity to support life.

Key Words and Phrases

Abortion
Abstinence
Acid Rain
Acquired Immunodeficiency Syndrome (AIDS)
Age-Structure Diagram
Baby Boom
Birth Control
Birth Rate
Carrying Capacity
Condom
Consumption-Based Resource Degradation
Contraceptives
Death Rate
Demographer
Demographic Transition
DepoProvera
Diaphragm
Doubling Time
Exponential Growth
Extinction
Family Planning
Human Immunodeficiency Virus (HIV)
Immigration
Intrauterine Device (IUD)
J-Shaped Curve
Less Developed Countries (LDCs)
Malthusian Overpopulation
More Developed Countries (MDCs)
Morning After Pill
Natural Increase
Net Migration
Norplant
Overpopulation
Percent Annual Growth Rate
Pill
Population Histogram
Population-Based Resource Degradation
Rate Of Natural Increase
Replacement Level Fertility
Replacement Rate
Rhythm Method
RU-486
Sterilization
Sustainable Economic Development
Technological and Consumption-Based Resource Degradation
Technological Overpopulation
Total Fertility Rate (TFR)
Vaginal Sponge
Zero Population Growth (ZPG)

Critical Thinking and Discussion Questions

1. Construct a simple graph that indicates (roughly) the buildup of the human population since humans appeared on the Earth. Explain why growth was so slow for so long, then began to skyrocket in the past 200 years.
2. What is the global population? What is its rate of natural increase and its doubling time?
3. What is the current population size of the United States? What's its growth rate and doubling time? What are the global birth, death, and natural increase rates?
4. How is the percent annual growth rate determined?
5. How do demographers determine how long it would take for a population to double?
6. List several reasons why birth rates tend to decrease when the standard of living rises.
7. Does a decrease in the total fertility rate of a nation to replacement level fertility necessarily mean that nation's population is decreasing as well? Why or why not?
8. Using your critical thinking skills, analyze the following statement: "The United States receives an influx of about 1,000,000 immigrants yearly. This influx is important to us and should not be stopped." Here are some questions to help you grapple with this issue: Do you approve of such massive immigration? Should it be stopped to stabilize population growth? Why or why not? What benefits might result from continued immigration? What might be the disadvantages to the United States?
9. Describe the shape of age-structure diagrams for (a) rapidly growing populations, (b) slowly growing populations, (c) stable populations, and (d) declining populations.
10. Would a study of age-structure diagrams of the U.S. population be of benefit to automobile manufacturers? The agricultural industry? School administrators? Why?
11. Give five examples of environmental stress caused by the technological overpopulation experienced by the United States today.
12. Using your critical thinking skills, analyze the following assertion: "Population growth is a problem only in the less developed nations."
13. Compare the population trends of the MDCs and LDCs.
14. Discuss the major features of China's population control program. Could any of these be used in the United States?
15. Discuss future population growth in terms of the carrying capacity of the Earth.
16. You are appointed head of a department in the government of a developing country facing rapid growth, poverty, and environmental destruction. Outline features of a population control policy aimed at reducing growth, stimulating environmental protection, and promoting an improvement in health and welfare of your people.

Suggested Readings

Brown, L. R., G. Gardner, and B. Halweil, 1998. *Beyond Malthus: Sixteen Dimensions of the Population Problem.* Worldwatch Paper 143. Washington, D.C.: Worldwatch Institute. A must read for all students. This aptly describes the many impacts of overpopulation.

Brown, L. R. and B. Halweil, 1999. "Where Death Rates are Rising," *World-Watch* 12(5): 20–29. A startling look at rising death rates.

Charlesworth, B. 1994. *Evolution in Age-structured Populations,* 2nd ed. New York: Cambridge University Press. A high-level study.

Ehrlich, P. R., and A. Ehrlich. 1990. *The Population Explosion.* New York: Simon and Schuster. Updated version of the 1971 classic, *The Population Bomb.*

Frazer, E. "Thailand: A Family Planning Success Story." 1992. *In Context* 31: 44–45. Provides information on the success of Thailand in promoting family planning.

Hardin, G. 1993. *Living Within Limits: Ecology, Economics and Population Taboos.* New York: Oxford University Press. A comprehensive analysis of the current population crisis by a world authority.

Jacobsen, J. L. 1992. *Women's Reproductive Health: The Silent Emergency.* Worldwatch Paper 102. Washington, D.C.: Worldwatch Institute. Examines the importance of family planning as a means of protecting the health of women.

Malthus, R. 1994. *An Essay on the Principle of Population.* New York: Oxford University Press. A classic; must reading for all students.

McFalls, J. A., Jr. 1998. *Population: A Lively Introduction.* Population Bulletin 53(3). Washington, D.C.: Population Reference Bureau. An excellent overview.

Gelbard, A., C. Haub, and M.M. Kent. 2000. *World Population Beyond Six Billion.* Washington, D.C.: Population Reference Bureau. A great survey of major population changes and projected changes to year 2050.

Population Reference Bureau. 2000. *World Population Data Sheet.* Washington, D.C.: Population Reference Bureau. A great source for information on populations of every country in the world.

Statistical Abstract of the United States. 2000. Washington, D.C.: U.S. Government Printing Office. Provides latest population data for the United States.

United States Bureau of the Census. *Current Population Reports.* Washington, D.C.: U.S. Government Printing Office. Published annually. Latest information on births, immigration, and changing age structure of our nation's population.

Web Explorations

Online resources for this chapter are on the World Wide Web at: **http://www.prenhall.com/chiras** *(click on the Table of Contents link and then select Chapter 4).*

Chapter 5

World Hunger: Solving the Problem Sustainably

Of all the problems created by the human population, undoubtedly that with the greatest potential for disaster is a shortage of food. Food shortages result in widespread hunger, malnutrition, and starvation. As noted in the previous chapter, these and many of the other consequences of overpopulation were first predicted in 1798 by Thomas Malthus, a British clergyman and economist. Malthus argued that the fixed land base imposed limits on food production. In addition, he believed that the human population and its demand for food would grow faster than food supplies. Without population control, he said, starvation, disease, and war would prevail to correct the ecological imbalance.

5.1 World Hunger: Dimensions of the Problem

Today, Malthus's glum predictions may be coming true. Worldwide, an estimated 12 million people, nearly half of them children, die each year from hunger and starvation or diseases worsened by hunger. This death toll is the equivalent of 65 jumbo jets, each carrying 500 passengers, fatally crashing every day of the year.

Undernutrition, Malnutrition, and Overeating

According to recent estimates of the World Health Organization, approximately 1.2 billion people—or one in five of the world's population—regularly consume less food than they need to stay healthy. Although the United Nations Food and Agriculture Organization estimates suggest that the number of undernourished people in the less developed countries has declined from the 1970s to 2000, current trends in population growth and declining cropland suggest that the improvements may prove to be short-lived. Most people experiencing food shortages live in Africa, Asia, and Latin America. The greatest concentration of chronically hungry are found in Africa and South Asia. India is also suffering from intense hunger. Hunger is endemic in pockets in the more developed countries, too. In the United States, for instance, an estimated 10% of the households are hungry, on the edge of hunger, or worried about hunger. Hunger affects one in five American children.

Food issues are really three fold. In areas characterized by food shortages, people suffer from two basic maladies: undernutrition and malnutrition. In many parts of the world there's another problem, overnutrition. In fact, worldwide, the number of overweight people now rivals the number that is chronically malnourished (Table 5.1). Let's consider each one.

Undernutrition is a quantitative phenomenon characterized by an inadequate amount of food—or calories—resulting in half-empty stomachs and gnawing hunger pains. Nutritionists assume that the average person requires a minimum of 2,200 to 2,400 calories daily. Western Europeans and North Americans consume 3,200 calories daily. They are among the fortunate. The average Ethiopian consumes only 1,600 calories daily—only half of the intake of an average American adult (Figure 5.1). The deficit of 600 calories per day undermines strength, causes severe mental and physical lethargy, and weakens resistance to a variety of diseases.

Malnutrition, on the other hand, is a qualitative phenomenon characterized by inferior food quality or a lack of certain important foods. This results in a deficient supply of one or more of such key nutrients as proteins, vitamins, and minerals.

Overnutrition results from excess intake of food. In the United States, 55 percent of all American adults are overweight. The percent of the population classified as obese (severely overweight) has climbed from 15 to 23 percent since 1970. Even children are afflicted, with one of every five now classified as overweight or obese. But the U.S. is not alone in this regard. Fifty percent of the adults in Russia, the United Kingdom, and Germany are classified as overweight. Even in less developed nations, overnutrition is becoming more commonplace among the wealthier classes. While many struggle to get enough to eat, the wealthy eat too much.

TABLE 5.1 *Overweight Adults and Underweight Children*

Country	Percent Adults Overweight
United States	55
Russia	54
United Kingdom	51
Germany	50
Colombia	43
Brazil	31
Country	**Percent Children Underweight**
Bangladesh	56
India	53
Ethiopia	48
Viet Nam	40
Nigeria	39
Indonesia	34

Source: Worldwatch Institute

Malnutrition and undernutrition often go hand in hand. In many less developed nations, poor children suffer from a lack of proteins and calories. As shown in Figure 5.2, children lacking the calories and protein their bodies require become thin and emaciated. They gnaw endlessly on their clothing to appease the insatiable hunger that plagues them day and night. This disease, called **marasmus** (pronounced mar-AZ-mus), most often afflicts infants who are separated from their mother's breast milk, which is rich in proteins and calories. Marasmus may result from maternal death, a cessation of maternal milk production, or the use of milk substitutes, which are often diluted by poor families and therefore often provide inadequate amounts of nutrients.

Millions of children under six years of age in the Third World suffer from the protein deficiency disease, **kwashiorkor** (pronounced kwash-ee-OR-core). Kwashiorkor is a West African word that means "the disease the child gets when another baby is born" because the mother can no longer feed the older child with her breast milk. Such children often receive a high-starch, low-protein diet. Although they receive adequate calories, they lack essential protein.

Kwashiorkor usually occurs in slightly older children, from one to three years of age. Although the disease was first discovered in tropical Africa, it has since been identified among children of Central and South America, the Caribbean, the Middle East, and the Orient.

Children with kwashiorkor lie in their mother's arms, weak and lifeless. Their limbs are thin and wasted. Their abdomens protrude because of the fluid that has accumulated inside (Figure 5.3). They suffer from skin ulcers and increased susceptibility to the infectious diseases common in many less developed countries.

Protein deficiency retards physical as well as mental development and, even if a proper diet is restored several years later, the brain damage cannot be reversed. As a result, protein deficiencies create a real impediment to the success of many less developed nations. Countries where protein deficiencies are common are not only eroding their many natural resources, they are eroding their human resources as well.

Kwashiorkor and marasmus are two clinically recognizable conditions. However, for every child diagnosed with one of these diseases, a hundred suffer from milder forms of malnutrition and undernutrition.

By the year 2025, the number of people afflicted by severe hunger could rise catastrophically in many parts of the world. The most severe problems will likely occur in Latin America and Africa, where many nations are simultaneously besieged by an explosion in human population and political turmoil. For residents of these countries, hunger and starvation could become even more commonplace.

Micronutrient Deficiency

Another problem that gets less attention than those already discussed is **micronutrient deficiency. Micronutrients** are chemicals required in small amounts. These include vitamins,

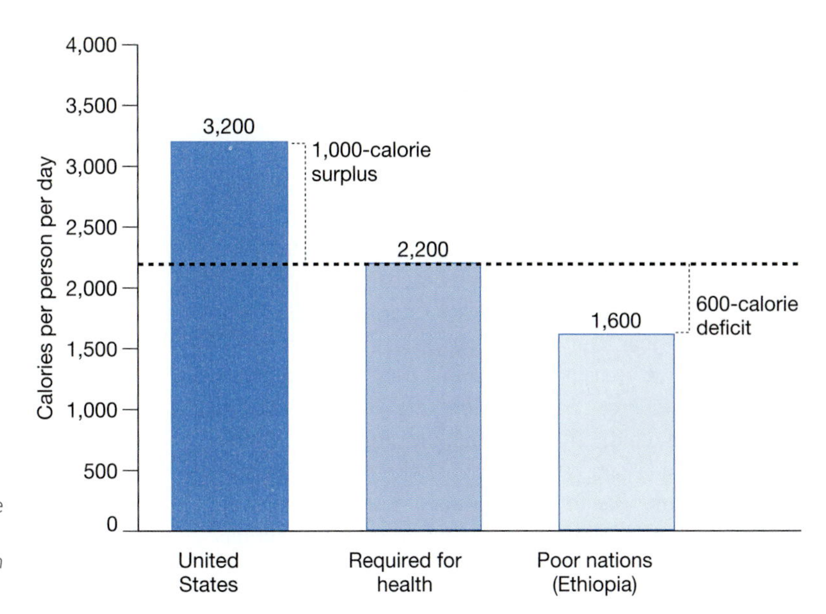

FIGURE 5.1 *The average American is overfed; the average person living in a poor nation is underfed. Hunger, diseases worsened by hunger, and starvation kill an estimated 12 million people a year.*

such as vitamin A, and minerals such as iron. **Macronutrients** are substances like protein and carbohydrate that are required in large quantity.

According to one estimate, approximately two billion of the world's people suffer from micronutrient deficiency. The three most prevalent are vitamin A, iodine, and iron deficiencies. All three can lead to serious medical problems. Iodine deficiency may result in mental retardation. Vitamin A deficiency can result in blindness and, in severe cases, death in children.

FIGURE 5.2 *Death from starvation, Parvati Pura, India. This village suffered from a local famine because of an extended drought. The two-year-old boy is almost dead from protein and calorie deficiency (marasmus).*

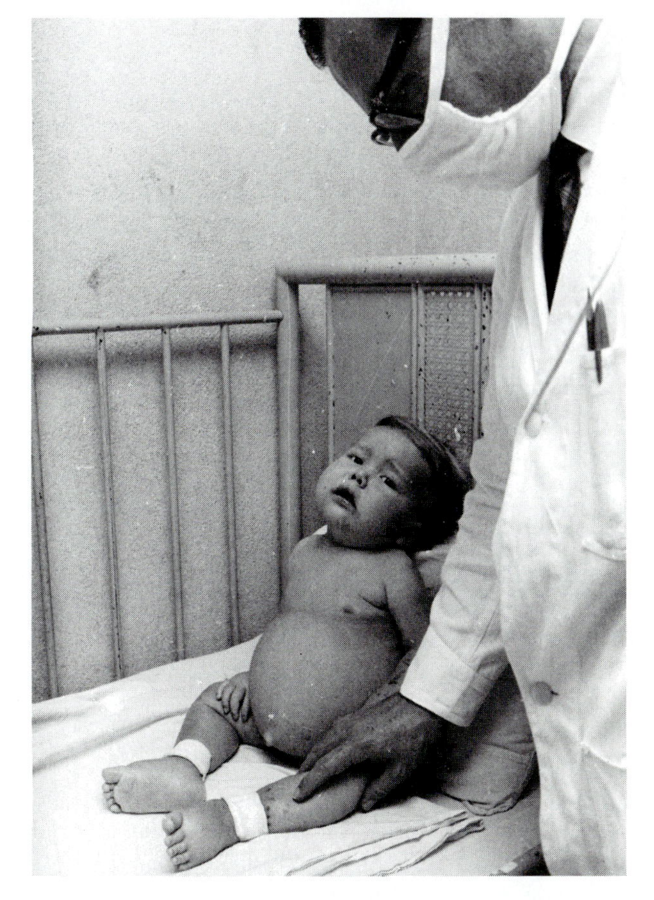

FIGURE 5.3 *Guatemalan child suffering from kwashiorkor (protein deficiency). Note the protruding abdomen.*

Food Trends and Challenges

The world food supply is at an all-time high, yet hunger and malnutrition persist at very high levels. Researchers often point to grain production statistics to characterize world food situation. They note that from 1950 to 1984, increases in grain production outstripped population growth, resulting in a dramatic increase in grain production per capita. But since that time, growth in grain production has fallen behind population growth, with grain production per person (per capita) dropping about 0.5 percent year. But averages like this always hide region trends. In some areas, such as China, food production per capita has improved greatly due to economic reforms and population control. In others, such as India, the situation has deteriorated in recent years with grain production per capita remaining the same due to massive uncontrolled population growth that countered increases in grain harvest. Pakistan has witnessed a recent drop of about 1 percent per year for the past decade. All in all, however, the majority of world's less developed nations have achieved impressive gains in grain harvests over the past 50 years. Solving food problems requires an understanding of the unique conditions in each part of the world.

Food aid has helped alleviate widespread starvation and hunger in Ethiopia, the Sudan, Chad, Somalia, and other countries, but the land may be paying the price. In Ethiopia, for example, people struggling to feed themselves on the parched landscape are hastening the spread of deserts and accelerating soil erosion on farmland. In fact, over 1 billion tons of topsoil is blown or washed away from Ethiopia's highlands each year. So severe is the population pressure that the land may never recover from the drought that has gripped the area for more than a decade. Ethiopia's agriculture has steadily expanded over the past two decades primarily as a result of an increase in the amount of land under the plow and dramatic increases in grain production. Despite the increased production, the annual harvest provides 15 percent less grain per person than it did in 1950.

The nations of the world face two major food-related challenges. First, in nations where hunger and malnutrition is commonplace, the most immediate challenge is to find ways to end them to end current suffering. The second challenge has to do with meeting food demands resulting from future growth. That is to say, all countries must find ways to provide more food to meet future demand. In the face of rapid population growth and rising costs, this task may be difficult.

There's a third challenge, too. That is, meeting demand for food—whether to solve hunger or meet future demand to prevent even greater hardship—in sustainable manner. That is, we must meet immediate and long-term demands while protecting the soil and water upon which agriculture depends. Rangelands must also be managed intelligently to ensure their long-term productivity.

The challenge facing the global community can be best placed in perspective by examining expected global population growth. According to the Population Reference Bureau, a Washington, D.C.-based nonprofit organization, the world's population of 6.1 billion people in 2000 is expected to increase by 1.7 billion people by 2025. Food production must climb accordingly just to maintain the woefully inadequate status quo.

To meet present demands and ensure a long-term supply of food for the world's ever-growing population, most experts agree that all nations must engage in sustainable agriculture and ranching. This chapter and the next three deal primarily with soils and farming. Chapter 12 covers rangeland.

Although there is some difference of opinion regarding the definition of **sustainable agriculture,** it is defined here as a means of producing adequate amounts of high-quality, affordable food while protecting, even enhancing, the soil and other natural resources essential to agricultural production, such as water supplies (for more details on sustainable agriculture see Chapter 7).

Meeting present and future needs and creating an enduring system of agriculture are essential to building a sustainable future. Without such efforts, actions in other arenas are meaningless. But how do we increase food supplies sustainably?

5.2 Increasing Food Supplies Sustainably: An Overview

To understand how we can meet the demands of the present and future generations while protecting the agricultural land and other resources vital to food production, we must first understand the problems that keep us from producing adequate amounts of food in the first place.

Strategies fall into six categories: (1) protecting existing soils from destructive forces such as erosion and conversion to nonagricultural uses, (2) increasing productivity (output per hectare) of existing farmland and restoring depleted soils, (3) reducing pest damage, (4) improving food storage and distribution, (5) developing new food sources, and (6) expanding the land under cultivation. The food crisis, however, cannot be solved in the farm fields of the world alone. Humane population stabilization strategies are essential. Without population stabilization, the technical measures described in this and other chapters will only postpone the day of reckoning predicted by Malthus nearly two centuries ago. See Ethics in Resource Conservation Box 5.1 for an exploration of issues regarding population stabilization.

Protecting Existing Farmland

One of the keys to sustainable development is prevention. Prevent pollution from ever being created so you don't have to deal with it later. Prevent the waste of energy so you won't have to scramble to find new sources. Prevent farmland from being lost from erosion and other forces, so it can remain productive forever.

Farmland is currently being lost at record rates in the United States and most other countries as a result of erosion, nutrient depletion, desertification, salinization, water logging, and farmland conversion. How big a problem are they? Consider a few examples.

ETHICS IN RESOURCE CONSERVATION 5.1

FEEDING PEOPLE OR CONTROLLING POPULATION GROWTH?

Rare is the year that a famine in some part of the world does not dominate the news. In 1993, it was Somalia. In recent years in the African nation of Somalia, political strife, drought, crop failure, and a massive, hungry population teamed up to produce an incredible famine that killed tens of thousands of people. Many countries and international relief organizations rushed to feed the starving Somalis. The United States and later the United Nations moved soldiers in to quell the civil strive caused by fighting warlords. These measures, combined with an end of the drought, seemed to work.

Important as these measures were, many people recognized that they were nothing more than stopgap measures. That is, they afforded a temporary relief to a chronic or recurring problem. Unless something permanent is done, starvation will very likely return in the near future.

Ending this repetitive cycle, say some observers, will require efforts to stabilize populations. But how? Can the developed nations ethically impose this mandate on other nations? Can the United States, for instance, agree to give food aid only if a nation embarks on a population-stabilization program?

Rather than answer the question ourselves, we ask you to make a list of points that support this viewpoint and another list of points that oppose it. Then, analyze each argument carefully and decide which side of the debate you agree with. Once you have done this, take a few moments to think about your position and where it has come from. In other words, how has your ethical viewpoint evolved? Were your parents or teachers influential in forging your ethics? Have your friends influenced your thinking? Have books helped shape your beliefs?

Each year, an estimated 24 billion metric tons of topsoil is washed from the world's farmland. To put this statistic into perspective, 24 billion tons a year equals 240 billion tons of lost topsoil, per decade, which is equivalent to about half of the topsoil on all of the farms in the United States. Eroded soil reduces agricultural productivity. In severe cases it makes farming impossible, thus permanently sidelining productive farmland. You'll learn more about it in the Chapter 7.

Farming can also deplete the topsoil of nutrients. Without proper fertilization, soil fertility declines over time, further decreasing agricultural productivity.

Desertification is the spread of desert typically in semiarid regions, largely because of poor land management. This trend robs us of much arable land and decreases our food supply and our potential for increasing it as the human population increases (Figure 5.4). Desertification results from a number of factors. Global warming may be causing it. Overgrazing and deforestation also contribute to the process that has robbed the world of millions of hectares of farmland and rangeland.

After population stabilization, the next most important tactic for meeting present and future demands for food is preventive—improving the way existing farms and ranches are managed worldwide. Sustainable management can be considered a form of first aid for ailing farms and ranches. Such measures are the most cost-effective and environmentally sustainable means of increasing food supplies.

You'll learn more about prevention in future chapters. Controls on soil erosion and nutrient depletion, for example, are discussed in Chapter 7. Ways to reduce salinization and water logging are described in Chapter 9. Desertification on rangelands and range management are outlined in Chapter 12. Deforestation and desertification are discussed in Chapter 14.

FIGURE 5.4 *Desertification caused by overgrazing and poor agricultural practices destroys millions of hectares of marginally productive land each year in the United States, South America, Asia, and Africa.*

Measures to reduce global warming are discussed in Chapter 22. This chapter tackles ways to reduce farmland conversion.

Reducing Farmland Conversion In less developed countries, growing cities, new villages, roadways, reservoirs, and other uses consume once-productive farmland at an alarming rate. Much of the same is occurring in the more developed countries (Figure 5.5). The loss of farmland to urban sprawl, shopping centers, new highways, airports, and other human uses is called **farmland conversion.** In the United States, estimates of farmland conversion vary. In a recent study, the National Resource Inventory, losses between 1992 and 1997 were estimated to be 6.4 million hectares (16 million acres)—an average of 1.2 million hectares (3 million acres) per year. On average, that included 0.4 million hectares (one million acres) of cropland and 0.8 million hectares (2 million acres) of pasture every year! Worldwide, between 1975 and 2000 an estimated 150 million hectares (370 million acres) were lost to nonfarm uses. That figure is nearly equal to all of the land currently farmed in the United States. The loss of farmland is especially troublesome given the ever—expanding human population.

Protecting farmland from this dangerous force is vital to ensuring a long-term food supply for the growing population. How can this force be reckoned with?

Population stabilization is essential to this effort. The more successful we are in curbing growth here and abroad, the more land will be saved for farming.

Growth management, controls on the land people develop, can also slow down the loss of agricultural land. Growth management strategies include restrictions on land use. Zoning laws, for instance, are used in the United States and other countries to determine which lands can be used for industry, housing, recreation, and so on. Direct growth management strategies are a more recent phenomenon in the United States. One of the best programs exists in Oregon, a progressive state that passed growth management legislation in the early 1970s. This legislation directs cities and towns to restrict urban sprawl to certain areas to protect farmland, forests, wetlands, and other ecologically important lands. With growing pressure from expanding populations, Washington, Florida, New Jersey, and Tennessee have recently passed similar legislation.

Unfortunately, growth management even in the rich, more developed nations is spotty and poorly applied. Expecting adequate planning in the poor, less developed nations may be unrealistic.

Farmland can also be protected by other means as well. Some states buy the development rights of the land, essentially paying farmers what they'd get to sell to a developer, for example. This ensures that the land remains in agricultural production and isn't threatened by a buyout.

Increasing the Productivity of Existing Farmland

So far, we have learned that the first line of defense in the battle to feed the world's people is population stabilization. Second on the priority list are efforts to protect existing farmland from destructive losses such as erosion and farmland conversion. The third item on the agenda involves measures to increase the productivity of existing and new farmland through irrigation, fertilization, and the development of new crops and strains of livestock that produce more food (Figure 5.6).

Increasing Irrigation and Irrigation Efficiency Irrigation has helped boost cropland production in the United States and elsewhere enormously in the past four decades. Between 1950 and 1990, for instance, the amount of land under irrigation nearly tripled (Figure 5.7). Since that time, however, the amount of irrigated land has only increased slightly. Furthermore, although the amount of land under irrigation continues to climb, the number of acres of irrigated land per capita (that is, per person) has declined since the late 1970s. Today, growth in irrigation is not keeping up with population increase and the demand for food.

FIGURE 5.5 *Cropland is destroyed by housing developments, highway and airport construction, new shopping malls, and other byproducts of urbanization. Prime agricultural land is often taken out of production because it is flat and suitable for building and is often located near expanding cities.*

FIGURE 5.6 *Irrigation made possible in Egypt by the Aswan Dam. An area of 1 million acres formerly dependent on the annual Nile floods for irrigation is now cultivated under a system of perennial irrigation based mainly on lifting water. The new irrigation system makes possible a 40 percent increase in crop production in upper Egypt, because at least one additional crop can be grown per year. Photograph shows a mechanized irrigation pump at Habu Hat has replaced the laborious methods of lifting water by a hand-operated or animal-driven waterwheel. Unfortunately, the increased food production can barely keep up with Egypt's population increase.*

One way to improve the situation is to improve irrigation efficiency—that is, use water more efficiently so that more acres can be irrigated with the available water supply. In the western United States, for instance, cement-lined ditches and pipelines are replacing open ditches to transport water from supplies to farm fields in some areas. These strategies can cut water loss by 50 to 90 percent, respectively. Drip irrigation systems that irrigate certain crops, especially fruit trees, use a fraction of the water of sprinkler systems. Computer-monitoring and computer-controlled irrigation systems determine the amount of soil moisture and help farmers avoid overwatering, which can lead to salinization, waterlogging, and waste. In recent years, many American farmers have adapted center pivot irrigation systems, shown in Figure 5.8, to reduce water use. These systems once sprayed water upward. By a simple and rather inexpensive modification, farmers can divert the spray downward, just above the crop. This change reduces water demand for crops by about 50 percent. Other, more costly solutions to increase water supplies are discussed in Chapter 9.

Improving Yields Through Selective Breeding: The Green Revolution Since the dawn of human existence, nearly 4 million years ago, humans have succeeded in domesticating only about 80 species of food plants. Nonetheless, most of our food comes from a handful of species, with three crops, wheat, rice, and corn, accounting for well over 50 percent of the world's cropland.

After the domestication of wild plants, scientists began to tinker with ways to increase their yield. The most significant advances began in 1943 with the establishment of the International Maize and Wheat Improvement Center in Mexico. Sponsored by the Rockefeller and Ford foundations, the center set out to develop high-yield varieties of wheat and rice under the leadership of Norman Borlaug, an agricultural geneticist (Figure 5.9).

After 25 years of intensive research, the center had amassed an impressive record of new seed varieties. Scientists at the center produced strains of wheat and rice that yielded three to five times as much grain as previous strains. Best publicized,

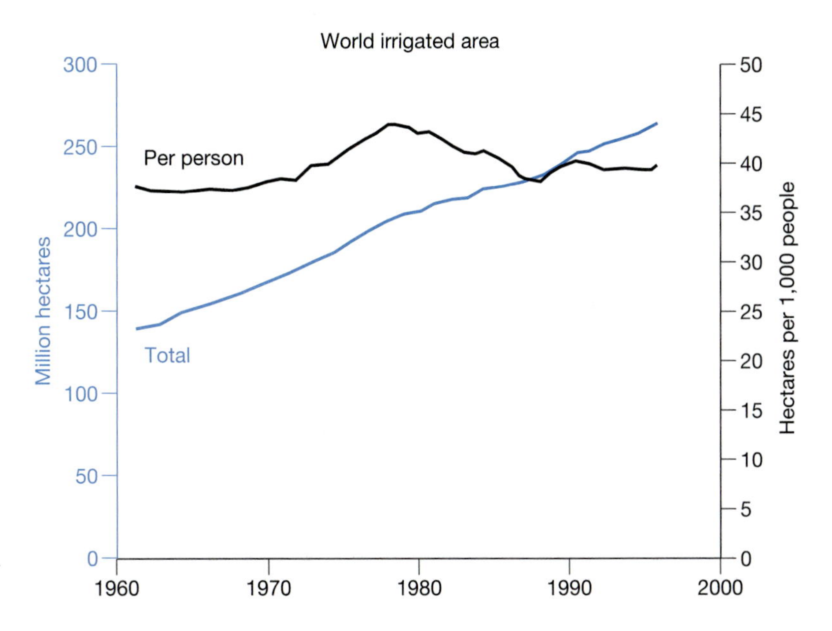

FIGURE 5.7 *Irrigated cropland has increased steadily since 1960, but irrigated land per capita has remained more or less constant because of population growth.*

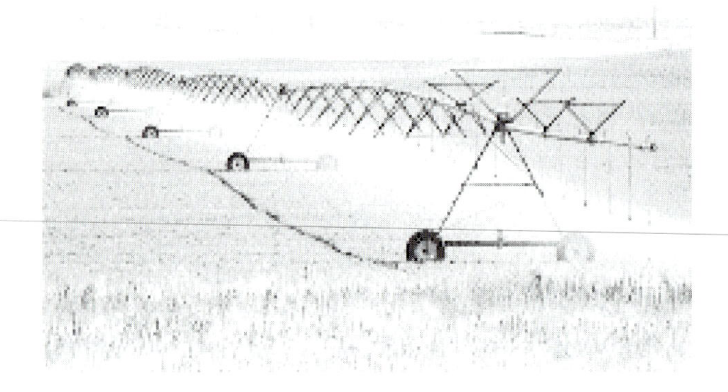

FIGURE 5.8 *Center pivot irrigation systems. (Top) Incredible amounts of water are sprayed into the air to irrigate crops. (Bottom) Fairly inexpensive modifications of these systems greatly reduce water loss to evaporation, protecting supplies.*

FIGURE 5.9 *One of the architects of the Green Revolution, Nobel Prize-winning Norman E. Borlaug is shown recording the vigor and stage of growth of wheat plants on a selective breeding plot in Mexico. Borlaug was successful in developing the so-called miracle wheats that, at least temporarily, greatly boosted wheat production in Mexico and other less developed nations around the world.*

however, was the new high-yield wheat that boosted Mexico's grain production from 780 kilograms per hectare (700 pounds per acre) to 4,700 kilograms per hectare (4,200 pounds per acre). Another important contribution was the development of new, high-yield varieties of rice. In addition to wheat and rice, scientists developed blight-resistant potatoes that outdid previous strains by an amazing 500 percent. Disease-resistant beans quadrupled yield. Efforts to develop high-yield grains with higher protein content than their predecessors are part of a movement referred to as the **Green Revolution**.

The new wheat and rice developed at the center were especially developed for use in tropical and subtropical countries. Extremely responsive to fertilizer and irrigation water, these plants not only grew faster and produced more grain, they were shorter and stouter and thus better able to withstand winds and harvesting than the traditional long-stemmed strains. Their short growing seasons ensured two or three harvests per year.

It took Borlaug and his associates 30 years to develop the new "miracle" strains, which won him the Nobel Prize in 1971, a fitting recognition of his service to humankind.

The high-yield varieties spread more quickly and more widely in the more developed countries than any other agricultural innovation in history. By the mid-1980s, nearly 50 percent of the wheat cropland and nearly 60 percent of the rice land in the more developed nations were sown with high-yield seeds (Figure 5.10). The amount of rice and wheat grown in the developed world shot up 75 percent between 1965 and 1980, even though the area planted in those crops increased only 20 percent. The benefits of the Green Revolution also spread to the less developed countries. India, once crippled by food shortages, became self-sufficient. Mexico and Indonesia boasted similar progress.

Despite the obvious benefits of the new plant crops, critics point out that the Green Revolution has not been a panacea for world hunger. The new high-yield varieties proved to be disappointing, for a number of reasons.

Perhaps the most common criticism is that the Green Revolution mostly benefited well-to-do farmers who could afford to irrigate their farmland and buy the fertilizer necessary for the new high-yield varieties. In rural Africa, where food is desperately needed, the Green Revolution had virtually no influence. Worldwide, an estimated 1.4 billion people live by subsistence farming. They grow barely enough food for them-

FIGURE 5.10 *A new rice strain for hungry India: IRS (left), a new high-yielding variety of rice under testing at a research station in Aduthurai. The plant to the right is the traditional variety.*

selves and their families, and they cannot afford the expensive seed or the costly fertilizer needed to make the new miracle seeds grow.

The high-yield varieties also proved inadequately equipped to ward off pests and disease. The geneticists had inadvertently eliminated the natural resistance of wheat and rice, and had produced a generation of genetic lightweights. To protect their crops, therefore, farmers needed large amounts of pesticides, which further drove up the cost of farming. A farmer who could afford the seed might not have enough money to pay for costly fertilizer and insecticide.

Another problem of the Green Revolution, and of modern agriculture in general, is that it fostered the use of a limited number of genetic strains, thereby fostering a reduction in genetic diversity (the number of genetic strains in use). Where dozens of strains had once been grown, huge fields containing one variety sprang up. The fewer the strains, the more devastating was the emergence of a plant disease or hungry insect pests.

Many critics claimed that the Green Revolution was a failure, but such criticism was premature. Learning from their errors, geneticists have developed new varieties of wheat and rice that grow under less favorable conditions. In Bangladesh, for example, one half of the wheat crop is a new high-yield strain that is grown without irrigation. Research is continuing on strains that can withstand drought and are more resistant to frost, pests, and crop diseases. A frost-resistant strain of winter wheat has been developed and is now used in Canada, helping to increase wheat production. Perhaps one of the most promising developments for the rural poor of Africa and Latin America is the relatively new genetic research aimed at increasing the yield of staples such as yams, potatoes, and various legumes. The Rockefeller Foundation, a key funder of the Green Revolution, recently announced plans to concentrate its agricultural program on genetic improvement of such crops, a great potential benefit to the 1.4 billion rural subsistence farmers who grow them.

Improving Yield Through Genetic Engineering Plant breeding, while successful in raising the productivity of food crops, is slow, tedious work. Until recently, 10 to 20 years was needed to develop desirable hybrids. Today, a new technology, discovered in 1973, may accelerate genetic enhancement. That technology is genetic engineering. You've probably heard about it in recent years as protestors line up to express their dismay over this new development.

Genetic engineering is a process in which scientists isolate **genes,** segments of the hereditary material of cells, the DNA, that determine various characteristics of organisms such as pest resistance, drought tolerance, fat content, and protein content. Isolated genes can be manufactured in one of several ways. These genes can then be transplanted from one strain to another, creating new strains that could potentially outperform their predecessors. It is also used to transplant genes from one species to another, crossing boundaries impossible in evolution.

Genetic engineering is proving to be a quick but controversial. In fact, many popular crops today are the result of genetic engineering. They are called **genetically modified, genetically engineered,** or **transgenic crops.** At this writing, more than 60 crops including corn, soybeans, strawberries, and apples, have been field tested in 45 different countries. In 1992, China became the first nation to grow transgenic crops commercially. The number of nations now permitting transgenic crop production is around two dozen with the United States leading the way. It accounts for about three-fourths of all transgenic crop production in the world. A large percentage of all soybeans and corn grown in the U.S., for instance, are now derived from genetically modified seeds. You've probably eaten products like corn syrup or soybean oil made from them without even knowing.

Desirable genes for transplantation may come from existing **cultivars,** strains that are currently under cultivation, often in remote parts of the world. Or they may come from the wild species from which modern species were derived many years ago. Many cultivars and wild ancestors of domesticated species are endangered by the relentless expansion of the human population but contain valuable genes that could enhance crop and livestock production. Consequently, many nations have launched aggressive programs to search out the vanishing cultivars and wild species that gave us corn, wheat, beans, and other commercially important plants—species that could benefit greatly from genetic boosts from their untamed cousins. The U.S. Department of Agriculture currently houses over 10,000 species with 450,000 genetically different samples. They're adding approximately 10,000 to 20,000 new samples per year from wild plant species and cultivars. Should either or both vanish from the wild, their genetic potential would theoretically be protected (Figure 5.11). We say theoretically because storage facilities are not a perfect answer. Seeds deteriorate in storage even under optimal conditions.

The importance of genetic improvements cannot be overstated. In the past 60 years, for instance, corn harvests have increased more than fourfold, from 20 bushels per hectare to 100 to 250 bushels, thanks to genetic infusions.

Geneticists are pursuing two basic approaches in crop modification through genetic engineering. They're tinkering with "input traits" and "output traits." Input traits refer to genetically controlled traits that affect tolerance to agricultural inputs such as pesticides or fertilizers. For example, much of the work in genetic engineering has focused on making crops resistant to herbicides. That way, farmers can use herbicides without affecting the crop. Much of the future work, some experts predict, will be in output traits, that is, genetic traits that have to do more with the plant's ability to produce food. For example, workers are developing high-oil corn designed to fatten beef cattle faster. Some alterations may affect flavor or nutrient content or even the color, making food more appealing.

Although transgenic crops have been fairly quickly been accepted by North American farmers and many consumers, this is not the case everywhere. In less developed nations, some farmers have protested against the use of transgenic crops and have even destroyed fields planted in them. Several European nations have even proposed placing a moratorium on transgenic food crops, while more research is done on their effects. What are the possible effects?

FIGURE 5.11 *Scientist at the National Seed Storage Laboratory at Colorado State University checks seeds stored in liquid nitrogen. This technique allows for better long-term storage than previous measures.*

Concerns are several. First, there are concerns about human health. The introduction of new genes into food crops could result in allergic reactions among those eating them. Some evidence of this is already occurring. There are environmental concerns, too. Scientists have documented examples in which nontarget species were harmed by pesticide residues in the soil, which were produced by crops that have been genetically modified to produce the chemicals. Even more startling, Canadian farmers have discovered that weeds had acquired herbicide resistance from nearby transgenic crops two years after the introduction of the transgenic crop. Herbicide-resistant weeds could outcompete other species, spreading uncontrollably.

Clearly, there is much to be learned about transgenic crops. Caution is advised as we launch this bold new experiment. In fact, in January 2000, an international agreement, known as the Cartagena Protocol, was signed in Montreal, Canada. This treaty regulates the international trade of food from transgenic crops. Its main focus is on sharing of information on existing regulations, risk assessments, and other agreements. It also calls for solicitation of consent from an importing country prior to the release of genetically modified organisms–for example, seeds from genetically modified crops. The treaty calls on the more developed nations to assist in helping other nations develop the skills and the capacity to perform risk assessments.

Increasing Fertilizer Use Japan has less than 0.07 hectare (0.166 acre) of arable land per capita, one-thirteenth that of the United States. Only with the most intensive agricultural methods has this tiny island nation been able to feed its 127 million people. Japanese farmers have achieved amazing success, producing 5,300 calories per cultivated hectare (13,200 calories per acre) per year, almost three times the productivity of American farmers. One key to Japan's accomplishments is the large amounts of fertilizer applied to the land: sardine-soybean-cottonseed cakes, animal wastes, green manure, human solid wastes, and artificial fertilizer.

Japan's success with soil enrichment can be repeated in many less developed countries, although on a smaller scale. Even without modifying any farming method, 9,500 artificial fertilizer trials conducted by U.S. agricultural specialists in 14 less developed nations have shown an overall average yield increase of 74 percent. In the next 25 years, however, fertilizer use must increase sevenfold if rapidly growing populations are to be properly fed. How likely is this?

Between 1950 and 1989, world fertilizer use increased from 14 million tons to 146 million tons (Figure 5.12). This was responsible for a dramatic rise in grain production—from 620 million metric tons in 1950 to nearly 1,700 million metric tons in 1989. Despite these impressive gains, fertilizer use has begun to taper off. In the 1970s, fertilizer use climbed an average of 6 percent per year. In the 1980s, growth slowed to

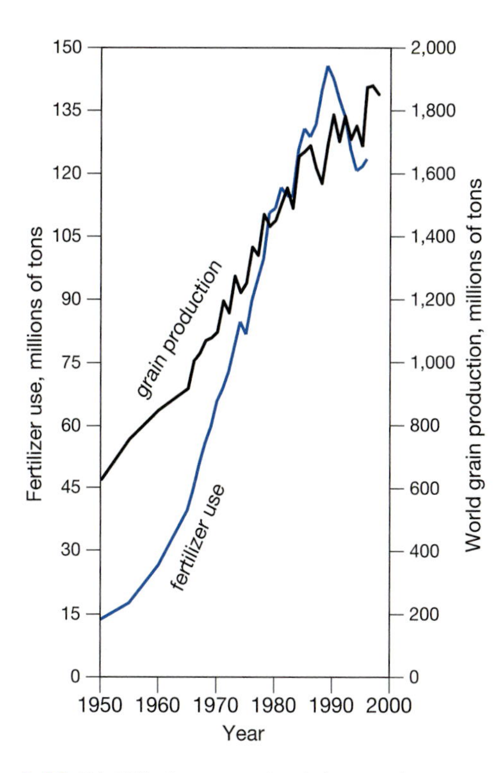

FIGURE 5.12 *World fertilizer use and grain harvested per capita, 1950–2000.*

3 percent per year, and from 1990 to 2000, total fertilizer use declined further (Figure 5.12). It is currently about 120 million tons. Total grain production has only increased slightly since 1990. When population growth is taken into account, it turns out that grain production per capita has actually declined in the past decade.

The long-term prognosis for fertilizer is not very good. Despite its many benefits, many farmers in less developed countries simply cannot afford it, nor can their much wealthier counterparts in the more developed nations. Heavily dependent on fossil fuel energy for its production and application, fertilizer is likely to become more expensive in the future as world oil supplies taper off. Further declines are possible. To solve this apparent dilemma, some experts argue that what is needed is more efficient application of fertilizer.

Edward Wolf of the Worldwatch Institute, for instance, points out that many countries overfertilize their most productive areas. They would achieve far more benefit from applying that additional fertilizer to marginal land. For instance, Chinese farmers apply most of their fertilizer to the most productive one third of their cropland. Applying that fertilizer to the marginal cropland—the remaining two thirds—would yield three to 15 times more grain per ton of added fertilizer than it does on the productive land.

In Africa, Latin America, and Asia, where farmers could benefit most from fertilizer use, farmers frequently cannot afford it. Instead of food aid, therefore, which is a stopgap solution, some observers believe that the industrial nations should consider donating fertilizer to help hungry nations become self-sufficient. Or they might assist in building fertilizer factories and the transportation networks needed to distribute fertilizer economically and quickly to rural areas. This would be an immensely costly and time-consuming task.

Recent evidence shows that low doses of artificial fertilizer applied to nitrogen-fixing plants actually enhance the plants' ability to fix atmospheric nitrogen. Heavier doses, however, impair plants' nitrogen-fixing capabilities. Thus, small donations of nitrogen fertilizer and technical assistance (training) to help farmers achieve a proper balance of artificial fertilizer and nitrogen fixation could help many nations improve production dramatically.

It is important to point out, however, that artificial fertilizer is made from fossil fuel—notably natural gas. Excess use alters nutrient cycles and pollutes waterways. **Organic fertilizer** such as cow manure, which is discussed in more detail in Chapter 7, may be a better choice. Organic fertilizers have been used in many parts of the world for thousands of years. They represent a more sustainable choice because they do not rely on a finite resource (natural gas) for fertilizer production. However, in many less developed nations one of the chief sources of organic fertilizer, cattle dung, is a valuable source of fuel. Where it once enriched farmland, it now is dried and packed in cakes that are used for cooking. Population pressures have reduced woodlands in many less developed nations, forcing peasants to use this alternative fuel. To return to the earlier, more sustainable way, some countries are promoting plans to develop sustainable forests near villages that could reduce the burning of dung and allow farmers to use it to enrich their fields. More efficient cooking stoves are also being widely distributed to reduce the amount of wood burned by villagers. Solar cookers are also being promoted in sunny climates to reduce the demand for wood.

World hunger will not be solved just by pouring more fertilizer on the land. Fertilizer is a costly alternative and bound to become more costly in the near future. But improvements in fertilizer application, the use of low doses in combination with nitrogen-fixing plants, better soil management, and the use of organic fertilizers can all help boost cropland production. These approaches are all part of a multifaceted plan needed to increase productivity of existing farmland.

Reducing Pest Damage

Roughly 40 kilograms out of every 100 kilograms of food grown throughout the world are destroyed by pests and disease organisms (mostly fungi), according to most estimates. Rats, insects, and fungi annually destroy enough food each year to feed one third of India's population. One in every 14 people in the world will starve because of food deprivation caused by agricultural pests. Let's examine the impact of pests more carefully.

Rodents The rat is one of humankind's greatest competitors. A single rat consumes 40 pounds of grain a year. The estimated 120 million rats in the United States alone destroy several billion dollars worth of food a year. In India, where rats may outnumber people by ten to one, up to 30 percent of the crops are lost to rodents.

Disease-Transmitting Insects In equatorial Africa, 23 varieties of tsetse fly, the transmitter of the protozoan that causes African sleeping sickness in humans and an equally serious disease in livestock, have effectively prevented livestock production in an area larger than the United States. In Southeast Asia, the microscopic malarial parasite *Plasmodium,* which is injected into the human bloodstream by mosquitoes, has been a scourge to the farmers. It incapacitates millions of rice farmers during the critical periods of transplanting and harvesting. To escape the malarial season in northern Thailand, for instance, farmers do not sow a second rice crop.

Locusts and Other Insect Pests Crop-destroying insects cause billions of dollars of damage worldwide each year. Huge crops inadvertently improve the prospect for harmful pests. Migratory hordes of locusts have plagued farmers since the time of Moses and continue to devastate crops. One locust swarm near the Red Sea, so thick it blocked out the sun, blanketed an area of over 5,000 square kilometers (2,000 square miles). Because of their great mobility, locusts may destroy crops more than 1,600 kilometers (1,000 miles) and several nations away from their hatching sites (Figure 5.13). Swarms of locusts have been known to travel from Saskatchewan, Canada, to Texas. (For more on pests and pest control see Chapter 8.)

Effective control of agricultural pests in the less developed nations would markedly boost food output. But such measures must be inexpensive, reliable, and environmentally safe to represent a sustainable solution to the problem. Of special interest are the biological techniques that allow farmers to minimize the use of chemical pesticides, long known to have serious environmental impacts. Relatively simple measures, such as altering the time of planting to avoid the emergence of harmful insects or increasing crop diversity to minimize pest population growth, can dramatically control pest populations without the damaging side effects of chemical pesticides.

Improving Food Storage and Distribution

As noted in the previous section, lots of food doesn't make it to the end user. It rots or is destroyed by pests. Improvements in storage, often simple changes in the types of storage bins, can drastically reduce food loss from rats and other pests. The more developed nations can assist the less developed nations in this simple, cost-effective means by providing financial assistance and technical assistance.

In many less developed nations, roads are inadequate and food distribution is problematic. Trucks transporting food may bog down in flooded roadways and never make it to the end user. Improving the distribution of food through the development of a better transportation system can also help increase the available food supply.

Developing New Food Sources

Much has been said about developing new food sources to feed the world's hungry people. Algae, yeast, and other food supplements rate high on the list. Despite the promise these new options offer, they are at best only of minor importance. Consider algae.

Algae have been grown for food in the United States, England, Germany, Venezuela, Japan, Israel, and the Netherlands for many years. One particular species, *Chlorella,* can be grown in small ponds. *Chlorella* is rich in proteins, fats, and vitamins and apparently contains all of the essential amino acids required by humans. Each hectare devoted to algal culture could produce nearly 90 metric tons of dried algae. The protein content of this algae crop may be 40 times greater than the protein yield of soybeans and 160 times greater than the yield of beef protein.

FIGURE 5.13 *Locust swarm threatens crops in Somalia. The operator of the spray plane (left in photograph) wished to spray this swarm with insecticides, but the engine would not start because it was clogged with locusts. The locusts are so thick that they blot out the terminal building (right in photograph).*

So why haven't food-producing nations gone berserk over algae the way they have over hamburgers? For one, algae are extraordinarily expensive to grow. Except perhaps in urban areas, where abundant sewage is available to nourish the microscopic plants, the economic costs are so high that one cannot justify widespread algae culture.

Aside from the formidable barrier of cost, new foods like algae face tremendous consumer resistance. Foods not integral to a culture often fail to catch on, making the financial risks of such ventures even greater. In the less developing countries, algae culture is an unlikely candidate for raising food production. However, more and more catfish, trout, and other tasty freshwater fish are grown in ponds in the United States. Carp are popular in many less developed nations. In Colorado, the Rocky Mountain trout on the menu is likely to have been reared in a nearby fish pond. Today, millions of tons of fish now come from commercial operations.

Native Grazers Another potential source of food lies in schemes to raise native grazers: herbivores indigenous to regions, such as Africa, that are adapted to the local climate and the local diseases.

A number of years ago, a team of wildlife biologists compared the meat production of domesticated livestock raised in Africa with native populations of antelope, zebras, giraffes, and even elephants maintained in the wild (Figure 5.14). Their conclusions are illuminating.

First, wild herbivores made exceptional use of the available plant food base. Cattle, on the other hand, were very selective, consuming only certain highly palatable grasses and ignoring other apparently less tasty forage. From the viewpoint of a range manager, cattle underutilize the available resource. Fences that protect them also result in overgrazing that could eventually destroy the productive capacity of the land. In sharp contrast, when several native species grazed a given area, they consumed many different plants, making fuller use of the available forage: antelopes, for instance, fed on grasses and low-level foliage, giraffes consumed foliage higher up on trees, and elephants dined on bark and roots.

Second, wild herbivores were much better adapted to drought than domesticated livestock. When a water hole dried out, they would move to another, often many miles away. Cattle, restricted by fences, cannot wander off in search of water. Native species also require less water per pound than livestock. A zebra, for instance, can get along without drinking water for three days. The gemsbok does not require drinking water at all but survives on **metabolic water** (water produced when glucose is broken down by cells to produce energy) and the water in the plants it eats. Also, native herbivores are much better able to avoid native predators. A cow is an easy target for a hungry lion. A native antelope has a fighting chance of escape.

Third, wild herbivores are immune to the potentially lethal sleeping sickness transmitted by the tsetse fly. Introduced livestock, however, are not.

For these reasons, ranches that raise wild species for commercial meat production can expect a profit margin fully six times that of the traditional livestock ranch. This is not to suggest that native grazers are a panacea. Several problems exist. First, because native herbivores migrate in response to available forage and water, ranching operations must include extensive land holdings. This system therefore naturally favors wealthy landowners. But to contain herds, extensive fencing might be required and that can be costly. In addition, if one is ranching wild animals, there must be a system to prove and maintain ownership.

Livestock Improvements Man does not live by bread alone. Many people acquire at least some of the protein they need from livestock. Solving world hunger, therefore, requires improvements in livestock production. Chapter 12 outlines methods to improve rangeland management.

In an effort to increase food production, scientists have developed new strains of cattle with meatier carcasses as well as new strains of pigs that grow faster and produce larger litters. In more developed nations, scientists have developed strains of dairy cattle with vastly superior milk production compared to dairy cattle of less developed nations. For example,

FIGURE 5.14 *Giraffes and zebras in Zimbabwe. Such animals could be raised on large game ranches more profitably and efficiently than traditional livestock such as cattle and goats.*

American test breeds of Holsteins produce up to 900 kilograms (2,000 pounds) of milk a year, compared to the 140-kilogram (300-pound)-per-year production from the yellow cattle of China.

A good part of the success in animal breeding results from artificial insemination, the injection of sperm from bulls into cows. By selecting genetically superior bulls and cows, scientists can develop strains that are capable of producing more food. Chickens with greater egg-laying capacity and more efficient feed-to-biomass conversions have been developed as well. Genetic engineering, discussed earlier, may also enhance scientists' ability to improve livestock.

Increasing the Amount of Land Under Cultivation: Tapping Into Farmland Reserves

Many countries have tried to boost agricultural production by farming new land. In an ambitious period from the mid-1950s to the mid-1970s, China, the former Soviet Union, and the United States, for instance, all expanded their farms onto previously untilled land. But much of the expansion was on marginal land, which was often difficult to farm, poor in nutrients, or highly erosive. In the 1980s in the United States, 50 percent of all the soil eroded from cropland came from a mere 10 percent of the arable land.

Starting in the late 1970s, the high cost of farming marginal land forced officials in these three countries to cut their losses. All began to withdraw marginal land from production. In the United States, the Food Security Act of 1985 provided a way to withdraw about 15 million hectares (35 million acres) of highly erodible farmland from production—over one tenth of the land then under cultivation. This goal was achieved by 1993. So successful was the program that Congress has extended the law at least to 2002.

The prospects for expanding food production in the Third World are mixed. In Southeast Asia, most of the cultivable land is already under cultivation. In Southwest Asia, more land may be farmed than is believed sustainable. In Africa and South America, however, only a small percent of the land thought to be suitable for agriculture is currently being farmed. Although expansion is feasible on both of these continents, the wisest strategy for increasing food output may be taking better care of existing farmland so that it remains productive indefinitely. It is also important to note that much of the land that agriculturalists view as potential farmland includes tropical rain forests, arid lands, and wetlands. Should they or can they be used?

Tropical Rain Forests: A Potential Farmland? Tropical rain forests are blessed with abundant sunshine, rainfall, and perpetually warm days—all conditions conducive to forest growth. To the untrained eye, the dense tropical forests might appear to be prime candidates for farmland conversion (Figure 5.15). Just the opposite is true, however: Most tropical rain forests, when stripped of trees and planted in crops, make some of the worst farmland known to humankind.

FIGURE 5.15 *Shifting type of cultivation in central Sumatra. Agricultural land is opened up by cutting and burning the forests. The cleared area will then be intensively cropped for a few years until the soil fertility has been exhausted, then it will be abandoned. Small plots revegetate and can be used again in 35 years; large plots tend to suffer from extreme erosion, making them permanently unsuitable for agriculture.*

Unlike deciduous forests of the eastern United States, China, and Europe, where leaves and other plant litter form a thick, spongy layer that decays over time, making the soil rich and productive, the tropical rain forest soils are bare and nutrient poor. Fallen leaves and branches are quickly decomposed by bacteria, insects, fungi, and earthworms. The nutrients released by decomposition into the soil, however, are rapidly absorbed by the roots of the large trees that tower over the forest floor. Thus, the soil that supports the most productive ecosystem on Earth is, paradoxically, among the poorest known to humankind. Chopping down trees and planting crops is a prescription for disaster. What nutrients are in the soil are quickly taken up by the crops or leached from the soil into the deeper layers, where they are inaccessible to crops. Nutrients are also washed away during the rainy season, impoverishing a land that has little recuperative ability.

Another problem with tropical forest soils is that many are rich in an iron compound that, when exposed to sunlight during dry periods, hardens the soil, creating a bricklike layer impervious to plants and farm equipment alike. Known as **laterites** (*later* is Latin for brick), these reddish brown soils may have caused the downfall of the Khmer civilization in Cambodia and the Mayas of Mexico. In more recent times, they caused the failure of an agricultural colony started by the Brazilian government in the heart of the Amazon basin.

Today, tropical rain forests fall at an alarming rate. Each year, by several estimates, an area of tropical forest the size of the state of Washington is cleared. This land is cleared for timber and other wood products. Farms and pastures are feebly started on the denuded landscape but, almost without exception, soon fail. Plans to expand agriculture at the expense of forests, most agree, must be stopped. (For more on tropical rain forests, see Chapter 13.)

Arid and Semiarid Lands Many an optimist looks to deserts and semiarid lands as a source of potential farmland. Few agriculturalists would deny that with adequate water and fertilizer, the sandy soils of arid and semiarid lands could be made to produce crops. For example, in Egypt the gigantic Aswan Dam on the Nile River has made it possible to increase crop production on the vast desert surrounding the project.

However promising many irrigation projects are riddled with problems. High costs and low returns are key stumbling blocks. Adding to the purely economic barriers are salinization, the buildup of salts on irrigated farmland, and waterlogging, the saturation of the topsoil in heavily irrigated land, described in Chapter 8. Salinization today threatens every arid region of the world where irrigation is used. Currently in intermediate or final stages of salinization, this land will soon fall into disuse. Thus, plans to expand agriculture into arid and semiarid lands are very likely a short-term answer with high costs and limited utility.

Wetlands Throughout the world, wetlands have been drained to provide living space and valuable farmland. In the United States, agriculture has been a major source of wetland losses. **Wetlands** are lands that are wet part or most of the year and include swamps, bogs, salt marshes, and mangrove swamps. The land alongside of rivers can be classified as wetlands if they are flooded part of the year. The agriculturally productive fenland of Britain was once a swamp. Flourishing Israeli settlements now occupy the site of the former water-logged Huleh marshes. Drainage has made crop production possible in Italy's Po Valley and in the Yazoo Delta of the Mississippi. Much of the fruits and vegetable crop produced in southern Florida is grown on drained swampland.

Wetlands are productive fish and wildlife habitat. Many species of waterfowl live and breed in wetlands. Many commercially valuable fish and shellfish depend on coastal wetlands. Destroying these areas endangers fish and wildlife populations. But wetlands are also water purifiers that trap sediment and other pollutants. They act as sponges as well, holding back rainwaters and reducing flooding and increasing groundwater recharge. (See Chapter 9 for more on wetland functions.) Converting wetlands to farmland robs us of the free ecological services they provide and may require costly engineering solutions such as water pollution control facilities to eliminate pollutants and dams to control floods.

Efforts to protect wetlands have increased in many nations. In the United States, wetlands in the lower 48 states once covered an area twice the size of California (90 million hectares, or 220 million acres): Today, half of all coastal and inland wetlands have been drained or filled. The greatest losses have occurred in California (91%), Ohio (90%), and Iowa (89%).

Fortunately, the loss of wetlands has slowed dramatically in recent years. New laws prohibit further drainage and filling. In fact, the federal government has strengthened controls over wetland drainage for farmland by refusing federal crop insurance and other economic support to farmers for crops grown on newly drained and filled wetlands. Local environmental groups are waging successful campaigns to prevent the loss of important swamps. The Federal government requires losses to be mitigated. That is, if wetlands must be lost due to development, steps must be taken to create new wetlands or prevent losses elsewhere. In Florida, a huge swamp along the Kissimmee River (pronounced keh-seh-mee) is being restored only a few years after a multi-million-dollar drainage project was completed, not as an effort to mitigate losses, but because the destruction had proved economically and ecologically costly.

Despite these changes, wetlands continue to be lost to development. Losses are particularly high in the less developed nations, but the United States and Canada still experience unacceptably high losses. Further wetland drainage here or abroad must be viewed with caution. The impacts on wildlife and fish, stream flow, and water quality often far outweigh the benefits realized by converting them to farmland or other uses. In fact, limiting farmland expansion onto wetlands and other ecologically sensitive regions in many African nations may prove to be more economically profitable than farming, for the wealth of wild species in these nations attracts tourists who pour millions of dollars into national economies. Revenues from such activities could help support programs of improved soil management and population control, reducing the present pressures on undeveloped land.

In summary, although there is some potential to expand the land under cultivation, it is not as great as the figures suggest. It can only be counted on as part of a mix of solutions designed to expand food output.

5.3 Poverty, Conflict, and Free Trade: Vital Strategies Needed to Feed the World's People

Feeding the world's people, like so many issues, requires a multifaceted approach—population control, protecting farmland from erosion and conversion, increasing productivity, and so on. The next three chapters outline many solutions, but one aspect they do not touch on is poverty. One of the key stumbling blocks to feeding the world's people is inability to pay. Even in the chronically food-short nations, the well-to-do can buy food while the poor starve to death. Huge numbers of people go hungry because they cannot buy or grow enough food. Some experts believe that not until the standard of living is raised in many less developed nations can hunger be eliminated. Even modest improvements in annual income could go a long way in feeding the world's people.

Thus, in addition to the measures described in the previous section, many experts believe that countries must find ways to raise the earning power of their poor. But any economic development strategy must be sustainable. It must provide livable wages and decent work. It must use resources efficiently, and so on. In other words, it cannot cause further deterioration of the environment. If possible, it should result in improvements.

Some proponents of sustainable development argue that one key may be the development of small, sustainable businesses in local villages and cities and towns. Sustainable businesses would use local resources, rather than imported ones, for local consumption, rather than export. In other words, such projects would promote local or regional self-reliance. Bicycle factories or factories that turn out efficient cookstoves, for example, could help peasants earn enough money to feed themselves and their children, and the bicycles and stoves could be used locally. Without such improvements in personal wealth, most experts agree, hunger may be with us for a long time.

Economic development by itself is not enough. Many experts argue that the less developed countries should strive to become self-sufficient in food production, rather than relying on food imports or using their own land to produce export crops (coffee, tea, fruits, and the like) for consumption in Western nations. Ironically, many LDCs were once self-reliant, but became indebted to more developed nations that financed many Western-style development projects. To generate money to pay back loans, many less developed nations turned their cropland, which once produced staples, into citrus and tea plantations. The products, of course, were bound for the West. This reduced the amount of food that LDCs could produce and made them even more reliant on the West, this time for basic food. Free trade, discussed shortly, has also thwarted local food production as farmers find it difficult to compete with large, industrial farmers in the MDCs.

The problem with world food problems, writes John Tanner, "has less to do with growing more food than growing it in the right places and getting it to the people who need it most." By growing their own food locally and in sufficient quantities and at a price suitable to local demand, LDCs can achieve self-sufficiency and ease suffering and death.

Self-sufficiency for the less developed nations may be an essential component of achieving a sustainable future. More developed nations can lend a hand in this important transition by providing technical assistance and measures to support locally based, sustainable agriculture. Together, these and other measures mentioned in this chapter could bring us closer to the goal of feeding a hungry planet and avoiding an ecological catastrophe already in the making.

Two additional factors also play a key role in hunger and malnutrition. Civil conflict and free trade. In numerous countries, warring factions often use food as a weapon. They cut off supplies to weaken an opponent or destroy crops or force farmers to abandon their farms (who can farm with bullets flying over their heads?). War also has negative impacts on the economy, for example, it eliminates jobs and makes it difficult for people to afford food. Destruction of roads and other elements of infrastructure like stores, railroads, storage facilities, and bridges also makes food distribution difficult, if not impossible. Even after a war has ended, it may take many months for agriculture to recover. In Afghanistan, land mines planted in a 1979 conflict remain buried in farmfields. It's estimated that somewhere between one half and two-thirds of the nation's farmland cannot be worked because of the threat of mines.

Another culprit has emerged in recent years, notably free trade. Although this may seem like a good thing, it does have a downside. In particular, farmers from industrial regions such as North America and Europe gain access to markets in LDCs where they sell government subsidized products, often undercutting local farmers. Many subsequently give up farming. Farmers who have provided for their families are no longer able to. They must seek employment elsewhere or go hungry.

Dependence on imported crops for staple foods can be a dangerous proposition for other consumers in less developed nations, too. Price fluctuations and currency devaluations, for instance, can cause dramatic increases in food prices. In Mexico, for example, consumers initially benefitted from cheap corn imported from the United States. However, in 1995, the peso was devalued the price of corn doubled in a year, making it difficult to afford.

Summary of Key Concepts

1. The eighteenth-century British economist Thomas Malthus advanced the theory that because of the limits set by finite land supply, human populations would tend to increase faster than the food supply. Eventually, therefore, the human population would experience a massive die-off that would bring the population back in line with food production.
2. Malthus's glum predictions may be coming true. An estimated 12 million people, nearly half of them children, die each year from hunger and diseases worsened by hunger.
3. According to recent estimates, 1.2 billion people are undernourished, 2 million suffer from micronutrient deficiency, and 1.2 billion are overfed.
4. Undernutrition is a quantitative phenomenon that results from an inadequate intake of food. Malnutrition is a qualitative phenomenon that is characterized by inferior food quality, usually resulting in a deficient intake of proteins and vitamins. Overnutrition is an equally dangerous condition caused by overeating.
5. Two clinically identifiable diseases are seen in many of the world's children. Marasmus is a disease that afflicts infants separated from their mother's milk, which causes a deficiency of protein and calories. Kwashiorkor occurs in slightly older children and results from an inadequate intake of protein. For every case of kwashiorkor and marasmus there are a hundred other children who suffer from milder cases of malnutrition and undernutrition.

6. World agriculture faces two major challenges: (a) feeding people who are starving today and (b) meeting the demands of future citizens. Both challenges must be met in a sustainable way.
7. World population is expected to increase by 1.73 billion people between 2000 and 2025. Food supplies must keep pace just to meet the demand of new world citizens. Fortunately, there are many ways to solve world hunger. The foremost strategy is population stabilization.
8. Another important strategy involves measures to protect existing farmland from such forces as erosion, desertification, nutrient depletion, and farmland conversion.
9. Efforts to increase productivity can also be helpful. These efforts include measures to irrigate more efficiently and to fertilize land with artificial and organic fertilizer. New high-yield plant and animal species developed through selective breeding or genetic engineering may also prove helpful, although there are many concerns about potential ecological and health impacts of transgenic crops (genetically modified crop species)..
10. Pest control is also an important strategy to curtail preharvest and preconsumption losses. Pests destroy an estimated 40 percent of the world's food supply each year. Through better pest control, farmers can raise food production. But pest control strategies must also be environmentally safe.
11. New food sources such as algae and yeast appear to be of limited value. Far more promising are plans to raise native grazers. Native grazers fully utilize the plant food base, tend not to overgraze, are mobile, and are often resistant to diseases.
12. Expanding the amount of land under cultivation is a strategy of limited value, for many countries are farming nearly all of their good land.
13. In the United States and other developed nations, much of the potential farmland is of marginal quality. This land is often too costly and environmentally harmful to farm.
14. In Africa and South America, large amounts of farmland lie in reserve, but much of it is ecologically sensitive land, including tropical rain forest. Before such "reserves" are tapped, more careful land management should be applied to existing farmland to avoid further losses. In addition, tropical rain forests, arid lands, and wetlands should probably be avoided whenever possible. Tropical rain forest soils are poor and highly erodible. Arid lands are easily damaged. Wetlands are an important biological resource better left undisturbed.
15. Reducing poverty through sustainable economic development is also a viable strategy, for many people are hungry simply because they cannot afford food.
16. Ending war is also important to ensuring food for the world's people as food is often used as a weapon in war and war itself disrupts farming or the distribution of food.
17. Free trade agreements may also threaten food supplies.
18. Despite the long list of potential ways to solve world hunger, little progress will be made without population growth stabilization strategies and measures to increase the standard of living of many of the world's poor people.
19. What is needed is an integrated approach that includes the many measures discussed in this chapter and emphasizes self-sufficiency.

Key Words and Phrases

Borlaug, Norman
Brown, Lester
Cultivar
Desertification
Economic Development
Farmland Conversion
Farmland Reserves
Free Trade
Genes
Genetic Diversity
Genetic Engineering
Genetically Engineered Crops
Genetically Modified Crops
Green Revolution
Growth Management
High-Yield Strains
Integrated Approach
International Maize and Wheat Improvement Center
Irrigation
Kwashiorkor
Laterites
Malnutrition
Malthus, Thomas
Marasmus
Micronutrients
Native Grazers
Obesity
Organic Fertilizer
Overnutrition
Salinization
Self-Sufficiency
Sustainable Agriculture
Transgenic Crops
Tropical Rain Forests
Undernutrition
Waterlogging
Wetlands

Critical Thinking and Discussion Questions

1. Debate the following statement: "Food surpluses from wealthy nations should be donated to poor, hungry nations, now and in the future. Through such donations, world hunger can be solved."
2. Thomas Malthus voiced dire warnings about world population growth nearly 200 years ago. What were they, and how valid are they today?
3. Refute this statement with statistics on world hunger: "The world's doing fine. There are very few hungry people."
4. Define malnutrition and undernutrition. Why are these conditions particularly harmful to infants?
5. Describe the connections between overpopulation and hunger. In what ways does overpopulation worsen hunger? What is the role of poverty in overpopulation and hunger?
6. How prevalent is overnutrition? Where is it most prevalent? Why?
7. What are the most immediate challenges facing the world in regard to food production?
8. Define the term sustainable agriculture.

9. Do you agree with the following statement? Why or why not? "The answer to world hunger is cropland expansion. We must plow all available land. There's plenty of good farmland left."
10. Give several reasons why the land on which tropical rain forests grow might appear to be a prime candidate for farmland conversion. Describe why these impressions are false. What critical thinking rules does this illustrate?
11. Of what value are swamps, salt marshes, lagoons, and other wetlands? Why should or shouldn't they be drained to make more farmland?
12. Describe the importance of protecting farmland from such forces as erosion and conversion as a strategy for meeting the demands of the world's population for food.
13. List ways to increase crop productivity.
14. What is the Green Revolution? In what ways was it initially successful and in what ways was it unsuccessful? What new developments will help the world's subsistence farmers? How could genetic engineering help increase plant production?
15. Do you agree with the following statement? "Fertilizer has helped farmers the world over increase food production. By applying more fertilizer to cropland, especially in the LDCs, farmers can boost agricultural production." What are the limitations to this strategy? Can you suggest any more effective strategies?
16. In what ways does war interfere with food production and distribution?
17. Sketch the broad outlines of a master plan to increase food production worldwide.
18. Using your critical thinking skills, debate the following statement: "Poor countries must eventually become self-sufficient in food production. They cannot rely on the wealthy nations of the world to bail them out of their troubles."

Suggested Readings

Brown, L. R., M. Renner, and B. Halweil 1999. *Vital Signs: The Environmental Trends that are Shaping Our Future.* New York: W. W. Norton. Great analysis of trends in agriculture and population.

Brown, L. R. "Facing Food Insecurity." 1994. In *State of the World.* L. Starke, (ed.) New York: W. W. Norton. An excellent discussion of recent trends in food production.

Conway, G. "Food for All in the 21st Century." 2000. *Environment* 42:(1) 8–18, Describes the need for a second Green Revolution and ways this can be achieved.

Durning, A. T. 1993. "Supporting Indigenous Peoples." In *State of the World.* L. Starke, (ed.) New York: W. W. Norton. Describes ways to protect indigenous people and their systems of sustainable agriculture.

Durning, A. T., and H. B. Brough. 1992. "Reforming the Livestock Economy." In *State of the World.* L. Starke, (ed.) New York: W. W. Norton. Full of ideas on ways to improve livestock production in environmentally sustainable ways.

Gardner, G. 1996. *Shrinking Fields: Cropland Loss in a World of Eight Billion.* Worldwatch Paper 131. Excellent update on the current loss of cropland and food trends.

Gardner, G. and B. Halweil. 2000. "Nourishing the Underfed and Overfed In *State of the World.* L. Starke, (ed.) New York: W. W. Norton pp 58–78.

Gupta, A. 2000. "Governing Trade in Genetically Modified Organisms: The Cartagena Protocol on Biosafety," *Environment* 42:4, 22–33. Describes the provisions of an important international treaty on trade in genetically engineered food.

LaBastille, A. 1999. *Jaguar Totem: New Wildlands and Wildlife,* Westport, NY: West of the Wind Publications. A delightful chronical of an ecologist's studies of tropical rain forests. Many insights.

Levidow, L. 1999. "Regulating BT Maize in the United Statesand Europe: A Scientific-Cultural Comparison," *Environment* 41:10, 10–21. Worthwhile reading for those interested in the controversy over genetically engineered foods.

Paarlbert, R. 2000. "Genetically Modified Crops in Developing Countries: Promise or Peril?" *Environment* 42:1, 19–27. Candid look at this important subject.

Postel, S. 1993. "Facing Water Scarcity." In *State of the World.* L. Starke, (ed.) New York: W. W. Norton. Describes water conservation measures for food production.

Reganold, J. D., R. I. Papendick, and J. E. Parr. 1990. "Sustainable Agriculture." *Scientific American* 262:6, 112–120. Excellent overview.

Rosegrant, M. W., and R. Livernash. 1996. "Growing More Food, Doing Less Damage." *Environment* 38:7, 6–11, 28–32. Good look at policy changes required to promote sustainable agriculture.

Rosson, P. "Will Erosion Threaten Agricultural Productivity?" 1997. *Environment* 39:8, 4–9, 29–31. Important reading.

Soule, J. D., and J. K. Piper. 1992. *Farming in Nature's Image: An Ecological Approach to Agriculture.* Washington, D.C.: Island Press. Essential reading.

Web Explorations

Online resources for this chapter are on the World Wide Web at: **http://www.prenhall.com/chiras** *(click on the Table of Contents link and then select Chapter 5).*

Chapter 6

The Nature of Soils

Civilization is only skin deep. In this chapter we examine the "skin"—the soil—that forms a thin layer around the surface of the Earth. It is this soil on which human survival depends.

6.1 Value of Soil

Although soil is far more complex than air or water, it is sometimes the forgotten part of the environmental picture. This is a dangerous oversight and not just because soils affect our future food supply. High-quality soils not only promote the growth of plants, they also prevent water and air pollution by resisting erosion and by degrading and immobilizing agricultural chemicals, organic wastes, and other potential pollutants. Most city dwellers equate soil with "dirt," but to the farmer, soil is the essence of survival. The farmer's economic well-being is inextricably linked with the quality of the soil.

Empires and nations, like individuals, are dependent on the soil. To the extent that a nation's soil resources are fertile and abundant, that state will have vigor and stability. When that resource is exhausted because of the mounting demands of a swelling population or mismanagement, the nation's survival is in danger. Some historians believe that the demise of many great civilizations was due to destruction of the soil resource base that had originally made possible the rise of those civilizations.

6.2 Characteristics of Soil

Soil is a natural system consisting of four components: **mineral matter, organic matter, water,** and **air.** Soil inherits mineral matter from its parent rock and organic matter from its living or decaying organisms. As shown in Figure 6.1, a healthy loam surface soil is usually composed by volume of about 50 percent solids (minerals and organic matter) and 50 percent pore space (occupied by water and air). Under optimal conditions for crop yields, the pore space should be occupied by equal amounts of water and air.

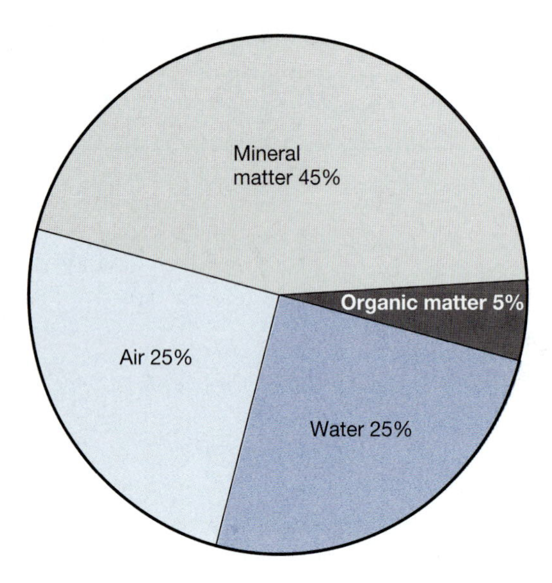

FIGURE 6.1 *The composition (by volume) of a healthy loam surface soil. Mineral and organic matter constitute 50 percent and pore space, containing air and water, makes up the other 50 percent.*

TABLE 6.1 *USDA Division of Soil Particles Into Fine-Earth and Coarse Fractions*

Particle Name	***Diameter Range***
Fine-Earth Fraction	
Sand	2.0–0.05 mm
Silt	0.05–0.002 mm
Clay	<0.002 mm
Coarse Fraction	
Gravel	2 mm–7.6 cm (3 inches)
Cobble	7.6–25 cm (3–10 inches)
Stone	25–60 cm (10–24 inches)
Boulder	>60 cm (>24 inches)

Major characteristics of soils are texture, structure, organic matter, living organisms, aeration, moisture content, pH, and fertility. An understanding of these characteristics is an essential prerequisite to the study of soil profiles, soil types, soil productivity, and soil management.

Texture

Mineral particles in soils are extremely variable in size, ranging from submicroscopic clay to large rock fragments such as stones and gravels. For convenience and efficiency, the U. S. Department of Agriculture (USDA) divides soil particles into two major groups based on size: the **coarse fraction,** consisting of particles greater than 2 millimeters in diameter, and the **fine-earth fraction,** consisting of particles less than 2 millimeters in diameter (Table 6.1). The fine-earth fraction, to which the term **soil texture** applies, has a greater effect on soil behavior because it is the most chemically and biologically active. The fine-earth fraction has been further divided into three main size groups or separates: **sand** (2.00 to 0.05 millimeter), **silt** (0.05 to 0.002 millimeter), and **clay** (less than 0.002 millimeter). The relative proportions of sand, silt, and clay in a particular soil determine its soil texture.

Since soils are composed of different percentages of sand, silt, and clay, specific terms are used to convey some idea of their textural makeup. For this, **textural classes** are used, such as sandy loam, loam, or silty clay. There are 12 major soil textural classes, which are defined by the percentages of sand, silt, and clay as shown in the textural triangle (Figure 6.2). These percentages are on a weight, not volume basis. The tex-

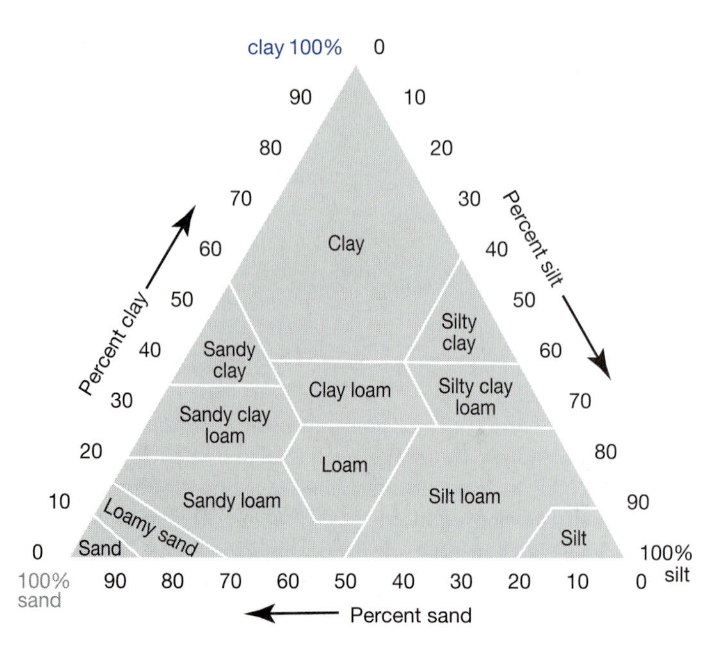

FIGURE 6.2 *Standard USDA textural triangle for classifying soil textures, based on percentages of sand, silt and clay.*

tural triangle is used to determine the soil textural name after the percentages of sand, silt, and clay are determined from a laboratory analysis. Organic matter and coarse fragments are not included in determining soil texture.

Soils that are dominantly clay are called *clay;* while those with a high sand percentage are *sand.* A soil that does not exhibit the dominant physical properties of sand or silt or clay may fall into the textural class *loam* or *silt loam.* A **loam** soil, the most desirable soil texture for many crops, does not contain equal percentages of sand, silt, and clay. It does, however, exhibit equal properties of sand, silt, and clay. It usually takes a large amount of sand or silt particles to exert as much influence on soil properties as a comparatively small quantity of clay particles.

Soil texture is one of the most important characteristics of the soil because it helps determine soil **permeability** (the ease with which air and water pass through a layer of soil), water storage, the ease of tilling the soil, the amount of aeration, soil fertility, and root penetration. For instance, a coarse sandy loam or loamy sand is easy to till, has plenty of aeration for good root growth, and is easily wetted, but it also dries rapidly and easily loses plant nutrients, which are drained away in the rapidly lost water. High-clay soils (over 30 percent clay) have very small particles that fit tightly together, leaving little open pore space, which means there is little room for water to flow into the soil. This can make high-clay soils difficult to wet, difficult to drain, and difficult to till.

In general, the larger the **surface area** of soil separates, the greater the possible chemical reactivity with the soil air and soil solution (water plus solutes). So, clay particles are the most chemically reactive, followed by silt, then sand particles. The soil separates vary significantly in their influence on air and water (Table 6.2).

TABLE 6.2 *Influence of Soil Separates On Air And Water Characteristics*

Soil Characteristic	Sand	Silt	Clay
Water permeability	Rapid	Moderate	Slow to very slow
Water-holding capacity	Low	Medium	High
Air movement	Rapid	Moderate	Slow to very slow

It should be emphasized that soil separates are relatively stable. Despite the dynamic nature of soil, despite the continuous physical, chemical, and biological activities that are continuously transforming it, and despite the soil management activities of the farmer, sand will not change to silt, nor silt to clay, within the average human life span.

Clay is an important reservoir of plant food, a function that largely depends on two characteristics of clay particles: (1) their very large surface area, and (2) their negative electrical charge. Soil scientists estimate that the aggregate surface area of the clay particles in the topsoil of only 2 hectares (5 acres) of an Iowa cornfield down to a depth of 30 centimeters (1 foot) is roughly equal to the entire surface area of the North American continent. The negatively charged surface of the clay particle attracts positively charged ions (cations) of nutrient elements, such as calcium, potassium, magnesium, zinc, and iron (Figure 6.3). These nutrients form a loose chemical bond with the clay particles. This process is called **adsorption.** From the standpoint of soil fertility, adsorption is of fundamental

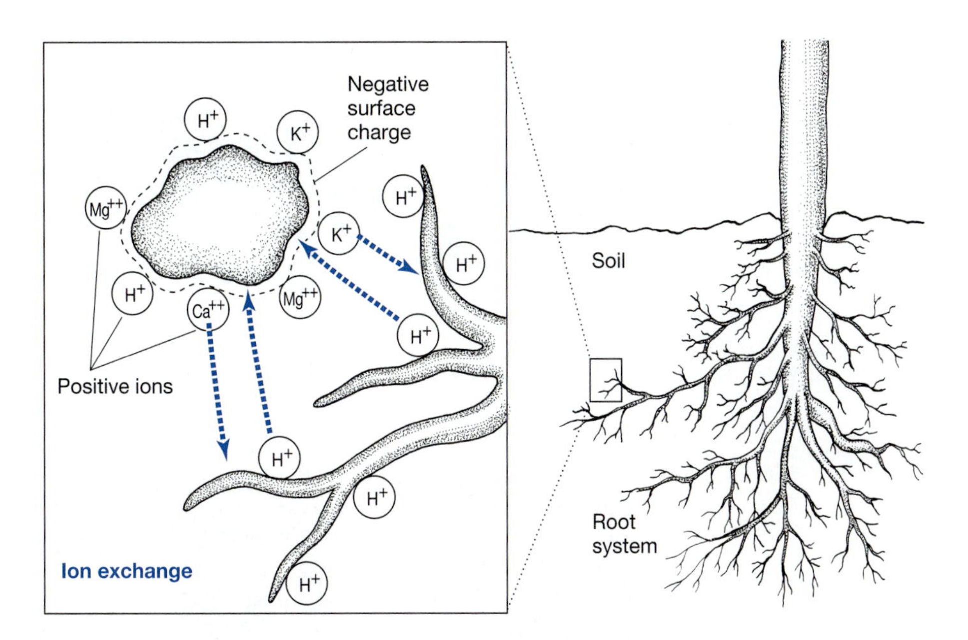

FIGURE 6.3 *Ion exchange between clay and plant roots. The positive ions of nutrient elements, such as potassium (K^+), calcium (Ca^{++}), and magnesium (Mg^{++}), are attracted to the negatively charged surface of the clay particle. These nutrient ions on clay particles are available to be taken in by the plant. Note that as these nutrient cations are absorbed by plant roots, H^+ ions are excreted into the soil solution from the root systems of plants (or more organic acid anions are produced inside the cell) to balance the absorbed cations.*

importance because it prevents the leaching (removal) of nutrients from the soil by the action of water. These nutrient ions on clay particles are available to be taken in by the plant. Just 10 grams of dry Iowa topsoil may have 1.2 quintillion ion-exchange sites that can hold nutrients for use by crops!

On the other hand, clay particles do not retain negatively charged nitrate ions (NO_3^{2}) ions very well. As a result, if too much nitrogen fertilizer is applied to a crop, nitrates may leach through the soil and into the groundwater, possibly causing contamination. A judicious use of nitrogen fertilizers by farmers and gardeners should ease this problem.

Structure

Soil structure is the arrangement or grouping of soils particles into clusters or **aggregates.** These aggregates may have a variety of shapes, such as granular, blocky, or prismatic, as shown in Figure 6.4. Aeration, water movement, heat transfer, and root growth in a soil are all to some degree dependent on its structure. Physical changes imposed by farmers in plowing, cultivating, liming, and manuring the land are with soil structure, not texture. Other factors affecting soil structure include alternate freezing and thawing, wetting and drying, plant root penetration, burrowing by animals like worms and pocket gophers, addition of slimy secretions from animals, bacterial decay of plant and animal remains, and compaction by farm equipment and off-road vehicles.

When farmland has good soil structure, like granular structure, crop production is enhanced. When you squeeze a handful of such soil, it has a crumbly quality. Soil with granular structure has an abundance of pores through which life-sustaining water and oxygen can move to plant root systems. Such soil allows water to infiltrate after a rainfall or snow melt (Figure 6.4). A soil with poor structure, like platy structure, has a minimum number of pore spaces for air and water because of the closely packed soil aggregates. Water permeability is greatly reduced in poorly structured soils.

Some soils lack structure, having no observable aggregation. Such structureless soils are either single-grained, as in very sandy soils, or massive, as in soils high in silt or clay, where the natural structure has been destroyed or become puddled. Soil structure can be improved by adding organic matter, such as crop residues, compost, and animal manures. The growth and decay of grasses and legumes stimulate aggregation and enhance soil structure.

Organic Matter and Soil Organisms

Soil organic matter is the living or dead plant and animal materials in the soil. It includes plant and animal residues at various stages of decomposition, small animals, microorganisms, and humus. The semistable, dark-colored material that represents the decomposition products of organic residues and materials synthesized by microorganisms is known as **humus.** Maintaining humus is indispensable to good soil management. Humus has many functions. It (1) improves soil structure, (2) increases pore space so that air and water can penetrate more readily, (3) buffers the soil against a rapid change in pH, (4) reduces erodibility of the soil, (5) minimizes leaching of nutrients, (6) increases the nutrient storage ability and water-holding capacity of the soil, and (7) provides a suitable habitat for valuable soil organisms, such as bacteria and earthworms.

Humus formation is favored in moist and cool climates, not in dry and hot climates. Soil humus is richer under native

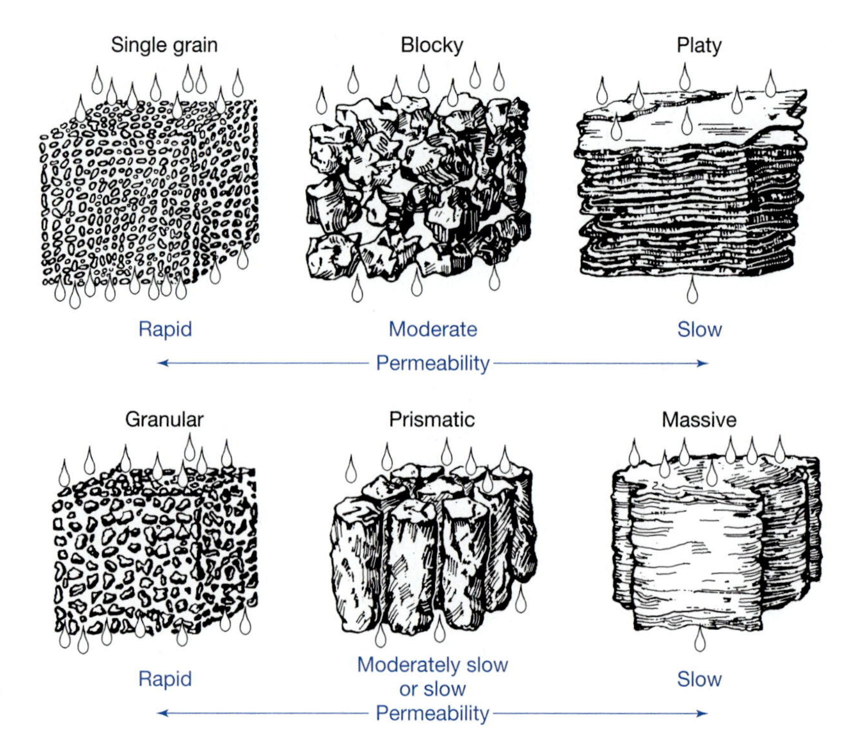

FIGURE 6.4 *Effect of structure on water permeability. Note the variety of shapes occurring in soil aggregates and the relative rates at which water can move through them.*

grassland than under forest vegetation. However, humus content in both native grassland and forest environments is higher than under cultivated conditions. This is due to the harvesting of crops, to soil erosion, and to the increased oxidation of organic matter as a result of tillage.

Three percent of organic matter in the soil averages **biomass** or living organisms. Soil organisms can be large or small: **soil macroorganisms** (such as insects and earthworms), or **soil microorganisms** (microbes such as fungi, bacteria, and protozoa) (Figure 6.5). Some soil organisms have little effect on plant growth, some are absolutely essential to plant growth in natural systems, and others cause damage by reducing or destroying plant yields and by spreading animal infections and disease. Though the detrimental soil organisms are many, overall, the beneficial ones exceed them in effect. Some soil organisms aerate and aggregate soils.

Soil microbes convert toxic carbon monoxide to carbon dioxide, fix atmospheric nitrogen (convert it to chemical compounds that plants can use), and are a major source of antibiotics used for disease control. Soil microbes are instrumental in decomposing organic matter and releasing available plant nutrient elements. There may be billions of soil microbes in 1 gram of soil (one-fifth teaspoon). They can weigh 680 kilograms (1500 pounds) in a hectare (2.5 acre) of healthy soil. This means that a steer grazing one hectare of rangeland could be outweighed by the microbes grazing within it.

Aeration and Moisture Content

Healthy soil "inhales" and "exhales" continuously. Oxygen, which is present in greater concentration in the atmosphere than in the soil, diffuses into the soil pores and is needed for the respiration of plant roots and microorganisms. Carbon dioxide moves from the soil into the atmosphere. This constant "breathing" action of the soil is dependent on the numerous soil pores, which serve as air reservoirs and passageways.

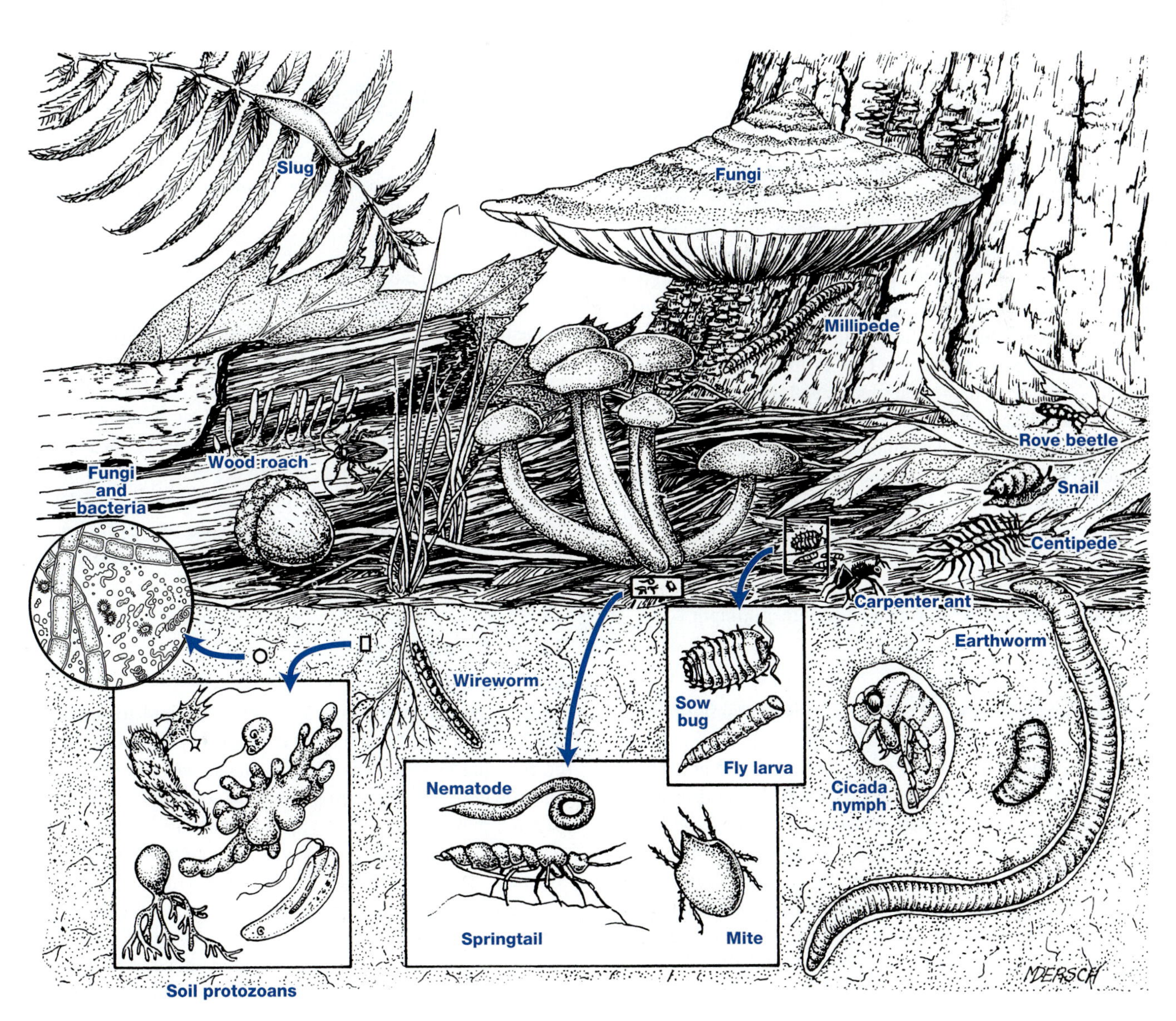

FIGURE 6.5 *Characteristic organisms found in the soil.*

The **pore space** of a soil is that portion occupied by both air and water. The amount of pore space is determined largely by texture and structure. With smaller surface area, as in sands, or with closeness of particles, as in compact subsoils, the total pore space is low. With greater exposed surface area or if the particles are arranged in porous aggregates, as is often the case in medium-textured soils high in organic matter, the pore space per unit volume will be high.

Two types of pores occur in soils: **macropores** and **micropores.** Macropores (larger than 0.06 millimeter in diameter) are readily drained of water and filled with air after a heavy rain. In contrast, with micropores (less than 0.06 millimeter), attraction forces in the soil retain the water within the fine pores and air is greatly impeded. Thus, in a sandy soil, in spite of the low total pore space, the movement of air and water is surprisingly rapid because of the dominance of the macropores. Fine (clayey)-textured soils have mostly micropores and, thus, allow relatively slow gas and water movement in spite of their high porosity. Aeration in micropores is frequently inadequate for satisfactory root development and desirable microbial activity. Soil conditions are generally ideal where macro- and micropores are in equal proportions.

Water serves several important plant functions. It is an essential raw material for photosynthesis; it is the solvent medium in which minerals are transported upward to the leaves and sugar is transported downward to the roots; and it is an essential component of protoplasm, forming 90 percent of the weight of actively growing organs, such as buds, roots, and flowers.

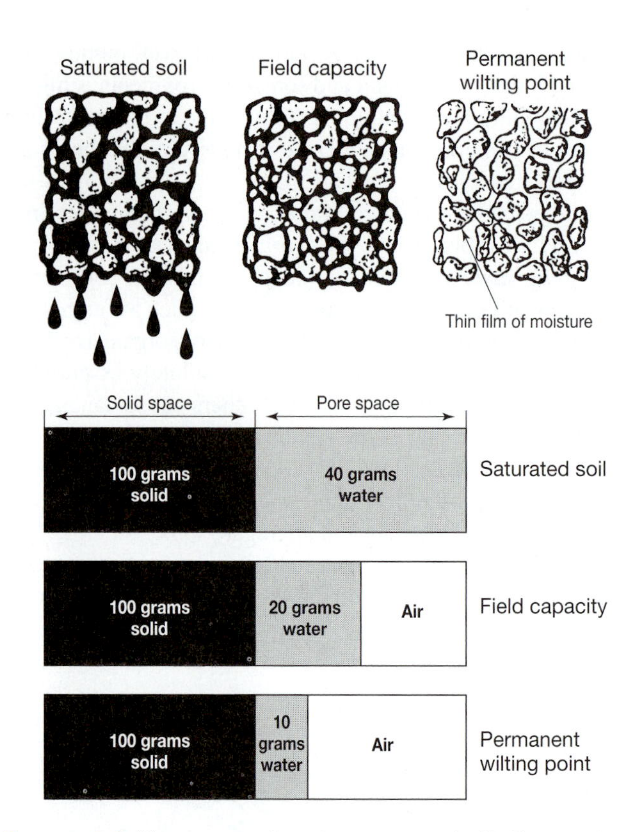

FIGURE 6.6 *Water content of a soil can vary considerably. Also note that as the water content of a soil decreases, the air content increases.*

The amount of water in the soil is critically important to crop health and survival. When the pore spaces are completely filled with water to the exclusion of air, the soil is said to be at **saturation** (Figure 6.6). This condition frequently occurs after a heavy rain or an intensive irrigation. Two to three days later, after water has drained from the macropores and is held in the micropores, the soil is at **field capacity.** This water is readily absorbed by plants. When plants have removed all the water they possibly can and have begun to wilt, the soil is at the **permanent wilting point.** The only moisture left in the soil is in the form of thin films that line the pores. These films are held so tenaciously by the soil that roots cannot absorb them—a condition experienced by many farmers and backyard gardeners who have seen their beans and lettuce shrivel during a midsummer drought.

Soil pH

An important property of the soil solution is its **reaction;** that is, whether it is acid, neutral, or alkaline (basic). The acidity, neutrality, or alkalinity of a soil depends on the relative amounts of hydrogen (H^+) and hydroxide (OH^-) ions. Soil scientists use the **pH scale** to determine a soil's reaction. The pH scale ranges from 0 to 14 (Figure 6.7). A pH of 7 is neutral, a condition in which the number of hydrogen ions equals the number of hydroxide ions. At a pH below 7, hydrogen ions outnumber hydroxide ions and the soil is acidic; at a pH above 7, hydroxide ions outnumber hydrogen ions and the soil is alkaline.

Soils vary in pH from 4.5 to a little above 7 in humid regions, and from a little below 7 to 9 in arid regions. Most vegetables, grains, trees, and grasses grow best in soils that are slightly acidic to neutral, having a pH between 6.5 and 7.0. Soils under hardwood (oak, maple, and beech) forests are usually more alkaline than soils under coniferous (spruce, fir, and pine) forests.

Soil pH is the soil chemical property most commonly measured by farmers and urban homeowners alike. It can be determined with color indicators purchased at a nursery or by electrometric methods in the laboratory. Lime may be applied if the soil is too acidic, as indicated in the reaction in Figure 6.8. This reaction illustrates that if lime is mixed with acid clay, the calcium ion (Ca^{++}) from the lime replaces two hydrogen ions (H^+) attached to the clay particle, with water and carbon dioxide being formed.

Soil Fertility

Soil fertility is the status of a soil with respect to its ability to supply nutrients in amounts and forms required for maximum plant growth. The fundamental components of soil fertility are the essential nutrients absorbed by plants and utilized for various growth processes. Most plants are known to need at least 16 essential nutrient elements to grow, although more than 90 elements can be absorbed by plants. An **essential element** is a chemical element required for the normal growth of plants. Each essential nutrient plays one or more special roles in plant growth.

Essential nutrients that are required in large amounts by plants are called **macronutrients.** These include carbon, hydrogen, oxygen, nitrogen, phosphorus, potassium, sulfur, calcium, and magnesium (Figure 6.9). The other essential

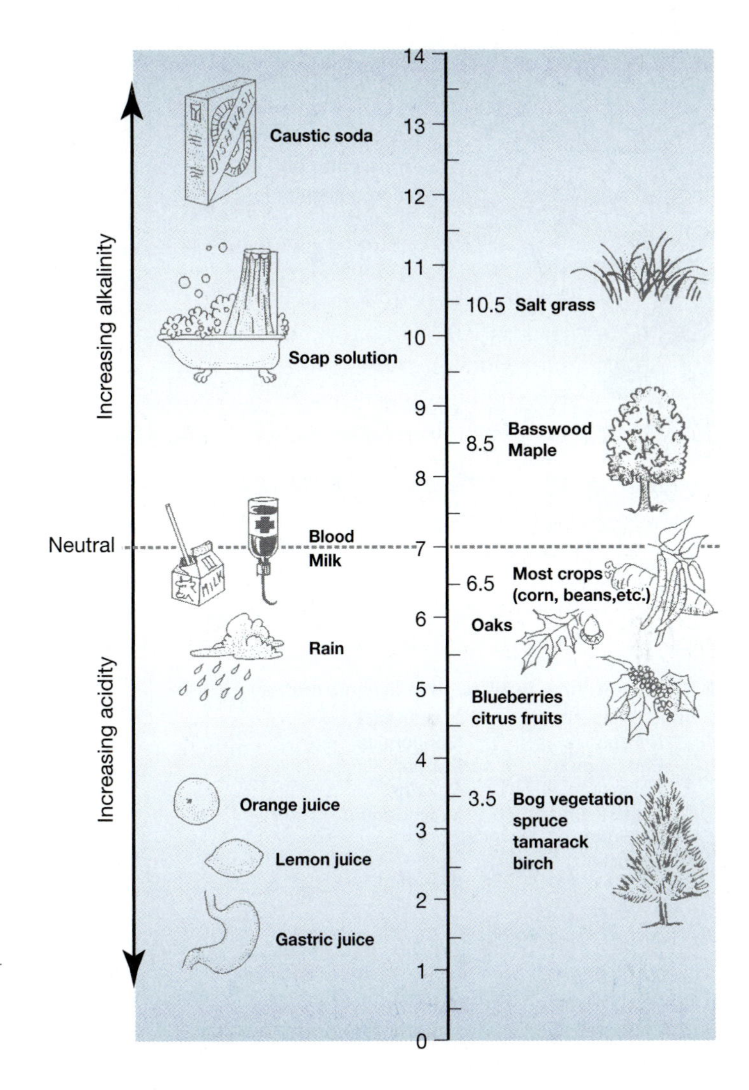

FIGURE 6.7 *The pH scale. The pH values of certain well-known substances are shown on the left. The preferred soil pH values of some plants are shown on the right.*

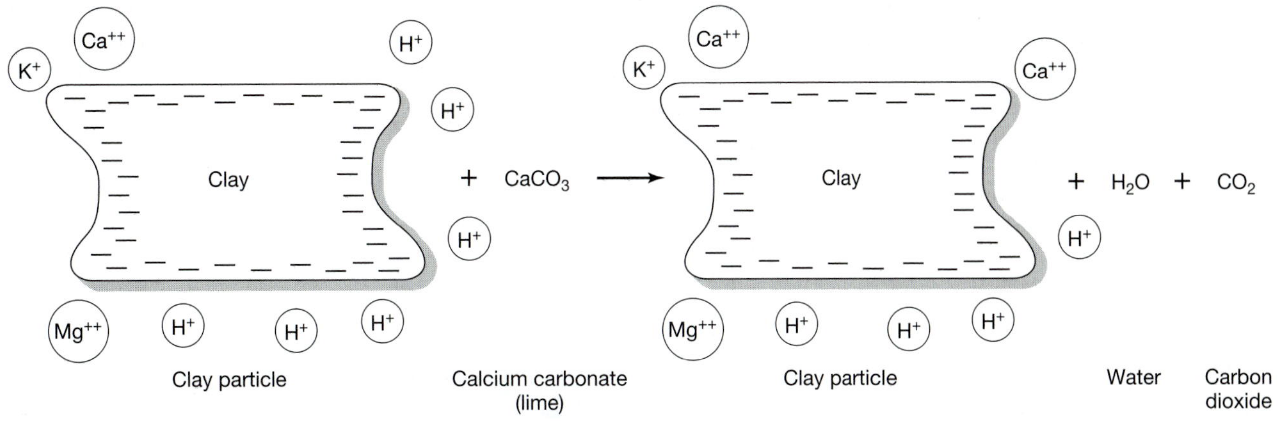

FIGURE 6.8 *How lime neutralizes soil acidity.*

nutrients (manganese, copper, chlorine, molybdenum, zinc, iron, and boron) are used in only small quantities by plants and are known as **micronutrients.** As an example of the trace amount of a micronutrient needed by a plant, only 70 grams of the micronutrient molybdenum is sufficient to satisfy the needs of 1 hectare (2.5 acres) of clover.

Research has shown that each essential element must be present in a specific concentration range for optimum plant growth (Figure 6.10). If the concentration of a given element in the plant root zone is too low, a deficiency of that element occurs and plant growth is restricted. Likewise, if the root zone concentration of that element is too high, toxicity occurs and plant growth is similarly limited. Only in a specific middle range of concentration (the sufficiency range) is optimum plant growth attained. One of the principal objectives of good farmers is to manage the fertility of their fields so that the concentration of each essential element is maintained in the proper range.

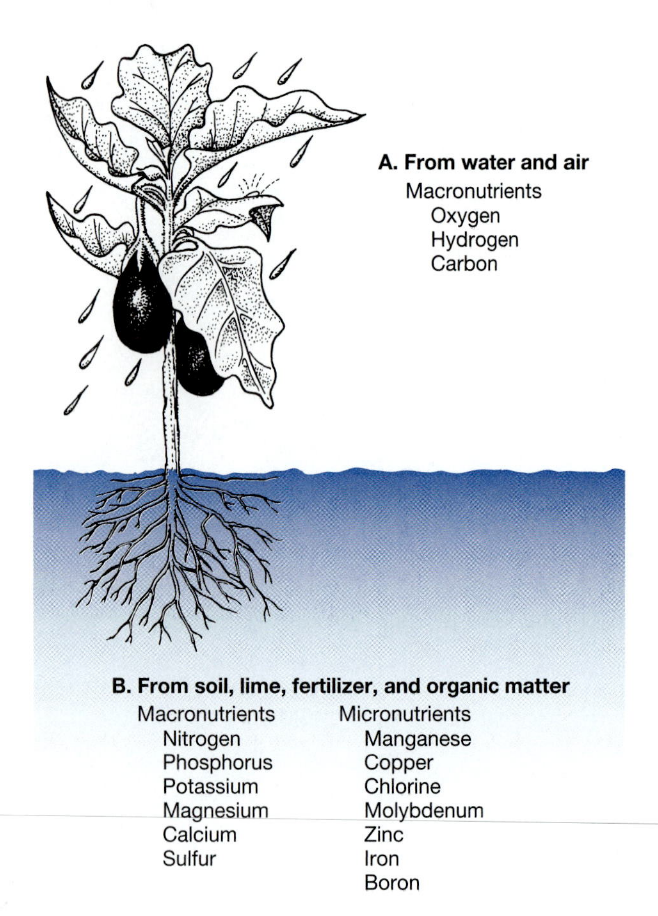

FIGURE 6.9 *Nutrients required by plants.*

The soil gains nutrients from (1) nitrogen fixation, (2) decomposition of plant and animal remains, (3) animal wastes, (4) weathering of parent materials, and (5) fertilizer (Figure 6.11). The soil loses nutrients mostly by (1) root absorption, (2) leaching due to the downward movement of water, (3) soil erosion, and (4) volatilization.

6.3 Soil Formation

The development of a mature soil is a complex process that may require centuries to a million years to complete. If we are to care for our soils so that they will last for future generations, we should understand how they formed and how they relate to their environment. Soil formation depends on five major factors, called **soil-forming factors.** These five factors are responsible for kind, rate, and extent of soil development. They are (1) climate, (2) parent material, (3) organisms, (4) topography, and (5) time.

Climate

Climate is an increasingly dominant factor in soil formation with increased time. The climate also influences soil development indirectly in determining the natural vegetation. The most important climatic influences that affect soil development are temperature and precipitation. They exert profound influences on the rates of biological, chemical, and physical processes within parent material and soil. For example, for every 10°C rise in temperature, the rate of chemical reactions doubles. In addition, many soil processes are controlled by the activity of soil organisms, which in turn is affected by temperature and moisture.

Mineral **weathering** occurs through physical and chemical reactions whose rates are influenced by temperature and precipitation. When precipitation and the other four factors remain constant, an increase in temperature causes an increased rate of weathering and clay formation. The rate of weathering increases by an increase in the presence of water (up to a point, of course). Thus, high average temperature and precipitation, as in tropical and subtropical areas, tend to encourage rapid weathering and clay formation. Leaching and water erosion are also more intense and of longer duration in warm and humid regions. Conditions that result in a minimum degree of weather-

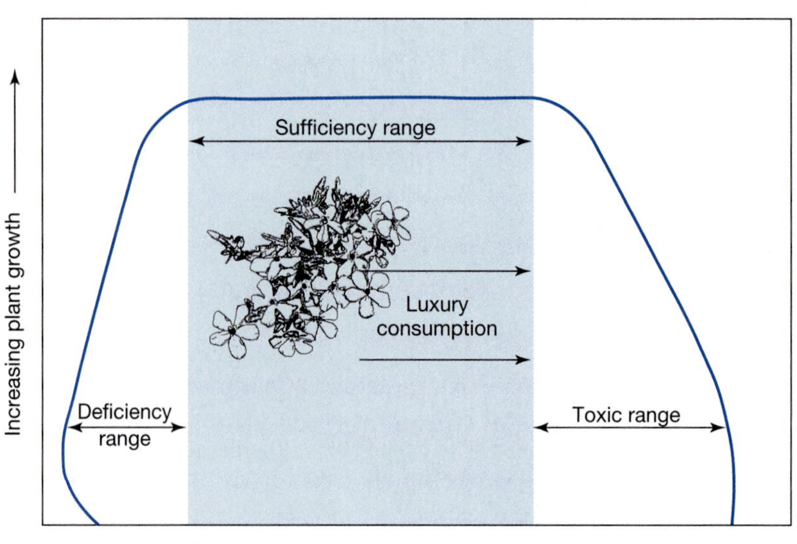

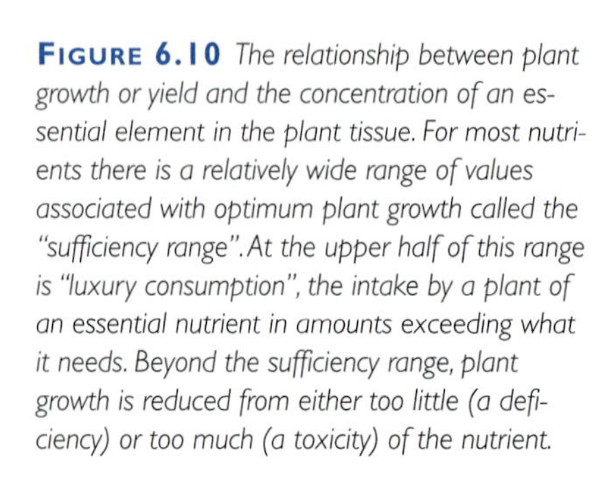

FIGURE 6.10 *The relationship between plant growth or yield and the concentration of an essential element in the plant tissue. For most nutrients there is a relatively wide range of values associated with optimum plant growth called the "sufficiency range". At the upper half of this range is "luxury consumption", the intake by a plant of an essential nutrient in amounts exceeding what it needs. Beyond the sufficiency range, plant growth is reduced from either too little (a deficiency) or too much (a toxicity) of the nutrient.*

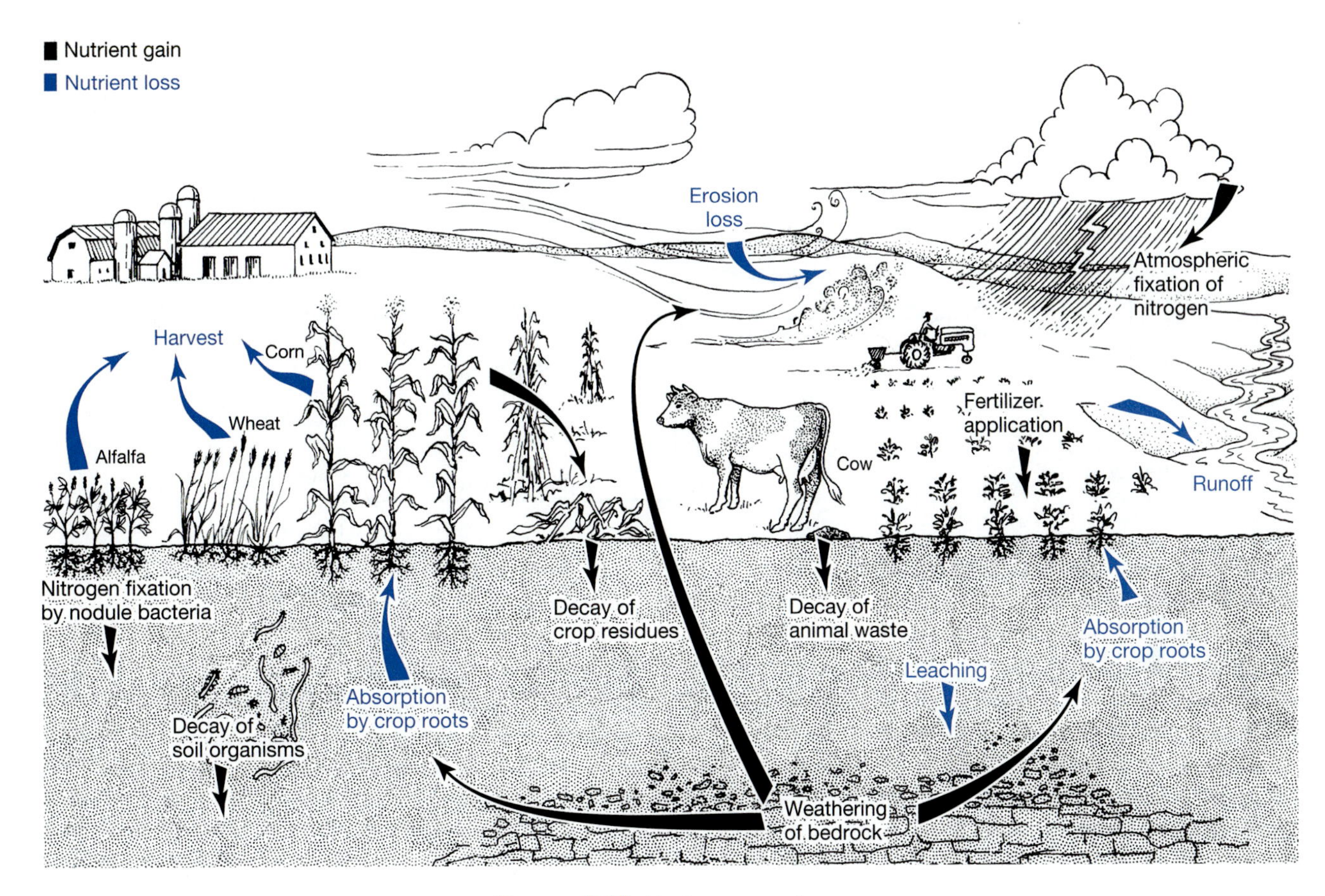

FIGURE 6.11 *Nutrient sources of crops.*

ing, leaching, and water erosion are found where the climate is cold and dry.

Parent Material

Parent material is the unconsolidated (soft and loose), weathered mineral or organic material from which soils form. Parent material might lie in its original position above bedrock for centuries or it might be moved to new positions by the mechanical forces of nature. Soils can form then from this unconsolidated material sitting on top of consolidated bedrock or from transported unconsolidated material. The formation of soils starts with and results from changes in the parent material.

Parent materials can be classified into three major groups: residual, transported, and organic. **Residual parent materials** formed from the weathering of rock in place. The rate of residual parent material formation is usually slow, requiring several tens of thousands of years to affect hard rock to appreciable depths. For example, in subtropical Zimbabwe, Africa, weathering granite was studied at two sites in 1978 and was estimated to form soil depths of 11.0 millimeters (0.43 inch) and 4.1 millimeters (0.16 inch) per 1,000 years. Because soil forms so slowly, it can be seen why it is important to protect soil from erosion (Chapter 7).

Transported parent materials are the most extensive materials on the Earth's surface. Transported parent materials were transported from their place of origin and redeposited in a new location. There are four transporting agents: ice, water, wind, and gravity (Figure 6.12). Glacial deposits are material carried by ice, such as glacial till, or material deposited by the meltwaters of glaciers (outwash plain). Water deposits can be alluvial (flood plains or deltas), lacustrine (lake deposits), or marine (coastal plain or beach deposits). Wind deposits include sand dunes, loess (mostly silt-sized particles), and tephra (volcanic ash). Colluvium is a deposit of rock fragments and soil material accumulated at the base of steep slopes as a result of gravity.

Most **organic deposits** accumulate in stagnant water (lakes or swamps) where decomposition of plant residues is retarded by the limited supply of oxygen (Figure 6.13). Stagnant waters in temperate climates allow good growth of adapted plants, which die or shed leaves into the water. The decomposition of the plant material in stagnant water is very slow because of the deficiency of oxygen. Over centuries, such organic accumulations may reach depths of many meters.

Organisms

The activity of living plants, animals, and soil microorganisms and the decomposition of their organic wastes and residues have marked influences on soil development. The development of a mature soil depends on the activity of a great number and diversity of **organisms.** Perhaps the most

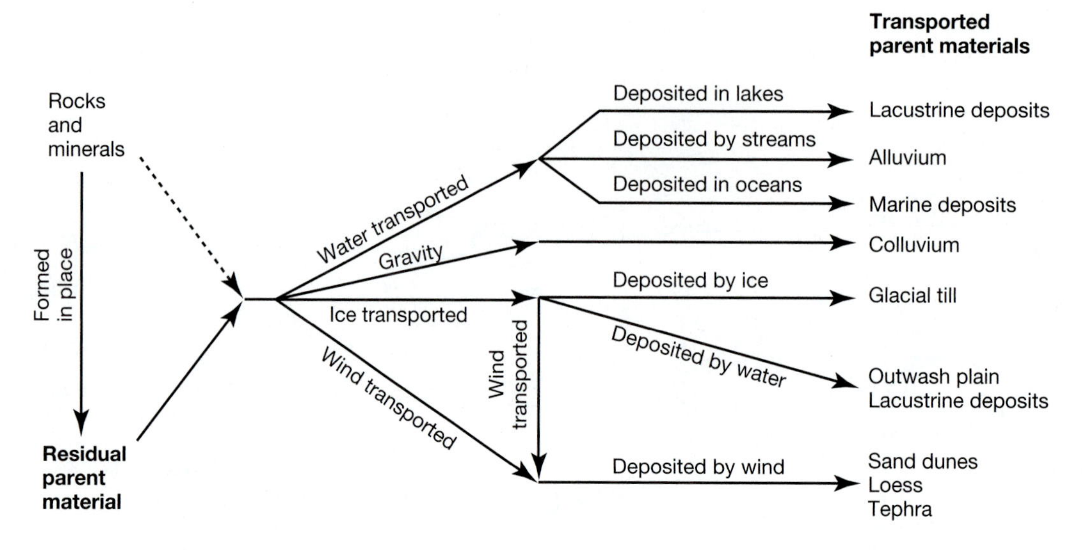

FIGURE 6.12 *How different parent materials are formed, transported, and deposited.*

important soil organisms are the microorganisms, especially the fungi and bacteria. They decompose organic matter and beneficially influence soil structure, aeration, and fertility. Microorganisms, like **lichen** (a symbiotic association between algae and fungi), secrete very dilute carbonic acid (H_2CO_3) that slowly dissolves rock, thus adding inorganic material to the developing soil. Upon their death, the lichens decompose, releasing nutrients that enrich the soil.

Rocks may be splintered by the roots of trees and other plants. Rooted vegetation absorbs mineral nutrients from lower levels. These nutrients become incorporated into plant matter, such as the leaves. When leaves, fruits, and nuts fall,

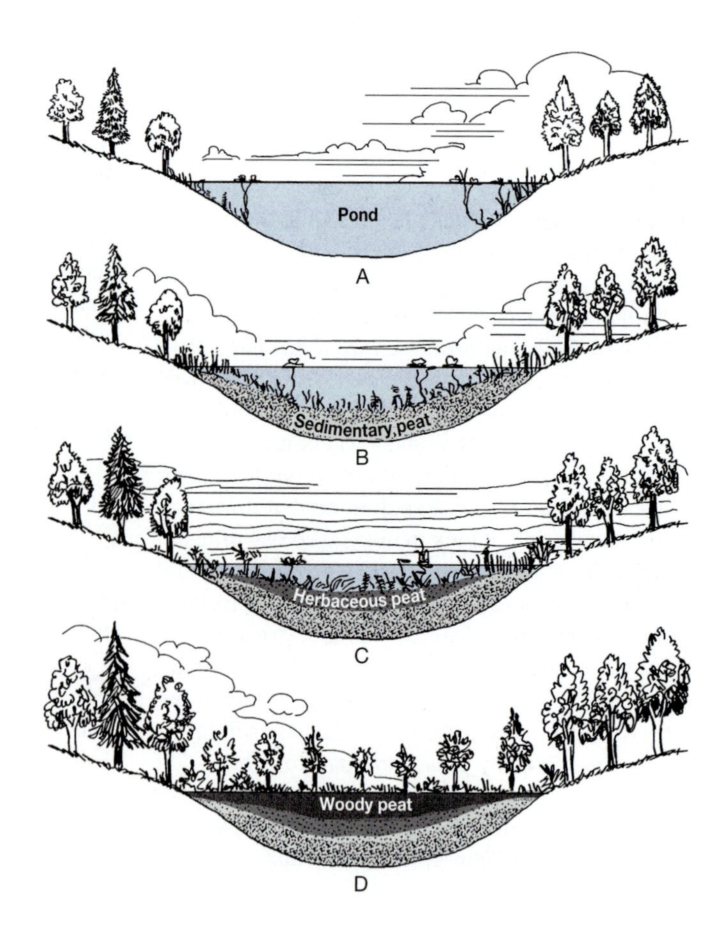

FIGURE 6.13 *Stages of development of a bog (an organic deposit): (a) Aquatic plants slowly grow around the edges of the pond; (b) water-loving plants continue to invade more of the pond; (c) with time different layers of organic debris fill the bottom of the pond; (d) trees and shrubs eventually cover the area forming a woody peat bog.*

and when the plant dies, the nutrients are released, thus recycling the nutrients and contributing to soil formation. Earthworms, beetle larvae, bull snakes, pocket gophers, moles, and ground squirrels burrow through the soil, aiding in the movement of air and water through the ground. Different vegetation types, especially the type of organic material they leave on or in the soil, has a marked effect of soil formation. For example, forest soils tend to develop acid, leached soils of moderate to low fertility, whereas grassland soils tend to produce thick topsoils rich in organic matter and fertility (Figure 6.14).

Topography

Topography, which refers to the shape or contour of the land surface, largely determines how water moves in a landscape and how susceptible the soils are to water erosion. For example, thin soils usually develop on steeply sloping surfaces because the newly formed soil is washed down the slope by runoff waters almost as quickly as it develops. Conversely, on the relatively flat valley floor, much thicker soils develop because the valley receives soil particles and organic material that erode from the hill above. However, many flat valley floors can have water-logged soils, in which water moves too slowly through the soil for proper drainage.

Time

When the Earth first formed, about 4.5 billion years ago, it was completely devoid of soil. With the passage of **time,** however, physical, chemical, and—when organisms appeared on Earth—biological processes resulted in the formation of the first thin envelope of soil on this planet. The length of time required for a soil to form depends on the combined effect of climate and organisms, as modified by topography, acting on parent material. A given period of time may produce much change in one soil and little in another (due to the other four soil-forming factors). For soil development from consolidated hard rock, the time may be very great (as in the previously mentioned Zimbabwe example, where less than one-half inch of soil derived from hard granite requires 1,000 years to develop). On the other hand, fresh alluvium may develop into crop-supporting soils in just a few decades.

6.4 The Soil Profile

When one looks at the exposed face of a road cut or the side of a stone quarry, it is apparent that many soils are organized into layers or **horizons.** Each of these horizons is characterized by a specific thickness, color, texture, structure, and chemical composition. A cross-sectional view of the various horizons is known as the **soil profile** (Figure 6.15). When people travel, they can make their trip more interesting if they pay attention to the changing soil profiles in the road cuts along the way.

The major layers from the ground surface downward to bedrock are called Master horizons and are designated as **horizons O** (organic layer), **A** (topsoil), **E** (subsurface), **B** (subsoil), **C** (parent material), and **R** (bedrock). These horizons may not exist in all soil types. For instance, in immature soils, where weathering has not fully progressed, some horizons will be missing.

The soil profile is the product of the actions of vegetation, temperature, rainfall, and soil organisms on parent material located on a specific land surface operating for many thousands of years. The soil profile, therefore, tells us a great deal about soil history

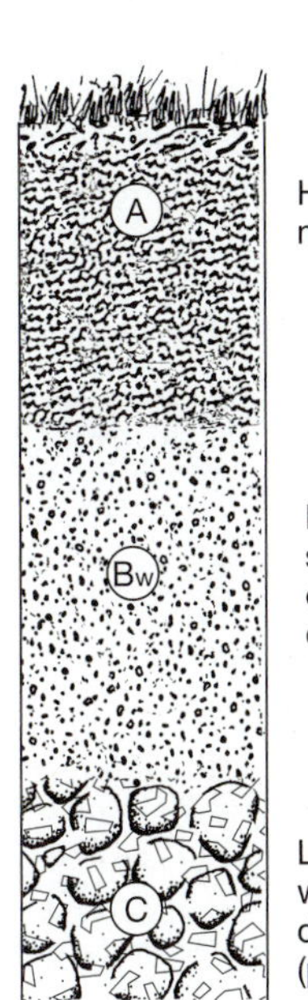

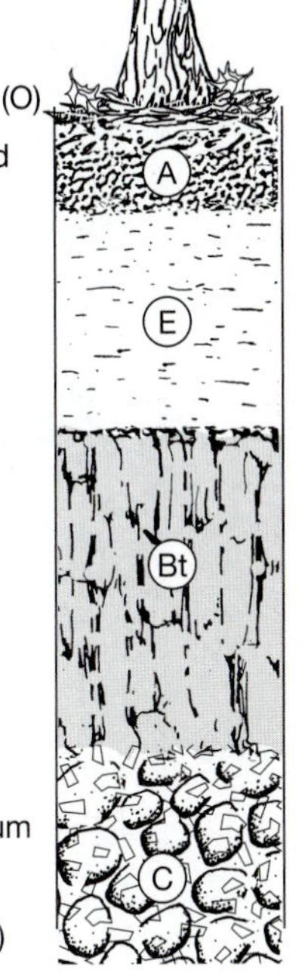

FIGURE 6.14 *Two soil profiles, the one on the left showing a soil from a subhumid grassland and the one on the right illustrating a soil from a humid forested region.*

or genesis. It represents a kind of soil autobiography. From a practical standpoint, the soil profile is of great importance, for it can tell the soil scientist immediately whether the soil is suited for agricultural crops, for rangeland, for timber, or for wildlife habitat and recreation. The profile also reveals the suitability of the soil for various urban uses, such as home sites, highways, sewage disposal plants, sanitary landfills, and septic tank fields.

We shall examine the basic characteristics of a soil profile, beginning with the uppermost horizon and moving downward to bedrock (Figure 6.15).

O Horizon

The **O horizon** is an organic horizon that forms above the mineral soil. It results from organic litter that came from dead plants and animals. O horizons are common in forest areas but generally absent from grassland areas.

A Horizon

Human survival depends on the thin layer of topsoil, or **A horizon,** that covers much of the Earth. In the United States, its thickness ranges from 2.5 centimeters (1 inch) on the slopes of the Rockies to more than 1 meter (more than 3 feet) in the Palouse region of Washington state. This layer is rich in humus. It is from the topsoil that crop roots absorb much of their water and nutrients. It is within this layer that most soil organisms live.

E Horizon

The **E horizon** is called the zone of leaching because here much material is dissolved and carried downward to the B horizon by water. The E horizon comes from the word *eluvial,* meaning washed out.

B Horizon

The **B horizon,** commonly called the subsoil, is a zone of accumulation that receives and stores clays, soluble salts, and humus that are leached downward from the A and E horizons. After many years of abusive farming, the topsoil may be completely eroded away. The B horizon, which now lies at the surface, usually reduces crop yields because of its poorer physical, chemical, and biological properties.

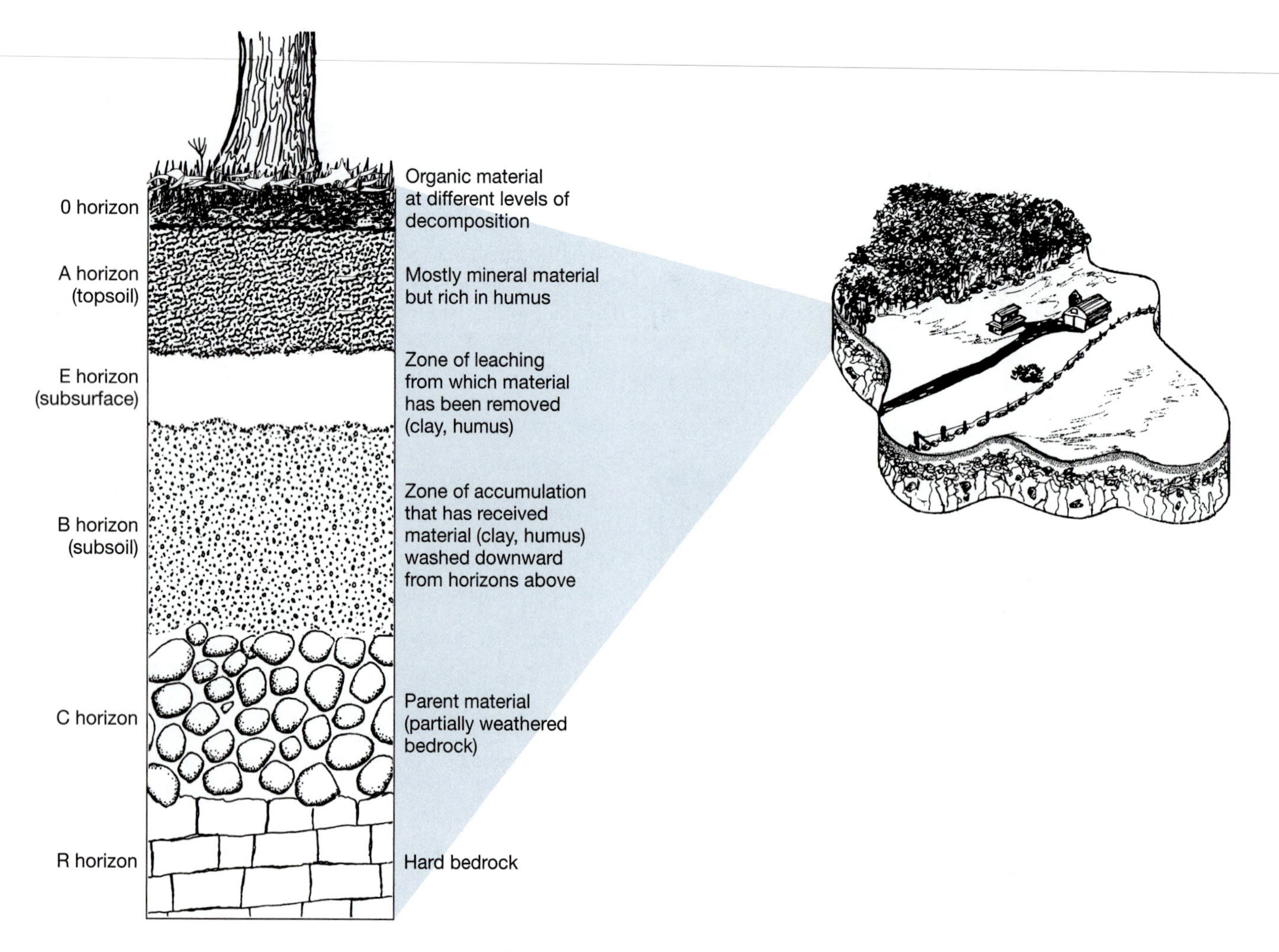

FIGURE 6.15 *A soil profile showing the major or Master horizons.*

C Horizon

The **C horizon** is composed of unconsolidated parent material. Usually this material has been transported to the site by glaciers, wind, or water. However, the parent material in this horizon can also be derived from underlying bedrock. The parent material in the C horizon determines many soil characteristics—its texture, water-holding capacity, nutrient levels, and pH. For example, soils formed from parent material derived from granite tend to mature slowly and to be acidic. The quartz mineral grains in granite are resistant to weathering and become sand particles in the soil. The less resistant feldspars and ferromagnesium minerals in granite can form clay particles. Thus, depending on the degree of weathering, soils formed from granite can have a wide range of textures. On the other hand, soils formed from parent material derived from limestone develop rapidly and tend to be alkaline. Soils that develop from limestone tend to be more productive than those derived from granite.

R Horizon

The **R horizon** consists of bedrock with little evidence of weathering. As bedrock weathers, it contributes parent material to the C horizon above. The R horizon is absent, of course, in loess and alluvial soils, where the parent material was transported by wind or water instead of being derived from bedrock.

6.5 Soil Classification

There are many thousands of kinds of soils in the United States and around the world. The study and understanding of their importance in supporting such renewable resources as crops, range grasses, forests, and wildlife would be severely hampered without some system of classification. Soil classification makes it possible to use each kind of soil safely and productively.

Soils are classified in order to remember their names and important properties and because it is inefficient to have to consider every soil to find information about one particular kind. Classifying soils into groups with similar properties minimizes the problem of locating information about any one soil. Classifying soils also supports soil surveys (mapping). Soil classification tells us which soils are the same, and soil surveys tell us where they are located. Soil classification and surveys permit the transfer of soils information from one place to another and from the present to future generations.

The soil classification system used in the United States (and some other countries) is **soil taxonomy,** which was developed by the U. S. Department of Agriculture. Other countries and international organizations have also developed soil classification systems. However, soil taxonomy is designed to include all soils of the world.

Diagnostic Horizons

Soil taxonomy recognizes twelve major soil groups in the world, known as **orders.** Most soil orders are defined on the basis of having **diagnostic horizons,** that result from distinctive soil-forming processes and have specific physical and chemical properties. Diagnostic horizons can be divided into surface horizons (epipedons) and subsurface and subsoil horizons. The most common diagnostic horizons are shown in Table 6.3.

Specific limits have been placed on the properties of these horizons to help the observer define and recognize them. Some of the properties, such as color and structure, can be determined by observation, while others, such as organic matter content, require laboratory analysis.

The Soil Orders

Each of the twelve major soil orders ends in *sol,* which is derived from the Latin word *solum,* meaning soil. Seven of the twelve soil orders occur in broad zones that depend largely on the two soil-forming factors, climate and organisms (mainly vegetation) (Figure 6.16). For example, **Aridisols** form in hot, dry climates under desert vegetation like cactus, mesquite, and sagebrush, whereas **Oxisols** form in warm, wet climates under tropical rain forests. Two more orders, **Entisols** (young soils) and **Inceptisols** (weakly developed soils), can be found under

TABLE 6.3 *Common Diagnostic Horizons. Lower case letters after Master horizon symbols designate subordinate distinctions within the master horizons; for example, a Bt horizon is a layer with an accumulation of silicate clays.*

Diagnostic Horizon	*Main Characteristics*
Mollic epipedon (A horizon)	Thick, well-structured, dark-colored surface horizon with base saturation >50%
Ochric epipedon (A horizon)	Light-colored surface horizon, or thin, dark-colored surface horizon
Argillic horizon (Bt horizon)	Accumulation of silicate clays
Cambic horizon (Bw horizon)	Weakly developed B horizon
Spodic horizon (Bhs horizon)	Accumulation of humus and Al and Fe oxides
Oxic horizon (Bo horizon)	Accumulation of Fe and Al oxides and kaolinite

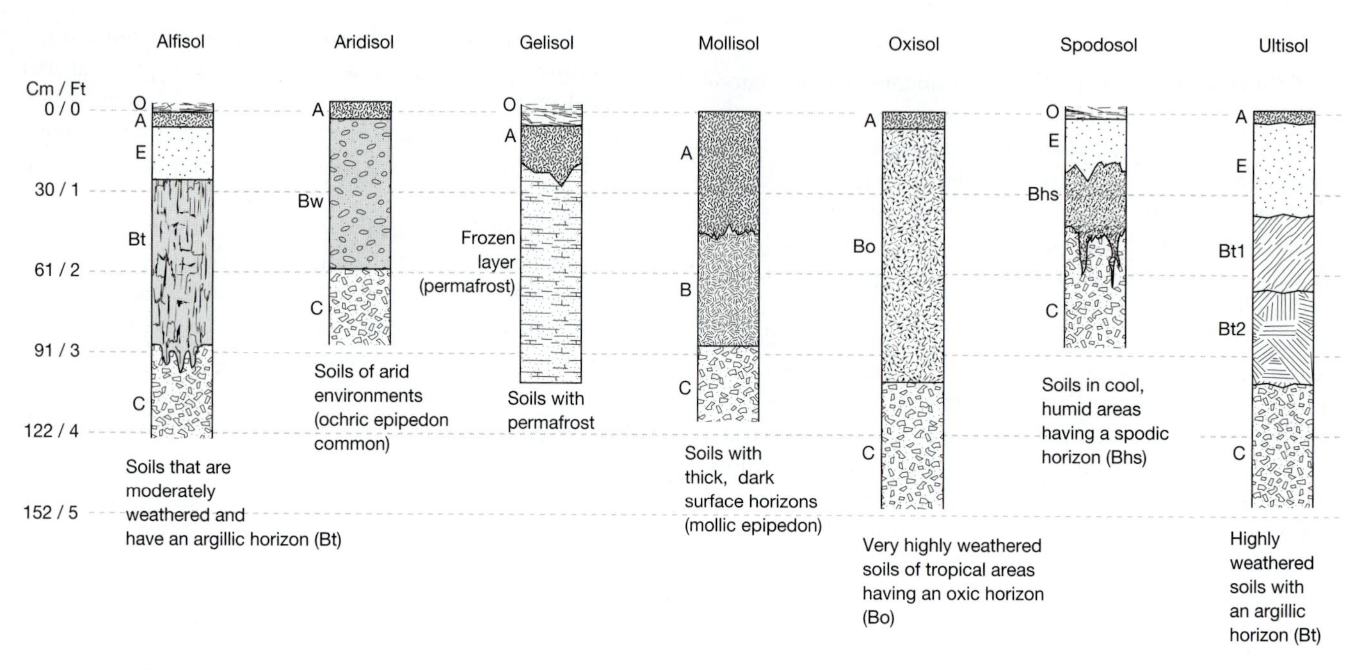

FIGURE 6.16 *Typical soil profiles of the seven soil orders that depend largely on the two soil-forming factors, climate and organisms (mainly vegetation).*

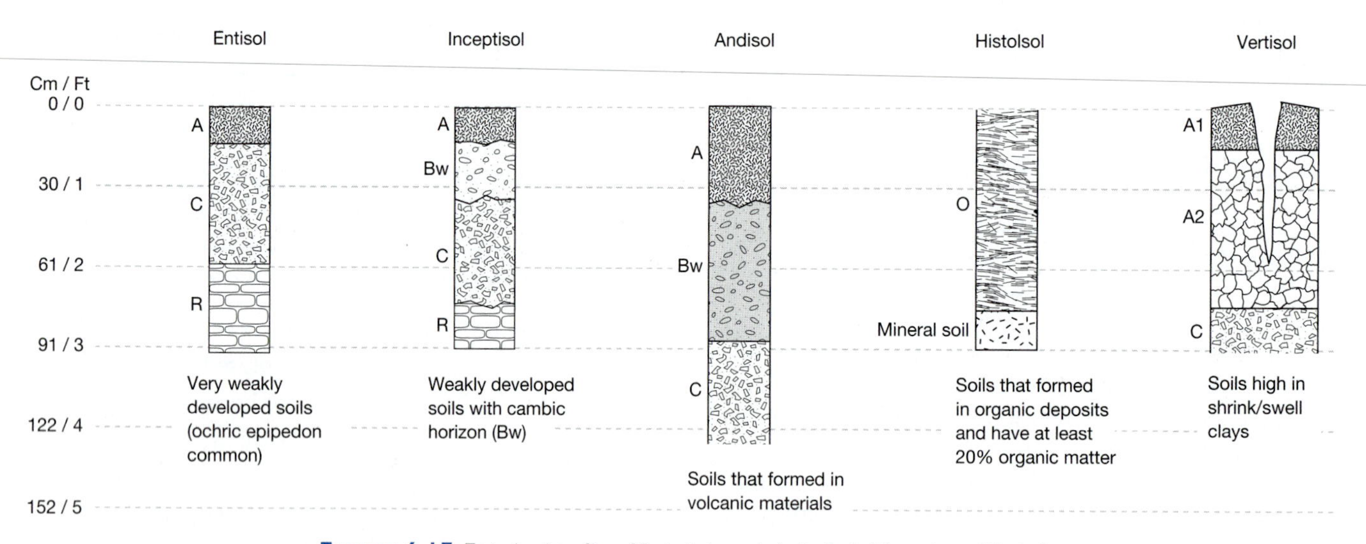

FIGURE 6.17 *Typical soil profiles of Entisols, Inceptisols, Andisols. Histosols, and Vertisols.*

any moisture and temperature conditions (Figure 6.17). The remaining three orders of **Andisols, Histosols,** and **Vertisols** are more related to distinctive parent materials (Figure 6.17); for example, **Histosols** form in organic materials that built up in stagnant waters.

Mollisols include some of the world's most productive soils. The United States is more than 22 percent Mollisols, whereas the world's land area is almost 7 percent Mollisols (Table 6.4). It would be fair to say that the United States is blessed with good agricultural soils. In comparison to the world average, the United States has less than the average share of Aridisols and Entisols but about the normal share of Inceptisols and Histosols. The United States has almost no Oxisols (tropical soils), compared to 7.5 percent in the world's land area (Table 6.4). Oxisols in the United States can be found in Hawaii.

The distribution of the major soil orders in the United States is shown in Figure 6.18. The extensive soil orders in the eastern part of the United States are Spodosols, Inceptisols, Alfisols, and Ultisols. **Spodosols** are found mainly on the sandy deposits in the northern Great Lakes states, New England, and Florida. They formed in a cool, relatively humid climate under forest vegetation, mostly conifers. They are characterized by a subsoil **spodic horizon** in which iron, aluminum, and humus have accumulated.

Inceptisols are weakly developed soils that occur mainly on sloping mountainsides extending from southern New York through central and western Pennsylvania, West Virginia,

TABLE 6.4 *Percent Land Area and Major Uses of Soil Orders. Miscellaneous land is made up mostly of rocky lands and shifting sands.*

Order	Percent Land Area of the United States	Percent Land Area of the World	Major Land Use	Fertility
Alfisols	14.5	9.6	Cropland, forest	High
Andisols	1.7	0.7	Cropland, forest	Moderate
Aridisols	8.8	12.1	Rangeland	Low
Entisols	12.2	16.3	Rangeland, forest, cropland	Moderate to low
Gelisols	7.5	8.6	Bogs, tundra	Moderate
Histosols	1.3	1.2	Wetland, cropland	Moderate
Inceptisols	9.1	9.9	Cropland, forest	Moderate to low
Mollisols	22.4	6.9	Cropland, rangeland	High
Oxisols	<0.01	7.5	Cropland, forest	Low
Spodosols	3.3	2.6	Forest	Low
Ultisols	9.6	8.5	Forest, cropland	Low
Vertisols	1.7	2.4	Cropland, forest	High
Miscellaneous land (Rocky land and shifting sands)	7.8	14.0	Wildlife, recreation	
Total	100%	100%		

and eastern Ohio (Figure 6.18). They also can be found on the floodplains of the larger rivers, especially the Mississippi. Inceptisols have more significant profile development than Entisols but less development than the other mineral soil orders.

Alfisols and **Ultisols** formed in humid climates and are characterized by a subsoil **argillic horizon** in which silicate clays have accumulated from the layer(s) above. **Alfisols** formed mostly under deciduous forests, although some formed under grasslands. A large area of Alfisols can be found in the Midwest in Ohio, Indiana, Wisconsin, Minnesota, and Michigan. They also occur in a narrow belt east of the Mississippi River, where they formed in loess. Alfisols seem to be more strongly weathered than Inceptisols but less weathered than Spodosols. **Ultisols** typically formed in forested subtropical and tropical areas. They are more leached, weathered, and acidic and thus less fertile than Alfisols. Most of the soils in the southeastern part of the United States fit in the Ultisol order.

The extensive soil orders of the western United States are Mollisols, Aridisols, Entisols, Alfisols, and Gelisols. Most of the **Mollisols** developed under grass vegetation, mainly in the Great Plains states in the central part of the United States and in the Northwest in Oregon, Washington, and Idaho. They also extend eastward through Iowa, Illinois, and eastern Indiana. Mollisols are characterized primarily by the presence of a **mollic epipedon,** a thick, humus- and nutrient-rich surface horizon (Figure 6.19). The mollic epipedon is intrinsically more fertile than any other surface soil in the United States. In part this is because of the dense mesh of roots extending down through the A horizon. When a grass plant dies, its root system decomposes in place and releases nutrients that are available to future plant generations.

Aridisols are dry soils found in desert areas where the natural vegetation is desert shrubs and short grasses. They commonly have an **ochric epipedon** that is generally light in color and low in organic matter. Large areas of Aridisols can be found in California, Nevada, Arizona, and New Mexico. Aridisols can be made highly productive for growing cultivated crops when irrigation water and fertilizers are made available. For example, the Aridisols of California's Imperial Valley produce a great variety of high-value crops ranging from citrus to celery, from walnuts to dates. However, they generally require large amounts of water and must be carefully managed to prevent the buildup of soluble salts.

Entisols are young soils with very little soil profile development. They can be found under various environmental conditions. They formed where the soil is very steep, as in the Rocky Mountains, very sandy, as in the Sand Hills of Nebraska, or very young, as in the California trough. In the western United States, Alfisols occur mostly in forested areas in California.

Gelisols, like Entisols, are young soils with little profile development. Unlike Entisols, however, the principal defining feature of Gelisols is the presence of a **permafrost** layer, a

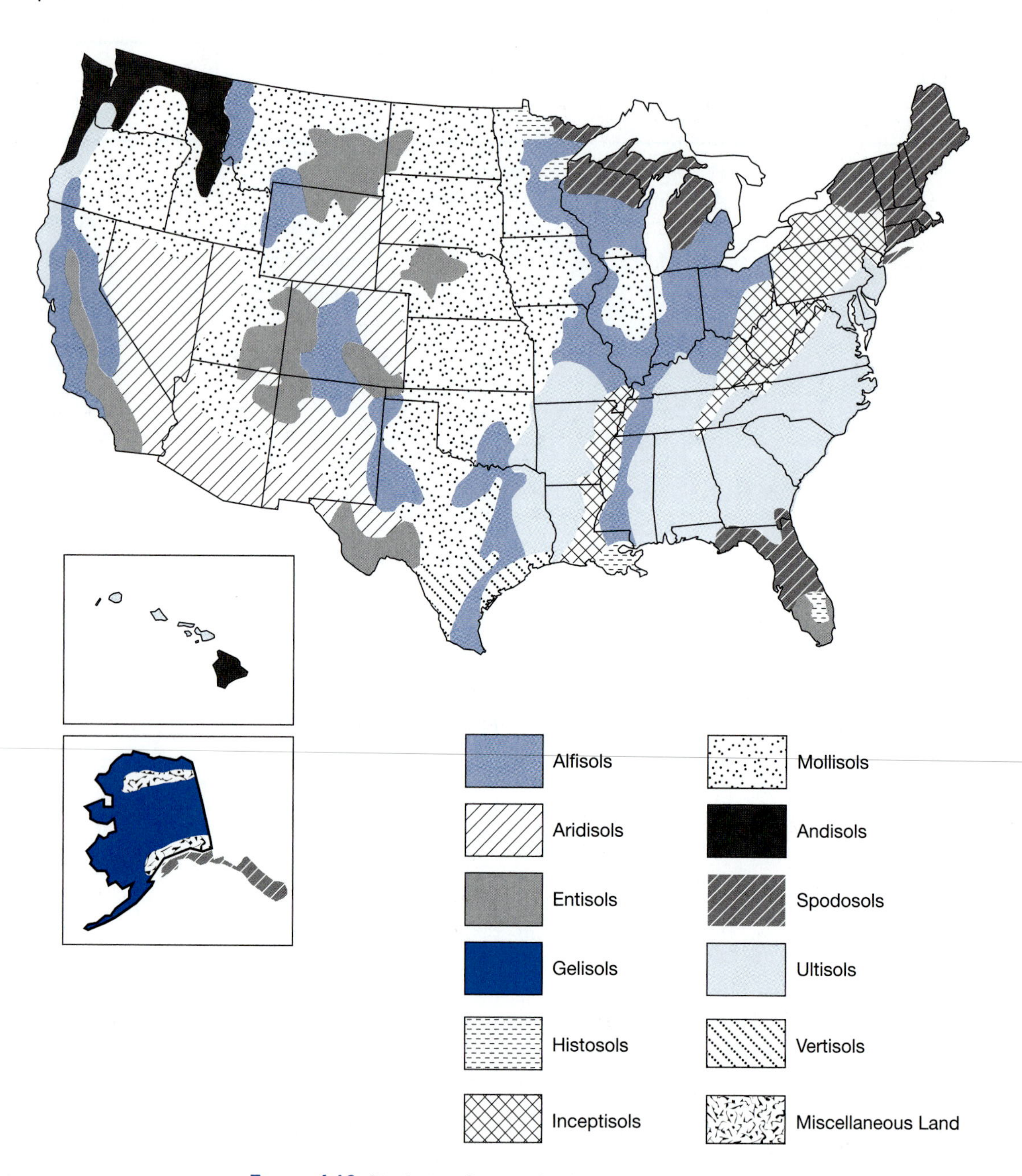

FIGURE 6.18 *Distribution of major soil orders in the United States.*

layer of material that remains at temperatures below 0°C (32°F) for more than two consecutive years. Gelisols cover most of Alaska, mostly supporting tundra vegetation of lichens, grasses, and low shrubs.

Histosols, Vertisols, and Andisols are less extensive in area because they formed in unique parent materials. **Histosols** are organic soils with at least 20 percent organic matter. They form in organic deposits that accumulate in shallow lakes, marshes, bogs, and swamps. They can be found in the northern Great Lakes states, Florida, and the Mississippi Delta. **Vertisols** are found where clayey materials are in abundance and are characterized by a high content (more than 30 percent) of shrink/swell clays. They are most extensive in Texas. **Andisols** formed on volcanic deposits in the Pacific Northwest. Recent eruptions, like that of Mount St. Helens in 1980, are giving rise to the formation of new Andisols.

Summary of Key Concepts

1. Soil is one of the most important components of terrestrial communities. High-quality soils not only promote the growth of plants, they also prevent water and air pollution by resisting erosion and by degrading and immobilizing agricultural chemicals, organic wastes, and other potential pollutants.
2. Soil is a natural system consisting of four components: mineral matter, organic matter, water, and air.

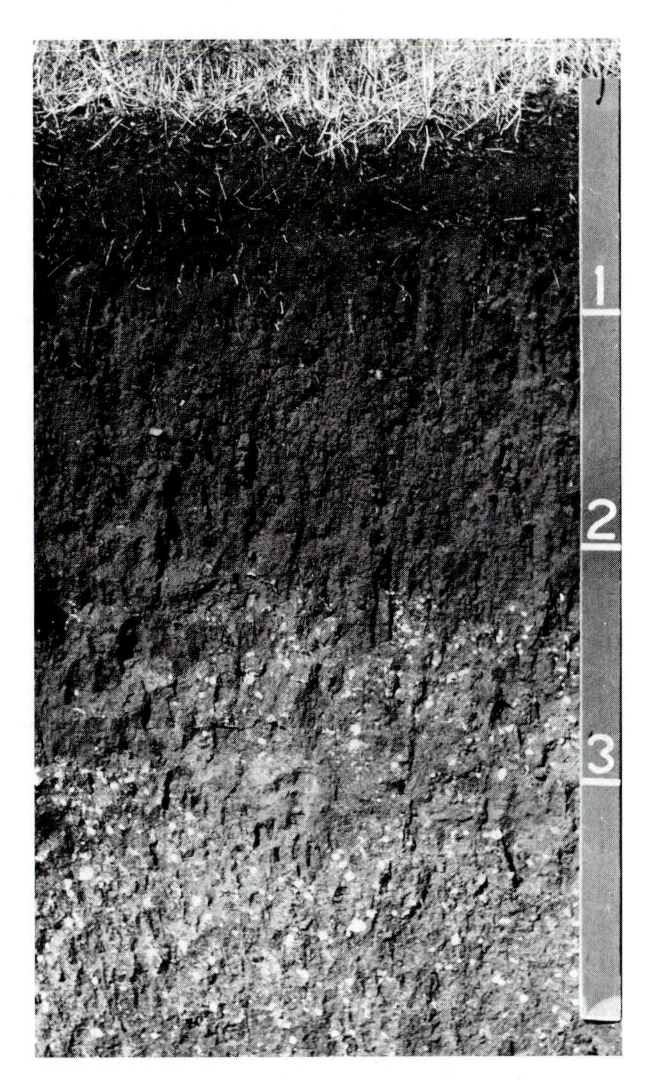

FIGURE 6.19 *Profile of Mollisol soil formed from glacial till.*

3. Soil texture is determined by the relative proportions of sand, silt, and clay in a particular soil. It helps determine soil water permeability and storage, the ease of tilling the soil, the amount of aeration, soil fertility, and root penetration.
4. A loam soil exhibits equal properties, not percentages, of sand, silt, and clay.
5. Clay particles have negatively charged surfaces, attracting positively charged ions (cations) of nutrient elements such as calcium, potassium, magnesium, zinc, and iron.
6. Soil structure is the arrangement or grouping of soils particles into clusters or aggregates.
7. Soil organic matter is the living or dead plant and animal materials in the soil.
8. Humus is the semistable, dark-colored material that represents the decomposition products of organic residues and materials synthesized by microorganisms. Humus improves soil structure, buffers the soil against a rapid change in pH, reduces soil erodibility, and increases the nutrient storage ability and water-holding capacity of the soil.
9. Soil microorganisms (such as fungi and bacteria) are instrumental in decomposing organic matter and releasing available plant nutrient elements.
10. The pore space of a soil is that portion occupied by both air and water.
11. Macropores (larger than 0.06 millimeter in diameter) are readily drained of water and filled with air after a heavy rain, whereas micropores (less than 0.06 millimeter) retain water within the fine pores and impede air movement.
12. Soil scientists use the pH scale to determine whether a soil is acidic, neutral, or alkaline. In general, soils vary in pH from 4.5 to a little above 7 in humid regions, and from a little below 7 to 9 in arid regions.
13. Soil fertility is the status of a soil with respect to its ability to supply nutrients in amounts and forms required for maximum plant growth.
14. An essential element is a chemical element required for the normal growth of plants. Most plants are known to need at least 16 essential nutrient elements to grow.
15. Soil formation depends on five soil-forming factors, which are responsible for the kind, rate, and extent of soil development. The five soil-forming factors are (1) climate, (2) parent material, (3) organisms, (4) topography, and (5) time.
16. When precipitation and the other four factors remain constant, an increase in temperature causes an increased rate of weathering and clay formation. The rate of weathering increases by an increase in the presence of water (up to a point, of course).
17. Parent material is the unconsolidated (soft and loose), weathered mineral or organic material from which soils form. Parent materials can be classified into three major groups: residual, transported, and organic.
18. Residual parent materials formed from the weathering of rock in place.
19. Transported parent materials were transported from their place of origin and redeposited in a new location. The main transporting agents are ice (glacial), water (alluvial), and wind (aeolian).
20. Most organic deposits accumulate in stagnant water (lakes or swamps) where decomposition of plant residues is retarded by the limited supply of oxygen.
21. The development of a mature soil depends on the activity of a great number and diversity of organisms.
22. Topography, which refers to the shape or contour of the land surface, largely determines how water moves in a landscape and how susceptible the soils are to water erosion.
23. The length of time required for a soil to form depends on the combined effect of climate and organisms, as modified by topography, acting on parent material. The major horizons from the ground surface downward to bedrock are designated as horizons O (organic layer), A (topsoil), E (subsurface), B (subsoil), C (parent material), and R (bedrock).
24. A soil profile is a vertical section of the soil extending through all its horizons and into the parent material.

25. The soil classification system used in the United States (and some other countries) is soil taxonomy, which was developed by the U. S. Department of Agriculture.
26. Soil taxonomy recognizes 12 major soil groups in the world, known as orders. Most soil orders are defined on the basis of diagnostic horizons that result from distinctive soil-forming processes and have specific physical and chemical properties.
27. Spodosols form in a cool, relatively humid climate under forest vegetation, mostly conifers. They are characterized by a subsoil spodic horizon in which humus and aluminum and iron oxides have accumulated.
28. Inceptisols are weakly developed soils, but have more significant profile development than Entisols, but less development than the other mineral soil orders.
29. Alfisols and Ultisols form in humid climates and are characterized by a subsoil argillic horizon in which silicate clays have accumulated from the layer(s) above. Ultisols are more leached, weathered, and acidic and thus less fertile than Alfisols.
30. Mollisols include some of the world's most productive soils. They are characterized primarily by the presence of a mollic epipedon, a thick, humus- and nutrient-rich surface horizon.
31. Aridisols are dry soils found in desert areas where the natural vegetation is desert shrubs and short grasses, whereas Oxisols form in warm, wet climates under tropical rain forests.
32. Gelisols are young soils with little profile development because they form in cold temperatures and frozen conditions for much of the year.
33. Histosols, Vertisols, and Andisols form in unique parent materials. Histosols are organic soils with at least 20 percent organic matter. Vertisols are soils characterized by a high content (more than 30 percent) of shrink/swell clays. Andisols form in volcanic deposits.

Key Words and Phrases

A Horizon
Acidic
Adsorption
Aeration
Alfisols
Alkaline
Andisols
Argillic Horizon
Aridisols
B Horizon
C Horizon
Clay
Climate
Coarse Fraction
Diagnostic Horizons
E Horizon
Entisols
Essential Element
Field Capacity
Fine-Earth Fraction
Gelisols
Histosols
Horizons
Humus
Inceptisols
Loam
Macronutrients
Macroorganisms
Macropores
Micronutrients
Microorganisms
Micropores
Mollic Epipedon
Mollisols
O Horizon
Ochric Epipedon
Organic Deposits
Organic Matter
Organisms
Oxisols
Parent Material
Permafrost
Permanent Wilting Point
Permeability
pH
Pore Space
R Horizon
Residual Parent Materials
Sand
Saturation
Silt
Soil Classification
Soil Fertility
Soil Formation
Soil Orders
Soil Profile
Soil Structure
Soil Taxonomy
Soil Texture
Soil-Forming Factors
Spodic Horizon
Spodosols
Time
Topography
Transported Parent Materials
Ultisols
Vertisols
Weathering

Critical Thinking and Discussion Questions

1. Is there any correlation between soil fertility and the power and influence of nations like the United States? Discuss your answer.
2. Discuss five major characteristics or properties of soils.
3. A farmer has two level fields, each 2 hectares in size. One field is a clay loam soil and the other field is a sandy loam. If she grows the same crop on both pieces of land, should she fertilize and irrigate the same amount on both fields or should she manage them differently? Discuss your answer.
4. What characteristics of clay are undesirable for crop production? What characteristics are desirable?
5. Is there a difference between humus and organic matter?
6. What are three beneficial effects of microorganisms on soil?
7. Why is water permeability faster in a sandy-textured soil than in a clayey-textured soil?
8. Give three reasons why a plant cannot survive without water.
9. What does pH mean? What pH is desirable for most trees and food crops?
10. Suppose that a farmer has soil with a pH of 5.0 but wishes to raise a crop that requires a pH of 6.0. What can he do?
11. What is an essential element? Is a macronutrient any more essential for plant growth than a micronutrient?
12. Describe the five soil-forming factors involved in the development of soils. What is the importance of each?
13. How could a climatic change eventually cause a change in the type of soil occurring in a given region?
14. List three forces that play an important role in the transport of parent materials for soil development.
15. Discuss the statement, "A soil profile represents a kind of autobiography." Is this statement valid? Why or why not?
16. What is the difference between an E horizon and a B horizon?

17. What is the basis for naming a soil order? Give an example.
18. Compare the general features of a Spodosol and a Mollisol. What are the major differences?
19. Compare Aridisols to Ultisols with respect to their distribution in the United States, crop-producing potential, appearance, and fertility.

Suggested Readings

Brady, N. C., and R. R. Weil. 1999. *The Nature and Property of Soils,* 12th ed. Upper Saddle River, N.J.: Prentice Hall. This classic work on soils provides a comprehensive and in-depth treatment.

McLaren, R. G., and K. C. Cameron. 1996. *Soil Science: Sustainable Production and Environmental Protection,* 2nd ed. Auckland, New Zealand: Oxford University Press. Excellent introductory soils text with emphasis on sustainability and environmental issues in New Zealand.

Miller, R. W., and D. T. Gardiner. 1998. *Soils in Our Environment,* 8th ed. Englewood Cliffs, N.J.: Prentice Hall. Very good comprehensive, introductory soils text.

Plaster, E. J. 1997. *Soil Science & Management,* 3rd ed. Albany, New York: Delmar Publishers. Easily readable and practical soils text with emphasis on management.

Rowell, D. L. 1994. *Soil Science: Methods and Application.* New York: Hallsted Press. Very readable, comprehensive text.

Singer, M. J., and D. N. Munns. 1999. *Soils: An Introduction,* 4th ed. Upper Saddle River, N.J.: Prentice Hall. Excellent, simplified beginning soils text.

Soil Survey Staff. 1998. *Keys to Soil Taxonomy.* Washington, D.C.: USDA Natural Resource Conservation Service. Describes the latest U.S. soil classification system.

Suggested Readings

Online resources for this chapter are on the World Wide Web at: **http://www.prenhall.com/chiras** *(click on the Table of Contents link and then select Chapter 6).*

Chapter 7

Soil Conservation and Sustainable Agriculture

The main reason for conserving soil is to maintain it as a permanent, useful resource for future generations. The rate of soil removal should be limited to an amount that will not compromise its future productivity. We are all dependent on conserving soil because we rely on products, like food, clothing, and shelter, that come from the soil. The demand for these products is rapidly growing because of an increasing population and a higher standard of living.

The safe way to protect soil from erosion is not to expose it directly to rain or wind. This becomes difficult and often not practicable when growing crops. A number of methods have been developed to protect soil against erosion. Proper soil conservation is only one indicator of a farm or agricultural system being sustainable. Other criteria include farming systems that produce adequate yields of high quality and, at the same time, are economically viable, environmentally safe, resource conserving, and socially responsible.

7.1 The Nature of Soil Erosion

During the three-century history of soil deterioration in the United States, erosion has played a dominant role. The word *erosion* is derived from the Latin word *erodere,* meaning "to gnaw out." **Soil erosion** is, thus, the process by which soil particles are detached from their original site, transported, and eventually deposited at a new location (Figure 7.1). The principal agents of erosion are wind and water. In humid areas, water is the main cause of soil erosion, whereas in semiarid areas, wind and water are responsible for most soil erosion.

Geological, or Natural, Erosion

Geological, or **natural, erosion** is a process that has occurred at an extremely slow rate ever since the Earth was formed 4.5 billion years ago. It wears down high places and fills up low places. It has worn down mighty mountain systems and filled sea basins

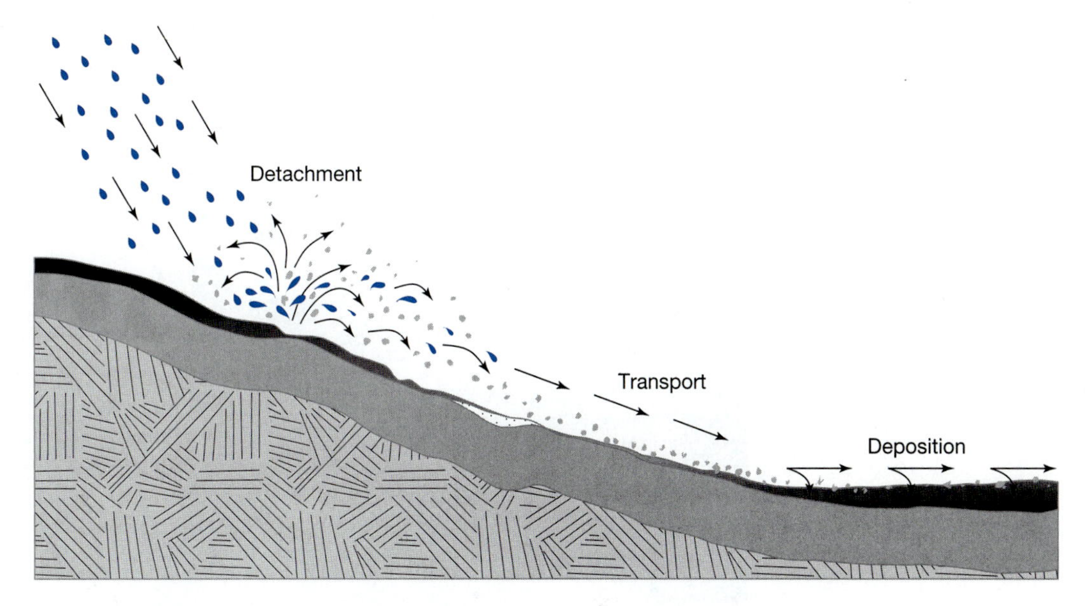

FIGURE 7.1 *The three-step process of soil erosion by water. Raindrop impact destroys soil structure, detaching soil particles. The detached soil particles are then transported and eventually deposited downhill.*

with sediment. Scientists estimate that water erodes a rocky surface at the rate of one-fourth to one-fiftieth of a millimeter per year—the rate depending, of course, on the force of the water and the nature of the rock.

The mountains, valleys, canyons, coastlines, and deltas on the Earth's surface have been sculpted by water and wind erosion working over the eons. The Appalachian Mountains were once as tall and rugged as the Rocky Mountains, but since their formation 200 million years ago, they have been gradually worn down by erosive forces. Were it not for geological erosion, New Orleans would be resting on the bottom of the Gulf of Mexico, for the delta on which it is built was formed by a deposit of soil transported by the Mississippi River from sites as far as 1,600 kilometers (1,000 miles) away. The Grand Canyon originated as a shallow channel 100 million years ago. It was ultimately scoured to its awesome 1-mile depth by rain and the churning waters of the Colorado River (Figure 7.2). As a final example of geological erosion, wind-blown silt and fine sand, known as loess, have been deposited in thick sheets on North America, China, and Europe, forming some of the Earth's most fertile soils.

Accelerated Erosion

Geological erosion, then, has continued to operate at a slow, deliberate pace for millions of years. However, with the appearance of humans, a different type of erosion, known as **accelerated erosion,** began and is often ten to 100 times as destructive as geologic erosion. Accelerated erosion occurs when people disturb the soil or natural vegetation by plowing hillsides, cutting forests, grazing livestock, or constructing buildings or roads. An example of gully erosion, one type of accelerated erosion, is shown in Figure 7.3.

Worldwide, most soil removal by wind and water erosion comes from agricultural land. It is with this accelerated erosion that the conservationist is primarily concerned, and for good reason. The removal of soil degrades arable land and can eventually render it unproductive.

Soil erosion rates are highest in Asia, Africa, and South America and lowest in Europe and the United States. However, the relatively lower rates in Europe and the United States exceed the average rate of soil formation or replacement. Some of the most destructive erosion the United States has ever experienced was caused by heavy winds and occurred on the Great Plains during the 1930s. The part of the Great Plains most affected by this cataclysm was in the Southern Plains, which are large areas of Texas, Oklahoma, Kansas, Colorado, and New Mexico. Because of the frequency and severity of the dust storms in the Southern Plains, this region was called the **Dust Bowl.**

7.2 The Dust Bowl

Throughout history, the Great Plains have experienced alternating periods of drought and adequate rainfall. Drought visited the Great Plains in 1890 and again in 1910. During each dry spell, crops withered and died. Farms and ranches were abandoned, only to be reoccupied during the ensuing years of adequate rainfall.

In 1931 there was no better place to be a farmer than in the Great Plains. Farmers grew wheat and turned the Plains of rich prairie grass into one of the most productive areas in the world. Then, drought and high summer temperatures began to set in the northern Great Plains. The drought spread to the Southern Plains, where the most spectacular storms occurred toward the middle of the decade.

Droughts had visited the plains before, so had windstorms. But never before in the history of the North American prairie had the land been more vulnerable to their combined assault. Gone were the profusely branching root systems of the buffalo grass, the grama grass, the big

FIGURE 7.2 *Geological erosion caused by water: the Grand Canyon of the Colorado River.*

FIGURE 7.3 *Accelerated erosion caused by humans resulted in these huge water-carved gullies on a North Carolina farm.*

bluestem, and the little bluestem, which had originally kept the rich brown soil in place. Gone were the high concentrations of decomposing organic material that aided in building stable soil aggregates and the soil cover of grass mat and sagebrush. On the ranches, soil structure deteriorated under the pounding of millions of cattle. On the wheat and cotton farms, soil structure broke down under the abuse inflicted by heavy machinery.

The Dust Bowl was the result of tillage and not enough rain. Many authorities believe that the Dust Bowl was the nearly inevitable clash of the drought with the agricultural practices that until then had been successful in the wetter east. The drought prevented the wheat from growing and there was no ground cover to blanket the soil. The stage was set for the "black blizzards", as the blinding winds of the Dust Bowl came to be called.

In 1932 there were 14 dust storms in the Great Plains. In 1933 the number of dust storms increased to 38. Farmers, confident that rain would return, continued to plow. But the storms in 1934 and 1935 kept coming with gale velocities and alarming frequency (Figure 7.4). In western Kansas and Oklahoma, as well as in the neighboring parts of Texas, Colorado, and New Mexico, the wind whirled soil particles upward into the prairie sky. Brown dust clouds up to 2,000 meters (6,500 feet) thick filled the air, with an upper edge almost 3.3 kilometers (2 miles) high (Figure 7.5). One storm on May 11, 1934, lifted about 275 million metric tons (300 million tons) of fertile soil into the air. (This roughly equals the total soil tonnage scooped from Central America to form the Panama Canal.) In many areas, the wilted wheat was uprooted by the strong winds. In the Amarillo, Texas, area during March and April 1935, 15 windstorms raged for 24 hours; four lasted for over 55 hours.

Dust from Oklahoma prairies came to rest on the deck of a steamer 330 kilometers (200 miles) out in the Atlantic. Dust sifted into the plush offices of Wall Street and smudged the luxury apartments of Park Avenue. When it rained in the blow area, the drops would sometimes come down as dilute mud. In Washington, D.C., mud-splattered buildings of the Department of Agriculture were a rude reminder of the problem facing it and the nation. A thousand miles westward, people stuffed water-soaked newspapers into window cracks, to no avail (Figure 7.6). The dust sifted into kitchens, forming a thin film on pots and pans and fresh-baked bread. Blinded by swirling dust clouds, ranchers got lost in their own backyards. Motorists pulled off to the side of the high-

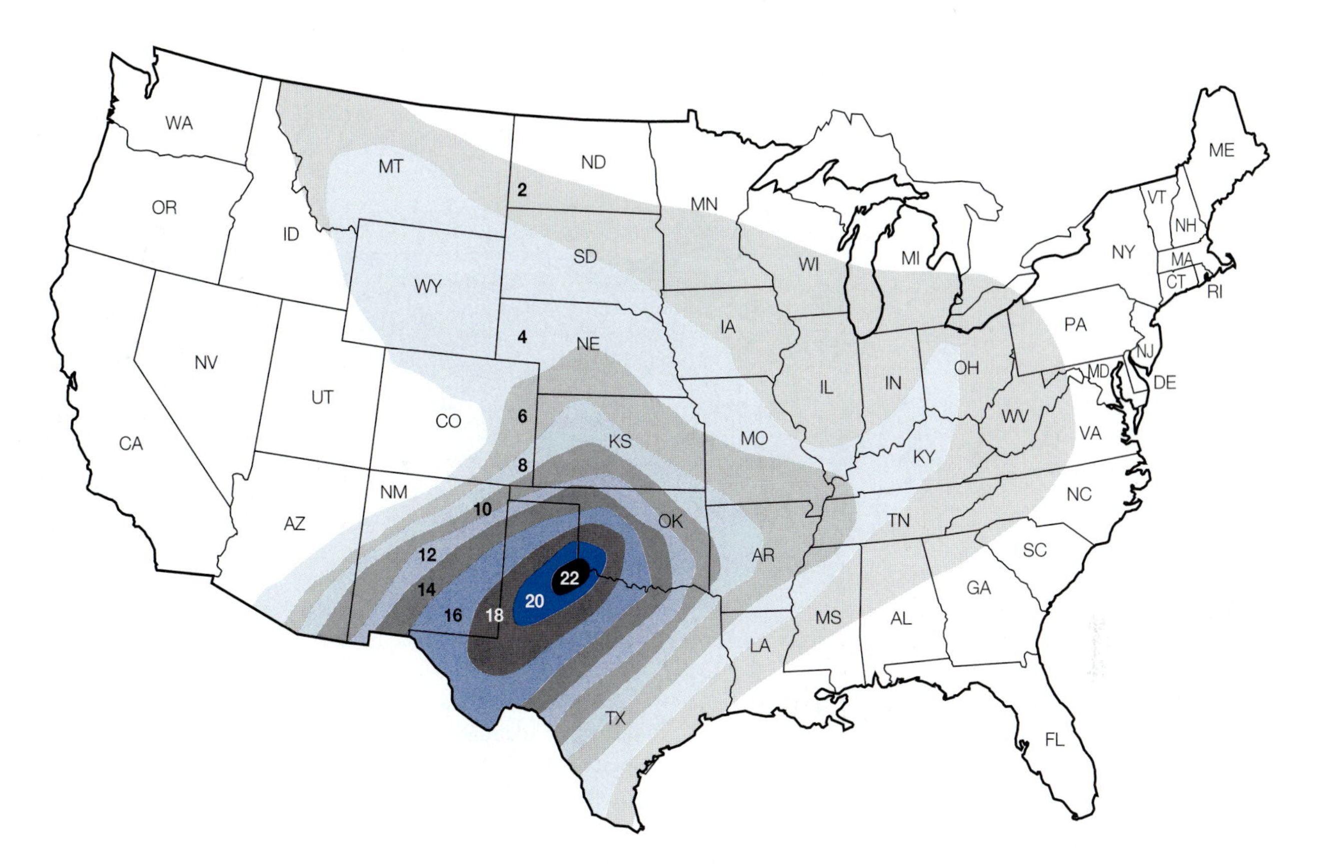

FIGURE 7.4 *Number of dust storms in March 1936 in the Great Plains. The greatest number, 22, was recorded in northern Texas.*

ways. Hundreds of airplanes were grounded. Trains were stalled by huge drifts. Hospital nurses placed wet cloths on patients' faces to ease their breathing. In Colorado's Baca County (March 1935), 48 relief workers contracted "dust pneumonia"; four of them died.

When the rains finally returned and the winds finally subsided, ranchers and farmers wearily emerged to survey the desolation. Five to 30 centimeters (2 to 12 inches) of fertile topsoil, made up mostly of silt and clay particles, had been carried to the Atlantic seaboard. Sand particles, too coarse and heavy to be airborne, bounced across the land, sheared off young wheat, and finally accumulated as dunes to the leeward side of homes and barns. Heavily mortgaged machinery became shrouded in sand.

The dust storms of the 1930s inflicted both social and economic suffering. Yet a few ranchers and farmers were philosophical about their misfortunes and even cracked jokes about birds flying backward "to keep the sand out of their eyes" and about prairie dogs "digging burrows 100 feet in the air." However, for most Dust Bowl victims, the dusters were not funny. Many victims were virtually penniless. The 275 million metric tons of topsoil removed in a single storm on May 22, 1934, represented the equivalent of taking 3,000 farms of 40 hectares (100 acres) each out of crop production. Up to 1940, Dust Bowl relief alone cost American taxpayers over $1 billion; $7 million (more than was paid for Alaska) was pumped into a single Colorado county. The only recourse for many of these ill-fated farmers was to find a new way of life. They piled their belongings into rickety cars and trucks and moved out—some to the Pacific coast, some to the big industrial cities of the Midwest and East. However, our nation was still in the throes of a depression, and many an emigrating family found nothing but frustration, bitterness, and suffering at the end of the road.

FIGURE 7.5 *Dust storm approaching Springfield, Colorado, on May 21, 1937. This storm reached the city limits at exactly 4:47 A.M. Total darkness lasted about one-half hour.*

FIGURE 7.6 *Abandoned Oklahoma farmstead, showing the disastrous results of wind erosion.*

7.3 The Shelterbelt Program

In an attempt to prevent future dust bowls, the federal government launched a massive **shelterbelt** system in 1935. More than 218 million trees were planted on 30,000 farms across the Great Plains from North Dakota south to Texas. The green checkerboard patterns formed by the 32,000 kilometers (20,000 miles) of windbreaks have added color and variety to the prairie landscape. On the Central Plains a typical shelterbelt consists of one to five rows of trees planted on the western margin of a farm in a north–south line, to intercept winter's prevailing westerly winds (Figure 7.7). Conifers, such as red cedar, spruce, and pine, provide the best year-round protection. By planting a few rows of grain between the rows of trees, farmers can further reduce wind erosion. A properly designed shelterbelt of adequate height and thickness can reduce a wind velocity of 50 kilometers (30 miles) per hour to only 13 kilometers (8 miles) per hour leeward of the trees.

Although **windbreaks** occupy valuable land that otherwise could be used for crop production, are relatively slow to grow, and must be fenced from livestock until the stands are well established, the accrued benefits far outweigh these minor disadvantages. In addition to controlling wind erosion, properly designed windbreaks provide aesthetic benefits, increase soil moisture by reducing evaporation and trapping snow, and provide a habitat for wildlife. Moreover, the fuel requirement for heating and cooling nearby homes is significantly reduced (Figure 7.8). Unfortunately, many

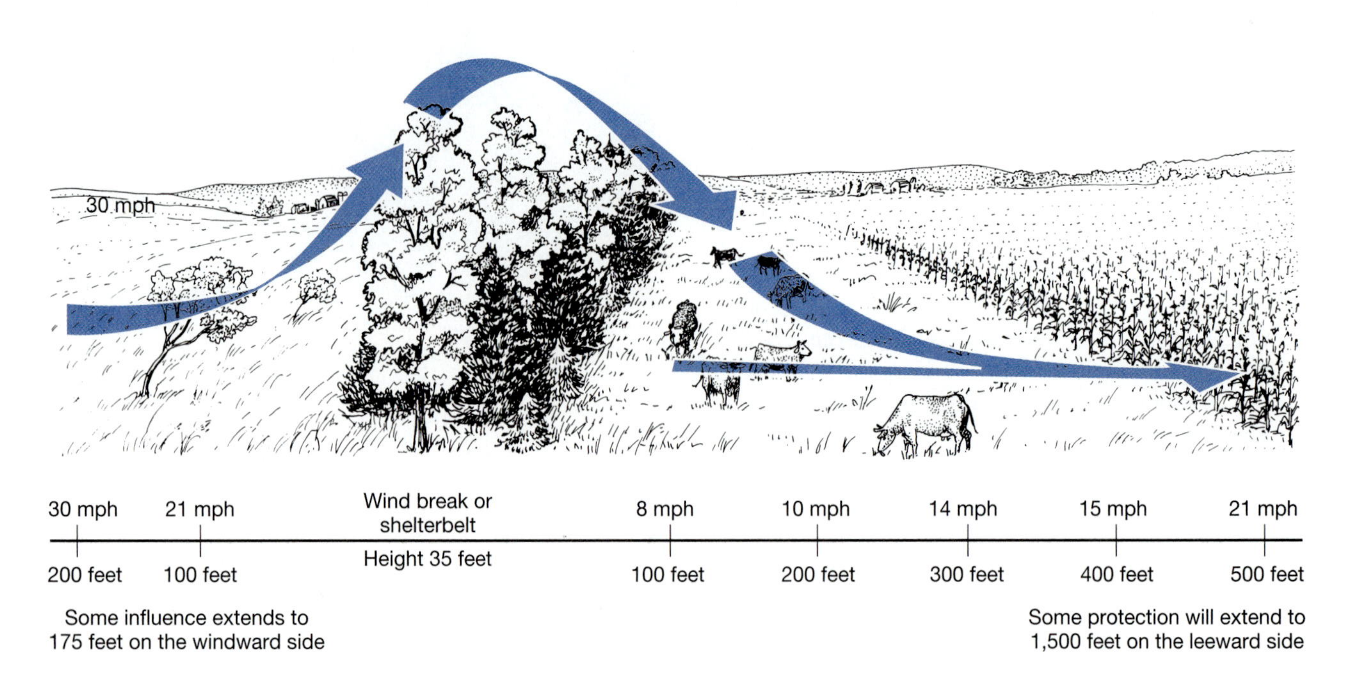

FIGURE 7.7 *The effect of a windbreak on wind velocity. When protecting farmfields, the windbreak is planted perpendicular to the prevailing winds.*

FIGURE 7.8 *This North Dakota farm is well protected from wind and snow by a 17-year-old windbreak of conifers, fruit trees, and shrubs.*

of the shelterbelts planted in the 1930s have been removed—in some cases so that the farmer could use the wood as fuel, in other cases to make room for crops and to facilitate the use of heavy farm machinery and sprinkler irrigation systems.

7.4 Soil Erosion Today

The Dust Bowl period was devastating to our nation's soil resource, to our economy, and to the emotional well-being of millions of Americans. In the more than 60 years that have passed since that critical period in American agriculture, the federal government has spent tens of billions of dollars to control erosion. Scientists at many major universities have conducted erosion-control research with tax-money support. Hundreds of scientific publications have been written on the subject. The U.S. Department of Agriculture (USDA) has established more than 3,000 conservation districts whose prime function is to assist the farmer with erosion problems. So, after devoting all this time, energy, and money to soil erosion control, we, as taxpayers, are justified in asking, "Is the American farmer doing any better in controlling erosion today than during the Dust Bowl years?" Generally speaking, the answer is "yes".

Since the 1930s, soil conservation practices have significantly reduced the rate of wind and water erosion on U.S. croplands. These practices include terraces, residue management, cover crops, crop rotations, contour farming, strip farming, windbreaks, and putting highly erodible farmlands into grass or trees for at least 10 years. However, although many farmers are doing a good job in controlling erosion, more need to be.

In the early 1930s soil erosion in the United States was estimated at 3.6 billion metric tons (4 billion tons) of soil removed each year. In 1982 the figure was 2.8 billion metric tons (3.1 billion tons) and by 1997 the figure dropped to 1.8 billion metric tons (2 billion tons). These estimates are more significant when you take into account that more land was being cropped in 1982 or 1997 than in the 1930s. About two thirds of annual soil erosion is by water and one third by wind. More than 50 percent of the soil lost to water erosion and about 60 percent of the soil lost to wind erosion occur on croplands that produce most of the country's food (Figure 7.9).

The good news is that not only has the total amount of soil erosion decreased, but annual rates of soil erosion (measured in metric tons per hectare) have declined. The bad news is that more than 34 percent of U.S. cultivated cropland is losing soil through water and wind erosion above the maximum sustainable rate of 11 metric tons per hectare per year. In some regions, soil loss is or has been shocking. In some farm fields of Washington, Oregon, and Idaho, 120 to 250 metric tons per hectare per year have been lost because of farming on steep slopes and highly erodible soil types (Figure 7.10). This is a serious problem that is not acceptable for our future sustainability.

Topsoil depth in the United States ranges from a few centimeters to more than 1 meter. The USDA has determined that soils with a thick layer of topsoil can tolerate erosion losses of 11 metric tons per hectare (5 tons per acre) per year without losing their ability to support crops economically and indefinitely. Although there is no sound basis for this **tolerable soil loss** value, this amount of soil, they say, is normally regained by the natural processes of soil formation. In fact, research estimates of soil formation rates average about 1 metric ton per hectare (0.5 ton per acre). Thus, critics argue that the tolerable levels of soil erosion accepted by the USDA, which actually range from 2 to 11 metric tons per hectare, depending on the particular soil, actually exceed

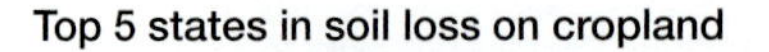

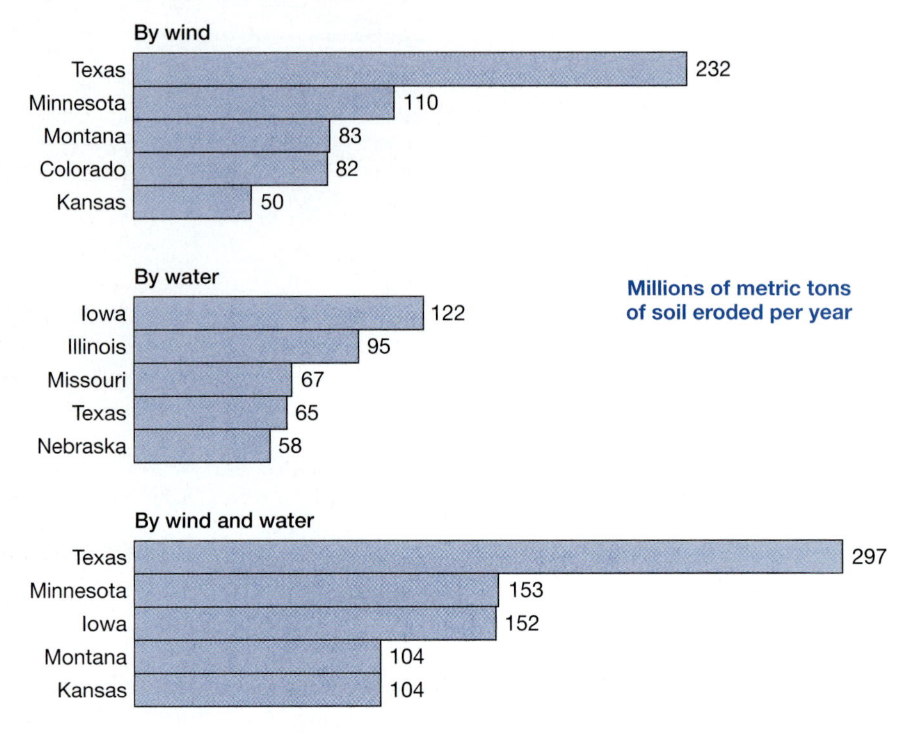

FIGURE 7.9 *Five states sustaining the greatest wind and water erosion of cropland in 1992.*

the recuperative or renewable capacity of agricultural soils. In other words, these so-called tolerable levels of soil erosion are not sustainable because soil is not being rebuilt (even under some of the best conditions) as fast as these rates require. To be sustainable, soil erosion should be far lower.

Soil erosion reduces crop productivity and removes valuable topsoil and water from a farmer's field. These are on-site damages from erosion. However, there are also off-site problems like pollution of air and water with soil particles. Dust in the air can be unhealthy. Sediment clogs rivers and lakes, reducing their potential for recreation and navigation (Figure 7.11). Can we put a value on the on-site and off-site damages caused by water and wind erosion? Some attempts have been made to do just this.

In one 1995 study in the United States, researchers found these costs to be exceptionally high. Using an average value for wind and water erosion in the United States of 17 metric tons per hectare per year (a 1992 USDA estimate of 14 metric tons per hectare per year could have been used instead), they estimated total annual on-site costs at $27 billion. This estimate is based on the fertility or nutrient value of a ton of eroded soil being worth about $5, which does not account for other soil components like organic matter and soil organisms. Since the researchers used an estimated 4 billion metric tons of soil lost each year in the U.S. (a 1992 USDA estimate of 2.2 billion metric tons of soil could have been used instead), they put the replacement value of nutrients in the eroded soil at $20 billion (of the $27 billion). The remaining $7 billion is for lost water (130 billion metric tons of water are lost annually) and lost soil depth. Using similar techniques, these researchers estimated the off-site costs for items such as dredging harbors and waterways, reduced water storage in reservoirs, loss of wildlife habitat, flooding, health costs, and cleaning up domestic water supplies. Off-site costs totaled $17 billion, which brings the grand total to $44 billion.

This erosion-cost study has been criticized for using higher than more accepted, average rates of erosion in the U.S. However, even if we cut the costs of erosion in half, the estimated annual cost of erosion to our nation is still a staggering amount ($22 billion). Since erosion rates are the lowest in the United States and Europe, one can only imagine what the costs of erosion are in Asia, Africa, and South America, where erosion rates are the highest on the planet and annually average 30 to 40 metric tons per hectare. The U.N.-affiliated International Food Policy Research Institute reported in 2000 that nearly 40 percent of global farmland is seriously degraded. This human-induced soil degradation has been caused by erosion, loss of organic matter, hardening of soil, chemical penetration, nutrient depletion, and excess salinity. The problem is the worst in developing countries, where populations are the highest.

7.5 Factors Affecting the Rate of Soil Erosion by Water

Rainfall

Annual precipitation in the continental United States ranges from almost nothing in some areas of Death Valley, California, to 360 centimeters (140 inches) in parts of Washington state. The amount of precipitation a region receives greatly affects erosion rates. However, even more important factors are rainfall intensity and the seasonal distribution of the rainfall. Hard-beating, intensive rains can cause serious erosion, even in areas of low annual rainfall. In contrast, gentle rains of low intensity may cause little erosion, even if the total annual precipitation is high.

A town in Florida once experienced a deluge of 60 centimeters (24 inches) of rain in only 24 hours. The soil loss resulting from runoff waters must have been severe. Had this 60-centimeter rainfall resulted from daily half-inch drizzles occurring over 48 consecutive days, the erosion threat would have been negligible because the soil would have had sufficient time to absorb the water. Surprisingly, even in the arid deserts of Nevada and Arizona, where annual rainfall averages 13 centimeters (5 inches), excessive erosion occurs because the entire annual precipitation occurs in a few torrential cloudbursts. As a result, the desert floor is dissected by canyons gouged out by runoff waters.

FIGURE 7.10 *Rill erosion (causing shallow, narrow channels) in the Palouse region of southeastern Washington, one of the most productive dry-farmed wheat areas in the world. Notice the sediment accumulating at the bottom of the slope.*

Soil Erodibility and Surface Cover

Soil structure, discussed in Chapter 6, greatly influences a soil's erodibility. To resist erosion, soil should be composed of **water-stable aggregates,** so that the soil particles are grouped into clusters that cling together even when they are submerged in water. Increasing the level of organic matter in soils makes them more resistant to erosion. The structure of a soil can be improved and the erosion hazard decreased by plowing under a crop of clover or alfalfa **(green manuring),** by using conservation tillage methods where more crop residue is left on the soil, or simply by adding organic material like barnyard manure. For example, in Iowa, soil loss from nonmanured land was over five times that from heavily manured land.

The addition of organic material also improves the soil's water-absorbing ability or infiltration capacity. When more water enters the soil, less runoff is available to transport soil particles. This trait, in turn, results in a more dense, vigorous growth of vegetation, which further protects the soil. A developing corn root system, for example, penetrates the soil particles and binds them in place, reducing erosion.

The influence of agriculture on soil erosion depends mostly on the amount of surface cover and the intensity of tillage operations. A dense cover of vegetation with plant residue on the ground, as found under bluegrass pasture or undisturbed natural forests, provides excellent protection from erosion. Clean tillage systems, where crop residues are turned under, leave the soil unprotected for considerable periods of time and can result in accelerated erosion. Thus, a cultivated field without crop cover or residues has a much greater potential for erosion than a dense pasture or woodland. Between these two extremes, there are numerous intermediate conditions of soil cover, some of which are shown in Figure 7.12.

FIGURE 7.11 *Sediment-laden river emptying into the clear waters of a lake. This is an example of off-site damage from erosion because the turbid water can clog fish gills and negatively impact submerged aquatic vegetation.*

Topography

The **slope** of the terrain greatly affects the intensity of surface runoff and soil erosion. Increasing the steepness of a slope greatly increases the velocity of runoff and the rate of

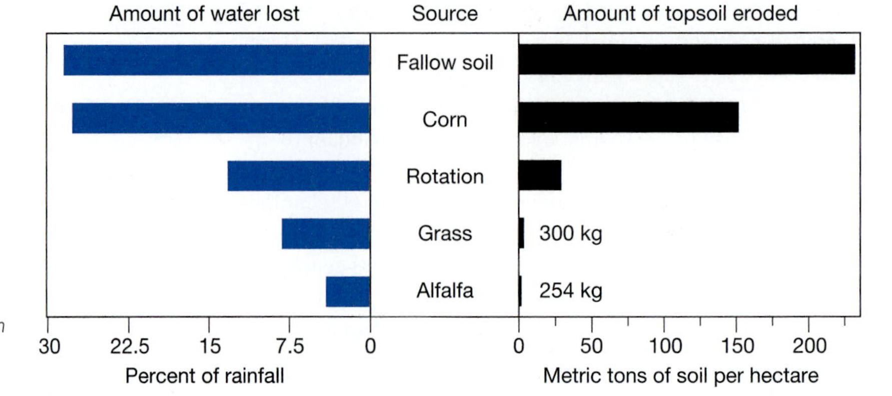

FIGURE 7.12 *The influence of plant cover on soil erosion and water runoff. Data from research conducted by the Soil Conservation Service (now the Natural Resource Conservation Service) at Bethany, Missouri, on Shelby loam on land with an 8 percent slope. Average annual rainfall in the area is 40 inches.*

FIGURE 7.13 *Contour farming in the Palouse region of Washington state. The pattern of farm conservation is reflected in the fields of this Palouse wheat farmer, who uses contour farming to reduce rainwater runoff and its erosive effect on soil. Besides slowing water so that it can soak into the soil better, such measures reduce siltation, the most common cause of water pollution in the United States, according to the U.S. Department of Agriculture.*

erosion. The steepness of a slope is indicated as a percentage. Thus, a 10 percent slope is one that drops 10 meters over a horizontal distance of 100 meters. In many parts of potato-growing Aroostook County, Maine, where slopes may be as steep as 25 percent, more than 60 centimeters (24 inches) of topsoil has already been removed by erosion since farming began. On farms planted to row crops like corn and cotton, a doubling of the slope results in a doubling of soil erosion by water. Also, as the length of slope increases, so does the amount of runoff water accumulating downslope and, thus, the erosion hazard.

7.6 Controlling Soil Erosion by Water

Erosion Control Practices

If the factors that contribute to soil erosion are altered, soil erosion can be controlled. Some important erosion control practices include (1) contour farming, (2) strip cropping, (3) terracing, (4) gully reclamation, (5) conservation tillage, and (6) removal of cropland from production.

Contour Farming **Contour farming** may be defined as plowing, seeding, cultivating, and harvesting at right angles to the direction of the slope, rather than down it (Figure 7.13). It was practiced by Thomas Jefferson, who wrote in 1813, "We now plow horizontally, following the curvature of the hills ... scarcely an ounce of soil is now carried off." Jefferson, however, was an exception. In the early days of American agriculture, the farmer who could plow the straightest furrows (usually up and down slopes) was considered a master plowman and was praised by his neighbors.

An experiment conducted on a Texas cotton field with a 3 to 5 percent slope revealed that the average annual water runoff from a noncontoured plot was 12 centimeters (4.6 inches), whereas that for a contoured plot was 65 percent less, or 4 centimeters (1.6 inches). The lower the water runoff, the lower the erosion rate.

Strip Cropping **Strip cropping** is the practice of growing crops that require different types of tillage, such as row and sod, in alternate strips along the contours or across the prevailing direction of the wind. On sloped land, planting crops on contoured strips is an effective erosion deterrent (Figure 7.14). When viewed from a distance, such farmland looks like a series of slender, curving belts of color. A row crop, such as corn, cotton, to-

CASE STUDY 7.1 A 100-YEAR STUDY OF THE EFFECTS OF CROPPING ON SOIL EROSION

Just imagine a scientific study that ran continuously for 100 years! It seems hard to believe. But that's exactly what has been done by soil scientists at the University of Missouri. Three generations of workers have collected research data on the rates of soil erosion from plots at Sanborn Field, the oldest agricultural experiment station west of the Mississippi River. The initial data were obtained in 1888, when Grover Cleveland was in the White House, the world's first skyscraper was being built in Chicago, and the nation was agog with the invention of the automobile. It was also a year when many thousands of primitive mule-drawn plows were slicing into the virgin prairie so that it could be planted to grain.

Of course, the University of Missouri research team, headed by Clark Gantzer, knew that the intrinsic nature of the soil and the degree of slope were factors in determining the degree of erosion. But they were primarily interested in finding out the role played by a row crop such as corn, which leaves much land exposed, and by a dense cover crop such as timothy grass. The soil on the study plots was a sandy loam, characteristic of more than 4 million hectares (10 million acres) in Missouri, Kansas, and Illinois.

The slope on the gently rolling plots varied from 0.5 to 3 percent. Gantzer and his co-workers compared the amount of erosion that had occurred on three different plots over the 100-year period. Plot A had been planted continuously to corn. Plot B was planted to a 6-year rotation of corn-oats-wheat-clover-timothy-timothy. Plot C was planted to timothy grass. The amount of vegetative cover was least in plot A, greater in B, and greatest in C.

The results of this highly important research were reported in the Journal of Soil and Water Conservation. The most severe erosion, about 46 metric tons per hectare per year (20 tons/acre/year), occurred on the corn plot. Indeed, this plot had 56 percent less topsoil than the plot planted to timothy. This degree of erosion cut corn yields by 60 percent. The rotation plot lost 21 metric tons of topsoil per hectare per year (9.4 tons/acre/year) and retained 30 percent less topsoil than the timothy plot. The scientists found that during the 100-year study period, timothy was 54 times more effective in controlling soil erosion than was corn. They further concluded that cover management was about 35 times more significant than soil erodibility and 28 times more significant than slope in explaining the differences in erosion between the plots.

bacco, or potatoes, and a cover crop of hay or legumes are alternated along the contours. Strip cropping is frequently combined with crop rotation, so that a strip planted to a soil-depleting corn crop one year will be sown to a soil-enriching legume crop the next. For a study of erosion, see Case Study 7.1.

Terracing **Terracing** has been practiced by humans for centuries. It was used by the Incas of Peru and by the ancient Chinese. Plagued with relatively dense populations and a scarcity of arable land, those civilizations were forced to till extremely steep slopes, even mountainsides, in order to prevent widespread hunger. The flat, steplike bench terraces that those ancient agriculturists constructed, however, are not amenable to today's farming methods. To be effective, terraces must check water flow before it attains a velocity of 1 meter per second (3 feet per second), which is sufficient to loosen and transport soil.

A variety of terraces are in use around the world today (Figure 7.15). Broad-based terraces permit the entire surface to be planted. The steep back-slope terrace allows cropping on most of the surface area except for the backslope which is planted to permanent grass. Flat-channel terraces permit large volumes of water to move off the soil without erosion. Bench terraces can keep water on the cropped area (no runoff) and are mostly used for rice production. Growers must be careful to maintain any of these terraces.

A channel terrace is formed by digging a channel across the slope. It is frequently used in the Tennessee and Ohio valleys, as well as the Southeast and Mid-Atlantic states, where rainfall is high but the water-absorbing capacity of the soil is poor.

Gully Reclamation **Gullies** are erosion channels too large to be erased by ordinary farm operations. They are active when their vertical walls are free of vegetation and inactive when they are stabilized by vegetation. They are especially common in the southeastern United States due to a long history of soil abuse and intense rains. Gullies are danger signals indicating that land is eroding rapidly and that the area may become a wasteland unless erosion is promptly controlled. Some gullies work their way up a slope at the rate of 5 meters (15 feet) a year (Figure 7.16A). In North Carolina, a 45-meter (150-foot)-deep

FIGURE 7.14 *Strip cropping in northwestern Illinois. Alternating contour strips of corn and alfalfa reduce erosion and pesticide demands (because of increased diversity) and helps enhance soil fertility.*

gully was gouged out in only 60 years, swallowing up fence posts, farm implements, and buildings in the process.

If relatively small, a gully can be plowed and then seeded to a quickly growing "nurse" crop of barley, oats, or wheat. In this way, erosion will be checked until sod can become established.

In cases of severe gullying, small check dams of manure and straw constructed at 6-meter (20-foot) intervals may be effective. Silt collects behind the dams and gradually fills the channel, allowing plants to take root. Dams may be constructed of brush or stakes held securely with a woven wire netting. Earth, stone, and even concrete dams may be built at intervals along the gully. Once dams have been constructed and water runoff has been restrained, soil may be stabilized by planting rapidly growing shrubs, vines, and trees. Willows are effective. Not only does such pioneer vegetation discourage future erosion, it also obliterates the ugly scars and provides food, cover, and breeding sites for wildlife (Figure 7.16B).

Conservation Tillage Before we discuss the revolutionary agricultural development known as conservation tillage, we must describe the traditional procedures involved in **cultivating** the soil (Figure 7.17):

1. One pass is made over the field with a moldboard plow. During this process, the crop residues remaining after the previous harvest are plowed under and the upper 15 centimeters (6 inches) of the soil is inverted and broken up.
2. One or two passes over the field are then made with a disk to break up any clods and to prepare a seedbed for the next crop; sometimes a field cultivator is used on the second of the two passes.
3. After row crops like corn or cotton have sprouted, a cultivator is used to remove weeds that compete with the crop for moisture and nutrients.

In contrast, **conservation tillage** restricts plowing of the soil to reduce erosion. Such restriction may vary from slight to complete. If tillage is omitted completely, the term **no-till** is applied. By definition, conservation tillage systems leave enough of the previous crop residues so that at least 30 percent of the soil surface is covered when the next crop is planted. Soil can be conserved, time and fuel consumption reduced, and comparable yields obtained with various forms of conservation tillage.

Conservation tillage has spread rapidly from only 2 percent of our nation's harvested croplands in 1962 to about 36.6 percent in 2000. No-till alone accounted for 17.5 percent of harvested cropland in 2000. It is projected that in the next decade or so conservation tillage could be applied to as much as three-quarters of U.S. cropland. This high estimate for these soil-saving systems already holds true in some farmland areas of the U.S.

In the no-till form of conservation tillage, special machinery is used that cuts narrow slits into the ground in which the seeds of the next crop are planted. This method is also called direct drilling or seeding. All operations are done in a single pass over the field. No seedbed preparation is necessary. Corn and soybeans, for example, can be sown directly in wheat stubble. Almost all of the residue from the previous crop harvest remains on the surface, compared to only 1 to 10 percent with conventional tillage using the moldboard plow. There is a trade-off, though. Weeds usually increase with no-till methods, resulting in greater use of herbicides.

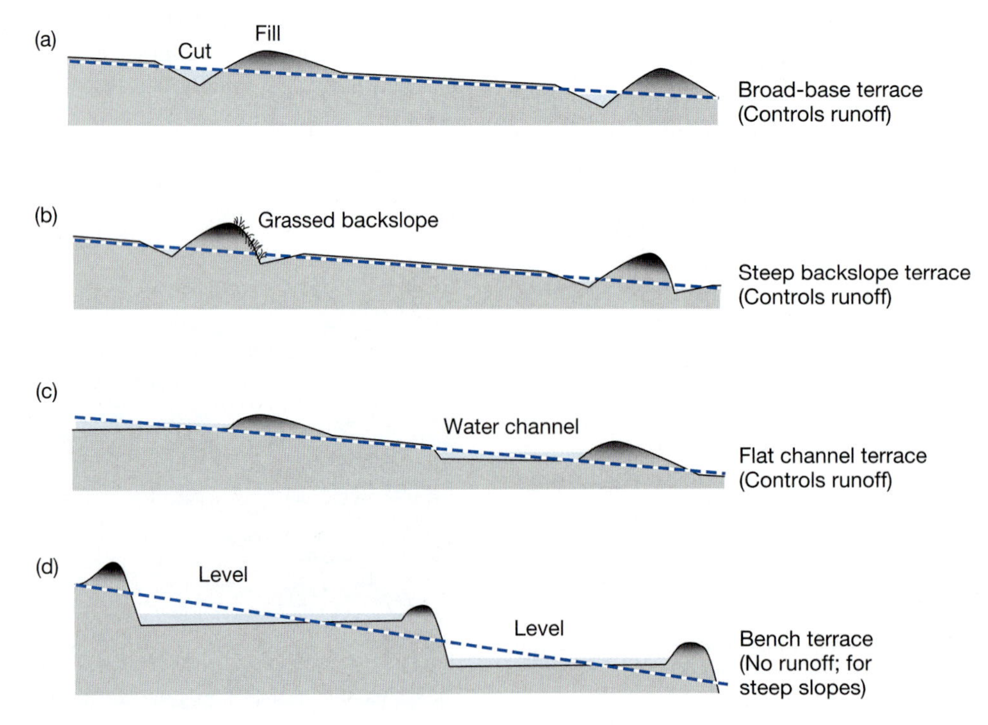

FIGURE 7.15 *Four common types of terraces in use around the world. Terraces are very effective at controlling water runoff and erosion. To save moisture, terraces are designed to cause ponding of water on the terrace, allowing water time to infiltrate the soil.*

FIGURE 7.16A *Gully erosion on a Minnesota farm.*

FIGURE 7.16B *To prevent further erosion the area was planted with protective vegetation, primarily locust trees. Five growing seasons later, the locust trees averaged 4.5 meters (15 feet) in height and not only served to control erosion, but provided wildlife cover and beautified the landscape.*

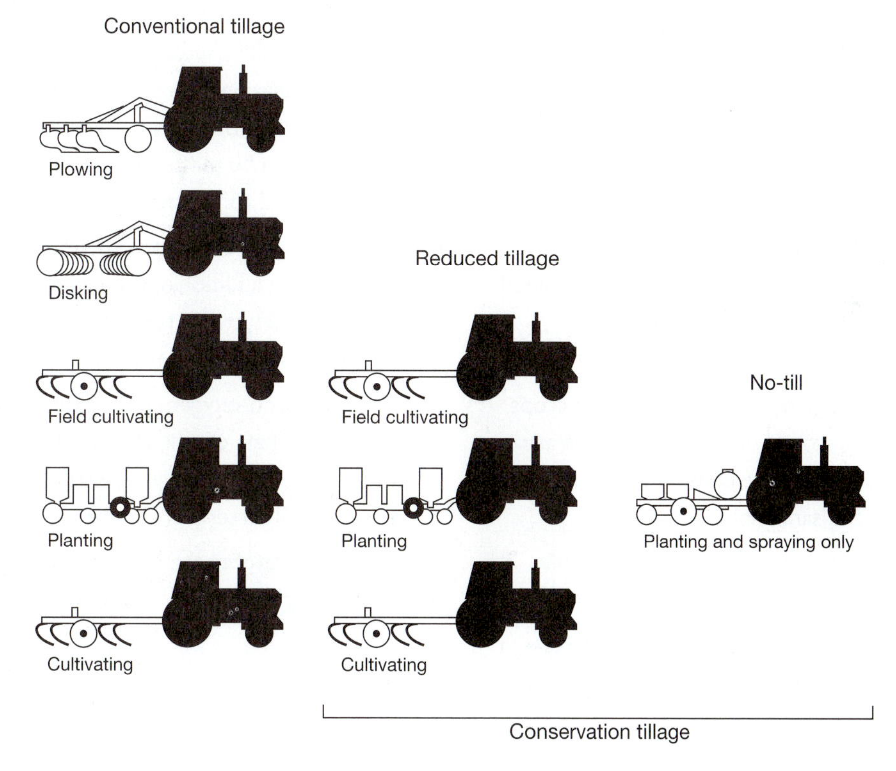

FIGURE 7.17 *Equipment differences between conventional tillage, reduced tillage, and no-till systems.*

Research conducted on farms in Illinois has shown that no-till farming controls erosion even on slopes as steep as 9 percent. On many farms, the layer of crop residues on the surface can reduce soil loss from erosion by 90 percent. In fact, a recent experiment conducted by the Natural Resource Conservation Service in Georgia showed that no-till farming reduced erosion losses from 58 metric tons per hectare (26 tons per acre) per year to only 0.2 metric ton per hectare (0.1 ton per acre)—more than a 99 percent reduction.

No-till farming provides a number of other benefits. There also are some disadvantages. The pros and cons of this revolutionary farming technique are summarized in Table 7.1.

Removing Cropland from Production In 1985 Congress passed the **Food Security Act** (or Farm Bill). It calls for the removal of 18 million hectares (45 million acres) of marginal cropland from the cropland base. The cropland taken out of production is that which is most vulnerable to erosion when cultivated. This land is to be planted with grasses, trees, and other long-term cover to stabilize the soil. This removal of cropland is termed the **Conservation Reserve Program (CRP),** which is only one of a number of programs of the Food Security Act.

The objective of the CRP is to control erosion on 18 million hectares of highly erodible cropland. Under the terms of this

TABLE 7.1 *Pros and Cons of No-Till Farming*

Pros	Cons
1. Labor may be reduced by 30 to 50 percent. The only passes over the field are for planting and fertilizing, pesticide application, and harvesting.	1. Farmers require greater management skills.
2. Use of diesel fuel is reduced by 30 to 50 percent.	2. Special equipment is needed to plant seeds directly in crop residues.
3. Wear and tear of farm equipment is reduced.	3. No-till farming is not suitable for all crops.
4. Soil erosion is generally reduced by 90 percent.	4. Seeds may not make contact with soil if the seed planter is not perfectly level. As a result, seed germination can be lowered.
5. Soil retains more moisture because of reduced rates of evaporation and surface runoff.	5. Plant diseases, such as fungi, may be more abundant because of the higher soil moisture levels.
6. Sediment pollution of lakes and streams is reduced.	6. Populations of crop-destroying insects and rodents are often greater.
7. Fertilizer pollution (eutrophication) of lakes and streams is lessened.	7. Weed populations are often greater; therefore, competition with crops for available nutrients and moisture can be more intense. More herbicides are often needed to control these weeds.
8. Three to 5 years after startup, crop yields often increase, except in level, fine-textured, poorly drained soils.	8. More energy is consumed (by manufacturers) to produce the additional pesticides. (However, in most cases, no-till systems require less total energy than do their conventional counterparts.)
9. Double-cropping, the growing of two different crops in the same growing season, is possible in some areas. Thus, soybeans can be planted immediately after a wheat harvest on the same field.	9. Anaerobic (no-oxygen) conditions may prevail in certain pockets of soil. This, in turn, increases the rate of nitrogen loss through denitrification because denitrifying bacteria are more active.
10. Crop residues provide food and cover for wildlife.	10. Soil temperatures are cooler in spring, often resulting in slower germination and more stand problems.
11. Air pollution is diminished because of the decrease in fuel consumption and potential dust from wind erosion.	

program, the farmer makes a contract with the USDA to withdraw erodible farmland from crop production for 10 years and to establish vegetative cover (such as grass or trees) on this land to stabilize the soil. The USDA (the taxpayers), in turn, makes "rental" payments to the farmer during this period. Not only are the plantings valuable in checking erosion, they are useful in providing food and cover for wildlife as well.

Although the Food Security Act called for the removal of 18 million hectares of highly erodible land, by the early 1990s final enrollment of land taken out of production was 14.7 million hectares (36.4 million acres), with the first contracts to expire in 1995 under an early-out provision. This impressive retirement cut annual erosion on U.S. farmland by 500 million metric tons per year. This is enough topsoil saved in a year to fill a convoy of dump trucks 58 wide stretching from Los Angeles to New York. The CRP also improves wildlife habitat, as birds and other animals take advantage of the newly restored habitat.

The CRP was to be terminated in 1995 but the 1996 Farm Bill extended it through 2002. In addition, no more than 14.7 million hectares of highly erodible land may be enrolled in the program at any one time.

The Food Security Act also established national incentives for erosion control on actively farmed lands. For example, conservation compliance, sometimes called "Sodbuster", requires growers to develop and implement erosion control plans for certain highly erodible lands (HEL) to remain eligible for federal price support programs, such as subsidies. These erosion control plans are developed in conjunction with the USDA Natural Resource Conservation Service (formerly the Soil Conservation Service). In addition to the "Sodbuster" provision is the "Swampbuster" provision, which protects wetlands by denying eligibility for other USDA programs to growers who drain and farm certain wetlands.

Enforcement of the erosion control provisions of the Food Security Act poses a formidable challenge for the USDA. However, this problem may be solved by the work of USDA scientists, who have recently developed regionally adaptable computer models of erosion rates (called the Revised Universal Soil Loss Equation or RUSLE). The models, based on the **Universal Soil Loss Equation (USLE)** (A Closer Look 7.2), will enable the USDA to know instantly how much soil erosion a particular farming practice on a certain soil type will cause in a specific type of climate anywhere in the United States. Suppose, for example, that the USDA learns, with the aid of computer models, that a particular farmer's uphill-and-downhill plowing method on a silty loam soil in northern Illinois is causing the erosion of 74 metric tons per hectare (30 tons of soil per acre) annually—62 metric tons per hectare (25 tons per acre) more than is considered acceptable. If the farmer refuses to shift to contour plowing, the USDA could reduce or even eliminate his or her federal subsidies, as required by the Food Security Act.

The Natural Resource Conservation Service and Its Program

The black blizzards of the 1930s alerted a hitherto apathetic nation to the plight of its soil resources more forcefully than thousands of urgent speeches. In 1934 the newly organized Soil Erosion Service set up 41 soil and water conservation demonstration projects. The labor force for these projects was supplied by Civilian Conservation Corps workers drawn from about 50 camps. The projects impressed Congress so much that it established the Soil Conservation Service in

A CLOSER LOOK 7.2 *The Universal Soil Loss Equation*

The Universal Soil Loss Equation (USLE) was developed by soil scientists after many decades of research to be able to predict soil losses from water erosion. The equation is:

$$A = RKLSCP,$$

where

A = number of metric tons (tons) of soil lost per hectare (acre) per year

R = rainfall erosivity

K = erodibility of soil

L = length of slope

S = steepness of slope

C = cover type (grass, wheat, forest, etc.)

P = practice used in erosion control (strip cropping, contour farming, etc.)

The USLE has been widely used since the 1970s. In the early to mid-1990s, the equation was revised into a modern, computerized tool called the Revised Universal Soil Loss Equation (RUSLE). RUSLE still uses the same factors of USLE shown above, although now some of the factors are better defined, which improves the accuracy of predicting soil loss from water erosion. RUSLE is a readily available computer software package. Farmers or soil scientists can use it to estimate soil loss for any farm in the United States.

Suppose that a farmer in southern Ohio, whose soil is a silty loam, would like to know her erosion losses. From information in the RUSLE software package or even from the original USLE tables available from the NRCS office, she determines the following about her farm:

$$R = 150$$

$$K = 0.33$$

$$LS = 0.40$$

Suppose now that the farmer's land is almost devoid of cover from the time of the harvest in autumn to the planting of the next crop the following spring. In this case, $C = 0.9$. Suppose further that she does not use any soil erosion control practices, such as terracing or strip cropping. In this case, $P = 1.0$. The expected soil loss on her farm can then be calculated by using the equation:

$A = (150)(0.33)(0.40)(0.90)(1.00) = 17.8$ tons per acre annually, or 40.2 metric tons per hectare.

Obviously, erosion is excessive on the farmer's land, the rate being roughly 3.5 times the tolerable limit of 11.2 metric tons per hectare (5 tons per acre) per year. Since factors *RKLS* remain fairly constant, the only practical way for her to reduce erosion losses on her farm would be to reduce the values of *C* (cover) and *P* (practices). She decides to substitute conservation tillage for traditional tillage on her farm. In this way, some crop residues are always on her land, even between the harvest and the next planting. As a result, the value for *C* drops to 0.1 in the USLE. She further decides to till and plant on the contour. Adoption of this practice, in turn, reduces the value of *P* to 0.4 in the USLE.

Now let us recalculate the soil loss on this Ohio farm after substituting the new values for *C* and *P* in the USLE:

$A = (150)(0.33)(0.40)(0.10)(0.40) = 0.8$ ton per acre (1.8 metric tons per hectare) per year.

It is clear that conservation tillage and contour farming have been extremely effective, reducing erosion losses on the farm from 40 to 1.8 metric tons per hectare (17.8 to 0.79 tons per acre) annually.

1935, renamed the **Natural Resource Conservation Service** (NRCS) in 1994. The major function of the NRCS has been to provide technical assistance to farmers and ranchers so that they can better utilize land with methods that are as consistent with the needs of the soil as with those of the landowner.

The administrative and operative unit of the NRCS program is the **conservation district,** which is organized and run by farmers and ranchers. Each district is staffed by a professional conservationist and several aides who work directly with farmers on their land. The highly diversified types of assistance provided by the NRCS to the farmer are indicated by the kinds of specialists on its staff: agricultural engineers, botanists, chemists, ecologists, foresters, irrigation engineers, land appraisers, land use specialists, soil scientists, and wildlife biologists. Any farmer in a particular district can request assistance to set up and maintain a sound conservation program on his or her farm. Participation in the NRCS program is voluntary.

Today nearly 3,000 conservation districts have been organized, embracing roughly 2 million hectares (5 million acres) and 96 percent of the nation's farms and ranchlands. In a typical year the NRCS assists more than 900,000 farmers and ranchers. This assistance includes:

1. Conducting and publishing soil surveys.
2. Promoting conservation tillage.
3. Controlling salination.
4. Identifying important farmlands, such as prime and unique farmlands.
5. Constructing thousands of terraces and farm ponds.

The NRCS has even helped to establish plant cover on the lava-covered slopes of Mount St. Helens in Washington.

The NRCS prepares **soil surveys** to meet not only the needs of farmers and ranchers, but also those of homeowners, highway engineers, and land-use planners. A soil survey is a systematic examination, description, classification, and mapping of the soils in a given area (usually a county). A soil survey can help in making interpretations for all kinds of soil uses, not just those uses that were intended at the time the survey was made. The NRCS uses aerial photographs as a base map which shows the boundaries or delineations of the different soil mapping units (Figure 7.18). A completed soil map becomes part of a soil survey report, which has four major features: (1) a set of soil maps, (2) map legends that explain the map symbols, (3) descriptions of the soils, and (4) use and management eval-

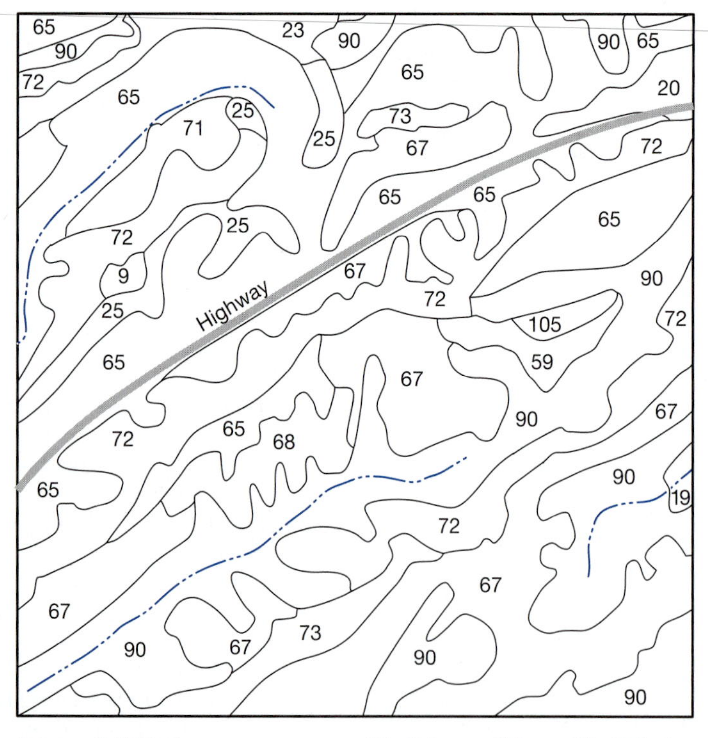

9	Athena silt loam, 7–25% slopes
19	Caldwell silt loam
20	Caldwell silt loam, drained
23	Caldwell silt loam, 5–25% slopes
25	Caldwell silt loam, 25–40% slopes, eroded
59	Naff silt loam, 7–25% slopes
65	Palouse silt loam, 7–25% slopes
67	Palouse silt loam, 25–40% slopes
68	Palouse silt loam, 25–40% slopes, eroded
71	Palouse-Thatuna silt loams, 7–25% slopes
72	Palouse-Thatuna silt loams, 25–40% slopes
73	Palouse-Thatuna silt loams, 40–55% slopes
90	Snow silt loam, 7–15% slopes
105	Thatuna silt loam, 25–40% slopes

FIGURE 7.18 *An example of part of a soil survey map from the U.S.D.A.* Soil Survey of Whitman County, Washington. *The different mapping units (designated by numbers) indicate specific soil types with unique physical, chemical, and biological characteristics, the details of which can be found in the survey report.*

uations for each soil. Soil reports are available free to the public from the NRCS offices or Land Grant Universities. Some survey reports are now being placed online.

One of the most significant accomplishments of the NRCS early in its history was the development of the **Land Capability Classification System.** This system evaluates land according to its limitations for agricultural use. It classifies land into eight categories, with Class I the least susceptible to erosion and Class VIII the most susceptible. Classes I through IV are suitable for cultivation, whereas Classes V through VIII generally are not. Class I land is most suitable for crop production, being flat, fertile, and not vulnerable to erosion. Classes II and III may be used for growing crops, but proper erosion control measures must be practiced. Class IV land is suitable only for limited cultivation. Classes V through VII are most suited for pasture, range, forest, wildlife, or recreation. Class VIII land, being stony, infertile, or very steep, is suitable only as wildlife habitat, wilderness, or recreational areas (Figures 7.19 and 7.20). The NRCS uses the Land Capability Classification System in preparing farm (conservation) plans for farmers.

The Development of an NRCS Farm Plan

If a farmer requests technical assistance from his NRCS district, four steps are followed in executing the **farm plan** for the farm.

First, the technician and the farmer make an intensive survey of the farm. On the basis of such criteria as slope, fertility, stoniness, drainage, topsoil thickness, and susceptibility to erosion, the technician maps each parcel of land on the basis of its capability. Each plot is given a land capability class symbol in the form of a Roman numeral or color. This map is then superimposed on an aerial photograph.

Second, the farmer draws up a farm plan with assistance from the technician. This plan involves decisions on how each acre will be used and how it will be improved and protected. For example, should a given acre be used for crops, pasture, forests, or wildlife? Usually alternative uses and treatments are considered.

Third, the treatment and uses called for in the plan are actually applied. Although much of this application can be completed by the farmer alone, he or she may find it helpful to enlist the aid of the NRCS technicians for more complex conservation measures such as terracing and strip cropping. Recommended management practices for a Georgia cotton farm (Figure 7.21) are shown in Figure 7.22.

The final and most important phase of the program is its maintenance from year to year with the assistance of conservation technicians. As time passes, agricultural geneticists might develop a new strain of rust-resistant wheat or a tick-resistant breed of cattle. A new variety of bluegill may be discovered that thrives in farm ponds. These new developments can gradually be incorporated into the conservation program.

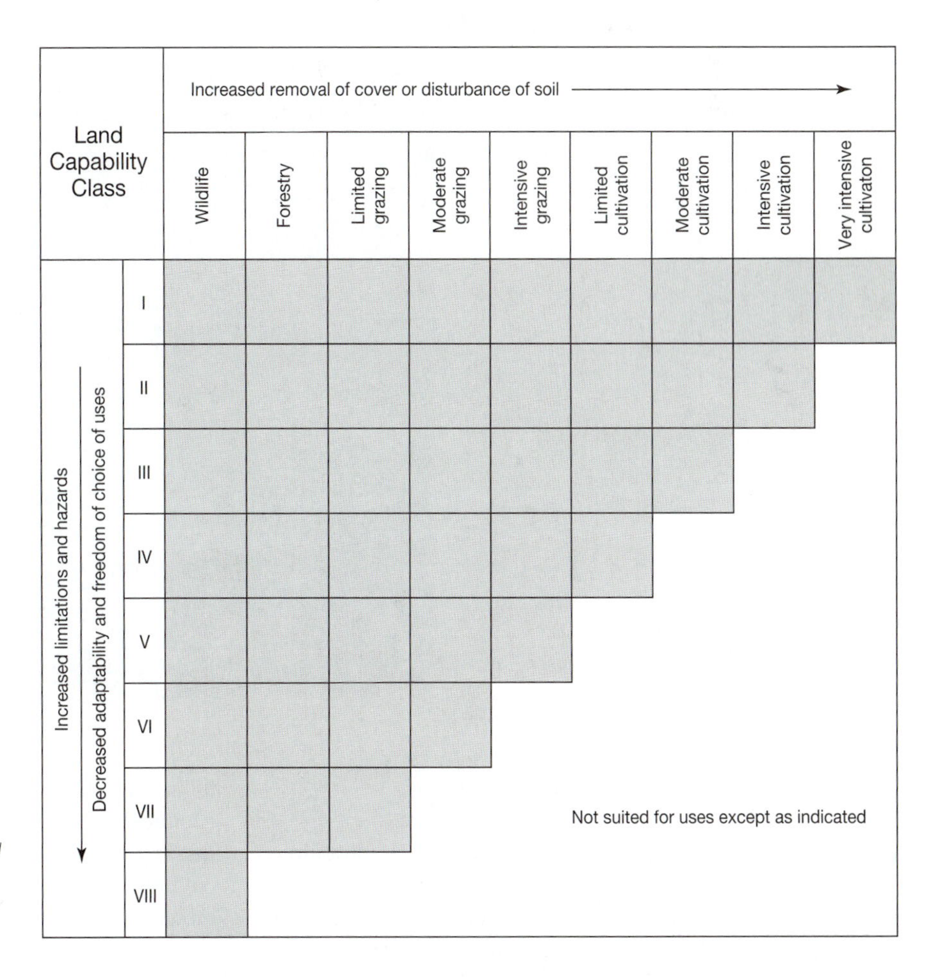

FIGURE 7.19 *Intensity of land use for each land capability class. Class I lands are the most versatile. The limitations and needed conservation practices increase from Class I lands to Class VIII lands.*

FIGURE 7.20 *The eight USDA land capability classes.*

FIGURE 7.21 *Harvesting a cotton crop on a Georgia farm.*

7.7 Alternative Agriculture

For nearly four decades after World War II, U.S. agriculture was the envy of the world, almost annually setting new records in crop production and labor efficiency. During this period U.S. farms became highly mechanized and specialized, as well as heavily dependent on fossil fuels, borrowed capital, and chemical fertilizers and pesticides. Today many of these farms are associated with declining soil productivity, deteriorating environmental quality, reduced profitability, and threats to human and other animal health. A growing cross section of American society is questioning the environmental, economic,

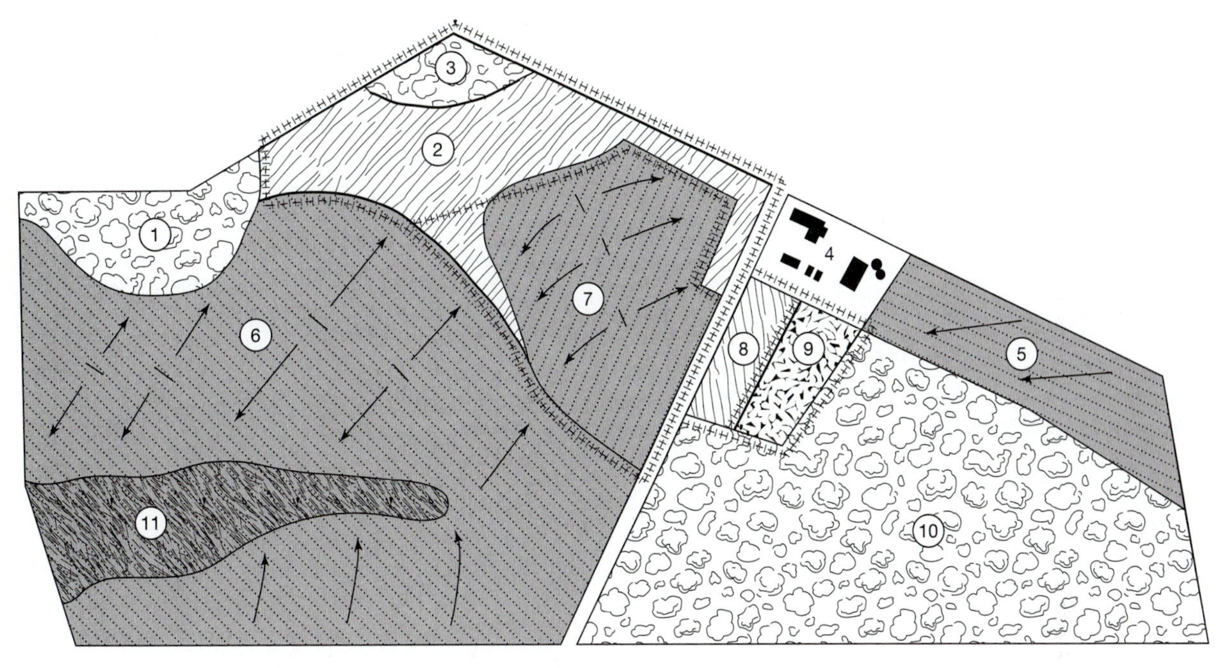

Field number	Acres	Recommended Use	Recommended Land Management
1	2.5	Woods	Thin and cull; cut annully for sustained yields
3	1.5		
10	22.0		
2	5.0	Costal Bermuda and Crimson Clover	Construct fence; lime and fertilizer as determined by soil analysis
8	2.0	Costal Bermuda and Crimson Clover	Lime and fertilizer
9	2.0	Fescue and Ladino Clover	Construct fence; lime and fertilizer as determined by soil analysis
4	2.0	Farmstead	
11	5.0	Permanent sod	Lime and fertilizer
6	39.0	Rotation cropland	Build up terraces; establish 4-year rotation of small-grain row crops
7	9.0	Rotation cropland	Fence to permit grazing and small grain; build up terraces; lime and fertilizer as needed
5	5.0	Rotation cropland	Build up terraces; establish 4-year rotation of small grain with row crops; lime and fertilizer as needed

FIGURE 7.22 *Conservation program for a Georgia cotton farm. Harvesting of cotton crop is shown in Figure 7.21.*

and social impacts of conventional agriculture (Table 7.2). Consequently, many individuals are seeking alternative practices that would make agriculture more sustainable.

Alternative agriculture embraces several variants of nonconventional agriculture that are often called organic, biological, biodynamic, integrated, low-input, natural systems, or no-till. **Organic** or **biological farming** excludes the use of agrochemicals, using instead naturally derived products as defined by organic certification programs and relying on crop rotations, biological pest controls, and modern technologies. **Biodynamic farming** is similar to organic farming but differs in that biodynamic farmers use a series of eight soil and plant amendments, called preparations, made from cow manure, silica, and various plant substances. **Integrated** and **low-input farms** place minimal reliance on the use of materials, like fertilizers and pesticides, from outside the farm. For example, an integrated farmer might use both organic and chemical fertilizers for soil fertility and control pests biologically or mechanically, with only limited use of pesticides. **Natural Systems Agriculture** is an ecology-based approach to agricultural production in which the vegetative patterns of natural ecosystems are used as guides to the composition and management of agricultural crops. For example, at The Land Institute, located in the North American tall-grass prairie region, researchers are working to develop systems of perennial, grain-producing crops grown in mixtures to mimic the prairie ecosystem. No-till farming has already been discussed in this chapter.

One of earliest landmarks of the alternative agricultural movement in the U.S. was *Farmers of Forty Centuries: Permanent Agriculture in China, Korea and Japan,* by Franklin King, published in 1911. It documents how farmers in parts of East Asia worked fields for 4,000 years without depleting the fertility of their soil. This text and others of the early twentieth

TABLE 7.2 *Concerns About Conventional Agriculture*

Increased costs and uncertain availability of energy and farm chemicals
Increased resistance of weeds and insects to herbicides and insecticides
Decline in soil productivity from erosion
Decrease in number of farms, particularly family farms
Pollution of surface water and groundwater with sediment and agrochemicals
Destruction of wildlife and beneficial insects
Hazards to human and animal health from pesticides and food additives
Depletion of finite reserves of plant nutrients

century focused on holistic aspects of agriculture and the complex interactions within farming systems. Yet, around this same time, U.S. agriculture was in the early stages of industrialization. New technologies and scientific methods were developed to help farmers meet the growing demands of expanding urban populations. By substituting mechanical power for horses, for example, farmers could increase their grain acreage by 20 to 30 percent, because they could plow more ground in less time and did not need to grow fodder.

Many groups and individuals continued to believe that biology and ecology rather than chemistry and technology should govern agriculture. Their efforts helped to give birth to the soil conservation movement of the 1930s, the ongoing organic farming movement, and considerable related research. Nevertheless, by the 1950s technological advances had caused a shift in mainstream agriculture, creating a system that relied on agrochemicals, new varieties of crops, and labor-saving but energy-intensive farm machinery. This system has come to be known as **conventional farming.**

As pesticides, inexpensive fertilizers, and high-yielding varieties of crops were introduced, it became possible to grow a crop on the same field year after year—a practice called **monocropping**—without depleting nitrogen reserves in the soil or causing serious pest problems. Farmers began to concentrate their efforts on fewer crops. Government programs promoted monoculture by subsidizing only the production of wheat, corn, and a few other major grains. Unfortunately, these practices set the stage for extensive soil erosion and for pollution of water by agrochemicals.

In the United States between 1950 and 1985, as a share of total production cost, the cost of interest, capital-related expenses, and manufactured farm inputs (such as chemical fertilizers, pesticides, and equipment) almost doubled, from 22 to 42 percent, while labor and on-farm input expenses declined, from 52 to 34 percent. During most of this period, relatively little research on alternative agriculture was conducted because of lack of funding and public interest.

By the late 1970s, however, concern was mounting that rapidly rising costs were endangering farmers nationwide. In response, the USDA commissioned a study in 1979 to assess the extent of organic farming in the United States, as well as the technology behind the farming and its economic and ecological impact. The study, *Report and Recommendations on Organic Farming,* published in 1980, was based heavily on case studies of 69 organic farms in 23 states. The USDA report concluded that organic farming is energy-efficient, environmentally sound, productive, and stable, and tends toward long-term sustainability. In addition, the report stimulated interest, nationally and internationally, in alternative agriculture. Its recommendations provided the basis for the alternative agriculture initiative passed by Congress in the Food Security Act of 1985, which called for research and education on sustainable farming systems.

The alternative agriculture movement received a further boost in 1989 when the Board on Agriculture of the National Research Council released another study, *Alternative Agriculture.* Although controversial, the report found that well-managed farms growing diverse crops with little or no chemicals are as productive and often more profitable than conventional farms. It also asserted that "wider adoption of proven alternative systems would result in even greater economic benefits to farmer and environmental gains for the nation."

Alternative farming systems do not represent a return to preindustrial revolution methods. Rather, alternative farming combines traditional conservation-minded farming techniques with modern technologies. Alternative production systems use modern equipment, certified seed, soil and water conservation practices, genetically improved crop varieties, and the latest innovations in feeding and handling livestock. Emphasis is placed on rotating crops, building up soil, diversifying crops and livestock, and controlling pests naturally. Whenever possible, external resources—such as commercially purchased chemicals and fuels—are appropriately reduced or replaced by resources found on or near the farm. These internal resources include solar or wind energy, biological pest controls, and biologically fixed nitrogen and other nutrients released from green manures, organic matter, or from soil reserves. In some cases, external resources (such as herbicides for no-till systems) may be essential for reaching sustainability. As a result, alternative farming systems can differ considerably from one another because each tailors its practices to meet specific environmental and economic needs.

Alternative farmers who practice soil conservation and reduce their dependence on fertilizers and pesticides generally report that their production costs are lower than those of nearby conventional farms. Sometimes the yields from alternative farms are somewhat lower than those from conventional farms, but they are frequently offset by lower production costs, which leads to equal or greater net returns. Organic farmers, in particular, are often more profitable than their conventional counterparts largely due to price premiums for organically certified products. Research comparing the agronomic, economic, and ecological performance of alternative and conventional farming systems confirms this. In fact, studies generally have found that organic and low-input systems have somewhat lower yields but less variability in production from year to year, are equally if not more profitable, cause less erosion and pollution, have better soil quality, are more energy efficient, and rely less on government subsidies than their conventional counterparts. In other words, the organic and low-input systems seem more sustainable.

7.8 Sustainable Agriculture

Principles and Practices

Just because a farm is alternative does not mean that it is sustainable. For a farm to be **sustainable,** it must produce adequate amounts of high-quality food, conserve resources, and be environmentally safe, profitable, and socially responsible (Table 7.3). So, if an organic farm produces high yields of nutritious food and is environmentally friendly and energy efficient but is not profitable, then it is not sustainable. Likewise, a conventional farm, that meets all the sustainability criteria but pollutes a nearby river with sediment because of soil erosion, is not sustainable. Thus, for any farm to be sustainable, whether it be conventional or alternative, it should meet all the sustainability criteria as listed in Table 7.3 or as deemed important by society.

Before we have sustainable farms, we must have good land that can be farmed. Between 1982 and 1997 about 240,000 hectares (600,000 acres) of cropland in the U.S., lying primarily on the edge of towns, were converted each year to urban use (Figure 7.23). We need to protect our best or prime agricultural lands for use as cropland, pasture, or forestland and build on our lower quality farmland.

Sustainable agriculture addresses many serious problems afflicting U.S. and world food production: high energy costs, groundwater contamination, soil erosion, loss of productivity, depletion of fossil resources, low farm incomes, and risks to human health and wildlife habitats. It is not so much a specific

TABLE 7.3 *Examples of Evaluation Criteria for Farm Sustainability*

Economic	*Environmental*	*Social*
Farm profitability	Energy efficiency	Adequate yields
Operating costs	Soil, water, and air quality	Food and fiber quality
Income variability	Soil and water conservation	Farmland protection from urbanization
Financial risks	Wildlife protection	Farmworker salaries
Food costs	Food and feed safety	Quality of life for farmers
Return on investment	Farm safety	Ethics of farming practice

FIGURE 7.23 *Prime farmland slowly being converted to highway and housing developments near Puyalup, Washington.*

farming strategy as it is a systems-level approach to understanding the complex interactions within agricultural ecologies.

A central component of almost all sustainable farming systems is the **rotation of crops**—a planned succession of various crops grown on one field. When crops are rotated, the yields are usually about 10 percent higher than when they grow in monoculture. In most cases monocultures can be perpetuated only by adding large amounts of fertilizer and pesticide. Rotating crops provides better weed and insect control, less disease buildup, more efficient nutrient cycling, and other benefits. Alternating two crops, such as corn and soybeans, is considered a simple rotation. More complex rotations require three or more crops and usually a 4- to 10-year cycle to complete. In growing a more diversified group of crops in rotation, a farmer is less affected by price fluctuations of one or two crops. This may result in more year-to-year financial stability. There are, however, disadvantages, too. They include the need for more equipment to grow a number of different crops, the reduction of acreage planted with government-supported crops, and the more time and information needed to manage more crops.

Besides a diverse assortment of crops, **diversity** also results from mixing species and varieties of crops and from systematically integrating crops, trees, and livestock. When most of North Dakota experienced a severe drought during the 1988 growing season, for example, many monocropping wheat farmers had no grain to harvest. Farmers with more diversified systems, however, had sales of their livestock to fall back on or were able to harvest their late-seeded crops or drought-tolerant varieties.

Regularly adding crop residues, manures, and other organic materials to the soil is another central feature of sustainable farming. Organic matter improves soil structure, increases its water storage capacity, enhances fertility, and promotes the **tilth,** or physical condition, of the soil. The better the tilth, the more easily the soil can be tilled or direct-drilled with seed (as with no-till), and the easier it is for seedlings to emerge and for roots to extend downward. Water readily infiltrates soils with good tilth, thereby minimizing surface runoff and soil erosion. Organic materials also feed earthworms and soil microbes.

The main sources of plant nutrients in some sustainable farming systems are animal and green manures. A **green manure crop** is a grass or legume that is plowed into the soil or surface-mulched at the end of a growing season to enhance soil productivity and tilth (Figure 7.24). Green manures help to control weeds, insect pests and soil erosion, while also providing forage for livestock and cover for wildlife. Some sustainable farming systems rely mostly on synthetic chemical fertilizers but use them in the most appropriate and efficient amounts. Other sustainable farming systems may take an integrated farming approach, in which synthetic fertilizers are considered a supplement to the nutrients made available from biological nitrogen fixation, organic fertilizers, crop residues, and soil organic matter decomposition.

Controlling insects, diseases, and weeds without chemicals is also a goal of sustainable strategies, and the evidence for its feasibility is encouraging. One broad approach to limiting use of pesticides is commonly called **integrated pest management** (IPM), which may involve disease-resistant crop vari-

FIGURE 7.24 *Sweet clover, a green manure legume, being turned under. Green manure crops, when plowed under or surface-mulched instead, enrich the soil and improve future crop productivity. Legumes are very good green manures because they contribute biologically fixed nitrogen to the soil, thus reducing the need for any synthetic nitrogen fertilizers.*

eties and **biological controls** (such as natural predators or parasites that keep pest populations below injurious levels) (Figure 7.25). Farmers can also select tillage methods, planting times, crop rotations, and plant-residue management practices to optimize the environment for beneficial insects that control pest species or to deprive pests of a habitat. Pesticides are used as a last resort and applied when pests are most vulnerable or when any beneficial species and natural predators are least likely to be harmed.

In practice, IPM programs are a mixed bag. They have dramatically reduced the use of pesticides on crops such as cotton, sorghum, and peanuts. More than 30 million acres (about 8 percent of U.S. farmland) in 1990 were being managed with IPM programs, resulting in annual net benefits of more than $500 million. On the other hand, IPM programs have also been reduced to "pesticide management" for many crops like corn and soybeans, for which pesticide usage has actually increased significantly.

Biological control techniques are some of the best ways to control pests without pesticides. They have been used for more than 100 years and have been commercially successful in controlling pests, especially insects, in more than 250 projects around the world. Yet USDA funds for studying them have declined.

Barriers to Farmers Adopting Sustainable Methods

What are the forces that inhibit farmers from adopting sustainable methods? One obstacle is the federal farm programs, which generally support prices for only a handful of crops. Corn and other feed grains, wheat, cotton, and soybeans receive roughly three fourths of all U.S. crop subsidies and account for approximately two thirds of cropland use. The lack of price supports for other crops effectively discourages farmers from diversifying and rotating their crops and from planting green manures. Suppose, for example, that a farmer was growing wheat on a steep, severely eroding hillside and replaced the wheat with erosion-controlling grass. The benefits from government subsidies, let alone the annual income, for that farmer from the grassed area would be nil.

Government subsidies have given farmers powerful incentive to practice monoculture or short rotations to achieve maximum yields and profits. However, farm legislation, called the "Freedom to Farm Act", passed in 1996 was supposed to snap the decades-old link between crop prices and government subsidies. The law was to end government-guaranteed prices for corn and other feed grains, soybeans, cotton, rice, and wheat. Instead, farmers were to get guaranteed payments that would decline over 7 years and were to have an immediate end to most planting controls. The payments were to total $36 billion over 7 years and terminate at the end of 2002. The idea behind "Freedom to Farm" was that farmers would flourish in an unfettered market. However, the subsidies have not been drying up. Since the Act was passed, Congress has repeatedly passed emergency increases in spending for agriculture. And the subsidies have been flowing faster than ever. In the first three years since the Act passed, government payments to American agriculture tripled. The subsidies have helped the biggest producers, packers, and processors, not the small family farmer.

The long-term economic benefits of sustainable agriculture may not be evident to a farmer faced with having to meet payments on annual production loans. Many conventional farmers are greatly in debt, partly because of heavy investments in specialized machinery and other equipment, and their debt constrains the shift to more sustainable methods. For example, contour farming requires much more tractor time (as well as costly diesel fuel) than straight row farming. It takes considerably more time (and skill) for the farmer to follow the curving topographical pattern than to plow straight up and down the hills. It is estimated that the application of soil conservation measures may add 10 to 20 percent to the farmer's operating expenses—at a time when profit margins on many farms are thin.

A significant portion of today's farmers lease farmland and practice what is called **tenant farming.** An owner-operator may be willing to invest time, energy, and money in effective soil conservation measures, even if the payoff in enhanced crop yields may not materialize for several years. However, a tenant farmer, who often rents a given farm on a year-to-year basis, does not have the same motivation.

More than half the land in the United States is farmed by people over age 55, and almost one fifth is farmed by people over age 65. There are now twice as many farmers over age 65 as there are under age 35. In short, there is a shortage of young farmers. Furthermore, a recent study showed that most farmers

FIGURE 7.25 *Ladybug beetles: natural predators of cherry aphids (shown above) and other insect pests. Integrated pest management (IPM) programs now in use on many farms take advantage of natural predator-prey relationships or other biological-control mechanisms to reduce the need for chemical pesticides. Farmers who practice biological control encourage the proliferation of beneficial microbes and insects, such as ladybugs. They also make their fields generally inhospitable to herbivorous pests.*

who practice sustainable agriculture adopted the practices when they were under age 40. In other words, young farmers are more apt to make changes.

Then, too, there is little information available to farmers on sustainable practices. Government-sponsored research has inadequately explored alternative farming and focused instead on agrochemically based production methods. Agribusinesses also greatly influence research by providing grants to universities to develop chemical-intensive technologies for perpetuating grain monocultures.

Legislative support for change in the U.S. agricultural system is growing, but financial support for sustainable agricultural projects is still only a small part of the total outlay for agriculture. The U.S. Congress created the **Sustainable Agriculture Research and Education Program** (SARE; formerly called LISA, or the Low-Input Sustainable Agriculture program) in 1988 to implement research and education programs in sustainable farming. SARE has a number of objectives: to reduce reliance on fertilizer, pesticide, and other purchased resources to farms; to increase farm profits and agricultural productivity; to conserve energy and natural resources; to reduce soil erosion and the loss of nutrients; and to develop sustainable farming systems. Since 1988, the amount of funding that U.S. Congress has allocated to SARE is less than 1 percent of the total USDA research and education budget during that same time. Shifting mainstream agriculture toward more sustainable methods will require a much larger research and education effort. Universities and the USDA are slowly putting more emphasis on sustainable agricultural research.

Future Research and Education Efforts

Cropland in the United States (Figure 7.26) and other industrial countries has been receiving more nitrogen than plants can effectively use. For example, nitrogen inputs (commercial fertilizer, biological fixation, crop residues, and manure) to all cropland in the United States in 1987 were about 20.9 million metric tons, while only 13.5 million metric tons of nitrogen were used by crops (crop harvest and residues). In other words, nitrogen inputs were more than 50 percent greater than intended outputs, leaving 7.4 million metric tons of excess nitrogen (mostly from commercial fertilizers) to leak into the environment through leaching, erosion, and runoff, or through gaseous losses to the atmosphere. This excess needs to be reduced if nitrogen pollution is to be controlled. A high research priority is the development of specific cropping systems that produce and consume nitrogen more efficiently. The use of slow-release fertilizers provides another means of reducing nitrogen losses from fertilized soils. More, however, must be learned about alternatives to fertilizers and the cycling of nutrients through the agricultural ecosystem.

Integrating modern technologies like global positioning systems with several other computer-based technologies (such as geographic information systems) shows promise in improving the efficiency of fertilizer and pesticide applications on large

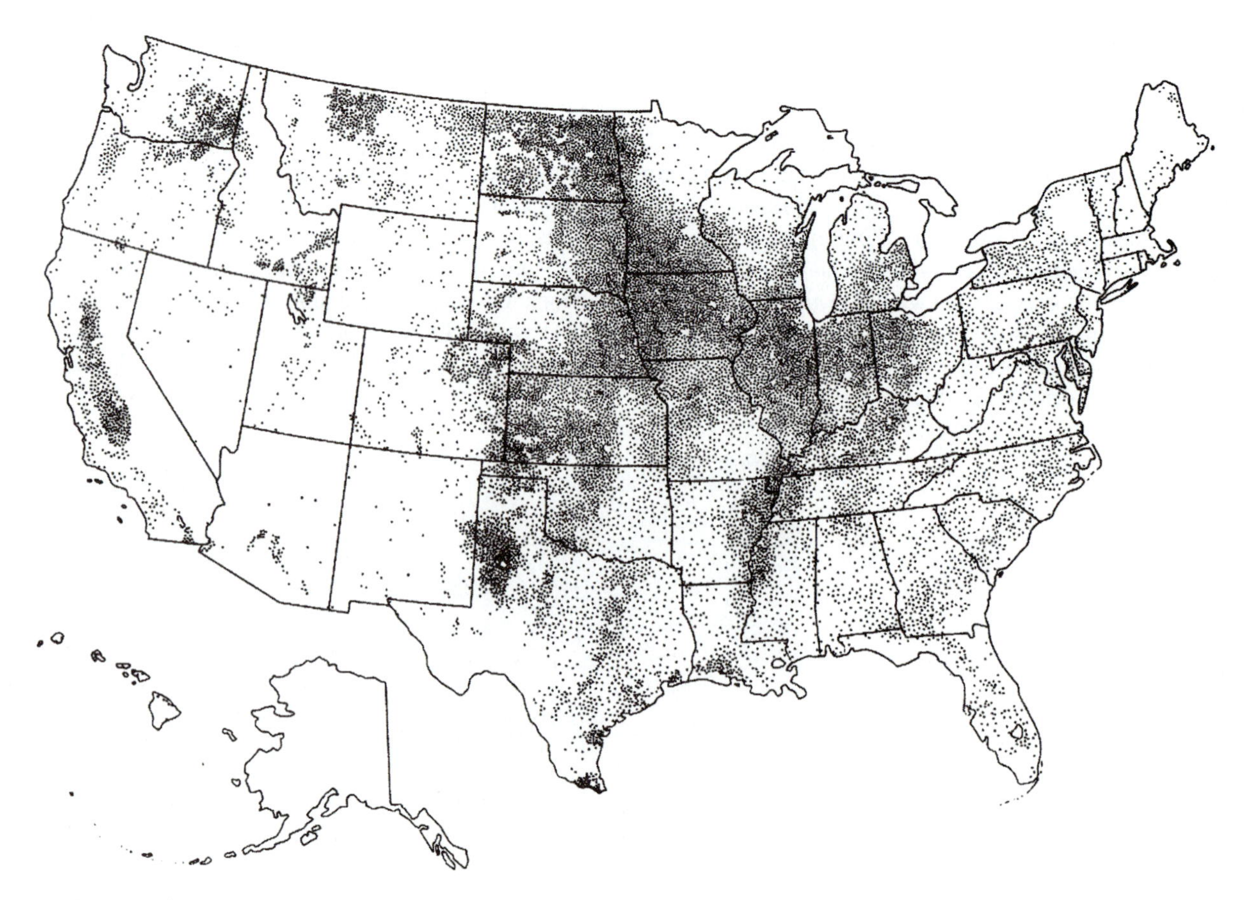

FIGURE 7.26 *Cropland distribution in the United States. Each dot represents 25,000 acres (slightly more than 10,000 hectares).*

agricultural fields. The resulting **precision farming** system attempts to treat each small soil unit (say, 1 hectare in size) of a larger farm field (say, 20 hectares in size) according to its particular properties (such as contents of clay, organic matter, and nutrients). Where considerable variation for relevant soil properties exists from each 1-hectare cell to the next, precision farming should be more efficient than the traditional method of averaging soil properties of all 20 1-hectare cells of a large field and treating the entire field with the same fertilizer and pesticide rates according to the average properties. It remains to be seen if the expense of all the computerized equipment and extra soil sampling and analysis of precision farming systems can be offset by the increase in crop yields and reduction in inputs. For more on precision farming, see GIS Application Focus 7.3.

Effective strategies must also be developed for controlling pests, weeds, and diseases biologically. The strategies may rely on beneficial insects and microorganisms, allelopathic crop combinations (which discourage weed growth), diverse crop mixtures, and rotations and genetically resistant crops. More research should be done on the relative benefits of various cover crops and tillage practices and on integrating livestock into the cropping system.

U.S. farmers now use only a fraction of the thousands of crop species in existence. They may benefit by increasing cultivation of alternative crops such as triticale, amaranth, ginseng, and lupine, which are grown in other countries. Yet in addition to diversification, germ plasm (seeds, root stocks, and pollen) from traditional crops and their wild relatives must be collected and preserved continually. Well-managed collections of germ plasm will give plant breeders a broader genetic base for producing new crops with greater resistance to pests, diseases, and drought. Today much of the germ plasm that U.S. plant breeders use to improve crops comes from developing countries.

New breeds of crops being developed by biotechnology, such as grains that fix their own nitrogen, may eventually be included in sustainable cropping systems. Since no-till practices commonly require increased herbicides, yet at the same time have so many other beneficial environmental aspects, can reduced-herbicide or even organic, no-till systems be developed with the aid of biotechnology? But neither biotechnology nor any other single technology can fix all the problems addressed by a balanced ecological approach. The success of sustainable agriculture hinges on a blending of conservation practices and modern technologies.

Better education is as important as further research. Farmers need to know clearly what sustainable agriculture means, and they must see proof of its profitability. The USDA and the Cooperative Extension Service should provide farmers with information that is up-to-date, accurate, practical and applicable to local farming conditions. Farmers and the public also

GIS AND REMOTE SENSING

GIS, REMOTE SENSING, AND PRECISION FARMING

The integrated use of field observations, computers, geographical information systems (GIS), global positioning systems (GPS), remote sensing (aerial photography), and farm implements is termed precision farming. It is allowing a growing number of farmers to practice site-specific management of their production operations which can improve efficiency, create cost savings, increase yields, and reduce environmental hazards. Site-specific management allows the farmer to selectively apply pesticides, fertilizers, and other agricultural chemicals within a field as opposed to a uniform application of these substances across a field. This means, for example, that a grower can prevent some parts of a field from being over-fertilized or under-fertilized. This alone translates into cost savings because fertilizers are not wasted. In addition, there is less potential of polluting surface or ground waters with nitrates from fertilizers.

The elements that allow all of this to happen are the following: (1) the installation of modern GPS receivers on farm equipment that enables the grower to establish positions within a field to about one meter (or less); (2) data collection devices that can be "location stamped" with geographic coordinates and that can even include on-the-fly digitizing capability such that a grower could sketch conditions on a computer map or digital aerial photo regarding pest infestations or yield variability; (3) point sampling in the field where technical measurements are made of soil nutrient content, soil moisture status, and other physical or chemical properties, all of which are precisely located in the spatial database; (4) the use of GIS in a mobile setting to display information, to derive and evaluate relationships among the data, and to conduct spatial modeling (including the use of remote sensing in the form of large-scale aerial photography), which incorporates data layers of spatial information about the environmental and agricultural properties of the field; (5) the technological development of variable rate farm implements, which allow farmers to incorporate and continuously locate a tractor's position in the field and on GIS maps, and to vary the application rate of field inputs, such as fertilizers, seed spacing, and seeding rates according to precise instructions for each location.

This combining of the technologies of GIS, GPS, aerial photography, ground information and farm implements allows the modern farmer for the first time to truly attempt to incorporate field variability into the day-to-day management of the farm. All problems have not been solved, however. Issues, such as how to incorporate the point sampling of technical data into continuous surface information and how to characterize and visualize error and uncertainty in all facets of the operation, are the focus of much theoretical and applied research today by geographers, soil scientists, and other specialists. Additionally, this technical system of data collection, data integration and management, spatial analysis, and output for decision-making is complicated and evolving rapidly. Since precision farming is a very dynamic enterprise, farmers and land managers need to be well-informed to keep abreast of the scientific and technological changes.

need to be better educated about the potentially adverse environmental and health consequences of soil erosion and the pollution created by certain agrochemical practices.

One of the most effective methods for communicating practical information about sustainable agriculture is through **farmer-to-farmer networks,** such as the Practical Farmers of Iowa. Farmers in this association have agreed to research and demonstrate sustainable techniques on their lands. They meet regularly to share information and compare results. Because such networks have aroused growing interest and have proved effective, the land-grant community should try to promote their development. There are also farmer-to-consumer networks, called **community-supported agriculture (CSA).** CSA networks have been developed throughout the United States; the consumers "subscribe" to farmers and buy their products directly; in turn, farmers bypass all the middlemen and get paid a higher price for their products.

Some scientists and environmentalists have recommended levying taxes on fertilizers and pesticides to offset the environmental costs of agrochemical use, to fund sustainable agricultural research, and to encourage farmers to reduce excessive use of agrochemicals. This approach is precisely how funding for the Leopold Center for Sustainable Agriculture was established by the Iowa State Legislature in 1987 as part of the Iowa Groundwater Protection Act.

Agriculture is a fundamental component of the natural resources on which rests not only the quality of human life but also its very existence. If efforts to create a sustainable agriculture are successful, farmers will profit and society in general will benefit in many ways. More important, the United States will protect its natural resources and move closer toward attaining a sustainable society.

Summary of Key Concepts

1. Soil erosion is a three-step process by which soil particles are detached from their original site, transported, and eventually deposited at a new location. The principal agents of erosion are wind and water.
2. Geological erosion is the wearing away of the Earth's surface by water, wind, ice, or other natural agents under natural environmental conditions, undisturbed by humans.
3. Accelerated erosion is erosion that is much more rapid than normal, geological erosion and results primarily from human activities and sometimes other animal activities.
4. Soil erosion rates are highest in Asia, Africa, and South America and lowest in Europe and the United States. However, the relatively lower rates in Europe and the United States exceed the average rate of soil formation or replacement.
5. The severe dust storms that ravaged the Dust Bowl in the 1930s resulted from a combination of factors, including severe drought, high-velocity winds, overgrazing, and poor soil management. When rain returned and the winds finally subsided, from 2 to 12 inches of topsoil had been removed. Millions of farmers and ranchers in states like Oklahoma, Colorado, Kansas, Texas, and New Mexico suffered severe economic hardship and were forced to seek employment in urban centers.
6. A shelterbelt or windbreak is a wind barrier of living trees and shrubs established and maintained for the protection of farm fields.
7. In the early 1930s soil erosion in the United States was estimated at 3.6 billion metric tons (4 billion tons) of soil removed each year. In 1997 the figure was half or 1.8 billion metric tons, but more land was being cropped in 1997 than in the 1930s.
8. A tolerable soil loss is considered as the maximum combined water and wind erosion that can take place on a given soil without degrading that soil's long-term productivity. More than 34 percent of U.S. cultivated cropland is losing soil through water and wind erosion above the maximum tolerable rate of 11 metric tons per hectare per year.
9. The rate of water erosion is influenced by such factors as (a) rainfall intensity and seasonal distribution, (b) soil erodibility, (c) surface cover, (d) tillage practices, and (e) topography.
10. Effective methods of erosion control include (a) contour farming, (b) strip cropping, (c) conservation tillage, (d) terracing, (e) shelterbelts, (f) gully reclamation, (g) removal of land from production, and (h) conservation tillage.
11. Conservation tillage systems restrict plowing of the soil to reduce erosion and leave enough of the previous crop residues so that at least 30 percent of the soil surface is covered when the next crop is planted. If tillage is omitted completely, the term no-till is applied.
12. The Conservation Reserve Program authorizes farmers to remove, in exchange for government payments, marginal cropland that is most vulnerable to erosion when cultivated. This land is to be planted with grasses, trees, and other long-term cover to stabilize the soil.
13. The Revised Universal Soil Loss Equation (RUSLE) is $A = RKLSCP$, where A = number of tons of soil lost per hectare (acre) per year, R = rainfall and runoff, K = erodibility of soil, L = length of slope, S = steepness of slope, C = cover type, and P = practice used in erosion control. By using RUSLE, fairly good estimates of soil loss can be made on any farm in the United States.
14. Almost 3,000 soil conservation districts, the administrative and operative units of the Natural Resource Conservation Service (NRCS), have been organized to provide technical and financial assistance (by developing farm plans) to farmers so that they can effectively manage each acre of land according to its capability.
15. The USDA Land Capability Classification System evaluates land according to its limitations for agricultural use. It classifies land into eight categories, with Class I the least susceptible to erosion, and Class VIII the most susceptible.
16. Alternative agriculture embraces several variants of nonconventional agriculture that are often called organic, biological, biodynamic, integrated, low-input, natural systems, or no-till. It combines traditional conservation-minded farming techniques with modern technologies.

17. Studies generally have found that organic and low-input systems have somewhat lower yields but less variability in production from year to year and are equally if not more profitable, cause less erosion and pollution, have better soil quality, are more energy efficient, and rely less on government subsidies than conventional farming systems.
18. Just because a farm is alternative does not mean that it is sustainable. For a farm to be sustainable, it must produce adequate amounts of high-quality food, conserve resources, and be environmentally safe, profitable, and socially responsible.
19. Sustainable agriculture addresses many serious problems afflicting U.S. and world food production: high energy costs, groundwater contamination, soil erosion, loss of productivity, depletion of fossil resources, low farm incomes, and risks to human health and wildlife habitats.
20. Central components of most sustainable farming systems are the rotation of crops, the addition of organic materials like crop residues, composts, and animal and green manures, diversity of crops and/or livestock, and integrated pest management, with an emphasis on biological control techniques. When commercial fertilizers are used, they are to be added in the most appropriate and efficient amounts.
21. Barriers to farmers adopting more sustainable methods include government subsidies; unwillingness of older farmers to change; more time, skill, and initial investment needed in applying sustainable farming methods; tenant farming; and little information available to farmers on sustainable practices.
22. The U.S. Congress created the Sustainable Agriculture Research and Education (SARE) program in 1988 to implement research and education programs in sustainable farming. Since 1988, the U.S. Congress has allocated less than 1 percent of the total USDA research and education budget to SARE.
23. Shifting mainstream agriculture toward more sustainable methods will require a much larger research effort. Some future research priorities include developing (a) specific cropping systems that produce and consume nitrogen more efficiently, (b) cost-effective precision farming systems, (c) new biological control strategies, (d) organic or reduced-herbicide, no-till systems, (e) more land in alternate crops, and (f) genetically improved crop varieties.
24. Better education of farmers and consumers through, for example, extension courses, farmer-to-farmer networks, and farmer-to-consumer networks is also needed to move us toward a sustaining agriculture.

Key Words and Phrases

Accelerated Erosion
Alternative Agriculture
Biodynamic Farming
Biological Control
Black Blizzard
Community-Supported Agriculture (CSA)
Conservation Districts
Conservation Reservation Program (CRP)
Conservation Tillage
Contour Farming
Conventional Farming
Crop Rotation
Cultivation
Dust Bowl
Farm Plan
Farm Subsidy
Fertilizer
Food Security Act
Geological Erosion
Green Manure
Gully Reclamation
Integrated Pest Management (IPM)
Land Capability Classification
Legumes
Low-Input Farming
Monocropping
Natural Resource Conservation Service (NRCS)
Natural Systems Agriculture
No-Till Farming
Organic Farming
Pesticide
Precision Farming
Rill Erosion
Runoff
Shelterbelt
Slope
Soil Erosion
Soil Loss Tolerance
Soil Survey Maps
Soil Tilth
Strip Cropping
Sustainable Agriculture
Sustainable Agriculture Research and Education (SARE)
Tenant Farming
Terracing
Universal Soil Loss Equation (USLE)
Water-Stable Aggregates

Critical Thinking and Discussion Questions

1. What is the difference between geological and accelerated erosion? Give one example of each.
2. Were the black blizzards of the Dust Bowl era caused directly by drought? Could they have been prevented despite the drought? Discuss your answer.
3. Make a list of the on-site and off-site effects of soil erosion.
4. Is the level of soil erosion in the U.S. higher or lower today compared to the Dust Bowl era? Discuss your answer.
5. What does it mean to have tolerable soil loss?
6. Discuss the advantages and disadvantages of conservation tillage.
7. What is the difference between contour farming and strip cropping? Can both be done at the same time?
8. What are the benefits of windbreaks? Why are they sometimes taken out?
9. Briefly list five benefits of no-till farming.
10. What is the Revised Universal Soil Loss Equation? How can the farmer use this equation to control soil erosion on the land?
11. What is the Natural Resource Conservation Service? What are some of its responsibilities?
12. Why is there such an interest in alternative agriculture?
13. What is monocropping? What is crop rotation? What are green manures, and can they be incorporated into a crop rotation?
14. Briefly list five benefits of conventional farming.
15. Is there a difference between alternative agriculture and sustainable agriculture? Discuss your answer.
16. What are the barriers that inhibit farmers from adopting sustainable practices?

17. Do you think that U.S. agriculture is on a sustainable path for the future? If so, discuss your answer. If not, discuss the types of changes that are needed to make it more sustainable.
18. Suppose that all commercial fertilizers were banned for agricultural use. Discuss the advantages and disadvantages of this development.
19. Suppose that all pesticides and herbicides were banned for agricultural use. Discuss the advantages and disadvantages of this development.
20. Discuss the statement: "The only energy an ear of corn on your dinner table represents is the solar energy involved in photosynthesis that made the development of that ear possible."

Suggested Readings

Bosworth, D. A., and A. B. Foster. 1982. *Approved Practices in Soil Conservation.* Danville, Ill.: Interstate Printers & Publishers. Practical guide to soil and water conservation for the farmer, rancher, camp manager, contractor, and urban dweller.

Croveto, C. 1996. *Stubble Over the Soil.* Madison, Wis.: Amer. Soc. Agron. Excellent book about no-till farming written by a farmer.

Francis, C. A., C. B. Flora, and L. D. King, (eds). 1990. *Sustainable Agriculture in Temperate Zones.* New York: John Wiley & Sons. Exhaustive review of all aspects of sustainable agriculture in temperate zones.

Gantzer, C.J., S.H. Anderson, A.L. Thompson, and J.R. Brown. 1990. "Estimating Soil Erosion after 100 Years of Cropping on Sanborn Field." *Journal of Soil and Water Conservation* 45: 641. This is the reference for the 100-year case study of the effects of cropping on soil erosion.

Hudson, N. 1995. *Soil Conservation.* Ames, Ia.: Iowa State University Press. Comprehensive coverage of the mechanics and types of soil erosion and their control measures.

Klinkenborg, V. 1995. "Farming Revolution." *National Geographic* 188(6): 60. Wonderfully readable article on sustainable agriculture in the U.S.

Lampkin, N. 1990. *Organic Farming.* Ipswich, England: Farming Press. Excellent book on the principles and practices of organic farming systems.

National Research Council, Board on Agriculture. 1989. *Alternative Agriculture.* Washington, D.C.: National Academy Press. Detailed coverage of the issues in alternative agriculture, including numerous case studies of commercial alternative farms in the United States.

Pimentel, D., et al. 1995. "Environmental and Economic Costs of Soil Erosion and Conservation Benefits." *Science* 267: 1117. This is the reference for the comprehensive study of what the on-site and off-site costs of erosion are and how much it would cost to remedy the effects of erosion.

Smolik, J. D., T. L. Dobbs, and D. H. Rickerl. 1995. "The Relative Sustainability of Alternative, Conventional, and Reduced-Till." *American Journal of Alternative Agriculture* 10(1): 25. One of the best scientific papers to analyze the major sustainability indicators of three different farming systems.

Statz, B. 1993. "The Landscape of Hunger." *Audubon* 95(2):54. Emphasizes the need for soil protection if world hunger is to end.

Trimble, S. W., and P. Crosson. 2000. "U.S. Soil Erosion Rates—Myth and Reality." *Science* 289: 248. Solid discussion of what the scientific data can tell us concerning soil erosion rates in the U.S.

Troeh, F. R., J. A. Hobbs, and R. L. Donahue. 1991. *Soils and Water Conservation,* 2nd ed. Englewood Cliffs, N.J.: Prentice Hall. Full coverage of soil erosion and conservation methods.

Web Explorations

Online resources for this chapter are on the World Wide Web at: **http://www.prenhall.com/chiras** *(click on the Table of Contents link and then select Chapter 7).*

Chapter 8

Pesticides: Protecting Our Crops, Our Health, and Our Environment

Worldwide, approximately 2.6 million metric tons of pesticides are applied to crops, golf courses, and other lands each year. As its name implies, a **pesticide** is a chemical substance that kills pests. Pesticides are sprayed on crops, orchards, swamps, pastures, forests, gardens, golf courses, lawns, trees, and even entire neighborhoods the world over in an effort to control potentially harmful pests, including insects, fungi, weeds, rodents, and others. Despite their initial popularity, pesticides have come under scrutiny worldwide. Initial concern in the United States began in 1962 after the publication of Rachel Carson's *Silent Spring*. Despite increased awareness of the dangers of pesticide use and efforts to reduce our dependency on chemical pesticides, total pesticide use has nearly tripled since the early 1960s in the United States. This increase is largely due to a rise in the use of **herbicides,** chemicals designed to control weeds, which has increased six-fold in that period, while insecticide use remained more or less constant and in recent years has begun to decline.

As the use of these chemical pesticides continues, so does debate over several key questions: How effective are pesticides? What risks do they pose? Are pesticides adequately regulated? Are there safer alternatives? This chapter examines these and a number of other important issues and presents material on alternatives that many experts believe are essential to building a sustainable system of agriculture.

8.1 Where Do Pests Come From?

In undisturbed ecosystems, naturally occurring regulatory mechanisms tend to keep populations of all organisms in a dynamic equilibrium (Chapter 15). This is especially true for the insect species that have become major crop pests, that is, pests that cause significant, costly damage each year to one or more crops. Troubles begin when humans alter natural systems by plowing up prairies and planting large fields of wheat, or leveling forests and planting a single commercial species (Figure 8.1). Such actions destroy the complex web of life in ecosystems, replacing them with

FIGURE 8.1 *Monotype—a pest's paradise. Modern agriculture is based on the planting of monotypes or monocultures, a vegetational plot made up of a single species of plant, such as the citrus trees shown in this photograph of the irrigated Gila River Valley in Arizona. Such monocultures often cover many hundreds of square kilometers.*

highly simplified ecological systems meant to maximize the production of "useful" species that serve human interests. In such instances, ecosystems that once housed several dozen species, perhaps hundreds, are replaced by artificial systems in which a single species dominates. This is called a predatory **monoculture.**

At least two changes of extraordinary importance occur in such instances. First is the loss of natural insect predators—for example, birds and predatory insects that keep potential pests in check. Second, monocultures, often present vast expanses of genetically identical plants that provide an enormous supply of food for pests, in a sense a kind of tinder for an outbreak of hungry pest species. In these simplified ecosystems, then, species that were once held in check by predators and by a limited food supply proliferate rapidly and can cause extensive damage. By reducing biological diversity, human civilization has inadvertently unleashed a force it now must struggle to control. In the United States alone, 19,000 agricultural pests exist. One thousand of these are considered major pests (Figure 8.2).

Pests may also arise from accidental or intentional introduction of insects or other organisms. In their new habitat, the aliens often find little environmental resistance. Populations explode. Consider some examples.

In 1869, the pupae of the gypsy moth, a native of Europe, were shipped from France to Medford, Massachusetts, at the request of a French astronomer, Leopold Trouvelot, of Harvard University, who was interested in developing a disease-resistant silkworm moth. Unfortunately, a few of the insects escaped from captivity and made off into the woods, where they lived in relative obscurity. Twenty years later, however, the town of Medford was crawling with gypsy moth caterpillars—and so are many other areas (Figure 8.3).

Released from density-dependent control agents such as predators and parasites, which had kept this species in check in its native Europe, the gypsy moth population exploded in its new environs. One observer wrote, "The street was black with them...they were so thick on the trees that they stuck together like cold macaroni...the foliage was completely stripped from all the trees...presenting an awful picture of devastation. On a quiet summer night one could actually hear the sound of thousands of tiny mandibles shredding foliage in the trees. Pellets of waste excreted by the larvae rained down from the trees in a steady drizzle."

A single caterpillar can devour a square meter of foliage in a day. Although the larvae prefer oak leaves, they will also consume birch and ash foliage, and when fully grown they will even eat pine needles. Deciduous trees can withstand a single defoliation. Several repeated seasons of defoliation will almost certainly kill them because the loss of leaves eliminates a tree's ability to photosynthesize—to produce the food it needs to grow and survive. Defoliation also makes trees more susceptible to fungi, winds, and drought.

Since its escape from captivity, the gypsy moth has spread throughout the northeastern states and westward into Michigan, Wisconsin, Colorado, and California (Figure 8.4). In recent

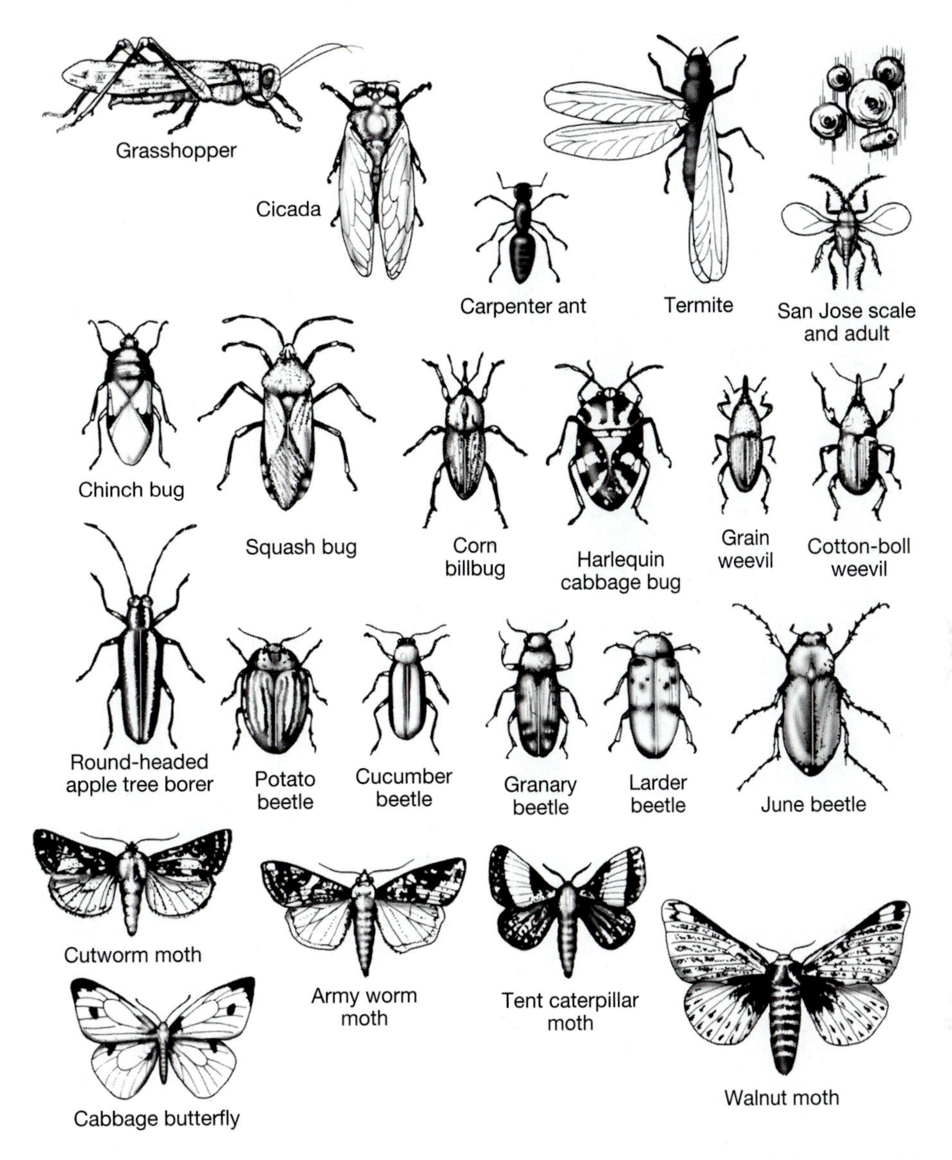

FIGURE 8.2 *Insecticides are employed against the harmful species of insects shown here. Only 0.1 percent of the 800,000 species of insects in the world are considered pests.*

years, the spread of the moth to distant forests has been facilitated by recreational vehicles, on which the moth frequently deposits its eggs.

In Colorado, several communities are now officially quarantined because of gypsy moths. Fearing further spread, California officials are now stopping Coloradans at the border to check their vehicles for eggs. California and other states that strictly enforce controls on gypsy moths require newcomers to be certified free of gypsy moths before they can enter the state. Such controls came in the wake of the severe outbreaks that occurred in 1980 in California and the northeastern United States. In the Northeast, trees were defoliated over a 2.5-million-hectare (5-million-acre) area stretching from Maine to Maryland.

Another exotic or alien species whose impact has been strongly felt is the fungus that causes Dutch elm disease. Accidentally introduced from Europe around 1933, the fungus thrives in the American elm, a tree that, unlike its Dutch counterpart, is not resistant to this organism.

The American elm is a stately tree that once graced parks, boulevards, college campuses, and suburbs throughout much of the eastern United States. Today, most elms are either dead or dying of Dutch elm disease. In one of nature's cruelest ironies, the fungus itself does not kill the tree; the tree kills itself in trying to fight off the organism. Specifically, elm trees produce chemicals to ward off the fungus, but these substances clog the vessels carrying water from the roots to the limbs and leaves. As a result, photosynthesis stops and the trees die.

The spores of the fungus are spread by bark beetles and also from the root of one tree to the roots of adjacent trees. Because it was once customary to plant elms in rows along residential streets, the fungus spreads rapidly from tree to tree. Attempts to stop it by killing the beetles with the insecticide DDT only

FIGURE 8.3 *Leaf-eating caterpillars of the gypsy moth damage hundreds of thousands of dollars worth of forest and shade trees in the northeastern states annually. They hatch in April from eggs laid the previous year.*

succeeded in killing many robins and other songbirds. By 1976, despite vigorous efforts to stop the disease, it had spread from Massachusetts south to Virginia and west to California. More than a million elms die each year. The lovely elms of Santa Rosa, California, once a tourist attraction, are now succumbing rapidly. Virtually gone are the once magnificent elms in St. Paul and Minneapolis.

Alien species are a major problem in the United States. In fact, more than half of the weeds in the United States and many of the most destructive insect pests, such as the cotton boll weevil and the Mediterranean fruit fly, are foreigners that were sometimes intentionally, sometimes accidentally, imported into the United States.

8.2 Types of Chemical Pesticides: An Historical Perspective

Chemical pesticides are not a new development. In fact, a variety of chemical pesticides such as arsenic, ashes, and hydrogen cyanide were used by farmers before World War II. These substances were often highly toxic to people or ineffective against pests. In 1939, the Swiss scientist Paul Müller discovered that a synthetic chemical known as DDT was a powerful insecticide and started a revolution in agriculture with far-reaching impacts on agriculture, people, and the environment.

Chlorinated Hydrocarbons

DDT fits within a class of chemicals called **chlorinated hydrocarbons.** These are organic compounds that contain chlorine atoms. DDT's insecticidal activity spurred research that led to the discovery of a string of chlorinated hydrocarbons, such as endrin, dieldrin, mirex, heptachlor, and Kepone—all of which have since been banned or severely restricted in the United States. They're all nerve toxins that kill pests by altering the function of their nervous system.

Before being banned in the United States, DDT swept the market. It proved to be an extraordinary ally, wiping out lice that infested many soldiers in Europe during World War II. It also proved to be an effective means of controlling the malaria-carrying mosquito in the tropics. So effective was it in this regard that the incidence of malaria worldwide dropped to almost zero. In India, for instance, the number of cases of malaria fell from 1 million per year in the 1950s to 50,000 in 1961.

DDT also proved to be an effective means of controlling a variety of insect pests, saving crops literally on the brink of destruction. Few people questioned the wisdom of the Nobel Prize selection committee when they announced that Müller would receive the award in 1944.

But all of this hoopla had a dark side. Biological studies of the effects of DDT and other chlorinated hydrocarbons caused some scientists and public policy makers to question whether the damage caused by this chemical might outweigh its benefits.

First and foremost, DDT and all other chlorinated hydrocarbons persist in the environment for many years because bacteria lack enzymes needed to break them down. Studies suggest that DDT and its harmful breakdown products remain for 15 to 25 years (Figure 8.5). Today, despite its banning in 1972, DDT and its breakdown product, DDE, which is equally harmful, can still be found in the mud on the bottoms of American lakes and rivers.

DDT also **bioaccumulates**—that is, it concentrates in body tissues, especially fat. Because it is fat soluble, it may remain in fat tissues for decades. Making matters worse, DDT and other chlorinated hydrocarbons build up in food chains, so that the highest-level consumers have levels many hundreds of thousands, sometimes millions, of times higher than the environment (Figures 8.6 and 8.7). This process is known as **biomagnification.**

These problems and others described in the section on the hazards of pesticides caused a furor the world over. As the evidence grew, it became clear that the chlorinated hydrocarbons were too risky to use. Consequently, chemists introduced a new variety of pesticides, the organic phosphates.

Organic Phosphates

Malathion and parathion are the two best known **organic phosphates.** These neurotoxins, which are much more quickly degraded in the environment than chlorinated hydrocarbons

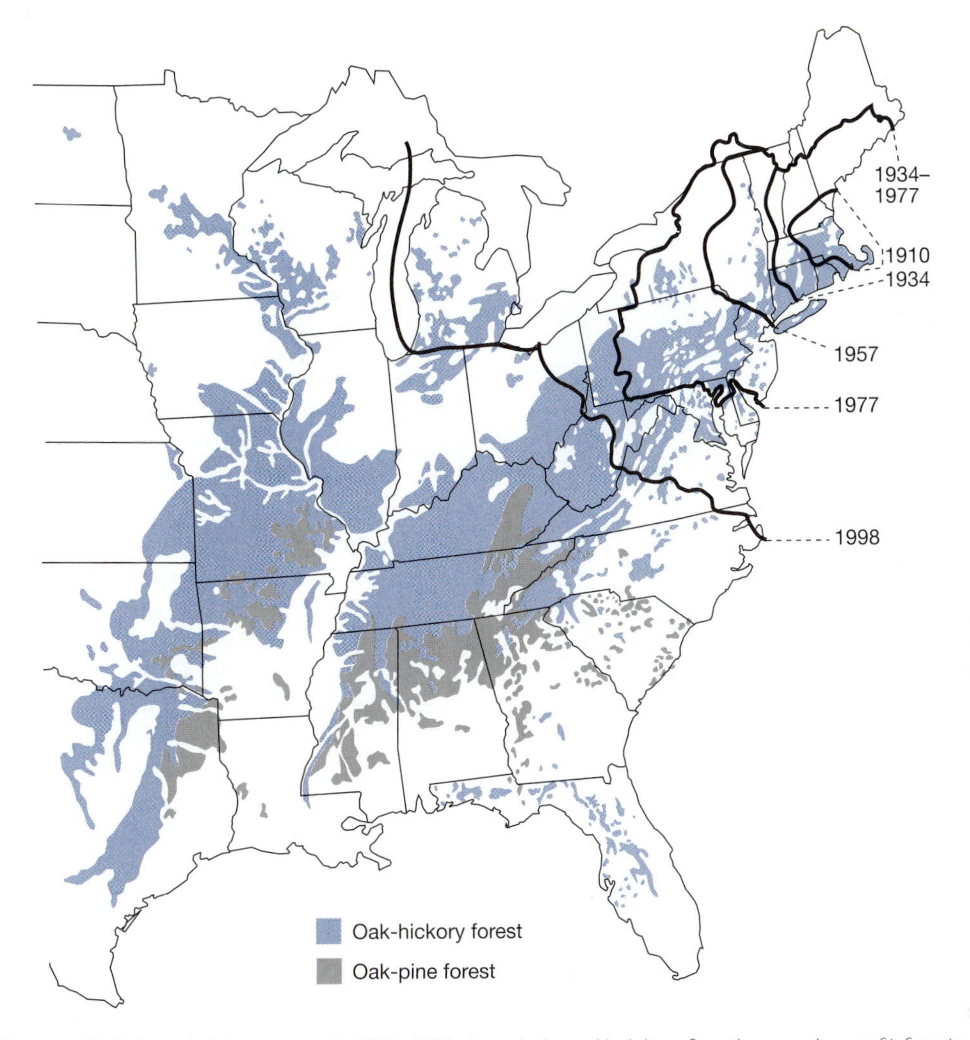

FIGURE 8.4 *Spread of the gypsy moth, 1969–1998, shown by heavy black lines, from the general area of infestation in the Northeast. Male moths have been trapped as far west as Wisconsin and as far south as Alabama. The oak forests shown on the map are potentially vulnerable to gypsy moth invasions.*

and are unlikely to biomagnify because they are water soluble, were originally considered to be safe substitutes for the chlorinated hydrocarbons. However, experience soon showed that even at low levels, organic phosphates were hazardous to humans. Low-level exposure often led to dizziness, vomiting, cramps, headaches, and difficulty breathing. Higher levels led to convulsions and death. Symptoms were most common in farm workers and people living around farms.

Like the chlorinated hydrocarbons, many of the organic phosphates have disappeared from shelves or have been severely restricted in the United States.

Carbamates

To create a safer class of chemicals that biodegraded even more rapidly, pesticide manufacturers developed an entirely new line of chemicals, the **carbamates.** Perhaps the best known is the commercial preparation called Sevin (carbaryl). Carbamates persist in the environment, but only from a few days to 2 weeks at most, and therefore are referred to as nonpersistent pesticides. They are nerve poisons, like the chlorinated hydrocarbons and organic phosphates.

8.3 How Effective are Pesticides?

Each year, weeds, insects, fungi, bacteria, birds, and other pests consume or destroy approximately 42 percent of the annual food production in the United States. By various estimates, pests annually destroy or consume 33 percent of crops in the field and 9 percent of food at various stages after harvest. Damage is worse in the tropics, where two or three crops are grown on a field in a single year and where conditions are ripe for insect growth. If this damage could be prevented, it would greatly increase the available food supply.

The most common form of pest control today entails the use of chemical pesticides. But how effective is it? Studies suggest that chemical pest control works, but not as well as one might suspect. One's perspective on the issue depends on whom one talks to. Farmers and pesticide manufacturers, for

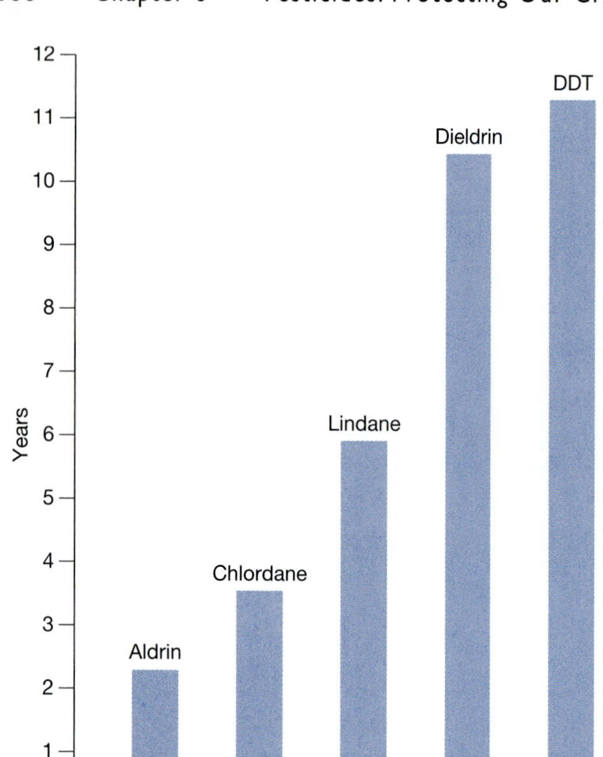

FIGURE 8.5 *Average persistence of pesticides in the soil.*

example, convincingly argue that chemical pesticides have helped farmers produce much more food than would have been possible without their arsenal of chemicals. According to the U.S. Department of Agriculture (USDA), crop production per hectare in the United States has increased more than 70 percent in the past 30 years. In reality, only part of that increase is due to pesticide use; irrigation, fertilizer, and genetic improvements have made much larger contributions.

According to agricultural economists, each dollar invested in pesticides results in about \$2 to \$4 in improved yields. The U.S. Office of Technology Assessment estimates that without pesticides, American farmers would lose an additional 25 to 30 percent of the annual crop, livestock, and timber production (Figure 8.8). The USDA estimates that without pesticides, food bills would be 50 to 75 percent higher. David Pimentel, a Cornell University expert on insect pest control, however, believes these figures are exaggerated. A complete ban on pesticides, he estimates, would increase preharvest losses in the United States only 9 percent—from 33 percent to 45 percent.

Pesticides have helped to increase crop yields in the United States. They have also helped save millions of lives worldwide by killing insects like mosquitoes, lice, fleas, and tsetse flies, which can carry a number of fatal diseases including the West Nile disease, a type of encephalitis (brain infection) transmitted by mosquitoes. This disease was recently unintentionally

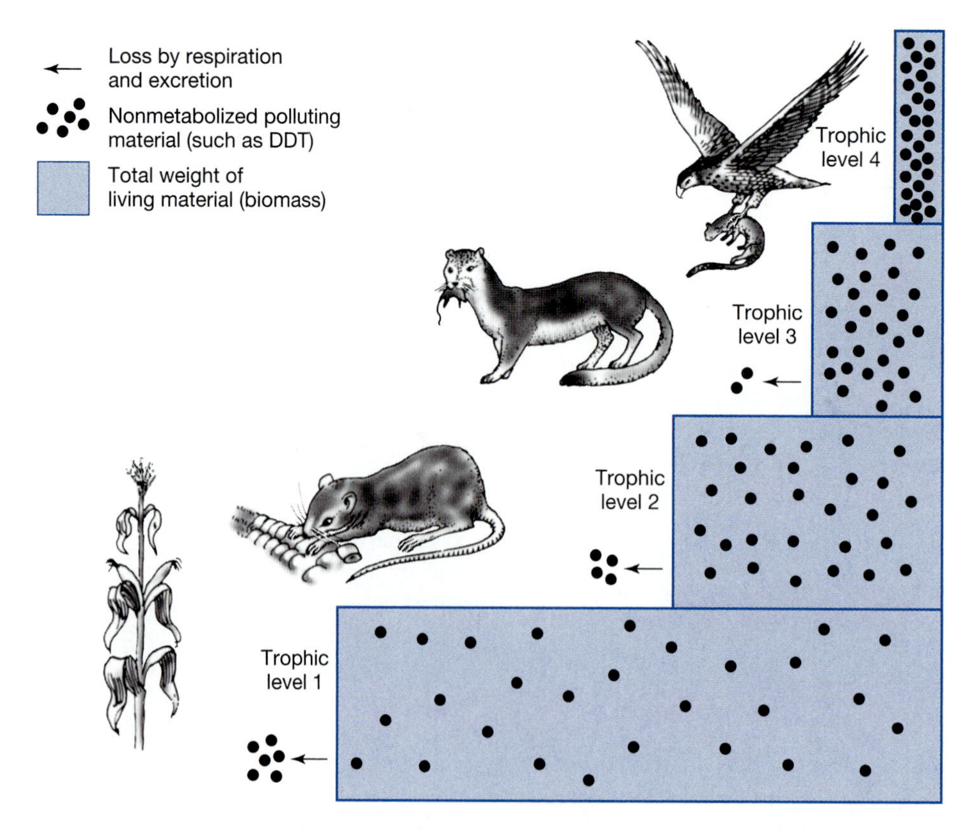

FIGURE 8.6 *Biological magnification: the increasing concentration of certain toxic chemicals, such as DDT, in the food chain. A given organism takes in large amounts of contaminated food. Much of the food may not be converted into protoplasm but may be burned up as fuel during respiration, or may be excreted as waste. However, pollutants, such as DDT, that are taken into the body along with the food may remain inside the cells of the organism. As a result, the concentration of the pollutant increases progressively in the food chain.*

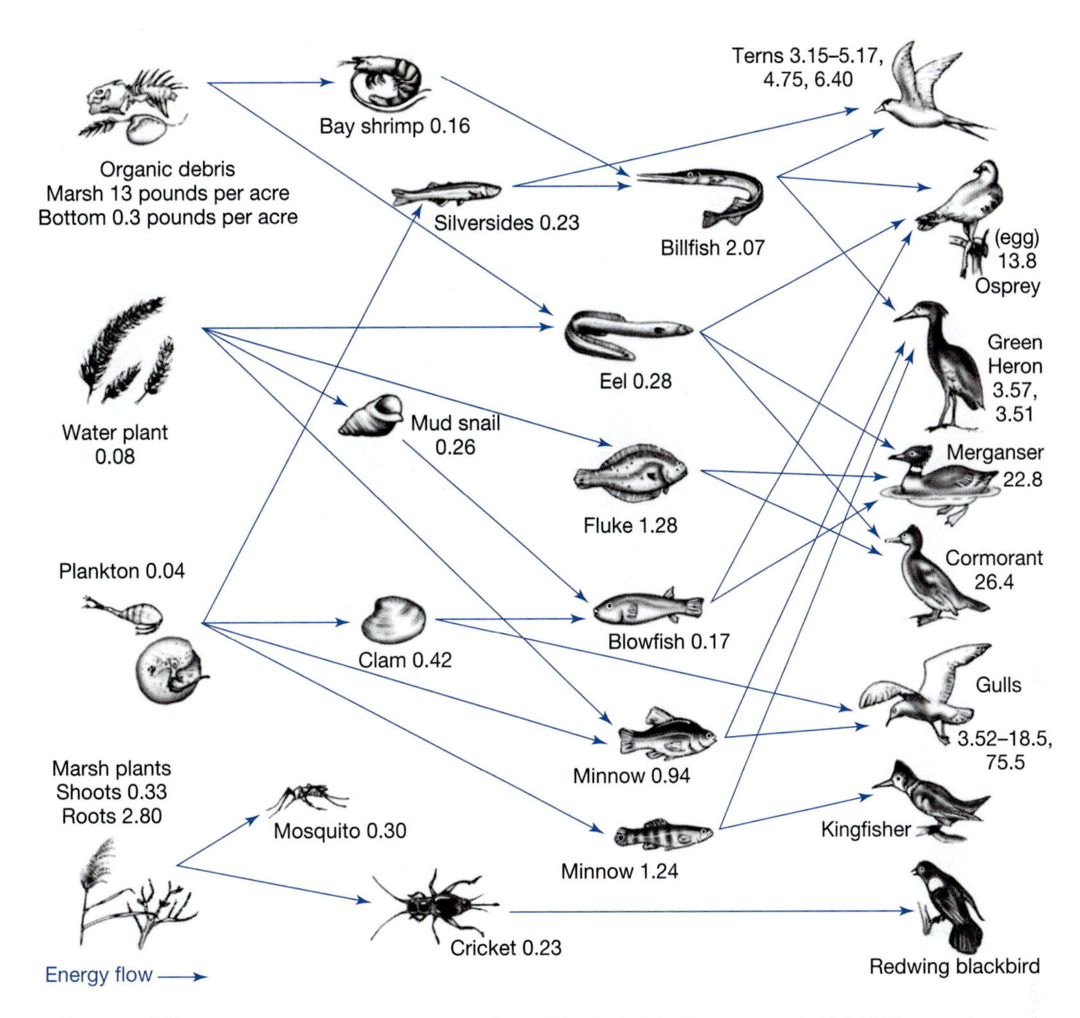

FIGURE 8.7 *Food web of the marsh ecosystem off Long Island, which had been sprayed with DDT for mosquito control. Note the biological magnification of DDT as it moved up the food web. The greatest concentrations were found in the fatty tissues of fish-eating birds, such as gulls and mergansers.*

introduced into the eastern United States. Caused by a virus that is spread from mosquitoes to birds to mosquitoes to people, this disease kills thousands of birds and has killed a number of people, although it is usually not fatal.

Unfortunately, in the past decade scientists have noted that the effectiveness of chemical pesticides has steadily dwindled. In fact, despite continued pesticide use, insect damage to U.S. crops has doubled in the past 30 years. Why?

The reason is twofold. First, many insect pests have become resistant to insecticides. Each time a field is sprayed, a small percentage of the insect population survives because it is genetically resistant to the insecticide. This subset of the original population flourishes in the absence of competition. To kill them and their offspring, farmers must spray again at higher doses. The next application kills off much of the once genetically resistant population, but again leaves behind a small number of even more resistant insects. With time, they may flourish, creating a more difficult-to-control pest. To combat this genetic strain, higher doses or more frequent applications may be necessary. The net effect is that new, resistant strains develop and pesticide use skyrockets. This phenomenon is often called the **pesticide treadmill.** Once on the pesticide treadmill, farmers find it difficult to get off. They may switch to different pesticides to control resistant species, but genetic resistance to the new formulation invariably starts them on the treadmill again. In the 1960s, Nicaraguan farmers sprayed cotton fields five to ten times a year. Today, these same fields are sprayed 30 times a year to control insect pests genetically resistant to malathion.

Today, 550 species of insects and mite species are resistant to at least one insecticide. More than 20 species are resistant to most types of insecticide. Farmers are also finding that weeds and a variety of organisms that cause crop diseases are becoming resistant. By one estimate, 230 plant diseases and 220 weed species have developed resistance to at least one chemical designed to control them. What is most alarming is the fact that the number of pesticide-resistant insect species is increasing rapidly.

The rise in insect damage noted above also results from the destruction of **beneficial insects**—that is, natural predators that help control populations of pests. Ladybugs, praying mantises, spiders, and wasps, are a few of the many beneficial organisms that rid our fields and lakes of potential pest insects free of charge. By poisoning them with pesticides, farmers are

FIGURE 8.8 *Benefits derived from insecticides. Untreated cotton on the left yielded only one-quarter bale per hectare. Cotton on the right, treated with insecticide, yielded 2.5 bales per hectare.*

destroying helpful natural allies that are a part of environmental resistance. Populations of pest species, now unleashed from their natural controls, may surge. In California the spider mite was once an innocuous insect held in check by its natural predators. Today it is the state's leading insect pest, causing over $116 million in damage each year. How did it become such a pest? The answer is, pesticide use destroyed its natural predators.

According to David Pimentel, American farmers spend an additional $120 million a year on pesticides to battle insects that have developed genetic resistance. They spend an additional $150 million per year to destroy pests whose natural predators have been destroyed. If that were not enough, pesticides annually destroy 400,000 bee colonies in the United States. Bees pollinate many commercial crops, such as fruit trees. The destruction of these pollinators reduces crop yields, costing farmers an estimated $135 million a year.

FIGURE 8.9 *Cotton boll weevil attacking cotton boll. Ten percent of the average cotton crop in the United States is destroyed by this weevil. Pesticides have been used to control its populations.*

8.4 How Hazardous are Pesticides?

Pests cause an enormous amount of economic damage, irritation, pain, sickness, and even death. By one estimate, rodents, weeds, and insects in the United States annually cause $2 billion, $5 billion, and $7 billion worth of damage, respectively. As noted earlier, in the United States pests of various sorts consume or destroy about 42 percent of the annual food production. Ten percent of the average annual cotton crop in the United States is destroyed by a single insect species, the cotton boll weevil (Figure 8.9). According to the U.S. Forest Service, pests destroy 5 billion board-feet of timber annually.

Controlling such pests is imperative, but as a study of the impacts of conventional pest-control strategies shows us, safer methods are needed. Let's examine some of the problems of chemical pesticides before considering a more sustainable approach.

Each year Americans apply 560 million kilograms (1,230 million pounds) of pesticides to their farms, forests, golf courses, roadways, rivers, lawns, and gardens (Figure 8.10). That is about 600 grams per hectare, if it were evenly applied to all land in the United States. In reality, however, not all land is treated. Sprayed fields, for instance, may receive 3 to 18 times that amount. In addition, according to the U.S. Environmental Protection Agency (EPA), homeowners often apply pesticides at a rate five to ten times greater than farmers.

Pesticides end up on American soils and in our water, our food, our wildlife, and our people. It is this widespread contamination that worries biologists and health officials. Are these concerns justified?

FIGURE 8.10 *Aerial spraying of sulfur, a fungicide, to control mildew on grapevines, 20 miles south of Fresno, California. Unfortunately, many pesticides applied in this manner drift to neighboring fields and houses.*

Human Health Effects

Western Colorado fruit-grower Dorsey Chism's health took a turn for the worse in 1984. Once a happy, optimistic man, he became irrational and depressed. His face and body bloated, and he was constantly short of breath. Local doctors thought he had emphysema.

In August 1984, after years of spraying his fruit trees with pesticides, sometimes without wearing a protective mask or clothing, Chism went into convulsions and was rushed to Aspen Valley Hospital, where he spent 14 days in intensive care. There a doctor familiar with the health effects of chemical toxins diagnosed his illness as chronic pesticide poisoning. In 1985 Chism died, after months of lingering near death hooked up to an oxygen tank.

Chism is not alone. According to Lewis Regenstein, author of *America the Poisoned,* at least 100,000 Americans are poisoned each year. The National Coalition Against the Misuse of Pesticides puts the number higher, at 300,000. The National Agricultural Chemicals Association, an industry group, claims that the figure is no higher than 20,000 a year. Most people affected are farm and factory workers who have been exposed to high levels of the agents.

Unfortunately, no precise figures on farm worker poisonings are available—the government does not keep them. Nonetheless, there is good reason to believe that the higher estimates are accurate. In California, for instance, the only state that keeps such figures, 1,400 people are poisoned seriously enough to be reported each year. These official figures may grossly understate the real rate. For example, in a 1976 incident in Madera, California, 118 workers were poisoned, but only six cases made the official list. In September 1978, pesticides sprayed on cotton fields drifted to nearby schools, causing respiratory difficulties in children. The incident was serious enough to merit closing the school for a week, but not one case made the official list. By some estimates, fewer than 1 percent of California's poisonings are reported.

All told, 200 to 1,000 Americans like Chism die each year from pesticide poisoning. Worldwide, half a million people are poisoned by pesticides. These poisonings result in 5,000 to 14,000 immediate deaths and numerous chronic and fatal illnesses.

One of the most vulnerable groups is farm workers in less developed nations. Lax or nonexistent laws and regulations, lack of training and safety precautions on the part of employers, and illiteracy, which renders safety instructions on packages useless, all contribute to high exposure levels.

Another problem of concern is the ingestion of pesticides in foods. Some doctors believe that general maladies, such as dizziness, insomnia, indigestion, and frequent headaches, may be caused by ingesting pesticides in food. This promoted a *U.S. News and World Report* journalist to write: "When an American sits down to a typical breakfast, chances are the menu includes bug spray, weed killer, an embalming agent, and arsenic." Dr. Marshall Madell summed up this situation aptly: "Future historians will say we were foolhardy. We sprayed our food with poison, then ate it."

A recent study by the National Academy of Sciences estimates that pesticides contaminating the most common American foods could cause as many as 20,000 cases of cancer per year in the United States, costing $1.1 billion in health care costs in 1980 dollars. Tomatoes, beef, potatoes, oranges, lettuce, apples, and peaches topped the list.

Pesticide residues on food may be minute, but these chemicals can accumulate in body tissues and could cause cancer, birth defects, and other problems. In 1984, for instance, a University of Hawaii researcher discovered that birth defects were higher in the offspring of women who ingested milk contaminated with heptachlor, a chlorinated hydrocarbon, than in women whose milk was free of the contaminant.

In 1970, a sampling of 1,400 Americans showed they had nearly 8 parts per million (ppm) of DDT in their body fat (Figure 8.11). In 1983, 11 years after DDT was banned, the levels were 1.67 ppm. What the long-term effects are, no one knows. Dieldrin, chlordane, heptachlor, and other pesticides long since banned or restricted persist in our bodies.

Herbicides, applied to fields to destroy weeds, have also been linked to a growing number of problems. For instance, the herbicides 2,4-D and 2,4,5-T have been linked to one form of cancer in farm workers. Shelia Hoar, a University of Kansas researcher, found that farmers and farm workers exposed to the herbicide 2,4-D 20 days or more per year were six times more likely to develop non-Hodgkin's lymphoma (a type of cancer) than nonfarm workers. She also discovered those who were most in contact with the chemical—for example, those who mixed the chemicals—were eight times more likely to develop this disease than nonfarm workers. In 1978 Bonnie Hixl, a young mother who lives in Alsea, Oregon, complained to the

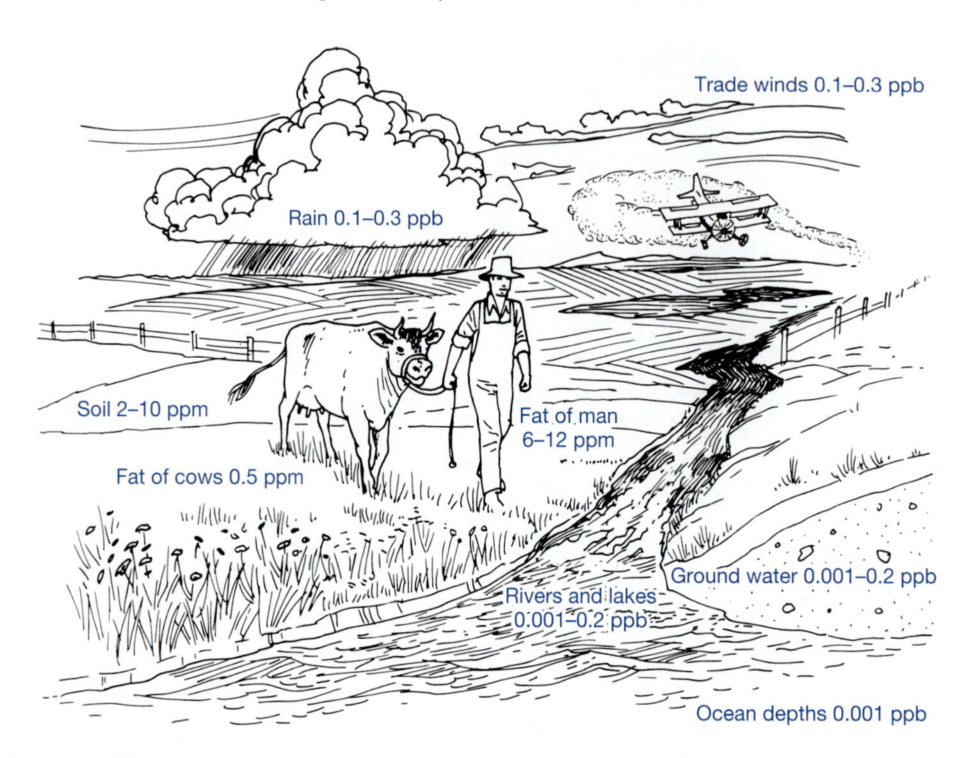

FIGURE 8.11 *Avenues for the dispersal of pesticides, such as DDT, once they are released. Average values for DDT concentrations are indicated in parts per million (ppm) and parts per billion (ppb).*

EPA that she and seven other women had aborted ten times in 5 years. Researchers sent to study the problem found that spontaneous abortions occurred most frequently in these women shortly after the forests were sprayed with the herbicide 2,4,5-T to control brush growth. Although it could not be sure that a cause-and-effect relationship existed, the government banned the use of this chemical for brush control.

Far better known are the multitude of health problems that have resulted from the use of Agent Orange, a 50:50 mixture of 2,4-D and 2,4,5-T, during the Vietnam War. These chemicals were used to defoliate trees along rivers, roadways, and around camps to reduce the chances of ambush of U.S. soldiers and their allies. In some cases, they were sprayed on crops that could have been used to feed enemy soldiers.

Millions of kilograms were sprayed during the war, and it was not too long after aerial sprayings commenced that problems began to arise. U.S. soldiers heavily exposed to Agent Orange complained of headaches, nausea, dizziness, and diarrhea. Many men were afflicted with an irritating skin rash, called chloracne, over large parts of their bodies. Others became uncontrollable and depressed.

Studies showed that Agent Orange was contaminated with a toxic substance known as **dioxin,** which is now believed to be largely responsible for the many health problems reported by soldiers and Vietnamese citizens. Dioxin is a potent toxic substance that causes birth defects and cancer in mice and rats.

In 1969, a Saigon newspaper reported that Vietnamese soldiers and villagers had developed serious health effects, and linked them to the chemical defoliants being liberally sprayed on their land. The newspaper report claimed that miscarriages and birth defects had increased in villages. Initially dismissed as propaganda, the report soon caused a furor in the United States, prompting steps to ban the use of Agent Orange in the war and some domestic uses of 2,4,5-T.

Soon, doctors found that certain cancer rates (for example, testicular cancer) were higher in Vietnam veterans than in the general public and that many men who had been exposed to Agent Orange fathered children with birth defects. For many years, though, health officials denied the validity of veterans' complaints of dizziness, attributing many of them to stress caused by the war. Growing evidence, however, strongly suggests that Agent Orange was indeed the culprit. One study of 40,000 Vietnamese couples, performed by Vietnamese doctors, showed that women whose husbands had fought in areas sprayed with the defoliant were 3.5 times more likely to miscarry or give birth to babies with birth defects than women whose husbands had been lucky enough to avoid the regions.

In 1984, the Veterans Administration reached an out-of-court settlement with U.S. veterans. The VA established a $180 million fund to compensate victims.

Interestingly, recent studies suggest that dioxin may also suppress immune functions. One form of dioxin, TCDD, suppresses the immune system in mice at least 100 times more effectively than corticosterone, one of the body's glucocorticoids. In addition, studies indicate that TCDD may bind to the cell membrane and cytoplasmic receptors that normally bind to hormones, upsetting body functions. Some researchers are calling dioxin an *environmental hormone.*

TCDD also appears to have a variety of direct biological effects—not necessarily involving hormones. In some cells, it causes rapid cell growth. In others, it may alter cellular differentiation. Nonetheless, its effect on the immune system may be far more important than its impact on cancer.

Studies suggest that a number of common herbicides (the thiocarbamates) may upset the thyroid's function and result in the formation of thyroid tumors. These herbicides are chemically similar to thyroid hormone and therefore block the secretion of thyroid-stimulating hormone, resulting in goiter. At higher levels they may cause thyroid cancer.

Pesticides are now becoming a major health concern in cities and suburbs, where they are used on trees, gardens, parks, and golf courses to control a host of insect pests. For years, pest control workers have sprayed suburban areas to control mosquitoes, but they did so usually at night, and only an occasional protest was raised. The lawn-care industry, now a lucrative endeavor worth several billion dollars a year, is also responsible for poisonings. One woman, for instance, was sprayed by a careless applicator as she rode by on her bicycle. Her tongue went numb and her head ached for 3 days after the incident. Children are generally more sensitive and sometimes experience extreme reactions to pesticides, including difficulty breathing. Playing on a recently treated lawn can cause severe chemical burns as well. Adults suffer from headaches, dizziness, and nausea after their lawns are sprayed with pesticides.

In 1982, Navy Lt. George Prior, an avid golfer, died after a severe reaction to a chemical pesticide (chlorothalonil) that had been applied to a golf course. In response to this incident and to reports of bird kills and groundwater contamination, the EPA launched a study of pesticides on golf courses. The EPA's report noted that the United States' 13,000 golf courses annually receive 5,500 metric tons (12 million pounds) of 126 different pesticides applied to control weeds and insects.

As these and other examples show, pesticides pose a risk to the health of the world's people, especially chemical workers and farm workers. People living near farms may also be exposed to toxic levels, because 50 to 75 percent of the spray applied by planes and other methods drifts away to contaminate surrounding fields and homes. Of all the people endangered by pesticide use, farm workers in the undeveloped nations suffer the most because of poor or insufficient worker protection practices.

Although these and other examples suggest that pesticides are highly toxic, many toxicologists, note that the overall impact on human health of toxic chemicals released into the environment is probably small, especially compared with the impact of tobacco smoke. Natural carcinogens may be far more harmful than chemicals applied intentionally to our food crops. Nevertheless, an impact on humans is only one part of the complex ecological equation. We must also consider impacts on other species.

Effects on Fish and Wildlife

Pesticides affect fish and wildlife in many ways, often profoundly. Consider tributyl tin, for example. TBT is a potent biotoxin added to paint, which is applied to ships to prevent the buildup of marine algae and barnacles. These organisms increase a ship's drag, and therefore decrease its speed and fuel efficiency. The U.S. Navy estimates that if the entire fleet were painted with TBT-containing paints, its fuel bill would fall by 15 percent, saving $150 million a year. Underwater cleanings would also be reduced, saving additional money.

Unfortunately, TBT is released from the paint and pollutes bays and harbors. In France, TBT from recreational watercraft caused massive reproductive failure in commercially important oyster beds in 1980 and 1981. Banning the use of this chemical on pleasure craft less than 25 meters long resulted in an abrupt turnabout. Oyster reproduction resumed in 1982 and has continued ever since. In sections of San Francisco Bay, TBT levels were once as high as 500 parts per trillion. Mussels, barnacles, and other marine organisms vanished from the waters as a result.

Pesticides and other chemical pollutants are also believed responsible for an epidemic of fish cancers in waters of the United States. In Puget Sound, for instance, 70 percent of the English sole have liver cancer. Similar findings have been made in a number of U.S. rivers.

Pesticides kill beneficial insects, like honeybees and praying mantises, and birds, as mentioned earlier. In the 1960s and 1970s DDT was responsible for drastic reductions in the populations of a number of flesh-eating birds. Hardest hit were peregrine falcons, ospreys, brown pelicans, and bald eagles (Figures 8.12 and 8.13). Biomagnified in the food chain, DDT reached high levels in fish and insectivorous birds. Predatory birds that preyed on fish and other birds, being the highest on the food chain, ended up with extremely high concentrations of DDT in their tissues. DDT levels in the predatory birds, however, were not high enough to kill adults, but high enough to decrease calcium deposition during eggshell formation. As a result, the eggshells of DDT-contaminated peregrine falcons and other species became thin and were easily broken by the parents (Figure 8.14). The fragile eggs cracked during incubation and killed the embryos. As a result, hatching decreased. By the time scientists had discovered the problem, none of the 200 breeding pairs east of the Mississippi River was able to produce young. Peregrine populations in Europe and the western United States had fallen by 60 to 90 percent. By 1970, the outlook for this regal bird was bleak. The bald eagle and osprey met similar fates.

Today, thanks to captive-breeding programs, hundreds of peregrine falcons have been released into the wild. Free of DDT, the birds may well be able to reestablish their population. Ospreys, bald eagles, and brown pelicans also appear to have weathered the storm and are now making remarkable recoveries.

DDT, which was used in a vain attempt to kill elm bark beetles thought to be responsible for the spread of Dutch elm disease, has also been linked to the death of thousands of robins and other insect-eating songbirds. Many birds died in convulsions shortly after eating insects taken from freshly sprayed trees. Others died from eating earthworms that had been contaminated by DDT. Researchers found that DDT remained on the lower sides of leaves throughout the summer despite rains. The leaves shed in the fall decomposed and released their DDT to the soil. This was taken up by earthworms, which feed on organic matter in the soil. Shortly after the summer spraying, scientists found that worms contained 4 to 400 parts per million (ppm) of DDT. A hungry bird returning in the spring would succumb from eating 11 worms—an hour's snack for a hungry robin. Up to 744 ppm of DDE has been found in the tissues of dead robins.

FIGURE 8.12 *Down and out? Pesticides are responsible for the drastic decline of the Peregrine Falcon population in the United States. A ban on DDT and captive-breeding programs have helped this magnificent bird recover.*

FIGURE 8.13 *Our nation's symbol, the Bald Eagle has been placed on the official list of endangered species. Its decline has correlated closely with the amount of chlorinated hydrocarbon pesticides, such as DDT, occurring in its environment. Now that DDT has been banned, the eagle is recovering nicely.*

In the United States, more than 30 million kilograms (67 million pounds) of chemicals are used each year to control fungi, insects, and weeds on our lawns and golf courses (update).

Birds have also been poisoned by granular carbofuran, a pesticide that eradicates nematodes and insects from corn, rice, and other crops. Although this pesticide apparently poses no threat to humans when applied to crops, it is lethal to songbirds even in minute quantities. In the late 1980s, approximately 2 million birds were dying from carbofuran poisoning each year, according to EPA records.

In summary, although chemical pesticides may reduce pests and save crops, the success of an application is frequently only temporary. Pest populations often rebound because of the destruction of predatory insects and because of genetic resistance. Such short-term successes are often costly to the ecosystem and people, leading to widespread contamination of insects, birds, mammals, and human beings.

8.5 Sustainable Pest Control

Pesticides pose many problems for people and the environment. But are there ways to reduce their use? And are there safe alternatives for farmers that meet the dual criteria of being reliable and economical?

Reducing and Eliminating Pesticide Use by Careful Monitoring

Farmers the world over are finding many ways to reduce pesticide use, saving enormous sums of money in the process. One discovery they have made is that they have been applying pesticides far in excess of what is needed to do the job.

One of the main reasons for such widespread overuse is that many farmers fail to perform careful scientific analyses prior to pesticide application. In other words, they do not determine pest population levels or the distribution of pests in their fields. A cursory examination of their crops may turn up a pest or some pest damage, which triggers an automatic response: a spraying of a whole field. In many cases, farmers apply chemical pesticides on the basis of predetermined schedules, often established by chemical pesticide salespeople. They do this without first checking to see if there is an insect problem.

With costly crop loss at stake, farmers could hardly be blamed for such cautious steps. However, with the cost of farming escalating and the environmental impacts of pesticide use becoming widely known, many farmers are first assessing whether the presence of an insect pest represents a real threat, farmers can apply pesticides much more sparingly, often reducing use by 50 percent with no increase in crop loss. They are also abandoning set schedules, too. Farmers may find that a pest infestation is limited to a small portion of the field, which is sprayed. Or, they may find that the pest populations are at a level that will not create problems, so they do not spray at all.

For a discussion of ways that GIS and remote sensing are being used to monitor pest damage in crops and forests, see GIS and Remote Sensing boxes in this chapter and chapter 7

Integrated Pest Management

Farmers are also learning a whole new repertoire of measures to control insect damage. Some of these are preventive in nature and others use natural pesticides to combat an outbreak or

FIGURE 8.14 *Newly hatched Bald Eagle in nest. One egg has not hatched. DDT contamination of the eagle's food chain has caused an eggshell thinning. Sometimes the embryos inside the abnormal eggs are crushed under the weight of the incubating female.*

GIS AND REMOTE SENSING

USING SATELLITE REMOTE SENSING TO DETECT PEST DAMAGE IN OREGON'S FORESTS

Each year, insects cause enormous damage to the forests of the world. Measuring this damage is vital to forest management, but is time-consuming and expensive. Recent developments in GIS, however, could make the task much simpler and much more precise. To understand the new techniques, let us look first at the old.

For years, forest managers relied on manual mapping of forests to indicate the extent and severity of insect damage. Information from field inventories was simply transferred onto paper maps. Forest managers then turned to a more sophisticated technique, known as aerial sketch mapping. In this technique, trained observers flew over forests and estimated the extent and severity of tree damage visually from the cockpit of planes or helicopters. This data was then transferred to paper maps.

Today, researchers and forest managers have begun to explore a more high-tech and precise solution, remote sensing and GIS. The process begins with infrared satellite images of forests. They're used to assess insect damage, which, in turn, is used to create forest management plans.

One way satellite images can be used to assess damage is by comparing photos taken from different times side by side. With side-by-side infrared photographs to work from, forest managers can examine changes in forest cover over large areas. A decline in green on the more recent satellite image, for example, indicates a loss of tree cover from insect damage. An increase in magenta (bright pink) indicates the possibility of a decrease in vegetative cover. From this qualitative examination, researchers can estimate the amount of change that has taken place over time.

Another way of assessing damage over time is by combining multitemporal imagery (images from different times) to create a single image, which scientists call a change detection image. A change detection image shows areas where vegetation is lost or gained or where it has not changed at all. Although more useful than the previous method, it still provides a qualitative look at change. In other words, no measurements of actual change are represented.

For a more precise, quantitative assessment of change, researchers have developed a software that generates a change detection map—an actual map of an area that shows the percent of change in the vegetative cover of the landscape. Although this technique may sound complicated, it is really basically fairly simple. Scientists use an automated image-processing technique that subtracts pixel values from one satellite image from the pixels of a second image taken at a later time. Land-cover changes are represented by differences in the pixel value changes, and are shown on a second image, the change-detection map. This map shows the approximate range of forest cover loss.

Using remote sensing (satellite images) and GIS equipped with computerized digital imaging processing software, forest managers can obtain fairly accurate depictions of the amount of forest cover lost to insect damage over time in large regions. This technique is cost effective and easy to update.

From this map, forest managers can devise forest management plans that stipulate possible salvage logging operations as well as forest fire prevention and suppression. This new approach also helps managers assess the impacts of insect damage on future harvests, wildlife habitat, and recreational uses of forests. It can also be used to assess other important aspects of the forest, such as habitat change, fire hazards, and forest health.

to keep pests under control. Many of these techniques are extremely effective and far safer to human health and the environment. Interestingly, the USDA's pest control program now spends nearly 70 percent of its money earmarked for pest control on nonpesticide approaches. Even some large chemical companies, like Monsanto, have begun to look for potentially less harmful ways to control pests.

Pest control today often takes an integrated approach, known as **integrated pest management,** which capitalizes on four major strategies: environmental, genetic, chemical, and cultural controls.

Environmental Controls **Environmental controls** are measures that alter the biotic and abiotic environment of the pest. The most important means of environmental control are (1) crop rotation, (2) heteroculture, (3) trap crops, and (4) the use of natural predators, parasites, and disease-causing organisms. These measures are often simple, highly effective, inexpensive, and environmentally benign, and they all are preventive in nature.

Crop Rotation In Chapter 7, we noted that farmers use crop rotation to reduce soil erosion and increase soil fertility. It also helps control pests. How?

Many pest species are highly specialized; they feed on one or a few species of crop plants. For example, the alfalfa weevil feeds mainly on alfalfa, whereas the corn rootworm feeds primarily on corn, and so on. If a farmer plants corn on the same plot year after year, the corn rootworm population grows. Damage can be costly. If, however, a farmer plants corn and oats in alternating years, rootworm populations are kept low. During the years in which the plot is oats, the corn rootworm's food supply is greatly diminished. Food therefore becomes a limiting factor. As a result, crop loss tends to remain low without the use of chemical pesticides.

Heteroculture Monoculture (single-crop) farming simplifies the ecosystem, creating something of a "Garden of Eden" for pests, as noted earlier in the chapter. **Heteroculture,** or planting several crops on a farm, reduces food supplies for pests and is therefore an excellent preventive measure. Heteroculture is often combined with crop rotation. Pests stand little chance against this simple, effective combination. Not only are their food supplies limited, but the supplies change from year to year, so that pest populations cannot get firmly established.

A novel approach to heteroculture is called **intercropping,** which entails planting two different crops in alternating strips. For instance, intercropping corn and peanuts has been shown to reduce corn borers by 80 percent. Researchers believe that this combination works because the peanuts provide habitat for predatory insects that feed on corn borers. Intercropping also reduces the amount of food available for specialized insect pests.

Intercropping can also dramatically increase crop production. In Nebraska, for example, farmers intermix strips of soybeans and corn. This opens up the corn patch, greatly increasing the amount of sun that strikes each plant and raising production by a remarkable 150 percent. The corn protects the soybeans from the drying effects of the wind and raises output by about 11 percent. So, not only do farmers reduce their pesticide use, which saves them money, they also increase their production, which makes them more money per hectare!

Trap Crops Farmers can lure insects from commercially valuable crops by planting low-value **trap crops** nearby. For example, alfalfa can be used to lure the lygus bug from cotton, where it can cause substantial damage. In Hawaii, fields of melons and squash are bordered by rows of corn, which attract melon flies. Insecticides can then be sprayed sparingly on the trap crop to kill pests. Or, to avoid pesticide use, farmers may plow under the insect-infested trap crop or burn it. In Nicaragua, a major exporter of cotton, farmers are required by law to plow under cotton stalks after harvest to reduce boll-weevil damage. Many also plant postharvest trap crops of cotton to attract the harmful boll weevil, which is then destroyed by insecticides. Insecticide use is greatly reduced and environmental contamination is minimized.

Introducing Predators, Parasites, and Disease-Causing Organisms By alternating crops, intercropping, and planting trap crops, farmers can greatly reduce, in some cases even eliminate, pesticide use. One reason is that such practices tend to restore the populations of predatory insects and birds that consume harmful insect pests.

Farmers can also intentionally introduce these biological control agents to help reduce pest problems. The deliberate introduction of natural control agents, such as predators, parasites, and diseases, is one form of environmental control called **biological control.** This form of pest control artificially alters the biotic environment of a pest, creating a more natural condition in which the environmental resistance on the pest population keeps it under control. Moreover, once a natural predator or parasite is established in the more complex farm ecosystem, it may survive and prosper from year to year, holding pest species in check without the use of costly pesticides (Figure 8.15).

Several hundred species of insects, viruses, and bacteria are currently used to control pests. The first major success occurred late in the nineteenth century when agricultural scientists introduced the vedalin beetle from Australia to California to control an insect, the cottony-cushion scale, which was destroying the state's citrus crop. Since then, several hundred other parasites and predators have been successfully introduced in the United States to protect a wide range of crops, including olives, alfalfa, apples, and corn. In California, for instance, entomologists from the University of California, Davis, introduced several parasitic wasps from Iran, Iraq, and Pakistan to control a pesty insect called olive scale, which once caused severe damage. Today, these natural enemies have managed to keep the olive scale completely under control. In Colorado, peach growers have long enjoyed the protection of a natural insect predator that is released in their orchards by state officials. The predator is grown at the state's insectary and released each year in orchards, where it feeds on the Oriental fruit moth, an insect that ruins peaches by boring into the

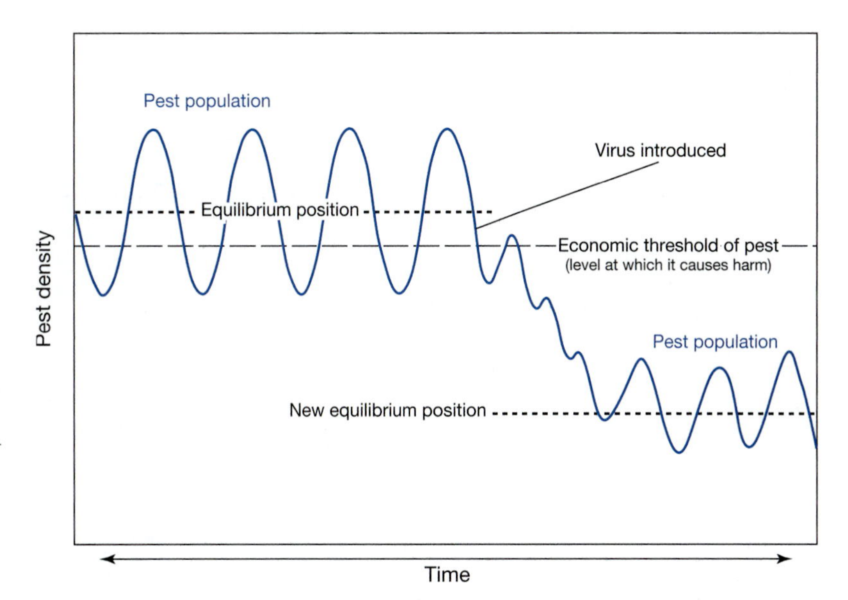

FIGURE 8.15 *Classic biological control resulting in total elimination of an insect pest as an economic problem. Biological control agents include viruses, predators, and parasites. Notice that the pest population increased and decreased initially in response to pesticides, but the introduction of a biological control agent (virus) caused levels of the pest to drop below the economic threshold—i.e., the level at which it causes economic damage.*

fruit's interior, where it is also safe from pesticides. Colorado's insectary also releases eight species of parasites to control the alfalfa weevil. Collectively, they control about 40 percent of the pest population, which, by one estimate, increases alfalfa hay production by as much as $12 million annually.

Bacteria and other microorganisms can also be used to control harmful pests. For example, farmers, gardeners, and foresters now use a bacterium called BT (short for *Bacillus thuringiensis*) to control leaf-eating caterpillars. BT spores are sold as a powder that can be applied to fields and forests. The spores are consumed as the caterpillars devour the leaves of trees and crops. Inside the caterpillars' stomachs, however, the spores hatch into bacteria that release a toxic protein that kills the hungry pests.

BT has been used successfully to control pine caterpillars and cabbage army worms in China. In the northeastern United States, it is being used to help control gypsy moths, which feed on leaves of deciduous trees but, when hungry enough, devour conifers as well. California has used BT for 25 years or so to control a number of caterpillars and mosquitoes, helping the state cut its pesticide use by 90 percent. In Australia, government officials and scientists used a virus to control its burgeoning rabbit population. Early in the twentieth century the European rabbit was introduced into Australia, apparently at the instigation of European immigrants, who longed to indulge once again in their favorite sport of hunting the elusive brushland "bounders." Unfortunately, their numbers sharply increased, for there were no natural predators to control their population size. As their numbers increased, the rabbits began to invade sheep ranges. Under the combined grazing pressure of both sheep and rabbits, the rangelands rapidly deteriorated. Grasses were clipped to ground level. The denuded earth became vulnerable. Dust clouds and sand dunes were the inevitable result. Faced with economic ruin, Australian ranchers banded together in an all-out effort to eradicate the rabbits. They tried all the conventional control methods. They poisoned. They trapped. They staged mammoth roundups. They launched huge rabbit-hunting parties. They even erected a fence several hundred miles long from Queensland to North Wales in an attempt to contain the rabbits. Their efforts were to no avail. Finally, in 1950 government biologists introduced the myxoma virus (mix-OH-mah), lethal to rabbits exclusively, into the target area. It is transmitted to healthy rabbits by virus-carrying mosquitoes, which bite only live rabbits. The immediate results were spectacularly successful. Only one year later, 99.5 percent of Australia's rabbits had succumbed to the virus. Unfortunately, the rabbits gradually developed resistance to the virus, and by 1958 the mortality rate from the virus had dropped to only 54 percent. Rabbit populations are once again on the rise.

Biological control, while effective in many instances, has at least two important limitations. First, it tends to be slower than conventional pesticides. When an insect population explodes, farmers may not have enough time to wait for natural predators to be introduced into their fields and to gain control. By the time they do, the field may be devastated. To compensate for this, farmers must carefully monitor pests and time the release of biological control agents to avoid outbreaks. In addition, farmers can create conditions conducive to beneficial insects to ensure their continued presence.

Second, alien species introduced as biological control agents can become pests in their own right, as described in Chapter 15. Years of research are needed to avoid creating additional pests. Regardless, biological control agents, combined with crop rotation, heteroculture, and other techniques, can provide an important measure of protection.

Genetic Controls Yet another method of controlling pests involves genetic manipulations. Called **genetic control,** it consists of at least two methods: genetic resistance and the sterile-male technique.

Genetic Resistance Chapter 5, on world agriculture, described ways in which scientists are developing high-yield crops through plant-breeding programs and genetic engineering. Scientists are also working on ways to make plants resistant to insect pests. Today, thanks to this research, much of the wheat planted in the United States is resistant to the Hessian fly, once a notoriously destructive insect that caused several hundred million dollars worth of damage a year. Scientists have also developed strains of cotton, soybeans, alfalfa, and potatoes that are resistant to leafhoppers.

Combined with other control measures, genetic resistance provides an environmentally safe method of reducing pest damage while protecting the environment. Through genetic engineering, described in Chapter 5, new strains could be produced to help facilitate the transition to a global sustainable system of agriculture.

Sterile-Male Technique Scientists have also been able to capitalize on a fluke in insect biology. Many female insects breed only once in their lifetime. If this mating is infertile, perhaps because the male is sterile, the female produces no young. Knowing this, scientists have devised an ingenious method of control, known as the **sterile-male technique,** which they have used to control several species of harmful insects. The most notable is the screwworm fly—an insect approximately three times the size of a housefly.

The screwworm fly is widely distributed in South America, Central America, and Mexico and ranges northward into the southern United States, including Georgia, Florida, Alabama, Texas, Arizona, New Mexico, and California.

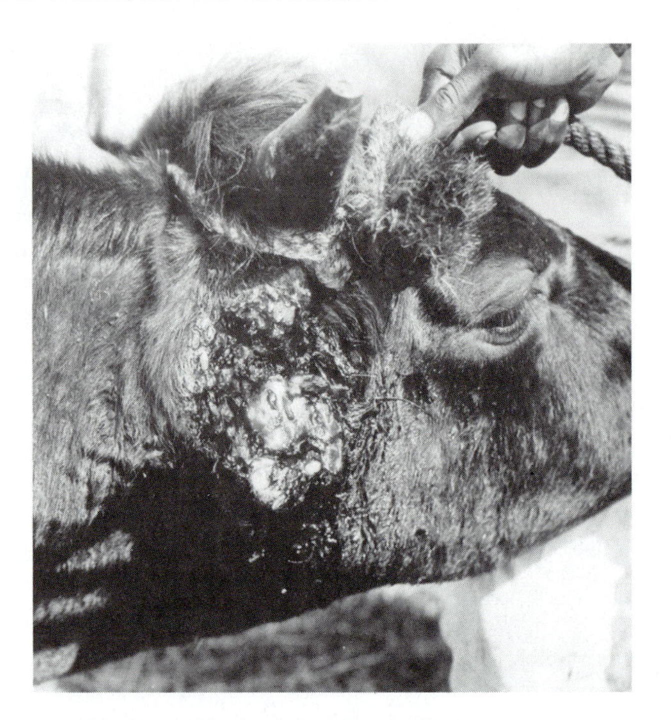

FIGURE 8.16 *Screwworm infestation in the ear of a steer. An untreated, fully grown animal weighing about 1,000 pounds may be killed by several thousand maggots feeding in a single wound.*

Shortly after mating, the adult female deposits about 100 eggs in open wounds of warm-blooded animals, such as cattle or deer. The eggs soon hatch into parasitic maggots that feed ravenously on the flesh and blood of the host (Figure 8.16). As the feeding continues, the wound discharges a fluid that attracts more adult flies. Eventually, in severe infestations, more than 1,000 may feed in a single wound, killing a full-grown, half-ton steer within 10 days. After 5 days of intense feeding, the larvae drop to the ground and pupate. Soon after, they emerge as adults. After mating with a male fly, the female lays its eggs in another open wound, thus completing the life cycle. In a single year, ten generations of flies may be born.

Screwworm flies are a sizable pest. In fact, experts estimate that livestock damage in the United States caused by the flies ranges between $40 million and $120 million per year. In the 1930s, Edward Knipling, chief of the USDA's Entomology Research Branch, had an idea. He reasoned that screwworm flies could be controlled by sterilizing and releasing captive-raised male flies. After highly successful preliminary tests on a Caribbean island, he decided to try this method in the United States. Starting in 1958, Knipling set up a sterilization factory in an old airplane hanger. There, he and other workers sterilized 50 million flies grown in captivity each week by exposing them to radioactive cobalt (Figure 8.17). Over 2 billion sterilized males were packed in boxes and dropped over target areas in the southeastern United States. Boxes dropped from planes popped open when they struck the ground, releasing the flies (Figure 8.18).

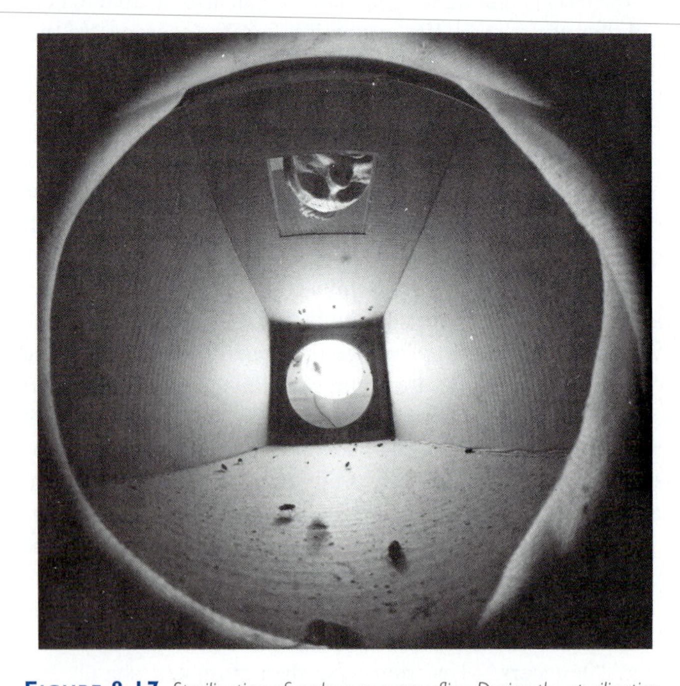

FIGURE 8.17 *Sterilization of male screwworm flies. During the sterilization process, canisters containing 30,000 flies are exposed to a cobalt-60 radiation source at the sterilization factory.*

The sterilized males competed with fertile wild males for mates. When the ratio of sterile to fertile males is 9:1, over 80 percent of the matings are infertile (Table 8.1). As a result, the screwworm fly population gradually declined. Eighteen months after the project was begun, the screwworm fly had been eradicated from the Southeast. The screwworm fly still persists in the Southwest, where it migrates in from Mexico. However, the continued release of sterile males in Mexico and the United States will likely keep the population of screwworms at manageable levels.

The sterile-male technique has been used in California to control the Mediterranean fruit flies, or medflies, which were introduced from Hawaii. The medfly lays its eggs in over 230 fruits, nuts, and vegetables. The larvae that hatch from these

FIGURE 8.18 *Nearly ready for release. Mexican technicians unload containers containing sterile screwworm flies in preparation for their release from an airplane over an infested area.*

eggs then consume the fruit, with potentially devastating effects. The program has worked well for a number of years when combined with other techniques.

This technique has also been used in Africa in the island of Zanzibar off the coast of Tanzania to control disease-carrying tsetse flies. These flies carry a parasitic disease known as trypanosomoisis that affects people and livestock in much of the African continent. For more details, see Case Study 8.1.

The sterile-male technique offers many advantages over traditional pest-control strategies. It is species specific, eliminates or greatly reduces the need for pesticides, and can work when the density of the pest population is low. Researchers note, however, that the technique has its problems. Most important, sterilized males may be less sexually active than fertile males, hence reducing the effectiveness of this approach.

Natural Chemical Controls Over millions of years, many plants have evolved a number of chemical substances to ward off potential enemies. Biodegradable and nonpersistent, many of these chemicals are now being considered for widespread use. For example, rotenone (RO-teh-known), derived from the roots of certain Asiatic legumes, is used today by gardeners against an army of insect pests. Pyrethrum, extracted from chrysanthemums and daisylike flowers by pesticide manufacturers, is also used in home gardens.

Aside from these chemicals, researchers are experimenting with two additional chemical groups: pheromones and insect hormones.

Pheromones Insects and other animal species release a number of chemicals into the environment that influence other members of the same species. These chemical substances are called **pheromones** (fair-uh-moans). One class of pheromones, and perhaps the most important, is the **sex attractants,** chemicals that attract males to females for mating.

Sex attractants help ensure the survival of various species and are an important biological adaptation. Consider how they help the gypsy moth. Although the male gypsy moth has strong, functional wings, the female is far too heavy-bodied for effective flight. After emerging from her pupa case, the virgin female flutters about near the ground or creeps up tree trunks. Soon after emerging she is ready to breed. To attract males, she secretes minute quantities of a sex attractant pheromone, called gyptol, from glands in her abdomen (Figure 8.19A). The sensitive antennal receptors of the male moth detect this scent and, after he locates the female, he mates with her.

Scientists at the USDA synthesized a chemically similar compound, called gyplure, which has proved as effective as gyptol as a sex attractant. This and two dozen other synthetic pheromones are now used in a variety of ways to control insect pests. One of the most common uses is in **pheromone traps** (Figures 8.19B and 8.20). A minute amount of the pheromone is placed in a trap laced with insecticide or some sticky substance that entraps unwary males. Deluded into thinking that a willing female waits in the trap, the male enters and is immobilized or killed by the insecticide. Each year, farmers install thousands of traps to control a variety of insects.

Traps can also be used to determine when pest species emerge in the spring, so that insecticide use can be carefully synchronized to do the most good. In addition, traps can be used to monitor the population levels, so that farmers know when to release predatory insects or if an outbreak is occurring.

TABLE 8.1 *Population Reduction of a Pest Population When a Constant Number of Sterilized Males Are Released in a Pest Population of 1 Million Males and 1 Million Females*

Generation	Number of Virgin Females	Number of Sterile Males Released	Ratio of Sterile to Fertile Males	Number of Fertile Females in the Next Generation
1	1,000,000	2,000,000	2:1	333,333
2	333,333	2,000,000	6:1	47,619
3	47,619	2,000,000	42:1	1,107
4	1,107	2,000,000	1,807:1	Less than 1

CASE STUDY 8.1 TSETSE FLIES BROUGHT UNDER CONTROL IN ZANZIBAR

Many residents of Africa are plagued by sleeping sickness, a debilitating disease caused by a parasitic microorganism transmitted by tiny tsetse flies (Figure 1). The parasitic microorganism, known as trypanosome, also infects cattle bitten by tsetse flies, causing anemia that sometimes leads to death.

FIGURE 1 *Tsetse fly.*

For years, government officials and researchers have attempted to control tsetse flies to reduce the risk of infection to both humans and livestock, which render huge chunks of the Continent uninhabitable or unusable for livestock grazing. This, in turn, reduces the potential output of useful products, among them milk, hides, meat, and fertilizer. Direct losses are estimated to range from $600 million to $1.2 billion per year, according to the United Nations Food and Agricultural Association.

Inhabiting 36 countries, Africa's 22 species of tsetse fly are found in humid and semihumid areas, as well as semi-arid regions. Their presence is an effective deterrence to farming (because land is uninhabitable) and livestock grazing (because cattle are affected). Accordingly, many efforts have been launched to control tsetse fly population, but most of them have a huge environmental cost. For example, to eliminate tsetse flies many African farmers have cleared brushlands in an attempt to destroy tsetse fly habitat. Many native species on which the fly feeds have also been killed. Workers have sprayed habitat from the ground and from airplanes, too. Although spraying proved effective, the ecological cost caused officials to look for other solutions.

One safer solution involved the use of insect pheromones, naturally occurring chemicals released by female insects to attract males of the same species for reproduction. Officials used these attractants to lure tsetse flies into traps containing poisons. Although effective and much safer to people and the environment than widespread pesticide use, this approach did not eradicate the insect as officials had once hoped.

On the tiny, densely populated island of Zanzibar, off the coast of Africa's Tanzania, eradication efforts in the 1980s included living targets, cattle dipped in insecticides. The cattle were essentially sacrificed to eradicate the fly. This technique worked in some areas but not others. New tsetse fly traps were introduced on Zanzibar, too. However, researchers found that the species of tsetse fly that infested the island, didn't enter traps like species on the mainland. Rather, they landed on them. By adding a sticky glue to the outside panels of the traps, and changing the color to blue and white, which seems to be the most appealing trap for the fly, researchers were able to produce an effective means of capturing and immobilizing tsetse flies.

In 1999, however, the program was terminated without achieving full eradication. It was at this time that the governments of Tanzania and Zanzibar launched a successful attempt using sterile males. Releasing tens of thousands of radiation-sterilized males over the landscape in biodegradable boxes, the researchers were—in a few years time—able to eradicate the insect over a wide area. The environmental impact of the project has been minimal, far less than previous campaigns. The elimination of pesticide spraying itself is a big plus.

Already benefits are beginning to accrue. Cattle herds in remote areas are growing, bringing greater prosperity to rural poor peasants. There's even talk of using the technique on mainland Africa. Ethiopia, in Northern Africa, has launched an ambitious program. Conservationists, however, worry that such efforts, while beneficial, may also exact a huge toll by opening up land for settlement and agriculture. Clearly, success will require a balanced approach of eradication followed by sustainable development of regions opened up in the process.

Unsuspecting male insects can also be duped by another ingenious technique (Figure 8.19C). Farmers impregnate wood chips or cardboard shards with sex attractants. Then, flying over infested fields and orchards, they drop their payload. Alerted by the sudden presence of female sex attractant, eager males fly off in search of females, but find only wood chips. Undaunted by the lack of physical similarity, many males mount the wood chips and try to mate with them. Alternatively, farmers may simply spray the sex attractant over the infested area. The males spend much of their time and energy tracking down nonexistent females.

In either case, pheromones decrease the likelihood that a male will find a female, and this technique is appropriately called a **confusion technique.** In one experiment, researchers found that gyplure applied at only 5 grams per hectare (12 grams per acre) reduced mating in gypsy moths by 94 percent.

Sex attractants are nontoxic, species specific, and biodegradable. Moreover, it is impossible for an insect to develop resistance to them without at the same time developing resistance to the very act of reproduction. Since its first use on the gypsy moth, the pheromone technique has been used successfully on

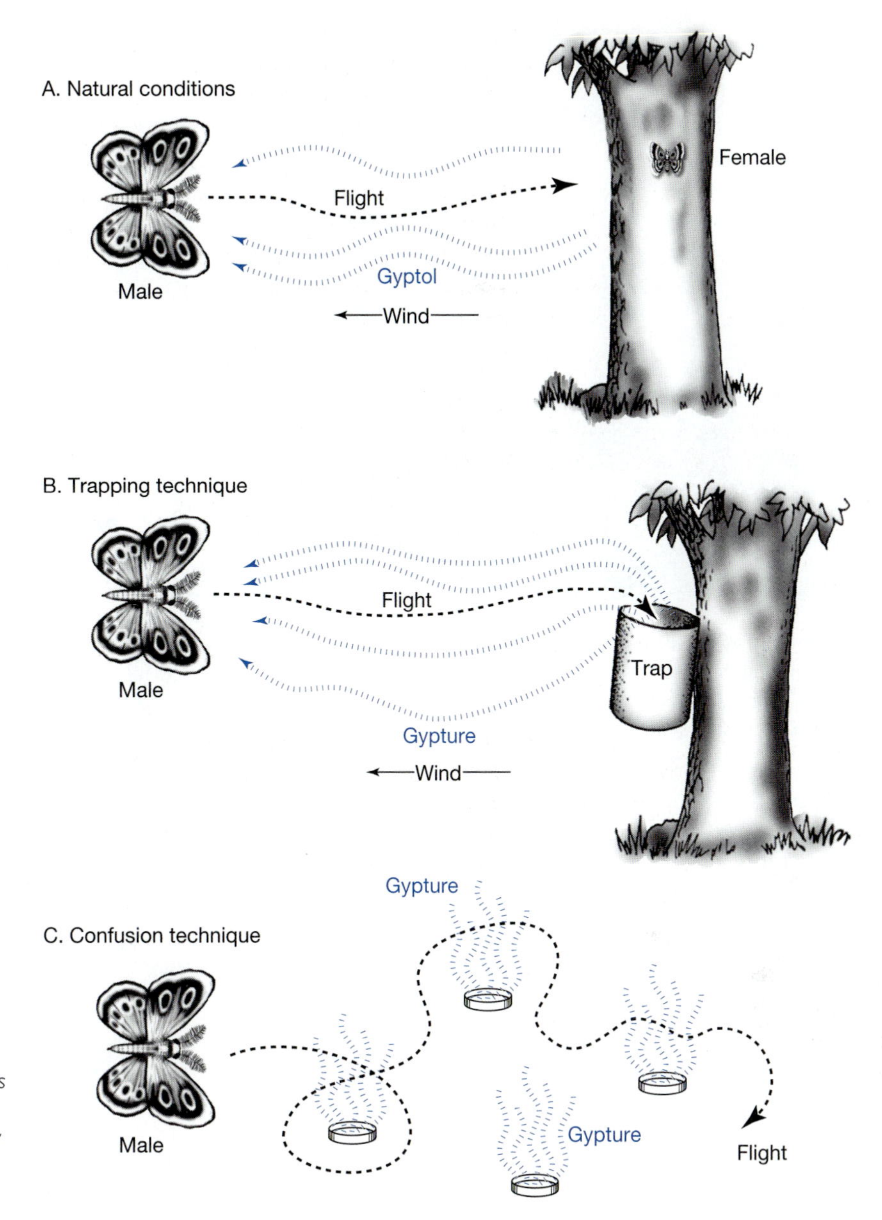

FIGURE 8.19 *Male gypsy moth attracted to a female by the sex attractant gyptol, which she emits (A). Male gypsy moth lured into a trap baited with synthetic gyplure (B). Male gypsy moth, confused by multiple sources of gyplure, is unable to find a female gypsy moth with which to mate (C).*

many other insect pests, including the cabbage borer, the European corn borer, the cotton boll weevil, the Japanese beetle, the tomato hornworm, and the tobacco budworm.

Insect Hormones Many insects hatch from eggs and pass through larval and pupal stages before maturing (Figure 8.21). Caterpillars, for instance, are larvae that develop into moths and butterflies. During the larval stage, the caterpillar produces a hormone, called **juvenile hormone,** that keeps it in its immature state. When levels fall, the insect pupates.

By spraying insects with juvenile hormone, farmers can prevent them from maturing. However, the larval forms of most insects do the greatest damage. Thus, the advantage of this method is that it prevents insects from maturing and prevents them from reproducing, allowing farmers to reduce pest populations over the long haul.

Juvenile hormone has been successfully used to control mosquitoes in Central and South America and could be used on a variety of other pests. Environmentally safe because it is biodegradable, nonpersistent, and nontoxic, juvenile hormone is unfortunately not as species specific as pheromones. As a result, it may kill predatory insects and nonpest species, seriously upsetting the ecological balance. In addition, juvenile hormone acts slower than traditional chemical pesticides. A week or two may be required before hungry crop-eating larvae die, and by that time they may have devastated a crop or forest. Furthermore, juvenile hormone is relatively unstable in the environment—so much so that it often breaks down before it can act. Chemists hope to find ways to make it last longer in the environment. Finally, juvenile hormone must be applied at the precise time that levels of the hormone in the caterpillar fall. Only if it is applied at that time will the larva remain immature.

To avoid these problems, scientists are now developing hormone inhibitors, chemicals that block the secretion of juvenile hormone by larvae. If they are successful, this new line of nontoxic chemicals may cause larvae to mature early, disrupting their life cycle and killing the insects.

FIGURE 8.20 *A typical gypsy moth trap. It contains captured gypsy moths lured into the trap by gyplure, a synthetic attractant that confuses male moths into thinking a female is inside the trap. Once inside, the moth becomes entangled in a sticky substance and is unable to extricate itself.*

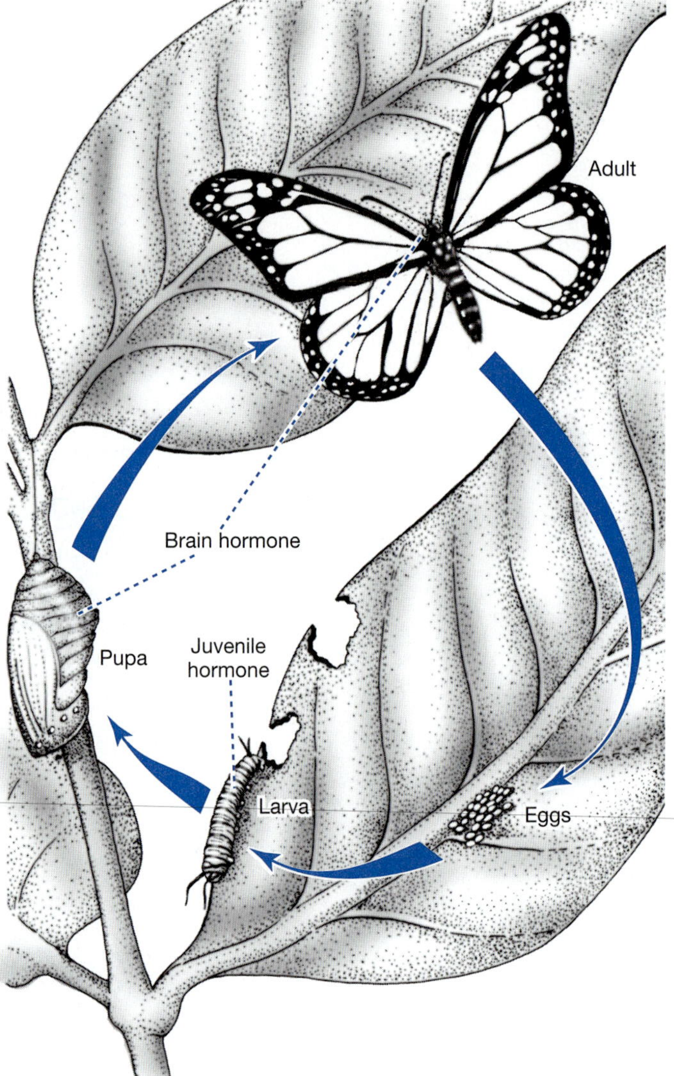

FIGURE 8.21 *Life cycle of insects and the role of hormones in insect metamorphosis. Juvenile hormone, secreted mainly during the larval stage, keeps the caterpillar in this immature state until it is ready to metamorphose into a pupa and adult. The hormones must be secreted in the right amounts at the right time.*

Cultural Control **Cultural control** is a catch-all term covering all of the techniques that do not fit into the other categories of pest control. Examples of cultural control methods include scarecrows and noisemakers that frighten birds away from crops, electrocution devices that zap unsuspecting bugs, and agricultural inspection stations that monitor the flow of fruits and vegetables across state and national borders. All of these are essential elements of integrated pest management.

Changes in our attitudes toward the fruits and vegetables we consume—another cultural control—could also go a long way toward reducing the contamination of our environment with chemical pesticides. Many thousands of tons of insecticide are sprayed on American orchards each year to prevent cosmetic damage to fruits. The San Jose scale, for example, is an insect that creates tiny blemishes on apples, pears, peaches, plums, and apricots, and is now controlled by insecticides. Orchards are sprayed every 10 to 14 days throughout the growing season to produce picture-perfect fruit. Eliminating the spray, at least in part, may yield less aesthetically appealing fruit that tastes just as good as its blemish-free counterpart. The hidden benefit would be a cleaner, healthier environment. Society must choose. Would it rather have perfect fruit or more birds overhead?

Integrated Pest Management: Combining Measures

Environmental, genetic, chemical, and cultural controls offer abundant opportunities to get off the pesticide treadmill and to clean up our environment. Which of these solutions is best?

In practice, the best strategy often involves a combination of methods, an integrated pest management approach. Farmers may plant pest-resistant crops, for example, and may rotate crops, alternate crops, use trap crops, and apply insect hormones or pheromones to reduce pests to economically acceptable levels. Integrated pest management emphasizes control, not the complete eradication of pests, which is often expensive and environmentally disastrous. The various strategies may be used singly or in combination, simultaneously or in sequence, depending on the particular pest. They call on our ingenuity and our knowledge of the life cycles of pests. No doubt insecticides will continue to be used to control insects, but integrated pest management, if successful, could greatly reduce their use.

Numerous studies show that integrated pest management works. In Texas, for example, researchers applied a variety of techniques to control the cotton boll weevil once controlled by heavy doses of pesticides. By planting an early-fruiting variety of cotton that was unappealing to the weevil and by reducing the spacing between rows, reducing fertilizer use, and reducing irrigation water, researchers were able to reduce pest populations and cut back on costly pesticide applications—from 12 per growing season to virtually none. Banks that once balked at lending farmers money for cotton because of average crop losses of $4.60 per hectare ($1.88 per acre) were suddenly thrilled to dole out funds to farmers who enjoyed a $900 per hectare ($364 per acre) profit.

8.6 Are Pesticides Adequately Regulated?

Pesticides are one of the pillars of the modern chemical- and energy-intensive system of agriculture and will be here for many years to come. In the meantime, society must be sure that pesticide manufacture and use are properly regulated. Unfortunately, critics argue that the current system of regulation is wholly inadequate in the United States and abroad.

Federal Regulation

In the United States, formal regulation of pesticides at the federal level began in 1947, when the Congress passed the **Federal Insecticide, Fungicide, and Rodenticide Act (FIFRA).** This law required manufacturers to register pesticides being shipped across state borders with the USDA, and required them to be labeled as well. No attempt was made to control the use of pesticides or to limit potentially harmful chemicals. In actuality, the law provided little protection.

Because of an outpouring of public concern regarding the harmful effects of pesticides in the 1960s, Congress amended FIFRA in 1972, 1975, and 1978. Broadening the scope of the act, the amendments required chemical companies to submit to the EPA information on all new pesticides intended to be sold and used in the United States. Each application specifies the crops and insects on which the pesticide is to be used, supported by research data. The EPA then analyzes the costs and benefits of each pesticide and approves pesticides that are deemed effective and safe. These pesticides are technically referred to as registered. Those that are not considered safe or worth the risk are denied registration and are effectively banned from use.

To gain some measure of control over the use of pesticides, the act also called on the EPA to classify pesticides as either general or restricted. **General pesticides** can be used by anyone. **Restricted pesticides,** however, can only be applied by state-certified applicators—special applicators, or farmers who have been certified. Farmers, for instance, not only need a permit to apply a restricted pesticide, but must keep records of the date, type of pesticide used, the crop, and the amount applied. In addition, the EPA or local agricultural department can inspect applicators' facilities to be certain that the pesticide is being used according to directions.

Although this may sound good, critics argue that certification is lax and inspections are rare. To become certified, applicators must take a course or read a book on pesticide use, then take a test, often an open-book test. Certification, say critics, does not ensure that applicators will use the product as instructed.

FIFRA also permits the EPA to cancel its registration of pesticides when new information suggests they are highly likely to threaten human health and the environment. Cancellations may require months or years of red tape. In the meantime, production and sale are permitted to continue. If, however, the EPA believes that a chemical being considered for cancellation is imminently hazardous to human health and the environment, it can suspend a chemical's use while the cancellation procedure lumbers on. In this process, the EPA weighs the economic, social, and environmental benefits against all costs.

The United States has taken the lead in pesticide management. On December 31, 1972, for instance, the EPA officially banned DDT for all uses except emergencies. (In early 1974, the EPA did permit the U.S. Forest Service to use DDT to control a highly destructive outbreak of the tussock moth in valuable coniferous forests in the Northwest.) In the years since the ban, the concentration of DDT in the soil, water, and wildlife has declined substantially.

In August 1974, after 2 years of hearings, the EPA also banned the general use of two additional chlorinated hydrocarbons, aldrin and dieldrin, considered by some authorities even more toxic than DDT. The EPA also suspended the use of heptachlor, endrin, lindane, Kepone, and toxaphene, all chemical cousins of DDT.

Is the Public Adequately Protected?

Despite laws and regulations on U.S. pesticide manufacturing and production, these substances continue to cause problems. As noted in the previous discussion, one of the biggest problems is the enforcement of safe use. The EPA dictates who can apply what chemicals, but enforcement is generally left to the states, and the level of enforcement varies widely from state to state. Poorly educated farm workers are especially at risk and difficult to patrol. Unable to read and understand labels, they often misuse products or are not provided with protective clothing and headgear. Farmers who ought to know better, like Dorsey Chism, mentioned earlier in the chapter, often apply chemicals excessively without protective gear, and often suffer the consequences. In 1996, two Louisiana men were arrested for applying malathion to dozens of houses for pest control. They did not have a permit and sprayed this highly toxic substance liberally in the homes, causing considerable sickness. What is to stop this from happening elsewhere?

Another problem comes with unanticipated effects. Invariably, products once deemed safe on the basis of toxicity data provided for EPA registration can turn out to have adverse effects on human health and wildlife that were not anticipated by EPA officials. In such instances, public outcry may be needed to get the EPA to reverse a pesticide's registration. Making matters worse, in 1976 the FDA found that Industrial Bio-Test, a leading testing facility that provided toxicity data for many chemical manufacturers, had falsified its results for a decade, providing a rosy picture of harmful substances. Several health-threatening pesticides (toxaphene, paraquat, and DBCP) were registered by the EPA on the basis of Industrial Bio-Test's screenings.

Since 1945, approximately 1,500 chemical pesticides and more than 35,000 different formulations have been introduced to the U.S. pesticide market. A U.S. Congressional investigation found that 60 percent of those in use lacked adequate information on their potential to cause birth defects, 80 percent lacked adequate cancer data, and 90 percent had not been sufficiently tested for possible mutations.

Insufficient testing remains a major stumbling block in the U.S. government's efforts to protect human health. Another major problem is accurately monitoring pesticide levels in the food Americans consume every year, 15 percent of which is imported. Studies have shown that small percentages of domestically produced and imported food are contaminated by unacceptable levels of pesticides.

Closing the Circle of Poisons Nine years after the EPA's ban on DDT, USDA officials turned back a shipload of beef headed for U.S. consumers from Central America because the meat contained unacceptable levels of DDT. This action pointed up a major problem in U.S. pesticide policy: Substances banned here were being sold abroad and reimported in food products, something called the boomerang effect or the circle of poisons.

Today, a dozen chemical companies supply about 90 percent of the world's pesticides. These European and American companies often manufacture and sell pesticides that are banned or restricted in their country of origin, such as chlordane and dieldrin, to less developed countries, which use 30 percent of the world's pesticides and are the fastest growing market. When the substances started showing up in food imported from these countries and began to be detected in migratory birds that wintered in the warm equatorial countries, many people grew alarmed.

Because of pubic interest, Congress amended FIFRA. Under the new amendment, the EPA is required to notify all governments and international organizations worldwide each time it cancels or suspends a pesticide's registration. The EPA also requires manufacturers and exporters of pesticides that have been banned or have failed registration in the United States to notify the purchaser of the status of the pesticide. The EPA also notifies the government through the U.S. State Department. Other countries have agreed to send out similar notifications.

Notification is only a small step, however, in protecting the farmers and farm workers of the undeveloped nations. Countries must have an intact system of internal regulation as well. However, a recent survey by the International Pesticide Industry Association found that only 51 countries have strict controls on pesticides. Forty-three have less stringent controls, and 41 have no controls whatsoever. Mistakes, carelessness, and ignorance resulting from lack of information or from poor training are responsible for the injuries to farmworkers and deaths. Even though container labels describe in the language of the importing country ways to apply the chemicals safely, farmworkers often ignore the warnings or may not be able to read them.

Controlling the Lawn-Care Industry Although agricultural pesticide use is viewed by many as a serious problem, homeowners and lawn care companies apply far more pesticide per acre than farmers. Throughout the nation, citizens have been pressing their representatives to regulate the lawn-care industry—and with growing success. Already, Maryland and a host of towns and counties throughout the United States have passed right-to-know laws and ordinances that require applicators to notify residents in advance of sprayings and to post warning signs after application. Rhode Island, Massachusetts, Minnesota, Iowa, and several other states have similar regulations. Some critics say that this is not enough. Signs are too small and placed at ground level, making them difficult to see. The information they contain is vague and not really a warning.

The lawn-care industry may find it advantageous to switch to integrated pest management, as Jay Kolby, president of Chem-Free Lawns, Inc., of Lancaster, New York (outside Buffalo), did. His thriving business relies on naturally occurring pesticides and biological controls. "It all works a little slower," says Kolby, "but in the long run it's actually better, you get a better lawn."

Modern society has come a long way in regulating pesticides, but much more needs to be done. Stronger laws, much better enforcement, and worker education are badly needed, especially in the Third World.

Many experts believe that we must reduce our use of pesticides, relying instead on integrated pest management. Natural predators and crop rotation, among other methods, hold promise for a pesticide-free world.

Summary of Key Concepts

1. Worldwide, 2.6 billion metric tons of pesticides are used per year. Despite increased awareness of the dangers of pesticide use and efforts to reduce our dependency on these substances, total pesticide use has nearly tripled since the early 1960s in the United States. Insecticide use remained more or less constant during that period

and has recently begun to fall, but herbicide use has climbed sixfold.

2. Insects become pests when humans simplify ecosystems, destroying the complex ecological control mechanisms that hold populations in balance. Planting large expanses of a single or a few crop species also provides insects with an enormous supply of food. Pests may also arise from the accidental or intentional introduction of insects (gypsy moths) or other organisms (rabbits) into environments in which there is little environmental resistance.
3. Each year, pests destroy or consume about 43 percent of the food produced by the United States. Damage is much worse in the tropics, where two or three crops are grown on a field in a single year and where conditions are ideal for insects.
4. Pesticides help control the damage caused by pests. According to agricultural economists, each dollar invested in pesticides results in about $2 to $4 in improved yields. By some estimates, abandoning pesticides would cause a 25 percent to 30 percent reduction in food production in the United States, and food bills would be 50 percent to 75 percent higher. David Pimentel from Cornell University, an authority on pests, however, argues that these figures are exaggerated. A complete ban on pesticides would, he says, increase preharvest losses in the United States by only 9 percent.
5. Pesticides also help control disease-carrying insects such as mosquitoes, which transmit malaria.
6. Scientists are finding that the effectiveness of pesticides is rapidly dwindling because hundreds of insects have become genetically resistant to insecticides. The widespread use of insecticides also destroys beneficial insects and other organisms, like birds, that help hold pest populations in check.
7. Today, at least 550 insects are resistant to insecticides. Twenty species are resistant to all insecticides, and that number is rapidly rising.
8. A variety of chemical pesticides were used by farmers before World War II, including ashes, arsenic, and hydrogen cyanide. In 1939, however, Swiss scientist Paul Müller discovered that DDT, a chlorinated hydrocarbon, was a powerful insecticide. He started a revolution in agricultural pest control.
9. DDT was followed by a string of chlorinated hydrocarbons. Unfortunately, chlorinated hydrocarbons persist in the environment for many years, bioaccumulate, and biomagnify. All of them have since been banned or severely restricted in the United States.
10. As a result of growing discontent with chlorinated hydrocarbons, chemical manufacturers created a new line of pesticides, the organic phosphates. They decompose more rapidly in the environment and are less likely to bioaccumulate and biomagnify. These substances, however, are potent nerve toxins.
11. Pesticide manufacturers produced a newer line of potentially safer insecticides called carbamates, which are even more quickly degraded than the organic phosphates and chlorinated organics.
12. Pesticides end up on American soils and in our water, our food, our wildlife, and our people. Each year approximately 100,000 to 300,000 Americans are poisoned by pesticides, mostly chemical or farmworkers. All told, 200 to 1,000 Americans die from pesticide poisoning. Worldwide, at least 500,000 people are poisoned by pesticides each year, and an estimated 5,000 to 14,000 die.
13. Some doctors believe that general maladies such as dizziness, insomnia, indigestion, and headaches may be caused by ingesting insecticides in food. Some pesticides may also cause cancer and birth defects.
14. Herbicides, applied to fields to destroy weeds, have also been linked to a growing number of problems, such as cancer, birth defects, and nervous disorders. Agent Orange, a herbicide used in the Vietnam War, was found to be contaminated with dioxin, which is believed largely responsible for many health problems reported by veterans and Vietnamese citizens.
15. Pesticides affect wildlife as well. Tributyl tin, a potent biotoxin added to paint to retard the buildup of algae and barnacles, is released from paint on ships' hulls and pollutes bays and harbors, killing algae and shellfish.
16. Pesticides have been implicated in the rash of fish cancers reported in American lakes and rivers. And each year, insecticides destroy an estimated 400,000 bee colonies.
17. DDT has been linked to the decline in Bald Eagle, Osprey, Brown Pelican, and Peregrine Falcon populations in the United States. DDT, scientists found, disrupted calcium deposition in eggshells, resulting in eggshell thinning and low embryonic survival.
18. Scientific research has yielded numerous alternatives to pesticides. Pest control today often combines many techniques, including the judicious use of pesticides. This approach is called integrated pest management. It capitalizes on four major control strategies: environmental, genetic, chemical, and cultural.
19. Environmental controls are measures that alter the biotic and abiotic environment of the pest. The most important ones are (a) crop rotation, (b) heteroculture, (c) use of trap crops, and (d) the introduction of natural predators, parasites, and disease-causing organisms. These measures are often simple, effective, inexpensive, and environmentally benign.
20. Genetic controls include (a) the sterile-male technique, in which sterilized males of the pest species are released into the wild in infested areas to breed with fertile females, which results in an infertile mating and greatly cuts down on pest populations, and (b) genetic resistance, or efforts to improve the genetic resistance of plants and animals to pests through genetic engineering and conventional animal and plant breeding programs.
21. Chemical controls include (a) natural chemical substances produced by plants to ward off insects, (b) pheromones, the sex attractants produced by females to attract males for mating, which can be synthesized and applied to infested fields to confuse males, (c) insect hormones, which can be applied to crops to alter the life cycle of pests and,

ultimately, reduce their number, and (d) the judicious use of chemical pesticides.

22. Cultural controls include all other techniques, such as scarecrows, noisemakers, electrocution devices, agricultural inspection stations, and changes in attitudes about blemish-free fruits and vegetables.
23. Formal regulation of pesticides at the federal level began in 1947 when Congress passed the Federal Insecticide, Fungicide, and Rodenticide Act. Since that time the act has been substantially strengthened to improve control. The amendments require all manufacturers to submit applications to the EPA produce all new pesticides to be sold and used in the United States. The EPA approves only those substances for which it believes the benefits outweigh the potential risks. The EPA also classifies registered pesticides as either general or restricted, and can cancel or suspend the registration of pesticides that experience shows are unsafe. To protect overseas users, the EPA must now notify all governments when it cancels or suspends a pesticide registration. It also requires exporters to notify customers as well if they are importing a substance whose use has been banned or restricted in the United States.
24. Despite strict laws and regulations, pesticides continue to cause problems. Enforcement of proper use remains one of the largest headaches. Additionally, many products deemed safe on the basis of toxicity data provided for EPA registration have caused unexpected damage when used in the field. Many chemicals in use today have not been adequately tested.

Key Words and Phrases

Agent Orange
Beneficial Insects
Bioaccumulation
Biological Control
Biological Diversity
Biomagnification
Cancellation of Registration
Carbamates
Chemical Control
Chlorinated Hydrocarbons
Confusion Technique
Crop Rotation
Cultural Control
DDT
Dioxin
Ecosystem Simplification
Eggshell Thinning
Environmental Controls
Federal Insecticide, Fungicide, and Rodenticide Act (FIFRA)
General Pesticide
Genetic Control
Genetic Resistance
Gyplure
Herbicide
Heteroculture
Hormone Inhibitors
Insect Hormones
Integrated Pest Management
Intercropping
Juvenile Hormone
Monoculture
Natural Predators
Nonpersistent Pesticides
Organic Phosphates
Persistence
Pesticide
Pesticide Treadmill
Pheromone
Pheromone Trap
Registration of Pesticides
Restricted Pesticide
Sex Attractants
Sterile-Male Technique
Suspension of Pesticides
Trap Crops
Tributyl Tin

Critical Thinking and Discussion Questions

1. In what ways does planting monocultures create problems with diseases and pests? How do they affect environmental resistance and the biotic potential of a species?
2. Why do alien species often become pests? Give some examples.
3. Describe the pros and cons of chemical pesticide use.
4. What is the pesticide treadmill?
5. Discuss why insect damage has increased dramatically in the past 30 years despite the use of insecticides.
6. Name the three types of chemical pesticides and give an example of each one.
7. Given what you know about the harmful effects of pesticides, describe a perfect chemical pesticide. What characteristics would it have? Can you think of any potential candidates?
8. Describe some of the health effects of pesticide use. Which sector of our population is most likely to be affected?
9. What is Agent Orange? Where was it used? Why? What are some of the consequences of its use?
10. Describe the effects of pesticides on wildlife. Give some examples.
11. Define integrated pest management. What are the major strategies of this technique? Explain each one and give an example.
12. "Insecticides are the only way to control insects," says one Midwest wheat farmer. Do you agree or disagree? What suggestions might you make to the farmer?
13. Describe how crop rotation and heteroculture keep pest populations in check.
14. What is a trap crop? How can it be used to reduce insecticide use?
15. Describe the pros and cons of biological control.
16. How can scientists alter the genetic resistance of crops and livestock to reduce pest damage?
17. Describe the principle behind the sterile-male technique.
18. Define the terms *pheromone* and *insect hormone*. How are they different and how are they similar? List some pros and cons of pheromones and insect hormones as pest control tools.
19. Describe the major provisions of the Federal Insecticide, Fungicide, and Rodenticide Act.
20. You are a farmer. You want to grow corn, alfalfa, and potatoes without using pesticides. How would you go about it?

Suggested Readings

Ames, B. N., and L. S. Gold. 1989. "Pesticides, Risk, and Applesauce." *Science* 244(4906): 755–767. Looks at natural carcinogens and important considerations for controlling human-produced pesticides.

Brady, C. 1992. "The Perfect Poison for the Perfect Banana." *Global Pesticide Campaigner* 2(3): 8–9. Alarming interview with a Honduran farmworker who applies herbicides.

Carson, R. 1962. *Silent Spring*. Boston: Houghton Mifflin. The book that raised worldwide alarm over the use of pesticides.

Curtis, J., L. Mott, and T. Kuhnle. 1991. *Harvest of Hope*. New York: Natural Resources Defense Council. Documents potential for reductions in pesticide use on nine major U.S. crops.

Di Silvestro, R. 1996. "What's Killing the Swainson's Hawk?" *International Wildlife* 26(3): 38–43. Describes the impacts of pesticide use on a migratory species.

Dreistadt, S. H., and D. L. Dahlsten. 1986. "California's Medfly Campaign: Lessons from the Field." *Environment* 28(6): 18–20, 40–44. A sober look at pest control.

Gardner, G. 1996. "Preserving Agricultural Resources." In *State of the World 1996*, L. Starke, (ed.). New York: W. W. Norton. Covers several important topics including pest management.

Goodstein, C. 1996. "Stood Up by the Birds and the Bees." *Amicus Journal* 18(1): 26–30. A look at the ecological consequences of disappearing birds and insects, both pollinators of flowering plants.

Gup, T. 1991. "Getting at the Roots of a National Obsession." *National Wildlife* 29(4): 18–20. Excellent coverage of the use of pesticides on lawns.

Halweil, B. 1999. "The Emporer's New Crops," *World-Watch* 12(4): 21–29. Examines the controversy over genetically modified crops.

Kinley, D. H. 1998. "Aerial Assault on the Tsetse Fly," *Environment* 40(7): 14–18, 40–41. Tells of a successful effort to control disease-carrying tsetse flies using the sterile male technique.

Lipske, M. 1990. "Natural Farming Harvests New Support." *National Wildlife* 28(3): 19–23. Highlights some alternative farming methods.

McGinn, A. P. 2000. "POPs Culture," *World-Watch* 13(2): 26–36. Examines the problems created by persistent organic pollutants, among them pesticides.

Siedenburg, K. 1992. "Philippine Pesticide Bans Show NGO Strength." *Global Pesticide Campaigner* 2(3): 1, 6–7. Excellent case study showing how nongovernmental organizations can have an impact on pesticide use.

Swezey, S. L., D. L. Murray, and R. G. Daxl. 1986. "Nicaragua's Revolution in Pesticide Policy." *Environment* 28: 6–9, 29–36, Jan.–Feb. Fascinating account of the evolution of Nicaragua's pesticide policy away from chemical insecticides.

Tuxill, J. 2000. "The Biodiversity that People Made," *World-Watch* 13(3): 25–35. Superb article on the loss of genetic diversity in crops.

Weber, P. 1992. "A Place for Pesticides?" *World-Watch* 5(3): 18–25. Superb overview of the pesticide issue, with a frank discussion of options.

Wilcox, F. A. 1983. *Waiting for an Army to Die: The Tragedy of Agent Orange*. New York: Vintage Books. A superb book on the troubles facing Vietnam veterans.

Youth, H. 1994. "Flying into Trouble." *World-Watch* 7(1): 10–19. Documents the startling decline in bird species and the role of pesticides in this tragic decline.

Web Explorations

Online resources for this chapter are on the World Wide Web at: **http://www.prenhall.com/chiras** *(click on the Table of Contents link and then select Chapter 8).*

Chapter 9

Aquatic Environments

Human uses of land and water have had far-reaching impacts on natural ecosystems. In order to manage ecosystems for sustainable use, one must understand the basic physical, chemical, and biological components and functions of those systems. The interrelationships between ecosystems must be understood as well. In this chapter, we introduce the ecosystem characteristics and processes of five important aquatic environments: wetlands, lakes, streams, coastlines, and oceans. We also discuss several human use conflicts associated with wetland and coastal environments. The health and continued productivity of aquatic habitats are essential for the survival of a number of organisms and imperative for successful fisheries management.

9.1 Wetlands

Definition

Many of us could easily recognize swamps, salt marshes, or bogs as wetlands. **Wetlands** are highly diverse and usually occupy a transitional zone within a landscape, lying between a well-drained upland area and a permanently flooded deepwater habitat. These transitional areas are characterized by a water table that is at or near the soil surface or land that is permanently or intermittently flooded with shallow water (Figure 9.1). The term *wetland,* however, is not easily defined and continues to cause controversies. The difficulty began in the late 1970s when regulatory laws relating to the preservation of wetlands were passed and it became necessary to **delineate** or define the boundaries of a wetland. Defining the exact boundary of a wetland on a piece of land now became an economic issue for owners of private lands and for managers of public lands.

Currently, the legal definition of wetland used by most people involved in the regulation and management of wetlands is the one adopted by the U.S. Army Corps of Engineers (USACE) and the Environmental Protection Agency (EPA) in the early 1980s and is worded as follows:

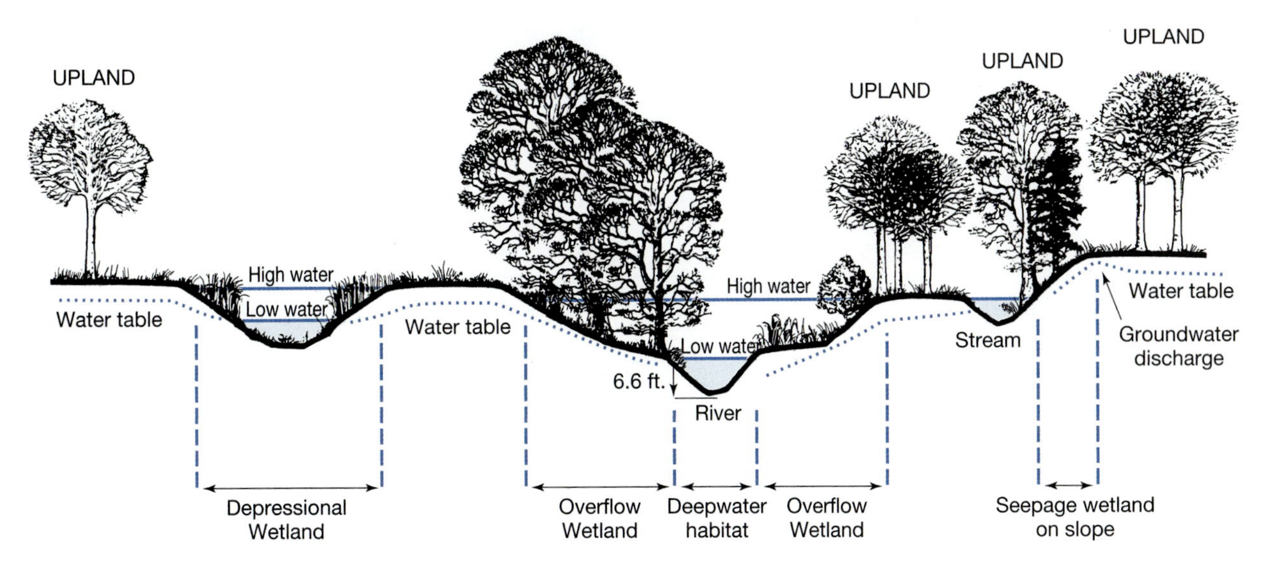

FIGURE 9.1 *Most wetlands occupy a transitional zone between well-drained uplands and deepwater habitats, represented by the overflow wetlands in this diagram. A permanent water depth of at least 2 meters (6.6 feet) is usually considered the boundary between a wetland and a deepwater habitat. Depressional wetlands can form between two upland areas. Seepage wetlands can form on hill slopes exposed to groundwater discharge.*

> The term "wetlands" means those areas that are inundated or saturated by surface or ground water at a frequency and duration sufficient to support, and that under normal circumstances do support, a prevalence of vegetation typically adapted for life in saturated soil conditions. Wetlands generally include swamps, marshes, bogs, and similar areas.

Professional wetland delineators define the legal boundaries of a wetland area using this definition and thorough field investigations to identify the presence of diagnostic environmental characteristics of wetland soils, hydrology, and vegetation. **Wetland delineations** are difficult for a number of reasons, including: (1) boundaries are inherently diffuse, almost never abrupt; (2) water levels vary from season to season; and (3) human land uses, such as farming and logging, alter vegetation, soils, and water regimes.

Classification

For the purposes of inventory, evaluation, and management, the U.S. Fish and Wildlife Service (USFWS) developed a wetland classification system in 1979. This system, known as the **Cowardin Classificaton System,** has become the standard for identifying and classifying wetlands throughout the world. Wetland types are divided into five major categories or systems: marine, estuarine, riverine, lacustrine, and palustrine. Each of these five systems includes a combination of wetlands and deepwater habitats (> 2 meters or 6.6 feet deep) that is influenced by similar hydrologic, geomorphic, chemical, and biological conditions. Wetland systems are further classified into subsystems, classes, and subclasses by identifying tidal water levels, freshwater depths and velocities, dominant vegetative form, and composition of substrate (Figure 9.2 and Table 9.1).

Marine system **Marine wetland systems** include the open ocean overlying the continental shelf and associated high energy coastlines that are subject to waves and currents of the open ocean. Water salinity exceeds 30 parts per thousand (ppt). Examples of marine wetlands include coral reefs, rocky shoreline cliffs, and sandy shorelines that lack appreciable freshwater influence.

Estuarine system **Estuarine wetland systems** include deepwater and wetland habitats that are adjacent to the open ocean but semi-enclosed by land. These areas are protected from high energy waves and currents of the open ocean yet are influenced by tidal fluctuations. Saltwater here is frequently diluted by freshwater runoff via coastal rivers. The lower limit for water salinity in an estuarine system is 0.5 ppt. Two examples of estuarine wetlands are tidal salt marshes and mangrove swamps.

Tidal salt marshes are found in the world's mid- to high-latitude coastal regions. These marshes form near river mouths, in bays, on coastal plains, around lagoons, or behind barrier islands. In these areas, the shoreline is gently sloped, is protected from high-energy waves and storms, and has an accumulation of sediment sufficient to function as the substrate for plant root growth. The majority of the salt marshes in the U.S. are found in states bordering the Atlantic Coast and the Gulf of Mexico and along Alaska's coastline. Fewer and narrower strips of salt marshes occur on the country's Pacific Coast, primarily due to the steepness of the shoreline.

Organisms that thrive in salt marshes are those tolerant of salinity, fluctuations in water levels (inundation at high tide and dry conditions during low tide), and extreme changes in daily and seasonal temperatures. Plants common to the salt marshes of the U.S. are salt-tolerant grasses and rushes and mud algae. A network of **tidal creeks** and shallow ponds throughout the marsh provides passage, spawning grounds, food, and shelter for fish and numerous other marine organisms (Figure 9.3).

Mangrove swamps dominate the coasts of the tropical and subtropical regions of the world. Mangrove swamps have a limited distribution in the U.S., the majority located in

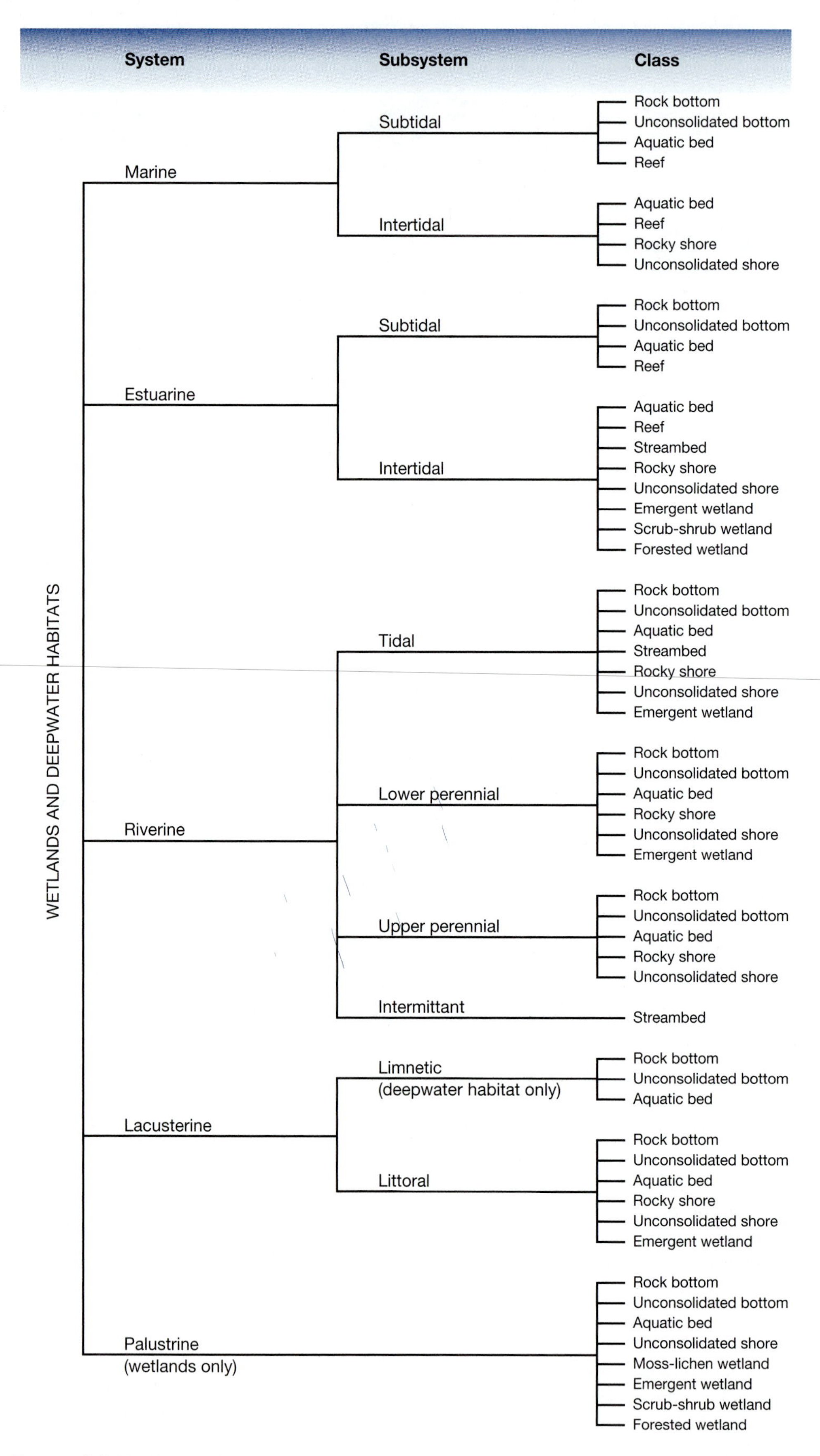

FIGURE 9.2 *The Cowardin Classification System hierarchy of wetlands and deepwater habitats divided into the first three categories: system, subsystem, and class.*

TABLE 9.1 *Descriptions of Classes and Associated Subclasses Used to Classify Wetlands and Deepwater Habitats in the Cowardin Classification System.*

Class	Brief Description	Subclasses
Rock bottom	Generally permanently flooded areas with bottom substrates consisting of at least 75 percent stones and boulders and less than 30 percent vegetative cover.	Bedrock; rubble
Unconsolidated bottom	Generally permanently flooded areas with bottom substrates consisting of at least 25 percent particles smaller than stones and less than 30 percent vegetative cover.	Cobble-gravel; sand; mud; organic
Aquatic bed	Generally permanently flooded areas that are vegetated by plants growing principally on or below the water surface.	Algal; aquatic; rooted vascular; floating vascular
Reef	Characterized by elevations above the surrounding substrate and interference with normal wave flow; they are primarily subtidal.	Coral; mollusk; worm
Streambed	Channel whose bottom is completely dewatered at low water periods.	Bedrock; rubble; cobble-gravel; sand; mud; organic; vegetated
Rocky shore	Wetlands characterized by bedrock stones or boulders with areal coverage of 75 percent or more and with less than 30 percent coverage by vegetation.	Bedrock; rubble
Unconsolidated shore	Wetlands having unconsolidated substrates with less than 75 percent coverage by stones, boulders, and bedrock and less than 30 percent native vegetative cover.	Cobble-gravel; sand; mud; organic; vegetated
Moss-lichen wetland	Wetlands dominated by mosses or lichens where other plants have less than 30 percent coverage.	Moss; lichen
Emergent wetland	Wetlands dominated by erect, rooted, herbaceous hydrophytes.	Persistent; nonpersistent
Scrub-shrub wetland	Wetlands dominated by woody vegetation less than 6 meters (20 feet) tall.	Deciduous; evergreen; dead woody plants
Forested wetland	Wetlands dominated by woody vegetation 6 meters (20 feet) or taller.	Deciduous; evergreen; dead woody plants

southern Florida and in Puerto Rico. Similar to salt marshes, mangrove wetlands require protection from high-energy wave action and need sediment deposits for root growth, although a few mangroves are found on rocky substrate.

Mangrove swamps are named after their dominant vegetation type, the mangrove tree, and are actually forested wetlands consisting of dense, impenetrable stands of these salt-tolerant trees and shrubs and associated understory ferns. Vegetation in mangrove swamps has adapted to thrive in a relatively wide range of salinity and water inundation frequencies. Tidal creeks are also a common feature of these wetlands.

Riverine Systems As the name implies, **riverine wetland systems** encompass the deepwater and wetland habitats associated with rivers and streams. A riverine system begins upstream where tributary streams originate or where the channel departs from a freshwater lake or pond and ends downstream where freshwater mixes with saltwater and salinity exceeds 0.5 ppt.

Riparian zone is a common term used to define the entire linear strip of land on either side of a river or stream that includes **riparian wetlands** as well as land that may not meet the criteria for a legally defined wetland (Figure 9.4). The riparian zone exhibits distinct soil, vegetative, and hydrologic characteristics and is considered an **ecotone,** or transitional area, between the aquatic (river or stream) and upland habitats. Riparian zones are characterized by high species diversity, high species densities, and high productivity and are considered conduits for the exchange of energy and nutrients. Many animals use this area for refuge, food, and travel.

All riparian zones exhibit a linear form but vary considerably in width, from broad alluvial valleys many miles wide in

FIGURE 9.3 *A tidal salt marsh off the coast of Louisiana. Note the branchlike patterns created by the tidal creeks.*

the southeastern United States to narrow strips of vegetation along stream banks in the arid regions of the West.

Geographic location also influences the vegetative community composition in riparian zones due to variations in climate, soils, topography, and geology. For example, large expanses of bottomland hardwood forests dominate the broad, flat floodplain of the Mississippi River and the floodplains of the Southeast's smaller rivers. Many of these riparian zones are connected hydrologically to deepwater swamps. But unlike the permanently flooded deepwater swamps, bottomland hardwood forests consist of species that tolerate shallower water depths and shorter periods of inundation including willow, cottonwood, oak, water hickory, green ash, maple, and birch. In contrast, riparian zones of the arid and semiarid western states are narrow and steep. Trees and shrubs of alder, elm, box elder, hawthorne, sycamore, walnut, and mesquite dominate these areas, as well as willows and cottonwoods, which are found throughout the country.

Lacustrine Systems **Lacustrine wetland systems** include deepwater (limnetic) and wetland (littoral) habitats of lakes, reservoirs, and large ponds (refer to Figure 9.8). These areas are situated in a topographic depression as in the case of a lake or a pond or formed by damming a river (reservoir). Lacustrine systems are greater than 8 hectares (20 acres) in size, experience permanent or intermittent flooding, consist of <30% vegetative cover, and can be tidal or nontidal as long as water salinity remains <0.5 ppt. Upland habitat or vegetation-dominated wetlands form the outer boundaries of lacustrine wetland systems (Figure 9.5).

Palustrine Systems The majority of all wetlands fall under the category of **palustrine wetland systems.** These systems are dominated by vegetation and are bound by upland habitat or any of the other four systems. Water depth in the deepest part of the basin is <2 meters (6.6 feet) at low flow and water

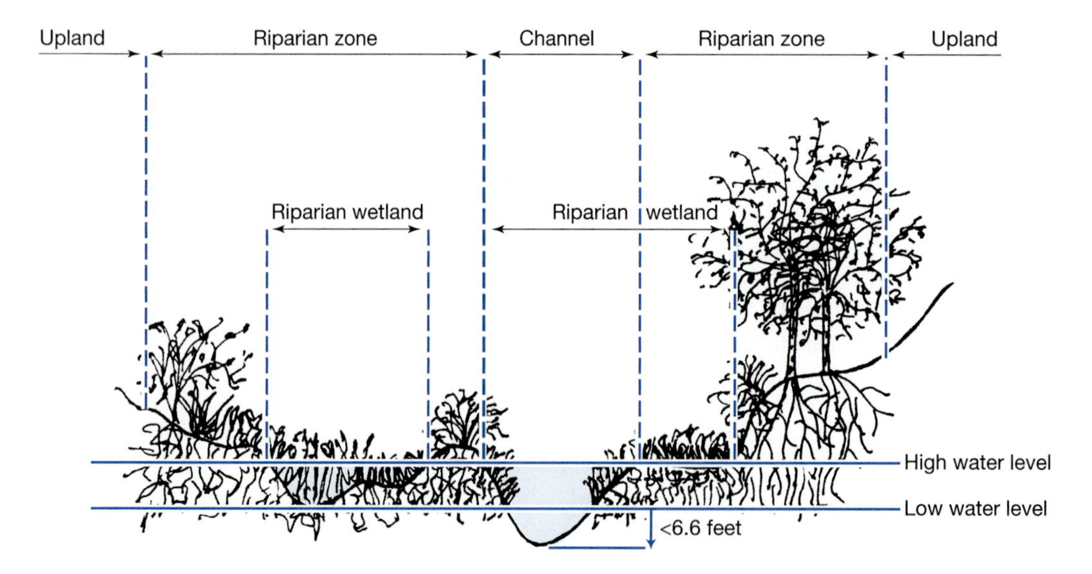

FIGURE 9.4 *Relationship between riparian zone and riparian wetland. The boundary of a riparian zone is not easily delineated but often includes drier sites that do not conform to the legal definition of a wetland.*

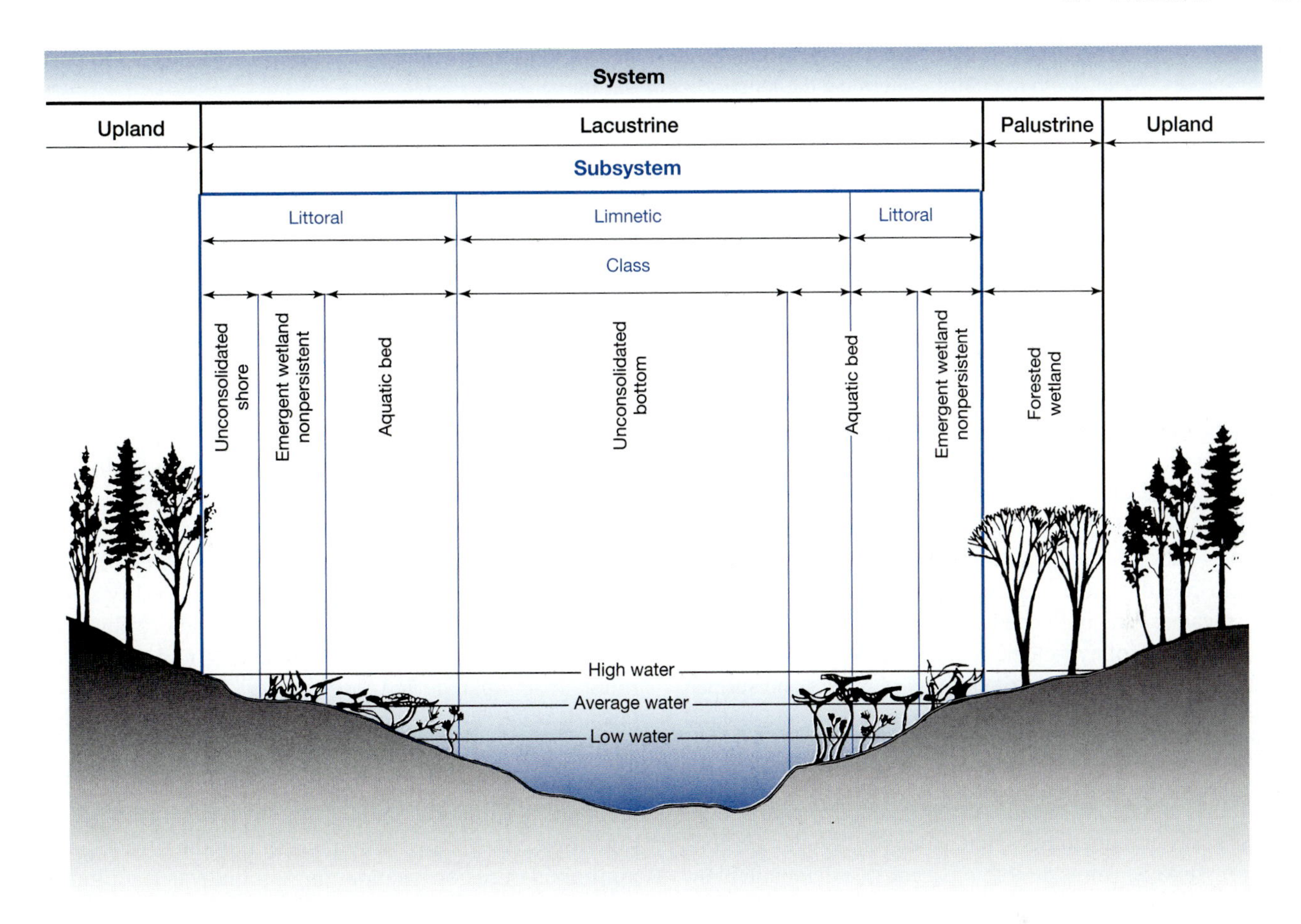

FIGURE 9.5 *A typical cross section of a lacustrine wetland system delineated into subsystems, limnetic and littoral, and classes within those subsytems. This particular palustrine system is bordered on one side by upland habitat and the other by a palustrine wetland.*

salinity is <0.5 ppt. Examples include bogs, fens, freshwater marshes, wet meadows, swamps, pocosins, and small shallow ponds. Some riparian wetlands that persist due to periodic flooding of a river channel may also be considered palustrine wetlands.

Bogs and **fens** are the major types of **peatlands** and are found primarily in the colder northern, humid regions of the Northern Hemisphere. In general, bogs are associated with relatively stagnant, water-filled depressions and support an abundance of acid-loving plants (such as mosses of the genus *Sphagnum*). Fens are considered a successional stage in the development of bogs and, therefore, act as a transition between shallow, open lake systems and bogs. In the U.S., most peatlands are located in Minnesota, Michigan, Wisconsin, Alaska, and Maine. Canada has the largest peat resources in the world. Extensive areas of peatlands also occur in Scandinavia, eastern Europe, western Siberia, and Labrador. **Peat** is a generic term used to describe organic soils. Organic soils form in cool, wet climates where an abundance of plant material and anaerobic soil conditions, due to standing water or lack of drainage, exist. This type of environment is conducive to extremely slow plant decomposition and can result in peat accumulations several feet thick. Peat is a valuable source of fuel in some parts of the world and is commonly added to garden and potting soils to improve soil structure and water-holding capacity.

Although numerous types of peatlands exist, their development will be described here using the classic example of a **quaking bog.** Quaking bogs are a natural successional stage in the "filling in" of small, shallow lakes. Mats of reeds, sedges, grasses and herbaceous wetland plants attach to the peat soil developing at the edges of the lake. The plant cover continues to proliferate from the edges toward the middle of the lake. Root systems are only partially attached to the lake bottom, resulting in the creation of a floating raft of vegetation. The floating mat soon becomes consolidated and dominated by sphagnum moss and other bog flora that is cushionlike and, when walked upon, bounces, or quakes. After the peat soil has accumulated for a period of time, the outer edges may be colonized by shrubs and trees able to thrive in the nutrient-poor, acidic conditions characteristic of late-stage bogs.

In North America, **freshwater marshes** are scattered throughout inland Canada and the U.S. The prairie pothole marshes of north-central U.S. and south-central Canada, the Great Lakes marshes, and the Everglades of Florida represent the majority of freshwater marshes. Vernal pools, playas, near-coast marshes, and the wetlands of the Nebraska sandhills are other types of freshwater marshes. The vegetative community within freshwater marshes is characterized by nonwoody **hydrophytes** (water-loving plants) such as grasses, reeds, cattails, sedges, rushes, broad-leaved plants, and various floating and submergent aquatic species. Water depth tolerances determine the longitudinal stratification of plant species thriving in an inland marsh (Figure 9.6). Generally, shallow water depths and shallow peat formation are characteristic of these marshes.

Distinctions between these different freshwater marshes are based on geologic origin and hydrologic regime. For example, the **prairie potholes** region of Minnesota, Iowa, and the Dakotas originated as millions of depressions formed by the scouring action of glaciers. These depressions later filled with meltwaters as the glaciers retreated. **The Everglades** are maintained by the periodic freshwater overflow of Lake Okeechobee onto the flat, limestone deposit that forms most of southern Florida.

Freshwater swamps occur along the Mississippi floodplain and Atlantic coastal plain of the southeastern U.S. and support woody vegetation, primarily various species of cypress, gum, and tupelo trees that have adapted to thrive under anaerobic conditions. Swamps remain flooded all or most of the year and can occur in a number of hydrologic settings, including: (1) isolated water-filled depressions fed primarily by rainwater and surface inflow, (2) edges of lakes fed by lake overflow and upland runoff, and (3) on broad floodplains permanently flooded by rivers or streams.

Although the canopy vegetation in swamps is dominated by cypress, gum, and tupelo trees, the understory vegetation consists of many other tree, shrub, herbaceous, and aquatic plant species. In the deep South, Spanish moss is frequently associated with these swamps and grows in abundance, hanging from tree branches and stems.

Pocosins are shrub-scrub wetlands usually located on relatively flat terrain between stream systems. These wetlands are found primarily along the southern Atlantic coastal plain. Coastal precipitation and seasonal flooding regimes of nearby stream systems influence the frequency and duration of inundation periods.

Functions and Values

Wetlands cover only about six percent of the world's land surface but are found in every climate regime from the tropics to the tundra and on every continent except Antarctica. Since the detailed study of wetland ecosystems began in the 1960s, scientists have discovered that wetlands perform many functions necessary for the maintenance of healthy ecosystems. Many of these functions provide benefits to human society as well, and therefore, have corresponding human-based values (Table 9.2). Keep in mind that not all functions are performed by every kind of wetland and that some wetlands will perform the same function better than other wetlands.

Wetland functions can be grouped into three broad categories: (1) hydrologic processes, (2) water quality improvement, and (3) wildlife habitat. Wetlands influence the dynamics of water flow by intercepting and storing storm runoff from uplands and "soaking up" floodwaters from adjacent waterbodies. Wetlands stabilize stream banks, lakeshores, and coastlines from the forces of stream flow and wave action. Additionally, wetlands contribute to the recharge and discharge of groundwater. These hydrologic functions are of value to humans in reducing flood damage, minimizing erosion, maintaining groundwater aquifers, and supporting base flows during the dry season for aquatic species.

Wetlands act as a sink for sediments, inorganic and organic nutrients, and toxic materials. Various physical, biological, and chemical processes remove these materials from the water column, enabling them to be transformed into nontoxic substances, absorbed by plants, or buried in soil or peat accumulations.

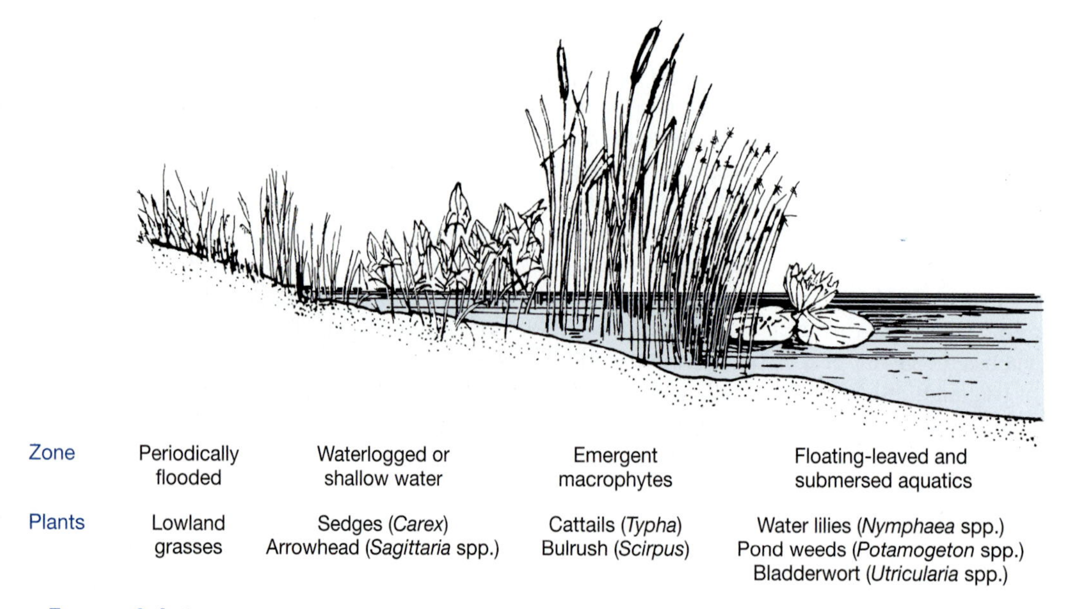

FIGURE 9.6 *A cross section through the edge of a freshwater marsh illustrates the distribution of plants according to particular water depth tolerances. Typical plant species are listed below each water inundation zone.*

These processes lead to water quality improvement, which is of extremely high value to humans.

Wetlands provide a number of unique habitats and feeding niches and therefore, are home to a wide variety of flora and fauna. Many fish, waterfowl, and other wildlife species require wetland habitats for all or part of their life cycles (Figure 9.7). Wetland ecosystems are characterized by high primary productivity compared to other ecosystems, producing an abundance of plant material that fuels an extensive food chain. Wildlife abundance and diversity are of value to humans for sporting activities such as hunting and fishing. In many countries, wetland plants and wildlife are harvested for food, fiber, and energy needs. Currently, a disproportionately high percentage of endangered and threatened species depend on wetlands for their survival.

Not related to a specific wetland function, but important to humans, are the cultural values wetlands provide such as aesthetics, open space, recreational opportunities, outdoor classrooms, and research and historical sites.

Wetland Protection

Once thought to be sinister wastelands and of no economic or ecological importance to humans or the environment, wetlands were drained, filled, dredged, and/or flooded for agriculture, navigation, or land development. At the time of European settlement (1780s) in the United States, the area now encompassing the lower 48 states contained approximately 89.5 million hectares (221 million acres) of wetlands. By the mid 1980s, less than 50 percent of the original wetland acreage remained. Wetland drainage for agriculture has been the major cause of inland wetland losses in the U.S. Along the nation's coastal areas, wetlands have been drained and filled primarily for urban and industrial development.

Recognition of the valuable benefits of wetlands has changed the way we manage wetlands. Wetlands are now federally protected. The U.S. has a **"no net loss" policy** that utilizes a stringent, permitting system under Section 404 of the **Clean Water Act** to regulate activities in and around wetlands and to protect wetlands from development. Five federal agencies share responsibility for wetland protection. The USACE protects wetland resources in relation to navigation and water supply. The EPA's authority is related to protecting wetlands for their contribution to the quality of our nation's water. The USFWS is concerned with protecting the high quality habitats that wetlands provide for fish and wildlife, including game species and threatened and endangered species. The National Oceanic and Atmospheric Administration's (NOAA) wetland authority is related to the agency's responsibility to manage the nation's coastal resources. And, the Natural Resources Conservation Service (NRCS) is charged with protecting wetland resources on agricultural lands. Together, these agencies administer a myriad of programs that provide various mechanisms for wetland protection. These mechanisms include acquisition, land-use planning, restoration and creation, mitigation, incentives to prevent degradation and destruction, disincentives to stop conversion to other land uses, technical assistance, education, and research. As of 1993, 29 states had their own wetland protection programs, many of which go beyond the federally enacted programs.

Between the mid 1950s and the mid 1970s, prior to wetland protection efforts, wetland acreage was converted at a rate of 185,400 hectares (458,000 acres) per year. Since then, however, reports indicate that wetland losses are slowing. From the mid 1970s to the mid 1980s, the estimated rate of wetland loss had declined to 117,400 hectares (290,000 acres) per year. An estimated 42.7 million hectares (105.5 million acres) remained in the

FIGURE 9.7 *Coastal wetlands provide valuable wildlife habitat. Great egrets feed in marshes along the New Jersey coast.*

TABLE 9.2 *Wetland Functions and Their Value to Humans*

Wetland Function	Corresponding Human Values
Hydrologic Processes	
Flood/Stormwater Control Wetland vegetation, with massive roots and rhizome systems, binds and protects soils. Wetlands intercept, store, and slowly release floodwaters, reducing the amount and velocity of surface runoff. Wetlands (particularly those immediately adjacent to rivers and streams) serve as floodway areas by conveying flood flows from upstream to downstream points. ***Base Flow/Groundwater Support*** Wetlands have the capacity to store water and then release it slowly to groundwater deposits. In this manner, wetlands also support water levels in streams and lakes.	• Buffers human coastal settlements from storm-driven waves. • Reduces flood damage to human property. • Minimizes soil erosion caused by wave action and stream flow. • Maintains groundwater aquifers. • Supports stream base flows during the dry season, necessary for the survival of many aquatic species.
Water Quality Improvement	
Pollutant Removal Wetlands act as settling ponds for sediments, inorganic and organic nutrients, and toxic chemicals. These pollutants settle out as water velocity slows down in wetland areas. ***Pollutant Detainment*** Wetland vegetation binds soil particles and retards the movement of sediment. Excess nutrients and other chemicals can be taken up by wetland plants or buried in wetland soils. ***Pollutant Transformation*** Various physical, biological, and chemical processes in wetland ecosystems can cause the breakdown of pollutants into nontoxic substances.	• Controls contamination of drinking water supplies. • Reduces the amount of pollutants in surface and groundwaters. • Maintains or enhances water quality for a number of domestic, commercial, and recreational uses.

conterminous United States in 1997. Between 1986 and 1997, the net loss of wetlands was 260,700 hectares (644,000 acres). During this same period, the annual conversion rate was 23,700 hectares (58,500 acres), representing an 80 percent decline compared to the previous decade's annual conversion rate. This slowing trend is due to implementing and enforcing wetland protection measures, eliminating several incentives for draining wetlands, improving public education and outreach, and monitoring and protecting coastal environments. Wetlands have also been created and restored as a result of both human activities and natural, long-term hydrologic cycles.

Wetland benefits have been recognized and wetland management policy has shifted from wetland drainage and conversion to protection, restoration, and creation schemes. Although the rate of wetland loss is slowing, these highly sensitive ecosystems will continue to be subject to alteration by human land uses and pollution. Controversy arises because conflicting interests still exist between landowners and the general public and between developers and conservationists. Nonetheless, through the continued enforcement of wetland protection regulations and the cooperative efforts of informed citizens, we can keep moving toward our "no net loss" wetland goal.

9.2 The Lake Ecosystem

Lakes can be formed as a result of tectonic activity, volcanic eruptions, landslides, glaciation, fluvial processes, meteorite collisions with Earth, or human activities. Tectonic activity has created a number of inland seas (saltwater lakes), such as the Caspian Sea, located between southeast Europe and southwest Asia. Lakes can form in a collapsed crater following a volcanic explosion, as in the case of Crater Lake, Ore-

TABLE 9.2 *Wetland Functions and Their Value to Humans (continued)*

Wetland Function	Corresponding Human Values
Wildlife Habitat	
High Primary Productivity Relative to other ecosystems, wetlands produce an abundance of organic material that supplies food for a diverse group of invertebrates, amphibians, fish, mammals, and birds. ***Nesting, Breeding, Resting, Migration Sites*** Many wildlife species require one or more of the habitat features unique to wetlands for all or part of their life cycles.	• Provides renewable sources of food, fiber, and energy. • Maintains plant and animal populations. • Preserves biological diversity. • Supports threatened and endangered species. • Supplies "game" for hunting and fishing and "subjects" for birdwatching and photography.
Ecosystem Interconnectivity Wetlands are transitional areas that can function as conduits between various ecosystems. For example, riparian wetlands provide food, water, and shelter for wildlife species that travel between upland and deepwater habitats. Coastal wetlands contribute nutrients needed by fish and shellfish to nearby estuarine and marine waters.	***Human Values Unrelated to a Specific Wetland Function*** Scenic beauty, aesthetic enjoyment, hiking/picnic areas, educational/research opportunities, archeological/historical sites, open space, and wilderness preserves.

gon. Ice scouring, ice damming, and moraine depositions from repeated glaciation created many of the water-filled depressions and valleys in our northernmost states. The power of flowing water (fluvial) may cause a river channel to shift direction, creating an oxbow lake. Reservoirs created behind dams are the largest type of human-made lakes. Lake Mead in Arizona and Nevada is the largest artificial lake in the world (637 square kilometers or 246 square miles) and was formed when Hoover Dam was built on the Colorado River.

The largest and deepest lakes are those formed by tectonic activities or major glacial processes. However, smaller, shallower lakes are most common. Essentially all lakes, regardless of size or origin, have several features in common.

Lake Zonation

Littoral Zone The **littoral zone** is the shallow margin of a lake. This zone is characterized by rooted vegetation (Figure 9.8). The rooted plants usually are arranged in a well-ordered sequence similar to that of freshwater marshes (Figure 9.6), from shore toward open water, as emergent, floating, and submergent. Typical **emergent plants**—that

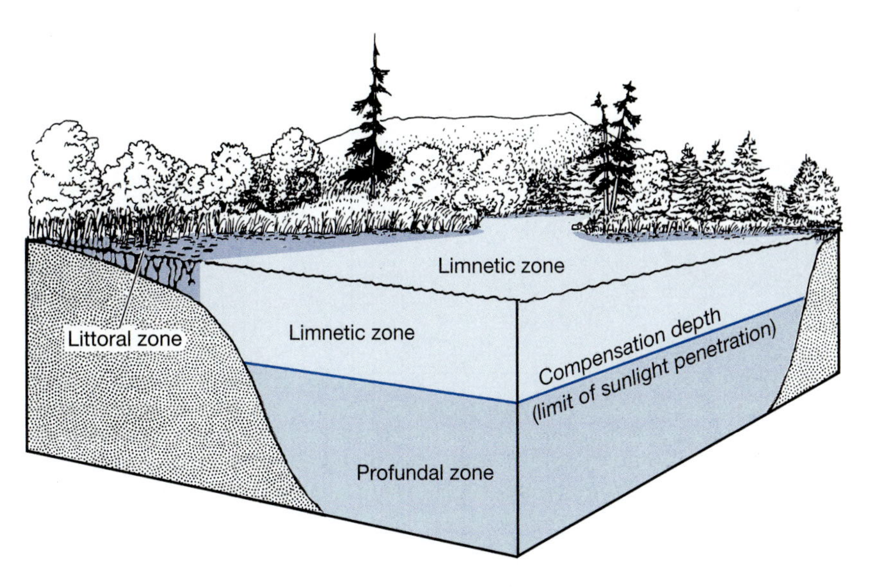

FIGURE 9.8 *The principal zones of the lake ecosystem.*

is, those that extend above the water—include cattails and bulrushes. Characteristic **floating plants** are water lilies and duckweed. Plants that are completely underwater, such as pondweed and milfoil, are called **submergents.**

Because sunlight penetrates to the lake bottom in the littoral zone, it sustains a high level of photosynthetic activity. In this zone, floating microorganisms known as **plankton** frequently give the water a faint greenish brown color. Plankton consist of two groups: plants (chiefly algae), known as **phytoplankton,** and animals (primarily crustaceans and protozoa), known as **zooplankton.** Plankton are largely incapable of independent movements and therefore are passively transported by water currents and wave action.

Limnetic Zone The **limnetic zone** is the region of open water beyond the littoral zone. This zone extends from the surface to the maximum depth at which there is sufficient sunlight for photosynthesis (Figure 9.8). At this depth, known as the **compensation depth,** photosynthesis bal-

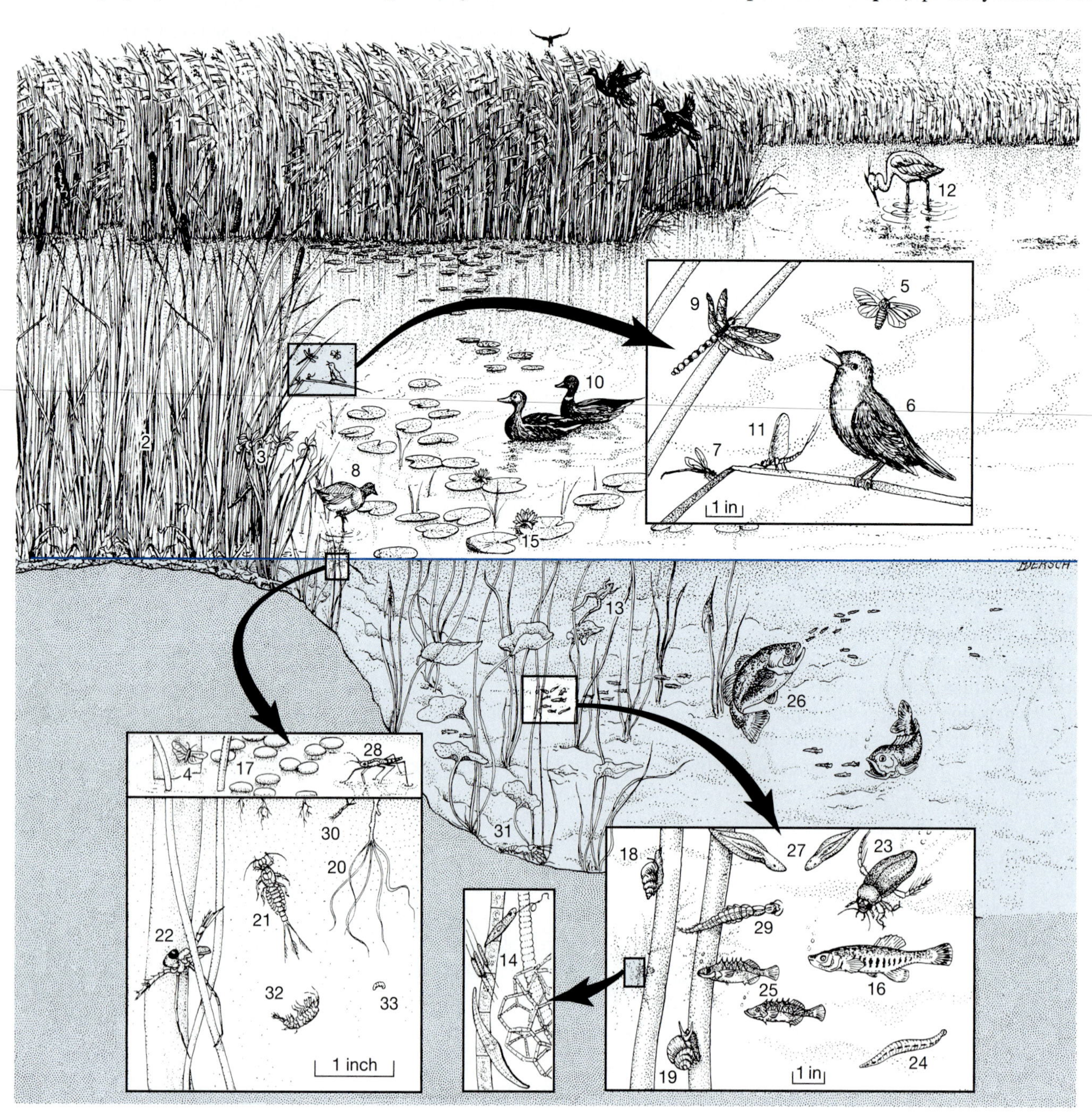

FIGURE 9.9 *Characteristic plants and animals of the lake ecosystem: 1. reed; 2. cattail; 3. yellow flag; 4. adult caddis fly; 5. moth; 6. warbler; 7. damselfly; 8. coot; 9. dragonfly; 10. mallard; 11. mayfly; 12. heron; 13. frog; 14. phytoplankton; 15. water lily; 16. minnow; 17. duckweed; 18. snail; 19. snail; 20. hydra; 21. mayfly (immature); 22. water boatman; 23. diving beetle; 24. leech; 25. stickleback; 26. black bass; 27. tadpoles; 28. water strider; 29. diving beetle larvae; 30. gnat larvae; 31. crayfish; 32. freshwater shrimp; 33. caddis fly larva.*

ances respiration. The light intensity here is only 1 percent of that of full sunlight. Although rooted plants are absent, this zone frequently contains numerous phytoplankton, mostly algae. In large lakes, the phytoplankton may actually produce more total biomass than the much larger rooted plants of the littoral zone. In spring, when nutrients and light are optimal, phytoplankton populations can "explode" to form **blooms.**

The limnetic zone derives its oxygen from the photosynthetic activity of phytoplankton and from the atmosphere immediately over the lake's surface. The atmosphere becomes a significant source of oxygen primarily when the water's surface is disturbed by wind and waves. Suspended among the phytoplankton are the zooplankton (animal plankton), primarily tiny crustaceans that form a trophic link between the phytoplankton food base and larger aquatic animals like fish (Figure 9.9).

The Profundal Zone The **profundal zone** lies beneath the limnetic zone and extends to the bottom of the lake (Figure 9.8). Because of the limited penetration of sunlight, green plants are absent. In north temperate latitudes, where winters are severe, this zone has the warmest water in the winter and the coldest water in summer. Large numbers of bacteria and fungi occur in the bottom ooze, sometimes up to 1 billion per gram. These bacteria constantly decompose the **organic matter** (plant and animal remains and wastes) that accumulates on the bottom. Nitrogen and phosphorus are released during decomposition and are put back into circulation as soluble salts. In winter, the metabolism of aquatic life is reduced and the colder water has greater oxygen-dissolving capacity. At this time, therefore, oxygen usually is not an important limiting factor for fish if the ice cover remains clear of snow. In midsummer, however, when the metabolic rates of aquatic organisms are high, the oxygen-dissolving capability of the warm water is relatively low, and the oxygen-demanding processes of bacterial decay proceed at high levels. Under these conditions, oxygen depletion, or stagnation, of the profundal waters may cause extensive fish mortality.

Thermal Stratification In temperate latitudes, lakes show marked seasonal temperature changes.

Winter As temperatures drop below freezing, ice forms. Since ice is less dense than liquid water, it floats on the surface of the lake. The water at increasing depth below the ice is progressively warmer than the surface layer. The heaviest water, at the bottom of the lake, has a winter temperature of 4°C (39°F). All winter the water temperature remains relatively stable (Figure 9.10).

Spring Following the ice melt, the surface water gradually warms to 4°C (39°F). At this point, all the water is of uniform temperature and density from the surface to the bottom. Strong spring winds stir the water, causing a complete mixing of water, dissolved oxygen, and nutrients from the lake surface to the lake bottom. This mixing is called the **spring overturn** (Figure 9.11). As spring progresses, however, the surface water becomes warmer and lighter than the water at lower levels. As a result, the lake becomes thermally stratified again.

Summer In summer, the upper stratum usually has the highest oxygen concentration and is characterized by a temperature gradient of less than 1°C (34°F) per meter (3.3 feet) of depth. This layer is called the **epilimnion** (upper lake). The middle layer of the lake, typified by a temperature gradient of more than 1°C (34°F) per meter, is the **thermocline.** The bottom layer of water, the **hypolimnion** (bottom lake), has a temperature gradient of less than 1°C (34°F) per meter (Figure 9.12). During the summer, the hypolimnion of many lakes becomes depleted of oxygen. This is caused by the biological oxygen demand (BOD) of bacterial decomposers, the lack of photosynthetic activity due to the absence of sunlight, and the minimal mixing with upper waters as a result of density differences (Figure 9.12).

Autumn The surface waters gradually cool in the fall. Eventually a point is reached where the lake temperature is uniform from top to bottom. Because the water is now also of uniform density, it becomes well mixed by wind and wave action during the **fall overturn** (Figure 9.11). Nutrients, dissolved oxygen, and plankton become uniformly distributed.

9.3 The Stream Ecosystem

Origin and Classification

Because the amount of water that falls as precipitation is greater than that lost by evapotranspiration to the atmosphere, the Earth experiences an excess of water. This excess water is carried to the ocean by streams and rivers. Water falling as precipitation flows into streams via two pathways: (1) flowing overland as surface runoff or (2) infiltrating the soil surface and then flowing underground into streams as groundwater. **Ephemeral streams** are those that flow only during the wet season. Streams that flow year-round are called **perennial streams** and are sustained by groundwater during dry periods.

A stream channel that flows into a larger channel is called a **tributary** of that channel. The area of land, confined by topographic divides, in which all tributaries drain into a single river system is termed a drainage basin, or **watershed** (Figure 9.13). Large watersheds are composed of several smaller **subwatersheds** that contribute runoff to different locations but ultimately converge into the common river system.

Stream size increases downstream as more and more tributary segments enter the main channel. Viewed from above, the network of water channels that form a river system displays a branchlike pattern (Figure 9.14). A classification system based on the position of a stream within the network of tributaries, called **stream order,** was developed by Robert E. Horton and later modified by Arthur Strahler. In general, the smaller and fewer the tributaries emptying into it, the lower the stream order number. For example, first-order streams

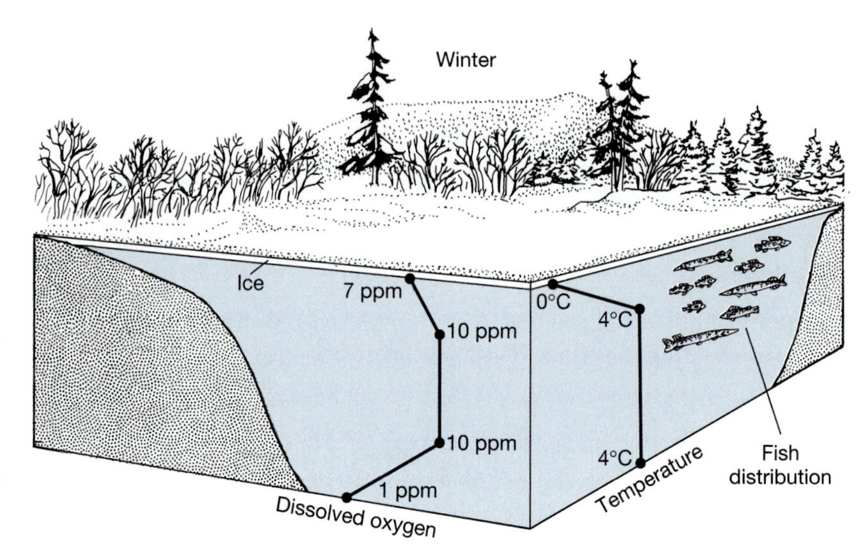

FIGURE 9.10 *The lake ecosystem in the northern states in winter. Because water becomes less dense as its temperature drops below 4°C (39°F), the ice which forms at 0°C (32°F) is relatively light and floats at the surface. Note that the concentration of dissolved oxygen drops off sharply in the deeper part of the lake from about 10 ppm to only 1 ppm. Fish remain where the level of dissolved oxygen is high.*

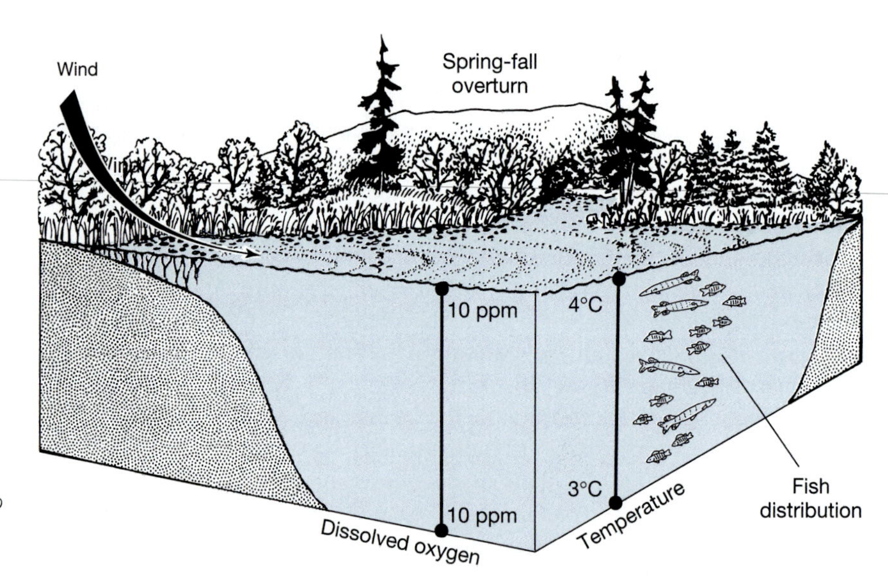

FIGURE 9.11 *The lake ecosystem in spring and autumn. Note the uniform distribution of temperature and dissolved oxygen from top to bottom due to "turnover" or thorough mixing of the water. Fish are also uniformly distributed in the vertical dimension.*

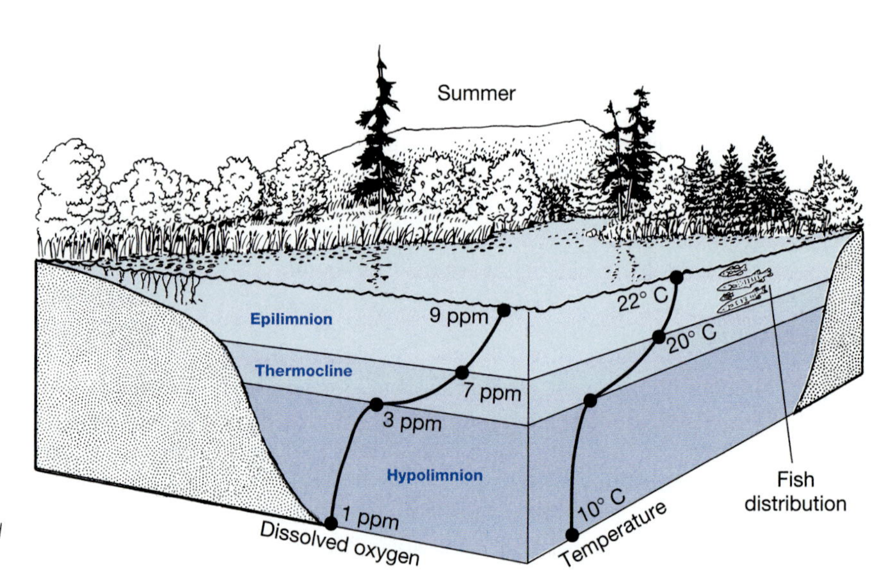

FIGURE 9.12 *The lake ecosystem in summer. Cross-sectional view shows thermal stratification and the vertical distribution of dissolved oxygen.*

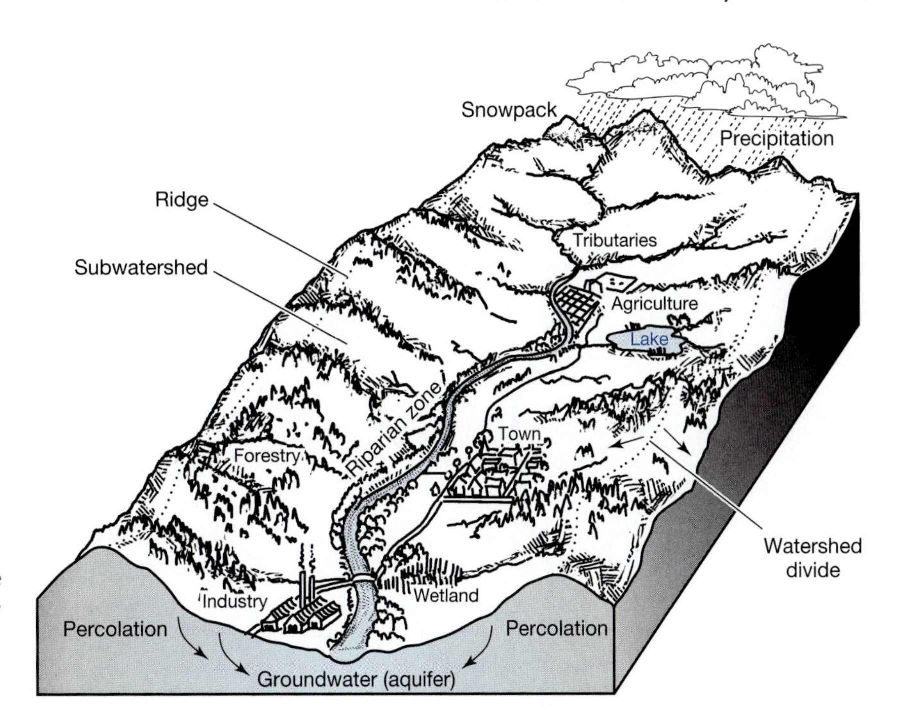

FIGURE 9.13 *Water sources, tributaries, and human land uses within a single watershed. Note the topographic divides that define the boundaries of the watershed. Numerous subwatersheds of tributaries drain the area between the secondary ridges situated at nearly right angles to the divides.*

(Order 1) are the smallest streams within the network, having no tributaries. Second-order streams (Order 2) are those that have only first-order streams as tributaries, and so on (Figure 9.14).

Another stream classification system, the Rosgen method, describes in detail short lengths of a stream channel based on gradient, bed material, bankfull width-to-depth ratio, extent of meandering, and composition of bank material. The proper use of this classification system requires extensive training and experience.

Physical Features

Channel Shape The cross-sectional shape of a particular reach of stream channel depends primarily on water discharge (erosional force), sediment load, and the composition of bed and bank materials (degree of resistance to erosional forces). In general, stream reaches with densely vegetated banks and cohesive, silt or clay bed and bank materials will be narrower than similar reaches with sparsely vegetated banks and noncohesive, sandy bed and bank materials.

During high water events, when the force of water flow is enough to erode banks and transport bed materials, stream channels take shape. Most river cross sections take on a somewhat trapezoidal shape in straight reaches but tend to be asymmetric in the curves and bends. As the river gets larger downstream, width increases faster than depth, resulting in a larger width-to-depth ratio than an upstream tributary.

Channel Pattern Viewed from above, a river channel can take on three different patterns: meandering, straight, or braided. The position of a stream in the watershed influences its channel pattern. For example, **headwater streams** (the beginning segments of a river system) are usually straighter, being confined by narrow valleys, shallow bedrock, and coarse sediments; **lowland streams** (higher order) have more freedom to meander across the finer sediments of floodplain materials.

The **meandering pattern** is by far the most common. In the meandering pattern, river channels are only straight for relatively short distances between curves, resembling a continuous string of S shapes. Meandering results when a channel naturally migrates laterally, eroding one bank and depositing material on the opposite bank. The depositional material forms what is called a **point bar.** The sinuosity of a channel, or lateral extent of channel meanders, is variable (Figure 9.15).

Pools and Riffles Along the length of a channel bed are regularly spaced deep and shallow areas that create an undulating topography. Storm flows scour out deep areas that become **pools** in nonstorm periods; sediments are deposited as mounds on the streambed, forming shallow-water areas called **riffles.** In meandering channels, pools are typically associated with the outside or concave bank of a bend where erosional forces are greatest. At high flow, the visible "white water" caused by a riffle is obscured. Steeper mountainous channels, having bed materials of boulders and large rocks, exhibit what are called step-pools, or several short, deep or shallow pools spaced between the boulders.

Floodplains and Terraces **Floodplains** are the flat areas of fine sediment deposition extending outward from the bank tops of lowland streams. Floodplains are formed by a river, in the present climate, and receive river overflow whenever water level exceeds the top of the banks, or **bankfull discharge.** As climate changes occur over relatively long time periods, rivers downcut into floodplains. Downcutting can continue until the floodplains are so far above the river that they no longer receive overflow discharge. These abandoned floodplains are called **terraces.** Terraces lie adjacent to but at a higher elevation than a current floodplain (Figure 9.16).

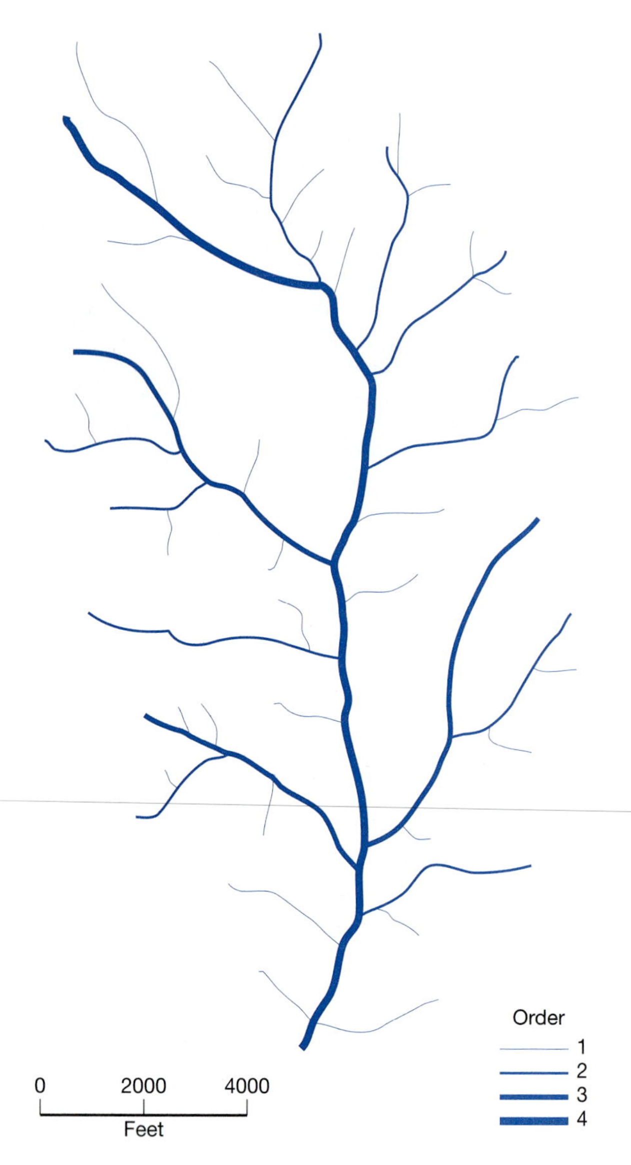

FIGURE 9.14 *A network of water channels forming the branchlike pattern of a typical river system. Network channels have been assigned a stream order number based on Horton's classification system.*

Biological Community and Energy Flow

Organisms that live in the stream environment can be divided into three categories based on the function they perform: producers, consumers, and predators. Aquatic plants are the **producers** that provide energy to the stream community through photosynthesis and include diatoms, algae, and macrophytes (cattails, bulrushes, water lilies, river weeds, and so forth). Microorganisms, such as fungi and bacteria, provide energy through the decomposition of organic matter. **Consumers,** including invertebrates (aquatic insects, snails) and fish, use the energy provided by these plants and microbes. **Predators** (fish, birds, mammals) feed on consumer groups for their energy requirements.

Streams are rather open ecosystems. In addition to instream sources of energy, stream food chains receive a considerable portion of their basic energy supply from materials originating on land. Materials are constantly being received from the terrestrial ecosystems that border them (riparian zones). Dissolved organic matter enters the stream via groundwater flow. Surface erosion of stream banks adds organic matter directly to the stream. Terrestrial animals transport organic matter to the streams, insects fall into the streams, and fish die and decay in the stream. Leaves fall into the water in autumn, and organic debris (stems, nuts, twigs, needles, seeds, dead weeds, cow manure, and the bodies of insects, worms, and mice) is washed into the stream during the spring runoff. Many of the primary consumers in a stream consume **detritus** (decomposing organic material) rather than living aquatic vegetation.

Riparian vegetation is the source of instream, **large woody debris.** Logs, branches, and root wads that have fallen into the stream, due to wind or erosional forces, shape channel structure by deflecting and slowing water flow. Large woody debris creates prime fish habitat by providing cover, creating pools, and trapping organic matter. Log jams can sometimes disrupt the migration of spawning fish, but can also form backwaters, sloughs, and marshes, which increase the amount of rearing habitat for juvenile fish.

Roots from riparian vegetation stabilize stream banks against erosional forces and maintain **undercut banks,** another excellent fish habitat feature (Figure 9.17). The overhead canopy, whether from trees or grasses, shades the stream from the sun, keeping stream temperatures down in the summer.

The composition of the biological community in any given reach of stream is determined by channel characteristics and energy sources. Slow-flowing, open streams support a different community of organisms than fast-flowing, shaded streams. And compared to narrow, headwater streams, lowland streams of wider widths depend less on energy input from adjacent riparian vegetation and more on that supplied by the transport of excess organic matter from upstream sources. The type, amount, and function of riparian vegetation are highly variable from one geographic location to another and are controlled by the climate, geology, and adjacent land uses.

9.4 The Coastal Environment

Structure of Coastlines

Coastal areas are naturally dynamic environments, continually subjected to the forces of wind, rain, waves, tides, currents, and seaspray. These forces, along with living organisms, shape the profile features of a coastal area through various erosional and depositional processes. The structural features of coastlines are highly variable and are dependent on climate, geology, and ocean dynamics.

Beaches and Dunes **Beaches** (sandy or gravelly) can be found on any coastline in the world where there is a source of eroding mineral or sediment deposits. **Sand dunes** develop where there is an active sand supply, relatively low precipitation, regular strong winds, and large tidal ranges. In the United States, the longest and widest stretches of sandy beaches are found on the Atlantic and Gulf coasts. Wind patterns and extensive glacial deposits have formed extensive areas of sand

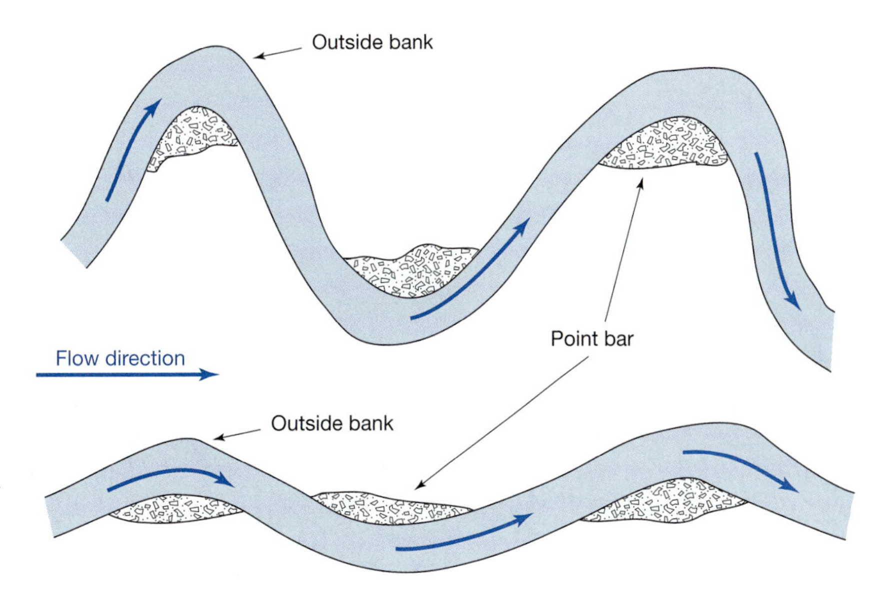

FIGURE 9.15 *The lateral extent of channel meanders is variable; however, the erosional and depositional processes are similar. The outside bank is eroded and point bars are created by the deposition of sediment on the opposite, inside bank.*

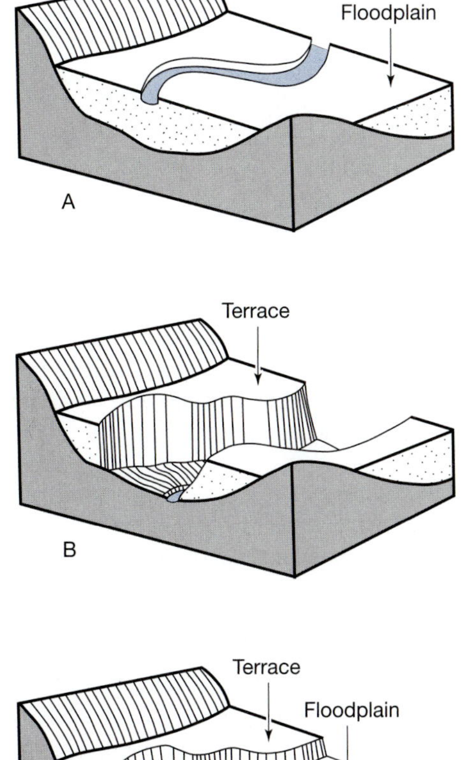

FIGURE 9.16 *The formation of a terrace, or abandoned floodplain. (A) Numerous flood events cause a of buildup of sediment, or floodplains, extending laterally from the tops of the riverbanks. (B) Changes in climate over time cause the river to downcut into the original floodplain, creating a terrace. (C) The number of flood events increases, causing new floodplain deposits.*

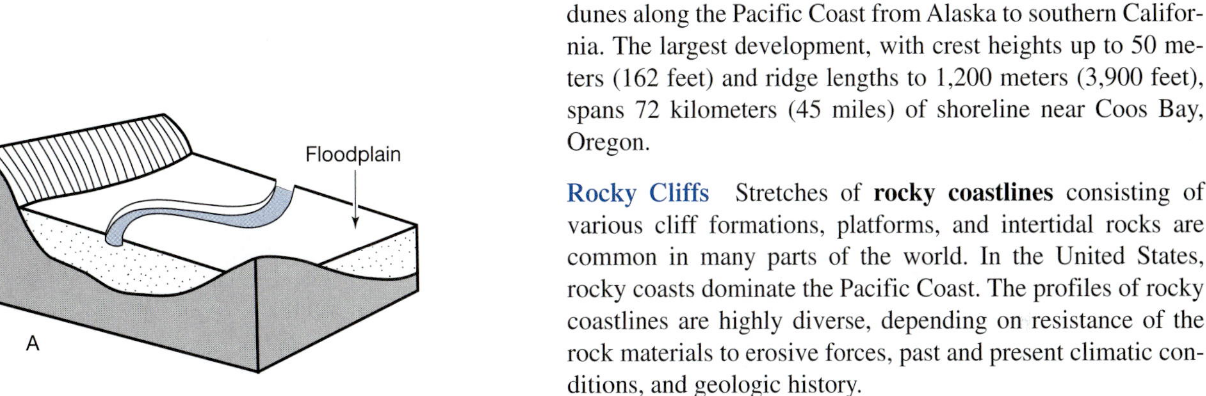

dunes along the Pacific Coast from Alaska to southern California. The largest development, with crest heights up to 50 meters (162 feet) and ridge lengths to 1,200 meters (3,900 feet), spans 72 kilometers (45 miles) of shoreline near Coos Bay, Oregon.

Rocky Cliffs Stretches of **rocky coastlines** consisting of various cliff formations, platforms, and intertidal rocks are common in many parts of the world. In the United States, rocky coasts dominate the Pacific Coast. The profiles of rocky coastlines are highly diverse, depending on resistance of the rock materials to erosive forces, past and present climatic conditions, and geologic history.

Hard rocks, such as granite or basalt, often form steep cliffs that plunge directly into the sea or that rise up sharply behind beach areas. Because these rocks are highly resistant to weathering, they have experienced very small erosion rates over time. In contrast, cliffs comprised of unconsolidated or weakly consolidated material, such as clays or glacial deposits, are receding at relatively high rates. These cliffs are prone to failure and highly susceptible to the erosive forces of waves and storms. Cliffs prone to failure frequently experience mass movements (landslides, rockfalls, and the like) and are more likely to display profiles with gradual or steplike slopes.

Vegetation thriving on hard rock cliffs consists of those species adapted to high salinity and rocky substrate. Inaccessible to humans and many predators, the cracks, ledges, and crevices of rocky cliff faces provide undisturbed and protected nesting habitat for many seabirds. In contrast, cliffs composed of weakly consolidated materials support vegetation for only short durations before it is torn away by erosional processes.

We have distinguished rocky cliffs in two simple categories: stable (hard rock) and unstable (weak, unconsolidated rock materials). Of course, the real world is more complicated. Cliffs are rarely made up of uniform rock materials but of complex combinations of strong and weak structured materials, and therefore, the variety of rocky coast profiles is enormous.

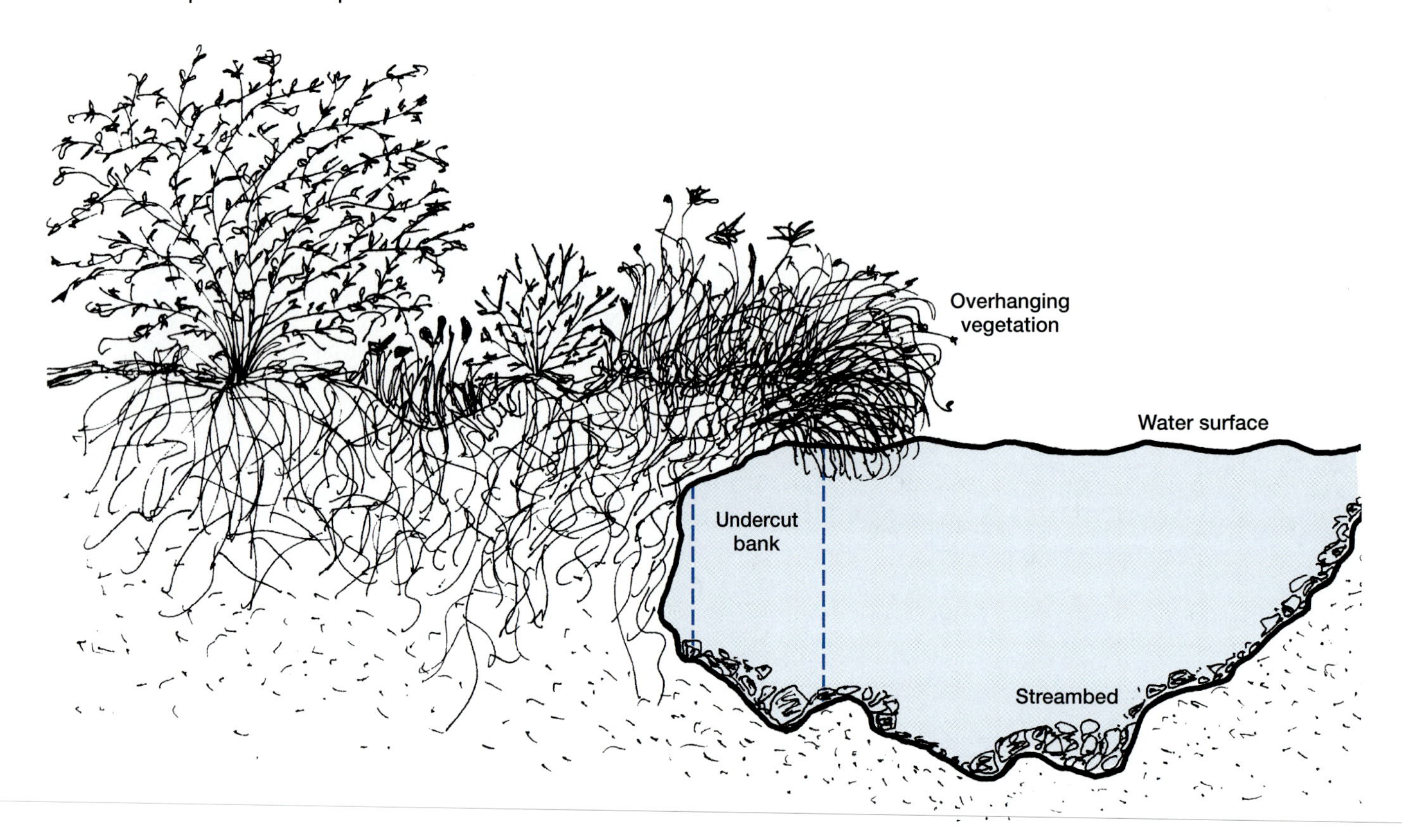

FIGURE 9.17 *Rooted riparian vegetation stabilizes stream banks and maintains undercut banks. The undercut banks and overhanging vegetation provide cover, shade, and resting places for fish.*

Barrier Islands Coastal wave, wind, and current action have created accumulations of coastal sediments parallel to and near the shore called **barrier islands.** These low-elevation islands are found worldwide and, in North America, are most common on the Atlantic and Gulf coasts. Structurally, these islands usually consist of sandy beaches and dunes and a few inlets, and are separated from the mainland by a salt marsh or lagoon area. Off the Atlantic and Gulf coasts there are 295 of these sandy islands covered with grasses, marshes, and sparsely distributed pine and oak trees (Figure 9.18). Most of these islands are long and narrow. The maximum width of these islands is 5 kilometers (3 miles); the maximum length is 100 kilometers (62 miles). Barrier islands are highly dynamic environments, constantly changing shape and size because of the erosion caused by wind, waves, and offshore currents.

Coral Reefs **Coral reefs** are formed from animals having calcareous skeletons (hard structures made from calcium materials) and certain species of algae that provide the sediment or "cement" to seal the coral framework. Coral animal offspring bud out and attach to the side of the parent. As animals die, new generations build onto the dead skeletons, often creating massive structures. Coral reefs are considered the largest biological constructions on Earth. The largest and most well-known chain is the **Great Barrier Reef,** off the eastern coast of Queensland, Australia, reaching a length of over 2,000 km (1,260 miles) and containing over 2,500 individual reefs. Coral reefs are primarily found in tropical and subtropical waters where temperatures never exceed 20°C (68°F) and depths are less than 100 meters (325 feet). In the United States, coral reefs are found off the coasts of Hawaii, Florida, and the Virgin Islands. Coral reefs are highly productive environments and home to an abundant and diverse group of plants and animals that live in, upon, and around them. These environments are also highly sensitive to natural and human-induced disturbances.

The Estuarine Ecosystem

Estuaries and their associated coastal marshes are a prominent feature of the U.S. coastline. **Estuaries** are semi-enclosed inlets (bays) that form transitional zones between coastal rivers and the sea (Figure 9.19). In one sense they represent a river–ocean hybrid, possessing some of the characteristics of each ecosystem. Nevertheless, the estuary has some distinctive properties and therefore must be considered a unique ecosystem.

Characteristics The water level in an estuary rises and falls with the tides. The water is a mixture of fresh water from the stream and salt water from the ocean. The salinity of estuarine water is highly variable, changing by a factor of 10 within a 24-hour period; salinity is higher when the tide comes in and lower when the tide moves out.

The concentration of dissolved oxygen is relatively high because estuaries are shallow and the water is generally quite turbulent. Because of the stirring action of the tides, **turbidity**

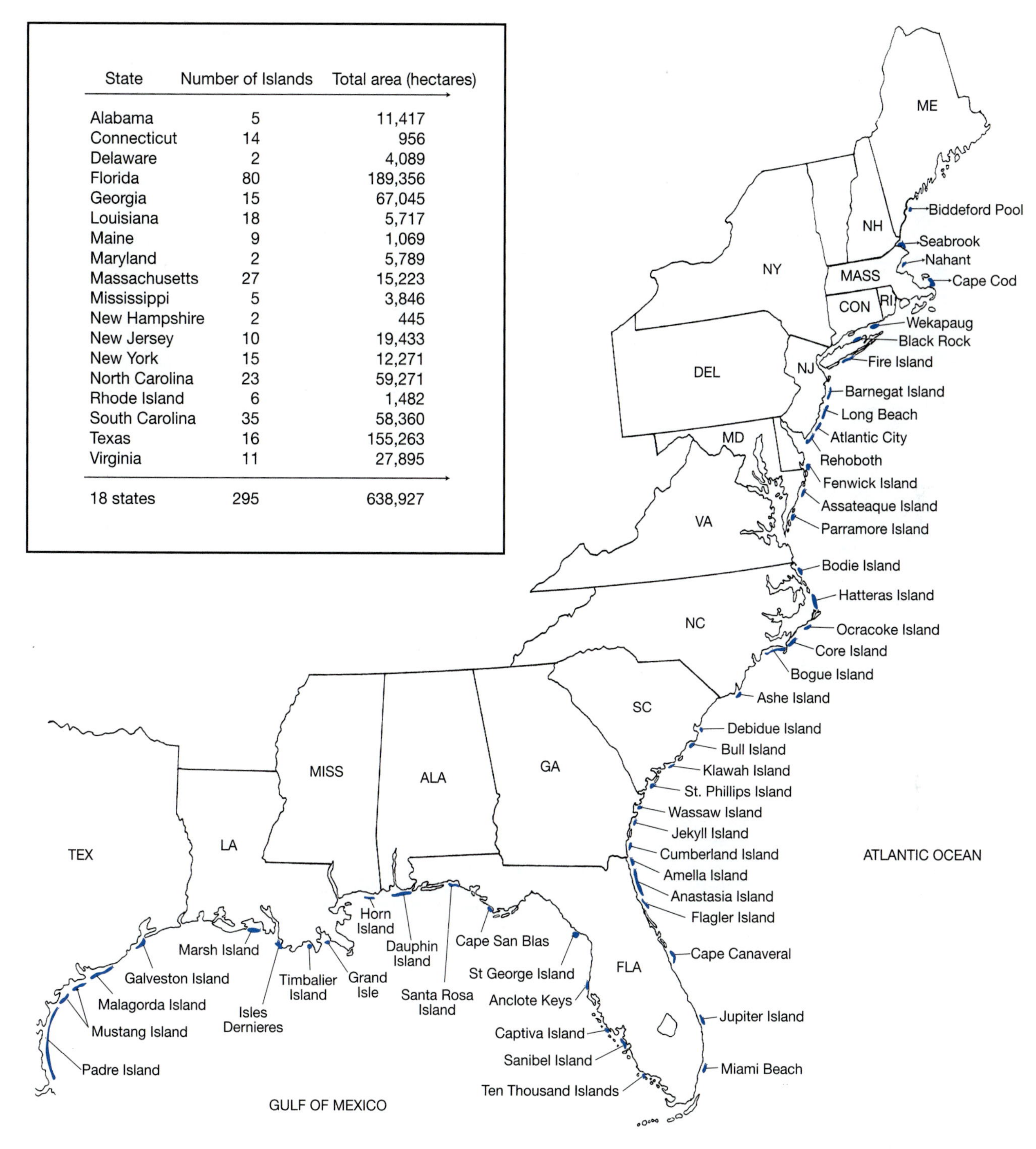

State	Number of Islands	Total area (hectares)
Alabama	5	11,417
Connecticut	14	956
Delaware	2	4,089
Florida	80	189,356
Georgia	15	67,045
Louisiana	18	5,717
Maine	9	1,069
Maryland	2	5,789
Massachusetts	27	15,223
Mississippi	5	3,846
New Hampshire	2	445
New Jersey	10	19,433
New York	15	12,271
North Carolina	23	59,271
Rhode Island	6	1,482
South Carolina	35	58,360
Texas	16	155,263
Virginia	11	27,895
18 states	295	638,927

FIGURE 9.18 *Major barrier islands along the Atlantic and Gulf coasts of the United States. These islands were built up from sediments by the action of waves, wind, and sea currents.*

or the concentration of fine sediment suspended in the water column, is characteristically high. The reduced penetration of sunlight, due to this turbidity, limits phytoplankton populations. However, nutrient levels are high. Nutrients carried down to the estuary by stream flow and those carried up by the incoming tides are concentrated in the estuaries, making them a highly productive resource.

Two food chains operate in the estuary. The first is the *grazer food chain,* where dissolved nutrients may be absorbed directly by phytoplankton and rooted plants, then passed into consumers. The second is the *decomposer food chain,* where inert organic material (decayed bodies of marsh grasses, crustaceans, worms, fishes, bacteria, and algae), known as detritus, is consumed directly by detritus feeders, such as clams, oysters, lobsters, and crabs (Figure 9.20).

Because of the abundant supply of nutrients and the high oxygen levels, the estuarine habitat produces more organisms than any other ecosystem except the coral reef.

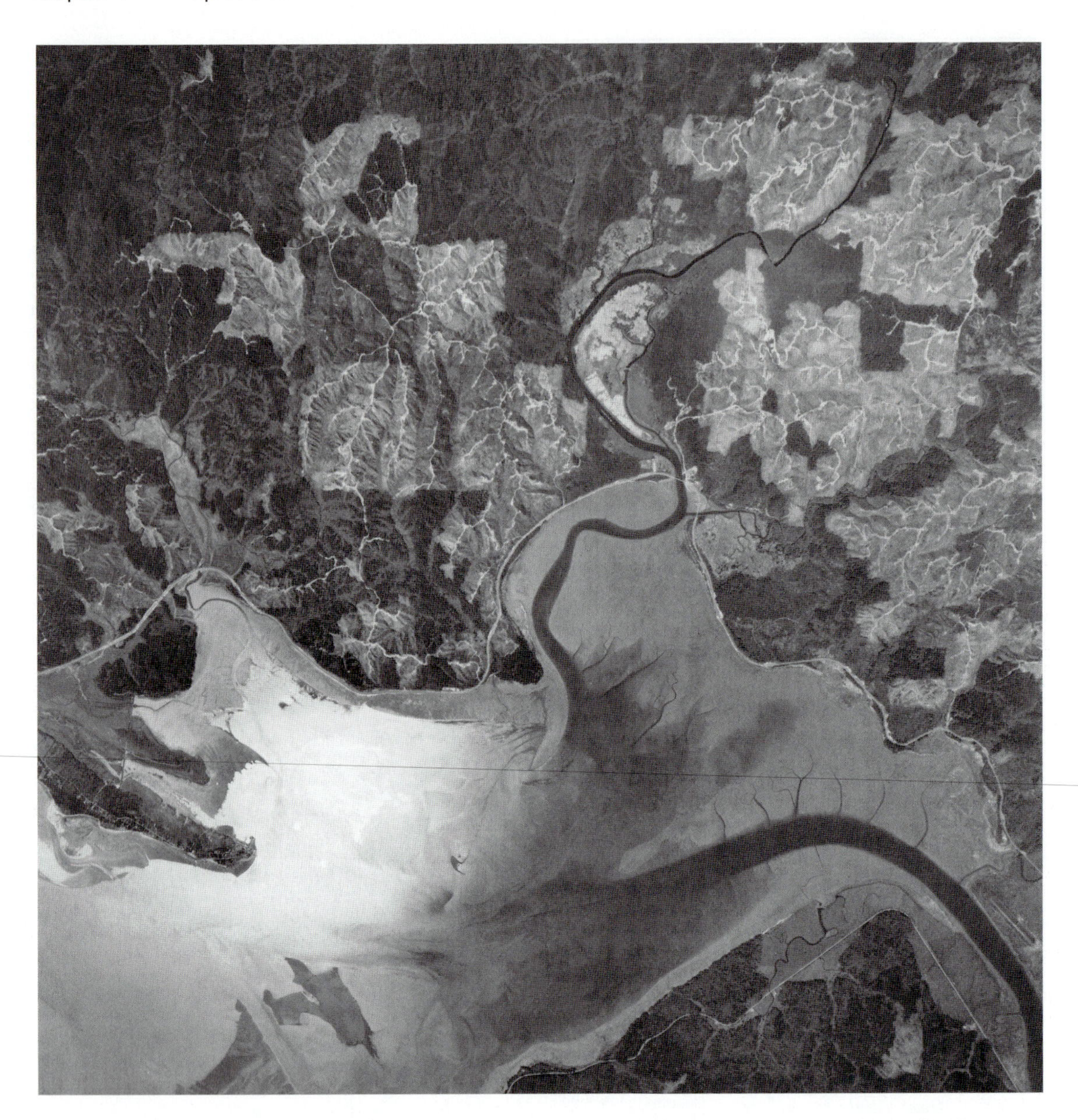

FIGURE 9.19 *Willapa Bay, off the coast of Washington State, is a transitional zone between the salt water of the Pacific Ocean and the freshwater rivers that drain into this bay. The two major coastal rivers forming this estuary are the Willapa River (lower right) and the North Lower Salmon River (upper center).*

Values Estuaries and their associated coastal marshes are indispensable to commercial and recreational fishing industries worth $56 billion annually. Sixty percent of the marine fish harvested by American commercial fishing interests spend part of their life cycle in estuaries. Of every 100 fish taken in the Gulf of Mexico, 98 are estuary and salt marsh dependent. Many marine species use the estuary as a "nursery" in which they spend their larval period immediately after hatching from the egg. Other species, such as the Pacific salmon, pass through estuaries twice during their stream-ocean-stream migration.

Estuarine ecosystems provide food, shelter, and breeding sites for millions of waterfowl and fur-bearing animals, such as muskrats. Almost 45 percent of all U.S. threatened and endangered species are dependent on coastal habitats.

Coastal wetlands play an important role in flood and erosion control. They absorb the shock of storm-driven waves before those waves rush inland and cause destruction of property and human life in heavily populated areas (Figure 9.21).

Estuaries and coastal wetlands are also natural pollution-filtering systems. They cleanse the water of industrial and domestic sewage delivered to the estuaries by rivers. An area of only 5.6 hectares (14 acres) of estuary has the same pollution-reducing effect as a $1 million waste treatment plant!

Human Development of Coastal Environments

The ocean shores are biologically rich and aesthetically stunning. Coastal areas have additional attributes that make them attractive for human development including mild climates, recreational opportunities, and prime locations for shipping, receiving, and transporting by sea. Population near the coast has grown faster than any other area in the United States. Today 53 percent of all

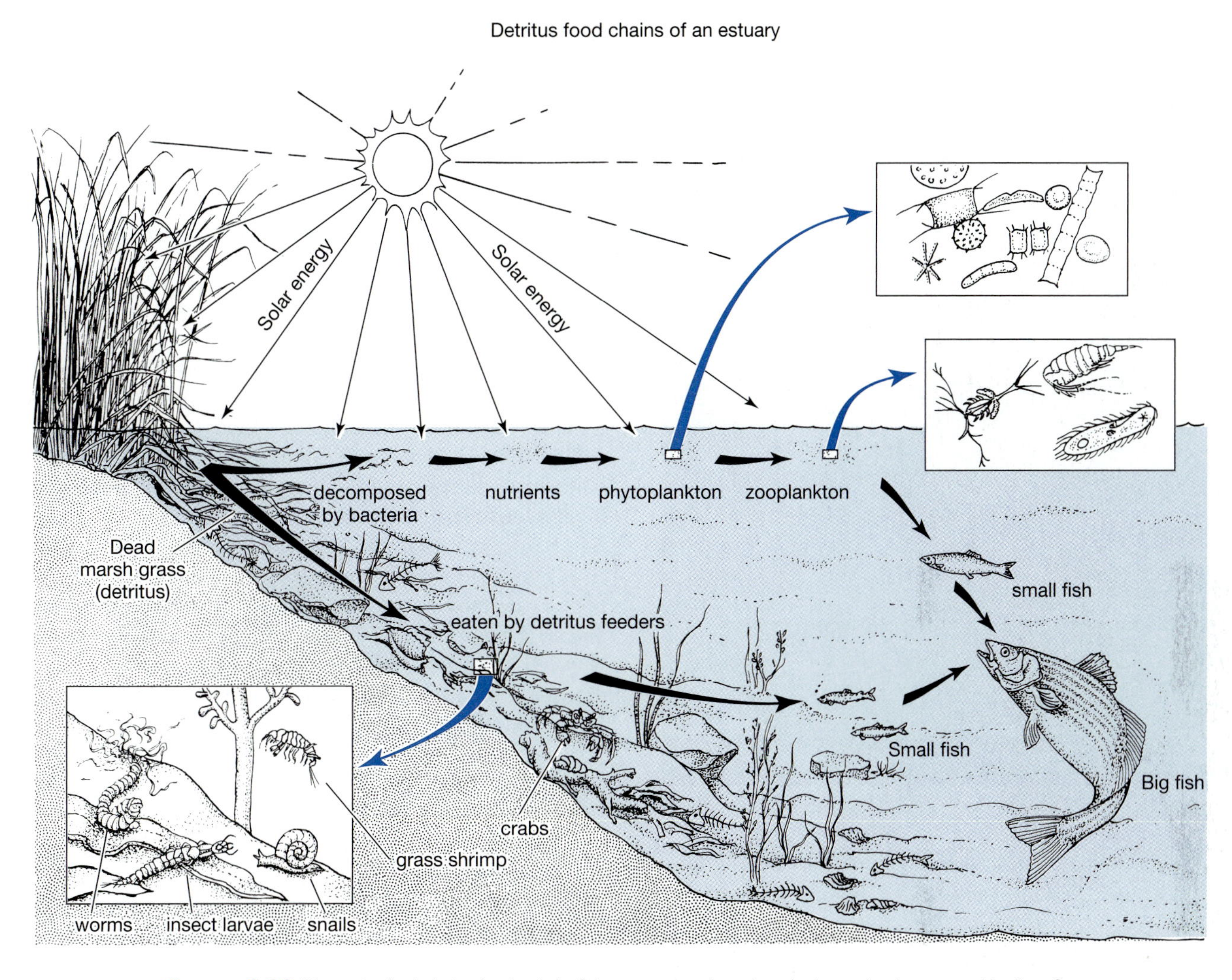

FIGURE 9.20 *The major food chains (and webs) of the estuary are based on detritus—the decomposed bodies of*

Americans live in this delightful region where the land meets the sea—a narrow fringe of land that comprises only 17 percent of the nation's contiguous land area. In 1994, the coastal population was growing at a rate of 3,600 people per day. Should growth continue at this rate, we will see a total increase of 27 million people by 2015. Worldwide, two thirds of the largest cities (population greater than 2.5 million) are located near estuaries.

Shortly after World War II, our coastlines were invaded by enterprising developers who soon launched a multi-billion dollar building boom. Many expensive beachfront homes, condominiums, hotels, and resorts were constructed. Ample federal subsidies were given for water supply systems, sewers, roads, bridges, and shoreline stabilization, spurring the feverish spate of construction. In addition to homes and resorts, human activities in coastal areas include port and industrial development, dredging for navigation, filling to increase land-use area, agriculture, sediment mining for sand and gravel, fishing, sewage disposal, and offshore drilling for gas and oil.

During the last decade, 17 of the 20 fastest growing counties and 19 of the 20 most densely populated counties were located along the coast. In addition, coastal areas include 16 of the 20 counties with the greatest number of new housing units under construction. Between 1994 and 2015, the largest increases in population are expected in the coastal areas of southern California, Florida, Texas, and Washington. Economic development, relocating retirees, and the proliferation of vacation homes are among the factors fueling this rapid population growth.

Coastal Problems

Coastal environments are fragile and highly sensitive to both human-induced and natural events. Habitat destruction, pollution, and accelerated erosion are examples of problems associated with human development (Figure 9.22). The hazards of natural events (tectonic activity, tsunamis, hurricanes, monsoons, storm surges) are exacerbated as human development increases in coastal areas.

Destruction of Estuarine Habitat and Coastal Wetlands

According to the U.S. Fish and Wildlife Service, dredging to deepen channels for navigation and filling to form solid land for construction sites destroyed 260,000 hectares (640,000 acres),

FIGURE 9.21 *Wetland along the North Carolina coast protects the coastline from erosion.*

or 4 percent, of our estuarine habitat between 1950 and 1969. Between the mid 1970s and the mid 1980s coastal wetland area decreased by 29,000 hectares (71,000 acres). From 1986 to 1997 estuarine and marine wetlands decreased by 4,200 hectares (10,400 acres). Although wetland destruction continues, this most recent estimate represents a dramatic reduction in coastal wetland losses compared to previous time periods.

Wave action produced by ship and boat traffic has caused substantial erosion of estuarine margins, destroying precious fish habitat. Freshwater fisheries in the estuaries have been severely damaged by dams upstream. These dams block the inflow of fresh water, resulting in saltwater intrusion from the ocean.

In addition to physical habitat destruction, human activities also increase nutrient loading and discharge pollutants into nearby estuarine ecosystems. Human activities contributing nutrients and other pollutants include sewage treatment plants, industrial effluents, urban storm runoff, agricultural runoff, and airborne contaminants.

Damage and Loss of Life from Storms Natural salt marshes and mangrove wetlands located in coastal areas act as storm buffers, absorbing the high-energy wind, rains, and waves before the ocean storms move further inland. Mushrooming development on the coastlines and barrier islands and the associated decrease in the areal extent of coastal wetlands have increased the vulnerability of both property and human life to hurricanes and other ocean storms. On average, five hurricanes strike the U.S. coastline each year. Of these five, two will be major hurricanes, reaching a category 3 or higher (defined as having winds 111 miles per hour or above). Locations of major landfalling hurricanes in the U.S. between 1899 and 1996 are depicted in Figure 9.23. Some examples of devastation from hurricanes hitting the U.S. coastline follow:

1. In 1969, Louisiana and Mississippi suffered $1 billion in damage from Hurricane Camille when a surge of water 7.5 meters (25 feet) high crumbled 2,822 homes and damaged 40,000 more.
2. In 1989, Hurricane Hugo struck the East Coast, killing 32 people and causing property damage of well over $4 billion. Fifty thousand people were left homeless in Charleston, S.C.
3. In 1992, Hurricane Andrew struck southern Florida, devastating homes in a wide path (Figure 9.24). All told, 75,000 homes were destroyed, leaving 250,000 people homeless. Total costs were estimated to be more than 22 billion dollars.

Erosion Coastal erosion is a natural process, given the inevitable forces of wave, wind, and storm events acting on

FIGURE 9.22 *Humans invade the beaches of Hatteras Island, a barrier island along the North Carolina coast.*

coastal land formations, especially those comprised of sand or other weakly consolidated materials. This natural process, however, continually threatens developed coastal areas with potential loss of life and billions of dollars in property damage. All 30 states bordering an ocean or the Great Lakes experience erosion problems. Net loss of shoreline is occurring in 26 of these states. The coastline on some of the barrier islands in the southeastern U.S. recedes an average of 7.5 meters (25 feet) per year. Erosion rates as high as 15 meters (50 feet) per year have occurred along the shores of the Great Lakes.

Examples. The following are just a few examples of erosion problems in some of our coastal states:

1. *New York.* Aerial photographs show that Long Island's shoreline retreated 30 meters (100 feet) in the last half of the twentieth century.
2. *Louisiana.* Since 1970, more than 762 square kilometers (300 square miles) of coastal areas have been flooded by waters of the Gulf of Mexico, washing away once-useful soil.
3. *North Carolina.* The Cape Hatteras lighthouse on the North Carolina coast was once at least 460 meters (1,500 feet) inland from the high-tide line, but it may be surrounded by the sea in the near future. So much beach erosion has taken place that the wind-whipped waves now surge to within 54 meters (180 feet) of this picturesque landmark.
4. *California.* More than 86 percent of California's 1,771 kilometers (1,100 miles) of exposed Pacific shoreline is receding at an average rate of 15 to 60 centimeters (6 inches to 2 feet) per year. Monterey Bay, south of San Francisco, loses 152 to 300 centimeters (5 to 10 feet) annually.
5. *Washington.* Cape Shoalwater, on Washington's Olympic Peninsula, has eroded 30 meters (100 feet) per year since the turn of the century. The Cape's sparsely settled sand dunes have retreated more than 3.2 kilometers (2 miles) since 1910.
6. *Massachusetts.* The resort town of Chatham nestles in the elbow of Cape Cod. The sand that has eroded from the shore has choked the harbor and impeded movements of boats. A number of recent storms have ravaged the Chatham shoreline. As a result, at least ten beachfront homes have collapsed into the sea and 30 more are in imminent danger.

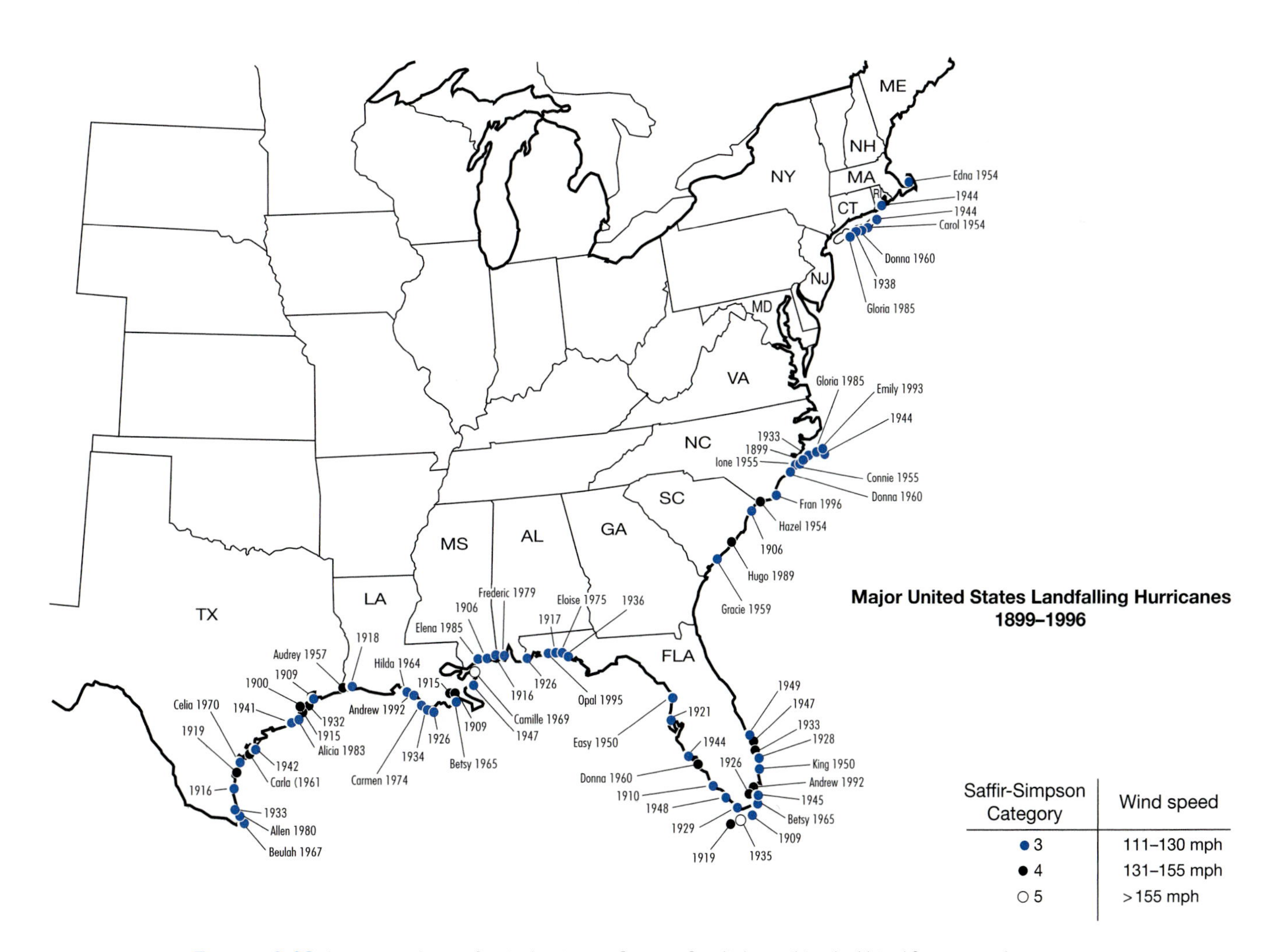

FIGURE 9.23 *Location and year of major hurricanes, Category 3 or higher, striking the United States coast between 1899 and 1996.*

FIGURE 9.24 *Hurricane Andrew (August 25, 1992) caused $22 billion in property damage and killed 54 people along the Atlantic and Gulf coasts. Here are a few trailer homes destroyed in Florida City, south of Miami.*

Accelerated Erosion Human activities are causing the world's shorelines to erode at a faster rate than that due to natural forces alone. We have altered the dynamics of coastal sediment deposition by either depleting or interrupting the transfer of supplies.

Numerous upstream dams have been constructed to generate hydroelectric power or to provide irrigation water for agriculture. These dams have blocked the flow of water that formerly carried millions of tons of sediment to river mouths to form deltas. Many deltas and beaches, therefore, are no longer replenished.

Minerals (such as diamonds), sands, gravels, and rocks have been "mined" from beaches for uses elsewhere. Beach sediment sources also decrease when human activities prevent or alter rockfall, natural downhill movement of unconsolidated materials, or water runoff from cliff areas.

Expensive structures built to lessen the force of wave action (breakwaters), prevent erosion (seawalls), or retain beach sediments (groins) often result in additional erosion problems. **Seawalls** protect the land behind them from further erosion but eventually cause the narrowing and lowering of the beach in front of them. Often, the scouring of the front beach continues until the seawall is undermined, causing failure and, ultimately, the need for a replacement wall. **Groins** are piers of stone spaced about 30 meters (100 feet) apart that extend into the sea at right angles to the shoreline (Figure 9.25). These structures trap sand washed against them by currents that move parallel to the shore, however; groins deprive shores that are immediately down-current of beach-replenishing sand. The only way to overcome this problem is to keep extending groins down-current. The construction of 15 groins by the U.S. Army Corps of Engineers in Suffolk County (Long Island) trapped large quantities of sand, but in doing so accelerated erosion down-current in the Westhampton (New York) area. Suffolk County was sued for $70 million by Westhampton residents.

The extraction of oil in coastal regions and the mining of water from coastal aquifers in Louisiana, Texas, and California have caused the surface of the land, which already was roughly at sea level, to sink or subside. This permits ocean waves to surge inland and wash soils away. In Louisiana the land has subsided 1 meter (3.3 feet) in the past 100 years.

Rise of the Oceans Ocean levels have been rising since the end of the last Ice Age, about 11,000 years ago. However, in the last century or so, the rate has quickened. In New Jersey, for example, since 1900 the sea level has risen about 5.1 meters (17 feet). Why? The primary cause is the *greenhouse effect* resulting from the increasing levels of carbon dioxide in the atmosphere. (See Chapter 19 for more details.) This increase has resulted from the burning of fossil fuels (coal, oil, and natural gas). Carbon dioxide has a warming influence on the atmosphere. The slight increase in temperature has caused the melting of some glaciers and ice caps. Also, when water warms up, it expands. As a result, the sea level has risen. Within this century, carbon dioxide levels should double. As a result, ocean waves might surge inland hundreds of meters in low-lying coastal regions, washing away beaches, causing serious property damage, and inflicting considerable mortality of both wildlife and humans.

Pollution Beach pollution and harm to the natural ecology of coastal environments increase with the increasing use of coastal areas for human developments and beaches for recreation. Sources of pollution include sewage pipe and stormwater drain outflows, industrial or agricultural effluent into rivers, ocean dumping from ships at sea, offshore oil drilling, and general litter left by beach visitors. In some areas of the country, beaches have been closed for public use because levels of sewage-related bacteria have exceeded recognized safe standards.

Sustainable Coastline Management

The need for management of our fragile and valuable coastal environments exists due to the inherent competition and con-

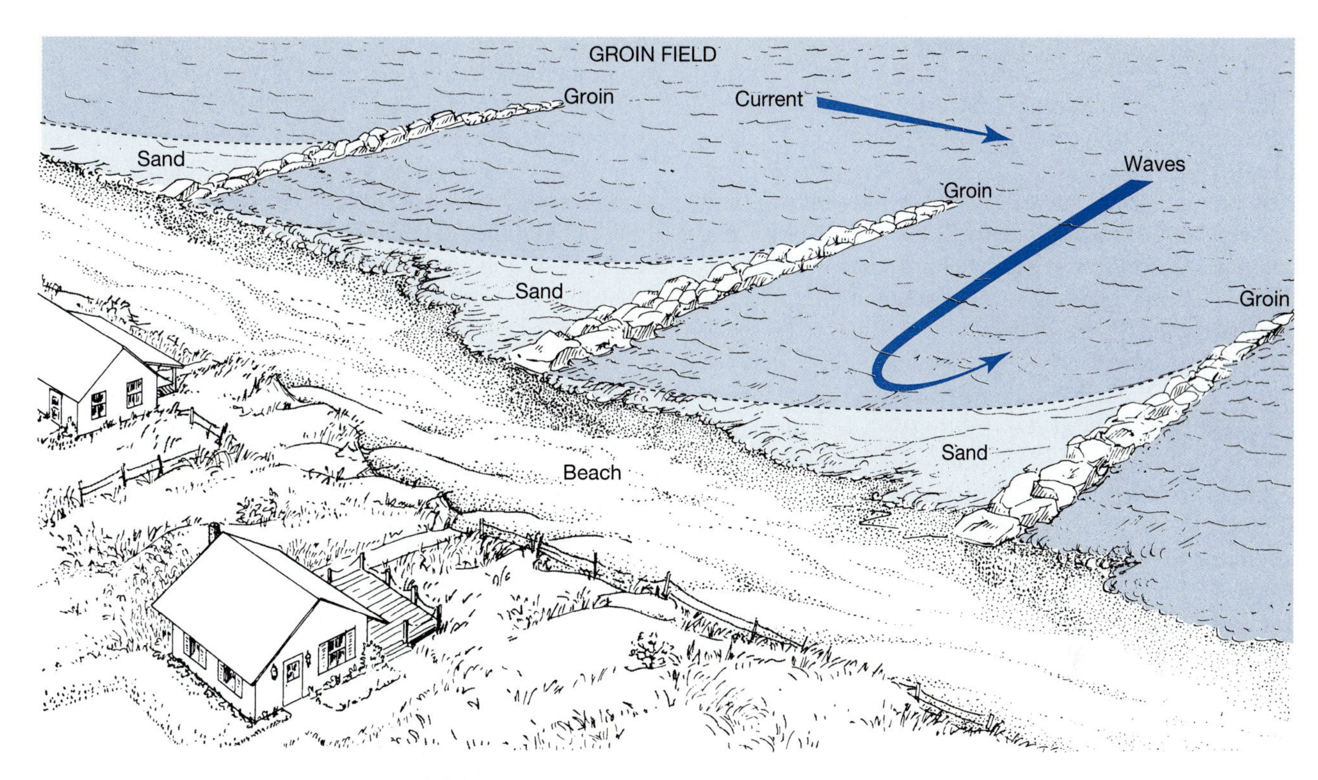

FIGURE 9.25 *Use of groins in an attempt to structurally control beach erosion.*

flict between human activities and healthy ecosystems. Sustainable coastline management seeks to guide the use of coastal environments in a manner that protects the value and function of the natural resources for future generations while allowing human communities and economies to thrive. In 1972, the U.S. Congress passed the landmark **Coastal Zone Management Act (CZMA)** to guide and support this endeavor by establishing voluntary partnerships between the federal government and state, territory, and commonwealth governments. With federal and state matching funds, states and local governments develop and implement coastal zone management programs to achieve the following national objectives identified in the CZMA:

1. Protect natural resources.
2. Manage coastal development to reduce the impact of natural hazards.
3. Protect and restore coastal water quality.
4. Provide public access to the coast.
5. Give priority consideration to coastal-dependent uses and orderly siting of major facilities.
6. Encourage urban waterfront and port redevelopment and historic and cultural preservation and restoration.
7. Support comprehensive planning and management for living marine resources.
8. Plan for the effects of land subsidence and sea level rise.
9. Coordinate and simplify government decision-making.
10. Encourage public participation in coastal management decisions.

Amendments in 1990 and 1996 broadened the scope of the CZMA including additional funding to address nonpoint source polluted runoff into coastal wetlands and waters and to establish a National Estuarine Research Reserve System. Currently, 33 of the 35 eligible states, territories, and commonwealths have federally approved, coastal management programs covering nearly 99 percent or 153,594 kilometers (95,439 miles) of the nation's shoreline (including the Great Lakes). Each state's program is unique due the diversity of coastal habitats, amount and spatial distribution of coast resources, and varying degrees of development and conflicting uses.

Between 1972 and 1997, more than $1.6 billion in federal and state matching funds were appropriated to support coastal management activities under the provisions of the CZMA. The Coastal Wetlands Planning, Protection, and Restoration Act (CWPPRA) of 1990 provided additional matching funds to states for the sole purpose of protecting, restoring, conserving, and enhancing threatened coastal wetlands. With the help of local jurisdictions, non-governmental organizations, universities, and private citizens, coastal management activities include estuarine and coastal wetland protection, restoration, and creation; erosion control and hazard reduction techniques in beach, dune, bluff, and rocky shore environments; and coral reef protection.

Estuarine and Coastal Wetland Protection States have a variety of techniques to protect, enhance, or restore wetlands. Most states have their own "no net loss" policy related to wetland acreage and support local zoning ordinances, land-use plans, and permit requirements that regulate or otherwise avoid alteration to remaining wetland areas. Twenty-six states also use acquisition or conservation easements to protect wetlands and 25 states have programs that actively restore, enhance, and/or create wetlands.

Touted as the "largest environmental restoration project in history", work to reverse the damage to the Florida Everglades has begun. In an effort to control flooding, a century of human modifications has converted the once more than 1 million hectares (2.5 million acres) "river of grass" into farmland, housing developments, and recreational areas. While the farmland is some of the most productive in the U.S., the hydrologic modifications had severe negative impacts on wading birds, fisheries, and the Florida panther. Restoration efforts include re-engineering the channelized Kissimmee River and associated waterways to recreate historical flooding patterns. Estimated costs are $3 billion to $5 billion over a 10 to 15 year period.

Louisiana contains a *whopping* 35 to 40 percent of the nation's coastal wetlands including ubiquitous old growth cypress swamps and vast stretches of salt marshes formed in the sediment deposits of the great Mississippi River delta. However, flood control, navigation, oil and gas exploration, and urban and agricultural land development projects have had devastating effects on the wetlands that border Louisiana's Gulf of Mexico coastline. Deprivation of river sediment supplies, accelerated coastal erosion rates, salt water intrusions, conversion to deep water habitats, and drainage for other uses have resulted in a 25 percent decrease of Louisiana's coastal wetlands since the early 1900s. In 1990, coastal erosion rates were estimated at 65 to 105 square kilometers (25 to 40 square miles) per year. The loss of Louisiana's wetlands is closely associated with loss of shellfish and finfish productivity and revenue as well as recreational, cultural, and ecological diversity and value. Oil and gas wells located in coastal wetlands are threatened by the conversion of the wetlands to open water. In response, the State and Federal governments have dedicated as much as $30 million annually on large-scale wetland creation and enhancement projects (Figure 9.26). In 1988, the U.S. Army Corps of Engineers launched a $25 million project to divert fresh water from the Mississippi River into Louisiana wetlands and estuaries to check saltwater intrusion. Sediment dredged from river channels has been used to create marshes in the coastal region of the Gulf of Mexico. Since 1993, funding has been allocated to restore nearly 20,235 hectares (50,000 acres) of Louisiana's coastal wetlands.

Erosion Control Techniques We established earlier that shoreline structures engineered to protect beaches from erosion have actually increased erosion problems. A couple of other erosion control strategies have been attempted with improved success.

Beach Nourishment Sand has been trucked into eroded beaches to replenish material lost during storms or as a result of normal beach erosion. Nineteen states have used this approach. The benefits of such **beach nourishment,** however, are extremely expensive and only temporary, most lasting from 2 to 7 years before returning to prefill conditions. In 1976, New York City began a beach replenishment project such as this for Rockaway Beach. The project was completed several years later. This ambitious program, funded by more than $50 million from city, state, and federal coffers, involved the transport of 9.5 million cubic yards of beach-replenishing sand. A $2 million nourishment project at Ocean City, Maryland, lost 60 percent of its sand to violent storms only 2 weeks after it was completed!

Vegetation Establishment Planting of native vegetative to control erosion can be successful if done properly. Patches of salt-tolerant cordgrass have been planted by Texas conservation workers in the shallow waters near the coastline of Galveston Bay. This grass serves as a living buffer against the sand-washing waves of the Gulf. However, planting schemes that fail to mimic the natural patterns of succession and spacing may have limited success. For example, overplanting can lead to overstabilization of certain areas, causing unnatural erosional and depositional processes elsewhere.

Hazard Reduction Techniques Coastal residents and vacationers are vulnerable to the hazards of long-term erosion and damaging forces of natural, catastrophic events such as hurricanes, tsunamis, and earthquakes. Coastal states are utilizing various approaches to prevent or lessen losses of life and property and protect shoreline resources.

Restriction of Coastal Development States have restricted development and other human activities on or near high energy beaches, dunes, and cliffs by requiring setbacks, eliminating subsidies for private development, regulating or prohibiting the construction of shoreline stabilization structures, and/or preventing vehicle access. In North Carolina, for example, new buildings must be located at least 40 meters (120 feet) inland from the first line of dunes. In Delaware, state law forbids the construction of any industrial plants within 3 kilometers (2 miles) of the seashore. New Jersey's Waterfront Development Act (1988) empowers the Department of Environmental Protection to fine waterfront owners $1,000 if they illegally place structures too close to the shoreline and to impose a $100-a-day fine for each day the violation continues. In addition to addressing human safety and potential property damage concerns, the measures serve to reduce erosion, preserve scenic beauty, and protect vulnerable coastal wetlands and estuaries.

FIGURE 9.26 *The Point au Fer Wetland Restoration Project off the Gulf coast of Louisiana. Boulders were placed along a 1,220 meter (4,000 foot) stretch of this narrow strip of beach to prevent storm overwash and potential breaching into the wetland situated behind the beach. The salt marsh is now protected from accelerated erosion, saltwater intrusion, and hydrologic alteration.*

Planning and Public Education Planning and educational outreach are valuable tools to lessen the loss of life and damage to existing coastal development from various natural and human-induced hazards. State and local governments are identifying areas susceptible to storm damage, developing improved hurricane evacuation plans and emergency communications systems, and ensuring public awareness of appropriate evacuation routes and shelter locations. In addition, education and public outreach programs aim to increase public awareness of the dangers associated with building in dynamic shoreline areas (Figure 9.27).

Coral Reef Protection Coral reefs have historically exhibited the capacity to recover rapidly from natural stresses including catastrophic tropical storms, predation, diseases, and freshwater runoff that causes sedimentation and upsets the natural nutrient and salinity balances. However, increases in human populations in coastal subtropical and tropical areas during the past 50 years have resulted in new stresses that have accelerated damage to coral reefs and valuable reef fisheries resources. These stresses include polluted stormwater runoff, agricultural and industrial point and nonpoint source pollution, sewage effluent, and over-fishing. Direct contact by humans via snorkeling and diving in popular coral reef destinations has led to physical damage to the reefs and removal of coral and other marine life for souvenirs or for sale. Recent coral-eating starfish plagues and coral diseases have been particularly devastating in some areas of the Caribbean, escalating speculations that both of these stresses may be exacerbated by human activities.

Many coral reefs are located in remote locations away from human development or are under good management, such as the Great Barrier Reef in Australia. Still, recent estimates categorize up to 56 percent of the world's reefs as threatened and another 10 percent severely damaged or destroyed. Several management techniques are in place in the U.S. and other countries to help lessen or reverse human impacts to these fragile and valuable resources. Protection techniques include prohibition of coral removal, restrictions on touching coral and anchoring boats in coral reef areas, limitations on recreational activities and boating traffic, and regulations that control polluted runoff. Public outreach and education are important components of many coral reef protection plans and include partnerships with dive shops and clubs to increase the awareness of activities that harm coral reefs. Proactive efforts such as installation of mooring buoys along the outskirts of some coral reefs allow visiting boaters to tie up without anchoring on the reef itself. Several coral reefs have been set aside for research, restoration, and/or established as marine protected areas such as the Florida Keys National Marine Sanctuary and the Gray's Reef National Marine Sanctuary off the coast of Georgia.

9.5 The Ocean

General Features

The ocean covers 70 percent of the Earth's surface and is divided into three major basins: the Pacific, Atlantic, and Indian oceans. The Pacific Ocean is the deepest and largest ocean basin. The deepest point is the 11-kilometer (6.5-mile) Mariana Trench southeast of Japan. The Atlantic Ocean is the warmest ocean and relatively shallow. The land area draining into the Atlantic Ocean is four times that draining into the Pacific Ocean. The Indian Ocean is the smallest of the major basins and lies primarily in the Southern Hemisphere.

The ocean is about 70 times as salty as a lake or stream. Ocean water is comprised of approximately 96.5 percent water, 3.5 percent salt, and tiny amounts of living things, dust, and dissolved organic matter. The salts come from volcanic eruptions and the chemical weathering of rocks on the ocean bottom and on exposed land surfaces. The four ions that account for 97 percent by weight of the ocean's salt are Na^+ (64 percent), Cl^- (29 percent), Mg^{2+} (3 percent), and SO_4^{2-} (2 percent).

The ocean is continuously circulating (Figure 9.28). The Alaskan Current brings cold water down the Pacific Coast; the Gulf Stream brings warm water upward along the Atlantic Coast. These currents modify the temperatures of nearby coastal regions. For example, New York City has a relatively moderate climate, even in winter, whereas summer evenings in San Francisco may be quite chilly. Vertically moving currents, or **upwellings,** bring nutrient-rich cold water from the ocean bottom to the surface (Figure 9.29).

The sea is relatively infertile compared to fresh water. Nitrates and phosphates are extremely scarce. Two exceptions are the areas of upwelling and estuaries, where streams discharge massive loads of sediment.

Ocean Zonation

Neritic Zone The **neritic zone** is the marine counterpart of the littoral zone of the lake. This zone is a relatively warm,

FIGURE 9.27 *This beach house, located on the North Carolina coast, is much too close to the sea. This damage was a result of the house's vulnerability to flooding and beach erosion.*

nutrient-rich, shallow region, that overlies the continental shelf, a submerged extension of the continent. Occurring along our Atlantic, Pacific, and Gulf coasts, the neritic zone is 16 to 320 kilometers (10 to 200 miles) wide and up to 60 to 180 meters (200 to 600 feet) deep. The neritic zone ends where the continental shelf abruptly terminates and the ocean bottom plunges to great depths (Figure 9.30).

The nutrients of the neritic zone are supplied primarily by upwelling and the sedimentary discharge of streams. Sunlight normally penetrates to the ocean bottom, permitting considerable photosynthetic activity in a large population of floating and rooted plants. Animal populations are rich and varied (Figure 9.31). Oxygen depletion is not a problem because of photosynthesis and wave action. *The total amount of biomass in the neritic zone is greater per unit volume of water than in any other part of the ocean.*

Euphotic Zone The **euphotic zone** is the open-water zone of the ocean that corresponds to the limnetic zone of a lake (Figure 9.30). The term *euphotic,* which means "abundance of light," is appropriate to this zone, for it has sufficient sunlight to support photosynthesis and a considerable population of phytoplankton. In turn, the phytoplankton support a host of tiny grazing herbivores such as the small crustaceans. The total energy available to animal food chains from euphotic phytoplankton is much greater than that produced by plants of the neritic zone; this is largely because of the vast area of the euphotic zone, which extends for thousands of miles across the open sea. The degree to which light penetrates, of course, depends on the transparency of the surface waters. Because sunlight cannot penetrate deeper than 200 meters (650 feet) in most marine habitats, this is frequently considered the lower limit of the euphotic zone.

Bathyal Zone Beneath the euphotic zone is the **bathyal zone** (Figure 9.30). This zone is a region of semidarkness. Photosynthesis cannot occur here and so no producer organisms can survive.

Abyssal Zone Beneath the bathyal zone is the **abyssal zone.** The abyssal zone is the cold, dark-water zone of the ocean depths that roughly corresponds to the profundal zone of the lake habitat. This zone lies immediately above the ocean floor (Figure 9.30). Animal life is rather sparse. Any animal living in the abyssal zone must be adapted to low light, intense cold (because the abyssal water frequently approaches the freezing point), extremely low levels of dissolved oxygen (because no photosynthesis can occur here), water pressures of thousands of pounds per square inch, and scarcity of food.

In many areas, the sediment of the abyssal zone is rich in nutrients. These nutrients come from the decaying bodies of organisms from the sunlit waters above and from excretions of living animals. For example, in certain regions of the western Atlantic, the phosphate concentration at a depth of 900 meters (3,000 feet) may be 10 times greater than the concentration at 90 meters (300 feet). Because neither phytoplankton nor herbivorous animals can exist in the abyssal zone, most consumers are either predators or detritus feeders. A number of the deep-sea fish of the abyss have evolved luminescent organs that may aid them in finding food and mates.

Marine Food Chains

Because the ocean covers 70 percent of the Earth's surface, it receives 70 percent of the Earth's solar energy. Except for the anchored green plants of the neritic zone, this solar energy is trapped primarily by the phytoplankton (producers) in the open water of the oceans. Scientists estimate that 18 billion metric tons of living plant matter (mostly phytoplankton) are produced annually. This plant matter, in turn, supports 4.5 billion metric tons of zooplankton. Marine zooplankton may be consumed by a variety of filter feeders, including shrimp, herring, anchovies,

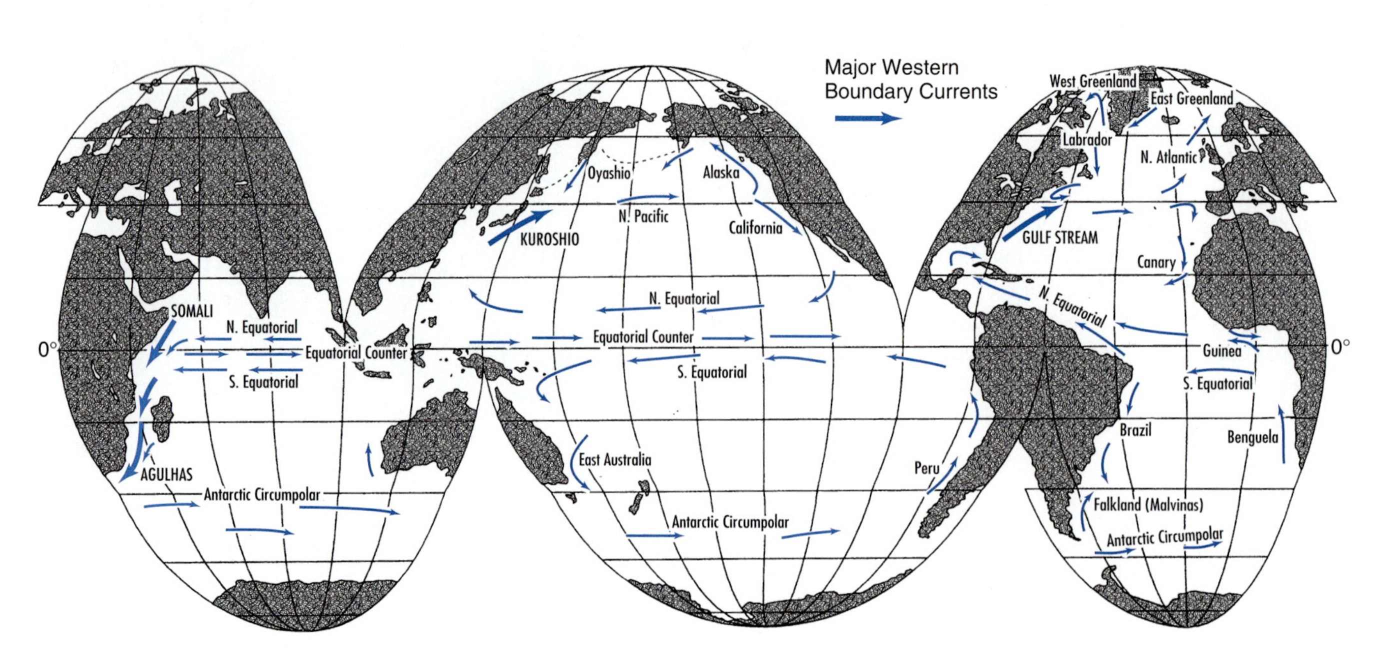

FIGURE 9.28 *Major ocean surface currents. The thicker the arrow, the stronger the current.*

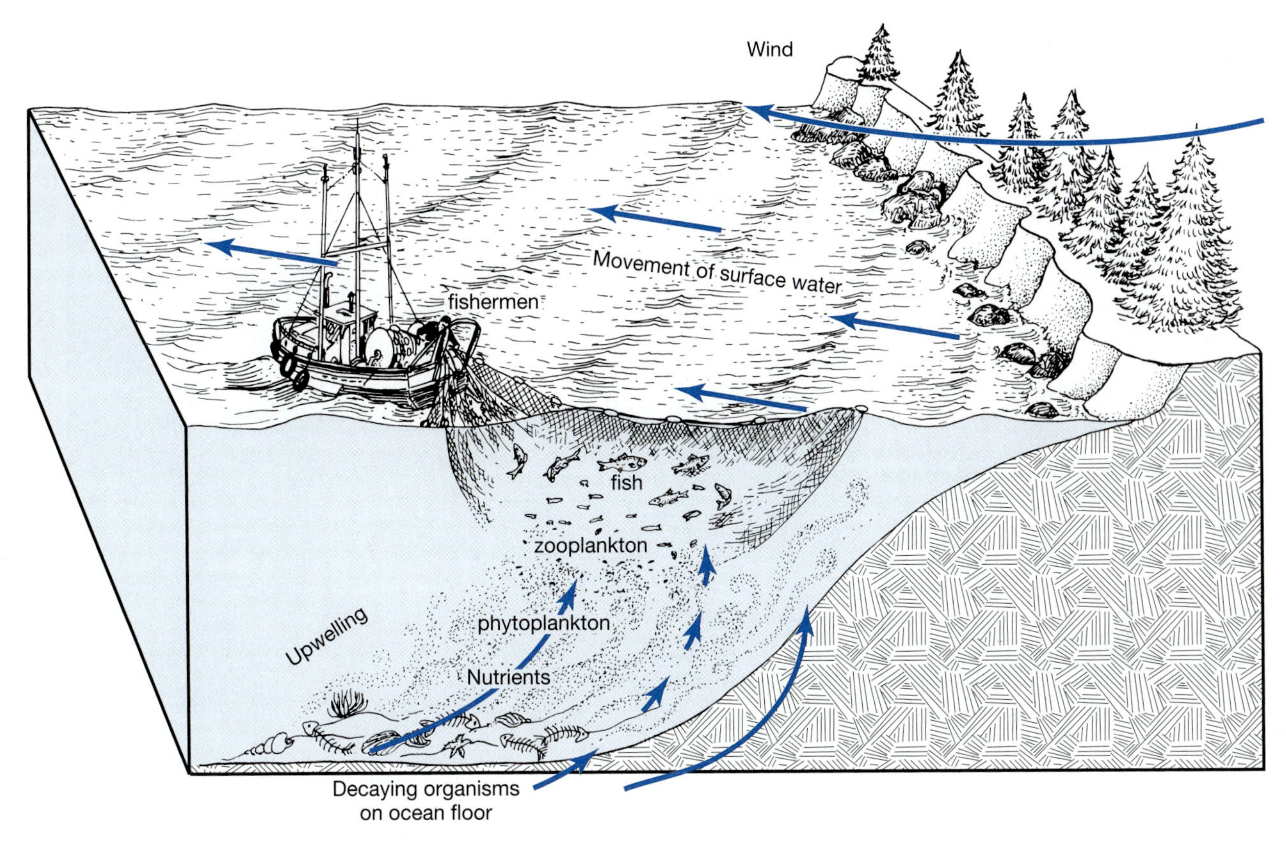

FIGURE 9.29 *The upwelling phenomenon. Wind moves the surface water outward. As a result, cold water from the ocean bottom moves upward and shoreward, carrying dissolved nutrients with it. Consequently the area of upwelling is biologically productive.*

and whales. Fish-eating predators such as the shark, barracuda, cod, salmon, cormorants, pelicans, and humans represent the terminal link of the marine food chain (Figure 9.31). As in the terrestrial food chains, the shorter the chain, the more biomass is available to the top-level organisms.

Ocean Resources

The ocean serves human needs in a great variety of ways. The ocean's 32-million-cubic-mile volume is a virtually limitless water supply for all organisms on this planet, including humans. Trillions of tiny algae in sunlit ocean waters have aided in replenishing the oxygen supply of the Earth's atmosphere, upon which the survival of all life depends. The ocean absorbs 20 times more carbon dioxide (responsible for global warming) than all the Earth's vegetation.

Ever since early humans scooped fish from its tidal pools with their bare hands, the ocean has provided people with abundant supplies of essential protein. The ocean serves as a highway for national and international transport and communication. And, the ocean is valued for recreational uses, including boating, sailing, recreational fishing, and cruising.

The ocean bottom is a treasure house of valuable minerals such as manganese, nickel, copper, and cobalt. Only a very small area of the ocean has been surveyed for these deposits. At this time, extraction technology is too expensive to make exploitation of these mineral reserves economically feasible.

Over 26 percent of all oil production is from offshore wells. Natural gas production from offshore wells accounts for about 17 percent of the world's production. Additional ocean oil and natural gas deposits have yet to be exploited. The ocean also provides the potential for renewable energy production if we can find an effective means to harness the energy from thermal underwater vents, waves, currents, tides, and thermal columns. Research sites in several parts of the world are experimenting with this type of renewable power generation.

These abundant ocean resources have the potential to provide for the long-term needs of a growing human population but only if regulated to prevent excessive exploitation and disruption to productive ecosystems.

Summary of Key Concepts

1. Wetlands are highly diverse and usually occupy a transitional zone between a well-drained upland area and a permanently flooded deepwater habitat.
2. Wetland and deepwater habitats are classified into five major categories by the Cowardin Classification System: marine, estuarine, riverine, lacustrine, and palustrine.
3. Marine and estuarine wetland systems comprise saltwater environments such as coral reefs, rocky shorelines, tidal salt marshes, and mangrove swamps.

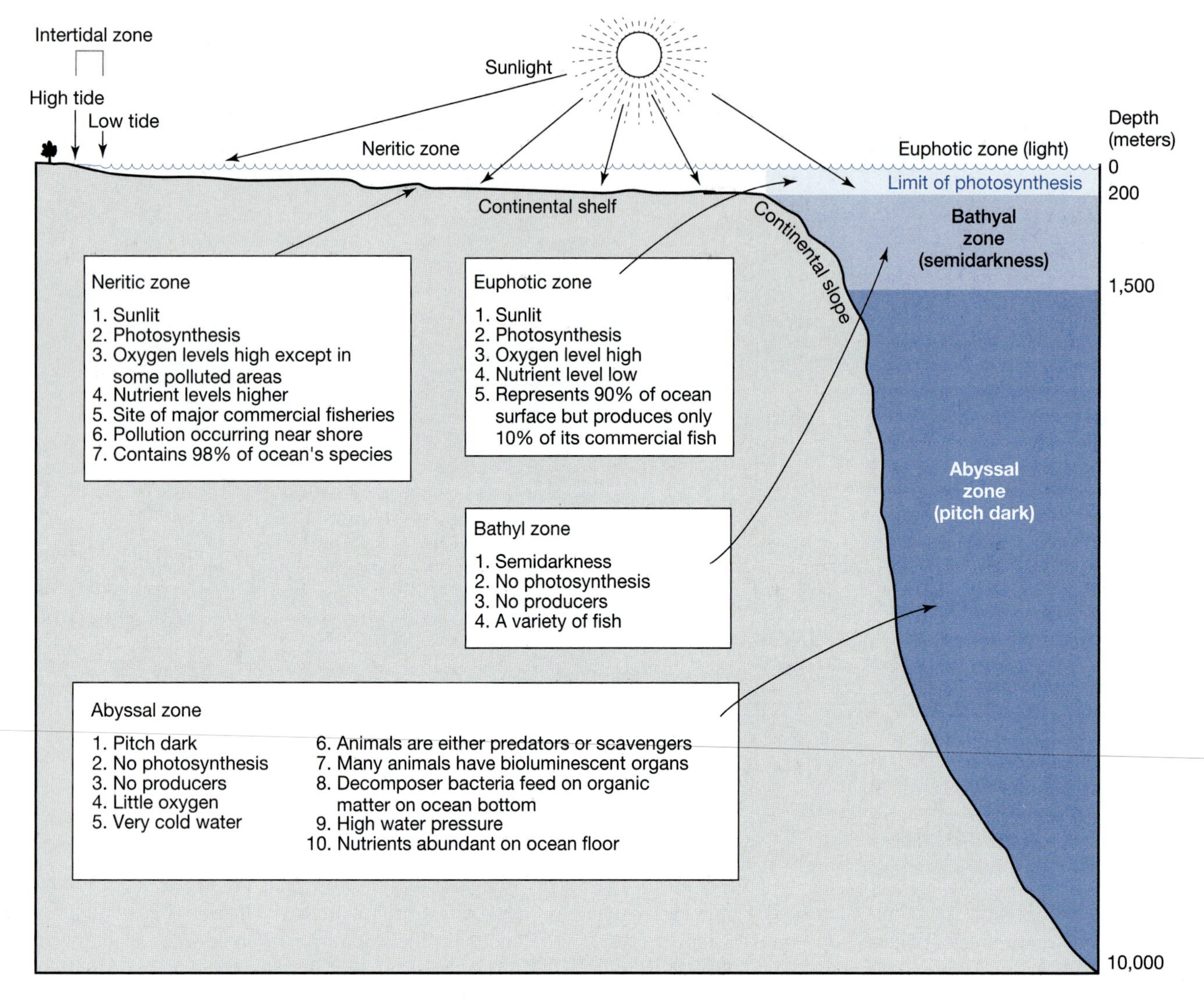

FIGURE 9.30 *Locations and characteristics of the ocean's life zones.*

4. Riverine wetland systems include riparian wetlands situated along rivers and streams; lacustrine wetlands are those bordering the edges of lakes, large ponds, and reservoirs.
5. Palustrine wetland systems contain the majority of all wetland habitats including bogs, fens, freshwater marshes, freshwater swamps, and small shallow ponds.
6. Wetland functions can be grouped into three broad categories: (a) hydrologic processes, (b) water quality improvement, and (c) wildlife habitat. Many human-based values are derived from the functions wetlands perform.
7. The United States has a "no net loss" wetland policy. Wetland management policy has shifted from wetland drainage to wetland preservation, restoration, and creation schemes. Activities in and around wetlands are regulated by Section 404 of the Clean Water Act.
8. Ecologists recognize three major lake zones: littoral, limnetic, and profundal.
9. The littoral zone is the shallow marginal region of a lake that is characterized by rooted vegetation.
10. Plant species colonize a lake margin according to specific hydrologic requirements, with those less tolerant of water inundation situated closest to the shore.
11. The suspended, floating microorganisms in a lake are known as plankton.
12. The limnetic zone is the region of open water beyond the littoral zone down to the maximum depth at which there is sufficient sunlight for photosynthesis. The profundal zone lies beneath the limnetic zone.
13. In spring and autumn, the waters of temperate zone lakes undergo a thorough mixing that is known as the spring and fall overturn.
14. Ephemeral streams dry up during the dry season, whereas perennial streams flow year-round.
15. A smaller channel that flows into a larger channel is called a tributary of that channel.
16. A watershed is the entire land area, confined by topographic divides, in which all tributaries drain into a single river system.

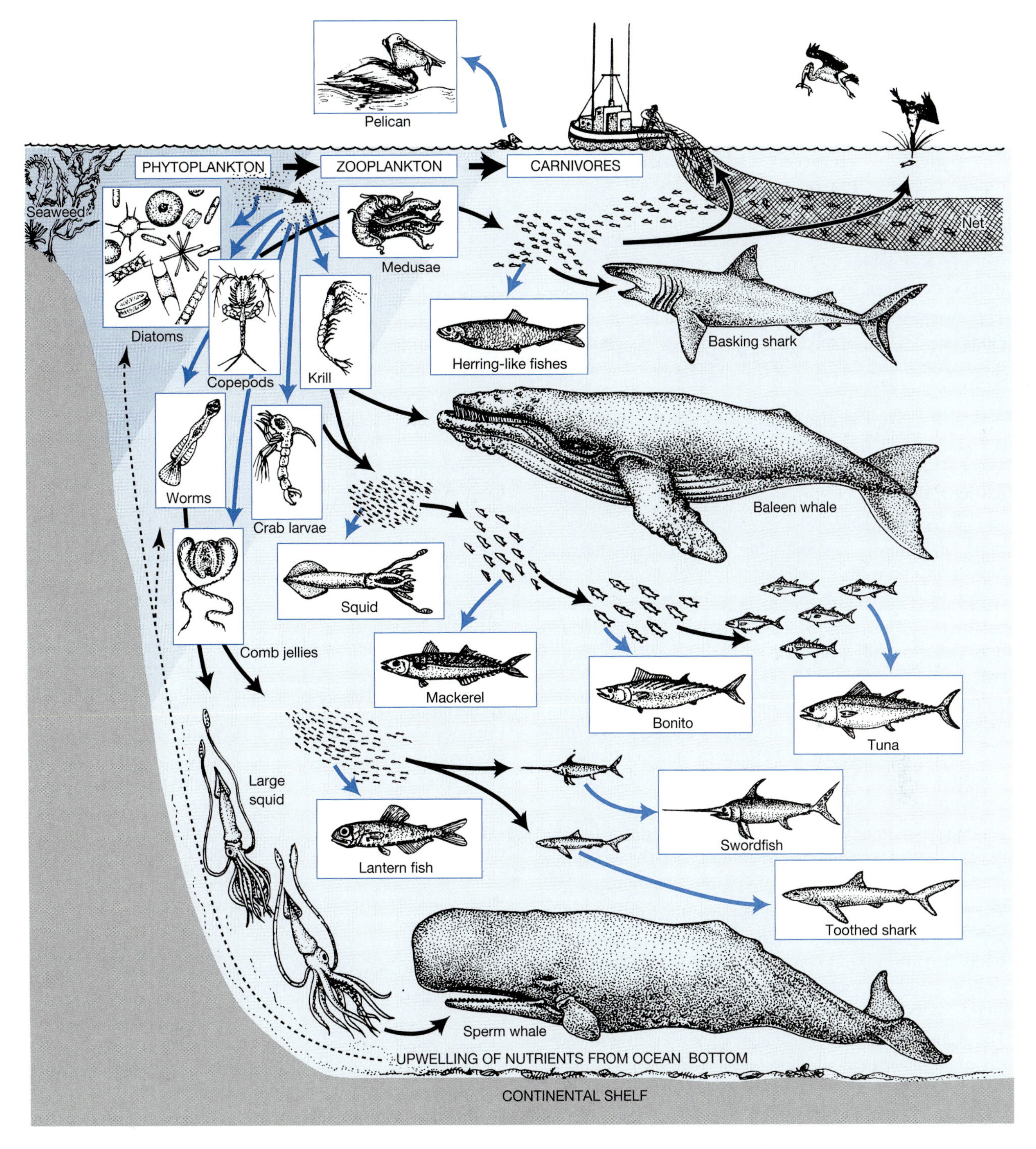

FIGURE 9.31 *Communities occupying the neritic zone. Seaweeds and algae are the producers, forming the base of the food chains and webs. Tiny zooplankton feed on the producers. In turn, zooplankton are consumed by fish, lobsters, crabs, and whales. The top consumers are the seabirds, sharks, tuna, swordfish, whales, and humans. The bottom-dwellers, which cling to rocks or are buried in the mud, include clams, snails, crabs, and bacteria. They feed on detritus. The dashed arrows on the left indicate upwelling of nutrients from the ocean bottom.*

17. A first-order (Order 1) stream is the smallest stream within a river system, having no tributaries. A second-order (Order 2) stream has only first-order streams as tributaries, and so on. In general, the smaller and fewer the tributaries emptying into a stream, the lower that stream's order number.
18. A stream channel takes shape during high water events, when the force of water flow is enough to erode banks and transport bed materials.
19. The pattern of a river channel can be meandering, straight, or braided. The meandering pattern is the most common.

20. Pools and riffles are regularly spaced along the length of a channel bed.
21. Overflow water and suspended sediment are received by a river's floodplain. Downcutting by a river results in an abandoned floodplain, called a terrace.
22. The stream ecosystem receives a considerable portion of its energy supply from materials like leaves and twigs that originate in the riparian zone.
23. Riparian vegetation is also important to streams and wildlife as a source of large woody debris, erosion control, shade, cover, food, and nesting grounds.
24. The structure of coastlines is highly variable and may include one or more of the following features: beaches, sand dunes, rocky cliffs, barrier islands, coral reefs, coastal salt marshes, and estuaries.
25. A typical estuary is characterized by (a) brackish water, (b) water levels that rise and fall with the tides, (c) high levels of dissolved oxygen, (d) high turbidity, and (e) high nutrient levels.
26. Because of the rapid development of the coastlines and barrier islands, hurricanes and other oceanic storms threaten human life and valuable property.
27. Accelerated coastal development has (a) degraded scenic beauty, (b) destroyed coastal wetlands, (c) accelerated beach erosion, (d) increased pollution, and (e) damaged fish and wildlife habitats.
28. Among the factors contributing to coastal erosion problems are (a) storm events, (b) the reduction of sediment discharge into the ocean due to dam construction, (c) subsidence, and (d) engineered erosion control structures.
29. Among the methods that have been used to control coastal erosion are (a) beach nourishment projects, (b) vegetative plantings, and (d) restriction of coastal development.
30. The National Coastal Zone Management Act (1972, 1990, and 1996) has provided financial assistance to coastal states, territories, and commonwealths to develop sound strategies for protection of the sensitive environments along coasts and barrier islands.
31. The ocean (a) covers 70 percent of the Earth's surface, (b) contains vertical and horizontal currents, (c) is about 70 times as salty as a lake or stream, and (d) is relatively infertile.
32. The four major zones of the ocean are (a) neritic, (b) euphotic, (c) bathyal, and (d) abyssal.
33. The neritic zone is a relatively warm, shallow, nutrient-rich region adjoining coasts.
34. The euphotic zone is a region in the open ocean that extends from the surface to the deepest area where there is sufficient sunlight for photosynthesis.
35. Below the euphotic zone is the bathyal zone, a region of semidarkness.
36. The abyssal zone lies immediately above the ocean floor. It is characterized by (a) darkness, (b) intense cold, (c) low levels of dissolved oxygen, and (d) a scarcity of food.
37. In areas of upwelling, fish production is highly efficient because many of the food chains have only two links: the producer phytoplankton and the consumer fish. In the deep waters of the open sea, however, fish production is much less efficient because the food chains may have as many as six links.

Key Words and Phrases

Abyssal Zone
Barrier Islands
Bathyal Zone
Beach Nourishment
Bog
Clean Water Act
Coastal Zone Management Act
Compensation Depth
Coral Reefs
Cowardin Classification System
Deepwater Swamp
Detritus
Ecotone
Ephemeral Stream
Epilimnion
Estuary
Estuarine Wetland System
Euphotic Zone
Everglades
Fall Overturn
Floodplain
Freshwater Marsh
Great Barrier Reef
Groin
Headwater Streams
Hydrophytes
Hypolimnion
Lacustrine Wetland System
Large Woody Debris
Limnetic Zone
Littoral Zone
Lowland Streams
Mangrove Swamps
Marine Wetland System
Meandering Stream
Neritic Zone
"No Net Loss" Wetland Policy
Palustrine Wetland System
Peat
Peatlands
Perennial Stream
Plankton
Point Bar
Pool
Prairie Potholes
Profundal Zone
Quaking Bog
Riffle
Riparian Wetlands
Riparian Zone
Riverine Wetland System
Sand Dunes
Seawall
Spring Overturn
Stream Order
Terraces
Thermocline
Tidal Creek
Tidal Salt Marsh
Tributary
Turbidity
Undercut Banks
Upwelling
Watershed
Wetland Delineation
Wetland Functions
Wetland Values
Wetlands

Critical Thinking and Discussion Questions

1. Define the key words and phrases in the list of key words and phrases.
2. Explain the difficulties in determining wetland boundaries.
3. Describe the general characteristics of each of the five wetland systems and give an example of each.
4. All riparian wetlands are included in the riparian zone, but not all land within the riparian zone is wetland. Explain.

5. Distinguish between wetland functions and wetland values and give two examples of each.
6. The annual rate of wetland conversion in the U.S. decreased by 80 percent compared to the previous decade. Does this mean we should "relax" our wetland regulation and protection policies?
7. Describe the characteristic distribution of rooted plants in the littoral zone.
8. Identify three insects, three fish, and three birds that are characteristic of the littoral zone.
9. Suppose that the light intensity on the bottom of a lake is 10 percent of that at the compensation depth. What percentage of full sunlight would that be?
10. Identify two sources of the dissolved oxygen present in a lake.
11. Give three reasons why oxygen might be a limiting factor for fish in northern states during the winter season.
12. What causes the fall turnover? What causes the spring turnover?
13. Do the fall and spring turnovers have any significance for the survival of aquatic organisms? Explain your answer.
14. Describe the stream characteristics that control its shape and pattern. Compare the shape and pattern of a mountainous headwater stream to a lowland, valley stream.
15. What defines the boundaries of a watershed?
16. How does a healthy riparian zone contribute to the health of the stream ecosystem?
17. In what ways does an estuary differ from an ocean? From a river? Why are estuaries so productive?
18. How do coastal wetlands protect the shore from erosion?
19. Loss of life and property from coastal storms and flooding is likely to increase in the future. Why?
20. What impact(s) will the predicted rise in sea level have on coastal and barrier island habitats and communities.
21. Engineered structures designed to control beach erosion have actually caused additional erosion problems. Explain.
22. Discuss the various strategies states have implemented to restrict coastal development in high energy or erosion-prone areas. Which one do you think has the greatest chance for success?
23. Characterize the neritic, euphotic, and abyssal zones of the ocean.
24. Describe at least four ways in which the ocean provides value to humans or satisfies human needs.

Suggested Readings

Cowardin, L. M., et al. 1979. *Classification of Wetlands and Deepwater Habitats of the United States.* Washington, D.C.: U.S. Department of the Interior, Fish and Wildlife Service, Publication FWS/OBS-79/31, U.S. Government Printing Office. The national and international standard for identifying and classifying wetland and deepwater habitats throughout the world. Report includes detailed descriptions of hierarchical systems and numerous photographs as examples.

Dahl, T. E. 1990. *Wetlands: Losses in the United States 1780s to 1980s.* Washington, D.C.: U.S. Department of the Interior, Fish and Wildlife Service, U.S. Government Printing Office. Report to Congress documenting historical wetland losses over a 200-year time span for each state in the nation.

Dahl, T. E. 2000. *Status and Trends of Wetland in the Conterminous United States 1986 to 1997.* Washington, D.C.: U.S. Department of the Interior, Fish and Wildlife Service, U.S. Government Printing Office. The most recent and comprehensive report on the status and trends of wetlands on a national scale.

Dahl, T. E., and C. E. Johnson. 1991. *Status and Trends of Wetlands in the Conterminous United States, Mid-1970s to Mid-1980s.* Washington, D.C.: U.S. Department of the Interior, Fish and Wildlife Service, U.S. Government Printing Office. The first update of the National Wetlands Inventory completed in 1982.

Frankl, E. 1995. *Ocean Environmental Management: A Primer on the Role of the Oceans and How to Maintain Their Contributions to Life on Earth.* Upper Saddle River, N.J.: Prentice Hall. Overview of current environmental problems relating to ocean resources and discussions of alternatives for sustainable use.

Gross, M. G. and E. Gross. 1996. *Oceanography: A View of the Earth.* Upper Saddle River, N.J.: Prentice Hall. Excellent introductory textbook on the ocean environment.

Kemper, S. "The Beach Boy Sings a Song Developers Don't Want to Hear." *Smithsonian* 23(7): 72–85, 1992. Nontechnical discussion of the beach erosion problem.

Leopold, L. B. 1994. *A View of the River.* Cambridge, Mass.: Harvard University Press. Author successfully transforms lengthy, mathematical, and inherently complex descriptions of the physical forms and processes of rivers into a compact, easy to read, interesting resource book.

Mitsch, W. J., and J. G. Gosselink. 1993. *Wetlands.* New York: Van Nostrand Reinhold. Wonderful, comprehensive text on all aspects of wetland science, restoration, and management.

Murphy, M. L., and W. R. Meehan. 1991. "Stream Ecosystems." In *Influences of Forest and Rangeland Management on Salmonid Fishes and Their Habitats.* Meehan, W.R., (ed.) American Fisheries Society Publication 19. Bethesda, Md. Extremely informative description of stream ecology and the implications for successful fisheries management.

Tiner, R. W., Jr. 1984. *Wetlands of the United States: Current Status and Recent Trends.* Washington, D.C.: U.S. Department of the Interior, Fish and Wildlife Service, U.S. Government Printing Office. First report of the results from the National Wetlands Inventory completed in 1982.

U.S. Army Corps of Engineers. 1987. *Corps of Engineers Wetlands Delineation Manual.* Technical Report Y-87-1. Washington, D.C.: Department of the Army. Describes technical guidelines and methods using soil, hydrology, and vegetation parameters to identify and delineate wetlands for purposes of Section 404 of the Clean Water Act.

U.S. Geological Survey. 1996. *National Water Summary on Wetland Resources.* Water-Supply Paper 2425. Washington, D.C. U.S. Government Printing Office. Excellent summary of wetland resources in the United States and individual states including overview sections on technical aspects; management and research; and restoration, creation, and recovery of wetlands. Beautiful color illustrations and photography.

Viles, H., and T. Spencer. 1995. *Coastal Problems: Geomorphology, Ecology and Society at the Coast.* London: Edward Arnold. Excellent descriptions of coastal formations and processes and current discussion of human activities and their effects on coastal environments.

Web Explorations

Online resources for this chapter are on the World Wide Web at: **http://www.prenhall.com/chiras** *(click on the Table of Contents link and then select Chapter 9).*

Chapter 10

Managing Water Resources Sustainably

Severe water shortage is one of the most serious long-range environmental prob lems facing the United States. How can this be? After all, the United States receives on average 16 trillion liters (4.2 trillion gallons) of precipitation per day—about 57,000 liters (15,800 gallons) a day for every man, woman, and child. Although this is an enormous amount of water, the impending shortage is real. Moreover, many other countries are facing similar problems. There are several reasons for this apparent paradox: (1) rapidly increasing population, especially in arid or semiarid areas, (2) rising demand by agriculture, industry, and cities, (3) continued inefficient use by many sectors, and (4) unequal distribution. Even water pollution, the subject of the next chapter, affects water supplies. Consider the following facts concerning the United States' critical water situation:

1. A rapidly increasing population. In the United States, population growth is 1.1 percent a year. Although that may not sound like much to worry about, it does mean that the demand for water is on the rise. Regional growth can be even higher. In the South, Southwest, and West, for instance, some cities are posting growth rates of 4 percent per year! Florida, Colorado, Utah, and Arizona are four of the most troubled areas. Such regional booms in population growth put tremendous strains on already overstrained water supplies.
2. Agricultural water use. Irrigated crop acreage in the United States has more than tripled during the past 30 years to about 24 million hectares (60 million acres). The capacity of one of the biggest groundwater supplies, the Ogallala aquifer, which extends from Nebraska to Texas, is being strained to the limit because of withdrawals—primarily by farmers. In some areas, the water table (the upper limit of the groundwater) is dropping 1.5 meters (5 feet) per year. Worldwide, irrigated cropland now totals over 260 million hectares, and many areas are suffering **groundwater overdraft**—a condition in which water is removed faster than it can be replenished.

3. Industrial water use. Water is used by industry to cool equipment, for use in products, and to make steam to heat buildings and to generate electrical power. In the United States, electrical power plants use half of all water drawn from surface and ground waters, a whopping 735 billion liters (190 billion gallons) a day. Industries use about one-seventh as much. Why? It takes water to grow food and make things. For example, it takes 53.2 liters (14 gallons) of water to make a pound of sugar, 570 liters (150 gallons) for the Sunday newspaper, and 247,000 liters (65,000 gallons) to produce an automobile. Population growth and rising affluence will very likely cause water demand from these activities to rise.

4. Water consumption and waste. Eighty percent of the United States population (220 million people) depend on 21,000 municipal systems for their water supply. The average American family uses 230 liters (60 gallons) per person per day. Although this may not seem like a lot, bear in mind that European families use only half as much. In places where summer lawn irrigation is common, such as Denver, Colorado, a family can use 733 liters (190 gallons) per day per person.

5. Unequal distribution. U.S. water problems are not caused by a shortage in the overall supply. They result from maldistribution—that is, the intensive concentration of water-using people and industry often in areas with marginal water supplies. The accelerated migration of people to the Southwest, for example, has reduced water supplies in this region to the vanishing point. As a result, water may have to be transported long distances from water-rich regions, or strict conservation measures and recycling programs may have to be started.

6. Pollution. Contamination of lakes, streams, and groundwater contributes significantly to the nation's water shortage problem. Lead from old pipes has made water in parts

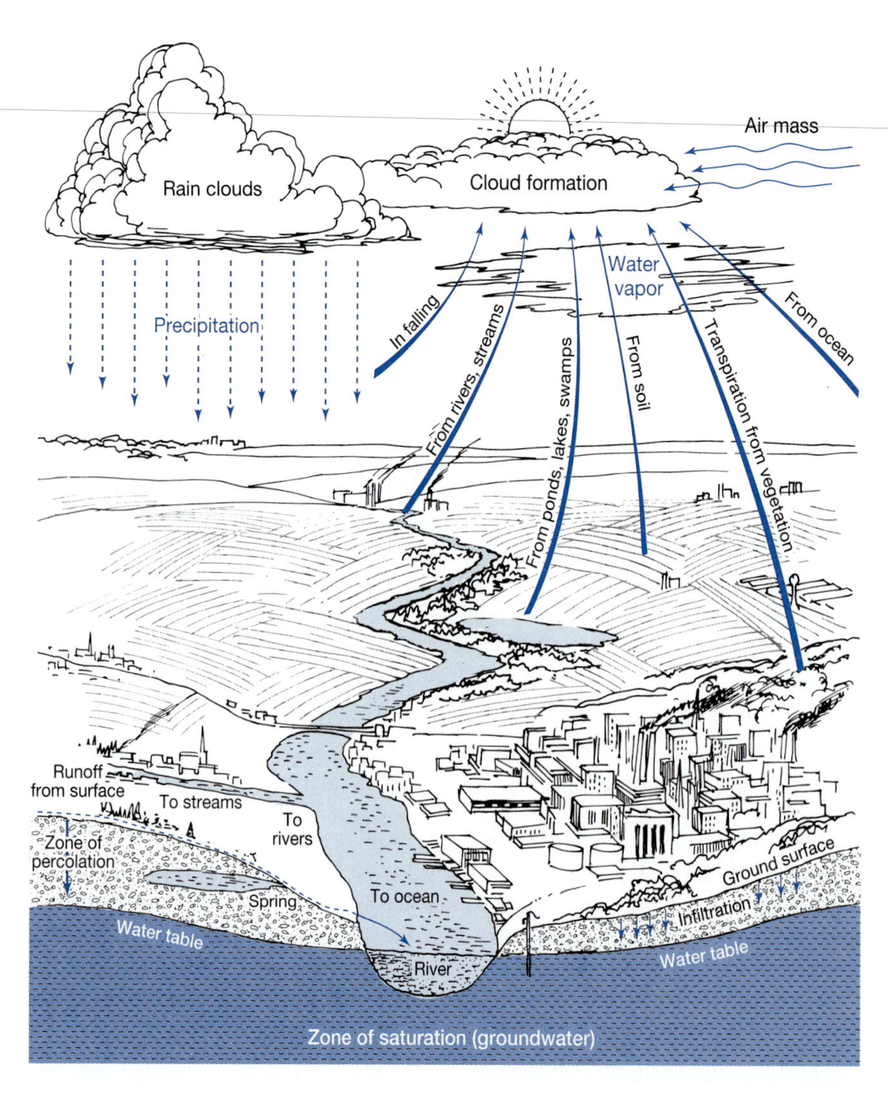

FIGURE 10.1 *The water cycle. This illustration shows the flow of water through the water cycle.*

of Boston unfit to drink. According to the Environmental Protection Agency (EPA), a number of our rivers, such as the Mississippi River below Minneapolis and St. Paul, are so polluted that public health may be jeopardized. Groundwater is increasingly becoming contaminated. This problem is especially severe in Florida, Wisconsin, and California.

These problems are already causing regional shortages that are bound to grow worse as the U.S. population increases. But they are not unique to the United States. Many other countries experience similar problems, so much so that supplying water could be one of the most difficult challenges of the coming decades.

TABLE 10.1 *Water Cycle Facts*

Location of Reservoir	Total Water Supply (%)	Renewal Time
On the Land		
Ice caps	2.225	16,000 years
Glaciers	0.015	16,000 years
Freshwater lakes	0.009	10–100 years (varies with depth)
Saltwater lakes	0.007	10–100 years (varies with depth)
Rivers	0.0001	12–20 days
Subsurface		
Soil moisture	0.003	280 days
Groundwater		
To half-mile depth	0.303	300 years
Beyond half-mile depth	0.303	4,600 years
Other		
Atmosphere	0.001	9–12 days
World's oceans	97.134	37,000 years
Total	100.000	

10.1 The Water Cycle

Water is an intensively used, life-sustaining resource that moves in a circular path through the ecosphere, a phenomenon called the **water cycle** or **hydrological cycle.** The water cycle is really just another nutrient cycle like the carbon and nitrogen cycles discussed in Chapter 3. Familiarity with the water cycle is basic to an appreciation of the nature and complexity of the serious water conservation problems confronting the world today. Before you read on, you may want to take a few minutes to study the water cycle shown in Figure 10.1.

The important thing to remember about water is that it is recycled over and over. Because of the cyclical nature of water movement, a given water molecule may be used thousands of times throughout the centuries. For example, the bath water used by Cleopatra over 2,000 years ago has flowed to the sea and been mixed with ocean water. Some has already evaporated and fallen on the continents as rain. A few molecules from her bath may be present in your next bath. Even the 50 kilograms (110 pounds) of water in the body of the average college student is replaced many times during the school year.

When thinking about the water cycle it is it important to differentiate between the components (oceans, rivers, lakes, and groundwater) and the processes (such as evaporation, precipitation, and so on).The water cycle is powered by solar energy and gravity. The daily solar energy input to the cycle is greater than all the energy utilized by human beings since the dawn of civilization. Solar energy causes water to evaporate from the soil, plants, oceans, lakes, and streams. Warm air rises and lifts the water into the atmosphere, where it forms clouds. Gravity pulls it down again as rain or snow. Snow melts and runs into streams or seeps into the ground. Rain water follows a similar route. All along the way, water can evaporate, once again becoming atmospheric moisture.

Although water moves rapidly and constantly, some water may be stored for varying lengths of time, either within the Earth's crust as groundwater, on the Earth's surface as polar ice caps or glaciers, in oceans and other water bodies, or in the atmosphere as clouds and moisture. The time required for the complete replacement of the water at a particular phase of the cycle is known as the **replacement period** (Table 10.1). Average replacement periods range from 9 days for water in the atmosphere to 37,000 years for water in the deep oceans. Table 10.1 shows renewal times for various water reservoirs. To understand the workings of the water cycle, we will look at the different components one at a time, beginning with the ocean. As you read each section, you may want to refer to Figure 10.1.

Oceans

When the astronauts peer down at the Earth from outer space, the planet appears blue, with just a few patches of brown and green. This is understandable because oceans cover 70 percent of our planet. If the Earth were a perfectly smooth sphere, the ocean water would be sufficient to submerge the entire globe to a depth of 242 meters (800 feet). The oceans are therefore a major reservoir of liquid water (Figure 10.2). It is from this reservoir that vast amounts of water evaporate.

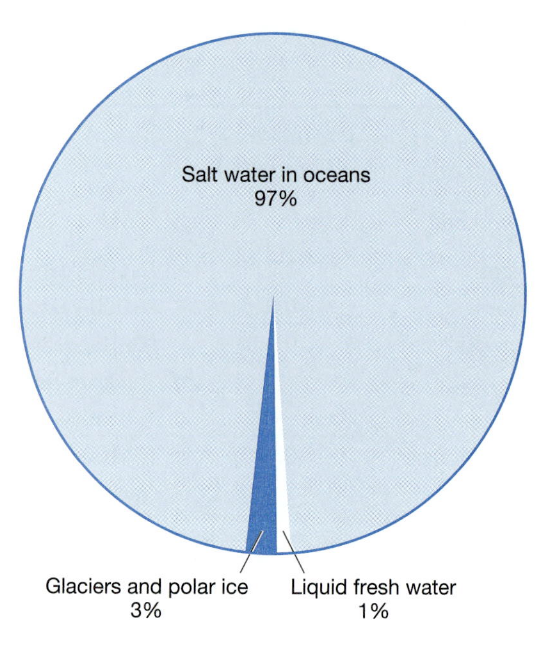

FIGURE 10.2 *The oceans represent an enormous reservoir of water, containing 97 percent of all the water passing through the water cycle.*

Precipitation

Water molecules at the ocean's surface are warmed by the sun; they rise into the atmosphere as a gas in a process called **evaporation.** As the water vapor rises, it gradually cools, condenses, and forms clouds that are transported many miles by wind currents. Eventually, the moisture falls as rain or snow. Water that evaporates from a Louisiana rice field may fall as rain on a college campus in Ohio or in Great Britain or France.

The United States' average annual rainfall would be sufficient to cover the entire country (if it were perfectly level) to a depth of 1 meter (3.3 feet). In the United States, rainfall is very unevenly distributed both in time and in space. The average annual precipitation per state is shown in Figure 10.3. Death Valley receives only 4.3 centimeters (1.7 inches) annually, whereas the western slope of the Cascades receives 350 to 400 centimeters (140 to 150 inches).

When rain or snow originally forms high above the Earth's surface, it is uncontaminated with foreign materials. However, as the raindrops and snowflakes fall earthward, they intercept various atmospheric pollutants, such as carbon dioxide, soot, dust, pollen, and bacteria. Where rain falls through air that is polluted with oxides of sulfur and nitrogen, as is the case in many industrial areas, it is converted into dilute sulfuric and nitric acids. The acidity of the rain may be so high that it is sufficient to corrode water pipes, accelerate the leaching of soil nutrients, and poison

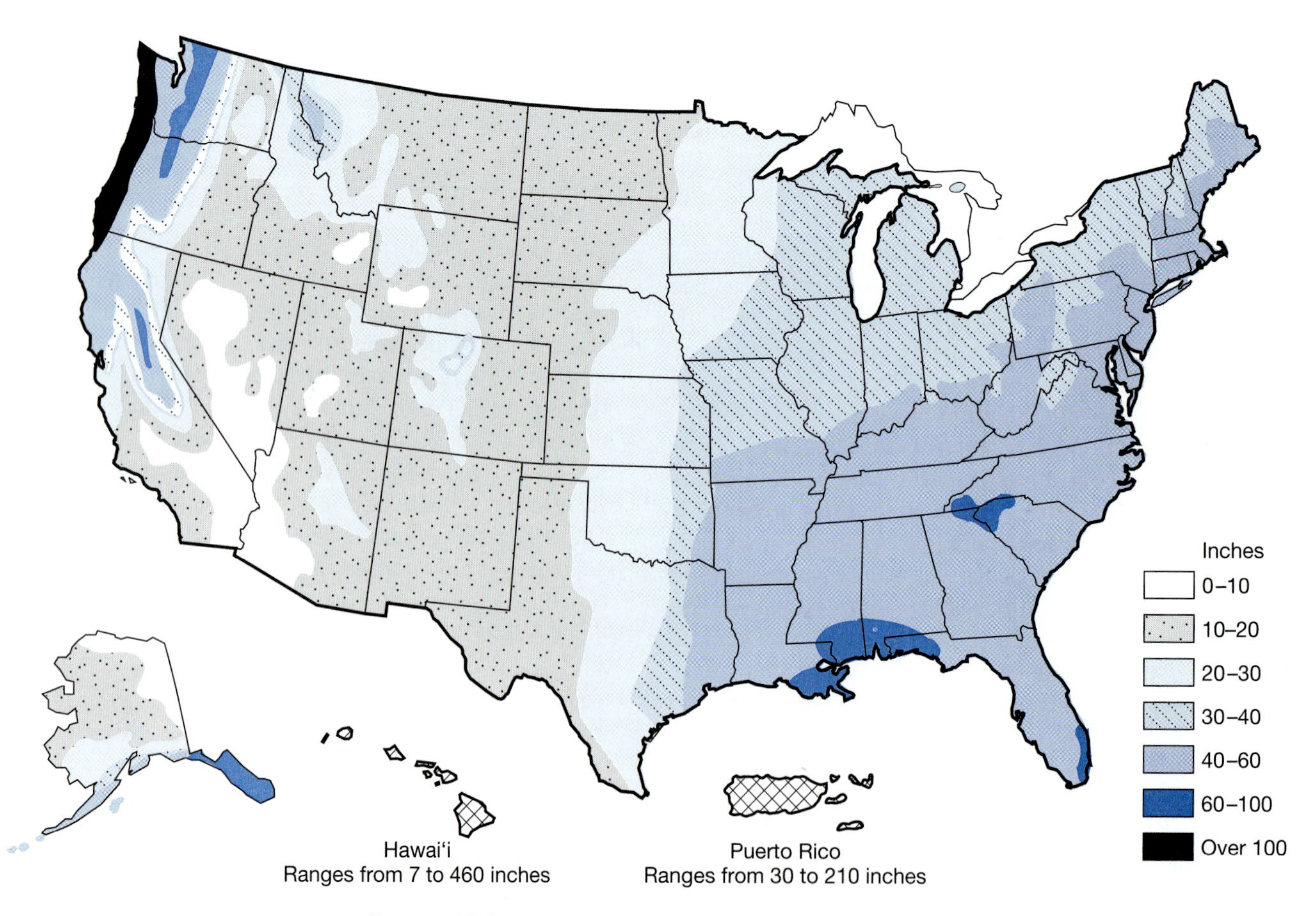

FIGURE 10.3 *Average annual precipitation in the United States.*

aquatic and terrestrial life. Eventually, much of the contaminants in the precipitation is carried to streams, lakes, and finally the ocean, which serves as the ultimate sink.

Bodies of Organisms

Each of the organisms on this planet, from the amoeba to the blue whale, must take water into its body to survive. In most species, including humans, 70 percent of the body is composed of water. The body of the average human adult contains 50 liters (110 pounds) of water. We get that water in the food we eat and the beverages we drink. Plants, on the other hand, must absorb water from the soil through their root systems. A large oak needs 400 liters (100 gallons) per day.

Water serves many functions. As a principal component of the blood, it is a transport medium for many substances, including hormones, enzymes, vitamins, oxygen, and minerals. Water distributes body heat in the body and it helps rid our bodies of metabolic waste as well as excess heat (as in perspiration). If the human body loses more than 12 percent of its water, death quickly follows.

Evaporation and Transpiration

What happens to water that falls on the Earth? Of the 0.75 meter (30 inches) of average annual rainfall, about 0.5 meter (21 inches) is released into the atmosphere by evaporation and transpiration. As noted earlier, evaporation may take place directly from streams, lakes, puddles, ponds, oceans, soils, vegetation, and the bodies of animals and their wastes (Figure 10.4). Plants lose water through transpiration, the escape of water from a plant through pores in its leaves. This process is essential to the plant's survival, for it draws dissolved nutrients from the soil up through the stem (or trunk) to the leaves. One mature oak tree may transpire 380 liters (100 gallons) per day—more than 150,000 liters (40,000 gallons) in a year.

Surface Water

About 22 centimeters (9 inches) of the 76 centimeters (30 inches) of annual rainfall in the United States ends up in ponds, lakes, and streams—bodies that contain surface water. Some may filter

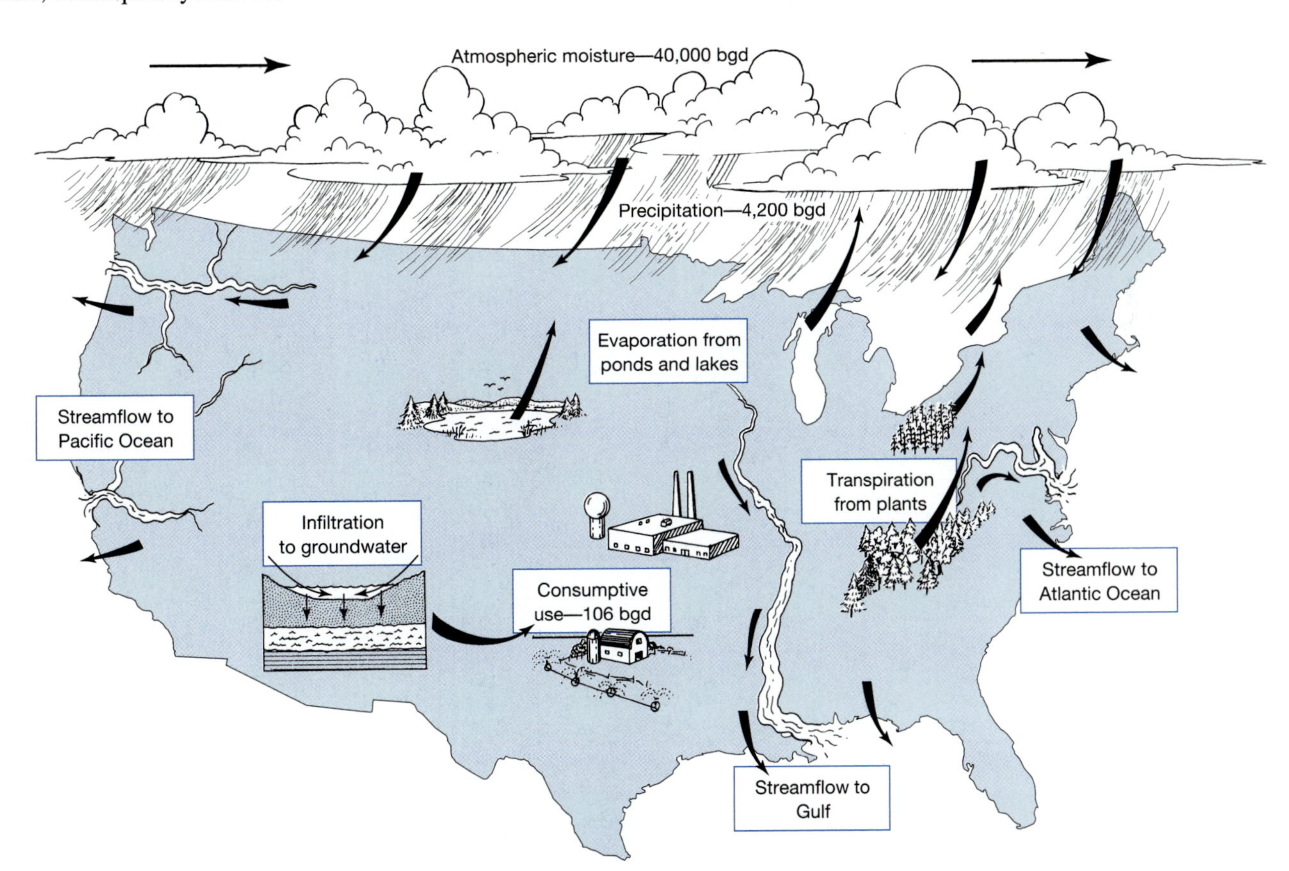

FIGURE 10.4 *What happens to the rain and snow? Water continuously evaporates into the atmosphere, most of it from the oceans. About 40,000 billion gallons per day (bgd) passes over the United States as water vapor, even in times of drought. Roughly 1 gallon in 10—4,200 bgd—falls to the surface of the coterminous United States. That works out to an average of 30 inches a year, of which 26 inches arrive as rainfall and the rest as snow, sleet, and hail. But few places receive the average precipitation. In the United States, precipitation ranges from less than 4 inches a year in the Great Basin to more than 200 inches a year along the Pacific Northwest coast. More than two thirds of the precipitation returns to the atmosphere, but 9 inches (1,300 bgd) either soaks down to the water table or runs into lakes or streams, from which it eventually moves to the ocean. Only a small fraction of the precipitation, 106 bgd, is consumed.*

down through the pores and channels of the Earth's crust to form groundwater (Figure 10.5). Surface water and groundwater are of great concern to conservationists, for it is these sources that are usable for domestic, industrial, and recreational purposes. Surface waters are also habitat for many species. Surface water and groundwater are often polluted with toxic chemicals, pesticides, human waste, and many other contaminants—a problem we discuss in the next chapter.

Stream flow in the United States averages 4,560 billion liters (1,200 billion gallons) a day. It may be in the form of a tiny mountain brook or a mighty river such as the Mississippi, which drains 40 percent of the land area in the United States and flows 3,800 kilometers (2,300 miles) across mid-America to the Gulf of Mexico. Surface water, in the form of streams, ponds, and lakes, satisfies about 75 percent of our water requirements.

Groundwater

Some of the rainfall and snowmelt gradually seeps down through the soil in a process called **infiltration** (Figure 10.5). This water first moves through the **zone of aeration.** This zone, which includes both the topsoil and subsoil, is characterized by pore spaces that contain water and air. The soil moisture in this zone is known as **capillary water.** Some of this water is absorbed by plant root systems. It then passes up plant stems and tree trunks to the leaves. Most of this water then transpires from the leaves into the atmosphere. A small amount of the water in the leaves is used as a raw material in photosynthesis.

Much of the soil water continues to filter down through the zone of aeration into the **zone of saturation**—so named because the pores are filled (or saturated) with water. The zones of saturation are also known as **aquifers.** As the water moves downward through the soil on its way to an aquifer, many pollutants adhere to the surfaces of the soil particles. As a result, the water quality is often improved. If there's enough soil overlying the zone of saturation, the water is completely purified. If not, pollutants can drain down into the groundwater.

Eventually, the downward movement of the water through the zone of saturation is stopped by a layer of impermeable rock (Figure 10.5). As a result, the water accumulates in the soil and gravel above the impermeable layer filling the spaces, pores, and cracks, creating an aquifer. The upper level of the zone of saturation is known as the **water table.**

Although it is now shown in Figure 10.5, the water table may intersect the surface and form marshes, ponds, or springs. In other cases, the water table may be more than a mile deep.

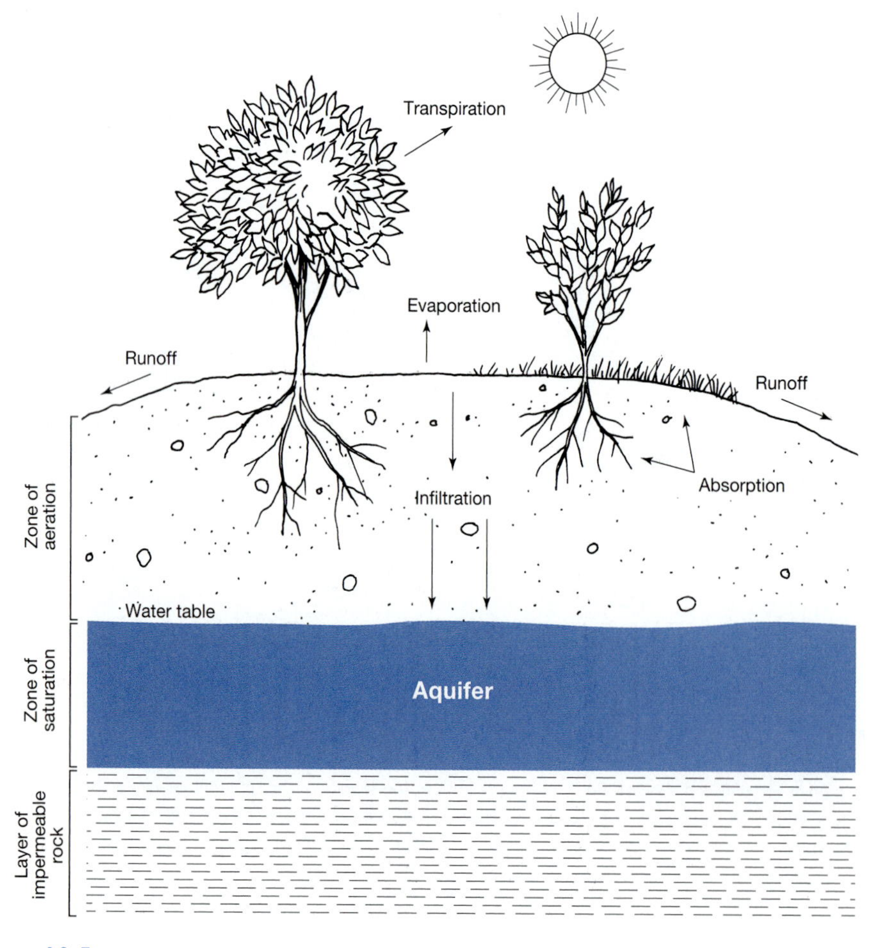

FIGURE 10.5 *Fate of water that has fallen on the ground as precipitation. This drawing also shows how an aquifer is formed.*

Since time immemorial, humans have tapped groundwater by drilling wells below the water table or siphoning off water from springs.

The water table may rise and fall depending on the amount of water leaving and entering the zone of saturation. After heavy rainfall, for example, the water table rises. An exception would be flash floods in deserts, where most of the water runs off instead of infiltrating. However, during periods of drought, the water table drops. In some cases, drought may cause wells to run dry (Figure 10.6).

Humans also affect the water table. Heavy withdrawals can cause the water table to drop. If the water in the aquifer is replaced rapidly at some location, the **aquifer recharge zone,** and the rate of withdrawal does not exceed the rate of replenishment, the aquifer can be counted on as a continual supply of water. If the aquifer is slowly recharged, then heavy water withdrawals can cause water tables to drop, leaving shallower wells high and dry. This problem is particularly acute in deep aquifers that contain water that took hundreds of thousands of years to fill, sometimes referred to as fossil groundwater. In cases where recharge is very slow and rates of withdrawal are high, ground water can be thought of as a nonrenewable resource.

It may be hard to believe, but most (97 percent) of the world's supply of liquid fresh water (2 million cubic miles) is held in aquifers, porous and permeable layers of sand, sandstone, and limestone. In the United States, the groundwater in aquifers in the upper 0.5 mile of the Earth's crust is equal to all the water that will run off into the oceans during the next 100 years! However, water in these great underground reservoirs moves very slowly and often comes from precipitation that fell hundreds of years earlier. For example, the aquifer that supplies a large amount of Chicago's water came from rain and snow that fell on the Great Plains far to the west a million years ago, and then slowly trickled through rocks at a rate of a few centimeters or meters a year. The Ogallala aquifer, mentioned earlier, lies under 572 square kilometers (225,000 square miles) of land that extends through eight states from Nebraska south to Texas (Figure 10.7).This aquifer contains 2.5 million billion liters (650 trillion gallons). It provides drinking water for several million people and is a major source of agricultural water, but it is being drained faster than it can recharge. It makes possible a multibillion-dollar economy based on irrigation farming that is fast running into trouble.

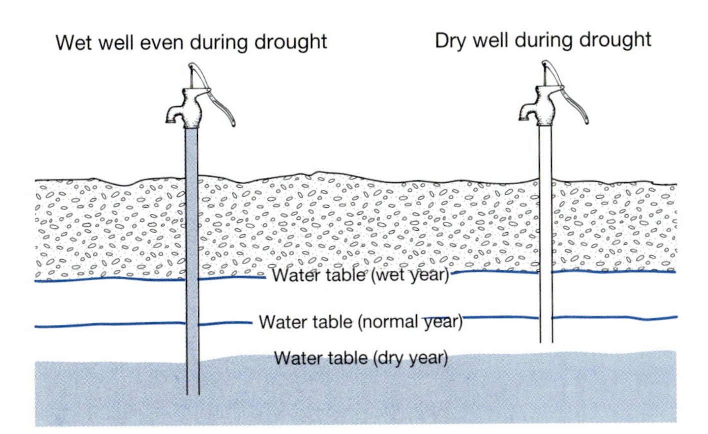

FIGURE 10.6 *An aquifer serves as a source of well water. Continued withdrawal of water would eventually deplete the supply if the rate of withdrawal exceeded the rate of recharge.*

10.2 Flooding: Problems and Solutions

When most people think about water problems in a nation, they think about water shortages. But many nations are plagued with the equally serious problem of too much water from time to time—flooding. Throughout history humans have suffered from destructive floods (Table 10.2, Figure 10.8). For a discussion of one of the latest and most costly floods in human history, see Case Study 10.1.

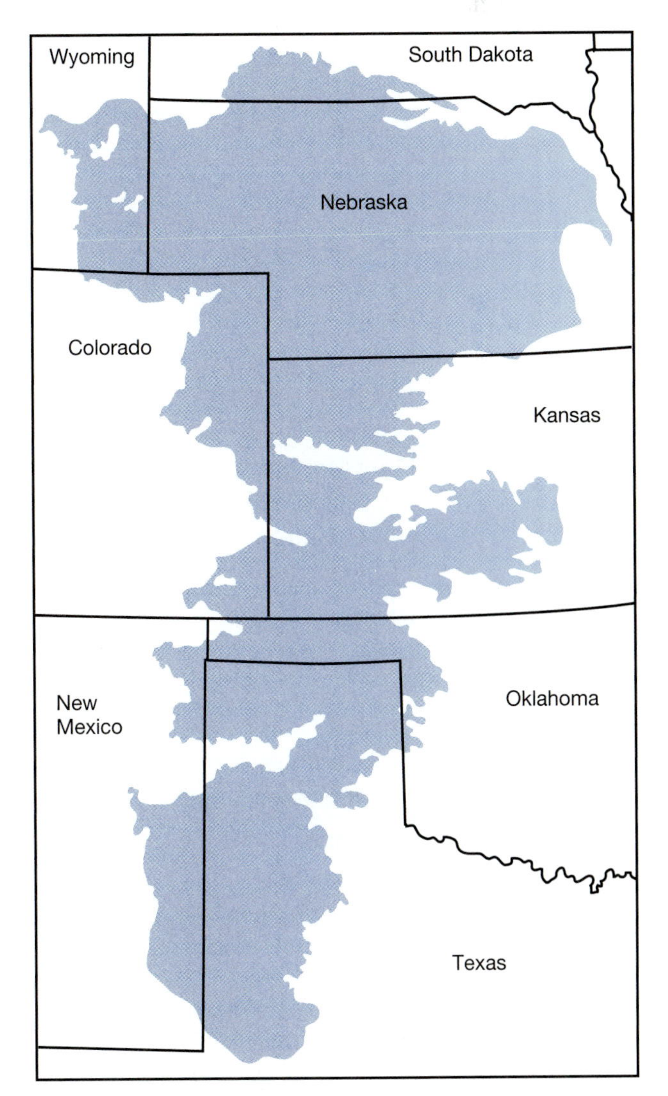

FIGURE 10.7 *The Ogallala Aquifer. This monstrous body of water is quickly being depleted primarily by farmers. It took lhundreds of thousands of years to fill and therefor is essentially being mined.*

Paradoxically, however, in some parts of the world, human survival may depend on floods. The flourishing agricultural economy of Egypt, for example, was sustained for millennia by the recurrent flooding of the Nile River. Each inundation was eagerly awaited by farmers, for when the floodwaters receded, they left behind extremely fertile topsoil carried from its massive upstream watershed, the area drained by a stream or river (Chapter 1.)

Flooding is a natural event. It happened long before humans appeared on the planet. But in recent years, flooding has grown worse, in large part because of the ways people affect the landscape. Changes in rainfall patterns resulting from a shift in climate caused by certain air pollutants may also be contributing to this problem. We'll consider the ways people affect landscapes here. Climate change is discussed in Chapter 19.

Humans alter the landscape in many ways. We cut trees, graze our cattle on the land, and plow meadows under to grow crops. Deforestation, overgrazing, and farming all decrease the vegetative cover. Vegetated land often acts as a sponge, sopping up water that falls on it. The water then percolates into the ground, where it feeds plants and replenishes groundwater. The loss of vegetation increases the amount of water flowing over the surface of the land—known as **surface runoff.** The increase in surface runoff causes streams and rivers to fill with water and flow over their banks, causing flooding (Figure 10.3).

This phenomenon is well illustrated by the following episode. Some years ago, merry celebrations ushering in the New Year at La Crescenta, California, were abruptly ended when floodwaters rushed down from the adjacent hills of the San Gabriel Mountains, inflicting $5 million in damage and killing 30 people. When flood control experts investigated the watershed above

TABLE 10.2 *Examples of Flood Disasters*

Date	Location	Deaths and Injuries	Property Damage
1811	Danube (Germany and Austria)	2,000 drowned	24 villages washed away
1861	Sacramento River (California)	7,000 drowned	300 villages destroyed; 2 million homeless
1889	Conemaugh River (dam burst) (Pennsylvania)	2,000 drowned at Johnstown, PA	$10 million
1900	Galveston, Texas (hurricane-spawned floodwaters)	6,000 dead	3,000 buildings destroyed
1936–1937	Mississippi River 800,000 injured	500 drowned	$200 million
1942	Columbia River		$100 million
1955	Atlantic Coast (Hurricane Hasel)		$1,6 billion
1965	Upper Mississippi River	16 drowned, 330 injured	$140 million
1979	Zambezi River (Monzambique)	45 drowned	250,000 homeless
1979	Brazos River	22 drowned	
1980	Southeastern Brazil	700 drowned	350,000 homeless
1981	Northern India	1,500 deaths	Extensive crop losses
1993	Mississippi River	40 deaths	$10 billion 42,000 home destroyed or isolated, 70,000 people temporarily homeless, farming on 70 million acres severely disrupted
1997	Red River	?	45,000 citizens in Grand Rapids, North Dakota evacuated

FIGURE 10.8 *Flooding is becoming a more common occurrence. It costs billions of dollars worth of damage and takes many lives each year. Much of the world's flooding is caused by humans.*

La Crescenta, they discovered that the flood waters originated from an 18-square-kilometer (7-square-mile) area in the San Gabriel Mountains that had been burned over only a short time before. However, the unburned portion of the watershed, with its vegetational sponge of chaparral, herbs, and grasses, thwarted the downhill rush of runoff waters; peak flows were roughly 5 percent of those of the burned areas. Whether it is California chaparral, Alabama alfalfa, or Wisconsin woodlands, any type of vegetational cover is useful in flood control.

Impervious surfaces such as roads, parking lots, and rooftops also increase surface runoff and flooding in watersheds. That's why cities are so prone to flooding. When heavy rains fall, water runs down streets and parking lots in a sudden torrent, filling streams with water. In an undisturbed ecosystem, much of the water would have been absorbed by the ground.

Flooding is an important resource management issue because it damages homes, cities, towns, and farm crops, causing billions of dollars worth of damage each year. It can also be quite lethal to humans, domestic animals, and wildlife. And, it can be controlled by better management of farms, forest cutting operations, grazing, and construction. For a map of the most flood prone areas in the United States, see Figure 10.9.

Controlling and Preventing Floods

Although humans cannot prevent all floods, we could prevent or reduce the magnitude of many of them, even some of the monstrous floods of the 1990s, experts say, could have been greatly reduced had certain measures been taken during development.

Flood control measures vary in complexity, cost, and effectiveness. Measuring the snowpack each winter helps in fighting floods, as it allows officials to predict future flooding and to take evasive action. Among the most popular approaches are levees, dredging, and dams. However, one of the simplest and most effective approaches is to restrict development in flood plains. Efforts such as this minimize human damage. Flooding can be minimized by protecting wetlands and watersheds and maintaining a normal vegetative cover to reduce surface runoff. Let's take a look at each of these measures, considering the environmental repercussions of each one.

Measuring the Snowpack to Predict Floods The U.S. Geological Survey, an agency of the federal government, measures the depth of the snowpack at more than 1,000 snow courses in the western mountains. Surveys can help warn officials of impending floods, allowing officials time to prepare for high waters. Several years ago, such a survey predicted that the imminent spring snowmelt in the Northwest would cause the Kootenai River to crest at 10 meters (35.5 feet), sufficient to cause extensive flooding at Bonner's Ferry, Idaho. Alerted by this forewarning, federal troops evacuated all residents and reinforced the dikes. On May 21, the river crested at 10 meters (35.5 feet), as predicted. Flood damage was minimized, however, and not one life was lost. In the 1997 flood in Grand Rapids, North Dakota, however, snow pack measurements caused officials to raise the levees (earthen structures built along rivers to reduce flooding) but officials underestimated the flood's potential. When the flood waters came, water flowed over the newly raised walls, causing much more damage than predicted.

Although measurements of snow pack can be helpful, they do not always work. Other factors may complicate matters. Snow may melt faster than anticipated or a heavy rain storm may occur when snows are melting. Further measures are therefore generally needed.

For a discussion of ways that GIS and remote sensing are being used to monitor snowpack and predict water supplies, see the GIS and Remote Sensing box on page 230.

Levees Levees are a popular choice. **Levees** are dikes constructed of earth, stone, or mortar at varying distances from the riverbank to protect valuable residential, industrial, and agricultural property from floodwaters (Figure10.10). Levees along the Arkansas, Red, White, and Ouachita rivers in Arkansas have given a measure of protection to more than 0.8 million hectares (2 million acres) of fertile alluvial land. During the past 150 years, a mammoth system of over 7,500 miles of levees and 29 dams has been constructed along the lower Mississippi River.

In the early part of this century, the Mississippi River and its tributaries flooded quite frequently. As a result, Congress passed the landmark Flood Control Act of 1936. Under this act, a $377 million program was authorized for the construction of a huge system of levees, flood walls, dams, and reservoirs. The U.S. Army Corps of Engineers was given the responsibility of doing the work. Since then this structural

CASE STUDY 10.1 THE GREAT MISSISSIPPI FLOOD OF 1993

Some people have called it the flood of the century. Others give it even more status, claiming that it was the biggest flood ever seen by the white man in America. No matter where the Great Flood of 1993 ranks among the deluges of history, it certainly is one that millions of Midwesterners would like to forget. The almost incessant thunderstorms that released torrential rains on Wisconsin, Minnesota, Iowa, Illinois, and Missouri swelled the Mississippi River to record heights. Indeed, the first 7 months of 1993 were the wettest in Iowa for the 121-year period that records had been kept. Some regions received up to ten times their normal amount of rainfall.

The rampaging waters of "Ol' Miss" swept downstream at a speed of 8 miles per hour. At Hannibal, Missouri, the river crested at 32 feet—2 feet higher than the 500-year flood mark. At St. Louis the river was above flood stage for a record-breaking 80 consecutive days, and in early August it reached an all-time high of 49.5 feet. Eventually the boisterous river spilled over into low-lying farmlands and swamped an area twice the size of New Jersey.

Much of the river from Minneapolis south to St. Louis was bordered by levees—walls of rock and concrete that were effective in containing the annual spring rise of the river from rain and snowmelt. Unfortunately, however, many levees were not designed to handle the flood of 1993. In fact, more than 800 of the levees either crumbled under the force of the swollen waters or were overrun. Paul Schloesser, Chief Petty Officer with the Coast Guard, was standing within a few yards of a levee at West Quincy, Illinois, when it broke. He recalled water pouring through the opening "with a raging roar like that of a freight train." The waves shot 5 feet into the air. Nearby workers ran for safety.

In many a river town where levees were either ruptured or nonexistent, the residents erected sandbag barricades to hold back the water. The dedication and energy expended on sandbagging were impressive. For example, in Cape Girardeau, Missouri, thousands of residents, in addition to hundreds of volunteers who came from a 400-mile radius, built a barricade of more than 500,000 bags. The sandbaggers ranged from schoolchildren to old-timers, from carpenters to lawyers, from farmers to college professors. In the process of defending the town from the river, these workers exhibited a magnificent degree of cooperation, self-sacrifice, and community bonding.

In countless spots along the upper Mississippi Valley, the floodwaters swirled over banks and then surged over and through levees and the frantically constructed sandbag barriers. The water kept getting higher and higher. It swept into fields of soybeans and corn. It raged into downtown shopping areas. Finally, it made its way into the living rooms, kitchens, bedrooms, and bathrooms of hundreds of riverfront homes, leading TV commentator Tom Brokaw to declare that "Mid-America was under siege."

In many of the inundated towns, grocery shopping could only be done with the aid of rowboat or canoe. One elderly native of Grafton, Illinois, who was proud of his tomato garden did his best to save his prized red beauties. He dug them up, put them in baskets, and strung them along his wife's clothesline, where they stayed "high and dry." One distraught woman, while wading in hip-deep water, came across a 3-foot water snake swimming in her bedroom. She tried to calm herself by imagining it was only a gigantic earthworm.

The flood raised havoc with transportation. Water swamped low-lying airport runways and severely disrupted flight schedules. Rail service on the transcontinental railways in areas gripped by flooding came to a halt. In Quincy, Illinois, the Bayview Bridge—the only Mississippi crossing within a 200-mile span of the river—was closed for a month. During the height of the flood, the Army Corps of Engineers closed the river to all commercial traffic from Dubuque to St. Louis. The shutdown forced at least 200 barges out of action, for an aggregate financial loss of $1 million per day.

At the peak of the crisis, Secretary of Agriculture Mike Espy made a helicopter survey of crop damage. Ironically, farm flooding was so extensive that in some areas there was very little land to see. The USDA estimated that at least 8 million farm acres either never got planted or were under water, and on another 12 million acres the corn and soybean crops were severely stunted. Eventually, several hundred counties in Wisconsin, Minnesota, Iowa, Illinois, and Missouri were declared official disaster areas and eligible for emergency federal aid.

The flood of 1993 left in its wake an estimated 40 dead, at least 42,000 homes destroyed or isolated by water, about 70,000 people temporarily homeless, the productivity of 20 million acres of cropland severely disrupted, and overall financial damage of at least $10 billion. No wonder that many authorities consider the flood of 1993 to be the greatest natural disaster in the history of the United States!

The flood, while devastating and costly, had some benefits, too. It has begun a movement to return floodplains to rivers. Rather than facing devastating floods again, some communities have opted to relocate to higher ground. This not only saves taxpayers money, as they're the ones who typically foot the bill to rebuild after devastating floods, land once occupied by people is being returned to wildlife. Natural habitat, nearly gone from the banks of many larger rivers, is being slowly restored. Wildlife biologists hope eventually to create a "string of pearls," that is, a series of natural habitats along large rivers, helping provide valuable habitat for ducks and other species. This will also reduce damage from floods, because it puts people out of harm's way, and provides safe places for water to go in case of floods. Restored wetlands could help reduce the intensity of floods, too.

method of flood prevention and control has been greatly expanded. Such structural control methods have increasingly come under attack. Indeed, a debate is currently being waged over the merits of expanding the levee system, especially along the upper Mississippi River. Let's briefly consider some of the arguments pro and con.

The Pros

1. The levees and dams built by the U.S. Army Corps of Engineers since 1938 have effectively controlled flooding along the Mississippi River system and have prevented at least $250 billion in damage—a ten-to-one return on the cumulative cost of $25 billion.

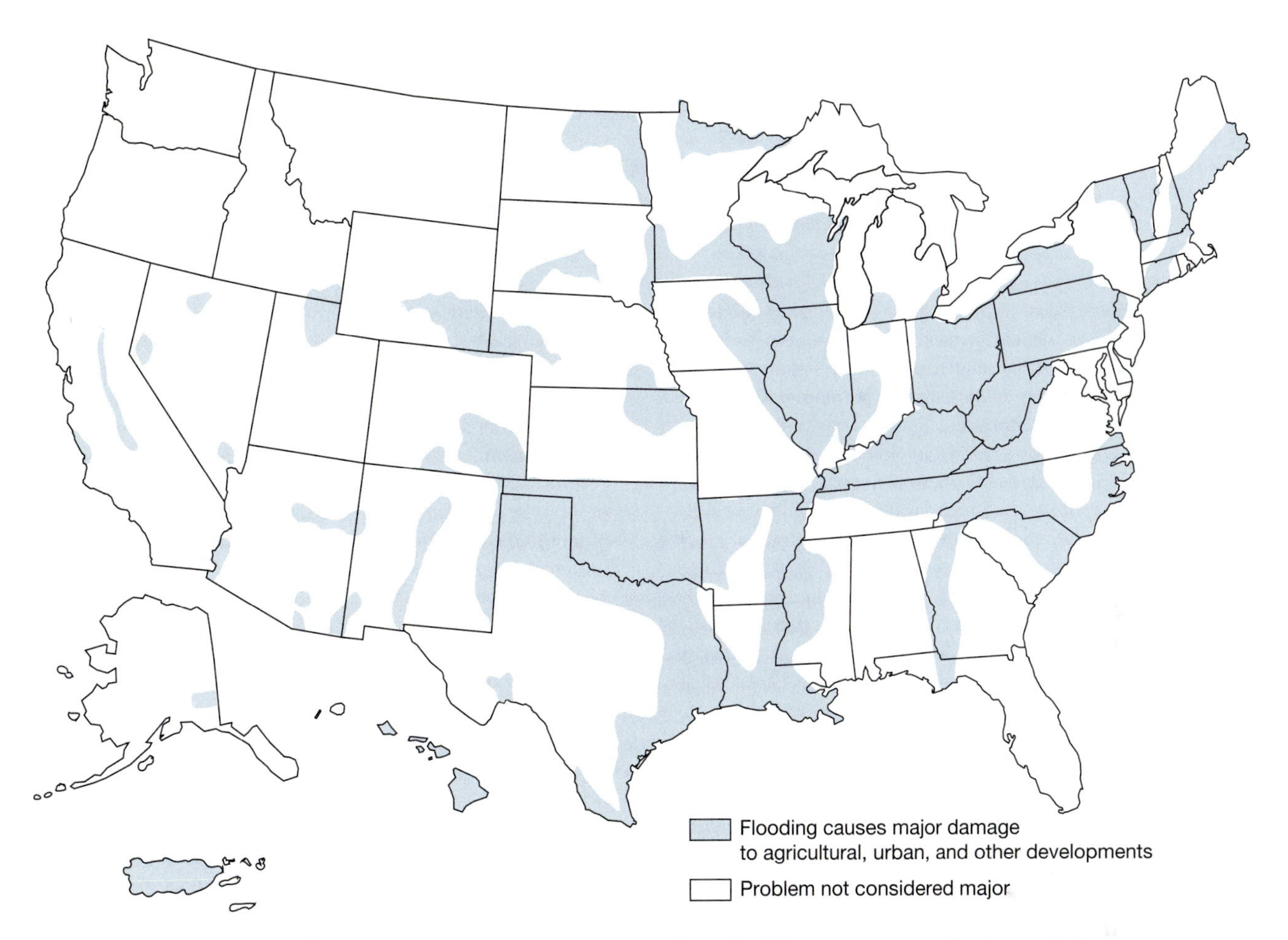

FIGURE 10.9 *Areas of the United States with flooding problems. Flood damages are expected to increase in the future on 175 million acres of land that is flood prone. (A flood-prone area is land adjoining rivers, streams, or lakes, where there is a 1 percent chance of flooding during any given year.) Forty-eight million acres of flood-prone land are cropland, 102 million acres are pasture, range, and forest, and 21 million acres represent other land, including built-up areas. Twenty-one thousand communities are subject to floods, including 6,000 towns or cities with populations exceeding 2,500. The potential damage caused by floods in any given year in the United States is about $4 billion. Because the number and value of buildings are increasing, flood damages will also increase.*

FIGURE 10.10 *Flooding along this river in Sedro Woolley, Washington often results when protective levees break.*

GIS AND REMOTE SENSING

GIS AIDS SNOW MONITORING AND MODELING AT THE NATIONAL WEATHER SERVICE

In the mountainous states of the western United States and in the western provinces of Canada, winter snows are a vital source of liquid water for many urban and rural areas. So massive is the snowmelt in western states, in fact, that the spring runoff provides a very large percentage of the annual streamflow of many rivers. In the spring when the snows begin to melt, huge reservoirs depleted by farmers, industry, and residences in the previous year begin to fill once again, creating a supply of water that often receives little additional input until the next spring snowmelt begins.

Unfortunately, things don't always go as we'd like them to. Sometimes the snowpack, the amount of snow accumulating in the mountains over the winter, is excessive. Other times, the snow melts much more rapidly than anticipated. The result of both occurrences is flooding—accompanied by loss of life and property damage. Sometimes the snowpack is below average and water shortages emerge.

In the United States, the National Weather Service (NWS) seeks to help those who are dependent on spring snowmelt or those who live in the floodplains of rivers. The NWS's National Operational Hydrologic Remote Sensing Center (NOHRSC), headquartered in Chanhassen, MN, uses remote sensing and GIS, combined with other techniques (such as mathematical snow modeling and field-gathered data) to provide vital information needed to estimate snowpack and to help predict water supplies and potential flooding.

NOHRSC monitors the extent of snow cover in the United States, its water content, and air temperature, then uses GIS to create detailed maps for display and analysis. NOHRSC obtains data from numerous sources. For example, it receives snow tube measurement data collected at fixed sites along hundreds of snow courses throughout the West from the Natural Resources Conservation Service, another government agency. Data is also supplied by a network of automated snow telemetry stations. Located at remote sites, these stations relay snowpack and weather data via satellites to ground-receiving stations which, in turn, transmit the data to the NOHRSC. The Center also obtains data on the water content and the depth of snow cover across the United States by conducting surveys from airplanes whose onboard instruments measure gamma radiation emitted by soils and snowpacks. (Researchers have developed mathematical equations that allow them to estimate the liquid water content of snow based on the amount of gamma radiation detected by the airborne sensors.)

The Center processes the data using computer programs and GIS. Much of the data is provided to an extensive network of river forecast centers operated by another branch of the National Weather Service.

Timely information about the quantity and location of winter snowpack is vital to many end-users. Farmers, irrigation district managers, hydroelectric power generators, wildlife managers, and flood managers, however, are the main benefactors of these new and exciting technologies. If floods are predicted, evacuation plans can be examined and fine tuned. If water supplies will fall short, water departments can implement water conservation programs. Given the lead time to take actions, GIS becomes a valuable tool in managing ourselves and our resources.

2. In 1973 and again in 1993, the Mississippi River reached record once-in-a-century flood levels. Nevertheless, many thousands of Americans remained safe behind the levees. Without this protection, the loss of life would have been considerable.
3. The levees help maintain a river depth that permits extensive barge traffic, which is invaluable to the nation's economy and to our position in world trade.
4. The protection afforded by the levees has made possible extensive agricultural, industrial, commercial, and residential development very close to the riverbanks.

The Cons

1. Since 1938, the per capita flood damage has actually increased by a factor of 2.5.
2. Despite the ambitious levee-building program of the Corps, the annual number of flood-related deaths has remained constant on a per capita basis from 1916 to the present.
3. The dam and levee system is not structurally perfect. During the 1993 flood, more than 800 of the 1,400 levees that had been built by the Corps or by state and local agencies either crumbled or could not withstand the surging waters.
4. The "straightjacketing" of a river between levees may actually increase flooding severity downstream. The reason is that restricting flow causes water level to rise more than if a river is allowed to spill over its banks. That is, the water level in a river increases more when its contained by levees than when it is left open. The Mississippi River near St. Louis is actually 3.3 meters (10 feet) higher today than it was in 1881 in large part due to the levees that have been built along its banks. In addition to causing rivers to rise, the levees result in an increase in velocity and water pressure. Therefore, if a levee does rupture, the property damage and loss of life will be much greater at that point than if the levee had never been built in the first place.
5. The levee system causes the drying out and death of any wetlands that once bordered the river. Of course, all the important functions of the wetland, such as a purifier of groundwater, as a habitat for wildlife, and as a living sponge that soaks up floodwaters, are lost forever.

6. The levee system prevents surrounding croplands from receiving nutrient-rich sediment from the river during the annual low-level flooding in spring.
7. Levees are unsightly barricades of earth or cement. Because many are 40 to 50 feet tall, they effectively block the view of the scenic river valley.
8. Once the levees are constructed, agricultural, industrial, commercial, or residential development often takes place nearby. In other words, development accelerates in areas once prone to flooding. As a result, if the levee is breached by a massive flood, the property damage and human misery could be both extensive and severe.
9. Levee maintenance is very expensive. For example, in just one 35-square-mile drainage district along the Rock River in northern Illinois, the 1993 disaster caused at least ten ruptures and the flooding of 10,000 acres of farmland. The repair of these breaks alone cost American taxpayers at least $500,000.
10. Finally, levees also tend to trap water behind them. Water falling on the watershed behind a levee may not be able to reach the river and water pools up behind the levee, causing extensive damage.

As in many issues, both sides present convincing arguments. Today, there is a movement underfoot to let the river have their flood plains—that is, to restrict farming and other activities. One town inundated in the 1993 flood actually moved to higher ground. In recent years, several federal agencies involved in flood protection like the Federal Emergency Management Agency and the Army Corps of Engineers have begun to spend huge sums helping communities along rivers locate to higher ground (Figure 10.11). This not only protects them from flooding, it reverts flood plains to their rightful owners, the rivers. Land once occupied by houses is now being preserved for wildlife and natural flood control.

Dredging Huge amounts of soil are washed into streams from watersheds disturbed by human activities, causing sediment to accumulate in channels. Called **streambed aggradation,** this buildup decreases the depth of waterways and thus increases the probability and severity of flooding. The size of this problem can be appreciated if we note that the Mississippi River transports roughly 2 million metric tons of sediment daily. To cope with this problem (as well as to deepen the channels for navigation), the Mississippi River and a host of others are periodically dredged by the U.S. Army Corps of Engineers.

The importance of dredging is emphasized by the 1852 Yellow River catastrophe in China. As the channel of this river became choked with silt, levees were built higher and higher, until the Yellow River was flowing above the rooftops. Eventually, a massive surge of floodwaters crumbled the retaining walls and drowned 2 million people.

Dams Despite serious criticisms of dams, the United States, until recently, had been committed to a vast program of dam construction. The largest was the Hoover Dam on the Colorado River (Figure 10.12). The Colorado drains almost one-thirteenth of the U.S. land area. This enormous concrete dam was built to control the river, prevent flooding, and provide irrigation water and electrical power for thousands of farmers in the arid Southwest. The Hoover Dam stands 220 meters (726 feet) high. The impounded water forms Lake Mead, the largest reservoir in the world; 115 miles long, it has an area of 640 square kilometers (246 square miles). Its storage capacity would be adequate to meet all the water requirements of New York City residents for 20 years.

Are dams effective in flood control? According to Brigadier General W. P. Leber, Ohio River Division Army Engineer, an Ohio River flood of several years ago would have caused additional damage costing $290 million had it not been for the coordinated system of 30 flood control reservoirs plus 62 flood walls

FIGURE 10.11 *Headed for higher ground. This church is being moved 500 feet away out of the flood zone.*

FIGURE 10.12 *Hoover Dam, one of the world's largest dams. It is located on the Arizona–Nevada border, about 25 miles southeast of Las Vegas. The electrical power generators at the dam supply most of the power needs of southern California, Arizona, and Nevada. The dam restrains the turbulent waters of the Colorado River and is useful in controlling downstream floods. Water from Lake Mead, the 125-mile-long impoundment behind the dam, irrigates 1 million acres and has increased crop production 120 percent in this region.*

and levees. He stated that flood crests were reduced by up to 3.2 meters (10.5 feet) by these flood control facilities. Several years ago, Los Angeles County, which has experienced repeated floods from rain-swollen rivers, established a coordinated complex of control structures costing nearly $600 million. It involves 60 headwater dams in the mountains, 15 retention reservoirs in the Los Angeles and San Gabriel rivers, and 6 major flood control dams. This system has proved to be very successful in protecting 130 million hectares (325 million acres) from flooding.

Despite the value of dams in flood control and in generating hydroelectric power, big dams create a number of problems (Table 10.3). One major drawback associated with the construction of big dams is the speed with which reservoirs fill with sediment. The rate of filling depends, of course, on the soil types in the drainage basin, the topography, vegetative cover, and the degree to which soil erosion is controlled. Because the Columbia River is relatively sediment free, dams such as the Grand Coulee and Bonneville might have a storage life of 1,000 years. However, the life span of dams constructed across muddier streams may be quite short. For example, the huge Lake Mead Reservoir behind Hoover Dam on the Colorado River in Arizona is filling with silt at a rate sufficient to destroy the operation of this multi-million-dollar structure in less than 250 years. The death of a Texas reservoir is shown in Figure 10.13. California's Mono Reservoir, which was designed to provide a permanent water source for the people of Santa Barbara, filled up with sediment in 20 years. In addition, biological succession proceeded so rapidly that a thicket of shrubs and saplings became firmly established. For all practical purposes, Mono Reservoir has been reclaimed by nature.

Evaporation losses from reservoirs in hot, arid regions where winds are prevalent can be considerable. The top 2 meters (7 feet) of water in Lake Mead, for example, evaporate

TABLE 10.3 *Disadvantages of Big Dams*

1. Extremely expensive: hundreds of millions to billions of dollars. (Example: Hoover Dam on the Arizona–Nevada border: cost $120 million.)
2. Prime agricultural land is flooded.
3. Scenic beauty is destroyed. (Example: The Rainbow Bridge National Monument in Arizona is threatened by the Glen Canyon Dam on the Colorado River.)
4. The resulting saltwater intrusion in coastal areas destroys cropland and pollutes freshwater aquifers. (Example: This has happened in both Florida and California.)
5. Drawdowns periodically eliminate the shallow-water areas where fish frequently spawn.
6. The natural habitat of endangered species in destroyed. (Example: The snail darter at the Tellico Dam.)
7. The upstream migration of adult salmon is blocked, interfering with reproduction.
8. There are excessive water losses from reservoirs because of evaporation. (Example: Lake Mead behind the Hoover Dam.)
9. The life of a dam is shortened because of siltation of the reservoir. (Example: Mono Dam in California.)
10. The collapse of the dam is possible as a result of faulty construction. (Example: Teton Dam in Idaho.)

FIGURE 10.13 *The death of the Lake Ballenger, Texas, reservoir. Although the original depth of this lake was 35 feet, it eventually had to be abandoned because of siltation.*

every year. About 6 million acre-feet are lost each year from 1,250 large western reservoirs, an amount sufficient to supply all the domestic needs of 50 million people.[1]

On rare occasions, a dam will collapse because of faulty design. Such was the case with the mighty Teton Dam in Idaho (Figure 10.14). It was built by the U.S. Bureau of Reclamation, an agency that has constructed more than 300 dams, including the world-famous Grand Coulee and Hoover dams. On June 5, 1976, the bureau's record was sullied. Only hours after the first fissure appeared in the Teton Dam, the monstrous earthen structure gave way with a deafening roar, and a mammoth wall of water surged down the valley. The resultant destruction was awesome. At least 14 lives were lost. Estimates of property damage approached $1 billion. Shortly afterward, the Secretary of the Interior and the governor of Idaho appointed a panel of distinguished engineers to investigate the cause of the Teton Dam's failure. As reported in *Science,* the panel concluded that "under difficult conditions that called for the best judgment and experience of the engineering profession, an unfortunate choice of design measures together with less than conventional precautions ultimately led to its failure."

Dams cause a host of other problems, too. They inundate fertile valleys and destroy streams that are not only vital habitat for fish and nonaquatic species, but are valuable for

FIGURE 10.14 *The death of a dam. An aerial view of the Teton Dam, Idaho, shortly after its rupture. We are looking upstream. The site of the break is roughly in the center of the photograph. The torrential waters released from the reservoir above the dam caused the deaths of at least 14 people.*

[1]An acre-foot is a measurement of water volume. It's one acre one foot deep.

recreational benefits bestowed on anglers, campers, rafters, canoeists, and others.

Stream Channelization The Soil Conservation Service (SCS), now renamed the Natural Resource Conservation Service (NRCS), has built up a good reputation during many years of valuable service in soil erosion and flood control. However, a storm of criticism has swirled around the Agency's channelization efforts. **Channelization** is the intentional deepening and straightening of streams to control flooding.

There are two main steps in channelization. First, all vegetation is bulldozed away on either side of the stream. The denuded area is then planted with a cover crop to stabilize the soil and prevent erosion. Second, bulldozers and draglines deepen and straighten the channel; in essence, the steam is converted into a water-filled ditch that moves water quickly out of a watershed.

The benefits of channelization are several. First, as noted above, deepened channels help remove water from a stream, reducing flooding. Second, because of this, cropland, towns, and buildings along the channelized portion of the stream are protected. Another benefit is that small lakes for recreation and wildlife were often constructed by the SCS as an adjunct to the main channelization process.

Unfortunately, the disadvantages of channelization are far greater:

1. A picturesque meandering stream is converted to an ugly eroding ditch.
2. Much valuable hardwood timber, already in short supply, is destroyed.
3. The diversity of the wildlife declines.
4. Water temperatures increase because of the removal of overarching trees that formerly intercepted the sunlight. This alters abiotic conditions of streams and rivers, making them unsuitable for many species of fish, a topic discussed in more detail in the next chapter.
5. Stream enrichment from leaf fall is considerably reduced.
6. Downstream flooding along nonchannelized sections may increase as a result of increased upstream flows.

In its long-range plans, the NRCS proposed to channelize several thousand small watersheds in the United States by the year 2000. That would have meant degrading nearly half of the nation's small watersheds. Fortunately, the controversy over channelizations has greatly slowed this process.

Protecting Watersheds One of the most economical and sustainable of all approaches to flood control is watershed protection. As noted in Chapter 1, a **watershed** is an area of land drained by streams and rivers. It is also sometimes called a **drainage basin.** A watershed delivers water, as well as sediment and nutrients, via small streams to a larger stream or a river. Watersheds vary in size from extremely small to extremely large. A knowledge of watersheds is important to many aspects of natural resource conservation, including wildlife management, pollution control, and water supply.

All watersheds, large or small, have the basic function of converting precipitation into stream flow and groundwater. Even during a light shower of only 0.25 centimeter (0.1 inch) of rain, a 2.5-square-kilometer (1-square-mile) watershed would produce 6.6 million liters (1.7 million gallons) of stream flow.

Watersheds vary considerably in vegetative type and topography and it is these characteristics, combined with precipitation, that determine how much water seeps into the ground and how much flows over the surface. Vegetative type and the extent of coverage determines how much sediment flows off the land during rainstorms. In the West, where vegetation is less dense, rain storms often result in considerable erosion. The lack of vegetation, even in undisturbed watersheds, often results in floods, even sudden flash floods. In the East, in undisturbed ecosystems, heavy vegetative cover reduces erosion in undisturbed watersheds and reduces stream flow, limiting flooding. When watersheds are disturbed, either by natural (fires, for example) or human factors (farming, for instance), two things typically happen: surface water flow increases and erosion increases. The result is flooding and the build up of sediment in streams and lakes. The effect of sediment on rivers and reservoirs was discussed earlier.

Watershed protection is the preventive medicine of flood control. It is perhaps the most sustainable activity humans can engage in to reduce flooding. In the United States, the **Watershed Protection and Flood Prevention Act** provides some degree of protection for watersheds. Under this law, protection of small watersheds is the duty of the Natural Resource Conservation Service of the U.S. Department of Agriculture (USDA). According to the USDA, 8,000 (61.5 percent) of the 13,000 small watersheds (smaller than 100,000 hectares or 250,000 acres) in this country have flood and erosion problems. Protection of large watersheds is assigned primarily to three agencies: the Bureau of Reclamation, the U.S. Army Corps of Engineers, and the Tennessee Valley Authority (TVA). Although the primary objective of the act is flood control, the law establishes a multipurpose concept and, where possible, it encompasses the problems of erosion, water supply, wildlife management, and recreation.

Within watersheds that are used for agriculture, soil erosion measures (discussed in Chapter 7) can go a long way in reducing flooding. Erosion control measures around mining operations and new construction also help. Controls on grazing that permit the emergence of a healthy vegetation also protect the land and reduce surface runoff. Such measures are often highly effective and extremely cost effective.

Zoning of Floodplains The U.S. Army Corps of Engineers, the Bureau of Reclamation, and the SCS have spent more than $15 billion on structural flood control projects (dams, levees, sea walls, etc.) since 1925. However, despite this enormous expenditure of tax money, property damage from flooding continues to rise. Annual costs increases considerably, from $3 billion in the early 1980s to more than $10 billion by 2000.

More and more experts now believe that nonstructural flood controls like watershed protection and floodplain zoning are the most effective and economical strategies.

The **floodplain** is the low-lying area along a stream that is subject to periodic flooding. As noted earlier, floodplains belong to rivers. Unfortunately, however, floodplains have been taken over by humans. The flatness of the land, their natural beauty, and the availability of the river for cheap transportation have attracted people since the beginning of time. Today, more than 2,000 cities in the United States, including Harrisburg, Pennsylvania, Des Moines, Iowa, and Phoenix, Arizona, are located, at least partially, on floodplains. Most of these cities experience flooding every 2 to 3 years. Interestingly, of the 42,000 homes destroyed or damaged by the great Mississippi flood of 1993, most were located on a floodplain.

In 1973, Congress passed the **Federal Flood Disaster Protective Act** to regulate floodplain development and control flood damage. This law encourages the zoning of floodplains for use as parks, golf courses, bicycle trails, nature preserves, and parking lots. The construction of buildings on the floodplains is discouraged. Under the terms of the act, any home, store, or factory built in a hazardous area on the floodplain is denied federal flood insurance. Unfortunately, the act has had little effect. People have continued to build in floodplains. Cities and towns have been repeatedly flooded. Only in recent years have federal agencies begun to encourage them to relocate out of flood plains, converting flood prone land into natural vegetation.

10.3 Water Shortages: Issues and Solutions

According to the U.S. Weather Bureau, a **drought** exists whenever rainfall for a period of 21 days or longer falls substantially below the average. The Great Plains, from Texas to Montana, average about 35 consecutive drought days each year and 75 to 100 successive days of drought once in 10 years. Up to 120 consecutive rainless days have been recorded for the southern Great Plains, or Dust Bowl, region.

J. Murray Mitchell, a research climatologist with the National Oceanic and Atmospheric Administration, studied drought patterns as revealed by tree rings. Going back to 1600 A.D. he found that the western states have experienced a prolonged, extended drought about every 22 years. Much to his surprise, Mitchell found that the drought cycle correlated with sunspot patterns. Such information is extremely valuable for national planning.

Drought is a far more common event than most people realize. It causes severe water shortages in cities and agricultural regions. This can cause dramatic loss in crops and severe financial hardship. Drought, of course, is accompanied by record heat, which takes a toll on people and livestock. In recent years, record temperatures are attributed to the death of 100 to 200 people a year in the United States. Most of the victims are the elderly unable to cope with or escape from the heat. Drought also affects wildlife populations from grazers to predators.

It is important to point out that water shortages during drought can be attributed to excessive water use and the inefficient use of water. Densely populated areas in semi-arid regions are also extremely vulnerable. Part of the problem, too, results from a failure of agricultural and municipal users to adapt to drought—that is, to conserve water during dry years.

Drought is a global phenomenon that plagues every continent. If predictions of global warming are correct, drought may occur more often, resulting in widespread damage and suffering.

Increasing Water Supplies

Even if per capita use of water remains the same, total water use in the world will increase substantially. In the United States, population will increase by approximately 115 million people between 2000 and 2050 (Chapter 4). In the less developed nations of the world, population is projected to increase by about 1.2 billion. Where will additional water to supply their needs come from?

Several strategies can be pursued, among them (1) water conservation (using water much more efficiently), (2) reclamation of sewage water, (3) development of groundwater resources, (4) desalination of seawater, (5) rainmaking, and (6) diversion of surface water to water-short regions. Changes in crops can also help ease the crunch. For example, efforts to develop salt-resistant and drought-resistant crops can permit farmers to continue producing crops needed to feed the expanding human population. The following sections outline the pros and cons of these solutions, and provide a good opportunity for you to practice your critical thinking skills and select the most sustainable approaches—that is, the solutions that make sense from social, economic, and environmental perspectives.

Conservation: Using Water Much More Efficiently

For years, water supplies have been expanded by building new dams that impound river water. This water was then stored and used during the year. Today, however, the era of big dam building is over, at least in the United States, for reasons described later. A far more cost-effective and environmentally sound approach is water conservation.

On The Farm Nearly 70 percent of all water used in the United States is used by farmers, mainly for irrigation. Crop growth requires enormous amounts of water—600 liters (150 gallons) of water for 0.45 kilogram (1 pound) of cotton and 60,000 liters (15,000 gallons) for a bushel of wheat. Conservation on the farm, mentioned in Chapter 5, can greatly expand water supplies throughout the world.

Reducing Seepage Losses Each year large amounts of irrigation water are lost by seepage—that is, they drain out of dirt-lined ditches. Seepage loss can be minimized by lining irrigation canals with concrete or plastic or by replacing ditches with closed pipes. For instance, the city of Casper, Wyoming, now has larger supplies of water available for domestic and industrial purposes because irrigation canals in the surrounding farmland were relined. These measures can cut water loss by half. Pipes can cut water loss by 90 percent.

Increasing the Use of Drip Irrigation At least 1.1 million acre-feet of water could be saved annually nationwide if farmers made maximal use of **drip irrigation.** In this method, water is slowly released at the base of plants by means of perforated plastic pipes. It then seeps into the ground, supplying the roots with ample water without much loss. In fact, water loss during irrigation can be cut in half. Unfortunately, however, not all crops can be irrigated by this method.

Using Heat Sensors Infrared sensors have been developed by the U.S. Water Conservation Laboratory in Phoenix to detect the amount of heat being given off by crops. This device indirectly determines soil moisture content because plants become progressively warmer as they dry out. Farmers could use this instrument to find out whether a given crop requires irrigation. Excess irrigation of the crop would then be avoided. Special computer monitors can also be used to measure soil moisture, helping farmers apply water when its needed (Figure 10.15).

In Industry For many years, industry has been flagrantly wasteful of water, especially in the water-rich eastern states, where it has been assumed that water is virtually inexhaustible. Contributing to the waste of water in this region is water's extremely low cost. For example, one study showed that when water costs only 1 cent per 4,000 liters (1,000 gallons), a coal-fueled electrical power plant uses 200 liters (50 gallons) of water for each kilowatt-hour of electricity produced. On the other hand, when the price of water rises to 1.25 cents per liter (5 cents per gallon), the same power plant reduces water use to 3.2 liters (0.8 gallon) per kilowatt-hour.

FIGURE 10.15 *Specialist taking soil moisture readings from a delicate instrument operated by the Bureau of Reclamation. With such information available, farmers irrigate only when and where needed. Such irrigation efficiency reduces not only the amount of water used but also the severity of salt buildup.*

With a blend of newly developed technology and creativity, a number of industries are making great progress in conserving water. As water-saving technology improves, the percentage of water saved will certainly increase.

On the College Campus and at Home As a student in a resource conservation class, you have by now developed a sensitivity to, and an understanding of, the overriding importance of saving the Earth's precious resources, among them water. Practical methods by which you, your fellow students, and your family can save water are listed in Table 10.4, entitled "Water Conservation Strategies."

Overcoming Legal Barriers to Conservation It may come to a surprise to you that certain laws actually impede water conservation efforts in the United States. In the western United States, for example, water in rivers and streams is allocated by a permit system. Two principles guide the allocation and use of this water.

The first, known as the Doctrine of Prior Appropriation, is colloquially known as *first in time, first in line.* What this means is that the oldest permit holders hold senior rights to water in a stream. Later permittees hold junior rights. Each year the water in all streams is allocated on the basis of how much is available and who has rights. But senior water right holders get their water first—they were the first in time and so get their water before anyone else. If there's any water left over, junior water right holders get their water. In drought years, then, farmers with junior water rights may not get any water at all.

The second and more important doctrine is colloquially known as the *use it or lose it* principle. What this means is that a water right holder must use his or her water each year or risk losing rights to that water. For example, if a farmer with 100 acre-feet of water finds a way to conserve his water, say by cutting use by half, he could lose the 50 acre-feet allocation. It will be assigned to another user. This archaic system obviously dissuades ranchers and farmers from conserving. Each year, they are essentially forced to use their full allocation, whether they need it or not.

Reclamation of Sewage Water

Sewage produced by homes contains human waste—solid and liquid waste. Huge amounts are generated each day. This waste is delivered to sewage treatment plants that remove many of the pollutants, then release the effluent into nearby waterbodies—lakes, streams, or the ocean. Sewage treatment plant effluent is 99 percent water. If the 1 percent pollution is removed, the final water product may be purer than the original substance.

Processed sewage water is already being used for a variety of purposes. The Bethlehem Steel plant in Baltimore, Maryland, uses 600 million liters (150 million gallons) of sewage effluent daily to cool steel. Golf courses are sprinkled with it in San Francisco, Las Vegas, and Santa Fe. Treated sewage water is used to irrigate crops in San Antonio, Texas, and Fort Collins, Colorado. Ornamental shrubs along highways in San

TABLE 10.4 *Water Conservation Strategies*

In the Bathroom
1. Take shorter showers.
2. Don't use the toilet as a wastebasket.
3. Don't let the water run while brushing your teeth.
4. Don't run the water while shaving. Plug and partly fill the basin to rinse your razor.
5. Repair leaks promptly.
6. If you replace your toilet, install a low-water unit.
7. A few drops of food coloring or dye tablets in your toilet tank can help you spot a leak. If color appears in the bowl, you have a leak. Fix it promptly.
8. Place bricks or plastic bottles in your toilet tank and save 1–2 liters every time you flush.
9. Install a low-flow shower head.
In the Kitchen
1. Wash only full loads in your washing machine and dishwasher.
2. If you wash dishes by hand, don't let the water run.
3. Cool your drinking water in the refrigerator, not by letting the water run.
4. Don't use water-wasting garbage disposals.
5. Stop those leaks (New York City alone wastes 800 million liters each day because of them). Leaking faucets and wasteful people are robbing Americans of scarce water supplies. A small leak (eighty drips per minute) wastes 26.5 liters (7 gallons) of water daily.
Outside Your Home
1. Use a broom, not a hose, to clean driveways, sidewalks, and steps.
2. Use an on–off spray nozzle on your hose.
3. Wash your car with a bucket of water; use a hose only to rinse.
4. Water your lawn and garden only during the cool of the day or during the evening. Older trees and shrubs often do not require irrigation. Plants are frequently overwatered.
5. Remove water-stealing weeds from the lawn and garden.
6. Use less fertilizer; it increases the need of plants for water.
7. Apply a mulch between rows in your garden to hold the soil moistur.

Bernardino, California, and the grounds at the Denver International Airport are watered with it (Figure 10.16).

Los Angeles daily discharges 68 million liters (17 million gallons) of processed sewage water over sewage-spreading beds at the edge of town. Eventually, this water seeps into aquifers that supply the town's wells. This water is of higher quality than the water from the Colorado River. Treated wastewater has also been used in southern California to stop saltwater intrusion from the ocean—that is, the penetration of salt water from the ocean into the land-based aquifers. By the early 1960s, water table levels had been dropping steadily in the coastal region. As a result, salt water encroached at the rate of 1.6 kilometers (1 mile) per year. To check this invasion, a freshwater barrier was formed by injecting treated wastewater into a series of coastal wells.

Waste water can also be recycled by homeowners. For example, some homeowners recycle **gray water** from sinks, showers, baths, and washing machines. This water, which actually appears gray, contains soaps and dirt, and is used to supply outdoor vegetation. Because the continued use of detergents for such purposes may damage soils, some homeowners use biocompatible soaps, that is soaps that contain substances that are actually good for plants. Others avoid detergents altogether by using laundry discs, small plastic devices containing ceramic beads containing copper and silver (Figure 10.17). These are placed in the wash and supposedly increase the ionization of the wash water, pulling dirt out of clothes and keeping it suspended in solution.

Many creative ways are being tried as well to treat **blackwater,** that is water from toilets. Constructed wetlands, for

FIGURE 10.16 *This golf course is irrigated with treated sewage water, a measure that conserves water and supplies valuable nutrients.*

example, are proving to be highly successful. In these systems, blackwater is diverted to small ponds, often filled with small rocks. Here, it is broken down by bacteria and other microorganisms. Soil may be placed over top of the rock bed. Plants incorporate nutrients from the waste. When the water comes out of the system, it is often pure as drinking water.

Developing Groundwater Resources

Conservation and water reclamation are two highly sustainable solutions to water supply problems. They promote key principles of sustainability, efficiency and recycling, respectively. The potential of these options is incredible.

The more traditional and perhaps less sustainable approach to meet rising demand involves the development of new water sources. In the United States alone, estimates show that there are approximately 53,000 cubic miles of fresh water in the aquifers located in the upper 0.8 kilometer (half-mile) of the Earth's crust. Wells drilled into these aquifers, 150 to 600 meters (500 to 2,000 feet) deep, could help alleviate impending water shortages.

It is important to point out, however, that these supplies must be carefully used to avoid depletion. In some situations, the proper decision may be to mine the water until the supply is exhausted; in other cases, it may be better to draw the water on a sustained-yield basis; that is, at a rate equal to the recharge. Some cities are looking at replenishing groundwater in years when surface flows are in surplus. The aquifer becomes an elaborate underground reservoir. Although this option sounds good, it does require a fair amount of energy for pumping and creates enormous pumping costs. Not all of the water would be recoverable either. Some experts assert that only a small portion of the water intentionally injected into groundwater from surface water supplies would actually be recoverable.

Desalination

Seventy percent of the Earth's surface is covered by oceans, in some places up to 10 kilometers (6 miles) deep, yet a water shortage plagues civilization from New York to New Delhi. The problem, of course, is that sea water is salty and thus not drinkable without treatment.

A number of U.S. communities are now operating desalination plants, facilities that remove salt from seawater to produce potable water. The nation's first desalination plant was built in California and produces 100,000 liters (28,000 gallons) of

FIGURE 10.17 *Laundry Disks. There's no need to use detergents to clean clothes, says the manufacturer of these devices. For really dirty clothes, just add a little special enzyme extracted from papaya.*

fresh water daily. Numerous desalination plants have been built on the west coast of Florida alone. According to recent statistics, there are currently nearly 1600 desalination facilities in the United States, producing 3.7 billion liters (950 million gallons) of water a day. Although desalination plants still produce only a small fraction of the United States' total freshwater, several countries in the Middle East rely much more heavily on desalination for drinking water.

Although desalination may seem like a viable option for meeting future demands, it does have some problems. For one, it is a rather expensive process. In general, it is much cheaper to pump fresh water—if it's available. However, if the freshwater source must be pumped more than 150 kilometers (90 miles) to the site of consumption, desalination becomes economically feasible. Desalination plants also require large quantities of energy and produce brine, salt waste that must be properly disposed of.

Developing Salt-Resistant Crops

After 6 years of research, two scientists from the University of California, Davis announced the development of a new strain of barley that will grow well even though it was irrigated with seawater. Barley plants have been grown on a tiny windswept beach at Bodega Bay in northern California (just north of San Francisco). They have achieved yields of 1,480 kilograms per hectare (1,320 pounds per acre), equal to the average global per acre yield of barley provided with fresh water. As one of the researchers, Emanuel Epstein, noted: "We have shown that sea water is not pure poison to crops." Their success is highly significant. Millions of acres of once prime agricultural land the world over (1.8 million hectares [4.5 million acres] in California alone) have been rendered worthless because of salinization, the accumulation of salts on irrigated cropland (discussed later in the chapter). Until now, farmers have been advised to cease cropping salinized soils or to flush out the salt, at considerable expense, with huge volumes of fresh water. However, the new saltwater barley and other crops now under development might do well in these soils. Such applications of this new genetic strain are probably more important than growing crops in salt water that will surely damage soil in short order.

Developing Drought-Resistant Crops

Food production could also be increased by developing new varieties of crops that are resistant to drought. George G. Still, a scientist for the USDA, is optimistic about the water-saving potential of such plant-breeding projects: "Sorghum is a very important grain crop in the arid portion of the Third World. There is something inherent in the plant, the germ plasm or the genes, that causes sorghum to put itself on idle during a dry spell and then go on and yield a crop when the rains come," he says. With the aid of recently developed techniques in genetic engineering, it may be possible to breed plants that require considerably less water.

Rainmaking

Rainmaking is a novel approach to increasing our water supply. One technique is cloud seeding, dispersing tiny crystals of silver iodide in moist air. The hope is that these crystals will serve as **condensation nuclei,** particles around which moisture will collect until raindrops are formed. The U.S. Bureau of Reclamation is confident that the weather modification techniques now available can increase the water supply of the San Joaquin River Basin by 25 percent, that of the Upper Colorado River Basin by 44 percent, and that of the Gila River Basin (Arizona) by 55 percent. In the United States, there are about 40 cloud seeding programs. Over half of them are in California and Nevada. Although the U.S. Congress eliminated federal support for weather modification research, it still supports many of the cloud seeding projects.

Cloud seeding could create some potential problems—some legal, some political, some economic, and some environmental. One drawback is the high cost. Another problem is the possibility that cloud seeding, which increases rainfall in one area, reduces rainfall in downwind areas. So, artificially enhanced rain in one region could result in drought in other areas. Another problem is the lack of control over the amount and precise location of the precipitation. For example, a late July rainfall might benefit corn but might damage the alfalfa that is awaiting the baler in a nearby field. Increased rainfall might improve forage for a rancher's cattle but might raise havoc with a nearby citrus grower's orange crop. Even more serious, lack of control over precipitation might lead to flooding, soil erosion, property damage, and even loss of life.

Long-Distance Transport: The California Water Project

Water diversion, the transport of water from one region to another through pipelines and canals, is viewed by some developers and government officials as a way to provide water to water-short regions. Because of costs and environmental impacts, however, the idea is frequently a controversial one. The California Water Project is an example of what can be done.

Looking down on the Earth, a bluish sphere far below, America's moon-bound astronauts could identify only two artificial structures: the Great Wall of China and the main aqueduct of the California Water Project (CWP). California has long been victimized by the curious fact that 70 percent of its potentially usable water falls on the northern third of the state (in the form of relatively abundant rainfall and the snowmelt of the High Sierras), while 77 percent of the demand is located in the semiarid southern two thirds, where only 12 centimeters (5 inches) of rain falls per year.

The CWP was built to rectify this problem. The most complex and expensive water diversion project in the history of the world, the CWP includes 21 dams and reservoirs, 22 pumping plants, and 1,140 kilometers (685 miles) of canals, tunnels, and pipelines (Figure 10.18). An expensive "faucet," the project cost well over $2 billion—enough money to build six Panama Canals.

Despite the obvious benefits derived from this colossal project, certain aspects have drawn criticism from environmentalists. They claim that the CWP was built at an excessive cost, not only in tax dollars but in energy costs, losses in scenic beauty, and the destruction of fish and wildlife habitat. Environmentalists also criticize the proposal to dam up other free-flowing wild rivers in northwestern California, such as the Eel, Klamath, and Trinity. In their view, too many of such unharnessed streams have already been lost to irrigation and power production.

Regardless of these criticisms and the enormous cost to the people of California, the CWP is an accomplished fact and is helping to alleviate southern California's recurring water shortages. Impressive as this project may sound, it is likely to be one of the last of its kind. Far more cost-effective—and more sustainable—means of meeting future demands must be employed, for example, water efficiency measures in homes, offices, and businesses, which have been discussed in this chapter. Also essential to creating an environmentally sustainable water supply system are measures to recycle water and recharge groundwater.

10.4 Irrigation: Issues and Solutions

The desert biome and the arid portions of the grassland biome (short-grass prairies) experience more or less permanent drought. Plant and animal residents that live in these ecosystems have evolved moisture-securing and moisture-conserving adaptations to survive. Humans, however, relative newcomers to this austere region, have not had to depend on long evolutionary processes to adapt to the environment. Instead, we have shaped the environment to fit our purposes. The most significant and dramatic example of our habitat-shaping talent in this region is modern irrigation.

Irrigation is an extremely complex and exacting operation that demands a high degree of planning, field preparation, and technical skill. It is also a rather costly process. Nevertheless, the high market value of many crops and the long growing season that permits several crops per year in some locations such as southern California make irrigation economically feasible. More than 10 percent of our nation's croplands, about 50 million acres, is now under irrigation. Four of every five irrigated acres are in the West (Figure 10.19).

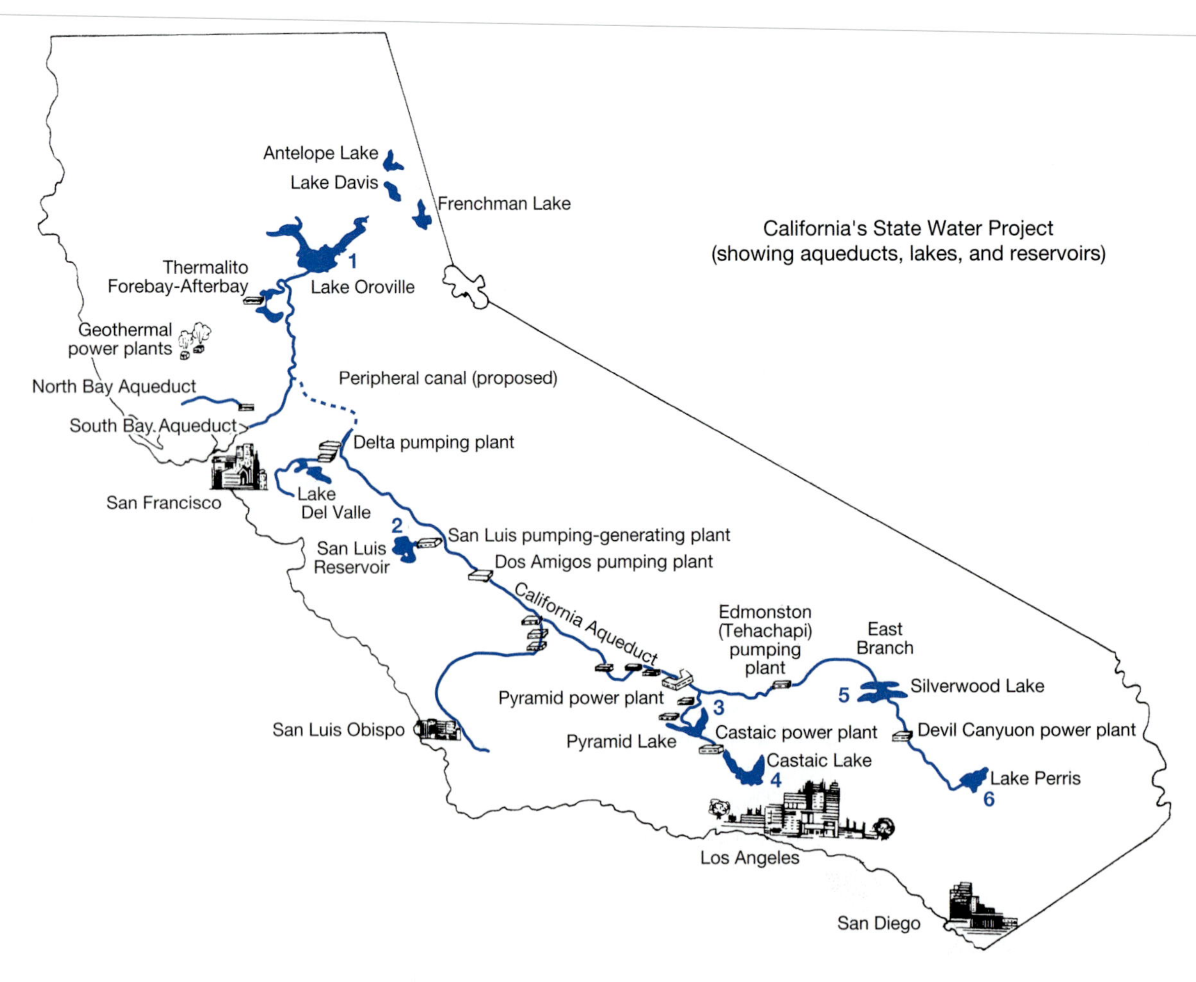

FIGURE 10.18 *California's State Water Project. Major lakes and reservoirs are indicated by numbers.*

Until recently, the amount of irrigated farmland increased steadily.

Methods of Irrigation

Irrigation water is either pumped or gravity fed through the main irrigation canal or pipe to laterals, which convey the water to individual farms. The laterals frequently follow field borders and fence lines. Up to 2.5 million liters (750,000 gallons) of water may be needed to irrigate a single acre in one growing season. The major methods are sheet, furrow, sprinkler, and drip irrigation.

Sheet and Flood Irrigation The sheet irrigation method, usually used for hay, grain, and pasture crops, is suitable on land with a slight grade. For proper sheet irrigation, the topsoil must be carefully prepared in advance so that water infiltrates properly. The water gradually flows downslope in the form of a sheet (Figure 10.20). Soil erosion and leaching of nutrients can result from this method.

Flood irrigation is also common, especially in the West, for growing rice. As its name implies, farmers flood fields with irrigation water. The fields are not prepared at all.

Furrow Irrigation In the furrow irrigation method, water may be drawn from laterals by siphon tubes that empty into furrows between crop rows (Figure 10.21). This method is used primarily on row crops such as corn, cabbage, and sugar beets. Erosion can be lessened by contouring the furrows.

Sprinkler Irrigation The sprinkler irrigation method may be used in places where the previous methods are undesirable because of erosion due to the steepness of the slope. It is also used in flatland where water is acquired from deep wells. It involves costly equipment and is restricted primarily to crops of high cash value. Considerable amounts of energy are required to power the pumps.

Sprinklers may be either stationary or rotary (Figure 10.22). Center pivot rotary sprinklers are very popular. In this technique, water is sprinkled from a raised lateral pipe that is in continuous slow rotary motion around a central pivot point. One model, known as the big gun, can propel a stream of water half the length of a football field. Most models spray water from numerous sprinkler heads spaced evenly along the length of the irrigation pipe. A circular area of 53 hectares (130 acres) can be irrigated in one revolution in 33 hours with a 400-meter (1,300-foot) pipe. Evaporation losses with all sprinkler systems may be considerable. Because of this, many center pivot irrigation models have been retrofitted with downward-facing spray heads that are suspended a relatively short distance above the crop. This greatly reduces water loss through evaporation.

Drip Irrigation Drip irrigation is used in orchards and vineyards where crops are permanent. With the drip method, water is delivered through perforated or highly porous plastic pipes (Figure 10.23). The pipes may either be placed on the surface or buried underground. Water drips from the pipes and seeps into the ground, watering the roots while minimizing evaporation. Since drip irrigation requires 20 to 50 percent less water than sprinkler systems, it can help save considerable amounts of water. Bear in mind, however, that drip irrigation is not suitable for all crops. It cannot be used for wheat or corn, because of the need to lay considerable amount of pipe year after year. It is quite suitable for permanent crops such as fruit and nut trees.

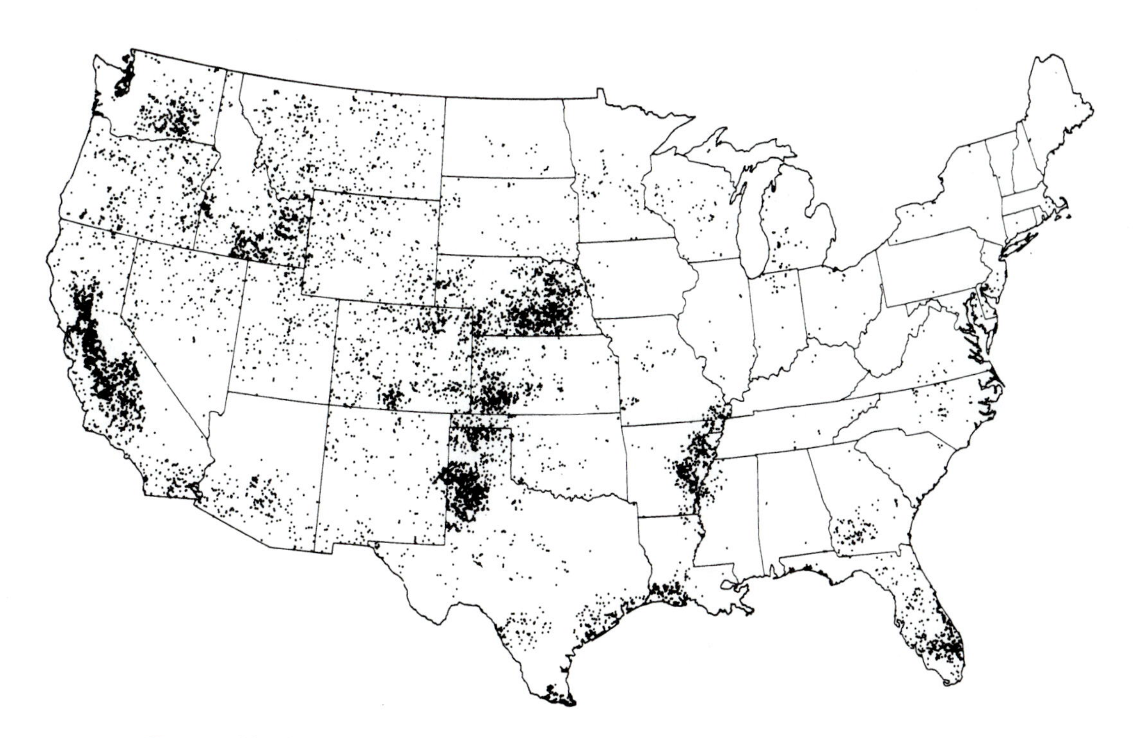

FIGURE 10.19 *Irrigated acreage in the United States. One dot equals 8,000 acres of irrigated land.*

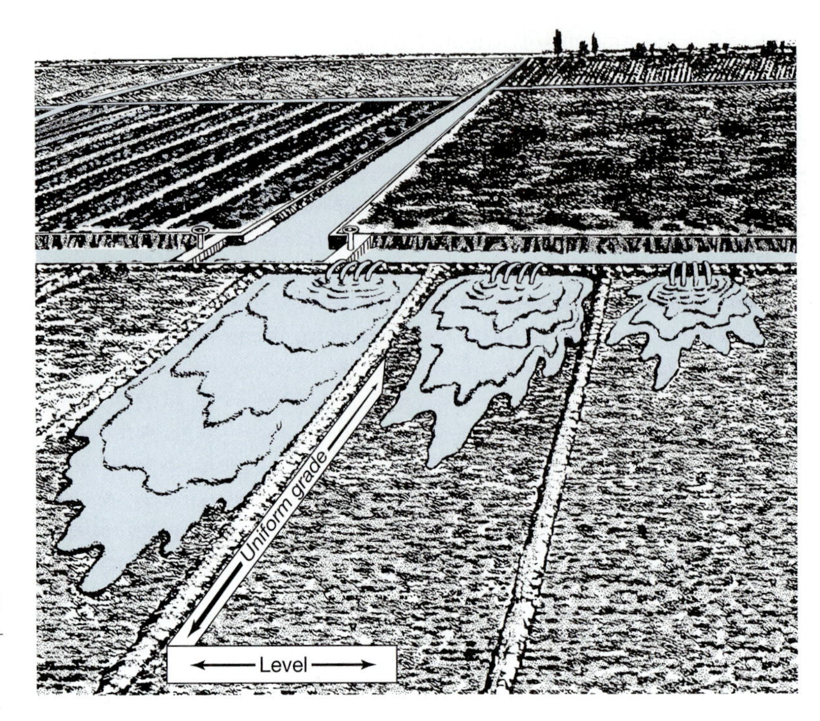

FIGURE 10.20 *Sheet or flood irrigation. This field has been prepared for irrigation by building small levees around each leveled area. The areas are then flooded in rotation to irrigate them.*

FIGURE 10.21 *Furrow irrigation. Water is transferred from the lateral in the foreground to a series of parallel furrows to irrigate this lettuce crop in the Palo Verde Valley.*

Irrigation Issues

Irrigation clearly helps increase productivity and is essential to meeting future demands (Figure 10.24). But irrigation is not without its problems.

Water Loss Many irrigated fields in the United States receive their water from reservoirs or streams located hundreds of kilometers away. The fruit-raising Central Valley of California, for example, gets its water from the Colorado River, 500 kilometers (300 miles) to the east. During transit, considerable amounts of water are lost. According to the USDA, only 1 of every 4 gallons drawn for irrigation is actually absorbed by crop root systems. The remaining 3 gallons are lost to evaporation, to water-absorbing weeds, or to ground seepage. In some areas, seepage may be sufficient to raise the water table and form a marsh. Water efficiency measures discussed earlier in the chapter can help solve this problem.

Salinization and Waterlogging Would you believe that bringing fresh water to a desert might be destructive to crops? Sounds incongruous, doesn't it? However, even fresh water is slightly salty, having acquired dissolved sodium, calcium, and magnesium salts as it flows down mountain slopes and through valley bottoms. In some cases, water flows through regions that greatly increase salt concentrations. When such water is brought by irrigation canals to hot deserts, where drainage down through the soil is very poor and the evaporation rate is very high, much of the water passes into the atmosphere. As a result, the salts are left behind on the ground as a white crust. Additional salt may be deposited because of the evaporation of groundwater that has been drawn to the surface by capillary action.

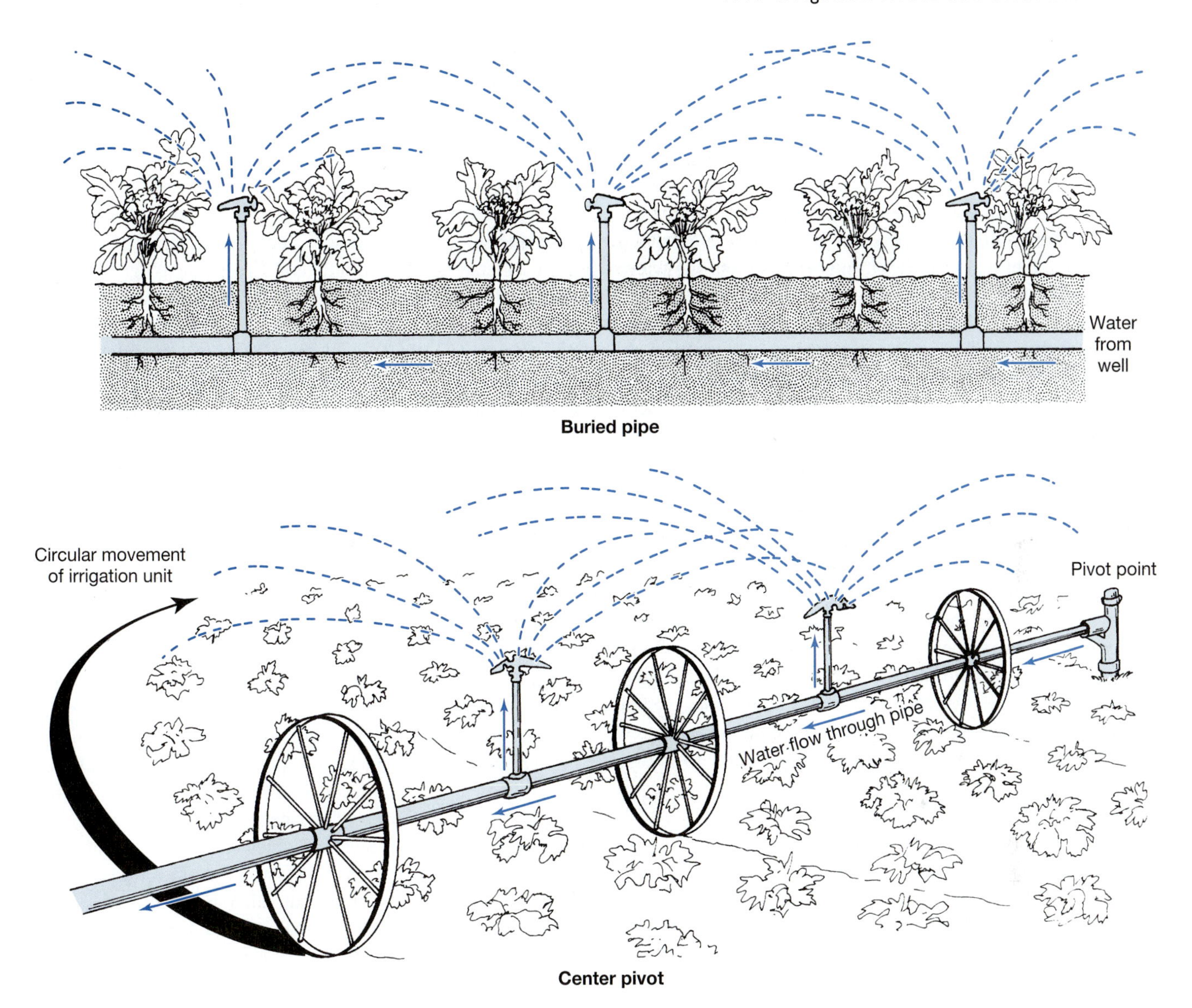

FIGURE 10.22 *Two types of sprinkler irrigation. In center pivot irrigation the perforated pipe moves slowly in a circle and may irrigate an area of about 150 acres.*

The buildup of salt in the soil is called **salinization.** As time passes, the salt deposits increase, becoming toxic to crops. Agricultural experts estimate that 30 percent of the West's irrigated land has salinity problems. After 20 years of irrigation, some land may have a salt load of more than 30 metric tons per hectare (80 tons per acre). Even in California's Imperial Valley, where crop harvests have been so bountiful, salinization has caused many farms to be abandoned. Salinization is also a problem in other agricultural nations where semiarid and desert lands are irrigated.

Salinization can be reduced by installing underground drainage pipes that collect excess salty groundwater. These porous pipes connect to a master drainpipe from other fields. This fairly effective measure, however, creates a waste disposal problem. The leftover water is rich in salts and potentially toxic heavy metals leached from the soil. If dumped in natural water courses, it could cause serious effects. Contaminated water from farms in the San Joaquin Valley of California caused extensive birth defects and mortality to waterfowl living in and around specially built evaporation ponds at the Kesterson National Wildlife Refuge and other locations in the 1990s.

Efficient irrigation, which reduces water use, can also reduce salinization. Salt-resistant crops that replace traditional strains could also be used. Finally, it may be necessary to convert areas plagued with recurring salt problems from cropland to grazing land—a solution to which many farmers are resistant.

Salts in soil also come from water that has already been used to irrigate crops. Water that flows off of fields into streams may be used further downstream as irrigation water, and then again even further along the course of the river. Each time it gets more salty. The Colorado River, for example, receives large amounts of salty water that has been flushed from thousands of irrigated farms in western Colorado, Utah, and Arizona. As a result, the river's water becomes progressively more salty as it flows southwestward across the Mexican border toward the Gulf of California. As a result of irrigation practices, the salinity of the lower Colorado has increased by 30 percent in the last 20 years. The Colorado's saltiness jeopardizes cotton production in the Mexicali district, where farmers have used the river as a source of irrigation water for decades. The economy of the region was threatened to such an extent that Mexican presidents frequently conveyed their concern to the U.S. government. The United States was eventually forced to construct a desalination plant near the Mexican border so that the water would be usable by Mexican farmers.

Another related problem is waterlogging. **Waterlogging** occurs when soil receives too much moisture. This may occur naturally or as a result of applying too much irrigation water. When soil becomes saturated with water, that is, when the air spaces are filled with water, oxygen flow to the roots is impaired. Plants suffocate and die. In addition, excess water moves up through the soil via capillary action. This water evaporates, leaving behind once-dissolved salts. In other words, excess irrigation, especially by water with excess salts, not only suffocates plants, it can result in salinization.

Depletion of Groundwater and Surface Water Yet another problem caused by our heavy use of irrigation water is groundwater depletion. Over the years, the number of hectares irrigated by farmers in the United States, Canada, and many other nations has increased sharply. On the high plains, a vast region extending from Nebraska south to Texas, most of the irrigation water is pumped out of the Ogallala aquifer (Figure 10.7). In 1946 there were only 2,000 irrigation wells in western Texas; today there are more than 70,000. The rate of withdrawal around Lubbock, Texas, is 50 times the rate at which the aquifer is naturally recharged by rain and stream flow.

As a result of this enormous overdraft for irrigation, groundwater levels have dropped significantly in some western states—more than 30 meters (100 feet) in portions of Kansas, Oklahoma, Texas, and New Mexico, and more than 120 meters (400 feet) in parts of Arizona. As a result, new wells must be drilled more deeply at considerable expense. And, of course, it takes more fossil fuel or electricity to pump water from the deeper wells. If this trend continues, irrigation farming will eventually become prohibitively expensive. West Texas could experience a 95 percent decline in irrigation and a 70 percent decrease in crop harvests by 2015. Serious declines in irrigation farming are expected over an 11-state area of the Great Plains and the Southwest—during an era of ever-increasing food demand.

Groundwater depletion or groundwater mining, as it is sometimes called, may worsen in years to come, according to some scientists. Projections of climate change caused by greenhouse gases and deforestation indicate that the mid-continent areas of North America, which are heavily dependent on groundwater for food production already, will become hotter and drier. If the water from the large Ogallala aquifer has been seriously depleted, the world renown breadbasket of the world, could face devastating losses. The effects on U.S. and global food supplies, famine and hunger, and economics could be substantial.

Groundwater withdrawals have other more immediate effects, too. When large volumes of water are removed from fine-grained, porous aquifers, the weight of the overlying soil and rock occasionally causes compression or collapse of the aquifer. As a result, the earth above the aquifer sinks, or subsides (Figure 10.25).

Groundwater overdraft causes significant land sinking or **subsidence** (sub-SIDE-ence). Subsidence has occurred in at least 11 states in the South and West due to groundwater overdrafts. Water mining caused the dramatic appearance of a sinkhole at Winter Park, Florida, in 1981. The huge pit was 37.5 meters (124 feet) deep and 120 meters (396 feet wide). It swallowed up a house, a swimming pool, six sports cars, and a camper. In the San Joaquin Valley of California, an 11,000-square-kilometer (4,200-square-mile) area sank more than 0.3 meter (1 foot). Some regions subsided more than 9 meters (30 feet).

Subsidence causes damage to irrigation facilities such as canals and underground pipes. The Department of the Interior spent $3.7 million in a single year to repair the damage caused by subsidence to federal irrigation projects.

In the Galveston Bay region near Houston, Texas, subsidence over a 10,000-square-kilometer (4,000-square-mile) area resulted in widespread flooding of shoreline properties. When subsidence occurs under urban areas, damage may be very extensive: sewer and water pipes burst, wells are destroyed, building foundations crack, and concrete highways buckle.

Groundwater overdraft can be reduced by efficient water distribution and use, methods described earlier in the chapter. Protection of groundwater recharge zones, areas where aquifers are replenished by water percolating from the surface, by restricting growth and other activities can also help (Figure 10.26). Wastewater applied to the ground in such areas can also help renew groundwater supplies.

Streams and rivers in many parts of the world are greatly depleted each year by water withdrawals for irrigation and other uses. Some streams run completely dry because of the massive withdrawal of water. For an aquatic organism or an organism dependent on the waterbody, the dewatering of its habitat can be quite devastating.

Some states have attempted to rectify the problem by maintaining minimum stream flows—that is, modest flow levels below which a stream or river cannot drop in order to protect aquatic species, especially fish. Scientists have developed numerous ways of determining minimum flows. However, minimum stream flows are often set very low—so low that they greatly reduce a stream's full potential to support life. Fish and other aquatic organisms often require more water to thrive and reproduce than is allowed, but powerful interest groups like farmers and water departments also need the water. Their interests often prevail over the interests of wildlife. Protecting wildlife is socially and politically difficult and further troubles lie ahead as populations expand and demand for water by people increases. Many states in the West are continuously involved in litigation over minimum flow, being sued by environmentalists and wildlife agencies. In this region, though water flow in

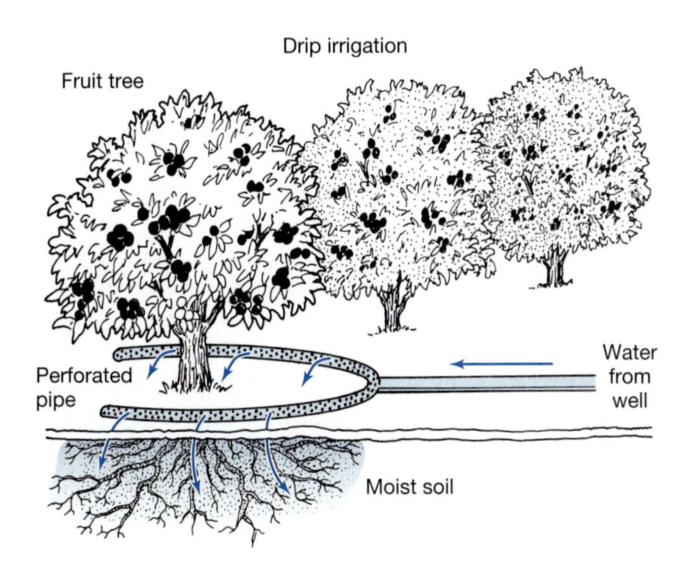

FIGURE 10.23 *Drip irrigation. This method is used primarily on fruit trees and high-cash crops. Water is applied directly to the roots of the fruit tree. The perforated plastic pipes may either lie on the ground surface or be buried.*

streams is determined by a Doctrine of Prior Appropriation, discussed earlier.

Sale of Water Rights Some farmers and ranchers in the arid states are discovering that their irrigation water is more valuable than the wheat, fruits, vegetables, and beef cattle they produce. As a result, they are now beginning to sell their land and water rights to nearby communities that need the water. Surface water is also growing more costly. For example, in Reno, Nevada, the cost of water from the Truckee River, which flows right past the town, surged from a mere $50 per acre-foot in 1976 to $3,000 per acre-foot in 1991—a 60-fold increase in 15 years.

Summary of Key Concepts

1. Water moves continuously from oceans to air, to land, to rivers, and back to oceans, in what is known as the water cycle. This cycle is powered by solar energy and gravity. Water evaporates from land and waterbodies, forms clouds, then is redeposited as rain or snow.
2. Water that falls on the ground as rain runs into streams or percolates down into the soil and rocks to become groundwater. The uppermost level of the zone that is saturated with groundwater is called the water table.
3. Ninety-seven percent of the world's supply of fresh water is held in porous layers of sand, gravel, and rock known as aquifers. These are important sources of fresh water for drinking, irrigation, and other uses.
4. Floods are a major problem the world over, costing billions of dollars a year and taking the lives of many people and livestock.
5. Floods are natural events but occur with greater frequency and in greater magnitude in disturbed environments—areas in which the natural vegetative cover has been stripped or reduced. Impervious surfaces such as roadways, parking lots, and roofs also increase the incidence and severity of floods. Damage from floods is more likely as people settle in floodplains.
6. Floods can be controlled by (a) measuring the snowpack to predict flood conditions, (b) protecting watersheds, (c) restricting development and other human activities in

FIGURE 10.24 *Irrigation transforms the desert! Water from the Colorado River conveyed by the All American Canal system makes it possible for crops to be produced in Imperial Valley, California. Note the stark contrast between the nonirrigated desert on the right and the irrigated fruit orchards on the left.*

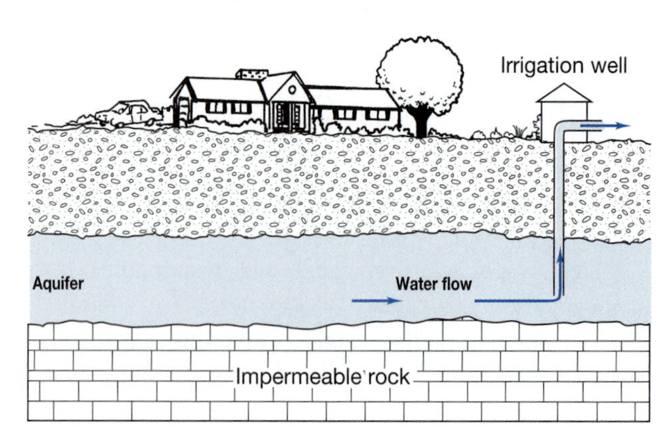

FIGURE 10.25 *The land-sinking phenomenon, called subsidence, is caused by water mining.*

floodplains, (d) building levees, (d) dredging rivers, and (e) building dams.

7. Big dams can control flooding but have several major drawbacks, among them (a) high economic costs, (b) endangerment of life because of possible collapse, (c) reservoir evaporation losses, (d) flooding of prime agricultural land, (e) siltation of reservoirs, which reduces their life span, (f) saltwater intrusion into coastal areas, (g) destruction of scenic beauty and recreational uses like rafting and kayaking, and (h) destruction of the habitats of species.
8. Stream channelization increases the speed with which water is removed from a watershed, thus reducing the potential for floods in the channelized section. Like other measures, it has disadvantages: (a) it destroys wildlife habitats, (b) it may increase flooding downstream, (c) it may lower the water table of groundwater as it reduces water absorption by the land, (d) it causes severe aesthetic losses, (e) it destroys many of the recreational functions of streams, and (f) it is costly.
9. Many experts believe that watershed protection and floodplain zoning are the most effective, economical, and sustainable methods of flood control.
10. Water shortages occur with great frequency throughout the world. As the world population increases, many experts predict more severe water shortages arising in many nations.
11. Drought is a natural occurrence but may also be worsening as a result of changes in global climate brought on by air pollution from human sources and deforestation.
12. Many factors contribute to the occurrence of water shortages in drought and normal times: (a) a rapidly increasing

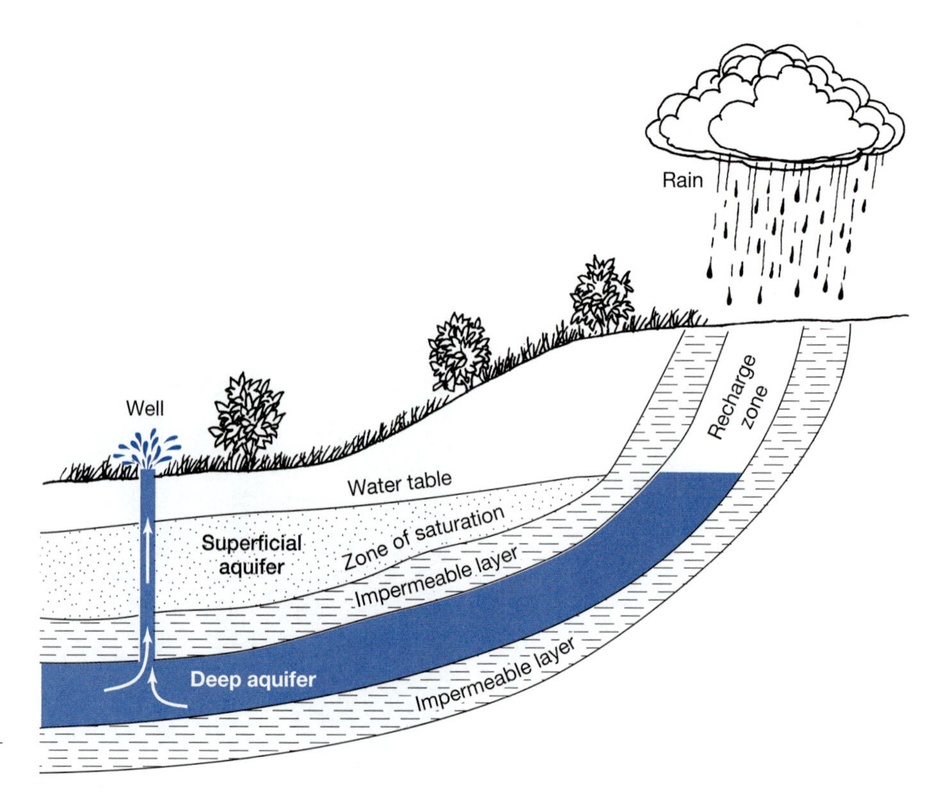

FIGURE 10.26 *An aquifer serves as a source of well water. Continued withdrawal of water would eventually deplete the supply if the rate of withdrawal exceeds the rate of recharge.*

population, (b) increasing demands by agriculture, cities, and industry, (c) inefficient use of water, (d) unequal distribution of water, and (e) pollution.

13. Satisfying present and future water demands requires a multifaceted approach with an eye toward sustainability. Possible methods include (a) using water much more efficiently, (b) purifying and reusing the aqueous effluents of sewage treatment plants, (c) developing new sources of groundwater, (d) desalinizing seawater, (e) developing drought-resistant and salt-resistant crops, (f) rainmaking, and (g) water diversion projects, that is, transferring surplus water to water-short areas.
14. Irrigation permits a flourishing agricultural economy in arid portions of the world and accounts for a very large percentage of the world's food output. Obviously, then, it is a very important way of producing food.
15. Four methods of irrigation are (a) sheet and flood, (b) furrow, (c) sprinkler, and (d) drip or trickle. These vary greatly in their efficiency, with the last two being the most efficient.
16. Serious problems associated with irrigation include (a) water loss during transit along canals, (b) inefficient practices during the application of water to the crops, (c) salinization of the soil, and (d) depletion of groundwater.

Key Words and Phrases

Aquifer
Aquifer Recharge Zone
Blackwater
California Water Project
Condensation Nuclei
Desalination
Drainage Basin
Dredging
Drip Irrigation
Drought
Drought-Resistant Crops
Evaporation
Federal Flood Disaster Protection Act
Floodplain
Floodplain Zoning
Furrow Irrigation
Graywater
Groundwater
Groungwater Overdraft
Hydrological Cycle
Imperial Valley
Infiltration
Irrigation
Irrigation Water Subsidies
Levee
Ogallala Aquifer
Open-Ditch Irrigation
Precipitation
Rainmaking
Renewal Time
Replacement Period
Rotary Sprinkler
Salinization
Salt-Resistant Crops
Saltwater Intrusion
Seepage
Sheet Irrigation
Snowpack
Sprinkler Irrigation
Stream Channelization
Streambed Aggradation
Subsidence
Surface Runoff
Surface Water
Transpiration
Upper Mississippi Flood
Water Cycle
Water Diversion
Water Mining
Water Table
Waterlogging
Watershed
Watershed Protection
Watershed Protection and Flood Prevention Act
Zone of Aeration
Zone of Saturation

Critical Thinking and Discussion Questions

1. What are the major water-related issues facing the United States and other countries?
2. Discuss four major factors that contribute to water shortages occurring throughout the world.
3. What is the water cycle? What powers the water cycle? What are the main water reservoirs? What are the renewal times for the atmosphere, rivers, lakes, glaciers, and oceans?
4. Using your critical thinking skills, discuss the statement, "The United States does not really have a water shortage; it is plagued with a water distribution problem. We can easily solve our problems by diverting water from areas blessed with abundant supplies to areas less fortunate."
5. Using your critical thinking skills, discuss the following statement: "Flooding is a natural occurrence. The best way to control it is by building more dams."
6. Using your critical thinking skills, analyze the following assertion: "To protect against flooding we should build more levees along river banks and channelize more rivers."
7. Describe the most sustainable means of reducing flooding. What are the features of these approaches that make them sustainable? What are the least sustainable measures? Why?
8. Describe some of the positive and negative effects of dredging operations.
9. Discuss the statement, "Big dams, like the Hoover, are engineering masterpieces that have been indisputable successes in boosting human welfare."
10. Suppose that the dams that now exist in the United States had never been built. What would have been the disadvantages? Would there have been any advantages? Discuss your answer in terms of the American economy, human safety, agricultural production, wildlife preservation, scenic beauty, and endangered species.
11. List the disadvantages of stream channelization to agriculture and to the stream ecosystem.
12. Describe the many ways we can expand the world's supply of fresh water. What the most sustainable approaches? Why? What are the least sustainable approaches?
13. What causes salinization? Briefly describe five ways in which salinization could be controlled.

Suggested Readings

Abramovitz, J. N. 1996, "Sustaining Freshwater Ecosystems." In *State of the World 1996,* Starke, L., (ed.) New York: W. W. Norton. Important reading.

del Moral, L. and D. Sauri, 1999."Changing Course: Water Policy in Spain." *Environment* 41(6): 12–15, 31–36. An in-depth look at alternative means of meeting demands for water in Spain.

El-Ashry, M. T. 1988. *Water and the Arid Lands of the Western United States.* New York: Basic Books. Examines the intense competition for scarce water between farmers, cities, commercial interests, and industries.

Brown, L. R. 1991. "The Aral Sea: Going, Going…" *World-Watch* 4(1): 20–27. Powerful story about the consequences of abusing one's water resources.

Dzurik, A. A. 1990. *Water Resources Planning.* Totowa, N.J.: Rowman and Littlefield. Comprehensive treatise on water resource planning and management.

Gardner, G. 1996. "Preserving Agricultural Resources." In *State of the World 1996,* Starke, L., (ed.) New York: W. W. Norton. Excellent in-depth view of agricultural problems, including water supply.

Gleick, P. H. 2000. *The World's Water 2000–2001: The Biennial Report on Freshwater Resources.* Washington, D.C.: Island Press. Very valuable resources with tons of useful information.

Howe, C. W. 1991. "An Evaluation of U.S. Air and Water Policies." Environment 33(7): 10–15, 34–36. Skillful analysis.

McGuane, T. 1993. "Wild Rivers." *Audubon,* Nov.–Dec. Celebrates our wild western rivers. A tribute to the 25th anniversary of the National Wild and Scenic Rivers Act.

Platt, R. H., P. K. Barten, and M. J. Pfeffer. 2000. "A Full, Clean Glass? Managing New York City's Watersheds." *Environment* 42(5): 8–20. Useful case study for those interested in learning more about watershed management.

Postel, S. 1996. "Forging a Sustainable Water Strategy." In *State of the World 1996,* Starke, L., (ed.) New York: W. W. Norton. Very important reading for students who want to learn more about this important topic.

Postel, S. 2000. "Redesigning Irrigated Agriculture." In *State of the World 2000,* Starke, L., (ed.) New York: W.W. Norton. Excellent overview of what can be done to improve the efficiency of irrigation.

Reisner, M. 1993. *Cadillac Desert: The American West and its Disappearing Water.* New York, Penguin. Fascinating book for anyone interested in learning about water and the political struggles that surround it in the arid American West.

Steinhart, P. 1993. "Mud Wrestling." Sierra, Jan.–Feb. Discusses the political barriers to saving our nation's valuable wetlands.

White, G. 2000. "Water Science and Technology: Some Lessons from the 20th Century." *Environment* 42(1): 30–38. Good overview of changes in water policy, many of which were discussed in this chapter.

Web Explorations

Online resources for this chapter are on the World Wide Web at: **http://www.prenhall.com/chiras** *(click on the Table of Contents link and then select Chapter 10).*

Chapter 11

Water Pollution

Widespread illness in Jackson Township, New Jersey, ranging from skin rash and kidney malfunction to premature death; warnings posted by Minnesota and New York state wildlife officials not to eat fish because of mercury contamination; a once-clear lake near Chicago converted to pea-green "soup"; the skeleton of a perch loaded with radioactive chemicals; a stream bottom near Baltimore blanketed with sludge worms; maggot-infested fish rotting on a Lake Erie beach; eight youngsters sick with typhoid fever after eating a watermelon they found floating in the Hudson River; 140 million fish deaths in U.S. waters in a single year; the Cuyahoga River, Ohio, bursting into flames; 18,000 people in Riverside, California, stricken with fever and vomiting; an outbreak of hepatitis in New York and New Jersey; thousands of fish floating belly up in the Potomac River just below our nation's capital; cholera spreading through South America—all these seemingly diverse events have one thing in common: They were caused by water pollution (Figures 11.1 and 11.2).

11.1 Types of Water Pollution

Water pollution can be defined as any contamination of water that lessens its value to humans and other species. Water pollution can be classified in two ways—by source or chemical type. We begin with a discussion of the sources. In this classification, scientists recognize two major types of water pollution: point and nonpoint (Figure 11.3).

Point Source Water Pollution

Point source water pollution has its source in a well-defined location such as the pipe through which a sewage treatment plant or a factory discharges waste into a body of water, either a surface water or groundwater. **Surface water** is a body of water in direct contact with the atmosphere, and includes lakes, rivers, streams, and springs. Groundwater, discussed in Chapter 9, refers to water in the ground, either saturated soil or deeper deposits known as aquifers.

FIGURE 11.1 *A sign of the times. Although considerable progress has been made in cleaning up our nation's waters, signs like this still appear.*

Although point source pollution may be very serious, it usually can be reduced or eliminated with current technology, notably **pollution control devices.** These devices capture the pollutant and then treat it so that it becomes less harmful, or concentrate it so that it can be disposed of in a more appropriate manner. More recently, companies are finding ways to eliminate waste entirely through **pollution prevention,** techniques that eliminate the production of pollutants in the first place. Pollution prevention is beneficial to the economy and the environment alike.

Nonpoint Source Water Pollution

Nonpoint source water pollution has its source over large areas such as farmland (pesticides, fertilizer, manure, sediment), grazing lands (animal wastes, sediment), stream banks (sediment), abandoned coal mines (acid drainage), construction sites (sediment), and roadsides (lead, sediment, de-icing salts). It may be surprising to learn that cities are nonpoint sources that release great volumes and varieties of water pollutants. Cities produce an incredible amount of runoff because of rooftops, streets, freeways, parking lots, shopping centers, and so on that retard water absorption, as noted in Chapter 9. This water flows across such areas, picking up pollutants, then flows into street gutters and storm sewers; it is eventually discharged into waterways and oceans. Some of the pollutants carried into

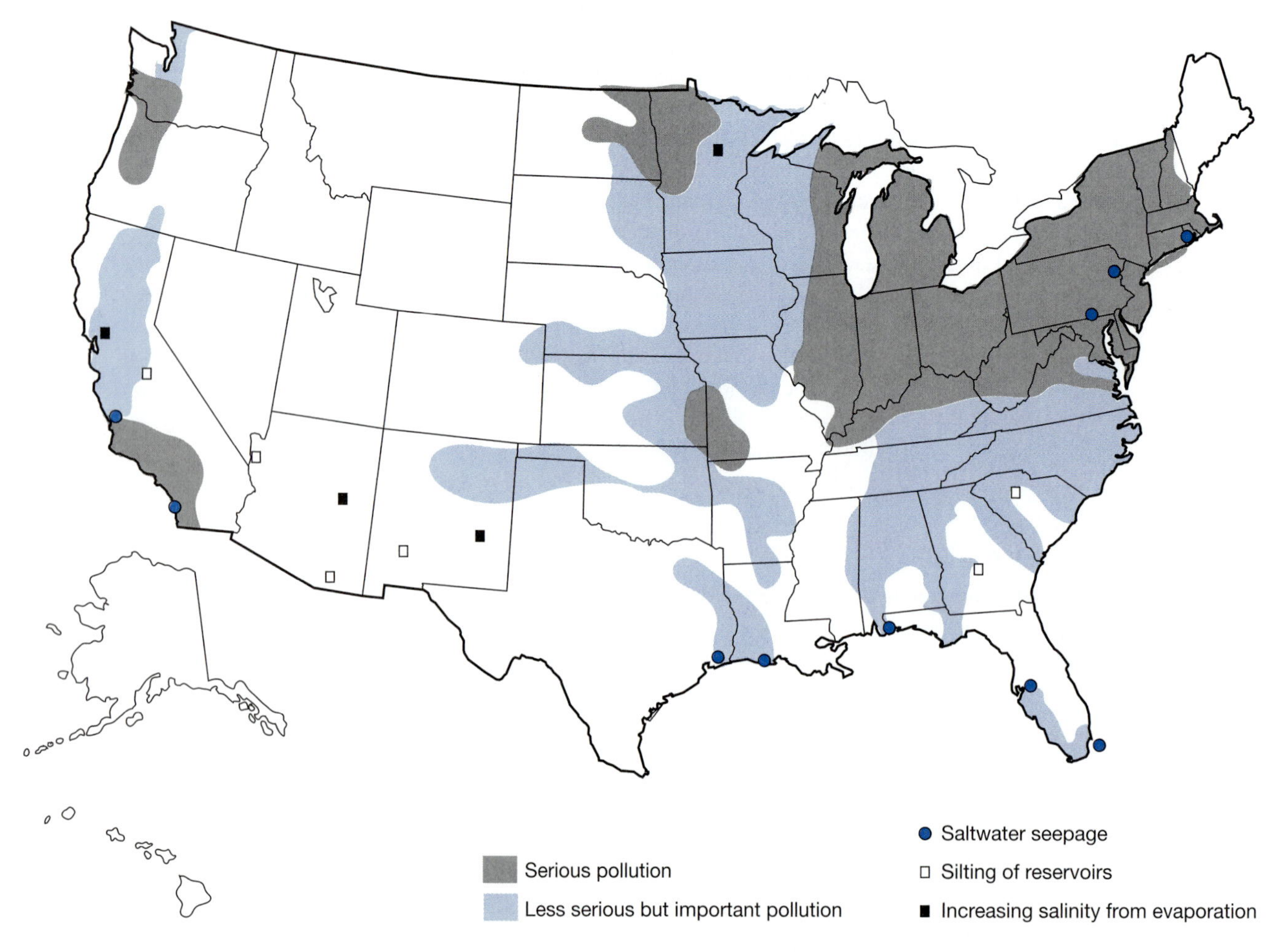

FIGURE 11.2 *Water pollution in the United States.*

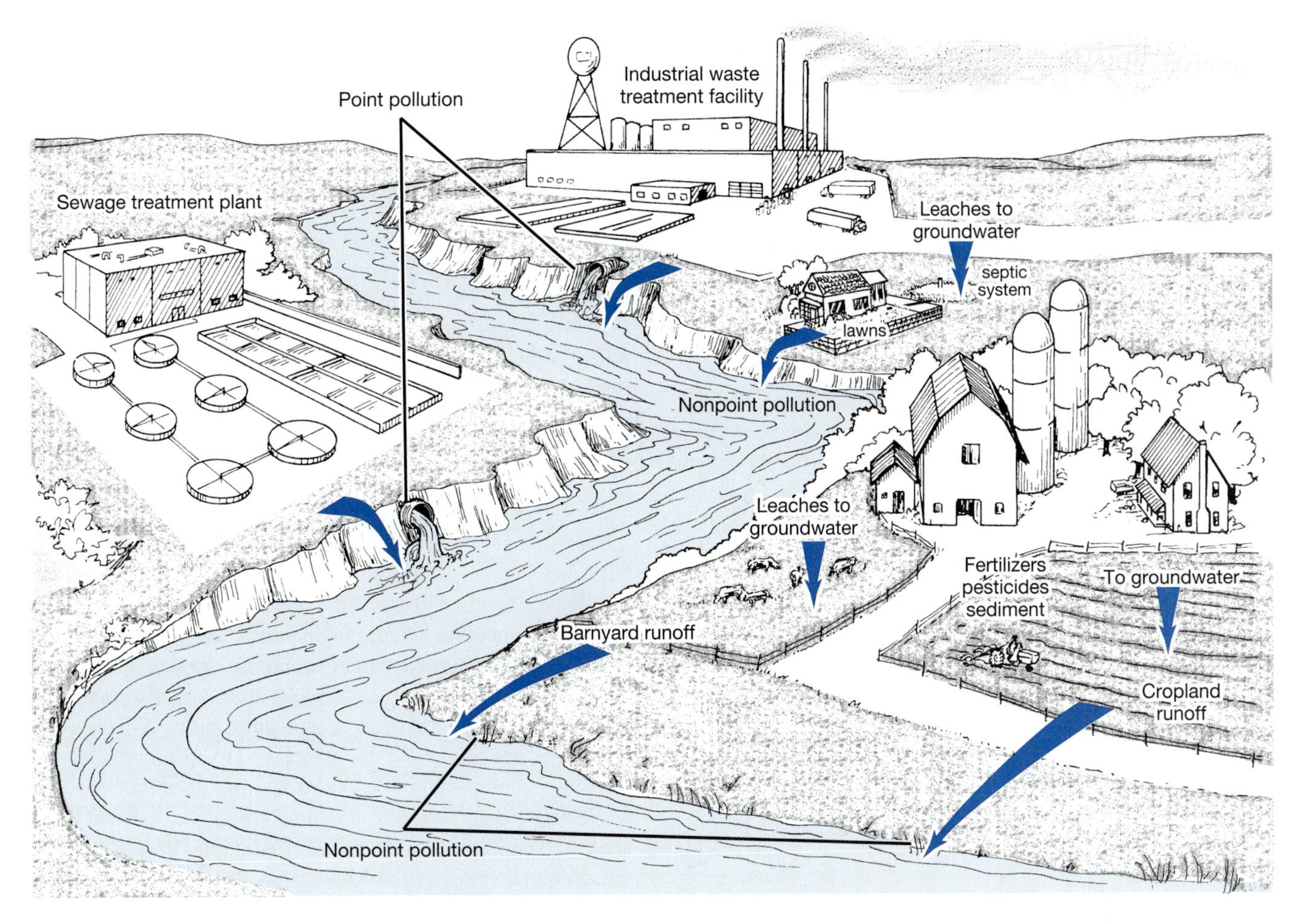

FIGURE 11.3 *Examples of point and nonpoint pollution.*

streams includes nutrients from pet wastes and lawn and garden fertilizers; heavy metals (lead, zinc, copper); bacteria; phosphorus leached from fallen leaves; salts from streets; toxic organic compounds; gasoline from service station spills; oil from roadways and parking lots; and sediment from construction sites for homes, schools, and freeway cloverleafs.

Storm drains empty into streams, rivers, and other surface waters and introduce a wide variety of pollutants from many nonpoint sources. In many cities, sewage from homes and storm runoff flow in the same system of pipes. Under normal conditions, this combined waste is delivered to sewage treatment plants. During heavy rain storms, however, sewage treatment plants may be overwhelmed, and excess drains directly into surface waters.

Pollutants enter waterbodies such as streams, rivers, lakes, bays, and oceans at many locations along their banks or shores from both point and nonpoint sources.

Awareness of the importance of nonpoint source pollution is relatively recent and many countries like the United States and Canada have therefore concentrated most of their efforts on point sources. Bob Adler, attorney for the U.S.-based Natural Resources Defense Council, once commented: "The irony is that Congress has spent $45 billion over the past 15 years to build sewage treatment plants but precious little to control nonpoint pollution—which the EPA and the states agree is half of our water pollution problem." Fortunately, the U.S. **Environmental Protection Agency (EPA)** and similar agencies in other countries have recognized the need to change and have begun to shift priorities to include non-point source water pollution control. Many states have also recognized the need to address nonpoint pollution and have begun to devote more resources to this problem.

Water pollutants can also be classified according to their chemical type. The major categories in this classification system are (1) sediment, (2) inorganic nutrients, (3) thermal pollution, (4) disease-producing microorganisms, (5) toxic organic chemicals, (6) heavy metals, and (7) oxygen-demanding organic wastes. The following section describes these pollutants, their impacts, and ways to control them.

11.2 Major Pollutants and Their Control

Before we examine the various types of water pollutants, we provide a few words on pollution control to help orient you and get you thinking about the most effective and sustainable approaches. As noted earlier, there are two major approaches for dealing with pollution—pollution control and pollution prevention. Under each of these categories there are many options with a range of costs and benefits.

Let's begin with water pollution control—technologies that remove pollutants from waste streams. The wastes, as noted earlier, can be treated to render them less toxic or nontoxic. Or, they can be disposed of in other ways. Pollution control technologies like sewage treatment plants are a form of **output control.** They deal with the pollutant *after* it has been produced.

As its name implies, pollution prevention involves techniques that eliminate the production of harmful substances. For example, a factory might find ways to alter its manufacturing processes so that it no long produces toxic substances. Or, it may find nontoxic substitutes. Or, it may recycle wastes, using them again. In all three instances, it has found ways to prevent pollution from being produced in the first place. Pollution prevention involves **input controls.** This is generally a more desirable approach from a social, economic, and environmental viewpoint than pollution control.

Unfortunately, most efforts in the United States and other countries throughout the world have focused on output controls. While important, output controls are really nothing more than a means of pollution diversion. That is, they capture wastes that were bound for one medium (such as water) and divert them to another. Much of the organic wastes removed from sewage treatment plants is converted to sludge, which is disposed of in some landfills. There, wastes may leach into the groundwater. Input controls are far better and more sustainable, for they eliminate problems in the first place. They're often a lot cheaper, too.

Sediment Pollution

America's aquatic ecosystems are being polluted with 1 billion tons of sediment annually. The Mississippi River alone carries 210 million metric tons of sediment into the Gulf of Mexico each year. To transport just 1 year's load would require a train of boxcars 37,000 miles long—sufficient to circle the Earth one and one-half times at the equator! It is a curious paradox that the very soil that makes the production of life-sustaining food possible suddenly becomes one of our nation's most destructive pollutants when washed into our lakes and streams.

Where Does Sediment Come From?

Sediment comes from natural and human sources or activities. In a well-vegetated, undisturbed watershed, sediment production is minimal. But in disturbed watersheds, erosion and sediment pollution can reach elevated levels. Much sediment comes from fields on which farmers have failed to use appropriate soil conservation measures. Such soil abuse has resulted in record-high concentrations of silt—up to 270,000 parts per million (ppm) in certain Iowa streams! Sediment also comes from construction sites—for homes, shopping centers, roadways, and the like. Because the land is laid bare in construction and not revegetated until the building is complete, erosion rates can be as much as ten times higher than from cropland (Figure 11.4). Although many construction companies practice better erosion control, they're sometimes inadequate. Improper logging activities that remove large blocks of trees from steeply sloping land contribute to sedimentation, especially in the Northwestern United States and southwestern Canada. Strip-mining operations, in which the soil mantle is scooped away by giant machines in order to expose the coal, also result in severe sediment pollution of streams, particularly in Appalachia.

Harmful Effects

Sediment pollution causes about $1 million a day in damage in the United States, according to one estimate. It damages the turbines in hydroelectric plants, clogs irrigation canals, and fills in navigable rivers. Turbine blades must be repaired and drainage ditches must be cleaned. Harbors and river channels are routinely dredged to remove sediment. Soil particles also carry nutrients and toxic chemicals, such as pesticides, which in turn cause additional damage in aquatic ecosystems.

Suspended sediment clouds surface waters blocking sunlight; the loss of sunlight in turn kills algae. Algae form the base for many aquatic food chains. This loss of algae, in turn, reduces the levels of dissolved oxygen in the water. The loss of oxygen harms other aquatic organisms such as fish and bacteria. Bacteria that live in water decompose organic material that enters from natural sources (leaves) and human sources such as canneries, slaughterhouses, and pulp mills. When oxygen levels fall, wastes from these plants tend to accumulate instead of being decomposed by oxygen-requiring bacteria.

Sediment also buries the breeding grounds of fish and buries shellfish habitat. Sediment has, for example, destroyed many valuable clam beds in the Mississippi River. Silt-induced fish mortality is extensive in many American streams (Chapter 12).

Sediment contaminates the nation's public water supplies as well. Over 7,000 billion liters (1,800 billion gallons) of silt-

FIGURE 11.4 *Unprotected soil can be washed away by runoff waters after heavy rainfall, as at this construction site.*

polluted water must be filtered annually so that Americans can have clean drinking water. The life span of thousands of reservoirs throughout the United States and other countries has been shortened as a result of sedimentation (Figure 9.13). Finally, as noted in Chapter 9, sediment fills streams and makes them more prone to flooding.

Control of Sedimentation

Input Control On croplands, sedimentation can be effectively reduced by the proper application of the erosion control strategies described in Chapter 7, such as conservation tillage, strip cropping, contour farming, terracing, gully reclamation, and the removal of marginal land from production. Highly erodible sites should be avoided altogether.

On construction sites, erosion can be reduce or eliminated by a host of measures—for example, if workers can be kept from bulldozing any more soil than is absolutely necessary. Protecting trees is also helpful. Careful site selection can reduce erosion, too—for example, by avoiding hilly, highly erodible terrain. Revegetation is also vital. Denuded areas can be seeded or sodded with grass as soon as possible or covered with straw mulch to prevent erosion. New trees can be planted to replace those lost during development. Water erosion controls like small dams made of straw bales can be placed across drainage ditches to slow the flow of water. In many sites, workers are now applying "sediment fences" made of a fabric that is erected around the perimeter of the site. The material is partially buried and extends above a foot or two above the surface, thus helping stop the flow of water and sediment from the site. Seeding on steep road cuts can be done with a hydroseeder, a machine that blows a slurry of seeds, fertilizer, straw mulch, and water onto the slope (Figure 11.5). Steep slopes can be regraded to reduce their slope and thus reduce runoff and erosion. All of these measures are preventive in nature. Additional measures are discussed at the end of the chapter in the section on watershed management.

Output Control Many output control methods are also possible, but less desirable. Muddy water, for example, can be directed to swamps and marshes, which can serve as natural sediment-removing filters, provided the runoff is not toxic to organisms that live in these habitats. However effective this is at keeping sediment from entering streams, it does not prevent the loss of sediment from the site. Sediment can also overwhelm wetlands and destroy them. Dredging rivers and harbors, mentioned earlier, is another possibility. It is costly and environmentally harmful, for the dredged sediment must be placed elsewhere. Drinking water is also removed by chemical coagulation followed by a rapid sand filtration at municipal water treatment plants. All in all, preventive measures appear far more cost-effective and desirable in the long run because they not only save us headaches and costs, they help protect valuable topsoil and the environment.

Inorganic Nutrient Pollution

All aquatic organisms require carbon, hydrogen, oxygen, nitrogen, phosphorus, sulfur, and a host of other elements to survive. While all of these are important, some of them play a particularly important role in determining growth of individuals and populations. An essential element that plays a key role in growth of individuals and populations is referred to as a **limiting factor.** It exerts its effect because it is typically found in low concentration. In aquatic ecosystems, nitrogen and phosphorus are the most important limiting factors. Both are inorganic nutrients. In small quantity, they're essential; in large quantity they can become pollutants, disrupting aquatic ecosystems. Nitrogen usually is available in the form of nitrate (NO_3) ions or ammonia (NH_3); phosphorus usually is available as phosphate (PO_4) ions. Because phosphorus is less abundant and is required in such minute quantities, it is far more likely to be the limiting nutrient than nitrogen.

FIGURE 11.5 *Worker uses hydroseeder to propel a stream of water, grass seeds, and fertilizer to establish grass cover on steeply sloping land that has been denuded by construction.*

When nitrogen and phosphorus become abundantly available, the productivity of an aquatic ecosystem often increases sharply. With the passage of time, therefore, most bodies of water become increasingly productive; however, as you will soon see, there's a price to pay for this.

Eutrophication The nutrient enrichment of aquatic ecosystems is known as **eutrophication** (you-trow-feh-KAY-shun). This process occurs naturally as a lake or river ages over a period of hundreds or even thousands of years. This is called **natural eutrophication.** The release of excessive amounts of nutrients into aquatic ecosystems as a result of human activities speeds up the process and therefore is called **cultural** or **accelerated eutrophication.** It is a serious pollution problem.

Classifying Lakes Based on Their Productivity Ecologists recognize three major types of lakes: **oligotrophic** (nutrient-poor), **mesotrophic** (middle-nutrient), and **eutrophic** (nutrient-rich). The characteristics of each lake type are summarized in Table 11.1.

The oligotrophic (awl-i-go-TROH-fic) type is represented by Lake Superior, Lake Huron, the Finger Lakes of central New York, and many glacial lakes in northern Minnesota, Wisconsin, and Michigan. Oligotrophic lakes are frequently beautiful, clear-water lakes, skirted by pine and spruce. Their waters are clear because of a scarcity of floating algae (phytoplankton), resulting from the low levels of dissolved nutrients. Food chains are largely based on bottom-dwelling producers (green plants) growing in shallow water near the lake margins. The total biomass per unit volume of water is much lower than in the other two lake types.

Mesotrophic lakes (mez-oh-TROH-fic) are characterized by a moderate amount of nutrients. Fertility, clarity, levels of dissolved oxygen, and total biomass are intermediate between those of oligotrophic and eutrophic lakes. These lakes are good for swimming, boating, and fishing. Fish like northern pike and walleyes may be plentiful in more northern lakes; bass may abound in southern lakes. Most of the lakes in the United States were probably mesotrophic at the turn of the century.

As time passes, the nutrient-enrichment process described earlier continues. As a result, a mesotrophic lake is gradually converted into an extremely fertile eutrophic lake (you-TROH-fic). When the average concentration of soluble inorganic nitrogen exceeds 0.3 ppm and the soluble inorganic phosphorus content exceeds 0.01 ppm, algal populations may "explode," creating an **algal bloom.** Occurring usually during the summer, algal blooms convert once-clear water into "pea soup" and restrict visibility to 0.3 meter (1 foot) (Figure 11.6). The effects of such a bloom are many:

1. It destroys the aesthetics of the lake, rendering it repulsive to swimmers and other sports enthusiasts. Canoe paddles, motorboat propellers, water skis, and fishing lines (as well as human arms doing the crawl stroke) become fouled up in the green slime.
2. The bloom gives water a bad taste and odor. If the lake is a source of drinking water, considerable expense may be involved in improving its quality (Figure 11.7).
3. As a result of wind and wave action, huge masses of algae (and even rooted plants that have been torn loose from the lake bottom) can pile up along the shore and decompose, giving off hydrogen sulfide (H_2S) gas, which smells like rotten eggs. At high concentrations it is toxic.
4. Some of the blue-green algae release chemicals that are poisonous to fish and humans.

TABLE 11.1 *Characteristics of Oligotropic, Mesotrophic, and Eutrophic Lakes*

Oligotrophic Lake	Mesotrophic Lake	Eutrophic Lake
Poor in nutrients	Intermediate	Rich in nutrients
Deep basin	Intermediate	Shallow basin
Gravel or sandy bottom	Intermediate	Muddy bottom
Clear water	Intermediate	Turbid water
Plankton scarce	Intermediate	Plankton abundant
Rooted vegetation scarce	Intermediate	Rooted vegetation abundant

FIGURE 11.6 *This lake is overgrown with algae, the growth of which is stimulated by pollutants.*

Eutrophic lakes have many other undesirable features, such as muddy bottoms (poor spawning surface for most desirable fish) and populations of less desirable fish such as carp. The dense blooms of phytoplankton (algae and other photosynthetic microorganisms) shade out bottom-dwelling plants that normally produce large amounts of oxygen during photosynthesis. Although phytoplankton are also photosynthetic, they usually occupy the upper levels of the lake. Therefore, much of the oxygen they release escapes into the atmosphere. As a result, levels of dissolved oxygen in eutrophic lakes are actually lower than in mesotrophic or oligotrophic lakes, despite the much greater abundance of phytoplankton. Carp are often abundant in eutrophic lakes because they are adapted to live in warm, shallow, turbid (cloudy), oxygen-poor waters. When conditions are bad, these fish often rise to the surface to gulp air.

Cultural Eutrophication Human activities are greatly accelerating eutrophication of lakes the world over. Scientists estimate, for example, that roughly 80 percent of the nitrogen and 75 percent of the phosphorus entering lakes and streams in the United States come from human activities. As a result, the eutrophication of many lakes is proceeding 100 to 1,000 times faster than it would under natural conditions. An estimated 33,000 medium-sized to large lakes, and about 85 percent of the large lakes located in urban areas in the United States are undergoing cultural eutrophication. The effects are summarized in Table 11.2.

FIGURE 11.7 *Biologist at the Amarillo, Texas, water treatment plant obtains water sample to determine algal population. If algal populations are high, water will have a bad taste. Much of Amarillo's drinking water comes from Lake Meredith, 45 miles northeast of town.*

Stream Eutrophication Although eutrophication can occur in streams, the effects are usually not quite so severe as in lakes. Many rivers are naturally oligotrophic at their headwaters, where the waters flow clear and cold. Such is the case of the small northern streams feeding into the Mississippi or the mountain streams of the western portion of the United States. However, by the time waters near the stream's mouth, they have received so many nutrients that they become turbid, warm, and muddy, harboring bullheads and carp rather than trout. Stream eutrophication is more easily reversed than lake eutrophication; because the nutrient input is often stopped, the current eventually purges the stream.

Sources of Nutrients The inorganic nutrients nitrogen and phosphorus come from three major sources: (1) agricultural fertilizers, (2) domestic sewage, and (3) livestock wastes (Figure 11.8).

Farmers fertilize their fields with commercial fertilizers (artificial fertilizers) and animals wastes. Fertilizers promote crop production because they are rich in nitrates and phosphates, two very important plant nutrients (Chapter 5). Regrettably, however, as farmers the world over strive to boost crop production with fertilizer to feed a rapidly growing human population, they inadvertently are promoting a population explosion of aquatic plants. The amount of agricultural fertilizer used in the United States has increased from less than 5 million metric tons in 1950 to 25 million metric tons in the 1990s, a fivefold increase. Much of the fertilizer, which is not absorbed by crop roots, is washed by runoff waters into lakes and streams. Scientists estimate that over 0.45 billion kilograms (1 billion pounds) of agriculture-generated phosphorus enters America's aquatic ecosystems yearly.

TABLE 11.2 *Adverse Effects of Nutrient Pollution on a Lake*

Lake aesthetics are destroyed.
The recreational values of a lake are destroyed.
Water quality is impaired by foul tastes and odors.
Gases that emanate from rotting algae have foul odors, tarnish silverware, and discolor painted houses.
Toxins given off by algae result in gastric disturbances if ingested.
Dense algal blooms at the surface reduce penetration of sunlight to the lake bottom.
Decomposing algae at the lake bottom reduce dissolved oxygen levels.
Pollution contributes to the winterkill of fish in northern lakes.
Rooted weeds interfere with navigation and recreation.
Game fish are replaced with less desirable fish.
Lake basin is gradually filled in, and the lake becomes extinct.

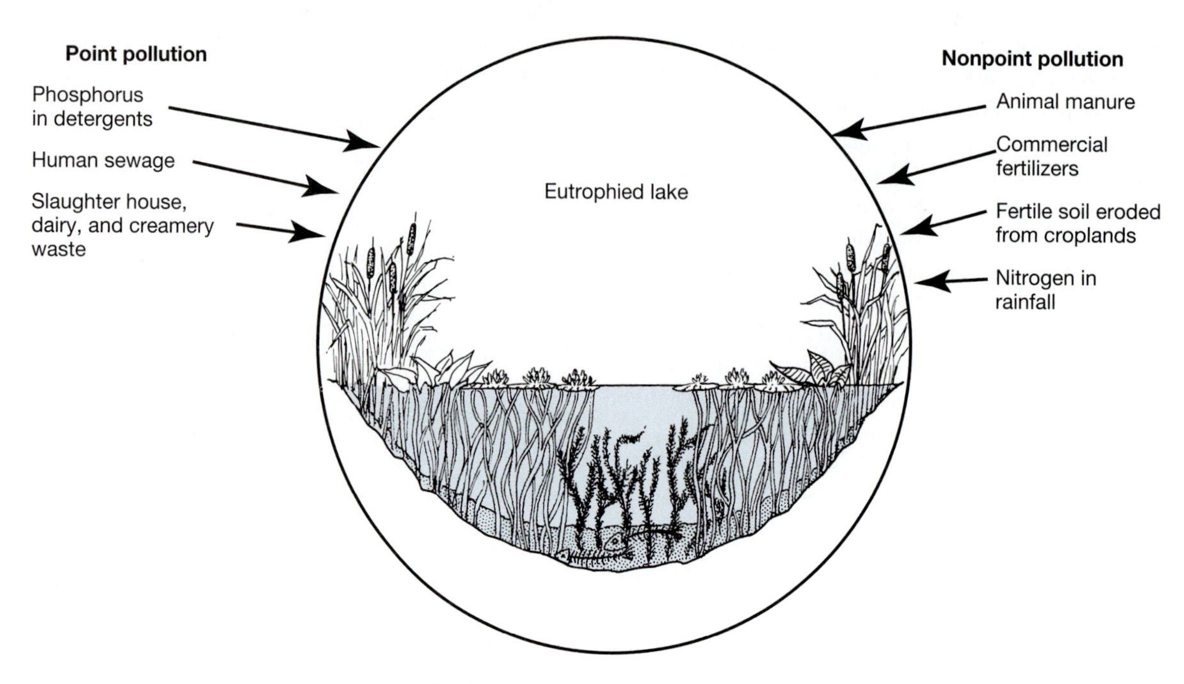

FIGURE 11.8 *Sources of nutrients causing eutrophication.*

Domestic sewage containing human wastes and household detergents contributes millions of kilograms of phosphorus to aquatic ecosystems yearly. Conventional sewage treatment plants are ineffective at removing nitrogen and phosphorus and are a major cause of cultural eutrophication. (It's only the more sophisticated sewage treatment plants that effectively remove these pollutants.)

The waste deposited on city lots, sidewalks, and streets by the United States' more than 110 million pet dogs and cats contributes significantly to eutrophication. In New York City alone, 500,000 dogs produce 18,000 metric tons (20,000 tons) of feces and 4 million liters (1 million gallons) of urine annually. It is not surprising, therefore, that urban runoff may carry up to 5 ppm of both nitrogen and phosphorus.

Each of the more than 100 million cattle in the United States produces ten times as much waste per day as a human being. In other words, our country's cattle alone produce the waste equivalent of more than 1 billion people's waste—nearly four times the U.S. population. That does not include the waste produced by other types of livestock, such as hogs, horses, sheep, pigs, goats, chickens, ducks, and turkeys. If all this animal waste were washed into lakes and streams, its eutrophication potential would be enormous. Hogs have become especially troublesome in the southeastern United States where the rapid increase in hog farming, more frequent hurricanes, and improper siting of facilities (they're often built in floodplains) result in flooding of facilities and detention ponds. Hurricane Denis in 1999 caused hundreds of detention ponds to break spilling millions of gallons of hog waste into flood waterways.

During the winter, it has long been the farmer's custom in the northern states to spread animal manure on the frozen ground. When spring comes, of course, some of the nitrogen and phosphorus is absorbed by crop root systems. Unfortunately, however, almost half of this manure may be washed by spring runoff into aquatic ecosystems. Because of this, the state of Vermont has banned this practice.

The problem has been accentuated by the recent practice of crowding livestock into feedlots, where food is brought to the animals, instead of permitting them to forage for their own food in the open pasture. A large portion of our nation's cattle are maintained in feedlots. One large lot, accommodating 10,000 cattle, yields 180 metric tons (200 tons) of manure *daily*. The total amount of waste generated by our nation's feedlots comes to 720 million metric tons (800 million tons) annually. Without proper control measures, much of this manure ends up in waterways.

So far, most of the discussion of inorganic nutrient pollution has focused on surface waters. Groundwater can also be contaminated by these nutrients. Septic tanks, discussed shortly, are household waste disposal systems used in many rural areas. They contribute to nitrate and phosphate contamination in groundwater, especially in areas with porous layers over aquifers and in regions with high water tables. Agricultural fertilizer can also contaminate groundwater as can animal waste.

Control of Eutrophication Eutrophication and its impacts, like excessive plant growth in aquatic ecosystems, may be controlled by both input and output methods. Consider some output methods first.

1. Output Measures and Controls
    A. Upgrading many wastewater treatment plants to the tertiary (advanced) level. These facilities would remove a higher percentage of the phosphorus and nitrogen. Wastes from facilities could be used to

fertilize rangeland, pastures, and farm fields, thus recycling nutrients back to the soil they came from.

B. Using detention basins on feedlots to collect animal wastes that are washed by runoff water into lakes and streams (Figure 11.9).

C. Employing weed-cutting machines to remove excess vegetation from aquatic ecosystems. Unfortunately, this would have to be done several times during the year and is rather costly.

D. Destroying plant growth with herbicides. Great care must be taken to prevent fish kills and the destruction of spawning beds.

E. Dredging the bottom sediment from lakes to remove the nutrients. This is rather costly. It would be impractical in deep lakes.

F. Avoid dumping treated sewage in lakes susceptible to eutrophication. At Lake Tahoe, Nevada and in Cambridge, Wisconsin officials successfully bypassed the sensitive lakes by pumping treated wastewater to less vulnerable waters downstream.

G. Applying treated wastewater (still rich in plant nutrients) to golf courses and other grassy areas.

H. Treating wastewater biologically in artificially constructed wetlands where aquatic vegetation and microorganisms utilize the nutrients and purifys the water.

I. Improve the activated sludge process to enhance bacterial absorption of phosphates.

2. Input Controls in Urban Areas

A. Banning the use of phosphate detergents or reducing phosphate levels in them. Bans have been imposed by several cities (Akron, Chicago, Miami, and Syracuse) and states (Indiana, Michigan, Minnesota, Wisconsin, New York, Maryland, and Vermont). Several states have limited the phosphorus content in detergents to less than 9 percent. Detergent manufacturers have subsequently lowered the phosphate content of their products, and many low- and no-phosphate detergents are now available in grocery stores in the United States and abroad.

These steps have resulted in immediate changes. Lake Onandaga in New York is an example. For many years it served as a liquid container for the sewage of Syracuse. The once beautiful lake gradually took on the pea-green color of eutrophication. The outraged city officials passed an ordinance banning the use of phosphorus-containing detergents. In only 1.5 years, the phosphorus level in the lake was reduced by 57 percent, the frequency of algal blooms decreased, and dissolved oxygen levels suitable for game fish were restored.

B. Using newly developed phosphate-free natural detergents. Perhaps the ultimate solution to the phosphate detergent problem is the use of phosphate-free natural soaps and detergents developed by the USDA. Because they are made from beef fats, they are to some degree similar to the old-fashioned soaps used before World War II. They perform as well as or better than phosphorus detergents in hot or cold, soft or hard water. Moreover, they are relatively inexpensive, are biodegradable, and, of course, do not cause eutrophication. The natural detergents have already seen wide acceptance in Japan. For some reason, however, American manufacturers were slow to produce a product that appeared able to solve the detergent-caused eutrophication problem once and for all.

FIGURE 11.9 *Control of feedlot runoff at Boys Town, Nebraska. This basin "catches" the manure and urine that would ordinarily be washed into a stream or lake after a rainfall. Catch basins, however, can release pollutants into groundwater.*

C. Imposing an excise tax on lawn and garden fertilizers to reduce sharply the volume of use.
D. Educating citizens to use less detergent and fertilizer.

3. Input Controls in Rural Areas
   A. Minimizing the use of fertilizers on cropland. Farmers, like homeowners, typically apply more fertilizer than is needed.
   B. Injecting liquid fertilizer directly into the soil rather than spreading it on the surface to be carried away by surface runoff.
   C. Postponing application of manure in northern states until after the spring melt of ice and snow.
   D. Planting vegetative barriers between fields and waterways to trap pollutants.
   E. Implementing soil erosion controls on farms (terracing, contour farming, etc.) to minimize surface runoff.
   F. Building hog farms and other areas of concentrated livestock production out of floodplains.

Take a few moments to compare the strategies and think about their social, economic, and environmental costs and benefits.

Thermal Pollution

Biscayne Bay, Florida, is an unusually productive ecosystem that supports many species of aquatic organisms, including lobsters, crabs, fish, and wading birds. More than 270,000 kilograms (600,000 pounds) of seafood is harvested annually from the bay. In the early 1970s, however, scientists discovered a 30-hectare (75-acre) region that was virtually lifeless—a biological desert caused by the discharge of heated water from a Florida Power and Light Company's power plant.

Thermal pollution is an increase in the temperature of water that adversely affects organisms that live there. Although thermal pollution may result from both natural causes (excessive heating by the summer sun) and human actions, the latter are far more significant. Many industries take water from a lake or stream to cool equipment or products. The electrical power, steel, and chemical industries are the most important users of cooling water. Electrical generating plants in the United States withdraw and use 733 billion liters (190 billion gallons) per day, nearly half of the total water withdrawal.

Remember the second law of energy? It states that whenever energy is converted from one form to another, a certain amount is lost as heat. There are several energy conversions involved in a power plant fired by fossil fuel. The chemical energy in coal or oil, or the nuclear energy in radioactive fuels, is converted into heat that generates steam to turn the blades of the turbine. That mechanical energy is then converted into electrical energy by the generator. In order to condense the steam back to water to return it to the boiler, the steam is passed over coils that carry cold water drawn from a stream or lake (Figure 11.10). As a result, the steam is condensed and heat is transferred to the cooling water. The temperature of the cooling water may be raised by 20°F. This warmed-up cooling

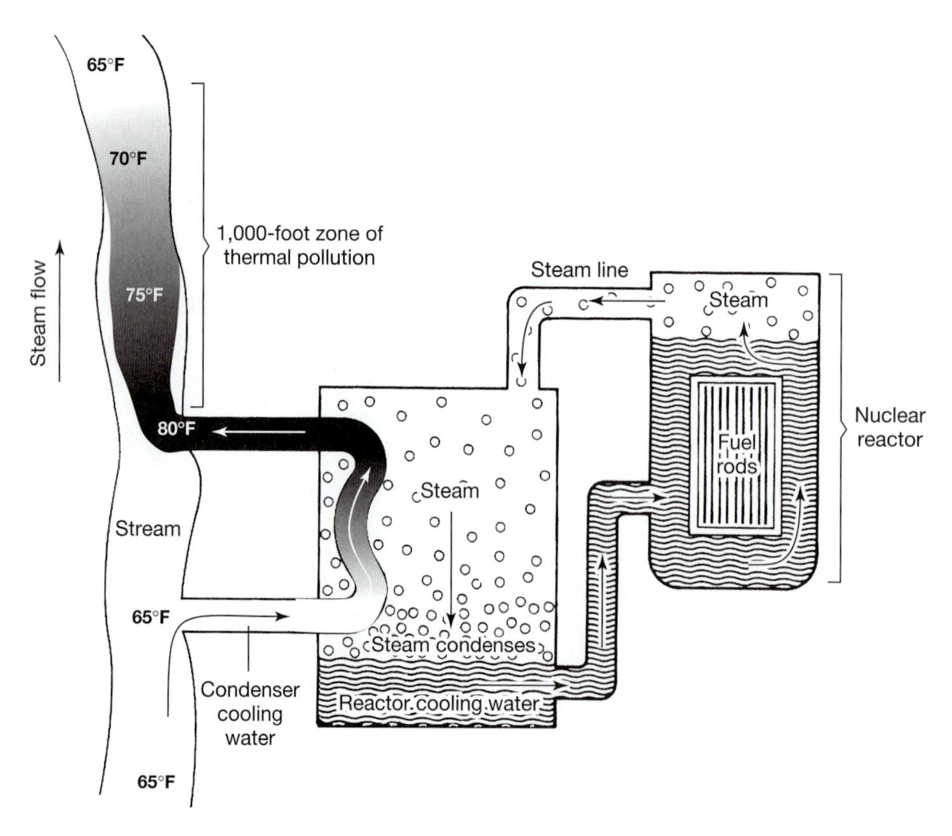

FIGURE 11.10 *Nuclear reactor cooling system. Note the 1,000-foot zone of thermal pollution caused by the discharge of cooling water.*

water may then be discharged into a lake or stream. The warmed-up area that results is called a **thermal plume.** Such plumes may extend 1 kilometer (3,280 feet) or more from the point of discharge (Figure 11.11).

The Biological Effects of Thermal Pollution

Thermal pollution has many adverse effects on aquatic ecosystems. Let us consider some of them.

Reduction in Dissolved Oxygen When water is warmed, its capacity to dissolve oxygen falls. For example, at 0°C (32°F) water that has been thoroughly mixed with oxygen has an oxygen content of 14.6 ppm, but at 40°C (104°F) it contains only 6.6 ppm. Unfortunately, as water temperature increases, oxygen requirements for fish increase. Thus, rising temperature decreases oxygen concentration in waters and increases a fishes demand for oxygen.

Fish vary in their need for oxygen. The carp, for example, requires water concentrations of only 0.5 ppm of oxygen at 0.6°C (33°F). Cold-water fish such as trout and salmon, which need about 6 ppm to survive, cannot tolerate the warm water that prevails in the thermal plume. If they remain in the plume, they will die from oxygen starvation.

Interference With Reproduction Fish are cold-blooded animals whose body temperatures and activity levels vary with the temperature of the external environment. Most fish are extremely sensitive to slight changes in water temperature. Many kinds of fish are instinctively "tuned" to certain thermal signals that trigger such activities as nest building, spawning, and migration. Changes in temperature can disrupt these activities and may kill eggs.

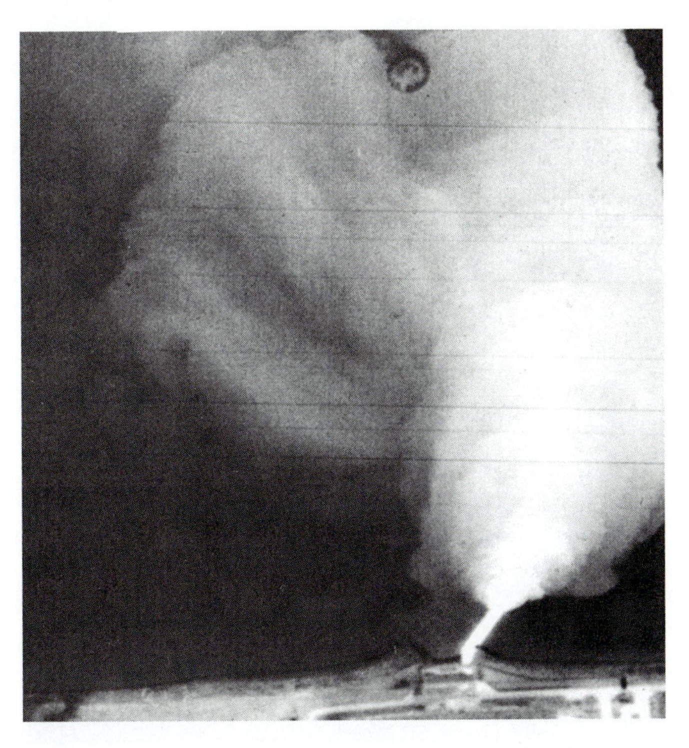

FIGURE 11.11 *Aerial photograph of thermal plume in Lake Michigan caused by the discharge of warmed-up cooling water from the Point Beach nuclear plant on the Wisconsin shore. The photograph was taken with infrared film, which is sensitive to heat.*

Increased Vulnerability to Disease Another problem with thermal pollution is that it may increase the vulnerability of fish and other aquatic organisms such as turtles to disease organisms. The ability of some bacteria such as *Chondrococcus* to penetrate the body of a fish is poor at 16°C (60°F), but it gradually increases with rising water temperatures. *Chondrococcus* is believed to be responsible for the massive kills of blueback salmon in the Columbia River in 1946. Thermal pollution of the Columbia was undoubtedly a contributing factor.

Direct Mortality The body temperature of a fish is determined by the temperature of the water in which it lives. A lake trout will perish if the water temperature is much higher than 10°C (50°F). Because the discharge of cooling water may raise stream or lake temperatures 11°C (20°F) above normal, cold-water fish caught in the thermal plume die from shock. Such species would have to emigrate to survive.

Invasion of Destructive Organisms Thermal pollution may permit the invasion of organisms that tolerate warm water and are highly destructive. A good example is the invasion of shipworms (highly specialized relatives of clams and oysters) into New Jersey's Oyster Creek where thermal discharges from the New Jersey Central Power and Light Company's power plant gradually warmed up the creek. One result was the shipworm invasion. They burrowed into wooden docks and the hulls of ships, wreaking considerable damage.

Undesirable Changes in Algal Populations Three major groups of algae live in freshwater ecosystems: diatoms, green algae, and cyanobacteria (once called blue-green algae) (Figure 11.12). Each has a distinct range of tolerance for water temperature. Diatoms prefer the coolest water (14°C or 58°F), green algae like it warmer (32°C or 90°F), and cyanobacteria like it even hotter, 40°C (104°F). The most valuable algae, as far as fish and human food chains are concerned, are the diatoms. Those that prefer warmest water—the cyanobacteria—are the least desirable as aquatic animal food. Many are too large to be eaten by small crustaceans such as water fleas. Not only that, cyanobacteria give off toxic substances and cause the multiple problems already discussed in the section on inorganic nutrient pollution. Along with nutrients, elevated water temperature is an important factor in promoting cyanobacterial blooms. It is easy to see how aquatic food chains could be disrupted by discharges of heated water effluents. As Richard Wagner, environmentalist at the University of Pennsylvania, states: "A water flea, for example, which might be able to tolerate the thermal extreme of 95°F, would probably starve to death if the diatoms on which it fed were unable to survive at that temperature. In turn, fish feeding on water fleas would be similarly hard pressed to survive, regardless of their tolerance or adaptability to high temperature."

Destruction of Organisms in Cooling Water The volume of water removed from a stream for cooling is enormous, sometimes involving a substantial part of a stream's total flow. Unfortunately, many of the plankton, insect larvae, and small fish that are sucked into the condenser along with the cooling water are

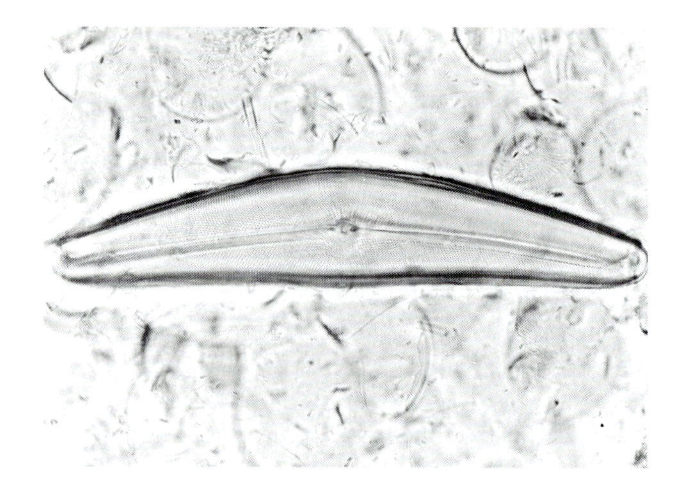

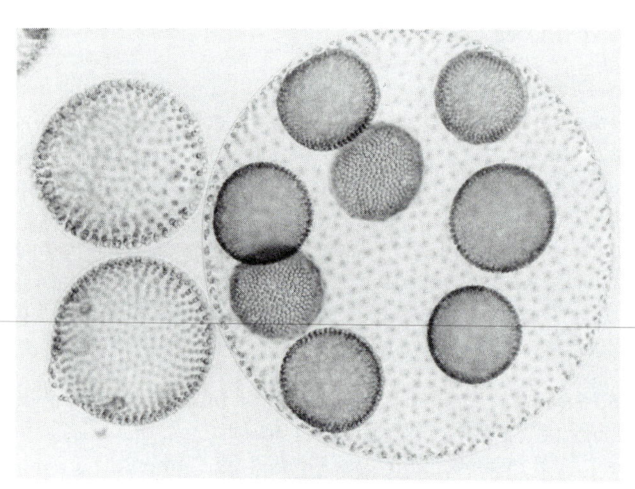

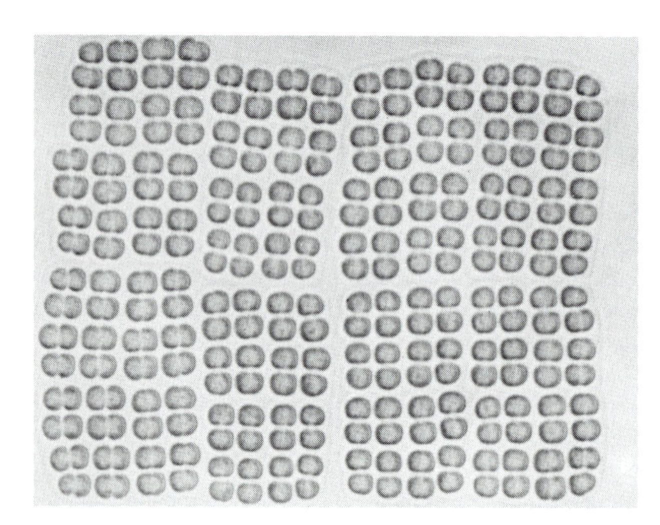

FIGURE 11.12 *Aquatic microorganisms. (A) Photomicrograph of a diatom. This alga prefers a water temperature of about 148°C (588°F). It is an important producer link in fish food chains. (B) Photomicrograph of a green alga. This is Volvox. It prefers water temperatures of about 328°C (908°F). (C) Photomicrograph of a blue-green alga. This colonial form prefers water temperatures of about 408°C (1048°F). An abundance of blue-green algae may be a sign of thermal pollution. In general, blue-green algae are not important producer links in fish food chains. Moreover, some species may be toxic to fish and other aquatic animals, as well as to humans.*

destroyed by thermal shock, as well as by water velocity and pressure.

Controlling Thermal Pollution After much prodding by state and federal environmental protection agencies, the power industry has tried to control the thermal pollution problem with **cooling towers.** These towers transfer heat from the water to the atmosphere. Cooling towers on some power plants, especially nuclear power plants, are mammoth structures about 30 stories tall (Figure 11.13). They are large enough at the base to cover a football field. In the **wet cooling tower** illustrated in Figure 11.14, the heated water is piped to the top of the tower and then flows downward over a series of baffle plates. During this time, about 2.5 percent of the water evaporates, a process facilitated by the upward flow of fan-propelled air. Smaller evaporative towers are used in most coal-fired power plants. The coolant water is then either (1) discharged into the stream, lake, or ocean from which it was drawn, (2) directed into a lagoon for further cooling,or (3) cycled back to the plant's condenser.

Although cooling towers reduce the thermal pollution, they create some problems. First, the towers generate a considerable amount of fog on cold winter days in colder regions. The fog may come into contact with solid surfaces like roads, forming a thin layer of ice and greatly increasing the likelihood of traffic accidents. Fog may also obstruct visibility. Second, the water that evaporates from cooling towers is lost to the aquatic ecosystems (rivers and lakes) from which it came. Third, toxic chemicals such as chlorine are used to prevent bacteria and other organisms from growing in the pipes associated with cooling towers. These chemicals can enter waterways, killing other organisms. Fourth, although the tall towers are remarkable engineering accomplish-

FIGURE 11.13 *Wet cooling towers in operation. Note water vapor being discharged into the air by one of the towers.*

ments, they are 121-meter (400-foot)-tall masses of concrete and steel that dominate the skyline, becoming eyesores. Fifth, the towers are very expensive. Such an expenditure, of course, is eventually passed along from the utility to the consumer in the form of increased electrical rates, adding perhaps 1 percent to a customer's annual bills. This cost, say environmentalists, seems reasonable when weighed against the damages caused by thermal pollution.

Beneficial Effects of Heated Water Artificially heated water is usually harmful to aquatic ecosystems. Nevertheless, under certain circumstances, it can be put to beneficial use. Let's consider some examples.

1. In Eugene, Oregon, heated water from a paper mill is sprayed onto fruit trees to prevent frost damage.
2. In Vineyard Haven, Massachusetts, heated cooling water is used by a hatchery to accelerate the growth of lobsters. The time required to produce a marketable lobster has been reduced from 8 years to only 2 years.
3. Researchers in Georgia have found that thermal pollution of the Savannah River resulted in an increased rate of growth in black bass. The heated water attracts fish to the area, resulting in record catches for anglers.

Some scientists and engineers have suggested that warm water from various factories and power plants be combined with the effluent of domestic sewage treatment plants and emptied into specially built ponds that are stocked with fast-growing fishes that are accustomed to warm waters—for example, the Asiatic milkfish. Thus, both the nutrient and thermal pollutants could be put to constructive use: producing food. During the winter in the northern states, the warm water would prevent ice from forming and permit the milkfish to feed and grow year' round. Furthermore, because the fish are plant eaters, not only would the nutrients be efficiently converted into fillets, but the weed problem commonly associated with eutrophic ponds would be under control.

Warm coolant water could also be used to heat homes. Because pipes required for conducting the heated water are very expensive, however, the power plants and the homes that are to use the water must be designed with this heating scheme in mind.

Disease-Producing Organisms: Causes, Impacts, and Controls

Water that is contaminated with infectious microorganisms is responsible for more cases of human illness worldwide than any other environmental factor. Diseases such as cholera, typhoid fever, dysentery, polio, and infectious hepatitis are all caused by microorganisms that are transmitted by water polluted with human or animal wastes. Humans contract the disease when they drink, swim in, or bathe in contaminated water. Fortunately, the death rate from these diseases has dropped dramatically in the last century in many countries. For example, in the 1880s the death rate from typhoid was 75 to 100 per 100,000 people per year. Today it is a mere 1 per 1 million people. Nevertheless, people are not exactly germ free. According to a report by the Natural Resources Defense Council, the bacteria, viruses, and protozoa that occur in U.S. drinking water cause 900,000 Americans to become ill every year. Even worse, these contaminants kill about 900 people annually, usually the very young, the elderly, and those people already weakened from diseases such as AIDS, cancer, pneumonia, or surgery. In 1993, a protozoan parasite, *Cryptosporidium*, causes an epidemic of sickness with flulike symptoms in Milwaukee. More than 312,000 residents became ill and eight died. Workers at Milwaukee's water purification plant found large numbers of the parasite in water that had already been treated—water that was being pumped to people's homes. *Cryptosporidium* comes from the intestines of land-dwelling animals and had been washed into the water source by heavy spring rains that year—an excellent example of nonpoint water pollution. Normally, a sophisticated filtration process would have removed all of the organisms, but the plant's filter system was not operating at peak efficiency for a few days, which permitted them to invade the water system.

Another disease organism that has been causing problems in recent years in parts of the United States is a dinoflagelate known as *Pfiesteria* (pronounced fee-STEER_ee-uh). One species has been associated with fish lesions (open sores) and massive fish kills in the coastal waters of the eastern United States ranging from Delaware to North Carolina (Figure 11.15). This microscopic

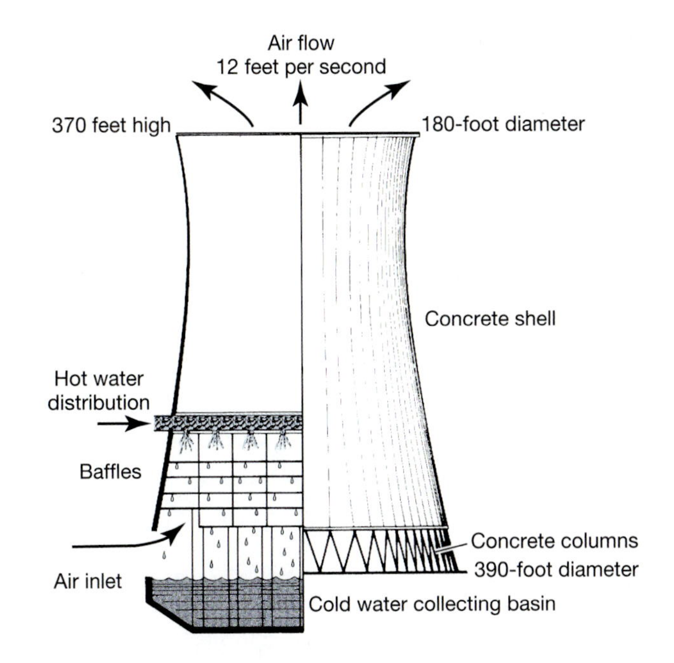

FIGURE 11.14 *Operation of a wet cooling tower. Heat from the cooling water is removed by evaporation on direct contact with air rising up through the hollow concrete shell. Its large size is necessary to provide sufficient surface area and draft to cool thousands of gallons of water each day. The distinctive shape channels the air flow and provides great structural strength with a minimum of material.*

single-celled organism is often classified as a type of algae and the majority of them are not toxic. Found in tidal rivers and estuaries, this organism feeds on algae and bacteria. Interestingly, when in the presence of fish, and in particular schooling fish such as Atlantic menhaden, Pfiesteria changes form and begins to release a powerful toxic chemical that stuns the fish, causing them to become sluggish. Other toxicants produced by the organism are believed to cause the fishes skin to break down, resulting in open sores or lesions. Pfiesteria then begins to feed on tissues and blood of the fish. Pfiesteria are not an infectious agent, like other bacteria and viruses. They are not transmitted from one fish to another and do not kill fish themselves. Rather, scientists believe that it is the toxic chemicals they produce or the open sores, which become infected by other organisms, that are fatal to the fish.

Pfiesteria is responsible for many fish kills, some very massive ones in tributaries of Maryland's Chesapeake Bay and in the brackish coastal waters of North Carolina, the latter of which killed millions of fish. There's also some evidence of human health damage as a result of exposure to the organism. Studies suggest that memory loss, confusion, a variety of other respiratory, skin, and gastro-intestinal problems may result from exposure to waters containing toxic forms of Pfiesteria. To date, there's no evidence that eating fish or shellfish exposed to this organism causes any health problems in people.

Pfiesteria outbreaks occur in warm, brackish, poorly flushed waters when schools of fish are present. Although research continues on this mysterious organism, there is growing evidence to suggest that high nutrient levels may also play a contributing role. Elevated levels of nitrogen and phosphorus, in particular, may cause outbreaks, either by stimulating growth of algae upon which Pfiesteria feeds or stimulating growth more directly. As noted earlier, these pollutants, which are common in coastal waters, come from a variety of sources, including sewage treatment plants, septic tanks, suburban runoff, and farms.

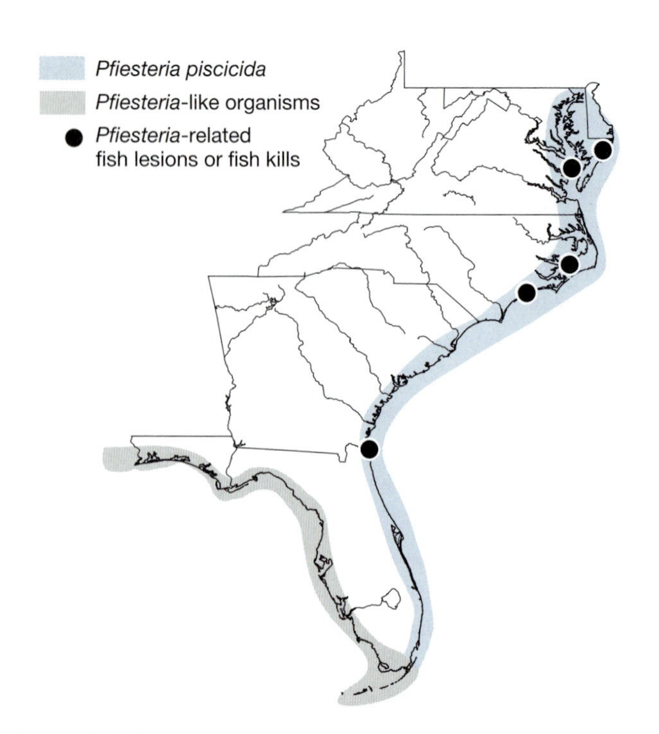

FIGURE 11.15 *Map of Pfiesteria-contaminated waters.*

Protecting Ourselves from Disease-Producing Organisms in our Water Supplies

Protecting ourselves from disease-producing organisms requires a variety of measures. Good sewage treatment is essential. As discussed later in the chapter, sewage treatment is designed to remove pollutants and kills potentially infectious agents contained in wastewater, which is often dumped into drinking water supplies. Water purification is also essential.

If you live in a densely populated region, the last glass of water you drank may already have passed through the bodies of eight other people living upstream from you. Of course, it has been cleansed over and over again after each use. In many countries, the use and reuse of water from a single river or stream by communities along the course of the waterway is very intensive and will become more so as the world's population grows. Safe reuse will depend on extremely effective fail-safe water-treatment methods to reduce harmful microorganisms to very low levels. Without such methods, the incidence of infectious waterborne diseases will very likely increase in years to come.

Drinking water is filtered at water treatment plants, often through elaborate sand filters. It is then chlorinated to kill potentially harmful microorganisms. Water suppliers, both private and public, take frequent samples of drinking water they distribute to ensure that disease-causing bacteria are held to an absolute minimum, as required by the Safe Drinking Water Act (1974), which is discussed shortly. Because the disease-causing bacteria or **pathogens** are so numerous, it is physically impossible to test for all of them. Instead, the law requires sampling of **coliform bacteria.**

Coliform bacteria, an extremely common form, are found in the intestines of humans and other warm-blooded animals. Domestic sewage contains 2 to 3 million coliform bacteria per 100 ml due to the high coliform content of human feces. Waterborne disease organisms also multiply in the intestines of infected individuals and can be spread to others through sewage when it is released into surface waters. Coliform bacteria become an indicator of levels of potentially harmful bacteria and viruses. A low coliform count suggests a low level of harmful bacteria. A high coliform count suggests the possibility of the presence of a large number of pathogenic bacteria. If water has more than two coliform bacteria per 100 milliliters, it is considered unsafe to drink. The city's water treatment plant then has to increase chlorination to destroy the bacteria. An alternative option might be to shift to another available water source, such as a well or lake. When the coliform count of a lake or stream exceeds

200 per 100 milliliters, it is considered unsafe for swimming and the beaches are closed (Table 11.3). For a discussion of the effects of another biological pollutant, see Case Study 11.1, on zebra mussels.

Studies suggest that as important as monitoring coliform bacteria is to protecting public health, it is not fail proof. Swimmers, in fact, have developed gastrointestinal disorders and flulike symptoms after swimming in waters deemed safe by virtue of low coliform levels. Further investigation suggested that the culprit was a group of noncoliform bacteria, known as fecal streptococcal bacteria. As a result, the U.S. EPA is suggesting that waters be monitored for coliform bacteria and fecal streptococcal bacteria to ensure safety.

Water Pollution by Toxic Organic Compounds

Chemical substances are divided into two groups, inorganic and organic. The inorganic are those that generally contain no carbon atoms. They are usually small compounds like sodium chloride, water, and oxygen. Nitrates and phosphates, discussed earlier, are two inorganic compounds. In contrast, organic compounds are compounds made primarily of carbon and hydrogen. Over a million organic compounds are known to science. Organic compounds such as protein, fats, and carbohydrates are important components of living organisms. Our interest here, however, is with a group of toxic organic compounds that are synthesized in chemical factories and often released into groundwater and surface water.

During this century, hundreds of thousands of such synthetic organic compounds have been produced. They form the basis of drugs, plastics, solvents, pesticides, and synthetic fibers. Unlike naturally occurring organic compounds, many of the synthetic organics resist decomposition by bacteria, sun, air, or water. As a result, once they are discharged into a body of water, they may persist as pollutants for several years, sometimes decades. Good examples of such persistent organics are the pesticide DDT and the industrial chemicals called polychlorinated biphenyls (PCBs).

The toxicity of synthetic organic compounds is largely a result of their ability to disrupt normal enzyme function in living cells. Enzymes are proteins in the cells of organisms that speed up the rate of important chemical reactions.

For years, toxic organic compounds have been released by factories directly into streams and other surface waters. Factories have also poured their toxic organic wastes into surface impoundments, ponds that are designed to let the organics evaporate. These substances then entered the atmosphere, only to rain down on the land downwind from the site. Some waste, of course, percolated into the ground, contaminating groundwater. Some companies have even injected toxic organics and other wastes, such as arsenic compounds, cyanides, and radioactive materials, all potentially harmful to humans, into deep wells in the ground that range from 90 meters (300 feet) to over 3.3 kilometers (2 miles) deep. For example, in order to dispose of the strong acid wastes that resulted from a steel-cleaning process, one company drilled a well 1.3 kilometers (0.80 mile) deep into a 540-meter thick(1,800-foot thick) layer of porous sandstone. In theory, the wastes would stay put because the waste-holding sandstone was completely walled off by impervious rock. However, if there were earthquakes, the wastes might be released from their sandstone "prison," move laterally, and eventually contaminate an aquifer used by a community as a source of drinking water.

Deep well injection is surprisingly common in some parts of the United States—and becoming more common as time passes. Most deep injection wells exist in the states of Texas, Oklahoma, Arkansas, Louisiana, and New Mexico. Unfortunately, pollution dispersal from these wells is not actively monitored.

There is good evidence to suggest that the deep well injection technique may actually cause earthquakes. Scientists have demonstrated a well-defined correlation between the volume

TABLE 11.3 *Diseases Transmitted by Water Contaminated With Fecal Matter*

Disease	*Organism*	*Symptoms*	*Comments*
Typhoid fever	Bacterium	Vomiting, diarrhea, fever, intestinal ulcers reddish spots on skin; may be fatal	500 cases in the United States annually
Cholera	Bacterium	Vomiting, diarrhea, water dehydration; may be fatal	Rare in the United States
Traveler's diarrhea	Amoeba	Diarrhea, vomiting	Not uncommon
Amoebic dysentery	Amoeba	Diarrhea, chills, fever, abdominal pain; death may occur	May be contracted by eating infected oysters and clams
Infectious hepatitis	Virus	Headache, fever, loss of appetite, enlarged liver	May be contracted by eating infected oysters and clams
Polio	Virus	Headache, fever, sore throat, weakness, paralysis; may be fatal	Rare in the United States since the use of the polio vaccine

CASE STUDY 11.1 THE ZEBRA MUSSEL: A WATER CONTAMINANT FROM EUROPE

The waters of North America are being invaded by a tiny but dangerous creature. It is not a fish or eel parasite, but rather a small mollusk, known as the zebra mussel, so named because of the striped pattern on many of their shells (Figure 1). Arriving in North America in 1988, as best as scientists can tell, the zebra mussel first appeared in Lake St. Clair, a small body of water connecting Lake Erie and Lake Huron (Figure 2). By 2000, this tiny mussel had spread widely throughout US and Canadian waters, and continues to move. If precautions aren't taken, it could spread to other waterways, even rivers and streams not connected to its present watershed. Studies have shown that the mussel attaches to boats and can live out of water for several days, so a recreationist boating in Missouri returning to California may be inadvertently helping the mussel spread even further than might otherwise be possible.

Zebra mussels are freshwater mollusks about the size of man's fingernail, but that can grow to a maximum length of about 5 cm (2 inches). They live four or five years. Females start reproducing in their second year and can produce up to a million eggs in a single spawning season. The fertilized eggs develop into free-swimming larvae, which are carried by currents from their origin to other sites, where they soon find a point of attachment, usually a solid substrate like a rock on the bottom of lakes and rivers. As adults, the zebra mussel attaches to rocks, piers, docks, boats, and water inlets.

Because this tiny creature is sedentary as an adult, its rapid spread seems amazing. Although the microscopic larvae are capable of swimming away from their parents, their dispersal is greatly facilitated by waves and water currents. The larvae may also be spread inadvertently by boaters and anglers, while they are in bait buckets, live wells (water containers for holding bait), or the cooling passages of outboard motors. Another method of dispersal is transport by air: the tiny larvae ride in small droplets of water on the feathers of migrating waterbirds, such as ducks, geese, herons, and terns. Even though

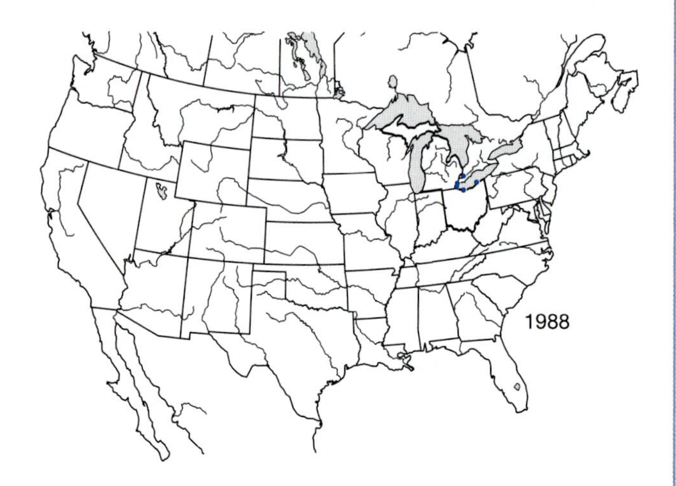

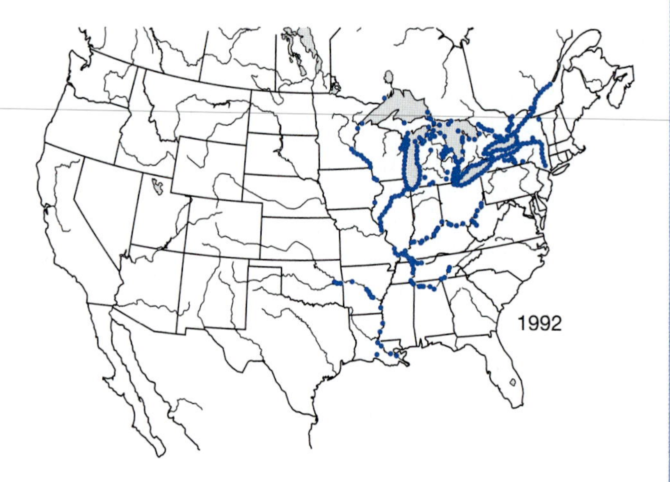

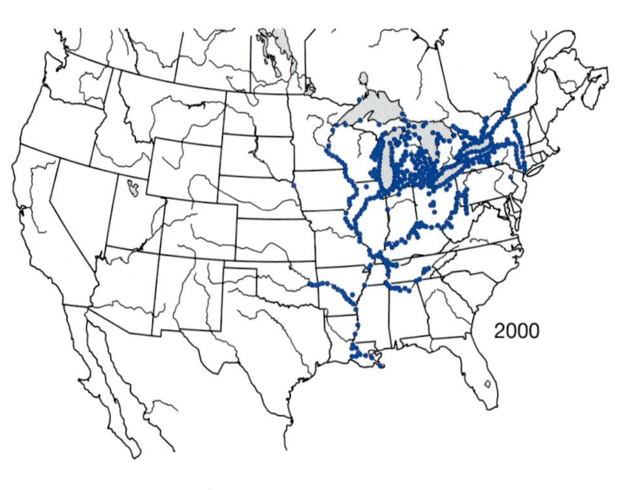

FIGURE 2 *Maps showing the spread of zebra mussels in the United States and Canada from 1988 to 2000.*

FIGURE 1 *The zebra mussel grows on just about any solid surface it can find.*

(continued)

sedentary themselves, adults may be moved about while attached to the backs of turtles and crayfish, or to the hulls of speedboats, barges, freighters, and even canoes. Although they disperse by floating in currents as larvae, the main means of transport appears to be boats. Adults attach to ships, boats, and barges and may be dislodged. Here, they set up new colonies.

Evidence of their ability to spread rapidly in North America became evident in 1993, 5 years after their first spotting in Lake St. Clair, when a weary rescue worker from Hannibal, Missouri happened to look up at some muddy tree trunks revealed after the flood waters had receded along the banks of the Mississippi. What he saw were hundreds of tiny, brown-striped "clams," holding tightly to the trees. It was a long way from Lake St. Clair!

Their invasive properties were clearly known to European scientists. From the mussels home in the Caspian Sea, the zebra mussel spread through all of Europe in the past century, creating incredible amount of damage, both economic and ecological. How the zebra mussel got to North America is not known. Its appearance in the Canadian waters of Lake St. Clair in 1988, however, is believed to be linked to shipping. The zebra mussel may have "stowed away" in the ballast water of a trans-Atlantic freighter. (Ballast water is carried in special tanks of an empty ship to steady it when at sea. After cargo has been loaded, the water is pumped out.)

Since its first appearance, the zebra mussel has spread relentlessly through many of the lakes and rivers of the United States. Margaret Dochoda, biologist with the Great Lakes Fishery Commission, predicts that within 20 years the zebra mussel will have spread throughout the entire East Coast of the United States. Indeed, in only 7 years it spread throughout all the Great Lakes, the Erie Canal, the New York State Barge Canal, the Hudson River, the St. Lawrence River, the Susquehanna River, and the Mississippi River. Some scientists believe that eventually it will be found in lakes and rivers throughout the United States east of the Rockies as well as in southern Canada. In 1999, it was spotted on the Missouri River near Sioux City, Iowa.

Zebra mussels feed by filtering green algae and organic material (detritus) from the water. As you may recall from previous reading and lectures, algae form the base of aquatic food webs that contain species valuable to humans such as fish, waterfowl, eagles, otter, and mink. Some scientists and conservationists believe that large colonies of zebra mussels pose a serious threat to these and other species. Some sport fish such as walleyes and lake trout breed mainly on lake bottoms that are covered with boulders. However, because the zebra mussel frequently colonizes such surfaces, the reproductive success of these fish may be jeopardized. This would be a serious financial blow to the walleye fishery, which is worth almost $1 billion in Lake Erie alone.

Zebra mussels often attach themselves to the inner surface of water intake pipes of municipal water systems, power plants, factories, agricultural processors, and irrigation systems. In 1990 this caused the entire city of Monroe, Michigan, to be without water for 2 days. At about the same time, Bill Kovalak, biologist with the Detroit Edison power plant, was startled when he found a zebra population density of 700,000 mussels per square meter blocking the company's intake pipes. This represented a formidable problem, for these pipes normally had a flow of 1.5 million gallons per second. Detroit Edison had to spend $500,000 to scrape the pests away. U.S. and Canadian water treatment plants along the shores of the Great Lakes that siphon off lake water, then filter it for domestic and other uses will probably spend billions of dollars to control the problem.

Zebra mussels have even been known to grow so heavily on their substrate that they can sink navigational bouys. Continued attachment to steel and concrete can cause corrosion and can lead to structural failure.

Scientists are working on a variety of strategies to control zebra mussels and stop the invasion. One major method for controlling the populations of this European invader is the manipulation of its reproductive activity. Male mussels are normally stimulated to release sperm by the scents they get from the explosive growth or bloom of billions of green algae. The female mussels, in turn, are triggered to produce eggs when they are exposed to mussel sperm in the water around them. This type of breeding behavior ensures that the young larvae have a superabundant supply of algal food. Jeffrey L. Ram of Wayne State University and his co-workers are trying to develop a false scent that would trigger the male mussels to release sperm in the absence of an algal bloom. The sperm would stimulate the females to release eggs. If successful, the resulting larvae would face mass starvation. Eventually, after repeated releases of such pseudoscents, the pest population might be brought under control.

Susan Fisher, a biologist at Ohio State University, recently discovered that a small amount of potassium released in the water prevents the zebra mussel from settling down and fastening itself to a hard surface. She suggests, therefore, that potassium could be constantly released at the openings of water intake pipes to prevent the mussels from forming colonies and blocking flow. Marine paint manufacturers have become interested in adding potassium to their paints to prevent zebra mussels from fouling boat hulls, buoys, and piers.

High-tech methods of control are also being investigated. They include acoustics to kill larvae, and the use of ultraviolet light and low-intensity electrical fields to repel them from intakes. Zebra mussels are consumed by natural predators, too, such as diving ducks, catfish, drum (a type of fish), sturgeon, and raccoons. However, studies suggest that these natural controls are currently no match for the reproductive capacity of the zebra mussel. There are simply not enough of them to make a dent in the population.

Surprisingly, not all of the environmental effects of this European invader have been bad. The zebra mussel has been found effective in clearing up water that for years had been the color of pea soup because of successive algal blooms. A single zebra mussel can filter 1 liter of water every 24 hours. A colony of 5 million could filter 1 million liters per day. Improvement in water clarity can be dramatic. A good example is the transformation in Lake Erie. Charles O'Neal, biologist at the State University of New York, commented, "The zebra mussels are like 'vacuum cleaners.' They suck all the nutrients out of the water. The West Basin of Lake Erie used to be green. You couldn't see your fingers. But now you can see down 30 feet!" Water clarity improves lake aesthetics and lakeside property values. It is a delight to swimmers and boaters as well. It has also increased sunlight penetration into previously murky waters, causing bottom dwelling plants to thrive. They serve as a source of shelter for many smaller fish.

(continued)

Zebra mussels also filter out considerable amounts of heavy metals such as lead, cadmium, and mercury. In so doing, they may lessen the degree of serious illness that humans may suffer from drinking metal-contaminated water or from eating fish in which such metals have accumulated.

Hardest hit by the zebra mussel are native mussel species. Already eliminated from the western basin of Lake Erie by zebra mussels, native species are simply outcompeted by their prolific European relative. Loss of plankton and dissolved nutrients caused by the thick masses of zebra mussels that often grow right on native species causes the latter to starve to death.

This case study on the zebra mussel is a classic example of how a natural ecological system, in balance for millennia, can be severely disrupted by the accidental invasion of an exotic organism. For better or for worse, despite strenuous attempts to control this pest, you can be sure that the zebra mussel will spread to most lakes and streams east of the Rockies. They will be here not only throughout your lifetime, but maybe forever!

of waste injected by the U.S. Army's chemical plant near Denver, Colorado, and the frequency of earthquakes in the immediate region.

Although deep injection wells in the United States are supposed to contain wastes for 10,000 years, not all critics believe that they will perform as hoped. Critics view this technique as a means of "sweeping pollution under the rug" and fear that it may be a stopgap measure that comes back to haunt future generations.

Let's consider groundwater contamination. Bear in mind that although we'll be examining toxic organic wastes, all forms of water pollution can contaminate our groundwater.

Groundwater Contamination Groundwater moves slowly through aquifers beneath our feet. Within only one-half mile of the Earth's surface, this vast water resource has four times the volume of the Great Lakes—about 160 quadrillion liters (40 quadrillion gallons)! More than one half of the people of the United States depend on this unseen resource for their drinking water, consuming more than 24 trillion liters (6 trillion gallons) yearly. At present, Americans pump 300 billion liters (80 billion gallons) of groundwater each day. This underground resource serves as drinking water for 95 percent of the nation's rural households and for 35 of our nation's largest cities.

Unlike a flowing river, groundwater has virtually no natural cleansing or diluting mechanisms. As a result, once it becomes contaminated, it may remain so for thousands of years. Furthermore, water withdrawn from wells draws groundwater in to replace what was taken out. This, in turn, tends to pull contaminants into wells. It is obvious, therefore, that pollution prevention is of the utmost importance in groundwater management.

The synthetic organic compounds that pollute groundwater come from a variety of sources: municipal and industrial landfills (sites where waste is buried), contaminated grounds of industrial facilities, evaporation ponds containing industrial wastes, agricultural and mining sites, places where sewage sludge is applied to the land, wells that inject industrial waste into the earth, septic tanks in rural and suburban areas that release household toxic organics, and underground gasoline storage tanks at service stations (Figure 11.16). Because water flows so slowly through most aquifers, it may take decades for a pollutant to move a mile or two from the site of contamination.

Adverse Effects on Health Obviously, groundwater—the source of drinking water for over 130 million people in the United States—is not free of impurities. As better instruments for detecting toxic organics become available, more and more toxic chemicals are found. The question naturally arises: "If I drink such contaminated water, will I eventually come down with some serious illness?" Unfortunately, the answer is not clear. Health officials simply cannot say that if you drink X liters of water contaminated with Y chemical at a concentration of Z ppm over a period of many years, you will develop cancer. At this point, medical knowledge is inadequate (Table 11.4). Of considerable interest in this regard, however, is a study conducted by a team of Harvard University scientists at Woburn, Massachusetts. Their investigation suggested a strong cause-and-effect relationship between the ingestion of certain industrial solvents (trichloroethylene, benzene, and chloroform) in contaminated drinking water and an abnormally high incidence of stillbirths, sudden infant deaths, and childhood leukemia.

It should be emphasized that polluted well water rarely contains only a single organic contaminant. Usually a given sample contains several. This means that it is possible for two or more organic compounds to interact and cause a synergistic effect on the body—an effect that is greater than the sum of the effects of each pollutant acting separately. Medical researchers have found, for example, that the industrial solvent TCE causes greater kidney damage when the individual is simultaneously exposed to PCBs (Table 11.5).

Another question that arises is whether the fetus is adversely affected by toxic organic compounds ingested by the mother. The answer is yes, at least in some cases. For example, PCBs are passed from the mother to the fetus and cause (1) delayed development, (2) reduced size at birth, (3) retarded reflexes, and (4) impaired memory.

In the early 1980s, a rash of illnesses occurred in Hardeman County, Tennessee. Among them were blurred vision, kidney malfunction, and liver damage. Health authorities concluded that the sicknesses resulted from drinking water that was contaminated with at least 15 different chemicals, including the organic solvents carbon tetrachloride and benzene. Further investigation showed conclusively that these toxic materials had leached into the source of drinking water from a "leaky" industrial landfill operated by the Velsicol Chemical Company, one of our nation's biggest chemical manufacturers. Finally, in 1986, after a series of legal skirmishes, the U.S. District Court

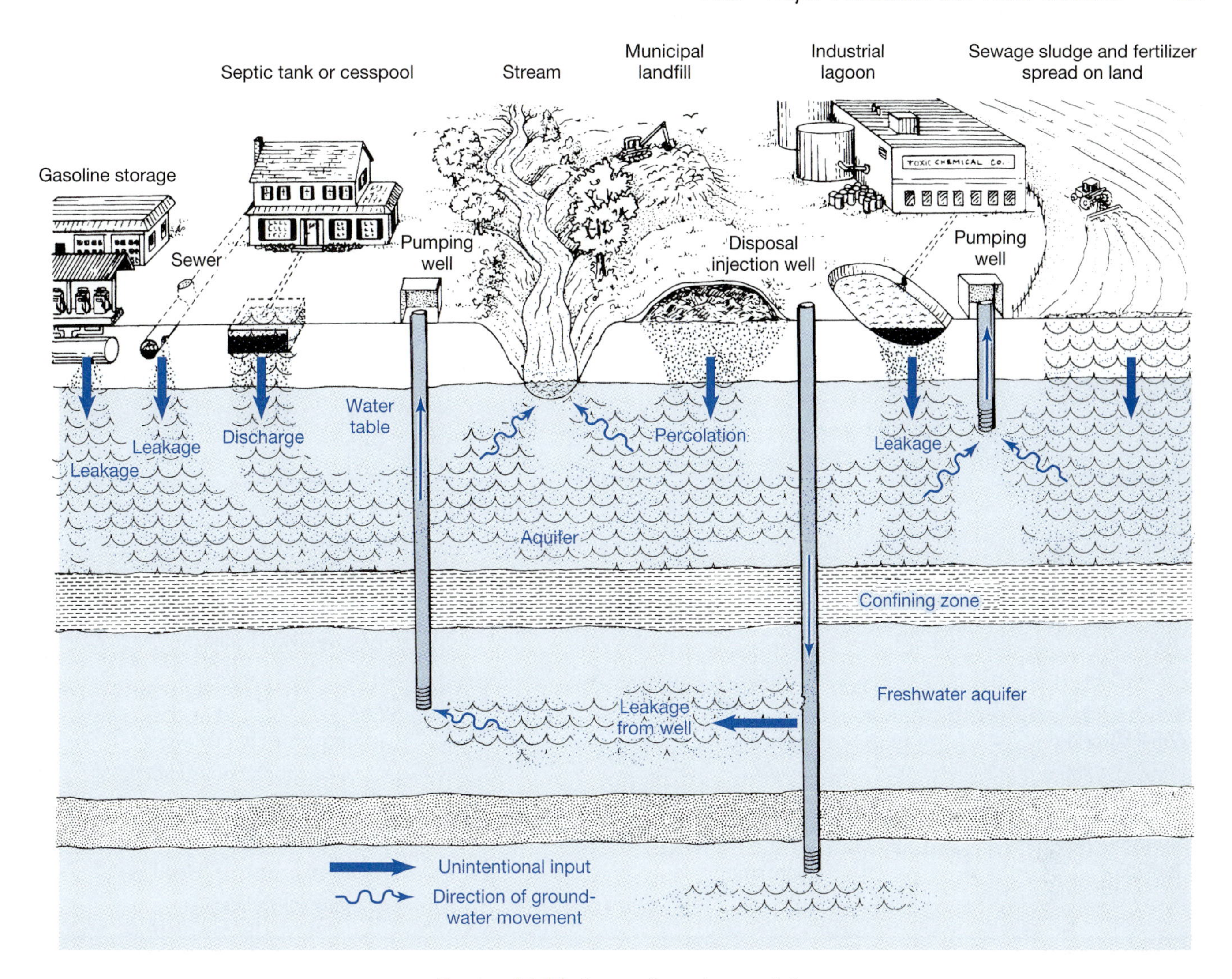

FIGURE 11.16 *Sources of groundwater pollution.*

ordered Velsicol to pay $12.7 million to about 100 people who had become ill from drinking the tainted water.

Groundwater contamination continues today as a result of improper and illegal disposal techniques. In fact, recent well-water testing by health officials in Gastonia, North Carolina, showed that 71 wells had levels of industrial chemicals that were three times higher than considered acceptable by the EPA. But Gastonia is just one of many areas in which this problem persists.

Similar, if not worse, problems exist in other countries. In many eastern European nations of the former Soviet Union, for example, many industrial sites were contaminated by hazardous materials. Improper disposal of wastes at these facilities and off site was commonplace. Although statistics on the extent of the problem are not available, the problem is considered very serious.

Controlling Toxic Organic Pollution Reducing the threat from toxic organic pollutants requires many different actions, many of them preventive. Preventing groundwater from becoming contaminated, for example, requires proper disposal or destruction (for example, incineration) of wastes. Bans on evaporation ponds and burial of toxic wastes in landfills are vital to meeting this goal, as well. Pollution prevention methods, such as hazardous waste recycling and substitution of nonhazardous materials in chemical processes for hazardous ones, which are discussed in Chapter 17, are even more effective. Such measures, which eliminate toxic waste production altogether, are now being pursued vigorously by businesses in many countries, including Canada and the United States. Water purification can also help rid toxic chemicals from waterways. Many homeowners have installed their own filters to ensure clean water.

Heavy Metal Pollution: Sources, Impacts, and Controls

Heavy metals are highly toxic elements such as lead and mercury. They come from a variety of sources. Mines are one example. Abandoned or poorly maintained operating mines, for

TABLE 11.4 *Effects of Some Toxic Organic Chemicals on the Health of Occupationally Exposed Workers*

Chemical	Exposure	Effects*
Carbon tetrachloride	Inhalation: absorption through skin	Liver and kidney damage Vomiting Abdominal pain Diarrhea Jaundice Red and white blood cells in urine Coma Death
Chloroform	Inhalation	Anesthetic effect Dizziness Mental dullness Kidney damage Liver enlargement Digestive disturbances Coma Death
Vinyl chloride	Inhalation	Chromosome abnormalities Increased spontaneous abortion (in the opinion of exposed workers)
Ethylene dibromide	Inhalation	Decreased fertility
Benzene	Inhalation	Prolonged menstrual bleeding Leukemia (blood cancer)

* All of these chemicals have caused cancer in laboratory rats and mice.
Source: Council on Environmental Quality: *Contamination of Ground Water by Toxic Organic Chemicals.* Washington, D.C.: U.S. Government Printing Office, 1981.

instance, may release groundwater that has leached heavy metals from the interior of the mine. Mine tailings, the waste material removed from the mine and sometimes discarded carelessly on the Earth's surface around the mine site, contain heavy metals such as lead, copper, zinc, and cadmium that may be washed into surface waters during heavy rains. Heavy metals are released by coal-fired power plants, garbage incinerators, and other industrial facilities and may be washed from the sky in rain or snow, falling on land and water. Those that land on the Earth may be washed into waterways. Several key point sources release heavy metals directly into waterways. They include metal-processing plants, dye-making firms, and paper mills. Heavy metals in our environmental can also arise from other sources. Lead, for instance, may be leached from lead pipes in homes built before 1930 and lead solder often used for joining copper pipes in newer homes, offices, and other facilities.

The problem with heavy metals is that, unlike organic pollutants, they are not broken down by bacteria. Consequently, they may persist in the water or bottom sediments for many years and eventually enter human food chains.

Metals are also poisonous. Their toxicity results from the fact that many heavy metals interfere with the normal function of enzymes—the proteins that facilitate many life-sustaining functions in our cells. Therefore, the intake of metals through contaminated food and water may cause serious illness. Lead poisoning, for example, may result in a broad spectrum of effects, ranging from decreased learning ability and gastric upsets to convulsions, coma, and death. Lead poisoning also causes mental retardation and stunts growth in children.

At present, the greatest concern among health officials and scientists is the subtle adverse effect of relatively low concentrations of lead in drinking water on embryonic development and on learning and memory in children. Children under the age of ten are especially vulnerable. The EPA estimates that the intellectual capacity of 143,500 American children is reduced by up to 5 IQ points annually because of lead-contaminated tap water. Unfortunately, lead in drinking water cannot be seen, smelled, or tasted. Consequently, its presence is not suspected until some time after the first symptoms of poisoning appear.

TABLE 11.5 *Some Toxic Chemicals in the Great Lakes: Use, Source, and Effects on Human Health*

Name	Use	Probable Source	Found In	Characteristics/ Health Effects
Asbestos in taconite tailings	Byproducts of iron mining.	Byproduct of iron ore mining. Secured on-land disposal ordered by court.	Lake Superior	Airborne effects may include asbestosis, lung cancer; water-borne effects not known, but cancer is implied.
DDT, chlordane, dieldrin, aldrin	Pesticides used widely in Great Lakes region to control insects and rodents. DDT banned in 1971; others now restricted.	Residues from previous widespread use; runoff from agricultural and forested areas, leaching from improper waste disposal sites; and atmospheric deposition.	All five Great Lakes	Bioaccumulation in fish, wildlife, and humans. Persistent in the environment. Long-range effects include reproductive disorders in wildlife; suspected cause of cancer in humans.
Heavy metals (mercury, lead, arsenic, cadmium, per, conchromium, iron, selenium, and zinc)	Wide variety of industrial uses, from antiknock agent in gasoline to paints, pipes, pesticides, glass and electroplating.	Industrial discharges, medical profession wastes via municipal discharges, agricultural runoff, disposal of waste products, mine tailings; urban nonpoint sources.	Lake Superior, Lake Ontario, Lake Huron, and Lake Erie	Excessive levels of heavy metals bioaccumulate in fish and wildlife. Human consumption of contaminated food may cause a variety of health problems. Mercury can cause brain damage, birth defects. Lead can cause anemia, fatigue, and irreversible brain damage, especially in children. Cadmium can cause kidney damage, metabolic disturbances. Arsenic can cause damage to the liver, kidney, digestive system bone marrow; suspected cause of cancer in humans. Copper, chromium, iron, selenium, and zinc are toxic to fish.
Mirex	Insecticide used to control fire ants; flame retardant; plasticizer.	Was produced and processed in Great Lakes region until ban in 1975; spills.	Buffalo River, Niagra River, Lake Ontario	Bioaccumulation in fish, wildlife and humans. Persistent in the environment. Suspected cause of cancer in humans.

TABLE 11.5 *Some Toxic Chemicals in the Great Lakes: Use, Source, and Effects on Human Health (continued)*

Name	Use	Probable Source	Found In	Characteristics/ Health Effects
PAHs (poly-aromatic hydro-carbons)	Variety of industrial uses.	Industrial oil and grease discharges; byproduct of all types of combustion; urban nonpoint sources; smelting.	All five Great Lakes	Persistent in the environment. Can induce cancer and cause chromosome damage in fish, wildlife, and humans.
PCBs (polychlorin-biphenyls)	Insulation for electrical capacitors, transformers; plasticizer, carbonless copy paper, wide industrial use. Total ban except by special EPA permit in July 1979.	Industrial discharges, municipal sewage treatment plant discharges, harbor sediments, low-temperature incineration of wastes; atmospheric deposition.	All five Great Lakes	Bioaccumulation in fish, wildlife, and humans. Persistent in the environment. Test monkeys developed reproductive failures, skin and gastrointestinal disorders. Probable human carcinogen.
Dioxins	No known technical use.	Microcontaminants in chlorophenols and banned pesticide 2, 4, 5, -T. Also bleach kraft paper process, and atmospheric deposition.	All five Great Lakes	Bioaccumulation in fish. Probable human carcinogen. Cause of birth defects and reproductive disorders in wildlife.

The U.S. EPA estimates that lead in drinking water is responsible for at least 680,000 cases of high blood pressure in adult men in the United States. In addition, about 560,000 children have unacceptably high lead levels in their blood, a considerable amount of which probably had its source in lead-tainted water. At least 350,000 Americans are drinking kitchen tap water with lead concentrations higher than deemed safe by the EPA.

Another heavy metal of grave importance is mercury. Mercury in waterways comes from direct industrial discharges and also from rain and snow. The latter comes from coal-fired power plants and incinerators that burn municipal garbage containing mercury batteries. Elemental mercury is fairly innocuous in aquatic ecosystems, but bacteria in sediments convert it to methyl mercury, a toxic form. Methyl mercury accumulates in body tissues and increases in concentration the higher one goes in the food chain. In lakes in the northern United States, concentrations in pike, a predatory fish, are 225,000 higher than they are in the water. This phenomenon is known as **biological magnification.**

Mercury is found in high concentrations in upper-level organisms in many parts of the world, including people. Three quarters of the U.S. states (virtually all that have examined the issue) have posted guidelines for eating fish contaminated by mercury (Table 11.6). In Michigan, health officials recommend that people eat no more than one meal a week of various fish taken from the state's inland lakes. Guidelines for the consumption of ocean-going fish such as shark and swordfish by pregnant women and women of childbearing age who may become pregnant have been issued by the U.S. Food and Drug Administration. Although health studies of the effects of low levels of mercury are not conclusive, some researchers warn that mercury may have significant developmental effects, including reduced cognitive abilities. Mercury can also affect wild species such as loons that live on fish. Studies now underway in the United States are designed to determine those effects. A summary of the effects of four heavy metals on human health is presented in Table 11.7.

Another source of potentially toxic heavy metals is agriculture. As noted in Chapter 9, irrigation water can remove toxic metals from the soil. Not long ago, officials at the Kesterson Wildlife Refuge in California were mystified by an extensive die-off of fish and waterfowl. In addition, many birds were hatched with deformities, such as crossed bills, which made feeding impossible. Furthermore, an abnormally high number of embryos died before hatching. Chemical

TABLE 11.6 *Health Advisory: PCB and Pesticide Contamination in Lake Michigan Fish*

Group 1	Group 2	Group 3
These fish pose the lowest health risk	*Women and children should not eat these fish*	*No one should eat these fish*
Lake trout up to 20"	Lake trout 20 to 23"	Lake trout over 23"
Coho salmon up to 26"	Coho salmon over 26"	Chinook salmon over 32"
Chinook salmon up to 21"	Chinook salmon 21 to 32"	Brown trout over 23"
Brook trout	Brown trout up to 23"	Carp
Pink salmon		Catfish
Smelt		
Perch		

analysis of the water in the artificially made marshes revealed very high levels of the heavy metal selenium (seh-LEAN-ee-um). The selenium was traced to runoff irrigation water from nearby croplands. Irrigation water leaches selenium from soils that naturally contain high concentrations of the potentially toxic element.

Controlling Heavy Metal Pollution Reducing our exposure to heavy metals requires many steps, for example, reductions in air pollution, pollution prevention and pollution control strategies at various factories. Changes in the chemical makeup of solder and the type of pipes used to convey water to our taps have already been made in the United States.

Interestingly, conventional municipal sewage treatment plants that receive wastes from homes and many businesses do not efficiently remove metals from the wastes. In fact, some metals may be toxic to the very bacteria the plant relies on to digest organic materials. It is therefore important that the amount of metal contamination in the incoming waste be reduced to a minimum. In the United States, local, state, and federal regulations require industries to pretreat their metal-laden waste before sending it on to a municipal sewage treatment plant. Under ideal circumstances, most of the metals are removed and transported to a certified hazardous waste dump. However, in most cases, the nation's industries have a long way to go. For example, the Milwaukee Metropolitan Sewerage District recently reported that 104 (74.8 percent) of 139 industrial waste discharge pipes failed to meet the district's standards.

Reducing our exposure to heavy metals requires actions at many levels. Separating industrial waste from municipal waste and requiring on-site treatment of industrial wastes have gone a long way in helping reduce heavy metal contamination of rivers and other surface waters. Changes in the type of pipe used in homes to supply water and changes in the type of solder used to attach pipe have helped as well. Tighter controls on air pollution emissions from factories and power plants have assisted, too, as many heavy metals in the environment originate as air pollutants, which may rain down on the land and water. This process, known as **cross-media contamination,** suggests the need for a holistic approach to controlling this and other forms of pollution.

TABLE 11.7 *Effects of Four Heavy Metals onHuman Health*

Mercury	Arsenic
Fatigue	Headache
Headache	Dizziness
Irritability	Fatigue
Loss of coordination	Vomiting
Numbness of hands and feet	Diarrhea
Shortening of attention span	Abdominal pains
Memory loss	Muscular pains
Kidney damage	Blood in urine
Death	Anemia
	General paralysis
	Heart malfunction
	Coma
	Death

Lead	Cadmium
Intestinal colic	Degenerative bone disease
Irritability	Severe crippling
Reduced resistance to infectious diseases	High blood pressure
Anemia	Heart malfunction
Blood in urine	
Brain damage	
Partial paralysis	
Mental retardation	

Oxygen-Demanding Organic Wastes

Most of us know of a friend or a child who has bought a goldfish, then found it floating belly up in its bowl a few weeks later. In many cases, the demise of the new pet was not due to neglect, but rather too much attention, more specifically, too much food. What happens in such instances is that an overzealous owner gives the fish too much fish food. The food which is not eaten consists of organic matter, biodegradable organic matter. The excess is consumed by bacteria in the water and as they consume the excess, they deplete the oxygen. (The decomposition of organic matter, as you will soon see, often requires oxygen.) The pet fish was actually asphyxiated because of a lack of **dissolved oxygen.**

Organic matter may accumulate in aquatic environments, as, for example, when an autumn leaf fall blankets a woodland stream, when a massive fish kill occurs, or when slaughterhouse debris is discharged. Streams contain a wide assortment of bacteria that decompose this and other organic material like sewage and animal wastes. These microorganisms are part of a self-purifying mechanism of streams.

The process by which such organic material is eventually decomposed by bacterial action may be summarized as follows:

high-energy organic molecules (fats, carbohydrates, and proteins) + oxygen	→	low-energy carbon dioxide + energy (used by bacteria to sustain life)

+ water + nitrate ions (NO_3^{-2})
+ phosphate ions (PO_4^{-3}) + sulfate ions (SO_4^{-2})

Note from this equation that oxygen in the water is essential to the process. In gold fish bowls and aquatic ecosystems, the bacteria actively compete with other oxygen-demanding aquatic organisms (fish, crustaceans, insect larvae, and so on). If sufficient organic material is present in the water, and if other conditions such as water temperature are favorable, the oxygen-consuming bacteria multiply rapidly. Levels of dissolved oxygen fall as the bacterial population increases. Levels may plummet from 10 ppm to less than 3 ppm, to the detriment of other aquatic organisms.

Because their decomposition consumes oxygen in aquatic ecosystems, naturally occurring organic wastes and human-produced wastes like sewage are called **oxygen-demanding organic wastes.** The federal government maintains a network of stream-monitoring sites at which dissolved oxygen levels are systematically checked. Of the thousands of measurements taken in the past few years, fewer than 5 percent were below 5 ppm of dissolved oxygen—the minimal level required for quality fish populations.

Water quality chemists measure the concentration of oxygen-demanding organic matter through a simple technique in which they mix water samples with a certain amount of oxygen, then measure the decline in oxygen over time. The amount of oxygen consumed by bacteria is dependent on the amount of organic material in the sample. The organic content is not expressed directly, but rather as the **biological oxygen demand,** or **BOD.** More recently, water quality personnel have switched to the term **biochemical oxygen demand.** Thus, it is customary to speak of the BOD of human sewage, of slaughterhouse wastes, and so on, since this material supports bacteria that require or "demand" oxygen.

If you live in a city or town served by a sewer system, every time you flush your toilet, you are making it a bit tougher for scrappy game fish in the stream or lake near your home to survive, for there are about 250 ppm BOD in the wastewater going down the pipe. Many of the wastes from canneries, cheese factories, dairies, bakeries, and meat-packing plants have BOD levels ranging from 5,000 to 15,000 ppm.

This does not mean that all biodegradable organic matter will cause surface water to become oxygen depleted. This outcome depends on the amount of organic matter. In fact, in streams and lakes fish, mussels, and vegetation eventually die. Aerobic (oxygen-requiring) bacteria will consume the organic compounds, producing carbon dioxide and water, using a proportional amount of dissolved oxygen. This is a natural recycling and cleansing mechanism. Problems arise when a organic matter is dumped into the water body in excess. In other words, it is only when the natural assimilation capacity is exceeded that we see oxygen depletion and damage. As you will soon see, a river's organic assimilation capacity is also determined by the rate or aeration—how fast oxygen is replenished. Cold turbulent water contains lots of oxygen and replenishes it more quickly than a slow-moving river. Rivers most vulnerable to oxygen depletion are the warm, flow-flow, stagnant waterways.

Effect of a High BOD on Stream Animals High BOD wastes from sewage treatment plants and other facilities like pulp mills and slaughterhouses can have a dramatic and devastating effect on aquatic animal populations. Results of studies of the kinds and numbers of organisms occurring immediately above and at several sites below the point of sewage discharge are shown in Figure 11.17. At point *A*, just above the outfall, the river is characteristic of an unpolluted stream. The high levels of dissolved oxygen (8 ppm) and the abundant food in the form of mayfly and caddis fly larvae make possible the survival of highly prized fish such as bass and trout. However, at point *B*, in the **zone of decline,** immediately below the outfall, dissolved oxygen levels drop rapidly because of the high organic component of the waste. In some streams the dissolved oxygen may drop to 3 ppm or less, which is insufficient to support the oxygen requirements of more desirable fish. Instead, only less desirable fish, such as carp and bullheads, which have low oxygen requirements, can survive. The larvae of mayflies, stone flies, and caddis flies, which require higher oxygen levels, are virtually absent, too. The dissolved oxygen concentration is so drastically reduced in the **damage zone,** from *C* to *D*, that even carp and bullheads cannot survive.

The most typical bottom-dwelling animals in the damage zone are reddish sludge worms, of which there may be 180,000 per square meter of stream bottom; bloodworms; and the red rat-tailed maggot (Figure 11.17). These animals are sometimes used as index organisms; their occurrence indicates that a particular stretch of stream is highly contaminated with organic waste. Unpolluted aquatic ecosystems usually have a much greater species diversity than their polluted counterparts. Rather surprisingly, however, the total biomass in severely deoxygenated areas may approach that of unpolluted water. The reason is that the few highly specialized species that can survive, such as the sludge worms, develop huge populations.

Beginning at point *D* in the **recovery zone,** the amount of oxygen removed by the sewage bacteria is more than counterbalanced by the oxygen entering the stream from the atmosphere because of wind action or photosynthesis of algae and stream-dwelling plants. As a result, the dissolved oxygen level rises, permitting the occurrence of carp and garpike. Finally, still farther downstream, at point *E,* most of the organic material discharged from the sewage plant has been decomposed; the level of dissolved oxygen rises to its original value. Fish and other organisms supported by the stream above the point of discharge can survive in the water below point *E.*

The characteristic dip of the oxygen curve at points *B* and *C* is known as the **oxygen sag.** The slope of the dissolved oxygen curve, which is highly variable, depends on the BOD of the sewage, the rate at which oxygen enters the stream, the water temperature, and the water velocity. If additional sources of pollution are present, recovery may be impossible.

The oxygen sag curve is not just of scientific interest. It is very practical information that is used to set BOD discharge standards for wastewater treatment plants. A wastewater treatment plant on a stream that is more prone to oxygen depletion will have a lower BOD discharge permit standard to maintain a minimum dissolved oxygen of 5 ppm in the receiving water than one that is swifter moving and better aerated. The discharge permit procedure, of course, is designed to protect aquatic life in the lake or stream into which treated sewage is released. The control of oxygen-demanding organic wastes is discussed in the next section.

11.3 Sewage Treatment and Disposal

Cities and towns produce enormous quantities of waste from nonpoint and point sources. In many parts of the world, especially the more developed nations like those of Europe and North America, the point sources discharge their wastes into sewers that transport them to **sewage treatment plants,** facilities that remove wastes. Sewage treatment plants receive numerous types of water pollution,

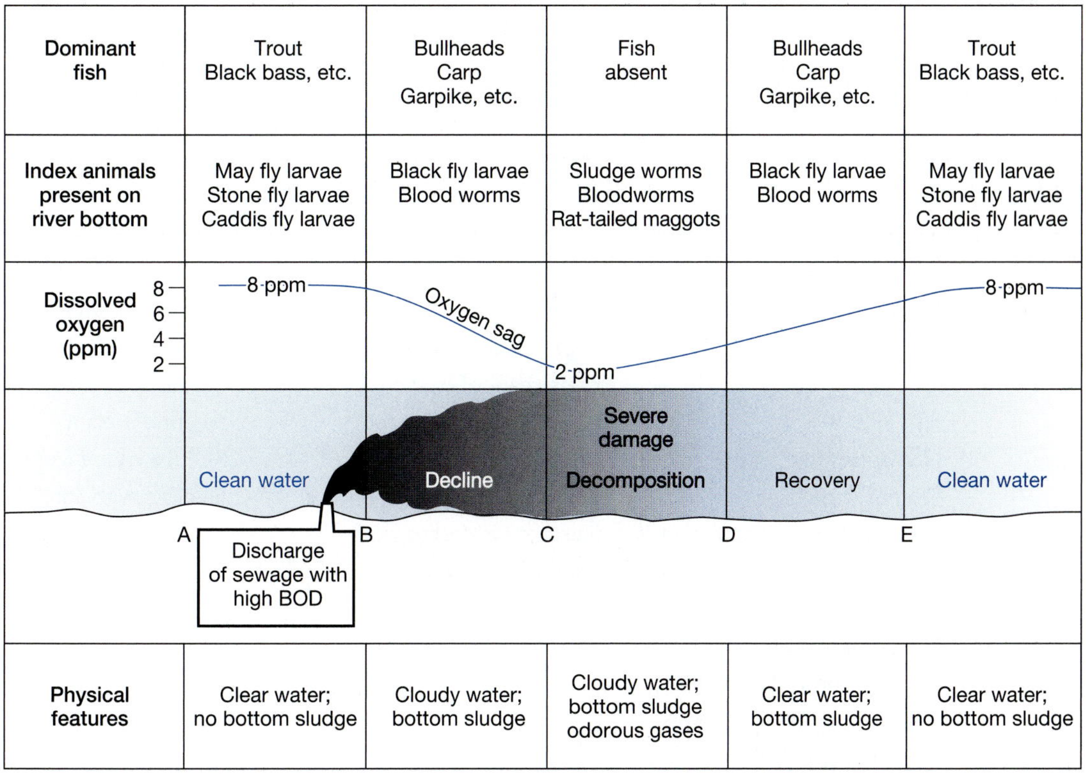

	A–B	B–C	C–D	D–E	Beyond E
Dominant fish	Trout Black bass, etc.	Bullheads Carp Garpike, etc.	Fish absent	Bullheads Carp Garpike, etc.	Trout Black bass, etc.
Index animals present on river bottom	May fly larvae Stone fly larvae Caddis fly larvae	Black fly larvae Blood worms	Sludge worms Bloodworms Rat-tailed maggots	Black fly larvae Blood worms	May fly larvae Stone fly larvae Caddis fly larvae
Dissolved oxygen (ppm) 8 6 4 2	8 ppm	Oxygen sag	2 ppm		8 ppm
Zone	Clean water	Decline	Severe damage Decomposition	Recovery	Clean water
Physical features	Clear water; no bottom sludge	Cloudy water; bottom sludge	Cloudy water; bottom sludge odorous gases	Clear water; bottom sludge	Clear water; no bottom sludge

Discharge of sewage with high BOD (at point A)

FIGURE 11.17 *Effect of sewage with a high BOD level on the amount of dissolved oxygen and the type of aquatic organisms in the stream.*

including sediment, infectious organisms, detergents (containing inorganic nutrients), human excrement, toxic chemicals, and organic material (Table 11.8). Many of the pollutants are removed by sewage treatment plants before the waste is discharged into a lake, stream, or ocean. These plants have done much to help clean up the waters of the United States, Canada, and other more developed nations. (For a discussion of the successes but hidden problems that remain see Case Study 11.2.)

Sewage Treatment Methods

Since 1880, when our nation's first sewage treatment plant was built in Memphis, Tennessee, more than 16,000 have been constructed. They serve 70 percent of the U.S. population. Domestic sewage treatment may be primary (rudimentary but rather inexpensive), secondary (more effective and of moderate cost), or tertiary (most effective but very expensive).

Primary Treatment

The wastewater entering a sewage treatment is highly liquid. But it does contain a few solid materials such as Band-Aids, sediment from storm water (discussed later), and toys flushed down the toilet by children. **Primary treatment** is mainly a physical process to remove these solids from the wastewater, as shown in Figure 11.18. Wastewater enters the plant and travels through a screen that removes large objects (gravel, garbage, leaves, and so forth). The stream of wastewater then passes into settling tanks, which are often called clarifiers, where suspended organic solids settle to the bottom. In primary-treatment-only plants, the fluid that remains is then chlorinated to destroy disease-causing organisms and discharged into lakes or streams.

The solids that have accumulated at the bottom of the settling tank are pumped to a sludge digester, where millions of bacteria "feed" on the organic waste, breaking it down in the absence of oxygen. One decomposition product, methane gas, is frequently burned to heat the digester to the temperature required for most effective bacterial action. In some plants, the methane is burned to produce electricity that can be used on site.

Primary treatment removes about 60 percent of the suspended solids and about 33 percent of the oxygen-demanding waste (BOD). Although primary treatment makes sewage look a lot better, it leaves a substantial amount of organic material, nitrates, phosphates, and bacteria, some of which may cause human disease. Fortunately, with the aid of cost-sharing grants from the federal government, thousands of American cities have been able to replace their primary plants with secondary plants.

Secondary Treatment

Virtually 100 percent of the municipal sewage in the United States receives secondary treatment (Figure 11.19).[1] Primarily biological in nature, **secondary treatment** uses aerobic bacteria to break down degradable organic materials and remove additional suspended solids. It may also remove ammonia by biological oxidation to nitrates. The two major methods available for secondary treatment are (1) the activated sludge process and (2) the trickling filter.

Activated Sludge Process In a sewage treatment plant with a secondary treatment capacity, fluid from the first settling tank (clarifier) is piped to another tank, the aeration tank, in which air is bubbled to provide a maximal supply of oxygen (Figure 11.19). In the aeration tank, aerobic (oxygen-using)

[1]Of the 16,000 sewage treatment plants in the United States, all but 68 are secondary treatment facilities. Those that are primary treatment only tend to be small.

TABLE 11.8 *Water Pollution: Sources, Effects, and Control*

Contaminant	Source	Effects	Control
Oxygen-demanding waste	Soil erosion Autumn leaf fall Fish kills Human sewage Domestic garbage Remains of plants and animals Runoff from urban areas during storms Industrial wastes (slaughterhouses, canneries, cheese factories, distilleries, creameries, and oil refineries)	Bacteria that decompose the organic matter will deplete the stream of oxygen Game fish are replaced by less desirable fish Valuable food for game fish (may flies, etc.) is destroyed Foul odors develop	Reduce runoff from barnyards Reduce BOD of sewage with modern secondary sewage treatment plants Reduce runoff from feedlots with catch basins

TABLE 11.8 *Water Pollution: Sources, Effects and Control (continued)*

Contaminant	*Source*	*Effects*	*Control*
Disease-producing organisms	Human and animal wastes Contaminated aquatic foods (clams, oysters)	High incidence of water-borne diseases such as cholera, typhoid fever, dysentery, polio, infectious hepatitis, fever, nausea, and diarrhea	Reduce runoff from barnyards and feedlots More effective sewage treatment Proper disinfection of drinking water
Nutrients (Phosphates and Nitrates)	Soil erosion Food-processing industries Runoff from barnyards, feedlots, and farmlands Untreated sewage Industrial wastes Household detergents Exhaust of motor cars	Eutrophication May cause methemoglobinemia in infants Foul odors Decreased recreational and aesthetic values	Tertiary sewage treatment Reduce use of commercial fertilizers Change detergent formula Control soil erosion with strip-cropping, contour plowing, and cover cropping Control feedlot runoff with catch basins
Sediment	Soil erosion from farmland, strip-mined land, logged-off areas, and construction sites (roads, homes, and airports)	Fill in reservoirs Clogs irrigation canals Increases probability of floods Impedes progress of barges Interferes with photosynthesis, reducing dissolved oxygen levels Destroys freshwater mussels (clams) Cause fish mortality due to asphyxiation Destroys spawning sites of game fish Necessitates expensive filtration of drinking water	Employ erosion-control practices on farms such as contour plowing, cover cropping, and shelterbelting Employ erosion-control practices at construction sites: sodding, use of catch basins, etc. Use mulching and jute matting on seeded road banks Establish temporary cover, such as rye and millet, at construction sites
Heat	Midsummer heating of shallow water by the sun Discharge of warm water from electrical power, steel, and chemical plants	Disrupts structure of aquatic ecosystems Causes shift from desirable to undesirable species of algae and fish Kills cold-water fish such as salmon and trout Blocks spawning migrations of salmon Interferes with fish reproduction Increases susceptibility of fish to diseases and to the toxic effects of heavy metals such as zinc and copper	Reduce the nation's energy demands for electricity Use closed cooling systems exclusively Instead of discharging heated water to streams, use it to heat homes, extend growing seasons on cropland, increase growth rate of food fish and lobsters and prevent frost damage to orchards

CASE STUDY 11.2 INVISIBLE THREAT: TOXIC CHEMICALS IN THE GREAT LAKES

The massive pollution of the Great Lakes in the 1960s was quite visible. Symptoms of gross eutrophication abounded: turbid bays, weed-choked shallows, floating mats of algae, rotting fish that fouled once-attractive beaches. Fortunately, however, those obvious signs of pollution gradually receded, thanks to a $9 billion Canadian–American investment in modern municipal and industrial wastewater treatment systems along the perimeter of these huge lakes. Today, the waters of the lakes once again, for the most part, are clear and sparkling.

Appearances can be deceiving, however. An invisible threat is present in these waters—posed by a chemical "broth" of toxic chemicals (Table 11.5). In fact, a joint study by American and Canadian scientists has indicated that human exposure (40 million people) to toxic pollutants in the Great Lakes region is greater than in other area on the North American continent.

One of the symptoms of the pervasive toxic contamination of this 162,500-square-kilometer (65,000-square-mile) expanse of fresh water is the appearance of skin lesions and cancers in fish. For example, liver cancers are frequently found in bullheads taken from Ohio's Cuyahoga River where it flows into Lake Erie. Bottom-feeding fish such as carp, suckers, and catfish are showing an increasing frequency of cancer—probably because they are ingesting toxic chemicals released from sediments. In the more contaminated tributaries of the Great Lakes, nine of every ten fish may have some form of cancer!

When the bodies of these fish are ground up and analyzed, they are usually found to contain relatively high levels of toxic chemicals; and the older and bigger the fish, the higher is the concentration of contaminants, including mercury and organic pesticides. As a result, health authorities have issued fish consumption advisories that instruct people to restrict their consumption of salmon, lake trout, and certain other species of Great Lakes fish (Table 11.6). Fetuses as well as young children apparently are especially vulnerable to the toxins.

Two questions naturally arise: What is the origin of these chemicals? And how did they get into the Great Lakes? Let's consider the case of toxaphene. This organic pesticide has been found in the tissues of fish taken from a lake on Isle Royale, an island in Lake Superior. Since toxaphene has never been used on Isle Royale, the only possible mode of entry to the island lake was by "fallout" from the atmosphere. Scientists believe that the toxaphene comes from southern states like Texas where it has been used to control an insect that attacks cotton, the cotton boll weevil. Northward-blowing winds carry the toxaphene to the lake. Other pesticides like DDT and chlordane, as well as PCBs and heavy metals like lead and zinc, apparently also enter the Great Lakes in substantial amounts. Other modes of entry for the more than 450 chemical contaminants of the Great Lakes ecosystem include (1) industrial wastewater discharge, (2) leaching from industrial waste storage lagoons, (3) municipal waste discharge, (4) agricultural runoff, (5) urban runoff, (6) mining site runoff, and (7) release from bottom sediments.

The health effects of many of these toxins, which are ingested in contaminated fish and drinking water, are only beginning to be understood. After all, it is one thing to feed laboratory animals high levels of a given toxic chemical and then observe the harmful effects, and quite another to determine precisely what will happen to human beings who ingest a few parts per trillion of a toxicant every day for 30 years. And since these contaminants in fish flesh or drinking water are not only invisible but tasteless and odorless as well, it is extremely difficult for a regulatory agency to convince legislative bodies and the public that something should be done about them.

In 1990, the United States and Canada formed the International Joint Commission to identify mutually important problems. As the water pollution problem in the Great Lakes became increasingly serious, the two nations entered into a Great Lakes Water Quality Agreement. Under the terms of this agreement, each nation has attempted to monitor the Great Lakes water under its jurisdiction, to identify areas of concern (hot spots), and to explore ecologically sound methods to control the pollutants. The United States and Canada have identified about 50 toxic hot spots that need immediate attention. One of them is the Detroit River, a tributary of Lake Erie. This waterway serves as a sewer for both municipal and industrial waste from scores of nearby cities. Even though this waste receives conventional treatment, the river is a diverse mix of toxicants, ranging from mercury to phenols, from PCBs to polyaromatic hydrocarbons (PAHs). Hundreds of contaminants are present in the bottom sediments of this stream. Other areas of concern include the harbors of Milwaukee (Wisconsin), Gary (Indiana), Muskegon (Michigan), Cleveland (Ohio), Toronto (Ontario), and Rochester (New York), as well as the mouths of many great rivers.

Under the terms of the Water Quality Agreement, the United States and Canada are using an ecosystem approach to deal with these focal points of contamination. The Remedial Action Plans drawn up so far are based on the premise that problems at a hot spot must be considered in their total ecological context before appropriate solutions can be developed. An important component of a given Remedial Action Plan, for example, is the examination and regulation of the modes of entry for a particular toxicant. For example, if pesticide contamination of the Buffalo River originates in agricultural runoff, farmers in the Buffalo River drainage will be urged to reduce their use of chemicals in pest control and to adopt more effective strategies to prevent soil erosion, such as contour farming and conservation tillage.

If the Remedial Action Plans developed by the United States and Canada for the detoxification of the Great Lakes are to succeed, the life-styles of many people living in this region may have to change. The use of low-phosphate detergents and more conservative application of lawn fertilizer could help. Moreover, great commitment and cooperation will be required from the industrial sector, from environmental agencies, and from government at all levels. Only in this way will the "invisible" chemical threat to the 40 million people of the Great Lakes be really made to disappear.

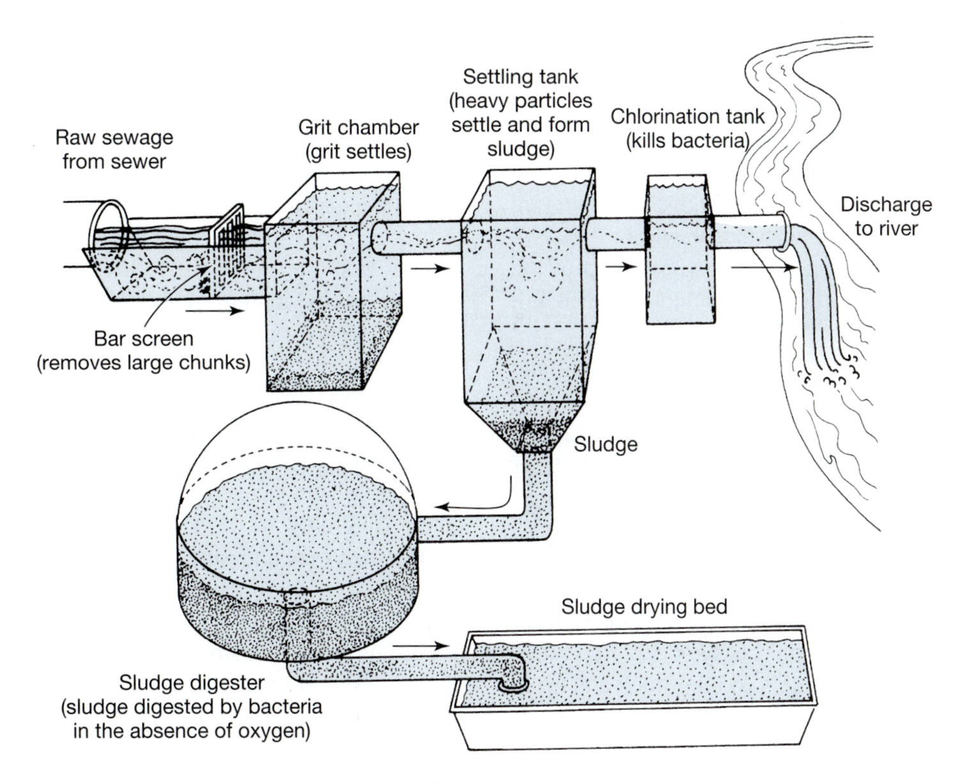

FIGURE 11.18 *Primary sewage treatment.*

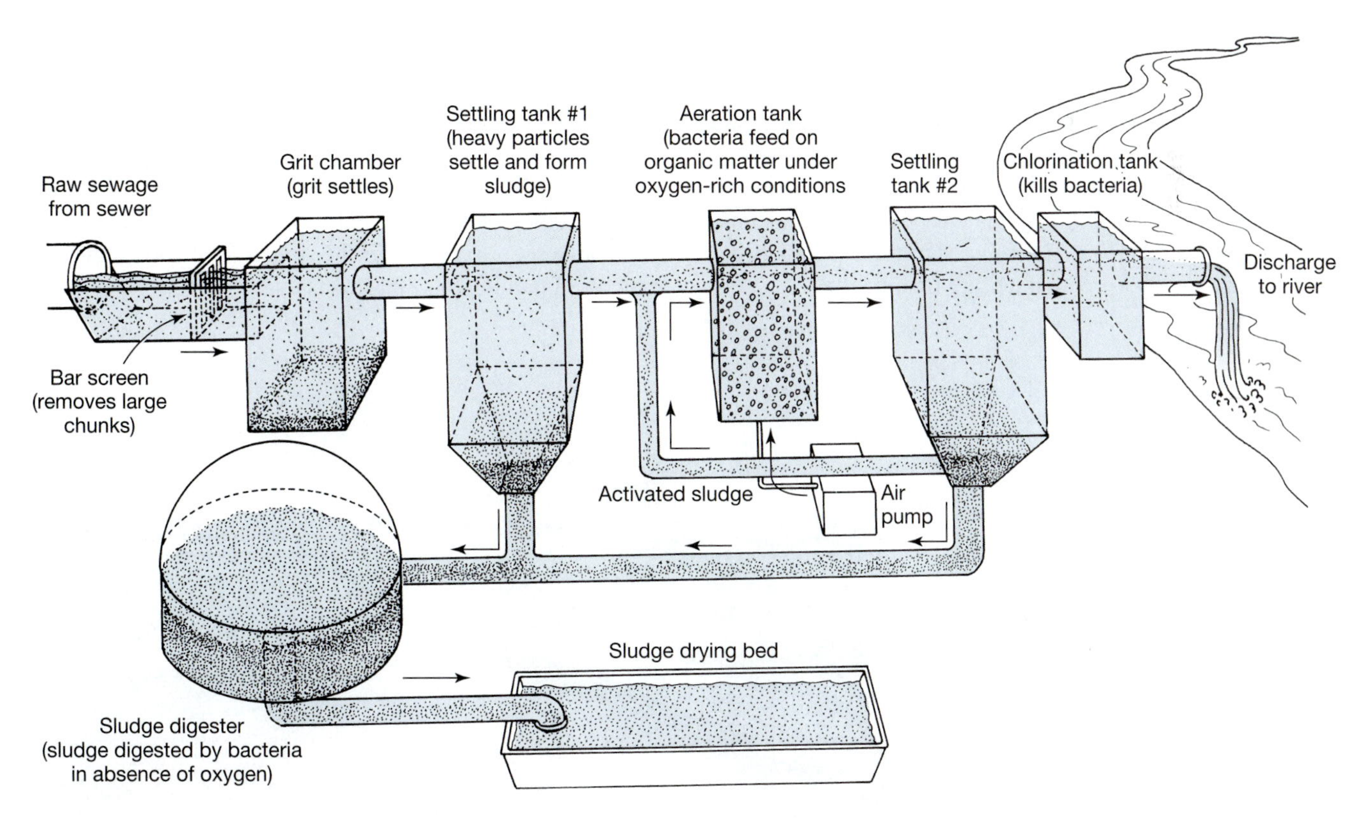

FIGURE 11.19 *Primary and secondary sewage treatment.*

bacteria decompose the organic compounds, and because there is plenty of oxygen, they do so at a fairly rapid rate. BOD drops rapidly because of the presence of oxygen and aerobic bacteria. The mix is then transferred to a settling tank. Here the remaining suspended organic matter and bacteria settle to the bottom, forming sludge. Most of the sludge from this process and also from primary treatment stages is then drained into a device known as an anaerobic digestor. Here it is further broken down in the absence of bacteria. Some sludge is recycled back to the aeration tank where in turn it provides a "seed" population of bacteria to act on the incoming waste. What's left at the anaerobic digester is dried and either incincerated, landfilled, or used as fertilizer. The term **activated sludge** is used to refer to the highly concentrated mix of aerobic bacteria and organic matter within the aeration chamber. The liquid that accumulates at the top of the second sedimentation tank is eventually chlorinated to kill bacteria and then discharged.

Secondary treatment reduces the BOD by 90 percent and removes 90 percent of the suspended solids. However, 50 percent of the nitrogen compounds and 70 percent of the phosphorus compounds (the chemical culprits responsible for eutrophication) still remain. To remove them, tertiary treatment is required.

Trickling Filter In the **trickling filter** process, the sewage is sprayed by the arms of a slowly rotating sprinkler onto a filter bed made up of stones or large chunks of bark (Figure 11.20). The filter bed may be about 2 meters (6 feet) thick and up to 60 meters (200 feet) in diameter. The stones or bark are coated with a slime of bacteria that has accumulated during the operation of the filter. The sewage, containing a load of dissolved organic compounds, trickles down through the stones, and the organics are consumed by the bacteria. Leftover solids are piped to a settling tank and later transferred to a sludge digester. About 80 to 85 percent of the dissolved organic compounds are removed by the trickling filter system. However, the wastewater still contains a high level of nutrients, such as phosphates, ammonia (NH_3), and nitrates, which could cause eutrophication of the lakes and streams into which they are discharged. The removal of these materials depends on still another process: tertiary treatment.

Tertiary Treatment The most advanced form of sewage treatment is **tertiary treatment.** It removes many more pollutants than secondary treatment. The water quality of our streams and lakes would be considerably better if all sewage were given tertiary treatment. Unfortunately, these plants are twice as expensive to build as secondary sewage plants and four times as expensive to operate. As a result, tertiary treatment is not used unless it is necessary to maintain a high level of purity in the receiving body of water. Currently, about 50 percent of the sewage is treated in this way in the United States.

The effects of primary, secondary, and tertiary treatment of sewage are summarized in Figure 11.21.

One final note about sewage treatment plants. Sewage treatment plants have traditionally been built by cities to accommodate only wastewater from homes and factories. But storm water, rain, and snowmelt often flow into the same pipes that carry wastes from our homes. It is then diverted to the sewage treatment plants. Consequently, after heavy thunderstorms the volume of storm water runoff to the treatment plant may be 100 times that of the wastewater alone. As a result, the incoming flow overwhelms the plant's capacity and the excess is allowed to overflow into a lake, stream, or bay. Regrettably, this overflow, which is a mixture of sewage and storm water, may pollute the receiving body of water so badly that it cannot be used for recreational activities. So, you ask, why don't communities build separate pipelines for storm runoff and sewage? The answer is that the cost, in many cases amounting to millions of dollars for large cities, is prohibitive.

Managing Storm Water Runoff

In newer cities, stormwater and sewage from homes is carried in separate set of pipes under the streets. That way, water running off of parking lots and streets during storms is kept sepa-

FIGURE 11.20 *Secondary sewage treatment. This rotary trickling filter device handles 4 million gallons of wastewater from Sacramento, California, daily. After chlorination, the effluent is discharged into the American River.*

Pollutant	Treatment		
	Primary	Secondary	Tertiary
Solids	Solids removed	Little removed	Little removed
Harmful bacteria	Bacteria removed	Little removed	Little removed
Dissolved organics	Little removed	Dissolved organics removed	Little removed
Harmful viruses	Little removed	Viruses removed	Little removed
Phosphorus	Little removed	Little removed	Phosphorus removed
Nitrogen	Little removed	Little removed	Nitrogen removed

FIGURE 11.21 *Relative effects of treatment on the removal of pollutants from sewage.*

rate from sewage. In older cities, however, **combined sewers** are more common. In such systems, storm water runoff and domestic sewage are combined. They both flow to the sewage treatment plant. This causes little problem, until a storm comes along, dumping millions of gallons of water on the city. Because sewage treatment plants can't handle the excess, storm water and sewage are diverted directly to streams and other water bodies. Waste water diversion during periods of heavy rainfall, in fact, is a major reason why many rivers do not meet federal standards set forth in the Clean Water Act.

The Clean Water Act, discussed in more detail shortly, established requirements for storm water discharges. Often managed by the states, these permits help to protect surface waters by requiring cities to reduce pollution reaching them through storm water. The permit system requires cities to adopt **storm water management programs.** These programs must seek to identify and eliminate illegal connections and illegal discharges of pollutants into storm drain systems—for example, companies that are illegally dumping waste into storm sewers. They also must reduce impacts of development projects by private and public entities on runoff and erosion, thus reducing nonpoint water pollution. And, they must help citizens learn more about storm water pollution (nonpoint pollution) and ways they can help minimize problems. Storm water management programs also require the cities and counties to periodically monitor water quality, identify problem areas that need addressing, and report their findings.

Septic Tanks

Many rural and suburban families in the United States and other countries have backyard septic tank systems that process their wastes (Figure 11.22). **Septic tanks** are underground sewage containers made of concrete or plastic into which all household wastewater flows. Solids settle to the bottom of the tank and form sludge. The fluids flow from the tank into a system of perforated pipes buried underground in a **leach field.** The waste stream then passes through the holes in the pipes and slowly percolates through the soil. The soil acts as a natural filter, removing bacteria, some viruses, and suspended materials. Phosphate binds chemically to the soil particles and hence is removed from the effluent. Organic material in the leach field is decomposed by soil bacteria. The sludge that accumulates at the bottom of the septic tank also undergoes bacterial decay but may have to be pumped out from time to time, especially if hard to digest materials like carrot peelings and toxic household cleaners are flushed down the drain.

Septic systems are an effective means of treating household wastes, but they do have drawbacks. For example, they cannot

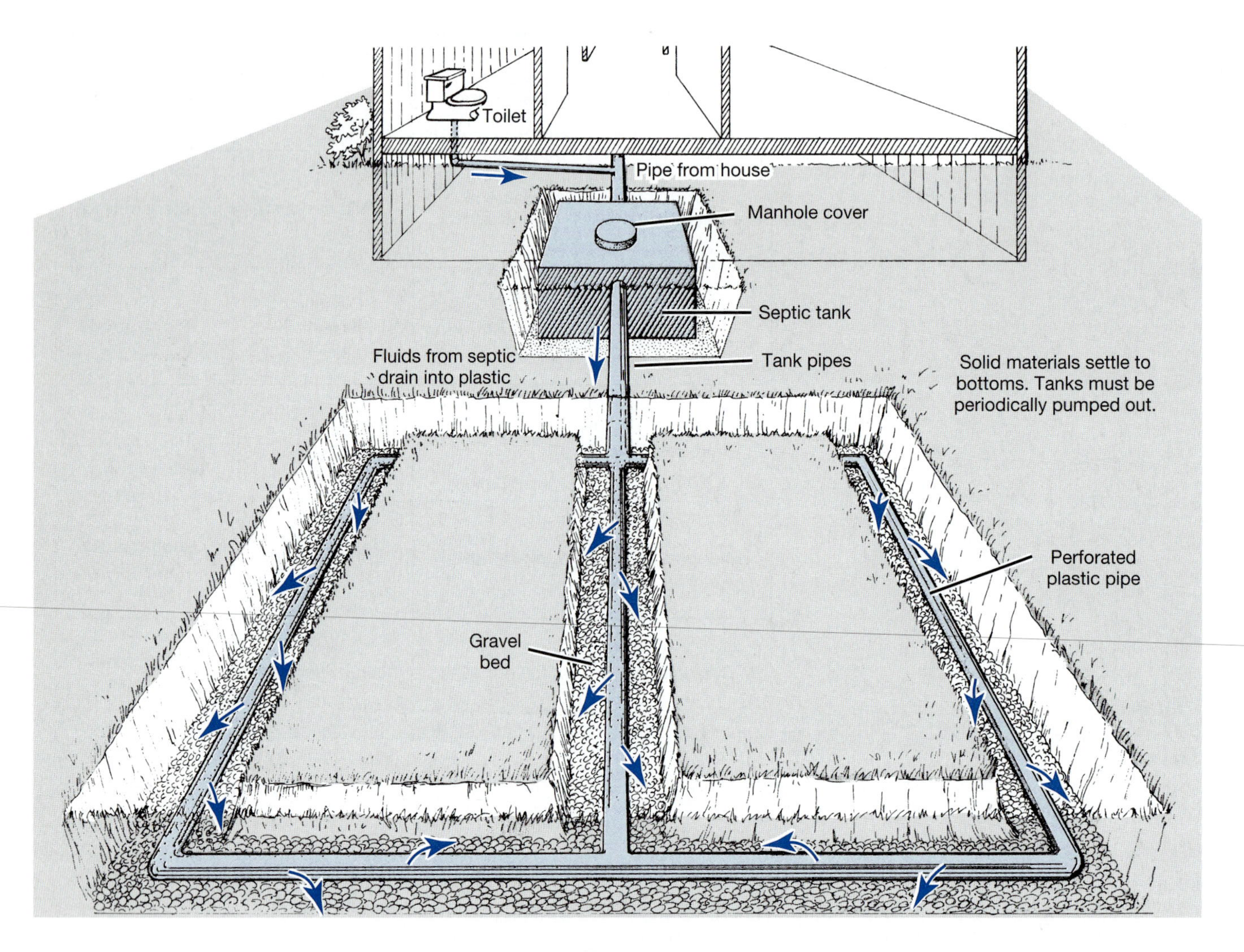

A. Septic tank and drain field

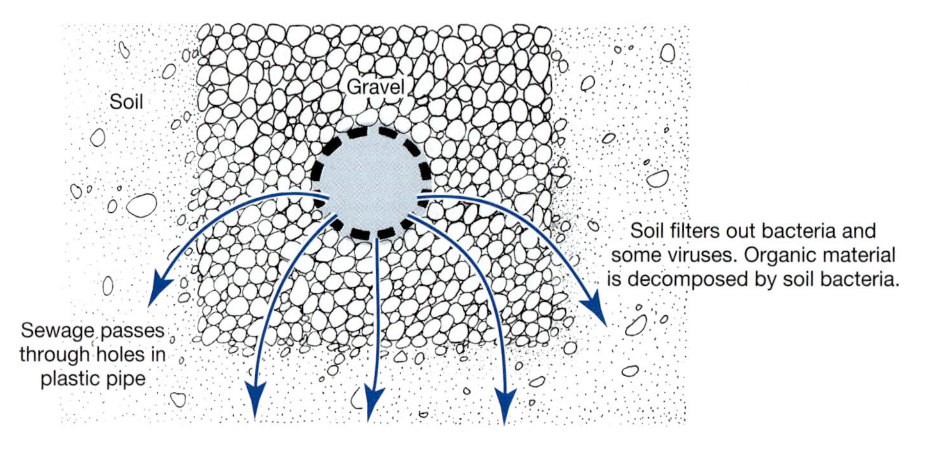

B. Cross section of drain pipe

FIGURE 11.22 *The septic tank and drain field system of sewage disposal.*

be used if the water table is too high, as it is in the southern United States, or if the soil is relatively impermeable. In such cases, they pollute the groundwater. If septic systems are overtaxed, their leach fields may become clogged with organic material. As a result, the partially decomposed waste may rise to the surface, causing visual pollution and generating foul odors. Septic systems also have a limited lifespan. Leach fields clog with undecomposed organic matter and must be replaced. Septic systems are the most frequent cause of bacterial contamination of groundwater, especially in areas where the population density is high. Septic systems servicing homes around lakes can also release substantial amounts of waste into the groundwater, which interfaces with surface waters. In crowded areas, nitrates and phosphates from the tanks can pollute surface waters, causing algal blooms.

Septic tank performance can be improved by filtering the liquid before it leaves the tank and enters the leach field. The performance and longevity of the leach field can also be enhanced by installing a small pump to force liquid from the septic tank through the field. This results in a more even distribution of liquid waste. Many leach fields fail when the first 10 meters (30 feet) of the system clog. The rest of the leach field is still in fine working order.

Alternative Treatment Methods

Although sewage treatment plants like those described in this section are the norm in most developed nations of the world, many countries are experimenting with alternative biological treatment facilities that are, quite gratifyingly for those interested in creating a more sustainable means of treating human waste, cheaper to build, much cheaper to maintain and operate, and more in harmony with natural cycles.

Holding Ponds, Indoor Biological Treatment Facilities, and Other Technologies In some cities and towns, sewage flows into a series of specially built ponds or marshes. In these facilities, various species of aquatic plants such as algae, cattail, water lilies, water hyacinths, and duckweed break down the waste. The tiny duckweed is a floating plant smaller than your little fingernail. In some ponds, duckweed grows profusely enough to form a solid green "living blanket" over the water. All these plants, of course, grow rapidly because they have access to the abundant supply of nutrients. Duckweed not only absorbs dissolved organics directly from the water but can be harvested and fed to livestock or even used as human food. In Thailand and Burma, the natives have long consumed duckweed, a highly nutritious plant that has six times the protein content of a soybean field of equal area! (The vegetation should be analyzed for the presence of metals before being eaten.) Such sewage holding ponds may even serve as wildlife habitats. Interestingly, the effluent from carefully built and well-thought-out sewage treatment ponds can be much cleaner than that coming from primary and secondary treatment facilities.

Indoor facilities are also being built. These generally consist of greenhouses that contain a number of tanks, each containing aquatic plants, microorganisms, and animals that degrade the waste material (Figure 11.23).

Sewage from homes can also be treated in specially built wetlands (Figure 11.24). One of the most popular designs are submerged wetlands. They consist of lined depressions filled with crushed rock or pumice, then covered with dirt. Sewage from the house empties into the system underground so there is no possible means of contact. Organic matter is then decomposed by bacteria in the rock bed. Plants growing in the soil absorb moisture and nutrients from the sewage, too. Effluent from such systems can be quite pure.

In areas where the topsoil is too shallow for septic tanks, many homeowners use composting toilets (Figure 11.25). This technology, invented in Scandanavia for such applications, has improved dramatically and several commercially available models are now available for home use. In these designs, wastes are deposited into a chamber. A handful of sawdust is used to cover the waste.Water in urine and in the feces evaporates through a vent pipe and the organic matter is converted into a rich, organic matter called humus. It can then be buried.

Land Application In the United States, nearly 7 million tons of sludge (dry weight) is produced each year. About 54 percent of it is returned to the soil or applied to the land. Sewage may be sprayed directly on land by means of tank trucks or conventional irrigation systems. In Lubbock, Texas, and Muskegon County, Michigan, this method is used extensively. Phosphates and nitrates are absorbed by the roots of crop plants and enhance yields. Any disease-causing microorganisms in the wastewater can be destroyed by preheating the waste before disposal. The organic material in the wastewater improves soil structure and increases its ability to absorb moisture and resist erosion.

For many years, Nassau County, Long Island, New York, dumped its sewage in the Atlantic Ocean. This practice is now illegal. As a result, the county sprays its treated sewage on land that overlies a severely depleted sandstone-limestone aquifer. As the wastewater percolates down through the soil, the impurities are gradually removed and the aquifer is recharged. The citizens of Nassau County are now taking showers and brewing their breakfast coffee with water that once suspended human waste—a remarkable example of the recycling of a precious resource.

Sewage Sludge: A Resource in Disguise?

For many years, sewage sludge has been buried in landfills or burned in order to get rid of it. Recently, however, many people have begun to see sludge as a valuable resource rather than a waste. They recognize that sewage sludge has many potential uses. It can be used as a fuel, livestock feed supplement, soil conditioner, and fertilizer. It can even be used as a building material. The wide use of sludge depends on efforts to prevent its contamination, especially efforts to ensure that industrial

FIGURE 11.23 *This greenhouse contains an artificial ecosystem constructed from plants, animals, and microorganisms that remove virtually all of the wastes from sewage. Such systems require few mechanical parts and are cheaper to operate than traditional sewage treatment plants. They also purify water without costly and toxic chemicals used in traditional sewage treatment plants.*

wastes containing heavy metals from factories is not comingled with municipal sewage or is pretreated to remove such pollutants. Today, only about 18 percent of the sewage sludge is currently landfilled in the United States. Let's consider some of the alternative uses for this material.

1. *Fuel.* During primary treatment of municipal sewage, much of the organic material in sewage is sent to a digester, where it is decomposed by bacteria in the absence of oxygen. During this process, a fuel called **biogas** is released. Biogas is mainly methane, the principal component of natural gas, a common household and industrial fuel. As noted earlier, biogas may be used at the sewage plant itself to heat the digester to the temperature required for proper operation and to power generators that provide electricity needed to run the plant. Even after removal from the digester, the treated sludge still has some fuel value left. If dried, it can be burned in an incinerator. The heat can then be recovered and used in industrial processes or to warm buildings.

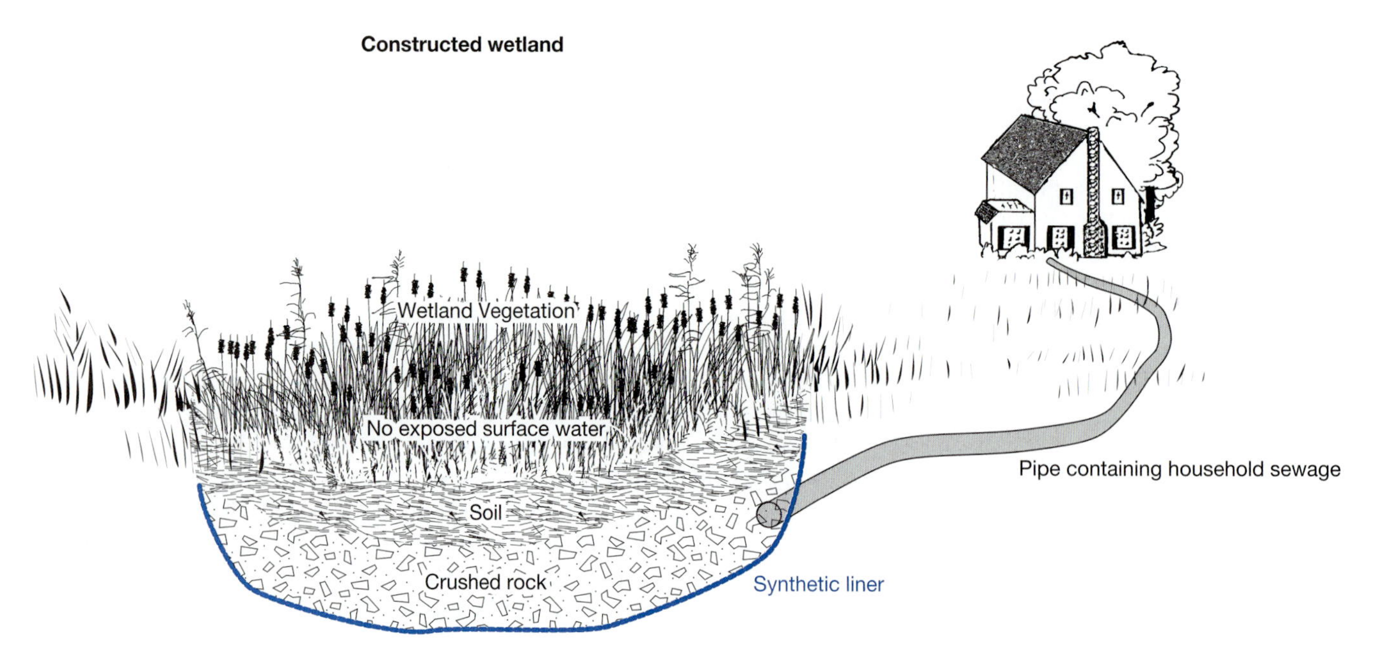

FIGURE 11.24 *Artificial wetland used to treat sewage from a house. Community-wide systems are also being used successfully in the United States and other countries.*

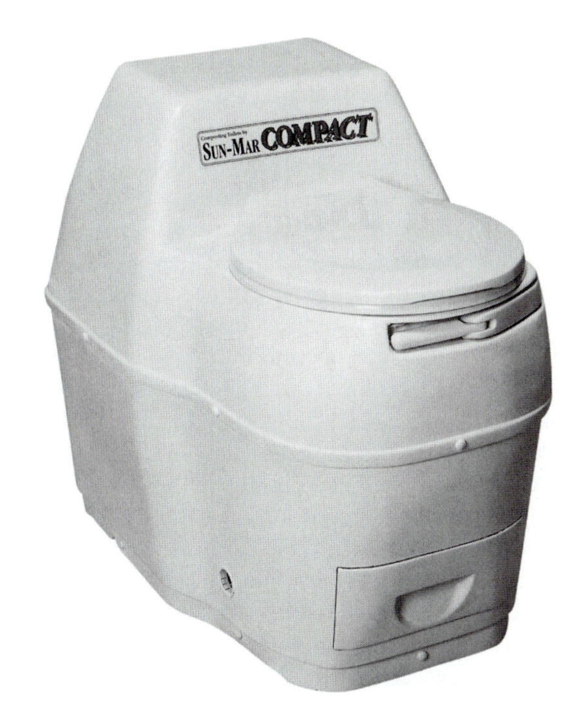

FIGURE 11.25 *Composting toilet. This SunMar composting toilet reduces toilet paper and feces to a fluffy organic matter rapidly and without odors.*

2. *Livestock feed.* Sludge has a fair content of nutritional proteins and fats and with proper treatment can be converted into a tasty feed supplement for cattle, pigs, and chickens.
3. *Soil conditioner.* Sludge has considerable value as a soil conditioner because it improves the ability of soil to retain nutrients, reduces erodibility, and promotes the ability of the soil to hold oxygen and moisture. Interestingly, the Metropolitan Sewage District of Milwaukee packages much of its sludge and markets it under the name Milorganite. Much of it is bought for use on lawns and gardens. At least 500 farms in the Milwaukee area regularly receive applications of Milwaukee sewage plant sludge.

FIGURE 11.26 *Application of sludge fertilizer on a Wisconsin farm.*

4. *Fertilizer.* Since sludge contains a fair amount of nitrogen and phosphorus, it has some value as fertilizer (Figure 11.26). However, the fertilizer value per pound is small compared to that of commercial artificial fertilizers. Nevertheless, Milwaukee area farmers who treat their land with sludge save almost $185 per hectare ($75 per acre) in fertilizer costs. As noted earlier, about 54 percent of all municipal sludge in the United States is now returned to the land as fertilizer.
5. *Building materials.* In Washington, D.C., stands one of the world's most unusual buildings: it is made in part of sewage sludge! The existence of this construction marvel is based on a technique developed by scientists at the University of Maryland. They succeeded in producing bricks from a mixture of sludge, clay, and slate. These so-called biobricks have no odors and look like ordinary bricks. The 750-square-meter (8,300-square-foot) sludge building was constructed by Washington's Suburban Sanitary Commission out of 20,000 biobricks. The widespread use of biobricks could have multiple environmental benefits, such as (a) reducing the cost of sewage disposal, (b) slowing the rate at which landfills are filling up, and (c) reducing both the soil erosion and visual pollution caused by the mining of clay.

11.4 Legislating Water Pollution Control

So far, we have examined the major forms of water pollution and discussed how many of them can be reduced through pollution control devices or, better yet, eliminated through preventive measures, the sustainable option. Water pollution control efforts in the United States and other nations have been inspired by numerous laws and regulations. In the United States, the **Federal Water Pollution Control Act** of 1972 is one of the most important environmental laws passed by Congress. The ambitious goal of the Water Pollution Control Act was to make the nation's waters fishable and swimmable by 1985. Toward this end, the U.S. Clean Water Act classifies surface waters according to their designated use: (1) drinking water, (2) swimming and fishing, and (3) transportation and agriculture. Most of the nation's surface waters are in the swimming and fishing category, and pollution control of these waters is meant to make such use possible. The nation has 120,000 miles of streams whose designated use is as a supply for drinking water. The transportation and/or agriculture category contains 32,000 miles of streams. The quality of these waters can be somewhat less than those in the swimming and fishing category.

The act also established minimal water quality standards for the nation's lakes, rivers, and streams and deadlines for industries and cities to reduce or eliminate their waste discharges. All U.S. cities were required to provide at least secondary sewage treatment. The law established fines of up to $50,000 per day and jail sentences of up to 2 years for violators.

In 1977, the Federal Water Pollution Control Act was amended under the name of the **Clean Water Act.** Amendments were also passed in 1981 and 1987. The 1987 amendments required all municipalities to have secondary sewage treatment plants in operation by July 1, 1988. The construction, operation, and maintenance of such plants were supported with $45 billion in federal funding and $15 billion in state and local funding during the period 1972–1986. Although an expenditure of $18 billion in additional federal funds was authorized by the 1987 amendments for the period 1987–1996, this represents a considerable reduction in the level of federal funding from that of the 1970s. In 1987, only about 70 percent of our nation's municipalities had constructed secondary sewage treatment plants due to cuts in federal spending for water pollution control. Since that time, secondary treatment has increased. In 1996, the latest year for which data were available, 40 percent of all sewage received secondary treatment, and nearly half of all sewage received tertiary treatment.

The Clean Water Act and its amendments requires water providers to drastically reduce or even eliminate chemical pollutants from drinking water that might cause serious human illness or even death. How effective has the Clean Water Act been?

The most inclusive information on water quality in the United States to date has been provided by the monitoring program of the U.S. Geological Survey. It is called the National Ambient Stream Quality Accounting Network (NASQAN). It includes 501 monitoring stations located on streams distributed throughout the country. In 1995, NASQAN reported that 4 percent of streams monitored exceeded the standard for phosphorus, and 1 percent of streams had unacceptable levels of dissolved oxygen, cadmium, and lead. In other areas, U.S. streams were not so well off. For example, 35 percent of all streams monitored exceeded the EPA standard for coliform bacteria, up from 25 percent 4 years earlier. What many cities have found is that water quality has not improved or has even declined, despite the construction of expensive treatment plants. Studies have shown that the reason for this is that proliferation of nonpoint sources. Although cities and towns have made remarkable progress in reducing pollution from sewage treatment plants, nonpoint sources have increased as populations expanded. Nonpoint water pollution has proved to be difficult to control, for reasons discussed earlier in the chapter. Nonpoint pollution has therefore thwarted progress in cleaning up our lakes and rivers.

The Clean Water Act (1987) authorized the expenditure of millions of dollars of federal money by the states to control nonpoint water pollution. In the first 5 years of this program, the U.S. Government spent $270 million. To receive the money, the states were required to match federal grants by 40 percent with their own money. One of the most popular approaches has been an "effluent trading policy" in which money has been spent to tighten controls on point sources to offset nonpoint pollution entering lakes and rivers. Over the years, hundreds of successful programs have been launched, but much more is needed to eliminate or greatly reduce this growing and serious problem.

The Clean Water Act is currently subject to intense debate. It was weakened by the 103rd Congress, and the 104th Congress tried to destroy the law, according to supporters of this important legislation. At this writing efforts are still underway to greatly weaken wetland protection. Another force that is threatening the law is the current shift from federal to state control. At the state level, clean water laws are being weakened by industries that oppose the legislation.

Another key piece of legislation aimed at protecting water is the **Safe Drinking Water Act** of 1974. This law, as its name implies, is designed to protect drinking water quality. It establishes the EPA as the main regulator of drinking water quality, one that oversees state and local governments, and other water suppliers. The Safe Drinking Water Act called on the EPA to establish regulations for pollutants in drinking water—that is, standards that tell us how much can be present in our water supplies. It also instructed the EPA to establish programs to protect groundwater drinking water supplies. Implementation of the 1974 law was slow in coming, so in 1986, the U.S. Congress passed a sweeping set of amendments to accelerate the pace at which the EPA was moving. The amendments called for the establishment of regulations for 89 specific contaminants by 1989. It also called on the EPA to establish regulations to ensure that drinking water from public supplies was filtered and disinfected. It banned the use of lead pipes and lead solder in plumbing systems and established a program to protect areas around wells supplying public drinking water.

Many of these goals have been met. Currently, nearly 80 National Primary Drinking Water Standards have been set. The EPA has also established standards for other pollutants, although compliance with these standards is voluntary.

In 1996, the act was once again amended. The newest amendment established a fund to help communities upgrade outdated drinking water filtration and purification systems. It established a reporting system, too, calling on all water suppliers who serve more than 10,000 customers to report violations of standards and to pinpoint sources of water pollution in their system, which then becomes public knowledge. The amendments also required the EPA to use sound science when establishing drinking water standards and to base decisions on risk and benefit calculations.

Another law that has helped in the battle against water pollution is the **Emergency Planning and Community Right-to-Know Act,** passed by the U.S. Congress in 1986. This law established a nationwide **Toxics Release Inventory (TRI),** a pollution accounting system. Under the law, major industrial facilities are required to publish data on the levels of pollutants that they have discharged into the air, water, and land or transferred to other sites for incineration, recycling, and disposal. These reports are open to the public via the Internet. TRI data, sometimes showing the release of huge amounts of toxic substances, has been used by local and national activists from national environmental groups to put pressure on companies to reduce toxic emissions. Many companies took steps to reduce their emissions because of such pressure. Others were astounded at the amount of pollution and the cost of getting rid of it and took steps to reduce emissions for financial reasons.

Watershed Management Plans

With nonpoint water pollution becoming such a major problem, and countering many gains in point sources, cities and towns have begun to explore other avenues to protect water quality. One approach gaining in popularity in the 1990s is **watershed management,** better management of entire watersheds. Watershed management involves actions by many individuals living and working within a watershed, among them farmers, homeowners, gardeners, city park managers, city street crews, and boaters. Watershed management involves many steps by many individuals with several key goals: to protect, even increase, vegetative cover; reduce impervious surfaces; and reduce nonpoint pollution sources in a watershed. The topic is so massive it could require a book to explain. Let's consider some examples to give you an idea of the depth and breadth of this approach.

In some cities and towns, officials are helping to establish vegetative **buffer zones** along streams, rivers, and lakes. Buffer zones help reduce surface flow into streams, cutting down on flooding and sediment pollution. Vegetative buffer zones may also help to remove nitrates, phosphates, and organic pollutants. Naturally occurring bacteria and other microorganisms in the zone decompose or incorporate organic and inorganic nutrients, respectively. Toxic metals are physically filtered out.

In other areas, officials are setting aside more open space, ranging from public parks to undeveloped wildlife habitat. In Boston, Massachusetts, officials have preserved land along the St. James river that runs into Boston Harbor. Consisting of parks and wetlands, this vegetative zone not only preserves that aesthetic qualities of an area and provides recreational opportunities, it helps to protect wildlife habitat. The protected areas reduce surface runoff, thus preventing flooding, erosion, and sediment deposition in the river.

Some cities and towns divert storm water into retention ponds, artificial structures that help reduce flooding by absorbing massive water flows from impervious surfaces such as streets, driveways, parking lots, and roof tops. Water is then allowed to flow into streams at more reasonable rates or soak into the ground, replenishing groundwater.

The City of Austin promotes responsible actions on the part of citizens through educational programs. Citizens are asked to wash cars on their lawns, rather than their driveways, so soaps don't run down into the sewer and into local waterways. Citizens are asked to apply fertilizer and pesticides sparingly and never before a storm, or to use natural, nontoxic alternatives. City officials offer advice on disposing chemical wastes, including motor oil and antifreeze. They even install "No Dumping, Drains to Creek" frog markers on stormdrains to remind citizens that a sewer is not a place to dispose of paints, oils, and other fluids. With boating a popular recreational activity, they promote cleaner, more environmentally sound ways of enjoying the sport. They, for instance, encourage boaters not to throw litter overboard, especially fishing line and six-plastic pack rings. They urge boaters to scrub their boats with a brush, and not use soap, or to use phosphate-free and nontoxic soaps.

Many cities and towns have also written formal **watershed protection plans.** They are usually derived after months of work by citizens, environmentalists, government officials, businesses, farmers, and other local stakeholders (people who have a stake in the outcome and will be affected by recommendations of the plan). A watershed protection plan outlines ways to protect watersheds for multiple reasons, maintaining water quality being one of the key ones. Citizens and public officials from Frederick County, Maryland worked with the Center for Watershed Protection, a nonprofit group, to develop an extensive set of recommendations to ensure that future development will be conducive to watershed protection. The plan called for shorter, narrower streets, fewer and smaller cul-de-sacs, and smaller parking lots to reduce impervious surfaces. It also called for increased stormwater treatment, to reduce stormwater flow into streams and other surface waters. And, it called for more community open space, increased vegetated buffers, limited clearing and grading of sites, and ways to enhance native vegetation.

Important as they are, watershed management plans often fall flat on their face, according to an analysis by the Center for Watershed Management. The reasons for their failure are many. But in generally, most are overly ambitious and prescriptive—that is, they outline what needs to be done, but don't result in the formulation of distinct regulations or mechanisms to fund such actions. As a result, they do little, if anything, to protect watersheds. According to the Center, many of them simply end up on the shelves of city officials, gathering dust with so many other plans that have been produced by similar processes in the past decade or two. Unless city officials and developers are required to take action, it won't happen. Unless there's money to monitor development and watershed conditions, there's little hope of anything worthwhile coming out of management plans. That said, we mustn't forget that watershed management programs do point out that many laws, zoning regulations, and ordinances actually work against watershed protection, permitting, sometimes encouraging activities that increase impervious surface, increase erosion, and decrease vegetative surface. Their true value, however, will only be achieved when plans are translated into new laws, zoning regulations, and ordinances. Their true value will be realized only when there are agencies to monitor watershed development. To be successful, these plans will need to produce desired, long-term outcomes of protecting streams and other resources from degradation.

11.5 Pollution of Oceans

Despite the many benefits humans derive from the ocean, we have treated it as if it were expendable. Up until 1970, oceans were generally viewed as vast and bottomless landfills for domestic, municipal, and industrial garbage. Disposal of waste materials in this manner appeared convenient, economical, and safe. Legislation related to the control of ocean oil pollution was enacted prior to 1970 but with little compliance. Lack of direct language in the laws and ineffective methods of enforcement were partially to blame. Also, prior to the 1970s, the full impacts and/or long-term effects of oil and other types of ocean pollution were unrecognized, misunderstood, or underestimated.

Since the 1970s, increases in scientific information, public awareness, and the number of hazardous incidents related to ocean pollution have influenced an onslaught of legislation designed to protect our valuable ocean resources from further degradation. The International Maritime Organization (IMO), an agency sponsored by the United Nations, has developed a number of programs and regulations aimed at reducing pollution of the ocean environment. Another U.N. agency, the International Seabed Authority, is responsible for regulating ocean mining in international waters. The **Federal Water Pollution Control Act (FWPCA)** sets water quality standards and regulates the discharge of pollution into U.S. waters. In addition, specific laws have been passed that ban or regulate disposal of U.S. waste products into the ocean, giving the job of enforcement primarily to the U.S. Coast Guard and the U.S. Army Corps of Engineers. These and other international and national attempts to share and protect ocean resources are encouraging; however, conflicts over use and pollution are ongoing problems due to the difficulty of enforcement.

Sewage

For decades the neritic zone bordering the Atlantic, Gulf, and Pacific coasts has been used as a dumping ground for a large volume of municipal sewage sludge, industrial wastes, and even household garbage. The harmful affects sewage dumping has on marine ecosystems are well illustrated by the problems at the New York Bight.

The bight is a relatively shallow area over the continental shelf opposite New York Bay. The site was used as a dumping ground for sewage sludge (and other wastes) for more than 60 years (Figure 11.27). In the early 1970s, for example, more than 7 billion liters (1.8 billion gallons) of sewage sludge from the New York metropolitan area spewed through 130 discharge pipes directly into the bight each year. More than 16 percent of this sewage had received no treatment whatsoever. In addition, raw sewage from 23 New Jersey towns entered the bight. As a result, the bottom of the bight is blanketed with a layer of black sludge that is 105 square kilometers (40 square miles) in area. This "blanket" is contaminated with a large variety of pollutants ranging from toxic metals and organic compounds (PCBs) to viruses and bacteria that can cause human disease.

The long-continued dumping of sludge and raw sewage in the New York Bight has had many adverse effects on the marine ecosystem:

1. Because of the high biochemical oxygen demand (BOD) of much of the waste, the concentration of dissolved oxygen in the region was often less than 2 ppm. Populations of microscopic algae and crustaceans fell sharply or disappeared altogether. This alteration, in turn, caused a decline in many commercially valuable species of plankton-eating fish.
2. The stomachs of bottom-dwelling fish like flounders that were caught near the dump site contained a number of abnormal items such as bandages, cigarette filters, and hair.
3. Some fish suffered from black gill disease, characterized by abnormally dark gill membranes and reduced respiratory function.
4. Toxic metals such as nickel, chromium, and lead reached unusually high levels in some fish.
5. A considerable number of harmful mutations resulting from chromosome damage were observed in young mackerel. Moreover, clam and oyster beds were so highly contaminated with disease-causing microorganisms that these creatures were unfit for human consumption. Apparently the bacteria and viruses were transported from the waste to the clam and oyster beds by shoreward-moving currents.

In 1986 the Environmental Protection Agency (EPA) directed New York City to dump its sewage sludge at a newly designated site at the edge of the continental shelf 170 kilometers (106 miles) from the coast. The city began dumping its sludge at the new site in 1987. In 1988, however, in an-

FIGURE 11.27 *Barge hauling refuse from New York City to dumping grounds.*

other amendment to what is referred to as the **Ocean Dumping Act,** the U.S. Congress agreed to ban all ocean dumping as of January 1, 1992. Severe fines can be imposed on any community in violation of the ban. The result of this action should finally phase out sewage sludge dumping by cities such as those in New York and New Jersey, which once amounted to 7 million metric tons per year. Today, nearly all ocean dumping that occurs in U.S. waters consists of dredged materials; however, other countries still dump sewage sludge and non-toxic industrial wastes.

Dredge Spoils

Eighty percent of the waste that has been dumped into U.S. coastal waters is dredge spoil. **Dredge spoil** is sediment (sand, silt, clay, and gravel) that has been scooped from harbor and river bottoms to deepen channels for navigation (Figure 11.28). More than 400 million cubic yards are dredged annually from U.S. channels and harbors—the equivalent of a four-lane highway, 6 meters (20 feet) deep, beginning in New York and ending in Los Angeles (Figure 11.29). This waste poses an enormous disposal problem. Annually, about 15 percent of this material, or 60 million cubic yards, is disposed of in the ocean. Where to dump can be a contentious decision. The urgently needed dredging of Baltimore Harbor was postponed for 15 years because no agreement could be reached on where to dump the spoil. Much of the spoil generated in the mid 1980s was dumped in 70 different ocean sites.

Unfortunately, about 1 in every 3 tons of dredge spoil is contaminated with both urban and industrial waste, as well as with pollutants resulting from urban and agricultural runoff. These contaminants (PCBs, heavy metals, and so on) eventually enter marine food webs and may harm not only ocean life but humans as well. Under the terms of the **Marine Protection, Research, and Sanctuary Act,** the U.S. Army Corps of Engineers, which does most of the dredging, was charged with finding suitable disposal sites beyond the continental shelf. At such sites, the water is deep enough that most of the pollutants should be greatly diluted, minimizing their adverse effects on the marine ecosystem.

Plastic Pollution

The world's oceans are seriously polluted with plastic (Figure 11.30). The enormous volume of plastic debris riding the waves is not appreciated until one hikes along a beach. During a 3-hour cleanup of a 260-kilometer (157-mile) stretch of the Texas coast, the following plastic objects were removed: 31,800 bags, 30,000 bottles, 29,000 lids, 7,500 milk jugs, 15,600 six-pack rings, 2,000 disposable diapers, and 1,000 tampon applicators. Even remote islands are not immune from an accumulation of plastic and other litter. The Ducie

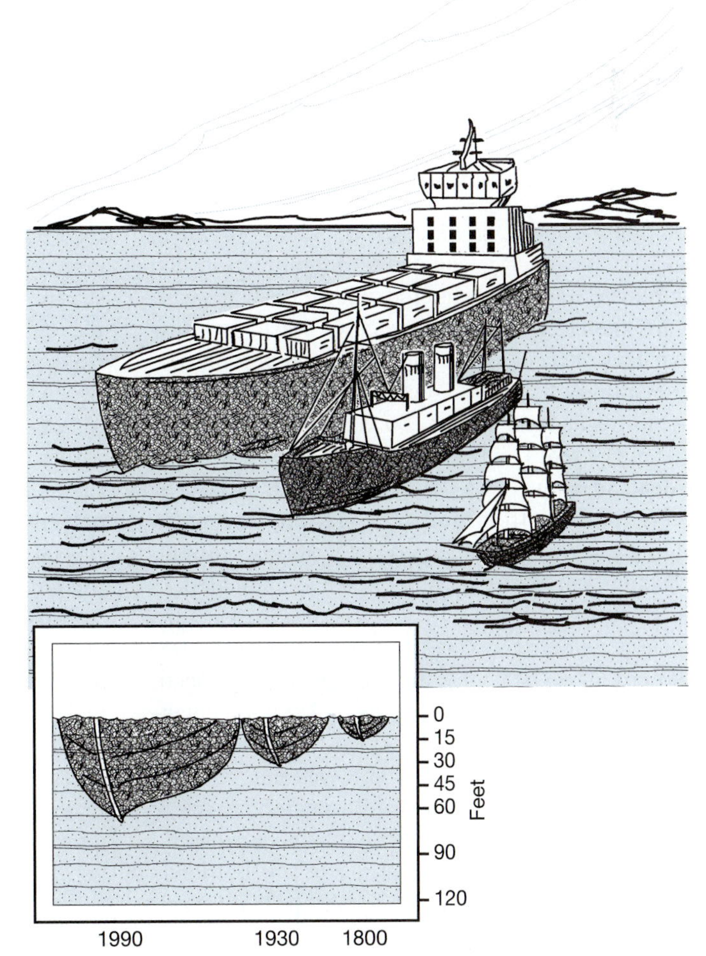

FIGURE 11.28 *Dredging harbors and channels to increase their depth is a continuing and growing necessity. Today an average ship needs 45 feet and a tanker needs 60 feet. In the early 1800s, an average cargo vessel needed only 15 feet; in the 1930s, an average steamship needed 26 feet.*

FIGURE 11.29 *Each year in the U.S., we dredge 400 million cubic yards of sediment from harbors and channels—enough material to build a four-lane highway, 20 feet deep from New York to Los Angeles.*

Atoll in the South Pacific, an uninhabited island 300 miles from the nearest inhabited island and 3,000 miles from the nearest continent, was found littered with human trash in 1991. In a 2.4 kilometer (1.5 mile) stretch of beach, 950 miscellaneous pieces of trash were collected, much of it plastic. Although seemingly harmless, such materials kill at least 1 million seabirds and more than 100,000 whales, porpoises, and seals every year (Figure 11.31). The gut of a sea turtle found dead in Hawaii was jammed with a variety of lethal objects, ranging from golf tees and bottle caps to bags and imitation flowers. Hundreds of seabirds, salmon, and marine mammals die when they become entangled in discarded fish nets.

Plastic causes wildlife mortality is several ways: (1) After being swallowed, it can be neither digested nor voided. Death is caused by blockage of the digestive tract. (2) Plastic entanglement may cause death by drowning. (3) Plastic entanglement may cause starvation because it prevents marine birds and mammals from searching for or swallowing food.

The sources of these lethal plastic pollutants are many. Every industrialized society lives in a plastic world. Manufacturers in the United States alone annually produce more than 6 million metric tons of plastic. Much of this plastic is discarded into streams by humans and then carried downstream to the ocean or dumped directly into the ocean from fishing boats and other commercial or recreational vessels or from garbage barges. The National Academy of Sciences once reported that more than 5 million plastic containers are tossed overboard from ocean-going vessels every day. Over a 3-year period, that would be one container for each person on Earth. In some regions of the Atlantic and Pacific oceans, polyethylene packing pellets reaching a density of 2,500 to 10,000 per hectare (1,000 to 4,000 per acre) have been found floating on the surface.

Most plastic cannot be broken down by bacteria. In other words, plastic is nonbiodegradable. Another problem is that most plastic items are quite buoyant. These characteristics make it possible for a golf tee from a Seattle golf course eventually to be ingested by a seabird in the South Pacific.

The amount of plastic floating and bobbing on the global seas will certainly increase. After all, population continues to grow and, therefore, so will demand for products. Commercial fishers lose more than 136,000 metric tons of plastic lines and nets annually. In the North Pacific alone, fishermen set out

FIGURE 11.30 *This "plastic mountain" represents only 1 week's accumulation of plastic and Styrofoam waste by an American middle-class family. The following items are included: six pieces of foam cushioning, five kitchen garbage bags, five sandwich bags, two toothbrush holders, two egg cartons, one wastebasket, two motor oil bottles, one soap tray, one mixing spoon, one fork, one ice cream bucket, one milk jug, one motor coolant jug, one plant food container, one cook's mixing bowl, and one bag for dry-cleaned clothes! Many plastic throw-aways are also accumulated at sea by fishermen, ship's crews, and passengers and are dumped overboard.*

FIGURE 11.31 *Plastic six-pack frame threatens a Western Gull on a California beach with death by strangulation.*

more than 32,000 kilometers (20,000 miles) of plastic nets nightly. Within a year, more than 4,800 kilometers (3,000 miles) of netting is lost, forming a considerable threat to marine life. Isn't it ironic that 10 times as many fur seals die from fish net entanglement as are killed by the Pribilof Island (Alaska) hunters, who have been sharply criticized for their fur seal butchery?

Control of Plastic Pollution

How can plastic pollution be controlled? At present, dumping in the ocean is regulated by the **London Dumping Convention** (1972) that has been signed by more than 85 nations. Annexes to the Convention regulate disposal from all trash-hauling ships. In the United States, this convention was implemented by the Ocean Dumping Act. Another international law that controls plastic pollution is the 1973 **Marine Pollution Convention,** or MARPOL Act. Annex V of the act bans the dumping of plastic by all ships other than trash ships. After several years of deliberation, Annex V was eventually ratified when both the United States and Great Britain signed the act in 1987. As a result, beginning in December 31, 1988, it became illegal for the ships of the 15 signatory nations and for the ships of any nation plying their waters to dump plastic at sea. Since the 15 signatory nations accounted for over 50 percent of the gross tonnage of the world's commercial ships, this agreement should help reduce plastic pollution. The Coast Guard enforces Annex V for the United States.

Another way to reduce this problem is to recycle more plastic materials. Some progress is being made. For example, in the United States, plastic soft drink bottles are reprocessed into paintbrushes, stuffing, and industrial straps. Some discarded plastic is recycled into building material used in the construction industry.

Perhaps the ultimate solution to plastic pollution is at the source of the plastic—the manufacturing process. Some manufacturers, for example, have recently developed a type of photodegradable plastic that will disintegrate when exposed to ultraviolet light from the sun. At least 11 U.S. states have laws that require photodegradable plastic in some products, such as the rings for soft drink and beer six-packs. A few manufacturers in the United States, Canada, and Italy are now producing biodegradable plastic bags. Should photo- or biodegradable plastics come into mass production, this potent threat to marine life may gradually disappear, assuming the breakdown products are not harmful.

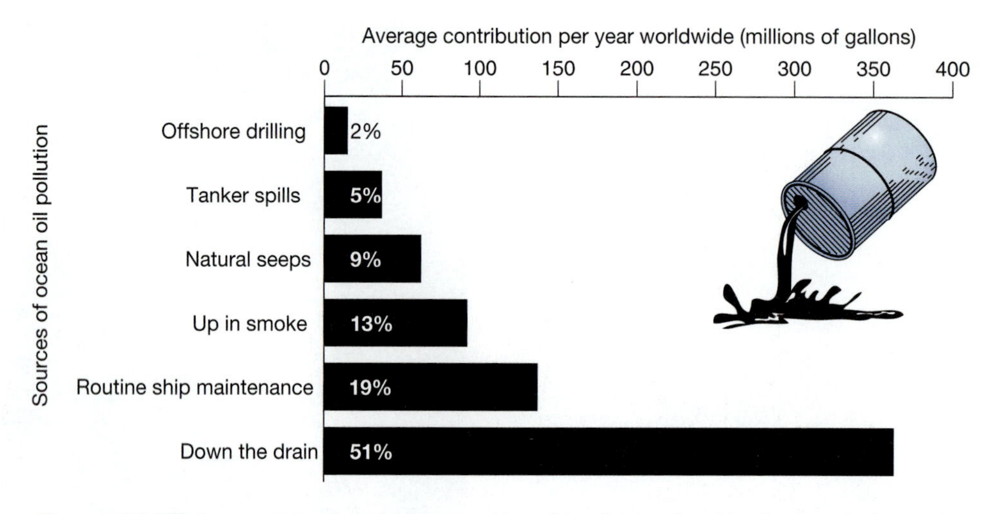

FIGURE 11.32 *Sources of oil pollution in the ocean. Most of the oil comes from inland sources via drain and sewer pipes and river runoff. Other human-related sources are tanker spills, routine ship maintenance, air pollution, and offshore oil well accidents.*

Oil Pollution

Oil has always polluted the sea. About half of the oil in the seas today has seeped through cracks in the ocean floor. This naturally occurring oil, however, is of little concern because the sources are widely distributed and contribute only 9 percent to the total oil input annually. The concern for most conservationists is oil from human activities that account for the remaining 91 percent of the annual input (Figure 11.32).

Oil Tanker Spills Although *oil tanker accidents* (Figure 11.33) are the most spectacular and newsworthy source, they account for only 5 percent of the oil that enters the ocean each year (Figure 11.32). Large spills can devastate several miles in localized areas since tankers usually run aground close to the sensitive and resource-rich coastal shores. Tanker accidents are widely publicized by the press, radio, and television. In fact, the breakup of the *Torrey Canyon* off the British coast in 1968, during which over 136 million liters (36 million gallons) of oil were lost, first alerted the general public to the problem. In 1976, the *Argo Merchant* broke apart on the Nantucket Shoals, 39 kilometers (24 miles) off the Massachusetts coast, and polluted the seas with 29 million liters (7.7 million gallons) of heavy fuel oil.

On March 17, 1978, one of the world's most serious tanker spills occurred. The Council on Environmental Quality described the episode as follows: "The oil tanker *Amoco Cadiz* went aground 2 kilometers (1.2 miles) off Portsall on the coast of France. Efforts to stop the spill or contain it were unsuccessful." The ship broke apart, and big winds and seas made it impossible to transfer oil to other tankers. As a result, the entire cargo of 228 million liters (60 million gallons) of crude oil was spilled, polluting the waters and shoreline along 198 kilometers (124 miles) of coast and 61 kilometers (38 miles) out to sea. Not only was there severe destruction of fish and seabirds, but the heavy oil ruined the beauty of the coastal area and impacted the economy of the shore-based villages. A number of citizens' lawsuits were filed against the Amoco Oil Company, owners of the tanker. In 1988, almost 10 years after the spill, the company was ordered by the courts to pay out millions of dollars in compensation.

On March 24, 1989, citizens of the United States were horrified by news of a 40 million-liter (11 million-gallon) oil spill in

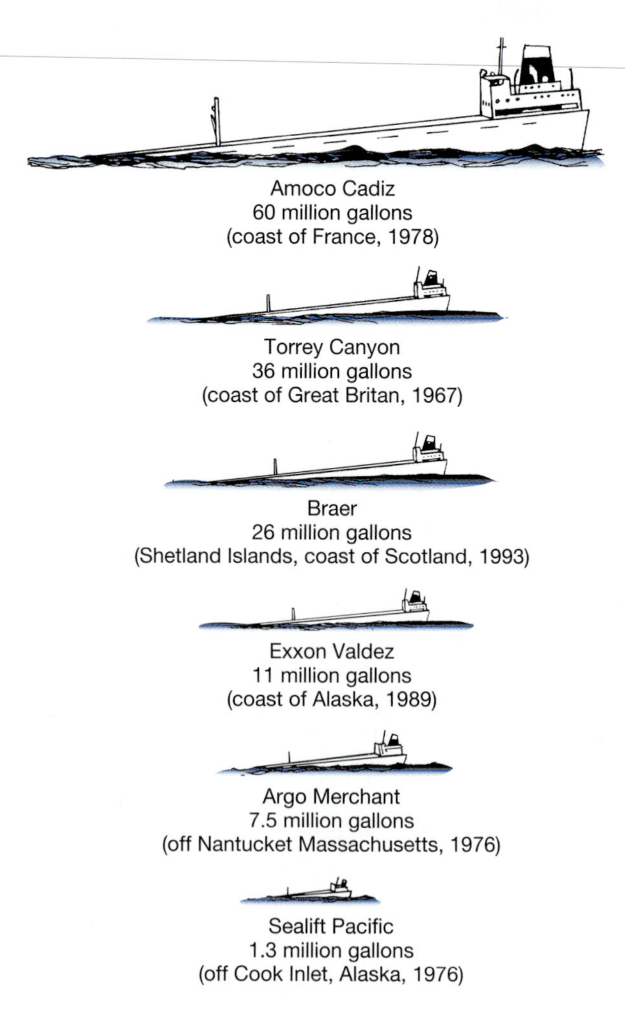

FIGURE 11.33 *A sample of major tanker oil spills.*

FIGURE 11.34 *The oil tanker* Exxon Valdez *ran aground in Prince William Sound, Alaska, March 24, 1989.*

Alaska's Prince William Sound near the port of Valdez (Figure 11.34). The supertanker, *Exxon Valdez,* ran aground on a reef in the sound, releasing much of its oil into the pristine, biologically rich waters. The oil slick spread quickly and by the end of the summer had polluted 2,300 kilometers (1,400 miles) of shoreline (Figure 11.35). Thousands of seabirds and otters perished in the oil, which formed a layer 1 meter (3.3 feet) thick in some places (Figure 11.36). Making matters worse, the spill occurred only 2 weeks before migrating flocks of waterfowl arrived. Many birds spend the summer in the waters of Prince William Sound; others merely stop there to feed and rest on their way to the Arctic tundra, where they breed.

The *Valdez* spill was not the largest in history but probably will go down as one of the most costly, economically and environmentally. The damage can be attributed to four key factors. First, the spill occurred in the relatively protected waters close to land. Second, cleanup was delayed for several crucial days. The special cleanup force stationed at Valdez had been all but abandoned by the oil companies operating there. Consequently, there were an insufficient number of oil skimmers close at hand. Third, the waters off the sound are cold, which retarded the biological degradation of the oil. Fourth, the waters were extraordinarily rich in sea life.

Routine Ship Maintenance The bulk of the spills caused by tankers and other ocean vessels occur during loading and discharging, bilge and fuel oil cleaning, dry-docking, oil ballast discharge, and other voluntary and routine shipping operations. Nearly 20 percent of the oil discharged into navigable waters each year are a direct result of millions of these routine operations that spill just a few gallons each (Figure 11.32).

FIGURE 11.35 *An Exxon worker cleans oil-covered rocks with a hot water spray in the aftermath of the massive oil spill.*

FIGURE 11.36 *An oil-soaked cormorant registers its protest to the atrocious oil spill off Alaska's coast in the once pristine waters of Prince William Sound. Thousands of birds and mammals died. Volunteers cleaned many birds and mammals that were caught in the oil, but the animals' prospects for survival were dim.*

FIGURE 11.37 *Offshore oil platform.*

Offshore Oil Well Accidents In 1969 a major oil well off the coast of Santa Barbara, California, accidentally released thousands of gallons of oil because of a faulty drilling technique. However, the spill was a mere grease spot compared to the 140 million gallon oil well blowout that occurred in the Bay of Campeche off Mexico's east coast in 1979. Oil escaped from the well for several months, threatening marine life along the Texas shore several hundred kilometers to the north. The world's largest oil spill occurred as a result of fighting in the Persian Gulf War during 1991. The spill involved a combination of wells, terminals, and tankers that amounted to a 240 million-gallon disaster in the Persian Gulf and surrounding area.

Offshore drilling accidents contribute less oil than tanker spills on an annual basis, but as in the examples above, can be even more devastating. Without tighter controls, oil well accidents could increase dramatically in the future. For example, American oil companies alone drill about 1,300 offshore oil wells a year (Figure 11.37). Four thousand new offshore wells are drilled worldwide each year.

Land Sources of Oil Pollution Over 50 percent of the oil input to the ocean each year comes from land sources, from both inland and coastal communities. The major sources are service stations, motor vehicles, and factories. This oil finds its way to the ocean via storm and sewage drains and as runoff in rivers and streams (Figure 11.32).

Remember the last time you drove up to a service station and asked for an oil change? Did you ever wonder what happened to the old oil? Until recently, it was probably discharged into sewers that drain into rivers, and eventually flowed to the sea. Three or five quarts of oil doesn't seem important. But how about billions of quarts? In fact, millions of motorists throughout the world are indirectly responsible for a much greater volume of oil pollution of the oceans than is caused by tanker breakups and oil well blowouts.

Air Pollution What about airborne hydrocarbons from factories, service stations, and vehicles? Maybe you fill up at a self-service pump. If so, you are well aware of the pungent odor of evaporating gasoline. Obviously, not all of the gasoline goes into your tank. Some of it escapes into the air. Some unburned gas also escapes from your exhaust pipe and becomes airborne. Evaporation of petroleum also occurs at thousands of industrial plants throughout the world. Combined, ocean-going vessels burn about 1.7 million tons of diesel fuel per day, spewing airborne effluent. These pollutants wash from the sky, polluting the oceans with at least 20 million metric tons of airborne petroleum hydrocarbons annually, contributing 13 percent of the oil inputs (Figure 11.32).

Adverse Effects of Oil Pollution Precisely how a given oil spill will affect marine life is difficult to predict. The effects depend on a number of highly variable factors, such as the amount and type of oil (crude or refined), proximity to organisms, season of the year, weather, ocean currents, and wind velocity. Adverse effects include: (1) a reduction in photosynthetic rates, (2) the concentration of chlorinated hydrocarbons, (3) death and abnormal behavior of marine animals, and (4) food chain contamination with carcinogens.

Because a heavy oil slick is opaque, it blocks sunlight. Photosynthetic activity of the marine algae below the oil barrier is sharply reduced, if not arrested completely. Such diminished rates of food production restrict the growth and reproduction of all organisms that are directly or indirectly dependent on marine algae for food.

An oil spill may also greatly increase the concentration of chlorinated hydrocarbons, such as the pesticides DDT, dieldrin, toxaphene, and the PCBs. These compounds are highly soluble in oil. As a result, marine organisms such as phytoplankton, as well as animals such as crustaceans and larval fishes, which migrate to the sea surface at night, are adversely affected. Even if those organisms are not killed directly, their physiology, growth, reproduction, and behavior may be impaired.

The heaviest influx of oil occurs in the neritic zone near the continental margins—the zone where virtually all of our shellfish (oysters, lobsters, and shrimp) and over half of our commercial fish crop are produced (Figure 11.38). (The *Amoco Cadiz* spill caused $25 million in damage to the French oyster industry alone.) The number of seabirds killed annually worldwide is enormous. In a single winter, more than 250,000 murres, eiders, and puffins were destroyed by oil pollution. In 1988, thousands of these birds were destroyed by oil spills in the North Sea.

Several of the hydrocarbons in crude oil mimic chemicals used by marine animals to guide them during mating, feeding, homing, and migrating. Flooding the ocean with pseudosignals from oil spills might derange the behavior of marine animals, causing them to expend valuable energy in nonadaptive pursuits. The eventual decline in vitality and numbers of the species would seem assured.

Crude oil is not just a single compound but a complex mixture of dozens of different hydrocarbons such as benzopyrene, an acknowledged cancer-inducing chemical, or carcinogen. These carcinogens may be concentrated in marine organisms such as shrimp, lobsters, and fish, and may eventually be consumed by humans.

Control of Oil Pollution In the U.S., the **Oil Pollution Act of 1990** was passed in response to public outcry of the devastation caused by the *Exxon Valdez* grounding. The law established provisions for new tanker structural design including double hulled construction; retirement or phase out of existing tanker fleets; financial responsibility, compensation, and liability for spills; improved spill response strategies; and inspection systems. With international cooperation, the number of oil spills from tankers has been reduced. But, tighter controls and improved enforcement methods are still needed.

Some strategies for oil pollution control after a spill has occurred include:

1. *Physical cleanup.* Oil that washes ashore may be cleaned up manually or with machines. In the *Valdez* spill, for

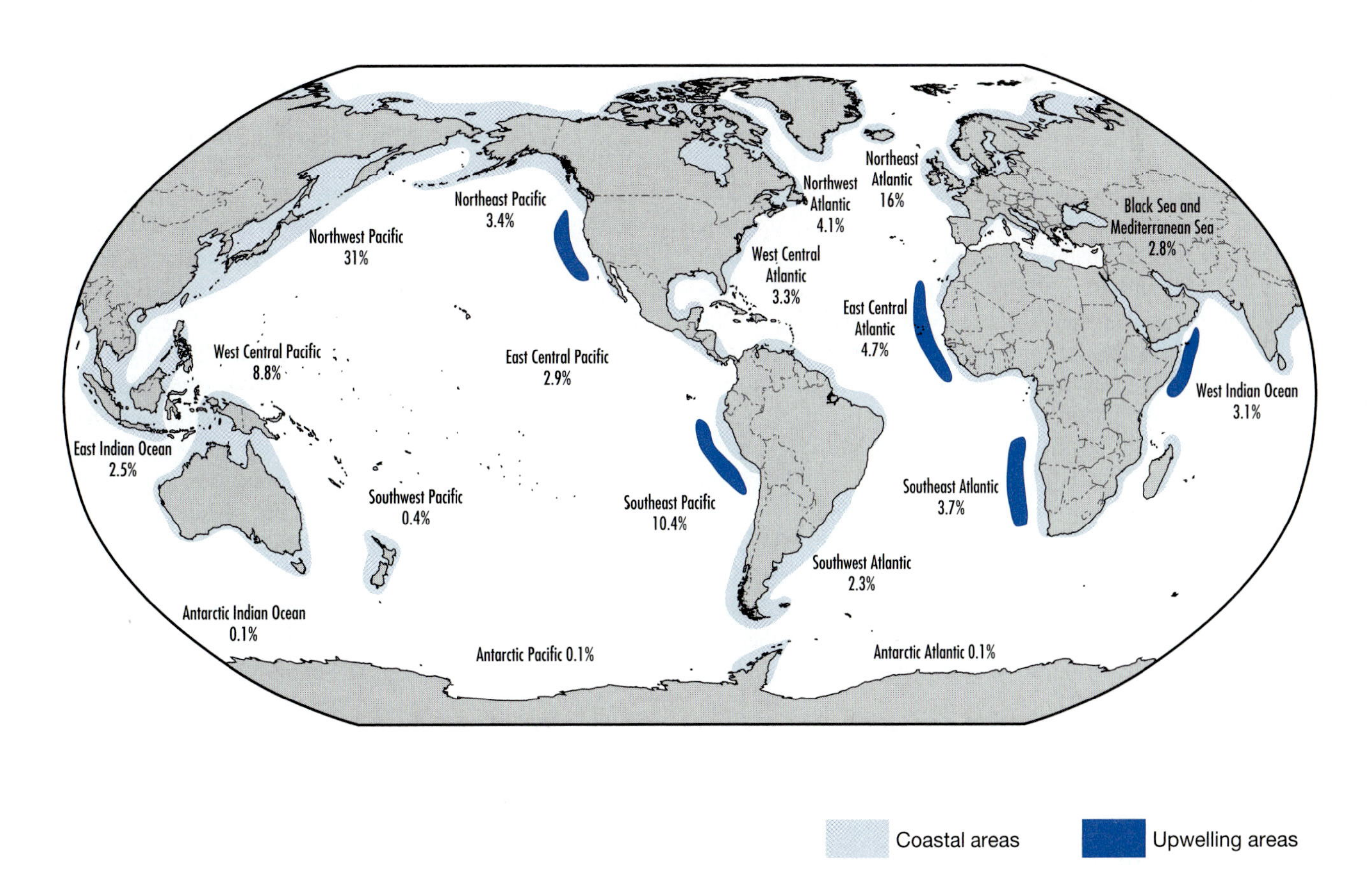

FIGURE 11.38 *Distribution of the world's fisheries. Coastal areas and upwelling areas together supply over 99 percent of world fish production. The deep ocean forms 90 percent of the ocean area but accounts for only 1 percent of the fish catch, if upwelling areas are excluded.*

example, workers sopped up oil with absorbent pads. Other workers scoured the beaches with hot water (Figure 11.35), washing the oil back into the sound where it was picked up by oil skimmers, vacuum-type devices that skim the oil off the surface and empty it into barges. Early in the spill, planes dropped absorbent material on slicks, which were later picked up by boats. In many spills, straw has been used to sop up oil that has washed ashore.

2. *Decomposition of oil by bacteria.* Two Israeli scientists developed a technique that uses bacteria to break down oil. Theoretically, an oil slick could be "seeded" by helicopter with the bacterial powder, accelerating the rate of decomposition and ultimate oil slick breakup. In the process of breaking down the oil, the bacterial population multiplies rapidly, to the point where it can be used as protein feed for livestock. The Israeli scientists estimate that hundreds of tons of animal food could be obtained from a spill. In the *Valdez* spill, Exxon applied oil-degrading bacteria to some beaches and found that the bacteria accelerated the destruction of oil.
3. *In-Situ Burning.* **In-situ-burning (ISB)** is a technique that can significantly reduce the amount of oil on the water, thereby minimizing adverse effects to the local environment. The controlled burn can be economical in certain instances, rapidly removing oil from the environment and reducing the need for a large amount of equipment and labor. The advantages of this method must be evaluated on a case by case basis. In-situ burning may not be advisable in some situations. Disadvantages include the generation of highly visible smoke, short-lived air pollution that increases health risks to humans and exposed populations downstream, localized temperature elevations that can harm or kill wildlife, and unknown long-term effects. The *New Carissa* tanker grounding off the coast of Coos Bay, Oregon in 1999 was an unusual event in which mitigation to avoid an oil spill included burning the ship's oil on board and sinking the ship out at sea.
4. *Fingerprinting.* The EPA's research laboratory at Athens, Georgia, is perfecting techniques by which a given oil sample can be identified by its "fingerprints." The distinctive fingerprints are recorded by a sophisticated instrument known as a **gas-liquid chromatograph.** With this technique, investigators trying to locate the source of an oil slick can determine with certainty whether the oil comes from a natural oil seep, a particular pipeline, a certain offshore oil well, or a tanker transporting oil from, say, Kuwait or Alaska. Once the source is identified, appropriate legal action can be taken against the parties responsible for the spill.

In most oil spill situations, a combination of these techniques is used. Appropriate methods should be chosen based on the unique conditions of each site. All of these methods are time-consuming and costly. Many of them are rather ineffective. Oil washed off the beaches near Valdez, for example, was spread back on by the rising tides. Cleanup methods also generate substantial amounts of waste that must be disposed of safely to avoid creating problems elsewhere. Given their shortcomings, these methods are clearly no substitute for prevention. To remind the public of the need for prevention, one environmental group sells T-shirts with the motto, "An ounce of prevention is worth 11 million gallons of cure."

The fact remains that the majority of oil and other sources of ocean pollution is generated from the result of small but numerous, individual land activities. Perhaps the best strategy for ocean pollution control is one that emphasizes and increases public education and awareness of everyday activities that contribute to this insidious form of ocean resource degradation. Educate yourself and then share your knowledge with others. The following list is a good beginning:

- Never pour oil, engine fluids, cleaners, or household chemicals down storm drains or sinks.
- Find government and industry-sponsored oil collection and recycling programs in your community and use them.
- Use lawn, garden, and farm chemicals sparingly and wisely. Before spreading chemicals or fertilizer, check the weather forecast for rain to prevent washing the substances into sewer and storm drains.
- Repair automobile and boat engine leaks immediately.

11.6 A World View of Water Pollution

Water pollution cleanup in the more developed countries has been moderate and, in many nations, is probably on a par with progress in the United States. The water pollution problems in the less developed countries in general are much worse than in the United States. There are several reasons for this: (1) lack of properly educated and technically trained personnel, (2) lack of funding for construction of waste treatment plants, (3) lack of tough pollution control legislation, and (4) lack of enforcement of such laws, if indeed they do exist. South American countries, for the most part, have relatively safe drinking water. However, many streams are seriously polluted with runoff from lead, zinc, and silver mines. As deforestation intensifies on this continent, river contamination with pesticides, fertilizer, and sediment is expected to increase accordingly. In Mexico, the drinking water has such high counts of infectious bacteria that American college students studying there have been advised to boil the water before drinking it lest they come down with diarrhea, fever, chills, and nausea, a complex of symptoms dubbed Montezuma's revenge. The scarcity of safe drinking water is even greater in Africa. For example, in rural Guinea, only 1 of every 50 people has access to it! Less than 10 percent of the rural populations of Madagascar, Mali, Sierra Leone, and Zaire have good drinking water available. In Pakistan, most of the human diseases, such as typhoid, diarrhea, dysentery, and infectious hepatitis, are caused by microorganisms that have contaminated public water supplies. In India, the

Yamuna River receives 200 million liters (54 million gallons) of untreated sewage from New Delhi every day. As a result, the coliform count in this stream is an almost unbelievable 24 million per 100 milliliters. (Recall that 200 per 100 milliliters is the standard for swimmable waters in the United States.) In Malaysia, 42 major rivers have been declared "ecological disasters" since they are virtually unable to support desirable aquatic life. Seventy percent of some stretches of Manila's Pasig River consists of untreated sewage.

In 1980 the United Nations launched the International Drinking Water and Sanitation Decade (1981–1990). The major objective of this program was to make all nations, especially the less developed countries, acutely aware of the importance of safe drinking water in the fight against disease. The ambitious goal of the program was to supply an additional 500,000 people with drinkable water every day of the decade! Unfortunately, this goal was not attained, but in the United States and other countries efforts are still underway to improve water quality—a goal that remains elusive today, in large part because of the continued increase in world population, resource demand, and industrial output. These three forces make it more important than ever to seek sustainable solutions—measures like erosion control and pollution prevention to avoid creating problems in the first place.

Summary of Key Concepts

1. Water pollution can be defined as any contamination of water that lessens its value to humans and nature.
2. Two broad classes of water pollution are point source and non-point source pollution.
3. Point pollution has its origin in specific, well-defined sources such as the discharge pipes of sewage treatment plants and factories. Non-point source pollution stems from widespread sources such as the runoff from agricultural lands or urban areas. Point sources are easier to control than nonpoint sources because they are so identifiable and are the source of concentrated releases of pollutants.
4. Pollution can also be classified by type. The major categories include sediment, inorganic nutrients, thermal pollution, disease-producing organisms, toxic organic compounds, heavy metals, and oxygen-demanding organic wastes.
5. Sediment comes from soil erosion from improperly managed and maintained farms, forests, construction sites, and the like. Better watershed management can go a long way in reducing sediment pollution with biological and economic benefits.
6. Sediment (a) fills up reservoirs, (b) damages hydroelectric plants, (c) clogs irrigation canals, (d) interferes with barge traffic along major rivers such as the Mississippi, (e) destroys fish spawning grounds, (f) reduces the photosynthetic activity of aquatic plants, and (g) makes necessary the costly filtration of water.
7. Inorganic nutrients include the phosphates and nitrates that arise from a variety of sources such as septic tanks, sewage treatment plants, and agricultural runoff.
8. Inorganic nutrients also come from natural sources. These compounds contribute to a phenomenon called natural eutrophication, a slow process of nutrient buildup in lakes that occurs over a period of thousands of years.
9. Human activities can accelerate the release of nutrients, causing cultural or accelerated eutrophication of lakes, and may age a lake 25,000 years in only 25 years.
10. Many factories and power plants use surface waters to cool various processes. This heated water is often released back into the source whence it came, creating thermal pollution.
11. Thermal pollution has multiple adverse effects. It kills fish because it lowers the concentration of dissolved oxygen. It also kills fish directly if the temperature of the water climbs outside the range of tolerance. Thermal pollution causes a shift in populations of microorganisms from the beneficial diatoms with undesirable blue-green algae. It also interrupts migration and enhances disease-causing organisms. Thermally enriched water, however, may be put to a variety of beneficial uses such as aquaculture.
12. Surface waters and groundwaters may become contaminated with disease-causing organisms, especially bacteria and viruses. Because it is costly to measure the many possible pathogens, water quality officials monitor fecal contamination and the potential for the presence of pathogens by measuring coliform bacteria, common but largely harmless bacteria found in the human intestine and in feces. A high coliform count indicates that the water sample may contain high levels of microorganisms capable of causing disease. Research, however, shows that fecal coliform levels may not be sufficient indicators and the U.S. Environmental Protection Agency is now promoting the use of an additional bacterium, fecal streptococcal bacteria, to monitor the safety of water.
13. Human society depends on a large number of potentially toxic organic compounds such as pesticides. These compounds can contaminate ground and surface waters. Groundwater contamination is especially troublesome because so many people throughout the world depend on groundwater for drinking water and because water flows so slowly in most aquifers, making natural purification a time-consuming process.
14. Toxic organic compounds cause many problems, including skin and eye irritation; brain and spinal cord damage; interference with normal kidney, liver, and lung function; cancer; and genetic mutations.
15. Sources of groundwater pollution from toxic organic compounds are (a) industrial landfills and lagoons, (b) municipal landfills, (c) septic tanks, (d) underground storage tanks for gasoline and other chemicals, (e) contaminated industrial sites, and (f) injection wells.
16. Heavy metals like mercury and lead come from many different sources. Most are toxic because they interfere with normal enzyme function.
17. Lead is one of the most troublesome of all heavy metals. It comes from solder used to join copper pipes and from old lead pipes, among other sources. Fortunately, new types

of pipe (copper) and new lead-free solder are being used in new construction.

18. The U.S. EPA estimates that the IQ of more than 140,000 American children has been reduced by 5 points because of lead in their drinking water.
19. Lead in drinking water is responsible for at least 680,000 cases of high blood pressure in American males.
20. Oxygen-demanding organic wastes come from (a) fruit and vegetable processing industries, (b) cheese factories, (c) creameries, (d) distilleries, (e) pulp and paper plants, (f) slaughterhouses, (g) bakeries, and (h) natural sources.
21. The discharge of organic waste reduces the dissolved oxygen in the stream and is lethal to many species of fish.
22. Sewage treatment plants primarily remove sediment, inorganic nutrients, and oxygen-demanding organic wastes.
23. Primary treatment of sewage is mainly a physical process in which solids are removed by sedimentation.
24. Secondary treatment of sewage is primarily a biological process in which organic wastes are decomposed by bacterial action. This process removes organic material as well as some of the nitrogen and phosphate, the inorganic nutrients.
25. The bacterial decomposition of organic waste during secondary treatment can be accomplished either by the activated sludge process or by trickling filters.
26. Tertiary sewage treatment, which is rather expensive, removes much of the nitrogen and phosphorus from the waste.
27. State and national laws promote measures to remove pollution from wastestreams and control nonpoint water pollution.
28. In the United States, the most important federal water pollution control law is the Federal Water Pollution Control Act of 1972 (now called the Clean Water Act) and its amendments (1977, 1981, 1987). Many other countries have similar laws patterned after this one. The Safe Drinking Water Act helps ensure that water sent to our homes is safe for consumption.
29. Toxic release inventory data have also been useful in the United States in pressuring major corporations to reduce their emissions of toxic substances into the air, water, and the soil.
30. Water pollution controls help reduce pollution, but in many instances water quality in and around major metropolitan areas has not improved because of nonpoint water pollutants resulting from the development of lands in the watershed. To address this problem, many cities and towns have begun to consider watershed management.
31. Watershed management involves steps taken by a wide variety of people, including government officials, homeowners, gardeners, city park officials, and farmers. They're designed to reduce the sources of pollution in a watershed and minimize the disturbance of vegetation to reduce surface runoff, and hence reduce the flow of pollutants into streams. Stormwater retention ponds and buffer zones are two of many measures called for in watershed management plans.
32. Water pollution is common in all countries, rich and poor. Although water pollution control efforts in more developed nations have been remarkable, there is much to be done to prevent further deterioration. In less developed countries, the challenge is even greater as a result of a lack of funds and technical expertise.
33. For decades the neritic zone bordering our nation's coasts has served as a dumping ground for raw domestic sewage, sewage sludge, industrial wastes, and dredge spoils.
34. Waste dumping at the New York Bight has caused (a) reduced levels of dissolved oxygen, (b) declines in plankton and plankton-dependent fish, (c) fish disease epidemics, (d) fish contamination with toxic metals, and (e) high rates of harmful mutations in fish.
35. Plastic pollution causes wildlife mortality by (a) blocking digestive tracts, (b) entangling-induced drowning, and (c) entangling-induced starvation.
36. A U.S. law that helps control plastic pollution is the Ocean Dumping Act, which bans plastic dumping from trash ships.
37. Today, nearly all ocean dumping that occurs in U.S. waters consists of dredged materials; however, other countries still dump sewage sludge and non-toxic industrial wastes.
38. The main sources of oil in the marine environment are (a) natural seeps, (b) oil well blowouts, (c) tanker spills (d) routine tanker and ocean vessel maintenance, (e) inland disposal via river and pipeline runoff, and (f) air pollution.
39. The often dramatic and highly publicized oil tanker spills contribute only 5 percent of the total oil inputs to the ocean each year.
40. The majority of ocean oil pollution is generated from the small but numerous, individual inland activities.
41. Oil pollution adversely affects the marine ecosystem by (a) reducing photosynthetic rates in marine algae; (b) concentrating chlorinated hydrocarbons such as pesticides; (c) contaminating human food chains with carcinogens such as benzopyrene, (d) disrupting chemical communication in marine organisms, which adversely affects such activities as feeding, reproduction, and escape from predators; (e) killing animals; and (f) causing long-term effects such as cancers due to chronic exposure to low levels of oil.
42. The Oil Pollution Act of 1990 requires (a) improved structural designs for new tankers; (b) phase out of the existing tanker fleet; (c) financial responsibility, compensation, and liability for spills; and (d) improved spill response strategies and inspection systems.
43. Strategies for clean-up of an ocean oil spill include (a) physical cleanup, (b) decomposition by oil-eating bacteria, (c) in-situ burning, and (d) the use of fingerprinting to determine the source of the spilled oil.
44. The best strategy for ocean pollution control may be one that emphasizes and increases public education and awareness of everyday activities that contribute to this insidious form of ocean resource degradation.

Key Words and Phrases

Accelerated Eutrophication
Activated Carbon Process
Activated Sludge Process
Algal Bloom
Amoebic Dysentery
Atmospheric Deposition
Biodegradable
Biological Magnification
Biochemical Oxygen Demand (BOD)
Biological Oxygen Demand (BOD)
Biogas
Biomass
Blue-Green Algae
Catch Basin
Chlorination
Clean Water Act
Closed-Cycle Cooling System
Cold-Blooded Animal
Coliform Bacteria
Combined Sewers
Cooling Tower
Cross-Media Contamination
Cultural Eutrophication
Detergent
Dissolved oxygen
Dredge Spoil
Dry Cooling Tower
Emergency Planning and Community Right to Know Act
Environmental Protection Agency
Eutrophic Lake
Eutrophication
Federal Water Pollution Control Act
Feedlot
Gas-Liquid Chromatograph
Great Lakes Water Quality Agreement
Groundwater
Heavy Metals
Holding Ponds
Hydroseeder
Input Control
In-Situ Burning
Industrial Sewage
Infectious Hepatitis
Injection Well
International Joint Commission
Land Application of Sewage
Landfill
Lead Poisoning
Leach Field
Limiting Factor
London Dumping Convention
Marine Pollution Convention
Marine Protection, Research, and Sanctuary Act
Mercury Poisoning
Mesotrophic Lake
National Ambient Stream Quality Accounting Network (NASQAN)
Natural Eutrophication
Nonpoint Pollution
Ocean Dumping Act
Oligotrophic Lake
Open-Cycle Cooling System
Output Control
Oxygen-Demanding Waste
Oxygen Sag Curve
Parts per Billion (ppb)
Parts per Million (ppm)
Pathogen
Point Pollution
Pollution Control
Pollution Prevention
Primary Sewage Treatment
Recovery Zone (of a River)
Safe Water Drinking Act
Secondary Sewage Treatment
Sediment Pollution
Septic Tank
Settling Tank
Sewage
Sewage Treatment Plant
Sludge
Sludge Digester
Sludge Worms
Species Diversity
Storm Water Management Plan
Surface Water
Synergism
Tertiary Sewage Treatment
Thermal Enrichment
Thermal Pollution
Toxic Organic Chemicals
Toxic Substances Control Act
Trickling Filter
Water Pollution
Watershed
Watershed Management
Watershed ManagementPlans
Water Pollution Control Act
Wet Cooling Tower

Critical Thinking and Discussion Questions

1. Define water pollution.
2. What is the difference between point source and nonpoint source water pollution?
3. List the seven basic types of water pollution.
4. List eight differences between oligotrophic and eutrophic lakes.
5. Distinguish between natural and cultural eutrophication.
6. Because photosynthetic levels are high in eutrophic lakes, you might suppose that the water would contain a relatively large amount of dissolved oxygen. Does it? If not, why not? Discuss your answer.
7. Using your critical thinking skills, discuss the following statement: "Eutrophication is a natural process, so there's no need to worry about human pollutants that contribute to this phenomenon."
8. List five adverse effects of algal blooms.
9. Describe the effects of a high level of BOD waste on the aquatic life of a stream.
10. Using your critical thinking skills and your knowledge of water pollution and ecology, discuss the following statement: "Lakes and rivers have naturally occurring bacteria that can cleanse water, removing organic wastes. Pollution from sewage treatment plants can therefore be released into them without harm."
11. Roughly what levels of dissolved oxygen does a particular stretch of stream have if one of the dominant organisms found in it is (a) a carp, (b) a trout, (c) a sludge worm, (d) a mayfly larva?
12. What does the term biological oxygen demand (BOD) mean?
13. Some manufacturers believe that thermal pollution should really be called thermal enrichment. Do they have a case for such a change in terminology? Discuss your answer.
14. Why are coliform bacteria levels monitored in lakes and streams?
15. Describe four important methods for controlling sediment pollution.
16. Summarize the main benefits derived from (a) primary sewage treatment, (b) secondary sewage treatment, and (c) tertiary sewage treatment.
17. Using your knowledge of ecology and water pollution, discuss the statement: "Groundwater contamination is more serious than contamination of surface waters."
18. Name six heavy metals that are toxic to humans when ingested.
19. What are the possible sources of lead in drinking water in a home?
20. Discuss the effect of plastic pollution on marine life.
21. List five sources of the oil that pollute the oceans and explain the contribution of each to the total oil input each year.

22. Discuss five adverse effects of oil on marine organisms.
23. Where should the U.S. focus efforts on controlling ocean pollution. Why?
24. What can you do to reduce the amount of ocean pollution?

Suggested Readings

Adler, R. W., D.M. Cameron, and J. Landman. 1993. *The Clean Water Act 20 Years Later.* Washington, D.C.: Island Press. Great summary of the Clean Water Act, including its strengths and weakenesses, by the Natural Resources Defense Council.

Canby, T. Y. 1991. "After the Storm". *National Geographic.* August: 2–33. Account of the devastation on the ground, in the water, and in the air from the largest oil spill in the world's history as a result of the fighting in the Persian Gulf War.

Carey, J. 1996. "Lessons from Loons." *National Wildlife* 34(5): 12–19. An excellent look at the effects of water pollutants on wildlife.

Cheremisinoff, P. N. 1993. *Water Treatment and Waste Recovery: Advanced Technology and Applications.* Englewood Cliffs, N.J.: Prentice Hall. Authoritative survey of latest methods of water purification.

Doppelt, B., et al. 1993. *Entering the Watershed: A New Approach to Save America's River Ecosystems.* Washington, D.C.: Island Press. Good overview of watershed protection.

Frankl, E. 1995. *Ocean Environmental Management: A Primer on the Role of the Oceans and How to Maintain Their Contributions to Life on Earth.* Englewood Cliffs, N.J.: Prentice Hall. Overview of current environmental problems relating to ocean resources and discussions of alternatives for sustainable use.

Garelik, G. 1996. "Russia's Legacy of Death." *National Wildlife* 34(4): 36–41. A telling story of the water pollution and other environmental atrocities caused by lax environmental enforcement.

Gleick, P. H. 2000. *The World's Water 2000–2001.* Washington, D.C.: Island Press. Contains a vast amount of information on the world's water resources, including waterborne diseases.

Hearne, S. A. 1996. "Tracking Toxics: Chemical Use and the Public's Right to Know." *Environment* 38(6): 4–9, 28–34. A description of an effective tool in control water pollution.

Riley, A. L. 1998. *Restoring Streams in Cities: A Guide for Planners and Citizens.* Washington, D.C.: Island Press. Examines a host of sustainable approaches to protecting watersheds and rivers.

Platt, R. H., P. K. Barten, and M. J. Pfeffer. 2000. "A Full, Clean Glass? Managing New York City's Watersheds." *Environment* 42(5): 8–20. Useful case study for those interested in learning more about watershed management.

Thompson, J. W. and K. Sorvig. 2000. *Sustainable Landscape Construction: A Guide to Green Building Outdoors.* Washington, D.C.: Island Press. Offers invaluable advice on landscape design in cities and towns that help protect watersheds.

UNEP. 1995. *The Pollution of Lakes and Reservoirs.* Nairobi, Kenya: United Nations Environment Programme. Documents pollution issues of the major lakes and reservoirs of the world and ways to address them.

UNEP. 1995. *Water Quality of the World River Basins.* Nairobi, Kenya: United Nations Environment Programme. Summarizes data on water quality from 82 major river basins around the world.

UNEP/GEMS. 1996. *Groundwater: A Threatened Resource.* Nairobi, Kenya: United Nations Environment Programme. Explains global groundwater pollution.

Williams, T. 1994. "Death in a Black Desert." *Audubon* 96(1): 24. Discusses California irrigation nightmare: toxic runoff is poisoning the land and killing thousands of birds.

Web Explorations

Online resources for this chapter are on the World Wide Web at: **http://www.prenhall.com/chiras** *(click on the Table of Contents link and then select Chapter 11).*

Chapter 12

Fisheries Conservation

Worldwide, there are at least 25,000 species of fish, making them the most abundant of the vertebrate animals. Over the past 200 years, however, human activities have had negative impacts on aquatic ecosystems and fish habitat. In the last few decades in particular, these activities have increased in number, variety, and intensity. The result of this accelerating human pressure has been a decline in the number and range of many fish species considered commercially and recreationally important to society.

Historically, both freshwater and marine fisheries management has emphasized the manipulation of fish populations and their environment for the sole purpose of increasing both sport and commercial fish harvest. Indeed, this is an important goal, justified by the ever-increasing demands for food and recreation of a growing world population. As evidenced by the continued decline of many fish populations worldwide, however, fisheries management has yet to attain strategies that produce sustainable fish harvests. Understandably, this ambitious goal will take time. To return the health of our declining fish stocks will require extensive commitment and cooperation on both national and international levels. There is general agreement that the health of our fisheries is dependent on the protection and restoration of aquatic habitats.

In order to protect and restore aquatic ecosystems for fish survival, the complex interrelationships between fish and their habitats must be understood. The fisheries manager must be able to accurately assess fish responses to alterations in their habitat, whether caused by natural or human-induced events, in order to make intelligent management decisions. In this chapter, we discuss both freshwater and marine fisheries, environmental limitations to freshwater fish productivity, problems facing the marine fishing industry, management options for sustainable freshwater and marine fisheries, and aquaculture.

12.1 Freshwater Fisheries

Most freshwater fisheries are managed for recreational anglers—about 9 percent of the global commercial fish harvest is caught in inland fresh waters. The freshwater game fish sought after by our nation's anglers include channel and flathead catfish; northern pike; muskellunge; chain pickerel; cutthroat, rainbow, brown, lake, bull, and brook trout; striped, smallmouth, and largemouth bass; bluegill; sunfish; crappie; perch; and walleye. Although most salmon are caught in marine commercial harvests, they are also considered an important freshwater fish, along with other **anadromous fish.** Anadromous fish begin life in fresh water, travel to and mature in the sea, and return to their native stream to reproduce and die (Figure 12.1). Nonanadromous fish are called **resident fish** and live their entire life in fresh water, sometimes confined to a lake, particular stream reach, or single tributary.

Habitat Requirements

All anadromous and resident fish species need relatively unaltered, or pristine, freshwater habitats during part or all of their life cycles to ensure optimum growth and survival rates. Fish migration, spawning, incubation, and rearing are examples of life-cycle stages. The successful completion of each of these stages is dependent on one or more of the following environmental conditions of the freshwater habitat: water temperature, depth, velocity, turbidity, dissolved oxygen, and salinity; substrate; cover; and food supply.

Temperature Stream temperature can delay **upstream migration** of fish by being either too cold or too warm. In addition, each native fish species has a specific time of year and temperature range in which **spawning** (egg deposition) occurs that ensures the maximum survival rate of their offspring. If water temperatures are too cold or too warm during incubation, embryos will not develop properly and will have a low chance of survival. Specific temperature ranges are required for normal behavior and optimal growth of juvenile fish during the rearing stage. Growth is retarded when temperatures become too low or too high. Many native fish species will move up- or downstream in response to adverse temperatures. Research has shown that several salmon and trout species prefer temperature ranges between 12°C and 16°C (54°F and 61°F).

Water Depth and Velocity Successful upstream migration depends on adequate water depths and flow velocities. Larger fish require a higher minimum depth and can tolerate higher stream flows than smaller fish. For example, during upstream migration, most salmon need a water depth of at least 0.24 meter (9.5 inches) and can tolerate a flow velocity of up to 2.4 meters per second (7.9 feet per second), whereas trout need only a depth of 0.12 meter (4.7 inches) but can tolerate a velocity of only 1.22 meters per second (4.0 feet per second). Spawning habitat is increased when the stream flow is high enough to cover suitable gravels but is decreased when velocity becomes too high for successful spawning activities. Juvenile fish need space for rearing—which is, in part, a function of water depth and stream flow.

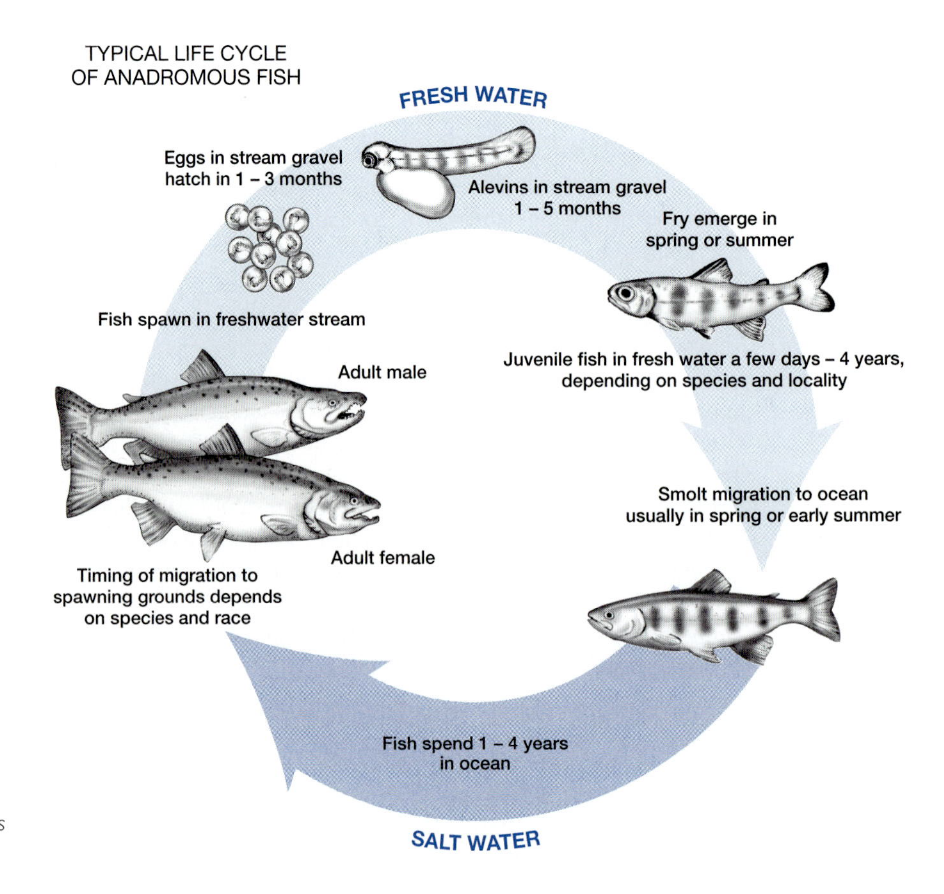

FIGURE 12.1 *Typical life cycle of anadromous fish.*

Turbidity **Turbidity** is a measure of suspended sediment in water. Research has shown that migrating fish will avoid or delay their migration when stream waters contain excessive amounts of suspended sediment. Both adults and juvenile fish can tolerate temporary episodes of seasonally high turbidity levels, such as storm events and spring runoff. However, fish will generally avoid streams that have chronically high levels of suspended sediments.

Dissolved Oxygen Streams are usually very well aerated. Flowing water, the relative shallowness of the stream, and the large surface area exposed to the atmosphere all contribute to oxygenation. All things being equal, the waters of shallow, fast-moving streams have much higher oxygen levels than those of deep, sluggish streams. The photosynthetic production of oxygen in a stream is not nearly as important as it is in a pond or lake. Because of the thorough mixing of stream water, oxygen depletion occurs less frequently than in lakes. However, stream fish are very sensitive to even slight reductions of oxygen levels. For example, a reduction in dissolved oxygen to below 5.0 parts per million (ppm) decreases the swimming speed of adult fish during migration and limit growth rate and food conversion efficiency of juvenile fish. **Hypoxia,** when dissolved oxygen levels are inadequate for aquatic life and slow-moving, trapped, or non-mobile aquatic life will perish, occurs at dissolved oxygen levels of 1-2 ppm. **Anoxia,** where no oxygen is available, occurs when dissolved oxygen levels get down to 0.5 ppm.

Salinity Salinity concentration is especially important for estuarine and coastal marsh organisms and can limit movement of both marine and freshwater organisms. Estuaries form in the transition zone where a river or rivers enter the ocean. They are characterized by tidal flux and a moving wedge of relatively dense, salty water under less dense freshwater. The size and movement of the salt wedge is dependent on the volume of freshwater flow from a river and can impact aquatic organisms, commercial fishing, and even drinking water supplies of an adjacent city. For example, if river flow is high, the salinity level in parts of the estuary can drop impacting shellfish beds.

Substrate The bed materials of a stream or river must consist of gravel particles of a suitable diameter for egg deposition. Spawning fish use the gravels to make a "nest," called a **redd,** for their eggs. In general, larger fish are able to use larger spawning gravels. Most eggs are deposited in spawning gravels ranging from 1.3 to 15 centimeters (0.5 to 6 inches) in diameter, with the largest proportion in the range of 1.3 to 3.8 centimeters (0.5 to 1.5 inches). The pores between the gravels must be large enough for water to pass through, providing the embryos with oxygen and carrying away waste materials. Fine sediment carried by stream waters during the incubation phase can be deposited on top of the spawning gravels. If this sediment layer becomes thick enough, the developing embryos will die from suffocation.

Cover Riparian vegetation and streambank and streambed structure determine the amount of cover available to spawning fish. Overhanging vegetation, undercut banks, aquatic vegetation, large woody debris, boulders, and deepwater pools are examples of natural structures that provide adult and juvenile fish with shade, resting areas, and protection from predation (see Figure 9.17).

Food Supply The abundance and growth rate of fish can also be food-limited. The productivity of a stream, or its ability to supply nutrients and energy, is variable from one stream to the next and between reaches of any one stream. In general, the greater the amount of organic matter (dead or living) in and adjacent to the stream environment, the larger the aquatic and terrestrial invertebrate populations that feed on these materials. In turn, the larger the aquatic and terrestrial invertebrate populations, the higher the numbers of adult and juvenile fish that feed on these invertebrates. Sources of organic matter in streams include leaf, needle, and twig litter; large woody debris; aquatic and streambank vegetation; organic detritus in various stages of decomposition; and dissolved organic matter.

12.2 Environmental Limitations to the Reproductive Potential of Freshwater Fish

Like most organisms, freshwater fish have great ability to reproduce. For example, a 16-kilogram (35-pound) female muskellunge may produce 225,000 eggs during a single breeding season. Some bass nests in Michigan lakes and streams contain more than 4,000 young per nest. In some species, like the bluegill, nests are built very close together. Such gregarious breeding behavior promotes reproductive success because an optimal spawning habitat can be shared by a large number of individuals. The reproductive characteristics of some species of fish are summarized in Table 12.1.

Fish, however, respond both physiologically and behaviorally to long-term or chronic alterations to stream and lake ecosystems that result in less than optimal habitat conditions. Were it not for the extensive mortality caused by **environmental limitations,** the lake basins and river channels would be choked with fish. Tagging studies have revealed that roughly 70 percent of a given fish population dies each year. Thus, if 1 million young of a given species hatch, 300,000 will be alive at the end of the first year, 90,000 by the end of the second, but only six fish by the end of the tenth. The major environmental limitations that cause mortality in fish populations can be divided into two categories: natural and human-induced (Table 12.2).

Natural Limitations

Storms and Soil Mass Movements A major storm event can have serious impacts on stream channel morphology and fish habitat. Flooding and high-velocity stream flow, resulting from high rainfall and overland water runoff, can alter pool and riffle distribution and structure, move large loads of sediment and debris, increase turbidity, and cause streambank ero-

TABLE 12.1 *Reproductive Characteristics of Major Species of Fish*

Common Name	Reproductive Age (or Length)	Spawning Time	Number of Eggs	Type of Reproduction
Bass, largemouth	2 years	Spring	2,000–100,000	Nest
Bass, smallmouth	2 years	Spring	2,000–20,800	Nest
Bluegill	1 year	May–August	2,300–67,000	Community nest
Carp	12 inches	Spring	790,000–2,000,000	Eggs, scattered
Channel catfish	12 inches	Spring	2,500–70,000	Nests
Muskellunge	3–4 years	Spring	10,000–265,000	Eggs, scattered
Northern pike	2–3 years	Spring	2,000–600,000	Eggs, scattered in marsh
Salmon, chinook	4–5 years	Fall	3,000–4,000	Eggs buried in gravel
Salmon, coho	3–4 years	Fall	3,000–4,000	Eggs buried in gravel
Trout, brook	2 years	Fall	25–5,600	Eggs buried in gravel
Trout, brown	3 years	Fall	200–6,000	Eggs buried in gravel
Trout, rainbow	3 years	Spring	500–9,000	Eggs buried in gravel
Trout, lake	5–7 years	Winter	6,000	Eggs scattered over gravel
Walleye	3 years	Spring	35,300–615,000	Eggs scattered

Source: *What Fish Is This?* Illinois Department of Conservation, June 1986.

sion. Impacts on fish habitat include disruption and siltation of spawning gravels, deposition of fine sediment over food sources such as bottom-dwelling flora and fauna, disturbance of side-channel rearing areas, and blockage of fish passage due to woody debris jams.

Mudslides, avalanches, slumps, and debris flows can have similar effects on fish habitat. Sediment, boulders, large organic debris, and other organic materials are transported into the stream that cause damming and obstruction of the channel, heavy siltation and turbidity loading, and local flooding.

Animal Activities Beaver dams are common on relatively undisturbed, lower-order streams (Order 4 or less) in the forested regions of North America. Construction of these dams alters fish habitat by diverting and ponding stream water, increasing local water temperatures, blocking fish passage at low flows, flooding side-channel rearing areas, and causing siltation of spawning areas. However, depending on the site, the benefits to fish habitat provided by these dams may outweigh the costs. For example, the flooding caused by beaver dams can provide important rearing and overwintering habitat. The silt that accumulates in the ponds behind the dams, with subsequent warmer water temperatures, can result in an increase in local biological productivity. And finally, beaver ponds may act as sediment traps, improving water quality to regions downstream.

Deer, elk, and moose browse and trample riparian areas. Heavy browsing reduces the amount of understory shrubs, grasses, and wetland plants growing along the shores of a stream or lake. The loss of this vegetation affects the quality of fish habitat by reducing cover and important nutrient and energy sources. Large concentrations of elk and moose trample the banks of lakes and streams, further reducing vegetation and causing soil erosion and bank instability.

Natural Barriers Waterfalls, excessive water velocities, and debris jams can impede upstream migration and, therefore, spawning activities. Research has shown that ideal leaping conditions for fish are when the depth of a pool below a waterfall is 1.25 times the height of the falls (Figure 12.2A, B). Of course, the success of any one fish to pass a given barrier also depends on other factors including the swimming velocity of the fish, the jumping ability of the fish, the horizontal and vertical distances to be jumped, and the steepness of the incline to be negotiated (Figure 12.2C, D).

Fish may be unable to continue their upstream migration if water velocities are too high. For example, large fish such as salmon and steelhead can swim for extended periods of time in water velocities of 2.4 meters per second (7.9 feet per second) or less, for a few minutes at time in water velocities of 3 to 4 meters per second (9.8 to 13.1 feet per second) or less, and are unable to navigate in velocities above 4 meters per second (13.1 feet per second).

Some woody debris jams can delay or prevent upstream migration. Removal of the jam, however, must be done with care to prevent siltation of active spawning and rearing areas downstream and harmful alterations in water levels and velocities.

Vegetation Disturbances We have previously established the crucial role riparian vegetation plays in high-quality fish habitat. In addition to vegetation losses caused by ungulates (deer, elk, and moose), major disturbances such as windthrow, wildfire, insects, and disease can also destroy riparian vegetation. The resulting negative impacts on fish habitat include reducing cover, altering stream temperatures, decreasing litter fall, increasing erosion and sediment delivery, decreasing bank stability, and increasing woody debris jams.

TABLE 12.2 *Major Factors That Limit the Reproductive Potential of Fish*

Cause of Limitation	Effects on Fish and Their Habitats
*Natural**	
1. Major storms	1. Altered pool and riffle distribution and structure, increased turbidity and sedimentation, streambank erosion, siltation of spawning gravels, burial of food sources, disturbance to rearing areas, blockage of fish passage
2. Soil mass movements (mudslides, avalanches, slumps, debris flows)	2. Damming and obstruction of stream channel, increased sediment load and turbidity, localized flooding, siltation of spawning gravels, burial of food sources, disturbance to rearing areas, blockage of fish passage
3. Animal activities (beavers, ungulates)	3. Diversion and ponding of water, localized temperature increases, flooding of rearing areas, siltation of spawning gravels, blockage of fish passage at low flows; trampling and browsing of riparian vegetation, reduction of cover and nutrient sources, alteration of stream temperatures, increased soil erosion, bank instability
4. Natural barriers (waterfalls, debris jams, excessive water velocities)	4. Impeded upstream migration, delayed spawning activities
5. Vegetation disturbances (windthrow, wildfire, insects, disease)	5. Reduction of cover, altered stream temperatures, decreased nutrient and energy sources, bank instability, increased erosion and sedimentation, increased woody debris jams
6. Predation by native animals	6. Fish mortality
7. Winterkill	7. Fish mortality
Human-Induced	
1. Water pollution (eutrophication, sedimentation, acid deposition)	1. Excessive growth of aquatic vegetation, oxygen depletion, increased sediment load and turbidity, acidification of water, alteration in chemical composition of water, interference with reproductive and feeding behavior, siltation of spawning beds, burial of food sources, decrease in spawning success, asphyxiation and mortality of fish in all stages of development
2. Alteration of stream temperatures	2. Disrupts timing of migration and spawning; causes abnormal growth rates, increased disease outbreaks, abnormal behavior patterns
3. Predation and competition by exotics	3. Effects are dependent on the behavior of the introduced species but can include loss of aquatic vegetation and food sources, loss of spawning grounds, increased turbidity, decreased dissolved oxygen levels, decreased spawning success due to predation of eggs, disease outbreaks
4. Predation by humans	4. Increased fish mortality and injury, population decline or elimination
5. Dams	5. Altered water levels and velocities, loss of rearing habitat, increased mortality from nitrogen intoxication and contact with turbines, population decline, delay and blockage of fish passage, spawning delay
6. Resource extraction (logging, grazing, mining)	6. Loss of riparian vegetation, altered stream temperatures, modifications of stream channel structure, abnormal fluctuations in water depth and flow, chemical pollution, siltation of spawning gravels, increased sediment loads and turbidity, streambank erosion, decreased dissolved oxygen levels, decreased spawning success, population decline
7. Channelization	7. Altered streams flows and depths, loss of spawning and rearing habitat, increased sediment load
8. Human recreation (swimming, boating, hiking, camping, horseback riding, mountain biking)	8. Riparian vegetation trampling, streambank erosion, soil compaction, decreased water quality, loss of cover, alteration in stream temperatures, loss of nutrient and energy sources

* Note that these natural events, in the long term, in some cases may actually increase productivity and improve habitat quality.

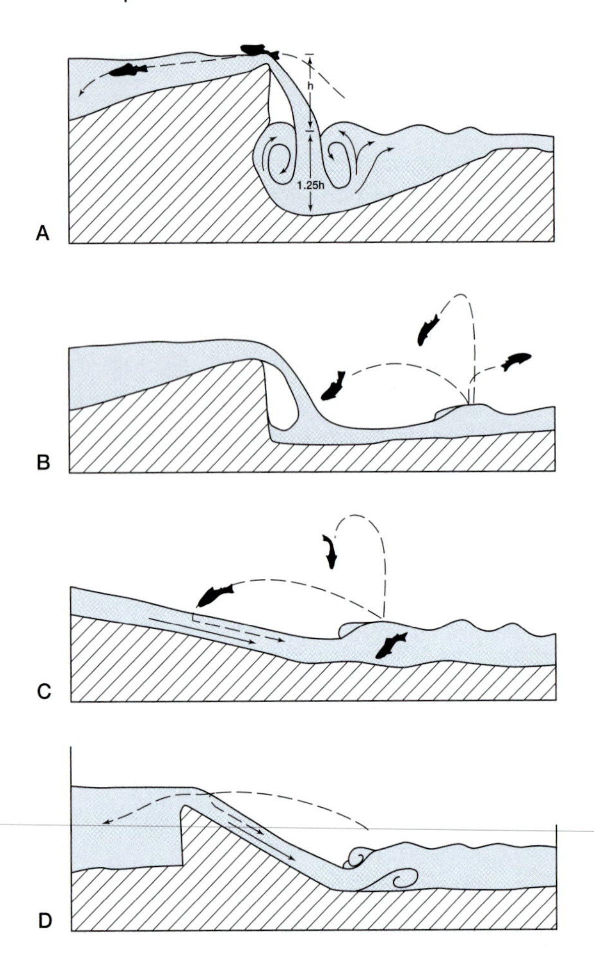

FIGURE 12.2 *Leaping ability of fish. (A) Water depth in the pool below the falls is 1.25 times the distance (h) from the top of the falls to the water surface of the pool. Fish use the upward momentum of the standing wave, formed close to the waterfall, to leap over the falls. (B) The height of the waterfall is equal to that in A, but the pool is shallower. The standing wave is formed too far from the falls to be of any use to the fish. (C) An incline too steep and too long for fish to negotiate. Fish can be thrown from the standing wave at the foot of the incline. (D) An incline steeper than in C, but shorter in length. Some fish, depending on their leaping ability and the amount of energy in the standing wave, may be able to pass over this barrier.*

Predation by Native Animals Fish are subjected to intense predatory pressure from other fish and from reptiles, birds, and mammals. A 6-inch muskellunge will consume 15 minnows a day! Three thousand fish have been eaten by a walleye by the time it is 3 years old. When other food is scarce, many fish resort to cannibalism. Smaller species may eat the eggs of larger species, which as adults regularly feed on the smaller species.

Wading birds and other waterfowl, such as egrets, herons, and ducks, consume large numbers of fish. A single merganser (a diving duck) may eat more than 35,000 fish annually! Bear, otter, fisher, and mink prey extensively on fish, especially during droughts, when water levels are low and the fish are easily caught. The Alaskan brown bear can easily eat 15 large salmon per day.

Winterkill Another environmental factor that controls fish populations is **winterkill.** During the long winters of the northern states, an icy barrier seals lakes off from atmospheric oxygen. As long as the ice remains clear of snow, sufficient sunlight may penetrate the ice to sustain photosynthesis (Figure 12.3). As a result, oxygen levels remain adequate. Snow, however, forms an opaque barrier that prevents sunlight penetration. If photosynthesis activity falls, so do oxygen levels, often resulting in heavy fish kills, especially if the lake is fertile and shallow. The decay of dead vegetation worsens the problem. As winter progresses, oxygen levels may drop to 5 ppm, at which point many of the more sensitive fish die; the more resistant fish, such as carp and bullheads, may die later if levels drop to about 2 to 3 ppm.

A classic example of winterkill occurred several years ago in a shallow southern Minnesota lake that had a dense population of bullheads. Although the ice cover that formed in November ultimately became 50 centimeters (20 inches) thick, oxygen levels initially were adequate. During the second week of January, however, a storm covered the ice with a 6-inch layer of snow. Two days later, oxygen levels fell sharply. After the ice melted in spring, thousands of dead bullheads littered the shore. Not one fish survived.

Human-Induced Limitations

Water Pollution More than 30 million fish are killed by water pollution in the United States each year. In Chapter 11, we described fish kills caused by industrial and municipal pollutants. Table 12.3 lists some pollution-caused fish kills. Here we shall describe mortality caused by eutrophication, sedimentation, and acid deposition.

Eutrophication As you recall, the enrichment of a body of water with nutrients that promote the excessive growth of aquatic vegetation is known as **eutrophication.** Dense algal blooms form near the surface, preventing sunlight from reaching the billions of algae at lower depths. As a result, algae and submerged vegetation sink to the lake bottom and form a dense organic ooze. Billions of bacteria in the bottom sediment decompose this material, consuming the oxygen dissolved in the surrounding water. As a result, the oxygen concentration in the deep water (hypolimnion) may fall rapidly from 7 to 2 ppm or less. As a waterbody becomes polluted with oxygen-demanding organic material, such as human sewage or the waste from slaughterhouses, pulp mills, and canneries, the oxygen reduction that results may trigger a massive fish kill. Eventually the dead fish float to shore, decompose, begin to smell, and attract flies.

Sedimentation Many tons of soil are washed into nearby lakes and streams by runoff waters as a result of abusive land practices, whether on a farm, a mine, or an urban construction site. Sediment depresses the photosynthetic activity of aquatic plants because it reduces sunlight penetration of the water. This, in turn, causes levels of dissolved oxygen to drop sharply. This is especially stressful to fish like trout and salmon, which require a minimum of 5 ppm of dissolved oxygen. Although fish may tolerate turbidities of up to 100,000 ppm for brief periods, concentrations of 100 to 200 ppm are harmful if they persist for any length of time.

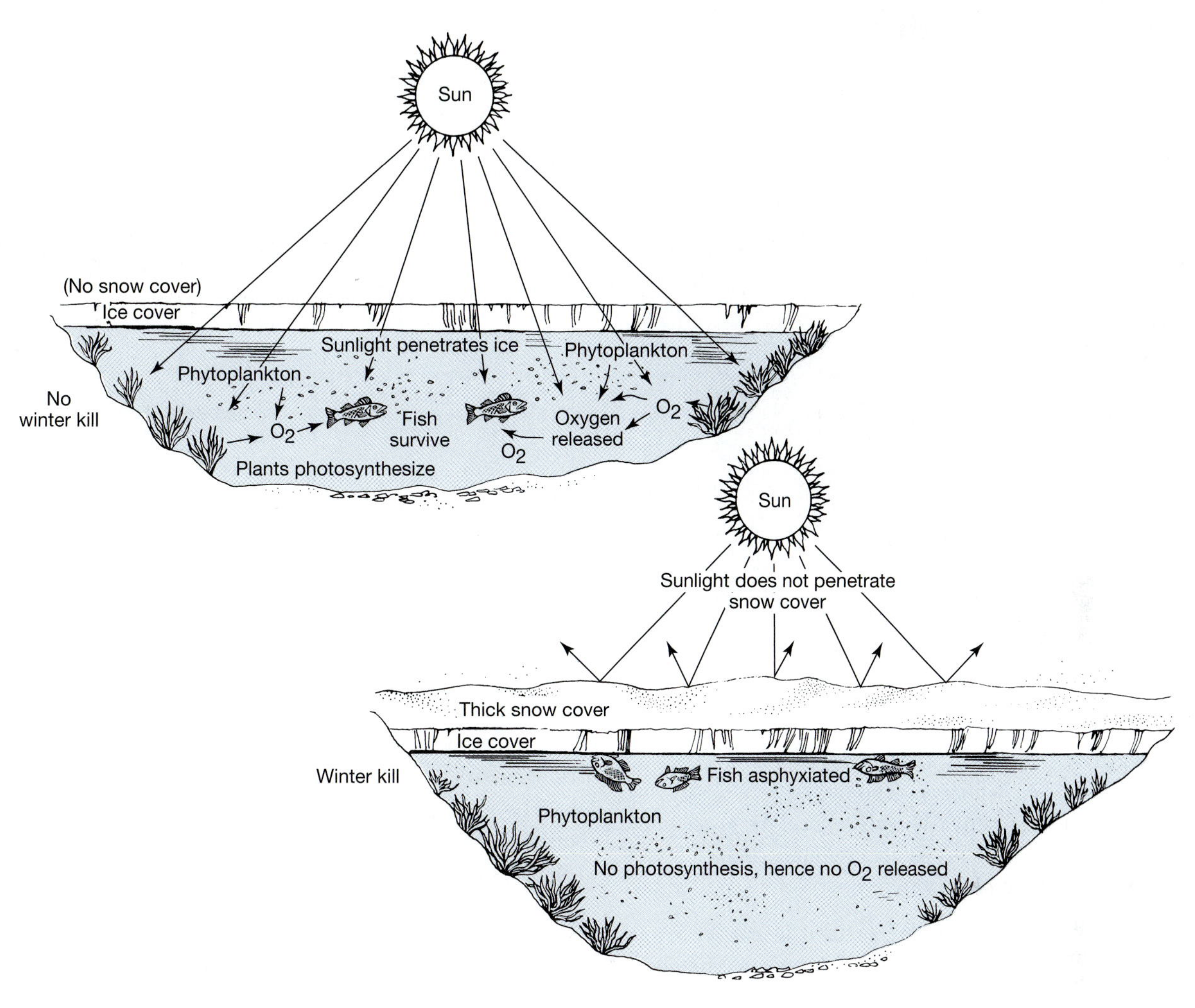

FIGURE 12.3 *Winterkill of fish.*

TABLE 12.3 *Fish Kills Caused by Water Pollution in Ohio, 1984*

Date	County	Name of Water	Number Killed	Suspected Pollutant
9/2	Montgomery	Great Miami River	158,234	Sewage and corn syrup
7/17	Daske	Greenville Creek	122,057	Hog manure
6/15	Fulton	Brush Creek	79,110	Ammonium nitrate
4/16	Crawford	Broken Sword Creek	57,237	Nitrogen fertilizer
9/1	Butler	Four-Mile Creek	41,335	Sewage
3/26	Columbiana	Beaver Creek	26,986	Gasoline
9/10	Coshocton	White Eye Creek	13,170	Cow manure
6/27	Morgan	Bell Creek	3,274	Cleaning chemicals
5/3	Marion	Riffle Creek	2,905	Herbicides

Source: *Water Pollution, Fish Kill and Stream Litter Investigation, 1984,* Columbus: Ohio Department of Natural Resources, 1985.

Thousands of fish die annually in American lakes and streams from asphyxiation caused by silt-clogged gills (Figure 12.4). **Suspended sediment** also interferes with the reproductive behavior of fish, which depend on visual cues provided by the gravel and sand of the stream or lake bed as well as on the color, shape, and behavior of the sex partner. Vast beds of inshore aquatic vegetation that were once important spawning beds for Great Lakes fish have been smothered by sediment. Mud may also cover fertilized eggs and reduce hatching success. A study of trout reproduction in Bluewater Creek, Montana, for instance, showed that egg hatching success was highest (up to 97 percent) where **siltation** was minimal (Figure 12.5). The larvae of aquatic insects, such as

FIGURE 12.4 *Fish kill caused by sediment. These fish suffocated when sediment clogged their gills during a flooding of the Iowa River in New Mexico.*

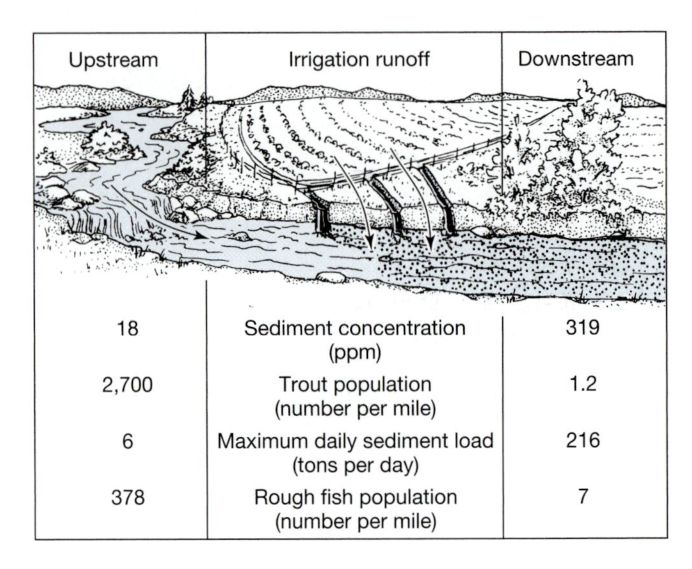

Upstream	Irrigation runoff	Downstream
18	Sediment concentration (ppm)	319
2,700	Trout population (number per mile)	1.2
6	Maximum daily sediment load (tons per day)	216
378	Rough fish population (number per mile)	7

FIGURE 12.5 *Effect of sediment from irrigation runoff on a Montana trout stream.*

mayflies and stone flies, which are favored fish foods, may also be destroyed by silt. Finally, mud sharply reduces the visual range of predatory fish, like bass, pike, and muskellunge, in their search for the smaller fish on which they feed.

Acid Deposition Can the burning of coal in an Ohio steel plant cause the death of trout high in the Adirondack Mountains of New York? Although it seems highly improbable, the answer is yes. The sulfur dioxide gas that is released from the smokestacks undergoes chemical reactions with oxygen to form sulfuric acid and water. Prevailing winds carry the acid droplets northeastward in clouds to a point high over the Adirondacks. Rain (or snow) then washes the acid into the lakes.

Normal, unpolluted rain has a pH of about 5.6—slightly acid, because carbon dioxide is dissolved in it to form carbonic acid. However, much of the acid rain in the eastern states has a pH of 4 or below—almost 100 times as acid as normal rain.

The effects of **acid rain** were not fully appreciated until the 1970s. Biologists have found that when the pH of lake water falls as low as 5, fish begin to die. The pH of many Adirondack lakes is below 5. Under such conditions, lake trout become deformed and embryos suffer high mortality.

The acidic water from rain and snowmelt also affects the chemical composition of the water. As the water drains off the land, the acids dissolve toxic metals from the soil and carry them into lakes and streams. Aluminum leached from the soil, for example, causes mucus to build up on the gills of fish. As a result, the fish suffocate. Aluminum is especially troublesome in the spring when the snows begin to melt, sending torrents of acidic water across the soil and eventually into lakes and streams.

A recent survey conducted by the New York State Bureau of Fisheries revealed that many of the 2,877 Adirondack lakes above 2,000 feet in elevation were devoid of fish. The Adirondack lakes are not exceptional. Many lakes in the northern parts of Minnesota, Wisconsin, and Michigan are also threatened.

The elimination of a species of fish from a lake may not occur suddenly. Instead it may develop gradually over a period of years owing to the inability of fish to spawn successfully. The effects of increasing acidity on fish populations in George Lake, Ontario, are shown in Figure 12.6. (For more on acid deposition, see Chapter 19.)

Alteration of Stream Temperatures Abnormal fluctuations in stream temperatures can adversely affect fish populations by disrupting the timing of migration and spawning activities, retarding or accelerating growth stages, increasing disease outbreaks, and altering behavior patterns. A variety of human activities alter stream temperatures. Although regulations now exist that prevent this practice, streambank vegetation has been cleared during forestry, mining, and cattle ranching operations. Water temperatures are altered when water is removed from streams and then returned after irrigation of agricultural lands and when water is released from deep reservoirs for flood control or power generation purposes. Cooling towers associated with nuclear power plants often discharge warm water into rivers and streams.

Predation, Competition, and Habitat Modification by Exotics The U.S. Fish and Wildlife Service is increasingly concerned with the "biological pollution" of native fish populations with **exotic (nonnative) species.** Some foreign fish were deliberately introduced by fisheries biologists to provide a desirable game or food fish or to aid in controlling an environmental problem. Many other exotics have been accidentally introduced. In either case, several have adversely affected the reproduction, growth, and survival of some of our most highly prized native fish.

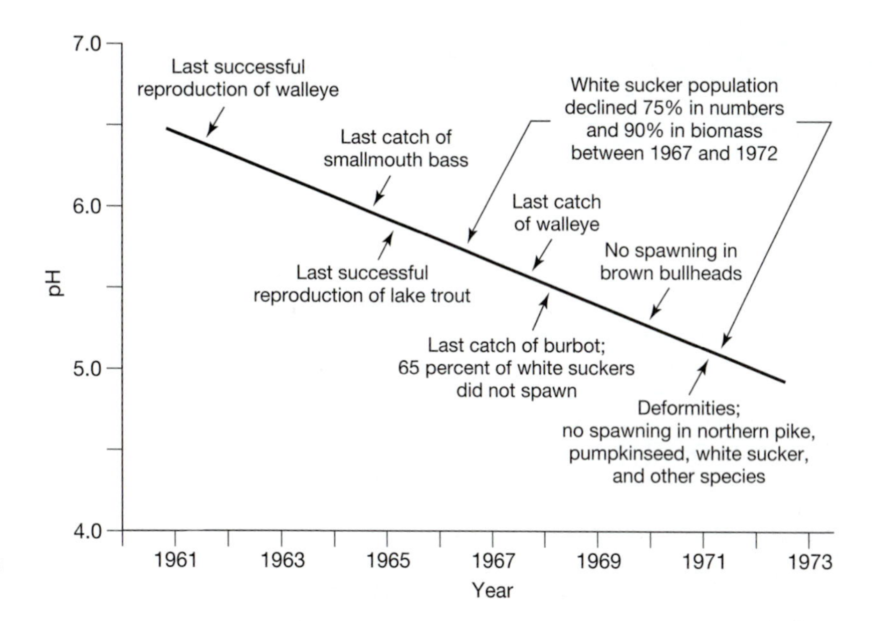

FIGURE 12.6 *Effect of increasing acidity (decreasing pH) of water on fish populations in George Lake, Ontario.*

The **European carp** is the most destructive exotic fish ever intentionally brought to the United States (Figure 12.7). It was originally introduced to California (1872), the Great Lakes (1873), and Washington, D.C. (1877). Hundreds of requests from all over the nation came to the U.S. Bureau of Fisheries to stock this so-called "wonder fish" in order to provide Americans with a valuable source of food.

The carp proved to be an extraordinarily adaptable fish. The introduced populations grew rapidly, following the characteristic S-shaped growth curve. Only 20 years after the introduction of European carp to Lake Erie, fishermen were able to harvest 1.6 million kilograms (3.6 million pounds) in a single year. Soon, however, the carp were affecting aquatic ecosystems in ways that had not been predicted. They uprooted aquatic vegetation during bottom-feeding. The results have been exceedingly harmful to game fish populations because (1) their spawning grounds were destroyed, (2) their food supplies were diminished, and (3) levels of dissolved oxygen were reduced as a result of interference with photosynthesis caused by the muddying of the waters.

One of the most recent import tragedies is the **river ruff,** another European species. The river ruff may have been accidentally introduced into Lake Superior in 1987 from a cargo ship that docked at Duluth, Minnesota. The ruff is a potential threat to the multi-million-dollar commercial fishery of the Great Lakes because it is a ravenous consumer of the eggs of the whitefish and other species. Since the ruff becomes sexually mature when only 1 year old, it has a reproductive edge on most species of native fish. In 1989, only 2 years after its introduction, the ruff's population had increased sharply in the harbor of Duluth. To control this problem, fisheries biologists planned to reduce the number of muskies and pike that anglers could take in Duluth harbor. They hope that these fish will prey on the ruff.

A nonnative or exotic fish species does not have to be introduced from another country. When intentionally or accidentally transplanted outside of its original drainage or water body, a

FIGURE 12.7 *The carp, an exotic introduced into American waters from Europe during the late nineteenth century. Note the mouth, specialized for bottom-feeding. This feeding habit causes this species to muddy waters and spoil habitat for game fish.*

native fish species from one area of the country becomes a nonnative species in another area. These introduced species can also compete with or prey on the native fish species of the new drainage system or waterbody, sometimes severely reducing populations (see Case Study 12.1).

Predation by Humans Fishing has been a major factor in the decline of many of our freshwater commercial and game fish. The elimination of the lake sturgeon, *Acipenser fulvescens,* from Lake Erie as a result of overexploitation by sport and commercial fisheries is an extreme example of **human predation.** A classic example of fishing pressure is offered by anglers on the

CASE STUDY 12.1 THE SEA LAMPREY—SCOURGE OF THE GREAT LAKES

Imagine a predator so efficient that it could destroy 97 percent of the lake trout population of the Great Lakes in only 21 years! That predator is the **sea lamprey**—an olive-gray, blood-sucking killer that completed its invasion of the Great Lakes in the 1950s. The lamprey is a primitive, jawless vertebrate with a slender, eel-like body. The muscular funnel around its circular mouth enables it to attach firmly to its prey (Figures 1 and 2). It moves its pistonlike tongue, armed with numerous hard, rasping teeth, back and forth through the lake trout's tissues, tearing flesh and blood vessels and causing severe bleeding. An anticoagulant prevents the blood from clotting. After gorging itself on a meal of blood and body fluid, the predator may drop off its host and permit it to swim weakly away. The trout may die from the direct predatory attack, or it may eventually succumb to bacterial and fungal infections that become established in the open wounds. During its short adult life of about 15 months, the average lamprey kills about 40 pounds of trout, salmon, and other Great Lakes fish. Even if a lake trout survives, the ugly scar left on its body would scarcely be admired by the grocery-buying homemaker.

The Great Lakes lampreys spend their entire life cycle in fresh water. When sexually mature, the adults swim up tributary streams to mate, spawn, and die. After hatching from the eggs, the larval lampreys are needle-thin and about 3 millimeters (one-eighth inch) long. They drift downstream until they come to a muddy bottom. Then they burrow tail-first into the mud, allowing only their heads to remain exposed to the current. During this time they feed on small algae, insects, worms, and crustaceans. Several years later, when they have grown to the size of a pencil, they acquire the muscular funnel and rasping tongue of the adult, emerge from their burrows, and swim into the open waters of the lake to prey on fish (Figure 3).

The lamprey originally occurred in the shallow waters off the Atlantic seaboard from Florida to Labrador, in the waters of the St. Lawrence River, and in Lake Ontario, at the eastern end of the Great Lakes chain. For many centuries, the westward extension of the lamprey's range into Lake Erie was blocked by Niagara Falls. However, in 1833 the Welland Canal was constructed to promote commercial shipping. Unfortunately, however, the canal also provided the lamprey with an invasion channel to Lake Erie (Figure 4).

The lamprey's colonization of Lake Erie was a slow process, probably because of a lack of suitable tributary spawning streams.

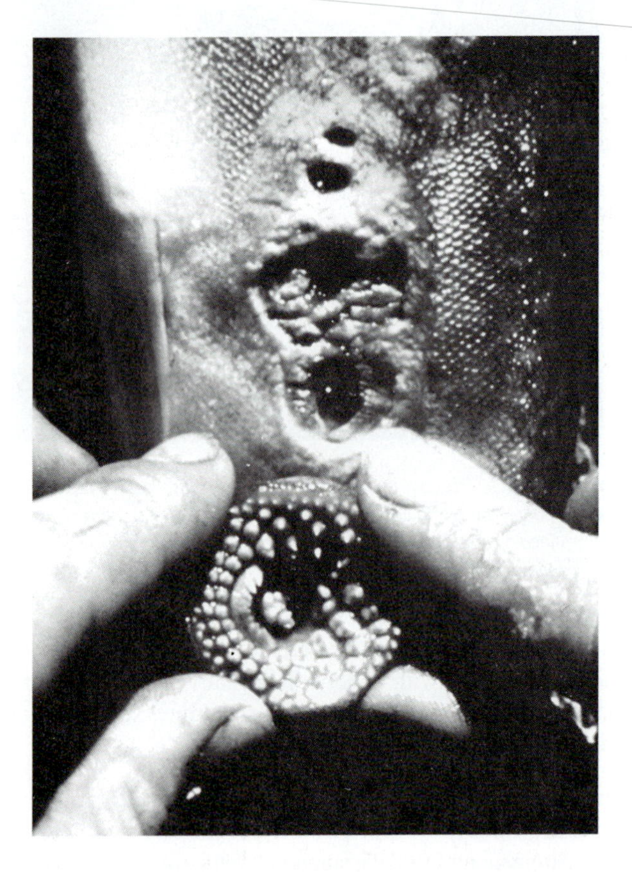

FIGURE 1 *Close-up of the muscular funnel, mouth, and rasping "tongue" of the sea lamprey. The end of the pistonlike tongue is visible inside the circular mouth in the center of the funnel. Note the wounds caused by the lamprey on the lake trout.*

FIGURE 2 *Lamprey adheres to netted lake trout.*

(continued)

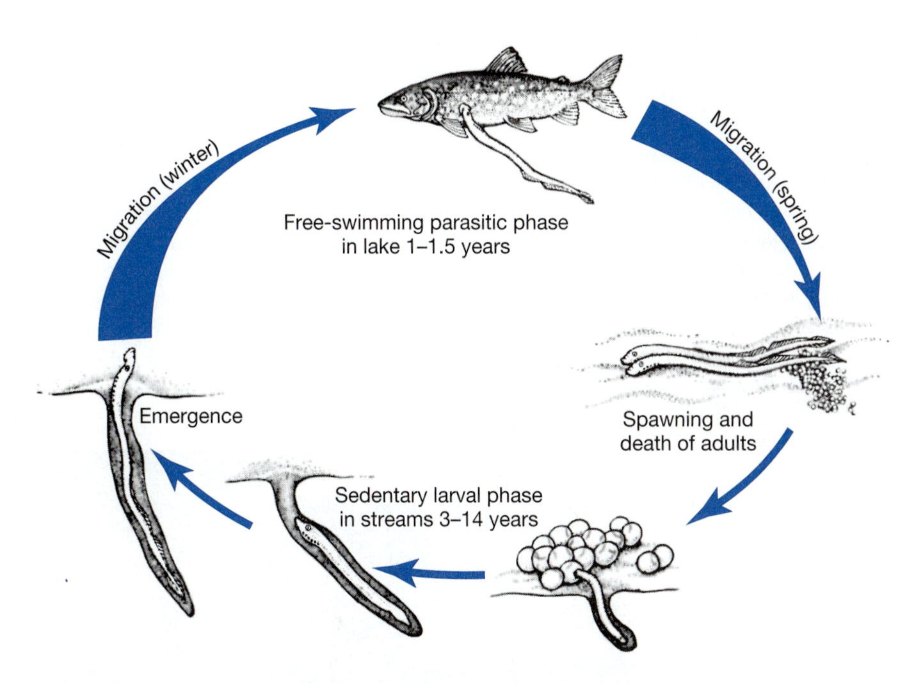

FIGURE 3 *Life cycle of the sea lamprey.*

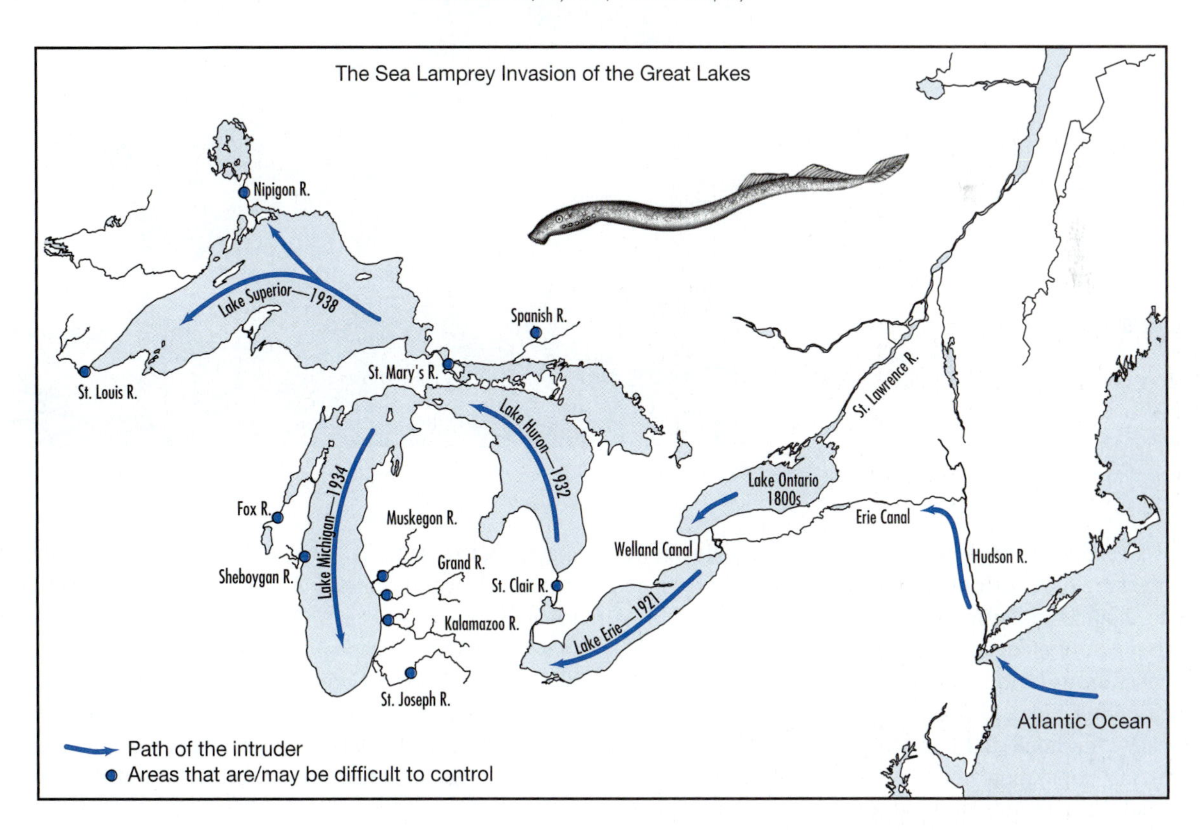

FIGURE 4 *Sea lamprey invasion of the Great Lakes.*

However, once it invaded Lake Huron, it spread rapidly into Lake Michigan and Lake Superior. By 1950 the lamprey was found in the western end of Lake Superior and had completed its Great Lakes invasion (Figure 4). Its predatory activity soon threatened the multi-million-dollar Great Lakes trout fishing industry with total collapse. The annual catch declined from 4,545 metric tons (5,000 tons) in 1940 to 152 metric tons (166 tons) in 1961, a 97 percent reduction in only 21 years! Idle nets rotted along the waterfront. Veteran fishermen, too old to acquire new skills, went on relief. Many of the younger men moved to Minneapolis, Milwaukee, Chicago, and Detroit in search of work.

(continued)

In 1955 the Great Lakes Fishery Commission was formed by treaty between the United States and Canada to control the sea lamprey. Members of the commission included all the Great Lakes states and the province of Ontario. A variety of control strategies were tried. Adult lampreys were netted and seined. They were even shocked with "electric fences" as they tried to move up their spawning streams (Figure 5). These methods, however, had only limited success. It was decided, therefore, to try chemical control—to use a **lampricide.** From 1951 to 1959, more than 6,000 compounds were tested as potential lamprey killers. Eventually an obscure poison, known as TFM, was selected. TFM was essentially nontoxic to humans as well as to game fish and their food organisms, such as minnows and aquatic insects. But it was lethal to lampreys. Larvae treated with low levels of TFM popped out of their burrows and quickly died from hemorrhage.

By 1960 the Great Lakes Fishery Commission had treated all of the lamprey-infested tributaries of Lake Superior with TFM. Only 2 years later, this chemical had reduced the number of spawning lampreys in these streams by 85 percent. In the ensuing years, the lampricide was also used on tributary streams of Lake Michigan (1963), Lake Huron (1970), and Lake Ontario (1972), with equally effective results. Ohio began chemical control of the lamprey on Conneaut Creek (Ashtabula County), a stream that flows into Lake Erie, in 1986. Because of the widespread success of TFM in reducing lamprey populations, the Great Lakes Fishery Commission has since been able to cut its use of the lampricide by 50 percent.

FIGURE 5 *Electric "fence" placed across Michigan stream near its entrance to Lake Michigan. As the lampreys swim through the "fence" they are stunned by an electric charge, float to the surface, and are easily removed.*

opening weekend of the trout season. During their enthusiastic quest for the king of American game fish, they frequently stand shoulder to shoulder along stream margins.

Fishing pressure is increasing. The number of licensed anglers has risen from 10 to more than 40 million within the last three and a half decades. One of every six Americans tries his or her luck with a hook and line each year. And with the nation's growing population, the increase in leisure hours and mobility, and the desperate need for a release from urban tension, the impact of the human predator on fish populations is bound to intensify.

Dams Dam building on river systems that eventually drain into the Pacific Ocean has had devastating effects on many anadromous fish species. For example, in the Columbia River on the Washington–Oregon border, many young salmon on their way out to sea are killed by **nitrogen intoxication** resulting from the high levels of nitrogen in the turbulent waters immediately below dams. Additional fish die as they are drawn into the dams' giant turbines. Only 10 percent of the salmon fry ultimately reach the ocean. After spending 1 to 4 years in the ocean, where they become sexually mature, the salmon swim back to the Pacific Coast. The fish then ascend their native streams, apparently recognizing them by their distinctive smell. (The hatching fish apparently learn how their native stream smells and remember it for the rest of their lives.) The psychological or behavioral process by which migratory fish assimilate environmental clues to aid their return to their stream of origin as adults is called **imprinting.** On their upstream migration to spawning grounds, the adult salmon must get around the dams. **Fish ladders** (Figure 12.8) and other fish passage systems were designed and built into dams for this purpose. However, many systems are considered ineffective and are in the process of being redesigned and upgraded on some of the older dams. Dams also alter anadromous fish behavior and habitat by causing abnormal fluctuations in water levels and velocities and, therefore, the natural dynamics of the river ecosystem.

Resource Extraction A variety of human land-use activities have affected fish habitat in similar ways. For instance, logging, road construction, mining, or livestock grazing near rivers or within a watershed all cause increased soil erosion and, therefore, sediment transport to streams. As a result, fish spawning habitat is degraded or lost and incubating embryos are killed as spawning gravels and redds are buried by the increased influx of sediment. Sedimentation also causes loss of stream insects, removing the forage base for most drift-feeding fish.

Logging Harvesting and log transport methods have changed over the life span of the timber industry on the west coast of North America, but each practice has altered natural river ecosystem processes and native fish habitat in harmful ways. In the late 1800s and early 1900s, instream debris, boulders, and riparian vegetation were cleared and water was impounded behind **splash dams** to facilitate floating the logs downstream to sawmills. Water was released from the splash dams when logs were ready for transport. These high-velocity surges of logs and water gouged and eroded stream banks, increasing the sediment load downstream. Streambed gravels were scoured out, and with them, spawning beds and developing embryos. Fluctuations in streamwater levels as these dams were opened and closed and secondary, unnatural hydrologic changes wreaked havoc on fish habitat, diversity, and productivity. Today, clear-cutting, road building, planting, thinning, burning, chemical application, mechanical site preparation, and milling practices continue to degrade fish habitat by altering sediment and nutrient delivery rates, water temperature levels, dissolved oxygen levels, and overall water quality.

Grazing Cattle graze heavily on riparian vegetation, if available, because of its succulence and variety. Unfortunately, however, after several years the heavy grazing pressure can eliminate plant cover, trigger erosion, and result in many harmful effects on the **carrying capacity** of the stream for fish. A study of a Montana stream showed that the portions with ungrazed banks had 27 percent more fish over 15 centimeters (6 inches) long than the grazed sections.

FIGURE 12.8 *King (chinook) salmon leaps up the top step of a fish ladder at the Red Bluff Diversion Dam during a late spring spawning run. This fish ladder enables salmon to move upstream beyond the dam and eventually spawn in California's Sacramento River.*

Mining Miners were once able to dredge, straighten, pollute, and strip vegetation from river channels on public lands without constraint. In most areas today, however, potential environmental damage must be reported, a restoration plan must be approved, and a reclamation performance bond must be posted before mining permission is granted. For the duration of the mining operation, strict water quality standards must be met. Unfortunately, monitoring and enforcement of these regulations are often lacking owing to personnel limitations.

Channelization Rivers and streams have undergone a myriad of hydrological alterations for the purpose of flood control, land drainage, and water diversion. A majority of these projects have resulted in fish habitat loss or alteration.

Recreation Other forms of human recreation, in addition to fishing, involve the use of rivers, lakes, and riparian areas. Future human population increases will no doubt increase the demand in these areas for recreational activities such as swimming, boating, hiking, camping, horseback riding, and mountain-bike riding. Overuse can have detrimental effects on riparian and aquatic habitats, including vegetation trampling, soil compaction, streambank erosion, and water pollution.

12.3 Sustainable Freshwater Fisheries Management

The fisheries manager has a challenging job. One reason the manager's job is so demanding is that each fishery, whether in a lake, stream, or artificial impoundment, is a unique system and should be managed according to its particular needs. Another reason the manager's job is taxing is that the successful production of just a single species depends on the interaction of a whole series of chemical, physical, and biological factors. Among these are lake (or stream) area, bottom, and depth; water currents; length of growing season; water temperature; levels of dissolved oxygen; water acidity or alkalinity; water pollution; water fertility (dissolved nutrients); shelter (cover); food availability; predator-prey relationships; reproductive potential; mortality rates; fishing pressure; fishing regulations; species composition; population size; population age structure; and growth rates. In addition, the fisheries manager must determine a suitable method to sustain fish populations on public lands that are managed for other resource uses (logging, grazing, mining, recreation) in addition to fisheries. And finally, a single river or stream ecosystem can flow through a large geographic area that includes both private and public lands. Applying consistent management practices throughout an entire watershed may be difficult when public and private land owners have different objectives for the resource. Some of the possibilities as well as difficulties faced by fisheries managers in attempting to restore fish populations in a given ecosystem are described in Case Study 12.2.

Humans have spent over 200 years destroying and degrading fish habitat—there is no quick fix. Fish resource managers must be able to develop and administer regulation, restoration, and protection schemes that will lead to the long-term health of aquatic ecosystems. With the exception of a few successes and the strategies acceptable in last-resort situations, there is much evidence indicating that the risks, negative impacts, and short-term nature of most artificial biotic manipulations to enhance or restore fish populations are, in general, poor management options. Currently, there is general agreement that the three components for sustaining freshwater fisheries are biota management (such as propagation and stocking), habitat management (maintaining adequate habitats), and human management (regulating the numbers of fish caught).

Population Enhancement Techniques

Artificial Propagation and Stocking In the early history of fish management, it seemed logical to biologists and anglers alike that if human beings could supplement the natural reproduction of a given fish species by artificial methods and introduce those artificially propagated fish into lakes and rivers, fish populations would be augmented and the angling success of fishermen virtually assured. In 1937 Wisconsin set a national record by stocking more than 1 billion fish!

Problems After intensive studies of population dynamics and reviews of numerous projects, it has become apparent that **artificial propagation** (also referred to as **captive breeding**) is more often a failure than a success. Moreover, the cost of artificial propagation in terms of facilities, maintenance, staff, and the rearing and eventual distribution of the young fish is almost prohibitive.

Artificial propagation is not well regarded by some fisheries biologists for a number of other reasons as well. Hatchery-bred fish are reared under more stressful environmental conditions and often are less well suited to the streams into which they are released. These fish tend to be physically weaker, generally more susceptible to parasites and diseases, and prone to more severe manifestations of a particular disease than wild or native fish. Therefore, artificial propagation and **stocking** of lakes or streams with these fish carries a high risk for the spread of disease to wild (transplanted but now self-sustaining) and native fish populations. In addition, captive breeding of *native* fish for the purpose of enhancing the population numbers in their native stream, carried out for more than one generation, puts the population at risk for inbreeding and loss of genetic diversity.

Whirling Disease An extremely pressing problem in the intermountain West is **whirling disease,** which is currently infecting wild trout populations. Whirling disease is caused by the fish parasite, *Myxobolus cerebralis,* believed to have been unintentionally introduced in 1955 from Europe to fish hatcheries on the East Coast of the United States. The mud bottoms of some of these hatchery rearing ponds presented an ideal habitat for the *Tubifex* worm, another host of the parasite. The transfer of the infected hatchery fish may have spread the disease to wild and native fish species in other regions of the country. Rainbow trout, especially those less than 2 years old, appear to be the most susceptible to whirling disease. Visible symptoms of the disease are whirling behavior, black tail, or cranial and skeletal deformity. There is no treatment for this disease; infected hatchery fish must be destroyed to prevent further spread.

In the upper Colorado River drainage, wild, young rainbow are severely infected. A 1994 study of an 8-kilometer (5-mile) reach of the river confirmed what may be interpreted as a catastrophic decline in population. Between 1991 and 1994, the rainbow trout population decreased so rapidly in this section of river that only a few fish remained that were less than 30 centimeters (12 inches) long. The lack of natural recruitment of young fish over several seasons may signal a local population collapse in the near future.

Whirling disease is also affecting an extremely popular trout fishing destination, the Madison River, in western Montana. After years of a stable wild trout fishery, 1991 population estimates in a section of the river indicated a significant decline in the numbers of wild rainbow trout. In 1993, a second section of the river also experienced an abrupt decline, and by 1994 both sections reported a 90 percent decrease in wild trout numbers, based on the averages reported in the 1970s and 1980s. Tissue analyses of fish specimens from these reaches confirmed whirling disease. Further sampling in the upper 88 kilometers (55 miles) of the river indicated that 75 percent of the young trout examined were infected. In 1995, population estimates showed further declines in the Madison River trout numbers, and whirling disease was discovered in other Montana waters.

Whirling disease is currently found in 22 states (Figure 12.9). Researchers are trying to determine whether unique conditions exist that may put wild trout populations in this region at risk. An increased understanding of the disease will also aid in the development of management strategies that will prevent the spread of whirling disease to uninfected waters.

Values Without the artificial propagation of trout, the thrill of hooking one of those scrappy fish would soon be nothing but a memory for most anglers. For instance, in Virginia, about 850,000 catchable trout are stocked annually in 185 streams and 20 lakes. Currently, both federal and state fish hatcheries rear brook, brown, cutthroat, rainbow, and lake trout. In mountainous areas of Wyoming and Colorado, trout fingerlings may be stocked by means of aerial drops. In Colorado nearly all of the trout stocked in its rivers are caught in just a few months!

Captive breeding and fish stocking can be valuable as a short-term measure in some cases. When used for restoring native species to their original habitat, however, this technique should be saved as a last resort. Fish stocking may be used to reestablish fish populations that have been destroyed by predators, drought, pollution, disease, or some other environmental factor. Fisheries biologists may stock southern farm ponds with tilapia to get rid of excess aquatic vegetation. They

CASE STUDY 12.2

REBUILDING FISH AND WILDLIFE POPULATIONS ON THE COLUMBIA RIVER DRAINAGE SYSTEM

Estimates based on tribal accounts and historical catch records from pre-industrial era canneries on the Columbia River indicate that somewhere between 10 and 16 million adult salmon and steelhead returned to the mouth of this river annually—the starting point for their long upstream migration to their native spawning grounds. Both commercial and sporting harvests of the native species that swam in these waters were bountiful. Some of the early anglers spoke of seeing salmon so numerous during their spawning runs that "you could almost walk across the stream on their backs." The brightness of the sun glinting off the backs of the silvery mass of migrating adult salmon was the inspiration for the name of White Salmon, a city located on the Washington side of the Columbia River, about 60 miles east of Portland. Huge chinook salmon, some weighing over 27 kg (60 pounds), were so big and fat they were referred to as "hogs" (Figure 1). But such sights are long gone. Populations of all the salmon species native to the Columbia River system have declined sharply in the past few decades. Both the sockeye and the chinook salmon of the Snake River in Idaho are listed as endangered. Moreover, the native coho salmon of the upper Columbia and Snake rivers have disappeared.

By any measure, the salmon situation in the Columbia River Basin is grim. In 1992, for example, only a single male Snake River sockeye completed the 900-mile upstream migration to its spawning grounds in Redfish Lake in central Idaho. Fisheries biologists dubbed him "Lonesome Larry." His sperm was removed and refrigerated, and will be used in artificial propagation of the species. In recent years, only about 1 million adult salmon and steelhead return to the Columbia River Basin system—less than 10 percent of the region's historic runs.

What factors caused the dramatic decline in the Columbia River Basin salmon and steelhead? One factor is the sediment pollution resulting from the building of logging roads near stream margins. Another is thermal pollution caused by the removal of riverside trees that used to shade the streams and keep them cool. Temperatures of 20°C (68°F) or above effectively block the upstream movement of salmon. However, when water temperatures drop below 20°C, migration is quickly resumed. Still another factor is urbanization, with its attendant housing developments, malls, and parking lots. The water runoff from thousands of acres of asphalt and concrete forms a "witch's broth" of pollutants, such as oil, grease, salt, and heavy metals, that eventually contaminates nearby salmon and steelhead streams. Overfishing has also been highly detrimental.

The most important factor in the salmon and steelhead decline, however, has been the construction of the vast hydropower system along the Columbia River and its tributaries. This huge complex—the greatest hydropower system in the world—consists of 66 major hydroelectric dams, including the 107-meter (350-foot)-tall Grand Coulee, as well as about 70 lesser ones (Figure 2). Of course, such structures constitute a daunting array of concrete barriers to successful upstream migration by adult salmon and steelhead. Juvenile fish that try to move downstream are often drawn into the giant turbines and die. Only about 10 percent of the juveniles finally make it to the ocean every year.

FIGURE 1 *Chinook (King) salmon.*

For many years attempts by state or federal agencies to rebuild the salmon populations of the Columbia River and its tributaries

(continued)

FIGURE 2 *Bonneville Dam on the Columbia River.*

were best described as piecemeal, haphazard, and ineffectual. In 1980, however, Congress passed the Northwest Power Act. The agency created by this act, known as the **Northwest Power Planning Council (NPPC)**, was formed by an interstate compact involving Washington, Oregon, Montana, and Idaho. Provision is made for influential input to the NPPC from federal fish and wildlife managers, the four Northwest states, Native American tribes (who have asserted their salmon fishing rights), and public and private utilities. The Northwest Power Act directs the NPPC to (1) prepare "a regional conservation and electric power plan" [Section 4.(d)(1)]; (2) "inform the Pacific Northwest public of major regional power issues" [Section 4.(g)(1)(A)]; and (3) "develop ... a program to protect, mitigate, and enhance fish and wildlife, including related spawning grounds and habitat, on the Columbia River and its tributaries" [Section 4.(h)(1)(A)].

The NPPC has authorized the expenditure of Bonneville Power Administration ratepayer money for the construction of several new hatcheries to augment the 100 state and federal hatcheries already in existence. At first glance, a program for the artificial propagation of salmon seems admirable. However, care must be exercised so that the natural diversity of native populations of salmon is not overwhelmed by genetically inferior, hatchery-reared fish. In fact, as of 1994, only 300,000 (12 percent) of the 2.5 million salmon in the Columbia River Basin were salmon that had hatched on natural spawning grounds. Increasing the number of artificially propagated fish may have harmful effects on the gene pool of a given species of salmon. The NPPC, in its Columbia River Basin Fish and Wildlife Program, has cited concerns over **genetic erosion.** It appears that the offspring of a hatchery fish bred with a native parent have a lesser chance of survival than pure native offspring.

Let us briefly examine the NPPC's program for mitigating the destructive effects of the hydropower system on migrating salmon. The fish ladders (Figure 3) and bypass systems originally built into the older dams by the U.S. Army Corps of Engineers left something to be desired; observers felt they truly did not work. The NPPC has upgraded ineffective fish ladders and bypass systems in older dams, such as those in the Dryden Dam of eastern Washington. In order to prevent the future construction of hydropower projects that would be destructive to salmon spawning habitat, the NPPC has declared 72,000 kilometers (44,000 miles) of salmon streams off-limits to future dam builders.

Another measure being investigated by the NPPC is the permanent drawdowns of mainstream reservoirs, so that the stream channel would assume once more its predam character. A report prepared for the NPCC by nine independent scientists suggested that this hydromodification would need to be permanent in order to restore the riverine habitat for fish and other wildlife.

FIGURE 3 *Fish ladder on the Bonneville Dam permits salmon to swim upstream to their spawning grounds.*

(continued)

Such a strategy would greatly accelerate the speed of water flow. Suppose, for example, that prior to dam construction, a salmon could move from its hatching site in some Columbia River tributary all the way to the Pacific Ocean in 10 days. Drawdowns of mainstream reservoirs would permit such a rate of migration. However, with the damming up of the water flow by numerous federal, public, and private hydroelectric projects, the downstream migration might well take over 30 days—three times as long. And, of course, this means that the young salmon would be exposed three times longer to pollution and predation than their ancestors were under predam conditions. Not only this, the juvenile salmon might undergo premature adaptations for a saltwater environment while still swimming in fresh water. The result could be highly adverse.

Many fisheries experts consider **hydromodification** the most important component of the NPPC's ambitious program to rebuild the fish and wildlife populations of the Columbia River Basin. This strategy, however, has not been universally accepted. Opponents of drawdowns worry about the potential impact on hydropower generation, the potential for higher electricity rates, diminishing supplies of irrigation water, and loss of jobs as the result of reduced reservoir levels and generating capacity at the dams. In addition, opponents argue that the technology should be proved effective or at least endorsed by scientists before being implemented. In contrast, proponents of drawdowns believe that fish and wildlife habitat cannot be restored without reverting to natural river dynamics and that salmon need a free-flowing river, not a chain of slow-moving lakes. Proponents further argue that drawdowns should be pursued despite the potential for higher electricity prices.

Even more controversial than drawdowns, the breaching of four Snake River dams has been considered by many as the best strategy to reverse the rapid decline of salmon stocks on the Columbia River's largest tributary. Because all four of the Snake River's salmon and steelhead populations are listed as endangered under the Endangered Species Act, the National Marine Fisheries Service (NMFS) has been charged with developing and implementing plans to reverse the river's salmon and steelhead population declines. Opponents to dam breaching argue that dam removal would likely cost more than $1 billion, reduce the region's power generating capacity, eliminate an important transportation artery for neighboring farmers shipping their products to markets via Snake River barges, and have an uncertain impact on salmon and steelhead populations. On the other hand, many others agree with a group of more than 200 scientists that argued in a statement to then President Clinton that dam breaching is a necessary prerequisite to restoring healthy salmon runs to the Snake River system. A 1999 NMFS plan, however, put dam breaching on the back-burner and instead called for focusing on immediate improvements to habitat, hatchery, and harvesting strategies. With opposition to dam breaching running strong amongst the region's political leaders, supporters of the NMFS's plan feel that focusing on immediate, less controversial strategies will further salmon recovery efforts more than waiting the decade or so likely required to overcome political opposition and carry out dam removal plans.

A serious obstacle to successful completion of either NPPC or NMFS efforts is the complex nature of the many stakeholders in the Columbia River and its fish populations. At least this is the view of Kai N. Lee, former member of the NPPC and director and professor of environmental studies at Williams College in Williamstown, Massachusetts. As Lee has noted, the Columbia River Basin Fish and Wildlife program is either implemented or significantly influenced by 11 state and federal agencies, 13 Native American tribes, over 10 public and private utilities that own and operate major hydroelectric projects in the Columbia drainage, and numerous organized interests ranging from agricultural groups anxious to protect water rights to fly fishers impatient for the return of native fish stocks. Lee commented that "if the river is to revive in any sustainable sense, it will have to be managed with a stability, a durability, and an awareness of biology which is exceedingly rare in human affairs."

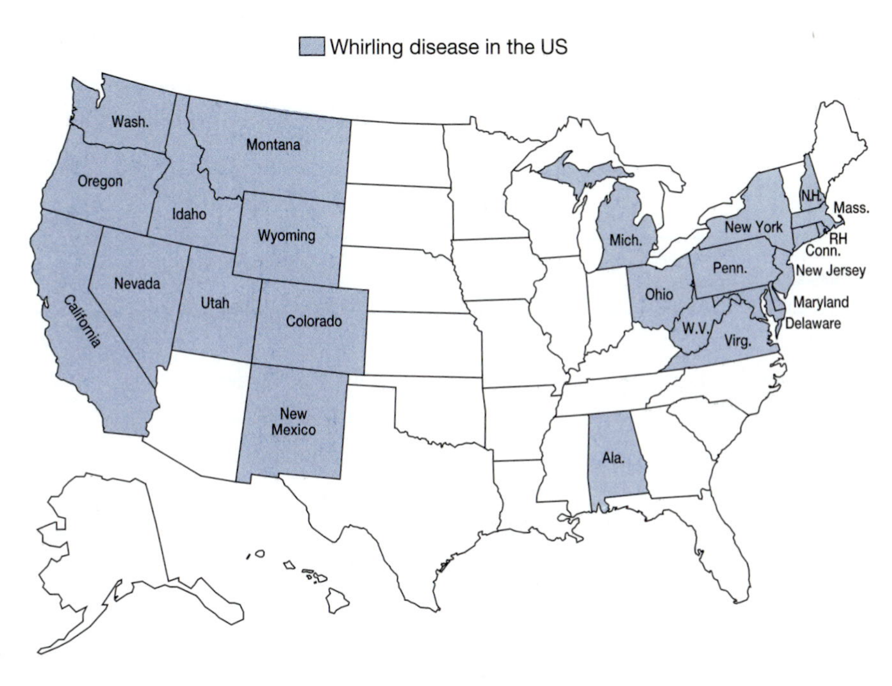

FIGURE 12.9 *Whirling disease is found in 22 states.*

may stock reservoirs with predatory species to reduce the number of bluegill. Reservoirs may be stocked with rainbow, brook, or brown trout if the water temperature is suitable. Largemouth bass may be stocked in reservoirs and farm ponds where water gets too warm for trout.

The official "trout policy" of federal hatcheries is to propagate trout to fill the following needs: (1) to stock trout in suitable waters in which they do not occur; such waters may be newly created reservoirs or may be water from which competitive nongame fish have been removed; (2) to stock trout in waters where conditions for growth are good but where natural spawning sites are inadequate; growth usually is rapid, but nevertheless, such streams must be restocked at intervals of 1 to 3 years; (3) to stock trout in waters where fishing pressure is heavy but there is no natural production. This is sometimes known as **put-and-take stocking.** The trout planted are of catchable size. Most of them are caught the same season they are planted. Put-and-take stocking is common in urban areas. For instance, thousands of legal-sized rainbow trout are planted annually in lakes in the metropolitan area of Denver, Colorado. Several of our nation's sportsmen presidents, such as Dwight Eisenhower and Lyndon Johnson, have boasted about the fishing potential of a particular trout stream on the basis of the lunkers they hooked only minutes after strategic stocking by publicity-sensitive conservation officials!

Introductions An **introduction** is the stocking of a nonnative fish. Although this practice was popular in the United States in the past, it can have some detrimental affects on the existing biological community. If the nonnative fish species becomes established, it can compete for resources with native fish, hybridize with genetically similar native species, prey on native fish species or their eggs, alter the community structure, and ultimately lead to the decline or extinction of some native species. The negative impacts due to the European carp and river ruff introductions were discussed earlier.

Some introduced species can survive without causing a major disruption by finding an ecological niche not fully utilized by the existing fish community. Coho salmon and chinook salmon, native to many drainages to the Pacific Ocean from Alaska to California and in Japan, have been successfully introduced into the Great Lakes (A Closer Look 12.3). Many lakes in Minnesota, Michigan, New York, and other states that had no walleye or muskie fishing now produce trophy-sized specimens, thanks to introductions from other waters in this country. New reservoirs are frequently stocked with species that were not originally found at the site. The great success of warm-water sport fishing in California has been possible largely because of introductions. Twenty-one of 24 warm-water species in California were introduced from states east of the Rocky Mountains, mostly in the late 1800s (Table 12.4).

The introduction of the European brown trout roughly 100 years ago is also considered highly successful in some areas of the country. This species has established itself in waters either too warm or too badly polluted for native trout. As a result, it has provided angling thrills for the anglers even in urban areas. The brown trout is able to survive in the relatively warm, somewhat muddy waters of Lowes Creek, for example, which is only a stone's throw from the city limits of Eau Claire, Wisconsin, a bustling city of 58,000.

Vibert Boxes Local sportsmen's groups can greatly increase trout populations in their favorite streams by using a simple but ingenious device called a **Vibert box** (Figure 12.10). This is a plastic box with slots on all sides to permit the free flow of stream water. About 100 trout eggs are placed in the box, which is then planted in the gravel bed of the stream. The Vibert box (1) permits the eggs to develop under natural conditions, (2) protects the eggs from predation, and (3) is inexpensive. (One trout fishermen's club planted 50,000 brown trout eggs by this method at a cost of only $300. Hatchery production methods would have been ten times as expensive.) Ninety percent of the eggs in a Vibert box actually hatch, compared to a natural hatching success of only 15 percent. The newly hatched fish (fry) are immediately conditioned to their environment of dissolved oxygen, water temperature, water chemistry, and stream flow. They therefore have greater ability to resist environmental stress than hatchery-reared stock.

Reservoir Stocking and Drawdowns Very few states are blessed with the number of natural lakes found in Minnesota

TABLE 12.4 *Some Native Fish Introductions to California's Inland Waters*

Species	Year	Source	Introduction Site
Smallmouth bass	1874	Lake Champlain, Vt.	Napa River
Channel catfish	1874	Mississippi River	San Joaquin River
Largemouth bass	1879	Eastern United States	Crystal Spring Reservoir (San Mateo County)
Yellow perch	1891	Illinois	Feather River (Butte County)
Lake trout	1894	Michigan	Lake Tahoe
Bluegill	1908	Illinois	Placer and Orange counties
White bass	1965	Nebraska	Lake Nacimiento (San Luis Obispo County)
Blue catfish	1969	Arkansas	Lake Jennings (San Diego County)

A CLOSER LOOK 12.3 *Salmon Fever in the Great Lakes*

The Great Lakes' fishery was almost destroyed 30 years ago because of overfishing, pollution, and the invasion of the predatory sea lamprey. Since then, however, the fishery has changed dramatically. Today, for example, the Great Lakes boast good salmon fishing. A key species that has made the salmon boom possible is the alewife.

The Alewife Invasion

A silvery, sardine-like fish, the alewife invaded the upper lakes from Lake Ontario. It first appeared in Lake Michigan in 1949. Lake trout fed on it ravenously, keeping its population down to moderate levels. However, a dramatic decline in lake trout stocks occurred soon thereafter, owing to heavy predation by the sea lamprey and overfishing. In the 1950s the alewife population exploded, probably due in part to the plummeting population of lake trout, its principal predator. By the 1960s the alewives formed 80 percent of the fish in Lake Michigan and over 50 percent of the lake's fish biomass.

During the spring of 1967, the alewives suddenly experienced a massive die in Lake Michigan owing to unusually cold weather and starvation. The carcasses of millions of these small fish washed up on shore, littering swimming beaches, decaying, releasing vile odors, and attracting flies. This was an unpleasant signal to fisheries biologists that the Great Lakes ecosystem was in trouble.

Salmon Stocking

In 1964, Michigan's Department of Natural Resources obtained 1 million coho salmon eggs taken from fish in the Columbia River on the Washington–Oregon border. This momentous event signaled the beginning of one of the great experiments in the history of American fisheries. In 1966, the 2-year-old salmon were planted in Lake Michigan and Lake Superior. By 1967, a few of the sexually mature coho, now 3 years old, began to make spawning runs up tributary streams. These sleek, muscular, heavy-bodied fish caused "coho fever" to rage among Michigan anglers. During the fall of 1967, more than 150,000 of them swarmed to the shores of Lake Michigan near Frankfort and Manistee. In their frenzied excitement, seven coho-crazy anglers ignored storm warnings and drowned when their boats sank in rough waters.

In 1968, coho were planted in the other Great Lakes. The success of these programs prompted the stocking of another Pacific salmon—the chinook. This species is often called the king salmon because of its unsurpassed fighting qualities and huge size.

Salmon stocking rates increased year by year. By 1983, 84 million coho and 108 million chinook had been stocked in the Great Lakes. These predators from the Pacific fattened up on their alewife prey. And the ultimate predators, the anglers, with their varied array of strategies and lures, were just as successful in catching the salmon.

Problems

The great Great Lakes salmon-stocking experiment, however, has not been without its problems. First, for some unexplained reason, the coho has not been able to spawn successfully except in the tributary streams of Lake Superior. Second, there is some concern that the coho of Lake Superior will interfere with the spawning activities of the long-established brown and rainbow trout—highly prized game fish in their own right. The most serious problem, however, has been the 86 percent reduction in the population of the alewife. Since 75 percent of the salmon's diet consists of alewives, fisheries biologists have become rightly concerned that salmon growth and reproduction might be sharply diminished, resulting in a disastrous collapse of the billion-dollar Great Lakes sport fishery.

Two strategies to prevent this problem have been considered. One is to reduce the tonnage of alewives that can be taken by commercial fisherman. In Lake Michigan alone in the mid-1980s, the annual commercial harvest of alewives for use in fertilizer and pet food amounted to about 18 million metric tons. The second strategy is to cut back on the salmon-stocking program. Accordingly, Wisconsin reduced its stocking rate by 10 percent several years ago.

What does the future hold? Conservation agencies of the Great Lakes states, as well as the U.S. Fish and Wildlife Service, are cautiously optimistic. Guided by computer models of predator-prey interactions, biologists hope to stabilize both the salmon and alewife stocks.

(22,000) or Wisconsin (8,000). However, damming rivers for hydropower production, flood control, and recreation has resulted in the formation of thousands of artificial lakes, or **reservoirs,** throughout the United States, especially in the South and West. Today they account for more than 25 percent of all freshwater fishing.

The temperature, chemistry, and biology of a reservoir are quite different from those of the portion of the river above the dam. The waters of a newly created reservoir dissolve nutrients from the newly submerged soil, which is of great significance to fish management. The sudden increase in water fertility that results is called a **nutrient flush.** The water-killed organisms (grass, shrubs, insects, worms, mice, and so forth) will eventually decompose and release nutrients as well. These nutrients are then channeled upward through the aquatic food chain. As a result, fish growth and reproduction are usually excellent. In one impoundment in Kentucky, anglers caught twice as many fish as they had at the same location prior to the construction of the reservoir.

Stocking Reservoirs Fishing success may improve dramatically if reservoirs are stocked. For example, as a result of stocking in Virginia's Smith Mountain Lake, it is now possible to catch trophy-sized muskies there. Fish stocking is most effective (1) in newly formed impoundments, (2) when introducing predators (bass) to control an overpopulation of stunted prey (bluegill), (3) when compensating for the severe reproductive failure of a game species, and (4) when stocking a forage fish (bluegill, threadfin shad) to provide food for desirable predatory species (spotted bass, northern pike).

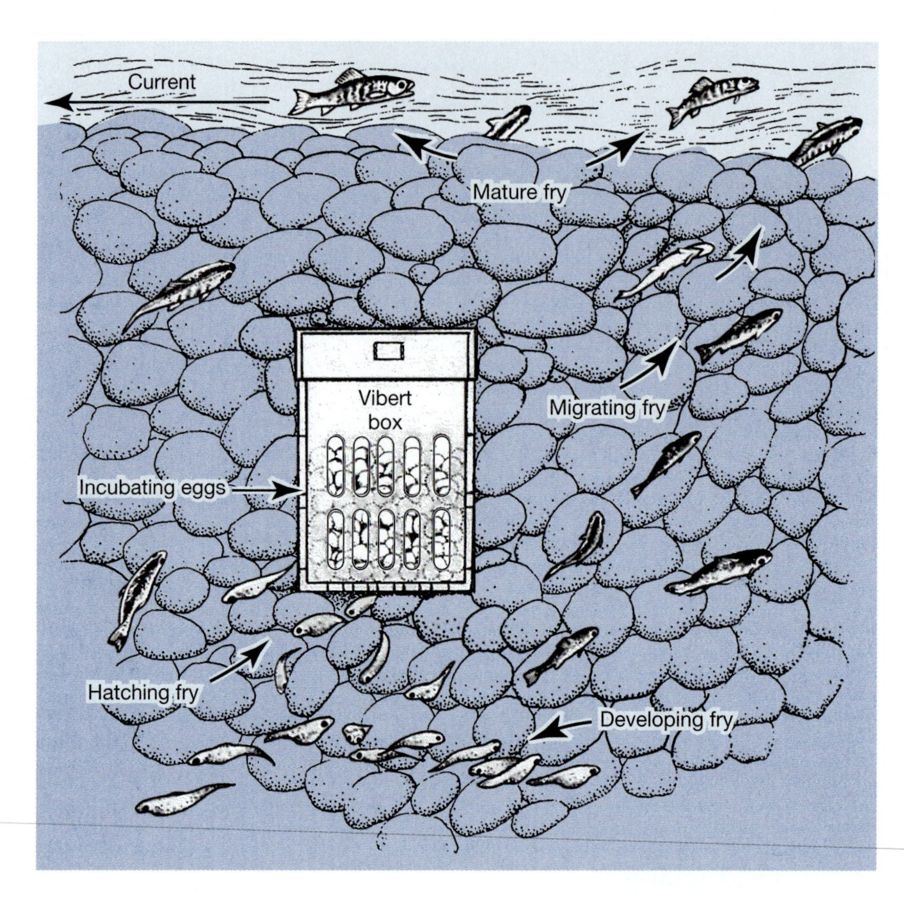

FIGURE 12.10 *Stream-planted Vibert box. View of section through gravel on stream bottom. All stages of fry development are shown.*

If the reservoir is sufficiently deep, it becomes thermally stratified during the summer, complete with epilimnion, thermocline, and hypolimnion. Such an impoundment is referred to as a **two-story reservoir.** At least 30 states stock their two-story reservoirs with both warm- and cold-water species: bass, catfish, and bluegill for the warm epilimnion and rainbow, brown, and lake trout for the cold hypolimnion.

Harmful Drawdowns If used for flood control or hydropower, the level of a reservoir can be purposely lowered or raised by the dam operator. Unfortunately, the fluctuating water level may seriously affect the fish (Figure 12.11). For example, a **drawdown** of just a few meters shortly after the lake trout have spawned could leave their eggs "high and dry." The entire spawn could be destroyed. On the other hand, if a drawdown occurs shortly before spawning, the lake trout are forced to spawn in an area where many of the eggs might be consumed by other fish, such as bullheads. Thus, the fish manager needs the cooperation of the owner and operator of the dam if the fish production potential of the reservoir is to be realized.

Beneficial Drawdowns Drawdowns can also benefit the fisheries of a reservoir. For example, in late summer, fish managers may request dam operators to draw down 10 to 80 percent of the water. The objectives include (1) aeration of the bottom muck, (2) acceleration of the bacterial decomposition of organic material and the release of nutrients, (3) restriction of forage fish to a small area where they can be more easily caught and eaten by predatory game fish, and (4) facilitation of nongame fish removal.

Some time after drawdown, the reservoir is refilled and restocked with game species. Due to the nutrient flush effect, a large quantity of food (diatoms, crustaceans, insects, minnows, and so on) become available to the fish of this artificial ecosystem. As a result, both growth and reproduction are enhanced. For example, in the 11,200-hectare (28,000-acre) reservoir in Beaver Creek, Arkansas, the average weight of pike increased 2 kilograms (4.4 pounds) annually during the first 3 years after drawdown. In an impoundment on the Rough River in Kentucky, the average weight of catfish progressively increased through the fourth postdrawdown year, at which time it was 473 percent greater than during preimpoundment years! Unfortunately, however, in most reservoirs, the positive effects of the nutrient flush diminish by the third or fourth year. At this time, therefore, another drawdown is required.

Translocations Replacing or enhancing native fish populations that have disappeared or diminished from their original waterbody can be accomplished with little threat to the existing species through **translocation** from another waterbody. The translocation process involves obtaining large numbers of fertilized eggs directly from spawning adults of the target species that are well established in an ecologically similar waterbody. The eggs are then immediately transferred to and placed in the original waterbody, thus creating a new population. The eggs may also be hatched in a captive breeding station (hatchery) and later introduced as young into the original waterbody. However, this second option is the least desirable owing to the aforementioned risks of stress, disease, and loss of genetic integrity. Prior to all translocations, however, the root cause(s) of the decline or ex-

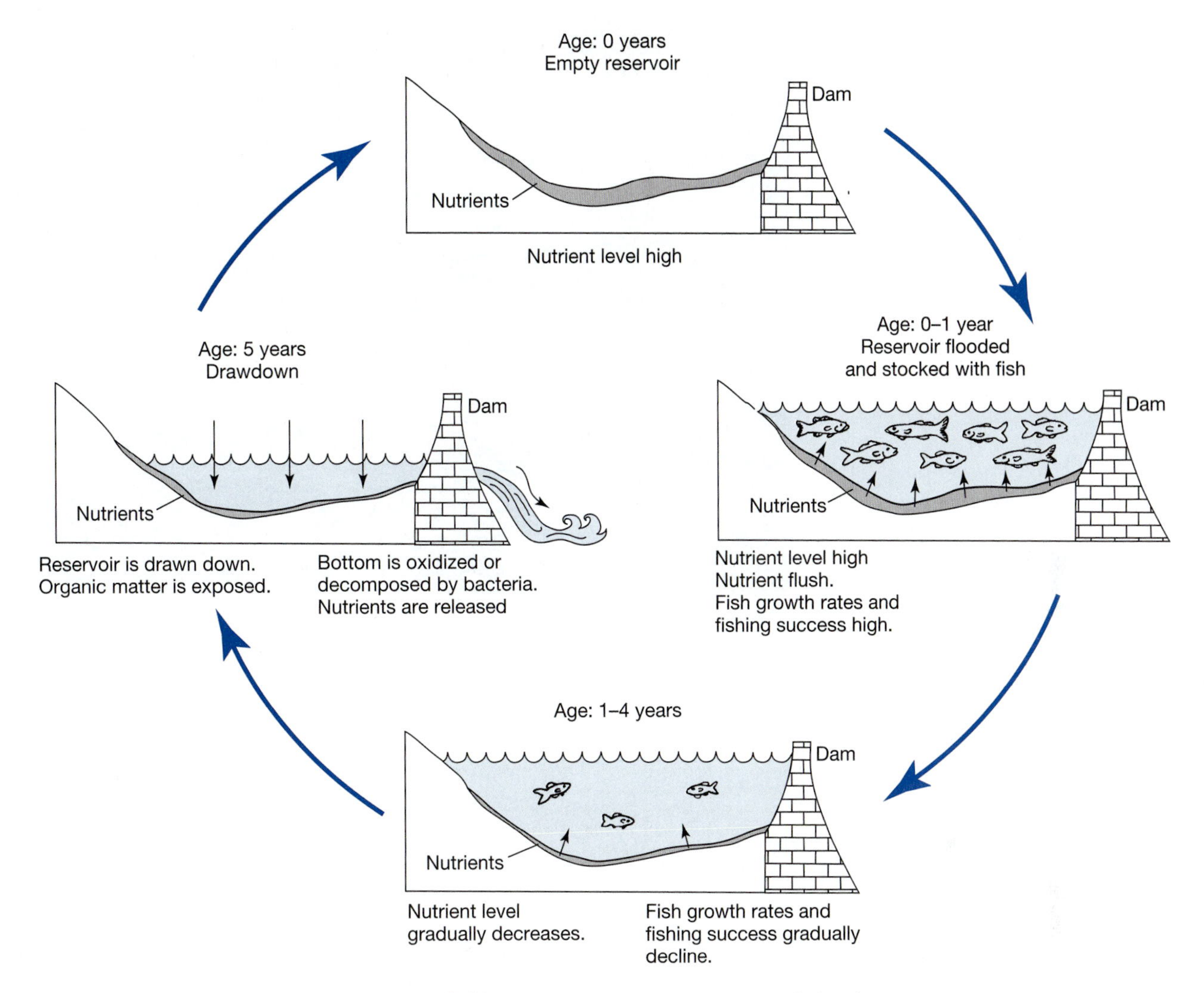

FIGURE 12.11 *Reservoir drawdown and the nutrient flush cycle.*

tinction of the target species in the original waterbody must be identified and ameliorated. Bypassing this first, important step will undermine the success of the translocation.

Removing Undesirable Fish Because of their destructiveness to game fish, large populations of **nongame fish** (gar, alewife, threadfin shad, common carp, quillback, spotted sucker, brook stickleback, mottled sculpin) and "stunted" **panfish** (black bullhead, white perch, rock bass, bluegill, black crappie, warmouth, redbreast sunfish, orange-spotted sunfish) are frequently the focus of intensive eradication projects. However, eradication of any of these species from any body of water is an enormously difficult task.

Various control methods under study involve chemicals, seining (netting), commercial fishing, manipulation of water levels, and fish-spawning control. Before state or federal biologists use a specific chemical, it must first be registered with the U.S. Department of Agriculture and approved by state health and pollution agencies and by the Federal Committee on Pest Control.

Rotenone, a chemical derived from the roots of an Asiatic legume, kills fish at a concentration of only 1 ppm within minutes at a water temperature of 21°C (70°F). Unfortunately, poisoning with rotenone is nonselective, resulting in the indiscriminate death of many species (Figure 12.12). The chemical control agent **antimycin** kills carp more readily than it does most other fish and does not appear to be deleterious to invertebrates. The long-term effectiveness of these techniques is questionable, however. In 1994 a group of researchers in Wisconsin reviewed 250 fish control projects located in 36 states and three counties. The study results revealed that less than 50 percent of the projects were considered successful, suggesting that improvements are necessary. The researchers concluded that many fish control projects are initiated without thorough investigation into the causes of the overpopulation of the undesired species. Many times habitat degradation, decreased water quality, or overexploitation of these species' predators by a fishery is the underlying cause of the problem. Therefore, chemical or physical removal techniques are only short-term treatments for the symptom rather than for the cause of the problem.

Controlling Oxygen Depletion in Winter Various methods are available to reduce winterkill of fish caused by oxygen depletion: (1) If the lake is small, the opaque snow blanket may be removed with plows. This will permit sunlight to penetrate

FIGURE 12.12 *Chemical control of undesirable fish. Dead fish by the thousands float belly-up in a small bay of Clear Lake near Watkins, Minnesota. The lake's entire population of undesirable (rough) fish, including carp and bullheads, was destroyed after the lake was treated with rotenone, a chemical lethal only to fish and other gill-breathers. This lake was later stocked with valuable species of game fish.*

to aquatic vegetation so that photosynthesis can occur and the water can be oxygenated. (2) Dynamite may be used to blast holes in the frozen lake to expose surface waters to atmospheric oxygen. (3) Oxygen can be introduced through ice borings by motorized aerators.

Selectively Breeding Superior Fish Fisheries biologists in Wisconsin have crossed northern pike with muskellunge to develop a hybrid known as a tiger muskie. The tiger muskie has been successfully stocked in reservoirs. In these artificial lakes, they fare much better than either of the parental species. Larger and higher-quality fish are being developed at the Federal Fish Farming Station at Stuttgart, Arkansas. A rapidly growing hybrid catfish, for instance, has been produced by crossing a channel catfish with a blue catfish. When 2 years old, the hybrids weigh 32 percent more than similarly aged blue catfish and 41 percent more than channel catfish of the same age. Another successful hybrid in the South is the hybrid striped bass, which has been widely stocked and exhibits rapid growth and good environmental tolerance.

Because it is essentially impossible to predict the impacts of a "new" fish on a habitat or native organisms, selectively breeding superior fish is highly controversial. Introducing genetically engineered or hybrid fish species should be done on a very limited basis, with considerable forethought of potential consequences, and only in artificial impoundments or small, "closed" waterbodies.

Protective Legislation for Freshwater Fish Populations

Fish populations can also be controlled by regulations that limit the take—the size and number of fish an angler can take home. Similarly, some species or fishing areas are regulated by **"catch-and-release-only" restrictions. Closed seasons** are also used to protect species at critical times (Table 12.5). Fisheries biologists have long recognized that when female bass or walleye are taken when swollen with eggs, anglers are removing much more than a single adult. They are also removing hundreds of future young fish. Over the years, certain fishing techniques have also been outlawed, such as seining, poisoning, dynamiting, spearing, and using multiple-hook lines.

In recent years, fisheries biologists have been experimenting with more liberalized regulations on many species of warm-water fish. In many states, size limits on panfish (sunfish, bluegills, rock bass, and crappies) have been lifted, permitting fish of any size, from runts to giants, to be taken. On the other hand, **minimum size limits** have been placed on predatory species such as bass, pike, walleyes, and muskies. The main objective of these regulations is to ensure the presence of large predators that can control populations of panfish. Such regulations provide more opportunities for anglers to land a lunker bass or pike. However, researchers have shown that size limit regulations do not affect the survival rate of northern pike (Figure 12.13). The effects of **creel (catch) limits,** varying fish methods and gear, open and closed seasons, and winter fishing on fish populations are continually being evaluated.

The most ecologically sound and hence most effective regulations are those that are tailored to a given body of water. Such regulations are formulated on a lake-by-lake or stream-by-stream basis. Unfortunately, however, their administration and enforcement are often difficult.

Habitat Restoration

Habitat protection and restoration, supported by legislation to control fishing pressure and potentially harmful population enhancement techniques, are the principal long-term measures by which natural fish populations can be sustained. The success of any **habitat restoration** depends on an effective man-

TABLE 12.5 *Typical Fishing Regulations for a Northeastern State*

Species	Season	Daily Limit	Minimum Size
Largemouth bass	May 7–March 1	5	None
Bluegill, sunfish, crappie, perch	Open all year	50 in all	None
Catfish	Open all year	10	None
Muskellunge	May 28–Nov. 30	1	32 inches
Northern pike	May 7–March 1	5	None
Walleye	May 7–March 1	5	None
Lake trout	Jan. 2–Sept. 30	2	17 inches
Trout (other than lake trout)	May 7–Sept. 30	3 in all	Brook trout—10 inches Brown trout—13 inches Rainbow trout—6 inches

agement plan that includes routine, long-term monitoring procedures for measuring vegetation establishment, water quality, and fish productivity. Managers must be able to evaluate changes in the habitat over time and adapt necessary strategies that address problems as they arise. Unfortunately, comprehensive programs of this sort are rare because of the financial and time obligations they require.

The first step in habitat management is to identify and mitigate the existing factors that are limiting production. Native fish populations may increase considerably when the carrying capacity of their environment is raised; that is, by providing adequate water flows, depths, and temperatures; space; substrate; cover; and food supplies. Various strategies can be used to improve the stream habitat for fish, including the following: adding spawning gravels, revegetating stream banks and lake edges, restoring the historic meanders of rivers, creating pools and riffles, and placing instream woody debris. Some methods of stream improvement are shown in Figure 12.14.

A series of **brush shelters** may be anchored along the inner margin of a lake's littoral zone with great effectiveness. In the winter season, brush piles can be set up on the ice cover in strategic areas and weighted with bags of stones; they will gradually sink to their proper place on the lake bottom as soon as the ice melts in spring.

In the case of heavy grazing, sometimes eliminating the culprits is all that is necessary. Research by the U.S. Forest Service has shown that if the sides of a severely grazed stream are fenced off from cattle for 5 to 10 years, the stream's original ability to support large trout populations will be restored, as shown in Figure 12.15.

The fish production of a body of water may be increased by improving natural **spawning sites** and by providing artificial spawning surfaces where suitable natural ones are lacking.

1. *Sites for bass.* Sand or small gravel can be spread on the muddy bottoms of lakes and streams to provide spawning sites for bass. Nylon mats can be used as an artificial spawning surface for largemouth bass. In one experiment, 90 mats were tested in several ponds. Over a 2-year period, the bass spawned on nearly 75 percent of the mats.
2. *Sites for northern pike.* Intensive study of the breeding behavior of the northern pike has shown that the shallow, marshy fringes of the littoral zone are the preferred spawning habitat. Regrettably, in recent years suitable spawning sites for this species have been greatly reduced as a result of real estate development, marina construction, and industrial expansion. State fish and

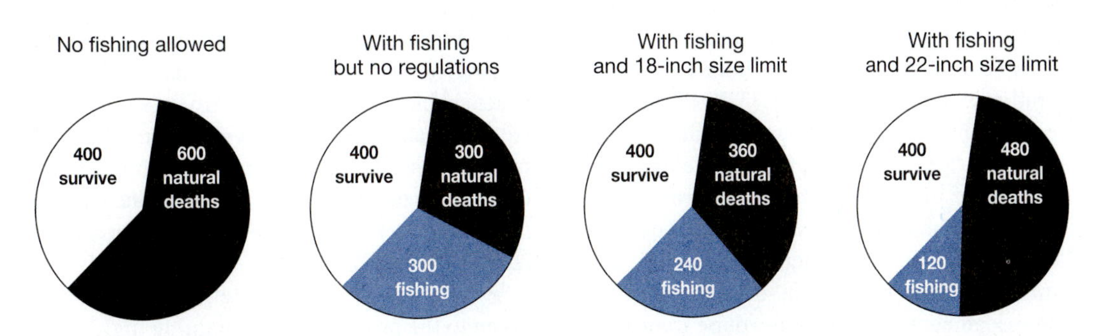

FIGURE 12.13 *Size limits on northern pike have little if any effect on the number of northerns that die during a given year. Even if no fishing at all is permitted, the mortality of a given northern pike population will be the same.*

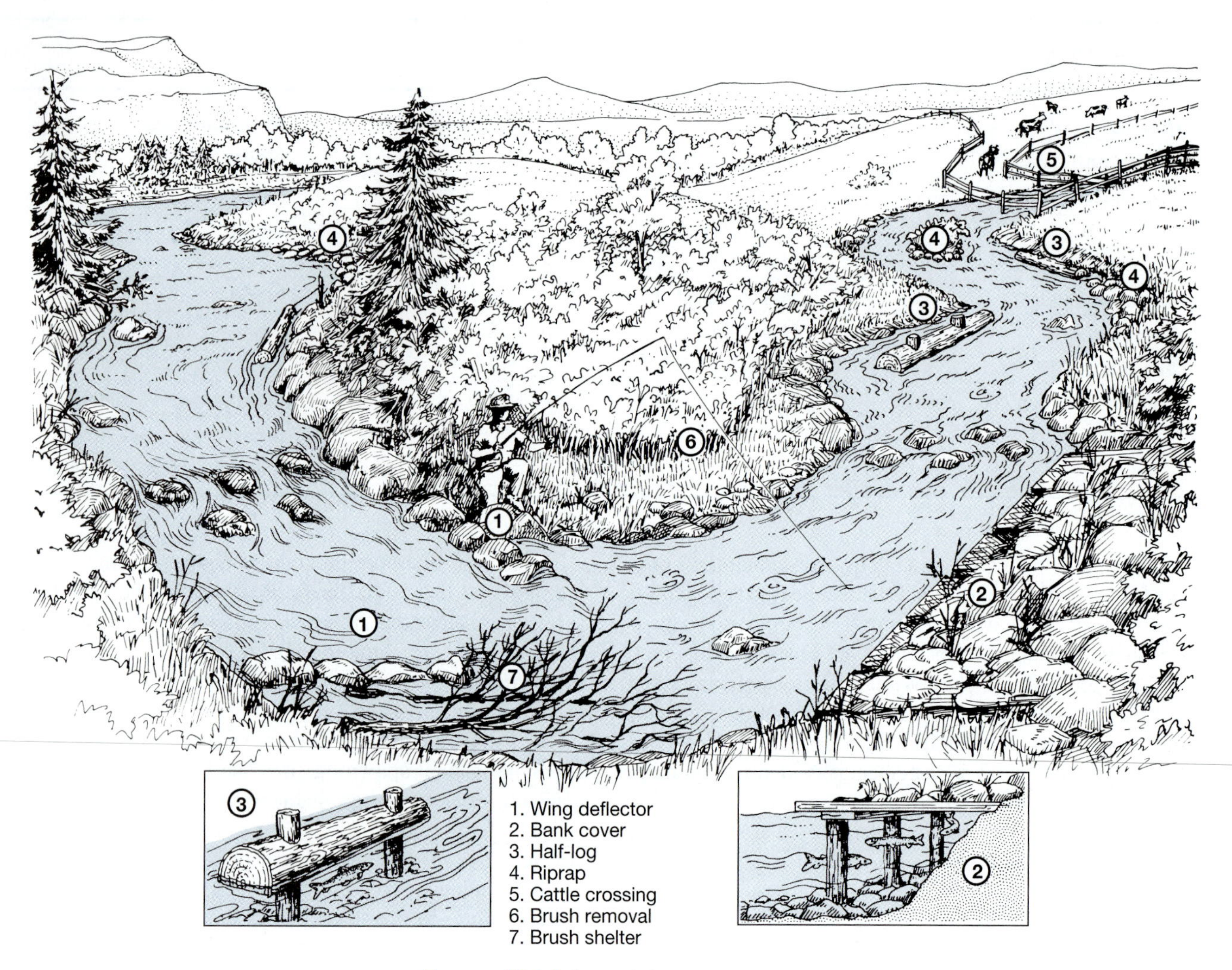

FIGURE 12.14 *Stream habitat improvements.*

game departments are attempting to rectify the situation. Wisconsin, Iowa, and Minnesota, for instance, have acquired thousands of acres of marshes to provide suitable pike breeding habitats.

3. *Sites for lake trout.* Hatchery-reared lake trout have been stocked in the Great Lakes for several decades. However, only a small percentage of the trout successfully spawn. Ross Horrall, a researcher from the University of Wisconsin—Madison, has recently located traditional lake trout spawning reefs (mounds of submerged rocks) that were successfully used long before the sea lamprey decimated this species. One of these reefs is located near the Apostle Islands in Lake Superior. The Horrall research team placed 273,000 fertilized lake trout eggs in huge **Astroturf "sandwiches"** (Astroturf is the plastic "grass" used on football fields). The sandwiches stabilize the eggs, protect them from sediment and predators, and make easy retrieval possible (Figure 12.16). After a 7-month incubation period, the Horrall team found that 88 percent of the eggs had hatched. The researchers hope that the young trout will remember the spawning reef and will return to breed when sexually mature.

Often a compounding effect is associated with restoration activities. Each habitat management method will have a positive effect that, in turn, may result in other positive effects. For example, suppose that cover is increased by the introduction of brush shelters (Figure 12.14). The cover protects the fish from predation by mink, otters, bear, fish hawks, and herons. As a result, fish become more numerous. Shelters have an additional positive effect. As shown in Figure 12.17, the twigs and branches of the shelter serve as attachment sites for insect larvae, snails, and crustaceans. Those organisms provide food for the growing fish populations. As a consequence, the growth rates increase, resulting in bigger fish. Such shelters also provide shade where fish may retreat during the heat of the day. The cooling effect on the water enables it to dissolve more oxygen. The increased levels of oxygen enable the fish to swim faster. Figure 12.17 shows the pathways of effect for several other stream management methods; among these effects are cover, an increased water flow, and improved water fertility. See Table 12.6 for a summary of **stream improvements** for fish.

Priority for restoration efforts and funding should be given to the protection of remaining high-quality sites and to the restoration of sites that are, at worst, moderately degraded yet still contain valuable, threatened, or endangered fish

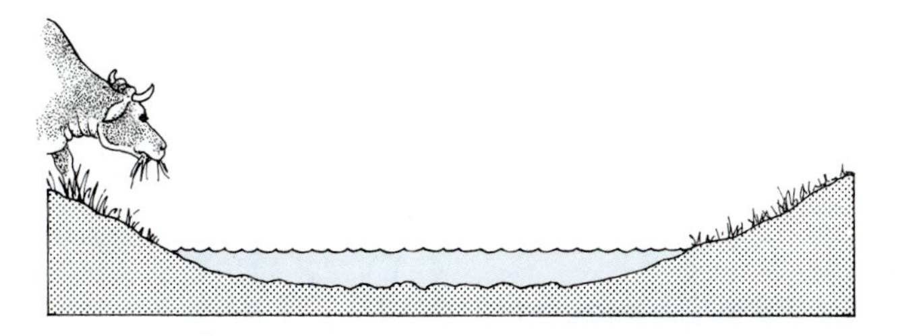

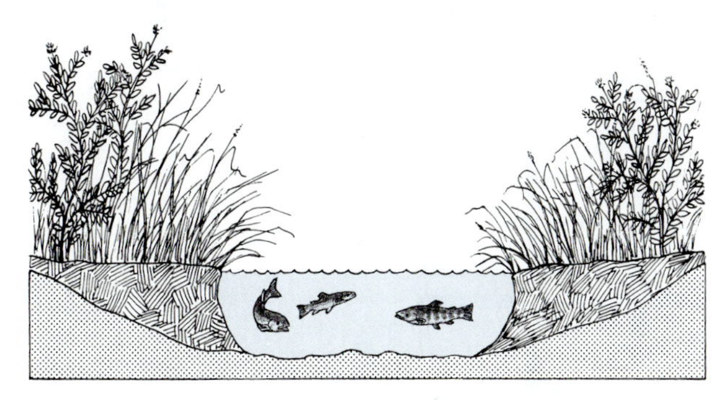

FIGURE 12.15 *Schematic representation of stream conditions subject to heavy grazing (top). Banks are grazed and trampled, leading to increased stream width, shallow water, and poor habitat for trout. Shallow water may be warmed enough by direct sunlight to limit a trout fishery. After 2 to 3 years without grazing, vegetation begins recovering as stream banks again develop their structure (middle). The habitat is improving for trout as food, cover, and spawning conditions become more favorable. Conditions after 5 to 10 years without grazing offer excellent trout habitat (bottom). Overhanging banks and deeper water have been restored as the vegetation recovered. Sedimentation also is reduced significantly.*

species. Restoration of habitats that are severely degraded or have lost most of their valuable fish may be prohibitively expensive and may have a much lower chance for success.

12.4 Marine Fisheries

The most important species to the U.S. commercial marine fishing industry in 1999 were the Alaska pollock, Pacific cod, flounders, hake, Pacific ocean perch, rock fish, anchovies, halibut, sea herring, mackerel, menhaden, butterfish, Atlantic cod, cusk, haddock, Atlantic pollock, Atlantic ocean perch, Pacific salmon, sablefish, tuna, clams, crabs, lobster, oysters, shrimp, scallops, and squid.

Most of the fish and shellfish that form the basis of the world's marine fishing industry are harvested near shorelines in the areas of the continental shelf and margin. In 1999 U.S. fishing fleets caught 4.2 million metric tons (4.6 million tons) of

FIGURE 12.16 *Astroturf egg sandwich! Ross Horall, a fisheries biologist at the University of Wisconsin—Madison, displays an Astroturf "sandwich," a recently developed device that holds and protects lake trout eggs during incubation.*

ocean fish and shellfish with a total value of $3.5 billion (Figure 12.18). Alaska led all states in both the volume and the value of fish caught. For example, Alaska's fisherman brought in almost one half of the total U.S. catch (2 million metric tons) (Figure 12.19). The most valuable U.S. catches for 1999 were shrimp ($561 million), crabs ($521 million), Pacific salmon ($360 million), lobsters ($353 million), and Alaska polluck ($163 million). In 1998, the U.S. marine fishery ranked third in catch quantity, behind China and Japan, with the Russian Federation fourth and Peru fifth.

Marine Commercial Fishing Techniques

Methods of Locating Fish In recent years, some highly sophisticated methods have been developed for locating commercial stocks of marine fish.

1. **Sonar** or echo-sounding systems, which use sound waves, help to locate fish schools and determine their relative abundance.
2. **Moored buoys** equipped with sonar or other devices detect the presence of fish schools swimming nearby and then radio the information to fishing vessels.
3. **Color enhancement** by either photographic or electronic means can detect color differences in the ocean

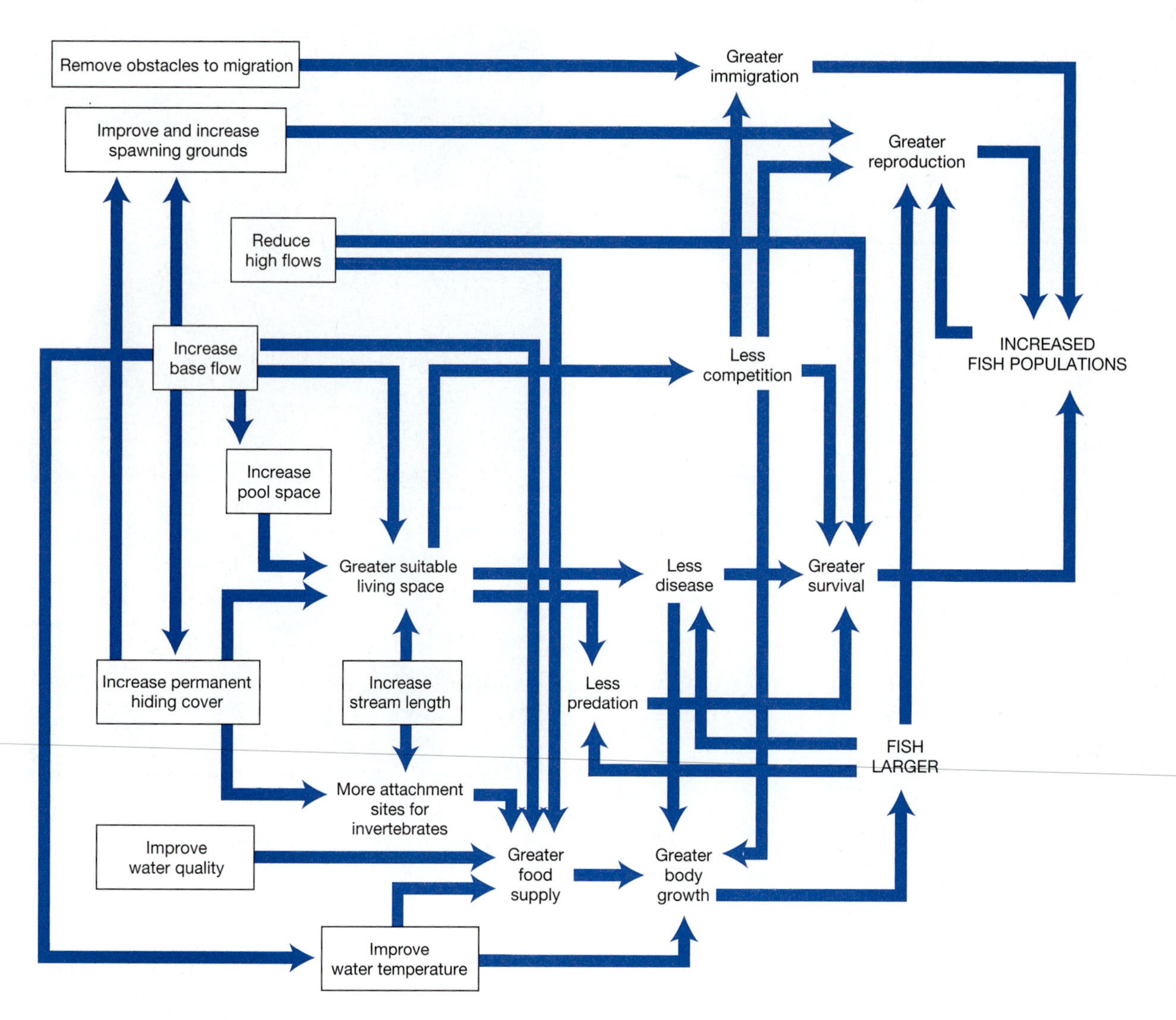

FIGURE 12.17 *"Chain responses" resulting from the application of various habitat improvement methods to streams. The methods are shown in boxes.*

TABLE 12.6 *Stream Habitat Improvements*

Problem	*Habitat Improvement*	*Result*
Not enough shelter or living space	Wing deflector Bank cover Half-logs	Channel deepens, pools form Cover increases, predation and competition decrease
Stream overgrown with trees and shrubs	Brush removal	Sunlight reaches stream, more food produced
Erosion of stream banks	Riprap Cattle crossing Fencing	Banks stabilize; water clears, channel deepens
Poor spawning success	Wing deflection	Silt scoured from gravel beds
Water too warm	Narrow and deepen channel	Colder water
Too much predation; lack of food	Brush shelters	Provides cover from predators, as well as habitat for food organisms

environment that are undetectable with the unaided eye. Such color-enhanced images provide information on the location of fish and on marine plant life and pollutants.

4. **Infrared sensors** borne by airplanes or spacecraft sense the ocean's temperature and detect fish movement. Temperature detection is valuable because some species of commercially valuable fish, such as tuna, have highly specific water temperature preferences.
5. **Ultraviolet sensors** can detect the presence on the ocean surface of oils given off by fish as they swim through the water. Since the oil rapidly dissipates when exposed to air, its occurrence indicates the relatively recent presence of a fish.
6. Airborne **electronic image intensifiers** can detect the flashes of light (bioluminescence) given off at night by microscopic marine organisms when disturbed by passing fish. The faint light is intensified 55,000 times and then projected on a television monitor.

Methods of Harvesting Fish Commercial fishing fleets use both passive and active fishing gear. **Passive gear** includes methods that are stationary and, therefore, depend on the fish or shellfish to come into contact with them. **Active fishing** gear includes methods that are mobile and are able to seek out and capture the fish or shellfish.

Passive Gear

1. **Pots and cages.** Pots and cages are used for catching crustaceans such as crabs and lobsters (Figure 12.20).
2. **Set nets, pound nets, and traps.** These nets are situated in such a manner that fish can easily swim in but can escape only with extreme effort.
3. **Hooks and lines.** A line with numerous baited hooks is set out close to the water surface; a long line can be over 1 km (0.62 miles) in length and contain more than 400 hooks (Figure 12.21).
4. **Gill net.** The fishing vessel bearing a gill net moves into a promising area shortly before nightfall. Up to 4.8 kilometers (3 miles) of net may be laid at right angles to the incoming or outgoing tide. The net is buoyed up by floats at the top and is weighted at the bottom with lead sinkers, thus forming a wall across the path of targeted fish stocks (Figure 12.22). The nets can be anchored (set nets) or allowed to free float (drift nets). As schooling species try to pass through the net on the way to their fishing grounds, their gills become entangled in the net and the fish die. The mesh is designed so that only fish above a minimum size are caught. The catch is hauled into the boat at dawn.

Active Gear

1. **Purse seine.** More fish are caught by purse seines than by any other type of net (Figure 12.23). The net is played out in a circle around schools of fish. When the school is completely encircled, the bottom of the net is drawn in to prevent fish from escaping. The purse line at the net bottom works like the drawstring on an old-fashioned purse. A large purse seine may be almost 1.6 kilometers (1 mile) long, have 16,000 floats, and weigh 13.6 metric tons (15 tons).
2. **Trawl net.** The trawl is a baglike net that is pulled by the fishing vessel. Most trawls are drawn over the ocean bottom and are used to harvest cod, flounder, haddock, ocean perch, and shrimp (Figure 12.24).
3. **Drift net.** In theory, drift netting is a passive technique, but the enormous size of some nets (sometimes called "walls of death") makes it impossible for many marine organisms to avoid them. Therefore, they are considered by some as an active technology.

12.5 Problems Facing the Marine Fisheries Industry

The annual marine commercial catch for the United States and the world is at a new high. This growth in marine harvest reflects an increase in the number of fishing vessels, advances in fishing gear technology, and the increasing protein demands of a growing world population. At the same time, however, the productivity of many important fish and shellfish continues to display serious declines. This biological stress has been caused by a number of factors, including the destruction and loss of

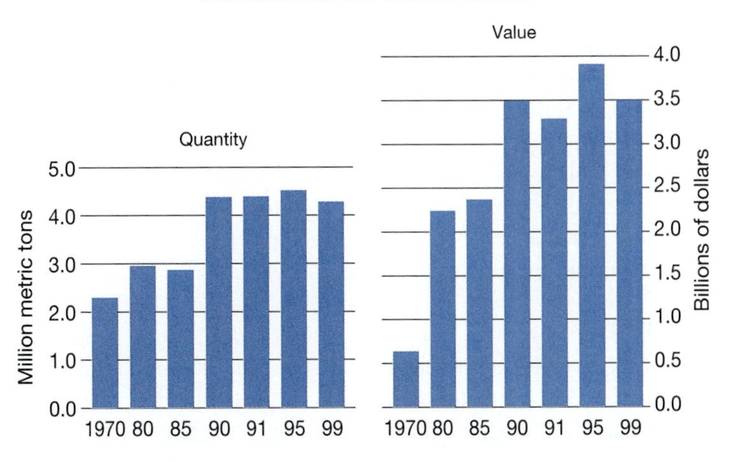

FIGURE 12.18 *The quantity and value of the U.S. catch of ocean fish and shellfish, 1970–1999.*

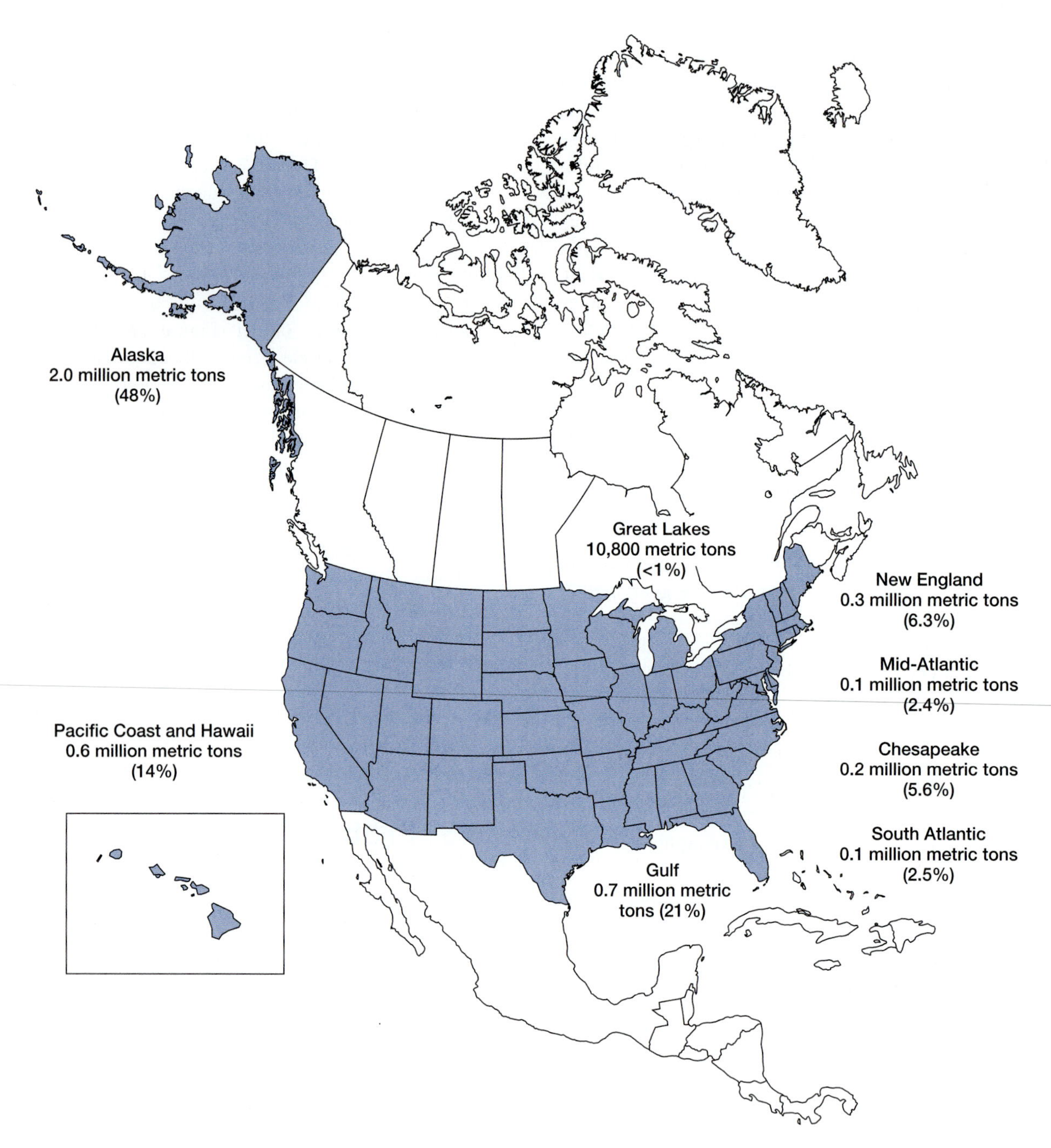

FIGURE 12.19 *U.S. ocean fish and shellfish catch by region, 1999.*

critical spawning, feeding, and rearing habitat that is found in coastal wetlands, estuaries, and bays. Marine and coastal habitat degradation from pollution caused by various human activities and losses caused by conversion to residential, commercial, and industrial development were discussed in Chapter 11. We continue the discussion here with at look at how overfishing and bycatch pose additional threats to the long-term health of our marine environment.

Overfishing

A large part of the **overfishing** problem is that the open ocean (all areas except the 322-kilometer or 200-mile strip extending out from each nation's shoreline) is the world's biggest **commons,** a public area over which no single nation has sovereignty. The tragedy of such a commons, however, as pointed out by Garrett Hardin of the University of California at Santa Barbara, is that its resources are ruthlessly plundered. After all, with no restriction on the take, it would seem foolish for a country with a marine fishing fleet not to get what it can as long as possible. This attitude has dominated the fishing industries of seaboard nations for years.

A classic example of the collapse of a once-major fishery caused by intense human predation is that of the Pacific sardine. In 1936–1937 the Pacific Coast sardine industry reached its peak. About 660,000 metric tons (727,000 tons) were netted (Figure 12.25), and sardine biomass was about 3 million

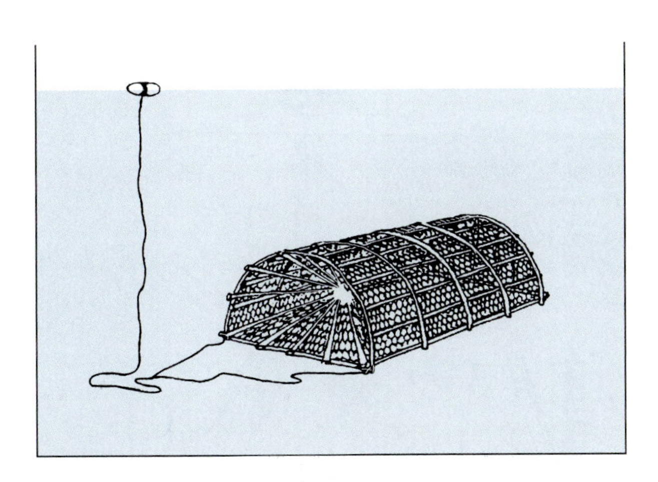

FIGURE 12.20 *Example of a pot used for crabs or lobsters.*

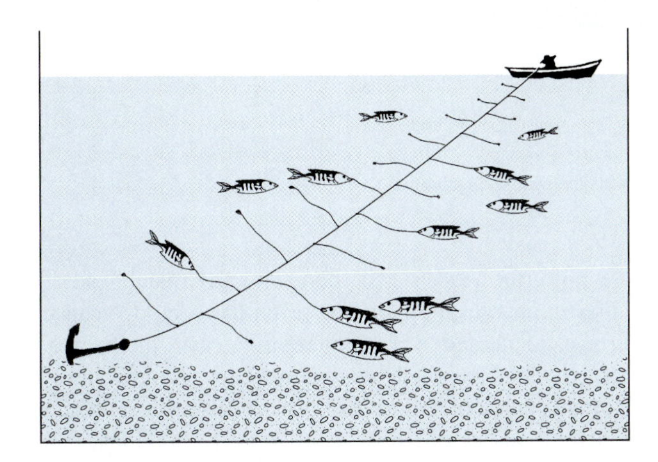

FIGURE 12.21 *Hook and line.*

metric tons (3.3 million tons). Biomass or fish weight is one way of expressing fish abundance or population size. This industry ranked first in the nation in metric tons harvested and third in value of catch. It grossed $10 million annually. The fish were used in many ways, from canned sardines to fish bait, pet food, and fertilizer. Unfortunately, the industry's prosperity depended on overexploitation. The fishing fleet was enlarged to make up for decreases in the harvest per boat. The industry rejected regulations based on the advice of fisheries scientists. As a result, sardine abundance and catches declined after World War II. The Washington–Oregon fishery collapsed in 1947–1948. In 1951, the San Francisco fleet returned with only 72 metric tons, well less than 1 percent of its take only a decade earlier, and the fishery closed down. In the late 1970s, Pacific sardine biomass declined to immeasurable low levels (a few thousand metric tons). Since 1986, sardine biomass has increased by 30–40 percent annually, and quotas have been allowed for commercial fishing. In 1997, biomass was about 600,000 metric tons.

Unfortunately, the sardine fishery is only one among many that have been going downhill. According to the Food and Agriculture Organization (FAO) of the United Nations, more than two thirds of the world's important marine fish stocks are being fished at or above their biological limit, or level of maximum productivity. Populations of many species, including such highly desirable food fish as Pacific perch, Spanish mackerel, grouper, flounder, bluefin tuna, Chilean seabass, Atlantic salmon and cod, orange roughy, rockfish (Pacific red snapper), and swordfish have been critically depleted or are overfished. Figure 12.26 illustrates exhausted fish stocks in the North Atlantic alone. David Crestin, deputy director of the U.S. Office of Fisheries Conservation and Management, recently commented on the critical nature of the problem: "At least 42 percent of the species in the American fisheries are being overharvested. In fact, some stocks are at their lowest level since we've been keeping records."

FIGURE 12.22 *Gill nets being used on the ocean bottom. These nets may also be suspended from the surface by floats. Fish become entangled in the net as they try to swim through it.*

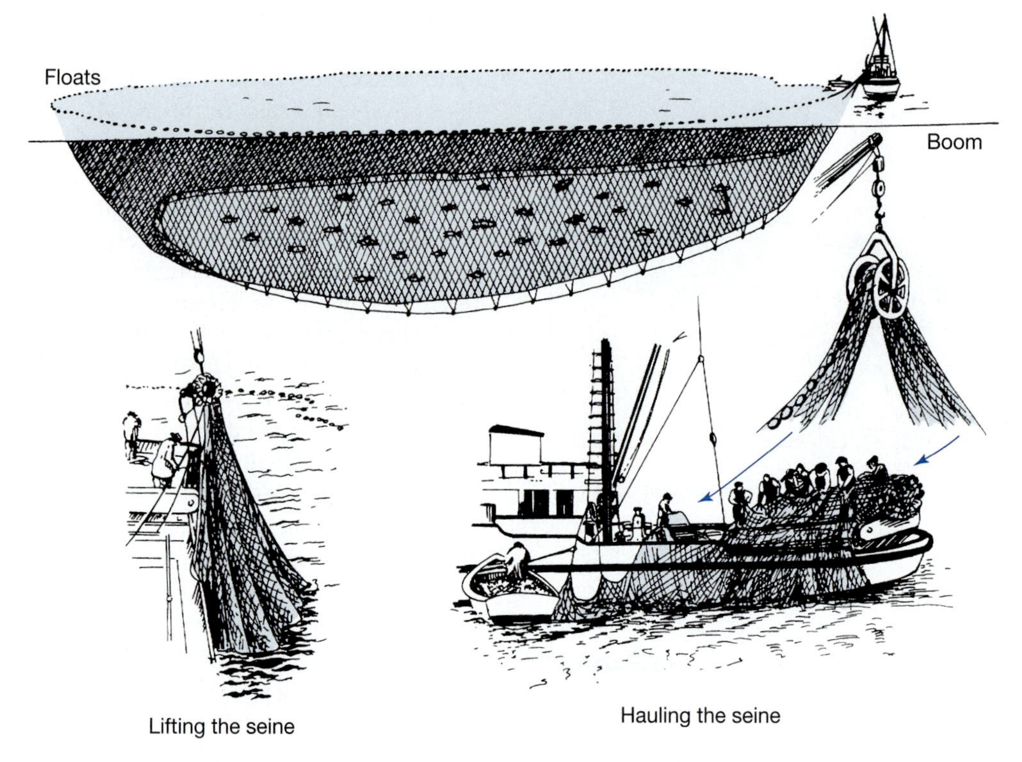

FIGURE 12.23 *Purse-seining for salmon: equipment and techniques.*

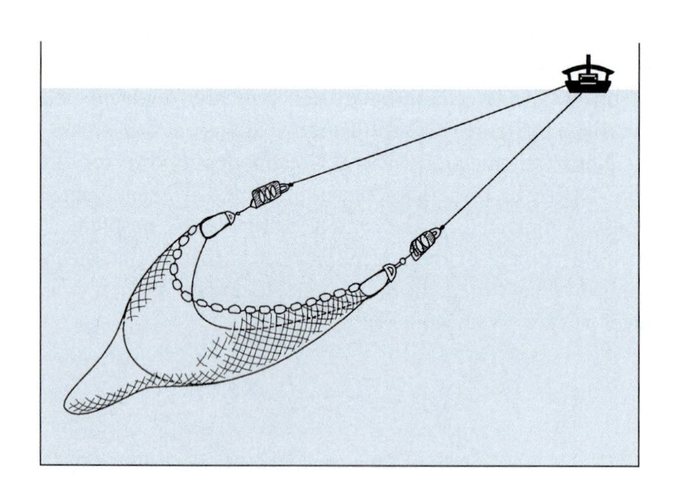

FIGURE 12.24 *Trawl net (otter type).*

Economics and legislation also play a role in overfishing. Each nation has economic jurisdiction over a 322-kilometer (200-mile) belt of sea space bordering its coasts. Each such area has been termed an **Exclusive Economic Zone (EEZ).** Scientists estimate that 95 percent of the world's living marine resources are contained within these EEZs. The intent of the EEZs was to reduce competition from foreign fleets and create a stronger incentive for each nation to manage its marine resources on a long-term, sustained-yield basis. What occurred, however, was a frenzied rush to expand national fishing fleets to take advantage of these new "noncompetitive" fishing opportunities. Governments subsidized construction of fishing vessels and fish-processing facilities, and industry developed more powerful vessels and more effective fishing gear for finding and harvesting fish. The result has been overcapacity, overinvestment, and overexploitation. Because the fishing industry employs and feeds a large number of people and is an important source of export earnings, governments continue to compensate losses (the FAO estimates that the global fishing fleet spends $50 billion more than it earns every year) through **subsidies.** In the presence of government subsidies, the fair market price does not accurately reflect the value or the scarcity of our marine resources.

Bycatch and Discards

Bycatch is the collective term used to describe captured marine organisms that are not the target species of the fishery. **Discards** are the bycatch that is thrown back for various reasons, including nontarget species, juvenile fish, endangered species, wrong size, inferior quality, or surplus to quotas. Bycatch also includes the capture of oceanic birds and marine mammals, the most famous example being dolphins that are caught in tuna purse-seine nets (Figure 12.27). Once thought to occur on a limited basis and otherwise considered an unavoidable aspect of the extensive use of nonselective fishing gear, bycatch is now recognized as a serious problem that may have far-reaching effects for the entire marine ecosystem.

Estimates for annual global discards in commercial fisheries are an astounding 27 million metric tons (29.8 million tons), more than one-third the weight of the total marine catch worldwide! The shrimp and prawn trawler fisheries have been singled

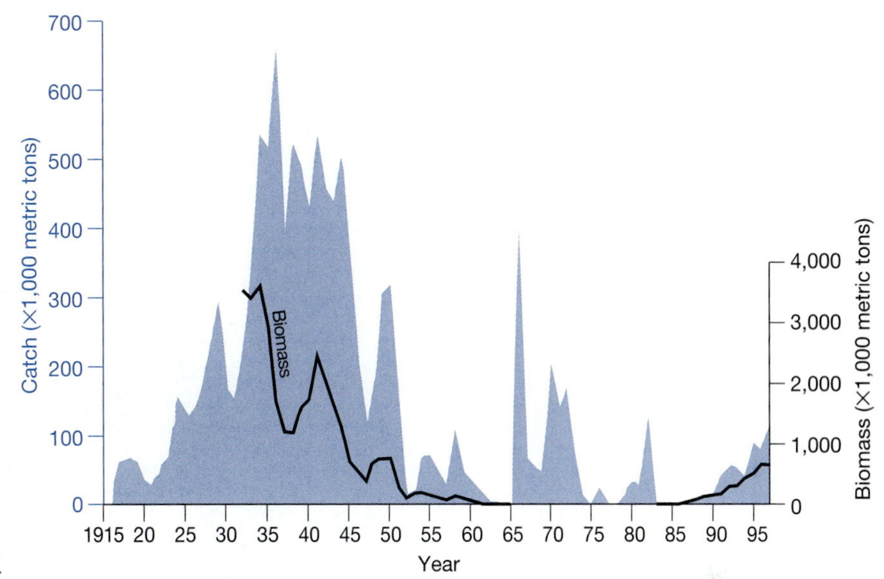

FIGURE 12.25 *Pacific sardine catch, 1916–1997.*

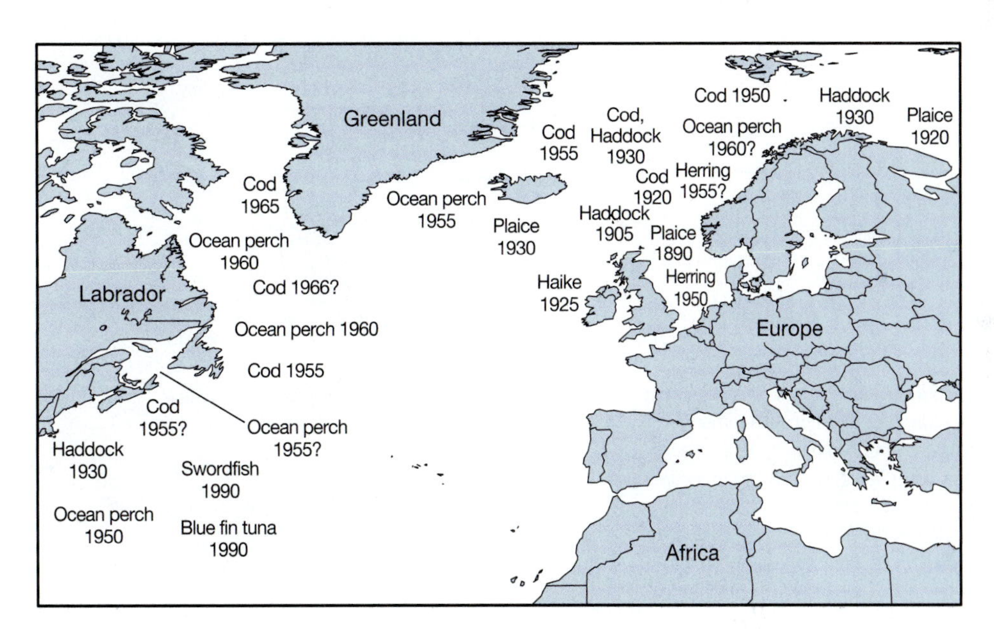

FIGURE 12.26 *Exhausted fish stocks in the North Atlantic. The depletion of the plaice (flounder) fishery in the North Sea became obvious already in 1890. Since that time many other fisheries have declined sharply. Note the depletion of the cod, ocean perch, and haddock fisheries off Labrador and Newfoundland.*

out as the largest contributors to this problem, accounting for more than one third of the world's bycatch discards.

The organisms hauled on board the vessel sustain trauma from exposure and handling, resulting in death or injury. Those organisms that are thrown overboard in an injured or weakened state become easy prey for hungry predators or have an otherwise poor chance of survival. The incidental capture and death of large numbers of juveniles of commercially and recreationally important species may reduce the potential biomass and yield of stocks crucial to other fisheries. Bycatch of endangered or threatened species is counterproductive to the recovery of these species. The bycatch of nontarget species that may not be of direct commercial value may represent an important food source for commercially valued or endangered fish, a source now removed from the ecosystem. So, as you can see, bycatch can not only affect a single species, it can cause a breakdown in the structure of entire biological communities.

12.6 Sustainable Marine Fisheries Management

Under the current marine management schemes, the overriding assumption appears to be that, if the fishing pressure is reduced, depleted fish stocks will bounce back. However, overfishing,

(A)

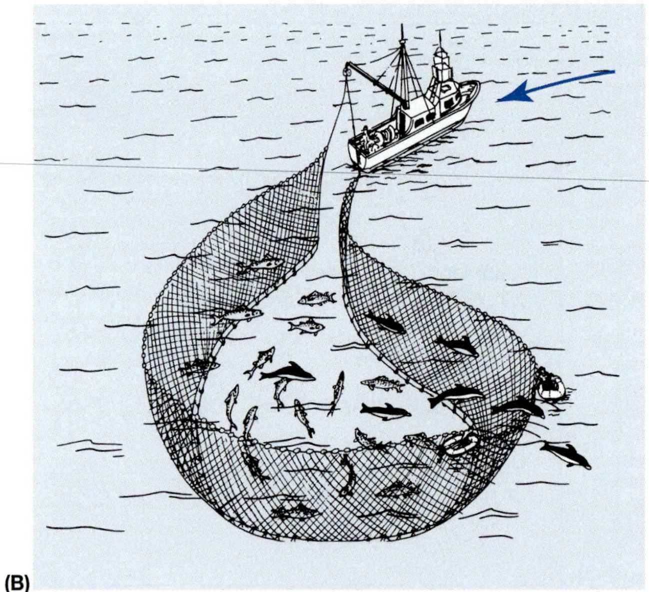
(B)

FIGURE 12.27 *(A) Tuna fishers accidentally net dolphins along with tuna. This happens rather frequently because dolphins swim underneath tuna schools. Fishers are now legally obligated to rescue as many dolphins as possible. (B) To permit the escape of the dolphins, the fishers first draw their net into an elongated shape. Then men on rafts help "spill" the dolphins over the end of the net into the open sea.*

bycatch, and habitat loss and degradation not only result in the decline in numbers and productivity of commercially important marine organisms, they also disturb nontarget species, disrupt predator-prey relationships, alter marine community structures, and reduce genetic diversity. Repair of all facets of the entire marine ecosystem will be necessary for the return of a healthy, naturally productive environment and a sustainable world fishery industry. Although a majority of the world's governments, scientists, and fisheries recognize the problems and acknowledge an urgent need to take action, this will be a formidable task requiring international cooperation and consensus.

Toward the goal of sustainable fish harvests, several management options have been suggested and debated over the years. Some strategies focus on regulations that limit harvests, some address the government subsidy issue, others aim at pollution cleanup and habitat restoration. Ultimately, the effort must include a combination of regulatory action, economic incentives, and habitat restoration.

Optimum Yield

There is an underlying premise for the success of any combination of the management options described below: In order to set quotas that ensure sustainable catches, we must obtain scientifically sound estimates of the optimum yield. In addition, it is crucial that a strict monitoring system be in place that discourages fishers from the temptation to catch more than their quota. The scientific research necessary to set quotas and the work force necessary to monitor catches will require considerable financial and technical support.

Most fisheries management has been operating under the principle of **maximum sustained yield (MSY)**. MSY is established for each species and is defined as the highest sustained fish or shellfish catch (number or weight) that can be allowed each season without affecting the reproductive capacity of that species. We know from the above discussion that, for a number of reasons, fisheries management is not meeting this goal. MSY is a mathematical calculation based on factors such as spawning mass, annual recruitment, annual growth in biomass, nonfishing mortality (natural death, predation), and fishing mortality. The problem with this calculation is that other factors such as marine pollution, habitat destruction, and declines in other marine organisms that have an impact on fish or shellfish breeding, feeding, and rearing success are not included in the equation. Therefore, MSY focuses on the productivity and quantity of a single species without regard for species interactions with each other and their habitats. In addition, the benefits and values that a fishery industry provides to a particular culture are not considered.

A more holistic view of fisheries management, and one some experts feel will achieve the requirement set forth in the 1976 **Magnuson Fishery Conservation and Management Act** for sustainable yields of important commercial fish and shellfish, is the principle of **optimum yield.** Optimum yield management considers the biological, economic, social, and political values of a given fishery in order to maximize benefits to society. It recognizes that fishing effort reductions alone will not improve the health of fish and shellfish stocks or generate greater social, economic, and nutritional benefits for society.

The optimum yield of a particular fishery must be considered on a case-by-case basis and involves an interdisciplinary team of scientists, managers, industry representatives, fishers, and laypeople. Figure 12.28 illustrates the theoretical concept of optimum yield. The intrinsic yield of a fishery to an ecosystem is the health of the ecosystem itself. An ecosystem achieves its highest intrinsic value when it is the healthiest; that is, in the absence of any human impacts. The extrinsic yield of a fishery is its sociocultural, economic, and nutritional benefits to society. As the extrinsic yield of a fishery increases, the intrinsic yield (aquatic ecosystem health) decreases. A maximum extrinsic yield is reached (at point *M* in Figure 12.28), after which the benefits to society decrease along with ecosystem health. The chosen optimum level of use established for any given fishery is dependent on the goals and resource objectives of the associated society. Point *M*, where extrinsic yield is the highest, might be chosen as the optimum level. However, by maximizing extrinsic yield, costs to the health of the ecosystem eventually outweigh the benefits to society. Ideally, an optimum yield would be one that allows the greatest extrinsic yield to society with minimum loss of intrinsic yield to the ecosystem (point *O* in Figure 12.28). In other words, the optimum yield is always less than the yield that would give maximum benefits to society.

Regulations and Economic Incentives

The following are selected methods that have been proposed or implemented to address the problems concerning the world's marine fishing industry:

1. Bycatch could be reduced significantly by banning some types of fishing gear (such as trawl nets and drift nets) in selected waters. In response, an economic incentive may be created for the development of more selective gear and fishing practices.
2. Some studies have estimated that sustainable harvests can be attained only by decreasing the intensity of fishing 30 to 50 percent—a considerable downsizing of the global fishing fleet. **Downsizing** could be accomplished by direct buyouts or by reducing or adjusting subsidies to encourage retirement of old vessels and discourage entry of new vessels.
3. A harvest tax or user fee for fishing could be imposed.
4. The current quota system encourages waste by forcing fishing vessels to rush and catch as much as they can in as little time possible before the "blanket" quota is reached. This "free-for-all" fishing causes investment in fishing equipment and practices that are efficient but not necessarily environmentally friendly. **Individual transferable quotas (ITQs)**, which give fishers a guaranteed portion of the catch, are currently being tried in various parts of the world. Evaluations of this strategy, however, have had mixed reviews. ITQs essentially transform a common property resource into a private one; fishers have an economic incentive to protect the resource in order to maintain the same quota in subsequent years. A drawback to this system may be that, over time, a small group of fishing fleet operators (the largest and wealthiest) will accumulate quotas and force smaller operators out of the

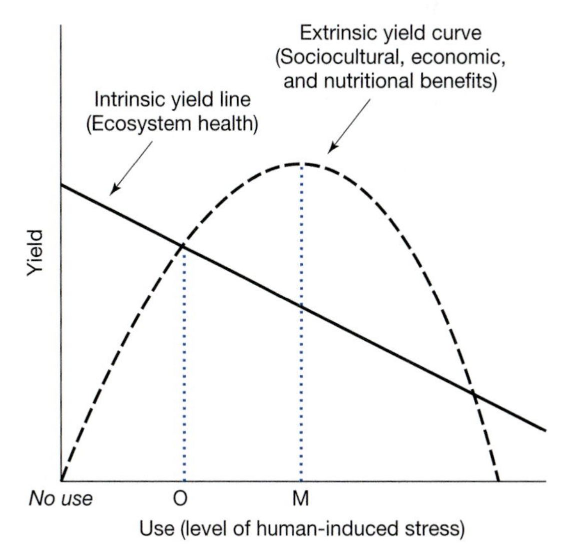

FIGURE 12.28 *Conceptual illustration of optimum yield related to fisheries management. The intrinsic yield (health) of an ecosystem declines from its pristine state as the extrinsic yield (sociocultural, economic, and nutritional benefits to a society) of a fishery increases. At point M, the maximum extrinsic yield is reached, after which benefits to society decrease along with intrinsic yield. Maintaining the maximum extrinsic yield comes at a high cost to ecosystem health that may eventually outweigh the benefits to society. In theory, point O is the optimum yield, which allows the greatest extrinsic yield to society with the smallest loss to ecosystem health (intrinsic yield). Note that optimum yield will be less than the yield necessary to maximize benefits to society.*

market. In addition, with less time constraints on their catch limit, fishers may actually exacerbate the bycatch problem by discarding smaller-sized fish caught earlier in return for larger, higher-valued fish caught later.

5. The problems of open access in international waters (beyond the EEZs) may be mitigated by the following regulations:
   - Establish size limits on fish that may be harvested.
   - Limit the number (or poundage) of a given species of fish that may be taken.
   - Restrict the number of times a particular vessel can fish.
   - Restrict the number of boats that can fish in a particular region.
   - Limit the number of fishing vessels that may be built during a given year.

All nations exploiting fish stocks in international waters would have to comply with mutually accepted restrictions to ensure a continued optimum yield year after year.

The Precautionary Approach to Fisheries

Agreements on harvesting fish in a commons are usually difficult, if not impossible, to attain. However, in December 1995, a United Nations–sponsored conference resulted in an international agreement that calls for a more conservative, or precautionary, approach to fisheries management. This agreement will become part of the 1982 U.N. Convention on the Law of the Sea. The **precautionary approach** requires increased monitoring, inspecting, and reporting in an effort to protect fish stocks *before* they show signs of decline instead of managing by reacting to the decline itself. The principles of the precautionary approach are outlined in Table 12.7.

Marine Habitat Restoration

Fish and shellfish need high-quality habitats in which to spawn, feed, and rear. Limitations on catch alone will not restore degraded habitats. As discussed in Chapter 11, we need to continue our efforts in the cleanup of polluted coastal wetlands, estuaries, bays, and oceans. Additional limitations on the development and conversion of wetlands should be also imposed. Projects to enhance, restore, and create these ecosystems should be encouraged and funded. Economic incentives should be put in place that will encourage the development of alternatives to the current fishing gear and techniques that destroy sea-bottom habitats, animals, and coral reefs.

Construction of Artificial Reefs Marine fisheries biologists are currently exploring the potential of artificial reefs to provide food and cover for both game and commercial species. Artificial reefs are especially helpful in raising the carrying capacity of flat, sandy coastal plains.

Marine scientists at the State University of New York at Stony Brook have constructed a reef from blocks of compacted sludge and fly ash—materials that create major disposal problems. These researchers have formed a 500-ton reef by dumping 18,000 blocks in the Atlantic Ocean 4 kilometers (2.5 miles) south of Saltaire, Long Island. It is hoped that such reefs will enhance sport and commercial fishing in an

TABLE 12.7 *The Precautionary Approach to Fisheries*

*Principle**	*Management Objective*
1. Fisheries stocks must be maintained at levels of abundance that are not substantially below their range of natural fluctuation.	1. Reduce fish stocks to no less than 75 to 80 percent of their unfished level. (Currently fish stocks are managed at maintenance levels below one-half their original abundance.)
2. New fishing gears and techniques must be evaluated before being introduced to a fishery.	2. Fishing gears would be widely tested before they are used on a commercial scale. Technologies that result in excessive levels of bycatch or substantial disturbance to the habitat would not be allowed unless modifications are made to reduce these effects to minimal levels.
3. Closed areas must be established to protect the marine habitat.	3. Set aside large areas where fishing gears that are destructive to the sea bottom are banned, protecting valuable habitat or allowing the recovery of a damaged habitat.

* These three principles are mutually supportive. Because abundant stocks require much less fishing effort for the same level of catch, there is less incentive to use destructive or nonselective fishing methods. Consequently, large parts of the habitat can be closed with only minimal impact on the fishery.

Source: Modified from M. Earle, "The Politics of Overfishing." *Ecologist* 25: 70, 1995.

area of high human population density and will not release toxins that contaminate the very fish they support. It could help solve an increasingly serious waste disposal problem for operators of coal-fueled power plants.

A series of reefs have been constructed along the New Jersey coast from Sandy Hook to Cape May by the New Jersey Department of Environmental Protection. Construction materials range from discarded tires and stainless steel drums to concrete bridge rubble and purposely sunken barges. More than 36,000 bald tires were placed on the Garden State Reef in 1 year alone (Figure 12.29). The reefs provide excellent recreational and commercial fishing for such highly regarded species as sea bass, cod, bluefish, mackerel, and tuna.

Choosing Seafood at the Food Market

When the Monterey Bay Aquarium, one the most prestigious aquariums in the world, buys seafood, they want to support sustainable fisheries, so that there will be plenty of fish for the future, so that marine habitats stay healthy, and so there is little bycatch. Their seafood choices (Table 12.8) for their own restaurant are based on the best available information, including data from the U.S. National Marine Fisheries Service, the U.N. Food and Agriculture Organization, and the Australian Bureau of Rural Sciences. They have three categories of choices: "best choices", "proceed with caution", and "avoid". You, as a consumer, play an important role in supporting or not supporting sustainable fisheries by the choices you make in purchasing seafood at the food market.

FIGURE 12.29 *Artificial reef being created off the New Jersey coast from tires that have been baled together and dumped overboard.*

12.7 Aquaculture

For thousands of years, human beings obtained their food by hunting animals and gathering eggs, fruits, berries, nuts, and seeds—a rather inefficient process that was unable to support more than a few million people the world over. Eventually, humans invented agriculture—the controlled rearing of plants and animals—a process that has been able to feed many more people. In a similar fashion, humans have for many years harvested fish from the sea by a relatively inefficient hunting-and-gathering technique: Fishing vessels move to fishing grounds where the harvest is unpredictable and then transport the catch back to market. The controlled culturing of fish and other aquatic food organisms is potentially a much more efficient process. **Freshwater aquaculture** is practiced in inland ponds. **Marine aquaculture** (also called **mariculture**) is practiced in shallow bays or estuaries.

Aquaculture has great potential to help meet the nutritional needs of a growing global population and relieve pressure on stressed ocean fisheries. Having increased aquaculture production by more than 120 percent between 1990 and 1995, China currently accounts for more than 67 percent of the nearly 31 million metric tons (34 million tons) of global production. India follows at a distant second in aquaculture production with a little over 2 million metric tons and the United States ranks eighth with nearly 450,000 metric tons. Global farmed fish production has more than doubled over the last decade and currently accounts for more than one-quarter of the fish directly consumed by the world's human population.

In addition to numerous crustaceans and seaweeds, over 200 species of fish and shellfish are farmed around the world. Important farmed marine organisms include shrimp, salmon, and bivalves (such as oysters, clams, and mussels). Freshwater-farmed fish include carp, talapia, and catfish, making up slightly more than 60 percent of global aquaculture production. In terms of trophic functioning, aquacultural organisms can be arranged on a food web scale ranging from seaweed at the bottom, herbivores (such as talapias) and omnivores (such as carps) in the middle, and carnivorous species (such as salmon, shrimp, and trout) at the top. These distinctions greatly determine production practices and their ecological impacts.

Production Methods

Mollusks (like oysters) are often raised in shallow, nearshore environments on rafts, in trays, or other structures to which the mollusks attach themselves (Figure 12.30). Crustaceans have traditionally been reared in saltwater ponds in coastal environments, although more recently they are also

TABLE 12.8 *Seafood Watch Chart of the Monterey Bay Aquarium. (For more detailed information, see the Monterey Bay Aquarium web site at* **www.mbay.aq.org***)*

Best Choices *(Wild population is abundant enough to sustain fishing; item is caught or farmed in ways that protect the environment.)*

Seafood Item	Where's It From
Albacore/Tombo Tuna	Pacific, Atlantic, Indian Ocean
Calamari strips & steaks (Pacific Squid	New Zealand, China, other Pacific
Calamari, whole (Market Squid)	U.S. Pacific Coast
Catfish	U.S. (farmed)
Clams	U.S., Canada, New Zealand (farmed)
Dungeness Crab	N. California
Halibut, Pacific	U.S. Pacific Coast or Alaska
Mahi-Mahi	Hawaii
Mussels, Black	U.S., Canada
Mussels, Green-Lipped	New Zealand (farmed)
New Zealand Cod	New Zealand
Oyster	U.S., Canada, New Zealand (farmed)
Rainbow Trout	Idaho (farmed)
Salmon	Alaska, California
Striped Bass	U.S. West Coast (farmed)
Sturgeon	U.S. West Coast (farmed)
Tilapia	Worldwide (farmed)

Proceed With Caution *(May or may not be environmentally-friendly, depending on how or where item is caught or farmed.)*

Seafood Item	Where's It From
American Lobster	East Coast of N. America
Bay Scallops	New England
Bay Shrimp	Pacific Northwest
English/Petrale Sole	U.S. West Coast
Imitation Crab/ Surimi (Pollack)	Alaska
Salmon	Washington, Oregon
Shrimp/Prawns	Georgia
Snow Crab	Bering Sea, Alaska
Spot Prawns	West Coast of N. America
Yellowfin/Ahi Tuna	Pacific, Atlantic, Indian Ocean

Avoid *(Wild population's survival is threatened by too much fishing, or item is caught or farmed in ways that damage the environment.)*

Seafood Item	Where's It From
Bluefin Tuna	Atlantic and Pacific Ocean
Chilean Seabass/ Patagonian Toothfish	Patagonian and Antartic waters
Cod, Atlantic	North Atlantic
Lingcod	West Coast of N. America
Monkfish	New England and mid-Atlantic
Orange Roughy	New Zealand
Rockfish (Pacific Red Snapper or Rock Cod)	Alaska, Washington, Oregon, California
Sablefish (Black Cod, Butterfish)	West Coast of N. America
Salmon	Pacific Northwest, Chile, Great Britain (farmed)
Sea Scallops	East Coast of N. America
Shark (al)	Worldwide
Shrimp/Prawns	Mainly tropical countries (farmed)
Swordfish	Pacific, Atlantic, Indian Oceans

being raised inland to expand production potential. Most marine fish are produced in near-shore floating cages. For anadromous fish, such as salmon, that hatch and develop in freshwater and subsequently mature in saltwater, both freshwater tanks and saltwater cages are used. Freshwater fish are commonly raised inland in earthen ponds or concrete pools (Figure 12.31).

Food sources used in aquaculture production are important considerations for producers and greatly determine the ecological impacts of the particular system. Low trophic level organisms, such as mollusks, rely on plankton and organic material provided by nutrient-rich seawater circulating through the production system. Mid-level herbivores and omnivores are most often maintained simply by providing nutrients to the naturally occurring aquatic organisms on which the fish feed. The high trophic level carnivores are reared on nutrient-rich diets requiring large amounts of fishmeal and fish oil derived primarily from ocean catches. The rising costs of feed inputs, particularly of fishmeal and fish oil, have helped to focus research efforts on developing lower cost feed formulations with greater portions of vegetable protein.

Ecological Impacts

The ecological impacts of aquaculture, as with terrestrial agriculture, vary depending on the intensity of production, the organism being produced, and the site of production. As previously mentioned, the factor that perhaps most determines the effect of an aquaculture system on the surrounding environment is the trophic level of the organism being produced. Mollusks, tilapia, and carp are often raised in so-called **extensive aquaculture systems** that rely on naturally occurring food sources and produce relatively small amounts of waste that are easily assimilated into the surrounding environment. Fish farming at this level, carried out on small scales and at low stocking densities, has traditionally been integrated within a complex polyculture system that includes multiple fish and plant species. In addition, nutrient-rich aquaculture effluent is often applied to surrounding crops that subsequently provide nutrient inputs to the aquaculture system. Although such production methods are still quite common throughout the world, greater market demand has encouraged replacement of extensive systems by more **intensive aquaculture**

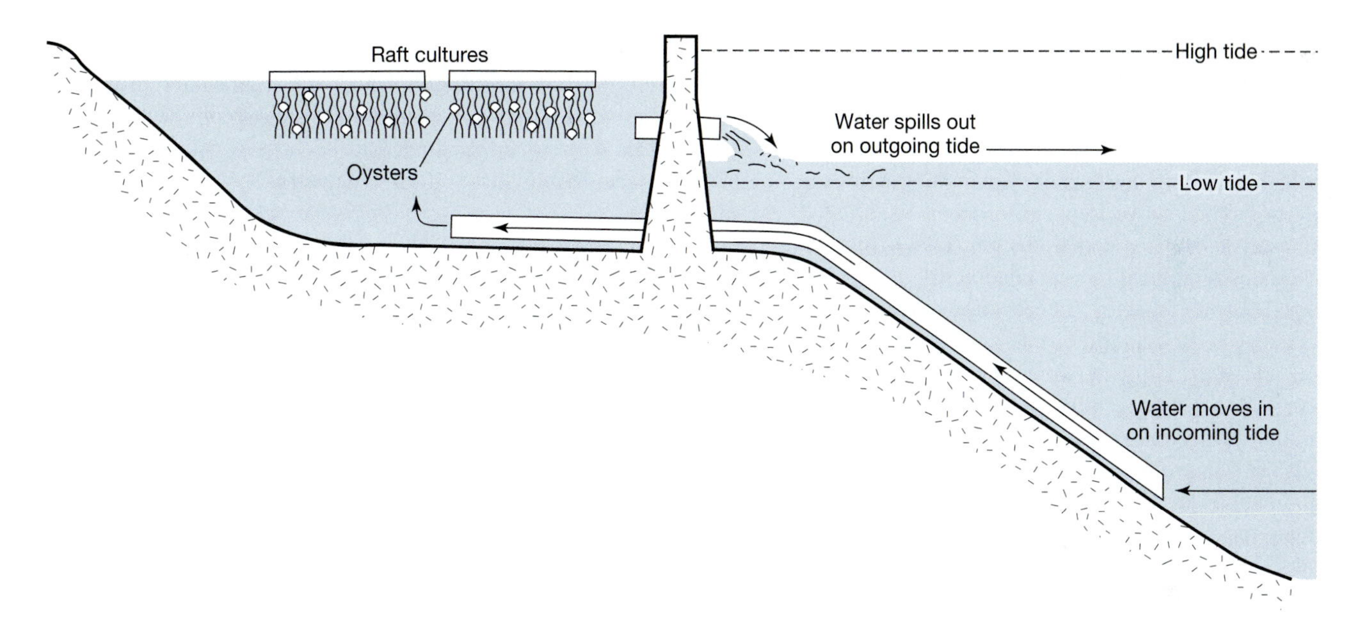

FIGURE 12.30 *Oyster culture. Deep, nutrient-rich water of the incoming tide enhances the growth of algae in the artificial impoundment. Because the oysters either feed on the algae directly or feed on crustaceans that feed on the algae, oyster production on these in-shore rafts is increased significantly.*

FIGURE 12.31 *Tilapia farming with cages in the Philippines.*

systems that focus on one species, have higher stocking densities, generate greater waste amounts, and are less likely to be involved in tight nutrient cycling systems.

Aquaculture systems involving high trophic level species are often intensive, high-input farms and are rapidly increasing in number. Global production of salmon and shrimp, for example, has increased six to seven fold since the early to mid-1980s. Relying heavily on wild-caught ocean fish stocks necessary for high-nutrient diets, these production practices can have dramatic and widespread ecological impacts, causing some fisheries experts to question the sustainability of intensive production of high trophic level species. For example, about 3 metric tons of wild-caught fish are required to produce 1 metric ton of farmed salmon or shrimp. Ocean stocks of small open-ocean (or pelagic) fish, such as anchovies, sardines, and herring, are particularly affected.

Regardless of the trophic level of the organism being raised, aquaculture systems can have other serious ecological impacts beyond those impacts to ocean fisheries. These include soil and water contamination, loss of natural ecosystems, and biological contamination of native species by escaped farmed species. Levels of waste discharge greater than ecosystems can filter or assimilate pollute surrounding water and soils. High stocking rates of farmed salmon in the North Atlantic have resulted in high discharges of both nitrogen and phosphorus into coastal waters. Inland saltwater ponds used to produce shrimp have resulted in salination and contamination of soil and water resources in China, Thailand, and Indonesia. Released wastes also encourage disease outbreaks such as infectious salmon anemia that forced the closure of 40 percent of Scotland's salmon farms in 1998.

The conversion of crucial ecosystems to aquaculture systems has also been a common problem as is seen by the displacement of important coastal mangroves in Asia by shrimp and milkfish farms. These mangroves provided important nutrient cycling services as well as habitat for plants and animals. Shrimp farming alone may account for as much as 10 percent of the total global loss of coastal mangroves. Conversion of coastal areas that serve as important nurseries for marine animals can also deplete ocean stocks. Researchers estimate that as much as 0.4 kilogram of fish and shrimp are lost from ocean fisheries due to habitat conversion for every kilogram of pond-raised shrimp produced in Thailand.

Biological contamination of native species by farmed species has also proven to be a widespread problem. Escaped, farm-raised Atlantic salmon account for as much as 40 percent of the salmon catch in the north Atlantic region and more than one-quarter million farmed salmon have escaped in the north Pacific Ocean in the last two decades. Escaped salmon may hybridize with native stocks irreversibly altering the genetic makeup of wild stocks, many of which are already endangered. Biological pollution also includes the spread of diseases from farmed populations to wild populations. Whitespot and Yellowhead viruses have spread from domestic to wild shrimp stocks in Asia and the United States, causing tremendous economic losses to shrimp farmers and fishermen alike.

To ameliorate some of these negative environmental impacts of the aquaculture industry, researchers recommend decreasing the production of high trophic level fish, reducing levels of fishmeal and fish oil in feed formulations, increasing the use of integrated polyculture systems, and improving aquaculture practices. By increasing production of herbivores and omnivores, or farming down the food web, demand for fishmeal and fish oil can be reduced. Integrated polyculture systems, where waste effluent is spread as fertilizer on adjacent cropland and multiple aquaculture species increase production efficiencies, can reduce pollution and input requirements. Careful site planning can also greatly mitigate damage to the surrounding environment with which an aquaculture system will interact. In order to provide a lasting source of protein to the world's growing population, such recommendations will need to be rapidly adopted to keep pace with the expanding aquaculture industry around the world.

Summary of Key Concepts

1. Anadromous fish begin life in fresh water, travel to the sea to mature, and return to their native stream to reproduce and die.
2. The successful completion of each stage in the life cycle of a fish is dependent on one or more of the following environmental conditions of the freshwater habitat: water temperature, depth, velocity, turbidity, and level of dissolved oxygen; substrate; cover; and food supply.
3. Fish respond both physiologically and behaviorally to degraded habitat conditions caused by long-term or chronic alterations to stream and lake ecosystems.
4. Roughly 70 percent of a given fish population dies each year as a result of environmental limitations.
5. The major environmental limitations that cause mortality in fish populations can be divided into two categories, natural and human-induced. Natural limitations include major storms, soil mass movements, animal activities, natural barriers, vegetation disturbances, native animal predation, and winterkill. Human-induced limitations include water pollution, stream temperature alterations, predation and competition by exotics, human predation, dams, resource extraction, channelization, and recreation.
6. Predation by the sea lamprey caused the annual lake trout harvest in the Great Lakes to drop from 4,545 metric tons in 1940 to 152 metric tons in 1961—a 97 percent reduction in only 21 years.
7. Lamprey populations in the Great Lakes have been controlled by treating their spawning streams with the chemical TFM.
8. Severe winterkills of fish caused by oxygen starvation occur in shallow lakes when the snow cover prevents aquatic plants from getting sufficient sunlight for photosynthesis.

9. The 1980 Northwest Power Act directs the Northwest Power Planning Council to (1) prepare "a regional conservation and electric power plan," (2) "inform the Pacific Northwest public of major regional power issues,"and (3) "develop … a program to protect, mitigate, and enhance fish and wildlife, including related spawning grounds and habitat, on the Columbia River and its tributaries."
10. Three components for sustaining freshwater fisheries are biota management (such as propagation and stocking), habitat management (maintaining adequate habitats), and human management (regulating the numbers of fish caught).
11. Controversial biotic manipulation techniques for enhancing or restoring fish populations include artificial propagation and stocking, introductions, chemical or physical removal of undesirable fish and predators, and selective breeding of superior fish.
12. Hatchery-reared fish tend to be physically weaker, more susceptible to diseases and parasites, more likely to contract severe manifestations of a particular disease, and less well suited to the streams into which they are released than their wild or native counterparts.
13. Whirling disease, caused by a fish parasite thought to have been introduced from Europe, is infecting several trout streams in 22 states in the United States.
14. Federal and state fish hatcheries stock trout in (1) suitable waters in which they do not occur, (2) waters where conditions for growth are good but where natural spawning sites are inadequate, and (3) waters where fishing pressure is heavy but there is no natural production.
15. Carp populations have been brought under partial control in some regions by periodic seining and the use of selective chemicals such as antimycin.
16. Many fish control projects are only short-term treatments for the symptom rather than for the cause of the problem. Thorough research into the causes for the overpopulation of the undesired species must be an initial step in the development of long-term solutions.
17. Levels of dissolved oxygen in snow-covered northern lakes can be increased by snow removal, opening the ice cover with dynamite, and using motorized aerators.
18. Coho and chinook salmon, native to the Pacific drainages, have been successfully introduced into the Great Lakes.
19. Two-story reservoirs are stocked with warm-water species in the epilimnion and cold-water species in the hypolimnion.
20. The water of a newly created reservoir receives a nutrient flush that greatly increases the abundance of fish food.
21. Replacing or enhancing fish populations that have disappeared or diminished from their original waterbody can be accomplished with little threat to the existing species through translocations.
22. Legislation that has been utilized to protect fish populations include (1) catch (size and number) limits, (2) catch-and-release-only restrictions, (3) closed seasons, and (4) outlawing destructive fishing techniques.
23. Habitat protection and restoration, supported by legislation to control fishing pressure and potentially harmful population enhancement techniques, are the principal long-term measures by which natural fish populations can be sustained.
24. Native fish populations may increase considerably when the carrying capacity of their environment is raised; that is, by providing adequate water flows, depths, and temperatures; space; substrate; cover; and food supplies.
25. Priority for restoration efforts and funding should be given to the protection of remaining high-quality sites and the restoration of sites that are, at worst, moderately degraded yet still contain valuable, threatened, or endangered fish species.
26. Among the techniques and instruments employed by commercial fishermen to locate marine fish are (a) sonar or echo-sounding systems, (b) moored buoys, (c) color enhancement of aerial photos, (d) infrared sensors, (e) ultraviolet sensors, and (f) electronic image intensifiers.
27. Fish and shellfish are harvested with both passive and active fishing gear. Passive gear includes pots and cages; set nets, pound nets, and traps; hooks and lines; and gill nets. Active gear includes purse seine nets, trawl nets, and drift nets.
28. Despite growth in marine harvests, the productivity of many important fish and shellfish continues to decline. Growth in marine harvests reflects the increase in the number of fishing vessels, advances in fishing gear technology, and the rising protein demands of a growing world population. The biological stress of the marine ecosystem is due to destruction and loss of critical spawning, feeding, and rearing habitat; overfishing; and bycatch.
29. Overfishing is the result of a combination of factors including fishing in a "common", government subsidies, economic and political forces, advances in fishing gear technology, and lack of harvest regulation enforcement.
30. Among the fisheries that have been depleted due to overfishing are the (a) Pacific sardine fishery, (b) northwestern Atlantic cod and herring fisheries, (c) northwestern Pacific salmon fishery, and (d) eastern Atlantic bluefin tuna.
31. Bycatch is the collective term used to describe captured marine organisms that are not the target species of the fishery. Discards are the bycatch that are thrown back because they are either a nontarget species, juvenile fish, endangered species, improper size, inferior quality, or surplus to quotas.
32. Many discarded fish and shellfish die or are injured, making them easy prey.
33. Overfishing, bycatch, and habitat loss and degradation not only result in population and productivity declines of

commercially important marine organisms, but also disturb nontarget species, disrupt predator-prey relationships, alter marine community structures, and reduce genetic diversity.

34. Recognition and concern for the state of the marine ecosystem is increasing throughout the international community. Several management options for sustainable fisheries have been proposed or are being initiated, including the optimum yield principle, economic incentives for downsizing the global fishing fleet and development of selective fishing gear, individual transferable quotas (ITQs), stricter harvesting regulations, the precautionary approach, and habitat enhancement and restoration schemes.
35. We play an important role in supporting or not supporting sustainable fisheries by the choices we make in purchasing seafood at the food market.
36. The practice of aquaculture is increasing and it holds promise as a protein source for a growing global population. However, if not managed correctly, aquaculture systems can destroy wetlands, pollute water, impact ocean stocks of small open-ocean fish, and cause biological contamination of native species by escaped farmed species.
37. To make the aquaculture industry more sustainable, researchers recommend decreasing the production of high trophic level fish, reducing levels of fishmeal and fish oil in feed formulations, increasing the use of integrated polyculture systems, and improving aquaculture practices

Key Words and Phrases

Active Fishing Gear
Anadromous Fish
Anoxia
Antimycin
Artificial Propagation
Astroturf Sandwich
Bycatch
Catch-and-Release-Only Restrictions
Closed Seasons
Color Enhancement
Commons
Creel (Catch) Limits
Discards
Drawdown
Electronic Image Intensifiers
Environmental Limitations
European Carp
Eutrophication
Exclusive Economic Zone
Exotic (Nonnative) Species
Extensive Aquaculture System
Fish Ladders
Freshwater Aquaculture
Genetic Erosion
Habitat Restoration
Hybrid Fish
Hypoxia
Imprinting
Individual Transferable Quotas
Infrared Sensors
Intensive Aquaculture System
Introduction
Magnuson Fishery Conservation and Management Act
Marine Aquaculture (Mariculture)
Maximum Sustained Yield
Minimum Size Limits
Moored Buoys
Native Species
Nitrogen Intoxication
Nutrient Flush
Optimum Yield
Overfishing
Passive Fishing Gear
Precautionary Approach
Put-and-Take Stocking
Redd
Reservoir
Resident Fish
River Ruff
Rotenone
Sea Lamprey
Siltation
Sonar
Spawning
Splash Dams
Stocking
Suspended Sediment
Translocation
Two-Story Reservoir
Ultraviolet Sensors
Vibert Box
Whirling Disease
Winterkill

Critical Thinking and Discussion Questions

1. How does water temperature affect fish behavior?
2. Discuss the importance of streambed materials (substrate) on fish spawning.
3. What factors limit food and energy supplies in the stream environment?
4. List three human-induced limitations to fish productivity and describe their impacts on fish habitat.
5. What is the source of acid rain? What is its effect on fish?
6. What causes the winterkill of fish in northern lakes? How can it be prevented?
7. Describe three activities of carp that are harmful to game fish populations.
8. What would you suggest as a solution for the increasing angler pressure on freshwater fish?
9. How have dams affected the survival of anadromous fish?
10. Discuss the pros and cons of artificial propagation and stocking.
11. What is the suspected cause of the spread of whirling disease to the intermountain West? What are the visible symptoms? Discuss some possible strategies to stop the further spread of this disease.
12. Why was the introduction of Pacific salmon in the Great Lakes considered successful? What role did the alewives play?
13. What three general strategies are available to fisheries managers to protect and restore freshwater fish populations?
14. How can the stream habitat be improved for fish?
15. Discuss the statement, "Predator control is an effective method for increasing game fish populations in the United States." Is it valid? Why or why not?

16. Describe the life cycle of the sea lamprey.
17. In what situation would you use translocation as a management option?
18. Describe efforts to control the sea lamprey in the Great Lakes.
19. Discuss the use of a Vibert box.
20. Why would a fish manager loosen regulations on the size and take of panfish?
21. Briefly list four benefits that may be derived from the drawdown of reservoirs.
22. Describe the types of protective legislation used to control freshwater fish populations.
23. Describe two methods used by commercial marine fishers to locate fish.
24. Why has the bycatch problem become an international concern?
25. Discuss the negative environmental impacts caused by modern fishing gear.
26. How has the establishment of Exclusive Economic Zones (EEZs) enhanced marine fishing pressures?
27. What is the "tragedy of the commons"?
28. List two methods proposed to address the problems of the world's marine fishing industry. Explain how each would reduce the overfishing or bycatch problem.
29. Explain the theoretical concept underlying the optimum yield principle.
30. What is meant by the "precautionary approach" to fisheries management?
31. Do you feel artificial reefs composed of human "garbage" are viable options for habitat restoration or enhancement? Explain.
32. Discuss the pros and cons of aquaculture.

Suggested Readings

Booth, W. 1994. "Turning the Tide on Dwindling Marine Resources." *Science* 263: 25–26. Discusses major political and environmental factors causing depletion.

Crivelli, A. J. 1995. "Are Fish Introductions a Threat to Endemic Freshwater Fishes in the Northern Mediterranean Region?" *Biological Conservation* 72: 311–319. Discussion of the controversies surrounding fish introductions.

Fairlie, S., M. Hagler, and B. O'Riordan. 1995. "The Politics of Overfishing." *Ecologist* 25: 46–73. Interesting explanation and discussion of the underlying reasons for the overfishing problem.

Fisheries Statistics and Economics Division. 2000. *Fisheries of the United States, 1999.* Washington, D.C.: U.S. Department of Commerce, National Oceanic and Atmospheric Administration, U.S. Government Printing Office. Annual report detailing the status and trends of our nation's commercial and recreational fishing industry.

Frankl, E. 1995. *Ocean Environmental Management: A Primer on the Role of the Oceans and How to Maintain Their Contributions to Life on Earth.* Englewood Cliffs, N.J.: Prentice Hall. Overview of current environmental problems relating to ocean resources and suggestions for sustainable use.

Hagler, M. 1995. "Deforestation of the Deep." *Ecologist* 25: 74–79. Warning that the effects of the world's overfishing may be causing permanent damage to the marine ecosystem.

Kohler, C. C., and W. A. Hubert. 1999. *Inland Fisheries Management in North America,* 2nd ed. Bethesda, Md.: American Fisheries Society. A mandatory reading assignment for current and future fisheries managers.

Lawren, B. 1992. "Net Loss." *National Wildlife* 30(6): 47–52. Well-written account of factors causing decline of American fisheries.

Maitland, P. S. 1995. "The Conservation of Freshwater Fish: Past and Present Experience." *Biological Conservation* 72: 259–270. Appeal for an increase in fish conservation programs in every country of the world.

Malvestuto, S. P., and M. D. Hudgins. 1996. "Optimum Yield for Recreational Fisheries Management." *Fisheries* 21: 6–17. Detailed explanation of the optimum yield management concept.

Mann, C. C., and M. L. Plummer. 1999. "Can Science Rescue Salmon." *Science* 289: 716–718. Excellent discussion off the Snake River salmon issue in the Pacific Northwest.

Meronek, T. G., et al. 1996. "A Review of Fish Control Projects." *North American Journal of Fisheries Management* 16: 63–74. Literature review assessing the "success" of 250 fish control projects.

Murphy, B. N., and D. W. Willis. 1996. *Fisheries Techniques,* 2nd ed. Bethesda, MD: American Fisheries Society. An excellent reference for fisheries managers.

Naylor, R. L., et al. 2000. "Effect of Aquaculture on World Fish Supplies." *Nature* 405: 1017–1024. Excellent article on the pros and cons of aquaculture.

Nehring, R. B., and P. G. Walker. 1996. "Whirling Disease in the Wild: The New Reality in the Intermountain West." *Fisheries* 21: 28–30. Interesting account of the history and extent of the whirling disease threat to trout in the rivers of the intermountain West.

Philippart, J. C. 1995. "Is Captive Breeding an Effective Solution for the Preservation of Endemic Species?" *Biological Conservation* 72: 281–295. Excellent discussion of the role, techniques, and constraints of captive breeding programs.

Potera, C. 1997. "Fishing for Answers to Whirling Disease." *Science* 278: 225–226. Good, short article on the cause of whirling disease.

Stegemann, E., and D. Stang. 1993. "The Herring of New York." *Conservationist* 47(5): 7–14. Popular account of several species of fishes in the herring family sought by anglers in the Hudson and other New York rivers.

Vincent, E. R. 1996. "Whirling Disease and Wild Trout: The Montana Experience." *Fisheries* 21: 32–33. Further reading on the incidence of whirling disease in trout.

Wilks, A. 1995. "Prawns, Profit, and Protein: Aquaculture and Food Production." *Ecologist* 25: 120–125. Pros and cons of aquaculture.

Web Explorations

Online resources for this chapter are on the World Wide Web at: **http://www.prenhall.com/chiras** *(click on the Table of Contents link and then select Chapter 12).*

Chapter 13

Rangeland Management

Rangelands are areas of the world that are a source of forage (such as grasses and shrubs) for free-ranging native and domestic animals, as well as a source of wood products, water, energy, wildlife, minerals, and recreational opportunities. They also produce intangible products such as natural beauty, open space, and wilderness that satisfy important societal values. They even store carbon and thus reduce atmospheric greenhouse gases. Most rangelands are unsuitable for cultivation because of physical limitations, such as low precipitation, rough topography, poor drainage, or cold temperatures.

The United States has about 312 million hectares (770 million acres) of rangelands, which include the wet grasslands of Florida, the desert floor of California, and the mountain meadows of Utah, to name a few of the diverse rangeland ecosystems. More than 99 percent of the nation's rangeland is west of the Mississippi River. Worldwide, rangelands occupy about half of the terrestrial surface (excluding permanently frozen lands).

This chapter is concerned with rangelands—their ecology, history, distribution, and condition. It concludes with a look at ways to manage these important lands more successfully.

13.1 Ecology of Rangelands

Types of Rangelands

Seven major categories of vegetation provide most of the world's grazing lands: grasslands, tropical savannas, tundra, desert shrubs, shrub woodlands, temperate forests, and tropical forests. **Grasslands,** as used here, occur in temperate areas, are made up of a mixture of grasses and some forbs (broad-leaved flowering plants), and are typically free of trees or shrubs. They usually develop in areas with 25 to 75 centimeters (10 to 30 inches) of annual rainfall. Grasslands occur on every continent in the world and represent some of the world's richest and most productive grazing areas. At one time, great herds of large, wild herbivores roamed the world's grasslands. Today, most of these wild animals have been killed or driven away. In wetter areas, these grasslands have been converted to cropland because of their deep, fertile soils. In the areas too dry to grow crops, grasslands are now used for livestock. There is, however, a growing movement to restore and preserve some of the native grassland areas around the world (see A Closer Look 13.1).

Tropical savannas, essentially tropical grasslands, occur mostly in Africa (Figure 13.1) and do not occur in the United States. They are characterized by a mixture of grasses, shrubs, and scattered trees. Savannas are called *campos* in Brazil and *llanos* in Venezuela. The crop-growing potential of tropical

A CLOSER LOOK 13.1 Prairie Restoration and the National Grasslands Story

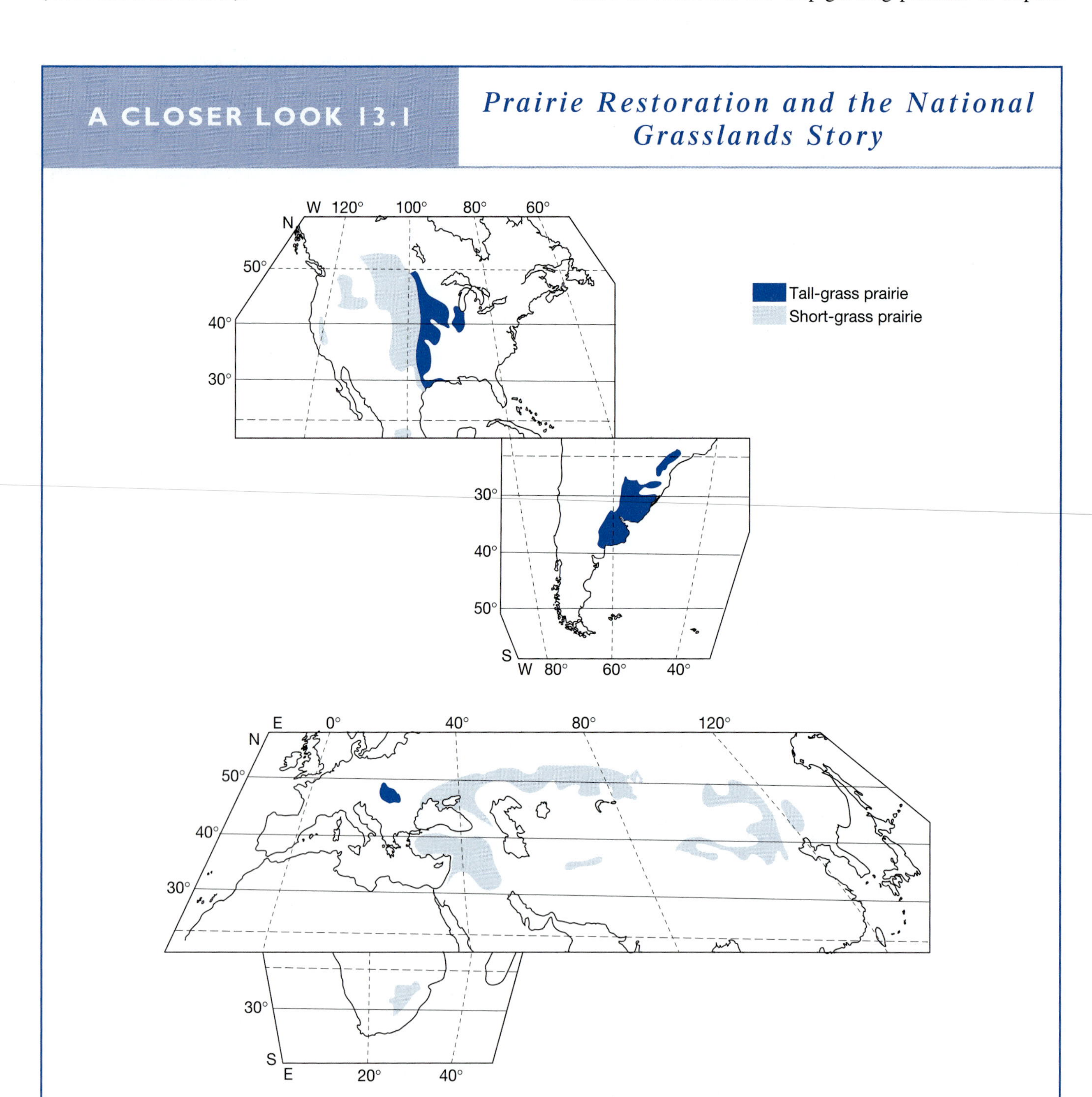

FIGURE 1 *World map of grassland ecosystems: the tall-grass prairie and the short-grass prairie.*

The grasslands of the Great Plains form the largest single expanse of true grass range in the world, extending from southern Canada to Texas. They include tallgrass prairies and short-grass prairies. Grasslands are called *pampas* in Argentina, *pastures* in Europe, *steppes* in Eurasia, and *veld* in Africa (Figure 1).

Prairies, especially tall-grass prairies which grow in moister areas than do short-grass prairies, have helped produce humus- and nutrient-rich soils. Before the European settlement in North

(continued)

America, these two types of prairie could be found in the Great Plains, southern Alberta, the northern border region of Montana, eastern Washington State, and western Idaho. These prairies teamed with abundant buffalo herds, elk, and other wildlife. They were also home to many tribes including Sitting Bull's Hunkpapa Sioux, Apache, Arapaho, Arikara, Assiniboine, Atsina, Bannock, Blackfeet, Cheyenne, Chippewa, Ojibwa, Bungi, Comanche, Cree, Crow, Hidsata, Kiowa, Klamath, Kootnei, Mandan, Metis, Modoc, Pawnee, Santee, Shasta, Shoshone, Teton, Wichita, Yankton and Yanktonia.

Since the time of European settlement, much of these prairies has been overgrazed and plowed up for growing crops. The settlers discovered that while these vast grasslands were productive in wet years, they were also subject to serious drought and bitter winters. In the Great Plains, land that should never have been plowed yielded its topsoil to incessant dry winds, resulting in the Dust Bowl in the 1930s (see Chapter 7 for more details). Today, most of these prairies are farmed and some are used for grazing. However, since the birth of the **National Grasslands** in 1960, 20 native grassland areas have been revegetated and restored to conserve the natural resources of the prairie ecosystem.

The 20 National Grasslands, totaling almost 1.6 million hectares (4 million acres), are important lands managed for sustainable multiple uses as part of the U.S. National Forest System. They have made important contributions to conserving grassland ecosystems while producing a variety of goods and services which, in turn, have helped to maintain rural economies and lifestyles. Wildlife, including many declining, threatened or endangered species, thrives in reborn habitats. With restored prairie vegetation, once wounded soil rebuilds itself. Construction of livestock ponds allow cattle grazing while simultaneously expanding the range of many wildlife species by providing water where none existed before. Private farmlands within the National Grassland boundary add diversity to the prairie habitat. The National Grasslands also provide diverse recreational uses, such as mountain bicycling, hiking, hunting, fishing, photographing, birding and just sightseeing.

Although most of the National Grasslands are located in Great Plains states, three are located in the Great Basin states of California, Oregon, and Idaho. The Little Missouri National Grassland in North Dakota is the biggest, with 416,215 hectares (1,028,051 acres). The Rita Blanca National Grassland, which includes 31,362 hectares (77,463 acres) in Texas and 6421 hectares (15,860 acres) in Oklahoma, has wildlife that varies as much as does the climate over the wide expanse of country.

There are other organizations and foundations responsible for restoring and preserving native prairie. The largest is **The Nature Conservancy,** whose mission is to preserve plants, animals, and natural communities by protecting the lands and waters they need to survive. The Nature Conservancy has helped to protect more than 4.5 million hectares (11 million acres) of habitat in the United States and nearly 24.3 million hectares (60 million acres) in Canada, Latin America, the Caribbean, Asia and the Pacific. They currently manage 1,340 preserves, including numerous native prairies. For example, the Oregon Chapter of the Conservancy recently purchased 11,000 hectares in northeast Oregon, part of the largest expanse of bunchgrass prairie remaining in North America. In addition to being home to one of the highest concentrations of nesting birds of prey in the country, Zumwalt Prairie Preserve is also habitat for elk, mule deer, bighorn sheep, bobcat, and the endangered Snake River steelhead.

FIGURE 13.1 *Rhinoceros grazing on open tropical savanna in Tanzania.*

grasslands is usually limited by water availability or poor soil. **Tundra** vegetation can be found in very cold areas in the Arctic or as meadows at high mountain elevations. The Arctic tundra is used mostly by wild animals, whereas the alpine tundra is used by both livestock and wild animals. Temperatures are too cold and soils too poor for tundra to be converted to crop production.

Desert shrublands make up the largest area of the world's rangelands and are characterized by an arid climate (usually less than 25 centimeters or 10 inches of rainfall), poorly developed soils, and sparse vegetation dominated by low-growing shrubs usually less than 2 meters in height (Figure 13.2). All arid regions of the world support some desert shrubland. Some are productive and of major grazing value, whereas others occur in areas of such low precipitation that they are little used.

Shrub woodlands usually occur in about the same annual rainfall belt as grasslands, but low-growing trees (usually less than 10 meters or 33 feet) and dense shrubs are the dominant vegetation. Where **temperate forests** occur in naturally open stands or where they are made open by timber cutting, considerable grazing may be provided. **Tropical forests** provide little herbaceous vegetation under their dense canopies and are of little relative importance for grazing.

Characteristics of Rangeland Vegetation

The vegetation on rangelands is mostly grasses, grasslike plants (sedges and rushes), forbs, and shrubs (Figure 13.3). These plants are known as **forage** and supply food and energy for domesticated animals such as cattle, sheep, and horses, and for wild animals such as deer, giraffes, and wildebeests. Some of these animals graze on grass, whereas others browse on leaves, twigs, and shoots. Humans, in turn, acquire food and

FIGURE 13.2 *Cattle grazing on desert shrubland in Idaho.*

energy from cattle and sheep in the form of veal, beef, mutton, or lamb.

A leaftip of rangeland vegetation like grass can be nibbled off without affecting growth as long as the **basal zone** (the lowermost portion of the leaf) remains intact. In a short time, the leaf can grow to its original length. In fact, the grass leaf can be grazed again and again without adverse effects, as long as the plant has some time to recover. Grasses can therefore provide a continuous food reservoir for the grazing animal.

Range ecologists generally regard the upper 50 percent of the grass shoot (stem and leaves) as a "surplus" that can be safely eaten by livestock or wild herbivores (deer, antelope, elk, and so on) without damaging the plant. The lower 50 percent, known as the **metabolic reserve,** is necessary for grass survival (Figure 13.4). This provides the minimum amount of photosynthesis needed to manufacture foods for the roots. Many root systems extend to a depth of 2 meters (6 feet) or more. Large amounts of nutrients are stored in them. This nutrient reserve enables range grasses to survive drought as well as brief periods of overgrazing. When a range is overgrazed, the herbivores bite into the metabolic reserve, frequently clipping the grass to the ground, starving and killing the root system and leaving the site vulnerable to erosion; hence the need for recovery between grazings.

Range ecologists and ranchers frequently classify the various range plants into three categories with respect to the dynamics of plant succession: decreasers, increasers, and invaders. **Decreasers** are highly nutritious, extremely palatable plants that generally decrease under even moderate grazing pressure (Figure 13.5). Representative decreaser species are big bluestem, little bluestem, blue grama, wheatgrass, and buffalo grass.

Increasers, on the other hand, are generally less palatable but are still highly nutritious climax species that tend to increase (at least temporarily) when a range is heavily grazed (Figure 13.6). Apparently this increase is the result of reduced competition from the decreasers. When severe grazing pressure continues over a long period, even the increasers begin to decline, apparently unable to withstand trampling by hoofs of grazing animals. They are replaced by invaders.

The **invaders,** such as ragweed, cactus, and thistle, may be considered undesirable weed species, are low in nutritional value, and are not very desirable for grazing animals (Figure 13.7). Some may be poisonous. The invader downy chess has sharp seeds that can harm animals by lodging in their throats or piercing their skin. Invaders frequently are annual plants that thrive best under sunlight intensities much higher than the 1 to 2 percent of full sunlight occurring in a dense stand of climax grasses. Because their roots are taproots instead of the dense fibrous roots of a typical grass, these invaders are not very effective in binding the soil.

A range in excellent condition has a high percentage of decreasers and almost no invaders. Conversely, a range in poor condition has a low percentage of high-forage-value decreasers and a large percentage of low-forage-value invaders (Figure 13.8).

Rangeland Carrying Capacity

The **carrying capacity** of a habitat is indicated by the size of the population of a species that can be sustained in that habitat. Grazing carrying capacity is the maximum number of animals (or quantity of herbivore biomass) that can graze each year on a given area of range, for a specific number of days, without causing a downward trend in forage production, forage quality, or soil quality. It is affected by annual climatic conditions, previous grazing use, kinds of grazing animals and how long they graze an area, and soil type. It varies from site to site and from season to season for each species. The carrying capacity or number of grazing animals for dry years should not be exceeded if a particular grazing site is to maintain a sustainable herd. That's because in dry years, the carrying capacity may be half that of a normal year.

The forage potential of a given range is usually described in terms of **animal unit months (AUMs)**. An AUM is the amount of forage needed to keep a 1,000-pound (454-kilogram) animal well fed for 1 month. Accepted animal unit equivalents are listed in Table 13.1. Notice that one steer is equivalent to five sheep or five goats grazing the same habitat. This means that five sheep

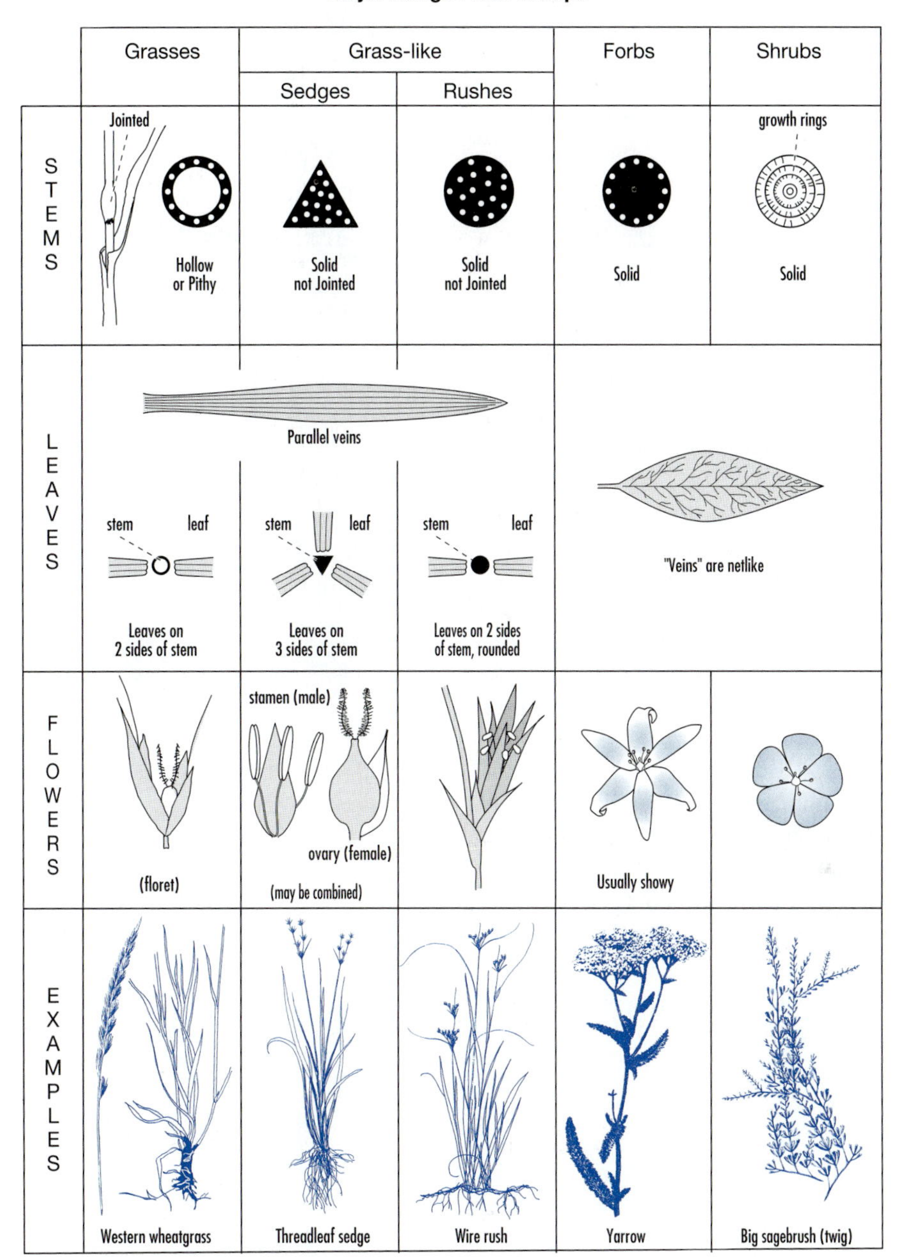

FIGURE 13.3 *Characteristics of the major groups of range plants.*

or five goats eat as much as one steer. In addition, a 1,000-pound horse typically eats 25 percent more than a 1,000-pound steer. When considering AUMs, range condition needs to be taken into account. For example, when a range is in excellent condition, 0.4 hectare (1 acre) may equal 1 AUM. However, on poor range, it may require 2 hectares (5 acres) to equal 1 AUM.

Effects of Human Activities and Overgrazing on Rangelands

When people plow up native rangelands and plant them with crops (called **sodbusting**), they destroy their ecological stability. When the vegetation is grasslands, the grasses are replaced with crops whose root systems are less effective in holding the soil in place. Without appropriate soil conservation practices, the soil may become vulnerable to wind or water erosion. A classic example of this is the Dust Bowl of the 1930s, which was associated with massive wind erosion on the Great Plains (see Chapter 7). A more recent example is the Palouse region in eastern Washington. Once a rich, rolling prairie, it now produces high yields of rain-fed wheat, barley, peas, and lentils. Because of a combination of steep hills and lack of proper soil conservation practices, the Palouse region is one of the most erodible areas in the United States.

Rangelands not converted to croplands can be properly grazed by domesticated livestock and wild animals. In fact,

TABLE 13.1 *Animal Unit Equivalents**

Number of Animals	Animal Unit Month (AUM)
1 steer	1.0
5 sheep	1.0
5 goats	1.0
4 deer	1.0
1 horse	1.25
1 bull	1.25
1 elk	0.67

* One steer is considered equal to five sheep grazing the same habitat.

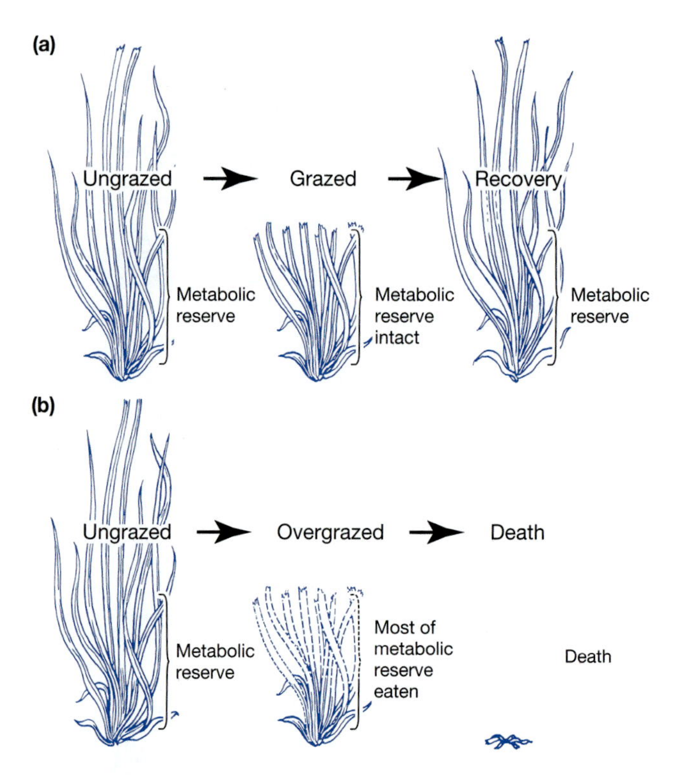

FIGURE 13.4 *Grasses can be grazed without harm as long as the metabolic reserve is left intact. Without it the plant dies.*

light to moderate grazing is necessary for the health of rangelands, especially grasslands. Normal grazing encourages healthier root systems and vigorous leaf growth, promotes nutrient cycling and the buildup of soil organic matter, and hinders soil erosion. Decreasing the number of grazers too much causes replacement of grasses by forbs and woody shrubs. Increasing grazers too much causes overgrazing.

Overgrazing is continued heavy grazing that exceeds the carrying capacity of the community and creates a deteriorated range. Although most overgrazing is caused by too many livestock feeding too long in a particular area, large numbers of wild herbivores can overgraze rangeland in long dry periods. Overgrazing changes the makeup and decreases the productive

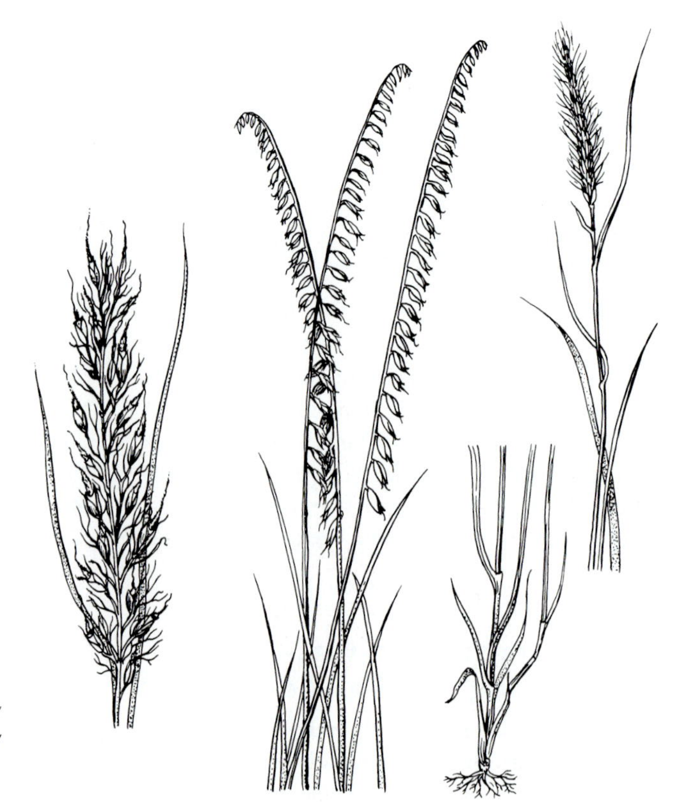

FIGURE 13.5 *Decreasers. These plants are usually the plants that grazing animals prefer. They are highly nutritious and palatable, and decrease under heavy grazing pressure.*

FIGURE 13.6 *Increasers. These plants increase in number (at least temporarily) under grazing pressure and replace the decreasers. Although they are fairly nutritious, they are less palatable and less preferred by cattle.*

potential of a plant community. More important, it can expose soil to water and wind, increasing its erosion potential.

Overgrazing can sometimes be so devastating that all vegetation disappears and the land becomes barren and highly prone to erosion. When combined with drought, severe overgrazing can cause desertification (see A Closer Look 13.2).

Undergrazing can also damage rangeland by allowing too much leaf and stem to get old, which in turn reduces the growth of grasses and increases the growth of forbs and woody shrubs. It can even lead to an increase in soil erosion and degradation. Thus, rangeland declines as a source of food for livestock and wild herbivores.

Effect of Drought on Range Forage

Drought is one of the greatest environmental problems ranchers encounter. They can, to some degree, control rodents, poisonous plants, brush and weeds, predators, insects, and unfavorable soil chemistry, but there is absolutely nothing they can do to control drought. They can adjust to it, but that is all. Moreover, drought is unpredictable. A severe drought results in a drastic deterioration of the rangeland plant community, even when grazing is light. In the Snake River region of southern Idaho, during the height of the Dust Bowl era in 1934, extended drought caused an 84 percent reduction in plant cover in ungrazed areas.

Once a dry spell has ended, range recovery may be fast or slow, depending on the amount of precipitation. Thus, a lightly grazed Montana pasture required 8 years to return to good condition after a severe drought because of limited rainfall. On the other hand, a Kansas range recovered very rapidly when rainfall was adequate (Figure 13.9).

13.2 Brief History of Range Use in the United States

American bison, pronghorn antelope, and elk roamed the ranges of North America before European settlers arrived. At one time as many as 34 million head of these species may have existed. The selective grazing and migration patterns of these animals and other wild herbivores helped prevent overgrazing. Their numbers were kept at the carrying capacity of the land because of competition and predation within and between species. Native Americans used their meat and hides for food, clothing, and shelter.

Introduction of Domesticated Animals

In the sixteenth and seventeenth centuries, European colonists, mostly Spaniards, were immigrating to the New World and brought a domesticated animal-based culture with them. Many wild herbivores were killed or driven from rangelands to reduce their competition with livestock. By 1800, cattle, sheep, and horses were a significant part of rangelands in the western United States and Canada. The cattle industry's phenomenal expansion during the 1800s was responsible for much of the settlement of the West. For this

FIGURE 13.7 *Invaders. These weeds replace the increasers if the range continues to be grazed heavily. Some of the invaders, like military grass (downy chess), have sharp seeds that can harm livestock by lodging in their throat or piercing their skin. Thistles have sharp spines that make them useless as forage.*

we can thank thousands of courageous, hard-working ranchers and farmers. However, in their zeal, many abused our once-bountiful grassland and caused widespread destruction that is still evident today.

On the range, cattle and sheep replaced the bison, which was nearly driven to extinction by the end of the nineteenth century. Between 1870 and 1890, cattle numbers jumped from about 5 million to almost 27 million. Sheep numbers rose from about half a million in 1850 to more than 20 million in 1890. As the numbers of domesticated grazing animals increased, the quality and carrying capacities of rangelands declined.

Cattlemen allowed their cattle to overgraze the ranges in many areas. If one person did not exploit the forage of an area, then a neighbor would. Some called this the "tragedy of the commons" (Figure 13.10). Where a given pasture could support 25 cattle, many grazed 100. Big bluestem, bluegrass, and buffalo grass were chewed off at the roots. Once the grass plant's metabolic reserve had been eliminated, the root system withered and died. Many of these stockmen were so obsessed with large numbers of range animals that they ignored the fact that four head of livestock, sick and scrawny from undernourishment, would sell for less than one head in prime condition after grazing on good forage.

Public Land Distribution and Abuse

The **Homestead Act** of 1862 sped up the development of the West. This act allowed the offering of 160 acres (64 hectares) of public land to anyone who would settle the land and reside on it for 5 years. In the eastern United States, 160 acres was

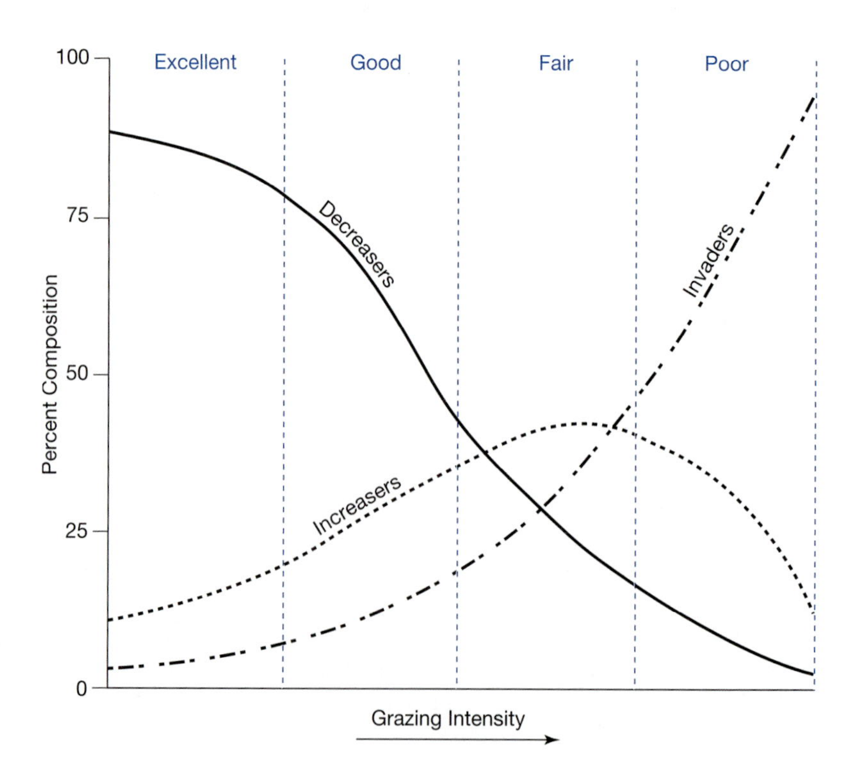

FIGURE 13.8 *Relationship of intensity of grazing to range condition and the relative proportion of decreasers, increasers, and invaders. Figure 13.8 is taken from L.A. Stoddart, A.D. Smith, and T.W. Box,* Range Management, Third Edition, *New York: McGraw-Hill, 1975, p. 191, which in turn was adapted from data from P.L. Sims and D.D. Dwyer, "Pattern of Retrogression of Native Vegetation in North Central Oklahoma,"* Journal of Range Management, *1965, 18:20–25.*

A CLOSER LOOK 13.2 Causes of Desertification

Desertification has been defined in many ways. The United Nations Desertification Prevention Treaty defines desertification as land degradation caused by natural (namely climatic changes) and artificial factors (human-induced activities) in dry, semidry, or dry and semihumid areas. The climate factor refers to large-scale global changes in climate in a dry area. Human-induced activities include overgrazing, overcutting, cultivation of unsuitable land, or inadequate irrigation. These destructive activities are intensified by high population density, poverty, and poor land management. The bottom line is that desertification results in a reduced productive potential and a diminished capacity to provide benefits to humans and other life on the planet.

Desertification is characterized by devegetation, brush invasion, groundwater depletion, salination, and severe erosion. Areas probably least affected by desertification are the naturally barren deserts of the world that are little used by humans. However, sparsely vegetated grazing lands adjacent to these deserts have, in many places, been severely affected.

Because of the different definitions of desertification, estimates of the degree of desertification vary. A recent detailed study on soil degradation worldwide found that some 680 million hectares of land have been degraded since 1945 due to overgrazing, most of which has occurred in Africa and Asia. This study is discussed in more detail in the text, in the section on range condition.

Desertification is probably most severe in **the Sahel** or Sahelian zone, just south of the Sahara Desert in Africa (Figure 1). The combination of long drought periods in 1969 to 1973 and again during the 1980s, overgrazing of common lands, overharvesting of fuelwood, and expanded cultivation on marginal lands reduced plant cover and left the soil bare in places. Hot, dry winds picked up the soil and moved it toward the Sahara Desert, resulting in an apparent advance southward of the desert. During the 1969-1973 drought, grain crops failed and cattle starved or had to be sold due to a lack of available forage. In 1973 alone, it has been estimated that 100,000 people died of starvation and disease and 5 million cattle perished.

Desertification also occurs in the United States, mainly in the Southwest. For example, the Sonoran and Chihuahuan deserts of the Southwest are perhaps a million years old, yet in places, they have become even more barren during the last 100 years. Their animal populations have diminished. Valuable grasses have declined. Invader species such as Russian thistle have multiplied. The original floodplain vegetation has changed beyond recognition in the Santa Cruz River Valley of Arizona.

Although there are differences of opinion about the steps needed to prevent and reverse desertification, scientists believe that this process can be stopped and desertified regions rehabilitated. They agree on the critical role that plants play in stabilizing the soil but differ somewhat on which soil, water, rangeland, and forest management strategies are most appropriate. The United Nations Environmental Programme estimates that the global cost would be close to $150 billion. However, the enormous expense of the rehabilitation effort would be more than offset by income from the increased agricultural productivity that would result.

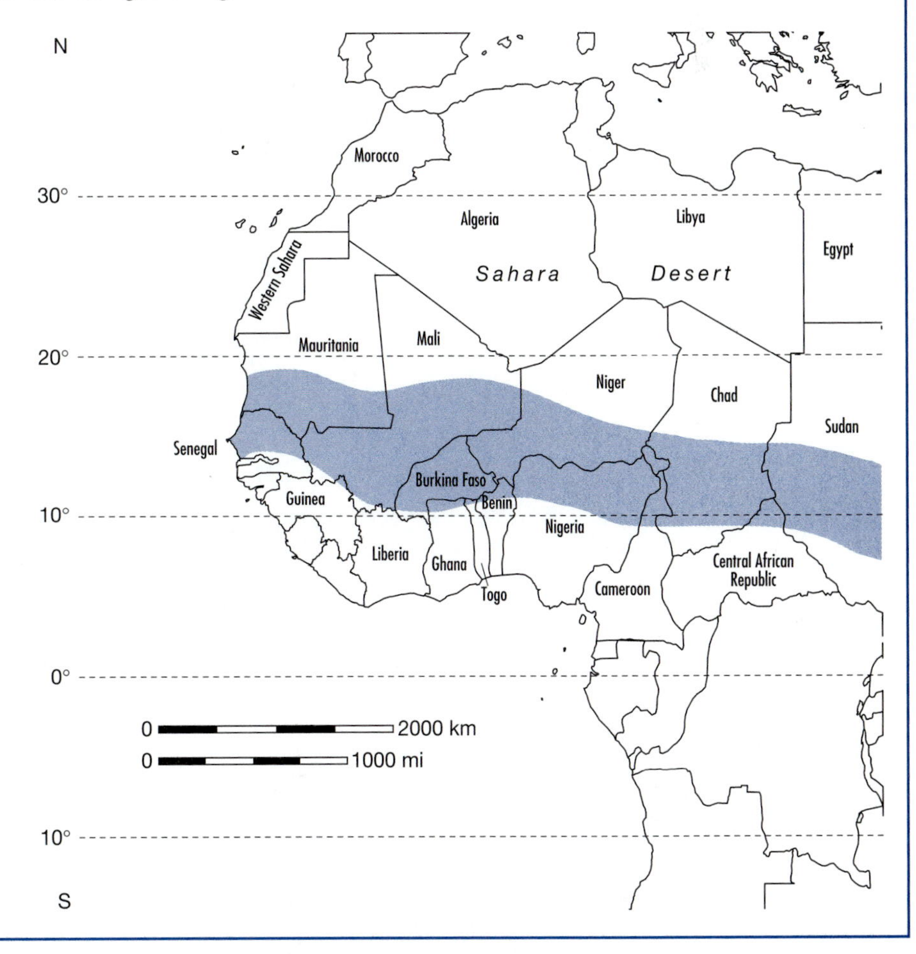

FIGURE 1 *The Sahel or Sahelian zone (shown in color). This zone has undergone periodic droughts throughout past decades.*

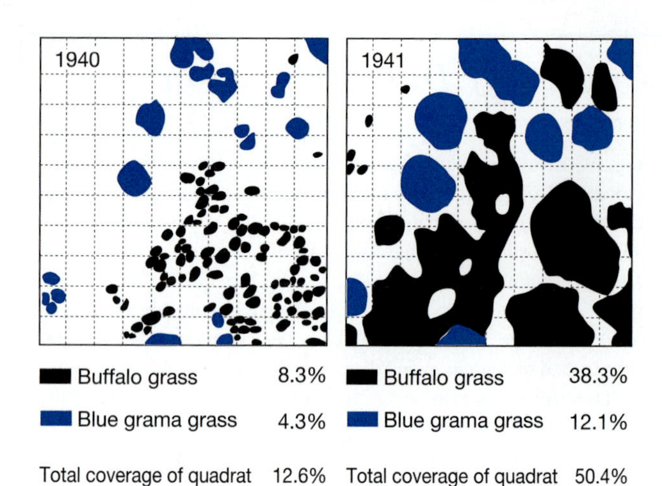

FIGURE 13.9 *Recovery of range from drought following a period of adequate rainfall. The basal cover of range vegetation near Ness City, Kansas, had been reduced to only 12.6 percent of the area by autumn of a drought year. Overgrazing had contributed to the deterioration. However, within 1 year after the return of adequate rainfall, range grasses responded sufficiently to cover 50.4 percent of the sample plot.*

plenty of land for a family to make a living. In the West, that was not true. Land in the arid West that should never have been cultivated was plowed and damaged by homesteaders trying to make a living.

In 1878, **John Wesley Powell,** an explorer who had studied the soil, water resources, plants, and animals of much of the arid West, submitted his *Report on the Lands of the Arid Region* to the U.S. Interior Department. He warned that the prevailing methods of distributing land and practicing agriculture would not work in the arid West. He recommended that the land was suitable for grazing sheep and cattle, not for growing crops. He felt that a single family ranch should be 2,500 acres (1,000 hectares) or larger and that the arid range could support far fewer cattle than cattle ranchers were raising. He also believed that if the arid land was to be farmed, farmers would have to be assured of irrigation water from building dams or diverting rivers, and that a single farm should be no more than 80 acres (32 hectares). Powell's report was ignored by government leaders. If his suggestions had been taken to heart, the health of our current rangelands would most likely be better today.

In 1873 barbed wire was developed, which allowed farmers, sometimes called sodbusters, to build fences to keep cat-

FIGURE 13.10 *Cartoon illustrating the "tragedy of the commons". Any individual user of a commons, a publicly owned rangeland, could profit by grazing as many of his or her cattle as possible, because the forage was free. But if all members of the community did the same thing, the commons would quickly be overgrazed and become useless for feeding animals and, thus, humans in the future.*

tle from trampling their crops. Barbed wire was also used to limit grazing and keep other livestock from trespassing. Farmers and ranchers fought and even killed each other because angry ranchers, wanting the land for grazing, cut and pulled down the farmers' fences. This dispute between farmers and ranchers is depicted in the famous western movie *Shane*.

By 1900, many rangelands had been degraded because of heavy grazing by cattle, sheep, and horses for more than 50 years. Other areas of rangeland had been degraded by plowing because they were too dry to support crops without irrigation. In 1905 the **U.S. Forest Service (USFS)** was formed and began to restrict livestock numbers and grazing seasons in national forests.

Although the condition of some rangelands in national forests began to improve, the remaining millions of hectares of public lands that were not controlled by grazing statutes continued to experience severe abuse from uncontrolled grazing. Finally, in 1932, the seriousness of the range problem prompted Congress to request the USFS to survey the range condition. The survey showed that rangeland productivity in the United States had been reduced by 50 percent. On some Utah ranges, rice grass (a valuable winter forage species when most rangeland is covered with snow) had been reduced by 90 percent by intense grazing (Figure 13.11). The survey further revealed that the extensive removal of grass cover had resulted in erosion on 80 percent of the range.

Taylor Grazing Control Act and Other Laws

In 1934, as a direct consequence of the Forest Service report, Congress enacted the **Taylor Grazing Control Act.** This was the successful culmination of a long struggle on the part of conservationists to place our ailing public rangelands under federal control. This act had three major objectives: (1) to halt overgrazing and soil deterioration, (2) to improve and maintain ranges, and (3) to stabilize the rangeland economy. Although much grazing land was under private ownership, the major focus of the Taylor Grazing Control Act was on rangeland owned by the public, which had been seriously abused by western ranchers. Under the provisions of the Taylor Grazing Control Act, the public range was divided into operational units called **grazing districts** and managed by the newly established Grazing Service. The Grazing Service, which became the **Bureau of Land Management (BLM)** in 1946, was not effective in administering this management system.

FIGURE 13.11 *A severely overgrazed range. Note the denuded land and the scrawny cattle.*

In 1976, Congress passed the **Federal Land Policy and Management Act,** which consolidated legislation related to public land management and gave the BLM authority to manage all public rangelands not in national forests or national parks. The BLM has the responsibility of preventing overgrazing and seeing that damaged rangelands recover. Also in 1976, the **National Forest Management Act** was passed, directing the USFS to develop and maintain a comprehensive inventory of all National Forest Service lands (including rangelands). In 1978 the **Public Rangelands Improvement Act** was passed; among its provisions are policies to manage and improve the condition of public rangelands. These three major pieces of legislation provide the framework for managing, inventorying, and improving U.S. public rangelands for both the BLM and the USFS.

In 1985 Congress passed the **Food Security Act** (or Farm Bill). One of the programs of the Farm Bill is called the **Conservation Reserve Program (CRP),** which calls for the removal of 18 million hectares (45 million acres) of highly erodible cropland from the cropland base. This land is to be planted back to grassland (and some trees) to control erosion. Under the terms of this program, the farmer makes a contract with the U.S. Department of Agriculture to withdraw erodible farmland from crop production for 10 years and to establish vegetative cover (most often grass) on this land to stabilize the soil. The USDA, in turn, makes "rental" payments to the farmer during this period. The CRP was to be terminated in 1995 but the program was so successful in reducing erosion that it has been extended through 2002. Not only are the grass plantings valuable in checking erosion, they are useful in providing food and cover for wildlife as well. What's so interesting is that most of these highly erodible lands were originally native grasslands before they were cultivated in the 1800s. It seems that Mother Nature knew what was best for her land all along. For more on CRP, see Chapter 7.

13.3 Rangeland Resources and Condition

Rangeland Resources

Worldwide, rangelands occupy about half of the planet's ice-free land surface and provide about 80 percent of the feed for domestic livestock (Figure 13.12). Most rangelands are in semiarid areas too dry for rain-fed crop production.

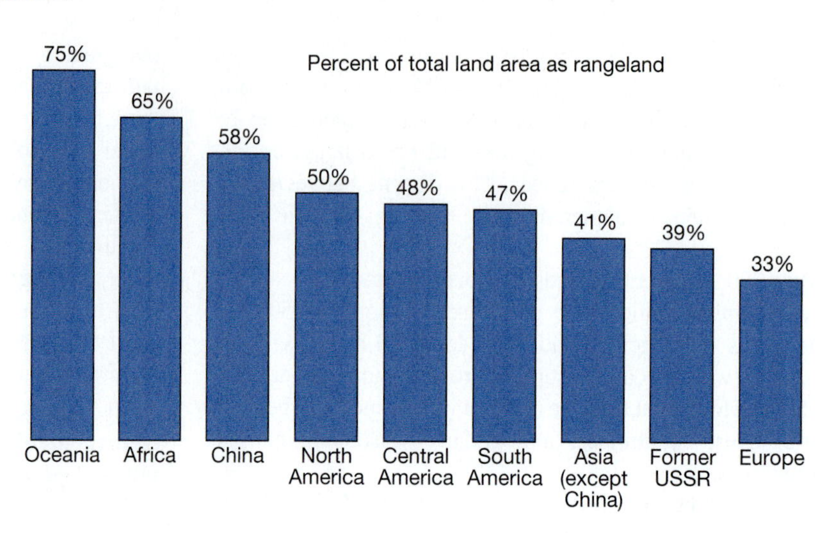

FIGURE 13.12 *Distribution of rangelands in the world (data from UN Food and Agriculture Organization).*

About 42 percent of the world's rangelands are used for grazing livestock. The remaining 58 percent are too cold, dry, or remote from population centers to be utilized by domesticated livestock. However, with good management, some of this land could be grazed by livestock.

Globally, more than 200 million people use rangelands for some form of livestock (pastoral) production. More than 15 percent are nomadic and pastoral people who are completely dependent on livestock grazing on rangelands. Most pastoral people live in the less developed countries in Africa and Asia. However, pastoralism can be found in industrialized countries like Australia, where rangelands cover more than 70 percent of the country. In Australia, pastoralism consists of mostly sheep and cattle grazing and uses 60 percent of the rangelands.

Rangelands occupy about 29 percent of the total land area in the United States. Most of this is short-grass prairies in the arid and semiarid western half of the country (Figure 13.13). More than half of these rangelands in the United States are privately owned, 43 percent are owned by the federal government, and the remainder are owned by state and local governments. Privately owned rangelands, compared to publicly owned rangelands, have relatively few restrictions placed on them by the government and are usually managed for the single use of grazing. The federal government (mostly the BLM and USFS) manages public rangelands according to the principle of **multiple use.** This means that, besides grazing, these lands are used for wildlife conservation, recreation, mining, energy resource development, soil conservation, and watershed protection.

About 2 percent of U.S. ranchers are issued permits by the BLM or USFS to graze their herds on public rangeland. How much ranchers should pay for **public grazing permits** has been a controversial issue (detailed in A Closer Look 13.3) between ranchers and environmentalists for many years.

About 75 percent of U.S. public and privately owned rangeland is grazed by livestock at some time during each year. However, America's rangelands have supplied only about 16 percent of the total amount of the forage that cattle consume prior to slaughter. The remaining 84 percent is provided by crops such as alfalfa, which are grown on fertilized pastures. This surprising statistic is explained by the fact that most cattle graze on the range only until they are mature. They are then shipped to **feedlots** (enclosed pens), where they are crowded together and fed with grain (largely grown on irrigated farms on the Great Plains) and special feeds to fatten them up for slaughter (Figure 13.14). Some feedlots in Nebraska hold as many as 10,000 cattle at a time.

The overriding importance of feedlots and the relatively minor role of rangelands in livestock nutrition is being modified somewhat because of America's awareness of the role that high-fat, high-cholesterol meat plays in heart disease and stroke. These health concerns have resulted in a dramatic decrease in annual per capita beef consumption from almost 39 kilograms (85 pounds) in 1975 to only 29 kilograms (64 pounds) in 1997. Studies have shown that the longer cattle are grazed on rangelands, the lower the fat content of their meat. Ever sensitive to consumer demands, the livestock industry is now keeping cattle on the range until they weigh 300 kilograms (660 pounds)—twice the weight at which they were once shipped to feedlots.

A few farsighted ranchers are replacing some of the traditional strains of cattle, such as the shorthorn and Hereford, with the leaner-meated longhorn—a favorite of the young livestock industry in the nineteenth century. Even more interesting, the American Water Buffalo Association established in 1986 promotes the scientific study of raising, breeding, and marketing water buffaloes (oxlike animals from Southeast Asia). Water buffalo cows and bulls are now for sale in Florida, and there is a market for water buffalo meat in the Washington, D.C. area. Water buffalo steaks are both tasty and lean, and should satisfy cholesterol-conscious Americans.

Rangeland Condition

U.S. Rangelands The term **range condition** may be defined as the current state of vegetation of a range site in relation to the potential natural plant community. In a sense, it is an estimate of how close a particular rangeland is to its productive potential.

What is interesting is that the three major U.S. federal land management and advisory agencies—the USFS, BLM, and NRCS (Natural Resource Conservation Service)—use somewhat different definitions and methodologies to determine range condition. In addition, methodologies and concepts with-

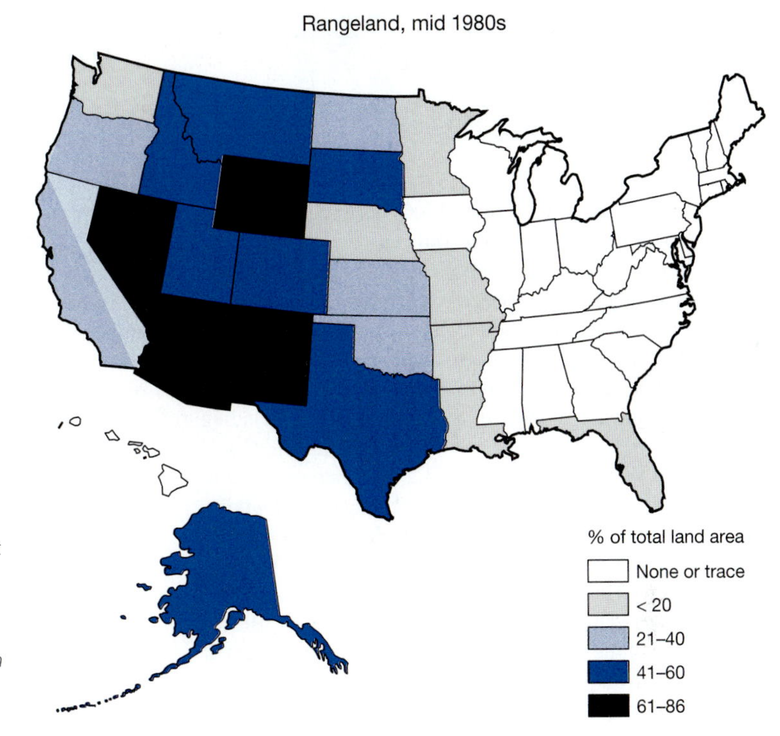

FIGURE 13.13 *Rangeland as a percentage of total area of each state in the mid-1980s. Note that the western United States is dominated by rangeland. Figure 13.13 is taken from D.A. Castillon,* Conservation of Natural Resources, *Dubuque, Iowa, Wm. C. Brown Publishers, 1992, p. 216, which in turn was taken from data from Enviromental Protection Agency,* Environmental Trends.

in agencies have changed somewhat over time. Yet these three agencies have traditionally classified range condition into four (or five) classes: excellent, good, fair, and poor. (Sometimes a fifth class of "very poor" or "depleted" has been used.) Although their terminologies have differed somewhat (and still do) as to what each class represents, their classifications have reflected comparisons between a site's existing vegetation (that is, which plant species are present and their relative amounts) and what the site could potentially support if natural plant succession progressed unimpeded through time. An excellent rating means the existing vegetation closely resembles its natural potential, whereas a poor rating means the existing vegetation is very dissimilar to its natural potential (Figure 13.15). Thus, although comparing range condition data from these agencies is difficult, we can still get an idea of the general trend of the condition of U.S. rangelands from examining past data.

Public rangelands in the United States were severely abused in the late 1800s and early 1900s due to improper livestock grazing. Soil and vegetation are still recovering in some places from these past abuses. Today, better grazing management

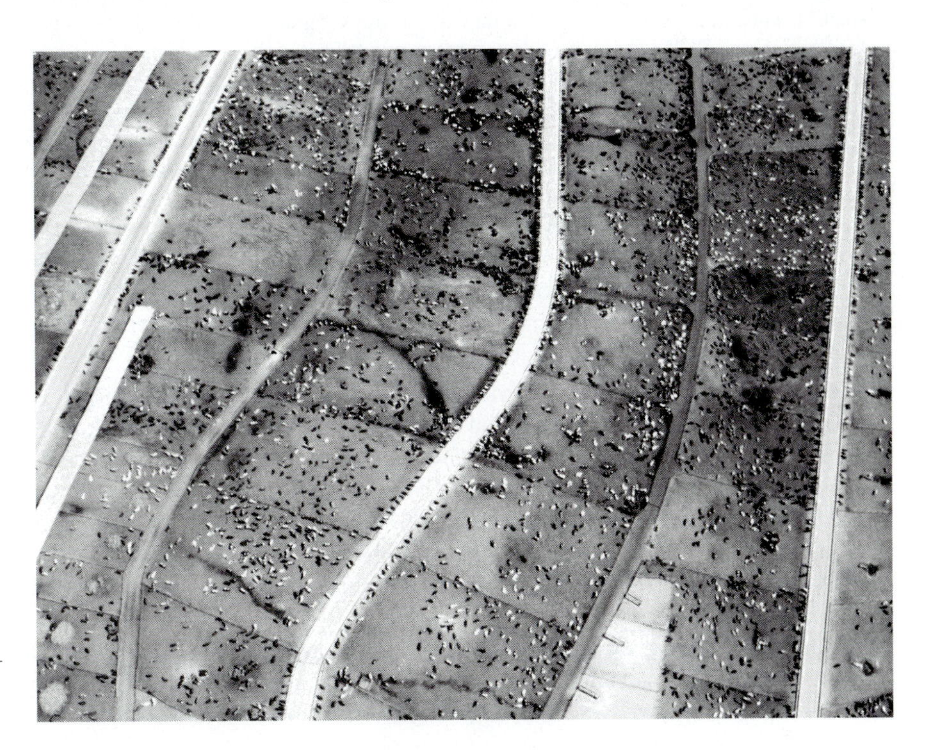

FIGURE 13.14 *Aerial view of a large western feedlot. Cattle are crowded together and food is provided so that they "fatten up" before being slaughtered.*

A CLOSER LOOK 13.3

Range Wars: Ranchers Versus Environmentalists

Cattle grazing in the West began in the 1500s, when Spanish longhorns were brought in from Mexico. The peak grazing year was 1884, when almost 40 million cattle munched the grasses of the open range. Sometimes the ranchers tried to flaunt federal laws. In 1896 some even tried to graze their livestock in Yosemite National Park. The U.S. Cavalry took charge and drove 189,000 sheep, 1,000 cattle, and 300 horses from the park.

Today, cattle grazing is legal in 45 national parks, including such celebrated ones as Grand Canyon, Rocky Mountain, Grand Teton, and Mesa Verde. Federal law even allows cattle grazing in 150 of our National Wildlife Refuges, as well as in official wilderness areas like the Big Blue Wilderness in Colorado and the Gila Wilderness of New Mexico. Additional grazing occurs in many of our national forests. However, the great majority of livestock graze on lands administered by the Bureau of Land Management (BLM). About 80 percent of the 110 million hectares (270 million acres) of public land in the West is now being used for grazing.

In recent years ranchers have bought permits to graze 2 million cattle and 2.3 million sheep annually on public rangelands. The cost of a grazing permit for cattle in 1996 was $1.35 per AUM. A permit to graze on private rangeland costs between $9 and $10, or more than seven times as much. U.S. taxpayers, who own the public grazing lands, are therefore subsidizing those ranchers who hold federal grazing permits. This **subsidy** (the difference between the fees collected and the actual value of grazing on federal land) has amounted to as much as $100 million annually, tax dollars used to maintain these public rangelands.

Some environmental groups, such as the Sierra Club, the Audubon Society, and the National Wildlife Federation, are very concerned with the western grazing situation. Their thinking goes something like this: Not only are these ranchers (and their livestock) feeding at the public trough, but they are destroying grassland ecosystems in the process. Consider streamside habitat, for example. It is here where 75 percent of the rangeland wildlife finds food, cover, and breeding sites. Heavy grazing pressure is destroying these areas. Indeed, a recent report of the Environmental Protection Agency stated that "riparian [streamside] areas throughout much of the West are in the worst condition in history."

Environmentalists believe that federal agencies tend to issue too many grazing permits, which results in our public rangelands inevitably getting overgrazing and degraded. Overgrazing causes the replacement of valuable forage grasses by worthless weeds and is the most important cause of grassland plant extinction in America. Livestock and wildlife compete for the same food supplies, but wildlife usually loses out in this competition. A good example is the situation in the Burns BLM district in Oregon, where such wild animals as grouse, deer, and antelope are able to get only 3 percent of the total plant food available.

Conservationists also argue that higher fees would provide more money for improvement of range condition, wildlife conservation, and watershed management. They contend that any rancher with a permit who cannot remain in business without government subsidies should not be in the ranching business.

Ranchers with permits are opposed to higher grazing fees because for most of them, profit margins are already razor thin. They argue that most federal land is extensive, steep, or otherwise difficult land on which to manage livestock, whereas private land usually is more productive and manageable. Most public rangelands remained in the federal domain because homesteaders found other areas more attractive. They say that overgrazing on public rangeland is also caused by increasing numbers of elk, deer, and other wild animals.

Private grazing leases are for single use, whereas public grazing leases have restrictions for multiple use. In addition, the restrictions imposed on federal grazing permits are more severe than those typically imposed on private grazing leases. These restrictions amount to extra costs for federal permittees. For example, federal grazing permits include range improvements and maintenance (such as fences), animal management not provided by the leaser, and forage limitations and withdrawals for watershed and wildlife habitat purposes. Almost all grazing permittees also incurred extra costs when they acquired their ranch properties, because they had to pay for the land's increased value that was gained from having a public land grazing permit assigned.

Some people argue that if public grazing fees are raised to the level of prevailing private grazing fees, many ranchers dependent on public rangelands will go out of business. In turn, rural communities where they buy their supplies will decline.

The grazing permit fee has become a perennial public policy issue. Maybe the debate should focus instead on the bigger issue: the uses of federal lands and their resources. With improved cooperation among all parties and organizations, these issues should be resolved.

practices are needed in many areas. Yet, with some exceptions, U.S. rangelands today are in their best condition in the last 100 years. For example, BLM data (Figure 13.16) on trends in range condition on their lands indicate that acreage in the combined category of excellent and good has more than doubled between 1936 and 1992, and that in the poor category has decreased by almost two-thirds. NRCS data (Figure 13.17) for private rangelands indicate similar improvement during a briefer period.

Despite this improvement, there is still a large body of evidence indicating that significant areas of U.S. rangelands are being overgrazed and in need of improvement. Let us not forget we still have large areas of fair and poor rangelands, both public and private. In all fairness, though, some rangelands classified as fair or poor may be in satisfactory condition under current management practices. Here's why: Remember that rangeland sites are each rated relative to their potential natural plant community. A site being managed for multiple uses (such as wildlife habitat, camping, hiking, and livestock grazing) may be most suited to vegetation that's very different from its natural potential composition. For example, deer forage is usually maximized on a site when shrubs and forbs, rather than perenni-

FIGURE 13.15 *This range near Miles City, Montana, is classified as in good condition. It must be emphasized that in comparing range conditions, a range in the arid sagebrush region of Wyoming may be in excellent condition and still be much less productive than a range in good condition in the semiarid shortgrass region of Nebraska.*

al grasses, are plentiful. But if such a site's natural potential were abundant perennial grasses, its existing vegetation (of abundant shrubs and forbs) would be judged dissimilar from its potential and the site would get a lower condition class rating (such as fair or poor). Yet the site may currently support the desired plant species in the desired relative amounts.

Rangelands Outside the United States Partial surveys indicate that rangelands in Canada have followed a similar upward trend as in the United States. Australia's rangelands have generally shown an improvement in recent decades, although some areas have certainly been overgrazed and degraded. Most of the rangelands in Europe are in excellent or good condition and are among the most productive in the world. This cannot be said for

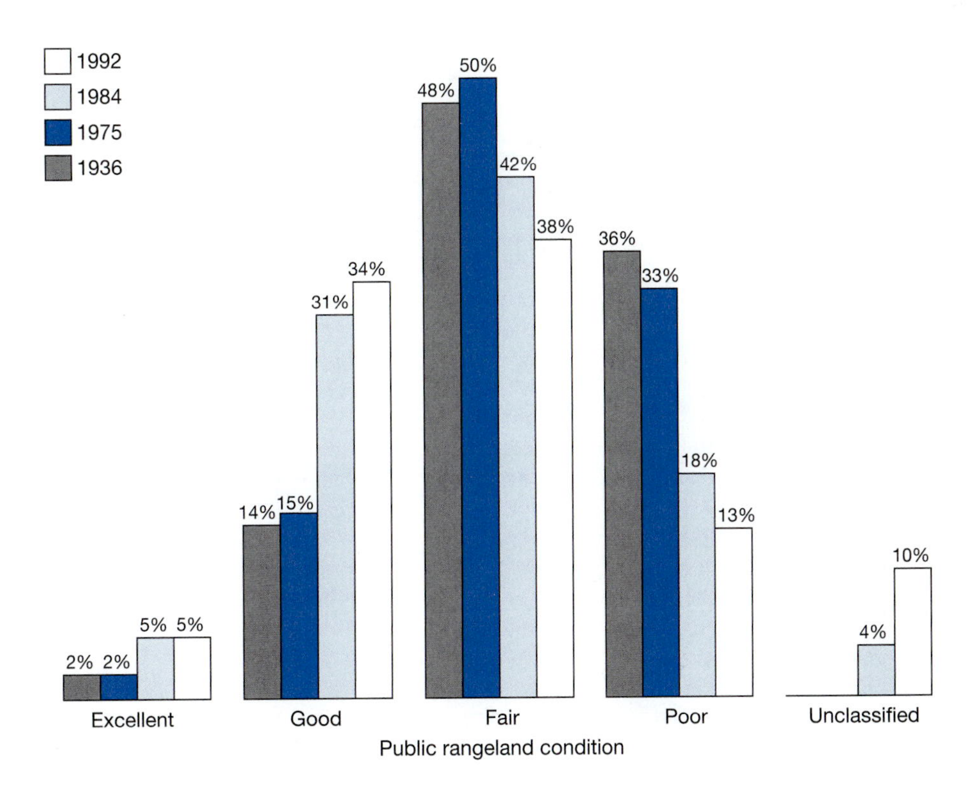

FIGURE 13.16 *Trends in range condition on lands administered by the Bureau of Land Management, 1936 to 1992 (data from the Bureau of Land Management and the U.S. Forest Service).*

most rangelands in other parts of the world. Limited data from partial surveys in Africa, Asia, and South America show that many of their rangelands have been degraded to some degree.

Rangeland deterioration is probably most visible in Africa and Asia, which together contain nearly half the world's rangelands. A 3-year U.N. study, *Global Assessment of Soil Degradation*, involving more than 250 scientists throughout the world, found that overgrazing has degraded some 680 million hectares since midcentury, of which almost 65 percent are in Africa and Asia (Table 13.2). This study suggests that 20 percent of the world's pasture and range is losing productivity and will continue to do so unless herd sizes are reduced or more sustainable livestock practices are put in place.

The nine countries in the arid Sahel region of West Africa, where livestock numbers commonly exceed rangeland carrying capacity by 50 to 100 percent, have reported severe overgrazing and an increase in desertification. The northern African countries of Morocco, Algeria, Tunisia, Libya, and Egypt are experiencing overgrazing and decreased forage production. Sudan shows signs of a rapidly deteriorating range and increasing desertification. Zambia has a serious overgrazing problem, where the carrying capacity of rangeland is far exceeded by livestock numbers.

In China, about one fourth of the rangeland has been severely degraded, most of it in the northern part. Much of the rangeland in the Middle East, particularly in parts of Iran, Iraq, Jordan, Oman, Pakistan, and Syria, is in poor condition. In India, much of the rangeland is degraded or overgrazed, although some is improving through rangeland improvement programs. Many rangelands in the South American countries of Brazil, Argentina, Uruguay, and Paraguay have been degraded from overstocking.

TABLE 13.2 *Worldwide Land Degradation by Overgrazing, 1945 to 1991*

Region	Land Overgrazed (million hectares)
Africa	243
Asia	197
Europe	50
North and Central America	38
Oceania	83
South America	68

13.4 Range Management

Range management is an interdisciplinary field that uses inputs from soil and plant sciences, animal and wildlife sciences, forestry, hydrology, economics, and other related fields. Rangeland ecosystems must be understood so that past changes can be explained and future influences predicted. Maximizing livestock or wild herbivore productivity without degrading rangeland quality is usually the major goal of range management.

The best strategy in range management is to prevent the range from deteriorating in the first place. One of the first

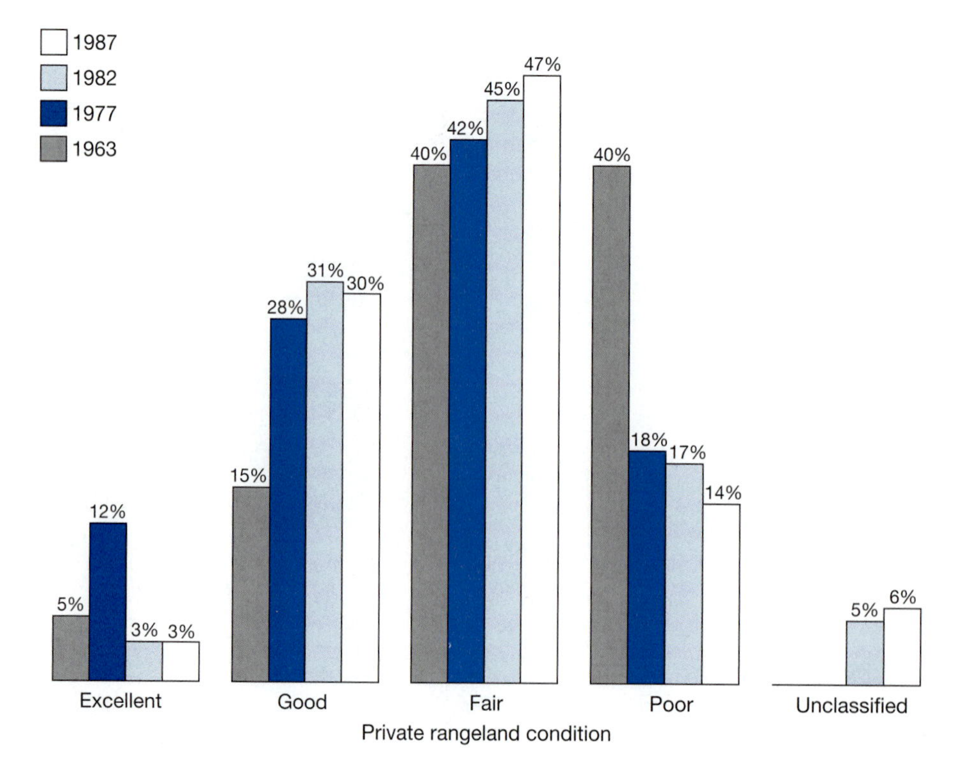

FIGURE 13.17 *Trends in condition of private rangelands, 1963 to 1987 (data from the Natural Resource Conservation Service).*

steps in good range management is determining the carrying capacity of the rangeland sites to be managed. This estimate is then used to manage the number and kinds of animals grazing the rangeland sites to avoid overgrazing. In addition, stock may need to be rotated from one site to another to prevent overgrazing. Other rangeland species, such as grasshoppers and jackrabbits, can compete with livestock for forage and can effectively decrease carrying capacity. Predators such as coyotes can also play a role in determining the carrying capacity of a range. Several range management options are available to increase carrying capacity, among them stock manipulation, artificial seeding, and the control of plant and animal pests.

Stock Manipulation

Distribution of Water and Salt Ranchers must make sure that livestock will graze their land uniformly. Because cattle tend to concentrate in wet meadows and along stream margins and to avoid ridges and slopes, part of a range may be severely overgrazed while another part may be ignored. Grazing can be directly controlled by barbed wire fencing and herding, both of which are rather costly methods. Ranchers can also use indirect methods that are much less expensive but highly effective. For instance, they can strategically locate water holes and salt blocks. Because cattle and sheep normally congregate around water sources, the salt blocks should be placed between 0.4 and 0.8 kilometer (0.25 and 0.5 mile) from the nearest water source, although this distance can vary depending on length of grazing time and other factors. Salt blocks should be put in ungrazed areas on ridges, gentle slopes, or openings in brush or forest, to induce livestock to frequent areas they would normally avoid (Figure 13.18).

Salt is essential to the vigorous health of range animals. Within 3 weeks after having been deprived of salt, cattle develop an unusual craving for it. When salt deprivation continues, the animals lose their appetite, become emaciated and weak, and may collapse. On the same ranges where the soil is naturally high in phosphate and sulfate salts, livestock may partially satisfy their salt requirements by grazing on salt-absorbing vegetation.

Grazing Systems There are a number of grazing systems commonly used in the United States and other parts of the world. These systems include continuous, deferred-rotation, rest-rotation, short duration, Merrill three-herd/four pasture, high intensity-low frequency, best pasture, and season-suitability. In selecting an appropriate grazing system, ranchers need to consider many factors, including climate, topography, vegetation, kind or kinds of livestock to be grazed, wildlife needs, watershed protection, labor requirements, fencing, and water availability. Only a few grazing systems will be discussed here.

Allowing livestock to graze in a pasture continuously throughout the grazing season is called **continuous grazing.** One problem with continuous grazing is that livestock have preferred areas of grazing that will often receive excessive use. In these preferred grazing areas, the more palatable (and usually more nutritious) range plants can be so seriously overgrazed that they lose vitality and nutritional value, resulting in a mosaic of overgrazed and undergrazed patches in an area.

The first specialized grazing system developed in the United States was **deferred-rotation grazing.** This system temporarily makes fields unavailable for use by moving livestock. The main features of this system are presented in Figure 13.19. Note that the ranch is divided into three pastures, *A*, *B*, and *C*, and that in each of them, grazing is deferred for 2 successive years within a 6-year period. During years 1 and 2, forage plants in pasture *A* are allowed to reach maturity and drop their seeds before livestock are permitted to graze on them late in the season. Even though the grasses become rather dry at this time, they still are highly nutritious. A certain amount of grazing after the seeds have been produced may be advantageous to the pasture, because foraging cattle scatter the seeds and trample them underfoot, which forces the seeds into the ground and enhances germination. Deferred-rotation grazing can increase the size, density, weight, vitality, reproductive capacity, and nutritional value of forage species. Note that in Figure 13.19, grazing in pasture *A* is deferred the first and second years, grazing in pasture *B* is deferred the third and fourth years, and grazing in pasture *C* is deferred the fifth and sixth years. Thus, for this hypothetical ranch, all three pastures, within a 6-year rotation plan, are temporarily removed from grazing pressure.

Short-duration grazing was developed in Zimbabwe by Allan Savory in the 1960s and introduced in the United States

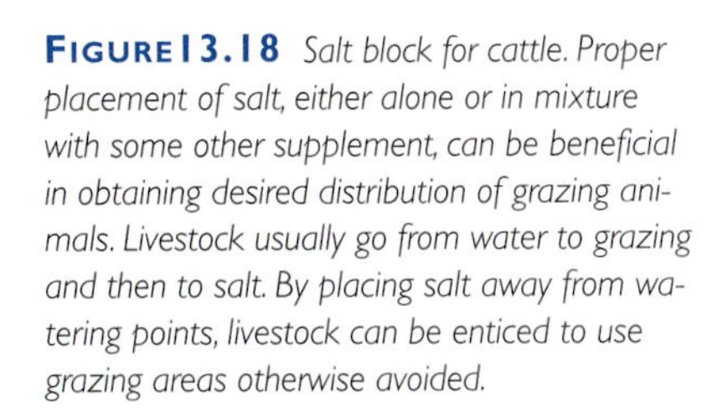

FIGURE 13.18 *Salt block for cattle. Proper placement of salt, either alone or in mixture with some other supplement, can be beneficial in obtaining desired distribution of grazing animals. Livestock usually go from water to grazing and then to salt. By placing salt away from watering points, livestock can be enticed to use grazing areas otherwise avoided.*

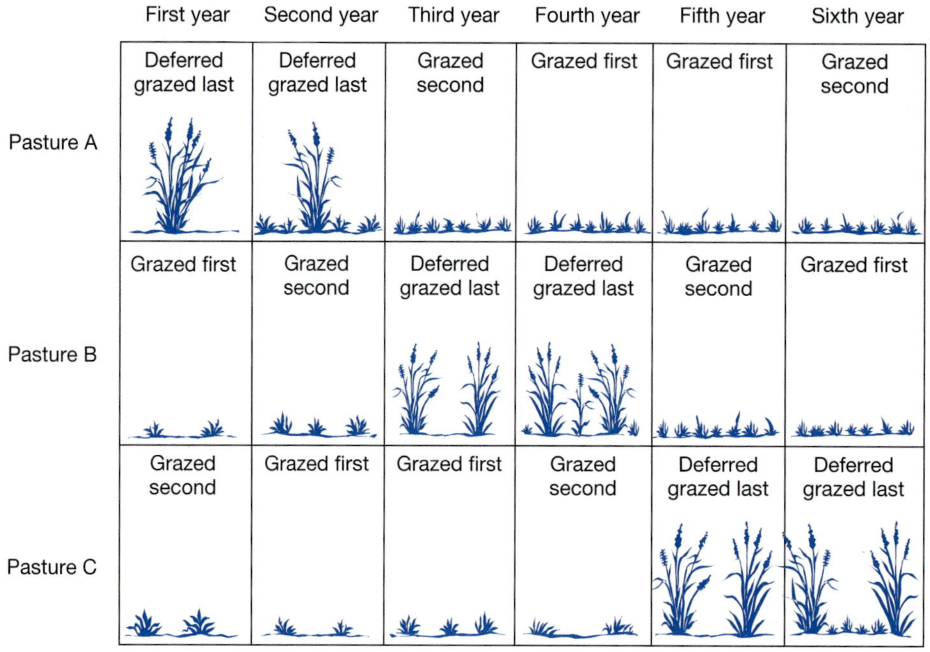

	First year	Second year	Third year	Fourth year	Fifth year	Sixth year
Pasture A	Deferred grazed last	Deferred grazed last	Grazed second	Grazed first	Grazed first	Grazed second
Pasture B	Grazed first	Grazed second	Deferred grazed last	Deferred grazed last	Grazed second	Grazed first
Pasture C	Grazed second	Grazed first	Grazed first	Grazed second	Deferred grazed last	Deferred grazed last

FIGURE 13.19 *Deferred-rotation grazing.*

in the 1970s. Savory has since made refinements to this system, which is now called **holistic resource management.** The system, which has gained popularity in specific areas of the United States, involves a high stock density (number of animals per unit area). Under good management it is claimed that labor costs are reduced, individual animal performance is increased, and range condition is improved. Although holistic resource management has been advocated for all rangeland types around the world, it is controversial mainly because of a lack of long-term research proving its effectiveness. For a complete discussion of Savory's views on grazing management, the reader is referred to Savory's book listed in "Suggested Readings" at the end of this chapter.

Artificial Seeding (Reseeding)

Ranchers can improve their rangelands by periodic seeding, also called **reseeding.** Reseeding helps restore severely degraded rangeland and increase its carrying capacity. Seeds may be broadcast by hand or airplane. However, unless some provision is made for covering the seeds, reseeding will probably fail. Uncovered seeds may be blown away by strong winds, killed by severe winter cold, eaten by birds and rodents, or washed away during heavy rainstorms. Ranchers can drive a herd of cattle over the area to trample the seeds into the ground. If seeding is done in recently burned-over timberland, the loose covering of ashes may ensure the success of reseeding. Similarly, if reseeding is synchronized with autumn leaf fall, the leaf litter may provide sufficient seed cover for successful germination.

Aerial broadcasting is the only feasible method in rugged mountainous regions. It can be highly effective. For example, several years ago, a burn in a fir and pine stand in the Cabinet National Forest of Montana was seeded by plane. Two years later, the area was cloaked with a dense stand of timothy grass and Kentucky bluegrass. This vegetation provided not only excellent protection from erosion but also 2,100 kilograms of food per hectare (1,900 pounds per acre) for grazing animals.

In general, ranges that have been properly reseeded support a greater number of livestock in better condition over a longer period than equivalent ranges that have not been reseeded. Many plantings in the West, for example, have been grazed for 15 successive years and still produce 3 to 20 times as much forage as they did before seeding. A classic example of how artificial reseeding can improve rangeland is provided by a 200-hectare (500-acre) plot in the Fish Lake National Forest in Utah. Before seeding, this area supported a vigorous cover of sagebrush and rabbit brush, which provided forage for only eight head of cattle. However, only 3 years after it was seeded to wheatgrasses (Figure 13.20) and bromes, it was able to sustain 100 head.

Control of Rangeland Pests

Another important aspect of rangeland management is the control of plants, such as weeds and woody vegetation, that compete with range grasses, the control of herbivores (grasshoppers, jackrabbits, rodents) that compete with livestock for food, and the control of predators, particularly the coyote.

Control of Plant Pests

Ranchers are periodically confronted with invasion of their grazing land by woody, low-value shrubs, such as mesquite, sagebrush, and juniper, that compete with range grasses for soil moisture, nutrients, and sunlight (Figure 13.21). Invasion of these shrubs is especially likely if the land is overgrazed. Unless effective control methods are established, the aggressive spread of these species may seri-

FIGURE 13.20 *Crested wheatgrass, an exotic bunch grass that was introduced from Russia. It thrives in the northern plains states, where summers are cool.*

ously lower the livestock-carrying capacity of the range. Consider the mesquite.

The **mesquite** is a thorny desert shrub with small leathery leaves. Like all members of the pea family, it produces large pulpy seed pods. The extensive root system may extend to a depth of 15 meters (50 feet).

Of all the woody plant invaders of southwestern grasslands, mesquite ranks first in distribution, abundance, and aggressive encroachment of rangeland. Plant ecologists believe that, for millennia, mesquite invasion was prevented by periodic fires that were ignited by lightning strikes or by Plains Indians (as an aid to hunting). Grasses were consumed in the fires along with mesquite. Many grass species, however, can mature and produce seeds in 2 years, whereas mesquite requires a longer period. For this reason, recurrent fires can control mesquite growth.

Ecologists believe that climax vegetation of the grassland biome, with the aid of deep, fibrous root systems, can compete successfully with mesquite for sunlight and limited soil moisture. However, when the white settlers drove off the Indians, introduced cattle and sheep by the millions, and instituted new methods to control fire, the main factor responsible for confining mesquite vanished. Ranchers frequently overstocked range herds, causing the deeply and extensively rooted climax species (decreasers) to decline. The decreasers, in turn, were replaced by increasers, which were then replaced by invaders. The shallow taproots of the first invaders were competitively inferior to mesquite, which gradually took over. The mesquite invasion was further facilitated by cattle. The late-summer-maturing mesquite pods, some up to 20 centimeters (8 inches) long, provide cattle with nutritious food. After digesting the pulp, cattle eventually void the seeds with their waste, frequently at a considerable distance from the parent plant. The seeds, still viable and well fertilized, show a surprisingly high germination rate.

FIGURE 13.21 *Woody shrubs like sagebrush (pictured here), rabbit-brush, and greasewood provide forage and cover for wild browsing animals like mule deer. Many of these shrubs invade rangeland that is overgrazed, and are indicators of range in poor condition.*

Dense, mature mesquite stands can be regulated by **controlled burning.** In this way, the rancher can duplicate the natural control by fire that operated for thousands of years before the coming of the white settlers. Controlled burning (Figure 13.22) encourages the growth of valuable forage plants such as grama, bluestem, and buffalo grasses and can even have beneficial effects on many wildlife species. It can also result in a more uniform distribution of grazing livestock. In addition, cattle "beef up" much more readily on control-burned ranges than on unburned ranges. This rapid weight gain is due not only to the increase in the protein, phosphorus, and moisture content of the forage plants after a burn, but also to their enhanced palatability and digestibility. All this said, in certain areas of the United States, the use of fires for vegetation control has been restricted in recent years because of the concern about air pollution.

Grubbing the mesquite plants out of the ground or plowing them up is effective but very costly. In the case of plowing, the whole area would have to be carefully reseeded with nutritious forage grasses. The control of extensive acreages of mesquite can be accomplished by the aerial spraying of herbicides. Great care must be taken, however, so that the chemicals have no harmful effect on valuable plants and animals of the rangeland ecosystem. Herbicides are not used that often in the United States to control range vegetation because of their expense and possible threats to human health.

Biological control of unwanted range plants includes the introduction of goats, camels, and predatory insects. Goats in particular are used in many countries as agents of brush control. Careful and complete ecological study should precede the introduction of any biological agent to be sure that the animal (or plant) concerned will not displace the desirable native species. Some range managers use periodic, short-term trampling by large numbers of livestock.

Control of Herbivores Insects can more severely overgraze ranges more than livestock. Although range caterpillars, blackgrass bugs, Mormon crickets, and harvester ants cause the greatest damage, grasshoppers are the number one insect in influencing range vegetation. Of the more than 100 species of **grasshoppers** collected in western range vegetation, the most destructive and widely distributed are the lesser migratory grasshoppers (Figure 13.23). During periods of peak abundance, they may gather in swarms and migrate several hundred kilometers.

FIGURE 13.22 *Controlled burning of a Montana range.*

When the weather is wet, grasshopper populations remain small. The rancher may not even be aware of their presence. Then, however, there may come a severe drought, with only 12.5 centimeters (5 inches) rather than the usual 37.5 centimeters (15 inches) of annual rainfall. With environmental factors now optimal, the grasshopper populations rapidly increase until it is almost impossible to take a single step through a grama grass pasture without flushing several of them. During a peak year, grasshoppers may so deplete forage that livestock must move to other pastures or starve. During a severe drought, grasshopper density may reach more than 30 per square meter; the insects may consume up to 99 percent of all vegetation.

One interesting facet of the grasshopper–rangeland relationship is that these insects are much more numerous in overgrazed ranges than in moderately grazed fields. This is because most grasshopper species prefer ranges with little grass and a high forb component. A study in southern Arizona revealed that the grasshopper population was 450,000 per hectare (180,000 per acre) on overgrazed lands, compared with only 50,000 per hectare (20,000 per acre) on range in average condition—a 9:1 differential. It may be, therefore, that an effective method of controlling grasshopper plagues is to ensure that the range is not being subjected to excessive grazing pressure.

During a drought, **jackrabbits** compete aggressively with cattle and sheep for high-quality forage (Figure 13.24). Between 75 and 150 jackrabbits consume as much forage as one cow, and 15 to 30 eat as much as one sheep. Feeding trials show that jackrabbits consume about 6.5 percent of their body weight per day on a dry-matter basis. Such heavy intake is more then three times the daily rate of most ruminants on range forages. A dense jackrabbit population, therefore, exerts heavy grazing pressure. This prevents the reestablishment of highly nutritious decreasers and favors the intrusion of lower-value increasers, as well as invaders like cactus and thistle.

The rangeland pest, whether grasshopper, jackrabbit, or even prairie dog, becomes a serious problem only during population peaks, and these peaks usually coincide with rangeland deterioration. It should be emphasized that these pests do not cause the initial depletion of the pasture. They are a symptom rather than a cause of range deterioration. We can compare range abuse to a wound. The "wound" was initially inflicted by excessive stocking, and the ensuing pest buildup merely irritated the wound and prevented it from healing properly.

Just why a range in good condition (well stocked with climax plants) should be an unsuitable habitat for certain rabbits and rodents has never been fully explained. It may be that the tall vegetation obstructs the vision of these relatively defense-

FIGURE 13.23 *Grasshopper outbreak. An insect specialist examines grasshoppers in his sweep net. By sweeping the range grasses a few times with his net, he can determine the severity of the outbreak. Appropriate control measures can then be applied.*

less animals and makes them more vulnerable to predators. In any event, most rangeland experts agree that shooting, trapping, and poisoning campaigns are only stopgap procedures and rarely worth the cost. Rabbits and rodents have high reproduction rates and their numbers can usually recover in a short time. The best long-term solution to the pest problem seems to be vegetation management, which often is simply a four-strand barbed-wire fence to keep excess cattle off the deteriorating range.

FIGURE 13.24 *The jackrabbit competes with livestock for forage. However, it becomes a serious pest only when the range has been overgrazed, as is the case in this photograph of the Santa Rita Experimental Range, Arizona.*

Control of Predators: The Coyote Predators, such as coyotes, black bears, golden eagles, bobcats, foxes, and mountain lions, can have considerable influence on range livestock. Of all rangeland predators, the **coyote** (Figure 13.25) poses the most serious problem and mainly to sheep. In one study on

FIGURE 13.25 *The coyote, a stealthy rangeland predator that has been accused, justly or unjustly, of killing many thousands of sheep annually.*

losses of sheep to predators in the Great Basin, coyotes accounted for 90 percent of the losses, bobcats for 2 percent, and undetermined predators for 8 percent. The wily "brush wolf" has become a thorn in the side of sheep ranchers, with some reporting sheep losses of more than 20 percent. Although herding of sheep greatly reduces sheep losses to predators, the number of sheepherders has been declining in the past three decades. Studies show that losses of lamb crops to predators are much lower for herded sheep than for unherded sheep.

When a rancher destroys a coyote that has been killing, say, 20 sheep per year, simple arithmetic might suggest that this rancher will be 20 sheep richer each year thereafter. However, things are not quite that simple and straightforward in nature. For one thing, the coyote feeds on animals other than sheep. Fifty percent of its diet is composed of range grass-consuming jackrabbits and rodents. Therefore, the value of the few sheep saved by destroying a coyote may be less than the value of the forage consumed by the hundreds of rodents and rabbits that the coyote would have removed from the rangeland community had it been allowed to live. Nevertheless, many ranchers appear irrevocably committed to predator control as a range management tool. In fact, where coyotes have become numerous, sheep ranchers have waged all-out extermination campaigns—poisoning, trapping, shooting, and even pursuing them to their dens.

The control of predators on grazing lands became a federal government responsibility by an act of Congress in 1931. Under the federal Animal Damage Control Program, 70,000 to 85,000 coyotes are destroyed annually in 13 western states. The annual kill represents 18 to 29 percent of the coyote population in this region. However, in light of the latest predation data, it should be emphasized that coyote control should focus on those relatively few ranches where sheep kills have actually occurred. A widespread, nonselective campaign to destroy all coyotes is expensive, time-consuming, and highly unwarranted. The pros and cons of various methods of coyote control are discussed in Case Study 13.4.

Summary of Key Concepts

1. Rangelands are areas of the world that are a source of forage (such as grasses and shrubs) for free-ranging native and domestic animals, as well as a source of wood products, water, energy, wildlife, minerals, and recreational opportunities.
2. Worldwide, rangelands occupy about half of the planet's ice-free land surface and provide about 80 percent of the feed for domestic livestock. Rangelands occupy about 29 percent of the total land area in the United States.
3. Seven major categories of vegetation provide most of the world's grazing lands: grasslands, tropical savannas, tundra, desert shrubs, shrub woodlands, temperate forests, and tropical forests.
4. The upper 50 percent of the grass shoot is a surplus that can be safely eaten by livestock; the lower 50 percent, known as the metabolic reserve, is necessary for the plant's survival.
5. Range ecologists and ranchers frequently classify the various range plants into three categories with respect to the dynamics of plant succession: decreasers, increasers, and invaders.
6. Grazing carrying capacity is the maximum number of animals that can graze each year on a given area of range, for a specific number of days, without causing a downward trend in forage production, forage quality, or soil quality.
7. Overgrazing is continued heavy grazing that exceeds the carrying capacity of the community and creates a deteriorated range.
8. Desertification is land degradation caused by climatic changes and human-induced activities, such as overgrazing, overcutting, cultivation of unsuitable land, or inadequate irrigation, in dry, semidry, or dry and semihumid areas.
9. The Homestead Act of 1862 allowed the offering of 160 acres (64 hectares) of public land to anyone who would settle the land and reside on it for 5 years.
10. In 1905 the U.S. Forest Service was formed and began to restrict livestock numbers and grazing seasons in U.S. national forests.
11. The major objective of the Taylor Grazing Control Act of 1934 was to improve the quality of U.S. rangelands.
12. About 2 percent of U.S. ranchers are issued grazing permits (subsidized by American taxpayers) by the BLM or USFS to graze their herds on public rangeland.
13. The term *range condition* may be defined as the current state of vegetation of a range site in relation to the potential natural plant community.
14. Public rangelands in the United States were severely abused in the late 1800s and early 1900s due to improper livestock grazing.
15. Although better grazing management practices are still needed in many areas, today U.S. rangelands are in their best condition in the last 100 years.
16. Many of the rangelands in Africa, Asia, and South America have been degraded to some degree.
17. Livestock may be better distributed to prevent overgrazing by the strategic distribution of salt and water.
18. There are a number of grazing systems, such as continuous grazing, deferred-rotation grazing, and short-duration grazing (also call holistic resource management), commonly used in the United States and other parts of the world.
19. Rangelands in poor condition may benefit from reseeding.
20. An important aspect of rangeland management is the control of plant pests, such as weeds and woody vegetation, that compete with range grasses, as well as the control of animal pests (grasshoppers, jackrabbits, rodents) that compete with livestock for food.
21. Various methods have been used to control coyote predation of sheep, including guard dogs, guard cattle, birth control chemicals, chemical repellants, and lethal poisons like Compound 1080, trapping, shooting and den hunting.

CASE STUDY 13.4 METHODS OF COYOTE CONTROL

Some methods for controlling coyote populations are nonlethal, such as using sheepherders, guard dogs or cattle, birth control, and chemical repellents. Lethal techniques include poisoning, trapping, shooting, and den hunting. Most of these control strategies are discussed below.

Use of Guard Dogs

For several centuries, European sheepherders successfully used guard dogs to protect sheep from predatory coyotes. Dogs with superior guarding instincts have been selectively bred at the Hampshire College Farm Center in Amherst, Massachusetts. Since 1978, sheep ranchers in 31 states have been using these dogs to protect their flocks (Figure 1). When placed in sheep flocks when still pups, the dogs soon consider themselves as a natural part of the flock. When a coyote approaches a sheep, the guard dogs react quickly, rushing fiercely at the intruder and causing it to flee.

The guard dog system is highly effective and can be a financial boon to the sheep rancher. For instance, a survey found that one of every three ranchers who had experienced heavy sheep losses from marauding coyotes reported not a single attack once guard dogs were used. Environmentalists firmly support the guard dog system. Nevertheless, many ranchers do not use guard dogs, either because they are unwilling to try a new approach to coyote control or because they mistakenly believe that the dogs themselves will kill some sheep.

Use of Guard Cattle

A recently developed predator control strategy involves the intermingling of lambs and young cows in pens for 1 month. During this period, the animals develop a strong attachment for each other. When released on the open range, the cattle protect the sheep from predators by kicking and butting the predators. According to a report by the USDA, sheep kills by coyotes can be sharply reduced by this method.

Birth Control Chemicals

With this control technique, carcasses of livestock are laced with birth control chemicals that reduce the reproductive ability of coyotes that consume the meat. Theoretically, coyote populations should then decline. Unfortunately, however, such a decline would probably be only temporary. The reason? Like many other species of wildlife, coyotes have tremendous reproductive resilience. When a population declines in a given year, the number of young the next year generally increases. The coyote is so resistant that, to eradicate it completely, 75 percent of the population would have to be destroyed year after year for half a century!

Chemical Repellents

This method, still in the experimental stage, involves the injection of a bad-tasting, nausea-inducing chemical, lithium chloride, into the carcasses of dead sheep. The carcasses are then left out for coyotes to feed on. When a coyote eats the tainted flesh, it becomes very sick, and may then avoid live sheep thereafter.

Lethal Poison: Compound 1080

The use of sodium monofluoroacetate, popularly known as Compound 1080, for coyote control has been highly controversial. The general public, environmentalists, and organizations like the Defenders of Wildlife, the National Audubon Society, and the National Wildlife Federation have strongly opposed its use. For one thing, it is extremely toxic; 28 grams (1 ounce) is sufficient to kill 20,000 coyotes.

In 1972, the use of Compound 1080 on all federal lands and by federal agencies anywhere was banned by an executive order of

FIGURE 1 *Sheepherder tending sheep on national forest land. The sheepherder can protect the sheep from predation by coyotes.*

(continued)

then-president Richard Nixon. A short time later, the EPA banned its use by state agencies and private individuals as well.

One major complaint concerning its use has been the accidental but fatal poisoning of nontarget species like golden eagles and bobcats, which had eaten Compound 1080-laced bait intended for coyotes or even the carcasses of poisoned coyotes. Some wildlife biologists estimate that before the ban on Compound 1080, about 9,000 bobcats were accidentally poisoned per year.

Nevertheless, sheep ranchers complained bitterly to their congressional representatives, saying that the ban on Compound 1080 deprived them of the most effective weapon in their coyote control arsenal. In 1985, the EPA yielded to pressure from the Reagan administration and approved the use of the poison in special collars that are worn around the necks of sheep. Since coyotes frequently lunge for the neck of their prey, Compound 1080-containing collars would seem to be an effective way to control these predators.

Is the political struggle to use Compound 1080 really worth the effort? The answer, ironically, is no. In one highly regarded study, sheep losses from predators and other causes prior to its use (1940–1949) were compared to losses from 1950 to 1970, when it was widely used. The study found no detectable difference in sheep losses from predation with or without Compound 1080. The present policy is for the U.S. Department of Agriculture to restrict predator control activities to those ranchers suffering high known predator losses.

Key Words and Phrases

Animal Unit Months (AUMs)
Basal Zone
Bureau of Land Management
Carrying Capacity
Conservation Reserve Program (CRP)
Continuous Grazing
Controlled Burning
Coyote
Decreasers
Deferred-Rotation Grazing
Desert Shrublands
Desertification
Federal Land Policy and Management Act
Feedlots
Food Security Act
Forage
Grasshoppers
Grasslands
Grazing District
Grazing Permit
Homestead Act
Holistic Resource Management
Increasers
Invaders
Jackrabbits
Mesquite
Metabolic Reserve
Multiple Use
National Forest Management Act
National Grasslands
Overgrazing
Powell, John Wesley
Public Rangelands Improvement Act
Range Condition
Rangeland
Reseeding
Short-Duration Grazing
Taylor Grazing Control Act
The Nature Conservancy
The Sahel
Tropical Savannas
Tundra
Undergrazing
U.S. Forest Service

Critical Thinking and Discussion Questions

1. What are the differences between grasslands, tropical savannas, and desert shrublands?
2. What is unique about the growth characteristics of grasses that enable them to survive despite moderate grazing pressure?
3. What is the relationship among the three groups of rangeland plants called decreasers, increasers, and invaders?
4. Briefly list three major objectives of the Taylor Grazing Control Act of 1934.
5. Name the two federal agencies that administer most of the public rangelands in the United States. When were they established?
6. Should fees for grazing on public rangelands in the United States be increased? Why or why not?
7. What is the function of the feedlot?
8. Describe the deferred-rotation grazing system.
9. Briefly describe some of the positive effects of rangeland burning.
10. What precautions must be taken when reseeding a range by broadcasting?
11. Why has mesquite become such a serious rangeland pest?
12. What would be an effective strategy for controlling grasshoppers on rangeland?
13. Describe the different methods, both lethal and nonlethal, of coyote control.
14. Define *range condition.* Describe the condition of U.S. rangelands today.
15. What are the major causes of desertification in Africa?
16. What are rangelands? Where are there significant areas of rangelands in the world?
17. What are the benefits of restoring native prairies?

Suggested Readings

Council for Agricultural Science and Technology. 1996. *Grazing on Public Lands.* Task Force Report No. 129. Ames, Iowa: CAST. Discusses and provides scientific information concerning livestock grazing on public lands in the western United States.

Dagget, D. 2000. *Beyond the Rangeland Conflict: Toward a West That Works.* Reno: University of Nevada Press. A book of ten real world examples of environmentalists and ranchers working together to deal with issues of open space, endangered species, functional ecosystems, environmental restoration, holistic management, and conflict resolution.

Heady, H. F., and R. D. Child. 2001. *Rangeland Ecology and Management,* 2nd ed. Boulder, CO: Westview Press. Rangeland management textbook that focuses on the ecology of rangeland grazing, practical management of animals, and vegetational manipulation.

Higgins, K. F., A. D. Kruse, and J. L. Piehl. 1989. *Effects of Fire in the Northern Great Plains.* Extension Circular 761. Brookings: South Dakota State University. Excellent discussion of the effects of fire on the grassland ecosystem of the Northern Great Plains, with special emphasis on the use of fire for wildlife management.

Holechek, J. L., R. D. Pieper, and C. H. Herbel. 2001. *Range Management: Principles and Practices,* 4th ed. Englewood Cliffs, NJ: Prentice Hall. Solid fundamental textbook covering almost all aspects of range science and management.

Knapp, A. K., et al. 1998. *Grassland Dynamics: Long-Term Ecological Research in Tallgrass Prairie.* New York: Oxford University Press. Comprehensive description of the long-term research studies on the ecology of the tallgrass prairie, focussing on the Konza Prairie in Kansas.

National Research Council. 1993. *Rangeland Health: New Methods to Classify, Inventory, and Monitor Rangelands.* Washington, D.C.: National Academy Press. Examination of methods used to inventory, classify, and monitor rangelands, with recommendations for evaluating the ecological health of U.S. rangeland ecosystems.

Rennicke, J. 1992. "Sacred Cows?" *Backpacker* 20(5): 47–51. Nontechnical account of the rancher-environmentalist confrontation over grazing subsidies.

Savory, A. 1988. *Holistic Resource Management.* Washington, D.C.: Island Press. Good discussion of Savory's views on grazing management and specifically his system called holistic resource management.

Stegner, W. 1954. *Beyond the Hundredth Meridian.* New York: Penguin Books. A fascinating look at the old American West in the late 1800s and at John Wesley Powell, who explored it and warned against the dangers of settling it.

Stoddart, L. A., A. D. Smith, and T. W. Box. 1975. *Range Management,* 3rd ed. New York: McGraw-Hill. A classic on characteristics, ecology, and management of rangelands.

Stubbendieck, J. L., S. L. Hatch, and K. J. Kjar. 1982. *North American Range Plants.* Lincoln: University of Nebraska Press. Comprehensive treatment of the habitats and forage values of American range grasses, with excellent drawings of each species.

Web Explorations

Online resources for this chapter are on the World Wide Web at: **http://www.prenhall.com/chiras** *(click on the Table of Contents link and then select Chapter 13).*

Chapter 14

Forest Management

Our nations forests range from the virgin stands of hemlock and Douglas fir in Alaska and the Pacific Northwest, to second-growth oak and hickory in the Mideast and East, to plantations of pine and black walnut in the South. As shown in Figure 14.1, six major forest regions exist in the conterminous United States. The occurrence of a particular forest region is the biological expression of the prevailing

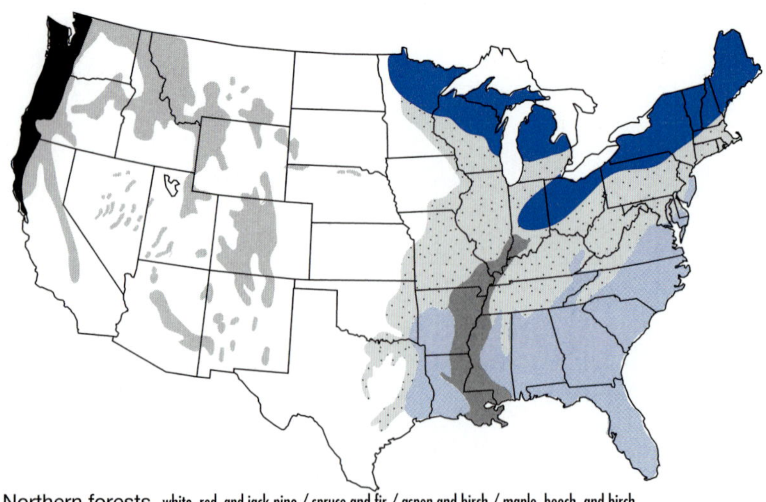

- Northern forests white, red, and jack pine / spruce and fir / aspen and birch / maple, beech, and birch
- Central forests oak and hickory
- Southern forests oak and pine / loblolly and shortleaf pine / longleaf and slash pine
- Bottom land forests oak, gum, and cypress
- West Coast forests Douglas fir /hemlock and sitka spruce / redwood / western hardwoods
- Western interior forests ponderosa pine /lodgepole pine / Douglas fir / white pine / western larch / fir and spruce / western hardwoods

FIGURE 14.1 *Distribution of the six major forest regions in the conterminous United States.*

environmental conditions, including rainfall, temperature, and soil.

The United States has more trees now than it did in 1920 on approximately the same amount of forestland. It also has the largest legally protected wilderness system in the world, while at the same time sustaining a highly productive and efficient wood products industry. Until the 1920s, forests were generally logged and abandoned. Today, an average of 1.7 billion seedlings are planted annually in the U.S. That's six seedlings planted for every tree harvested. In addition, billions of additional seedlings are regenerated naturally.

14.1 Forest Ownership

Forestland is defined by the U.S. Forest Service as land which is at least 10 percent covered by forest trees of any size. Our nation has more than 298 million hectares (737 million acres) of forests, covering 33 percent of the total land area of the United States. America's forests are owned by private individuals (54%), public agencies (37%), and private industries (9%).

About two thirds of U.S. forestland (198 million hectares or 490 million acres) is classified by the U.S. Forest Service as commercial timberland—of high enough quality to be used by the timber industry (Figure 14.2). Private owners other than the timber industry, such as farmers and owners of estates, whose major source of income is not from producing timber, own 58 percent of this commercial timberland. Such private owners are classified as **nonindustrial private forest owners.** The forest industry, which includes companies operating wood-using plants, owns 14 percent of the commercial timberland, of which more than half is located in the South, an important region in regards to the timber economy of the nation. These companies range from a small open sawmill with a few thousand hectares of land to a multinational conglomerate operating several dozen mills and owning millions of hectares of productive forestland. Eighteen percent of commercial timberland is located in the National Forest System. Other federal agencies, such as the National Park Service, and local and state governments own 10 percent.

14.2 The U.S. Forest Service

Four federal bureaus are charged with administering and managing our nations forests. Among them are the Natural Resource Conservation Service, which is concerned with farm management–associated forests; the Tennessee Valley Authority, which is charged with timberland management near numerous reservoirs along the Tennessee River and its tributaries; and the Fish and Wildlife Service, which is interested in improving the forest habitat for wildlife and fish. The fourth, the U.S. Forest Service, a bureau of the U.S. Department of Agriculture (USDA), has the primary responsibility for managing our nations forests to promote the greatest good for the most people over the long run.

The U.S. Forest Service was established in 1905. President Theodore Roosevelt appointed **Gifford Pinchot** to be its first chief forester (Figure 14.3). A forestry professor at Yale University, Pinchot promoted the use of several forest management methods he had learned in Europe. Pinchot was a zealous crusader for conservation, which he defined as the wise use of natural resources.

The Forest Service divides its attention among three major areas: (1) administering and protecting the national forests; (2) researching forest, watershed, range, and recreation management, wildlife habitat improvement, forest product development, and fire and pest control; and (3) cooperating with the state and private forest owners in the 50 states, Puerto Rico, and the Virgin Islands to promote sound forest management.

The Forest Service protects and manages 155 national forests and 20 national grasslands (Figure 14.4), embracing 77 million hectares (191 million acres). About 18 percent of this land is protected as wilderness areas. The remaining 82 percent is managed by the Forest Service according to the principles of multiple use and sustained yield.

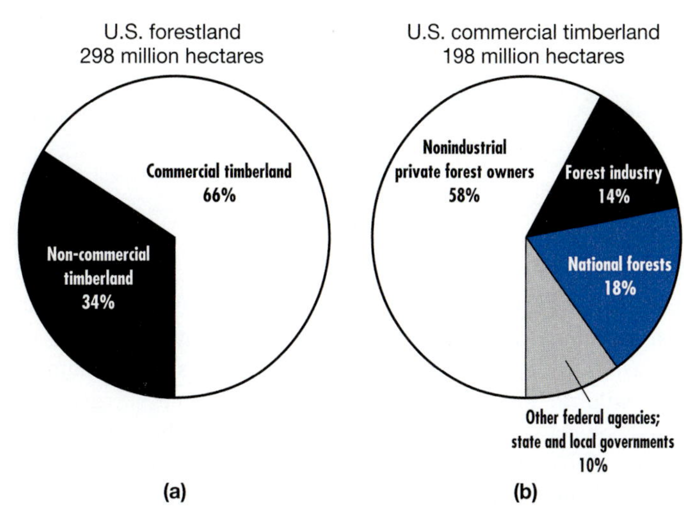

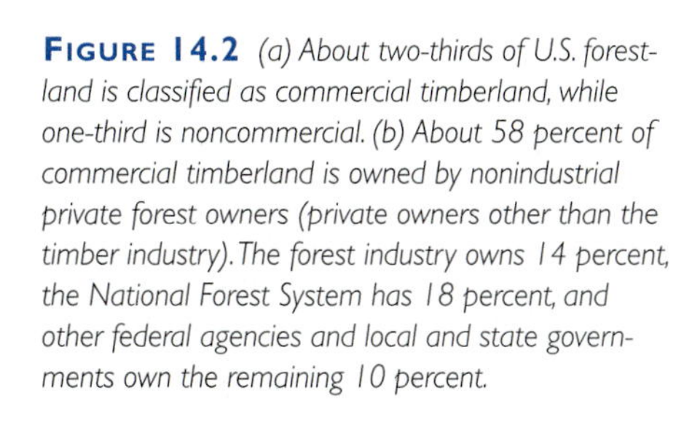

FIGURE 14.2 *(a) About two-thirds of U.S. forestland is classified as commercial timberland, while one-third is noncommercial. (b) About 58 percent of commercial timberland is owned by nonindustrial private forest owners (private owners other than the timber industry). The forest industry owns 14 percent, the National Forest System has 18 percent, and other federal agencies and local and state governments own the remaining 10 percent.*

FIGURE 14.3 *Theodore Roosevelt and Gifford Pinchot (to the left of Roosevelt) standing at the base of a giant redwood called Old Grizzly. Pinchot was the first chief of the U.S. Forest Service.*

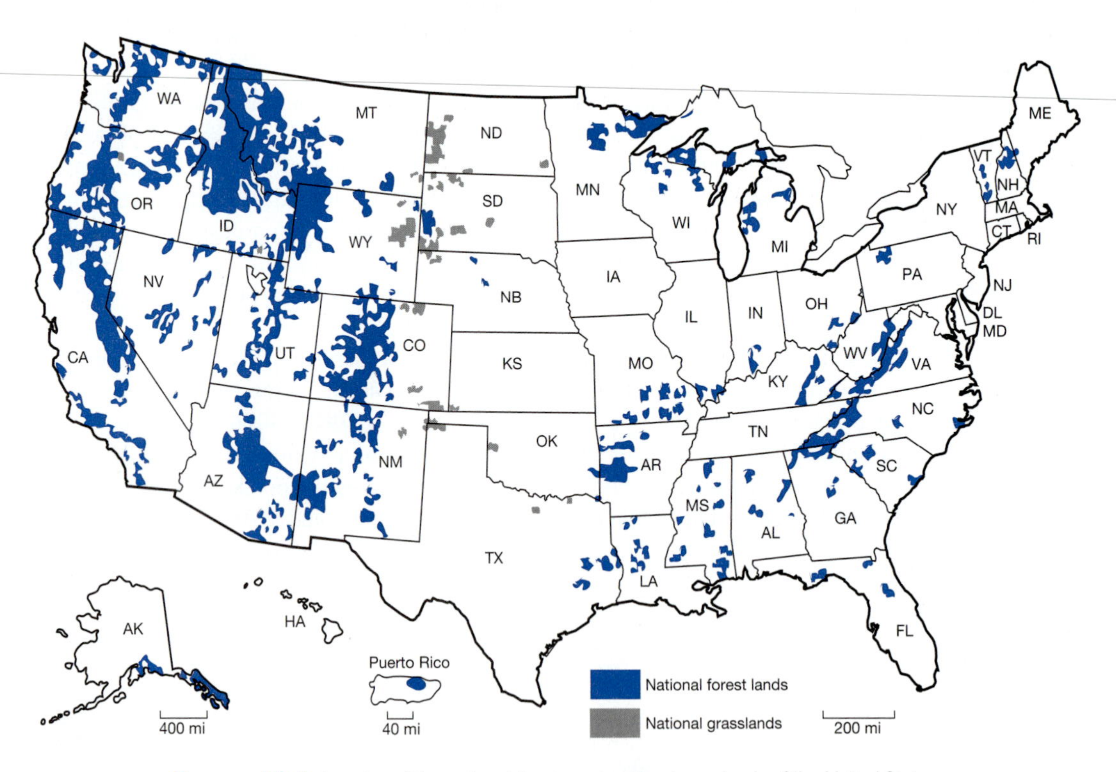

FIGURE 14.4 *Location of the national forests and national grasslands of the United States.*

Multiple Uses

Multiple-use management requires meeting a number of different needs, including timbering, grazing, agriculture, mining, oil and gas leasing, hunting and fishing, recreation, soil conservation, wildlife protection, and watershed management. A primary objective of the Forest Service is to make the greatest number of forest resources available to the greatest number of Americans, a goal mandated by the **Multiple Use–Sustained Yield Act of 1960.** The multiple-use management of forests looks simple on paper. In operation, however, it is an extremely complex ecological problem. For example, the Forest Service is frequently forced to use a given forest primarily for one purpose, thus sacrificing its potential use for others. A forest cannot be all things to all people. If a stand of Douglas fir, for example, is developed for high-quality timber, clearcutting may be the best way to harvest it. However, the wholesale removal of timber may impair natural flood and erosion control and may eliminate wildlife and recreational opportunities.

Sound multiple-use management must weigh the needs of many people, and these needs vary. Thus, timber production may have top priority in the Douglas fir and Western hemlock stands of Washington and Oregon, but in the low-value second-growth forests of populous New York, where many city dwellers go for a dose of wilderness tonic, recreational values have high priority (Figure 14.5).

Forests as a Source of Wood Products From early colonial days, when the straight, sturdy trunks of New England spruce and pine were fashioned into masts for the Royal Navy, to the present, three centuries later, our forests have been the source of a variety of valuable products. Today our commercial forests provide the raw materials for more than 10,000 products. They support an industry that ranks among the top ten employers in 40 of the 50 states.

The global average consumption of wood per person is about 0.7 cubic meters (almost 25 cubic feet) per year. To maintain the worlds highest standard of living, the United States uses more wood per capita than any nation on earth—about 2.3 cubic meters (81 cubic feet) of lumber per person per year. (A considerable amount is imported from Canada and Scandinavia.) In a typical year, each American uses the equivalent of a 31 meter (100 foot high), 40 centimeters (16 inches) in diameter tree for their wood and paper needs. Americans eat, sleep, work, and play in a world of wood. Whether it is toothpicks, telephone poles, photographic film, maple syrup, acetic acid, or structural timbers, we depend heavily on wood and wood-derived products (Figure 14.6).

FIGURE 14.5 *Use of the forest as an "outdoor laboratory". A scientist with the Audubon Environmental Education Project explains the ecological aspects of the forest to a nature class.*

The types of forest products used have changed dramatically in the past few decades. The harvest of pulpwood, as a source of paper, has increased threefold since 1950. In the 1980s about 50 percent of all the wood harvested in the United States was used as fuel—most of it by the forest products industry itself in its manufacturing processes. Even now, about 50 percent of all the wood harvested worldwide is used for fuel by direct combustion (Figure 14.7).

Forests in Flood and Erosion Control Forest vegetation reduces flooding and soil erosion. This was demonstrated in Davis County, Utah, on the eastern edge of the Great Salt Lake, a region frequently plagued by flash floods. Forest Service investigators discovered that much of a flood-triggering runoff originated from areas that had been depleted of vegetation. These denuded parts of the watershed had either been burned, overgrazed, or plowed up and converted into marginal croplands. In some areas, the runoff waters carved gullies 21 meters (70 feet) deep. During one rainy period, runoff was 160 times greater on an abused plot than on a nearby undisturbed one. With the aid of bulldozers, the gullies were filled in, slopes were contoured, and the bare soil was carefully prepared as a seedbed and planted with rapidly growing shrubs and trees. Only 11 years later, severe August rainstorms put the rehabilitated watershed to the test. An investigation revealed that fully 94 percent of the rainfall was retained by the newly forested area. Moreover, soil erosion was reduced from the pretreatment figure of 80 metric tons per hectare (36 tons per acre) to a mere trace.

Forests as Rangelands In addition to timber, U.S. forests frequently include considerable areas of high-quality livestock forage. Thus, of the 77 million hectares (191 million acres) of national forests and national grasslands, 40.5 million hectares (100 million acres) provide forage for 6 million cattle and sheep belonging to 19,000 farmers and ranchers. (Most of this is in the West. In the lake and central states, most forest grazing occurs on farm woodlots.) Ranchers pay fees for the privilege of grazing their livestock in national forests.

Forests as Wildlife Habitat The U.S. national forests, as well as many private woodlands, offer excellent wildlife habitat. More than 60 percent of the elk in the Rocky Mountain region find food, cover, and shelter in national forests.

The Forest Service tries to manage the national forests to provide the best possible wildlife habitat. Sometimes the best management involves increasing the amount of forest edge. Such habitat occurs between the forest and adjoining fields, meadows, and marshes. It frequently includes a considerable number of shrubs that provide food and cover for wildlife. Development of such habitat can be integrated with timber harvesting and the construction of fire breaks and logging roads.

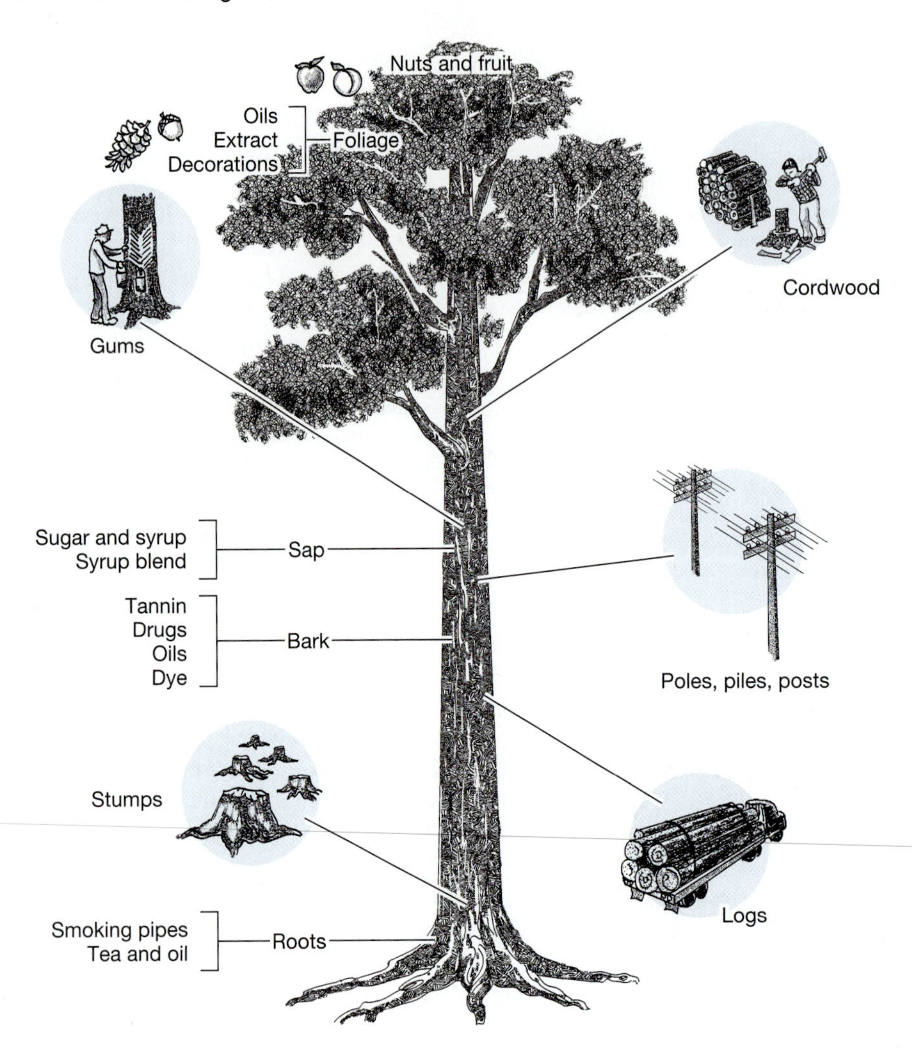

FIGURE 14.6 *Useful wood and wood-derived products.*

FIGURE 14.7 *A child of Bali, Indonesia, carrying firewood, which has been cut in the nearby woods and will be used for cooking.*

Because food and cover for elk and deer are more abundant in early stages of a succession than in climax stages, setting back the succession by periodic controlled burns may be very beneficial to wildlife.

Sustained Yield

Today's many lumbermen are a different breed from the cut-out-and-get-out loggers of the late nineteenth and early twentieth centuries. After studying German silvicultural techniques,

American foresters learned that a forest can be managed in such a way that a modest timber crop can be harvested indefinitely, year after year, if annual decrements are counterbalanced by annual growth. This is the **sustained-yield concept.** Under the terms of the Multiple Use–Sustained Yield Act of 1960, foresters have a mandate from Congress to employ the principle of sustained yield in their operations of our national forests.

If a forest is managed for sustained yield, the wood produced in a given year should equal the volume removed. Let us see how this might work with clearcutting, a harvest method in which all the trees in a given area are cut. Suppose that a Georgia farmer owns 40 hectares of pine woods and that the trees are harvested when they are 40 years old. If this pine stand has a normal distribution, as foresters say, this stand would have 40 age classes, ages 1 to 40, each covering 1 hectare (Figure 14.8). One 40-year-old hectare of forest could be harvested each year. This operation could be carried out indefinitely as long as the clearcut hectare is properly reseeded or replanted.

The length of the cutting cycle, or rotation, depends on the species of tree and on its intended commercial use. For aspen and birch to be used as pulpwood, it varies from 10 to 30 years; for pine pulpwood it is 40 years. On the other hand, the rotation for Douglas fir to be used as lumber may be up to 100 years.

14.3 Harvesting Trees

Preparation for Harvest

Once a forest matures, it can be harvested as efficiently as possible. Before harvesting a forested area, foresters estimate the volume and grade of standing timber on the site. Such an on-site survey is called a **cruise.** Individual trees to be cut as well as the boundaries of areas to be cut are marked. Data from the cruise are used by the forest manager to prepare a detailed **logging plan.**

In some states, such as California and Massachusetts, the plan must be submitted to a state board for approval. A typical plan might include the following items:

1. A map showing the stand to be logged.
2. A map that shows the location, distribution, age, and volume of the species to be logged. The distribution of species is shown on areas that are to be selectively cut.
3. Harvesting method (clearcutting, strip cutting, or selective cutting) to be employed.
4. The most suitable access roads.
5. An estimate of the amount of time required to complete the logging operation.
6. The cost of the operation and the probable gross income from the sale of the saw timber and/or pulpwood.

Harvest Methods

Managing a stand of trees requires serious planning and involves a silvicultural system. A **silvicultural system** is a long-range harvest and management program designed to optimize the growth, regeneration, and administrative management of a particular forest stand, usually with the goal of obtaining a perpetual and steady supply of timber. Such a stand is managed on a sustain-yield basis.

Silvicultural systems are generally classified by the method used to harvest and regenerate the stand. Several harvest methods are available to timber companies. The choice of a particular harvest method depends on many factors, both biological

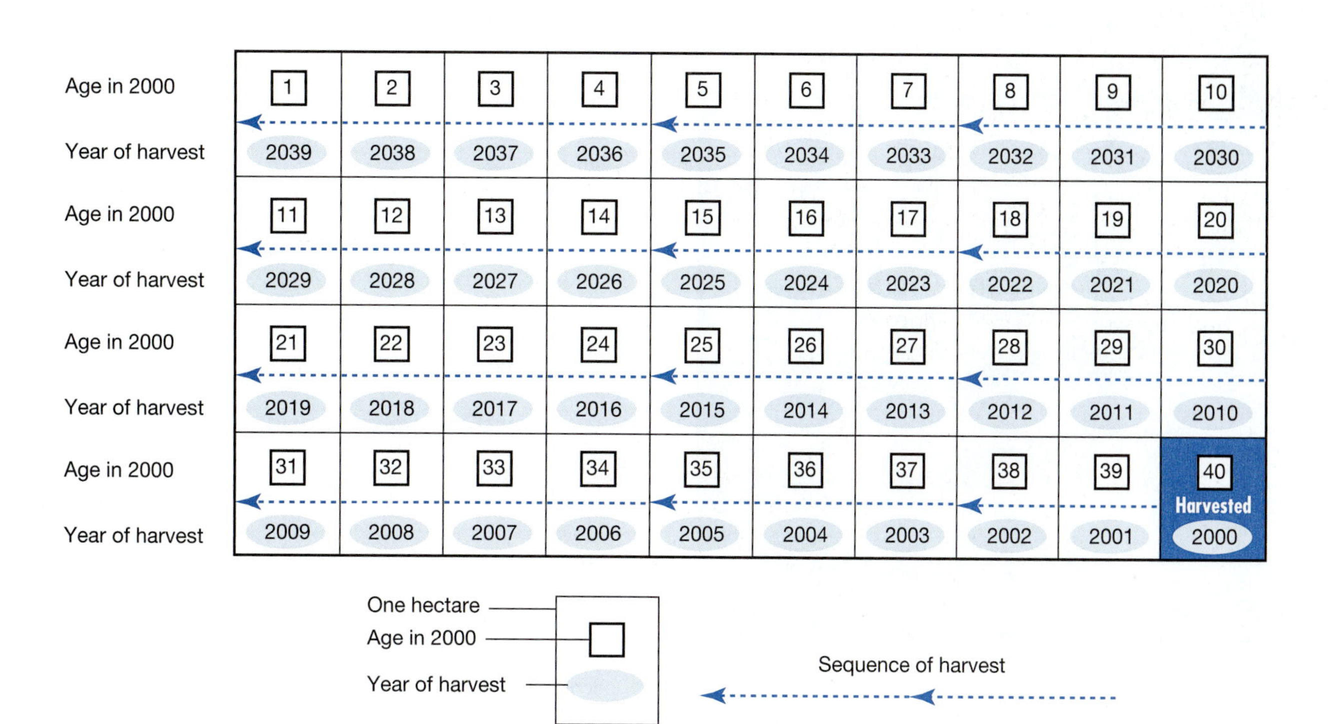

FIGURE 14.8 *A 40-year rotation harvest in a 40-hectare (100-acre) forest.*

and economic. Harvest methods may be grouped under two categories: even-aged and uneven-aged stands. Clearcutting, seed tree, shelterwood, and coppice methods are used to produce even-aged stands. Selective cutting and strip cutting are used to produce uneven-aged stands.

Even-Aged Stand Methods The most widely used method of timber harvesting is **clearcutting.** The clearcutting method, which is the standard logging practice in the Northwest and other areas in both private and public forests, is employed on **even-aged stands** composed of one or two species and is applicable only to trees whose seedlings thrive in full sunlight. Even-aged stands of trees are maintained about the same age and size and harvested all at once (clearcut). Growing a single-species stand of trees is called **monoculture,** which is a controversial practice (see A Closer Look 14.1).

The Douglas fir on the Pacific Coast is harvested by clearcutting. Perhaps the most valuable timber species in the

A CLOSER LOOK 14.1 *The Monoculture Controversy*

Monoculture is the practice of growing trees, much as the farmer grows a crop of corn or oats. It involves planting and raising a single-species stand of trees, with all individuals in the stand being of the same age and size (Figure 1). Monoculture is practiced on many tree farms and some public lands around the world. A **tree farm** is a private land area used to grow timber for profit. Tree farms make up 2 to 3 percent of the world's forest area. There are more than 74,000 tree farmers in the United States who in aggregate manage 38.5 million hectares (95 million acres) of woodland.

FIGURE 1 *A monoculture. This pine plantation occurs in the southern coniferous forest region. It is located in the Tennessee River Valley. These trees will be used as a source of pulpwood for the paper industry.*

Monoculture is advocated by some forest managers, forest-products corporations, and economists, but it has been criticized by many professional foresters and ecologists. Since it is controversial, it is instructive to list the major points both for and against monoculture.

Arguments for Monoculture

1. It is an efficient method of growing and harvesting a large volume of timber. Because growth is rapid, harvesting (by clearcutting) can be done on a relatively short rotation.
2. It increases short-term profits for timber companies and private land owners.
3. Planting a forest monoculture is the quickest way to reforest degraded land and prevent soil erosion and desertification.
4. It is amenable to the intensive application (frequently by air) of fertilizers, herbicides, fungicides, and insecticides.
5. It makes possible maximal use of such recent technological developments as machine seeders, tree-planting machines, the tree monkey (which climbs and prunes trees simultaneously), the chip harvester, the one-man logger, and the crusher (which can clear 240 hectares [600 acres] of forested land in 1 month).
6. It makes possible the establishment of sun-loving (shade-intolerant) seedlings of such valuable species as Douglas fir, redwood, longleaf pines, ponderosa pine, yellow poplar, red oak, cherry, and black walnut.
7. It can help increase wood production to meet growing demands.
8. High yields from forest monocultures can reduce pressure to clear large areas of old-growth forests.

Arguments Against Monoculture

1. A forest under monoculture is an artificial, simplified ecosystem. As such, it lacks the built-in balancing mechanisms found in the more complex natural ecosystem represented by the multiage, multispecies forest.
2. Although monoculture admittedly grows wood faster, the wood is inferior to the more slowly growing wood in a natural forest.
3. The intensive use of fertilizers and pesticides can pollute aquatic ecosystems. Pesticide contamination of food chains

(continued)

may have adverse effects on wildlife and humans (Chapter 8). Continuous use of insecticides can result in the development of insecticide-resistant strains of forest insects.

4. Monoculture depends on the intensive use of energy derived from fossil fuels. This energy may be used directly, as in the consumption of gasoline by tree planters, pruners, chain saws, helicopters, and air planes (used in seeding and in applying fertilizers and insecticides), or indirectly, in the manufacture of the heavy forest-planting and harvesting machinery or in the production of fertilizers and pesticides.
5. The single-species, single-age forest primarily serves one function: wood production. But other functions, such as erosion and flood control and maintaining wildlife habitat and diversity, scenic beauty, and recreational opportunities, are often better served by the naturally developed multispecies, multiage forest.
6. Because of the relative scarcity of moisture-absorbing organic material on the floor of the monoculture forest, the forest floor tends to be drier and warmer than that of the natural forest. As a result, the monoculture forest is more susceptible to fire.
7. A forest monotype is very vulnerable to destructive outbreaks of insects and disease organisms.
8. Eventually, of course, the monoculture forest will be clearcut, a harvesting method that can result in environmental problems.

One solution to the monoculture controversy is to limit forest monocultures on public lands and regulate them on private lands to prevent soil erosion and water pollution from runoff of sediment, pesticides, and fertilizers. What do you think?

world, it has been exported to Europe, where it has proved superior to native species. Some Douglas fir trees in Washington and Oregon are more than 60 meters (200 feet) tall and are over 1,000 years old. A Douglas fir, unlike a beech or a maple, is not a climax tree and is not shade tolerant as a seedling. Its seeds do not germinate in the shade of the forest floor. Therefore, the species is not amenable to selective cutting. If it were, its place in the forest would rapidly be appropriated by shade-tolerant species. In addition, a 30-meter (100-foot) Douglas fir weighing several tons could not be removed without badly bruising and killing younger growth.

With the clearcutting technique, an entire patch of evenly aged mature trees, 16 to 80 hectares (40 to 100 acres) in area, is removed, leaving an unsightly scar in the midst of the forest. (In national forests, the maximal size of a cut is 16 hectares [40 acres]). Because a large number of such blocks may be removed, a clearcut forest may resemble a giant green-and-brown checkerboard when viewed from the air (Figure 14.9). In addition to its use on Douglas fir in Oregon and Washington, the clearcutting method has been used effectively in harvesting even-aged stands of southern pine; aspen forests in northern Minnesota, Wisconsin, and Michigan; and coniferous forests in the West.

Clearcutting is done on a **rotation** basis. A rotation is the cycle between planting and harvesting. If saw timber is wanted, the rotation may be 100 years. The reason for this relatively long rotation is that the trees must be quite mature before the wood has the desirable density and durability. On the other hand, if pulpwood is desired, a rotation of only 30 years is satisfactory, since at that age pulpwood species such as pine, aspen, and birch have optimal characteristics. Moreover, if birch and aspen get much older, they become highly susceptible to diseases and insect attacks. Rotations of 100 years for

FIGURE 14.9 *Checkerboard clearcut area in Olympic National Forest, Washington.*

saw timber and 30 years for pulpwood are the most efficient from the standpoint of harvest volume. In other words, the trees are harvested before their growth rates sharply decline.

Large-scale clearcutting is controversial in many areas, especially the United States, Canada, Australia, and the tropical forests of Latin America, Indonesia, Asia, and Africa. The clearcutting practices of the U.S. Forest Service became a storm center of controversy in 1971. Much of the criticism focused on the ponderosa pine logging in the Bitterroot National Forest of Montana.

The clearcutting practices of private industry in the United States have also been the subject of criticism. Consider the Pacific Lumber Company of California, for example. In 1988 it began to clearcut magnificent, centuries-old redwoods (Figure 14.10) near Eureka, California, at an accelerated pace, allegedly to pay off junk bonds (bonds issued with little or no collateral) that paid for the takeover of a company with a long track record of sustainable harvest. Since one 500-year-old redwood has a market value of more than $50,000 dollars, the accelerated clearcutting made financial sense to Pacific Lumber. But it did not make aesthetic or ecological sense to an environmentalist group called the Coalition to Save the Redwoods. They protested the clearcut vigorously, even to the point of staging a sit-in on pulley-suspended platforms high up in the doomed trees. The advantages and disadvantages of clearcutting are listed in Table 14.1.

Other even-aged methods are designed to overcome some of the problems inherent in clearcutting with natural regeneration. For example, the **seed-tree** method is a silvicultural system in which all timber is removed in one cut, except for a scattered number of mature trees left to provide a source of seed for the new stand. The trees remaining need to be spaced to insure uniform seed distribution. The seed-tree method is most suitable in situations where intensive site preparation is feasible and the remaining trees are reasonably stable from the wind.

In the **shelterwood** method, part of the timber stand is removed in a series of cuttings during a relatively short time. Like the seed-tree method, this method also allows seed trees to be left standing but in adequate numbers so that protection and shade are provided for the new tree seedlings. In a typical shelterwood method, the first major cut leaves enough temporary partial overstory so that tree crowns shade 30 to 80 percent of the ground surface, depending on species and local conditions (Figure 14.11). Once the seedlings become firmly established (usually after several years), the remaining trees are completely cut down so that they do not reduce the growth of the saplings (young trees). The shelterwood method works well with tree species whose seedlings are not expected to germinate well in open conditions.

The **coppice** method depends on vegetative regeneration by stump sprouts instead of on stands developing from seed. Although this makes the coppice method unique, coppice stands are usually harvested by clearcutting and are therefore discussed here with other even-aged methods. The coppice method is used

FIGURE 14.10 *Large redwoods in Big Basin Redwoods State Park, California.*

TABLE 14.1 *Advantages and Disadvantages of Clearcutting*

Advantages	Disadvantages
1. Clearcutting is the quickest and simplest method of harvesting, requiring less skill and planning than other harvesting methods. It also reduces road building.	1. Clearcutting accelerates surface runoff and increases erosion on sloping land because the trees that protected the soil from rain and wind are removed. This erosion of topsoil often leads to siltation and sedimentation of stream channels which may have downstream effects on dams.
2. A few years after the area has been clearcut, sun-loving shrubs and saplings usually become established on the logged-off site, providing cover, food, and breeding sites for a great variety of wildlife such as rabbits, grouse, deer, and many songbirds.	2. It replaces an old growth forest that has a diverse, uneven-aged stand of trees with a monoculture more vulnerable to attack from disease, insects, and fire. In addition, high-quality, old growth timber is replaced with faster-growing, lower-quality timber.
3. It is the best way by which forests of highly desirable species, such as Douglas fir, can be regenerated.	3. It promotes the blowdown of trees. In a solid stand, most of the trees are protected from a windstorm. However, when a forest is clearcut, the trees bordering the open areas are left unprotected.
4. It is the only effective method for controlling some disease and insect outbreaks. To save the life of an infected stand, the forester uses the surgery of clearcutting, just as a surgeon amputates infected limbs to save the life of the patient.	4. It reduces biological diversity and greatly diminishes the carrying capacity (for some species) and habitat quality of an area, at least temporarily. How many grouse or deer can be supported by a bare patch of ground?
5. It usually gives the maximum economic return.	5. It reduces the recreational value of a forest and destroys the scenic beauty of a region, converting it into ugly, desolate scars.
6. It increases the volume of timber harvested per hectare and shortens the time needed to establish a new stand of trees. It often permits reforesting with genetically improved stock.	6. It creates a fire hazard. A large amount of debris (loose bark, branches, sawdust, broken logs) is often left behind after a forest is clearcut. Such slash could easily be ignited by lightning and start a wildfire.

for species, such as aspen and oak, that sprout vigorously and have sprouts with the potential to reach commercial size.

Uneven-Aged Stand Methods Clearcutting will not work in timber stands composed of **uneven-aged trees** (different ages and sizes) or in mixed stands composed in part of valuable timber species and in part of commercially unattractive species. Under such conditions, trees are harvested by **selective cutting,** a hunt-and-pick method in which mature trees of quality species are harvested at repeated intervals after being marked in advance with spray paint or some other method. Deformed trees and trash species are removed to upgrade the stand.

The selective-cutting method can be used to harvest single species such as maple, beech, and hemlock, whose seedlings can germinate in the shade of the forest floor. Selective cutting has been used extensively in mixed coniferous–hardwood stands and in deciduous forests (oak, hickory, butternut, and walnut) (Figure 14.12). It is more costly and time-consuming than clearcutting but has many advantages over the latter method. These advantages include the following:

1. It minimizes environmental abuses such as land scarring, accelerated runoff, soil erosion, and wildlife habitat destruction.
2. It reduces blowdown.
3. It decreases the fire hazard because of the volume of slash left after the harvest is reduced.
4. It results in a high rate of natural reproduction.

Selective cutting on a given stand may be done on a 10-year rotation. The entire stand is never completely cut. Trees may be cut individually or in small groups. Sustained-yield management is practiced, the volume of wood harvested during a given year being equal to the volume grown since the previous cutting. A small number of trees are removed per hectare. They are eventu-

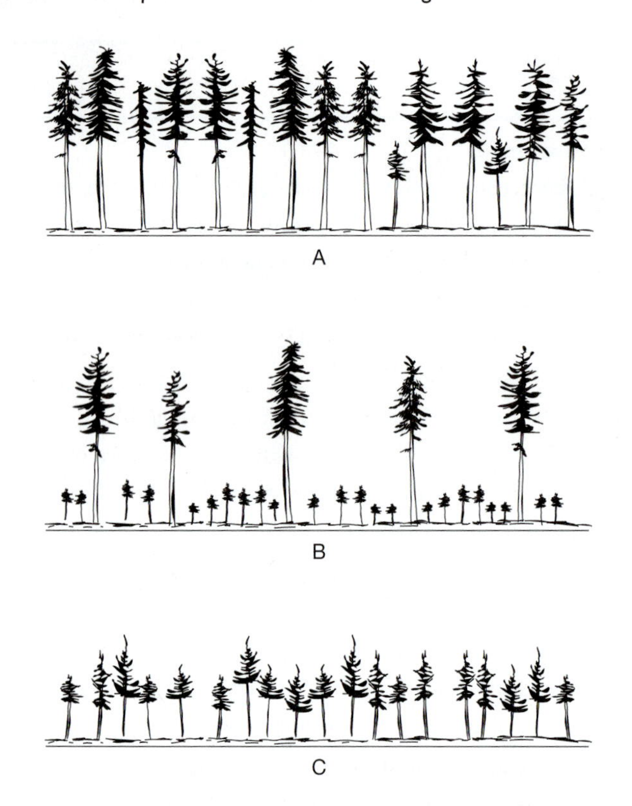

FIGURE 14.11 *Typical shelterwood method. (a) This is a mature tree stand before harvest. (b) The first major cut leaves enough trees as a temporary shelterwood overstory to provide 40 percent crown cover. (c) After tree saplings have become established, the remaining mature shelterwood is cut down.*

ally replaced by natural reproduction, an on-going process because openings are creating indefinitely.

The **strip cutting** method of harvesting timber has been used effectively in the forests of the northeastern United States as well as in tropical rainforests of Central and South America. In our discussion of erosion control on farmland (Chapter 7), strip cropping on the contour was mentioned as a highly desirable technique. Strip cutting is somewhat similar. It is usually used in hilly terrain where clearcutting might result in massive pollution of streams just below the logging site.

During strip cutting, loggers remove narrow strips of forest. Strip cutting has been successfully used in southern pines. A typical technique involves cutting a strip about 80 meters (250 feet) wide. Residual forested strips are left between the cut strips to serve as seed sources, as shown in Figure 14.13.

Strip cutting has several advantages over clearcutting: (1) it minimizes the loss of soil nutrients from the forest; (2) it curbs the pollution of mountain streams with sediment (thus preventing the destruction of spawning sites for trout and other species); (3) it minimizes the ugliness associated with much larger clearcut areas; and (4) it permits more effective reforestation by natural mechanisms.

The Logging Operation

The logging operation includes felling the trees, removing the limbs, and cutting the trunks into logs. The logs are dragged across the forest floor or air-lifted to a central point, where they are loaded onto trucks or train cars. The wood is then transported to a pulp processing plant or sawmill.

Using a chain saw, a logger can cut about 14 cubic meters (500 cubic feet or 7 cords) of saw timber of pulpwood per day. Large companies frequently use **hydraulic shears** and **whole tree chippers.**

The hydraulic shears resemble gigantic pincers. This machine grabs a tree near the base of the trunk and shears it off, cuts the trunk into logs, and loads the logs on a truck or a flatbed.

A whole tree chipper, used by the pulpwood companies, can chew up an entire 50-centimeter (20-inch)-diameter pine trunk, branches and all, into thousands of small wood chips in less than a minute (Figure 14.14). The chips are then blown into a waiting van and hauled to a paper mill.

The chipper has several advantages: (1) it enables the forester to utilize much more of the trees biomass than the 60 percent harvested by conventional methods; and (2) it leaves the forest floor relatively clean, facilitating the

FIGURE 14.12 *Selective cutting of an unevenly aged northern hardwood-hemlock stand. Top: Before selective cutting. Trees to be felled are those with a line drawn through the trunk. Bottom: Same stand 10 years after selective cutting.*

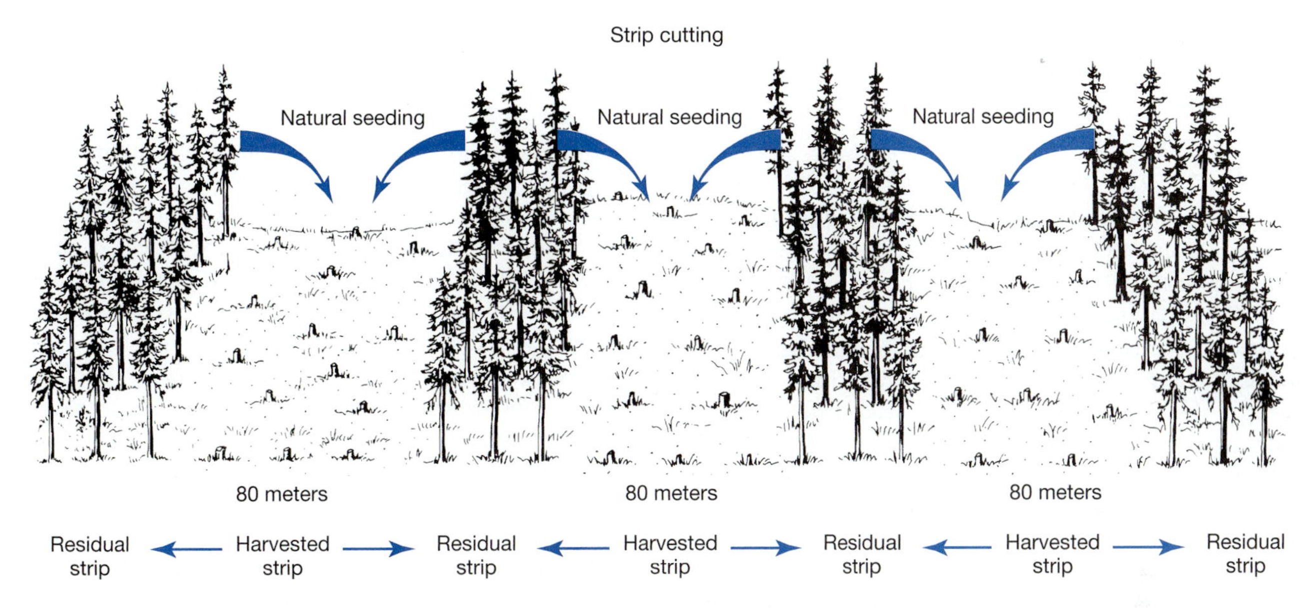

FIGURE 14.13 *Typical strip-cutting pattern. Residual forested strips are left so that natural reseeding can occur in the harvested strips.*

FIGURE 14.14 *A single worker operates equipment that cuts and removes pulpwood trees and transfers them to a chipping machine. The chips can then be trucked to a plant and converted into paper pulp or chipboard.*

growth of new trees and reducing the fire hazard posed by logging-accumulated slash. Unfortunately, this also reduces nutrient replenishment and soil formation, which may come back to haunt companies in the long term.

14.4 Reforestation

Whenever timber is removed, either by clearcutting, strip cutting, or selective cutting, the denuded area must be reforested to ensure a sustained yield. This may be done by natural or artificial methods. Similarly, any forested land that has been destroyed by fire, insects, disease, hurricanes, or strip mining also should be reforested, even though timber may not be its ultimate primary use. In recent decades, millions of hectares in the U.S. have been either seeded or planted for whatever the purpose intended.

Natural Reseeding

After clearcutting, a few mature, wind-firm trees may be left intact as a seed source within the otherwise logged-off site. Scattered by wind and, to a lesser degree, by birds, rodents, and runoff water, the seeds are eventually dispersed throughout the denuded area. **Natural reseeding,** however, is usually not completely adequate. One reason is that some tree species, such as loblolly pine, may have only one good seed-producing year every 2 to 5 years. (In a good year, seed production may be ten times that of a poor year.) Another reason is that the dispersed seeds must reach bare ground to develop properly so that the seedlings can absorb moisture and nutrients from the soil. If they fall on bark, logs, or stones, survival is greatly reduced. Because of these drawbacks, natural reseeding is usually supplemented by aerial, hand, or machine seeding.

Seeding by Foresters

In rugged terrain, **aerial seeding** is the best method. Seeds are sown from planes flying slowly just above the treetops. A helicopter can seed 1,000 hectares (2,500 acres) per day. Unfortunately, many of these seeds fall on infertile soil or are consumed by birds, mice, and squirrels. To minimize losses to animals, the seeds are frequently coated with a toxic deterrent. Except in the case of unusually small seeded trees, such as hemlock and spruce, rodent eradication is virtually a prerequisite to successful seeding.

If a logged-off site is flat, power-driven seeding machines may be used, as has been done in the cutover land of Wisconsin and Michigan. These machines plant up to 3.3 hectares (8 acres) per day, and simultaneously fertilize the soil and apply a herbicide to prevent weed encroachment.

Planting

In addition to bird and rodent problems, a major disadvantage of seeding is the high number of first-year seedlings killed by frost, drought, hot weather, insects, and autumn leaf fall. As a result, seeding, even by artificial methods, is less successful than planting young trees from plantation stock. Moreover, no rodent control is needed. Seedlings are produced under nursery or greenhouse conditions. Of course, successful regeneration requires good seed source, nursery or greenhouse practices, and planting practices. In the South and in the Great Lakes states, trees can be planted at a rate of 150 per worker-hour. On flat land, three workers, a tractor, and a planting machine can set 1,000 to 2,000 trees per hour.

Developing Genetically Superior Trees

The development and use of genetically superior trees is another important forest management technique used in reforestation. Crossing two species of trees, known as **hybridization,** may result in offspring that combine the best traits of the parents. For example, in northern California, plantations of Jeffrey pine were formerly very vulnerable to the attacks of the pine weevil. Economic damage was severe. The problem has been somewhat reduced, however, by crossing the cold-resistant Jeffrey pine with the weevil-resistant Coulters pine. The resultant hybrids were resistant to both cold and the pine weevil. Some other possibilities through genetic engineering are discussed in A Closer Look 14.2.

Research shows that in southern pine stands, selective breeding techniques may increase wood volume by 10 percent, straightness of the trunk by 9 percent, wood density by 5 percent, and resistance to rust (a fungal disease) by 4 percent. The economic gains resulting from selective breeding may be considerable. A forest breeding program in California, for example, resulted in an increased return of $68 per hectare ($27 per acre).

Another way to produce better trees is through **seed orchards,** which produce large quantities of high-quality seeds. Rootstocks are formed by removing the tops of young trees with superior traits. Small branches are then cut from other trees of high quality and grafted onto the rootstocks. The tree that develops from this graft is then crossed with still other trees that are commercially valuable. The seeds from this cross may then be saved for planting in commercial forests.

Another method to produce better trees is through **tissue culture.** In this technique, seeds are collected from superior trees and grown into seedlings. Tiny shoots from these seedlings are then cut off, chopped up, then placed in nutritive

A CLOSER LOOK 14.2

Genetic Engineering: The Key to Tomorrows Superforests

The rapidly developing field of **genetic engineering,** described briefly in Chapter 5, holds promise for dramatic improvements in our nations forests. Hereditary traits are determined by genes. These genes, in turn, are present in the DNA molecules of the chromosomes. In one technique, known as direct DNA transfer, the gene responsible for a given trait, such as rapid growth, can be transferred directly to the DNA of single, isolated tree cells. Each of these cells may then develop into a tree with the ability to grow at an accelerated rate.

What does the future hold? On the basis of recent research in the United States, the following is already or soon may be attainable:

1. Trees that will be distasteful to potentially destructive browsing herbivores, such as deer.
2. Spruce and hardwood (oak, maple) planting stock with a whole series of genetically engineered traits such as accelerated growth, reduced need for fertilizer, improved ability to compete with weeds, and higher survival rates.
3. Pulpwood species (spruce, aspen, birch) with traits that make them much more valuable to the paper industry.
4. Trees that will actually produce their own insecticides to ward off insects, as well as their own herbicides to reduce or eliminate weeds that compete with them for soil moisture and nutrients.

solutions. The cells grow into baby trees complete with roots, shoots, and leaf buds. Because all these individuals are derived from parts of the same parent tree, they have identical hereditary material and are called clones.

14.5 Control of Forest Pests

The most serious agents of mortality and growth loss for most forests are disease and insect pests (Figure 14.15). Under the authority of the **Forest Pest Control Act of 1947,** surveys are conducted annually in both private and public forests to detect pest population buildups so that they can be arrested before they reach disastrous levels.

Diseases

One of the natural causes of mortality in trees is disease. The most numerous and important diseases of forest trees are incited by fungi. These diseases vary in the species and parts of tree affected, in symptoms, and in the type of damage they cause. For example, fungi can cause decay in wood, most of which

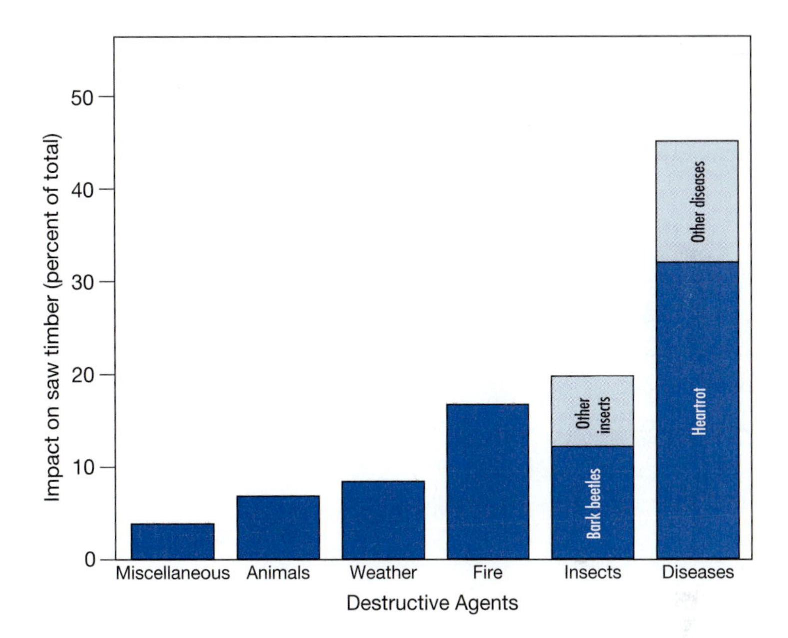

FIGURE 14.15 *Annual impact of destructive agents on saw timber. Saw timber or sawlogs are logs of sufficient size and quality to produce lumber or veneer.*

FIGURE 14.16 *A young western white pine stand in the St. Joe National Forest in Idaho is being sprayed with phytoactin, a blister rust-killing chemical.*

occurs in the central core of dead wood, or "heartwood", of the tree. Such heartwood decay is called **heartrot,** which may make wood unmarketable or lower its quality. Heartwood decay has the greatest effect on the growth of forest trees of all destructive agents, accounting for more than 70 percent of the total loss attributed to diseases of forest trees. Many different species of fungi cause heartrot. (Heartrot fungi, however, are beneficial as important agents in the decay of fallen logs, dead stubs, and slash, the recycling of elements, and the removal of flammable debris.) The remaining disease damage can be attributed primarily to white pine blister rust, dwarf mistletoe, and Dutch elm disease (Figure 14.16). The most injurious diseases are exotics, accidentally introduced into the United States, that have suddenly been released from environmental factors that ordinarily kept them in check in their native habitat (Table 14.2).

Insects

Insects account for about 20 percent of all timber destroyed, ranking second to diseases as agents of forest damage. Each species of tree has its own unique assemblage of insect pests. An oak tree may be eaten by more than 100 species. No part of a tree is spared (Table 14.3). A healthy tree can withstand the nibbling of insects, but if trees are already stressed because of drought, crowding, or pollution, or a combination of the three, they may be seriously damaged by insects.

Bark Beetles **Bark beetles** are among the most devastating insects of forests in the world. Adult beetles attack a tree by boring through the bark, then tunneling out egg chambers and galleries with their powerful jaws. The tiny grubs that hatch from the eggs consume the soft inner bark, and, if sufficiently numerous (1,000 per large tree), may actually girdle the tree

TABLE 14.2 *Common Fungus Diseases of Forest Trees: Damage and Control*

Common Name	Host	Symptoms	Control
Root and butt rot	Species of pine, mature stands, and plantations 15–25 years old	Decline in vigor; needles short, becoming yellow and then dying; cones appear prematurely; fruiting bodies of fungus at base of tree or on ground surface arising from roots	Avoid planting white or red pine on poorly drained or high-pH soils
Western red rot	Ponderosa pine	None, except presence of decay and red discoloration in heartwood	Periodic pruning of dead branches on crop trees
Brown spot	Longleaf pine	Small spots on needles causing needle dieback; retards growth of seedlings	Controlled burning in third year of seedling's life or spray with prescribed chemicals
Oak wilt	Oak species, particularly those of the black oak group	Leaves crinkle and become pale green, later turning brown or bronze; mature leaves shed at any symptom stage; lower branches affected last; trees usually die after the summer symptoms appear	Some promise of control has been obtained by cutting or poisoning healthy oaks for 50–100 ft around spot infections Mechanical or chemical root barriers
Dutch elm disease, associated with attacks by scolytid beetles	American elm	Progressive dwarfing and yellowing of leaves, accompanied by various degrees of defoliation; followed by death of branches or entire tree	Mainly through destruction of infected trees and protection against bark beetles by insecticidal sprays and maintenance of tree vigor by appropriate pruning and feeding practices Lignasan injection

TABLE 14.3 *Common Insect Pests of Forests: Damage and Control*

			Control Method		
Name	**Principal Species Affected**	**Type of Damage**	**Prevention**	**Direct Control**	**Control Season**
Bark beetles	Pines	Girdles tree by killing cambium layer under bark	Salvage green blown-down timber	Feel, peel, and burn bark; salvage infested logs; burn slabs	Spring, summer, fall
Gypsy moth	Oaks, birch, aspen	Defoliation		Spraying insecticide	May–June
Sawflies	Eastern and southern pines and tamarack	Reduces growth by defoliation; epidemics kill stands	Cut mature and overly mature stands	Aerial insecticide spray	Early summer in period of greatest activity
Spruce budworm	True firs, Douglas fir	Reduces growth by defoliation; kills older trees extensively	Cut mature and overly mature stands	Aerial insecticide spray	Early summer as insects emerge
White pine weevil	Eastern white pine, Norway spruce	Kills leaders, causes forked and crooked boles	Maintain shade where possible to 20 ft	Hand spray upper stem and new growth with sodium arsenate or cut and burn infected tip; leave only best side branch on first lower whorl	Early summer as insects emerge

and kill it within a month. The bark beetle group includes a large number of destructive species.

Pine beetles kill trees by transmitting a fungus that develops inside the tree. The fungus plugs up the vessels that transport water to the leaves, branches, trunk, and roots. The western pine beetle killed 59 million cubic meters (more than 2 billion cubic feet) of ponderosa pine along the Pacific Coast between 1917 and 1943. The mountain pine beetle has done extensive damage to sugar pine, western white pine, and lodgepole pine in California. The southern pine beetle, the most damaging insect in the southeastern U.S., is responsible for killing large areas of southern pine.

Overharvesting of trees can result in thick regrowth. These trees are less able to ward off insects like bark beetles. Thinning trees, therefore, effectively controls the beetles.

On a long-term basis, perhaps the best method of control is to prevent infestations from occurring in the first place. This may be done by using **sanitation techniques:** burning all potential bark beetle breeding sites, such as senile and wind-blown trees, and logging-accumulated slash and debris and the broken-off trunks of lightning- and fire-killed trees. (In removing those trunks, however, the forester is also removing the potential nesting cavities of beetle-eating woodpeckers and is removing nutrients that would be returned to the forest floor, so a cost–benefit analysis would have to be made.) Insecticides can be applied by planes that fly low over the infected area.

Integrated Pest Management In the late 1970s, certain environmental groups became concerned over the large amounts of chemical pesticides used by the Forest Service to control insect outbreaks. This criticism reached its peak when the Forest Service used DDT (banned except for emergencies) to control an explosion of tussock moths in conifer stands of the Northwest. Ecologists considered this strategy particularly questionable because ordinarily the tussock moth population is held in check by a death-causing virus. Eventually the Forest Service, stung by criticism from many quarters, decided to adopt the **integrated pest management (IPM)** policy, a concept described more fully in Chapter 8.

With IPM, methodologies from several disciplines are integrated in an ecological manner to control a pest or pests. Among the IPM strategies now used by the Forest Service are the following:

1. Use of biological control agents, such as natural predators and parasites.

2. Use of selective cutting methods rather than clearcutting methods (because even-aged trees are highly vulnerable to insect attacks).
3. Removal of bark-damaged trees because they serve as focal points of insect infestation.
4. Growing heterotypes rather than monotypes whenever feasible (see A Closer Look 14.3).

14.6 Fire Management

Wildfires

In the early part of the twentieth century, there were single years when there were more than 100,000 forest wildfires in the United States. They burned more than 27,500 square kilometers (10,450 square miles) of forest, an area the size of Maryland. However, because of better fire control techniques, the total acreage of destroyed timber in the U.S. has been reduced in recent years. Nevertheless, occasional serious wildfire outbreaks do occur, especially during periods of extended drought (Figure 14.17). Consider 1988, for example. In that year, the most severe drought in half a century transformed much of our nations timber into kindling wood. Lightning strikes set the kindling ablaze. By midsummer, scores of forest fires were raging in Alaska, Idaho, California, Oregon, Colorado, Utah, Wyoming, and Wisconsin. The Forest Service called the summer of 1988 the worst fire season in 30 years. The fires charred at least 1.5 million hectares (3.7 million acres), an area larger than Connecticut. The largest burn in Colorado's history destroyed more than 7,300 hectares (18,000 acres) of prime habitat for elk and deer. The Black Hills blaze in South Dakota scorched 7,000 hectares (16,500 acres), forced the evacuation of 1,000 campers, and smoke-shrouded the presidential shrine at Mt. Rushmore.

Most of the publics attention, however, was focused on the devastation in **Yellowstone National Park.** Fire boss Fred Roach, who had battled such blazes for two decades, told reporters that he never saw anything as awesome as this! Twenty-five hundred army personnel were flown to the scene to help contain the fire. President Ronald Reagan became so concerned that he sent Agriculture Secretary Richard Lyng, Interior Secretary Donald Hodel, and Deputy Defense Secretary William O. Taft to Yellowstone to assess the destruction. Eventually, in early September, as a result of cooler weather,

FIGURE 14.17 *Wildfire in the Ochoco National Forest, Oregon.*

A CLOSER LOOK 14.3 Controlling Insect Outbreaks with Heterotypes

Insect pest populations sometimes build up to impressive numbers. Charles Kendeigh, an ecologist at the University of Illinois, studied a severe spruce budworm outbreak in Ontario. The larvae, which were feeding on coniferous foliage, occurred in such vast numbers that their excreta sounded like drizzling rain as it fell on the forest floor. More recently, an explosion of forest tent caterpillars virtually destroyed 11,000 hectares (26,850 acres) of water tupelo in the region of Mobile Bay, Alabama. Aerial photographs revealed almost complete defoliation of extensive stands of once-healthy trees.

Is it possible to control such outbreaks without using expensive insecticides that may have adverse effects on the forest ecosystem? One possible alternative method is to establish forest **heterotypes,** as suggested by Kenneth Watt, an ecologist at the University of California–Davis. When monoculture is practiced in a large area, populations of forest pests tend to fluctuate widely. Over an evolutionary period of millions of years, certain species of insects have specialized to feed on a particular species of tree; examples are spruce budworm on balsam fir, larch sawfly on larch, gypsy moth on oak, and pine-back beetle on pine. Many also feed on a particular kind of tree at a particular stage (seed, seedling, sapling, or mature tree) in its life cycle. In natural forest ecosystems, the particular food tree to which a species of insect has become adapted may be widely dispersed. An insect, such as a pine-bark beetle, for example, might have to creep, crawl, or fly a considerable distance from one food tree to another. In the process, of course, such an insect becomes vulnerable to predation, windstorms and rainstorms, fire, and other mortality factors. Even if it survives, it may eventually starve to death before it is able to find another food tree. Conversely, in an artificial forest ecosystem, as represented by a monotype, the pine-bark beetle is surrounded not only by the right species of food tree, but also one of the precise age that the beetles prefer. The severity of forest insect outbreaks can be substantially reduced simply by breaking up the large contiguous stands of single-age, single-species trees into small, isolated stands interspersed with trees of different kinds and ages.

snow, reduced wind velocities, and the heroic efforts of thousands of firefighters, the fires went out.

In the end, eight separate fires had burned about 45 percent of the park's 890,000 hectares (2.2 million acres). In nearly half of the burned area, the trees were not killed. Often not discussed is the fact that more than half of the burned area resulted from three fires started by human beings on which suppression action was taken from the very beginning. In addition, two other fires had burned into the park from adjacent lands.

Since the Yellowstone event, fires have continued to ravage the West, in large part because of a prolonged drought and previous fire-prevention programs that protected forests yet permitted dead timber and debris to build up on the forest floor, which greatly increases the severity of fires.

About 15 percent of our nations forest fires are caused naturally (for example, by lightning). The rest are caused by people, either deliberately or accidentally through carelessness.

Firefighting

The actual suppression or attack pattern used by firefighters varies greatly, depending on the size of the fire, terrain, type of fire, wind direction, location of roads, availability of water, and relative humidity. A variety of fire suppression methods are used, including firebreaks, back fires, fire-retarding chemicals, and infrared systems (Figure 14.18).

Components of the infrared system include computers, satellites, satellite dishes, and video cameras. This system can pinpoint the location of a wildfire with considerable accuracy. Moreover, it can track the movements of a windblown fire. Because such movement can be highly erratic and could trap unsuspecting firefighters, infrared systems have the potential to save lives.

In rugged mountainous country, smoke jumpers may parachute into the burn area. The national forests of mountainous regions in Washington, Oregon, California, Montana, Idaho, and New Mexico are protected by Forest Service smoke jumpers. With their help, many remote blazes that in 1930 might have burned out of control may now be extinguished within hours.

FIGURE 14.18 *A U.S. Forest Service plane releases a fire-suppressing chemical on a wildfire in the Ozark National Forest, Arkansas.*

Use of Controlled Fires

Since the 1930s, the Forest Service has used the symbol of Smokey Bear to alert the American public to the highly destructive effects of wild forest fires. In recent years, however, the service has come to realize that some wildfires are ecologically beneficial. In fact, it is now believed that fire plays an important role in maintaining many economically important timber stands. Examples include the old growth, even-aged stands of Douglas fir in the Northwest, red pine in Minnesota, and white pine in Pennsylvania and New Hampshire. Certain less valuable species, such as pitch pine on sandy soils near the mid-Atlantic Coast and jack pine in the lake states, are also considered to be **fire types** or **fire climaxes.**

Today foresters in many countries use controlled or prescribed burns to improve the quality of timber, livestock forage, and wildlife habitat (Figure 14.19). A **controlled burn** is a fire that is purposely ignited by highly trained foresters for specific purposes. It has a low flame and moves slowly along the forest floor. Forest managers must be extremely cautious when performing controlled burns. They must be certain that the wood is not too dry and that the wind is neither too strong nor headed in the wrong direction.

About 0.8 million hectares (2 million acres) of pine stands in the South are control-burned annually. At the Alpha Experimental Range in Georgia, controlled burning is conducted only in the afternoon, under damp conditions, and the relatively cool, creeping fire is usually extinguished by nightfall. Any pine stand 2 to 4 meters (6.5 to 13 feet) high can be control-burned because trees of this size are protected from flames by their corklike bark. The highly resistant longleaf pine (whose terminal bud is protected by a group of long needles) can be burned without ill effect when the seedlings are only 15 centimeters (6 inches) high.

In the longleaf slash-pine stands of the South, controlled burning (1) reduces the crown-fire hazard by removing highly combustible litter, (2) prepares the forest soil as a seedbed, (3) increases the growth and quality of livestock forage, (4) retards a forest succession leading to a low-value scrub oak climax and maintains the high-timber-value pine subclimax, (5) promotes legume establishment and resultant soil enrichment, (6) increases the amount of soluble mineral ash (phosphorus and potassium) available to the forest plants, (7) stimulates the activity of soil microorganisms, (8) controls the brown-spot needle blight, a fungus that is highly destructive to longleaf pine seedlings, and (9) improves food and cover for wildlife and quail.

FIGURE 14.19 *A controlled burn in a forest in Nepal. Note the low flame.*

The "Let-It-Burn" or "Prescribed Natural Fire" Policy

In 1972 the National Park Service established a policy that utilizes wildfires in 17 national parks, including Grand Teton (Wyoming), Rocky Mountain (Colorado), Sequoia (California), Yosemite (California), and Yellowstone (Wyoming). Under this policy, wildfires are permitted to burn under careful surveillance. This **"let-it-burn"** (or let-burn) **policy** is also implemented by the Forest Service in the wilderness areas of our national forests. Since the phrase "let-it-burn" more appropriately describes what the fire does and not what the management strategy entails, many people prefer to use **"prescribed natural fire"** to reflect the monitoring and decision processes associated with these naturally ignited fires. This policy is quite a turnabout from the Park Service's policy between 1920 and 1972, when wildfires were considered the forests prime evil and all were to be fought. Periodic natural fires help revitalize old growth forests by burning up dead trees and debris and killing off diseased trees. In turn, this encourages new plant growth.

The massive Yellowstone fires in 1988, the most severe in the parks 112-year history, forced the National Park Service and Forest Service to reevaluate their prescribed natural fire policy. They admit, for example, that when the present fire management plans were established in the early 1970s, no provision was made for a firestorm of the 1988 variety. After all, data from fire-scarred trunks that are centuries old indicate that a conflagration on such a scale would occur only once in 150 to 200 years (Figure 14.20)! In the aftermath of the Yellowstone fires, then-Interior Secretary Donald Hodel suggested that the prescribed natural fire policy be scrapped. However, after reappraisal, it was decided that the policy would be retained, except under conditions of extreme drought like that of 1988. Still, some feel that the natural fire policy procedures now are too conservative, because of the greatly expanded checklists that in too many instances result in a decision to suppress a fire.

Even if the natural fire policy had not been in place, it is estimated that 25 to 30 percent of Yellowstone National Park would have burned anyway. Today, almost 14 years after the big fires, forests are reborn and wildlife blooms in the part of Yellowstone once described as destroyed. Many ecologists and biologists now believe that the unusually large fires of 1988 are in the normal range for a lodgepole pine ecosystem and that these fires will lead to a significant increase in the number of Yellowstone's plant species.

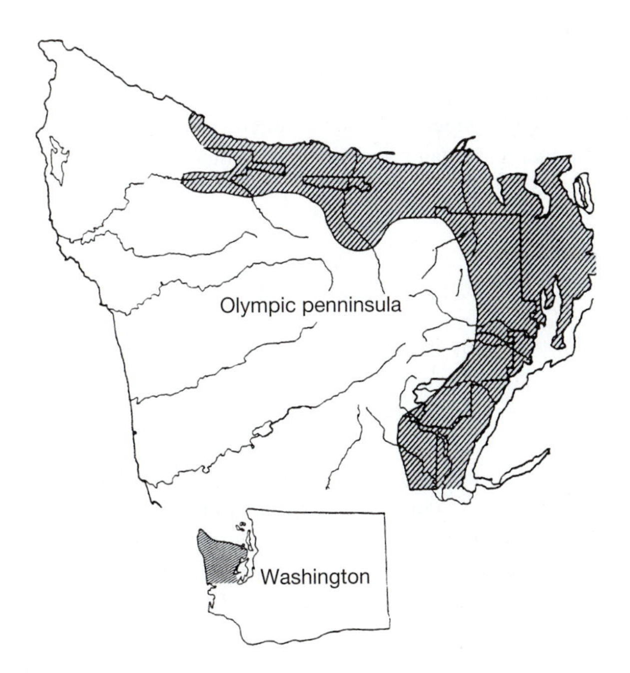

FIGURE 14.20 *Estimated area of a single fire or series of fires around 1700 in the eastern Olympic Mountains of Washington. This fire (or fires) burned from 1.2 to 4 million hectares (3 to 10 million acres) of forest land. Although we really do not know, such widespread fires were probably weather-driven by large lightning storms rather than fuel-driven. Other large fires in the western Olympics appear to have occurred around 1230 and again in 1480.*

14.7 Meeting Future Timber Demands

Humans currently consume about 5 billion cubic meters (177 billion cubic feet) of wood annually from a total world forest growing stock of 350 billion cubic meters (12.4 trillion cubic feet). With the anticipated growth in the human population, demand for wood is expected to increase to 6.6 billion cubic meters per year by 2025, an increase of almost 33 percent. In the next 20 years this global demand for wood corresponds to an average increase of about 80 million cubic meters annually, about half of the total annual timber harvest in Canada. And Canada accounts for about 10 percent of the world's forests, and supplies 50 percent of the world export trade in softwood lumber and more than 50 percent of the world trade in newsprint.

It is questionable whether this rate of increase in forest harvesting can continue. In fact, some people believe that the world will face conditions of timber famine within the next 20 years. Although this dire prediction has been made many times before, it has not been fulfilled. The increasing demand for wood has always been satisfied by technological developments, exploitation of areas of previously unexploited forest, and utilization of tree species previously considered unusable. However, because of increasing world population, deforestation in developing countries, and the establishment of new plantations lagging far behind the rate of forest depletion in developing countries, this prophecy may finally come true.

Turning our attention to home, by the year 2020 Americans may be using 50 percent more timber than they do now. Because our projected annual supply of timber for the year 2020 may be only half a billion cubic meters (18 billion cubic feet), we could face an annual deficit of one-quarter billion cubic meters (9 billion cubic feet) (Figure 14.21).

Theoretically, there are several ways in which the increasing demand for timber may be satisfied. Unfortunately, some of the solutions are fraught with economic or ecological problems. Among them are the following:

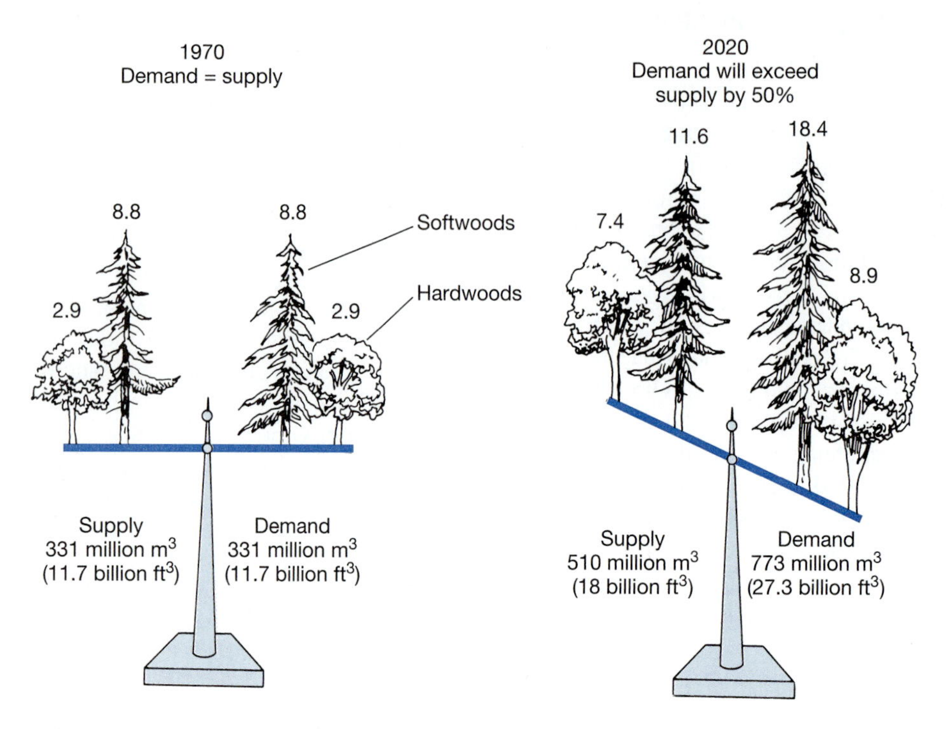

FIGURE 14.21 *In 1970, timber demand equaled supply in the United States. However, by 2020, timber demands may exceed supply by 50 percent. Several options are possible. We can cut demand by building smaller homes. We can recycle paper and cardboard. We can better manage forests to improve yield. We can import more timber.*

1. Upgrading and extending forest management in all forests, public and private, large and small. Where can the greatest gains be made? Primarily on small, privately owned woodlots. Of the 4.5 million small forests in the United States, over 50 percent are under 12 hectares (30 acres) each. Yet these small-forest owners control over three times as much timberland as the U.S. Forest Service. Until now, few small-forest owners have cut trees on their land for commercial sale.
2. Making more effective use of wood residues and weed species of trees. In a given year, more than 30 million cubic meters (1 billion cubic feet) is left as unused residue. (Of course, eventually this material will decompose and the resulting nutrients will serve to enrich the soil for the growth of future generations of trees.) Another 30 billion cubic meters (1 billion cubic feet) of wood is left as milling residue. Both types of residue could be used to manufacture wood products (see A Closer Look 14.4). It may well be, however, that the value of such residues and weed species as a source of fuel will exceed their value as wood products.
3. Developing superior (faster-growing, better-grained, and disease-, insect-, fire-, and drought-resistant) trees through the techniques of grafting, hybridization, and genetic engineering.
4. Increasing the use of wood substitutes. Plastic and other materials might be substituted for wood in packaging. The fibrous waste left over after the juice has been extracted from sugarcane can be used to manufacture paper. It must be emphasized, however, that the manufacture of some of these substitute materials is dependent on the intensive use of energy from oil fuels—a source that is gradually being depleted. Moreover, the extensive use of plastics is fraught with problems. Plastics do not decompose easily and therefore pose a serious waste disposal dilemma.
5. Increasing imports.
6. Recycling paper to reduce demand.
7. Reducing demand by building smaller houses or reducing paper packaging.

14.8 Preserving Wilderness

From time to time in our nations history, far-sighted environmentalists have urged that the federal government set aside large acreages of pristine forest as wilderness preserves. One of the most influential of these was **John Muir.** An immigrant from Scotland, he fell in love with the forests of America. He tramped through them with endless fascination. In 1892 he founded the **Sierra Club,** one of our nations most active environmental organizations. A magnificent redwood stand just outside San Francisco was named the John Muir National Monument in his honor.

For several decades Muir's crusade met only apathy. Truly, his seemed to be a "voice crying in the wilderness." However, in the mid-1900s a surge of environmental awareness and responsibility moved through the American public. This awareness was fostered by the writings of such people as Aldo Leopold and Rachel Carson. At long last the U.S. Congress passed the **Wilderness**

A CLOSER LOOK 14.4 *Forest Conservation by Efficient Utilization*

During the cut-it-and-get-out logging operations of the 1890s, lumbermen were interested only in logs. The rest of the tree—the stump, limbs, branches, and foliage—was left in the forest, where it frequently served as tinder for a catastrophic fire. (Of course, much of this biomass decomposed, and the nutrients were released into the soil for new growth.) Further waste occurred at the sawmill, where square timbers were fashioned from round logs. Slabs, trimmings, bark, and sawdust were hauled to the refuse dump and burned. To help meet the growing demand for wood and wood products, it is necessary to practice even more conservation after trees are harvested and removed from the forest.

A definite trend toward more efficient utilization of our timber resource by the U.S. wood and wood-products industry is under way. Whereas early in the timber industry wood had only two primary uses, as lumber or as fuel, through the efforts of the Forest Products Laboratory and similar research centers, a number of ingenious methods have been developed for utilizing almost every part of the tree, including the bark.

The forest industry has become much more diversified. Whereas in 1890 almost 95 percent of the forest harvest was converted into lumber, today only about 30 percent of the harvest consumed by Americans is fashioned into boards and timbers. In the last few years, techniques have been developed for making extremely useful products from scrap boards, shavings, wood chips, bark, and sawdust. For example, years ago, extremely short boards went to the scrap heap. Today, however, thanks to superior waterproof glues, such boards can be joined to form structural beams of almost unlimited length. Much wood that was once considered worthless is now converted into thousands of small chips. These chips are then compressed into sturdy and durable chipboard or hardboard.

Act of 1964, one of the most effective tools in the legislative arsenal for preserving the quality of our nations environment.

The purpose of this act was to preserve primitive areas in their natural state. The act designated such primitive areas in national forests, national parks, and national wildlife refuges as possible candidates for inclusion in a **National Wilderness Preservation System.** Congress defined its concept of **wilderness** in the following words:

> A wilderness, in contrast with areas where man and his own works dominate the landscape, is hereby recognized as an area where the earth and its community of life are untrammeled by humans, where humans themselves are visitors who do not remain.

Congress described a wilderness as having four major characteristics:

> (1) It generally appears to have been affected primarily by the forces of nature, with the imprint of the work of humans substantially unnoticeable. (2) It has outstanding opportunities for solitude or a primitive type of recreation. (3) It has at least 5,000 acres of land, except in the case of islands, which might have a smaller area. (4) It may also contain ecological, geological, or other features of scientific, educational, scenic, or historical value.

Access to an official area within the National Wilderness Preservation System can be made only by trail or canoe. All motorized vehicles or boats are prohibited, as are roads, buildings, and any commercial activities. Grazing, timber cutting, mining, and drilling for oil are also prohibited except for those parties who had filed claims by 1983. Human recreation is restricted to such quiet forms as hiking, camping, bird-watching, studying rock formations, identifying wildflowers, canoeing, and sport fishing.

The 1964 act designated 54 wilderness areas with 3.6 million hectares (9 million acres) of Forest Service land. Congress started expanding the wilderness system in 1968. As of the year 2000, the National Wilderness Preservation System included 633 wilderness areas (Figure 14.22), totaling 42.5 million hectares (more than 105 million acres), of which 56 percent is in Alaska. They are managed by four government agencies: the National Park Service (42 percent), the U.S. Forest Service (33 percent), the Fish and Wildlife Service (20 percent), and the Bureau of Land Management (5 percent).

Why is our wilderness system important? Today, when urban dwellers from Seattle to Miami and from San Diego to Boston are being crowded together shoulder to shoulder, when urban air is polluted with industrial gases and automobile fumes, when drinking water tastes of chlorine, and when noise assaults the ears in so many areas, it is reassuring to know that somewhere in the great forests of America is wilderness. And once you get there, you can hike or canoe for miles, free of modern civilization.

Of the 77 million hectares (191 million acres) of national forest in the United States, about 40 million hectares (51.5 percent) are roadless. About 13 million hectares of the roadless forests have been included in the National Wilderness Preservation System. However, an ongoing controversy has been raging concerning the designation of the remaining 27 million roadless hectares (see A Closer Look 14.5, The Wilderness Controversy). Environmentalists, led by organizations such as the Sierra Club, the National Audubon Society, and the Wilderness Society, would like most or all of the roadless acreage to be included in the National Wilderness Preservation System. In the name of progress, resource developers lobby elected officials and government agencies to build roads in these areas to prevent them from being designated wilderness in the future.

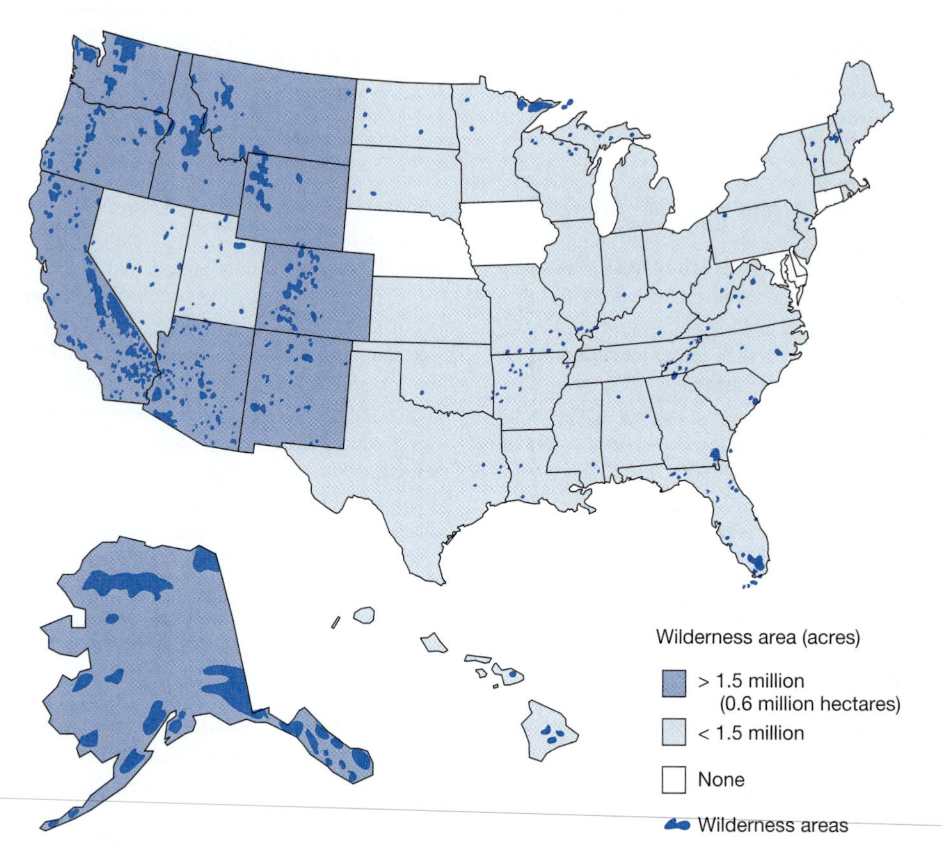

FIGURE 14.22 *Distribution of wilderness areas in the National Wilderness Preservation System in the United States.*

Under the terms of the Wilderness Act, obtrusive activities such as logging, mining, and the use of automobiles, motorboats, and snowmobiles are prohibited in designated wilderness. In some areas, even aircraft are not permitted to descend below a designated altitude. In an officially designated wilderness area, every attempt is made to permit the forest ecosystem to operate without human interference. If an overly mature pine riddled with bark beetle blows down during a windstorm, it remains where it falls. Barring a major catastrophe, such as a crown fire, the official policy is to let nature take its course. As a result, these areas have considerable scientific as well as recreational value. In such an area, university researchers can make observations, collect data, and formulate hypotheses, and college students can acquire valuable field experiences in geology, entomology, mammalogy, ornithology, ecology, soil science, field natural history, game management, and forestry, as well as other fields.

14.9 Protecting Natural Resources: National Parks

Much of our nation's forestland is located in **national parks** and administered by the National Park Service. In addition to the preservation and appropriate use of these woodlands, our national park system is committed to protecting other national resources such as wildlife and scenic beauty as well.

Brief History

The hunters and trappers that brought back game and furs from the rugged wilderness country of northwestern Wyoming also brought back tales of wondrous sights and sounds: exploding geysers that shot a stream of water 200 feet in the air; an awesome gorge almost a quarter-mile deep; dozens of plunging waterfalls; multicolored springs that bubbled water hot enough to heat your morning coffee; vast forests of spruce, fir, and pine; snow-crowned mountain peaks; and huge herds of antelope, elk, and deer. These accounts seemed almost unbelievable. However, U.S. Army General Henry Washburn led an expedition to Wyoming and found them to be true. Interest in the region grew rapidly. Ultimately, a Montana judge by the name of Cornelius Hedges suggested the area be converted into a national park so that all Americans could thrill to the sights and sounds of this natural wonderland. On March 1, 1872, President Ulysses S. Grant signed legislation that made **Yellowstone** the first national park in the United States—and in the entire world.

The national park concept seemed so attractive to Congress that several more were added in the 1890s, including Yosemite (Figure 14.23), Sequoiah, and Mt. Ranier. In 1916 the **National Park Service** was established as a bureau within the Department of the Interior. Congress gave the new agency the following mandates:

1. National parks are to be maintained in absolutely unimpaired form for the use of future generations.

A CLOSER LOOK 14.5 *The Wilderness Controversy*

Nearly 42 million hectares (104 million acres), or 4.6 percent of the total land area of the United States, has been officially designated as part of our National Wilderness Preservation System. However, more than 55 percent of this wilderness is in Alaska. In the contiguous 48 states, less than 2 percent of the land area is so designated. Government agencies, in reviewing remaining roadless areas in public lands, found that an additional 5 percent could qualify for wilderness designation. A spirited controversy now rages concerning the advisability of expanding our National Wilderness Preservation System. Conservationists say we don't have enough wilderness areas, whereas many ranchers and officials of the timber, mining, and energy industries operating on public lands say we already have too much. Lets examine arguments for and against wilderness expansion.

Arguments For Expansion

1. Access to wilderness provides temporary escape from the hustle and bustle of modern society. The awesome grandeur of the wilderness replenishes the human spirit.
2. Wilderness is an insurance policy that protects natural ecosystems against the type of disruption and degradation that occurs in areas not so designated.
3. Wilderness preserves biological diversity. It prevents the extinction of endangered species of plants and animals. It provides essential habitat for mountain lion, bear, moose, elk, bighorn sheep, and deer, and it protects the spawning beds of salmon and trout.
4. It preserves the inherent right of all creatures, both human and nonhuman, to survive.
5. It supports thriving outdoor gear, camping supply, and tourist industries.
6. Supplies of cattle forage, timber, copper, oil, and natural gas that are available in nonwilderness areas are completely adequate to support our economy and ensure our national security. In addition, potentially new wilderness areas (in the 5 percent that qualify) would not add much to the country's timber, mineral, and energy resources.

Arguments Against Expansion

1. The current interest in wilderness is a fad and will soon dissipate.
2. Wilderness use by visitors is only a fraction of the visitor use in our national parks.
3. Wilderness is just a playground for the wealthy members of our society.
4. The designation of wilderness lands has caused a substantial employment decline in the cattle, logging, and mining industries.
5. Resources like grasslands, timber, coal, copper, oil, and natural gas should not be locked up in wilderness areas. They should be fully utilized to promote our nations economic health and security. After all, the original vegetation quickly becomes reestablished soon after grazing, logging, and mining have ceased.
6. All of our public lands, including so-called wilderness, should be managed according to the multiple-use concept.

After reflecting on the arguments pro and con, do you think our nations wilderness system should be expanded?

---

2. They are to be set apart for the use, observation, health, and pleasure of the people.
3. The national interest must dictate all decisions affecting public or private enterprise within the parks.

Today the National Park System comprises 383 diverse units (Table 14.4). Besides national parks, they include national monuments, national historical sites, national recreation areas, national memorials, the White House, and more. This system embraces almost 34 million hectares (more than 83 million acres). Units of the national park system are located in every state (except Delaware), as well as in the District of Columbia, American Samoa, Guam, Puerto Rico, the Mariana Islands, and the Virgin Islands (Figure 14.24). Lets briefly examine some of the unit categories:

1. *National parks.* Our national parks are usually extensive acreages whose major function is the preservation of scenic grandeur, wilderness, and wildlife. Examples include Yellowstone, Yosemite, Glacier, and Grand Teton national parks. (Worldwide, there are more than 1,000 national parks today in more than 120 countries.)
2. *National monuments.* These units range in size from the Statue of Liberty in New York Harbor to the 800,000 hectare (2-million-acre) Death Valley National Monument in east-central California. Death Valleys Badwater region, at 85 meters (280 feet) below sea level, has the lowest elevation on the North American continent.

TABLE 14.4 *Categories of the National Park System. As of the year 2000, there were a total of 383 units in the National Park System.*

International historic sites	National parks
National battlefields	National parkways
National battlefield parks	National preserves
National battlefield sites	National recreation areas
National historic sites	National reserves
National historical parks	National rivers
National lakeshores	National scenic trails
National memorials	National seashores
National military parks	National wild and scenic rivers
National monuments	Other designations

FIGURE 14.23 *Yosemite Falls plunges down a rock wall to the valley below. The scenic crown of Yosemite National Park in California is its 739-meter (2,425-feet) plunge, which ranks it among the tallest falls in the world.*

3. *National recreation areas.* The beautiful Snake River that flows through Oregon and Idaho has cut Hells Canyon—the deepest gorge on the North American continent. For many years the utility industries did their best to get permission to build hydropower dams across the river. These strenuous attempts, however, were forever thwarted in 1975 when Congress established the Hells Canyon National Recreation Area. Not only does this region boast unsurpassed scenic grandeur, but sizable populations of trout swim in the Snake River itself.

   Two examples of national recreation areas that are located in heavily populated urban areas, bringing parks closer to people, are the Golden Gate National Recreation Area, near San Francisco, and the Gateway National Recreation Area, near New York City.
4. *National lakeshore and seashore areas.* One of our most popular national seashores is Padre Island, off the cost of Texas. It provides more than 109 kilometers (68 miles) of white sand beaches along the Gulf of Mexico. Among the varied recreational activities provided are marine bird-watching, swimming, surfing, sailing, and horseback riding. Padre Island is inundated with college students every year during their spring break.
5. *National historical sites.* These units include both national historic sites and national historical parks. The major function of these sites is to commemorate some significant event in the history of our nation. For example, Valley Forge National Historical Park was established when President Gerald Ford signed a bill on July 4, 1976, to celebrate the bicentennial of the American Declaration of Independence. It includes the house where George Washington lived in 1760 as well as the site of the colonial encampment in 1777–1778.

How Is Acreage for National Parks Acquired?

Early in the history of the national park system, new parks were established on lands already owned by the federal government. These parks were usually quite large. They had as their major function the preservation of scenic beauty, wilderness, and wildlife. However, in the past few decades most units added to the national park system have been purchased from private individuals, have been located east of the Mississippi River, are relatively small, and serve as monuments, historical sites, and recreational areas. In 1965 Congress passed the **Land and Water Conservation Act.** Under this act federal money is provided for the purchase of lands that will be designated as units of the national park system. More than 128 units have been purchased in this way, at a cost of more than $2 billion.

Sometimes land for the national park system is donated by private individuals, a corporation, or even a state. For example, in 1984 the state of New York donated 5 hectares (12.5 acres) of an attractive beach to the Fire Island National Seashore. The wealthy Rockefeller family donated the major portion of what is now Acadia National Park in Maine. They also donated the exquisite ocean beaches of what is now the Virgin Islands National Park on St. Thomas Island.

Preservation of Natural Beauty

A major function of our national parks is preservation of the natural beauty of our mountains, forests, canyons, waterfalls, wetlands, lakes, and streams (Figures 14.25). However, the concept underlying our national park system has changed considerably from that visualized by Congress when it established Yellowstone Park back in 1872. In that era the concept pretty much signified hands off, letting alone, or protecting completely from interference by humans. However, in those days our knowledge of ecology and the principles of biological succession and carrying capacity of habitat was quite limited.

Today, scientists know considerably more. They realize that if the National Park Service is to preserve plant and animal communities in their original form, some human manipulation

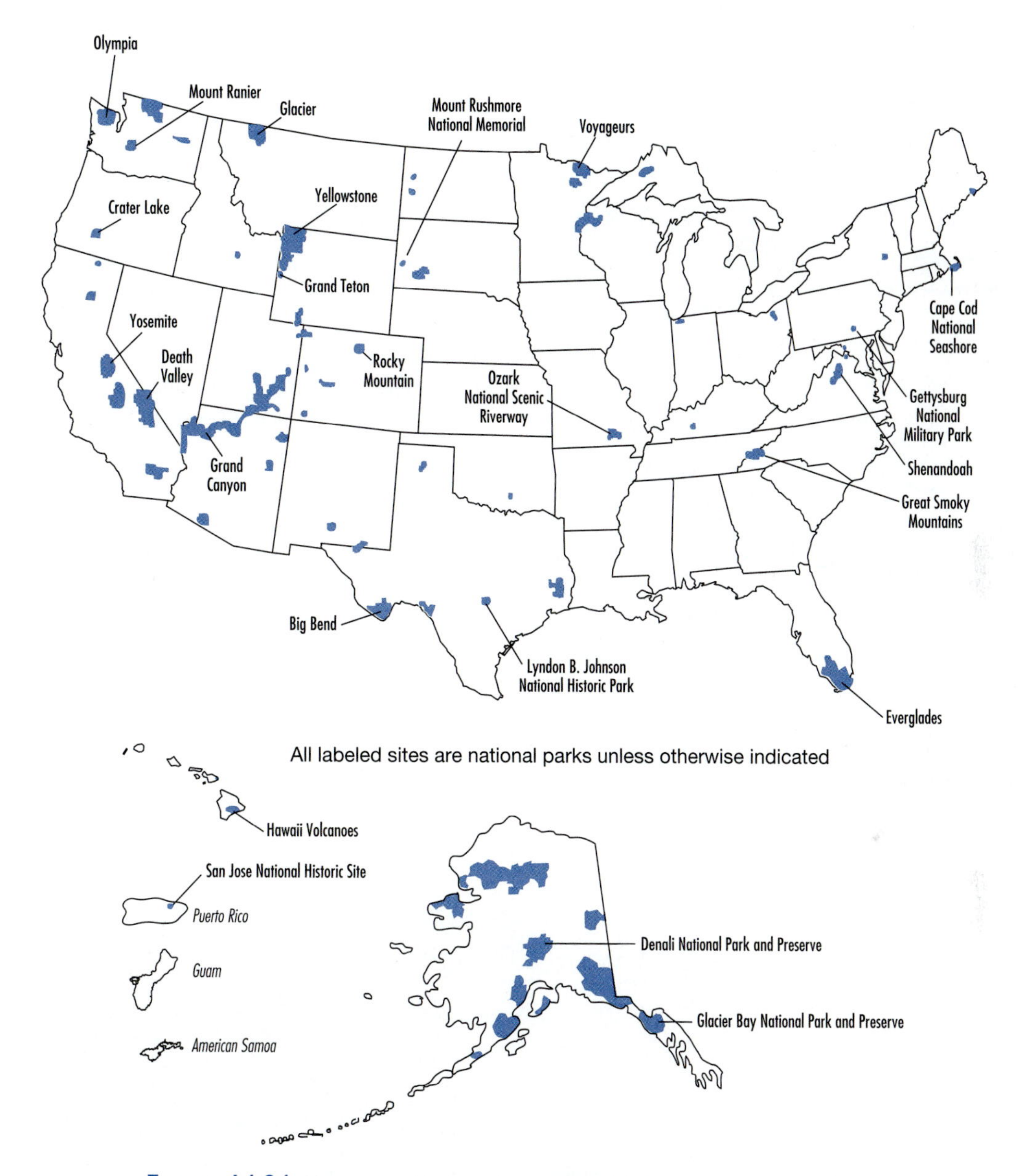

FIGURE 14.24 *Major national parks and other national designations in the United States.*

of succession and carrying capacity is absolutely essential. Of course, such manipulation must be based on research that is both sound and comprehensive.

Overcrowding: A Major Stress on Our National Park System

In recent years the national park system has been subjected to increasing stress. Much of it has been caused by attempts of private interests to engage in logging, mining, oil drilling, grazing, urban development, and other activities in nearby areas. Such human activities can threaten wildlife and recreational values of the park system. However, one of the most serious problems, ironically, has resulted from the very success of the system in attracting people. For example, consider the following statistics.

Since 1916, when the National Park Service was established, the annual number of visitors has soared from a mere 400,000 to more than 287 million in 1999—a greater than 700-fold increase! In fact, today there are more annual visits to our national park system than there are people living in the United States! By 2020, annual visits to our national park system are expected to reach 600 million people.

The people pollution problem in Yosemite National Park in midsummer is one example of current overcrowding. Yose-mite Village is grossly misnamed. In midsummer it certainly is not a village but a bustling city of more than 60,000 temporary inhabitants. It has suffered from all the environmental aches and pains of a major city, such as traffic snarls and pollution of air, water, and land. Motorists to Yosemite have agonized and cursed as they leaned on their horns in

FIGURE 14.25 *Canoeists in Montana. Such recreational activity is just one of the multiple uses provided by our national parks.*

modern urban shopping center located in what was once a wilderness. In an historic effort to restore the park, officials announced on November 4, 1997 a proposed ban on cars by the year 2001, requiring vistors to walk, bike, or take buses to get around Yosemite Valley.

Unfortunately, Yosemite is not alone in suffering from human impact. The list of modern developments in the 5 percent of Yellowstone National Park (Figure 14.26) that is most frequently visited by tourists includes 1207 kilometers (750 miles) of roads, 2,100 permanent buildings, seven amphitheaters, 24 water systems, 30 sewer systems, 10 electrical systems (with 150 kilometers or 93 miles of transmission lines), numerous garbage dumps, 54 picnic areas, 17,000 signs, and hotel-cabin accommodations for 8,586 people per night. Other parks that are often overcrowded in the summer include Mt. Ranier, Sequoiah, Crater Lake, and Rocky Mountain.

Just as in any urban center, crime occurs frequently in our most popular national parks. It includes drug peddling, sexual assaults, and wallet-snatching. Overzealous tourists often make off with illegal souvenirs such as endangered species of cacti and alpine flowers as well as redwood bark and slivers of petrified wood. Even in the backcountry of our parks, people pollution is in evidence. For example, in Rocky Mountain National Park, backpackers caused so much trampling of wildflowers and erosion that some wilderness trails had to be paved over with asphalt!

FIGURE 14.26 *Line of cars jams road in Yellowstone National Park.*

"beep-and-creep" traffic. Pollutant gases from thousands of exhaust pipes cause vile odors. What's more, eventually they contribute to a haze that sullies the beauty of Yosemite Falls and the mighty granite dome of El Capitan. Sewage systems are overtaxed. Soiled paper, candy wrappers, pop cans, paper plates, orange peelings, beer cartons, and other assorted litter desecrate once lovely campsites. Lines of tourists wait impatiently at checkout counters in Yosemite Mall—a virtual

A Blueprint for the Future of the National Park System

A number of conservation and environmental organizations are concerned with the ongoing pressures on our national park system. They have made a number of suggestions as to how these problems might be managed in the next few decades:

1. *Restrict commercial activity.* All logging, mining, oil drilling, and grazing now permitted on federal lands adjacent to any national park should be banned. All commercial ventures like restaurants, hotels, and camping equipment sales should be reduced to a minimum.
2. *Expand the park system.* To alleviate crowding and resultant pollution and deterioration of our present parks, the system should be greatly expanded. The Wilderness Society proposes the enlargement of many of the 383 units in the national park system. In addition, it suggests that many new parks be added within the near future.
3. *Control traffic congestion and pollution.* Bar all private motor vehicles from the park interior. All transportation within the park should be performed exclusively by federal shuttle buses (if needed), powered preferably by electric motors, which are not only quiet but nonpolluting as well.
4. *Reduce crowdedness.* Crowds will be reduced if the system is expanded. However, another strategy must be employed: greatly restrict the number of visitors by issuing a restricted number of entry permits. For example, a given park might issue only 1,000 permits per day, to a maximum of 180,000 during the 6 months the park is open to the public.
5. *Greatly increase park budgets.* The federal money available to the national park system should be boosted considerably. A park's nature education facility provides exciting and important information on the kinds of biological communities in the park and how their ecological systems work. These educational centers are in urgent need of improved equipment as well as additional staff. Money should also be available to provide summer research grants to college students who are interested in pursuing a professional career in wildlife biology, ecology, nature education, or park administration. More funds are required so that deteriorating roads, shelters, and visitors centers can be upgraded. A minimum of $2 billion is urgently needed for road repair alone. And, of course, billions of dollars will be required in the near future to acquire land for the proposed expansion of the national park system.

14.10 Reversing Tropical Deforestation

Researchers disagree about how much tropical forested land has been and is being logged. Most believe that humans have already cut about 50 percent of the worlds closed-canopy tropical rainforests. A reasonable average of current annual removal is about 105,000 square kilometers (42,000 square miles), an area roughly the size of Tennessee. By some estimates, present rates of removal will leave only scattered remnants of tropical rainforests by the year 2025, except for parts of the Amazon Basin and Central Africa. Figure 14.27 illustrates the percentage of tropical rainforests lost between 1981 and 1990 from the top five countries, with the largest acreage being removed in Indonesia and Brazil.

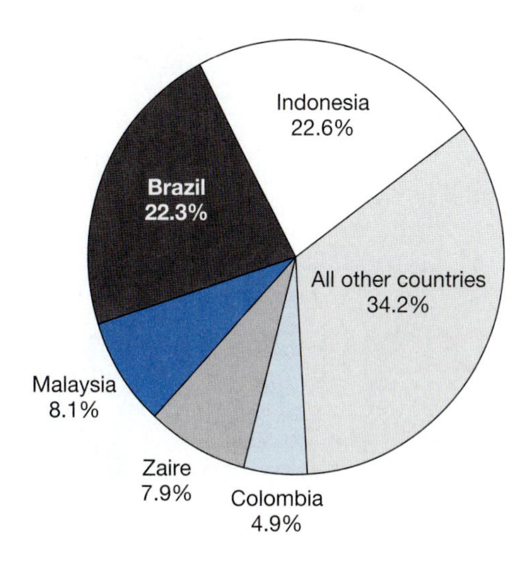

FIGURE 14.27 *Percent of tropical rainforests lost between 1981 and 1990 from the top five countries with the highest losses coming from all other countries combined.*

Madagascar provides a good illustration of forest removal in the tropics. Using satellite imagery to calculate the amount of forest remaining in Madagascar, scientists found that only 19 to 34 percent of the original forest remains. On the other hand, recent evidence from a satellite study of the Amazon suggests that estimates of deforestation that were made during the 1980s were too high. However, the same study found that rates of fragmentation (that is, the breakup and isolation of forest by farms and logging) were higher in the Amazon Basin than was previously thought.

By definition, a **tropical rainforest** is a forest with 200 centimeters (79 inches) of annual rainfall spread evenly enough through the year to support broad-leaved evergreen trees, typically arrayed in several irregular canopy layers dense enough to capture more than 90 percent of the sunlight before it reaches the ground. The ground is so dark where the sunlight does not hit that a flashlight is needed to closely examine the insects.

The remaining tropical rainforests are in Latin America, Africa, and Southeast Asia. They form a broad band around the equator about 7.8 million square kilometers (3 million square miles)—roughly the size of the United States (Figure 14.28). Some have called tropical rainforests a green hell, others a green cathedral. Whatever its name, the worlds tropical rainforests, which once covered the earth like a thick green blanket, are rapidly declining. And you say: So what? After all, what is a tropical forest good for? Snakes and squawking parrots and monkeys, maybe. But what does it have to offer me? The answer is, plenty. From the medicine you took for your last illness to the rubber tires on your car, tropical forests have a lot to do with you. And they have significance for the well-being of nations as well as individuals.

Value of Tropical Forests

Tropical forests (of which two thirds are rainforests) contain almost half of the world's growing wood. More than 1 billion cubic meters of wood are removed from the world's tropical

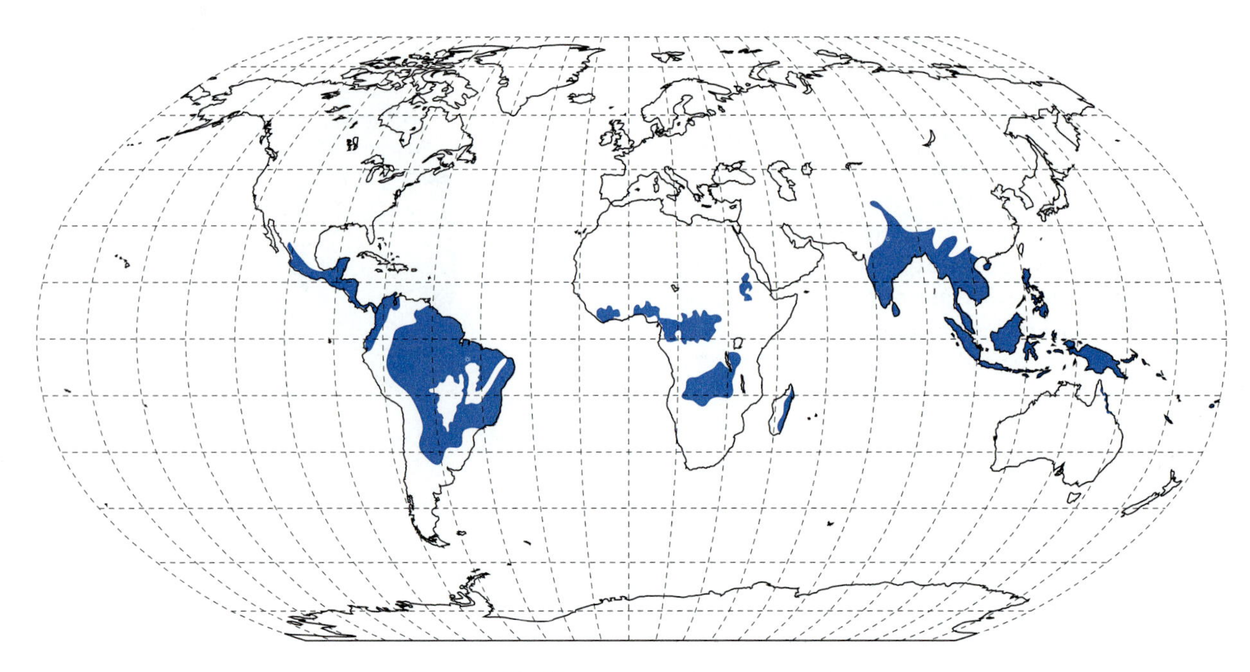

FIGURE 14.28 *Tropical moist forests of the world.*

forests for human use each year. Half of the world's annual harvest of hardwood is supplied by tropical forests. Tropical nations receive billions of dollars annually for the export of wood from the forests. Brazil, for example, exports much of its wood to the United States.

Rice production and irrigation farming in the tropics depend on water supplies slowly released from tropical forests. Tropical forests produce food products such as coffee, cocoa, nuts, tropical fruits, spices, and sweeteners. They also produce other materials, such as gums, dyes, waxes, resins, and oils, that are used in ice cream, shampoo, deodorant, sunscreen lotion, tires, shoes, and many other products.

Tropical forests provide habitat for an estimated 3 to 7 million species of plants and animals—the greatest variety of life occurring anywhere on this planet (Figure 14.29). They are home to at least half of all the earth's species of plant and animal life. On a global basis, the tropical forest biome has unsurpassed scientific and educational value. For example, the raw material for productive studies of plant and animal ecology is almost unlimited.

Many modern medicines, such as those that control disease (quinine), relieve mental stress (reserpine), or show promise to cure cancer, are derived from plants that grow in tropical forests. We've examined only a very small portion of the millions of plant species found in these forests as potential sources of medicines.

Causes of Deforestation

The main direct causes of tropical deforestation are slash-and-burn agriculture, fires, livestock grazing, fuelwood gathering, and commercial logging. Two major indirect causes are overpopulation and poverty in many tropical areas.

FIGURE 14.29 *The toucan, a colorful resident of the tropical forests of Central and South America. Scientists believe that the enormous bill, which is multihued, is used during courtship display.*

Slash-and-Burn Agriculture

Much forest destruction can be blamed on slash-and-burn or shifting agriculture. **Slash-and-burn agriculture** is practiced on about 1 of every 5 hectares in the tropical rainforests. The farmer cuts the trees, tills the land for a few years, then abandons it—leaves it fallow—for perhaps 3 to 5 years. The fallow period needs to be much longer, from 10 to 25 years, to regenerate the forest and restore soil fertility. In theory, at least, this system appears to be ecologically sound, as long as clearcuts are kept small and the fallow period is long enough. On the other hand, if the clearcut areas are too large, the effects can be devastating. Because of the increasing food demands imposed by increased population growth, the fallow period is often too short for soil fertility to be restored.

Fire Fire is used as a tool in slash-and-burn farming to dispose of felled trees and to prevent the invasion of weeds and shrubs into the crop area. Millions of hectares of tropical forests have been destroyed when such fires have raged out of control. Forest destruction by fire in Brazils Amazonian rainforest is accelerating.

Cattle Ranching In order to raise the cattle to supply the U.S. demand for hamburgers, hot dogs, and luncheon meat, cattle ranchers in Central America are clearing away tropical forests and converting the newly opened areas to grazing lands. Next time you bite into a hamburger, you should momentarily reflect on the fact that in a very small but very real way, you may be contributing to one of this hemispheres most serious environmental problems.

Gathering Fuel Wood More than 1 billion cubic meters of wood are harvested for fuel in the tropics. This wood is either used directly for firewood or is converted to charcoal. In Haiti the production of charcoal is big business. Unfortunately, the resulting forest destruction in this country is also immense.

Industrial Logging Commercial logging in tropical regions is frequently wasteful and inefficient. For example, during a selective logging operation in Malaysia, more than 55 percent of the trees were either severely damaged or destroyed. Until recently, commercial logging employed selective cutting. The only trees taken were mature specimens of the most valuable species, such as teak, mahogany, and rosewood. Today, however, any tree has potential value if it can be reduced to wood chips. As a result, clearcutting is being more widely used. Unfortunately, most clearcut areas do not become reforested by natural means. They must be replanted by people. Frequently, however, forestry personnel either are not knowledgeable about effective reforestation or simply refuse to use it. Moreover, land-starved squatters move in to carve out an existence based on slash-and-burn agriculture.

Effects of Deforestation

Firewood Scarcity Because of the severe firewood shortages in India, Haiti, and Nepal, hot meals are almost a thing of the past. Eric Eckholm, a former research ecologist with the Worldwatch Institute in Washington, has pinpointed the problem for the natives: "For one-third of the worlds people, the energy crisis does not mean. . . high prices of petroleum. . . . It means something much more basic—the daily scramble to find the wood needed to cook dinner." As a result of increasing wood scarcity, the poor are now shifting from wood fuel to animal dung. Unfortunately, however, dung is also valuable as fertilizer. A lot more grain could be produced if the cow dung being burned as fuel in Africa, Asia, and the Near East were used as fertilizer instead.

Climatic Changes Tropical forests influence the weather and the climate in important ways that scientists do not as yet fully understand. For example, the Amazon rainforest receives roughly twice as much rainfall as can be accounted for by moisture moving in from the ocean. By means of transpiration and water kept in circulation in these densely forested areas, the Amazon rainforest generates much of the rain that supports the forest. It's possible, as suggested by computer models, that the Amazon would become too dry to support tree growth if the rainforests were replaced by grasslands as the major vegetation type. Furthermore, it seems possible that there is a critical point at which deforestation will alter the climate of the Amazon basin sufficiently so that the remaining forests could not persist.

When a tropical forest is removed, the radiation of heat from the denuded area is greatly increased. As a result, the local climate and even the global climate may be adversely affected. The burning of wood releases carbon dioxide into the atmosphere. Trees also absorb carbon dioxide, so their loss accounts for an additional increase in global carbon dioxide levels. The result of the increased heat radiation and increased levels of carbon dioxide may contribute to a progressive warming of the earth (Chapter 19).

Loss of Gene Pools and Extinction of Species The irreplaceable gene pools of many kinds of organisms are rapidly being diminished. The tropical forests contain the parent species from which many of our present agricultural crops were derived. Those species may be needed as genetic reservoirs from which new varieties of disease- and pest-resistant crops can be developed.

The removal of the tropical rainforests also threatens the survival of many birds and mammals. Species diversity is greater in this biome than anywhere else on earth. That is, tropical rainforests possess more species of animals and plants than any other ecosystem. Extinction of such species is occurring at alarming rates.

Saving the Tropical Forests

How can the tropical forests be saved? The issue boils down to the following: How can people living in these tropical areas achieve a decent living from the land without destroying it? Forests need to be saved in a manner that improves local economies. This is called **sustainable development.** For example, tree planting programs have been shown to be most successful when local people are not only involved in their planning and implementation but also own the land or are given **ownership** of any trees grown on commonly owned land. With ownership comes the needed incentive for local villagers to plant and protect trees for their own use and for sale. It may even help to privatize a small portion of the tropical forests, to give the protection of ownership or contract control. Where a government cannot afford to protect a forest at risk, a well-regulated timber company, given the right long-term incentives, might provide good management and reforestation. National forests and wildlife preserves should certainly be kept and protected. The World Bank recommends that countries with rainforests convert 15 percent of their forests to national parks and preserves.

Some ecologists believe that the best method of harvesting timber in rainforests profitably with little loss of biodiversity is strip cutting or logging. With this method, a strip is harvested on the contour of a slope and logs are hauled out on a road along the upper edge of the strip. The area is then left for a few years until seedlings begin to grow in the harvested strip. Another strip is clearcut but this time above the road. One advantage of strip cutting is that nutrient runoff from the freshly harvested second strip is used by the regenerating vegetation downslope in the first strip. In addition, the recently cut second strip catches the seeds rolling down from the mature forest above. In contrast, these benefits do not occur in a clearcut system.

Agroforestry shows promise of increasing the production of food and fuelwood in the tropics. **Agroforestry** is the practice of growing trees among crops or on pasture land. Trees can supply timber, fuelwood, fruits, nuts, and food for livestock. They also provide fertilizer for crops, reduce soil erosion and weeds, and protect crops from winds. Agroforestry programs are operating successfully in a number of tropical countries in Asia, Africa, and Latin America.

The extraction of nontimber products from rainforests has been shown to yield similar financial returns as logging and farming in Peru and Brazil. In the Amazon, half a million rubber tappers, called *seringueiros,* make their living today from rubber, Brazil nuts, tonka beans, palm hearts, and other wild products. They also hunt, fish, and practice small-scale agriculture in forest clearings. The tappers preserve the rainforests because they depend on the forests' biological diversity for their livelihood. In 1987, the Brazilian government established *seringueiro* extractive reserves on government land with 30-year renewable leases and a prohibition on clearcutting timber.

So far, we have given some helpful but modest strategies. Governments and local people will need to be persuaded to adopt such innovations as agroforestry, strip cropping, and extractive reserves. As stated earlier, private ownership of some kind should help. Just as important, sustainable development will depend on education and social change.

Countries with rainforests should provide financial grants (with United Nations financial support) to support research and education on the ecology and management of their tropical forests. Science is already demonstrating its potential to preserve tropical forests through its advances in increasing tree yields on forested land and crop yields on agricultural land (so that less forestland has to be cleared for crop production). The capability and knowledge also exist to increase the efficiency of forest product harvest and use.

More developed countries must help because the loss of species anywhere in the world diminishes wealth everywhere. More developed countries already provide and need to continue to provide funds to help plan and implement successful strategies. The funds should be used to provide financial incentives to villagers and village organizations for reforesting areas from which trees have been removed for use as fuel. Incentives also include managing these fuel plantations so that harvesting is on a sustained-yield basis. Funds should also provide incentives to improve management practices in natural forests to increase yields and reduce forest degradation. Helping raise the standard of living of the people of tropical less developed countries is also important because people who worry little about food, clothing, and shelter can afford to care about forest preservation.

Summary of Key Concepts

1. Our national forests are administered and managed by four federal agencies: the Forest Service, the Natural Resource Conservation Service, the Tennessee Valley Authority, and the Fish and Wildlife Service.
2. President Theodore Roosevelt appointed Gifford Pinchot to be the first chief forester of the newly created U.S. Forest Service in 1905.
3. Under the terms of the Multiple Use–Sustained Yield Act of 1960, national forests must be managed for many uses in addition to timber harvesting. Among them are flood and erosion control, grazing land, wildlife habitat, biomass fuel, mining, oil and gas leasing, and scientific, educational, wilderness, and recreational uses.
4. Sustained-yield management ensures that annual forest harvests are balanced by annual timber growth. It also dictates that whenever timber is removed, the denuded area must be reforested, either by natural or by artificial means.
5. As of the year 2000, the National Wilderness Preservation System included 633 primitive areas. They are managed by the National Park Service (42 percent), the U.S. Forest Service (33 percent), the Fish and Wildlife Service (20 percent), and the Bureau of Land Management (5 percent).
6. Much of our nations forestland is located in national parks and administered by the National Park Service. As of the year 2000, the National Park Service administered 383 diverse units, such as national parks, national monuments, national historical sites, national recreation areas, and national lakes and seashores.
7. Our national park system has a number of stresses today, including the attempts of private interests to engage in logging, mining, oil drilling, grazing, and urban development in nearby areas, as well as overcrowding by visitors in summertime.
8. The logging plan includes (a) a map of the forest on which the location, species, age, and volume of the species to be logged are indicated, (b) a discussion of the harvesting method, (c) selection of haul roads, (d) consideration of the scientific, recreational, and wildlife functions of the forest, and (e) an estimate of gross income from the operation.
9. Harvest methods may be grouped under two categories: even-aged and uneven-aged stand methods. Clearcutting, seed tree, shelterwood, and coppice methods are used to produce even-aged stands. Selective cutting and strip cutting are used to produce uneven-aged stands.
10. Clearcutting is where the entire timber stand in a designated area is cut.
11. Opponents of clearcutting criticize it for the following reasons: (a) it accelerates surface runoff, flooding, and soil erosion, (b) wildlife-carrying capacity is sharply reduced,

at least temporarily, (c) scenic beauty is destroyed, (d) fire hazard is increased, and (e) tree blowdown is facilitated.

12. Supporters of clearcutting claim that it has many advantages: (a) it is the only way to regenerate forests in some areas, (b) it can help to control certain disease and insect outbreaks, (c) it is cheaper than selective cutting, and (d) timber grows faster on clearcut areas than on areas that are selectively cut.
13. The seed-tree method is a silvicultural system in which all timber is removed in one cut, except for a scattered number of mature trees left to provide a source of seed for the new stand.
14. The shelterwood method is a silvicultural system in which the mature timber is removed, leaving sufficient numbers of trees standing to provide shade and protection for new seedlings.
15. Strip cutting is the harvesting of wide strips (for example, 80 meter-wide strips), allowing natural regrowth.
16. Selective cutting is the harvesting of scattered trees or small groups of trees usually on mixed-species stands of unevenly aged trees. Selective cutting is a more costly and time-consuming method than clearcutting but reduces land scarring, runoff, soil erosion, and wildlife depletion.
17. Arguments for monoculture practiced on tree farms and in national forests are that it (a) takes optimal advantage of fertilizers and pesticides, (b) is rapid and efficient, (c) increases short-term profits, (d) makes maximal use of modern tree planting and harvesting methods, (e) is needed to satisfy Americas growing demand for lumber and wood products, and (f) is appropriate for the sun-loving seedlings of timber-valuable species such as Douglas fir, lodgepole pine, and ponderosa pine.
18. Arguments against monoculture are (a) it is an artificial system made by humans that lacks the built-in balancing mechanisms of a multispecies, multiage forest, (b) the fast-growing wood is inferior, (c) intensive fertilization causes the eutrophication of lakes and streams, (d) intensive use of pesticides may contaminate wildlife and human food chains, (e) it requires intensive inputs of costly energy derived from fossil fuels, (f) it is more susceptible to fires than the natural forest, and (g) it is highly susceptible to destructive disease and insect outbreaks.
19. A tree farm is a private land area that is used to grow trees for profit under sound management principles.
20. The logging operation includes felling trees, removing them from the forest, and transporting them to a sawmill or pulp mill. If trees are to be used as pulpwood, they are often harvested with hydraulic shears and whole tree chippers.
21. Genetic engineering holds promise for dramatic improvements in our nations forests, such as trees with desirable traits like accelerated growth, reduced need for fertilizer, and built-in chemical resistance against weeds and insects.
22. Ingenious methods have been developed to use almost every part of a tree. Useful products are being made from scrap boards, shavings, wood chips, and bark.
23. Major agents of forest destruction are disease (such as heartrot fungus), insect pests (such as bark beetles), and fire.
24. Insect pests can be partially controlled by (a) using insecticides, (b) clearcutting the infested area, (c) using biological control agents such as viruses, parasites, and predators, and (d) replacing monotypes with heterotypes.
25. Fires are controlled with the use of back fires, fire lanes, fire-retardant chemicals released from planes, and water pumped from tank trucks and released from planes.
26. The idea that every forest fire is destructive has been replaced by the concept that occasional fires are a natural and beneficial feature of the forest ecosystem and are needed to maintain the forest's health.
27. Controlled fires, purposely started by highly trained personnel, serve many functions. They (a) reduce the crown-fire hazard, (b) prepare the soil as a seedbed, (c) maintain subclimax species that are valuable as timber and as a wildlife habitat, (d) control insect and disease outbreaks, (e) stimulate soil microbial activity, and (f) increase the quality of livestock forage.
28. Projections indicate that by the year 2020, our nations timber demand will exceed the supply by roughly 50 percent. This increased demand may be met by (a) upgrading our small private forests, (b) more effectively controlling the agents of forest destruction (diseases, insects, and harmful wildfires), (c) developing superior (faster-growing, better-grained, and insect-, disease-, and drought-resistant) trees through the techniques of grafting, hybridization, and gene splicing, (d) increasing the use of wood substitutes, (e) increasing imports, (f) recycling paper, and (g) reducing wood use.
29. Tropical forests have multiple values: (a) erosion and flood control, (b) timber and forest products, (c) medicines, (d) wood fuel, (e) outdoor laboratories for scientific research, (f) the habitat for at least half of all the earth's species of plant and animal life, and (g) recreational functions.
30. Tropical forests have been greatly reduced because of (a) slash-and-burn agriculture, (b) conversion to grazing land, (c) industrial logging, and (d) fuel-wood gathering.
31. The effects of deforestation include firewood shortages, climatic changes, loss of gene pools, and extinction of species.
32. The issue of saving tropical forests boils down to the following: How can people living in these tropical areas achieve a decent living from the land without destroying it? Some modest strategies include private ownership of land or of trees, agroforestry, strip cropping, extractive reserves, research and education, and donations from more developed countries.

Key Words and Phrases

Agroforestry
Bark Beetle
Clearcutting
Coppice Method
Controlled Burning
Cruise
Douglas Fir
Even-Aged Stands
Fire Climax
Fire Control
Forest Conservation
Forest Pest Control Act

Genetic Engineering
Heartrot
Heterotype
Hybridization
Incendiary
Infrared System
Integrated Pest Management (IPM)
Land and Water Conservation Act
"Let-It-Burn" Policy
Logging Plan
Monoculture
Muir, John
Multiple Use Management
Multiple Use–Sustained Yield Act
National Parks
National Park Service
Natural Reseeding
Nonindustrial Private Forest Owners
Pest Control
Pinchot, Gifford
Prescribed Natural Fire
Reforestation
Rotation
Sanitation Techniques
Seed-Tree Method
Selective Cutting
Shelterwood Method
Shifting Agriculture
Sierra Club
Silvicultural System
Slash-and-Burn Agriculture
Smoke Jumpers
Strip Cutting
Sustainable Development
Sustained Yield
Tree Farm
Tropical Rainforest
Uneven-Aged Stands
U.S. Forest Service
Wilderness Act of 1964
Wilderness Area
Wildfire
Yellowstone National Park

Critical Thinking and Discussion Questions

1. Discuss the economic importance of our nation's forests.
2. Describe the functions of our nations forests in terms of (a) flood and erosion control, (b) rangeland, (c) wildlife habitat, (d) wilderness area, and (e) source of fuel.
3. What is meant by the term "sustained yield"?
4. Discuss the statement: "Clearcutting is the best harvesting method for our nation's forests."
5. Describe six advantages and six disadvantages of monoculture.
6. What type of forest is most suitable for clearcutting? For shelterbelt harvesting? For strip cutting? For selective cutting?
7. List four factors that make reforestation by seeding difficult.
8. What is the promise of genetic engineering for forest ecosystems in the twenty-first century?
9. Can you think of any reason why so many of our forest pests are exotics?
10. Discuss the factors that make heterotypes resistant to violent outbreaks of insect pests.
11. Suppose a forest fire breaks out in the Rockies. Wind velocities are high. Discuss methods that might be used to bring this fire under control.
12. List five effects of controlled burns in the longleaf pine stands of the South.
13. Describe several methods by which our nations future timber needs might be satisfied.
14. Why is it important to have wilderness areas?
15. The National Park Service was established in 1916. It administers 383 units, including parks, monuments, historical sites, recreation areas, and lakes and seashores. A blueprint for the future should include expansion of the system, control of traffic congestion, restriction of commercial activity, reduction of crowdedness, and major increases in budget. Why? Discuss your answer.
16. Suppose that tomorrow all of the tropical forests in the world were suddenly destroyed. Discuss the aftermath of such a catastrophe. What might be the long-range climatic effects? The biological effects? The social and economic effects? The effect on world peace?
17. What are some strategies for saving the tropical forests from deforestation? Discuss such strategies.

Suggested Readings

Brown, C. D. 1993. Mapping Old Growth. *Audubon* 95(3): 130–133. Well-written account of how environmental groups determined the amount of old growth acreage in some of the national forests in the Northwest.

Burton, L. D. 1998. Introduction to Forestry Science. Albany, New York: Delmar Publishers. Solid introductory text on forests and forestry.

Jordan, C. F. 1980. "Amazon Rain Forests." *American Scientist* 70: 394–401. Good discussion of strip cutting as a sustainable industry in the Amazon.

Kimmins, J. P. 1997. *Forest Ecology,* 2nd ed. Upper Saddle River, NJ: Prentice Hall. Comprehensive and in-depth treatment of forest ecology and management.

King, G., and M. Mardon. 1990. "Last Stand for the Redwoods." *Sierra* 75(4): 55–57. Popular account of the political, commercial, and environmental factors that promote redwood depletion.

Knickerbocker, B. 1993. "Forest Managers Learn How to Grow Green Lumber." *Christian Science Monitor,* 11–13. How a private logging company harvests timber on a sustainable basis while protecting the natural forest ecosystem.

Kohm, K. A., and J. F. Franklin, (eds.) 1997. *Creating a Forestry for the 21st Century.* Washington, D.C.: Island Press. Authoritative examination of the current state of forestry and its relation to the emerging field of ecosystem management.

Mitchell, J. G. 1997. "Our National Forests." *National Geographic* 191(3): 58–87. Excellent account and pictorial of the management and health of U.S. National Forests.

Parfit, M. 1996. "The Essential Element of Fire." *National Geographic* 190(3): 116–139. Wonderful article and pictorial of the blessings and curses of forest fires.

Perry, D. A. 1994. *Forest Ecosystems.* Baltimore, Md.: Johns Hopkins University Press. Comprehensive, high-level material for the serious student.

Repetto, R. 1998. *The Forest for the Trees: Government Policies and the Misuse of Forest Resources.* New York: Basic Books. Discusses the manner in which governments that are committed in principle to conservation are aggravating the losses of their forests through ill-conceived policies.

Wilson, E. O. 1992. *The Diversity of Life.* New York: W. W. Norton. Fascinating, well-written book, especially the sections on tropical forests.

Young, R. A., and R. L. Giese. 1990. *Introduction to Forest Science,* 2nd ed. New York: John Wiley & Sons. Good, general introduction to the science and practice of forestry.

Web Explorations

Online resources for this chapter are on the World Wide Web at: **http://www.prenhall.com/chiras** *(click on the Table of Contents link and then select Chapter 14).*

Chapter 15

Plant and Animal Extinction

Many of us live our lives seemingly apart from the rich and varied biological world. In our cities and suburbs, there's often little evidence of nature. To many people, nature seems abstract and inconsequential. But nothing could be further to the truth. Wild plants and animals—and the ecosystems they inhabit—enrich our lives in numerous ways (Figure 15.1). If you hike, hunt, fish, bird-watch, or photograph, your sensory rewards are many and varied. You may have been lucky enough to be graced by the eerie cry of a pack of coyotes, the graceful flight of the white-throated swift above a desert canyon, the cool, dark solitude of a primeval redwood forest, or the elegant outline of a pintail duck on a pond in the early morning sun.

Beyond the purely aesthetic benefits of wild things are the many substantial economic benefits we reap from nature: skin ointments, antibiotics, anticancer drugs, and a host of other medicines to combat heart disease, cancer, malaria, and numerous others. Today one of every two prescription and nonprescription drugs originates from a wild plant. These drugs not only reduce human suffering they contribute significantly to our economy, resulting in billions of dollars worth of revenue to pharmaceutical companies and retailers.

The economies of the United States and many other nations derive further economic benefits from nature. The hordes of hunters, anglers, birders, campers, and river runners annually spend billions of dollars on travel, lodging, food, and supplies. And nowadays, many people are flocking from richer nations like Canada, the United States, England, and Germany to view wildlife in remote corners of the world, spending billions of dollars on **ecotourism,** defined here as a form of tourism focused around adventures in the wild with the intent of doing little harm in the process.

Wild species are also an important reservoir of genes that aid agricultural scientists who are looking for ways to improve food crops and livestock. Plants and animals provide food for many of the world's people, from the Chicago stockbroker who sips white wine after a shrimp dinner in an elegant lakeside restaurant to the hunter who, crouched by a fire in the grasslands of Africa, tears meat from the bones of the gazelle he has killed and washes it down with water from a gourd.

Finally, the rich and varied plants and animals that form ecosystems provide innumerable benefits—free of charge—to humankind. Species in the web of life control

FIGURE 15.1 *Courtship dance of the whooping crane, an endangered species of North America.*

pests, recycle water, reduce flooding, and control erosion. Many, like plants and algae, provide oxygen and remove carbon dioxide from the air. Plants also help to maintain climate, and they and others recycle nutrients essential to the survival of humans and all other species. Just because these services come without a price tag, they are not without value. In fact, quite the opposite is true. Nature's services are priceless. Those services that we have lost, like the flood-control function of wetlands in many river basins, cost society billions of dollars a year.

Clearly, nature serves us well. And although we humans have seemingly isolated ourselves from nature in our urban centers, we have not emancipated ourselves from the bonds that link us to the environment. Few people recognize that nature is the **biological infrastructure** of society.

Despite these important benefits, humans are systematically destroying the Earth's life support systems. Economist Herman Daly argues that most nations are "treating the Earth as if it were a corporation in liquidation." This chapter looks at evidence of one aspect of the Earth's impending bankruptcy—plant and animal extinction. It examines the causes of extinction and ways to prevent it. It also provides background information on plant and wildlife populations to help you better understand the natural forces at work in a population—information that is essential to understanding ways to preserve our vanishing biological resources.

15.1 Extinction: Eroding the Earth's Biological Diversity

Extinction is the disappearance of a species. It is a biological fact of life. In fact, biologists estimate that 500 million species have made the Earth their home since life began over 3.5 billion years ago. To date, approximately 1.4 million species have been catalogued by biologists. Just a few years ago, biologists believed that the total number of species alive was somewhere between 3 and 5 million. Recent estimates, however, put the world total at well over 30 million species, perhaps as high as 80 million. At least half, perhaps as many as two thirds of those species live in the rich tropical rain forests of the world.

Despite the impressive number of fellow travelers on spaceship Earth, at least 90 percent of the species that once lived on this planet have vanished—become extinct. Why have so many species gone extinct?

Biologists believe that many prehistoric organisms such as the dinosaur became extinct because they were unable to adjust to changing environmental conditions and thus perished. In some cases, species that went extinct because of natural causes left behind only fossil remains. That is, their line disappeared entirely. In other instances, they evolved into new species. Their descendants carry on the line.

What caused the environmental conditions to change for dinosaurs? No one knows for sure, but theories abound. One popular theory to explain the disappearance of the dinosaurs some 65 million years ago says that a giant asteroid struck the Earth, producing a cloud of dust that shrouded the Earth and brought about a temporary but biologically devastating cooling in global temperature. Alternatively, some scientists believe that worldwide, intense volcanic activity may have produced conditions hazardous to life. Gases from volcanoes may have formed harmful acids that fell on the land and waters, wiping out plants and animals. And dust from the towering volcanoes may have blocked out the sun, shading the Earth and spawning a period of bitter cold intolerable to these magnificent creatures.

Changing conditions today, brought about by human actions, continue to drive species to extinction (Figure 15.2). Many biologists warn that we are on the brink of another mass extinction. Today, one vertebrate (backboned) species disappears every nine months. Add to that the plants, microorganisms, and inver-

FIGURE 15.2 *The rate of animal extinctions is increasing. These are just a handful of the planet's endangered species. Top: key deer, black rhinoceros; center: Florida panther,* golden lion tamarin; bottom: *gorilla, manatee.*

tebrates that disappear from the face of the Earth, and the rate of extinction is an alarming 40 to 100 species every day! With continued population growth and economic development, especially in the biologically rich tropics, biologists fear that the rate of extinction could climb even more.

Paul Ehrlich, ecologist and author of perhaps one of the most important books of modern times, *The Population Bomb,* argues that the human population has reached its present size by exploiting, degrading, and depleting the Earth's fossil fuels, minerals, soils, water, and biological diversity. "It is the loss of biological diversity," he argues, "that may prove the most serious." Harvard's eminent biologist E. O. Wilson contends that the consequences of widespread loss of species are potentially far more devastating than fossil fuel depletion, economic collapse, and limited nuclear war. The erosion of Earth's rich biological diversity, says Wilson, will take millions of years to repair. It is "the folly [for which] our descendants are least likely to forgive us."

Preserving biological diversity has become one of the most important and pressing environmental concerns of modern times. Before we discuss the reasons why species are vanishing from the Earth and study ways to prevent this dangerous occurrence, we will first examine some fundamental principles of population dynamics—that is, factors that contribute to increases and decreases in wildlife populations. This information will help you understand extinction and supplements the ecological principles discussed in Chapter 3.

15.2 Understanding Population Dynamics

Suppose that all American robins lived to be ten years old and that each adult female annually produced (fledged) eight young. If the robin population started in 2000, a single breeding pair and its offspring would produce 1.2 billion robins by the year 2030. This population would require 150,000 additional planets the size of Earth to accommodate it.

Biotic Potential and Environmental Resistance

The theoretical maximum rate of growth for a population, given unlimited resources, is called the **biotic potential.** In nature, few organisms reproduce at their full biotic potential because of environmental resistance. **Environmental resistance** refers to all those factors that reduce reproduction and suvival—for example, unfavorable weather, predation, disease, and competition. When environmental resistance is minimized, populations increase rapidly.

One of the greatest concerns of biologists tracking the steady decline in biological diversity is that the widespread elimination of species may remove environmental resistance that currently keeps many less desirable species under control. For example, the loss of species may remove key predators that hold pest populations in check. Edward C. Wolf of the Worldwatch Institute notes that "extinction's survivors tend to be ecological opportunists." Rats, sparrows, starlings, cockroaches, weedy plants, and other hardy species will proliferate as human society destroys the intricate checks and balances of nature that reduce their biotic potential. The proliferation of these results in a dramatic shift in ecosystems. Starlings, for instance, are known to reduce song bird populations.

Environmental resistance consists of many factors, which are generally grouped into two broad categories: density independent and density dependent.

Density-Independent Factors Any factor that limits the growth of a population of organisms, irrespective of the number of organisms present in a given habitat, is known as a **density-independent factor.** The most common of density-independent factors are drought, heat waves, cold spells, tornadoes, storms, floods, and natural contamination such as silt. A rare freeze in the tropics, for example, kills most members of a population of exotic butterflies, whether the population in a given area is large or small.

Human activities can also impose density-independent limits on the growth of the Earth's organisms. For example, heated water discharged from power plants may kill all of the fish in a stream, regardless of the population density. In such instances, all organisms perish when the range of tolerance is exceeded. Likewise, many of the chemicals accidentally or intentionally released into the environment—pesticides, air pollutants, water pollutants, and toxic wastes—affect populations regardless of their density.

Habitat destruction also acts independently of population density. Roadways, farms, suburbs, factories, cities, and villages all obliterate vital habitat, often destroying all of the species that once lived there, regardless of their density.

It is important to note, however, that some human interventions result in population increases rather than declines. In other words, they unleash biotic potential. For example, the house sparrow population has flourished in the United States because of the abundance of suitable nesting sites and food (waste grain and seeds in manure) provided by human settlements. Barred owls, once restricted to the eastern United States, have successfully moved westward into the Midwest and the West Coast, largely because of cities and towns. Research hypothesize that this successful expansion of habitat has been facilitated in large part by trees planted in and around cities and towns of the previously treeless Midwest.

The predicted increase in global temperature, caused by carbon dioxide and other factors, may shorten winters in North America, making conditions more favorable for insects. As a result, many insect populations could explode, creating even more damage to crops and forests. Disease organisms from the tropics could spread northward, and, in fact, already appear to be. In a report on global climate change issued by the National Science Foundation in 2000, researchers predict that over the next 100 years average temperatures in the United States will increase by 5 to 10°F (3 to 6 °C) with a major shift in climate zones. Temperatures in the northern United States will be similar to those in the middle tier and climate in the middle tier, they predict, will be more like that of the southern tier. The southern states will be "treated" to temperatures characteristic of the tropics. The effects on plants and wildlife could be devastating.

Density-Dependent Factors Many species have evolved mechanisms that allow them to survive Winter. Birds, for instance, may migrate to warm climates when Winter approaches. Some animals grow a thick coat of fur when Winter approaches to protect themselves from the cold. Other animals hibernate. Perennial plants, species that grow year after year from the same root structure, such as deciduous trees and grasses, enter a state of dormancy. Trees like the maple, oak, and aspen, for instance, shed their leaves in the fall and greatly reduce their metabolic activity over the winter to survive the harsh conditions.

For these species, density-dependent factors tend to play a larger role than weather in controlling population growth. **Density-dependent factors** are those whose influence in controlling population size increases or decreases depending on the density of a population. Ecologists recognize at least four major density-dependent factors: competition, predation, parasitism, and disease.

Predation A host of animals hunt and kill their prey. Known as **predators,** these animals include strictly carnivorous (meat-eating) species, such as mountain lions, coyotes, and cheetahs; herbivores (vegetation-eaters), such as cows, deer, elk, and grasshoppers (which "prey" on grass); and omnivores ("all-eaters," eaters of both plants and animals), such as bears and human beings.

Predation is a density-dependent factor. As a general rule, as the population density of a prey species increases, so does the percentage of organisms killed by predators. Why?

It is generally thought that as the density of a prey species increases, individuals become easier to find and attack. Furthermore, increased competition in the dense prey population may result in a larger number of weakened organisms that become easy targets for predators. Prey may also be forced into less suitable habitat and may become weakened or diseased, and again are more likely to be captured and killed by a predator. Ecologists have also found that when the density of a prey population increases, predators that eat a variety of prey tend to shift their attention to the most concentrated ones.

Competition Competition is another density-dependent variable. Competition occurs at two levels—within a given species and between species. Consider an example of competition within a single species: ponderosa pine trees in the Rocky Mountains. If the number of ponderosa pine trees increases in a given region, creating a more dense stand, competition for sunlight, soil nutrients, and water increases proportionately. Trees with shallow, less well-developed root systems receive less water than others. As a result, they are more vulnerable to insects like the pine bark beetle, which carries with it a deadly virus that clogs the water-transporting vessels in the wood and kills the tree.

Population increases in animal species intensify competition for food, water, cover, nesting sites, breeding dens, and space. Competition generally affects the physically unfit—the aged or diseased. In their weakened condition, some may perish from disease or injuries inflicted during territorial battles. Some weakened individuals may be forced to colonize poorer habitats, where death from starvation, disease, or predation awaits them.

Parasitism **Parasites** like tapeworms are organisms that feed on the bodies of other living organisms, called **hosts,** but generally without killing them. Most parasites have limited mobility but are easily spread from organism to organism. The higher the host population density, the more readily they are transmitted. As a result, parasitism is a density-dependent factor. In Canada, for example, a species of parasitic wasp attacks webworm moths. When the moth population density is low, deaths from parasitic wasps are negligible. But when the population density increases, parasitism increases.

In Colorado in the late 1970s, bighorn sheep offspring died in record numbers because of a parasite called the lungworm. The lungworm weakens the sheep and makes them susceptible to pneumonia and other diseases. Passed from mother to fetus, the lungworm killed 97 percent of the newborn. The lungworm is a common parasite that, under normal conditions, lives in harmony with the bighorn. But because of human encroachment, the bighorn sheep habitat was reduced and the population density increased. Animals were forced to calve and feed on the same land. Lungworm eggs are deposited in the sheep's feces, and because they were forced to occupy a smaller habitat, the number of lungworm eggs they ingested increased. Record numbers were found in females. The worms passed through the placenta and infected bighorn fetuses. The newborn, too weak to withstand the parasite, perished. Fortunately, wildlife biologists found out about the problem and have captured sheep, treated them, and relocated them to more suitable habitat.

Disease Infectious diseases also increase when population density increases. The incidence of infectious diseases in human populations, for instance, increases with increasing population density. This was demonstrated during World War I, when American soldiers crowded shoulder to shoulder in the barracks and trenches suffered massive mortality from the influenza virus. Because the organisms that cause infectious diseases (viruses, protozoa, and bacteria) can be transmitted by contact, through food, and by animal vectors (insects), the greater the density, the more likely it is that the disease will spread. Today's population is extremely vulnerable to infectious disease. In fact, according to one estimate, within two years of the emergence of a new influenza virus half the world's population has been exposed. High density is one of the reaons. Of course, modern transportation facilitates the spread of infectious diseases.

Disease organisms that attack crops represent another example of density-dependent factors. When large plots are planted with a single crop species such as wheat or corn, farmers increase the likelihood that disease organisms will wipe out their crops. The population-reducing effects of wheat rust, corn smut, and other plant diseases are well known.

Although we have grouped the population-regulating mechanisms into two groups, density-independent and density-dependent, in reality these factors operate together on a given population. For example, disease and predation may wipe a species out during a hard winter. Heavy rains may flood land and reduce the population of a species afflicted with a disabling parasite, causing the population to plummet. In this instance, the density-independent factor (flooding) combines forces with the parasite, a density-dependent factor.

Wildlife Population Dynamics

The size of a population of any organism results from the interaction of factors that contribute to environmental resistance with factors that contribute to its biotic potential. In general, when environmental resistance climbs, a population declines. When environmental resistance falls, a population increases. Stability may be achieved in ecosystems when the two opposing sets of factors are in balance. Don't expect to see a perfect balance, however, conditions in nature shift constantly and populations rise and fall in response to them. We say that populations are in a dynamic equilibrium, a concept that will discussed shortly.

This section looks at plant and animal **population dynamics**—the changes that occur in populations—and explains the factors that affect it. The study of population dynamics is a key tool in wildlife management, the subject of the next chapter.

The S-Curve Whenever a species is established in a new habitat with adequate resources, its population grows in a characteristic way. The plot of population growth is called an **S-shaped,** or **sigmoidal, curve** (Figure 15.3A). This curve has four distinct phases, listed in chronological sequence: (1) the establishment phase, (2) the explosive (logarithmic) phase, (3) the deceleration phase, and (4) the dynamic equilibrium phase.

In the first phase, the population grows slowly as its new members establish themselves in their new habitat. The population soon begins to expand and fill the unexploited habitat and begins growing at an explosive rate. The rapid growth generally occurs because of the availability of untapped resources (for example, abundant food and shelter) and the lack of environmental resistance such as predation. During the second phase, however, growth of the population begins to reduce the supply of resources. This, in turn, will reduce the rate of growth, resulting in the deceleration phase. Finally, once the population is well established and competition and other factors come into play, the species reaches an equilibrium. At this point, populations shrink and expand, but remain more or less constant over the long term. Thus, during years with adverse weather, populations may decrease, but during years with more favorable weather populations rebound (Figure 15.3B). Such a system is said to be in a **dynamic equilibrium.** At this point, no further population surges are possible, for the population has reached the carrying capacity of the habitat. As noted in Chapter 1, the **carrying capacity** is the number of organisms a habitat can support indefinitely.

A number of species introduced into the United States from abroad, such as the English sparrow, European starling, and the German carp, have experienced population growth that follows the S-shaped curve. For instance, a few pairs of house sparrows were introduced to the United States in 1899, and within ten years the sparrow population had increased to many thousands of birds. Although these and other species have been introduced with good intentions (the sparrow was brought over from Europe to control insect pests), many **exotic** or **alien species** have had an enormous impact on native plants and animals. The sparrow, for instance, outcompetes songbirds for nesting and food both in the United States and Canada. Other alien species have destroyed wildlife directly, disseminated diseases, or aggressively outcompeted the more valuable species for food, cover, and breeding sites.

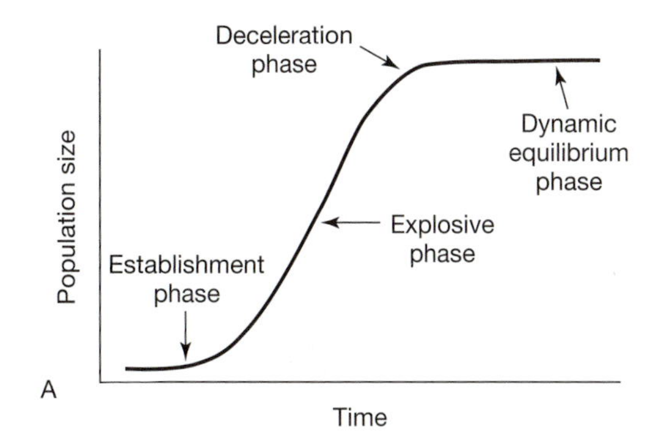

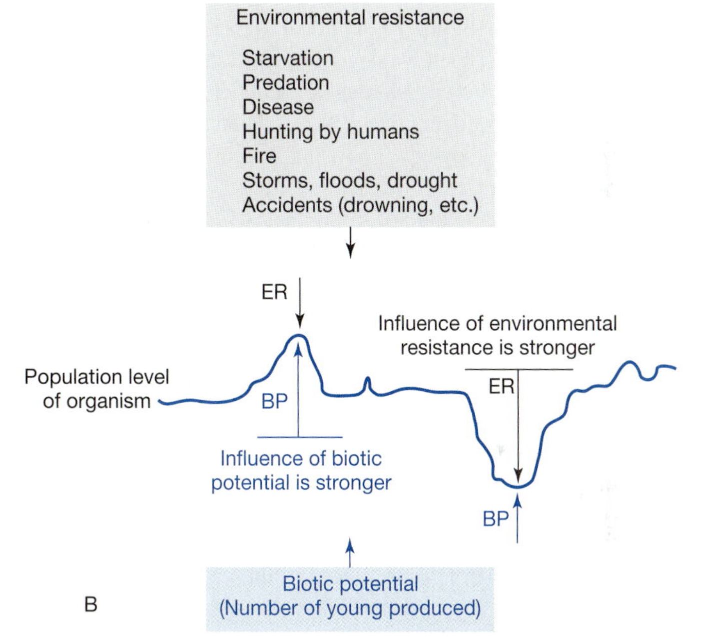

FIGURE 15.3 *(A) Sigmoidal curve illustrates the growth of a population of organisms in a new environment. After an initial establishment phase, growth explodes. As environmental resistance rises, growth decelerates, then stops. (B) A dynamic equilibrium occurs in this phase. The population may increase as a result of a decrease in environmental resistance or an increase in the number of offspring (an increase in biotic potential). A decrease in biotic potential or an increase in environmental resistance may cause the population to decline. Despite these shifts, the population remains more or less constant over time.*

Wildlife managers have used their understanding of the S-curve and population dynamics for years to manage commercially valuable species, and will continue to do so for years to come. Wild populations of deer, cod, pine trees, or even African elephants, for example, may be periodically harvested, much like a commercial crop. Perhaps the most important challenge facing scientists in the fields of forestry, wildlife management, agriculture, ranching, and fisheries is to find ways to secure the **maximum sustainable yield,** or **optimum yield**—the greatest yield possible that does not harm the population's long-term survival by upsetting the ecological balance of an ecosystem. Optimal harvests are generally achieved at the point on the S-curve between the explosive and the deceleration phases. Research has shown that in most species, this point occurs when a population is at roughly 50 percent of the habitat's carrying capacity. Harvests that drive the population below this level may severely reduce numbers and cause a further population decline, possibly driving the species to extinction.

Unfortunately, managers are discovering that what may appear to be an optimum yield may not in fact be optimum or sustainable over a long time frame. Factors not accounted for, such as soil deterioration caused by periodic harvesting of forests, have encouraged a re-examination of proper harvest levels.

Most plant and animal populations become fairly stable once they reach the plateau of the S-curve—the carrying capacity of the ecosystem. Even so, predators, parasites, competition, climate, and other factors may cause slight upward or downward swings in the population. Thus, the line in the graph in Figure 15.3a which is drawn straight actually shows upward and downward oscillations.

Irruptive Populations After fluctuating mildly for many years, some populations may suddenly increase sharply. A sudden increase in population size is called an **irruption.** It may occur after a period of unusually favorable weather that results in an unusually large supply of food or exceptional conditions for the survival of offspring. Expanded beyond the habitat's carrying capacity, the population typically plummets to a much lower level. Wide oscillations in the population size may continue for years if environmental conditions continue to vacillate. In some instances, the sudden upswing in a population may cause severe damage to the ecosystem—so severe, in fact, that the population crashes.

Cyclic Populations The brown lemming, a tiny rodent the size of a hamster, lives in the Arctic tundra of North America, a vast, treeless biome above the northern coniferous forest biome (Chapter 3). Every three or four years, the lemming population peaks, then crashes. A study of this predictable cycle showed that the peak lemming population overgrazes the vegetation that protects it from its prey, making the lemming highly susceptible to its main predators, the Arctic fox and snowy owl. As you might expect, the populations of these predators vary with the lemming's, increasing in times of abundance and declining in times of scarcity. Without alternative prey, many snowy owls starve during winters or migrate southward into the United States in times of scarcity.

Lemmings, the popular myth goes, migrate en masse to the sea when their populations top out, leaping from cliffs into the perilous waters, where they drown. Although researchers have found that lemmings do indeed migrate down mountainsides from crowded breeding sites, they now know that lemmings do not migrate far, and they disperse individually, not in great masses. Furthermore, the tiny rodents do not throw themselves into the sea, but they will occasionally cross streams, where some lose their lives. Stream crossings are made, however, only when they are absolutely necessary. In short, there is no suicidal mass march to the sea.

A number of other species experience regular three- to four-year cycles, among them the red-tailed hawk, the meadow mouse, and the sockeye salmon. Protecting these and other cyclic species requires special attention on the part of wildlife managers. Overharvesting of commercially important species in crash years could reduce numbers to levels that cannot be sustained.

The populations of snowshoe hares, found in the northern coniferous forests, undergo a ten-year boom-and-bust cycle (Figure 15.4). The lynx, which preys largely on the snowshoe hare, also has a ten-year cycle that lags just behind the hare's. The lynx cycle was first discovered by scientists studying the Hudson Bay Company's lynx pelt return records. The cycle is also characteristic of muskrats, grouse, and pheasants.

The original explanation given for the cycles goes as follows: When hares are abundant, food for the lynx is abundant. That makes it easier for lynx to raise their young and increases the survival of young and adults. But the success of the lynx population results in heavier predation on hares, which in turn reduces the hare population. With its food supply reduced, the lynx finds it more difficult to survive, and its numbers dwindle.

A look at the big picture, so essential to critical thinking, suggests that other factors may play a role in controlling the hare population size. Hares living on islands with no lynx also undergo a ten-year cycle. One explanation is that the rising population of hares reduces its own food supply. Hares begin to die off, and as the population declines, the lynx population plummets. A number of other hypotheses have been advanced to explain the ten-year cycle, but as yet scientists have been unable to determine which, if any, are valid. Among the causative agents in this puzzle are variations in weather, fluctuations in solar radiation, outbreaks of disease, and changes in the nutrient levels of plants.

15.3 Causes of Extinction

With this brief introduction to wildlife and plant populations, we now turn our attention to the question, What causes extinction? We will focus on anthropogenic causes—that is, human actions. "If a species becomes extinct," wrote David Day, author and conservationist, "its world will never come into being again. It

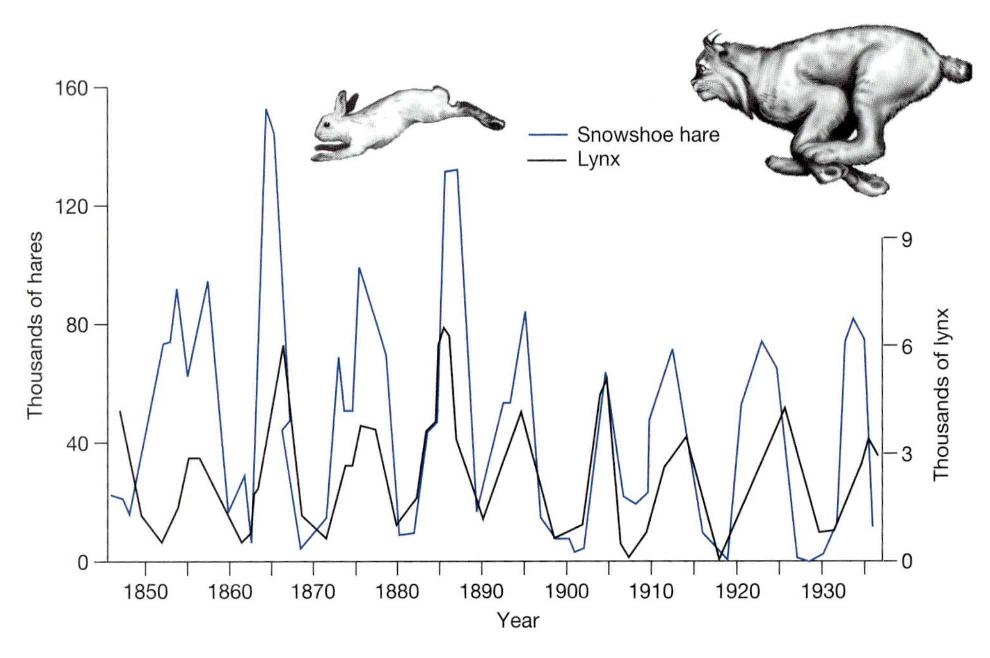

FIGURE 15.4 *An example of a wildlife population cycle, based on pelt records of the Hudson Bay Company. Note the approximate ten-year interval.*

will vanish like an exploding star. And for this we hold direct responsibility." Most of the species that have become extinct since the beginning of the Industrial Revolution have done so not because they have been unable to adapt to natural changes in the conditions on Earth, but rather because they have been unable to coexist with another species, *Homo sapiens.*

Plants and animals become extinct for a number of reasons. The greatest number vanish because their habitat is destroyed or altered beyond habitability. Put another way, habitat destruction or alteration is the number one cause of species extinction today.

Other species disappear because they are overharvested by commercial interests to supply food or other valuable resources for human society. Still others succumb to alien species that have been accidentally or intentionally introduced into their habitat. Alien species may kill a native species outright or simply outcompete it. Sport hunting has contributed to the decline of some species, as have pest and predator control programs. Pollution and the pet business have also contributed to the steady decline in the world's species. In some cases, several factors combine to drive a species to extinction. This section describes the three main causes of extinction: habitat destruction and alteration, hunting (commercial and sport), and the introduction of alien species.

Habitat Destruction and Alteration

Habitat is the area a population occupies. It provides food, water, shelter, and many other requirements of life. Destroying or altering that habitat can have profound impacts on the plant and animal species living there.

Habitat is altered and destroyed by many human activities: expanding urban areas, farming, logging, mining, and the construction of highways, railroads, pipelines, and dams. Today, these and other activities are growing at an alarming rate. The destruction or severe alteration of habitat has claimed innumerable species and threatens millions of others. In the United States, for instance, 500 species are known to have become extinct since Europeans arrived. Approximately 360 animal and nearly 600 plant species are listed as endangered (imminent danger of extinction). Another 126 animal species and subspecies and 140 plant species are listed as threatened (likely to become extinct), bringing the total of endangered and threatened species to about 1,200. Three hundred and fourteen species are awaiting listing. The story of one species, the snail darter, is presented in Case Study 15.1.

In many places only tiny islands of natural habitat, called **ecological islands,** now exist amid a sea of crops, pasturelands, towns, and mines (Figure 15.5). In the eastern United States, for example, small patches of deciduous forest are all that remain of once lush and vast forests. These islands exist today among farm fields, towns, and cities. **Habitat fragmentation,** the progressive destruction of habitat that results in the formation of isolated ecological islands, reduces populations of native species, making them vulnerable to extinction. Chandler Robbins of the U.S. Fish and Wildlife Service found that fragmentation of the forests in Maryland, New York, and Pennsylvania has caused a drastic reduction in the number of birds that once bred in them. The long-distance migrants such as vireos, tanagers, and orioles that winter in South America have been most severely affected.

Princeton biologist Robert MacArthur studied species diversity on naturally occurring islands and found that, all other things being equal, the smaller the island, the fewer species it supported. Other scientists found that this conclusion was also valid for terrestrially based ecological islands cut off from similar habitat by human activities. Thus, a 10 hectare plot of deciduous forest cut from a continuous forest supports far fewer species per hectare than a 10,000-hectare plot.

At least three reasons explain the reduced species diversity on ecological islands. First, small plots may not contain enough

FIGURE 15.5 *Farms, towns, roads, and other forms of development often leave only small islands of natural habitat, which support far fewer species than larger, undisturbed tracts of land.*

CASE STUDY 15.1 DAM VERSUS DARTER: A CLASSIC CONFRONTATION

In 1973 David Etnier, a fish expert at the University of Tennessee, discovered a new species of fish in the Little Tennessee River. The tiny fish, only 7.5 centimeters (3 inches) long, was given the name snail darter (*Percina tanasi*). According to Etnier, the entire world population of this species, numbering about 1,400 individuals, was confined to a 2.5-kilometer (15-mile) stretch of the Little Tennessee River. Obviously, the species qualified as endangered under the criteria of the Endangered Species Act. It was most unfortunate, however, that the fish's habitat would soon be destroyed because the Tennessee Valley Authority (TVA), a federal agency, was constructing the $116 million Tellico Dam on the Little Tennessee, a short distance from where the darter was discovered. The resulting reservoir, impounded behind the dam, would replace the shallow, fairly rapidly flowing water essential for snail darter survival with deep, quiet water—a completely different aquatic habitat.

The discovery of the tiny fish set the stage for a confrontation between technology and a powerful federal agency (the TVA), on the one hand, and a vanishing species, supported by a few dedicated environmentalists, biologists, and nature lovers, on the other. In a letter to us, Etnier described the confrontation:

The story is an interesting one that involves a small fish and a group of people with virtually no resources attempting to force a governmental agency to comply with federal law. The number of participants has been immense, on both sides. Our side fought with money from the sale of snail darter T-shirts. TVA's efforts to thwart us have probably cost that agency well in excess of a million dollars in lobbying and expenses associated with their staff of lawyers, biologists, and administrators working on the case. Virtually every newspaper, press service, and TV network has devoted some time or space to the issue and magazines such as *People, The New Yorker, Time,* etc. have carried lengthy articles.

Eventually the case was brought to the courts. In early 1977 a federal appeals court ruled that the TVA would have to terminate construction on the Tellico Dam. But the pro-dam people did not give up easily. In April 1978 no less an official than U.S. Attorney General Griffin Bell asked the U.S. Supreme Court to spare the dam and scrap the darter. But to no avail. In June 1978, our nation's highest court ruled in favor of the fish.

However, legislators began pondering the question: "When we formulated and enacted the Endangered Species Act, were we really concerned about saving tiny fish, or were we thinking of eagles and moose?" As one scientist wrote: "Congressmen are now finding themselves confronted with a Pandora's box containing infinite numbers of creeping things they never dreamed existed."

Nevertheless, in 1978, when the Tellico Dam was 90 percent finished, the Supreme Court ruled that the project had to be stopped because it violated the provisions of the Endangered Species Act. The court ruled that the act made any federal construction project illegal if it jeopardized the survival of any organism that had been formally classified as endangered under the provisions of the act.

A storm of controversy raged in Congress for many months concerning the act. Verbal battles were waged. Congressmen were heavily pressured by lobbyists. The construction industry naturally was interested in weakening the Endangered Species Act so that federal projects could be exempted. Environmentalists, on the other hand, were steadfastly opposed to any modification of the act that would lessen its influence in preserving endangered organisms.

Eventually the act was amended. Under the amendment, any requests for exemptions from the act were to be considered by a special high-level review committee. In 1979 the committee ruled to block any further construction of the Tellico Dam on the grounds that the economic benefits resulting from its construction did not justify its costs—in addition to threatening the survival of the snail darter. However, special interest groups with a financial interest in such multi-million-dollar projects as dams, levees, reservoirs, and highways refused to concede defeat. They succeeded in convincing legislators to amend a public works bill that would permit the completion of the Tellico Dam.

The pro-darter people were understandably dismayed. However, they did the best they could, under the circumstances, to preserve the fish. The entire snail darter population was removed from the Little Tennessee and introduced into the Hiwassee River nearby.

The Hiwassee River population survived the transplant and appears to be doing well in its new home. Recently, scientists have discovered snail darter populations in four additional creeks in Tennessee and Alabama.

As opponents projected, however, the TVA's Tellico Dam and reservoir have not fared as well. Built to stimulate industrial development and increase recreational fishing, the dam appears something of a failure. So far it has brought no industrial development to the region, and the projected recreation fishery is generously described as "average." Many critics think of it as a "pork barrel" project—built to please local and state politicians at a huge federal expense.

room and food for some species. The grizzly bear, for example, requires a 5,200-hectare (20-square-mile) habitat to find the food it needs. Anything smaller is inadequate. Second, small habitats may reduce the number of organisms in a given population below the critical size needed to reproduce. For instance, the now-extinct passenger pigeon once roamed in flocks that contained many millions of birds (Figure 15.6). Commercial hunters eliminated the huge flocks in mass slaughters to provide food for urban dwellers (Case Study 15.2). Excessive hunting accompanied by heavy deforestation spelled doom for the bird. By 1878, only 2,000 birds remained, in flocks too small to reproduce successfully (Figure 15.7). The population had fallen below the **critical population size,** the size needed to support a breeding population, and could never recover. Habitat destruction could have the same effect on other species. Third, tiny habitats may promote extensive inbreeding—that is, mating with close relatives. Inbreeding often results in inferior offspring.

FIGURE 15.6 *Shooting "wild pigeons" in Iowa, an illustration copied from* Leslie's Illustrated Newspaper *of September 21, 1867. Note the gunner firing into the densely massed birds. Over 100 birds are resting on the bare branches of the oak in the background.*

Yet another problem that occurs when a population reaches low level is that genetic diversity is decreased. **Genetic diversity** is a scientific term that refers to the amount of genetic variation in a population. **Genetic variation,** differences in the genetic makeup of the members of the population, results in variation in the structure, function, and behavior of organisms. Genetic variation means that not all members of a population are identical. For example, in a population of plants, some may be better able to survive drought. Or, they may produce more seeds. The greater the genetic diversity, the greater a species' ability to survive and reproduce. When a population is reduced, its genetic diversity declines and its options for survival may decline, too. In the late 1800s, for example, the northern elephant seal was hunted extensively, nearly to the brink of extinction. When hunting pressure stopped, the 20 or so remaining seals bred and produced all 150,000 of the modern descendants. Needless to say, they're all fairly similar genetically. This, in turn, makes them more vulnerable to environmental change.

Habitat fragmentation leads to **faunal collapse,** a decrease in animal species. Human activities may undermine the survival of

CASE STUDY 15.2 THE PASSENGER PIGEON: THE MANY CAUSES OF EXTINCTION

The passenger pigeon (*Ectopistes migratorius*) was once the most abundant bird on Earth. Early in the nineteenth century the renowned ornithologist Alexander Wilson observed a migrating flock that streamed past him for several hours. Wilson estimated the single flock to be 1 mile wide and 240 miles long and composed of about 2 billion birds. (The population of this flock was roughly ten times the total North American waterfowl population today.) Yet not one passenger pigeon is left.

What factors contributed to the passenger pigeon's extinction? First, many potential nest and food trees were chopped down or burned to make room for farms and settlements. The pigeon fed extensively on beechnuts and acorns; the single flock observed by Wilson could have consumed 17 million bushels per day.

Second, disease may have taken a severe toll. The breeding birds were susceptible to infectious disease epidemics because they nested in dense colonies.

Third, many pigeons may have been destroyed by severe storms during the long migrations between the North American breeding grounds and the Central and South American wintering region. Cleveland Bent cites a record of an immense flock of young passenger pigeons that descended to the surface of Crooked Lake, Michigan, after becoming bewildered by a dense fog. Thousands drowned and lay a foot deep along the shore for miles.

Fourth, their low biotic potential may have been a factor in their extinction. Although many birds, such as robins, lay four to six eggs per clutch, and ducks, quail, and pheasants lay 8 to 12 eggs, the female pigeon produced only a single egg per nesting.

Fifth, the reduction of the flocks to scattered remnants possibly deprived the birds of the social stimulus requisite for mating and nesting.

Sixth, one of the most important factors in the bird's demise was intense pressure from market hunters. They slaughtered the birds in their nests. Every imaginable instrument of destruction was employed, including guns, dynamite, clubs, nets, fire, and traps. Over 1,300 densely massed birds were caught in one pass of the net. Pigeons were burned and smoked out of their nesting trees. Migrating flocks were riddled with shot. Over 16 tons of shot were sold to pigeon hunters in one small Wisconsin village in a single year. Pigeon flesh was considered both a delectable and a fashionable dish in the plush restaurants of Chicago, Boston, and New York. Sold for two cents per bird, almost 15 million pigeons were shipped from a single nesting area at Petoskey, Michigan, in 1861.

The last wild pigeon was shot in 1900. Martha, the last captive survivor, died on September 1, 1914, at the age of 29, in the Cincinnati Zoo.

FIGURE 15.7 *Extinction is forever. When the last living passenger pigeon, Martha (shown here), died at the Cincinnati Zoo on September 1, 1914, a unique organism was removed from the human ecosystem forever.*

species living in fragmented habitat. Chemical contaminants in the air, water, and soil, for example, may hinder reproduction or kill organisms outright. Global warming may alter the distribution of plants and eliminate the animal species dependent on them. Increased ultraviolet radiation may destroy other plants or increase rates of cancer and mutations among animals. "The cumulative effects of such changes," notes Edward C. Wolf of the Worldwatch Institute, "can alter ecosystems in ways that increase the vulnerability of plants and animal species to extinction."

To protect biological diversity, human society has embarked on an ambitious program of park designation. Today, 425 million hectares of land are protected the world over. However, park protection is woefully inadequate. By some estimates, three times as much parkland as currently exists would be required to protect samples of each of the Earth's major ecological zones. Obviously, as the last paragraph pointed out, strict measures to control pollution are needed to enhance the survival of species living in these protected regions.

Recent studies have shown that many national parks in the United States, long seen as the last hope for America's vanishing wildlife, inadequately protect species diversity. Ecologist William Newmark studied the loss of mammal species in national parks and found an alarming drop in the number of species in all but the largest parks (Table 15.1). Bryce Canyon National Park, one of the smallest, lost 36 percent of its species. Yosemite, nearly 20 times larger than Bryce, lost 23 percent. Only the mammoth parks like Yellowstone suffered minor losses.

What lessons can we draw from these observations? Many parks are simply too small to support the diverse array of species that lived in the area before it became a park. If a park is cut off from neighboring areas by development in surrounding tracts, the park becomes an ecological island too small to support the diversity it once enjoyed. What is more, researchers point out, the faunal collapse in the world's parks may be continuing today.

Nowhere is the problem of extinction as critical as in the tropical rain forests, which contain an estimated one half to two thirds of the world's species. The dire predictions of extinction presented at the beginning of this chapter were based on a deforestation rate of two percent per year. Recent work by the United Nations Food and Agricultural Organization (FAO), however, showed that deforestation may be occurring at a much slower rate than once thought. Instead of 2 percent per year, the rate may be 0.6 percent per year globally. Although the revised estimate is still significant, it suggests that extinction may not be as rapid as many ecologists once predicted.

The FAO study also showed that the two largest areas of continuous forest—the great forests of Central Africa and Amazonia—are being cut, but at very slow rates. The Central African forests of Zaire, Gabon, Congo, and Cameroon, for instance, are in equilibrium—that is, new forests are being planted at a rate nearly equal to deforestation. Critical thinking suggests the need to consider this fact carefully. When you do, you find that the newly replanted forests are often monocultures, consisting of a single species grown for commercial use. Monocultures fail to support the biological diversity of the pre-

TABLE 15.1 *Faunal Collapse in America's National Parks*

Park	Area (Square Kilometers)	Share of Original Species Lost (%)
Bryce Canyon	144	36
Lassen Volcano	426	43
Zion	588	36
Crater Lake	641	31
Mount Ranier	976	32
Rocky Mountain	1,049	31
Yosemite	2,083	25
Sequoia–Kings Canyon	3,389	23
Glacier–Waterton	4,627	7
Grand Teton–Yellowstone	10,328	4
Kootenay–Banff–Jasper–Yoho	20,736	0

Source: Based on W. D. Newmark. "A Land-Bridge Island Perspective on Mammalian Extinctions in Western North American Parks." *Nature,* Jan. 29, 1987.

vious forest. In Amazonia, the rate of destruction is only 0.3 percent a year. At this rate, it would take 300 years to destroy the forest—still significant for the long-term well-being of the planet, but less pressing than some would have you think.

But global averages can be misleading, the FAO study showed. For instance, the tropical forests of the Ivory Coast of Africa are being destroyed at a rate of 6.5 percent per year. In Nigeria, the rate is 5 percent. In Paraguay, it is 4.7 percent. In Nepal, it is 4.3 percent. It is in these areas and other priority areas that conservationists must concentrate their efforts.

Throughout the world, large human populations live near estuaries, bays, and other coastal wetlands. Roads, highways, cities, homes, and airports now occupy space that once supported wetlands and an abundance of plants and animals. Inland wetlands have not fared any better. From Florida to Wisconsin, farmers have drained swampland and plowed it under. Thus, many of the world's coastal and inland wetlands have already been destroyed. In the Philippines, for instance, 50 percent of the mangrove wetlands have been filled in or dredged. In southern California, 90 percent of the salt marsh wetlands have met a similar fate. In the United States as a whole, nearly two thirds of the wetlands have vanished. Although the rate of loss has decreased dramatically in recent years (partly because of new laws and regulations, but also because many of the prime wetlands have already been destroyed), we still continue to lose wetlands.

Wetlands provide many benefits to people and the species that live in them, many of which have been touched on in previous chapters. (See Table 9.1 for a summary of them.) As the benefits of wetlands are many, so also are the dangers of destroying them. At least half of the biological production of the world's oceans occurs in coastal wetlands and estuaries. Sixty to 80 percent of the world's commercially important marine fish either spend time in estuaries or depend on them for food. And 60 percent of the fish caught in the ocean by commercial fishermen depend on the estuarine zone—the mouths of rivers and coastal wetlands—at some point in their life cycles. By filling in wetlands and polluting them with wastes from our homes and factories, we do a disservice to the other species that share this planet with us, and to ourselves.

The loss of wetlands increases the frequency and severity of floods. In seven states in the upper Mississippi River watershed, nearly 80 percent of the wetlands have been destroyed. Scientists calculate that if only a portion of those wetlands had been left intact, they could have contained the devastating floodwaters that caused nearly $16 billion in damage in the flood of 1993.

Filled wetlands also make poor sites for building, as many a flooded homeowner will tell you. Not only are they prone to flooding, because they're located in flood plains, but homes built on filled wetlands tend to sink. Loss of wetlands worsens flooding and results in poorer water quality, necessitating more costly water treatment.

Sir Edmund Hillary once noted that environmental problems are really social problems. They begin with people as the cause and end with people as the victims. Deforestation and wetland destruction show this relationship very clearly. Unfortunately, millions of species suffer along with us.

Hunting for Profit and Sport

Commercial hunting has a long history of causing species extinctions and near extinctions. The passenger pigeon, mentioned earlier, became extinct in the early 1900s because of widespread clubbing and shooting by commercial interests—in addition to habitat destruction. The great auk, a large penguinlike bird that once lived along the North Atlantic coast, became extinct in 1884 as a result of overharvesting. It had been slaughtered by sailors in search of meat. The heath hen, a bird similar to the prairie chicken, once lived in an area that stretched from New England to Virginia but was severely depleted by commercial hunters.

At the turn of the century, women flocked to stores to buy fashionable hats adorned with the elegant plumes of the snowy egret. Hunters gunned the birds down unmercifully in Florida, nearly wiping out the population (Figure 15.8). Fortunately, however, when restrictive laws were passed to protect the bird early in the twentieth century, the population rebounded.

The bison (commonly known as the buffalo), whose herds once blackened the prairies, remains today in tiny remnant populations. Commercial hunting and habitat destruction caused its decline and contributed to the decline of Native American populations, which were highly dependent on the bison, as well (Figure 15.9). Many species of whale, among them the blue whale and the humpback whale, were driven to the brink of extinction by overzealous whalers. Protected today, some whale species still remain in jeopardy. Many fear that the blue whale population may have been so severely depleted that it will not be able to recover. Today, numerous populations of some commercially important fish species have fallen to dangerously low levels because of overharvesting (Chapter 12).

Not all species are harvested for their food. The big cats of Africa, for instance, have been severely depleted by commercial hunters to provide furs for fashionable coats. Despite protective

FIGURE 15.8 *The snowy egret. This bird was nearly hunted to extinction to provide plumes for ladies' hats.*

FIGURE 15.9 *The bison is a multiple-use species. It formed an important base for the culture of the Great Plains Indians. By destroying the bison, commercial hunters reduced Native American populations.*

laws, poachers still hunt jaguars, cheetahs, tigers, and other furbearers.

Several species of African rhinoceros have also been hunted to near extinction. Although they are protected by law, rhinos are still killed by poachers for their horns which are sold to Yemen and China. The Yeminis, wealthy from oil, carve the horns into dagger handles for businessmen. The Chinese grind the horns to produce an aphrodisiac and a fever-depressing drug that is reportedly useless.

Sport hunting may contribute to the decline of animal populations if their populations are not well managed. For the most part, however, hunting has helped wildlife populations. The United States has over 400 national wildlife refuges, and thousands of state wildlife areas were paid for, in large part, by the sale of duck stamps and hunting licenses. Regulations on hunting have helped protect many species from overhunting and have become a valuable tool. Hunting also helps control the population size of deer and other game species, because predators have long been eliminated from most regions.

The Introduction of Alien Species

Humans have all too often introduced alien species of animals and plants only to discover, too late, that the anticipated benefits never really materialized or were greatly offset by their negative impacts. A classic example is the introduction of the water hyacinth into Florida. This South American flowering plant was brought in to adorn private ponds, but it was accidentally released into the waterways of Florida. It spread wildly throughout the southern states, clogging rivers and lakes, killing native plants, and making navigation impossible. Several southern states now spend millions of dollars each year to clear waterways of this fast-growing species. Another example is the mongoose, imported from India into the islands of Hawaii and Puerto Rico. A fierce, quick-moving, weasel-like predator, the mongoose was brought to these islands to control rats, which caused extensive damage to sugar cane. Unfortunately, the people in charge of this introduction had not studied the animal very carefully. Shortly after the mongoose arrived, they found that it hunts primarily during the day. The rat, on the other hand, is nocturnal. As a result, mongoose–rat encounters were rare, and few rats were killed. Unfortunately, the mongoose soon began to prey on ground-nesting birds. Some, like the Newell's shearwater and the dark-rumped petrel, were eradicated from the island of Molokai. Others, such as the Hawaiian goose, were driven to the brink of extinction.

As this example shows, islands are extremely vulnerable to alien species. The reason is that native species are often ill-equipped to cope with new species, especially predators, and the habitat is too limited for them to escape the pressure exerted by the invaders. The Hawaiian islands, for instance, have been particularly hard hit by alien species. Before humans settled there, the islands lacked any natural mammalian predators. Many bird species that had lived on these volcanic islands for hundreds of years had lost their ability to fly. Of what use are wings if there's plenty of food and there are no natural predators? Flightless birds have no other choice but to nest on the ground as well. After humans settled the islands their dogs, pigs, and goats decimated the populations of the flightless birds. The birds could easily be clubbed to death, and pigs raided their nests.

Alien species are sometimes assimilated into a new ecosystem without a wrinkle in the ecological fabric. In other instances, they perish because their new environment lacks essential resources or because environmental conditions differ too much from their native habitat. Many hardy species, however, tend to thrive in new environments. The zebra mussel that is spreading through the waterways of the United States and Canada, which was discussed in Chapter 10, is a good example. Without predators, competitors, disease, or parasites, such species proliferate, interfering with native plants and wildlife—literally reweaving the ecological fabric, turning a rich and varied cloth into a threadless, often colorless one.

For a discussion of ways in which GIS and remote sensing are being used to monitor and control noxious alien weeds in the Rogue River National Forest, see the GIS and Remote Sensing box.

The Many Causes of Extinction

As noted earlier, habitat destruction, commercial hunting, and the introduction of alien species are only three of a handful of factors that influence populations. Pollution, the pet trade, pest control, and predator control can have dramatic impacts on wildlife populations. As an example, scientists have recently

GIS AND REMOTE SENSING

MAPPING NOXIOUS WEEDS WITH GIS

GIS and remote sensing technologies are helping state and federal land managers combat an invasion of weeds that is damaging and displacing native vegetation all across our country. Many of these species have come inadvertently from other countries. Seeds reach North America's shores in seed stock or on livestock. In a new land without natural controls, such as diseases, the newcomers frequently out-compete native plants, spreading widely and causing considerable economic damage.

A weed is defined as a plant that is useless, undesirable, or detrimental. Weed species tend to restrict or interfere with the use of the land. Examples of important weed species in our country are ragweed, vetch, chicory, kudzu, sesbania, Russian thistle, black mustard, lambsquarters, crabgrass, horseweed, wild buckwheat, wild carrot, bull thistle, wild parsnip, Canada thistle, leafy spurge, quackgrass, johnsongrass, horsetail, St. Johnswort, horsenettle, sheep sorrel, wild rose, and sweetclover.

Weeds are commonly classified as either common or noxious. Common weeds are species of weed plants that are readily controlled by ordinary good-farming practices. Noxious weeds are those that are difficult to control because of an extensive perennial root system, effective means of propagation, and adaptability (meaning they do well under a variety of conditions). These and other features make them hardy and invasive.

The U.S. Department of Agriculture (USDA) defines noxious weeds as species of plants that cause disease or injure crops, livestock or land, and thus are detrimental to agriculture, commerce or public health. An example of a very invasive weed species that is causing ecological havoc in much of the United States is a stunningly beautiful wetland plant known as purple loosestrife. This beautiful plant aggressively displaces native plants, disrupting fish and wildlife populations and adversely affects agriculture and public recreation.

Noxious weeds have become well established on much public and private land. Because of this, federal and state agencies have implemented numerous programs to control the spread and eliminate these foreign invaders wherever possible. Until recently, most efforts involved field surveys that required visual assessment of weed infestation. Maps were then prepared to show the extent of weed growth. Such techniques were crude and inaccurate, depending on a subjective analysis by surveyors.

Today, GIS and remote sensing technologies are helping state and federal agencies understand the extend of infestation and mount strategies to control them. One example of this work is the management of the Rogue River National Forest (RRNF) in southern Oregon and northern California by the U.S. Forest Service.

In the past, USFS staff of the RRNF identified, located, and treated weedy species that were of particular interest to their land management objectives as the problem arose or was discovered. The approximate locations of infestations were marked on forest district maps, with only sparse recording of data describing the plant, its location, and density. Field workers often reported their findings quite differently. The result was a somewhat confusing, often inefficient way of plotting weed infestation.

Knowing the potential benefits of GIS and digital mapping techniques to their land management mission, the RRNF worked with a private company to develop a standardized noxious weed database to assist them in developing efficient, effective control strategies for noxious weeds. Because they needed to locate and map weed infestations with sub-meter accuracy, Forest Service workers and consultants decided to use GPS technology in their field mapping. GPS stands for Global Positioning System. It uses military navigation satellites orbiting the Earth to triangulate the position of GPS receivers on the ground. In this case, workers used handheld GPS receivers to pinpoint infestation.

An advantage of the computer technology aspect of GPS is that the receiver system can be programmed to input data in specific formats in a very standardized and user-friendly way. These programs, called *data dictionaries,* allow workers to enter data on location and other features (density of weed species) in a digital format that is readily transferable to a GIS. Digital photographs were taken at each site as well.

Standardizing data options helped workers establish a uniform system of reporting. After each day of mapping, field workers would download the GPS data, which was then exported to their GIS for integration with their existing database and map analysis system. Aerial photos of the forest already existed and were used to identify the locations of large noxious weed populations.

GIS, GPS, and remote sensing helped the RRNF develop a Noxious Weed Strategy. For the first time in a decade, the Rogue River National Forest is battling noxious weeds with organized, accurate data. They're using integrated weed management strategy that minimizes the use of potentially harmful herbicides.

discovered that 4 percent of polar bears are contaminated with PCBs, a toxic chemical used to clean nuclear reactors in Russian submarines. This toxicant passes up through the food chain and today about 5 percent of all polar bears are unable to successfully breed because of high levels of PCBs in their bodies.

As the human population grows and as more and more countries become industrialized, pollution could take an even larger toll on wildlife populations. Especially harmful to plants and animals will be the changes in weather and ultraviolet light penetration caused by an increase in carbon dioxide and the destruction of the ozone layer, respectively (Chapter 19).

It should also be emphasized that many species become extinct not as the result of a single action, but of many. The California condor once soared above much of the southern United States from California to Florida. This magnificent bird succumbed to habitat destruction, ingestion of lead in its scavenged prey (pollution), and pesticides (pest control). The bald eagle, the symbol of our great nation, was pushed to the brink of extinction as a result of habitat destruction, shooting, and pesticides. Fortunately, protective measures have helped the species recover. (Some of the ethical considerations that arise in extinctions due to human pressures are discussed in Ethics in Resource Conservation Box 15.3.)

Traits of Vulnerable Species

Making matters more complicated, some species have attributes that make them more vulnerable to extinction than others. Some of the major characteristics of concern are (1) specialization, (2) low biotic potential, and (3) nonadaptive behavior.

Specialization **Specialists** are organisms that have rather narrow requirements for reproduction and survival. Because of this they are extremely vulnerable to extinction.

A good example of a specialist that is highly vulnerable to extinction is a bird known as Kirtland's Warbler (Figure 15.10). A tiny bird with a powerful song, Kirtland's warbler has an extremely small breeding range. It is found in 13 counties in Michigan. Within this area, the warbler lives only in jack pine

ETHICS IN RESOURCE MANAGEMENT 15.3

DO OTHER SPECIES HAVE A RIGHT TO EXIST?

A few years ago a television news reporter interviewed a bulldozer operator regarding the imminent destruction of a prairie dog colony in Denver, Colorado, to make room for a new subdivision. The bulldozer operator rationalized his action by saying it was "survival of the fittest." Another construction worker shrugged his shoulders and said, "What's a few prairie dogs? They're in the way of progress."

Some animal rights advocates were also interviewed and argued that it was wrong to destroy the prairie dogs. At the very least, they said, the animals should be humanely trapped and transplanted elsewhere, even if it cost a little money.

This scenario is typical of what happens when humans and wildlife conflict. It hinges on an issue of rights. What rights do other species have, if any? Do human rights supersede the rights of all other species?

A survey of the thinking on this subject reveals a wide range of ideas. Some people think that other species have value and that humans ought to respect that. But just saying we ought to respect it isn't enough to ensure that we will. We ought to exercise more and eat better and respect the feelings of others, but that doesn't mean we will.

At the other end of the spectrum are individuals who believe that living and nonliving things have moral standing in their own right. That is to say, they have a right to exist, a viewpoint that many people have trouble with. When proponents of this point of view articulate it they are often met with disbelief.

In his book, *The Rights of Nature,* historian Rod Nash points out that incredulity met the first proposals to grant independence to American colonists, to free the slaves, to respect Native American rights, and to permit women to vote.

Historically, according to Nash, certain groups of people have benefited from the denial of rights to other groups or to nature. Today, in fact, many people see the idea of the rights of nature as a threat to human prosperity and progress.

If you poll environmentalists, you will find diverging views on this issue. To many, it is right to protect and wrong to abuse nature because it could harm people and their way of life. It denies us the goods and services of nature.

At the other end of the spectrum are those who maintain that nature has intrinsic value and intrinsic rights irrespective of human needs. In other words, other species possess a right to exist just as humans do, and the same rights that we attach to people in general should be attached to nature's cornucopia of living things.

Of course, as Nash points out, nature does not demand rights as people do. Wolves and redwoods don't petition for their rights. They can't. Thus, it is up to people to act as moral agents for nature. That is, we humans have a responsibility to articulate and defend the rights of the other occupants of the planet.

This seemingly radical notion of the rights of nature is an extension of the idea of liberty to nature, Nash points out. Liberty is an American tradition. The U.S. Constitution proclaims that we have certain inalienable rights: life, liberty, and the pursuit of happiness. Advocates of the rights of nature simply want to extend these American ideals to other living things.

Radical?

Believe it or not, these ideas are not new. Greek and Roman philosophers spoke of natural rights, called *jus naturale.* Jus naturale

(continued)

held that people had rights based solely on the fact that they existed. The Romans found it logical to assume that other animal species had rights, too, which they called *jus animalium.*

But after the decline of Greece and Rome, Nash points out, nature did not fare so well. Increasingly, people assumed that nature, animals included, had no rights, and that nonhuman beings existed to serve human beings. The relationship of people to nature emphasized expediency and utility.

Several centuries later the rights of animals once again came to the forefront, this time in England, over the issue of vivisection—the dissection of living animals. Renè Descartes (1596–650) was called on to support vivisection because he believed that animals could not feel pain. Moreover, he argued, animals did not think and therefore could not be harmed. Others, of course, disagreed.

In 1641, the Massachusetts Bay Colony passed the first law respecting the rights of domestic animals. The law read, in part, "no man shall exercise any Tirranny or Crueltie toward any bruite Creature which are usually kept for man's use."

John Locke, the seventeenth-century English philosopher, became an important source of thinking on animal rights. He argued that people have certain natural (inalienable) rights by simple virtue of our existence. For example, we share a natural right to continue existing. Interestingly, however, Locke did not assert that nature or animals had natural rights. He argued against cruelty toward animals for the way it affected people, maintaining that cruelty to animals would harden people toward people.

In the early history of the English humane movement, though, it was argued that animals were part of God's creation, and so people had the responsibility for being good trustees on God's behalf. All nature existed because of and for the glory of God, the Creator. Proponents of this view argued that God cared as much about the welfare of the most insignificant being as about human beings.

Additional support for the rights of nature came from the philosophy of animism, which held that a single and continuous force permeated all beings and things. The philosopher Benedict de Spinoza (1632–1677) put forth the notion that every being or object—wolf, maple tree, human, rock, star—was a temporary manifestation of a common God-created substance. When a person died, the matter in his body became something else: soil and food for a plant, which might nourish a deer and, in turn, a wolf or another person.

Spinoza's understanding of these interrelationships made it possible for him to place ultimate ethical value on the whole rather than on any single transitory part, such as a human life.To Spinoza, there were no higher and lower organisms; his idea of community held no bounds either. A tree or a rock had as much value and right to exist as a person.

Alexander Pope (1688–1744), the gifted British poet, summarized thousands of pages of animist philosophy when he wrote that living things "Are all but parts of one stupendous whole, whose body Nature is, and God the soul."

Things gradually got better for animals and for nature as people began to see that animals could think and feel pain. In 1789, England's Jeremy Bentham, who believed that animals could feel pain, said that pain was bad and pleasure good. He understood how the maximization of happiness could be extended from colonists to slaves to nonhuman beings. He rejected the ability to reason or to talk as an ethical dividing line between people and other forms of life. The question, he said, is not can they reason or can they talk, but can they suffer? Bentham wrote that the time will come when humanity will extend its mantle over everything that breathes.

John Lawrence, a little-known Englishman, argued that "life, intelligence, and feeling necessarily imply rights." He believed that in these respects animals were like people and that the "essence of justice" was not "divisible."

Edward Nicholson, a librarian at the London Institution, wrote in 1879 that saying animals have no powers of reason is inconsistent with common observations of household pets. Granted, an animal's "functions of mind are fewer and its feeling more limited than that of a man," but so are those of a human idiot. And, he maintained, no ethical person proposes to deny such a person's rights to life and liberty.

Nicholson argued that animals are capable of experiencing pain and pleasure and so have the "same abstract rights of Life and Personal Liberty" as humans.

You may have noticed that our discussion began with animals, and then gradually focused on domesticated animals. Limited as it is, this humanitarian view is the fundamental building block that is leading to a growing environmental ethic.

One can see the extension of human ethics to the environment in the writings of Henry David Thoreau. Thoreau referred to nature and its creatures as his society, thus transcending the usual human connotation of the word. "What we call wildness," he wrote in 1859, "is a civilization other than our own." He regarded sunfish, plants, skunks, even stars as fellows and neighbors. "If some are prosecuted for abusing children, others deserve to be prosecuted for maltreating the face of nature committed to their care."

This kind of thinking is prevalent today. Michael Fox of the Humane Society of the United States has said, "If a human has a natural right by virtue of his very being, to be free then surely this right should be accorded to all other living creatures."

"Firms that destroy the integrity of an ecological system are the same as individuals who make cash withdrawals from 7-Eleven with a shotgun," said Tom O'Leary, not an environmental radical, but president of a Seattle energy resources company. His sentiments are echoed in other high places.

What can one conclude from these observations?

First, there is no basis for the rights of nature except that which we assign. The whole idea of rights is a human construct. We call the shots. We can deny rights to other species, or we can assign them on the basis of existence, pain, thinking ability, or God spirit. They can be partial rights or complete rights or something in between. The choice is ours.

What do rights imply? Again, that's up to us. They can be sentiment without action or they can call for the complete negation of humanity. We believe that the goal should be to find some place on the continuum that works for us and for the planet and its millions of creatures.

Adapted with permission from R. F. Nash, *The Rights of Nature.* Madison: University of Wisconsin Press, 1989.

FIGURE 15.10 *Kirtland's warbler at nest. Nests are located in the protective lower branches of a jack pine that is 5 to 20 feet tall.*

habitat, among trees that are 6 to 15 years old and 2 to 7 meters (6 to 20 feet) high. Although it is a ground nester, the bird's survival depends on trees of this size because their branches extend to the ground, providing protection while nesting and at other times as well. In younger trees, the lower branches do not provide adequate nesting cover; in pines older than 15 years, the bottom branches become shaded out and die and are no longer suitable for nesting. Before humans intervened, pine growth was held in check by naturally occurring forest fires. Fires wiped out sections of the forest and created new growth, which in a few years provided trees of a suitable age and size. Well-intentioned forest fire protection, however, upset the natural cycle of renewal and caused the forests to age, wiping out the warbler's nesting ground.

Today, the U.S. Forest Service mimics nature by periodically burning sections of the forest in prescribed burns (Chapter 14). These small, well-managed fires destroy the mature pines and cause the jack pine tree cones to pop open, releasing seeds. The new seeds grow in the burned patches, ensuring a constant supply of trees of the right age for nesting. The Forest Service as well as other state and federal agencies are also planting new areas. Wildlife officials are also trapping cowbirds, a bird that poses a threat to warblers. The cowbird lays its eggs in warbler nests. Unaware of the addition, the warbler raises the cowbird young at the expense of its own young.

Clear cutting, burning, replanting, and cowbird control are now working wonders. The Kirtland's warbler population reached approximately 1600 in 1998, the highest recorded since the first census was taken in 1951, and up from 300 in 1974. Most of the increase has occurred in areas specifically planted for warblers. This effort not only provides habitat for the warbler, it also helps provide forest products and habitat for a variety of other species of songbirds, plants, and mammals.

Unfortunately, the Kirtland warbler faces another danger: the deforestation of its wintering ground in the Caribbean. Without it, the bird remains in peril, a fact that illustrates the importance of international cooperation in protecting a species.

The restricted habitat of some species makes them targets for extinction. So do restricted diets. China's panda, for instance, eats the leaves of certain bamboo trees and little else. If bamboo is destroyed, the panda will vanish (Figure 15.11).

Another example of an endangered specialist is the graceful hawk known as the Florida Everglade kite, now called the snail kite, one of the rarest species of bird in the United States. This graceful bird lives in central and south Florida and in parts of the panhandle of Florida. An important factor contributing to its

FIGURE 15.11 *The panda is a specialist that feeds only on bamboo, shown here. Destruction of the bear's only food source threatens to wipe out this magnificent creature.*

falling numbers has been its highly specialized diet. The kite, so named because it hovers like a kite above the swamplands where it feeds, subsists almost exclusively on apple snails. The snails are dependent on emergent vegetation (aquatic plants that protrude above the surface) such as spike rush, sawgrass, and cattails. These plants enable the snail to climb near the surface to feed, breathe, and lay eggs. The kites pick them off the vegetation, then extract the snail from their shells.

Snail kite populations began to decline in the 1940s as farmers and real-estate developers drained and filled in marshland, in the processing wiping out the snail's once vast habitat. To date, the Everglades habitat has been cut in half from these and other activities. Further declines have resulted from rising demand for water for irrigation and domestic uses, which lowers water levels in remaining swamplands and reduces snail habitat. Pollution of the waters from dairy and vegetable farms has resulted in a massive die-off of snails, too. In 1972, the snail kite population had plummeted to 65 birds. Although numbers have climbed nicely since then, there are only about 1,000 individuals left. (Fortunately, the species is well represented in Mexico and South America.)

Generalists are organisms that occupy a variety of habitats and eat a number of different foods. Because of their ecological versatility, they generally fare better. If their habitat is destroyed, they can move elsewhere. If they lose a food supply, they can shift to another. The coyote of North America is an excellent example. This marvelously adaptable canid is expanding throughout much of the United States, especially the Northeast, filling the empty niche created by the extinction of wolves.

Low Reproductive Rates Some animal species are extremely vulnerable to environmental stress, such as storms, drought, and disease, because of their low reproductive rates. The female polar bear, for example, breeds only once every 3 years, and then gives birth to only two cubs. The female California condor lays only a single egg every other year. The problem is further complicated by the fact that condors require six to seven years to reach reproductive age. The slow-moving orangutan of tropical rainforests of Borneo and Sumatra, breeds once every seven years.

Nonadaptive Behavior The Carolina parakeet, the only parrot native to the United States, became extinct in 1914, when the last survivor died in a zoo. The parakeet was hunted extensively by fruit farmers because the birds descended on their orchards in huge flocks, ravaging the trees. However, this exquisite red, yellow, and green "paint pot" might still be with us if it were not for one peculiar trait: When one member of a flock was shot, the remaining birds would hover above it, becoming easy targets for gunners.

Of more recent interest is the red-headed woodpecker, which ranges over two thirds of the United States. The woodpecker's population has declined in the last few decades in part because the bird has a curious tendency to fly along highways directly ahead of automobiles. Unfortunately, the latter usually win the fatal race.

15.4 Methods of Preventing Extinction

Three major methods are presently used to protect wildlife and plants, not just rare and endangered species but all species. They are (1) the zoo-botanical garden approach, (2) the species approach, and (3) the ecosystem approach.

The Zoo-Botanical Garden Approach

Wood's cycad, a tree from South America, and Cooke's kokio tree, from the Hawaiian islands, cannot be found in the wild today. Were it not for botanical gardens, the species would be extinct. Nurtured in climate-controlled facilities, the trees today hang on by a thread. Many animal species face a similar future. Pere David's deer, originally from China, continues its existence in zoos throughout the world. California condor's have been given a boost thanks to efforts of the Los Angeles Zoo and The San Diego Wild Animal Park. Once widely dispersed over much of the southern United States, the condor had suffered enormous losses. By the early 1980s, there were only 21 to 22 condors left in the wild and in zoos. Wildlife officials took steps to solve the problem. They trapped the remaining condors and sent them to the two zoos and began a captive breeding program. By late 1995, the condor population had increased to 105 birds. In January, 1992 wildlife officials began releasing captive-raised condors into the wild in carefully selected sites in California. Their hope was that the birds would establish wild breeding populations that will expand to other habitat. Additional birds are now also being introduced into a region near the Grand Canyon in Arizona.

Much of the work done to protect condors depended on state and federal government officials working jointly with zoos. More recently, The Peregrine Fund, a nonprofit organization based in Boise, Idaho, has become a partner in the costly efforts to save the condor. Their facility, The World Center for Birds of Prey, now houses 20 breeding pairs of condors. Thanks to the efforts of these organizations, at this writing (September, 2000) there are 39 California condors in the wild and numerous breeding pairs in captivity. The condor, still rare by any standard, is facing a much brighter future.

However important they are in preserving the rich biological diversity of the world, zoos and botanical gardens have some major drawbacks. First, they are a last-ditch effort, saving species before the final extinction. Second, many organisms simply do not do well in captivity. They may not breed, or may succumb to disease. Animal species accustomed to roaming over many square kilometers may become bored and restless when confined to a few square meters. Some refuse to care for their young. And caring for the offspring of these temperamental creatures is often costly and time-consuming. Third, captive-raised animals may be difficult to release into the wild.

Some zoos have taken important steps to mimic a species's habitat and have provided areas that allow animals to range more widely. Their reward: healthier, more productive animals that may help save some of our endangered species.

While harboring and breeding endangered animals in zoos may help save species in the short run, this approach is of limited value—a bit like saving a few of Renoir's and Monet's paintings for the sake of art. In the long run, however, saving species requires a more permanent solution: rebuilding wild populations in protected habitats. Realizing the importance of habitat, zoos are also beginning to play a more significant role in this effort, too. The San Diego Zoo's condors, as noted earlier, have bred successfully and are now being released into the wild in small numbers. Numerous zoos throughout the world are also cooperating in a program to breed golden lion tamarins that will be released into protected jungles in South America. In recent years, many progressive zoos have begun taking an active role in protecting habitat.

The Species Approach

Most species protection work has been the protection of those that are on the brink of extinction. Species not yet reduced to the critical level may be protected in the wild through special management programs. Scientists carefully study their niche to determine their habitat, critical food requirements, and other biological imperatives. They then design programs to enhance or expand the resources that the species needs to survive and prosper. This may require that human activities be altered. For instance, a species protection program may call for a ban on habitat destruction or controls on dangerous pollutants. In other instances, it may require measures that improve the habitat, such as fertilizers or prescribed burning and predator control. Stream improvements that protect spawning grounds and provide shelter may be needed to protect fish populations. Today, the California sea otter and the bison owe their survival to such actions.

One problem with this method is that it tends to overlook the needs of other species. By narrowly concentrating on one species, society may overlook other species that, in the long run, are more valuable. It is quite possible, for example, that a species that might be of more value to society would be neglected in a program aimed at protecting one that happens to be visually more appealing. For example, many people who would enthusiastically support protecting the grizzly bear would balk or laugh at similar efforts in behalf of the Furbish lousewort or some insect of far greater value to society.

Also, a narrow approach overlooks the fact that species are part of a complex ecosystem. To protect a species requires protection of its ecosystem and all the members of it, the subject of the next section.

The Ecosystem Approach

Perhaps the most significant outcome of the science of ecology is the concept of the ecosystem—an interacting and interdependent network of biotic and abiotic factors. A knowledge of ecosystems suggests that to preserve the Earth's species requires measures to preserve its ecosystems—a step that will benefit us as well. The ecosystem approach to preservation is perhaps the most effective and least costly means of saving plants, animals, and microorganisms. The concept of ecosystem management was introduced in Chapter 1. We've talked about a form of ecosystem management in Chapters 10 and 11 under watershed management. One of the key component of ecosystem management is habitat protection.

Habitat Protection By setting aside large areas of habitat that are populated with a sufficient number of species and letting nature take its course, biologists believe that we can prevent the steady decline in species diversity. This important measure, however, requires an understanding of the habitat requirements of a species and the species it depends on. It also requires an understanding of where the most critical wildlife habitats are.

The drawbacks to habitat protection, biologists are now finding, are significant. First, as discussed earlier in this chapter, unless the habitat is extremely large or connected to another natural habitat that is also relatively undisturbed, many species will be lost. A recent study of the tropical rain forests showed that small plots, 1-hectare (2.5-acre), lose all of their primates and other mammals and about half of their birds. Slightly larger islands, 10 hectares, do not fare much better. Even 100-hectare plots lose nearly half of their bee species and a few of their primates.

The ecosystem approach works best if special care is taken to set aside habitat sufficient to retain the original biological diversity. These are called **core preserves.** It also requires protection and careful management of surrounding lands, called buffer zones. A **buffer zone** is a region around a preserve in which limited human activity is permitted. The buffer, as its name implies, protects the core reserve from outside influences.

Another important element of ecosystem management is the use of **wildlife corridors**—connecting corridors that link similar habitats. Although not a panacea, wildlife corridors permit animals to escape predatory pressure and to expand into new territory. They increase food options and permit an intermixing of populations to preserve genetic diversity in neighboring populations.

Several extremely ambitious projects of this nature are now under way. One project, if successful, will create a series of connecting corridors in Florida that permit the endangered Florida panther freer access to its historical range. Another project in the midwestern United States hopes to eventually create a 362,000-square-kilometer (140,000-square-mile) bison preserve across ten Great Plains states. In China, officials concerned about the endangered panda bear are expanding preserves, creating additional ones, and connecting them by corridors. In the tropics, many countries have set aside huge tracts of land for protection. Called **extractive reserves,** these lands are also used by indigenous people for harvesting rubber, fruits, nuts, and other forest products from the forest. Such activities are thought to have little, if any, impact on the forests themselves and therefore serve human needs for food and other materials while protecting native species. Although extractive reserves are a good idea, scientists are finding vast areas of tropical rain forest that are practically devoid of animal life, having been overhunted by native peoples.

Combined with human population stabilization and reductions in pollution, habitat protection remains one of the sustainable answers to the problem of vanishing species. It may even surprise readers to learn that some parks, like New York City's massive Central Park, play a key role in protecting species. In this case, many birds use the park during the annual migration. It provides an important resting spot. Efforts are now underway by nonprofit conservation organizations and government entities to identify and protect such areas, known simply as important bird areas.

Important birds areas (IBA) refer to any habitat that is valuable to birds. It may be important nesting area or an important resting spot for migrants. The first IBA program emerged in Europe in 1985. Established by an international conservation organization, BirdLife International, the IBA program has spread to over 100 countries in Asia, Africa, the Middle East, and the Americas. Canada has an extremely active program with dozens of IBAs spread throughout its vast territory. The United States' program is growing, too.

Interestingly, IBAs are almost always private ventures, involving nonprofit organizations and private landowners. Individuals nominate sites that are essential to conserve bird diversity—that is, they are essential to the long-term survival and reproduction of naturally occurring bird species. In Canada, submissions are made to Bird Studies Canada, located in Port Rowan, Ontario. If the land is selected, it is added to the list. Nonprofit organizations and other interested parties in the region work together to determine the type of protection and management as well as those who will assume responsibility for the site. Conservation plans are drawn up in many cases.

Recent studies have also shown that many important wildlife areas in the United States are on private property. Vast government-owned tracts of land in the West, for example, while important, may not be nearly as species rich as private holdings. These biologically rich areas need to be protected, though, to ensure the survival of many species. Some conservation groups such as the Nature Conservancy often purchase such lands. Individual landowners are often inspired on their own to protect their lands, but in many cases some form of financial incentive is needed to ensure protection. Some state governments, for instance, pay for **conservation easements.** The farmer or rancher still owns the land and can pass it on to his or her heirs, but the land cannot be developed. Thus, the farmer or rancher sets the land aside in perpetuity.

Habitat Restoration Another element of the long-range plan is restoring damaged lands. In the Amazon Basin, for instance, at least 15 to 17 million hectares of forest have been converted to cropland and pasture. Approximately half of this land has already been abandoned because the poor soil of tropical forests lasts only 4 to 8 years when planted in crops or grazed by livestock. Refurbishing this land and others like it throughout the world could greatly slow the rate of loss of biological diversity.

But forest regeneration may be lengthy. The larger the disturbance, the slower is the recovery. A large clear-cut in the tropics may take 150 years to fully restore itself. Already a number of studies suggest that humans can accelerate forest regrowth and that careful work can reestablish a forest's full ecological diversity. But the costs could be exorbitant. Severe erosion must be stopped to prevent the soils from vanishing. Native species will have to be reintroduced to start natural succession.

In 1985 Rajiv Ghandi, then India's prime minister, established a program of tropical reforestation. Nearly 100% of India's large, intact forests have been cut down. The government hopes to replant 5 million hectares a year, but will limit its plantings to a few species of trees that can be used as food for livestock and fuel to supply the needs of India's rural poor. This effort could help reduce soil erosion and rural despair and would benefit wildlife, but it will not come close to restoring the full ecological diversity of the natural forests.

Some observers point out that India could easily broaden its project to restore some of its abandoned land to natural forest. Research in a variety of locations indicates that native populations can restore forests to near-natural conditions and that these forests can provide food for rural populations on a sustainable basis. For example, researchers in Mexico found that descendants of the Mayans protected and cultivated forests containing a variety of fruit- and nut-bearing trees. These forests, while not identical to native forest, were in many ways similar and supported a variety of species. Similar practices have been observed in Brazil, Colombia, Java, Sumatra, Tanzania, and Venezuela.

Habitat protection and restoration is occurring throughout the world. In the United States, much of our grassland biome has been plowed under for farms. Much of the deciduous forest biome has been leveled for pastures, farmland, towns, and cities. Conservation organizations, such as The Nature Conservancy, have taken an active role in setting land aside to protect plants and animals. And efforts are being made to restore native prairie, although on a small scale. Scientists believe that native prairie vegetation may prove to be one of the best long-term rotation crops for farmland that has been severely compacted, eroded, and depleted of its nutrients.

Protecting Keystone Species

In a stone archway, one stone located in the middle of the arch, the keystone, holds the others in place. In ecosystems, scientists are finding that single species have extraordinary influence on the well-being of others. These are called keystone species. Technically, a **keystone species** is one that, if lost, results in the loss of many others. It may, for instance, be a key food source, as in the case of the fig trees in tropical rain forests, which are a steady source of food throughout the period when fruits are not available for monkeys and birds. Or a keystone species's home may serve as a home to others. In the southeastern United States, the gopher tortoise digs burrows in the sandy soil that many other species live in or seek shelter in. So important is the gopher tortoise that if it vanishes, at least 37 other species will disappear in part because of the loss of shelter.

Understanding ecosystems and the importance of various species is therefore very important to proper management. Many little-known species may turn out to be extremely important for the ecological integrity of a region. Special efforts must be made to identify and protect such species.

Improving Wildlife Management and Living Sustainably

Yet another component of a successful plan to protect the Earth's fading biodiversity is improving wildlife management. As noted earlier, wildlife managers take steps to improve habitat and protect individual species. More and more managers are practicing ecosystem management. Efforts to better regulate commercial harvesting of fish and other species are also badly needed. Good scientific knowledge of the ecology, population dynamics, and sustainable harvest levels of commercially important species is vital to this effort, as are strong means of enforcing quotas. Cooperation on the part of companies that harvest fish and other species would go a long way, too.

Individuals, businesses, and governments can also contribute to the protection of plants and animals by creating a more ecologically sound society. Efforts to increase recycling, to use resources with much greater efficiency, and to tap into renewable energy supplies all reduce the pressure on natural systems. Population stabilization and growth management must also occur. Only by scaling back human impact can we ensure the well-being of the millions of species that share this planet with us.

15.5 The Endangered Species Act

The United States has long been a leader in protecting endangered species. America's first concerted effort to protect endangered species came in 1973, when the U.S. Congress passed the **Endangered Species Act.** This monumental act has helped to thwart the loss of species in the United States and abroad. It has also become a model for other countries.

The act requires the U.S. Fish and Wildlife Service to identify species that are **endangered,** that is, in imminent danger of going extinct, or **threatened,** that is, species likely to become endangered in the foreseeable future. These classifications are made based on population size of species under consideration and the rate of decline of their population. Table 15.2 shows the number of species officially listed as endangered in the United States and in foreign countries.

After a U.S. species is listed as endangered, it is afforded full legal protection under the act. It cannot be hunted, killed, or harrassed legally. Individuals cannot be exported. Violators can be fined up to $20,000 and can be imprisoned for one year.

The act also bans the importation of endangered species or their products from outside the United States. Recognizing the importance of habitat protection, Congress also directed the Department of Interior's Fish and Wildlife Service to identify the habitats of endangered species. Money was provided for habitat purchase.

The Endangered Species Act also promotes the protection of the habitat of endangered species by prohibiting federal projects (dams and highways, for instance) or federally funded projects on areas deemed to be critical to the survival of endangered species. Since the act was passed, thousands of projects have been modified to protect endangered species, with little or no problem. The exception to the rule is the Tennessee Valley Authority's controversial Tellico Dam, whose construction was temporarily stopped when scientists discovered the tiny snail darter in the stream to be dammed. The ensuing controversy was described in Case Study 15.1. Logging of old growth forest in the Pacific Northwest was curbed (not stopped) because of concerns over the spotted owl, which was listed as an endangered species in 1989.

The Endangered Species Act is considered by some to be a very effective tool in protecting species from extinction. Had the law not been enacted, a number of species might not have survived. Russell Peterson, formerly president of the National Audubon Society, agrees in part. He notes that the Endangered Species Act has been reasonably successful in protecting endangered species, but that it lacks the funding needed to restore and set aside habitat to protect species in the United States. Furthermore, he contends, the act "fails to

TABLE 15.2 *Endangered Species (1999)*

Item	Mammals	Birds	Reptiles	Amphibians	Fishes	Snails	Clams	Crustaceans	Insects	Arachnids	Plants
Total listings	336	274	114	26	121	29	71	20	41	5	706
Endangered species, Total	312	253	79	17	80	19	63	17	32	5	569
United States	61	75	14	9	69	18	61	17	28	5	568
Foreign†	251	178	65	8	11	1	2	-	4	-	1
Threatened species, Total	24	21	35	9	41	10	8	3	9	-	137
United States	8	15	21	8	41	10	8	3	9	-	135
Foreign†	16	6	14	1	*-	-	-	-	-	-	2

*Dash represents zero.

†Species outside the United States and outlyng areas as determined by the Fish and Wildlife Service.

Source: U.S. Fish and Wildlife Service, *Endangered Species.* Technical Bulletin 15(4), April 1991.

address the threat of extinction where it is greatest: in the undeveloped world." That threat is now being addressed by numerous agencies and private organizations. However, the specter of species extinction will not go away easily, and much needs to be done here and abroad to protect the world's rich diversity.

In recent years, the Endangered Species Act has come under attack by pro-development interests. They have tried to make economics a factor in classifying a species. They argue that if protecting an endangered species could result in economic hardship, the listing should be denied. Attempts are also under way to gut protection of internationally endangered species. When development of private property is prevented because of the presence of an endangered species, some argue that the private land holder (such as a developer) should be reimbursed for the legal "taking" of his or her property.

In response to criticism of the act, the National Research Council, which provides scientific advice to Congress, established a committee of scientists to examine the issues and make recommendations. They outlined numerous changes in the Act to make it more scientifically sound and economically responsible. For example, they called for faster development of recovery plans. Such plans, they said, should spell out which human activities are likely to harm recovery and which ones are not—a step that would allow for better economic planning in and around protected areas.

The committee also recommended that a core of survival habitat be established as an emergency, stopgap measure when a species is first listed as endangered. This habitat would be able to support the population for 25 to 50 years. After more careful study, scientists could determine the exact dimensions of the critical habitat needed for the species to recover. This might result in either a downsizing or an increase in the protected habitat.

The federal government, under the Clinton administration, also reacted to criticism of the Endangered Species Act. A previously rarely used provision of ESA permits private landowners, corporations, state or local government, or other nonfederal landowners who wish to conduct activities on their land to harm a threatened or endangered species can do as long as they have a permit. They're know as **incidental take permits.** To obtain a permit, the interested party must submit a Habitat Conservation Plan to the Secretary of the Interior or the Secretary of Commerce. **Habitat Conservation Plans** are documents that outline what parties will do to ensure that their activities won't affect the survival or recovery of an endangered or threatened species. They also outline ways that any adverse impacts would be mitigated or offset. If these conditions are met, they are approved.

As of August, 2000, over 300 Habitat Conservation Plans had been approved, covering approximately 20 million acres and protecting 200 endangered or threatened species. Although this sounds great, some critics say that the plans have serious shortcomings. Although activities outlined in the plan could have benefits, submission and approval of a plan does not mean that the permittee will necessarily carry through on it. Critics such as the National Wildlife Federation also point out that the plans authorize activities that destroy significant amounts of habitat of endangered or threatened species. What happens, they say, is that the responsible agency (the U.S. Fish and Wildlife Service) doesn't adequately assess a species recovery needs or doesn't assess the habitat destruction that is occurring elsewhere in a species range. Approval of a plan without a full understanding of these factors results in significant losses. The American Lands Alliance, a nonprofit group in Portland, Oregon, argues vigorously that most forest Habitat Conservation Plans in the West fail to offset the losses of habitat and do little, if anything, to restore degraded habitats. They say that the federal government is, in essence, giving exemptions to wood products companies and other major landowners to build roads, log, and damage habitats of threatened or endangered species on millions of acres of forestland for commercial gain. One of the many problems they cite is the failure of plans to provide habitat prior to eliminating existing habitat. As a result, many species will fail to colonize the new habitat, which is established after the original habitat is destroyed. In some instances, late successional forests are replaced with new forests that will eventually grow into buffer strips along rivers and lakes. Most often, however, they say that plans fail to provide any mitigation habitat.

Another criticism is that the Clinton Administration gave assurances to those whose plans were approved that there would be no additional conservation efforts besides those required by their plans for up to 100 years. This guarantee, sometimes called the "no surprises" policy, closes the door on necessary changes in management that may be required as further information becomes available.

The Foundation for Habitat Conservation argues in favor of Habitat Conservation Plans, noting that they result in restrictions that go beyond what is required by current law. They note that they provide benefits for unlisted species, too. They point out that the plans are based on **adaptive management,** the management of resources that changes if scientific data suggests needs. This principle, now being used more widely in resource conservation, is vital to our long-term success. It gives managers the opportunity to shift policies and actions when evidence suggests that previous ones were not appropriate. Despite their support, evidence is mounting that Habitat Conservation Plans fall short of their lofty goals and, as currently constructed, will undermine efforts to protect biodiversity. Many changes are needed to make them work. You can learn about them by visiting our Web site, http://www.prenhall.com/chiras, for links to other sites where the issue is discussed in more detail.

The Convention on International Trade in Endangered Species

Protecting wildlife requires efforts by all nations. Many have passed laws similar to the United States' Endangered Species Act. Despite this, species are still vanishing at a remarkable rate. One reason is the legal and illegal trade of plants and animals. The sale of animals skins, live animals, and plants is worth billions of dollars to those who engage in it, both legally and illegally. Recognizing that threat posed by trade, the

United Nations Environment Programme, the World Conservation Union, and the World Wide Fund for Nature launched an effort to ban international trade in endangered or threatened species. The result, a treaty known as the Convention on International Trade in Endangered Species of Wild Fauna and Flora, or simply CITES, went into effect on July 1, 1975. The treaty is currently endorsed by 150 nations. It bans the hunting, capture, or sale of endangered and threatened species. CITES, the organization that administers the agreement keeps a list of species that is currently threatened by extinction as a result of international trade. This list is known simply as Appendix 1.

Although efforts by CITES and its member nations are on going, illegal trade continues. Documents are falsified by traders so that it appears as if species being shipped have been taken from areas where they are relatively abundant. Species are often mislabeled. Wildlife officials are limited and can be bribed to look the other way. Enforcement is often lax. Penalities are mild.

As you have learned in this chapter, protecting species, especially those that are already threatened or endangered, will require a mix of actions by citizens, business people, and government officials. The task may not seem important to those who fail to understand the importance of ecosystems to human health and well-being.

Summary of Key Concepts

1. Wildlife and plants enrich our lives in many ways. Beyond the countless aesthetic benefits are purely economic ones derived from the sale of medicines, ointments, and foods. Sports enthusiasts and nature lovers spend billions to hunt, fish, and photograph nature's rich offering. Wild plants and animals provide food and innumerable ecological benefits, such as flood protection, erosion control, and nutrient recycling.
2. An estimated 3 to 80 million species share this planet with us, but this rich and varied biological world is fast disappearing. Currently, estimates suggest that 40 to 100 species go extinct every day. Most species are lost from the tropics.
3. Preserving biological diversity has become one of the most important and pressing environmental concerns of our time and is essential for creating a sustainable society.
4. To understand plants and animals and ways to protect them, one must understand ecology, especially the ways in which populations are affected by outside forces.
5. The biotic potential of a species is the theoretical maximum rate of growth for a population, given unlimited resources. In nature, few organisms reproduce at this rate for any appreciable length of time because of environmental resistance.
6. Environmental resistance refers to all of those factors that reduce population growth, for instance, adverse weather, food shortages, predation, and disease. In general, when environmental resistance climbs, a population declines. When it falls, the population increases. Ecologists group factors that contribute to environmental resistence into two categories: density-dependent and density-independent.
7. Density-independent factors are those that operate irrespective of the population density—for example, drought, cold spells, and storms. Habitat destruction by human populations is another example.
8. Density-dependent factors are those whose influence in controlling population size increases or decreases, depending on the density of a population. There are at least four major density-dependent factors: competition, predation, parasitism, and disease.
9. The size of a population results from the interaction of environmental resistance and the factors that create its biotic potential.
10. Whenever a species is introduced into a new habitat with adequate resources, its population grows in a characteristic way, following an S-curve. During the second phase, explosive growth occurs because little environmental resistance exists. During the final phase the population stabilizes as a result of environmental resistance. At this point, the population is said to be in a dynamic equilibrium—that is, it changes somewhat from year to year but remains fairly constant over the long term. The population has reached carrying capacity, the population size a habitat can support on a sustainable basis.
11. Wildlife managers use their knowledge of population growth curves to manage wild populations and to achieve the maximum sustainable yield.
12. Not all populations remain in dynamic equilibrium. Outside forces such as highly favorable weather can cause populations to increase abruptly. Such irruptive growth may extend the population past the habitat's carrying capacity and may result in a population crash. Human activities can easily upset the balance, resulting in irruption.
13. A number of species, such as the lemming, naturally experience 3- to 4-year cycles. Others, like the lynx and snowshoe hare, experience 10-year cycles.
14. Most of the species that have become extinct since the beginning of the Industrial Revolution have done so not because they have been unable to adapt to natural changes in the conditions on Earth, but rather because they have been unable to adapt to changes wrought by humans.
15. The major causes of extinction are (a) habitat alteration and destruction, (b) commercial overhunting and overharvesting, (c) the introduction of alien species, (d) pest and predator control, (e) pollution, and (f) the pet trade. Controlling these causes, especially the first three, could greatly improve the chances of survival for many of the world's vanishing species.
16. In many places only tiny islands of natural habitat, called ecological islands, now exist amid a sea of crops, pasturelands, towns, and mines. The smaller the island, the fewer species it can support. Thus, the progressive fragmentation of the Earth's biomes is gradually eroding its biological diversity. Ecologists are especially concerned with the damage now occurring in tropical rain forests.
17. Biologists once believed that tropical rain forests were being cut at a rate of 2 percent per year, but recent work by the Food and Agriculture Organization of the United Nations shows that the global average is probably only

about 0.6 percent. However, the average hides the rapid rate of destruction in some regions, now in desperate need of protection.

18. Biologists are also concerned about the impact of destroying wetlands. Already many of the world's wetlands have been destroyed by human development. The wetlands provide many direct and indirect benefits to human society and are an important habitat in need of protection.
19. Commercial hunting has a long history of causing species extinctions and near extinctions. Sport hunting may contribute to the decline in animal populations if their populations are not well managed. For the most part, however, hunters do more good than harm.
20. Humans have frequently introduced alien species of animals and plants only to discover, too late, that the benefits never really materialize. In some cases the introductions backlashed as the alien species proliferated and displaced native plants and animals. Islands have been particularly hard hit by the introduction of alien species.
21. Many species become extinct not as a result of a single action, but of many. And some species have traits that make them more vulnerable than others, such as specialization, low biotic potential, and nonadaptive behavior.
22. Three major methods are now used to protect plants and wildlife: (a) the zoo-botanical garden approach, (b) the species approach, and (c) the ecosystem approach.
23. The zoo-botanical garden approach was once narrowly confined to raising endangered species in captivity, but has been expanded to include programs in which endangered species that are bred in captivity are released in the wild.
24. The species approach involves plans developed to protect individual species by controlling human activities that threaten them and by improving habitat.
25. The ecosystem approach is perhaps the most effective and least costly means of saving endangered species. By setting aside large areas that are populated with a sufficient number of each natural species and letting nature take its course, biologists believe, we can reduce species extinction. Recent studies indicate, however, that large tracts must often be set aside to preserve species diversity.
26. The ecosystem approach also entails plans to restore damaged lands and waters, to better manage entire ecosystems, to establish preserves surrounded by buffer zones, and to create wildlife corridors.
27. Extractive reserves are also a valuable tool in protecting wildlife habitat. Extractive reserves allow humans access to natural areas to sustainably harvest nuts, fruits, and other plant products. Such uses can safeguard wildlife and plant populations.
28. Keystone species are often key components of their environment, and their loss can result in the loss of many other species. Protecting keystone species is obviously very important to the health and welfare of the world's ecosystems.
29. The Endangered Species Act of 1973 represents the United States' first concerted effort to protect endangered species. The act requires the U.S. Fish and Wildlife Service to identify endangered species and protect their habitat through a number of means. It also bans the importation of endangered or threatened species and has helped saved many species from extinction.
30. CITES, the Convention on the International Trade of Endangered Species, is another valuable tool. This international agreement bans the import of endangered species, although it is not uniformly or vigorously enforced in all countries.

Key Words and Phrases

Adaptive Management
Alien Species
Biotic Impoverishment
Biotic Potential
Biological Infrastructure
Buffer Zone
Carrying Capacity
Conservation Easement
Convention on Trade of Endangered Species (CITES)
Competition
Core Preserves
Critical Population Size
Cyclic Populations
Density-Dependent Factors
Density-Independent
Disease
Dynamic Equilibrium
Ecological Island
Ecosystem Approach to Species Protection
Ecosystem Management
Ecotourism
Endangered Species
Environmental Resistance
Extinction
Exotic Species
Extractive Reserve
Factors
Faunal Collapse
Generalists
Genetic diversity
Genetic variation
Habitat
Habitat Conservation Plan
Habitat Destruction and Alteration
Habitat Fragmentation
Habitat Restoration
Host
Important Bird Areas
Incidental Take
Irruptive Populations
Keystone Species
Maximum Sustainable Yield
Optimum Yield
Parasite
Parasitism
Population Dynamics
Predation
Predator
Prescribed Burn
S-Curve (Sigmoidal Curve)
Specialists
Specialization
Species Approach to Species Protection
Threatened Species
Wetlands
Wildlife Corridor
Zoo-Botanical Garden Approach to Species Protection

Critical Thinking and Discussion Questions

1. Suppose that a classmate of yours made the statement, "I could care less about the extinction of some weed or bug in Africa." How would you respond?
2. Describe the value of plants and animals to modern society.
3. Discuss the major hypotheses that attempt to explain why so many species have become extinct since life began on Earth.

4. Using your critical thinking skills, analyze the following statement: "Extinction is a natural phenomenon, so we really shouldn't worry about it."
5. Define the terms *biotic potential* and *environmental resistance*. Give some examples of factors that increase the biotic potential of a species.
6. Describe and give examples of density-dependent and density-independent factors.
7. A new rodent is introduced to an island with no natural predators. Describe the likely course its population will take. Draw a graph to show the shape of the curve, and explain what happens at each stage. What factors will bring the population into dynamic equilibrium?
8. List the causes of extinction. Which ones are the most important?
9. What is habitat fragmentation, and why is it so harmful to species diversity?
10. Suppose that a population of 100 passenger pigeons was discovered in Illinois. Would you expect this population to be able to survive, knowing what you do about the pigeon's reproductive requirements?
11. Give some examples in which commercial hunting has driven a species to extinction or near extinction.
12. Why are alien species such a threat to native populations?
13. Some species are vulnerable to extinction. What makes them so?
14. Suppose that fire was excluded from the Kirtland warbler's habitat. Would this promote or hinder the species's survival? Explain your answer.
15. In what ways can zoos help preserve endangered plants and animals?
16. What is meant by the ecosystem approach to species protection? Give some examples.
17. How would you respond to these questions put to you by a frustrated citizen: (a) "Who's more important, humans or alligators?" (b) "Wilderness is something I'll never be able to visit and don't care to, so why should we save it?"
18. Using your critical thinking skills and your knowledge of ecology, debate the following statement: "The Endangered Species Act is an impediment to progress and should be severely watered down. Environmental protection will only harm our economy."

Suggested Readings

Brandon, K., K. H. Redford, and S. E. Sanderson (eds.) 1998. *Parks in Peril: People, Politics, and Protected Areas.* Washington, D.C.: Island Press. Analyzes trends in park management and implications for protection of biodiversity.

Chester, C. C. 1996. "Controversy over Yellowstone's Biological Resources." *Environment* 38(8): 10–15, 34–36. Interesting insights into the value of biological resources.

Cox, G. W. 1999. *Alien Species in North America and Hawaii.* Washington, D.C.: Island Press. Valuable resource on invasive species.

DiSilvestro, R. 1996. "What's Killing the Swainson's Hawk?" *International Wildlife* 26(3): 38–43. An examination of the effect of pesticide use on migratory birds.

Freese, C. H. 1998. *Wild Species as Commodities: Managing Markets and Ecosystems for Sustainability.* Washington, D.C.: Island Press. Valuable reading.

Harris, T. 1991. *Death in the Marsh.* Washington, D.C.: Island Press. Tells the story of the Kesterson National Wildlife Refuge, illustrating how pollution affects wildlife.

Hudson, W. E. (ed.) 1991. *Landscape Linkages and Biodiversity.* Washington, D.C.: Island Press. Discusses innovative conservation strategies to protect biodiversity.

Kohm, K. (ed.)1990. *Balancing on the Brink of Extinction: The Endangered Species Act and Lessons for the Future.* Washington, D.C.: Island Press. Important collection of writings.

Laycock, G. 1996. *The Alien Animals.* Garden City, N.J.: Natural History Press. Superb account of the folly of introducing alien species into the United States.

Lipske, M. 1994. "Animal, Heal Thyself." *National Wildlife* 32(1): 46–49. Excellent article on the value of naturally occurring medicinals.

Margoluis, R., and N. Salafsky 1998. *Measures of Success: Designing, Managing, and Monitoring Conservation and Development Projects.* Washington, D.C.: Island Press. A guide for development projects that protects the environment.

Monks, V. 1996. "The Beauty of Wetlands." *National Wildlife* 34(4): 20–27. An excellent article with new insights into wetlands and the value of protecting them.

Mooney, H. A., and R. J. Hobbs. 2000. *Invasive Species in a Changing World.* Washington, D.C. Describes how changing patterns of global commerce are resulting in the spread of alien species and how climate change and other factors are influencing this spread.

Noss, R. E., and A. Y., Cooperrider, 1994. *Saving Nature's Legacy: Protecting and Restoring Biodiversity.* Washington, D.C.: Island Press.

Plotkin, M., and X. Famolare (eds.) 1992. *Sustainable Harvest and Marketing of Rain Forest Products.* Washington, D.C.: Island Press. Outlines ways of sustainably harvesting tropical rain forest products while protecting native species.

Posey, D. A. 1996. "Protecting Indigenous Peoples' Rights to Biodiversity." *Environment* 38(8): 6–9, 37–45. A look at the role of indigenous people in protecting biological resources.

Quammen, D. 1997. *Song of the Dodo: Island Biogeography in an Age of Extinctions.* New York: Touchstone. A massive book with much to offer on ecological science and extinction.

Reisner, M. 1991. *Game Wars: Undercover Pursuit of Game Poachers.* New York: Viking Press. Riveting account of poaching in the United States and efforts to stop it.

Rosenthal, D. 1996. "Showdown in Zimbawe." *International Wildlife* 26(6): 28–35. Shows what people can do to save their wildlife.

Ryan, J. C. 1992. "Conserving Biological Diversity." In *State of the World 1992.* Starke, L., (ed.) New York: W. W. Norton. Describes many ways to protect biological diversity.

Sunquist, F. 1988. "Zeroing in on Keystone Species." *International Wildlife* 18(5): 18–23. Good reference on keystone species.

Temple, S. A. 1998. "Easing the Travails of Migratory Birds." *Environment* 40(1): 6–9, 28–32. Describes the need for coordinated efforts among countries to protect migratory bird species.

Tudge, C. 1992. *Last Animals at the Zoo: How Mass Extinction Can Be Stopped.* Washington, D.C.: Island Press. Describes how captive breeding programs and restoration of natural habitat can be used to save endangered animals from extinction.

Tuxil, J. 1999. "Appreciating the Benefits of Plant Biodiversity" in *State of the World,* L. Starke, (ed.) New York: Norton. Great information on the value of plants.

Wilcove, D. 1990. "Empty Skies." *Nature Conservancy Magazine* 40(1): 4–13. Excellent overview of factors causing the decline of songbirds in the United States.

Web Explorations

Online resources for this chapter are on the World Wide Web at: **http://www.prenhall.com/chiras** *(click on the Table of Contents link and then select Chapter 15).*

Chapter 16

Wildlife Management

Wildlife management may be defined as the planned use, protection, and control of wildlife by the application of ecological principles. One major function of wildlife management is to protect endangered species, as described in the previous chapter. But wildlife management has other important functions. Consider the use of wildlife in the following scenario:

> High in the Adirondacks of New York, a weary but jubilant deer hunter, clad in blazing orange, drags out a 200-pound buck, lashes it to the top of his van, 12 points and all, and speeds home to show off his trophy kill. There will be deer steaks to grill for a long time.

This episode represents the **consumptive use** of our wildlife resources, which means the removal and alteration of natural resources by humans. Now consider a second scenario:

> In predawn darkness, a wildlife photographer creeps into his blind on a wildlife refuge. After a few minutes of cramped suspense, the salmon flash of dawn ushers in the courtship display of prairie grouse. The camera whirs, and one of the most spectacular breeding rituals in the world of birds is captured on film.

Wildlife photography represents the **nonconsumptive use** of wildlife, which is the use, without removal or alteration, of natural resources.

For much of this century, the management of wildlife for the hunter has been emphasized by wildlife managers. In recent years, however, management for nonconsumptive uses such as wildlife photography and bird-watching has received more attention.

Under some conditions, wildlife, ordinarily a desirable resource, can be harmful to society. There are a wide variety of animals, ranging from birds to elk, that can cause damage to agricultural crops, habitat, and other organisms. In such cases, the wildlife manager must develop strategies to control the destructive populations.

It is apparent that the effective management of our nation's wildlife is highly challenging and demands a great range of knowledge and skills. The involvement of well-trained professional wildlife biologists in decision making and action in many diverse arenas is essential to the wise use, protection, and control of our nation's wildlife resources.

FIGURE 16.1 *A red fox peers out from behind a rock in Kettle Moraine State Forest, Wisconsin. In order to survive, this species requires a habitat that provides cover, food, water, and adequate breeding sites.*

16.1 Wildlife

What Is Wildlife?

The term **wildlife,** in its most comprehensive sense, includes all animals on earth that have not been domesticated by humans. However, for most professionals in wildlife management, the term is largely restricted to wild vertebrate animals: birds, mammals, amphibians, and reptiles. In addition, wildlife does not include only game or commercially important species. **Game animals** are species that are harvested for recreational purposes, whereas **nongame animals** are, by subtraction, the majority of species that are not harvested for sport purposes. This includes songbirds, many rodent species, and most species of amphibians and reptiles. In reality, game animals are designated as such by legislation. For example, mourning doves are game birds in some states, where their hunting is legal. However, they are nongame birds in states where their hunting is illegal.

Wildlife Habitat

Habitat is the general environment in which an organism lives—its natural home. It provides the essentials for survival: cover, food, water, and breeding sites (den, nest, or burrow).

Cover **Cover** protects animals from adverse weather. Good examples are the dense cedar swamps that protect whitetail deer herds from winter winds and drifting snow and the leafy canopies of apple trees that shield nestling robins from the heat of the midday sun. Cover may also protect wild animals from predators (Figure 16.1). Representative of this function is the thicket into which a cottontail plunges when eluding a fox, or the marsh grasses that conceal a teal from a hawk. Even water may serve as cover, as for a muskrat or beaver, providing relative security from landbound predators ranging from wolves to humans.

FIGURE 16.2 *Parent bluebird bringing bill full of insects to hungry young. A young bluebird can consume half its weight in insects in one day. Nesting boxes provide additional protection for the young, helping boost the survival rate of this species.*

Food Within a single species, **food** preference often varies widely. It also varies in individuals, depending on the health and age of the animal, season, habitat, and food availability.

Birds and mammals spend a great deal of time searching for food. An animal's access to food is influenced by many factors, including population density, weather, habitat destruction (by fire, flood, or insects), plant succession, and type of habitat (Figure 16.2). Occasionally, when a food is abundant, an animal will exploit the source, even though it is not usually a dietary item. Consider some examples: Even though the green-winged teal's diet is 90 percent vegetation, it avidly consumes the maggoty flesh of rotting Pacific salmon. Although the lesser scaup (duck) is not normally a scavenger, the stomachs of these ducks feeding at the mouth of a sewer have been found to be filled with slaughterhouse debris and cow hair (as well as rubber bands and paper).

A house wren, normally an insect eater, fed its nestlings large quantities of newly hatched trout from an adjacent hatchery.

Animals that consume a great variety of foods are **euryphagous.** The opossum is a good example. It consumes fruits, blackberries, corn, apples, earthworms, insects, frogs, snakes, lizards, newly hatched turtles, bird eggs, young mice, and even bats. When its usual foods are scarce, this euryphagous animal is well adapted to survive by shifting to others that are available.

A **stenophagous** animal, on the other hand, maintains a specialized, or limited, diet. Such species are more vulnerable to starvation when their usual foods are scarce. For example, an early freeze that kills off insects frequently causes many swifts and swallows to die of starvation. In 1932 a disease caused a 90 percent destruction of the eelgrass along the Atlantic Coast. As a result, the wintering population of brants (small geese), which depend almost exclusively on eelgrass for food, was reduced by 80 percent.

Water Roughly 65 to 80 percent of wild animal weight is **water.** Water serves many functions. It flushes wastes from the body. As a major blood constituent, it transports nutrients, hormones, enzymes, and respiratory gases.

Animals can survive for weeks without food but only a few days without water. In the nineteenth century, buffalo herds living in the arid western grasslands of the United States traveled many kilometers to find water holes. Mourning doves may fly 50 kilometers (30 miles) from their nest site to a watering place. Dove and quail populations have been increased in the Southwest by the installation of "guzzlers"—devices that collect rainwater (Figure 16.3). Birds and mammals may get their water from dew or may drink it as it drips from foliage and tree trunks after a shower. During the northern winter, when liquid water is scarce, house sparrows and starlings will eat snow. Desert carnivores, such as the rattlesnake, fox, and bobcat, may derive water from the blood of their prey. Another desert animal, the kangaroo rat, may not need to ingest water during its entire life! It can use the water formed during cellular energy production.

The Edge Effect

The habitat essentials (cover, food, water, and breeding sites) for a given species are rarely all found in a single type of plant community. Usually an animal must rely on two or more plant communities to satisfy its needs. The region where two different ecosystems, such as marsh and oak woods, come together is called an **ecotone.** Ecologists also refer to it as an **edge.** As a general rule, the greater the amount of edge, the greater the population densities of many species. For deer, a clump of evergreens provides cover from storms and predators, a forest margin provides adequate food, and a dense thicket of shrubs provides a fawning site. However, increased edge may lead to more nest predation on forest interior nesting bird species.

Let us assume that a hypothetical animal with limited mobility requires an interspersion of four types of vegetation: grassland, shrubs, cornfield, and woods. The vegetation shown in Figure 16.4A, with minimal edge, supports only one covey (flock) of quail. On the other hand, the vegetation in Figure 16.4B, with much more edge, supports nine coveys, even though the total area of each vegetational type is the same in both cases.

Corridor

Corridors allow wildlife populations to expand habitat availability. A terrestrial corridor is a narrow strip of land that differs, usually in terms of dominant vegetation (such as forest or grass-

FIGURE 16.3 *Wildlife habitat improvement. Chukar partridge (introduced from Asia) have been attracted by a "guzzler," a watering device used in desert country for wildlife.*

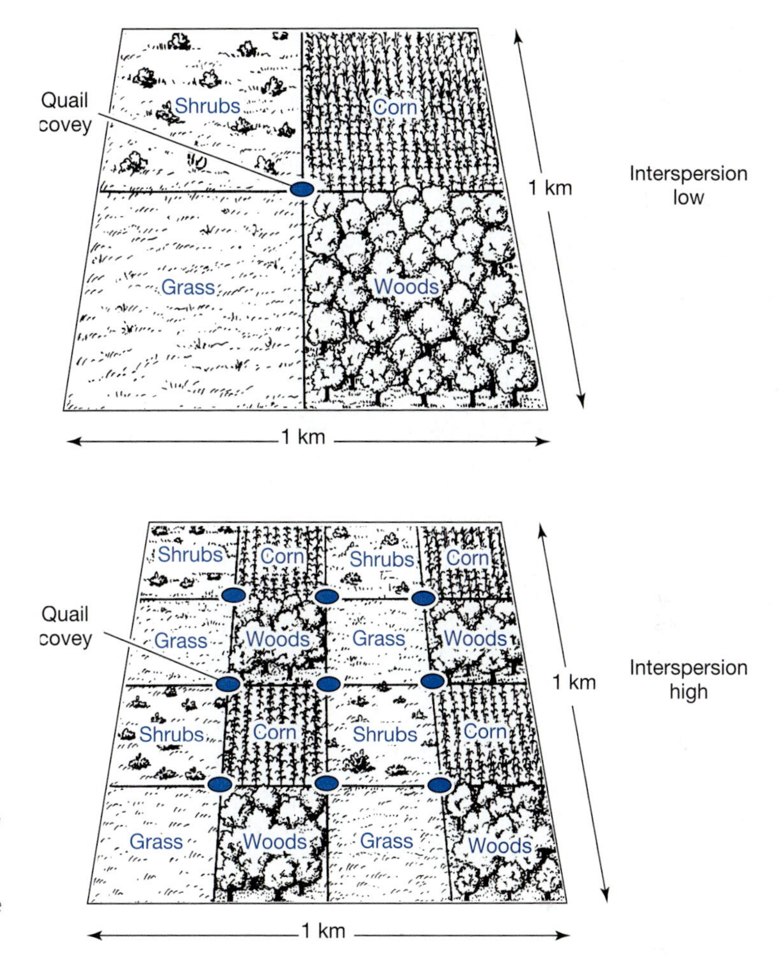

FIGURE 16.4 *(A) A square-kilometer area with minimal interspersion of four habitat types. This area can support only one covey (flock) of quail. (B) The same square kilometer with much greater interspersion of four habitat types. This area can support nine coveys of quail.*

land), from the surrounding areas. A good example of a corridor is a narrow strip of trees that extends across a large meadow and connects extensive patches of forest on either side of that meadow. A corridor can also be a narrow stretch of grassland running through a forested region. Disrupting corridors can have serious consequences on the movement of some animals. For example, in prairie or agricultural regions, wooded corridors next to rivers may be important for migratory birds and other animals.

Home Range

Ecologists define a **home range** as the area over which an animal habitually travels while engaged in its usual activities. The size of a home range can be determined by marking, releasing, and recapturing an animal. Animals can also be tracked with Geiger counters after having been fed radioactive materials. They can be fed dyed foods that result in colored feces; the home range can then be determined by studying the distribution of droppings. Birds can be individually marked with colored leg bands or spray paint. Small mammals can have their ears notched or toes clipped. Large animals (buffalo and elk) can be tattooed or marked with plastic collars so that visual identification is possible at a distance. Today, a wide variety of animals are tracked by radio waves (Figure 16.5). Animals are captured, fitted with a radiotransmitter, and then released. Scientists can then track the animal by remote sensor.

Herbivores usually have smaller home ranges than carnivores. A plant-eating moose, for example, may have a home range of only 40 hectares (100 acres or 0.4 square kilometers) of swamp. The omnivorous grizzly bear, on the other hand, requires 52 square kilometers (20 square miles) of habitat. Timber wolves have home ranges of at least 100 square kilometers (39 square miles).

Territory

A **territory** is defined as any defended area. Territories are usually defended against individuals of the same species. In many kinds of birds, threat displays (gaping, crouching, fanning the tail) and/or songs are used to defend territory rather than fighting.

Territorialism in birds may serve many functions. It can (1) ensure adequate food, (2) establish and maintain the pair bond, (3) spread birds out to control infectious disease, (4) reduce interference with breeding (mating, nest building, incubation), and (5) reduce predation because the territorial birds become familiar with refuge sites.

The size of a bird's territory varies widely, from 0.3 square meter (3.3 square feet) for the black-headed gull to 9,300 hectares (23,000 acres) for the golden eagle (Figure 16.6). The majority of songbirds (like the robin) have a territory size of 0.1 to 0.3 hectare (0.25 to 0.75 acre).

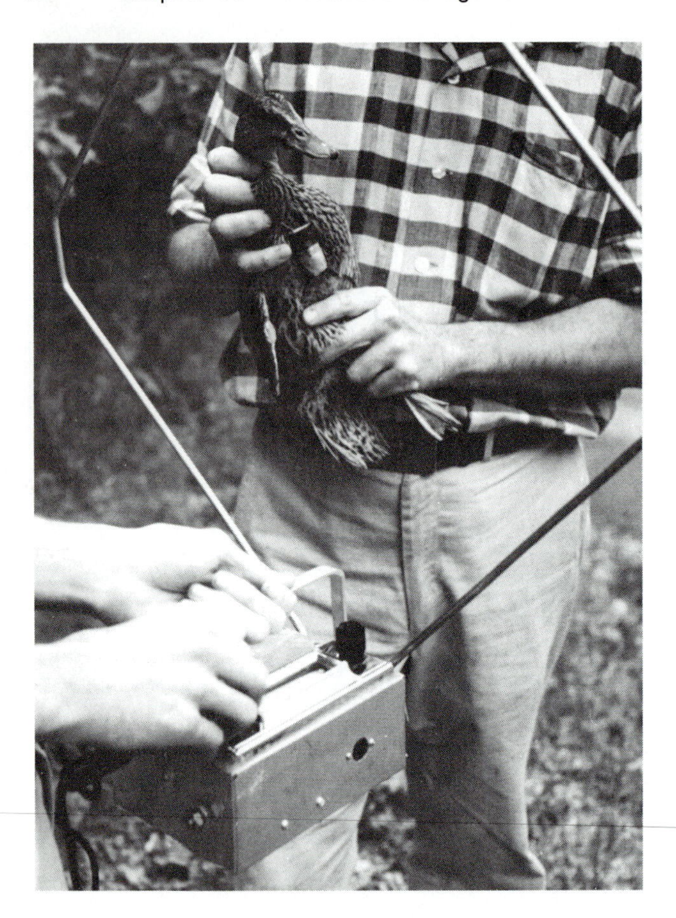

FIGURE 16.5 *Female mallard has been fitted with radio transmitter. After the bird is released, the radio waves generated by the transmitter will be picked up by the receiving set shown in the foreground. This is part of a University of Minnesota project conducted at the Cedar Creek Wildlife Area, Minnesota.*

16.2 Types of Animal Movements

For the greater part of their lives, birds and mammals occupy a relatively small area represented by their home ranges or territories. However, under certain conditions, a given species may move considerable distance from the original home range or territory. Such movements promote the survival of the species. Three basic types of movements are (1) dispersal of the young, (2) mass emigration, and (3) migration.

Dispersal of the Young

The phenomenon of dispersal occurs in the young of many birds (gulls, herons, egrets, grouse, eagles, and owls) and mammals (muskrats, fox squirrels, and gray squirrels). In a pine–oak habitat in central Pennsylvania, up to one half of the juvenile ruffed grouse leave their nesting areas, some traveling a distance of up to 7.5 miles (12 kilometers). Young bald eagles in Florida move north immediately after fledging, some arriving 2,400 kilometers (1,500 miles) distant in Maine and Canada (Figure 16.7). Up to 40 percent of a wintering muskrat population may disperse in spring. They are primarily young animals that have been ejected by the established, more aggressive adults. Such dispersals control population densities. Many of the dispersed young move into marginal habitats, where they incur heavy mortality from predation and accidents.

Mass Emigration

Mass emigrations frequently occur when a population has peaked because of extremely favorable conditions (such as food or weather) and has then experienced a greatly reduced food supply. Under such conditions, the alternatives to starvation are summer dormancy, hibernation, or emigration. Snowy owl emigrations into the United States from the Canadian tundra are correlated with the population crash of their lemming prey. Ornithologists recorded 13,502 snowy owls during the 1945–1946 emigration, which extended as far south as Oregon, Illinois, and Maryland. Twenty-four were observed over the Atlantic. Some were even seen in Bermuda. It is believed that very few of these owls live long enough to make the return flight to the Canadian tundra the following spring. Many are shot illegally and wind up stuffed with cotton on someone's mantelpiece.

Migration

Two types of migration occur with wildlife species. **Latitudinal migration** (north–south movements) involves changes in latitude, whereas **altitudinal migration** involves changes in elevation.

Latitudinal Migration Winter bird densities in the southern United States are high because many birds that breed in more northern latitudes temporarily join the permanent residents (Figure 16.8). Foods such as insects, fruits, and seeds are more available in the south than in the snow-covered lands to the north. In spring, however, the increasing day length eventually triggers hormonal secretions that stimulate migration. Presumably the northern habitats have a higher carrying capacity for the migrants and their future offspring. In far northern latitudes during the summer, there is more daylight in one 24-hour cycle for feeding the young. Biologists believe that the exploitation of two different habitats (winter and summer) may ensure a more balanced supply of vitamins and minerals.

Altitudinal Migration Latitudinal migrants move thousands of kilometers to find warmth and food. In contrast, altitudinal migrants achieve the same result simply by moving a few kilometers down the mountainside. The elk herds of the Rocky Mountains ascend the mountains in spring, keeping pace with the receding snow line, and spend the summer at the relatively cool upper levels. When the first snows cover their food supplies, the elk move down to the valleys for the winter. Herds of bighorn sheep make similar migrations.

16.3 Mortality Factors

In the previous chapter we saw that the population level of any species at a given time is the expression of two opposing forces—the biotic potential, which tends to push the popula-

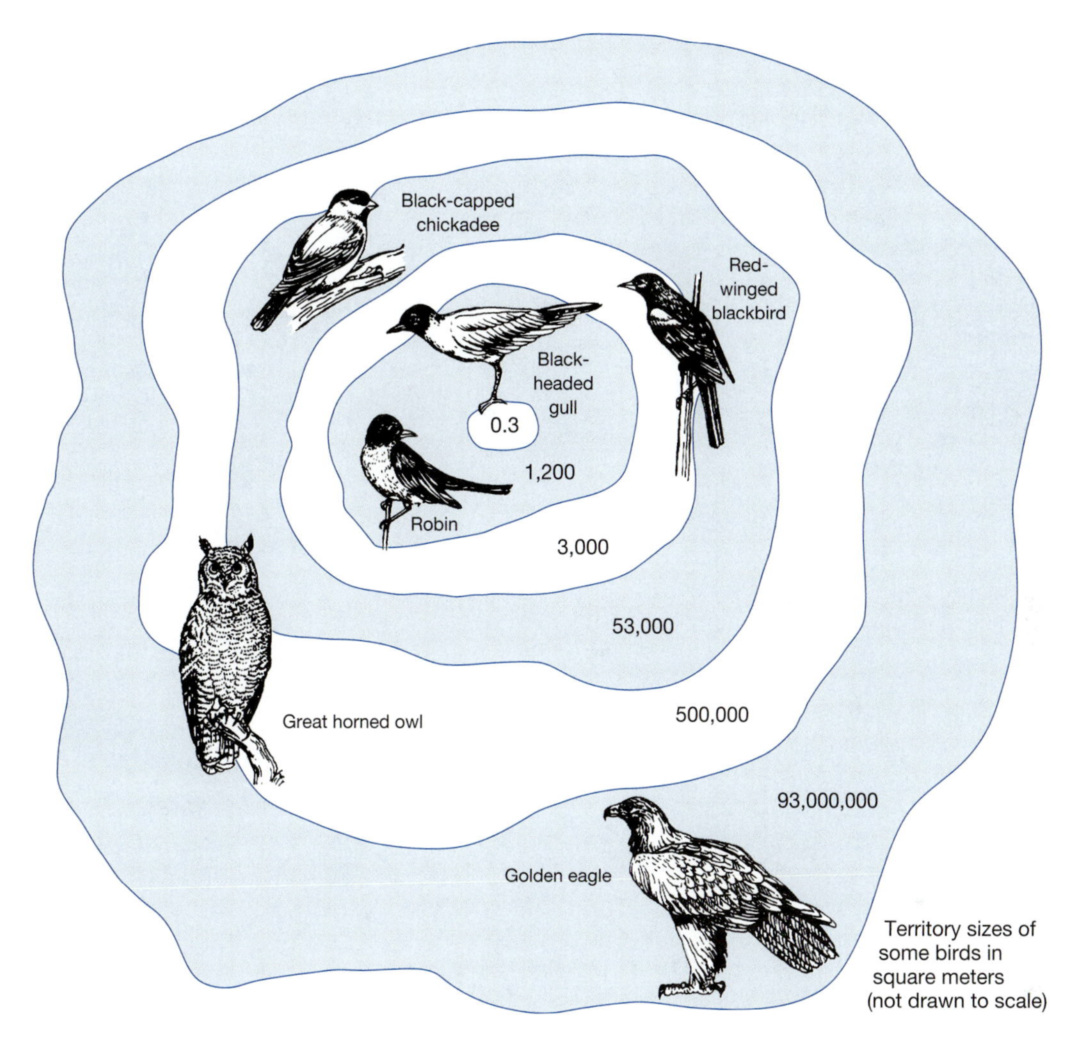

FIGURE 16.6 *Territorial sizes of several species of birds, ranging from 0.3 square meter for the black-headed gull to 93 million square meters for the golden eagle.*

tion up, and the environmental resistance or mortality factors, which tend to force it down. We shall now consider a variety of mortality factors affecting populations of deer and waterfowl.

Mortality Factors Affecting Deer

Starvation Winter is a critical season for deer in the northern states because available food is extremely limited. Herbs, mosses, fungi, seedlings, and stump sprouts are often covered by snow. Under such conditions, the only available plant materials are buds, twigs, and the foliage of conifers, such as white cedar and pine. If the deer populations grow too large, they will consume all the available browse up to the height they can reach when rearing up on their hind legs (Figure 16.9). As a result, a conspicuous **browse line** will form at a height of about 1.5 meters (5 feet), a definite warning to the wildlife biologist that the deer herds have overtaxed their food supplies.

Heavy snowfall in the Rocky Mountains may confine mule deer to 10 percent of their normal winter range. Under these conditions, the only available food is on the sunny south slopes, where snow melts rapidly. Of course, as deer crowd onto these slopes, food supplies can be depleted. One wildlife biologist once counted the carcasses of 381 starved deer in only 5 square kilometers (2 square miles) of a heavily used range in the Colorado Rockies.

Emergency feeding of starving deer is not considered sound management by most wildlife biologists. They argue that it permits the survival of deer whose future progeny will exert even greater demands on the available natural browse, thus aggravating the problem. Artificial feeding also may facilitate the spread of disease by promoting concentrations of highly susceptible animals. Moreover, it is expensive.

Predation A number of predators feed on deer, including wolves, cougars, bobcats, coyotes, and dogs. In Superior National Forest (Minnesota), wolves have been known to kill 1.5 deer per 2.6 square kilometers (1 square mile) annually. One year the wolves killed 6,000 (17 percent) of the 37,000 deer in the forest. Because hunting pressure in the remote backwoods country of Superior National Forest is extremely light, accounting for only 0.65 deer per 2.6 square kilometers, wolf predation theoretically might help control a deer herd that is often on the verge of exceeding the carrying capacity of the range. Because there are only 2,000 to 2,300 wolves in the entire state of Minnesota, the impact

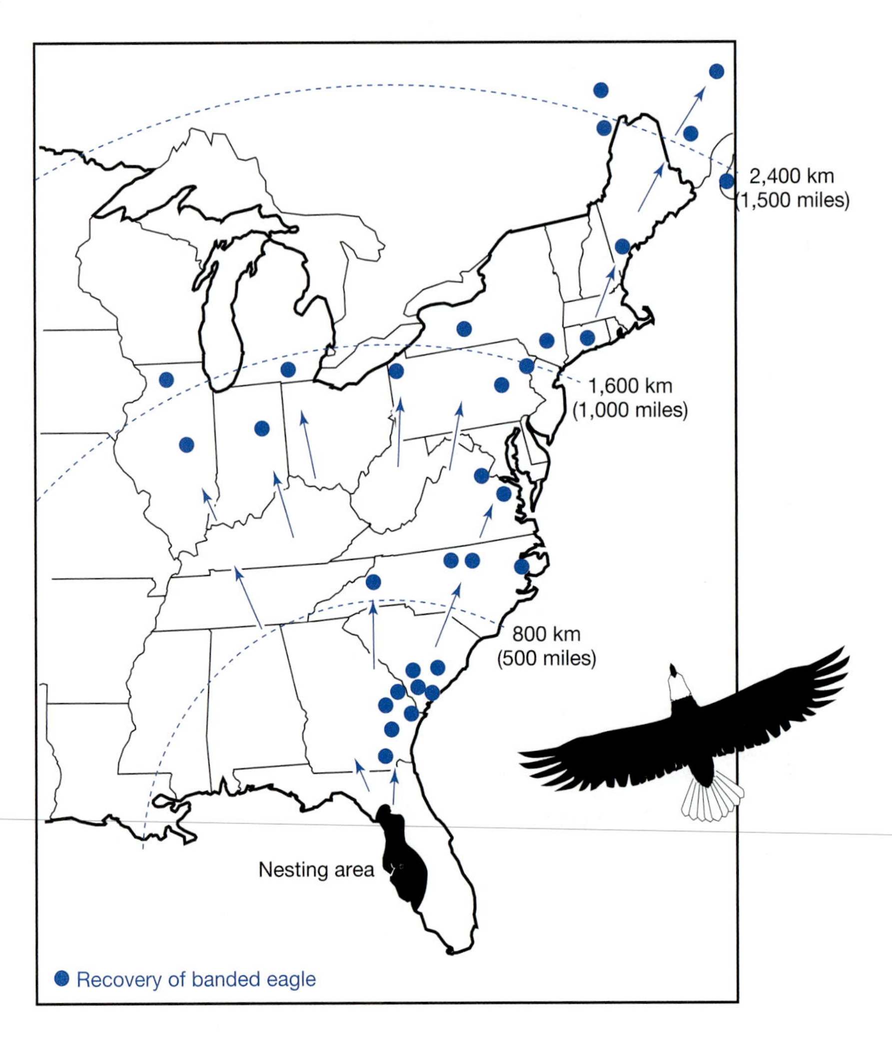

FIGURE 16.7 *Juvenile dispersal of young bald eagles that were leg-banded as nestlings in western Florida. Many of these birds moved up the coast, some to as far as 2,400 kilometers (1,500 miles) away! The function of this dispersal is unknown. After spending their first summer in the northern states, they return south to their breeding grounds.*

of wolf predation on deer in that state is probably negligible. In fact, wolves improve the deer herd and do not compete for the same deer that hunters want. A 1971 study conducted in Minnesota by the US Forest Service found that the vast majority of deer killed by wolves were at least five years old and in poor health, while the majority of deer killed by hunters were two years old or less and in good health. In addition, predator species populations are delicately balanced with their prey species. When deer populations decline, the wolf population declines soon after.

Although one cougar may kill 50 or more deer annually, cougars are unimportant as regulators of deer populations because of their scarcity, except in localized areas of the Southwest. Free-ranging domestic dogs can be a problem for deer, especially does and young fawns, but mostly in local cases.

Hunting A major source of mortality for some deer populations, especially the white-tailed deer, is legal, sport hunting. Illegal hunting and poaching can be important in some areas of the United States, but estimates are hard to acquire. Because deer populations can reproduce rapidly, effective control by natural predators and human predators (hunting) is necessary so that populations do not exceed the carrying capacity of their habit. However, some people and conservation groups oppose hunting and would like to see it banned or reduced. The hunting controversy is discussed in A Closer Look 16.1.

Disease Disease can be a concern for deer populations. For example, white-tailed deer are susceptible to outbreaks of epizootic hemorrhagic disease (EHD). This disease is caused by a virus. Outbreaks are common in the southeastern United States and sporadically in other regions and Canada. Outbreaks are usually associated with high white-tailed deer population densities. There is no effective control of this disease.

Accidents Collisions between automobiles and deer can be a significant problem in some states. For instance, Pennsylvania has about 45,000 reported deer–car collisions every year. The Pennsylvania Game Commission suspects that the true number is more than twice that reported.

Mortality Factors Affecting Waterfowl

The major mortality factors affecting waterfowl populations include loss of habitat, oil and chemical pollution, hunting and lead poisoning, disease, and acid rain (Figure 16.10).

Loss of Habitat Since habitat provides cover, food, water, and breeding sites, its loss is a major threat to migratory birds. Their breeding sites are in areas with wetlands, small ponds, and lakes. More than half of our nation's original inland and coastal wetlands have been lost to farming and other develop-

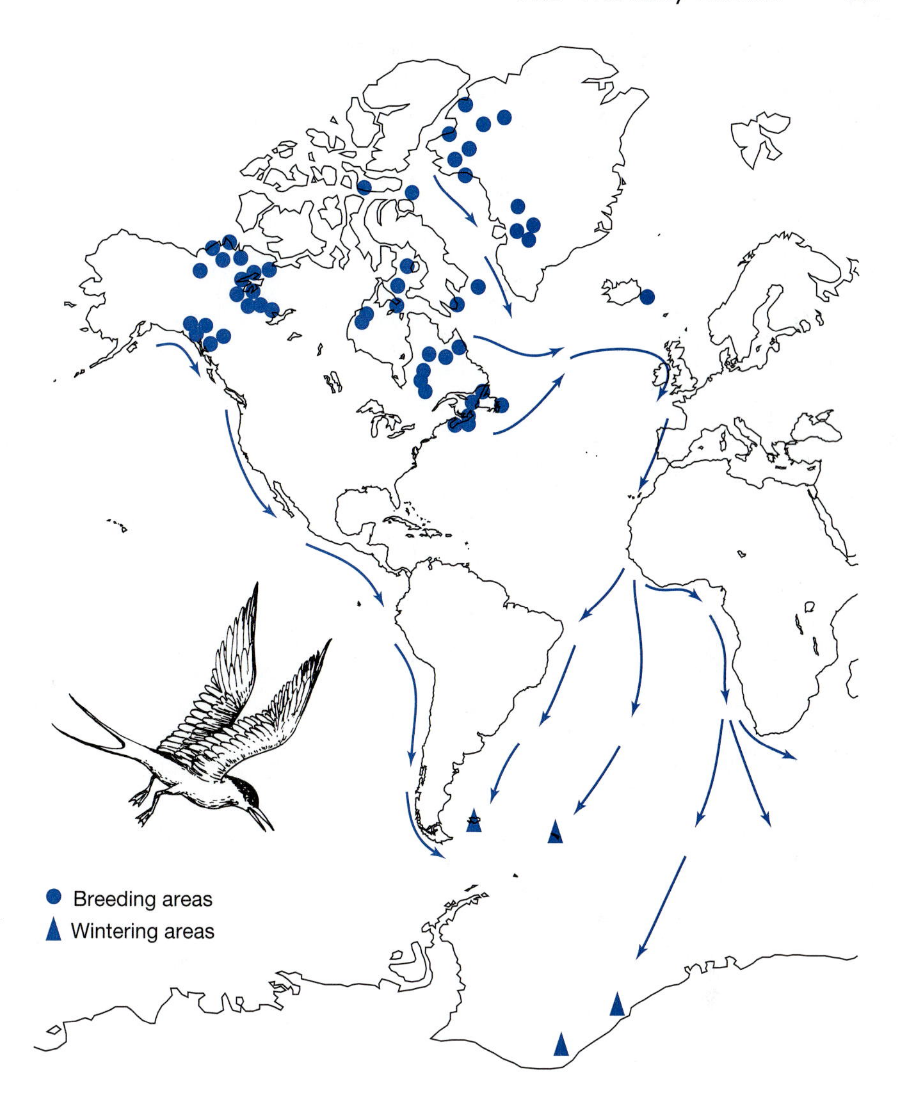

FIGURE 16.8 *Latitudinal migration of the Arctic tern. Only the southward movement is shown. Note that many of these birds nest in the Arctic and winter in the Antarctic. During their southward migration, some of these birds will cross the Atlantic Ocean twice and will complete an annual migration of about 40,000 kilometers (25,000 miles)—the longest migration of any organism in the world!*

ment (see Chapter 9 for a discussion of wetlands). For example, more than 1.5 million hectares (3.8 million acres) of wetland in Louisiana, Mississippi, and Arkansas have been destroyed in the past 25 years, largely because of conversion to soybean fields. Only about 20 percent of the original area, about 2 million hectares (5.2 million acres), remains. Wetland loss has serious adverse effects on thousands of waterfowl, like Canada geese, mallards, black ducks, wood ducks, and teal, that use wetlands as wintering grounds.

The most productive "duck factory" on the North American continent is located in the grassland biome of Manitoba, Saskatchewan, Alberta, the Dakotas, western Minnesota, and northwestern Iowa (Figure 16.11). This region produces more than one half of this continent's waterfowl. The ducks are raised primarily in tiny potholes, 0.5 to 1 hectare (1.25 to 2.5 acres) in size, where all their requirements for food, cover, water, and nesting sites are usually met. The density of the potholes may reach 50 per square kilometer (125 per square mile). An estimated 10 million potholes once existed in the prairie provinces of Canada alone.

Unfortunately, thousands of potholes in the United States have been drained by farmers, seriously threatening waterfowl populations. Much of this drainage has been subsidized by the U.S. Department of Agriculture (USDA) so that more corn, wheat, and soybeans could be produced. In Iowa alone, the number of potholes has been reduced by at least 94 percent since the late 1930s. Despite the seriousness of the problem, drainage continues. As a result, the number of ducks produced in the prairies of North America has dropped from 15 million to 5 million per year.

Drought can be as destructive to wetlands as agricultural drainage. For example, about 1 million potholes in the duck factory of Canada's prairie provinces, the Dakotas, Iowa, and Minnesota went bone dry during the drought of 1988. This was the most severe dry spell since the Dust Bowl days of the 1930s. As a result of the sharply reduced acreage of prime breeding habitat, the total North American duck population in autumn, just prior to the hunting season, was about 66 million—8 million less than in 1987 and the second lowest ever recorded.

Carp is a notorious destroyer of waterfowl habitat. It can eradicate dense growths of sago pondweed, water milfoil, and coontail, all favored duck foods. Lake Koshkonong in southern Wisconsin was once almost blanketed with rafts of canvasbacks, which consumed the abundant wild celery buds and pondweed nuts. Late in the nineteenth century, however, carp were introduced to the lake. In a brief time, the fish uprooted

FIGURE 16.9 *A deer rears up on its hind legs for browse in a Michigan forest, a possible sign that the deer herd is overtaxing the carrying capacity of the range.*

the choice waterfowl food plants, and the thrilling panoramas of ducks quickly vanished. To make matters worse, young carp compete directly with ducklings for the protein-rich crustaceans so essential for growth and development.

Oil and Chemical Pollution More than 100,000 waterfowl are killed annually by oil pollution (Figure 16.12). The timing of the spill greatly affects the mortality. On January 2, 1988, a storage tank collapsed near Pittsburgh, Pennsylvania, and released a tidal wave of diesel fuel into the Monongahela River. The 14-mile-long slick caused some waterfowl mortality. The death toll would have been much higher had the spill occurred during the spring, when thousands of ducks move up the Monongahela during their migration.

Why does oil kill ducks? There are several reasons: (1) Oil mats the feathers of waterfowl and reduces their ability to keep warm in ice-cold water. Death then results because of rapidly dropping body temperature. So dangerous is oil that an oil-soaked area the size of a quarter is sufficient to kill some waterfowl. (2) Waterfowl with oil-matted feathers may also starve to death because they lose their ability to swim or fly in search of food. (3) The accidental swallowing of toxic oil while feeding, drinking, or preening feathers may also cause kidney and liver failure.

Pesticides and other chemicals in irrigation runoff have polluted wetlands near farmed areas and endangered waterfowl. One of the more famous examples is in the Kesterson National Wildlife Refuge, which sits adjacent to California's heavily farmed San Joaquin Valley. Here, toxic selenium leached from the soil by irrigation water has caused birth defects in young birds hatched in the refuge.

Hunting and Lead Poisoning About 20 million ducks are killed each year by hunters in the United States and Canada, of

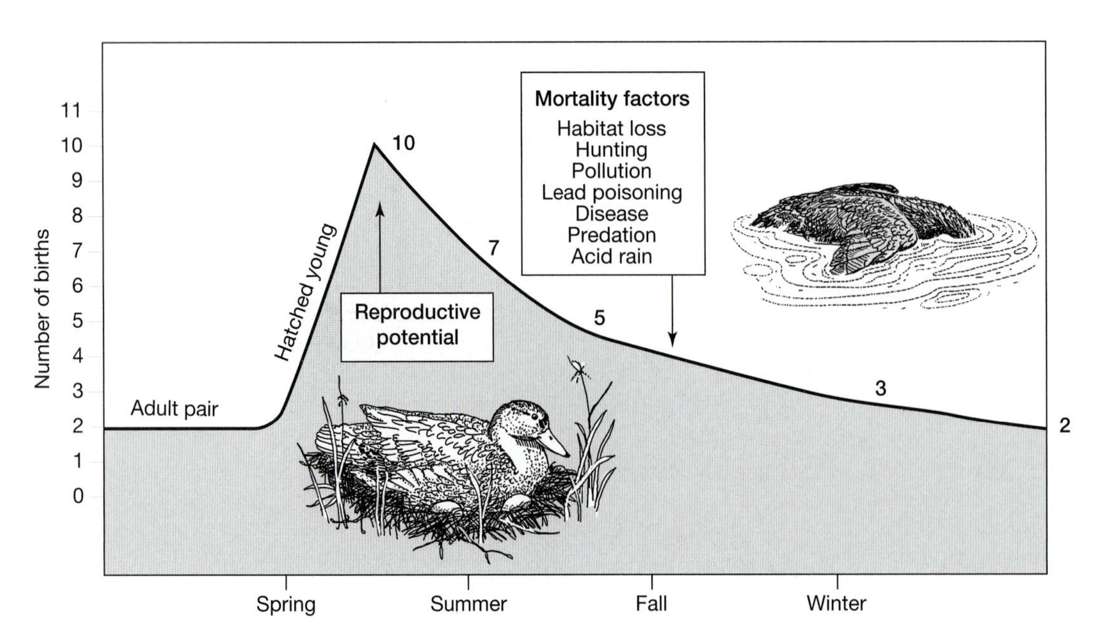

FIGURE 16.10 *Survival curve for a mallard family. The population of two adults and ten young was reduced by a variety of mortality factors to only two birds by the beginning of the next year's breeding season.*

A CLOSER LOOK 6.1 *The Hunting Controversy*

The hunter spots an elk herd edging into a clearing in the Colorado Rockies. He raises his rifle, picks out a big bull with a magnificent rack, squints down the sights, and squeezes the trigger. The crack of the rifle shot echoes through the valley. The bull staggers and slumps to the ground, blood spurting from the wound. He twitches his legs a few times and then lies still. Once again, the human predator has made a kill.

In 1996 about 22 million Americans hunted some form of wildlife. About 87 percent of the hunters in the United States are males. One half of them are quite young, ranging from 16 to 34 years of age. One of every two hunters has attended college. Their occupations are highly diverse. Almost every walk of life is represented, from farmers to surgeons, from factory workers to business executives. Many are still in college. Regardless of their occupation, however, most American hunters grew up in a rural or semirural environment and received hunting instructions from their fathers when they were teenagers.

In the rural environment of the past century, hunting had much greater acceptance than it does today. Probably this was because the slaughter of livestock for food was a common farm activity. As a result, the death of animals was accepted as necessary for human survival.

Today, however, our nation is highly urbanized. Most Americans have never seen a farmer butcher a chicken, pig, or cow. Many consider the killing of wild game unnecessary and cruel. About 50 percent of Americans today oppose hunting. They believe that hunting causes animals to suffer unfairly. They also draw a distinction between the behavior of hunters and wild predators. For instance, wolves, cougars, and other wild predators usually take prey that is weak, sick, crippled, aged, or suffering from disease. On the other hand, the human hunter prefers to bag trophy-sized animals in prime condition. This may weaken the species. Many people against hunting believe that state game commissions set hunting quotas not to keep wildlife in balance, but to satisfy the desires of hunters.

A number of citizens groups have been organized to outlaw hunting. Among them are the Friends of Animals, based in New York City, and the Humane Society of the United States, with headquarters in Washington, D.C. A common viewpoint of such groups is aptly expressed by Joseph Wood Krutch, a nature writer from New York City: "When a man wantonly destroys one of the works of man we call him Vandal. When he wantonly destroys one of the works of God we call him Sportsman."

Hunters believe that, since we have eliminated most of the natural predators of deer and other large game animals, properly regulated hunting by humans is needed. Without effective control by natural and human predators, these game animals will exceed the carrying capacity of their habitat. Sport hunters and hunting groups further argue that hunting is enjoyable recreation and brings in income to local economies. Taxes on firearms and ammunition and sales of hunting licenses provide states with funds for wildlife management, habitat work, and wildlife research.

Hunters have also been supported by many local sportsmen's groups as well as state and national organizations. Nationwide support for hunters is given by the National Rifle Association, the National Wildlife Federation, and the Wildlife Management Institute. Their pro-hunting philosophy is well represented by N. Adams in an article entitled, "Hunting: An American Tradition": "You show me a person who doesn't directly or indirectly kill on a regular basis, and I'll show you a bleached, well-weathered pile of human remains. Every living creature takes life to stay alive, and if it doesn't, it quickly starves and dies a slow agonizing death."

which 12 million are in the United States. At one time, hunters deposited more than 3,000 tons of lead shot on our nation's lakes, rivers, and marshes annually. Because there are 280 pellets of No. 6 shot in one shotgun shell, and the average hunter needs six shots to kill one duck, about 1,400 pellets were deposited on waterfowl habitat for each bird taken. In one study researchers counted 150,000 pellets per hectare (60,000 per acre) in the San Joaquin River marshes of California; 300,000 per hectare (120,000 per acre) were found on the bottom of Wisconsin's Lake Puckaway.

Until recently, lead poisoning from spent shot killed an estimated 2 to 3 percent of our waterfowl each year, an amount that nearly equals the combined duck production of North and South Dakota. (Secondary poisoning of bald eagles may occur when they eat the carcasses of lead-poisoned ducks.) The heaviest duck mortality from lead poisoning has occurred along the Mississippi flyway, especially in Illinois, Indiana, Missouri, and Arkansas.

To reduce this problem, the U.S. Fish and Wildlife Service began phasing out lead shot and replacing it with steel shot in 1987. Beginning with the 1991 season, lead shot was banned in the United States for waterfowl hunting. Starting with the 1999 season, lead shot accordingly was declared illegal in Canada. The advantage of steel shot is that it saves the lives of more than 2 million ducks per year. The disadvantages of steel shot are that it is more expensive than lead shot, does not kill at a distance of more than 40 meters (125 feet), and ruins outdated shotgun barrels made of soft steel.

Disease Waterfowl are subject to a wide variety of diseases. **Botulism,** a disease caused by the toxic metabolic wastes of the anaerobic bacterium *Clostridium botulinum,* can cause severe disease outbreaks in waterfowl. Although most prevalent in the West, it has been recorded from Canada to Mexico and from California to New Jersey. During the summer of 1910, this microscopic organism was responsible for millions of waterfowl deaths. Even today, botulism may kill 100,000 waterfowl a year in California and Utah (Figure 16.13).

Clostridium thrives in stagnant alkaline mudflats, where there is an abundance of trapped organic material (such as dead aquatic vegetation) and high water temperatures. These conditions are most likely to occur in the late summer during extended drought. Ducks become ill after eating contaminated organic material (decomposing plants and animals) or maggots and

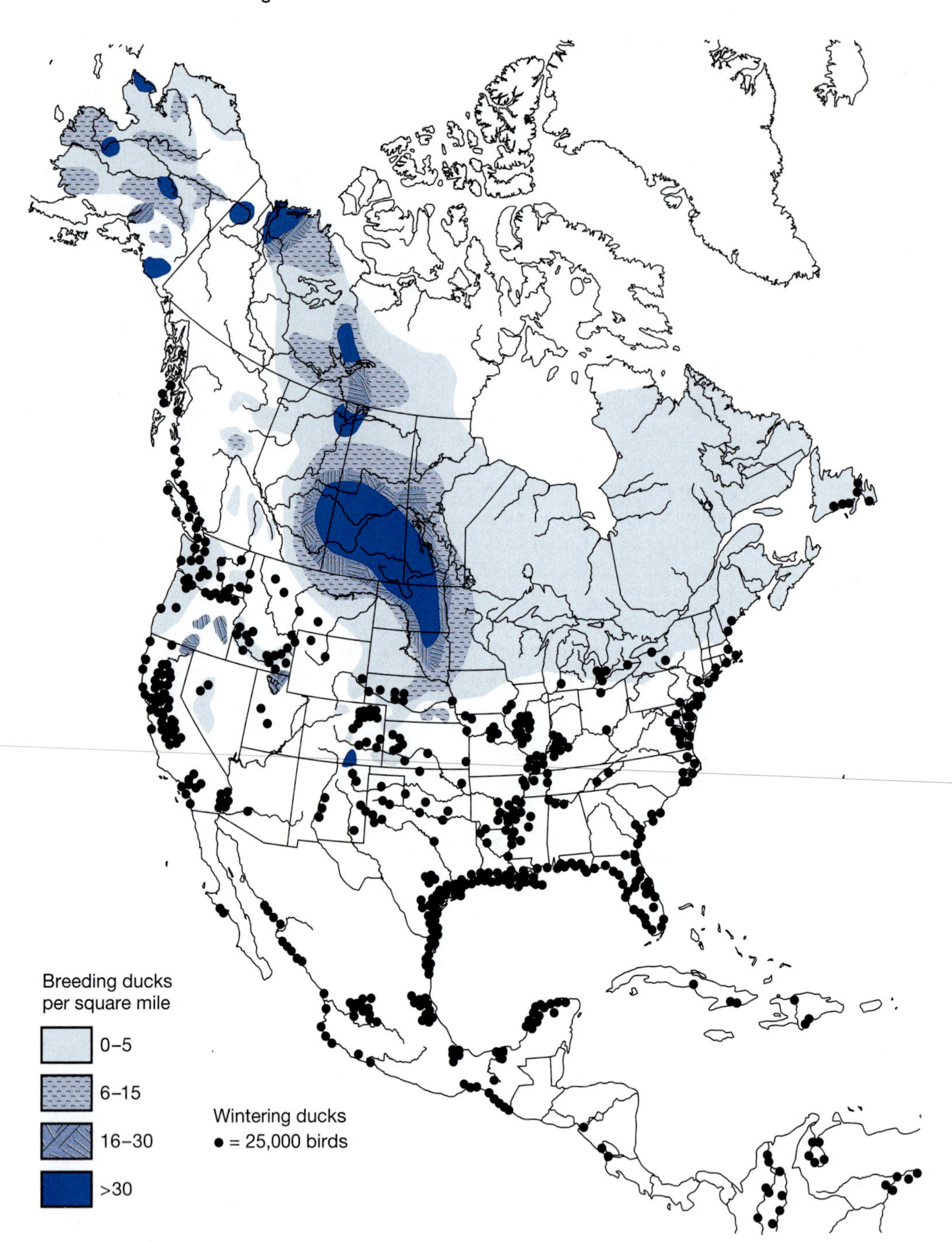

FIGURE 16.11 *Distribution of North American breeding and wintering ducks. More than 50 percent of North American ducks are produced in the "duck factories" of Manitoba, Saskatchewan, Alberta, the Dakotas, western Minnesota, and northwestern Iowa. The major wintering areas are the Atlantic, Gulf, and Pacific coasts, the lower Mississippi River Valley, and California. Because many ducks that migrate through the United States winter in Mexico, effective waterfowl management depends on the cooperative actions of Canada, the United States, and Mexico.*

other insects that harbor the bacteria. Apparently, insect larvae are a specialized microhabitat for the bacteria. After being absorbed by the bloodstream of the waterfowl, the toxin produced by the bacteria kills the birds by paralyzing their breathing muscles. Sick waterfowl cannot hold their necks erect, so the disease is also called "limber neck."

Other common diseases in waterfowl are avian cholera (a bacterial disease) and aspergillosis (a fungal disease). A viral disease that primarily attacks waterfowl is duck virus enteritis.

Acid Rain Acid rain in the United States and Canada creates another problem for waterfowl (Chapter 19). The black duck is highly prized by hunters, especially along the Atlantic Coast. It breeds in the northeastern United States and southeastern Canada, a region plagued by acid deposition. The mayfly, a common aquatic insect, is a valuable food for young black ducks. Unfortunately, however, mayfly populations have been sharply reduced in acidified lakes (Figure 16.14). Biologists wonder if future generations of young black ducks will be able

FIGURE 16.12 *Waterfowl mortality caused by oil spill. Oil-coated ducks pile up on the shore of the Mississippi River near Spring Lake, Minnesota. They were victims of a combined petroleum and soybean oil spill that killed 20,000 ducks.*

to shift to alternative food sources. It is possible that such substitute food will be either unpalatable or too difficult to catch in sufficient quantities. In addition, acidity may also kill alternative food sources, causing even greater problems for the black duck.

16.4 Wildlife Management

We have defined wildlife management as the planned use, protection, and control of wildlife using sound ecological principles. One basic approach to the management of wildlife is the acquisition and development of quality wildlife habitat.

Acquiring and Developing Habitat for Terrestrial Wildlife

Currently, the best prospect for increasing wildlife populations is to increase the amount and quality of habitat. Many wildlife biologists consider habitat development to be indispensable.

The best management programs are meaningless without habitat protection. If plenty of high-quality wildlife habitat is available, wildlife populations will remain relatively high, regardless of the lack of all other management methods. Federal, state, and private agencies are trying to reduce wetland loss by habitat purchase. Federal funds for this purpose come from an excise tax on the sale of hunting equipment. This excise tax originated from the 1937 Federal Aid in Wildlife Restoration Act, popularly known as the **Pittman-Robertson Act** and perhaps the most important legislative measure promoting wildlife management in the United States. This legislation originally levied an excise tax of 10 percent on the sales of sporting arms and ammunition; the tax was later increased to 11 percent. Revenues from the federal tax are distributed to the states for habitat purchase and for wildlife management and research. In addition to the Pittman-Robertson excise tax, other monies are derived from the sale of federal duck stamps, which must be purchased by waterfowl hunters at the beginning of each season.

Unfortunately, federal funds have been woefully inadequate. To help, Congress passed the Wetlands Loan Act in 1961, which has enabled the Fish and Wildlife Service to borrow funds from federal sources for wetland purchases.

Since federal and state budgets for the development of parks and sanctuaries are tightening, private lands play an increasingly important role in supporting wildlife. Among the organizations that have had great success in acquiring private land for the benefit of wildlife are the Nature Conservancy and the National Audubon Society.

The Nature Conservancy does not make a lot of headlines. Quietly and effectively, it goes about its business of purchasing or leasing essential habitat to preserve plants, animals, and nat-

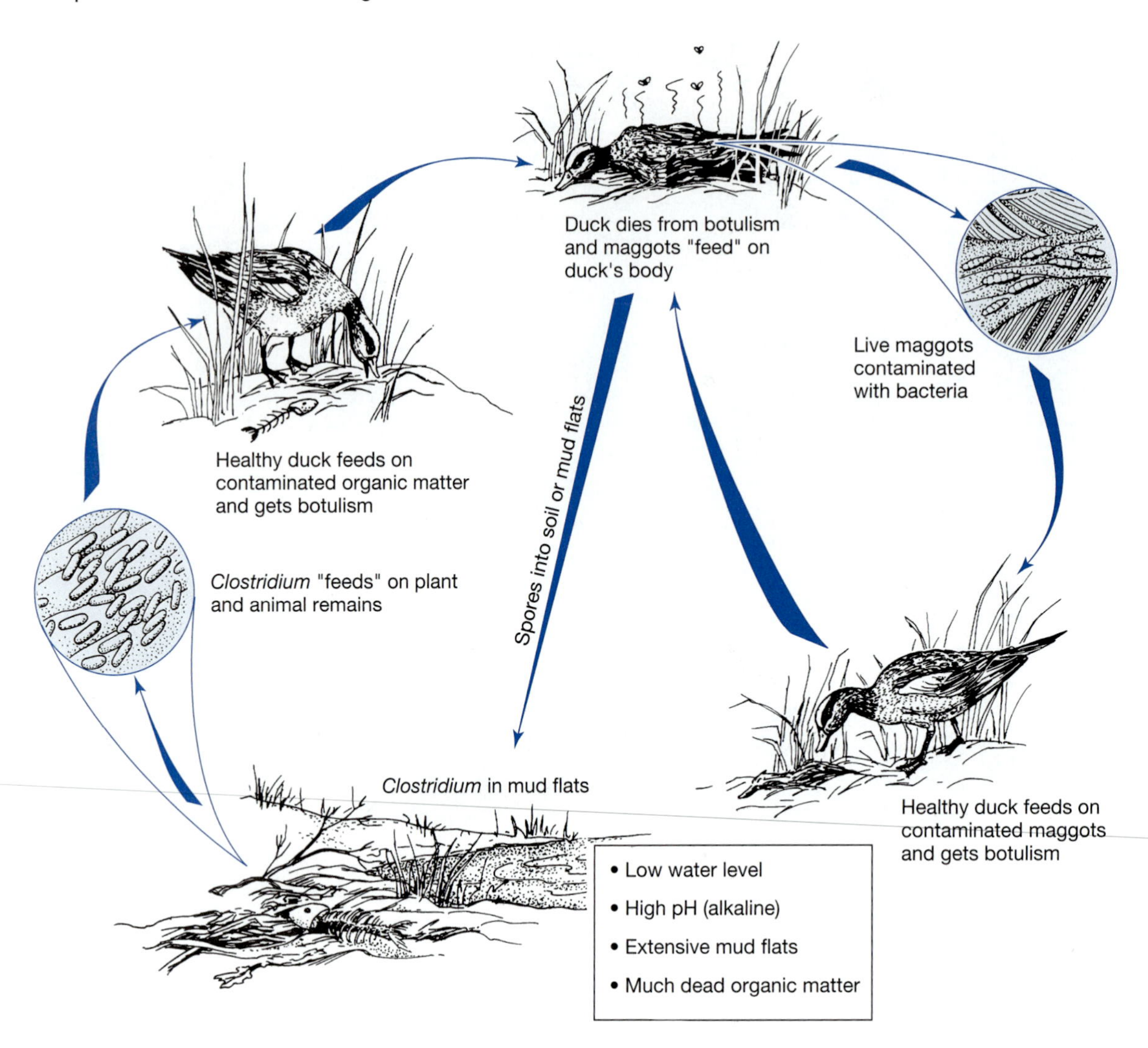

FIGURE 16.13 *Cycle of events leading to a botulism outbreak.*

ural communities. It has helped to protect more than 4.5 million hectares (11 million acres) of habitat in the United States and nearly 24 million hectares (60 million acres) in Canada, Latin America, the Caribbean, Asia and the Pacific. With the aid of a huge army of volunteers, it manages hundreds of wildlife sanctuaries in the United States. The National Audubon Society has leased or purchased about 61,000 hectares (150,000 acres) and has established more than 100 wildlife sanctuaries on this land.

Developing Habitat on the Farm

More than 85 percent of the hunting land in the United States is privately owned or controlled. Private farms, ranches, and woodlots in fact produce most of our nation's grouse, quail, doves, pheasants, and rabbits. Therefore, the biggest contribution to an abundant and varied game resource (as in the case of forest development) can be made by the private citizen. Fortunately, many soil and water conservation practices (Chapter 7), such as shelter belting and conservation tillage, also improve the habitat for wildlife.

During the Dust Bowl era of the 1930s, the USDA planted more than 30,000 kilometers (18,000 miles) of shelterbelts in the Great Plains. These narrow belts of trees on farms from the Canadian border down to Texas were planted primarily to control soil erosion. However, they also provided food, cover, and breeding sites for dozens of species of nongame birds like thrushes and warblers, as well as game such as grouse, quail, pheasants, squirrels, rabbits, and deer.

Conservation tillage has been discussed (Chapter 7) as a superb method for the control of soil erosion. The stubble and other vegetative debris from the harvested crops is left on the land. The new crop is planted the following spring directly in the stubble with special seed drills. Tillage is reduced to a minimum. The acreage of farmland that is being tilled in this way is increasing rapidly throughout the United States. Conservation tillage has benefited a great variety of wild animals, from pheasants in Nebraska to prairie chickens in Texas. Other ground-nesting species, such as quail, partridge, and early-nesting waterfowl, also benefit, since the stubble and other crop residues provide cover, breeding sites, and some food in the form of waste grain.

In 1985 Congress passed the **Food Security Act** (or Farm Bill). An extremely important provision of the act with regard to wildlife habitat improvement is the **Conservation**

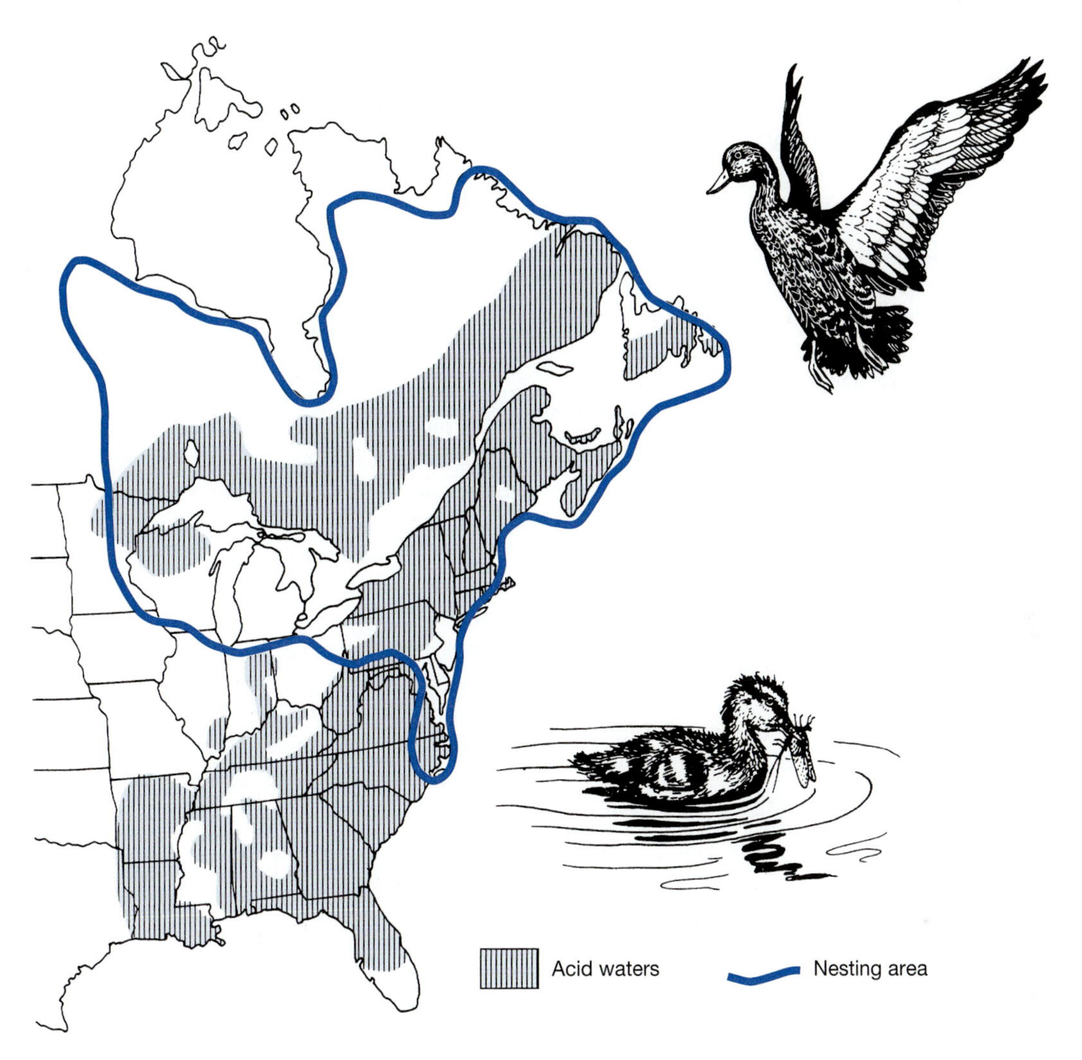

FIGURE 16.14 *The breeding areas of the black duck population of eastern North America broadly over-lap with regions that have acid waters. The acid conditions have caused a sharp decline in the mayfly population—an aquatic insect that is an important source of protein for young black ducks.*

Reserve Program (CRP), which started in 1986 and continues today. This provision enabled the U.S. Department of Agriculture (USDA) to make contracts with farmers to control soil erosion. In essence, the farmers received payments for discontinuing farming on highly erodible land. Instead of raising crops, the farmers planted grasses or trees. The program is responsible for setting aside 14.7 million hectares (36.4 million acres) of CRP land, which has greatly increased wildlife habitat.

The 1985 Farm Bill (and 1996 Farm Bill) created other USDA programs important to wetlands and wildlife protection. For example, the **"Swampbuster" provision** protects wetlands by denying eligibility for other USDA programs to growers who drain and farm certain wetlands. The **Wetlands Reserve Program** is a voluntary program offering landowners the opportunity to protect, restore, and enhance wetlands and their values (including optimum wildlife habitat) on their property. Landowners receive financial incentives to enhance wetlands in exchange for retiring marginal agricultural land. This program offers landowners an opportunity to establish long-term conservation and wildlife practices and protection beyond that which can be obtained through any other USDA program. The **Wildlife Habitat Incentives Program** also provides financial incentives to develop habitat for fish and wildlife on private lands. Participants agree to implement a wildlife habitat development plan and the USDA agrees to provide cost-share assistance for the initial implementation of wildlife habitat development practices. This agreement generally lasts a minimum of ten years from the date that the contract is signed.

In addition to all of these USDA programs, the U.S. Fish and Wildlife Service has a **Partners for Fish and Wildlife Program,** formerly named the Partners for Wildlife Program, which offers technical and financial assistance to private landowners to voluntarily restore wetlands and other fish and wildlife habitats on their land. The program emphasizes the reestablishment of native vegetation and ecological communities for the benefit of fish and wildlife in concert with the needs and desires of private landowners. Since the program began in 1987, these partnerships have generated significant habitat restoration accomplishments on private lands, primarily focused on the restoration of wetlands, native grasslands, stream banks, riparian areas, and in-stream aquatic habitats.

Manipulating Ecological Succession

In our discussion of succession in Chapter 3, we said that both plant and animal communities change as the physical environment changes. Thus, different organisms occupy a region as it goes through succession. Ecologist Raymond Dasmann has classified a number of species according to the successional stage: climax species (bighorn sheep, caribou, and grizzly bear); mid-successional species (antelope, elk, moose, deer, and ruffed grouse); and early-successional species (quail, dove, rabbit, and pheasant).

Wildlife biologists can regulate the abundance of these species by manipulating ecological succession. Thus, they can permit a succession to proceed on its natural course to a climax or, by employing such artificial devices as controlled burning (Figure 16.15), controlled flooding, plowing, and logging, can retard the succession or even return it to the pioneer stage.

The **early-successional species,** such as the rabbit, quail, and dove, depend heavily on major disturbance of the ecological succession by humans. These species prefer the weedy pioneer plants that invade an area denuded by human activity. Such vegetation may become established when farmland is abandoned and a pioneer community of invading weeds and shrubs becomes established.

Because **climax-associated species** such as caribou, bighorn sheep, and grizzly bear flourish only in relatively undisturbed climax communities, their survival depends largely on the establishment of state or national refuges and wilderness areas. Without such protected "islands" in the "oceans" of successional disturbance caused by humans, these climax-associated species will decline and become extinct.

As an example of manipulating ecological succession to encourage the population growth of a desired species, let's look at improving habitat for the ruffed grouse. The ruffed grouse is a highly prized game bird in the northern United States. It is named after the feathered ruff, or collar, it displays during courtship. When flushed by the hunter, it bursts from cover with a thunderous whir. A typical mid-successional species, grouse occupy a given region only temporarily.

Gordon Gullion, the nation's leading grouse expert, has found that young aspen stands provide virtually all of the food, cover, and breeding requirements of the grouse in the Great Lakes forests. Winter aspen buds provide an abundance of high-quality nourishment during a season when other foods are scarce. Note in Figure 16.16 that the highest population density (81 birds per square kilometer or 210 per square mile) occurs in aspen forests that are 12 to 25 years old. However, as the succession proceeds, the aspen are gradually replaced by other hardwoods, such as oak, and the grouse population declines. Gullion found that the carrying capacity of aspen forests in northern Minnesota can be increased by more than 600 percent when blocks of aspen are clear-cut at 10, 20, and 30 years of age. This harvesting pattern gives grouse access to aspen stands of variable age and staves off the invasion of other tree species.

Managing Habitat for Waterfowl

Waterfowl habitat can be improved by creating openings in dense marsh vegetation, constructing artificial ponds and islands, developing artificial nests and nest sites, and establishing waterfowl refuges.

Creating Openings in Marshes Although waterfowl require cover for protection from both weather and predators, they

FIGURE 16.15 *Wildlife biologists developing suitable habitat for prairie chickens in Minnesota. The controlled burn will prevent plant succession and the growth of shrubs and trees. As a result, the grassland habitat, which the prairie chickens require, will be maintained.*

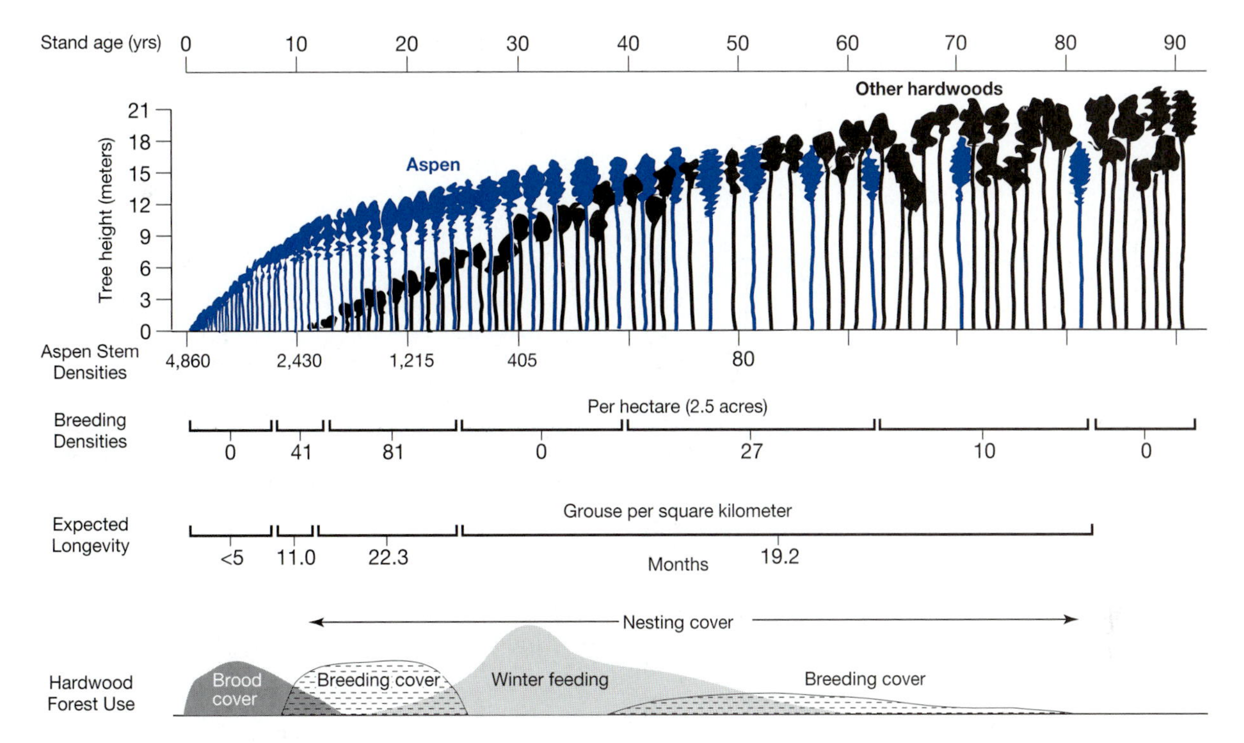

FIGURE 16.16 *Forest succession and grouse habitat. Note that the carrying capacity of the forest for ruffed grouse changes as the forest succession proceeds. Use of the forest as brood and breeding cover is high during the early successional stages, when young aspen dominate the woods. When the aspen have attained an age of about 35 years, the grouse no longer use the forest for breeding purposes but use it primarily as a source of winter food. Grouse feed extensively on winter aspen buds. By the time the stand is 80 years old, the aspen have been completely replaced by other hardwood species that are of no value to grouse. The highest grouse densities (81 per square kilometer) occur in a forest that is 12 to 25 years old and is dominated by aspen.*

also need channels and openings through which they can paddle or waddle between nesting sites and feeding areas and between feeding and loafing areas. Channels also provide areas where the birds can feed. These essential openings result from natural causes such as hurricanes and lightning-triggered fires, or they may be made by humans.

Constructing Artificial Ponds Where sloughs and potholes are scarce, waterfowl habitat can be constructed using artificial ponds. Between 1936 and 1994, the USDA assisted farmers in building more than 4 million farm ponds. Roughly two thirds of those ponds are used by waterfowl, either as nesting, feeding, and loafing areas for resident birds or as resting areas where migrating waterfowl can touch down for a brief respite before resuming their strenuous journey. Farmers can increase the carrying capacity of these ponds by erecting artificial nest boxes for mallards and wood ducks; by dumping piles of rocks or anchoring logs and bales of hay in the open water, where birds can preen and sun; and by seeding the pond with choice duck-food plants.

Constructing Artificial Islands In recent years, society has destroyed much valuable wildlife habitat in its attempts to promote its own welfare. As a result, wildlife populations have frequently been decimated. And for years it has been assumed, by at least a few biologists, that virtually all human-induced changes of the natural environment were detrimental to wildlife. The error of this type of thinking is shown by the effect of artificial islands in boosting waterbird populations. For example, in 1977 the Bureau of Reclamation (frequently assailed by environmentalists because of its obsession with big dam construction) used fill material to form 62 artificial nesting islands for Canada geese in Canyon Ferry Lake near Townsend, Montana (Figure 16.17). Since the construction of these artificial islands, biologist Robert Eng of Montana State University has noted a threefold increase in Canada goose production in the region.

For a number of years, the U.S. Army Corps of Engineers has constructed coastal "dredge islands" from the sand, mud, and shells it removes during dredging operations. More than 2,000 of these islands are scattered along the nation's coast from Long Island, New York, to Brownsville, Texas. Some have also been formed in the Mississippi River, in the Great Lakes, and along the Pacific Coast. They are usually located far enough from shore to afford protection from such predators as raccoons, foxes, and free-running dogs. Moreover, because the dredge islands are frequently about 2 to 3 meters (6.6 to 9.8 feet) high, they are not often flooded during high tides, as are many of the low-lying natural islands near shore. Sidney Island, now administered by the National Audubon Society, was formed from spoil resulting from the dredging of the ship channel at Orange, Texas. Clouds of herons and ibises leave their nests and circle above when visitors set foot on

FIGURE 16.17 *Artificial nesting islands for Canada geese. View of the Westside Pond in Canyon Ferry Lake near Townsend, Montana. Fill material was used to create about 62 artificial nesting islands. Note the Canada geese nesting on the island in the foreground.*

the island. According to Audubon Society counts, Sidney Island is home to 20,000 egrets, almost 8,000 herons, 2,000 ibises, 1,400 roseate spoonbills, and 380 cormorants. Several of these species have declined in other areas.

Developing Artificial Nests and Nest Sites Through the process of natural selection operating over millions of years, each species of waterfowl has evolved its own unique instinct for nest-site selection and nest construction. It would appear almost impertinent, therefore, for humans to attempt to improve on nature by constructing artificial nests and sites for waterfowl. However, wildlife biologists have done precisely this, and with encouraging results (Figure 16.18). These artificial nests promote reproduction and may be even more effective in minimizing mortality caused by mowing machines, predators, and nest-site competitors than natural nests.

Establishing National Wildlife Refuges America's system of national wildlife refuges was launched in 1903 by Theodore Roosevelt. He established the Pelican Island Refuge in Florida's Indian River to protect the brown pelican. From this modest beginning, the federal refuge system has grown to more than 530 individual refuges, wetlands, and special management areas, covering about 38 million hectares (93 million acres). Every state has at least one refuge, and most of the refuges were established primarily for use by waterfowl. In 1934 Congress passed the **Migratory Bird Hunting Stamp Act,** which provides funds to acquire, maintain, and develop waterfowl refuges through the sale of duck stamps.

Our national wildlife refuges provide over 1.2 billion waterfowl-use days. (One waterfowl-use day is one day's use by one duck, coot, swan, or goose.) Our refuges produce over 500,000 ducklings each year. The Tule Lake (California) and Agassiz (Minnesota) refuges each produce 30,000 ducks yearly, and the Malheur (Oregon) Refuge produces 40,000 annually. Huge concentrations of ducks and geese use many of the refuges during the fall migration (Figure 16.19). For example, in the Klamath Basin Refuge on the California–Oregon border, where considerable acreages of wheat and barley are grown exclusively as waterfowl food, a peak of 3.4 million ducks and geese has been recorded. Nearly 150,000 Canada geese have stopped over at the Horicon National Wildlife Refuge in southern Wisconsin—the greatest concentration of this species ever recorded in the United States. The geese fatten up or "refuel" at these stopover refuges before continuing their migration (Figure 16.20).

Many refuges have the specific mission of preserving one or more of our nation's endangered species. For example, the Aransas Wildlife Refuge in southern Texas provides a safe wintering ground for the whooping crane, while the Red Rocks Lake Refuge in Montana serves as a secure nesting area for the trumpeter swan. Some refuges are devoted primarily to saving endangered species of plants. Thus, in 1980 the Antioch Dunes Refuge was established in California to ensure the survival of the Antioch Dunes evening primrose and the Contra Costa wallflower. Although the national wildlife refuges have the primary mission of conserving wildlife, the U.S. Fish and Wildlife Service, which administers the refuge network, does permit a limited amount of farming, logging, sheep and cattle grazing, and even oil drilling and mining, as long as the wildlife is not jeopardized in the process. Unfortunately, on some refuges, these extracurricular activities have indeed destroyed some feeding areas and breeding grounds. Because of this, a number of environmental groups, such as the Audubon Society, the

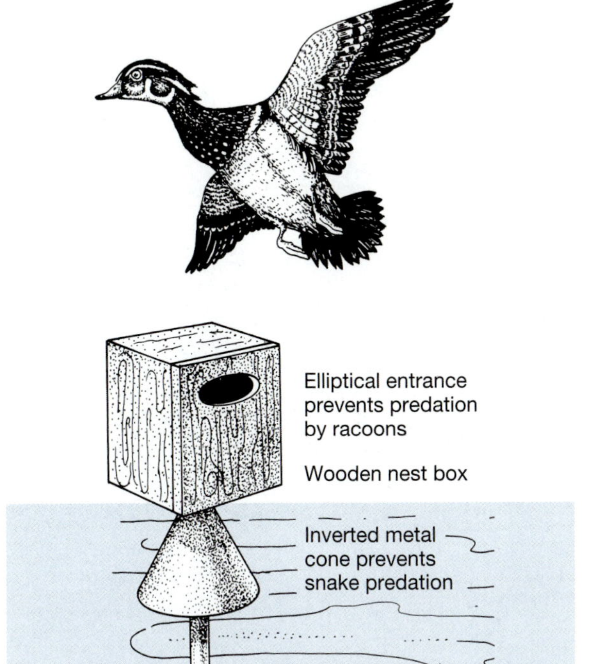

Use of nesting boxes

State	Number of box years	Number of nests	Percent use
Ohio	26,084	7,363	28
Connecticut	6,225	4,102	66
Illinois	3,218	1,579	49
Mississippi	2,475	1,747	71
Louisiana	1,229	416	34
Massachusetts	483	415	86

FIGURE 16.18 *Construction of nesting boxes for wood ducks has been an important factor in the population increase of this beautiful species from a very low level earlier in this century.*

Sierra Club, and Ducks Unlimited, have urged Congress to enact legislation that would ban such activities permanently.

The history of the Everglades, Florida, and the subsequent creation of Everglades National Park, which has some refuge-like qualities is discussed in Case Study 16.2.

16.5 Regulating Populations

In addition to habitat acquisition and development, wildlife managers use another basic approach to accomplish their objective—the regulation of populations. Most of our discussion will be concerned with deer and waterfowl.

Controlling the Harvest of Game Populations

Game managers may use hunting to control the populations of well-established species. Populations of upland game birds (quail, pheasants, grouse) and small mammals (rabbits, squirrels) usually recover rapidly after being reduced by hunting. Such populations are said to be **resilient.** Such resilience is due in part to their high biotic potential. For example, a hen partridge may lay 20 eggs per clutch; a doe rabbit may have six young per litter and rear several litters per year. Even deer have resilient populations. Under optimal conditions, a deer herd of six individuals can build up to 1,000 head in only 15 years!

In all species of animals, mortality factors, or environmental resistance, counteract the biotic potential. As a consequence, the population of a species remains about the same from year to year. Wildlife biologists distinguish between two types of mortality—additive and compensatory.

Additive mortality simply adds to the deaths caused by other factors (Figure 16.21a). For example, if predation accounts for 15 percent mortality of a population and an ice storm accounts for 15 percent, then the total mortality for the year is 30 percent. If in the next year predation takes 20 percent and an ice storm takes 25 percent, for a total mortality of 45 percent, the effects of the two factors are said to be additive.

Compensatory mortality denotes the replacement of one kind of mortality with another kind of mortality in animal populations (Figure 16.21b). Compensatory mortality occurs if increased mortality from, say, hunting results in decreased mortality from other factors, such as starvation, disease, and predation, so that total mortality remains the same. Note that with or without hunting in a properly managed ecosystem, the total mortality from all causes remains the same. For example, about 70 percent of a northern bobwhite quail population will die in a normal year, regardless of whether hunting mortality occurs. Furthermore, a **harvestable surplus** of animals, within limits, can be safely removed. A harvestable surplus is that portion of the population that can be taken by humans (hunters) without adversely impacting subsequent populations of that animal.

The number of animals taken by hunters depends on hunting regulations. The hunting kill can be adjusted by extending or shortening the hunting season, increasing or decreasing bag limits, permitting the kill of both sexes or limiting it to males, and regulating the use of weapons (bow, rifle, or shotgun). Wildlife managers should set hunting season regulations by annually surveying the land under their management to determine habitat conditions and numbers, reproduction rates, and age structures of game animals.

Regulating the Deer Harvest

Several decades ago, when deer were relatively scarce, state legislatures restricted hunting by closing or shortening the season, by timing the seasons to ensure an absence of tracking snow, by restricting firearms to shotguns, and by restricting the kill to one per hunter. The doe was afforded special status. The deer herd buildup was further promoted by winter feeding, introductions, predator control, and the establishment of refuges.

FIGURE 16.19 *Sky darkened with thousands of pintail ducks at the Sacramento National Wildlife Refuge in California. Our National Wildlife Refuge System includes more than 530 refuges covering about 38 million hectares (93 million acres).*

In response to these measures, as well as to the great abundance of edge and food available in the wake of extensive fires and logging, the herd increased rapidly—too rapidly. In only 13 years, the whitetail population in 45 states increased from 3.2 million (1937) to 5.1 million (1949). It soon exceeded the range's carrying capacity. Browse lines appeared, winter starvation became commonplace, and the range rapidly deteriorated.

Many state game departments advised legislators to reverse the trend by liberalizing hunting regulations. After much prodding from wildlife biologists, herd reduction was implemented by opening and extending seasons, timing the season to coincide with the occurrence of tracking snow, legalizing the use of rifles, lifting the ban on does, establishing bow seasons, and removing bounties on predators. Roads were built to facilitate hunter access in the back country.

The overpopulation problem is far from solved. Despite liberalized laws, hunters rarely harvest more than 10 percent of herds. One reason is the deer's secretive behavior: The animals rarely emerge from protective cover during daylight hours of the hunting season. Hunters sometimes do not see a single deer even in an area with large populations. A 10 percent annual harvest is simply not enough to appreciably check herd increase. In some states at least one third of the autumn deer herd could be taken during the hunting season year after year without affecting deer herd size.

Deer populations frequently vary from region to region within a state, often being highest in semiwooded agricultural regions and low in climax forests and urban areas. (Today, however, it is not uncommon to see deer on the outskirts of major cities like Milwaukee and Chicago.) Therefore, a state may be divided into a number of zones, each with its own set of regulations. (One state has had 60 zones.) In zones where herds are small, the season may be closed completely or may be open to bow hunters only. In overpopulated zones, the season may be opened on bucks, does, and even fawns.

In some states, wildlife managers are not permitted to practice what they preach because their technical knowledge in game management is far in advance of a receptive public or political climate. Too often the framers of our hunting laws yield to pressures exerted by hunters and resort owners, to whom the essence of game management is "more deer." Only when regulations are formulated in accordance with the advice of professionally staffed conservation departments will hunt-

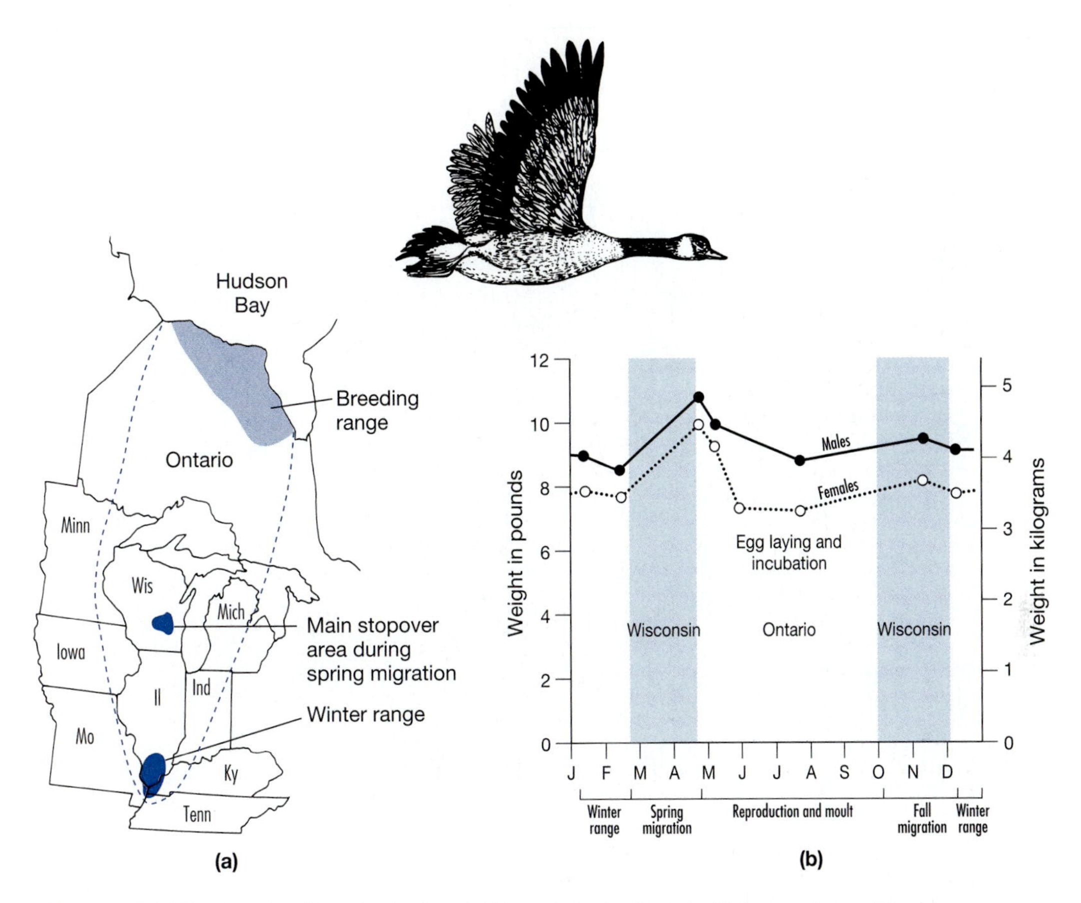

FIGURE 16.20 *The value of "stopover" refuges in Wisconsin for the Mississippi Valley population of Canada geese. These geese fly 725 kilometers (450 miles) nonstop to stopover refuges like Horicon Marsh in southeastern Wisconsin. There they find an abundance of food, which enables them to acquire the fat needed to fuel their 1,375-kilometer (850-mile) nonstop flight to their breeding ground in northern Ontario.*

ing regulations serve as an effective management tool. Some of the ethical considerations in regulating deer populations are discussed in Ethics in Resource Conservation Box 16.3.

Regulating Destructive Deer Populations

Deer can become pests in rural, suburban, and urban environments. Farms in particular provide sources of food for many kinds of wildlife, including deer. Deer frequently cause excessive losses of soybeans and other crops. Wisconsin provides a good example of crop depredation by deer.

In 1994 the Wisconsin deer population stood at more than 1 million. This was a population increase of 162 percent for the state as a whole compared to the deer population in 1962. However, when agricultural lands alone are considered, the increase was a dramatic 488 percent. And these farming country deer were taking a huge $37 million annual bite out of Wisconsin agricultural production. Fields of corn and soybeans were damaged when deer invaded them with crops still in the seedling stage. In response, the Wisconsin Department of Natural Resources liberalized its hunting regulations in areas where deer-inflicted crop damage was high. For example, during Wisconsin's nine-day hunting season in 1993, about 255,000 deer were taken—slightly over 25 percent of the total herd. Despite the huge annual harvests, however, deer damage to Wisconsin fields and orchards continues to rise. In the year 2000, Wisconsin's deer population stood at 1.7 million, the highest on record. The Wisconsin Department of Natural Resources has liberalized its hunting rules and hours even more in hopes that hunters will curb the deer population growth that Wisconsin has been experiencing.

Regulating the Waterfowl Harvest

The mallard is one of the most highly prized ducks sought by hunters on the North American continent. Each autumn, 12 to 18 million mallards wing their way south from their breeding grounds in southern Canada and the northern United States. And each autumn, several million duck hunters crouch in their blinds, hoping to lure these swift-flying birds to within shooting distance with their real-as-life decoys.

But how many mallards may the hunter take each day? One? Three? Five? Moreover, how many days may he or she hunt? The answers to these questions may vary from one state to another, and from year to year as well. Let's find out how, and on what basis, such waterfowl hunting laws are made.

Much of the information on which waterfowl hunting regulations are based is derived from population censuses conducted

CASE STUDY 16.2 THE EVERGLADES: WATER TROUBLES IN A WILDLIFE PARADISE

FIGURE 1 *A crocodile at the Florida Everglades National Park.*

The Everglades is a gigantic, 13,000-square-kilometer (5,000-square-mile) sea of grass that extends south from Lake Okeechobee almost to the tip of the Florida peninsula. This magnificent swamp is roughly 40 miles wide and more than 161 kilometers (100 miles) long. The characteristic vegetation is sawgrass, some of which grows 3.6 meters (12 feet) tall. The Everglades is dotted with raised oblong areas, called tree islands, that are vegetated with willow and myrtle. Wildlife is highly varied and of great interest, ranging from huge turtles, alligators, and crocodiles (Figure 1) to deer, cougars, and long-legged wading birds. The latter include the wood stork, great egret, tricolored heron, great blue heron, white ibis, and roseate spoonbill. In the 1930s more than 200,000 of these birds nested in the Everglades.

For many centuries water moved slowly through this great swamp from its source in Lake Okeechobee. The water flowed southward through two grassland rivers, the Shark River Slough, which empties into the Gulf of Mexico, and the Taylor Slough, which flows into Florida Bay. Water movement was so slow it took almost a year to move from Lake Okeechobee to the ocean. During the rainy season the sloughs would overflow and large numbers of fish would "spill out" into the grasslands. The result: a food bonanza for the huge population of wading birds and other fish-eating wildlife.

Such was the water flow pattern in the Everglades for millennia. Then, in 1905, Governor Napoleon Bonaparte Broward instituted a massive campaign to convert the northern part of the Everglades to agricultural and urban use. Since that time many thousands of acres have been drained, diked, and channeled to control flooding and to provide sites for crop production as well as housing developments. Much of the work was done by the U.S. Army Corps of Engineers. American taxpayers footed the $30 million bill for this grandiose adventure in swampland "plumbing."

Regardless of the heightened concerns of professional ecologists and such groups as the Audubon Society, the plumbing of the Everglades continued. Additional water was diverted from the Everglades by the South Florida Water Management District to supply the city of Miami. Because of these and other "plumbing projects," the natural rain-driven sheet flow of water through the Everglades was severely disrupted. In effect, the entire northern half of the Everglades was converted into a series of diked, man-made pools connected by canals.

These Everglade plumbing projects have caused a lowering of the water table. During the dry season the parched grasses are highly vulnerable to fire. Encroachment of ocean water has resulted in an increase in salinity. In essence, the original natural condition of the vast Everglades ecosystem has been greatly altered. Wildlife populations have suffered. For example, the number of wading birds nesting in the Everglades has decreased dramatically. In 1910 the number was 500,000. Today, only 25,000 nest here, a decline of 95 percent. The populations of crocodiles and alligators have sharply declined as well. Indeed, the numbers of some species are so low that the U.S. Fish and Wildlife Service has placed them on its official List of Endangered Species. They include the American crocodile, wood stork, snail kite, Cape Sable seaside sparrow, Florida panther, and manatee—all native to the Everglades. On the other hand, the populations of some foreign species of plants have exploded under the artificially altered conditions. Back in the 1920s, then-Governor Broward introduced the Australian eucalyptus. The purpose was to help dry out the Everglades so that sugarcane and other crops could be raised. The plant thrived where water levels had been reduced by drainage. Today this shrubby plant infests more than 40,000 hectares (100,000 acres). It forms dense stands and steals moisture and nutrients from more desirable vegetation that had served as good nesting and feeding habitat for wildlife.

(continued)

Environmentalists have worked for decades to restore the biological health of the Everglades. They got a big boost in 1947, when Congress created Everglades National Park in the southern portion of the marsh. Unfortunately, however, by that time the northern Everglades had been desecrated by powerful agribusiness and real-estate interests. A maze of canals, dikes, dams, and levees had replaced the natural sawgrass ecosystem. However, in the 1980s an emerging consensus developed on two critical points: first, that the original Everglades ecosystem had to be restored, and second, that this could only be done by reestablishing the natural, rain-driven sheet flow of the water from Lake Okeechobee southward through the marsh. Among the parties endorsing this policy were the National Park Service, the U.S. Fish and Wildlife Service, the National Audubon Society, concerned conservationists nationwide, and even the U.S. Army Corps of Engineers and the South Florida Water Management District.

This objective was given impetus by the Everglades National Park Protection and Expansion Act, which was passed by Congress in 1989. It provided for a 43,000 hectare (107,000-acre) addition to the park (today covering about 6000 square kilometers or 2,300 square miles) that was absolutely essential for reestablishing the original sheet flow of water from the Shark River and Taylor sloughs. The gigantic restoration project received additional assistance in 1996, when Vice President Al Gore signed a controversial act limiting the activities of the sugarcane industry, which allegedly damages the fragile Everglades with its industrial waste and refinery byproducts. Although these two acts aided in restoring the natural flow to parts of the Everglades, many believed that they would only temporarily prolong its life.

On Dec 11, 2000 the Water Resources Development Act of 2000 was signed into law by President Bill Clinton. Part of the Act contained the **Comprehensive Everglades Restoration Plan,** a $7.8 billion federal-state project intended to restore the natural flow of water in the national park. The goal of the Comprehensive Plan is to restore, protect, and preserve the defining ecological features of the original Everglades and other parts of south Florida. Led by the Army Corps of Engineers, the project will span 30 years. It will remove some of the manmade structures that have obstructed natural water flow and have diverted it to serve the needs of people, including drinking water and irrigation. The money will also pay for the construction of facilities to recycle wastewater and storage areas to hold fresh water that currently flows unused into the ocean.

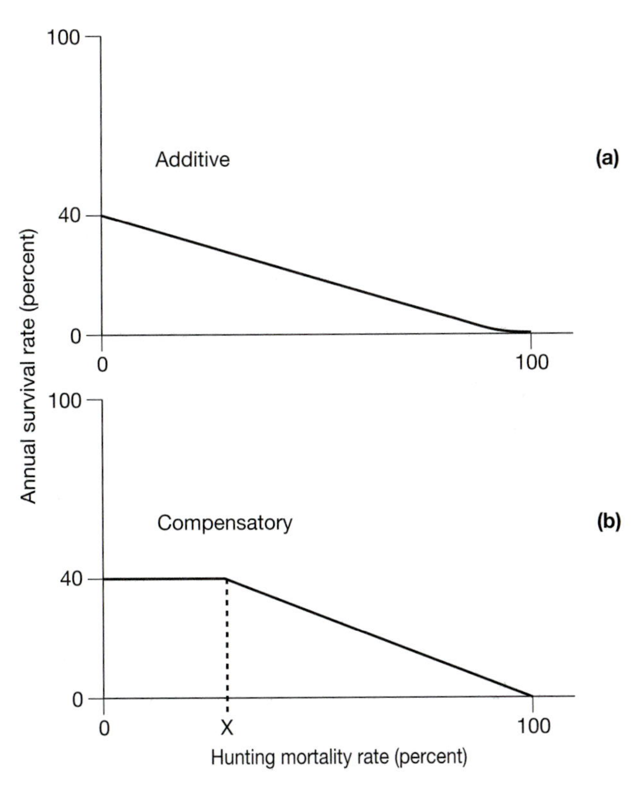

FIGURE 16.21 *Comparison of the influence of hunting mortality on annual survival rates of a theoretical bird population under additive and compensatory mortality conditions. (A) With additive mortality, an annual survival rate of 40 percent declines linearly beginning with the death of the first bird shot during hunting season; that is, the survival rate diminishes whenever a bird is killed. (B) With compensatory mortality, the annual survival rate of 40 percent is unaffected by hunting mortality until a certain point (X) is reached, after which hunting mortality becomes additive.*

on the breeding grounds several months before the fall migration. Regulations are also based on population data taken in the major migratory pathways of the ducks, which are determined from studies of banded birds.

In 1995 alone, almost 300,000 waterfowl (ducks, geese, and swans) were leg-banded in the United States and Canada with serially numbered bands. These banding operations are conducted by the U.S. Fish and Wildlife Service and the Canadian Wildlife Service. Records of the species, age, sex, weight, date, and banding locality are computerized. In 1995, more than 50,000 waterfowl bands were recovered. Although most recoveries are made by hunters, a considerable number are also recovered by birdwatchers and amateur naturalists who retrieve bands from birds killed by storms, pollution, predation, and disease.

Analysis of recovery data provides waterfowl biologists with information concerning growth rate, life span, and mortality, as well as the length, speed, and route of migration. For example, from such banding studies, we now know that some snow geese may travel 3,200 kilometers (2,000 miles) nonstop from James Bay, Canada, to the Texas coast in only two days!

From the practical standpoint of waterfowl population management, the most significant information derived from banding studies is that waterfowl, such as mallards that breed in Canada and the northern states, migrate along four (rather poorly defined) **flyways** on their way to southern wintering grounds. Known as the **Pacific, Central, Mississippi,** and **Atlantic flyways,** they have served as administrative units in the development of hunting regulations (Figure 16.22). These flyways have corridors that connect them (Figure 16.23). For example, a number of mallards that nest in the prairie provinces of Canada begin their fall migration by moving south along the Central flyway into South Dakota. Eventually, however, they swerve southeastward, joining the Mississippi flyway in Minnesota and Illinois,

ETHICS IN RESOURCE CONSERVATION 16.3

TO KILL OR NOT TO KILL?

Few areas of natural resource management are as heavily infused with (and confused by) ethical issues as wildlife management. One of the major issues that crops up from time to time has to do with managing wild species, especially deer and elk.

Individuals trained in wildlife management generally arrive at management decisions based on scientific facts. Basically, wildlife managers assess population size and growth rate, as well as habitat conditions, then determine the level of hunting needed to keep populations within the local or regional carrying capacity. Their goal is to keep deer and other wildlife from eating themselves out of house and home.

One thing that is rarely realized by outsiders is that the scientific approach to wildlife management is ultimately geared toward ecosystem protection. In other words, to the wildlife manager the ecosystem is of paramount importance. He or she manages wildlife populations in order to protect the entire system upon which all populations depend.

Critics of wildlife management professionals usually come at the issue from an entirely different vantage point. Their criticism of wildlife management, often presented in scientific rationale, is frequently underpinned by a profound belief in animal rights—that is, the right of individual species to flourish without human intervention. Some people call this the Bambi syndrome, a derogatory appellation meant to imply that people are emotionally involved, not intellectually engaged in the process of managing our wild species.

Varying from passionate zealots to ordinary citizens, animal rights advocates often pit themselves against wildlife managers, who are seen as cruel and manipulative. Time and again, conflicts between these two groups generate more heat than light. Why?

We believe that the chief reason for the degree of conflict is that the two groups are talking on different planes. Wildlife managers talk about the science of population management and ecosystem management, or so it seems; but animal rights advocates are talking moral issues. The former argue that we need to control wildlife populations, such as deer and elk; the latter argue that killing animals is inhumane. It smacks of human control over nature. It is arrogant and morally repugnant.

But in the debates that ensue, infrequent are the times when both groups talk on the same level. Rare is the individual who can dissect the arguments, teasing out moral or ethical concerns from scientific ones.

Animal rights advocates say it is wrong to kill deer or wild horses that are overrunning a region; wildlife management officials argue that it is absolutely essential for the health and long-term future of the land and other species as well.

An important point is often missed in the acrimonious debates: Wildlife managers do have an ethical position. They have aligned themselves on the side of ecosystems. They have taken the moral stance that ecosystem health is more important than an individual species's right to be free from human manipulation. You could argue that the ecosystem ethic (for lack of a better term) is really a scientific position, but it is not. It is ethics based on science.

Can there be any reconciliation? Not until both parties begin to sort out the ethical from scientific considerations and agree to talk a similar language. As a wildlife manager, you may want to meet with people who oppose your actions and lay out your ethical position. It may help to outline the argument for the primacy of ecosystems in scientific terms as well as ethical terms. In addition, it may be helpful to note that you care about humane treatment. Although a few animals may be killed to control population size, the benefit to the rest of the population and to the ecosystem is substantial.

With ethical positions laid out, you can debate scientific discrepancies and make real headway. You may find legitimate points that bear consideration or may force you to alter your plans. Even then, differing ethical viewpoints may make reconciliation impossible.

where they continue to their wintering grounds along the Gulf of Mexico.

Each of the four flyways is administered by a flyway council made up of directors of the various state conservation departments or their representatives. The flyway councils meet in August to frame regulations for the fall hunting season based on the size and distribution of the waterfowl population expected to be moving through their respective flyways. The councils then submit their recommendation to the U.S. Fish and Wildlife Service. The service also receives suggestions from various national conservation organizations, such as the National Audubon Society, Ducks Unlimited, and the Wildlife Management Institute. Armed with all this input, the service decides on regulations that it considers appropriate.

Let's examine a specific example of how waterfowl hunting regulations are tailored to the population level of the target species. In 1962, the estimated mallard population on the North American continent was extremely low—about 7.6 million. Highly restrictive hunting regulations were therefore established. In the state of Washington, for example, the hunting season in 1962 was shortened to 75 days, and the daily bag limit was reduced to four. As a result, the mallard harvest in that state in 1962 was only 177,000 birds. Such restrictive laws were continued. The mallard population responded to this protection. By 1970 it had reached 11.6 million on the North American continent.

Regulations on the mallard harvest were, therefore, liberalized. Washington extended its season to 93 days and increased the bag limit to six birds. The result was a kill of

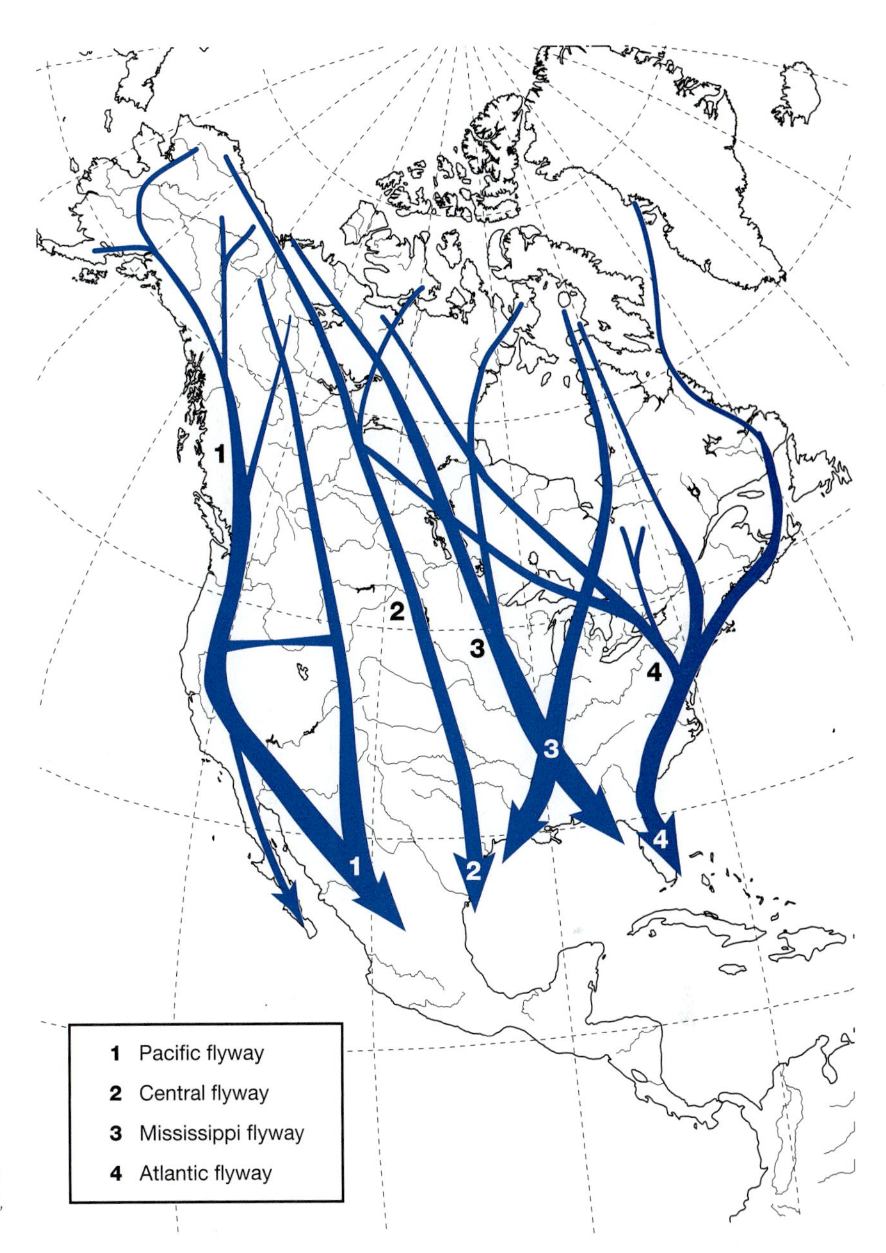

FIGURE 16.22 *Major waterfowl flyways. Recovery data on many thousands of waterfowl led biologists to believe, earlier in this century, that the birds migrated along four major flyways: (1) Pacific, (2) Central, (3) Mississippi, and (4) Atlantic.*

311,000 birds, 76 percent greater than in 1962. True, liberal hunting regulations (bag limits of three to seven per day and long seasons) increase the number of mallards shot compared with the number shot in restrictive seasons. However, a study done in the late 1970s analyzing band recoveries found that the survival rates for mallards, from September 1 to August 31 of the following year, were no different whether hunting regulations were liberal or restrictive. This indicates that other forms of mortality compensate for hunting mortality.

It is important to have population studies of mallards and other waterfowl that analyze whether a specific mortality factor like hunting is additive or compensatory. Such information is vital for determining bag limits and hunting season lengths. For example, if the hunting mortality is additive, then restrictive regulations should be adopted to help depressed populations recover. On the other hand, if it is compensatory, then liberal regulations (up to a point) may be adopted.

Regulating Destructive Waterfowl Populations

Waterfowl can cause damage to farmlands. The most serious example of wildlife-inflicted crop damage involves grain consumption by waterfowl in Canada. Losses to Canadian grain farmers may reach 380,000 metric tons per year. Financial setbacks have approached $40 million annually, although the annual average is about $14 million.

Although the long-term solution to crop destruction by waterfowl in Canada may involve some type of habitat modification, either in the cropped fields or in the nearby wetlands, wildlife managers are not even close to solving the problem. However, there are several management strategies currently

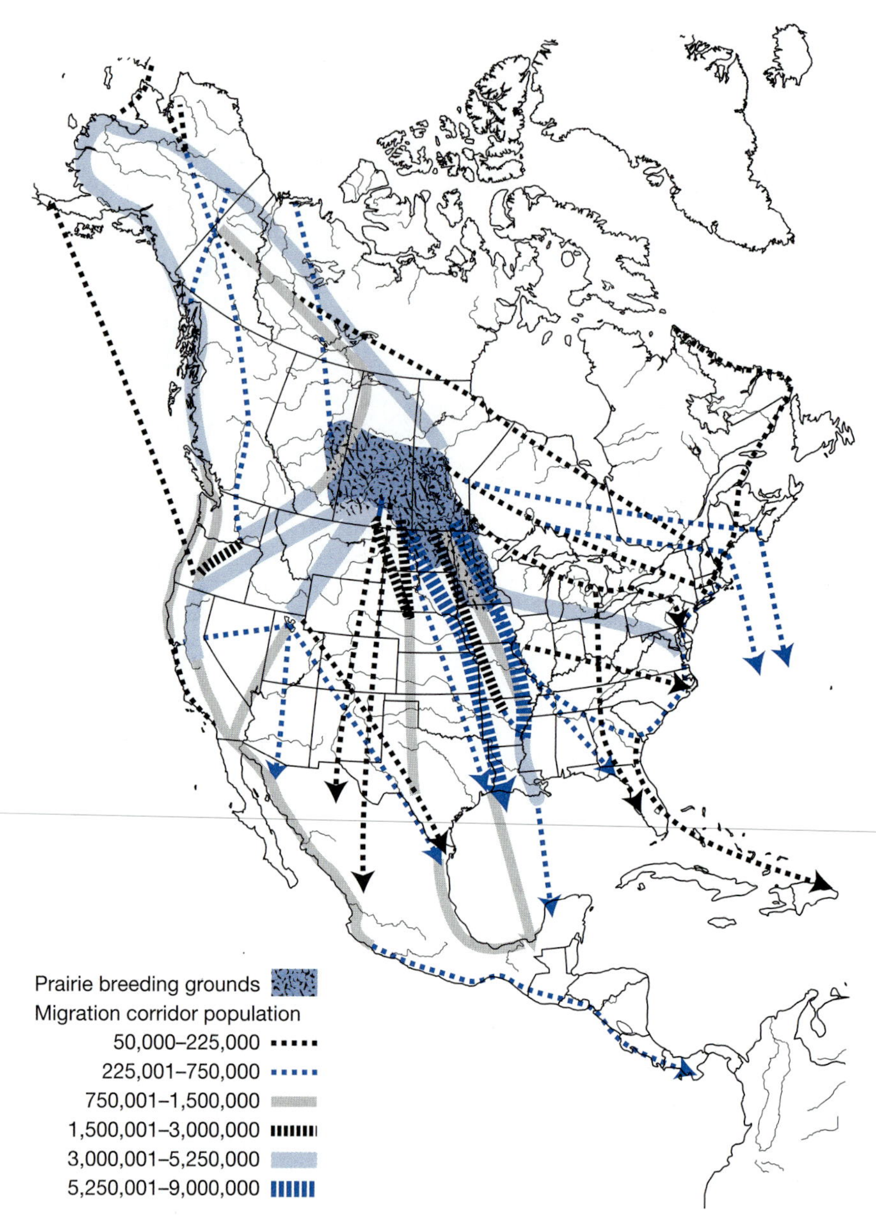

FIGURE 16.23 *Duck migration corridors. Wildlife biologists now know that the "flyway" concept was an oversimplification of an extremely complex migration phenomenon. The routes of many species of waterfowl may actually cross each other, as shown here. Moreover, many waterfowl may fly east or west during migration rather than north or south, as suggested by the flyway concept. Frank Bellrose, of the Illinois Natural History Survey, identified migration corridors for southward flying ducks and was able to estimate the approximate number of ducks using a particular corridor.*

available that lessen the damage. A variety of noisemakers, including propane gas exploders (Figure 16.24), are used to scare waterfowl away from their "wheaties." Special fields are planted in "lure crops" to attract ducks and geese from privately owned wheat fields. Feeding or bait stations are even deployed to attract ducks away from crops.

The Canadian government offers farmers insurance programs that partially offset the financial losses caused by marauding waterfowl. It's an interesting dilemma, to say the least. On the one hand, farming destroys waterfowl habitat when wetlands are converted to cropland. On the other hand, waterfowl destroy the crops that are planted on the original wetlands.

Canada geese have recently established breeding populations in some areas of northern Illinois. Unfortunately, however, huge flocks of these birds have become a serious nuisance on golf courses. Not only do they damage the greens, they cause problems at water hazards as well. In recent years hundreds of these geese have been trapped alive and transplanted to rural areas in southern Illinois, far from the golfing scene.

Wildlife Diseases Affecting Humans

Wildlife populations can be carriers of some diseases affecting humans. Tularemia (mostly transmitted by ticks or other arthropods) and sylvatic plague (transmitted by wild rodents) are two wildlife diseases of concern to humans, but rabies strikes the most fear in humans.

Rabies is a disease caused by a virus that lives in the saliva of the host (carrier) animal. The virus is transmitted to humans by the bite of an infected animal. It moves along nerve cell pathways and eventually reaches the brain. Unless the infected person is vaccinated against the virus, he or she will sink into a coma and die. Rabid animals will snap and bite at animals and humans. Many different species of wild animals

FIGURE 16.24 *A propane gas exploder in a cornfield in Ohio. Such noise-makers have been used to scare blackbirds, waterfowl, and other wildlife from farm fields. Such strategies often lose their effectiveness when the animals get used to the disturbance and continue eating the crops.*

can harbor the rabies virus, including skunks, bats, raccoons, and foxes. Extreme measures, such as poisoning and shooting, must sometimes be used to control rabid animals. Several years ago, strychnine-treated baits were used to control rabid foxes in Kentucky. Unfortunately, however, in addition to destroying 65 foxes, the poison killed 135 dogs. Moreover, the control measure was prohibitively expensive, costing $208 for every poisoned fox. More recently, an effective oral vaccine bait has been developed. It is dropped from airplanes over rabies-prone regions and confers immunity to the carnivores that consume it.

16.6 Nongame Wildlife

In recent years, state and federal wildlife agencies have developed a new focus of attention: nongame management. This is certainly appropriate, since Americans now spend more than $7 billion annually on recreation based on nongame wildlife. Their activities range from whale watching to winter bird-feeding to wildlife photography.

The status of nongame populations, such as those of amphibians and reptiles, is not well known. On the other hand, population trends in birds have been followed very closely for a number of years through annual surveys conducted by the U.S. Fish and Wildlife Service and the National Audubon Society. These surveys have shown that the populations of most nongame birds are fairly stable. Indeed, the annual Christmas bird censuses conducted by the National Audubon Society have shown that hawk populations are on the rebound after being decimated by pesticide contamination of their food chains in the 1950s and 1960s. On the other hand, populations of many species of vireos, tanagers, and flycatchers, which winter in Central and South America, are declining. Many biologists attribute this to the destruction of habitat in the United States as well as tropical rain forests (Chapter 14), the wintering range of these species.

Several species of fish-eating birds of the Great Lakes have suffered dramatic population declines in the last two decades. The suspected cause is their consumption of fish contaminated with toxic chemicals such as PCBs and dioxin. Female birds transmit the poisons to their eggs. A survey conducted by John Giesy, professor of wildlife management at Michigan State University, revealed toxic chemicals in the eggs of fish-eating birds throughout the Great Lakes. Among the species most strongly affected are cormorants, terns, and gulls. Many of the poisoned eggs never hatch. Many of the chicks that do emerge from the egg cannot survive owing to gross deformities. The abnormalities are crossed bills, club feet, dwarfed wings, and lack of eyes and skull. The incidence of these deformities has increased 30-fold in the last 40 years.

The restoration of the populations of these threatened species poses a formidable and ongoing challenge to state and federal wildlife managers. The eventual solution to the problem, of course, depends on cleanup of the chemical pollution of the Great Lakes.

Summary of Key Concepts

1. Wildlife management may be defined as the planned use, protection, and control of wildlife by the application of ecological principles.
2. The habitat of a wild animal provides certain essentials, chiefly shelter, food, water, and breeding sites.
3. Wildlife population densities tend to be higher in areas where there is a large amount of interspersion of plant communities, or edge.
4. A home range is the area over which an animal habitually travels while engaged in its usual activities.
5. A territory is any area that is defended. It is usually smaller than the home range.
6. Four types of movements in wild animals are (a) dispersal of young, (b) mass emigration, (c) latitudinal migration, and (d) altitudinal migration.
7. The movements of animals promote survival by providing (a) supplies of food and water and (b) a hospitable climate.
8. Additive mortality simply adds to the deaths caused by other factors, whereas compensatory mortality is the concept that one kind of mortality replaces (up to a point) another kind of mortality in animal populations.
9. The major mortality factors affecting deer populations include starvation, predation, hunting, disease, and accidents.
10. Emergency feeding of starving deer is not a sound management practice because it facilitates the spread of disease and is very costly.
11. Waterfowl populations are adversely affected by loss of habitat, oil and chemical pollution, hunting and lead poisoning, disease, and acid rain.

12. More than 50 percent of the waterfowl in North America are raised on tiny, 0.5- to 1-hectare (1.25- to 2.5-acre) potholes.
13. The Comprehensive Everglades Restoration Plan, part of the Water Resources Development Act of 2000, is a $7.8 billion federal-state project intended to restore, protect, and preserve the defining ecological features of the original Everglades and other parts of south Florida.
14. Waterfowl destruction by oil pollution may increase as more offshore oil wells are drilled and tanker traffic increases.
15. Lead shot, which can cause lead poisoning, was completely phased out in the United States in 1991.
16. Botulism, which can cause severe disease outbreaks in waterfowl, may kill 100,000 waterfowl in California and Utah per year.
17. Two major wildlife management techniques are (a) habitat development and (b) regulation of wildlife populations.
18. Wildlife biologists can regulate the abundance of wildlife species by manipulating ecological succession.
19. Current hunting and trapping regulations place restrictions on (a) the species and numbers of individuals taken, (b) the season and time of day, and (c) the type of firearm or trap used.
20. The 1937 Federal Aid in Wildlife Restoration Act, popularly known as the Pittman-Robertson Act, levies an excise tax on the sales of sporting arms and ammunition; revenues from the federal tax are distributed to the states for habitat purchase and wildlife management and research.
21. A number of voluntary government programs, such as the Wetlands Reserve Program, the Wildlife Habitat Incentives Program, and the Partners for Fish and Wildlife Program, help private landowners protect wildlife and improve their habitat.
22. The Federal Wildlife Refuge System includes more than 530 refuges. Most of them are designed for waterfowl.
23. Waterfowl habitat can be improved by creating openings in dense marsh vegetation, constructing artificial ponds and islands, developing artificial nests and nest sites, and establishing waterfowl refuges.
24. The mallard is one of the most highly prized ducks sought by hunters on the North American continent.
25. Analysis of recovery data from leg-banded waterfowl provides waterfowl biologists with information concerning growth rate, life span, and mortality, as well as the length, speed, and route of migration.
26. The most serious example of wildlife-inflicted crop damage involves grain consumption by waterfowl in Canada.
27. Many people have recreation activities based on nongame wildlife, such as whale and bird watching or wildlife photography.

Key Words and Phrases

Acid Deposition
Additive Mortality
Altitudinal Migration
Artificial Islands
Artificial Nest Sites
Artificial Ponds
Bird Banding
Botulism
Browse Line
Climax Species
Compensatory Mortality
Comprehensive Everglades Restoration Plan
Conservation Reserve Program (CRP)
Consumptive Use
Corridor
Cover
Disease
Dispersal of Young
Duck Stamps
Early-Successional Species
Edge Effect
Everglades National Park
Euryphagous Species
Food Security Act
Game Animals
Habitat
Harvestable Surplus
Home Range
Hunting
Latitudinal Migration
Lead Poisoning
Mass Emigration
Mid-Successional Species
Migratory Bird Hunting Stamp Act
National Audubon Society
Nature Conservancy
Nonconsumptive Use
Nongame Animals
Partners for Fish and Wildlife Program
Pittman-Robertson Act
Pothole
Rabies
Shelterbelt
Stenophagous Species
Succession
"Swampbuster" Provision
Territory
Waterfowl Flyway
Wetland Reserve Program
Wildlife
Wildlife Habitat Incentives Program
Wildlife Management
Wildlife Refuge

Critical Thinking and Discussion Questions

1. Compare stenophagous and euryphagous species. Give an example of each type.
2. Discuss the functions of territories.
3. Would you say that your college campus has a lot of edge? How does this edge increase survival for wildlife such as robins, squirrels, and rabbits?
4. What are the advantages of altitudinal and latitudinal migrations?
5. How can botulism be prevented or controlled in waterfowl?
6. What causes lead poisoning in waterfowl? How can it be prevented?
7. What is the difference between additive mortality and compensatory mortality?
8. How can ecological succession be manipulated to improve wildlife habitat?
9. Discuss the importance of cooperation between forestry officials and wildlife managers in developing high-quality habitat for ruffed grouse.
10. Discuss the ecological and economic impacts of the American hunter.
11. Give an example of restrictive hunting regulations for deer. Give an example of liberal hunting regulations for deer.

12. Give an example of destruction caused by waterfowl. How may such destruction be controlled?
13. Why is our national wildlife refuge system important?

Suggested Readings

Bellrose, F. 1976. *Ducks, Geese and Swans of North America.* Harrisburg, Penn.: Stackpole Books. The "Bible" on North American waterfowl. Excellent coverage of distribution, feeding habits, nesting behavior, migration, and mortality factors. Detailed migration maps.

Bolen, E. G., and W. L. Robinson. 1999. *Wildlife Ecology and Management,* 4th ed. Englewood Cliffs, N.J.: Prentice Hall. Informative introductory text on the management of wildlife populations and habitats.

Bookhout, T. A, ed. 1994. *Research and Management Techniques for Wildlife and Habitats.* Bethesda, Md.: Wildlife Society. Comprehensive manual of wildlife management that wildlife professionals use in their daily work.

Brandt, E. 1993. "How Much Is a Gray Wolf Worth?" *National Wildlife* 31(4): 4–12. The emerging science of resource economics proposes that wildlife preservation can be "good business" to entrepreneurs.

Caughley, G., and A. R. E. Sinclair. 1994. *Wildlife Ecology and Management.* Cambridge, Mass.: Blackwell Science. Intermediate text provides comprehensive overview of wildlife ecology and management.

Chadwick, D. H. 1996. "U.S. National Wildlife Refuges." *National Geographic* 190(4) 2–35. Popular account and beautiful pictorial of our National Wildlife Refuges.

Leopold, A. 1949. *A Sand County Almanac.* New York: Oxford University Press. A classic written in almost poetic prose. The founder of game management provides intriguing insights into the relations between humans and wildlife.

Morrison, M. L., B. G. Marcot, and R. W. Mannan. 1998. *Wildlife-Habitat Relationships: Concepts & Applications,* 2nd ed. Madison: University of Wisconsin Press. Provides a broad but advanced understanding of habitat relationships applicable to all terrestrial species.

Scalet, C. G., L. D. Flake, and D. W. Willis. 1996. *Introduction to Wildlife and Fisheries: An Integrated Approach.* New York: W. H. Freeman. Detailed introductory text on wildlife (and fisheries).

Web Explorations

Online resources for this chapter are on the World Wide Web at: **http://www.prenhall.com/chiras** *(click on the Table of Contents link and then select Chapter 16).*

Chapter 17

Sustainable Waste Management

Waste is a byproduct of all living organisms. We humans though hold the prize for waste among the animal kingdom. Worldwide, the human community produces billions of gallons of sewage, millions of tons of solid waste, and millions of tons of hazardous waste each day! This chapter examines two types of waste, municipal solid waste and hazardous waste. (Chapter 11 examined sewage.) Here you will learn where our waste comes from, how we deal with it, and how to deal with it more satisfactorily. Its aim is to outline sustainable waste management strategies.

17.1 Municipal Waste: Tapping A Wasted Resource

In 1997, American cities and towns produced a mountain of garbage, or **municipal solid waste,** from our homes and businesses (not industries). That year, in fact, U.S. solid waste amounted to nearly 200 million metric tons (217 million tons). This translates into a daily output sufficient to fill the Superdome in New Orleans 2.5 times every day! Add to that the over 70 million metric tons of solid waste that is produced by industries each year, and it is clear that Americans have a major problem on their hands. Further complicating matters, municipal solid waste production is increasing by 1.5 million metric tons (1.7 tons) per year.

Although Americans are the undisputed leader in total annual solid waste production, on a per capita basis, the Canadians are not far behind. While Americans produce about 2 kilograms (4.4 pounds) per day, on average, Canadians produce about 1.7 kilograms (3.8 pounds) per person per day (Figure 17.1). The Germans, French, Spaniards, and Italians produce about half as much on a per capita basis.

As shown in Figure 17.2, most of the waste from U.S. cities and towns consists of paper and yard wastes—grass clippings and leaves—with lesser amounts of metals, glass, and plastics making up the balance. Despite a marked increase in recycling, only about 26 percent of the municipal solid waste produced in the United States is currently re-

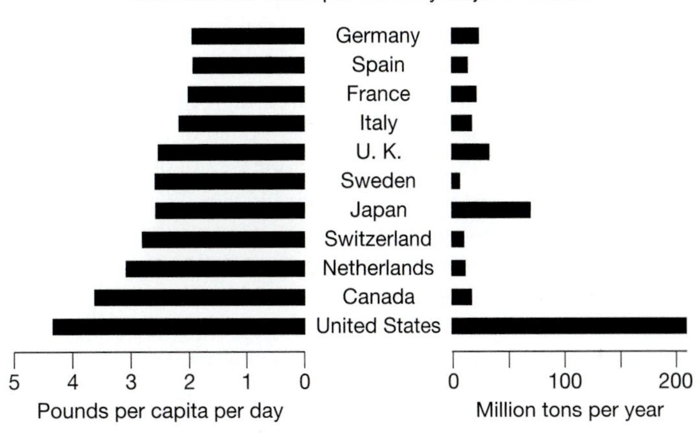

FIGURE 17.1 *Comparison of municipal solid waste production by country. Note that on a per capita basis the United States and Canada are leading producers.*

covered for reuse and recycling—up from 17 percent in 1990. Incineration, which many view as a less than desirable option, has skyrocketed from 4 percent in 1988 to 17.3 percent in 1997. The remainder, approximately 55 percent of the nation's waste, is dumped in landfills, where it is covered with dirt and forgotten.

Solid waste is a problem from several perspectives. First, it is expensive. Americans, for instance, currently spend over $300 million a year just to landfill disposable diapers. In many cities waste disposal is the second largest expenditure, preceded only by education.

Second, solid waste disposal takes up valuable and often costly land that could be put to better use. In crowded metropolitan areas, landfill space is often in short supply, and waste haulers must ship garbage many miles at huge expense. Many states, especially those in the Northeastern United States, are facing a shortage of landfill space. Some states have taken to transporting their wastes to other states, often many hundreds of miles away, or even other countries.

Third, garbage disposal and incineration represent a waste of valuable resources. (The term *waste,* which we use to define this valuable resource that is currently being squandered, is wonderfully appropriate.) David Morris of the Washington-based Institute for Local Self-Reliance wrote that "A city the size of San Francisco disposes more aluminum [each year] than is produced by a small bauxite mine, more copper than a medium copper mine, and more paper than a good-sized timber stand." Our trash is a gold mine of wastes, offering a wealth of resources, if only we were smart enough—or farsighted enough—to tap into the steady stream of recoverable resources pouring out of our cities, towns, and homes each day.

Fourth, because solid waste disposal and incineration squanders valuable resources, more minerals must be mined, more trees must be cut, and more oil must be drilled (to make plastic or to power mining equipment). Each of these activities produces enormous amounts of waste and has serious impacts on the environment.

Sustainable solid waste management, which includes recycling and other measures, can eliminate these problems and help foster a sustainable society. The next section describes various approaches to waste management from the conventional and unsustainable to the sustainable.

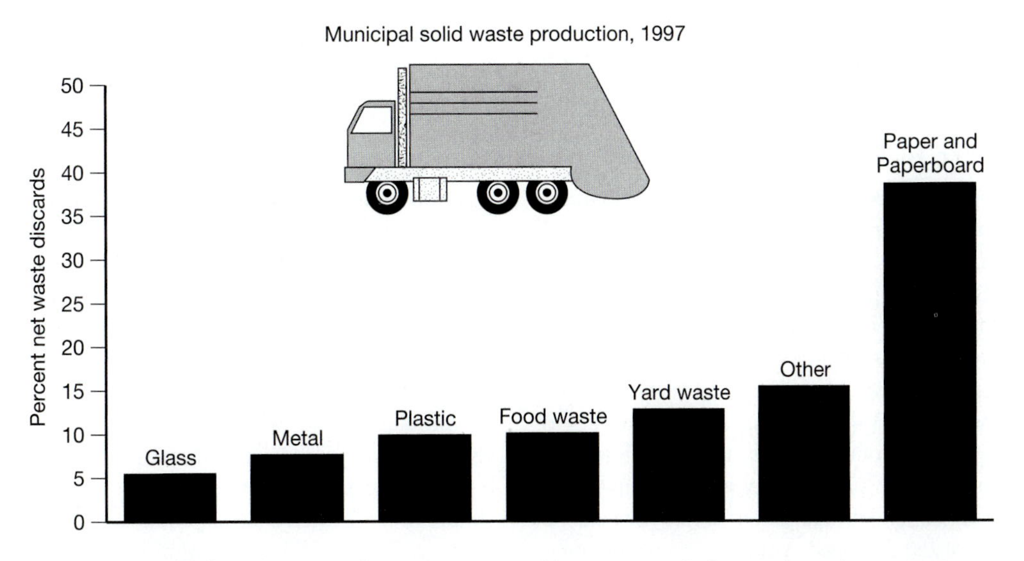

FIGURE 17.2 *Composition of America's municipal sold waste stream (before recycling and composting).*

17.2 Managing Our Municipal Solid Wastes Sustainably

Most industrial societies have traditionally viewed municipal waste as something to get rid of—to dump in the ground or at sea as far away as possible—with little regard for the wealth of materials wasted in the process. When one dump site was full, city planners or private companies found another. Faced with shortages of suitable dump sites and growing transportation costs to haul the mountains of trash to outlying sites, however, many city planners have come to look skeptically at the discard approach. As conservationists and environmentalists have long said, this approach is ecologically unwise and unsustainable. This view was expressed by the World Commission on Environment and Development in a 1987 report entitled *Our Common Future.* "Many present efforts to guard and maintain human progress, to meet human needs, and to realize human ambitions," said the commissioners, "are simply unsustainable—in both the rich and poor nations. They draw too heavily, too quickly, on already overdrawn environmental resource accounts to be affordable far into the future without bankrupting those accounts." Looking at the impact of our actions on future generations, they wrote, "We borrow environmental capital from future generations with no intention or prospect of repaying. They may damn us for our spendthrift ways, but they can never collect on our debt to them." Some arguments for the present generation shouldering its debt, rather than passing it on to succeeding generations, are presented in Ethics in Resource Conservation Box 17.1.

Two approaches can be used to create a sustainable waste management system and a more sustainable society. These approaches extend the planet's limited resource supplies, save energy, reduce pollution, and cut back on habitat destruction—in short, they reduce our destruction of environmental capital and create a more sustainable way of life. Some measures such as recycling also create jobs and can save money.

The first strategy in a sustainable society is the reduction approach, which calls for lower levels of material consumption in society. The second is the reuse and recycling approach, which maximizes the life span of a material in the production-consumption cycle.

The Reduction Approach

By reducing their per capita consumption of natural resources, modern societies can make tremendous inroads into the solid waste problem—and greatly reduce the world's demand for natural resources. But this approach is often unattractive to individuals and businesses. "Why cut back on our pleasures and our profits to save a few birds?" some opponents ask.

In the discussion of sustainable ethics in Chapter 2, you learned that many people think that other species have a right to exist, too. Protecting the environment helps ensure a rich and diverse biological world. Cutting back today also benefits us by protecting the environment and its many free services that can be lost when land is harvested or mined. Voluntary frugality, however, is also very likely inevitable. Advertizing is geared to promote maximum material consumption. Modern capitalistic economies promote maximum production and consumption, too.

Will frugality be forced upon those of us in the materialistic more developed countries? As you shall see in Chapter 20, the supply of certain minerals is limited. More important, however, oil supplies are finite and fast on the decline. As demand for oil begins to outstrip production, an inevitability projected to occur within the next ten years, oil prices will rise. Many observers believe that frugality will be forced upon us. To postpone this day, and the hardship that may result, why not begin now by using less now? "Sustainable global development," according to the World Commission on Environment and Development, "requires that those who are more affluent adopt lifestyles within the planet's ecological means."

Individuals can reduce their overall consumption level in many ways. One strategy is to purchase more durable items. Durable clothing, tools, furniture, computers, and calculators outlast their cheap counterparts and greatly reduce demand for energy, minerals, water, land, and so on. Buying sturdy goods casts a vote for environmental protection that will not escape the attention of business leaders.

Individuals can also hold on to products longer. A car that is driven for ten years rather than the typical three to five years, for instance, can greatly reduce your consumption of steel and other materials. If you live to be 70 and buy your first car at age 20, you would purchase five automobiles in your lifetime, if each one were held for ten years. If, on the other hand, you purchase a new vehicle every three years, you would need 17 cars.

Faced with rising energy and mineral prices in the 1970s, many manufacturers found another route to reduce material consumption: miniaturization. Computer manufacturers, for instance, have greatly reduced the size of their computers over the past two decades, requiring less material per unit. (For more on this subject, see Chapter 20.) Buying or building smaller homes is one way people can help. A 2,000 square foot home requires fewer building materials and less energy to heat and cool than a 4,000 square foot home.

Yet another approach to waste reduction is simply to consume less. We can lead simpler, less resource-intensive lifestyles. Do we need all of the modern conveniences? Do they really enhance our lives? Or are they what actor-environmentalist Dennis Weaver describes as simply something else to paint, repair, and insure?

By buying less, curbing our consumer passions, we can make tremendous inroads into the solid waste dilemma and promote environmental sustainability. But swimming against the mainstream will not be easy in any consumeristic society. U.S., Canadian, and Japanese societies have become cultures of mass consumption. They have made a virtue out of consumption and materialism. In the United States, bumper stickers proclaim, "He who dies with the most toys wins" or proudly advise citizens to "Shop till you drop." Others gleefully announce that the owners of the vehicles were "born to shop." In the United States, people are no longer referred to as citizens. The evening news reminds us over and over again that we are "consumers." Despite all of this consumeristic

ETHICS IN RESOURCE CONSERVATION 17.1

DO WE HAVE AN OBLIGATION TO FUTURE GENERATIONS?

Every action you take has an impact on the environment and will affect the welfare of future generations. Although the impact of any one action may be insignificant, when combined with actions of others we find that individual acts can mount up. But so what? All organisms have an impact on the environment.

Although it is true that all organisms affect their environment, humans have a much greater impact. Those of us who live in advanced technological societies have the greatest influence on the health and well-being of the environment. Given our disproportionately large impact, we face an important and difficult ethical question: Do we have an obligation to future generations to protect the planet? In other words, are we obliged by any ethical tenets to act in ways that respect the future of the planet, or is life a free-for-all?

Robert Mellert, who teaches philosophy at Brookdale Community College in New Jersey, argues that we do indeed have an obligation, which he bases on four ideas.

First, future generations will have many of the same needs that we have today. Although their priorities may differ, their needs for food, clothing, shelter, and recreation, among others, will remain. Mellert goes on to say that future generations, if they could be heard today, would claim a right to exist and to live well. To give them life without providing the means of meeting their needs would be cruel. To give them life without the means because of our carelessness and greed would be thoughtless and selfish.

Second, Mellert points out that an individual is born into a particular generation without his or her consent. None of us choose when to be born. He says, then, that because we have no claim to the time and place of our birth, justice would require that we have no more rights over the world and its resources than anyone else.

Third, Mellert contends that the survival of our species is more important than the survival of an individual member. Wildlife managers operate under this basic assumption. Their goal is to protect species and their habitat. That may require periodic culling—say, by sport hunters. No one is proposing this action, just that individual rights are of lesser importance than the rights of many—now or in the future.

Fourth, he argues that even after we die, the effects of our lives continue. The present generation is the product of its ancestors—all that they did, died for, and believed in—but we are also the product of our own decisions. Future generations will be the result of what we are now and how they use what we leave them.

Mellert concludes by saying that if we accept these four truths, it is clear that we have an obligation to future generations. Our obligation, he says, is based on the fact that we are part of a much larger whole. We are inhabitants of a planet that will be used by many others who follow in our footsteps. We owe the future as much as we have received from the past.

Ponder these thoughts for a while. Do you agree with them? Would you modify the case Mellert makes in any way? How? Why? Make a list of obligations we have to future generations.

language, the signs are suggesting that it may be the person with the fewest toys who actually wins!

The Reuse and Recycling Approach

You have worn a pair of pants for a year and, even though they are in pretty good shape, you have grown tired of them. What do you do? If you are like many Americans, you throw them away. But why not drop your usable goods off at one of the many charitable groups like Goodwill or Disabled American Veterans or a local used clothing store that will resell them?

Advocates of the reuse strategy point out that many products can be **reused**—put back into service in the same or different application. Boxes, appliances, clothing, furniture, grocery bags, and so on can all be reused in many ways. Newspapers, for instance, can be pulped and used for ceiling and wall insulation. They can be used to make animal bedding. Boxes can be used for moving or storing things. Glass can be crushed and mixed with asphalt and used for paving highways (Figure 17.3). Used office paper can be donated to schools for art projects. Even old eyeglasses are donated to the less fortunate by some outlets. Some cities and towns have public programs and private companies that reclaim construction waste. Old timbers from barns or wood floors from old buildings are removed and used in new buildings.

By diverting trash-bound items to collection centers for reuse, we can greatly extend the useful lifetime of the products and reduce our demand for new materials. Such efforts, therefore, help reduce resource consumption, pollution, and land disposal.

Reuse follows closely behind the reduction approach in its ecological appeal. For materials that cannot be reused, however, the next best approach is recycling. **Recycling** is another form of reuse, but usually involves some kind of conversion. For example, glass is crushed and melted, then used to make new glass, is said to be recycled. A glass container that once held applesauce but is now used to hold nails in the garage is said to be reused.

Recyclable materials can be extracted from municipal trash at central stations, which is convenient for the citizen but often costly. This process is called **end-point separation.** Recyclables can also be separated at the source—at homes and factories—and picked up by recyclers or delivered to recycling centers by producers (Figure 17.4). This option is called **source separation.** It involves considerably more citizen participation than end-point separation but reduces the costs of separation facilities.

Recycling reduces our demand for virgin materials and thus cuts down on mining and other resource extraction activities.

FIGURE 17.3 *Glassphalt. (A) Ground-up glass becomes an ingredient of a road paving material known as "glassphalt" (glass and asphalt). (B) Workers spread glassphalt along a busy street in Toledo, Ohio.*

Doubling the rate of paper recycling worldwide would meet increased demand and dramatically reduce the need to cut timber. Recycling a 3-foot-high stack of newspapers saves a single tree. Recycling a whole ton of paper saves 17 trees. As the statistics in the previous paragraph suggest, individual efforts added with those of others can have an enormous impact. If, for instance, the United States increased paper recycling 30 percent, an estimated 350 million trees could be saved annually. Enough electrical power would be saved to supply homes for 10 million people.

Paper recycling saves other resources and reduces pollution. Each ton of paper recycled reduces our water demand by 230,000 liters (60,000 gallons) and saves 255 kilowatt-hours of electricity—enough to run a refrigerator for a year. Recycling paper produces only 25 percent of the air pollution that comes from manufacturing paper from trees, and cuts water pollution as well (Table 17.1).

Similar, even larger, savings are available from recycling other materials. As shown in Table 17.1, producing an aluminum can by recycling uses 95 percent less energy than making it from aluminum ore, or bauxite. Put another way, we can produce 20 aluminum cans from recycled scrap with the same energy it takes to make one can from raw ore. Quite a savings!

Another form of recycling is called **composting.** Composting occurs when organic matter such as kitchen waste, yard waste (leaves and branches), and even paper and cardboard are allowed to decompose to a stable, humus-like material. Aerobic (oxygen-requiring) bacteria and fungi in the waste

FIGURE 17.4 *Garbage, garbage everywhere. Workers at Ecocycle in Boulder, Colorado, sort recyclable materials that will be shipped to market and used to make new products at a fraction of the energy cost of their counterparts made from virgin materials, and with much less pollution.*

TABLE 6.1 *The Benefits of Recycling*

Material	Energy Savings (%)	Solid Waste Reduction (%)	Air Pollution Reduction (%)
Paper	30–55	130*	95
Aluminum	90–95	100	95
Iron/Steel	60–70	95	30

* 130 percent is possible because 1.3 pounds of waste paper is required to make 1 pound of recycled paper.
Source: William Chandler, *Materials Recycling: The Virtue of Necessity*. Worldwatch Paper 56. Washington, D.C.: Worldwatch Institute, October 1989.

decompose the organic matter. The humus-like material can be applied to flower beds, lawns, gardens, and even farm fields. This material, which contains inorganic nutrients as well, is essentially recycled back to the earth. Homeowners can compost their own organic matter to create soil conditioners for their lawns, flower beds, and vegetable gardens. Cities and towns can compost waste. Because landfill space is limited, many states (22 as of 1997) have banned the disposal of yard waste (grass clippings and branches) in landfills. In 1999, there were 9,350 municipal yard waste composting programs in the United States, according to the U.S. Composting Council, most of which were in states where yard waste disposal had been banned. Some zoos have even gotten into the act, most notably Seattle's Woodland Park Zoo, which produces a compost from animal manure and sells it as Zoodoo (Figure 17.5).

Compost can be mixed with sewage sludge—a process called **co-composting.** Bacterial decomposition releases heat, and in a few days the compost pile may reach 150°F inside, sufficient to destroy pathogenic (disease-causing) bacteria that might be present.

In light of the progressive deterioration of American soils and the increasing demand for food, it would seem that composting would be a popular disposal method in this country. However, because the final product is relatively low in nitrogen and phosphorus, many farmers prefer to use commercial fertilizers. Another problem is the cost of transporting compost to farm field.

Composting itself has problems. One of them is that to be successful nonorganic material (for example, cans and glass) must be removed from the waste stream. Source separation could greatly reduce this problem and make composting more feasible. Still another problem is that large tracts of land are

FIGURE 17.5 *Zoodoo. Manure and bedding of zoo animals at Seattle's Woodland Park Zoo is composted and turned into soil supplements that are sold to area residents. This program cuts down on waste, reduces pollution, and makes the city a small profit.*

needed to compost organic wastes, and these areas must be situated away from people to avoid problems with insects and odor.

Composting is much more popular in Israel and in frugal European countries—in particular Italy, England, the Netherlands, and Belgium. In Holland, for example, a single company produces 180,000 metric tons of compost annually from the waste generated by 1 million people.

Model Recycling Programs

The United States recycles about 26 percent of its solid municipal waste, but the Environmental Protection Agency (EPA) believes that we could nearly double that amount. A 60 to 80 percent recovery rate may eventually be achieved as energy prices rise and resource supplies fall. Many other nations have a tremendous lead on the United States. Japan, the Netherlands, Mexico, and South Korea lead the world in paper recycling. South Korea and Mexico, in fact, now import wastepaper from other countries for recycling.

Japan, a leading consumer of paper, intensified its paper recycling program in the mid-1960s. Short on landfill sites, this tree-poor nation launched a major recycling program with prodding from environmentally concerned citizens. In Hiroshima, citizens separate their wastes and carry paper to local collection centers for recycling. Bottles, cans, and other items are also recycled. In Machida, Japan, an astounding 90 percent of the city's waste is recycled, thanks to citizens' efforts to separate trash, a highly computerized recycling system, and heavy fines.

The Japanese need to recycle paper and other goods because they import much of their energy, have little land for waste disposal, and lack the great forests of nations like the United States and Canada. But the success of their programs stems from more than necessity. The unity of the Japanese and their willingness to cooperate to solve a problem are crucial to their success. Their cultural ethic of frugality, their foresight, and strict laws also play a big role in their success. The government has helped out in many ways. In Fuchu City, a Tokyo suburb, the government purchased the costly recycling equipment, then turned it over to a private company, which operates it with profits earned from the sale of recycled materials.

The Netherlands is also a leader in paper recycling. Like Japan, it is short on land and forests. To reduce waste, the government established a waste exchange, a service that matches buyers and sellers of waste. The government also established a way to stabilize prices. Normally, the cost of recyclables vacillates wildly, according to supply and demand. In periods of low demand, recyclers may find few markets and prices too low to stay in business. To buffer against these destabilizing cycles, the government of the Netherlands buys recyclables at a set price when demand or market values drop. When prices increase, the government sells off its supplies and replenishes the funds that keep this system and the country's recyclers in business.

The Netherlands also promotes source separation. But, unlike Japan, where source separation is merely promoted, in the Netherlands it is the law. The state of New Jersey and the town of Islip, New York, have similar mandatory recycling programs.

The United States is slowly becoming a recycling nation. In 1996, the most recent year for which data were available, there were 8,817 curbside recycling programs, serving 134.6 million people—about half the population. Eight states—Maine, New York, Minnesota, New Jersey, South Carolina, Virginia, South Dakota, and Minnesota—had achieved recycling rates over 40 percent. Fifteen states had recycling rates of 30 percent to 40 percent.

The success of the United States' recycling efforts is largely the result of increased public awareness, the work of private companies, and can and bottle bills that require customers to pay a deposit on all beverage containers, which is returned to them when they return the container to a store. Eleven states now have container deposit bills in place, among them Oregon, Vermont, Maine, Michigan, Iowa, Connecticut, Massachusetts, Delaware, and New York. **Container deposit bills,** or more commonly **bottle bills,** require vendors to charge a small deposit for each bottle or can they sell, usually about five cents. The deposit is redeemed when the container is returned to a center or a machine in a grocery store, which scans the bar code and prints out a receipt which can be redeemed for cash.

Bottle bills have played an important part in the U.S. recycling strategy. In Oregon, the state with the first bottle bill, in 1972, 95 percent of the refillable bottles and 92 percent of the aluminum cans are returned. In Michigan, 96 percent of beverage bottles and cans are returned. Similar rates are found in other states as well. In states without aggressive recycling programs, the return is substantially below the rate in states with formal bottle bills.

The benefits of the container deposit programs are enormous. In states with such laws, the volume of roadside litter has been reduced by 35 to 40 percent. In New York State, where 400 million cases of beverages are sold each year, the Beer Wholesalers Association estimates that, in a two-year period, the deposit law saves the state $50 million on cleanup, $19 million on solid waste disposal costs, and $50 to $100 million on energy. The program created 3,800 net jobs—that is, 3,800 more jobs than were lost. In Michigan, 4,600 net jobs were created. A nationwide bill would create 100,000 net jobs, according to the U.S. General Accounting Office.

Commercial interests have responded well to the need for aluminum recycling. Coors, a Colorado-based brewery, has recycling centers throughout the United States. Other private companies have installed automated aluminum recycling machines, called reverse vending machines (Figure 17.6). Customers feed their aluminum cans into the machine, where they are weighed and crushed. The machine then pays the consumer for recycled aluminum.

In 1986, once again leading the nation in its efforts to promote recycling, Oregon passed the Recycling Opportunity Act. This law requires all cities with over 4,000 people to start curbside recycling programs that pick up materials at least once a month. For smaller communities, the law requires city officials to establish recycling centers at landfills. The Recycling Opportunity Act also ranks the waste management options available to the state in their order of desirability. First on the list are efforts to reduce the amount of waste generated—the reduction approach. Next comes reuse, followed by recycling. Finally, all leftovers must be buried in approved

FIGURE 17.6 *Reverse vending machine.*

landfills. Diligent efforts in the first two areas could greatly reduce the need for landfilling.

Deepening Our Commitment to Sustainable Waste Management

Most conservationists applaud the remarkable gains in recycling and composting throughout the world. They argue, however, that modern society has only begun to come to grips with waste and that much more can be done to create a sustainable system of waste management. Clues as to what needs to be done can be gained by looking at the commonly used recycling emblem. As you probably know, it consists of three arrows. The first stands for collection, the second for remanufacturing (the use of recycled materials in the production of new products), and the third for purchase (that is, the purchase of products made from recycled materials). All elements are essential to the success of recycling here and abroad.

Individuals, businesses, and governments can all play a significant role in promoting recycling by contributing to all three phases of this cyclical process. For the most part, however, efforts have been focused primarily on collection. Thousands of collection programs have been started in the past 10 to 20 years. Much less attention has been paid to **remanufacturing,** that is, making sure companies manufacture new products with recyclable materials. Even less attention has focused on **procurement,** ensuring products made from recycled materials are purchased by individuals, businesses, and governments. We will consider each one separately.

Promoting Collection Collection programs depend on all three key sectors of society: private business, government, and the citizenry. In some cities and towns, local government is responsible for picking up solid waste pickup and disposal. In others, private waste collection companies are responsible. Both can play a key role in promoting recycling. State and local governments as well as businesses, for example, can promote reuse and recycling campaigns through television advertisements, billboards, and publicly distributed pamphlets. Governments can also make recycling mandatory. Others may want to invest in the recycling equipment or in land for recycling facilities and either turn it over to private interests to run or enter into joint ventures with for-profit businesses. Governments and private industry can also help by setting up waste exchanges, that is, clearinghouses where consumers of reusable and recyclable materials find out what is available from various producers.

Governments can provide tax breaks to private trash haulers to help promote recycling. They can also remove subsidies from raw materials and energy, which will make the reuse and recycling option more attractive to commercial interests. Unfortunately, many current policies discriminate against recycled materials. For instance, the U.S. Forest Service sells timber each year at a substantial loss, which reduces the price of wood to consumers. The Wilderness Society estimates that the Forest Service loses about $265 million a year from below-cost timber sales. The Forest Service disputes this figure, but admits that its accounting practices do not reflect all costs like road-building into areas to be harvested.

This system sends erroneous signals to consumers—notably, that wood is much cheaper than it is. This discourages the efficient use of wood and wood pulp which is used to make paper. This system shifts the cost of wood products to the taxpayer, who essentially subsidizes the cost of raw wood and wood products. Charging companies more realistically would raise the price of wood, wood pulp, and other wood products and would be a shot in the arm to the recycling industry.

Cheap energy, a major factor in the production of goods from raw materials, is also heavily subsidized, and that hinders recycling efforts. According to one estimate, subsidies to U.S. energy producers—in the form of low-interest loans and tax breaks—cost U.S. taxpayers about $21–$36 billion per year. Oil currently costs about $35 per barrel (January 2000), but the real cost—calculated by adding in subsidies—is about $100 to $200 per barrel. Gasoline, if priced fairly, would cost U.S. citizens $6 to $12 per gallon. Recycling would be far more widespread if companies that produce materials and products paid the real cost of energy. "By underpricing energy and other natural resources," says the Worldwatch Institute's Cynthia Pollock, "governments subsidize the continuation of a throwaway society and the disruption of ecosystems."[1]

Recycling and reuse could also be stimulated by reducing the complexity of our waste stream. For example, manufacturers can eliminate containers made of several different types of plastics, which are difficult to recycle. On the South Pacific Island of Fiji, soft drinks come in one type of refillable glass container. Different brands simply have different paper labels. Therefore, glass bottles may be returned to any manufacturer, to be washed and refilled. The governments of Denmark and Norway allow fewer than 20 different returnable containers for beer and soft drinks. Standardization facilitates reuse and eliminates the cost of transporting bottles to distant manufacturers.

But before we select one packaging over another, it is important to examine how much energy each requires. Not all

[1]The quotations from Cynthia Pollock in this chapter are from *Mining Urban Wastes: The Potential for Recycling,* Worldwatch Paper 76. Washington, D.C.: Worldwatch Institute, 1987.

containers are created equal. An aluminum can, for instance, requires an enormous 7,000 British Thermal Units (BTUs) of energy to manufacture initially and 2,500 BTUs to recycle. On the other hand, a glass bottle requires 3,700 BTUs to manufacture and 2,500 BTUs to recycle. But glass is heavier to transport, which requires additional energy that must be taken into account.

Increasing Remanufacturing and Procurement The strategies we have examined would surely help to increase the supply of recyclable materials by increasing recovery rates. But increasing the recovery rate is only one third of the battle. Businesses, citizens, and governments must also find ways to incorporate recycled materials into their products.

To promote remanufacturing and stimulate demand, state governments and the U.S. federal government either require their agencies and branches to buy recyclable materials or permit the purchase of such materials. In some instances, recycled products must be economically competitive with products made from virgin materials. In other instances, state agencies are permitted to purchase recycled goods as long as they are within 5 to 10 percent of the cost of products made from virgin materials. Such actions could greatly increase the demand for recycled products, with spillover effects on the general economy, in large part because government purchases account for 20 percent of the gross national product of the United States. The gross national product, as explained in Chapter 2, is the value of all goods and services produced by an economy.

Recognizing the importance of recycling and the potential influence it could have on recycling, the U.S. Congress passed the **Resource Conservation and Recovery Act (RCRA)** in 1976. This law assigned to the EPA the task of drawing up guidelines for recycled materials, which was to be completed within two years of the act's passage. The guidelines were intended to establish recycled material content that would be acceptable for use by various government agencies. However, much to the dismay of many conservationists, the EPA dragged its feet for over a decade and did not issue guidelines until it was brought to court by the Environmental Defense Fund, a national environmental group. Since then, the EPA has issued numerous guidelines for recycled content.

RCRA also permits governmental agencies to purchase products that are made from recycled materials, such as paper, but requires that they be of reasonable cost. Unfortunately, "reasonable cost" has been widely interpreted as meaning the cheapest. Many offices and departments are currently purchasing recycled products.

In response to the federal government's slow pace, Maryland and all other states that have passed similar laws have, over the years, greatly increased the use of recycled materials in manufacturing. Interestingly, in 1993, the state of Colorado passed a law requiring all attorneys filing documents with the state government to use recycled paper.

Private business has also joined in by opening new plants that will incorporate recycled materials in their products. For instance, in Canada and the United States, dozens of mills have opened in the past decade to make paper and paper products from secondary (recycled) materials.

Individuals can also make a difference by purchasing recycled products whenever possible and by writing manufacturers to let their views be known.

Recycled products are becoming more and more common the world over. Numerous building products are now made from recycled material. Recycled milk jugs for example is used to make a safe, durable plastic wood that's used for decks and other purposes. Rubber from automobile tires is used to make durable roof tiles. Carpeting is made from recycled plastic pop bottles. Numerous directories are now available to direct builders and homeowners to sources. *GreenSpec* by the Environmental Building News and the *Green Building Resource Guide* published by Taunton Press are two examples.

Impressive as these efforts are, products made from recycled materials are still only a fraction of the total goods produced by the global economy. In a sustainable economy, just the opposite will very likely be true.

17.3 Waste Disposal: The Final Option

Ecologists envision an ideal world—some call it Ecotopia—in which there is no waste. In Ecotopia, virtually all plastics are either recycled or reused. Those that cannot be recycled are banned altogether. Waste paper, aluminum, steel, glass, and other materials are separated at their source, then shipped off to local recycling facilities, baled, and transported to regional manufacturers, who turn them into useful products once again. Yard wastes—grass clippings and leaves—are either composted by homeowners and used to enrich soils or are shipped to Ectopia's composting facilities, where they are piled in huge windrows and left to decay. Eventually, the rich organic materials are combined with wastes from municipal sewage treatment plants and sold as soil conditioners to gardeners and farmers. With all of this recycling and composting, there's no waste to speak of. Nothing needs to be landfilled or incinerated.

Ectopia may be years, even decades, away. Until that time modern industrial societies will undoubtedly continue to rely on landfills and incinerators.

Dumps and Sanitary Landfills

In the year 500 B.C., the Greeks and Romans hauled their trash outside the city walls and dumped it downwind, so as not to offend the residents. Flies and rats invaded the debris and, when the wind shifted, few people were spared the odorous onslaught.

Over 2,000 years later, in the United States, the most technologically advanced nation on Earth, many cities and towns still followed the same tradition. Open sores in the landscape, where rotting garbage swarmed with flies and rats, were common. Making matters worse, officials periodically burned the accumulating garbage to reduce its volume. Black smoke, filled with the toxic byproducts from burning rubber and plastic, billowed out of dumps everywhere. Rain and snowmelt trickled through the garbage, carrying sometimes hazardous liquids into the underlying groundwater, threatening groundwater and drinking water supplies. Troubled by their own garbage, Americans called for changes.

In 1976 their demands were met when the U.S. Congress passed the Resource Conservation and Recovery Act, already discussed. Among its many key provisions, RCRA called for an end to open dumps by 1983. What replaced them was the **sanitary landfill,** an excavation or hollow in the ground in which garbage was dumped, compacted, and covered daily with a fresh layer of dirt (Figure 17.7).

Used today throughout the developed world, the sanitary landfill reduces odors caused by rotting garbage and problems with insects and rodents. Because a soil layer is placed over the trash, compacted, and generally sloped to reduce water percolation into the garbage, groundwater contamination can be greatly reduced or eliminated. The protective layer of soil also reduces insects and other pests that could carry disease.

Landfills offer many other advantages over the open dump. Besides being cleaner, they can be "reclaimed" after they've been filled to capacity—that is, they are covered with topsoil and returned to some previous use, or given over to some new use when they are filled. In Evanston, Illinois, city officials built a park with baseball fields, tennis courts, and toboggan runs on Mount Trashmore, a hill 30 meters (60 feet) high and underlain by garbage. In Maryland, wastes from 87 roadside dumps were hauled to abandoned coal strip mines. After the gullies were filled with waste, the site was covered with soil and replanted.

Despite their advantages over the open dump, landfills are still a primitive method of waste management. If the soil or rock below a landfill is permeable, pollutants may drain into underlying aquifers, polluting water needed by municipalities, farms, or industry. Rotting debris also produces methane, a potentially explosive gas. Ordinarily, methane production is greatest during the first two years of operation but is not high enough to be hazardous. However, if large concentrations build up, methane can pass through the soil into nearby buildings. If it reaches sufficient concentration, it can explode.

More important, though, landfills are a waste of valuable resources that could be reused or recycled. They also require large tracts of land. A town of 10,000 people, for instance, produces enough trash each year to cover a 0.4-hectare (1-acre) site 3 meters (10 feet) deep. Landfills are also expensive—and are growing more expensive by the day—especially in more populated states and countries. Because of the growing shortage of suitable sites, city officials are often forced to haul their trash farther and farther from the site of production. As a result, the cost of landfilling garbage has skyrocketed. Philadelphia, for instance, has used up all of its landfill sites and must now transport its garbage to Ohio and Virginia. In eastern states, the cost of landfilling a ton of waste has climbed from \$20 in 1980 to over \$100 a ton today. In Western states where more open space is available, tipping fees (the cost of landfilling trash) are still on the low end of the scale, which is one reason why recycling efforts in many western states is way below the national average.

Worldwatch Institute's Cynthia Pollock notes, however, that in many areas landfill fees are held artificially low by local governments. Because of this, trash removal companies and local governments have little incentive to make better use of their wasted resources.

Incineration

Another strategy now being used in the United States and abroad to reduce waste and capture part of its intrinsic value is **incineration**—burning municipal solid waste. Incinerators can burn unseparated trash—containing plastics, metals, paper, yard waste, and glass—or separated trash that is relatively free of noncombustibles such as glass and metal. The heat produced during combustion is often used to generate steam for industrial processes, heating building, or, more commonly, electrical power generation. Virtually all municipal solid waste incinerators in the United States generate some electricity. Many of those that don't soon will.

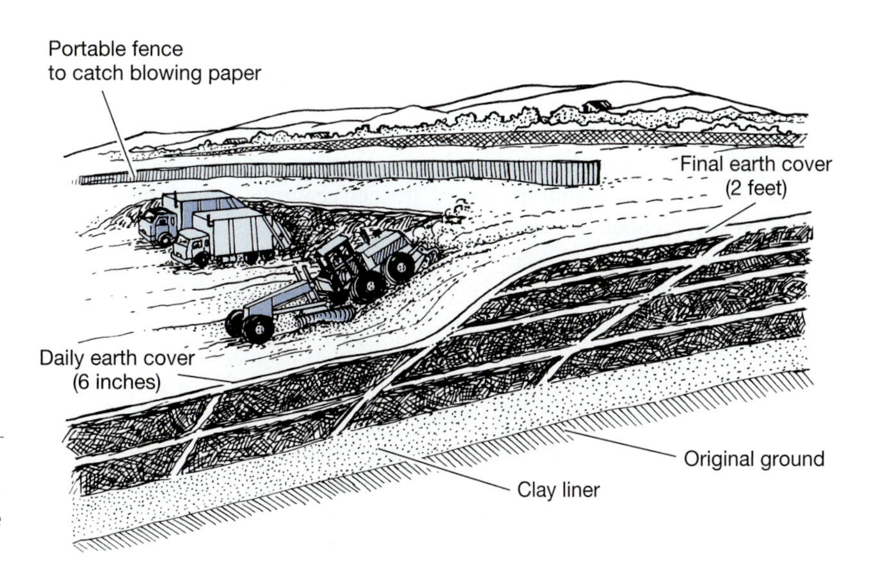

FIGURE 17.7 *Sanitary landfill operation. The bulldozer spreads and compacts solid wastes. The scraper (foreground) is used to haul the cover material at the end of the day's operation. Note the fence that catches any blowing debris.*

Incineration became popular during the oil crises in the 1970s, but proponents soon found that the technology had many problems—most notably, toxic air pollutants. Refinements in the technology have eliminated some of the problems, leading many U.S. cities to turn to this option. Currently (1997), there are about 150 incinerators operating in the United States, down from 170 in 1991. Incineration now handles 17.35 percent of the municipal solid waste output of the United States.

By comparison, there are over 350 incinerators in Western Europe, Japan, the former Soviet Union, and Brazil. Over half of the municipal solid waste in Japan, Sweden, Denmark, and Switzerland is burned.

Incineration is desirable for a number of reasons. First, it captures energy that would otherwise be lost. It also fits nicely with existing waste management practices and requires less land than landfills. No modifications of the pickup system are needed, as they are for recycling and composting. Incinerators can be designed to scale, providing flexibility to serve towns that produce as few as 100 metric tons of garbage a day, up to larger cities with over 3,000 metric tons.

Despite their mushrooming popularity, incinerators are still viewed skeptically by many. In fact, at least 300 proposals for municipal solid waste incinerators have been stopped in the United States because of public outcry. Why? Despite improvements in design and controls on air pollution, incinerators that burn plastics and other materials containing chlorine emit a dangerous class of compounds called dioxins. Dioxins have been linked to cancer, birth defects, and other problems. Recent evidence indicates that they may also weaken the immune system, making individuals more susceptible to cancer. Concerned about potential health problems, Sweden and Denmark have halted the construction of additional incinerators until their scientists can determine how substantial a risk they pose.

Toxic heavy metals such as mercury (contained in batteries) and acidic substances like hydrochloric acid (produced when plastics are burned) are also emitted from waste incinerators. The U.S. EPA requires pollution control devices on all municipal solid waste incinerators to eliminate several toxic pollutants. The cost of retrofitting some facilities will probably put some incinerators out of business. At this writing only a handful (six or so) of the United States' incinerators actually meet emission standards. Requiring residents to separate nonburnables and plastics could help reduce toxic emissions.

Critics also note that pollution control devices generate a hazardous residue called **fly ash**—materials that were removed from the smokestack gases. The ash in the bottom of the incinerator, called **bottom ash,** is also potentially toxic. It contains heavy metals and other pollutants. Consequently, some countries now classify ash from incinerators as a hazardous material that must be properly disposed of to protect human health. In the United States, however, the EPA has chosen not to classify municipal solid waste incineration ash as a hazardous material, meaning that many tons of potentially hazardous waste are currently being landfilled alongside garbage each year.

17.4 Hazardous Wastes

From a sustainable perspective, incinerators represent a less than prudent investment. Creating a sustainable society within the biophysical limits of the Earth will require a system of solid waste management based on reduction, reuse, and recycling, with minimal disposal through landfilling or incineration. Growing discontent in the 1960s and 1970s with solid waste disposal techniques brought about sweeping changes in governmental policy on solid waste disposal. But few people anticipated that another problem lay on the horizon, a problem brought to life by Love Canal, in Niagara Falls, New York. The canal had been the repository for over 20,000 metric tons of hazardous wastes, including dioxin, from 1947 to 1952 (Case Study 17.2). Troubles began in the late 1950s after the city pressured Hooker Chemical Corporation to turn over the land it had used as a hazardous waste dump to build a school and residential community. As construction on the project began, workers broke through the clay cap that Hooker had placed over the dump site. Homes and a school were constructed nevertheless.

In the late 1950s, toxic chemicals began to ooze out of the ground from the rusted steel drums. Children playing in the toxic ooze became ill and suffered skin burns; some even died. Subsequent health studies showed a significantly higher incidence of birth defects, respiratory difficulties, and other illnesses in people living near the site. State and federal officials evacuated hundreds of families and efforts began to clean up the site.

But Love Canal was not an isolated case. The Netherlands, Austria, Hungary, and Sweden all witnessed similar instances. Many additional examples came to light after the downfall of the Soviet Union. Eastern European nations, it seemed, contained many horribly polluted industrial sites. One incident after another revealed the seriousness of the problem (Figure 17.8). As people began to realize that highly toxic materials had long been carelessly discarded on the land and in the water, they began to call for action. Governments and international agencies drafted rules and regulations to clean up the thousands of potentially harmful toxic waste dumps already in existence and to prevent further tragedies. Despite many years of work on the problem, progress has been exceedingly slow.

How Big Is the Problem?

Hazardous wastes are broadly defined as chemical substances that adversely affect a wide array of organisms in a variety of ways from nonthreatening symptoms to debilitating illnesses, even death. Expose to hazardous wastes results from improper storage, disposal, and handling as well as accidents during transportation. Estimating the global production of hazardous waste is nearly impossible. Less developed countries, for instance, have few laws governing hazardous materials, and government officials often have no idea how much is produced. Other countries may classify as hazardous a substance that another country does not consider hazardous. Even states in the U.S. may classify materials hazardous that federal governments do not consider hazardous.

CASE STUDY 17.2 THE CHEMICAL TIME BOMB AT LOVE CANAL

In the 1880s, industrialist William T. Love began constructing a canal that would provide water and electrical power to the growing city of Niagara Falls, New York. Connecting the Niagara River just above the falls to a point below the falls, he envisioned as a focal point of industrial development. But because of economic troubles, the project was abandoned. Many parts were filled in as the city expanded, and by the 1900s only a 900-meter section of the ill-fated canal remained—a mute testimony to one man's dream gone sour.

In 1942, Hooker Chemical Company entered into an agreement with the canal's owner, Niagara Power and Development Corporation. Their agreement would allow the company to use the remaining section as a hazardous waste dump. There were no regulations for waste disposal at the time, and very little concern over the impacts of burying steel drums containing hazardous wastes directly in the earth.

In 1946, Hooker purchased the canal from Niagara Power, and for the next six years dumped thousands of steel drums containing 20,000 metric tons of hazardous materials into the canal. Then, in 1952, the city of Niagara Falls, which wanted land on which to build a school and residential community, began condemnation proceedings, which would allow them to take the land from the company. Hooker caved in to the pressure and signed over ownership of the site for a nominal $1 payment. It then sealed off the site with a clay liner and turned it over to the city, allegedly with a warning not to build on the dump site itself. In turn, the city signed an agreement releasing the company from any damages that might occur from use of the land.

In 1954, the city began building the 99th Street Elementary School right over the dump. Two hundred and thirty-nine homes went up nearby and another 700 soon skirted the chemical time bomb. During construction, though, workers broke through the protective clay cap and apparently did nothing to repair it.

A few years later troubles began. Rusty and leaking barrels began to emerge in low spots (Figure 1). Chemical wastes pooled on the surface. Residents noticed harsh chemical smells, so powerful that they killed grass and vegetables and took the bark off trees. Children playing near the pools of toxic ooze suffered serious chemical burns; some became ill, and a few died.

Then, in 1977—20 years later—unusually heavy rain and snow converted the site into a sea of mud. Through the years, corrosion had turned the steel drums into leaking sieves. Toxic wastes came bubbling up to the surface, pooling in people's backyards and filling basements with a black, smelly goo. Residents complained of strange odors. Pets began to die mysteriously. Many residents complained of severe headaches and rectal bleeding. Children playing in the area were seriously burned by chemicals. Concerned with the many complaints it was receiving, the New York State Health Department initiated a health study. To their surprise, they found a high incidence of liver, kidney, and respiratory disorders as well as epilepsy and cancer. In addition, the rate of miscarriages and birth defects among residents was found to be three times greater than the national average. In July 1978, the Health Department strongly recommended that pregnant women and children under two years of age move out of the area.

FIGURE 1 *Toxic chemicals and rusted drums worked their way to the surface in Love Canal, forcing the evacuation of hundreds of families.*

In an investigation of the site, more than 80 toxic chemicals were identified. At least one dozen were known carcinogens. As the evidence grew, city and state officials closed down the school, fenced it off, and evacuated several hundred families. On May 21, 1980, then-President Jimmy Carter declared Love Canal a disaster area. The federal government evacuated over 780 more and provided housing for them at a cost of $30 million.

In the fall of 1978, officials began a massive cleanup of the dump site. Some homes were removed, others bulldozed; still others were left intact, their doors and windows boarded up (Figure 2). As the cleanup proceeded, additional studies by the EPA revealed that the immediate vicinity was badly polluted but that the toxic wastes had not migrated much past the first two rows of houses on either side of the canal. The EPA concluded, therefore, that the 1980 evacuation was unwarranted. The study also showed that the chemical wastes had not migrated to deep aquifers and were unlikely to move much further.

Today, Love Canal and the schoolyard where children once played is surrounded by a chain link fence with bright yellow signs warning passersby of the hazardous wastes. The houses nearest the canal have been bulldozed. The dump site itself has undergone extensive work. Officials decided that the best strategy

(continued)

FIGURE 2 *One family took their home with them, leaving only the foundation as an eerie reminder of what had happened. Abandoned houses in the background are boarded up.*

was to leave the hazardous waste in place and to attempt to contain it—that is, to restrict its movement. To this end, a huge ditch was dug around the site. In the ditch, porous tile was installed. It's supposed to contain the waste—that is, keep it from migrating outward. The leachate collected by the drain system is pumped to an on-site treatment plant. Here, filters remove chemicals from the collected leachate. Water produced by the process is then released into storm water sewers. Critics point out that while this may seem like a good strategy chemicals, such as mercury and other heavy metals, that are not removed by the filters end up in the nearby Niagara River.

The site was also covered with a plastic liner and clay cap was to minimize the amount of water seeping into it from rainwater or snowmelt and to prevent chemicals from the site from evaporating into the air. The clay cap also prevents direct contact with contaminated soil.

In September 1998, the New York State Department of Health announced results of a five-year habitability study. Their conclusion: portions of the Love Canal neighborhood are "as habitable as other areas of Niagara Falls." They did not declare these areas safe, however. The U.S. EPA weighed in on the issue, too, and deemed the homes outside the immediate site suitable for habitation. A public corporation, the Love Canal Revitalization Agency, soon took ownership of the abandoned properties. They renamed the area Black Creek Village and repaired the homes, then sold them.

Critics argue that this new development is foolhardy at best because the "habitability" certification was improperly performed. In fact, the area was erroneously compared to two other badly contaminated sites. Lois Gibbs, a housewife turned activist, who has been instrumental in forcing the government to take action at Love Canal, notes that "the homes that will be reinhabited are still contaminated, still unsafe." She notes that there have been no cleanup measures taken around the homes, which were found to have several toxic chemicals in and around them. Only the creek and sewer systems were cleaned.

The canal itself and the region immediately surrounding it will probably remain fenced off forever, serving as a sort of national monument to our carelessness and as an impetus for a greater commitment to prevent future problems of this nature from ever happening again. In 1990, the first family moved into the outer region but sales were slow at first because local banks were unwilling to lend money. In 1992, however, the Federal Housing Administration agreed to provide mortgage insurance to families who wish to purchase Love Canal homes. Today, virtually all of the 239 homes have been sold and while the EPA considers the area safe and new residents have no fear of toxic chemicals, only time will tell if the decision to resettle the area was a wise one. Costing over $227 million to cleanup and other expenses, Love Canal is a symbol of the cost of shortsightedness.

FIGURE 17.8 *Workers clad in protective clothing and face masks pump toxic materials into a tank truck from barrels at an abandoned paint factory.*

Difficulties aside, experts estimate annual world production of hazardous materials to be about 600 million to 700 million metric tons. The United States, one of the few countries for which good information is available, produces the lion's share of these potentially harmful materials. In 1991, hazardous waste production peaked at about 275 million metric tons (306 million tons). Since then, production has declined dramatically, thanks to pollution prevention efforts. In 1997, the U.S. produced about 37 million metric tons (41 million tons).

For a century or more, hazardous wastes have been indiscriminately strewn about the landscape in the industrial nations of the world. Until quite recently in the history of these nations, wastes were frequently placed in large steel drums that were dumped in landfills or on vacant lots and left to rust. Over time, leaks developed, releasing a steady trickle of harmful chemicals into the soil that percolated into the underlying groundwater that drained into lakes and streams or was drawn to the surface in wells. In other instances, hazardous wastes were pumped into deep wells, into municipal sewage systems, or directly into lakes and streams. Some companies poured their wastes into sandy pits, creating evaporation ponds. From here hazardous materials seeped into the ground. What is interesting, is that all of these techniques were considered acceptable practice. Little concern was raised over the potential health effects or impacts on the environment. As environmentally unsound as the acceptable practices were, some practiced even worse disposal methods. Some unethical waste haulers, for example, pulled their trucks up to streams and discharged wastes under the cover of darkness. Others opened the spigots and drove along highways at night, spilling tons of toxic wastes.

Years of careless disposal have left a legacy of contaminated sites and polluted lakes, streams, and aquifers. In Europe and North America, hazardous waste dumps have caused the uprooting of entire communities. Every country in Europe is plagued with toxic waste sites—both old and new—needing urgent attention. In the United Kingdom alone, 5.5 million metric tons of hazardous wastes were discarded in 1980; three quarters of this material was dumped into landfills without adequate liners to prevent them from leaking from the site. In Holland, an estimated 8 million metric tons of hazardous materials are buried in the soil, mostly in leaky steel drums. As noted earlier, with the fall of the Soviet Union and the liberation of Eastern Europe, reporters and scientists have discovered a toxic nightmare—thousands of abandoned dumps and contaminated sites.

Many countries have spent enormous amounts of money to locate and begin to clean up leaking landfills and contaminated industrial sites. In the United States, estimates vary as to the number of sites in need of cleanup, and the eventual cost. The U.S. EPA currently includes nearly 1,400 sites on its priority list, 600 of which have been cleaned up. But these sites may just be the tip of the hazardous waste iceberg.

Soon after the Love Canal episode, the EPA estimated there were probably 10,000 additional sites in need of cleanup. In 1989 it increased its estimate to 39,000. The U.S. General Accounting Office currently estimates that the number could be much higher—as many as 100,000 to 300,000 sites. That estimate does not include the 17,000 toxic "hot spots" on military bases throughout the United States. The cleanup cost could come to hundreds of billions of dollars, a stern reminder of the economic sense behind prevention.

Alarmed at the number of hazardous waste sites in need of cleanup, the U.S. Congress passed the **Comprehensive Environmental Response, Compensation, and Liability Act** in 1980, commonly called the **Superfund Act,** or **CERCLA.** It called on the EPA to identify and clean up hazardous waste sites with the assistance of state governments (which were required to chip in 10 percent of the cost). The Superfund Act initially created a $1.6 billion fund garnered from taxes levied on the petroleum refining and chemical manufacturing industries between 1981 and 1985. During that period, however, only 13 sites were cleaned up, and some of them, say critics, inadequately so. Regulators soon found that cleaning toxic waste dumps was much more difficult and more costly than anticipated. Just analyzing the chemicals in a single site can cost upward of $800,000. Simple steps to stabilize a leaking site can cost $500,000.

The Superfund Act was renewed in 1986. The amendments expanded the fund, bringing the grand total to $16.3 billion, with money coming from federal and state coffers and an even broader business tax. Much of this money was earmarked for cleanup efforts, but the 1986 law also provided money for research on new hazardous waste treatment technologies.

The Superfund Act empowers the EPA to collect the cost of cleanup from the hazardous waste dump or hazardous waste site owners and operators, as well as from companies that paid to have their wastes dumped in them. Under the law, then, all participants are required to pay their portion. As one person put it, a company is liable for its waste forever.

While this provision may sound fair and reasonable, it has resulted in a costly legal nightmare. Millions of dollars in Superfund money have been spent to determine who is responsible for the hazardous waste at various sites. Because of the multiple responsibility provision, cleanups are occurring much more slowly than originally hoped.

Nevertheless, as of March 1992, the EPA had cleaned up 71 sites and had taken remedial action on 1,171. By 1996, 410 sites had been cleaned up. Because of widespread criticism, the EPA found ways to cut the time for cleanup by two years, and by the year 1999 had 600 sites completely restored. Twice as many sites have been cleaned up in last five years than during the previous 12 years

The Superfund Act is an important law that has helped address an important problem, but it has been criticized for several reasons. One criticism is that provides money for cleanup and financial compensation for property damage, but nothing to reimburse people for health effects. Critics also argue that toxic residues and contaminated soils that are excavated from hazardous waste sites and transported to new landfills equipped with liners of clay and synthetic material will eventually leak, no matter how carefully the new landfills are planned and constructed. This, they say, will very likely create a costly problem for future generations. Finally, critics point out that while CERCLA helps us clean up our polluted lands, many additional billions will have to be spent to purify the thousands of contaminated aquifers throughout the United States.

The Superfund program remains today, but the law has not been updated since 1986. The Superfund tax expired in 1995. No longer does the U.S. government levy a tax on chemicals to clean up sites. At this writing (September, 2000) efforts are underway to overhaul the act, but progress has been slow. Some organizations are calling for a complete overhaul; others prefer a piecemeal approach, revising one section at a time. Some groups are seeking to weaken the bill; others are calling for a strengthening of the legislation.

Brownfields: Converting Polluted Landscapes into Productive Ones

Within cities and many rural areas are abandoned industrial or commercial facilities, some of which are contaminated. Rather than convert them back into productive use, however, new sites are often selected. Why?

Lenders, investors, and developers fear that if they become involved in the redevelopment of a site and discover contamination they will become liable for cleaning up chemicals that already exists on the property. They could have to spend millions of dollars removing someone else's mistake. It's far cheaper to develop a pristine site, a **greenfield,** than a **brownfield**—a site that is contaminated or thought to be contaminated by hazardous materials.

The result of their fear is that many thousands of abandoned industrial sites are currently sitting idle, while companies bulldoze fields and cut down forests to build new facilities. Seeking to prevent this problem, the Environmental Protection Agency has launched a **Brownfields Economic Redevelopment Initiative.** It is designed to empower cities, states, tribes, and others to redevelop abandoned sites, working together with community members and other stakeholders. The initiative seeks ways to clean up and sustainably reuse brownfields within the existing Superfund Law.

To this end, the EPA has published a list of 31,000 properties in the United States taken off the Superfund site inventory. Assessments of the properties have convinced the EPA that these properties are not subject to Superfund cleanup. The EPA will also enter into agreements with prospective purchasers of brownfields not to sue them for cleanup if a site is found to be contaminated. The U.S. government also provides substantial tax incentives to companies that redevelop brownfields that are hoped to bring thousands of abandoned or underutilized sites back into productive use. Further economic incentives come in the form of revolving loans offered by the states.

Brownfield development helps reduce pressure on undeveloped land and puts a valuable resource (abandoned sites) back into service. Society and the environment, say proponents, are best served by developing these sites, not leaving them vacant. In most instances, contamination is slight to moderate. And many properties are owned by cities; many brownfields have been passed on to cities by banks and businesses who simply stopped paying taxes on them in a deliberate act to force their legal transfer. (Cities and counties can legally seize title to property if taxes are not paid within a certain period.) The upshot of this is that land is often cheap. Even if some clean up is required, it may be offset by the inexpensive price tag.

Brownfield development is occurring throughout the United States, with dozens of pilot projects underway thanks to federal and state assistance. In Portland, Oregon, for instance, there's a program underway to develop abandoned sites along the Willamette River. The City of Chicago is also successfully using this strategy.

Managing Hazardous Wastes: The Unmet Challenge

Cleaning up past mistakes is only half of the solution to our hazardous waste problem. The other half is preventing this kind of thing from occurring again. "Unless the wastes currently produced are better managed," argues Sandra Postel of the WorldWatch Institute, "new threats will simply replace the old ones, committing society to a costly and perpetual mission of toxic chemical cleanups." To prevent the indiscriminate and illegal disposal of hazardous wastes, Congress added special hazardous waste provisions to the Resource Conservation and Recovery Act of 1976. These provisions require all producers, transporters, and disposers of these materials to register them with the EPA. The materials can then be tracked from the site of production to the site of their disposal—from "cradle to grave," in the words of waste managers. It is more formally known as the **manifest system,** a manifest being a piece of paper used to denote the destination of cargo. Many other nations now have similar policies. RCRA also ordered the EPA to set standards for packaging, shipping, and disposal of wastes. To prevent further contamination, it required waste disposal companies to obtain licenses. Only licensed facilities could legally accept hazardous wastes.

Like municipal wastes, hazardous wastes can be dealt with in three basic ways, listed in order of desirability: the reduction approach, the reuse and recycling approach, and the discard approach. Hazardous wastes, however, are amenable to a fourth approach as well, called detoxification. The discard approach has historically been the most widely used method; until recently. Comparatively little recycling and reuse was practiced. Source reduction was almost unheard of.

Reducing Hazardous Wastes

Manufacturers have a number of options available to reduce hazardous wastes at their source, a process also called **source reduction.** The first line of attack in the source reduction mode, and often the cheapest, is **process manipulation** or **process redesign.** By modifying or redesigning the manufacturing processes that create hazardous wastes, companies can significantly reduce waste production. For instance, the Borden Chemical Company of California redesigned an equipment-cleaning procedure that once used toxic organic solvents and produced a dangerous sludge. The redesign reduced their discharge of toxic organic solvents by 93 percent and cut sludge wastes generated each year from 350 cubic meters to 25 cubic meters. As an added benefit, the changes save the company about $50,000 a year.

The Minnesota Mining and Manufacturing Company (3M), a leader in waste reduction since 1975, has cut its waste production in half and, over a 20-year period, has saved an estimated $827 million in the process (Figure 17.9). A large chemical company in the Netherlands has installed a new manufacturing process that has cut its waste production by 95 percent.

Companies throughout the United States such as Borden Chemical and 3M are learning that pollution prevention is not just good for the environment, it makes extraordinary economic sense.

Interestingly, the U.S. EPA has also taken a lead in pollution prevention, primarily by its 33/50 program. This voluntary program called on manufacturers to cut their output of 17 hazardous wastes by 33 percent by 1992 and by 50 percent by 1994. Numerous companies signed on and made significant progress toward reducing hazardous waste production. In fact, the program was so successful that it reached its 50 percent reduction goals one year ahead of schedule and the next year cut production of the targeted hazardous wastes another 10 percent.

To reduce their hazardous waste output, companies can also substitute safe materials for more harmful ones. A company based in Indianapolis, for instance, has introduced a new line of nonhazardous, biodegradable industrial cleaners called "Worksafe." These could replace highly toxic organic cleaners. Numerous companies produce environmentally and people-friendly paints, stains, and finishes.

Conservatively, nationwide efforts to modify manufacturing processes and substitute biodegradable compounds for toxic ones could reduce our waste stream by 15 to 30 percent. Many companies have pledged to reduce their output by 80 to 90 percent by process manipulation, substitution, and other methods.

Waste reduction represents a new way of thinking for businesses. For this idea to become mainstream, however, more companies must realize that waste is a sign of inefficiency that results in higher costs. Furthermore, top-level management must commit itself to a program of waste reduction, as 3M and other companies have. A surprisingly small effort can result in huge reductions and economic savings. USS Chemicals, for example, is a company whose management is committed to waste reduction. It has established a reward system for employees who suggest implementable waste-saving techniques.

Source reduction is extremely popular outside the United States, too. The governments of Canada, Japan, Sweden, Germany, Denmark, and the Netherlands, for instance, actively promote nonwaste and low-waste technologies.

Reusing and Recycling Hazardous Wastes Manufacturers can also make significant inroads into hazardous wastes by reusing or recycling them. For example, some companies have found that they can use relatively pure chemicals produced as a waste in one process in another process on site or at another facility. Or they can sell the "waste" to a willing buyer or even give it away, rather than paying exorbitant disposal fees. In some instances, hazardous wastes require some purification to make them reusable, but either way the savings can be substantial.

To facilitate the exchange of hazardous wastes, the Netherlands put into operation a hazardous waste clearinghouse or waste exchange. Established in 1969, it keeps track of as many as 150 different chemical substances produced by industry and links buyers and sellers. Numerous private and nonprofit clearinghouses now exist in many U.S. cities and throughout the rest of the developed world. The Northeast Industrial Waste Exchange in Syracuse, New York, operates a computerized network listing wastes from five different regions. Anyone with a computer and modem and the proper password can gain access to the files to find out what is available or to list wastes for sale.

As in other issues, the world may look to Japan for guidance. The Japanese produce about 200 million metric tons of industrial waste, including both nonhazardous and hazardous waste. They recycle over half of that material. Another 30 percent is incinerated. What is left, about 18 percent of the industrial waste stream, is disposed of, mostly in landfills.

Detoxification Some hazardous wastes cannot be reused or recycled but can be detoxified, that is, chemically altered to less toxic or nontoxic chemicals. **Detoxification** may be effected by

FIGURE 17.9 *The 3M corporation has been a leader in pollution prevention since 1975.*

biological, chemical, and physical treatment. For example, organic wastes such as PCBs, DDT, and even dioxin can be burned in high-temperature incinerators. Incineration converts harmful organic substances into relatively harmless carbon dioxide (a global pollutant in its own right, but not responsible for adverse health effects) and water. Long criticized by many in the environmental community because they do not completely eliminate toxic emissions, incinerators are growing in popularity.

The EPA owns and operates a mobile incinerator that destroys 99.999 percent of the dioxin wastes in soils and liquids. The incinerator can be transported to waste sites, thus avoiding the transport of hazardous materials to distant incineration facilities. Another option is to burn wastes at sea. At least six European nations currently ship some of their toxic wastes to huge oceangoing vessels equipped with high-temperature incinerators. These ships then head out to sea to burn the wastes (Figure 17.10).

Critics argue that land- and sea-based incineration would be acceptable only with much tighter controls on emissions. A newly developed plasma arc incinerator, for instance, burns toxic wastes at 45,000°F, destroying all traces of PCBs and other organic wastes. For comparison, the fire in a wood stove burns at 400 to 900°F. Plasma arc incineration may be the wave of the future.

Another promising development is the combustion of hazardous organic wastes in existing cement and lime kilns—huge furnaces in which cement and lime are heated. Typically powered by oil, kilns may be an efficient and cost-effective alternative to traditional incinerators. In Sweden, for instance, kiln operators burn a mixture of oil and hazardous organic materials as fuel. Most kilns already have state-of-the-art pollution control equipment. Basic materials in the kiln also neutralize acidic emissions. Cement companies are paid to incinerate wastes and are able to cut down on their own fuel consumption in the process, saving money on operations.

On another front, geneticists have developed strains of bacteria that decompose chemical solvents such as benzene, toluene, and xylene, converting them into carbon dioxide and water. Scientists have also found naturally occurring bacteria that successfully degrade oil wastes in soil and water.

Numerous chemical methods are also available. For instance, ozone can be used to destroy organic compounds. Special ion-exchange columns can efficiently separate out toxic heavy metals, and various bases can be used to neutralize acids.

Proper Disposal of Hazardous Waste

In an ideal world, hazardous wastes could be reduced by 60 to 75 percent (perhaps more) by process manipulation, reuse, recycling, and detoxification. Some toxic substances would still remain, however. Heavy metals are an example. The remaining material must be disposed of by any one of a half-dozen or so techniques such as in secured landfills, deep geological salt beds, surface impoundments, warehouses, and deep injection wells. Secured landfills and deep injection wells are the preferred methods and are discussed here.

Secured Landfills The most popular discard approach today is to deposit the waste in a **secured landfill**—a clay-lined pit designed to hold hazardous wastes (Figure 17.11). The thick, supposedly impermeable clay liners are often supplemented with synthetic liners and monitoring wells. Groundwater samples can be taken from the monitoring wells to determine if wastes are leaking out of the site. Special drain systems also pump liquid wastes from the bottom of the pit to treatment facilities where they are detoxified, thus minimizing the migration of these substances out of the site. Careful siting (locating a site in a dry region where the water table is not too close to the landfill) also minimizes the risk to groundwater and surface water. Grading and compaction of the soil over the site minimizes the penetration of rain and snowmelt, thus reducing leaching.

Although extraordinary precautions are taken to prevent the escape of materials from landfills, critics are unconvinced that they can contain wastes over the long term. Cracks in the liner and clay seal, for instance, could emit wastes that drain into groundwater and contaminate aquifers. Earthquakes could tear asunder the careful controls. Lax monitoring could unleash a local environmental catastrophe.

FIGURE 17.10 *The Dutch incinerator ship* Vulcanus *burns hazardous waste at sea. Is this a suitable way to get rid of the many hazardous organic wastes industrial societies generate, or will it lead to widespread pollution of the air and water?*

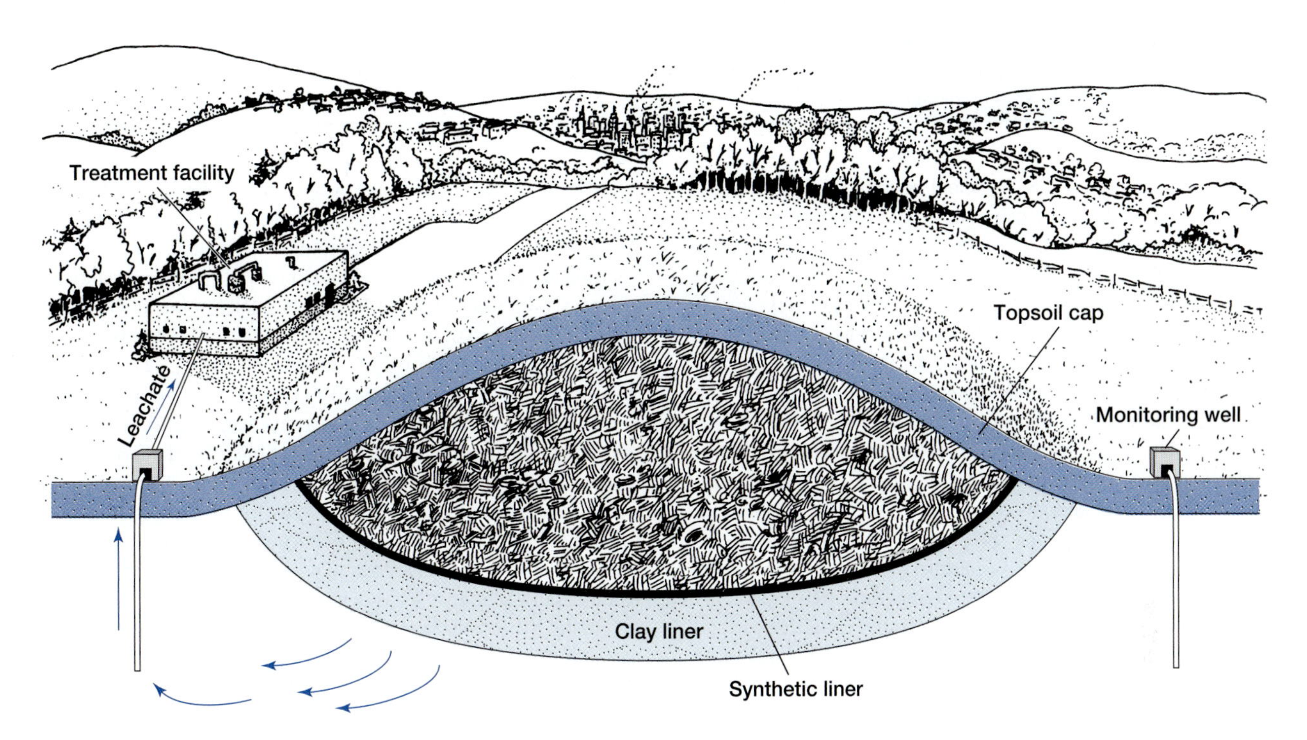

FIGURE 17.11 *Drawing of secured landfill showing monitoring wells, clay and synthetic liner, and treatment facilities for leachate.*

Deep Injection Wells Perhaps the least talked about but most important problem in hazardous waste management is that much of our hazardous waste is highly diluted in water. Separating out the hazardous materials is costly and expensive. Because of this, companies frequently dispose of their liquid wastes in deep wells or illegally in lakes and sewage systems.

Deep wells are drilled into the Earth's crust to porous zones sandwiched between impermeable rock layers. In theory, the hazardous material remains in place forever. In practice, though, this is not always what happens. Liquid wastes can migrate through unexpected fissures in the "impermeable" rock strata and contaminate aquifers. Cracks in the well casing can also result in leakage. Injecting large quantities of liquids into the Earth can destabilize rock layers, greatly increasing the frequency of minor earthquakes.

According to the latest statistics from the EPA, about 10 percent of all hazardous waste produced in the United States is injected into the ground. Because of the problems that can arise, many people would like to eliminate it entirely, preferring instead ways to separate out the hazardous materials for detoxification, incineration, or some other disposal method. (See Case Study 17.3.)

The Nimby Syndrome: Taking Personal Responsibility Besides growing awareness of the widespread nature of the problem, action to reduce hazardous waste production has been prompted by the unpopularity of hazardous waste facilities. Few people want a hazardous waste landfill or a deep injection well near their homes, farms, or schools.

Public policy makers have dubbed this the **NIMBY syndrome**—Not In My Backyard. Ironically, most people want the amenities provided by manufacturers but want nothing to do with the waste that comes from them. Let some other community or some other country take it, they say.

This situation is not likely to change in the near future, but citizens can take actions to reduce the production of hazardous chemicals. One of the most obvious is to cut back on the use of such things as pesticides, herbicides, solvents, and cleaners. This simple step helps to reduce the volume of hazardous waste churned out by the many factories in the world. Nontoxic substitutes can be used. Numerous nurseries and retail outlets offer safer alternatives to the potentially toxic group of chemicals they sell to control pests.

Because hazardous waste is produced in the manufacture of virtually all products, efforts to live with less or to buy more durable goods, discussed earlier in this chapter as a means of reducing solid waste, will also reduce hazardous waste output.

Individuals can also help by properly disposing of **household hazardous wastes**—batteries, paint, paint thinner, pesticides, cleaning agents, and so on. Many cities have hazardous waste programs designed to pick up household hazardous. In some locations, these wastes are recycled. Used paint, for instance, may be given to companies that combine like colors, mix it with virgin paint, then sell it. In other locations, you can actually acquire materials directly. That is, there are waste exchanges. Whatever you do, don't dump hazardous materials down the drain—they'll poison a septic tank or contaminate sewage at sewage treatment plants. And don't dump them into sewers, either.

CASE STUDY 17.3 EXPORTING TOXIC TROUBLES

In the 1970s and 1980s, U.S. laws and regulations for hazardous wastes were strengthened. Businesses found it increasingly more expensive to dispose of their toxic wastes. Some unscrupulous companies began to dispose of their wastes illegally to avoid the high cost of waste disposal. Many others turned to overseas markets to dispose of their wastes. Wastes were often shipped to the cash-hungry less developed nations of Africa or to eastern bloc nations, neither of which had adequate laws requiring proper hazardous-waste disposal. European companies also joined in.

The problem with exporting waste is that many of the countries that received waste didn't know what was in the shipments, didn't know how toxic the materials really were, and didn't have facilities to store it or dispose of it properly. Hazardous wastes were more often than not disposed of improperly. In trying to avoid the high costs of proper disposal these companies, often knowingly, were contributing to the contamination of other countries.

As the problem became widely known, U.S. lawmakers took action. In 1986, for example, the U.S. Congress amended RCRA, establishing procedures that require the U.S. to notify importing countries of shipments of hazardous waste and to obtain prior written consent. These regulations, however, were insufficient. EPA officials claimed that hundreds of tons of hazardous wastes were still being exported illegally.

Because exporting hazardous waste to a nation without its full consent goes against principles of international law, numerous African nations passed laws banning the import of hazardous wastes. In some countries, importing hazardous wastes is punished by stiff jail terms and multimillion-dollar fines. In Nigeria, an importer can be put to death. The Organization of Eastern Caribbean States and 22 Latin American countries joined forces to stop the dumping of hazardous wastes on their soils.

In 1990, the European Economic Community (a coalition of European nations) agreed to ban exports of toxic and radioactive waste to 68 former European colonies. Many of the less developed nations who were part of the accord also agreed not to import hazardous wastes from non-EEC members. Today, 121 nations including Canada, Mexico, and 13 European nations have signed an agreement (the Basel Convention) that bans the transfer of hazardous wastes to less developed nations. To date, the United States, has, refused to sign the agreement.

Although these are important steps forward, many less developed nations are still open to exports, representing a potentially huge repository for hazardous wastes from the industrial nations. Signing an agreement will not stop the illegal flow.

In the United Nations, talks are under way for the development of global standards to regulate the shipment of hazardous wastes from one country to another. But what is and what is not a hazardous substance is a contentious issue. Global standards could be important, but some critics argue that the action may be a step in the wrong direction, for it could end up promoting export and discouraging waste reduction, a sustainable solution. Furthermore, regulating and enforcing such a program could prove to be a nightmare. Thus, some proponents believe that a complete ban on the international movement of waste is the best answer. This would help protect the environment from inadequate disposal methods and would help nations develop long-term solutions that reduce hazardous-waste production—an essential element of a sustainable industrial design and a sustainable future.

Beyond that, citizens can join environmental groups such as the Worldwatch Institute, the World Resources Institute, the Institute for Local Self-Reliance, and the National Coalition for Recyclable Waste that are working on this and other issues. Citizens can write letters to government officials asking for more efforts to reduce wastes and reuse and recycle them. Without greater source reduction, reuse, and recycling, the hazardous waste problem could worsen, overwhelming future generations as surely as it has overwhelmed us.

Summary of Key Concepts

1. U.S. cities and towns generate 200 million metric tons of municipal solid waste each year. Only about 26 percent of the solid waste is recovered for reuse and recycling. Approximately 17 percent is burned to generate energy.
2. Many modern industrial societies have traditionally viewed municipal waste as something to be rid of—to dump in the ground or at sea as far away as possible—with little regard for the wealth of materials it contains. This is the discard approach
3. Far more sustainable are the reduction approach, which calls for lower levels of material consumption, and the reuse and recycling of waste materials.
4. Reducing per capita consumption of natural resources may become an economic fact of life in materialistic societies as resource supplies, especially energy, decline. Individuals can reduce consumption by buying durable items and avoiding unnecessary purchases. Individuals can help by consuming less. Miniaturization has successfully helped reduce resource consumption and will continue to be useful in the years to come.
5. Reuse and recycling reduce solid waste generation and conserve valuable resources. Advocates of this strategy point out that there are many products that can be reused. For materials that cannot be reused, however, the next best approach is recycling.
6. Recycled materials can be extracted from municipal trash at central stations or separated out of the trash at the source.
7. Many nations now have successful recycling programs. Japan, the Netherlands, Mexico, and South Korea lead the world in recycling paper. The United States is on its way to becoming a leader in recycling, too, with can and bottle bills and numerous curbside recycling and composting programs.
8. A successful recycling effort so essential to creating a sustainable system of waste management requires ways to promote remanufacturing and procurement.

9. Governments can play a key role in promoting all phases of recycling. Television ads, billboards, and pamphlets can encourage the public to take a more active role. New recycling laws, recycling centers, waste exchanges, and tax incentives can stimulate increased recycling as well.
10. Governments and businesses can also find ways to improve the market for recycled goods. Governments, for instance, can provide tax incentives for companies that use recycled materials in their manufacturing processes. They can also require their agencies to purchase recycled materials and can stipulate minimum recycled material content for products.
11. In an ideal world, wastes would first be reduced by the methods described above. What is left would need to be discarded. For years, Americans discarded their wastes in open dumps that were periodically burned to reduce their volume. Then, in the 1960s, Americans began to revolt. In 1976 Congress passed the Resource Conservation and Recovery Act that, among other things, called for a complete end to open dumps by 1983. The sanitary landfill replaced the dump.
12. Sanitary landfills are excavations or natural depressions in which garbage is dumped, compacted, and covered daily with a layer of soil to reduce pests.
13. Landfills offer many advantages over the open dump. Besides being cleaner, they can be reclaimed—returned to some previous use, or given over to some new use. But they do have their problems. Pollutants may drain into underlying aquifers, contaminating groundwater. Rotting garbage also produces methane, a potentially explosive gas. More important, landfilling squanders valuable resources that could be recycled or reused and also requires large tracts of land, which are in short supply in and around cities and towns, where huge quantities of waste are being produced.
14. Many local governments reduce their solid waste disposal by composting, a form of recycling in which organic matter like leaves and other yard wastes is piled, kept moist, and periodically turned. Aerobic bacteria decompose the organic matter to produce a stable humuslike material that can be used to condition and fertilize soil. Backyard composting is a popular strategy among some homeowners.
15. Another strategy now being used in the United States and elsewhere is incineration. Incinerators burn unseparated trash or glass- and metal-free trash to generate heat for electrical production, industrial processes, and heating.
16. Incineration is growing in popularity but is a source of air pollutants and ash, a hazardous waste. It also wastes perfectly recyclable materials.
17. Industrial nations also produce enormous amounts of hazardous waste. Hazardous wastes are substances that adversely affect a wide array of organisms, causing death or debilitating disease if not properly stored, transported, disposed of, or handled.
18. Careless disposal of such wastes have resulted in thousands of hazardous waste sites strewn about various nations.
19. Most nations face two problems in relation to hazardous wastes: first, cleaning up the tens of thousands of abandoned waste dumps and contaminated industrial sites, and second, dealing with the millions of tons of hazardous waste produced each year by factories and other sources.
20. Locating and cleaning up contaminated sites will be among the highest-priced items on the environmental agenda of many industrialized nations.
21. Although no one knows for sure, it is believed that about 600 million to over 700 million metric tons of hazardous waste are produced worldwide each year. The United States alone produces about 200 million metric tons per year.
22. In 1980, the U.S. Congress passed the Comprehensive Environmental Response, Compensation, and Liability Act, commonly called the Superfund Act. This new and important law calls on the U.S. EPA to identify and clean up hazardous waste sites in the United States using money largely from taxes levied on petroleum and chemical manufacturing industries. The fund is reimbursed by culpable parties, that is, individuals responsible for the hazardous waste site (producers and disposal companies).
23. Although Superfund cleanups occurred slowly, the pace has picked up and 600 of the nearly 1400 sites on the National Priorities List have been cleaned up to date.
24. The Resource Conservation and Recovery Act also includes measures to prevent further toxic contamination by establishing a system to monitor hazardous wastes from "cradle to grave." RCRA also gives the EPA the power to set standards for packaging, shipping, and disposal of wastes, and to license hazardous waste disposal facilities.
25. Like municipal wastes, hazardous wastes can be dealt with in three basic ways, in order of desirability: reduction, reuse and recycling, and disposal. Unfortunately, the last one, the discard approach, is the most widely used method.
26. To reduce hazardous wastes, manufacturers can modify or redesign their processes. They can also substitute safe materials for more harmful ones.
27. Manufacturers can also reduce the hazardous waste output of their factories by reusing and recycling wastes. Hazardous waste clearinghouses can help facilitate the exchange of wastes between businesses.
28. Some hazardous wastes that cannot be reused or recycled can be detoxified. For instance, organic wastes can be incinerated or decomposed by bacteria.
29. Some hazardous waste will inevitably be produced. This must be disposed of safely for hundreds, perhaps thousands, of years.
30. Secured landfills, the most popular approach today, are seen by few as a permanent solution. Many critics believe that it is only a matter of time before they begin to leak, creating problems for future generations.
31. Deep injection wells are also popular for the disposal of liquid wastes, but their use is riddled with problems.

Key Words and Phrases

Aseptic Containers
Basel Convention
Bottle Bill
Bottom Ash
Brownfield
Brownfield Development
Brownfield Economic Redevelopment Initiative
Cocomposting
Composting
Comprehensive Environmental Response, Compensation, and Liability Act
Container Deposit Bill
Deep Injection Wells
Detoxification
Dioxins
End-Point Separation
Fly Ash
Greenfield
Hazardous Waste
Household Hazardous Waste
Incineration
Manifest System
Municipal Solid Waste
NIMBY Syndrome
Open Dumps
Procurement
Process Manipulation
Process Redesign
Reduction Approach
Remanufacturing
Recycling
Recycling Opportunity Act
Resource Conservation and Recovery Act
Reuse
Reverse Vending Machine
Sanitary Landfill
Secured Landfill
Source Separation
Superfund Act
Waste Exchanges

Critical Thinking and Discussion Questions

1. Describe the three main techniques for managing municipal wastes. Discuss the pros and cons of each one. Give specific examples of each. What are the most sustainable measures and why do they contribute to sustainability?
2. Outline a plan to reduce your family's trash. What obstacles stand in the way of reducing the volume by 50 percent? How can they be overcome?
3. Why is it theoretically possible to recycle only 60 to 80 percent of America's aluminum?
4. Debate this statement: "In countries with large resource supplies it is more economical to use raw ore than recyclable materials."
5. Outline a plan for your city or town to reduce its solid and hazardous wastes. Would you involve private citizens, and if so, how? Contact local officials and ask them whether reuse and recycling programs exist and what the obstacles are to further waste reduction.
6. Describe ways to increase the demand for recycled goods.
7. Using your knowledge of this issue and of ecology in general, debate the following statement: "Governments have an obligation to help create markets for recyclable materials."
8. List the pros and cons of sanitary landfills, compost facilities, and incinerators.
9. What is hazardous waste? How can hazardous wastes best be reduced?
10. What problems does the Superfund Act address, and how does it address them?
11. What steps can individuals take to reduce hazardous waste production at home and at factories?
12. What hazardous waste problem(s) does the Resource Conservation and Recovery Act address, and how does it address it?
13. Describe ways that manufacturers can reduce hazardous wastes.
14. List and describe several methods of hazardous waste detoxification.
15. Debate the statement: "Secured landfills are the safest way to dispose of hazardous wastes."
16. Using your critical thinking skills and your knowledge of this issue, discuss the following statement: "It's up to manufacturers to control hazardous waste. They produce it and they should be responsible for reducing it."
17. Review the arguments presented in the Ethics in Resource Conservation box in this chapter. Using your critical thinking skills, analyze and discuss each one.

Suggested Readings

Municipal Solid Waste

Abramovitz, J. N., and A. T. Mattoon. 2000. "Recovering the Paper Landscape" In *State of the World 2000*. New York: Norton. Examines sustainable ways to produce paper, including recycling..

Andersen, M. S. 1998. Assessing the Effectiveness of Denmark's Waste Tax, *Environment* 40(4): 10–15, 38–41. Examines experience in Denmark with waste taxes and explores additional options to promote recycling.

Carless, J. 1992. *Taking Out the Trash: A No-Nonsense Guide to Recycling*. Washington, D.C.: Island Press. A practical guide showing how individuals, businesses, and communities can help alleviate the solid waste crisis.

Chiras, D. D. 1992. *Lessons From Nature: Learning to Live Sustainably on the Earth*. Washington, D.C.: Island Press. See Chapter 10 for a discussion of ways to build a sustainable waste management system.

EPA. Municipal Solid Waste Factbook–Internet Version. Washington, D.C.: EPA, 1995. An extremely valuable resource, though slightly out of date. Available on-line at www.epa.gov/epaoswer/non-hw/muncpl/factbook/internet/

Gardner, G., and P. Sampat. 1999. "Forging a Sustainable Materials Economy," In *State of the World 1999*. New York: Norton. Great look at consumption and ways to lessen it.

Kane, H. 1996. "Shifting to Sustainable Industries." In *State of the World 1996*. Starke, L. (ed.). New York: W. W. Norton. Describes important aspects of recycling.

Kates, R. W. 2000. "Population and Consumption: What We Know, What We Need to Know," *Environment* 42(3): 10–19. An important report on the subject.

Pollock, C. 1987. *Mining Urban Wastes: The Potential for Recycling*. Worldwatch Paper 76. Washington, D.C.: Worldwatch Institute. Informative coverage of the vast untapped potential of recycling.

Renner, M. 1992. "Creating Sustainable Jobs in Industrial Countries." In *State of the World 1992.* Starke, L. (ed). New York: W. W. Norton. Discusses the job potential of a sustainable economy based in part on recycling.

Rosenblatt, R. (ed.) 1999. *Consuming Desires: Consumption, Culture, and the Pursuit of Happiness.* Washington, D.C. Collection of writings on one of the most challenging aspects of modern times, consumerism.

Young, J., and A. Sachs. 1995. "Creating a Sustainable Materials Economy." In *State of the World 1995.* Starke, L. (ed.) New York: W. W. Norton. Addresses important issues regarding solid waste and recycling.

Hazardous Wastes

Carlin, A., P. F. Scodari, and D. H. Garner. 1992. "Environmental Investments: The Cost of Cleaning Up." *Environment* 34(2): 12–20, 38–45. Summary of U.S. EPA report to Congress. Center for Neighborhood Technology. 1990. *Sustainable Manufacturing.* Chicago: Center for Neighborhood Technology. Excellent discussion of efforts needed to reduce hazardous waste.

Fischhoff, B. 1991. "Report from Poland: Science and Politics in the Midst of Environmental Disaster." *Environment* 33(2): 12–17, 37. Describes the dimensions of the hazardous waste problem in Poland.

French, H. 1990. "A Most Deadly Trade." *Worldwatch* 3(4): 11–17. Documents the movement of hazardous materials to the developing countries and Eastern Europe.

Frosch, R. A. 1995."Industrial Ecology: Adapting Technology for a Sustainable World." *Environment* 37(10): 16–24, 34–37. Looks at ways to reduce waste production by factories.

Frosch, R. A., and N. F. Gallopoulos. 1989. "Strategies for Manufacturing." *Scientific American* 261(3): 144–152. Outlines the concept of the industrial ecosystem.

Gibbs, L. M. 1998. *Love Canal: The Story Continues.* Gabriola Island, B.C.: New Society Publishers. Fascinating account of the containment of wastes and resettlement of part of the contaminated area and more.

Krueger, J. 1999. "What's to Become of Trade in Hazardous Wastes?: The Basel Convention One Decade Later," *Environment* 41(9): 10–21. Examines the effects of an important international agreement on hazardous waste export to less developed countries.

McGinn, A. P. 2000. "Phasing Out Persistent Organic Pollutants." In *State of the World 2000.* Starke, L. (ed.). New York: W.W. Norton. Covers a variety of strategies for reducing and eliminating the production and release of persistent organic chemicals including those in hazardous waste.

Probst, K. N., and Bierle, T. C. 1999. "Hazardous Waste Management: Lessons from Eight Countries, " *Environment* 41(9): 22–30. Valuable resource.

Renner, M. 1994. "Cleaning Up After the Arms Race." In *State of the World 1994.* Starke, L. (ed.). New York: W. W. Norton. Discusses hazardous waste problems at federal facilities.

Russell, M. E., W. Colglazier, and B. E. Tonn. 1992. "The U.S. Hazardous Waste Legacy." *Environment.* 34(6): 12–15, 34–39. Discusses the cost of cleaning up America's hazardous wastes.

Steinhart, P. 1990. "Innocent Victims of a Toxic World." *National Wildlife* 28(2): 20–27. Discusses effects of toxins on wildlife.

Stigliani, W. M., et al. 1991. "Chemical Time Bombs: Predicting the Unpredictable." *Environment* 33(4): 4–9, 26–30. Shows how chemical contamination builds, then surpasses critical threshold levels, suddenly creating poisonous conditions.

Web Explorations

Online resources for this chapter are on the World Wide Web at: **http://www.prenhall.com/chiras** *(click on the Table of Contents link and then select Chapter 17).*

Chapter 18

Air Pollution

Human beings breathe in and out about once every four seconds, 16 times a minute, 960 times an hour—nearly 8.5 million times a year. Every year, we breathe nearly four million liters (1 million gallons) of oxygen-containing air from the Earth's atmosphere (Table 18.1).

In addition to being a vital source of oxygen, the Earth's atmosphere is of value to us in many other ways. The atmosphere, for instance, insulates the Earth. Without this capacity, the Earth would be subjected to drastic day–night temperature changes that would very likely be too extreme for living things. The atmosphere also helps to distribute heat, so the planet is more uniformly heated. Without the atmosphere, sound vibrations could not be transmitted; the Earth would be silent. There would be no weather, no spring rains for crops and lawns, no snow, hail, or fog. Without its atmospheric shield, our planet would be more heavily bombarded with meteorites and would be exposed to potentially lethal ultraviolet radiation from the sun. Without an atmosphere, Earth would be as lifeless as the moon.

As valuable as the atmosphere is, though, we humans have treated it with almost total disregard until recently. This chapter examines our unwitting assault on the atmosphere, notably the problem of air pollution. You will see how we impact the atmosphere and how we can live and conduct the affairs of business while protecting this vital element of the Earth's life support system.

TABLE 18.1 *Composition of Clean, Dry Air at Sea Level*

Gas	Volume Percent
Nitrogen	78.08
Oxygen	20.94
Argon	0.9340
Carbon dioxide	0.0310
Neon	0.0018
Helium	0.0005
Methane	0.0002
Krypton	0.0001
Sulfur dioxide	0.0001

Note: Gases such as carbon dioxide, methane, and sulfur dioxide are normal constituents of clean air. However, they often reach much higher concentrations in polluted air, and may have adverse effects on the environment and/or human health.

18.1 Pollution of the Atmosphere

Pollutants in the atmosphere arise from two sources: natural and human. This section describes each type.

Natural Pollution

Long before humans evolved, the atmosphere was to some degree polluted, not from artificial sources but from natural causes. Smoke from lightning-triggered forest fires billowed darkly across the land. Volcanoes spewed noxious gases into the atmosphere.

Natural pollution continues to be a problem. In May 1980, the massive eruption of Mount St. Helens in Washington released thousands of tons of dust and ash into the air and briefly caused breathing problems for both humans and wildlife downwind from the blast (Figures 18.1 and 18.2). Subsequent eruptions of other volcanoes like Mt. Pinatubo in 1991 had a similar effect.

A given sample of today's atmosphere may contain a host of natural contaminants, ranging from ragweed pollen to fungal spores, from disease-causing bacteria to minute particles of volcanic ash and salt to a host of harmful gases from many different sources. Interestingly, the release of pollutants from natural sources may sometimes exceed emissions from human sources.

Pollution Caused by Humans

Homo sapiens has been fouling the atmosphere ever since Stone Age people first roasted a deer over an open fire. The smoke smudged some of the magnificent cave-wall paintings in southern France—perhaps the first serious property damage caused by air pollution. In 1306, the English Parliament passed a law making it illegal to burn coal in a furnace in London; at least one violator was actually tortured for his offense. However, it was not until the Industrial Revolution that air pollution began to seriously affect the health of large segments of society. In 1909, over 1,000 people died in Glasgow, Scotland, as a result of polluted air. In conjunction with this incident, the word **smog** was coined as a contraction of smoke and fog.

So, if the Earth is naturally polluted why do we worry about pollutants coming from human sources? The reasons are several. First, human pollutants come from concentrated sources, usually cities or major industrial areas. Even though natural sources may exceed human sources, these concentrated release of pollutants results in elevated levels, locally and regionally. High concentrations can have serious impacts on people and the environment. Second, some human pollution sources can

FIGURE 18.1 *Mount St. Helens erupts. Generally, pollution from natural sources is far less harmful than pollution from human sources. That's because natural sources tend to release small quantities over huge areas, which results in low ambient levels. The eruption of Mount St. Helens was an exception.*

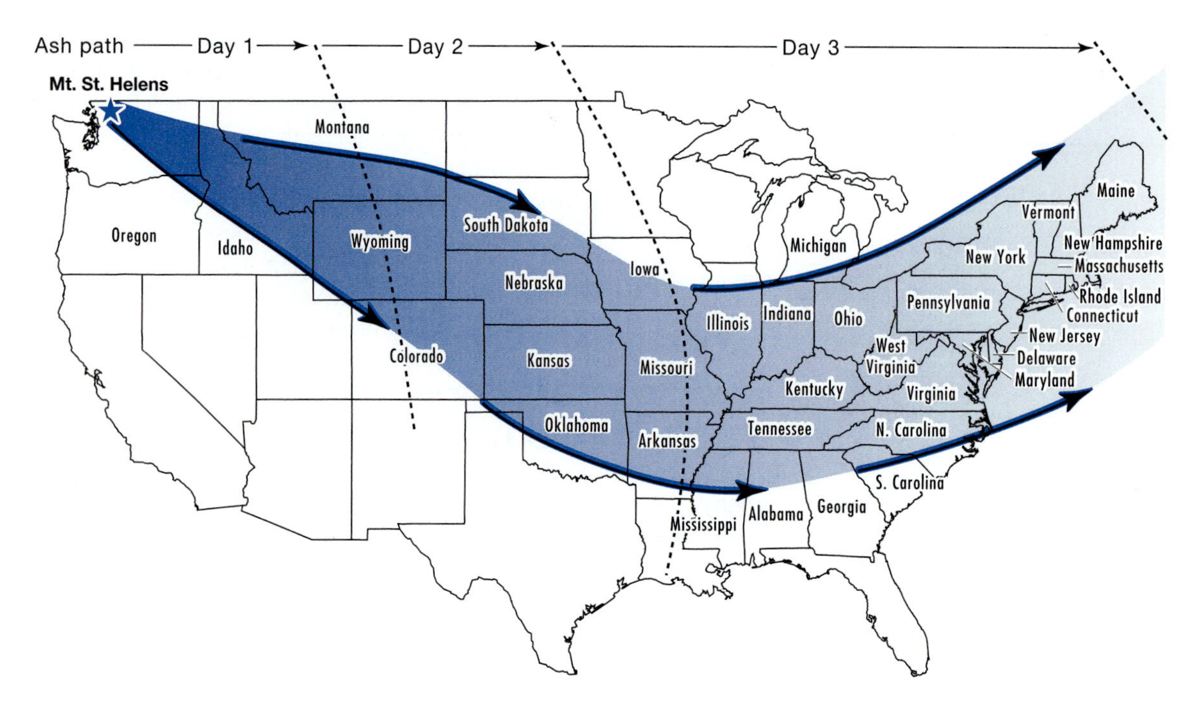

FIGURE 18.2 *Shaded area shows approximate path of ashes emitted into the air by the eruption of Mount St. Helens.*

be quite significant. Carbon dioxide releases from power plants, factories, homes, and automobiles, during the combustion of coal, oil, and natural gas, for example, have reached such high levels that they appear to be upsetting the Earth's energy balance, causing a global warming, a phenomenon that could have devastating effects on our environment, our economy, and our lives. To understand the scope of the problem, let's begin by examining the major pollutants, that is, those produced in largest quantity.

18.2 Major Atmospheric Pollutants

The major air pollutants in the United States are carbon dioxide, carbon monoxide, oxides of sulfur, particulate matter, hydrocarbons, and oxides of nitrogen (Figure 18.3, Table 18.2). Carbon dioxide will be discussed in the next chapter. These pollutants are all known as **primary pollutants.** They're produced by various sources; but some of them can be chemically converted into other, sometimes more troublesome chemicals, by natural processes. The result of these chemical reactions are known as **secondary pollutants.**

Carbon Monoxide

Carbon monoxide (CO) is a colorless, odorless pollutant released during the incomplete combustion of organic material such as wood, coal, oil, natural gas, and gasoline. It is common in sometimes rather high levels above city streets and freeway systems.

Rather surprisingly, roughly 93 percent of the CO in the global atmosphere is derived from natural sources, such as the oxidation of methane (marsh gas), which is formed by the decay of marshland organisms. However, this carbon monoxide does not build up to harmful concentrations because it is produced from widely dispersed sources and because it is quickly converted into carbon dioxide. (In this instance, carbon dioxide is considered a secondary pollutant.)

It might reasonably be asked, then, why we are so concerned with carbon monoxide as an atmospheric pollutant. The answer is that the seemingly insignificant 7 percent of the carbon monoxide generated by human activities, largely as the result of the incomplete combustion of fossil fuels, is concentrated in a relatively small volume of air in the world's major cities. In fact, the CO concentrations of urban areas are about 50 times greater than the worldwide average. The health effects can be substantial and are discussed shortly.

The amount of carbon monoxide released by U.S. sources decreased substantially during the 1980s. For example, between 1980 and 1997 it dropped from 105 million metric tons to 79 million metric tons annually. Much of this decrease can be attributed to the improvements in car engines that were mandated by the Clean Air Act of 1970. The **Clean Air Act** and its amendments, which are discussed in A Closer Look 18.1, address a wide range of air pollution issues. Since 1997, CO levels have risen dramatically to 87.5 million metric tons, in large part because of growing economic prosperity and widespread disregard for conservation. In the United States, a leading producer of carbon monoxide, cars are getting bigger by the day, as are our homes. With growing affluence, fuel consumption increases, and so does carbon monoxide.

Oxides of Sulfur

Oxides of sulfur are gaseous pollutants that form whenever sulfur-containing fuels, such as coal and oil, are burned. During combustion, the sulfur combines with oxygen in the air to

TABLE 18.2 *Sources and Effects of Major Air Pollutants*

Pollutant	Description and Major Anthropogenic Sources	Human and Enviromental Effects
Total suspended particulates	Solid or liquid particles produced by combustion and other processes at major industrial sources (e.g., steel mills, power plants, chemical plants, cement plants, incinerators).	Respiratory irritant, aggravates asthma and other lung and heart diseases (especially in combination with sulfur dioxide); many are known carcinogens. Toxic gases and heavy metals adsorb onto these particulates and are commonly carried deep into the lungs.
Sulfur dioxide	Colorless gas produced by combustion at power plants and certain industrial sources.	Respiratory irritant, aggravates asthma and other lung and heart diseases, reduces lung function. Sulfur dioxide damages plants and is a precursor to acid rain.
Carbon monoxide	Colorless gas produced by motor vehicles and some industrial processes.	Interferes with the blood's ability to absorb oxygen; can cause dizziness, drowsiness; impairs motor reflexes; may bring on angina.
Nitrogen dioxide	Brownish orange gas produced by motor vehicles and combustion at major industrial sources.	Respiratory irritant, aggravates asthma and other lung and heart diseases.
Ozone	A colorless gas formed from a reaction between motor vehicles emissions and sunlight. It is the major component of smog.	Respiratory irritant, aggravates asthma and other lung and heart diseases, impairs lung functions. Ozone is toxic to plants and corrodes materials.
Hydrocarbons	Small quantities of hazardous pollutants emitted from industrial processes and diesel motor vehicle exhaust.	Linked to organ damage, serious chronic diseases, and various types of cancer.
Lead	Very small particles emitted from motor vehicles and smelters.	Toxic to nervous and blood-forming systems; can cause brain and organ damage in high concentrations.

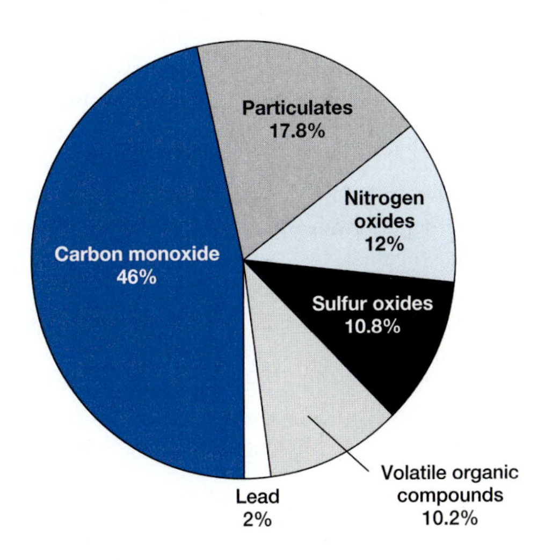

FIGURE 18.3 *Major pollutants released into the air over the United States.*

form sulfur oxides, denoted by the chemical formula SO_x. (The x can be a 2 or a 3.) About 20 million metric tons of sulfur oxides are released by U.S. sources into the atmosphere each year. Colorless sulfur dioxide stings the eyes and burns the throat. About 1 percent of the U.S. population will develop chronic weariness, tortured breathing, sore throat, tonsillitis, coughing, and wheezing when exposed for lengthy periods to the concentrations of sulfur dioxide normally occurring in polluted urban air. Sulfur dioxide slows down or even halts the cleansing mechanism of the lungs. It also contributes importantly to such chronic diseases as bronchitis and emphysema.

As you shall see in the next chapter, sulfur oxides combine with water in air to form sulfuric acid, a potentially harmful secondary pollutant. It falls to the earth in rain and snow and is deposited as sulfate particles. Chapter 20 discusses this problem in more detail.

A CLOSER LOOK 18.1 *The Clean Air Act*

The federal **Clean Air Act** is one of the most successful pieces of environmental legislation in U.S. history. But the Clean Air Act is not one law, but several. First passed in 1963, the law was fairly week and ineffective. However, over the past four decades, the Clean Air Act has been amended three times, strengthening it considerably, in response to our growing air pollution problems and our improved understanding of the best ways to solve them.

The first amendment to the Clean Air Act occurred in 1970. These amendments resulted in the establishment of (1) emissions standards for automobiles, (2) emissions standards for new industries, and (3) ambient air quality standards for urban areas. The **national ambient air quality standards** established by the EPA covered six pollutants: carbon monoxide, sulfur oxides, nitrogen oxides, particulates, ozone, and hydrocarbons. These standards were designed to protect human health and the environment.

The 1970 amendments were successful in reducing air pollution from automobiles, factories, and power plants. In addition, they stimulated many states to pass their own air pollution laws, some with regulations even tighter than federal ones. Despite these gains, the amendments created some problems. For instance, in regions whose air exceeded national ambient air quality standards, the law prohibited the construction of new factories or the expansion of existing ones. As you might suspect, the business community objected vociferously. In addition, some of the wording of the 1970 amendments was vague and required clarification. Of special interest were provisions dealing with the deterioration of air quality in areas that had already met federal standards. Environmentalists worried that clean air areas would deteriorate because of federal standards.

Because of these and other problems, the Clean Air Act was amended once again in 1977. To address the limits on industrial growth in areas that were violating the air quality standards, called **nonattainment areas,** lawmakers devised a strategy that allowed factories to expand and new ones to be built, but *only* if they met three provisions: (1) The new sources achieved the lowest possible emission rates, (2) other sources of pollution under the same ownership in that state complied with emissions-control provisions, and (3) unavoidable emissions were offset by pollution reductions by the company in question or other companies in the same region.

The last provision, known as the **emissions offset policy,** forces companies to make reductions in their own facilities and required newcomers to request existing companies in non-compliance areas to reduce their pollution emissions. In most cases, the newcomers pay the cost of air-pollution control devices.

The emissions offset policy is also used to produce an overall decrease in regional air pollution—because the air pollution emissions permitted from both the new and existing facilities are set below preconstruction levels.

The 1977 amendments set forth rules for the **prevention of significant deterioration (PSD)** of air quality in **attainment regions,** regions where air quality meets federal standards. However, PSD requirements apply only to sulfur oxides and particulates. Many air pollution experts think that the PSD requirements should be expanded to include other pollutants, such as ozone.

Another benefit of the 1977 amendments is that they strengthened the enforcement power of the EPA. In previous years, when the EPA wanted to stop a polluter it had to initiate a criminal lawsuit. Violators would often engage in protracted legal battles, knowing that legal costs were often lower than the cost of installing pollution control devices. The 1977 amendments, however, allowed the EPA to levy **noncompliance penalties** without going to court. The logic behind this new power is that violators have an unfair business advantage over competitors that comply with the law. Penalties equal to the estimated cost of pollution control devices eliminate the cost incentive to pollute.

In 1990, the Clean Air Act was amended once more to address other important issues, among them acid rain. The 1990 amendments, for example, set deadlines for establishing emission standards for 190 toxic chemicals from factories, a subject that had not been previously addressed. More important, it established a system of **pollution taxes** on toxic chemical emissions, charging manufacturers a tax on those chemicals they used that were potential toxic air pollutants. These provide a powerful incentive for companies to reduce their use of them.

The 1990 Clean Air Act amendments tightened emission standards for automobiles and raised the average mileage standards for new cars, a step that improved automobile efficiency and helped attack the pollution problem at its roots. In addition, the 1990 amendments established a market-based incentive program to reduce sulfur dioxide emissions, a primary contributor to acid deposition (Chapter 19). The law set up a system of **tradeable** or **marketable permits** to companies throughout the United States. Each company is granted a certain number of allowances for sulfur dioxide release. (One allowance is equal to one ton of pollution.) These permits stipulated lower than present emission rates to improve air quality and could be bought and sold. Therefore, companies that found innovative and cost-effective ways to reduce pollutants below their permitted allowance could sell their unused credits at a profit. (These were bought and sold on the commodities market starting in 1995.) This system encourages companies to develop cost-effective ways to prevent pollution. Thus, if a company can find inexpensive means of reducing its output of pollutants, it could benefit economically from the sale of its pollution credits. Most companies did. They switched to low-sulfur coal, which cost less than the coal they were using. So they not only saved money, cut sulfur dioxide emissions, but were able to sell their allowances to other companies. Finally, the 1990 amendments called for a phaseout of ozone-depleting chemicals, a topic discussed in the next chapter.

The Clean Air Act has worked admirably well and has grown more flexible as time has passed, giving companies greater leeway in how they control or eliminate pollution. It will no doubt change over the coming decade to reflect new demands and more creative and economically and environmentally beneficial responses.

Particulate Matter

Particulate matter refers to small solid particles and liquid droplets suspended in the air of varying size from very large to extremely small (Figure 18.4). These particles may remain suspended in the air for periods ranging from a few seconds to several months, depending on their size and weight. The finer the particulate, the longer it remains suspended in the atmosphere.

Most of the particulate matter in the atmosphere from human sources is emitted by facilities that use coal as fuel, such as power plants, iron and steel mills, and foundries. Automobiles and other motorized vehicles also contribute to particulate pollution. Diesel trucks and trains are an even greater source. In many major metropolitan areas that sand roads in the winter much of the atmospheric particulate pollution comes from sand on roadways that moving tires eject into the atmosphere.

Other sources of particulates include agricultural activities such as plowing, cultivating, and harvesting; slash-and-burn farming in Asia, Africa, and South America; debris burning by loggers; and strip-mine operations. In the United States, approximately 33.5 million metric tons of particulate matter are introduced into our atmosphere yearly from these and other sources.

Studies show that fine particulates may cause a significant increase in mortality due to respiratory and heart malfunction. Extrapolating from studies conducted in the areas of Detroit and St. Louis, researchers estimate that high levels of particulates could be responsible for more than 60,000 deaths nationwide per year. Especially vulnerable are the elderly, asthma sufferers, and people recovering from recent heart attacks.

Particulates are dangerous because they may contain sulfates or nitrates (discussed shortly), which may enter the lungs

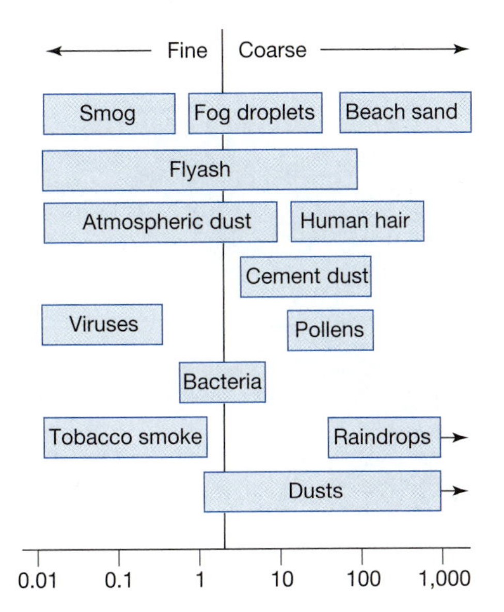

FIGURE 18.4 *Size ranges of some airborne particulates in microns. Some particles in smog, fly ash, dust, and tobacco smoke are only 0.01 micron in diameter, too small to be seen under an ordinary microscope.*

and combine with water, forming sulfuric acid and nitric acids in the tissues of the lung. Both can cause considerable damage.

Particulates also contain heavy metals such as lead. For years, lead was added to gasoline in many more developed nations to enhance its octane rating, that is, the efficiency with which it burns. We humans took this lead into our systems when inhaling air polluted with motor exhaust. A cumulative poison, lead is also ingested with food or water. It can damage the kidneys, blood, and liver. Moreover, it can damage the brains of children.

On the basis of lead concentrations in snows at high elevations in the Rockies, Clare C. Patterson, a California Institute of Technology geochemist, has suggested that lead concentrations in humans are 100 times the level of two centuries ago. In 1973 the U.S. Environmental Protection Agency (EPA) began restricting lead in gasoline, imposing a 90 percent reduction by 1985. The EPA eliminated lead from gasoline in 1995. Today, there are only about 4 tons of lead released into the atmosphere of the United States each year, down significantly in the past 20 years. Other industrial countries have followed suit. Unfortunately, lead additives are still widely used in less developed countries. Another particulate of grave concern is mercury, which was discussed in Chapter 10.

Hydrocarbons

A **hydrocarbon,** as the name suggests, is simply an organic compound that is composed of hydrogen and carbon. Examples are methane, benzene, and ethylene. In urban areas, humans may generate more than 200 kinds of hydrocarbons. Hydrocarbons are **volatile organic compounds** (or **VOCs**); that is, they evaporate readily. Many VOCs are chemically reactive. For example, some react with nitrogen oxides in the presence of sunlight to form **photochemical smog,** a potentially harmful mix of secondary pollutants (described below). Much hydrocarbon pollution results from the evaporation of gasoline from carburetors, crankcases, and gas tanks of cars and trucks. In addition, unburned hydrocarbons are released from the exhaust pipes of motor vehicles. The United States produces over 19 million metric tons of them per year.

Oxides of Nitrogen

The atmosphere is 79 percent nitrogen. When air enters a combustion chamber—say, the cylinder of a car—the nitrogen reacts with oxygen to produce **nitrogen oxides** (NO_x). **Nitric oxide** is the first nitrogen oxide formed. Its chemical formula is NO. Nitric oxide is relatively harmless at ordinary concentrations. At unusually high concentrations, nitric oxide can be lethal, causing death by asphyxiation because it combines with hemoglobin in the red blood cells, blocking their ability to transport oxygen. In fact, it binds 300,000 times more readily with red blood cells than does oxygen.

Nitric oxide also combines with atmospheric oxygen to form **nitrogen dioxide** (NO_2), a reddish brown gas with a pungent, choking odor. Nitrogen dioxide is a major component of photo-

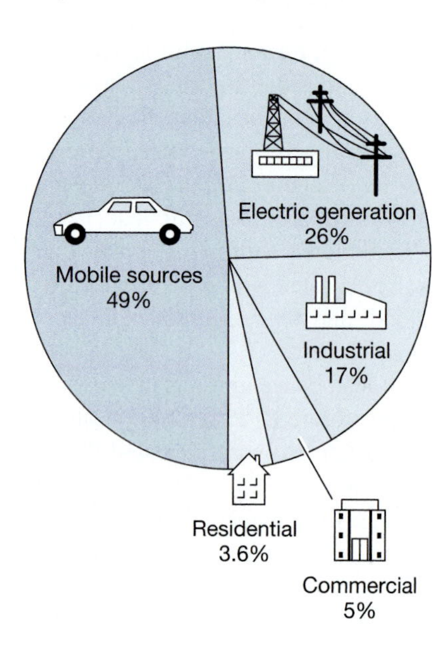

FIGURE 18.5 *Major sources of nitrogen oxides in the United States.*

chemical smog. It causes a variety of human ailments ranging from gum inflammation and internal bleeding to emphysema and increased susceptibility to pneumonia and lung cancer. Nitrogen dioxide is considered four times as toxic as nitric oxide. Major sources of the oxides of nitrogen are shown in Figure 18.5. Like carbon monoxide, nitrogen dioxide emissions have begun to increase. It is now about 24 million metric tons per year.

Ozone and Photochemical Smog Ozone is a secondary pollutant. This gaseous pollutant is a major component of photochemical smog—the brownish haze that shrouds many urban areas during the hot, sunny days of summer. **Ozone** (O_3) is produced by chemical reactions between hydrocarbons and nitrogen oxides. Because these reactions are powered by sunlight, the chemical soup produced in them is called photochemical smog. The greater the intensity of sunlight and the warmer the day, the larger the amount of ozone produced. Because hydrocarbons and oxides of nitrogen are mainly generated by motor vehicles, the ozone levels in cities gradually rise in the morning and peak between noon and 4 p.m. On cloudy days, ozone production is reduced; after sunset, it stops altogether.

Photochemical smog has traditionally been associated with Los Angeles, where it was first recognized and has been a severe problem. However, during the 1970s and 1980s, it proved to be a daunting challenge to health officials in Denver, Salt Lake City, Milwaukee, Chicago, New York, and Boston as well. During the extremely hot, dry summers of the 1990s, many East Coast cities were plagued with record high ozone levels. The EPA reported that in 1998, 32 areas failed to comply with the federal ozone standard (Figure 18.6). Standards are discussed shortly.

As many summer visitors to Los Angeles well know, the ozone in photochemical smog irritates the eyes, nose, and throat and makes breathing difficult (Figure 18.7). Recent scientific studies, however, have shown that ozone pollution may have much more serious consequences. For example, even low-level, short-term exposure to ozone can cause weight reduction and chromosome damage in rodents. Other animal studies have shown that chronic exposure to ozone causes permanent lung damage, including stiffening of the wall of the lung, which is normally associated with aging. Dr. Morton Lippmann, professor of environmental medicine at New York University, reported recently that jogging in an ozone-polluted urban environment may do the runner more harm than good. In fact, it can be as dangerous to one's health as smoking. The

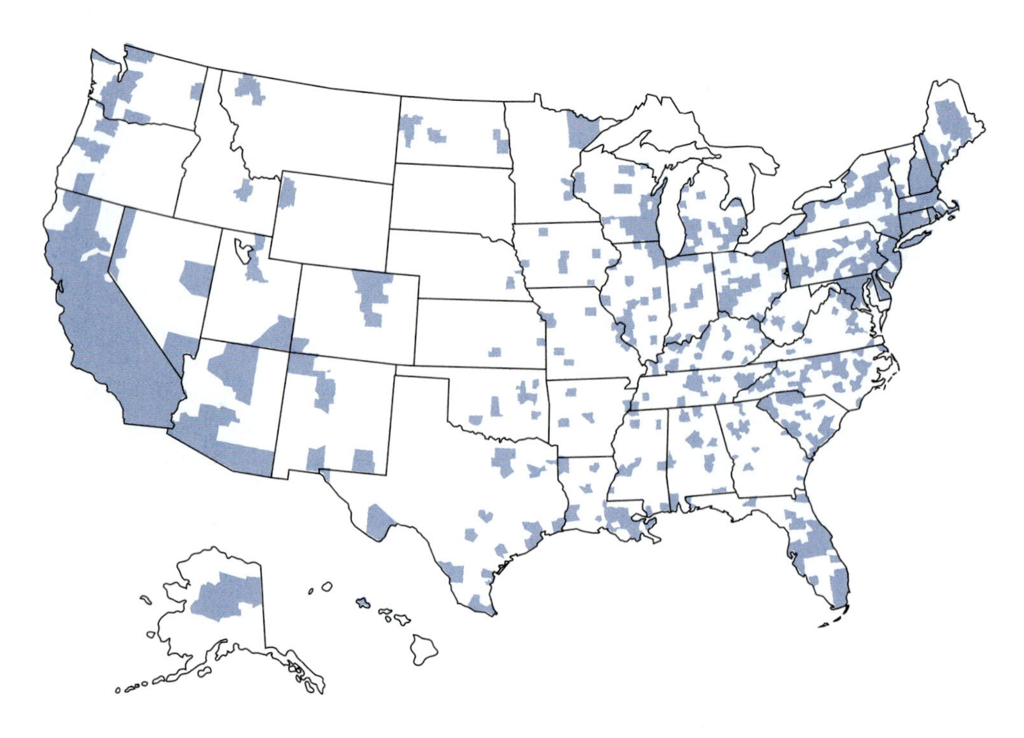

FIGURE 18.6 *Regions that at some time did not meet national ozone standards in 1998.*

FIGURE 18.7 *(Left) Aerial view of Los Angeles on a clear day. (Right) Los Angeles under smog. The smog in this photograph is trapped by a temperature inversion. Inversions are present over the Los Angeles Basin about 320 days of the year!*

effect is cumulative. With each breath, the runner's lungs are damaged a little more. Said Lippman, "People won't fall over from exercising one or two days, but you may have breathing difficulties later on."

But this is not all. The health of farm crops is adversely affected by ozone as well. For example, the U.S. Office of Technology has reported that an ozone-induced yield reduction of only four major crops—wheat, corn, soybeans, and peanuts—results in an estimated $3.2 billion annual loss to American agriculture (Table 18.3).

In the United States, the maximum concentration of ozone permitted under current EPA regulations is 0.08 parts per million (ppm) as an average over an eight-hour period. If the ozone in a city's air exceeds that level more than once in a given year, the city is subject to economic sanctions, such as the withdrawal of federal funds for highway construction.

18.3 Factors Affecting Air Pollution Concentrations

The concentration of air pollution in the air we breathe is dependent on the amount produced, but also on certain factors as well. This section discusses two of them, thermal inversions and urban heat absorption.

Thermal Inversion

Have you ever noticed that some days the air is cleaner than others? One reason for the buildup of atmospheric contaminants to high levels on some days is a meteorological condition known as a **thermal inversion** or **temperature inversion.** Let us explain how this forms, before we define it. It's easier to understand this way. Under normal daytime conditions, the air temperature gradually decreases with altitude from ground level to a height of several miles above the Earth's surface (Figure 18.8, top). This pattern permits hot air from the surface to rise and disperses pollutants at ground level as well. However, there are times when the temperature pattern deviates from this one. In such cases, temperature decreases for a ways, then begins to climb again. This is called a temperature inversion. During a thermal inversion, there's a layer of warm air over a region, which is often referred to as a warm air lid or thermal lid. In such instances, dispersion becomes impossible. Pollution becomes trapped near the surface and levels increase. Before we show you why, let's learn a little bit more about thermal inversions. First, meteorologists have identified two basic types of inversions: radiation and subsidence.

Radiation Inversion At night, heat radiates from the Earth's surface into the atmosphere. The Earth is a better radiator than the atmosphere. As a result, both the ground and the air layer next to it cool off more quickly than the air above them.

TABLE 18.3 *The Effect of Ozone on Plants*

Plant	Ozone Concentration (ppm)	Duration of Exposure	Reduction in Weight or Height
Alfalfa	0.10	7 hr/day/70 days	51% total dry wt.
Soybeans	0.10	6 hr/day/133 days	55% seed wt.
Sweet corn	0.10	6 hr/day/64 days	45% seed wt.
Wheat	0.20	4 hr/day/7 days	30% seed wt.
Beets	0.20	2 hr/day/38 days	40% root wt.
Ponderosa pine	0.10	6 hr/day/126 days	21% stem wt.
Hybrid poplar	0.15	12 hr/day/102 days	58% height
Red maple	0.25	8 hr/day/6 weeks	37% height

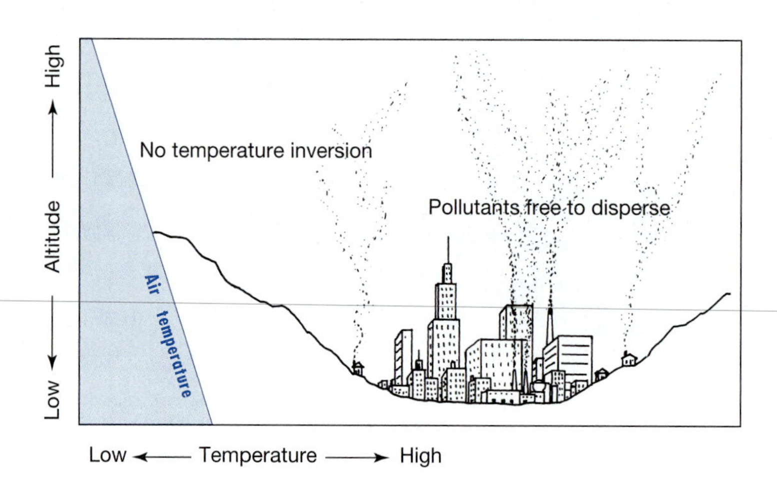

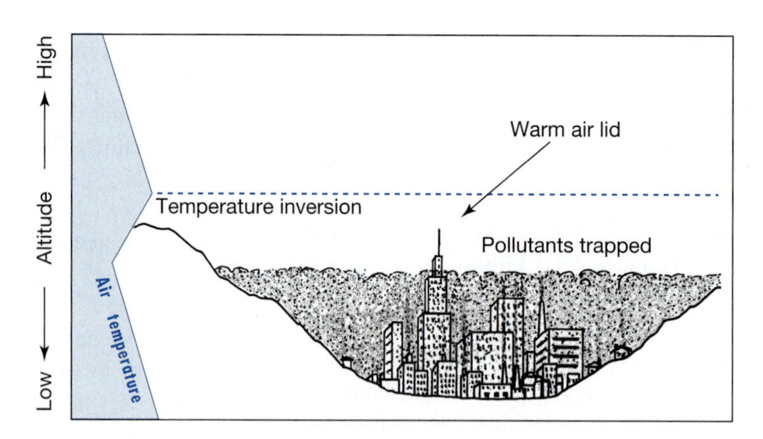

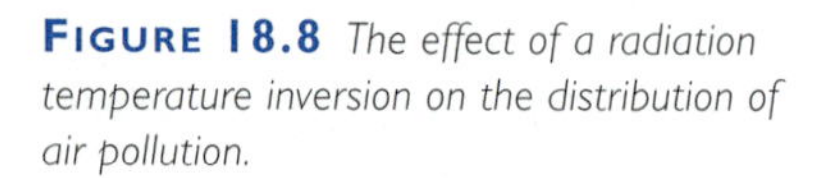
FIGURE 18.8 *The effect of a radiation temperature inversion on the distribution of air pollution.*

Consequently, a layer of cool air forms immediately above the Earth's surface. It may extend 300 meters (1,000 feet) above ground. Above this cooler layer is a warm layer, a thermal lid. The result is a **radiation inversion**—an inverted temperature profile caused by the radiative cooling of the Earth's surface (Figure 18.8, bottom). Because cool air does not rise, pollutants are trapped in this layer and remain relatively close to the ground. If winds are present, however, pollutants may disperse horizontally. When pollutants are not blown horizontally, they may build up to dangerous levels.

Radiation inversions are common in many areas of the United States, especially in mountainous regions. However, they are usually confined to small areas and they usually dissipate by late morning, when the Earth's surface is warmed by the sun. The air immediately above the Earth then warms up as well. As the day advances, the inversion gradually disappears. Pollutants disperse vertically and ground-level concentrations fall.

Subsidence Inversion A **subsidence inversion** is formed when a warm, high-pressure air mass stalls over an area and sinks toward the ground, at times as low as 600 meters (2,000 feet) above the ground, forming a huge warm air lid. Although less common than radiation inversions, subsidence inversions usually last longer and may be much more extensive. Sometimes the subsidence inversion forms a canopy over several states. This type of inversion worsens the air pollution problems of Los Angeles in the summer (Figure 18.9). During the summer months, a warm, high-pressure air mass is constantly present above the Pacific Ocean off the California coast. This air mass occasionally moves inland over Los Angeles, Oakland, and other coastal cities and puts a lid on the pollutant-laden air near the ground that has been cooled by ocean currents moving along the coast. Coastal California experiences this type of inversion on nine out of every ten days in summer, which is one of the reasons why air pollution in San Diego and Los Angeles can be so bad in the summer.

Dust Domes and Heat Islands

Any motorist speeding toward the outskirts of Chicago, St. Louis, Des Moines, or any other large city has observed the haze of smoke and dust that frequently forms an "umbrella" over the town. This shroud of pollutants, known as a **dust dome.** The dust dome is caused by a unique atmospheric circulation pattern that, in turn, depends on the marked temperature differences between the city proper and outlying regions. Let us explain. Cities tend to be warmer than the outlying areas due to the massive amounts of cement and pavement they contain. Although the average annual temperature of a city might be only 0.98°C (1.7°F) higher than that of the surrounding rural areas on a given day, occasionally a city may actually be 20°C (27°F) warmer. The city therefore forms a **heat island** (Figure 18.10).

Contributing to the warmth of the city are such heat-generating sources as people, kitchen stoves, industrial furnaces, utility boilers, and motor vehicles, as well as the heat-absorbing and heat-radiating surfaces of streets, parking lots, and buildings. In rural areas, on the other hand, heat-generating and heat-radiating structures are much less numerous. Moreover, there is more evaporative cooling in rural areas because of the vegetation. Evaporative cooling, you may recall, is the loss of water from transpiration. It cools the general vicinity.

Here's how the dust dome is formed. The heat generated in a city causes air laden with pollutants to rise. Cool air from the countryside moves into the city to replace warm air rising from the urban center. As a result, smoke dust, nitrogen dioxide, and other aerial "garbage" tend to concentrate above the city, creating a dust dome. One thousand times as much dust may be present immediately over an urban industrial area as in the air of the nearby countryside.

When the air is calm, the dust dome persists, but when winds as slow as 12.8 kilometers (8 miles) per hour or more develop, the dome is pushed downward and horizontally into an elongated dust plume. Such plumes originating in cities such as Chicago are occasionally seen a distance of 240 kilometers (150 miles) away.

18.4 Effects of Air Pollution on Local Climate

Just as certain climatic factors like inversions affect air pollution, growing evidence shows that air pollution can alter climate. Some forms of pollution (notably particulates) do so by reducing sunlight penetration, which decreases air temperature. Particulates also affect cloud formations and can alter precipitation levels. Such dramatic climatic changes may disrupt terrestrial and aquatic ecosystems. Chapter 19 provides evidence that suggests that carbon dioxide may be altering the climate on a global scale. This chapter examines the well-documented local climatic effects.

Air Pollution and Precipitation

Local climate is influenced primarily by particulates. Particulates (soot and dust) in urban air from factories, power plants, and roadways help to seed clouds in the atmosphere downwind of cities. That is, they serve as **condensation**

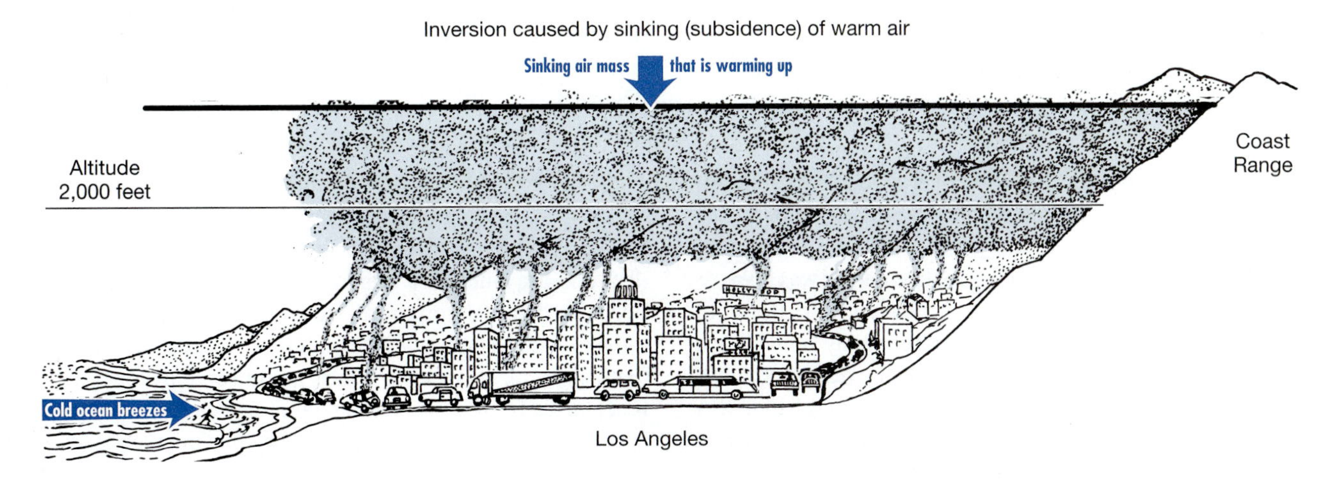

FIGURE 18.9 *The nature of a subsidence inversion. In a subsidence inversion warm air sinks down on an area, trapping cooler air, and causing pollution levels to increase.*

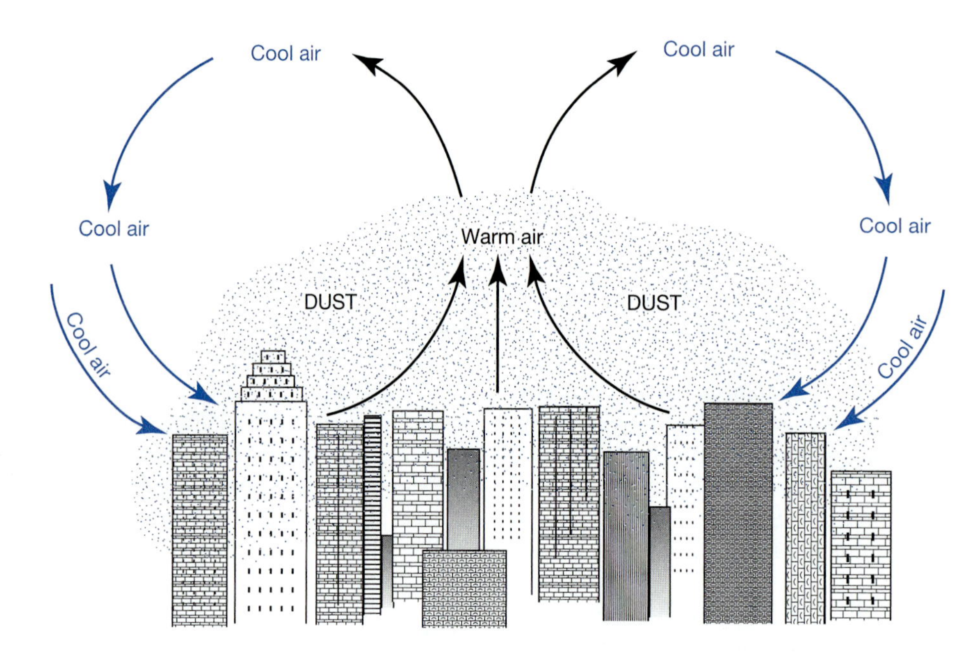

FIGURE 18.10 *Heat island effect of a city. The air is warmed up in the city because of large numbers of industrial furnaces, car motors, human bodies, heat-absorbing surfaces, and so on. This warm air moves upward and eventually cools. It may then sink, because of its greater density, move back into the city along with other cool air from the countryside, and be warmed again. As a result, a circular flow of air occurs. Any dust in the air will remain suspended above the city in a mushroom-like formation known as a dust dome.*

nuclei, small particles suspended in the air that absorb moisture in the atmosphere, forming tiny droplets. These droplets form clouds.

Because particulates promote cloud formation, they can result in an increase in rainfall downwind of cities and other sources of particulates. If human-generated air pollution induces rainfall, one would expect more rainy days from Monday to Friday, when factories are operating and pollutants are generated, than on weekends, when plants are closed. This is precisely the case in many instances. For example, in Paris, France, the average daily rainfall on weekdays is 31 percent higher than on weekends. Industrial contaminants, such as particulates, generated in Chicago are blown to the general region of LaPorte, Indiana, 80 kilometers (50 miles) to the east, where they trigger considerable amounts of rainfall.

Air Pollution and Decreased Average Temperature

The total annual load of particulates generated directly or indirectly by human activities the world over amounts to 800 million metric tons. Because particulates can block sunlight, an increase in particulate matter may have a cooling effect on a large area. The cooling effect of particulate matter was impressively demonstrated in 1883 when a volcano on the island of Krakatoa in the Dutch East Indies injected many tons of fine dust particles high into the atmosphere. Over a period of years, these particles eventually circled the globe several times. A short time after the eruption, the United States experienced a cooling trend; Bostonians, for example, had the rare privilege of throwing snowballs in June. More recent volcanic eruptions have had cooling effects, too, but much more modest.

Pollutants released into the atmosphere from natural and human activities continue to have a cooling effect, but overall, the Earth and its atmosphere are warming. Scientists believe that other forces are at play, for example, increased carbon dioxide pollution and deforestation, which appear to be offsetting the effect of particulates.

18.5 Effects of Air Pollution on Human Health

Scientists estimate that 40,000 to 50,000 Americans die each year at least partly because of air pollutants. That is about the number of American soldiers who were killed in the Vietnam War. Let's take a look at health effects, starting with some of the oldest known effects resulting from air pollution disasters.

Air Pollution Disasters

When many people die from air pollution in a short time, the episode is called a **disaster.** When most of the air pollution disasters are studied, a common pattern is revealed: (1) they occur in densely populated areas; (2) they occur in heavily industrialized centers where pollution sources are abundant; (3) they occur in valleys, which might serve as topographical receptacles for receiving and retaining pollutants; (4) they are accompanied by fog (it appears that the minute droplets of moisture are adsorbed on the surfaces of the pollutants); and (5) they are accompanied by a thermal inversion, which traps pollutants causing levels to increase substantially.

The Donora Disaster Forty-eight kilometers (30 miles) south of Pittsburgh, Pennsylvania, in a horseshoe-shaped bend of the Monongahela River lies the industrial community of Donora, Pennsylvania, with a population of 12,000. Almost encircled by hills rising to a height of 116 meters (350 feet), Donora is home to factories that manufacture steel, wire, sulfuric acid, and zinc, all crowded along the river margin for 5 kilometers (3 miles).

On October 26, 1948, a thermal inversion occurred. Soon afterward, a fog closed in on the valley. There was hardly a breath of air stirring. The black, red, and yellow smoke fumes that belched from the Donora smokestacks merged to form a multicolored blanket over the valley town. In a short time, the pungent odor of sulfur dioxide permeated the air. In addition to sulfur dioxide, the air over Donora contained high levels of nitrogen dioxide and hydrocarbons, which resulted from burning coal to provide heat and electricity for shops and homes. A sluggishly played football game between the Donora and Monongahela high schools was canceled in midplay when several of the players complained of chest pains and tortured breathing. Streets, sidewalks, and porches were covered with a film of soot. Motorists had to pull off to the side of the road because they couldn't see. People who had lived in Donora for over half a century got lost. The smog was so dense that it was extremely difficult to see from one side of the street to the other. The Donora fire department hauled oxygen tanks around the clock to people experiencing breathing difficulties.

Of the total population, roughly 43 percent—5,910 people—became ill, the most prevalent symptoms being nausea, vomiting, and severe headaches; nose, eye, and throat irritation; and labored breathing and constriction of the chest. Even pets and wildlife suffered. A veterinarian reported that the dense smoke caused the death of seven chickens, three canaries, two rats, two rabbits, and two dogs.

The complete death toll for Donora's "Black Saturday" was 17. Two more deaths occurred on Sunday. Then, on Sunday night, climatic conditions changed. A heavy rain washed some of the pollutants from the air. A breeze drove much of the smoke away. Visibility improved, and breathing became easier. The worst air pollution disaster in American history was over. Although the smog lasted for only five days, it left 20 people dead in its wake.

Incidents such as the one in Donora are dramatic examples of the effects of air pollution on humans. Similar examples have occurred in heavily industrialized cities, such as in London in 1952. In this episode, 4,000 people died as a result of pollution. Fortunately, such events are rare. Pollution controls have helped to reduce these events. More common though are chronic health effects resulting from long-term exposure to air pollutants. The next section discusses this problem.

Chronic Health Effects of Air Pollution

Many of us live in polluted areas, taking the pollution we breathe in stride. However, over the years scientists have found that air pollution in modern cities and towns is not something to be nonchalant about. It can kill people. But urban air pollution usually kills slowly and quietly, making the relationship between cause and effect difficult to detect. Thus, instead of stating that the deceased breathed in too much carbon monoxide, sulfur dioxide, nitrogen oxides, or particulates, the death certificate simply states that death was caused by a heart attack, lung cancer, or emphysema.

Chronic Bronchitis Many people, both smokers and urban residents, suffer from a chronic irritation of the bronchial tubes, which carry air into the lungs. This persistent disease, caused by pollutants in tobacco smoke and urban air, is called **chronic bronchitis** (bron-KITE-is). Symptoms include a persistent mucus buildup and cough. In some instances, individuals have difficulty breathing. Although cigarette smoking is the primary cause, urban air pollution is clearly a causative factor, even among children. Chronic bronchitis results from exposure to sulfur dioxide and nitrogen dioxide, which are found in both tobacco smoke and urban air, as well as ozone, which is found in urban air. These chemicals irritate the respiratory passageways, causing mucus secretion.

Emphysema To understand emphysema, you must first understand how the lung's function. When the air we breathe enters the millions of tiny air sacs (alveoli, pronounced al-vee-OH-lee) in the lung, oxygen passes through the ultrathin membranes of the tiny air sacs into blood capillaries (Figure 18.11). Oxygen is then distributed in the blood to the cells of the body. As oxygen flows into the lungs, carbon dioxide flows out, passing from the blood in the capillaries surrounding the air sacs into the alveoli. This helps the body get rid of excess carbon dioxide, a byproduct of cellular energy production. The total respiratory membrane surface presented by each lung's 300 million air sacs is about the size of a tennis court. Each breath we take draws air into the lungs. Expelling that air, however, is a passive process most of the time that depends on elastic connective tissue in the walls of the alveoli. Like inflated balloons, the lungs recoil after filling with air. This forces air out of the lungs.

Air pollutants can damage the alveoli and decrease the elasticity of the lungs. Sulfur dioxide, nitrogen dioxide, and ozone in urban air and in tobacco smoke, for instance, damage the alveoli. As the walls break down, the alveoli become larger and larger. This, in turn, decreases the surface area for exchange. This condition is known as **emphysema** (em-fi-ZEEM-ah).

The elastic tissue also progressively deteriorates in people with emphysema. In some cases, almost 50 percent of the lung's elastic tissue may be destroyed before the victim is aware of the problem. The decline in elasticity in emphysemic patients does not affect inhalation, but makes exhalation very difficult indeed. With each incoming breath, the air sac becomes overinflated. This process is repeated many times until the sac "pops" like a burst balloon, resulting in further destruction of the respiratory membrane and blood capillaries. Eventually, therefore, the area available for the exchange of respiratory gases is greatly reduced (Figure 18.11, bottom).

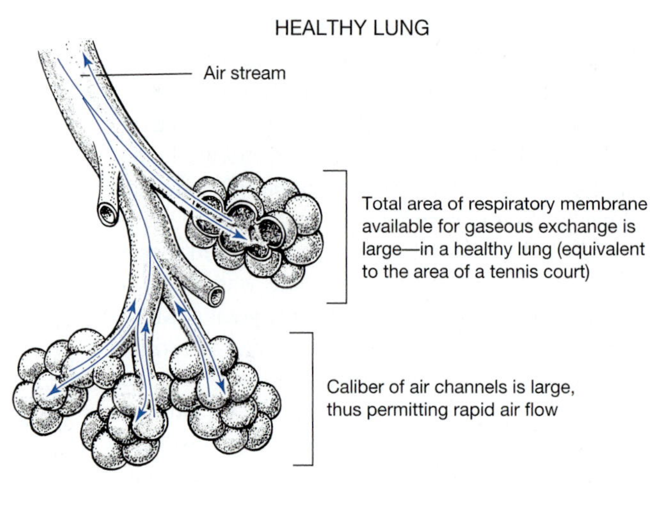

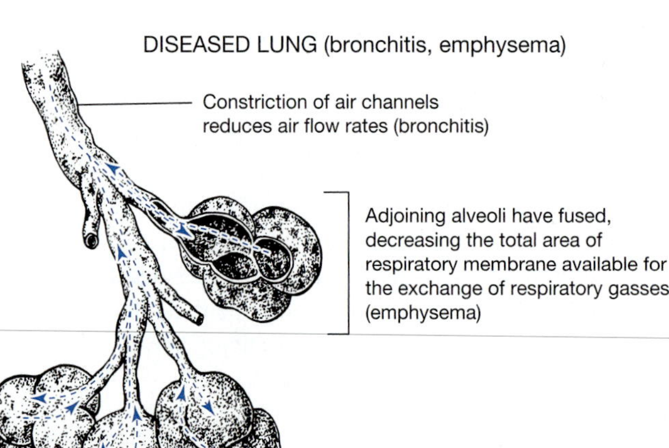

FIGURE 18.11 *Bronchioles and air sacs of a healthy lung compared to a lung diseased by air pollution. Note how air sacs break down in a diseased lung.*

In a severe case of emphysema, the cells throughout the body suffer from oxygen starvation. To counteract this, the victim's breathing accelerates in a vain attempt to aerate the blood properly. The heart speeds up to propel the blood more rapidly. Because the surface area for gaseous exchange is reduced, carbon dioxide levels in the blood remain higher than in healthy people. The skin of some sufferers turns slightly bluish (cyanosis) as a result.

Emphysema is most common in smokers. In fact, the rate of this disease is 13 times greater in smokers than in nonsmokers. However, studies suggest that air pollution also contributes to emphysema. The death toll from emphysema in the United States rose dramatically from a mere 1,500 in 1950 to more than 15,000 in 1990—a ten-fold increase. This mortality increase coincides with increased levels of atmospheric pollution in the areas where the fatalities occurred. Death rates have leveled off in the past decade. It is currently nearly 18,000 per year.

Lung Cancer Several air pollutants have been strongly implicated as **carcinogens**—cancer-producing agents. Among them are benzopyrene, asbestos, nickel, and beryllium. The hydrocarbon benzopyrene can get into the lungs from several sources: from the coal smoke issuing from industrial smokestacks or from the smoke issuing from cigarettes. The smoke generated when fat drippings from a charcoal-broiled steak spatter on the hot coals is also a little-suspected source of benzopyrene. Just one smoked steak may contain as much benzopyrene as 600 cigarettes!

Researchers have shown in laboratory experiments on rats and mice that, although two air pollutants may not induce cancer independently, cancer is induced when an animal is exposed to both pollutants simultaneously. Such an effect is called **synergistic.** For example, lung cancers resembling those in humans have been induced in laboratory animals by first exposing the animals to influenza virus and then to artificial smog. Tumors have also been generated in laboratory animals by forcing them to inhale a combination of benzopyrene and sulfur dioxide.

Researchers have found that the lung cancer rate of men over 45 living on Staten Island, New York, is 155 per 100,000 in the smoggiest area, compared to only 40 per 100,000 in the less smoggy region. To research biologists, such data strongly indicate a cause-and-effect relationship between atmospheric pollution and lung cancer in humans. Studies of other cities, however, do not show a clear relationship between urban living and lung cancer. The conclusion from this is that it is the

CASE STUDY 18.2 ASBESTOS: THE DANGERS OF A USEFUL PRODUCT

Asbestos is a naturally occurring mineral that has found extensive industrial use, especially since World War II. Roughly 30 million metric tons were used in the United States during the period 1900–1990. Asbestos has been used in an estimated 3,000 to 5,000 products. It's been used as pipe and boiler insulation and ironing board pads. It's been used to manufacture brake linings, protective clothing for firefighters, and talcum powder. In the construction industry, it is used to strengthen cement and plastics and to fireproof schools and skyscrapers (Figure 1). It's widespread use was made possible by the fiber's flexibility, great tensile strength, and resistance to heat, friction, and acid.

Dispersal Through the Environment

Wherever asbestos is mined or processed, or wherever asbestos products undergo wear, extremely minute fibers, asbestos dust, are released into the atmosphere. For example, a woman sets her iron down on her ironing board pad, inadvertently sending asbestos dust into the air. The driver of an older model automobile puts his brakes on, sending asbestos dust into the air. What happens to the material worn from brake linings? It's still around, in pulverized form, some of it as asbestos dust, possibly floating in the air, possibly forming a thin film on roads and highways, or possibly adhering to the soft, delicate lining of human lungs. Even food, water, and beverages may contain some asbestos fibers. In Rockville, Maryland, large areas of the city became contaminated by asbestos-containing crushed rock applied to school playgrounds and city streets. One wonders how many millions of asbestos fibers were eventually inhaled by schoolchildren during recess or by motorists driving by with their windows open.

FIGURE 1 *Abestos fibers. They have been widely used in industry because of their flexibility, great tensile strength, and resistance to heat. However, when inhaled they may cause serious illness.*

Some people claim that asbestos "time bombs" may be ticking away in thousands of schoolrooms throughout America. The reason? Starting in the 1940s, asbestos was mixed with paint and sprayed on ceilings and walls for fireproofing and sound insulation. Asbestos was also used to insulate pipes, boilers, and structural beams in schools. This practice continued into the early 1970s. (Such uses were banned by the EPA in 1973.) Dr. Lyman Condie, a toxicologist with the EPA, comments on the problem: "When sprayed surfaces are exposed to student activities—bouncing basketballs off gymnasium ceilings, or children running their hands along stairway ceilings—the asbestos can flake off into the air. Because the fibers are very small and light, they can move throughout the building, even though only a very small area was originally disturbed." Of our nation's 87,000 school buildings, at least 30,000 contain asbestos in their walls and ceilings. In the worse cases, levels of airborne asbestos inside these buildings materials may be 100 times greater than ambient levels (levels you'd see in the air we breathe) but are three to four times lower than historic workplace levels associated with the well-documented occurrence of asbestos-related disease.

Human Illness Caused by Asbestos

Exposure to asbestos causes a disease known as asbestosis, lung cancer, and mesothelioma (another form of cancer). Currently 65,000 Americans suffer from **asbestosis,** the symptoms of which include breathlessness, coughing, chest pains, barrel-shaped chest, club-shaped fingers, and bluish discoloration of the skin. These symptoms may not appear until 20 to 30 years after the exposure to asbestos. Over 50 percent of the people suffering from asbestosis eventually die from lung cancer.

Health experts estimate that 3,000 to 12,000 asbestos-induced cancer deaths occur in the United States annually. Asbestos poses the greatest threat to cigarette smokers. In fact, a cigarette-smoking asbestos worker has 90 times the chance of developing lung cancer as a nonsmoker who has no contact with asbestos! As might be expected, surveys have revealed a relatively high incidence of asbestos-related lung cancers among the 120,000 people (miners, asbestos product processors, and so on) in the United States who worked directly with asbestos.

Mesothelioma (pronounced mez-oh-theel-ee-OME-ah) is a cancer of the chest cavity lining, Although this disease was formerly quite rare, it has become much more common in the United States, especially among asbestos workers. Many of these cancer-stricken people or their families have sued the companies where their occupational exposure to the asbestos occurred. In Virginia alone, for example, asbestos product manufacturers were sued by 100 people for $300 million during a single year.

At this writing, it is still not clear whether children exposed to asbestos in school face any increase in health risk. Only time and further research will tell.

Control of Asbestos Emissions

In the 1970s, the EPA launched an aggressive program to remove asbestos from the walls and ceilings of schools, convention halls, theaters, and other public buildings. However, as of 1994,

(continued)

discouragingly little progress has been made. Part of the problem is monetary. It would cost at least $2 billion for a comprehensive, nationwide cleanup in schools alone. Improper removal can pose a hazard to workers and can contaminate the building even more. Accordingly, many experts have argued in favor of stabilizing asbestos, rather than removing it. Exposed asbestos can be kept from flaking off by applying paint or some other type of sealant.

Sites where old buildings are being razed are also significant sources of asbestos emissions. Moreover, they far outnumber such sources as factories. The problem is complicated by the fact that demolition contractors frequently ignore federal regulations on asbestos removal. Federal law requires that all asbestos fibers be wetted down before removal. This material then must be placed in leakproof containers, such as plastic bags, and conspicuously marked.

In 1986, the U.S. Congress passed the **Asbestos Hazard Emergency Response Act,** which is part of the Toxic Substance Control Act. This new law required schools to inspect their buildings for asbestos-containing building materials and then required officials to prepare management plans that recommend the best way to reduce the asbestos hazard, including containment. Options include repairing damaged asbestos, spraying it with sealants, enclosing it, removing it, or keeping it in good condition so that it does not release fibers. Schools were also required to notify parents of any problems as well as actions they were taking. Many schools immediately began removing asbestos because of parents' concerns. In the hands of experts, asbestos is carefully removed, sealed, then trucked off to an EPA-approved landfill. Unfortunately, the cost of removal and dangers from improper removal caused many school districts to find alternatives to removal, especially stabilization. There are no laws requiring asbestos to be removed from schools. It is only when the asbestos cannot be maintained in good condition or when measurements show unacceptable levels in the air or when demolition of a building is imminent, that it must be removed.

The asbestos industry, under EPA regulations, has also taken steps to reduce occupational exposure to the life-threatening fibers. The asbestos concentration in the air inhaled by the workers must be less than one fiber per cubic meter of air. The problem has been further diminished by the use of vacuum devices, by the mandatory use of masks, and by automating many manufacturing processes.

In addition, since the early 1970s, asbestos has been eliminated from many products, including drum brake linings, thanks to a ban imposed by the EPA.

According to the EPA, perhaps the most disturbing feature of the asbestos problem is that control measures have not prevented the gradual long-term development of cancer in persons who come into occasional, slight, or temporary contact with asbestos. Families of asbestos workers, for example, who inhaled asbestos fiber dust inadvertently brought into the home on the workers' clothing and shoes have an increased risk of disease. Such slightly exposed people also include those who live within 1.5 kilometers (0.93 mile) of an asbestos plant. Of the almost 2,000 autopsies on such people in New York City, one half revealed asbestos fibers in their lungs. (If you're interested in learning more, see our Web site.)

type of air pollution in urban air that is important in determining whether a city's air is carcinogenic.

Asbestosis Asbestosis, a disease caused by accumulation of asbestos fibers in the respiratory tract, is discussed in Case Study 18.2.

Carbon Monoxide Poisoning Carbon monoxide combines 210 times more readily with the hemoglobin in red blood cells than does oxygen. It therefore tends to replace oxygen in the bloodstream. At high doses, carbon monoxide is lethal. It is responsible for many fatalities—both intentional and unintentional—each year. Exposure to lower levels, say 80 ppm of carbon monoxide for 8 hours, as might be encountered in a tunnel or a line at a toll booth, causes cellular oxygen starvation equivalent to losing one pint of blood. This can cause headaches and other symptoms.

Carbon monoxide is produced by the combustion of all fossil fuels, especially if the combustion device is inefficient. Cars are a major source of carbon monoxide, as are power plants. Furnaces, water heaters, wood stoves, fireplaces, and gas cooking stoves are common sources in our homes. CO is also produced by the combustion of tobacco in cigarettes, pipes, and cigars. Carbon monoxide from tobacco smoke affects the smoker and those nonsmokers around him or her. Researchers have shown that cigarette smoke may contain 300 ppm of carbon monoxide. Carbon monoxide levels in a room of smokers can reach high levels, sufficient to deactivate roughly 10 percent of the victimized nonsmoker's hemoglobin.

The presence of carbon monoxide in the bloodstream of a pregnant woman has been suggested as a possible cause of stillbirths and deformed offspring. Certain conditions may render some people especially susceptible to carbon monoxide poisoning. They include heart disease, asthma, diseased lungs, high altitude, and high humidity.

Carbon monoxide may be the indirect cause of many fatal traffic accidents in the United States yearly. The effects of low-level carbon monoxide poisoning are similar to those of alcohol or fatigue; they impair the motorist's ability to control the vehicle (Figure 18.12). Because carbon monoxide is colorless and odorless, harmful levels may build up within the car without the driver being aware of them.

The Effects of Air Pollution on Other Organisms and Materials

Because it is so ubiquitous and damaging, air pollution affects many different species as well as many materials, even marble statues. Pollutants damage a wide assortment of plants from trees in the city to forests to farm crops, sometimes hundreds of miles downwind from major urban centers. The four main culprits are ozone, sulfur dioxides, sulfuric acid, and nitric acid. Studies of farms in California and along the East Coast of the United States show that ozone damages vegetable crops. Sulfur dioxide damages plants directly as well. Sulfuric and nitric acid damage plants either directly or through changes in the soil.

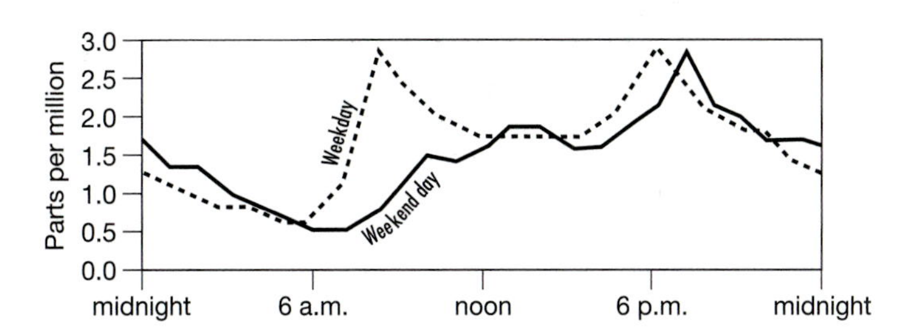

FIGURE 18.12 *Carbon monoxide levels on weekdays and weekends in Washington, D.C. Notice the difference in patterns due to the lack of early morning commuting during the weekend.*

Air pollutants damage metals, stone, concrete, clothing, rubber, and plastics. The four main culprits—once again—are sulfur dioxide, sulfuric acid, ozone, and nitric acid. Pollutants in cities often reach extraordinary levels and can cause considerable damage to buildings and valuable statuary. The Taj Mahal in India and the Statue of Liberty are two of the many victims of urban pollution. In fact, damage to the Statue of Liberty from sulfur dioxide and acids was so extensive that it recently required extensive repair, costing $35 million. Damage to crops and materials is estimated to be nearly $10 billion per year in the United States alone.

18.6 Air Pollution Abatement and Control

Because of its many impacts on people, plants, animals, ecosystems, and materials, air pollution has long been under attack. This section describes efforts to reduce and eliminate air pollution.

Pollution Control in Factories and Power Plants

Pollution can be reduced or eliminated by either pollution controls (devices that remove pollutants from smokestacks and other sources) or by preventive actions. This section will examine both strategies, beginning with the more traditional and often more costly pollution control methods. Let's begin by examining the control of particulates.

Controlling Particulates Three standard types of equipment for controlling particulates at **stationary sources,** such as power plants and factors, are available. The fabric **filter bag house** physically removes airborne particles from smokestacks via huge cloth bags suspended inside the device (Figure 18.13). A large filter bag house may consist of more than 1,000 elongated filter bags, each of which is several meters long. Up to 99.9 percent of the dust particles may be removed from the stack gases.

Another means of removing particulates is the **electrostatic precipitator.** This device is designed to remove small diameter solid particles (dust, fly ash, asbestos fibers, and lead salts) less than 1 micron (one-millionth of a meter) in diameter. They're most commonly used in coal-fired power plants to remove particulates from gases in the smokestack (Figure 18.14). In the electrostatic precipitator, pollutants pass between pairs of positively and negatively charged electrodes. The particles become negatively charged as they flow through the device, and are then attracted to a positively charged collector electrode in the wall. From time to time, the device is switched off, so particulates that have accumulated on the wall can drop to the bottom and be removed. Although the initial cost of a large precipitator can be more than $1 million, the power and maintenance costs are small. One unfortunate feature, however, is that the precipitator is more effective when high-sulfur rather than low-sulfur coal is burned.

The **cyclone filter** removes heavy dust particles with the aid of gravity and a downward-spiraling air stream (Figure 18.15). Particulate-laden air enters the device and travels through it. As it does, the particles strike the walls and fall to the bottom of the cylinder, where they can be removed.

Controlling Sulfer Oxide Emission The release of sulfur oxides into the atmosphere from transportation, industrial, and residential sources adversely affects human health, wildlife, forests, and farm crops, as well as irreplaceable paintings, monuments, stonework, and statuary.

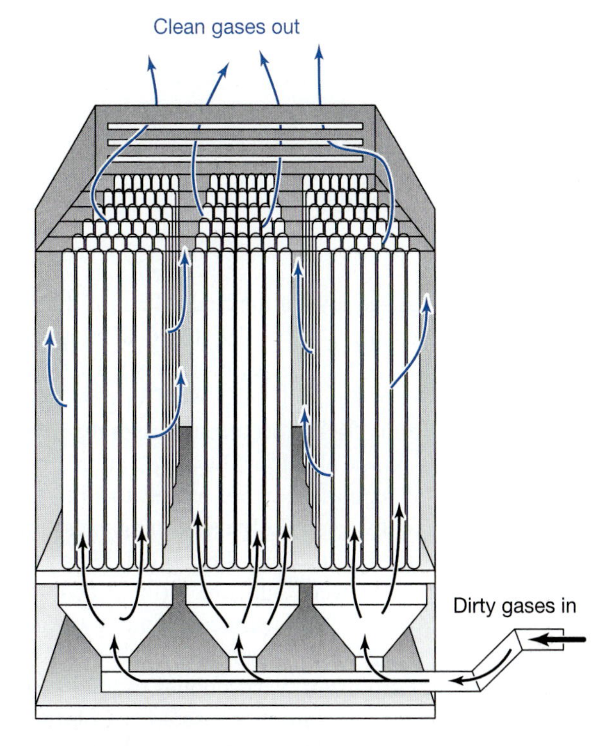

FIGURE 18.13 *Filter bag house. Solid particles are removed from exhaust gases by long vacuum cleaner-type bags that pick up particulates as air is passed through them.*

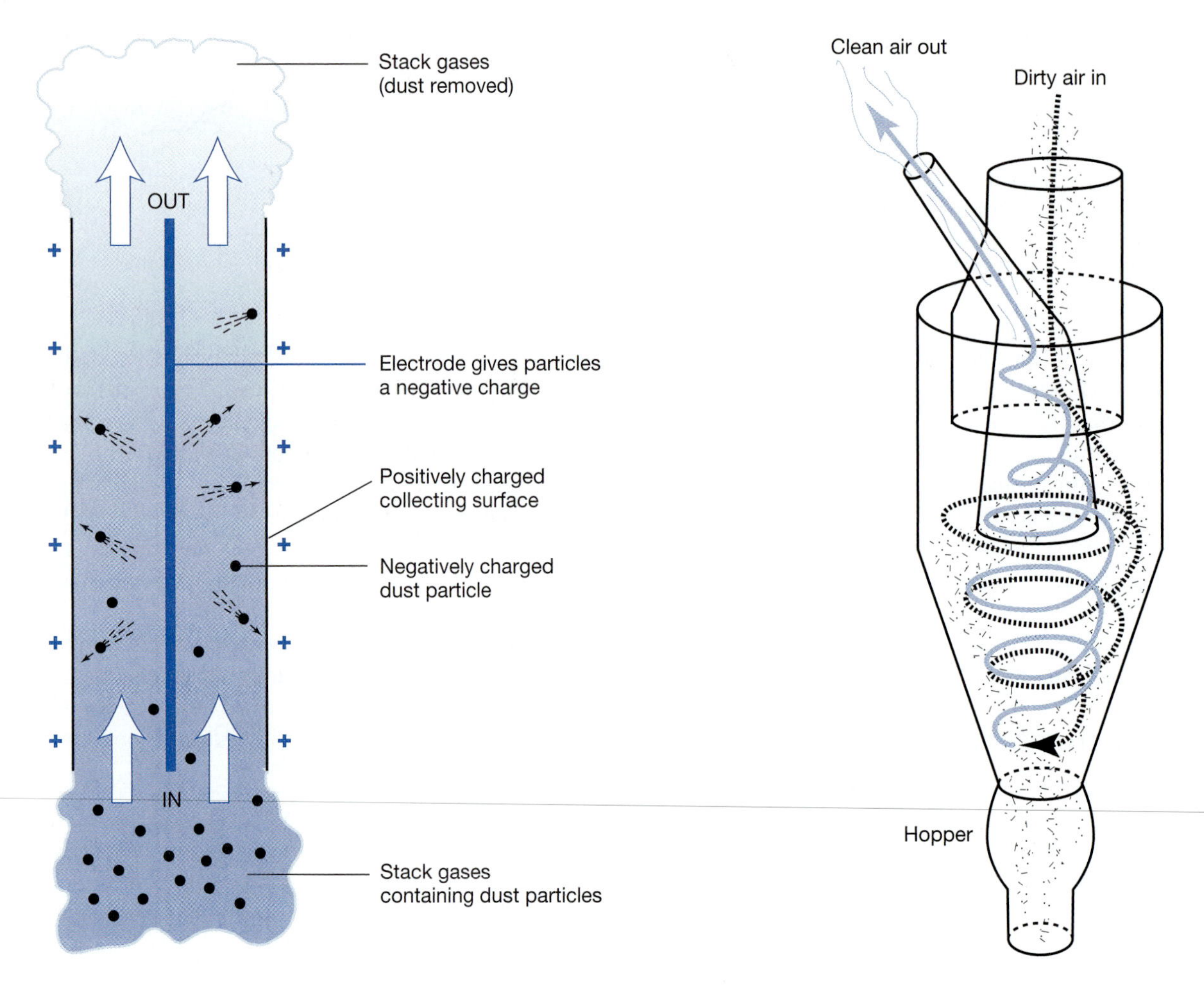

FIGURE 18.14 *Electrostatic precipitator. As the soot, dust, and other particulates pass through the precipitator, they are given a negative charge. They are then attracted to the positively charged wall of the precipitator. After accumulating on this collecting surface they are periodically released to a collecting chamber.*

FIGURE 18.15 *Cyclone filter. Large solid particles are removed by centrifugal force and are collected in a hopper.*

Sulfur oxide emissions can be reduced in several ways. Although no method will be satisfactory by itself, if the following approaches are used in combination, sulfur oxide emissions can be greatly reduced.

1. Shifting from high- to low-sulfur coal. Much of the industrial coal consumed before 1970 had a relatively high sulfur content (up to 3 percent or more). This coal came from mines in Pennsylvania, West Virginia, and Illinois. Once health officials become aware of the problems posed by oxides of sulfur, however, municipal, state, and federal regulations were passed to limit the burning of high-sulfur coal. Still, many companies continued to burn high- and medium-sulfur coal. However, when U.S. regulations were tightened further to control acid rain in the early 1990s, many companies switched to low-sulfur coal to avoid having to install pollution control devices. Fortunately, vast supplies of low-sulfur coal occur in many western states, such as Colorado, Montana, and Wyoming. These huge deposits can be removed by **strip mining,** a process in which the rock and soil overlying deposits of coal are removed in progressive strips. Strip mining can be very destructive to the environment, and thus precautions must be taken to restore the sites to their original condition (see Chapter 20 for a discussion).
2. Coal cleaning. Much of the sulfur can be removed from high-sulfur coal before it is burned. Transporting low-sulfur coal a distance of more than 1,600 kilometers (1,000 miles) to eastern industrial plants would add greatly to the cost of the coal. An alternative is to mine the high-sulfur coal available in the Midwest and East, where it is close to the industry that will use it, and then remove much of the sulfur before burning it. Much of the sulfur in high-sulfur coal comes from an impurity known as iron pyrite (fool's gold). Because iron pyrite is heavier than the coal, it sinks to the bottom of a tank containing a mixture of pulverized coal and water and can easily be removed (Figure 18.16).
3. Removing sulfur from smoke stack gases. Flue gas desulfurization process, more commonly known as **scrubbing,** is widely used in the United States and other more developed countries to remove sulfur from smokestack gases of power plants and factories (Figure 18.17). A scrubber is a device through which smokestack gases pass. In the scrubber, a mist consisting of ground limestone and water

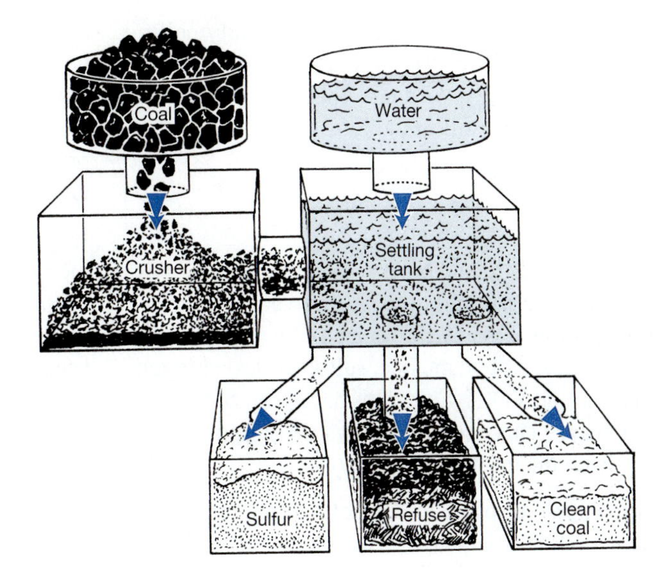

FIGURE 18.16 *Coal cleaning process.*

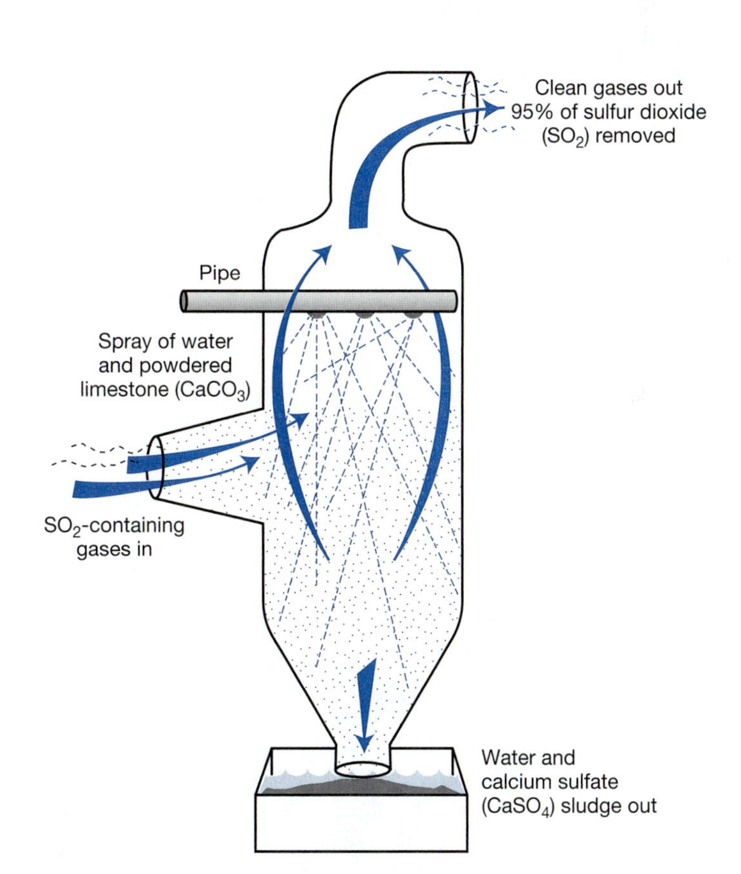

FIGURE 18.17 *Flue gas desulfurization. Downward-streaming water and powdered limestone ($CaCO_3$) "scrub out" sulfur dioxide from stack gases. A sludge of calcium sulfate forms and must be removed.*

is sprayed into the sulfur oxide—laden stack gases. The calcium in the lime reacts with the sulfur to form calcium sulfate.

Scrubbers can remove up to 95 percent of the sulfur dioxide in stack gases. The calcium sulfate sludge may be used in road beds or drywall production, but is most commonly disposed of in landfills. A big advantage of scrubbers is that they can be added to existing plants, as well as being used on new ones. The process is costly, however, and solves an air pollution problem by creating a solid waste problem.

Although pollution control devices work, they typically result in other problems. Scrubbers, for example, produce toxic sludge. Electrostatic precipitators remove particulates from smoke stack gases, but contain an assortment of toxic heavy metals. (Smaller units can also be purchased for in-home use.) More and more, policy makers and businesses themselves are seeking alternatives to these types of controls, notably pollution prevention measures. Why not, they say, avoid pollution altogether? If you can achieve the same goal without making pollution you're much better off in the long run.

The Control of Automotive Emissions

Automobiles, trucks, busses, and other types of motor vehicles are major sources of air pollution the world over. These **mobile sources** release an assortment of air pollutants, among them carbon monoxide, carbon dioxide, sulfur oxides, nitrogen oxides, particulates, and VOCs. Emissions have been controlled in a variety of ways. This section surveys these approaches with an emphasis on the most sustainable ones.

Catalytic Converters

In order to reduce the pollutant emissions from automobiles, as required by the Clean Air Act, U.S. auto manufacturers have relied heavily on a device known as the catalytic converter (Figure 18.18). The **catalytic converter** is a muffler-shaped device that is incorporated into the exhaust system of the car. One type of converter, used by the Ford Motor Company, has a honeycomb-like interior. The cells of the honeycomb are coated with a **catalyst,** a material that speeds up chemical reactions. The one used in automobile catalytic converters is platinum or palladium. The products of incompletely burned gasoline such as carbon monoxide and VOCs pass through the cells of the catalytic converter. In the catalyst, these gases react with oxygen and are converted into carbon dioxide and water.

One of the drawbacks of the catalytic converter used in vehicles today is that they don't remove nitrogen oxides. (That's one reason why it has been so hard to control nitrogen oxide emissions in most countries, and why this pollutant is increasing worldwide.) A three-way converter has been developed to reduce nitrogen oxides in automobile exhaust, as well as carbon monoxide and hydrocarbons. In this converter, nitric oxide oxidizes carbon monoxide to carbon dioxide and converts hydrocarbons to carbon dioxide and water. During this process, the nitric oxide is broken down to nitrogen and oxygen. Unfortunately, these devices are not in use.

Inspection and Maintenance Programs

Poorly tuned automobiles burn gasoline less efficiently than their well-tuned counterparts and thus produce more pollution. But how does one monitor how well millions of individual automobile owners are keeping their vehicles in tune? The 1977 amendments to the U.S. Clean Air Act requires those regions of the United States in which air quality standards for carbon monoxide and ozone were not met by 1982 to set up **inspection and maintenance (IM) programs.** Under such programs, automobile emissions are checked each year by state-licensed inspectors. When levels of pollutants in the exhaust, such as carbon monoxide and hydrocarbons, are too high, the owner must have the engine adjusted or repaired to meet emissions requirements, which usually requires only minor adjustments to the carburetor, a device that mixes fuel and air before it enters the cylinders where combustion takes place. The most common repairs, such as spark plug replacements and carburetor adjustments, may reduce pollutant emissions by 25 percent. In Portland, Oregon, an IM program resulted in a 40 percent reduction in emissions the very first year. The IM program in California, known as Smog Check, has cut smog-forming nitrogen oxides and VOCs by 18 and 90 metric tons per day, respectively, and emissions of carbon monoxide by 1,350 metric tons daily for the state as a whole.

Increasing Fuel Efficiency

Tuning automobiles makes them more efficient and cleaner-burning. But engines can also be redesigned to make them even more efficient and less polluting.

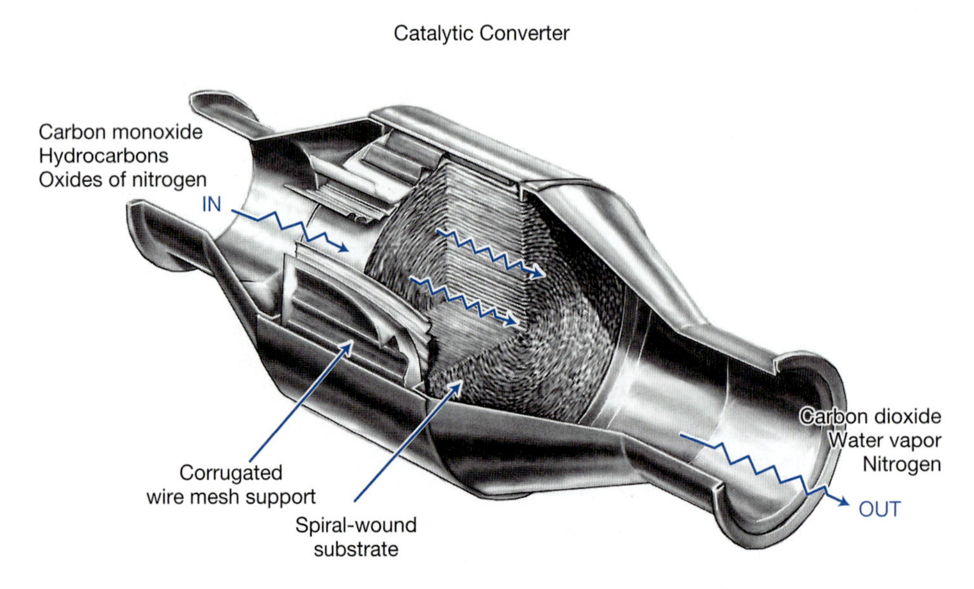

FIGURE 18.18 *Cutaway illustration of a catalytic converter used by Ford Motor Company. Resembling a small muffler, the emission-control device converts hydrocarbons, carbon monoxide, and oxides of nitrogen into nitrogen, carbon dioxide, and water. The chemical reaction depends on the platinum and rhodium catalysts that coat the internal surfaces of the converter.*

In 1975, a U.S. federal law placed a federal excise tax on all cars of a particular model that did not attain a minimum standard of efficiency set by the EPA. It was commonly referred to as the gas-guzzler tax. The size of the tax was inversely related to the car's fuel efficiency. For example, a few years ago, the biggest "gas guzzler" on the American market was an Italian import, the Lamborghini Countach, which achieved only 2.5 kilometers per liter (6 mpg) in the city—the lowest fuel efficiency ever recorded by the EPA. The tax on this car was a hefty $3,850. The gas-guzzler tax was designed to discourage consumers from buying energy-inefficient, pollution-producing vehicles. Unfortunately, President Clinton repealed the gas-guzzler tax in 1996.

Another attempt to encourage the manufacturing and sale of energy-efficient vehicles in the United States is the **CAFE standards—Corporate Average Fuel Efficiency standards.** These standards, established by the United States, set average efficiency goals for new autos sold in the United States, including imports and domestically manufactured vehicles. In 1994, the minimum efficiency required in models was a city—country average mileage of 11.5 kilometers per liter (27.5 mpg). Unfortunately, heavy lobbying by auto manufacturers has stalled any improvement in fuel efficiency. Fuel efficiency in many new cars has dropped steadily since that time (Figure 18.19). In 1999, light vehicles (passenger cars, sport utility vehicles, vans, pickup trucks) averaged 23.8 miles per gallon.

Some experts believe that in the not too distant future, a fuel efficiency of 33 kilometers per liter (80 mpg) will be attained by American cars. The Japanese already have a model that seats five and gets 33 kilometers per liter (80 mpg) on the highway. Honda's VX series, now discontinued, achieved gas mileage of 56 mpg. Chevy's Metro, formerly the Geo Metro, boasted similar mileage.

Today, however, automobile manufacturers are producing larger, more energy-consumptive cars. Sport Utility vehicles, vans, and light trucks, all which achieve gas mileage in the mid 20s, account for half of all new car sales. Ford and other auto manufactures have also begun selling supersize RVs like the Ford Excursion, which gets only 8 miles per gallon. Ford just announced the production of a new 10-cylinder vehicle.

Managing and Reducing Traffic Volume City officials, working with university researchers and departments of transportation, are working on ways to better manage traffic to prevent or reduce congestion. The goal behind this effort is to make traffic flow more evenly, that is, to prevent grid lock. Idling cars produce lots of pollution.

Computer systems that monitor traffic and flash messages to motorists are currently being studied. It is hoped that they'll alert drivers of potential jams, allowing them to take alternative routes. On-ramp controls are also being used in many cities to produce a more seemless integration of oncoming traffic with existing traffic flows. Traffic lights on the on ramp stop cars, then let them proceed two at a time at spaced intervals.

All of these measures, from more efficient cars to traffic control, are effective, but a far more effective means to urban air pollution is to reduce automobile traffic volume on highways (Figure 18.20). The EPA has made several proposals to reduce traffic volume in large urban areas. Among them are (1) terminating free parking facilities for employees by their employers; (2) placing an added tax on downtown parking fees; and (3) prohibiting commuters from driving to work one day per week, thus reducing commuter traffic by 20 percent. All of these could help promote alternative means of transportation such as buses, commuter trains, walking, carpooling, vanpooling, and bicycling.

The overall pollution-reducing strategy in New York City includes (1) banning all street parking in the main business section of Manhattan; (2) restricting cruising by Manhattan taxis; (3) placing several Manhattan bridges on the toll system;

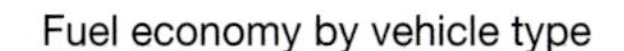

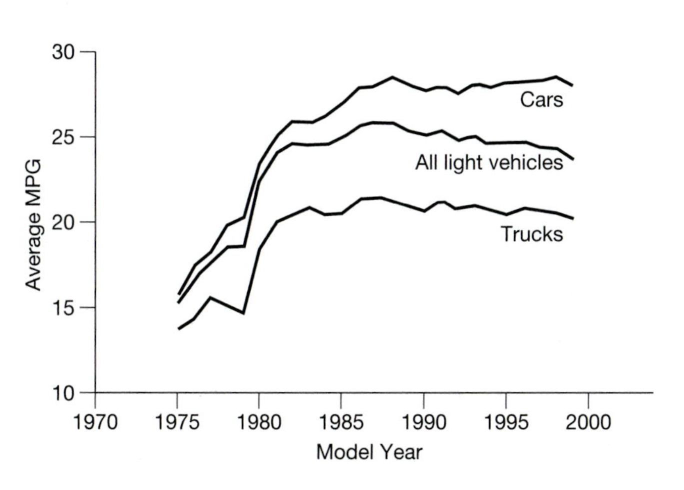

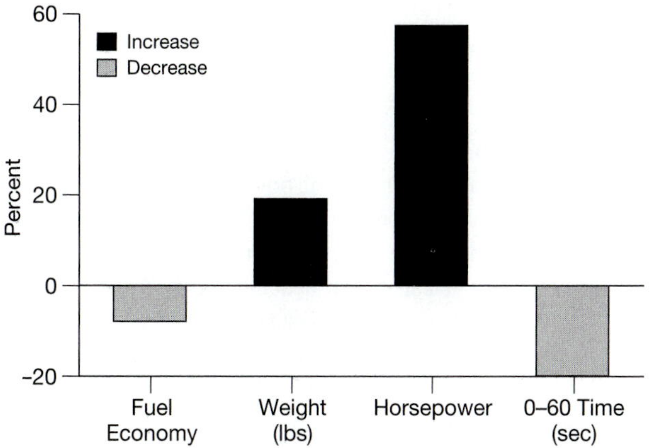

FIGURE 18.19 *Declining gas mileage. Fuel efficiency in passenger cars has remained fairly constant over the past decade, as shown in the graph on the left, but overall fuel efficiency in light vehicles (cars, vans, sport utility vehicles, and trucks) has decreased substantially as they have gotten heavier and more powerful. Note that acceleration of light vehicles has declined.*

FIGURE 18.20 *Traffic comes to a standstill in San Francisco. Like many cities, San Francisco faces a traffic crisis. Cars back up on the freeway during rush hour, making the commute a major ordeal and creating large amounts of air pollution.*

FIGURE 18.21 *In Manhattan and some other major urban centers, special lanes on busy thoroughfares are reserved for the exclusive use of express buses. This strategy has proved effective in reducing commuter traffic volume and air pollution.*

and (4) setting aside one lane on busy roads for the exclusive use of buses (Figure 18.21).

Replacing Gasoline With Ethanol Alternative fuels can also reduce pollution. **Ethanol,** the same chemical found in alcoholic beverages, for example, is capable of powering motor vehicles. Ethanol can be produced from a renewable source, too—from fermented grains such as corn or from sugar cane. Ethanol can be added to gasoline, or it can be burned in pure form, although some engine modifications are necessary for this. Ethanol not only comes from a renewable source, unlike gasoline, it also pollutes much less than gasoline, producing less carbon monoxide, hydrocarbons, and sulfur oxides. Ethanol produces no net carbon dioxide because all carbon dioxide it produces would be incorporated into plants to create more fuel for future generations. Therefore, ethanol could help reduce the buildup of carbon dioxide in the atmosphere, which evidence suggests could be causing a global warming trend (Chapter 19).

Ethanol is used widely by motorists in Brazil today. Although ethanol is currently more expensive in the United States than gasoline, researchers have developed less expensive means of producing it. If progress continues, it could become cost competitive in the near future. (For more on ethanol, see Chapter 22.) As fossil fuel reserves decline over the next decade or two, ethanol could become more attractive.

In 1994 the EPA ordered that ethanol must be mixed with any gasoline used in the nation's nine smoggiest cities—Baltimore, Chicago, Houston, Los Angeles, Milwaukee, New York, Philadelphia, San Francisco, and Hartford (CT). This fuel change, which started in 1995, is expected to reduce smog in these cities by 16 percent. California has instituted a similar program that began in 1996. Its program, which requires a greater percentage of ethanol in the fuel mix, is expected to cut California's smog by 40 percent.

Developing Alternatives to the Internal Combustion Engine Since 1970, when the Clean Air Act was passed, emissions of hydrocarbons and carbon monoxide by the average car have been reduced by 90 percent and emissions of oxides of nitrogen by 75 percent. Recently, however, federal analysts have been increasingly skeptical that further substantial overall reductions could be made, even with the use of the emission control strategies described earlier. The reason? The rapid increase in the number of motor vehicles on our highways. The number of motor vehicles has increased from 156 million in 1980 to 208 million in1999. Some people think that what is really needed is a radical shift from the traditional internal combustion engine to an engine powered by solar electricity or even steam.

Electric Cars In Europe, Japan, and the United States, 6,000 electric cars are produced each year. Several American cities, including San Francisco and New York, are making limited use of electric buses and delivery vans. Some experimental cars recently developed have a top speed of 128 kilometers (80 miles) per hour. However, although the electric car itself is relatively pollution free, electricity has to be generated to charge the batteries. Most electricity is generated by centrally located power plants, most of which burn coal. Thus, instead of having numerous, widely dispersed mobile sources of pollution (automobiles), the pollution problem is merely transferred to a few large stationary sources (power plants). It turns out that compared to a gas-burning automobile, the electric car isn't really that much leaner, if one takes into account the combustion of fossil fuel to generate electricity. The advantage of the electric car, then, is that emissions are restricted to a single source. If located outside the urban airshed, they can effectively reduce pollution in and around cities. Another big disadvantage of the electric car is that the batteries required to power the car are rather bulky. Electric cars are usually good for distances of up to 100 miles before needing recharging, so they are good only for commuting.

Electricity generated from clean, renewable sources such as wind generators and special solar electric cells (photovoltaics) could make the electric car a viable alternative to gas-powered vehicles. For more on the electric car, see Case Study 18.3.

Hybrid Cars One strategy that would eliminate some of the disadvantages of the electric cars now on the road would be to develop hybrid cars—automobiles equipped with both an electric engine and an internal combustion engine. At least two options have been explored. First is a hybrid that would rely on its electric motor to power it on short trips through urban areas—for example, on a shopping trip or a visit to the theater. The motor power for longer highway trips would be provided by an alternative source, a small conventional internal combustion engine. The gasoline engine would also kick in when additional power is needed, for example, when accelerating to pass another car.

Another perhaps more promising hybrid consists of a very small gasoline-powered engine that produces electricity that is transmitted to four small motors associated with each wheel. Amory Lovins, an internationally renowned expert on energy who has helped design such a car, believes that a fuel efficiency of 150 to 200 miles per gallon could be achieved by this type of hybrid.

Despite their general lack of concern for making and selling smaller cars—there's more profit, a lot more profit, in big cars and Americans seem to have an insatiable desire for such vehicles—the automobile manufacturers are making some headway. Two leaders are Honda and Toyota. Both have produced a **hybrid car** (Figure 18.22). These cars have small gasoline engines and electric motors. The engine and the motors work at different times. When cruising in the city at low speeds, the electric motors run. When you need to speed up or travel more rapidly, the gas engine starts up. Unlike electric cars, discussed in case study 18.2, the hybrid car doesn't require huge battery banks. It has batteries but they are recharged when the car is running on gasoline or when it is slowing down. When the car

FIGURE 18.22 *Two of the newest entries into the automobile fleet, both hybrid cars. (A) Toyota Prius, and (B) Honda Insight.*

decelerates, the electric motor produces electricity that helps to keep the batteries full.

Hybrid cars are an important step into a world of more efficient vehicles, although they do not improve gas mileage much. Honda's two-seater, the Insight, gets about 70 miles per gallon. Toyota's four-seater, the Prius, gets 52 mpg. On the open highway, when driven at 65 miles per hour, gas mileage falls to 46 mpg.

Hybrid cars can do better, says Amory Lovins. But they'll need to be made out of space age materials that are light but durable. Current hybrids are no more than conventionally made cars with steel bodies and frames with a hybrid power drive system. Unless radical changes are made in the construction of the hybrid, they won't achieve the mileage needed to make them a real contender for the car of the future.

Electric Cars Powered by Fuel Cells Most analysts believe that the hybrid car will be a transitional vehicle. The car of the future will be electric cars powered by fuel cells. **Fuel cells** look like batteries, but function quite differently. They take in oxygen and hydrogen, and produce electricity. The electricity is then used to run the car. Where will the hydrogen come from?

Hydrogen can be generated from water, for example, in a process called electrolysis. **Electrolysis** is the breakdown of water into oxygen and hydrogen that take place when an electrical current is passed through it. Electricity to make hydrogen can be generated from wind and solar sources, as discussed in Chapter 22. Hydrogen can then be stored and used to power the fuel cell to make electricity to run the electric motors of the car.

Hydrogen storage, you'll learn in Chapter 22, is very difficult, so car manufacturers are currently developing other sources. The main choice right now is gasoline. Car manufacturers are experimenting with a device known as a reformer, which strips the hydrogen from the hydrocarbon molecules in

CASE STUDY 18.3

GETTING CHARGED UP OVER ELECTRIC CARS

Electric cars whizzing down America's streets? It may be not too uncommon a sight even by the year 2010, especially on the Pacific coast and in the Northeast. In fact, in Europe, Japan, and the United States more than 6,000 electric vehicles are already being produced each year. Several American cities, among them San Francisco and New York, are making limited use of electric buses and delivery vans (Figure 1).

In 1993 California legislators passed a law mandating that, as of 1998, 2 percent of all cars a manufacturer sells in the state must be absolutely free of toxic emissions. They're called **zero emission vehicles.** Since that time, they've increased their goals, requiring 10 percent of all cars sold in California by 2003 to be zero or low emissions—that is about 170,000 cars. By 2010, they hope to have 800,000 ZEVs or LEVs on the road. At this writing (September, 2000) there are about 2,300 electric vehicles on the road in California as part of a demonstration fleet of ZEVs the automakers were required to produce prior to 2003. Unfortunately, automakers satisfied their demonstration fleet requirements and then stopped making ZEVs. They claim there isn't enough interest in them.

California, however, continues its pursuit and has developed partnerships with auto manufacturers to promote the development of several types of low emissions vehicles, including fuel cell powered cars and hybrid cars discussed in the text. They even offer a special ev (electric vehicle) loan program for state and local government agencies who want to lease electric vehicles.

Several auto manufacturers produce electric vehicles. In late 1993 Ford unveiled an EV prototype, dubbed the Ecostar. This modified minivan has a payload of 1,000 pounds and can carry two passengers. Its sodium-sulfur battery provides a range of 130 miles and can be recharged within six hours.

One problem with the Ecostar is its battery. It powers the car for only half the distance provided by a full tank of gas in a conventional vehicle. The battery has only a two-year life span. The biggest drawback to the Ecostar, however, is its price tag—it costs a lot more than a conventional car and offers much less, except of course in the environmental area.

Today, there are at least 18 electric vehicles being manufactured, ranging from sedans to pick ups to vans. But will motorists open their wallets and actually buy a General Motors EV1 or a Hundai Accent or a Toyota RAV 4EV? (For a list of Evs currently being sold see our Web site.) As you learned in the text, numerous auto manufacturers are also selling hybrid cars and are working on fuel cell powered electric vehicles. Paul A. Eisenstein, writing in the *Christian Science Monitor,* stated that the Big Three of the auto industry may have to spend millions of incentive dollars to persuade Californians—and other Americans—to buy a car they would rather do without.

It just may be that the electric car of America's future won't be powered by a chemical battery like the sodium-sulfur battery of the electric vehicles now making their way onto the world's highways. In 1994, the American Flywheel Systems (AFS) of Bellevue, Washington, announced the development of a radically different power storage medium—the flywheel battery. Flywheel batteries, according to their manufacturer, contain disks that rotate at high speed to store energy. Energy put in to them, causes the flywheel to turn. Developers claim flywheels can outperform chemical batteries in many ways. Flywheels currently being developed will be light weight, compact, and able to recover energy from a car when brakes are applied. Interestingly, flywheel batteries have been used in satellites and they can run for up to 20 to 30 years without maintenance. Some companies are now installing flywheel batteries to provide backup power to computer systems.

Eduard Furia, chief executive officer of AFS and a former EPA official, firmly believes the electric car's time has come. In reference to the gasoline-powered engines now in vogue, Mr. Furia said, "For 80 years we have been trying to make up for a very poor design." AFS announced that its luxury electric car will be

FIGURE 1 *General Motors introduces its new electric car. Dubbed the Impact, this car has a top speed of 100 miles per hour and a 125-mile range between charges.*

(continued)

powered by 20 flywheels, each with a diameter of 9 inches and weighing about 30 pounds. The AFS battery has several major advantages over the sodium-sulfur or lead-acid batteries used in other electric cars. It has a range of 350 miles instead of 80, it can be recharged in only 15 minutes instead of six hours, and it has a lifetime of 25 years instead of 3. Moreover, it accelerates from zero to 60 miles per hour in only 6.5 seconds.

Will the electric car replace the conventional gas-burner on the roadways of America? A number of skeptics don't think so. One of them is Marc Ross, a physics professor at the University of Michigan. According to Ross, "the electric car is only a special-purpose car for short distances and good climate. In any unusual driving situation, you'd always be in danger of running out of energy."

gasoline. Such vehicles, already being tested, appear to be more efficient than normal cars, but they do produce carbon dioxide. Given the inevitable decline in oil supplies, which many analysts believe will begin within a decade or two, it may be wiser to pursue hydrogen generated from water. Hydrogen is a renewable fuel. When it is burned, it combines with oxygen in the air to form water.

Mass Transit and Urban Growth Management In the long-term future, sustainable cities will depend more on **mass transit,** modes of transportation that carry large numbers of people from their homes to work or other activities. One of the reasons mass transit will likely be more popular in the future is that buses and commuter trains or light rail, which are the most common forms of mass transit, carry passengers four to five times more efficiently than automobiles. As a result, they produce four to five times less pollution per passenger mile. Mass transit will inevitably grow in popularity not just to reduce pollution and reduce energy consumption but to quell growing traffic congestion.

To be acceptable, mass transit systems must be designed to move people quickly to and from work. This will require special mass transit lanes on major highways going in and out of cities, which are already common in many urban areas. Light rail tracks could be laid in highway medians or alongside major rail lines. Some cities use existing rail lines. It is not inconceivable that over time, as pressure for mass transit grows, highway lanes, perhaps even entire highways, could be converted to rail lines.

Another component of improving transportation is growth management—efforts to restrict growth to certain areas often to create denser human settlements as opposed to less dense sprawl, which is inefficient. Efforts to locate new businesses in certain areas such as existing central business districts and to concentrate people, rather than spread willy nilly over the landscape, create high-density nodes that can be efficiently serviced by buses and light rail.

Growth management is now being practiced in several states, among them Oregon. For more information on this topic, see Chapter 5.

18.7 Indoor Air Pollution

When you think of air pollution, the sources that come to mind are belching smokestacks and the exhaust fumes of motor cars. But how about your family living room or kitchen? Recently, scientists have been finding that the air we breathe in our own homes, as well as in schools, stores, and offices, may be more dangerous to our health than the smog-ridden air outside. **Indoor air pollution** comes from a variety of sources—gas ranges, gas furnaces, tobacco smoke, new carpeting, paint, and even furniture.

Concern about indoor quality is justifiable, first of all, because people spend much of their time indoors—for most people, 90 percent of their time is spent inside. Even construction workers and home builders who spend a lot of time outdoors, still are inside about 60 percent of their life.

Indoor air pollution is a result of two factors. First, in an attempt to reduce their home heating bills, many homeowners have insulated their homes and caulked cracks to prevent warm air from leaking out, but these efforts can lock indoor air pollutants in. Before the energy crisis, the average residence time for a given molecule of gas inside the home was about one hour. However, in an air-tight home, the residence time is about 400 percent greater. Because many homes are more tightly sealed, fresh air exchange occurs less frequently. Pollutants once purged from the house now can build up inside.

Scientists at the Lawrence Berkeley Laboratory in California measured concentrations of pollutants inside a well-insulated house in which a gas stove and oven were used in ordinary meal preparation. What did they find? Concentrations of carbon monoxide and nitrogen dioxide in the kitchen, bedroom, and living room from the gas stove actually exceeded the levels of these gases outdoors! In fact, in some homes, the level of nitrogen dioxide may be two to seven times higher than is considered acceptable for outside air. Thus, for many pollutants such as particulates, oxides of nitrogen, carbon monoxide, and hydrocarbons (benzene, chloroform), indoor pollution is much more serious a health threat than outdoor pollution (outdoor pollution must be controlled to protect crops, forests, buildings, and other species).

The second major reason why indoor air pollution has come to the forefront in recent years is that new homes and buildings are made from products or furnished and decorated with products that release pollutants. Most brands of carpeting, for example, release formaldehyde long after you move into your new home. Wood products like cabinets and oriented strand board, a sheathing material that is used for floors, exterior walls, and roof decking, also release formaldehyde. Paints release volatile organic compounds. Latex paint releases mercury.

Many types of pollutants contaminate indoor air, too many for discussion here. (If you are looking for more information on the subject, see our Web site for references that may prove helpful to you.) This section will cover two of the major ones—formaldehyde and radon. Tobacco smoke is covered in Case Study 18.4.

CASE STUDY 18.4 TOBACCO SMOKE: THE DEADLIEST AIR POLLUTANT

Although many of us are concerned about air pollution, it is a relatively minor problem compared to tobacco smoke, which exposes both the participant and nonparticipants who are near smokers. Tobacco smoke is a witch's brew of toxic chemicals (about 200 of them) that kills nearly 1,000 Americans per day or about 325,000 a year. In North America, South America, and Central America, the death toll is believed to be about 625,000 people a year—about 2000 per day!

Some of the chemicals in tobacco smoke, such as carbon monoxide, sulfur dioxide, oxides of nitrogen, and particulates, occur as serious pollutants of the air in our major cities. Smoking greatly increases the amounts of these substances that a smoker takes into his or her lungs. In addition, the smoker inhales large quantities of cancer-causing pollutants like cadmium, nickel, benzopyrene, and radioactive polonium.

Medical research on the deadly effects of tobacco smoke is now firmly established. Today, more than 40,000 books and articles have been published on the effects of tobacco smoke. In 2000, the American Cancer Society estimated that approximately 157,000 people died of lung cancer. Approximately 85 percent of all lung cancers are believed to be caused by smoking (Figure 1).

Recent studies by the National Institute of Health have provided overwhelming evidence that the likelihood of a smoker dying from lung cancer is 22 times greater for males and 12 times higher for females than for people who never smoke. Many who smoke die in their fifties of lung cancer. What's more, tobacco causes cancers of the mouth, throat, larynx, esophagus, pancreas, uterus, kidneys, and bladder. Smokers also increase their chances of dying from strokes and heart attacks.

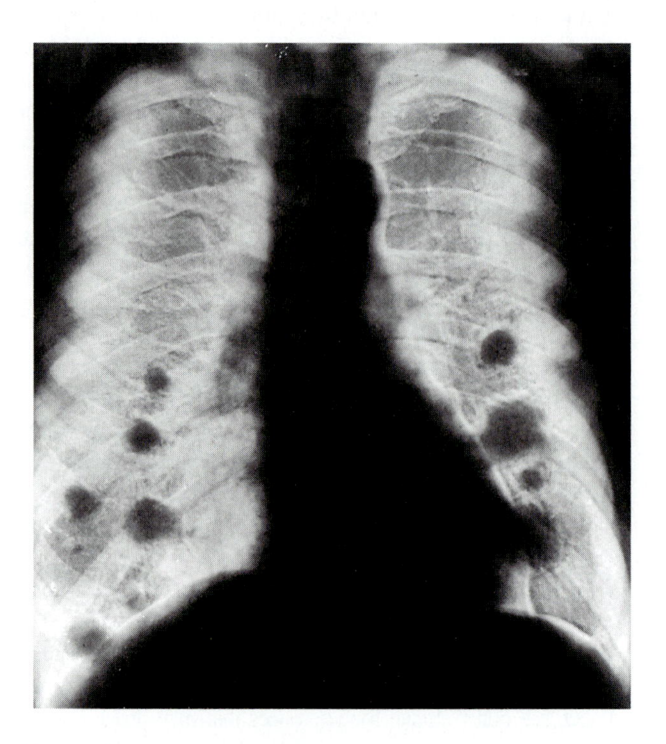

FIGURE 1 *Frontal chest x-ray showing lung cancer. The dark areas in the lower half of each lung are cancerous masses.*

Smoking has other adverse health effects, too. A pregnant woman who smokes passes on the tobacco chemicals in her body to her developing fetus. Although much lower in prevalence, the effects are similar to those of cocaine. The effects on pregnant women and their offspring include spontaneous abortions, a slightly higher rate of sudden infant death syndrome (SIDS), reduced number of brain cells, behavioral problems, and learning difficulties.

Smokers are not the only ones who suffer ill effects from this action. Nonsmokers can be severely impaired if they inhale **second-hand smoke,** that is, smoke exhaled by smokers or smoke that issues from the tip of the cigarette itself. The EPA estimates that second-hand smoke causes the death of at least 3,000 Americans every year. They also estimated that it causes 225,000 cases of bronchitis, pneumonia, and other respiratory infections in children up to 18 months old; and increases the frequency of asthma attacks in 600,000 asthmatic children each year.

Rate of Smoking

Smoking has declined markedly in the United States over the past four or five decades, but still one out of every four Americans age 18 years or older, or about 50 million people, smoke. One of the biggest concerns in the last decade has been the increase in teenage smokers. Because smoking is addictive, many of these people will be hooked for life. In addition, smoking in other parts of the world is on the rise. China, for instance, is experiencing a rapid increase in cigarette smoking.

Control of Tobacco Use

Smoking is a private decision with public ramifications. Not only are nonsmokers exposed, the cost of health care for many people who have smoked their whole life is paid through taxes—medicaid. Smoking costs countries millions of dollars a year in lost wages, too. Several strategies can be pursued to reduce smoking.

Smokers can be encouraged to use nicotine patches to help them kick the habit. When applied to the skin, these patches slowly release small amounts of nicotine into the bloodstream. The urge to smoke is therefore reduced. Twice as many smokers are able to end their nicotine addiction in this way as compared to those who do not use the patch technique.

The federal tax on tobacco products could be substantially increased to help defray public health costs. An added benefit of such a tax would certainly be a reduction in the rate of tobacco use. After all, this tax would probably price many teenagers and children out of the cigarette market. (Note that tobacco companies have raised prices of cigarettes quite substantially in recent years to pay huge class action lawsuits, which has driven the cost of cigarettes up considerably.)

Smoking could be restricted or banned in public places. Most states and hundreds of cities in the United States have done just that. In 1993, the city of Los Angeles was one of the first to ban smoking in all of its 7,000 enclosed restaurants. Any restaurant owner who violates the ban is subject to a six-month jail term and a $1,000 fine. Many others have followed suit. In addition, over the past decade or two all of the airlines have banned smok-

(continued)

ing on national and international flights. The government of Thailand recently banned actors from smoking in movies, a measure designed to reduce smoking among children and teens.

Yet another way to reduce tobacco addiction would be to restrict or even ban all tobacco advertisements. For years, the industry has advertised its products with highly effective ads in newspapers, magazines, and on highway billboards at an annual cost of more than $4 billion. Drastic restrictions in advertising have gone into effect as a result of state and federal lawsuits against cigarette manufacturers. Campaigns directed to younger individuals have been eliminated.

Litigation: Will the Tobacco Industry Die?

Tobacco companies are facing extreme pressure from two types of lawsuits, class action suits filed on behalf of smokers or their surviving relatives and lawsuits from states. In 1999, the tobacco companies lost their first class action law suit. Jurors in Miami found cigarette manufacturers were guilty of defrauding smokers and could be forced to pay billions of dollars in damages to smokers and their families. The Miami jury concluded that cigarette makers "engaged in extreme and outrageous conduct," concealing cigarettes' dangers, conspiring to hide their addictiveness and making a product that caused more than a dozen deadly diseases from heart disease to lung cancer.

The states of Florida, Minnesota, Michigan, and Texas also successfully sued the tobacco companies to seek compensation for state funds that have gone into the medical treatment of smokers. To avoid an avalanche of similar suits, the Federal government stepped in and came to an agreement with tobacco companies on behalf of the remaining 46 states, creating a monstrous settlement. The settlement will provide $206 billion to states, bans advertisement to youth and at sporting events, and has many other provisions. Full details can be obtained at our Web site.

Tobacco companies are also facing dozens of lawsuits in the United States and abroad. Lawyers have filed lawsuits against the tobacco industry on behalf of citizens in Argentina, Canada, Finland, France, Germany, Ireland, Israel, Italy, Japan, the Netherlands, Norway, Sri Lanka, Thailand and Turkey.

What will happen to the tobacco industry is uncertain. Only the future will tell whether this industry that causes so much death and disease will persist.

Formaldehyde You probably remember formaldehyde as the fluid used to preserve frogs and fetal pigs in high school and college biology classes. You may be surprised to learn that it is commonly found in foam insulation, furniture, carpets, particle board, and plywood. Formaldehyde causes a variety of human health problems, including eye irritation, nausea, respiratory problems, and cancer.

The Occupational Safety and Health Administration (OSHA), the federal agency that establishes rules to protect workers and sets standards for new products, has set 3 ppm as the maximum formaldehyde concentration permitted inside industrial plants. Studies in Europe and the United States have shown that formaldehyde levels inside homes are often higher. In Mission Viejo, a large development in California, for instance, formaldehyde concentrations in a research house having no furniture were relatively low. However, when furniture was added, formaldehyde levels increased almost threefold.

The cancer risk from breathing formaldehyde fumes is especially high in the United States' more than 5 million mobile homes due to (1) their relatively small air volume, (2) their poor air circulation, and (3) the relatively large amounts of formaldehyde-containing materials they contain, such as particle board and plywood.

Fortunately, many manufacturers are trying to reduce formaldehyde. You can, for example, purchase low or no formaldehyde oriented strand board. (We'll discuss other ways of reducing indoor air pollution, shortly.)

Radon Radium is a radioactive element that decays spontaneously to radon gas. Although radium is widely distributed in the rocks and soils of the United States, its abundance varies considerably from place to place. The radon gas from the decay of radium diffuses from the soil directly into the outside air. However, since the radon gas quickly disperses, its concentration in the atmosphere is negligible, only about 0.5 percent of the EPA's recommended safe level of 4 picocuries per liter of air.

By contrast, radon levels inside homes and other buildings are much higher because most homes have a lower atmospheric pressure than the outside air. The reason is that the inside air is "pumped" outside by clothes driers, fireplaces, and furnaces, especially during cold winter weather. As a result, radon gas is sucked up through cracks in the cement foundation from the underlying soil and rocks. Residents of the home are generally unaware of the presence of radon because it is tasteless, odorless, and colorless. The radon gas eventually decays into such products as radioactive lead and polonium. These radioactive pollutants may then be inhaled, lodge in lung tissue, and eventually cause cancer. In fact, the EPA estimates that 20,000 of the 130,000 lung cancer deaths in the United States each year are caused by radon.

Radon accounts for 55 percent of the annual radiation dose sustained by the average American from all sources, both natural (radon, cosmic rays) and artificial (television, nuclear power plants, x-rays for medical purposes, and so on.) In fact, Anthony Nero, a nuclear scientist with the Lawrence Berkeley Laboratory, believes that hundreds of thousands of Americans living in homes with high radon levels are exposed to as much health-threatening radiation as the Russians who were living in the vicinity of the Chernobyl nuclear plant in 1986, the year of the disaster.

The radon levels in some American dwellings can reach alarmingly high levels. Take a home in Boyerstown, Pennsylvania, owned by the Watras family, as an example. The Watras' home may have the dubious distinction of having the highest radon level of any dwelling in the United States—about 675 times higher than is considered acceptable even in a uranium mine. The lung cancer risk to the four occupants was equal to that of smoking 220 packs of cigarettes per day! Radon levels in some Wisconsin homes exceed 100 picocuries per liter—a higher level of radiation than one would get from 20,000 chest x-rays.

The EPA regulations for controlling levels of outdoor pollutants are usually set to keep the estimated risk of premature death below 1 in 100,000 (0.001 percent). In contrast, however, the estimated radon risk for most Americans in the United States is 4 in 1,000 (0.4 percent)—400 times higher.

For several years, it has been known that a region extending from Reading, Pennsylvania, into New York and New Jersey has very high levels of radon. However, a ten-state survey recently conducted by the EPA revealed other "hot spots" as well. For example, 63 percent of the homes surveyed in North Dakota and 46 percent of those in Minnesota had radon levels above the EPA guideline.

In late 1988, Lee Thomas, then head of the EPA, announced that residential radon pollution in the United States was both sufficiently widespread and sufficiently serious that every home should be tested. Fortunately for the home owner, this can be easily done with a radon-detection kit that can be purchased for about $15 to $30 from most hardware stores.

Control of Indoor Air Pollution

How do you reduce or eliminate indoor air pollution? Indoor air pollution is addressed in many ways. Experts on the subject use a three-part rule: eliminate, isolate, and ventilate.

To eliminate means to find ways to get rid of potential sources in an existing structure and not to use such things in a new home or building. Leaky furnaces, for example, can be replaced with new, more efficient models that bring in fresh air and exhaust all pollutants. To isolate means to find ways to seal off polluters. Oriented strand board, for example, can be coated with a sealant before use. Asbestos pipes can be painted to stabilize the material so it doesn't flake off. To ventilate, means just that, to find ways to provide controlled ventilation to ensure a steady fresh air supply. Special systems can be installed to bring fresh air into homes. Below are some specific suggestions:

1. Installation of an **air-to-air heat exchanger.** This device, costing $1,000 to $2,000, expels the polluted inside air to the outside and replaces it with fresh outside air. A large portion of the heat from the inside air is added to the fresh air coming into the house to save energy. For this reason, these devices are also called heat recovery ventilators. They're effective in controlling a wide number of indoor air pollutants.
2. The use of a subbasement vent system. In homes overlying rock or soil that emit radon gas, this system is probably an absolute necessity (Figure 18.23). It consists of a network of perforated pipe set in a gravel bed under the basement floor. The radon diffuses from the soil and rock, passes through the perforations, and accumulates inside the pipe. The radon is then vented to the outside air.
3. The establishment of regulations to protect lot and home buyers from high radon levels. Laws can be passed to require the seller of any lot to inform the prospective buyer

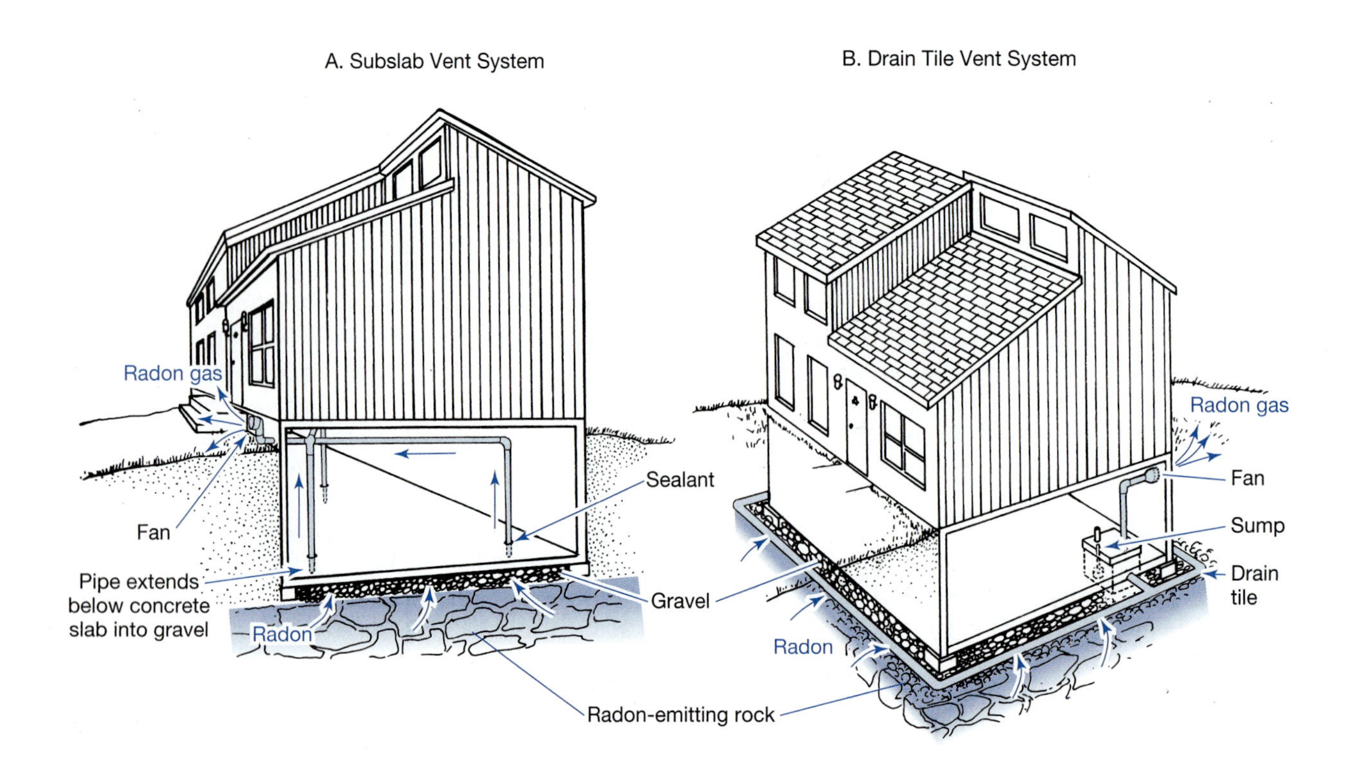

FIGURE 18.23 *A subbasement system for venting cancer-causing radon gas from the home. Several options are available to the home builder. Both types require a porous gravel bed laid down before the basement is poured. (A) Radon is drawn out of the gravel through pipes that penetrate the floor slab. (B) Porous pipes laid in the gravel collect radon, which is pumped outside by a small pump in the basement.*

of the level of radon gas in the air immediately above the property. Similarly, the contractor should be required to inform the future owners of a new home about precisely what levels of radon, formaldehyde, and other pollutants might be expected inside the home during normal living on a day-to-day basis.

4. Reductions in the use of toxic substances like formaldehyde in household products. In the United States and other more developed countries, laws and regulations have forced manufacturers to reduce or eliminate formaldehyde from building products and furniture. Fortunately, there are numerous sources of information on these new products like Environmental Building News' *GreenSpec*.
5. Use of nontoxic paints, stains, and finishes. A growing number of nontoxic or low-toxicity paints, stains, and finishes are now available to builders and home owners.

Our knowledge concerning indoor pollutants is still very incomplete. Much more research is needed.

Summary of Key Concepts

1. Air pollutants come from natural and human sources.
2. Although air pollutants from natural sources are often produced far in excess of those from human sources, it is the latter that are of greatest concern because they tend to be produced in limited areas, which results in local buildup of potentially harmful substances.
3. Pollutants are generally classified as either primary or secondary with primary being those directly produced by some activity or natural event, and secondary being those produced from primary pollutants through chemical reactions occurring in the atmosphere.
4. Hundreds of air pollutants are released into the atmosphere, but the major pollutants are carbon monoxide, carbon dioxide, sulfur oxides, nitrogen oxides, particulates, volatile organic compounds, and ozone.
5. Carbon dioxide is a greenhouse gas discussed in Chapter 19. Carbon monoxide is a colorless, odorless gas that is released from the combustion of fossil fuels and organic materials. It binds with hemoglobin in red blood cells and reduces the oxygen-carrying capacity of the blood.
6. Sulfur oxides are produced when naturally occurring sulfur in fuels such as coal and oil combine with oxygen. Much of the sulfur oxide generated by humans results from the burning of sulfur-containing fossil fuels in power plants and other industries.
7. Solid and liquid particles suspended in the air are known as particulates. Major sources are coal-burning facilities such as power plants, steel mills, fertilizer plants, jets, diesel trucks, and foundries.
8. Lead is one of the particulates of concern. Lead was once added to gasoline in the United States to improve its octane rating. It is released from auto exhaust into the air. Lead can damage the kidneys, blood, and liver. Furthermore, it can impair proper brain development in children.
9. A volatile organic compound is an organic compound composed of hydrogen and carbon. VOCs are released from the incomplete combustion of fossil fuel and from evaporation. Good examples of hydrocarbon pollutants of the atmosphere are methane and benzene. VOCs react chemically to produce photochemical smog.
10. Photochemical smog is produced by the reaction between VOCs, nitrogen oxides, and atmospheric oxygen in the presence of sunlight. Ozone is one of components of photochemical smog.
11. Concentrations of air pollution depend on the amount released by local sources, but also by certain climatic factors such as temperature inversions.
12. Radiation inversions occur primarily at night as heat radiates from the Earth's surface into the atmosphere, rapidly cooling both the ground and the layer of air next to it. This results in the formation of a warmer layer of air lying over a cooler layer containing pollutants that cannot escape by vertical mixing.
13. A subsidence inversion is formed when a high-pressure air mass sinks down and warms up. Here again, the warm air forms a lid on the cool air nearer the ground.
14. Cities also affect pollution patterns. Heat-producing processes and heat-absorbing structures usually translate into slightly warmer temperatures inside cities, known as the heat island effect. The heat island effect causes pollutants to concentrate in a dust dome above a city.
15. Air pollution is affected by climate but also affects climate—locally, regionally, and perhaps even globally. Particulates, which act as condensation nuclei, can enhance local precipitation.
16. The major illnesses caused by air pollution include emphysema, chronic bronchitis, and lung cancer.
17. Pollution levels can reach very high levels during inversions, causing death and illness in large segments of the population. The first major air pollution disaster in the United States occurred in Donora, Pennsylvania, in October 1948; it caused the deaths of 20 people. A more severe incident occurred in London several years later.
18. Asbestos is used in pipe and boiler insulation, brake linings, paint, plastics, and many other products and materials. People who inhale asbestos fibers may develop asbestosis, characterized by coughing, chest pains, and bluish discoloration of the skin. Over 50 percent of the people who have asbestosis eventually die from lung cancer.
19. Occupational exposure to asbestos has been reduced by the use of vacuum devices, the mandatory use of masks, and the automation of many manufacturing processes.
20. Two basic strategies exist for dealing with air pollution—the end-of-pipe approach and the preventive approach.
21. Many industries capture pollutants from smokestack gases. They are generally discarded in landfills. In some cases, though, air pollutants can be converted into useful products. For example, sulfur dioxide can be converted into sulfuric acid and high-carbon fly ash can be used

directly as fuel or can be converted into cinder blocks, paving materials, abrasives, and cement.

22. In some cases, end-of-pipe devices convert pollutants into relatively harmless substances. Catalytic converters on automobiles, for instance, increase the rate at which carbon monoxide and hydrocarbons are oxidized to carbon dioxide and water, respectively.
23. Pollution control devices employed by utilities and industries include electrostatic precipitators, cyclone filters, fabric filter bag houses, and flue gas desulfurization systems, popularly known as scrubbers.
24. Flue gas desulfurization is the most widely used method for controlling sulfur dioxide emissions by industry and utilities.
25. Companies can also reduce air pollution by modifying processes or switching to nontoxic materials. These constitute a preventive approach and are desirable from the standpoint of sustainability.
26. Five general strategies for controlling auto emissions are: (a) reducing traffic volume, (b) modifying the internal combustion engine, (c) replacing the internal combustion engine with electric and steam engines, (d) developing a hybrid car (combining an internal combustion engine and an electric engine), and (e) increasing the use of mass transit.
27. Another problem of growing concern is indoor air pollution. Recent attempts to insulate homes have worsened indoor air pollution problems from carbon monoxide, formaldehyde, radon gas, and tobacco smoke. Many new products and furnishings in our homes also contain toxic substances that are released into the air of our homes.
28. Addressing indoor air pollution requires a three-step process: eliminate, isolate, and ventilate. Pollutants may be controlled by (a) the installation of an air-to-air heat exchanger, (b) the use of vegetation as living air purifiers, (c) the use of a subbasement vent system in homes overlying rock or soil that emit radon gas, (d) the establishment of regulations to protect lot and home buyers from high radon levels, (e) reductions in the use of toxic substances like formaldehyde in household products, and (f) the use of nontoxic paints, stains, and finishes.

Key Words and Phrases

Air Pollution Disaster
Air-to-Air Heat Exchanger
Asbestos
Attainment areas
Carbon Monoxide
Carbon Monoxide Poisoning
Carcinogens
Catalyst
Catalytic Converter
Chronic Bronchitis
Clean Air Act
Condensation Nuclei
Corporate Average Fuel Efficiency Standards
Cyclone Filter
Donora Disaster
Dust Domes
Dust Plumes
Electric car
Electrolysis
Electrostatic Precipitator
Emissions Offset Policy
Emphysema
Ethanol fuel
Fabric Filter Bag House
Flue Gas Desulfurization
Flywheel Battery
Formaldehyde
Fuel Cell
Heat Islands
Hybrid Cars
Hydrocarbons
Indoor Air Pollution
Inspection and Maintenance Program
Lead Poisoning
Lung Cancer
Mass Transit
Mesothelioma
Mobile pollution source
Natural Pollution
Nitric Oxide
Nitrogen Dioxide
Nitrogen Oxides
Nonattainment areas
Ozone
Particulates
Photochemical Smog
Prevention of Significant Deterioration (PSD)
Primary Pollutants
Pollution Disaster
Reformer
Radiation Inversion
Scrubber
Secondary Pollutants
Smog
Smokestack scrubber
Source Reduction
Stationary pollution source
Steam Car
Strip Mining
Subsidence Inversion
Sulfur Dioxide
Sulfer Oxides
Synergism
Temperature Inversion
Thermal Inversion
Volatile Organic Compounds

Critical Thinking and Discussion Questions

1. Briefly describe the biological significance of the major components of unpolluted air.
2. List three sources of natural pollution.
3. Using your knowledge of air pollution and your critical thinking skills, discuss the following statement: "Natural air pollutants are produced far in excess of those from human sources, so we really don't need to worry about pollution from cars and such."
4. Discuss the health effects of each of the following air pollutants: (a) carbon monoxide, (b) oxides of sulfur, (c) oxides of nitrogen, (d) hydrocarbons, (e) ozone, and (f) lead.
5. List the major human sources of carbon monoxide, oxides of sulfur, oxides of nitrogen, hydrocarbons, ozone, and lead.
6. List the pollutants responsible for the formation of photochemical smog.
7. Cite an example of how severe air pollution actually caused the "death" of a town.
8. List five conditions that are usually associated with an air pollution disaster.
9. Describe three conditions that contribute to the heat island phenomenon.
10. Which climatic conditions would be most conducive to the rapid dispersal of atmospheric pollutants? To their concentration in a localized area?
11. Discuss the advantages and disadvantages of the electric car.
12. Using your critical thinking skills, analyze the following statement: "Electric cars are pollution free and therefore are a major boon to cities that are interested in cleaning up their air."

13. Discuss the advantages and disadvantages of replacing gasoline with ethanol.
14. What is a hybrid car?
15. Discuss hydrogen cars. You may want to do some more research on this.
16. Now that you have studied this material on air pollution, what changes, if any, do you plan to make with respect to the use of your automobile?
17. Discuss the problem of indoor air pollution. Why is it a problem? Where do indoor air pollutants come from? What are the general ways of addressing them?

Suggested Readings

Amdur, M. O. 1993. "Air Pollutants." In *Toxicology: The Basic Science of Poisons,* Amdur, M. O., J. Doull, and C. D. Klaassen (eds.), 4th ed., New York: Pergamon Press. Detailed account of air pollution and its effects on the body.

Beatly, T. 2000. *Green Urbanism: Learning from European Cities.* Washington, D.C.: Island Press. This book shows, among other things, how to revamp our cities to reducing air pollution and other environmental problems.

Bower, J. 1997. *The Healthy House: How to Buy One, How to Build One, and How to Cure a Sick One.* Bloomington, IN: The Healthy House Institute. Detailed reference.

Bower, J., and L. M. Bower. 1997. *The Healthy House Answer Book: Answers to the 133 most commonly asked questions.* Bloomington, IN: The Healthy House Institute. Great introduction to the subject of healthy homes.

Buck, S. J. 1998. *The Global Commons.* Washington, D.C.: Island Press. Examines management of global commons such as the atmosphere.

Clayton, T., G. Spinardi, and R. Williams. 1999. *Policies for Cleaner Technology: A New Agenda for Government and Industry.* London: Earthscan. A ground-breaking treatise on new ways to make industrial systems more sustainable by the business themselves.

Environmental Protection Agency. 2000. *National Air Pollutant Emission Trends, 1900–1997.* Washington, D.C.: U.S. EPA. Although this annual publication is always three years behind the time, it is a great resource.

Homes, D., et al. 1999. *GreenSpec: The Environmental Building News Product Directory and Guideline Specifications.* Brattleboro, VT: E Build, Inc. One of five books on alternative building materials that are good for the environment and easy on our health.

Kotov, V., and E. Nikitina. 1996. "Norilsk Nickel: Russia Wrestles with an Old Polluter." *Environment* 38(9): 6–11, 32–37. Excellent view of air pollution in Russia and efforts to control the problem now.

Lents, J. M., and W. J. Kelly. 1993. "Clearing the Air in Los Angeles." *Scientific American* 269(4): 32–39. Considerable progress has been made due to tough emission control standards.

Newman, P. and J. Kenworthy. 1999. *Sustainability and Cities: Overcoming Automobile Dependence.* Washington, D.C.: Island Press.

Roome, N. J. 1998. *Sustainability Strategies for Industry: The Future of Corporate Practice.* Washington, D.C.: Island Press. Great overview of the subject.

United Nations Environment Programme. 1999. *Global Environment Outlook 2000.* New York City: Earthscan. Great reference book on global environmental trends and policy changes needed to address them.

Web Explorations

Online resources for this chapter are on the World Wide Web at: **http://www.prenhall.com/chiras** *(click on the Table of Contents link and then select Chapter 18).*

Chapter 19

Air Pollution: Global Problems

In the summer of 1999, record heat and lack of rainfall in the United States raised havoc among farmers along the eastern seaboard. Many stood by idly as they watched their crops wither and die. Late that summer and throughout the fall, however, numerous hurricanes and tropical storms battered the coast of the southeastern United States, causing flooding and killing dozens of people and drowning countless livestock. The next year, in the Spring of 2000, drought set in among the western states while the Northeast was inundated with water. Farmers in Western New York, for instance, couldn't get their crops in because the soils were so wet, or could only plant on high ground. In the early summer of the year 2000, long before the traditional dry season in the West, wildfires broke out in California, New Mexico, Colorado, and Montana, burning millions of acres of forest land. That same year, researchers plying the waters of the Arctic Circle noticed the dramatic loss of ice. They were able to navigate the polar waters easily. Decades earlier, the area had been choked with ice, making ship travel nearly impossible.

At first glance, it would appear that these events have little in common. It might seem as if they were random fluctuations in weather. Nevertheless, these seemingly unrelated episodes may have a common thread: global warming resulting from a phenomenon called the *greenhouse effect.* Most scientists who've studied the subject believe that the planet has entered into a major period of global temperature increase, caused in large part by certain types of air pollution and deforestation.

Global warming, or more appropriately, global climate change, is just one of three global environmental problems resulting from human activities. We'll examine each one in this chapter, looking at impacts and solutions.

19.1 Global Climate Change

Atmospheric scientists have long known that several air pollutants trap heat escaping from the Earth's surface, causing the atmosphere to warm. One of the main culprits is carbon dioxide, which comes from the combustion of organic materials, including fossil fuels and vegetation. Without naturally occurring carbon dioxide, the Earth's temperature would be about 30°C (55°F) cooler, so cold that few, if any, forms of life could live here.

Carbon dioxide is aided by water vapor, ozone, nitrous oxide, and methane. The one thing these pollutants have in common is that they come from both human and natural sources. Over the past five decades, however, humans have also released millions of tons of a chemical known as **chlorofluorocarbons,** or **CFCs**, an air pollutant with no natural source.

The trapping of heat by gaseous pollutants is known as the **greenhouse effect.**[1] The gases responsible for this phenomenon are called **greenhouse gases.** To understand how these gases operate, consider a familiar example. If you park your car in the campus parking lot on a sunny day, even in the winter, the temperature inside the car rapidly increases, becoming nearly ovenlike. This rapid increase in temperature is due to the greenhouse effect. What happens is that sunlight passes through the car's windows and is absorbed by various dark surfaces like the seats. Here, the solar energy is converted into heat (infrared wave) energy. Since infrared radiation cannot readily escape through the windows, it is trapped inside the car. The same thing happens in a greenhouse or a sealed glass jar.

Carbon dioxide, methane, and the other greenhouse gases act like the glass in a greenhouse. They, in fact, form a vast global "greenhouse" around the Earth. These gases trap heat (infrared waves) escaping from the Earth's surface, radiating it back to Earth. Although the concentration of all the Earth-warming gases is increasing rapidly, the most significant one at the present time is carbon dioxide.

In your study of the carbon cycle in Chapter 3, you may recall that carbon dioxide is continuously removed from the atmosphere by green plants and algae (during photosynthesis). They use the carbon dioxide to make glucose and other molecules. Carbon dioxide is also absorbed by surface waters like lakes and oceans where it is taken up by plants and algae or simply dissolved in water.

Carbon dioxide is gradually released back into the atmosphere as plants and animals breakdown organic matter to make energy or when the organic matter decays. It is also released back into the atmosphere when forests, grasslands, or any organic material burn. Dissolved carbon dioxide may also be liberated from lakes, rivers, and oceans.

For thousands of years, the production of carbon dioxide and the release have been in balance. Thus, the amount of carbon dioxide removed from the atmosphere equaled the amount entering it (Figure 19.1, top). How do we know this?

Scientists have determined the concentration of carbon dioxide in the atmosphere of long ago by examining air bubbles trapped in glacial ice and by analyzing the wood of centuries-old trees. Such investigations have shown that carbon dioxide levels remained at about 270 parts per million (ppm) from the end of the last ice age (about 10,000 years ago) until

[1]The term *greenhouse effect* is a bit of a misnomer. The glass in a greenhouse warms the interior space because it lets light in, but physically traps hot air, not all infrared radiation.

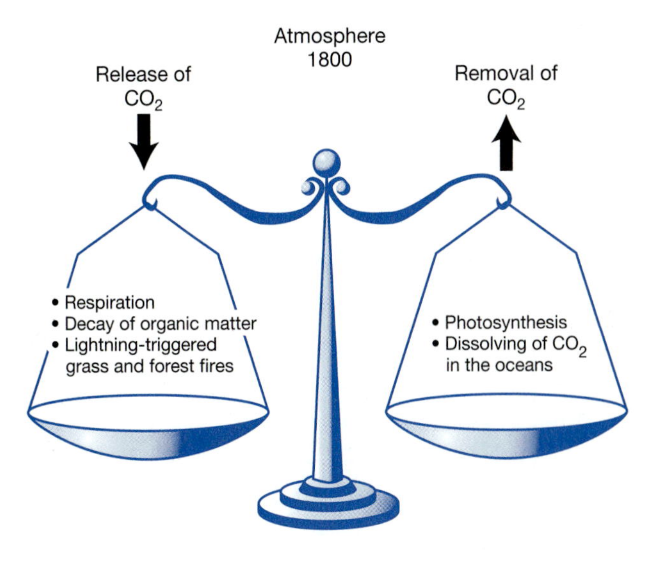

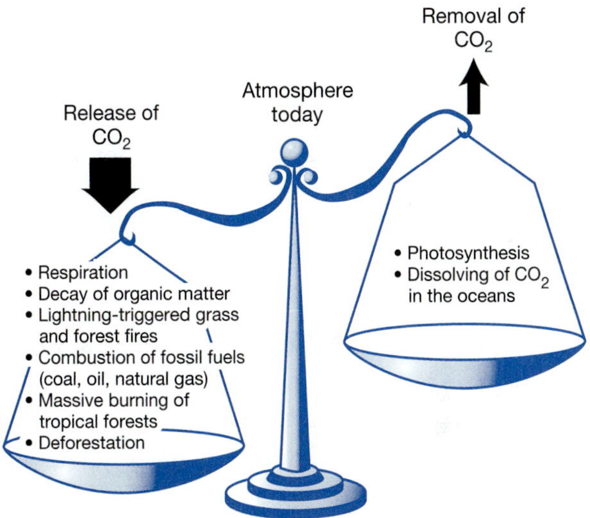

FIGURE 19.1 *Release and removal of atmospheric carbon dioxide today as compared with 1800.*

the nineteenth century. However, since 1860, these studies have shown that the concentration has risen from 290 ppm to about 361 ppm in 1995. Carbon dioxide levels in the atmosphere have been closely monitored by sensitive instruments atop Mauna Loa, an extinct Hawaiian volcano, since 1957. Records here show that carbon dioxide levels have risen from 315 ppm in 1957 to 368 ppm in 1999 (Figure 19.2). Overall, scientists report a nearly 27 percent increase in carbon dioxide levels in the atmosphere since 1860.

The increase in atmospheric carbon dioxide has largely been caused by human activities, two in particular stand out: the combustion of fossil fuels and the clearing and burning of forests the world over. In recent times, tropical deforestation to make room for cattle ranches, farms, and human settlements has been a major cause of the rise in global levels of carbon dioxide.

According to one estimate, global fossil fuel consumption is responsible for the annual release of nearly 6 billion metric

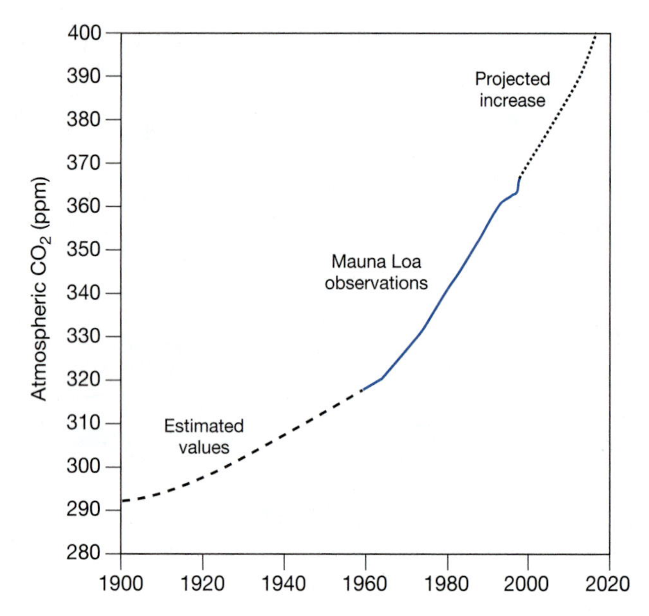

FIGURE 19.2 *The trend in atmospheric CO_2 concentration since 1900 and projected to 2020.*

tons of carbon into the air—about 1 ton for each person on Earth. The clearing and burning of tropical rain forests annually releases about 1.5 billion metric tons into the air, about one-fourth the amount released by the consumption of fossil fuels.

Although not definitive proof of cause and effect, the rise in the concentration of carbon dioxide in the atmosphere has been accompanied by a dramatic increase in global temperature from 1900 to 2000, which is shown in Figure 19.3. No one knows for sure if the continued release of carbon dioxide and the loss of carbon dioxide-absorbing vegetation will result in further increases in global temperature, but most atmospheric scientists think that it is very likely. Estimates of how much temperature will rise vary and seem to change every few years as our understanding of the climate and as computer models used to predict climate change improve. The most current computer projections by the **Intergovernmental Panel on Climate Change,** a group of climate researchers convened by the United Nations, called for an average increase in global temperature of about 2°C from the year 1990 to 2100, nearly twice the increase since 1950. Their models suggest that the warmup at the poles will be considerably greater than elsewhere. Certain continents may experience greater than average warming, too. In 2000, a report by National Research Council predicted an increase in North America of 3° to 6°C (5° to 10°F). In this report, scientists predicted a major shift in climate zones in North America. The authors suggest that the tropical climate of equatorial regions will shift northward into the lower-tier states. Their climate, in turn, will shift to the mid-tier states and the climate of the top-tier states would move into Canada. The effect of such a rapid shift on vegetation and wildlife could be devastating, with many trees simply dying as temperatures climb outside of their range of tolerance.

Although carbon dioxide is the major greenhouse gas, as noted earlier, other pollutants contribute to greenhouse warm-

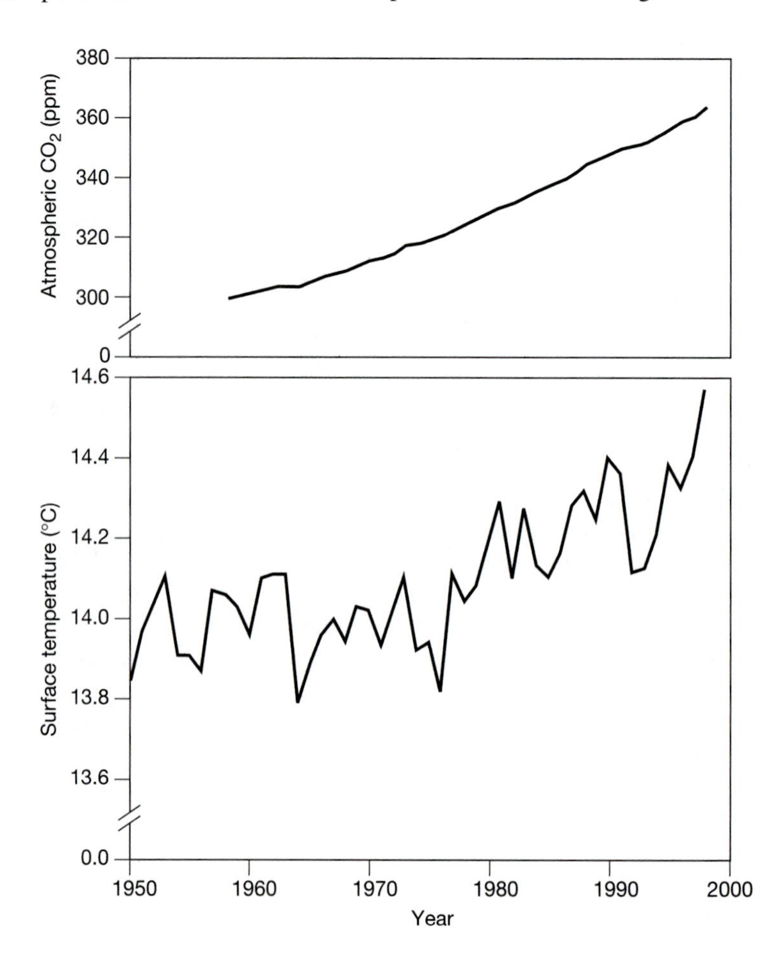

FIGURE 19.3 *Global surface air temperature from 1900 to 1995, showing a gradual increase.*

ing as well. Many of these are increasing at rates much higher than carbon dioxide. And many like the CFCs remain in the atmosphere for long periods—over 100 years.

Many factors influence global temperature and climate. Solar storms, which emit huge amounts of radiation, for instance, can cause the climate to warm. Slight changes in the tilt of the Earth also contribute to warming. Further complicating matters, after the last ice age the Earth entered a warm interglacial period. During this period, the Earth has slowly warmed. Skeptics of the global warming phenomenon are quick to point out that at least some of the increase in temperature we've experienced in the past 100 years may be due to a natural warming trend. They argue that other natural forces may be at play, too, in an attempt to discount their colleagues who argue in favor of a human-induced warming. The vast majority of scientists, however, believe that the majority of the global warming that has occurred since the Industrial Revolution is the result of human activities. Numerous scientific studies back up this assertion.

Further Evidence of Global Warming and Global Climate Change

The rise in carbon dioxide in the atmosphere and the increase in temperature are just two pieces of evidence suggesting that global warming is occurring. It is not enough to support the hypothesis fully, however. What other evidence is there that the Earth is warming up and that our climate is shifting?

Hardly a day goes by without some new evidence suggesting that the Earth's atmosphere is warming and that the climate is changing. Seventeen of the hottest years in the past 100 years, for instance, have occurred since 1980. Although a couple consecutive hot years are not unusual, the probability of having this many in a short period is astronomically small.

Not only is the Earth getting warmer, precipitation is changing. Studies show that over the past century precipitation has increased by about 1 percent over the World's continents. But the increase is not uniform. Rainfall appears to be on the decline in the tropics, while it is increasing in higher latitude areas. In addition to changes in rainfall, scientists are beginning to detect changes in the frequency and severity of storms. Hurricanes and tornadoes appear to be on the rise, often with devastating consequences.

Warmer temperatures are also wrecking havoc on glaciers and the polar ice caps. Antarctica is shrinking rapidly. In the past decade, numerous sections of Antarctic ice, some as large as Rhode Island, have broken loose. Glaciers are retreating, too, as is the Arctic ice cap.

One consequence of the melting of land-based ice is a rise in sea level. In the last century, sea level has risen worldwide approximately 15–20 cm (6–8 inches). According to the EPA, approximately 2–5 cm (1–2 inches) of the rise has resulted from the melting of mountain glaciers. Another 2–7 cm has resulted from the expansion of ocean water that resulted from warmer ocean temperatures. Whatever the cause, many islands throughout the world, especially those low-lying islands in the South Pacific, are facing almost certain ruin. Their beaches are being flooded by seawater and, if waters continue to rise, they could eventually be lost to the sea.

Not only are the seas rising, but they are also warming. Studies of the Pacific Ocean, for example, show a slight, but statistically significant increase in deep-water temperatures, consistent with the global warming hypothesis. Heat from the ocean is released into the atmosphere and is the engine behind the formation of hurricanes, which form in equatorial waters. The warmer the planet, say climatologists, the more frequent and more violent hurricanes will become. Evidence suggests this may indeed be occurring.

Warmer climates are also resulting in the spread of insects that carry malaria and dengue fever to higher altitudes and higher latitudes. In Africa malaria is spreading to higher altitude regions in five countries, scientists believe, because of the warming climate. In Mexico and Costa Rica, dengue fever, is spreading into the highlands as well. "The control issue looms largest in the developing world, where resources for prevention and treatment can be scarce," writes Dr. Paul Epstein, Associate director of the Center for Health and the Global Environment at Harvard Medical School in an article in *Scientific American.* "But the technologically advanced nations, too, can fall victim to surprise attacks—as happened last year (1999) when the West Nile virus broke out for the first time in North America, killing seven New Yorkers. In these days of international commerce and travel, an infectious disorder that appears in one part of the world can quickly become a problem continents away if the disease-causing agent, or pathogen, finds itself in a hospitable environment."

Projected Effects of Global Warming

Growing evidence suggests that the planet is warming and that the climate is shifting. What will happen to us and our planet? No one knows for sure, but many scientists have projected impacts. Below is a list of some of the social, economic, and environmental changes that might be expected. As you will notice, some impacts could be positive or beneficiial.

BENEFICIAL EFFECTS

1. One benefit of a warming climate is that the cost of heating homes, offices, and commercial buildings during the winter in the high latitudes regions might decline. Fossil fuel demand in these areas would be reduced, as would heating costs.
2. The far northern latitudes (tundra biome) might become more suitable for human settlement. Similarly, the ranges of many species of fish, birds, and mammals would be extended northward in North America, Europe, and Asia. Such has already been the case for the mockingbird, cardinal, and opossum.
3. Global warming could increase rainfall in some areas. Increased rainfall and/or a longer growing season could increase food production in certain areas of the world such as Canada, Mexico, Europe, northeastern Africa, and Southeast Asia. The political and economic clout of these nations may increase substantially.

4. For every 1 percent increase in atmospheric carbon dioxide, the rate of photosynthesis will increase 0.5 percent, provided that sufficient water and nutrients are available. As a result, the yields of such crops as rice, corn, and wheat may increase.

HARMFUL EFFECTS

1. Although heating costs may decrease in northern latitudes, summer cooling demand is likely to increase. With more and more warm regions, summer cooling demand could far outweigh any gains in this area. To keep buildings cool, more fossil fuels may be consumed, increasing carbon dioxide emissions and further increasing global warming.
2. Many scientists theorize that the increase in global temperature will further increase melting of the polar ice caps and glaciers. Scientists predict that further melting of the land-based ice—that is, glaciers and the massive Antarctic ice pack—will cause sea levels to continue to rise. The warming of the Earth will also cause ocean waters to expand, further adding to the rise in sea levels. According to the most recent scientific estimates, the oceans will rise 50 centimeters (20 inches) by the year 2100.

   A rise in ocean levels would cause the seas to move inland along many portions of the world's coastlines. Coastal properties worth billions of dollars would be lost. The city of Charleston, South Carolina, alone would suffer hundreds of millions of dollars in damage by the year 2035. Similar flood damage would probably be experienced by Boston, Philadelphia, New York City, Baltimore, Washington, Norfolk, Miami, St. Petersburg, New Orleans, and many other coastal cities throughout the world.
3. Many of the world's most productive rice-growing areas (low-lying coastal areas) would be destroyed by floodwaters, including the fertile, highly populated deltas of the Indus (Pakistan), Ganges (Bangladesh), and Yang Tze (China) rivers.
4. Rising sea level could increase damage caused by hurricanes and tropical storms. When these violent storms strike the shoreline, water levels rise and often flood coastal properties. A rise in sea level would therefore increase the distance that storm surges penetrate. As the seas rise, more and more inland property will be flooded during such events. Large tracts of agricultural land would be destroyed because of salts left behind by the receding waters.
5. Some scientists predict that global warming could also change weather patterns. Some areas of the world may be wetter and others may become drier. Computer models predict that the midwestern United States, a rich agricultural land, will be warmer and drier. Crop production would fall drastically. As a result of changing climate, corn production would actually shift northward into northern Minnesota, Wisconsin, and Canada. But the shift in growing regions may not offset the loss in production because northern soils are much less fertile. Overall, corn yields would very likely decline. The U.S. wheat industry could suffer a $506 million yield decline annually because of increased heat and drought.
6. Extensive drought on the Great Plains of the United States and other areas of the world would trigger huge dust storms like the "black blizzards" of the Dust Bowl.
7. Many aquifers along the Gulf and Atlantic coasts of the United States, which now serve as a source of drinking water for many American cities, would become contaminated by salt water from the ocean.
8. The increased salinity of many American estuaries, such as the Chesapeake Bay and Delaware River estuaries, would destroy their value as breeding and nursery grounds for many valuable species of game and commercial fish.
9. The carbon dioxide buildup would inflict enormous damage on the global economy. Consider, for example, just one factor—the displacement of agricultural production to the north (in the Northern Hemisphere). Such a dramatic shift would require hundreds of billions of dollars for the construction of new flood control, irrigation, drainage, and grain storage systems. For example, if new irrigation systems were required for just 15 percent of the existing irrigated area in the world, the price tag would be $200 billion—$15 billion for the United States alone. Then there would be the cost of new dikes, dams, levees, and so on to prevent the flooding of coastal areas.
10. Many plants and animals could become extinct because of the rapid change in temperature. Although species can adapt to climate change, the problem with human-induced change is that it occurs much faster than natural change that has occurred in the past.
11. As the Northern Hemisphere warms, diseases that are restricted to the tropics and warm climates in general could spread northward into major population centers. Evidence of this is already starting to mount up.
12. "Floods and droughts associated with global climate change could undermine health," writes Paul Epstein. "They could damage crops and make them vulnerable to infection and infestations by pests and choking weeds, thereby reducing food supplies and potentially contributing to malnutrition. And they could permanently or semipermanently displace entire populations in developing countries, leading to overcrowding and the diseases connected with it, such as tuberculosis."

As you can see, global warming has its benefits, but the social, economic, and environmental costs of the adverse impacts very clearly outweighs any gains we might make. In fact, most scientists believe that the consequences will be devastating. Already, hundreds of people die each year in heat waves. Crop production in many parts of the world is suffering. Islands are being inundated as sea levels rise, and coastlines are being battered by violent hurricanes that kill dozens of people and leave billions of dollars in property damage. Flooding is becoming more common in North America and Europe. Fires are raging in many parts of the world, burning millions of acres of forest land.

Reducing or Eliminating Global Warming

Despite the repeated warnings of scientists that the greenhouse effect could eventually have highly adverse effects on human society, the general public in the United States, as well as policy makers, was largely unconcerned with the threat until 1988. In that year, Americans experienced unusual drought, searing heat, floods, and hurricanes—the precise events scientists predicted would occur with increasing frequency and severity as the levels of greenhouse gases rise. The climatic disasters of 1988 may or may not have been valid signals that the greenhouse effect was finally in evidence. Nevertheless, some segments of the public, at least, have been aroused from their apathy. In 1992, many nations of the world signed a voluntary agreement, known as the **United Nations Framework Convention on Climate Change,** to reduce greenhouse gas emissions in an effort to stabilize the release of greenhouse gases and prevent dangerous warming of the Earth's atmosphere. The Convention on Climate Change, however, was weak and largely ignored. Many see it, however, as a good beginning. Since then, the nations of the world have met numerous times to hammer out goals and strategies for reducing greenhouse gas emissions, In 1997, meeting in Kyoto, Japan, they signed an agreement known as the **Kyoto Protocol,** which legally obliges Western nations and former Eastern bloc nations to cut emissions of greenhouse gases such as carbon dioxide to 5.2 per- cent below 1990 levels by between 2008 and 2012. How are we doing?

In 1997, the emission of carbon dioxide were 4.7 percent below 1990. Although progress appears good on the surface, much of the decline in carbon dioxide emissions can be traced to the economic faltering of the former Soviet Union where emissions fell 32.5 percent decline from 1990 to 1998. Carbon dioxide emissions in the United States increased nearly 12 percent during that period. In Europe, Germany, the United Kingdom, and France posted declines but these were offset by increases in other European nations. Overall, in Europe, emissions rose 3.1 percent during this period. In Japan, carbon dioxide releases climbed nearly 6 percent. In less developed nations, emissions rose nearly 40 percent. Clearly, much more work is needed if we're going to reach our goals. This section considers some of the strategies to help slash carbon dioxide emissions.

Reducing Automotive Emissions Carbon dioxide levels released into the atmosphere by motor vehicles could be substantially reduced by the emission-controlling strategies discussed in the previous chapter. As you recall, they include (1) reducing traffic volume, (2) instigating engine inspection and maintenance programs, (3) increasing fuel efficiency, (4) increasing the use of mass transit (buses and trains), (5) adopting alternative fuels such as hydrogen and ethanol, and (6) replacing gasoline engines with those powered by electricity. Below are some specific strategies that expand on the ideas presented in this paragraph,

Convert From Gasoline To Ethanol Motor vehicle fuel could be shifted from gasoline to ethanol (grain alcohol). Ethanol is a renewable resource produced from plants grown on "fuel farms." Corn, for instance, can be converted to ethanol. The use of ethanol has several advantages. Most important, it burns cleanly and it produces no net carbon dioxide because the amount of carbon dioxide that is released when it is burned equals only the amount removed from the air by photosynthesis in fuel crops. Most experts think that it will be needed sooner or later anyway to replace oil, the supplies of which are dwindling rapidly. Moreover, ethanol can be burned in cars mixed with gas. This requires no modification. Only slight modifications are required to burn 100 percent ethanol.

Improving Fuel Efficiency: Taxing Fuels Economists have long known that the price of fuel, whether it is home heating oil or gasoline, greatly affects the efficiency of its use. The cheaper the fuel, the less efficiently we use it. For this reason, some conservationists have proposed increasing the tax on gasoline in the United States and imposing a special carbon tax on all fuels that release carbon dioxide when burned. The greater the amount of carbon dioxide released from a given fuel, they suggest, the higher the fee. Such a tax would encourage industries and utilities to conserve fuel or switch from oil and coal to natural gas or solar energy. Motorists might be inclined to buy more efficient vehicles or choose other options for transport, for example, commuting by bus or light rail. Revenues from the tax could be used to develop and promote energy conservation and alternative fuels (Chapter 22).

Early in his first term, President Clinton proposed a carbon tax, but the idea was soundly rejected by senators from western states, those that produce coal and other fossil fuels. Other nations have imposed taxes. In Europe, it is not unusual to pay between $3 and $5 per gallon for gasoline at the pump. Most of the fee is for taxes. As the impacts of global warming become better known, and the costs become evident, policy makers in the United States and Canada may become more open to the idea, but at this writing it remains highly unpopular.

Use More Methane Fuel Methane gas generates only about half as much carbon dioxide as coal in producing the same amount of energy. Methane can be obtained from sanitary landfills by driving a pipe into the depths of the decomposing garbage. This is already being done at many sites in the United States, including the Fresh Kills Landfill on New York City's Staten Island.

Use Genetic Engineering Methods To Develop "Gasoline Plants" Through genetic engineering, plant geneticists could develop varieties of plants that produce hydrocarbons from which energy-rich fuels might be derived (Chapter 22). Even though carbon dioxide would be released into the atmosphere when these fuels are burned, the amount would only equal that removed from the air by photosynthesis during the life of the plants. (Burning such fuels would be quite different from burning coal derived from plants that carried on photosynthesis, and hence removed carbon dioxide from the air, more than 200 million years ago.)

Halt the Deforestation in the Tropics As described in Chapter 14, many square kilometers of tropical rain forests in Central and South America and Southeast Asia have been cut down and burned to clear areas for cattle ranches and farms.

Much wood is also used as fuel. Some experts believe that 40 percent of the closed canopy (solid stands) of tropical rain forest has already been removed. At present rates of deforestation in Nepal (4 percent per year), the tropical rain forest will have been completely cleared away by the year 2012. Much of the tropical rain forest that was cut down in Central America was removed so that ranchers could raise beef that winds up in the billions of hamburgers eaten by Americans. The United States currently imports about 200 million pounds of beef from Central America, much of which is raised on converted rain forest. Removing these vast stands of tropical rain forest prevents the uptake of more than a billion tons of atmospheric carbon by photosynthesis, thus contributing to the buildup of atmospheric carbon dioxide. Halting this activity and deforestation for other reasons could have a major impact on global carbon dioxide levels. Individuals can help reduce deforestation by planting trees, cutting back on all forms of waste, recycling,

Reforestation Many millions of hectares of tropical rain forest have been destroyed in the name of progress. At least two thirds of this forest land has never been replanted. Replanting this land and other devastated forests could go a long way toward helping reduce the threat of global warming. The American Forestry Association encourages the planting of 100 million trees yearly in the United States. Such projects will certainly help. Nevertheless, scientists estimate that to remove the 6.4 billion tons of carbon dioxide released into the air each year as a result of the combustion of fossil fuels, the new forests would have to cover an area larger than that of the continental United States!

It is encouraging to note that some major industries in the United States are cooperating vigorously in the reforestation effort. A good example is Applied Energy Services (AES). In 1989, this giant utility built a mammoth power plant in Uncasville, Connecticut. This plant will spew many tons of carbon dioxide into the air every day. However, to help nullify the greenhouse effect caused by these emissions, the company donated $2 million to a reforestation project in Guatemala. With this financial support, the Peace Corps, the Guatemalan Forestry Service, and 40,000 farmers planted 92 million tree seedlings. Scientists estimate that the net carbon dioxide taken in by the trees will roughly balance that emitted from the new AES power plant during its 40-year life span.

President Clinton's Climate Change Action Plan In 1993, President Clinton developed a program to slow down the buildup of greenhouse gases. It is largely based on 50 actions on the part of utilities, car makers, and other industries that would reduce the release of carbon dioxide, methane, and other gases responsible for climate warming. These actions would largely be voluntary rather than mandated by legislation. The Action Plan would cut annual greenhouse gas emissions by the equivalent of 109 million tons of carbon dioxide. Among the suggested actions of the plan are the following:

1. Plant enough fast-growing trees to absorb 10 million tons of carbon dioxide from the air every year.
2. Develop more energy-efficient electric motors. This would account for 8 percent of the carbon dioxide emission reductions proposed. This initiative has been endorsed by 27 different companies.
3. Foster programs that will reduce carbon dioxide emissions at electric power utilities.
4. Increase the energy efficiency of household appliances such as television sets, air conditioners, conventional ovens, and microwave ovens.
5. Expand the Green Lights program, under which more energy-efficient lighting is installed in businesses, through federal assistance.
6. Enact stricter federal controls on the emission of methane—one of the greenhouse gases—from landfills.

If the Climate Change Action Plan reduced carbon dioxide emissions as stipulated by the year 2000, the United States would have been able to satisfy a basic objective of the Convention on Climate Change, ratified at the 1992 Earth Summit in Brazil. Many environmentalists, however, are disappointed with the plan. They believe that its success depends too much on the voluntary cooperation of industry. Alden Meyer, of the Union of Concerned Scientists, complains that the plan doesn't commit the United States "to maintaining 1990 emission levels beyond 2000." He worries that "it could be only a one-shot deal and then business as usual."

Sustainable Solutions The previous strategies rely primarily on conservation (efficiency) and restoration efforts. But other strategies are also essential, to perhaps even eliminate the threat of global warming. Population stabilization is an obvious first step. Growth management strategies in urban areas are also important, for they can greatly reduce vehicle miles traveled and fossil fuel combustion. Recycling and tapping into the Earth's very generous supply of renewable energy are also vital components of curbing population growth. The last section in this chapter discusses efforts to reduce ozone-depleting chemicals, which also alter global temperature.

For a discussion of ways geographic information systems and remote sensing are being used to combat flooding caused in part by global climate change, see the GIS and Remote Sensing box.

19.2 Acid Deposition

"April showers bring May flowers"—or do they? Many scientists now believe that at least some April showers may bring death, not only to plants but also to fish, birds, mammals, and even humans. The problem is the acid in those April showers—in other words, **acid rain.** Acid rain is actually only part of the problem. Acids are falling from the sky in snow and in dry form (described shortly). Together, these various forms of acid fallout are referred to as **acid deposition.**

What is Acid Deposition and How is it Measured?

Acid deposition originates from two major atmospheric pollutants—sulfur dioxide and nitrogen dioxide. These gases combine with water and oxygen to produce sulfuric acid and nitric acid, respectively. Globally, sulfur dioxide is responsible for about 60 to 70 percent of the acid deposition. Nitrogen

GIS AND REMOTE SENSING

GIS AIDS EMERGENCY RESPONSE AND SURVIVAL STRATEGIES IN BANGLADESH

Bangladesh is a tiny country with huge problems. This tropical nation lies along the coastline of south Asia and is periodically devastated by floods associated with the seasonal monsoon rains and cyclones (hurricanes in North America). Flooding and storm surges cause frequent death in this densely populated land.

Because much of the nation lies at sea level, flooding during the monsoon season covers almost 20 percent of the country one year out of two. One year out of ten, 37 percent of the country is inundated. During the catastrophic flood of 1988, 67 percent of the land was under water. Problems could worsen as the populations of such nations expand and as the Earth's atmosphere warms.

Over the past decade, hundreds of thousands of lives have been lost in Bangladesh and other countries as a result of such extreme events. With no end to the disasters in sight and with millions of people crowding the low lying areas like the mouths of rivers where they eke out a living, an ambitious international research effort was recently launched. Known as the Bangladesh Flood Action Plan (FAP), the program is designed to create an advanced flood warning system for the country. It will help get news of impending disaster to villages in time so people can evacuate to higher ground. The program, when fully operational, will provide rapid emergency assistance to people stranded by flood waters. It will also develop plans to prevent future floods or, more likely, lessen their impact.

At the heart of this effort is GIS, which is being used as a planning and decision-making tool. GIS education, training, and technology transfer are supported by the International Council of Scientific Unions (ICSU) and U.S. researchers, working with faculty and students of the Geography Department at Dhaka University in Dhaka, Bangladesh.

One of the first projects was to create flood maps using handheld GPS (Global Positioning System) technology. Field surveys using hand-held GPS devices have been used to plot the extent of flooding, to locate higher points in the terrain where people escape floodwaters, and to identify escape routes. Such work provides data that is fed into a GIS and overlaid on Landsat satellite images. Knowing refuge locations and the numbers of displaced refugees will help emergency personnel deliver appropriate amounts of food, resources, and services during floods. Contrast this to the catastrophic flood in 1988 that affected the entire infrastructure of the country, including the international airport in the capital city of Dhaka, roads, and bridges. A lack of up-to-date maps and poor spatial knowledge about the nature of the disaster's effects on the landscape made it difficult for emergency response personnel to know what type of relief to provide and where to deliver it

One of the most important objectives of the Bangladesh FAP was to research the two major proposals on ways to deal with monsoon-caused flooding. One plan calls for costly, technological solutions: dams, dikes, and levees, such as those in the United States along much of the length of the Mississippi River. The other plan calls for mitigation efforts. For example, it would require efforts to plan and manage flood response better to warn people of impending storms and floods and to deliver assistance to people in refuges more quickly.

GIS, GPS, and satellite imagery are all being used to train personnel in government, non-governmental relief agencies, and academic institutions to better plan for and prevent future monsoonal flooding, hurricane-caused flooding, and death. Bangladesh not only faces the dangers of future monsoonal flooding, but also must prepare for possible sea level rise due to global warming. Efforts to use GIS and remote sensing could help in this effort as well.

dioxide is responsible for 30 to 40 percent of the rest. However, the contribution of these two gases varies from one region to the next. In the eastern United States, for example, sulfur dioxide has for years caused most of the acidity because the major pollution sources are coal-fired power plants in the upper Ohio River Valley. However, in the Rocky Mountains and along the Pacific Coast, oxides of nitrogen are the main chemical culprits. These pollutants arise primarily from automobile exhaust.

Acid deposition can be either wet or dry. Acids deposited in rain, snow, dew, fog, frost, and mist represent the **wet deposition. Dry deposition** occurs when dust particles containing sulfate or nitrate settle on Earth. Later those chemicals react with water to form sulfuric and nitric acids. Dry deposition also may occur when gases like sulfur dioxide and nitric oxide come in contact with soil, trees, lakes, and streams and then react with water to form acids.

Acidity is measured on the **pH scale,** which ranges from 0 to 14. As shown in Figure 19.4, a pH of 7 is neutral—that is, neither acidic nor basic. Numerical values above 7 indicate nonacidic or basic substances. For example, lye, with a pH of 13, is extremely alkaline. Substances that are acid have pH values below 7. Thus, battery acid, with a pH of 1, is extremely acidic. The pH scale is a special one. It is logarithmic. In logarithmic scales, a one-unit change results in a 10-fold change in the thing being measured. Thus, rain with a pH of 4 is 10 times as acidic as rain with a pH of 5 and 100 times as acidic as rain with a pH of 6. Rather surprisingly, normal unpolluted rain is weakly acidic, with a pH of 5.6. The reason is that carbon dioxide from the air reacts with the water to form a relatively weak acid known as carbonic acid.

Areas Affected by Acid Deposition

In 1980, the federal government established the **National Atmospheric Deposition Program.** It consists of acid deposition monitoring sites scattered throughout the United States. Through this program a comprehensive assessment of acidity trends can be made and maps can be drawn.

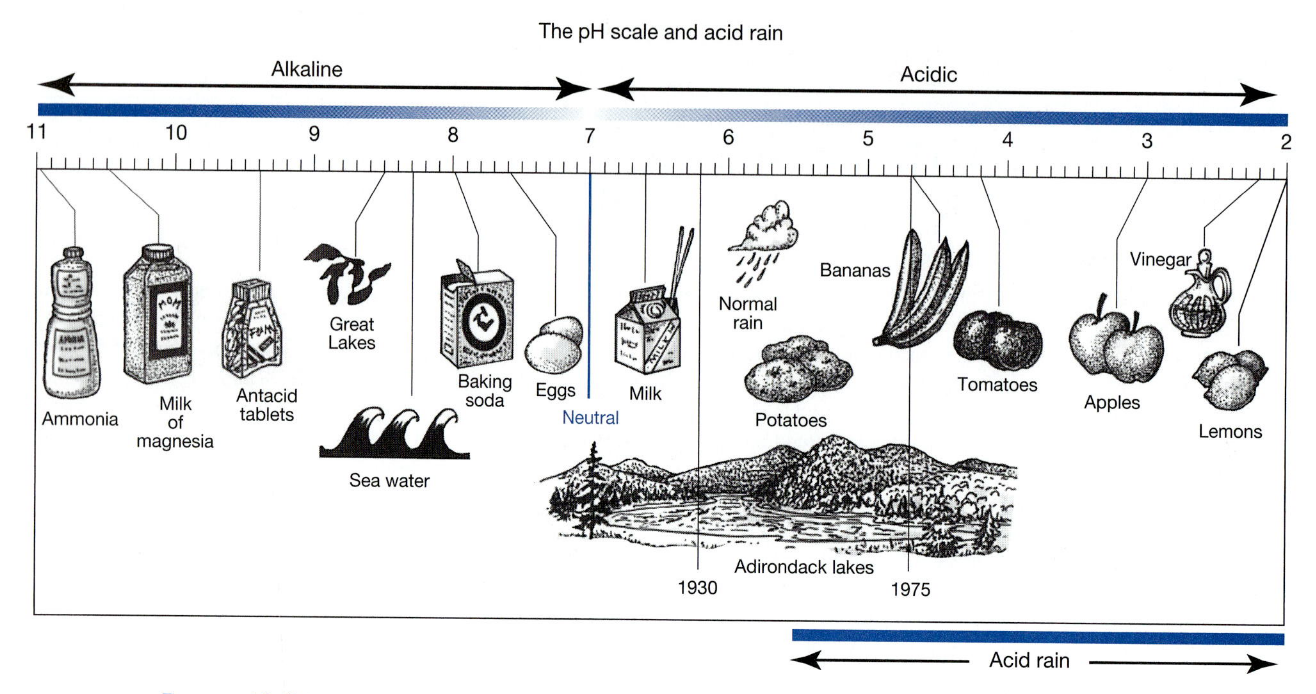

FIGURE 19.4 *The pH of acid rain compared with that of fruits, vegetables, and common household substances. Note the sharp decrease in the pH of New York's Adirondack lakes between 1930 and 1975.*

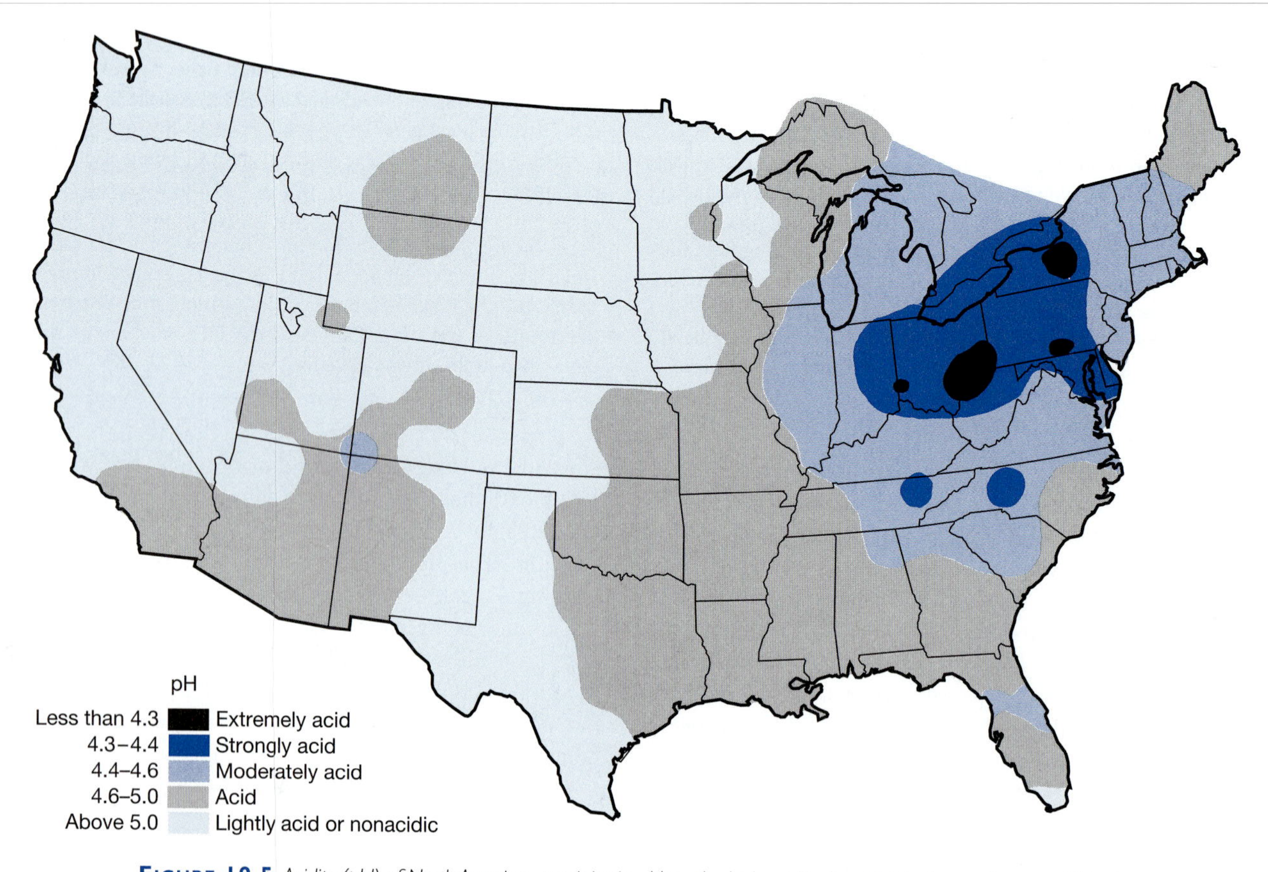

FIGURE 19.5 *Acidity (pH) of North American precipitation. Note the high acidity (low pH) in the northeastern United States and southeastern Canada.*

Figure 19.5 shows the pH of rainfall in the United States. As illustrated, the rain in the densely populated and highly industrialized eastern United States is quite acidic Over the past four decades, two disturbing trends have become evident. First, the level of acidity has been increasing. Second, the area exposed to acid deposition has grown considerably. This trend toward increasing acidity in the northeastern United States appears to be correlated with the increasing volume of sulfur and nitrogen oxides released into the air in the upper Ohio River Valley.

The average pH of rain in western New York, northern Pennsylvania, and lower Ontario, Canada, was 4.2, many times more acid than normal rain. In Los Angeles, a pH of 3 was recorded for a dense fog. The most acidic rain ever recorded in the United States fell at Wheeling, West Virginia. It had a pH of 1.4, making it considerably more acidic than lemon juice (pH 2.2) and almost as acidic as battery acid (pH 1)!

The Northeast has been the focus of many acid deposition studies. Recent work shows that many other states are now experiencing acid deposition. The California Air Resources Board, for example, has documented precipitation in the state that is among the most acidic found anywhere in the world. The pH of rain in California ranges from a low of 4.4 in San Jose and Pasadena to a high of 5.4 at Big Bear. Fog in California's Central Valley had a pH of 2.6—1,000 times more acidic than nonpolluted rain. Of considerable interest is the fact that dry acid deposition is 15 times more prevalent than acid rain in parts of southern California where rainfall is scant and the nitrogen oxide emissions from motor vehicles are high.

The United States is not the only country suffering from acid deposition. In fact, virtually every area in world downwind from a major industrial or population center experiences acid deposition. Scandinavia and Europe have been particularly hard hit, and much of what is known about the phenomenon comes from studies in these areas. Thousands of lakes in Sweden and Norway have been destroyed by acids deposited in rain and snow. The **acid precursors,** the substances that gave rise to the acids, were released by power plants and factories in England and the rest of Europe.

Canada has also been badly damaged by acid deposition, coming from sources within the nation and from the United States. At least 15,000 lakes in southeastern Canada have been destroyed because of this problem, first identified in the early 1900s.

Sources of Acid Precursors

Like most pollutants, acid precursors come from natural and human or **anthropogenic sources.** Sulfur dioxide, for example, is released from volcanoes, swamps, and power plants. Human activity now accounts for about as much sulfur dioxide emission as does nature. However, most human-generated sulfur dioxide is released from only five percent of the Earth's surface—primarily the industrialized regions of the United States, Canada, Europe, and eastern Asia. Roughly 18 million metric tons (20.4 million tons) of sulfur dioxide are emitted into the atmosphere of the United States annually: 64 percent from electric power plants, nearly 29 percent from industrial sources, and 7 percent from motor vehicles. Studies show that about 50 percent of the acid deposition in eastern Canada has its origin in the United States. In Ohio and nearby states, high-sulfur coal is king. As a result, Ohio smokestacks alone emit more sulfur dioxide into the air than those in New York, New Jersey, and the six New England states combined.

Nitrogen oxides also come from natural sources—for example, forest fires. But the major sources of concern are once again anthropogenic—primarily by cars, trucks, buses, and power plants. In the United States, these sources release 21 million metric tons (23.5 million tons) a year.

Sensitive Areas and Long Range Transport of Acids

Not all areas are equally vulnerable to acid deposition. Figure 19.6 maps those areas in the United States and Canada that are the most sensitive. As you can see, this includes much of southeastern Canada, the eastern United States, northern Minnesota, Wisconsin, and Michigan, certain areas in the Rockies, the Northwest, and the Pacific Coast. The vulnerability of a given region to acid deposition depends on the ability of the rocks and soils in the watershed (or the rocks on a lake bottom) to neutralize, or **buffer,** the acid. Soils derived from limestone, which are rich in calcium, are much more capable of neutralizing acids. Soils derived from granite, which are low in calcium, are highly vulnerable.

Researchers at Colorado State University have found that a given molecule of sulfur dioxide may remain in the atmosphere for up to 40 hours, and a sulfate particle may remain aloft for 3 weeks. Because of their relatively long residence in the atmosphere, these molecules may be carried hundreds of kilometers from their release point (Figure 19.7). Sulfur dioxide molecules that originate in an Arizona copper smelter may eventually fall in acid rain on a sensitive mountain lake in New York or Vermont. A molecule of sulfur dioxide that originated in Ohio may be transported to southern Ontario (Figure 19.8). It is estimated that 87 percent of the sulfate in New York and New Jersey and 92 percent of the amount in New England is delivered from outlying areas—most likely the upper Ohio River Valley (Figure 19.8). On the other hand, California generates all its own acid rain, receiving almost no inputs from outside its boundaries. This situation provides California legislators with a strong motive to control acid deposition, knowing that any reductions achieved in sulfur dioxide and nitrogen oxide emissions will result in a reduction in acid deposition in their state.

As noted in the previous chapter, superstacks, tall smokestacks that sent pollutants high into the atmosphere, were once thought to be an effective solution to air pollution. Although they did reduce ground-level pollutants near major sources, they simply transferred the problem to ecosystems and people downwind from the sites. Many superstacks served as mammoth sources of acid precursors. The superstack at the giant smelter at Sudbury, Ontario, towers to a height of 380 meters (1,140 feet) and releases 2,300 metric tons of sulfur dioxide into the air *daily.* It emits twice as much sulfur annually as did Mt. Saint Helens during its most active eruption year! It is the largest stack in the world and serves as a symbol of the acid rain problem in North America. Incredibly, this one stack gives off 1 percent of all the sulfur dioxide released worldwide!

Harmful Effects of Acid Deposition

Acids have many adverse effects on buildings and materials, soils and crops, ecosystems and the organisms within them, and human health. This section outlines the major impacts.

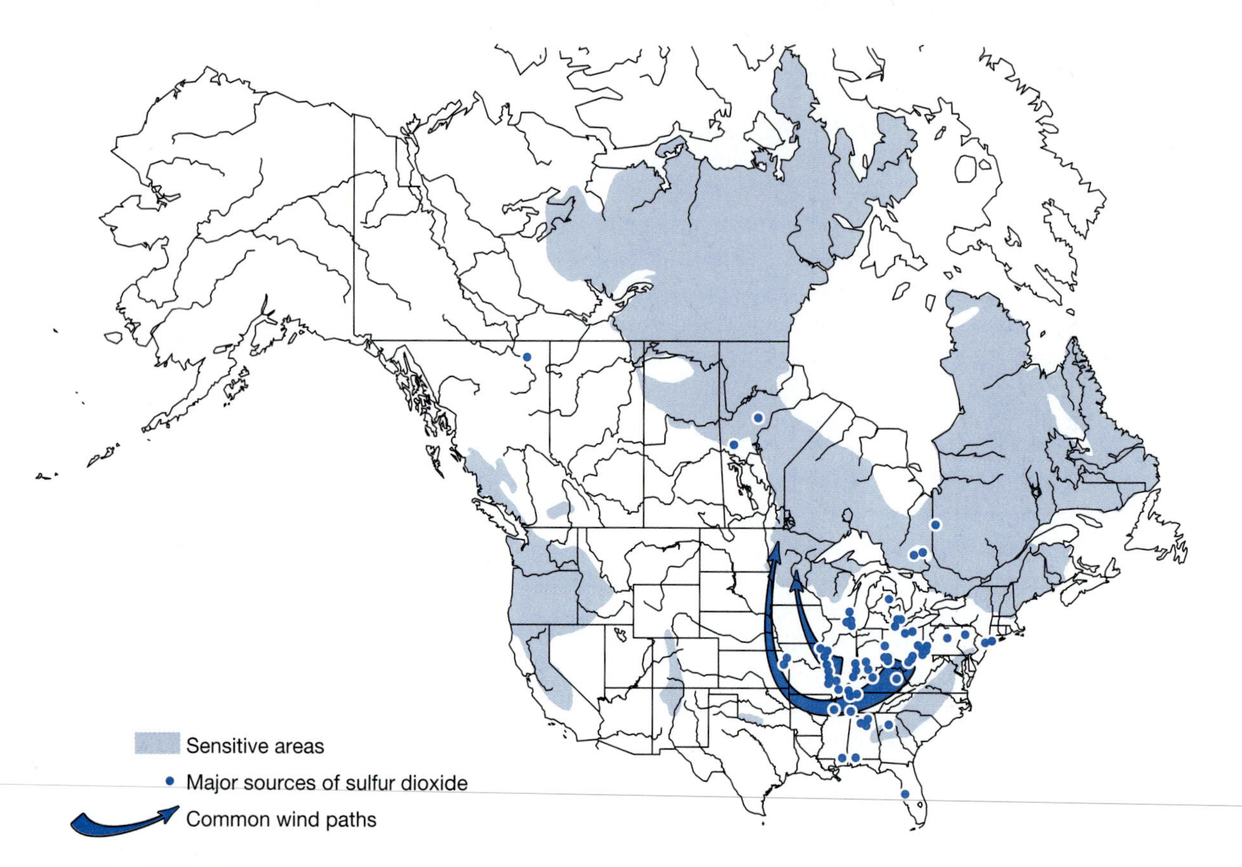

FIGURE 19.6 *Areas of acid deposition in North America. Each dot represents an area that emits more than 100,000 tons of sulfur dioxide annually. Arrows indicate direction of major wind movements. Shaded areas are low in natural buffers such as limestone and therefore are particularly susceptible to acidification.*

Effects on Buildings and Materials Acids erode the surfaces of structures made from sandstone, limestone, marble, and other materials (Figure 19.9). Priceless monuments of ancient Rome and Athens have been seriously damaged by acids, as have the Statue of Liberty, the Washington Monument, and the United States Capitol building. The marble statues in front of the Field Museum in Chicago have undergone rapid deterioration in only the past 30 years. Numerous metallic structures such as bridges, rails, and industrial equipment have been corroded by acid rain. The U.S. Air Force spends millions of dollars annually to repair acid rain–inflicted damage to their B-52 bombers and other planes. The overall damage to buildings and materials by acid rain has exceeded $5 billion annually in 17 eastern states alone. No estimates of global destruction are available, but the figure must be incredible.

Effects on Soils and Crops Acid deposition adds hydrogen ions to the soil. Hydrogen ions displace nutrient elements like calcium, potassium, and magnesium from the soil particles to which they are bound. Such nutrients may then be leached from the soil and carried away by runoff water. Acid deposition also inhibits the activity of nitrogen-fixing bacteria. As a result of these two effects alone, acid deposition may substantially reduce fertility in certain soils within a decade. Heavy metals, such as aluminum, that normally are harmlessly bound to soil particles are also displaced by the hydrogen ions in acid deposition. These metals then become soluble in water and can be absorbed by the roots of farm crops or trees, often with harmful results (Table 19.1).

Changes in the soil chemistry affect plant growth. But plants can be directly affected by acids. Acids can leach minerals directly from leaves. They can also kill buds and stunt growth. Acids in soil can reduce germination of seeds and impair growth of seedlings. Crop damage in the United States is estimated to be at least $5 billion per year.

In forests, acids deposit on needles and kill them, effectively reducing the photosynthetic capability of trees (Figure 19.10). Acids may also stimulate the growth of acid-tolerant mosses that grow on the forest floor. These impair tree growth by trapping moisture in the surface layers, drowning the feeder roots. Entire forests in the northeastern United States and Europe have perished because of acid deposition.

Effects on Aquatic Ecosystems Lakes and streams vary in their ability to cope with acid deposition. The most vulnerable are the soft-water lakes that contain relatively few dissolved minerals, such as calcium salts, that can neutralize acid. Their watersheds also have little or no soil, and what exists has little buffering capacity. Many lakes in the northeastern United States and in the Rocky Mountains and Sierra Nevadas are of this type. On the other hand, in the southern

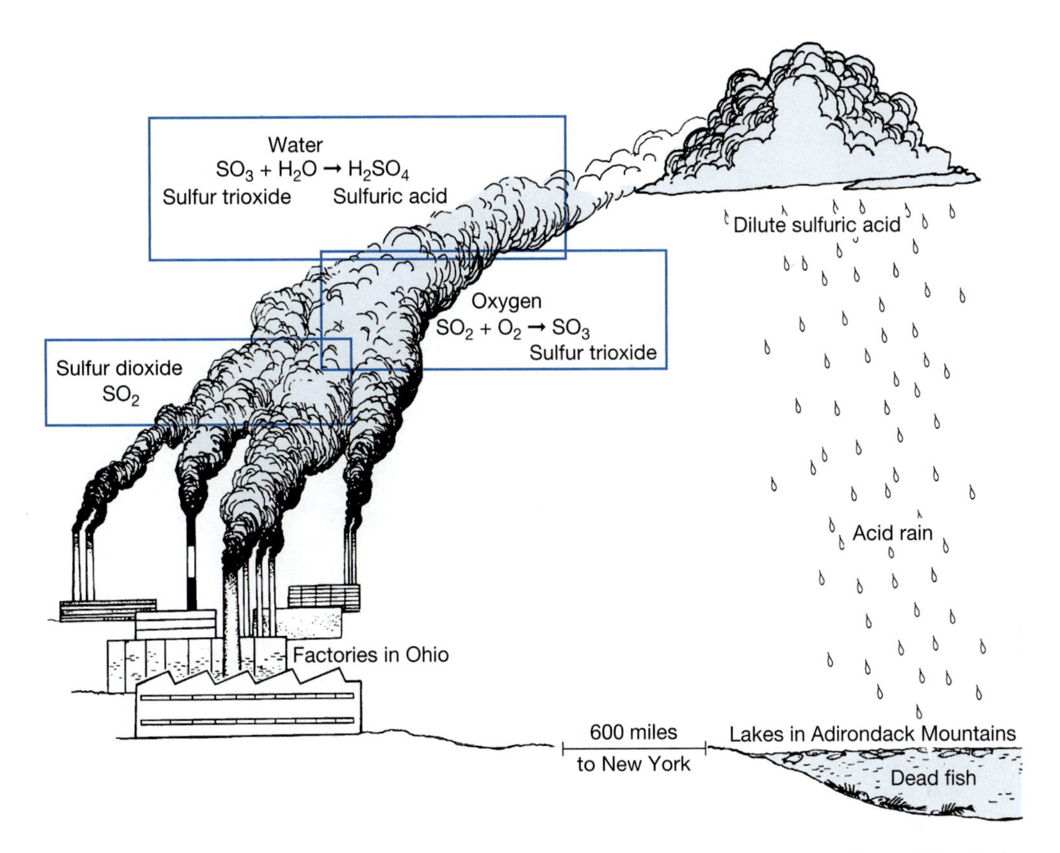

FIGURE 19.7 *Much of the acidity of the rain that falls on the Adirondack lakes in New York and destroys fish and other aquatic life has its origin in sulfur and nitrogen oxides (not shown) emitted from industrial smokestacks in the Midwest.*

and central regions of the United States, most lakes are hard-water lakes and contain acid-buffering minerals. Their watersheds tend to have well-developed soils with good buffering capacity. In Canada, many lakes in the Canadian shield, a region in the southeastern portion of the nation, are underlain by granite. The lack of acid-neutralizing capacity of these lakes makes them vulnerable to acid deposition. Mountainous regions in the Northern Rockies and central British Columbia, on the other hand, are more resistant to acid deposition.

Acid deposition has poisoned thousands of lakes in Scandinavia, the province of Ontario, southeastern Canada, and the northeastern United States. In the Adirondack Mountains of New York, 237 lakes have a pH of less than five—an acidity level lethal for many species of fish. In Canada, 300,000 lakes are vulnerable to acid deposition. At least 5,000 have become so acidic that they have suffered extreme fish loss. Furthermore, according to Environment Canada, Canada's equivalent of the U.S. EPA, in Nova Scotia, acidification of rivers has resulted in a dramatic decrease in Atlantic salmon.

Although a recent survey of lakes in the United States show that nationwide very few lakes have suffered from acid deposition, some areas are under extreme pressure. The U.S. Office of Technology Assessment conducted a survey of the acid rain problem in 27 eastern states. Their conclusions are as follows:

1. Eighty percent of the lakes and streams in the Northeast and upper Midwest are vulnerable to acidification.
2. Seventeen percent of the lakes and 20 percent of the streams in 27 states have been damaged by acid deposition.
3. Of the 17,059 lakes studied, 2,993 have already been acidified.
4. Of the 187,876 kilometers (117,423 miles) of streams investigated, 39,500 kilometers (24,688 miles) have already been damaged.

Harmful Effects of Acid Deposition on Lakes and Streams

Acids have many impacts on aquatic ecosystems:

1. They alter reproduction in many species, including fish (Figure 19.11).
2. Acidified waters may disrupt the homing ability in sexually mature salmon. Research conducted at the University of New Hampshire has shown that when the pH of water is between 5.0 and 5.5, salmon apparently lose their ability to detect certain odors in the water. Since adult salmon depend on chemicals to guide them to their native streams so that they can spawn, the survival of the species in acidified streams is threatened.
3. Increased numbers of fish embryos and newly hatched young develop abnormally and eventually die in waters in which the pH is below 5.5.
4. Populations of important fish food organisms, such as crustaceans and insect larvae, rapidly decline.
5. Game fish like bass and pike are replaced by more acid-tolerant but less desirable species such as bullheads and suckers.

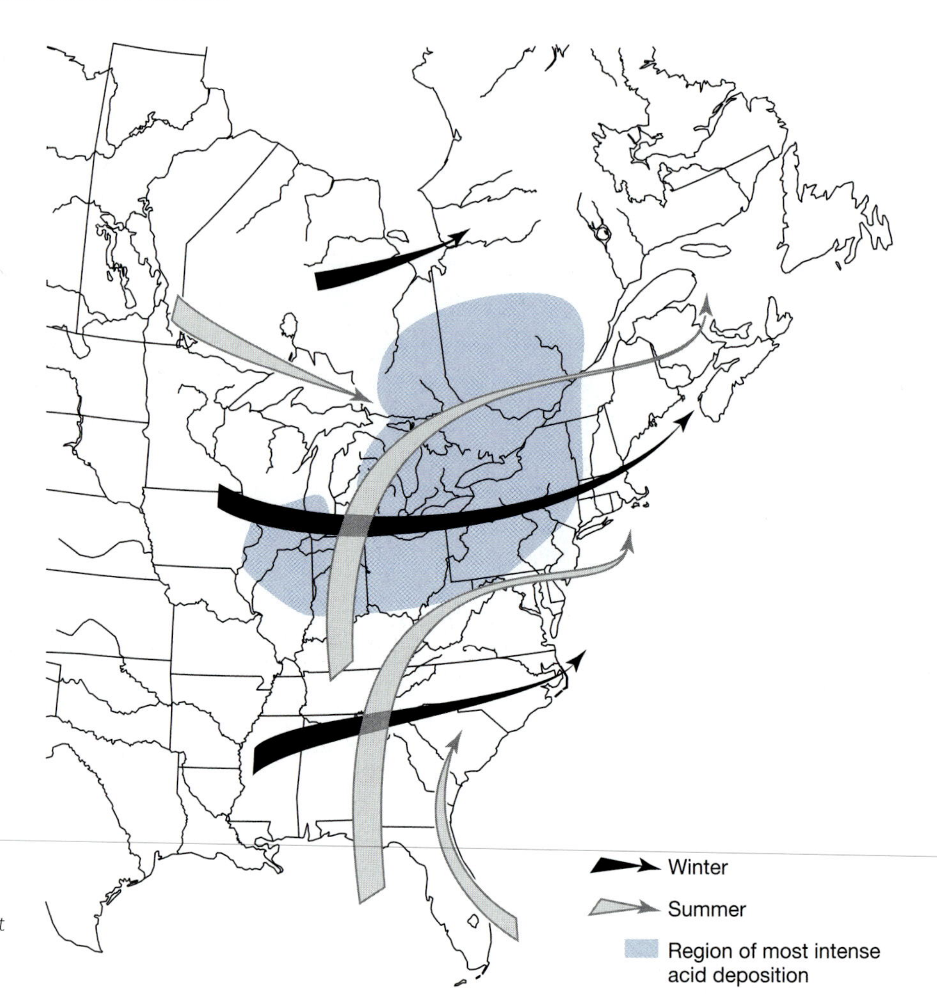

FIGURE 19.8 *Wind transports acid to distant areas. Southeastern Canada and Northeastern United States are particularly vulnerable.*

FIGURE 19.9 *Acid rain is slowly causing the erosion of well-known statues in Europe and North America. Among them is this statue located in Versailles, France. The damage costs society billions of dollars a year.*

6. Acids leach heavy metals such as aluminum, mercury, copper, zinc, and nickel from soils. These toxic chemicals are then washed into streams and lakes. Many researchers believe that heavy metals, more so than acids themselves, are a major cause of fish deaths. Aluminum is toxic to fish at concentrations of less than 1 ppm. At this concentration, it causes mucus buildup on gills and fish die by asphyxiation. Mercury builds up in the bodies of fish as a result of biological magnification in the food chain. In some acidic lakes in Wisconsin, predatory fish such as salmon, lake trout, and northern pike may have mercury concentrations above 2 ppm—more than twice the level considered safe for human consumption.
7. The skeletons of fish may become weakened by decalcification caused by acids. As a result, their bodies become distorted and the fish lose their ability to swim. Death from starvation, disease, or predation soon follows.
8. Bacterial decomposition is inhibited. As a result, essential nutrients like nitrogen and phosphorus stay locked up in plant and animal remains and are not available to terrestrial and aquatic plants, the producer base of food webs.

FIGURE 19.10 *Pines killed by acid rain in Czechoslovakia. Note defoliated and dead trees. Scientists believe that acid deposition may have contributed to this die-off.*

TABLE 19.1 *Pollution-Related Forest Declines of the Last 50 Years*

Widely assumed major role
Massive die-off of forests in Europe (*Waldsterben*)
Decline of ponderosa and Jeffrey pines in the San Bernardino Mountains of California
Regional decline of white pine in the eastern United States and Canada
Possible major role
Decline of red spruce, and balsam, and Frasser firs at high elevation in the Appalachian Mountains from Georgia to New England
Growth decline without other visible symptoms in loblolly, shortleaf, and slash pines in the Piedmont regions of Alabama, Georgia, North and South Carolina
Growth decline without other visible symptoms in pitch and shortleaf pine in the Pine Barrens region of New Jersey
Widespread dieback of sugar maples in northeastern United States and southeastern Canada
Declines related to biological or physical factors
Tannensterben (white fir decline) in Europe
Kiefersterben (Scotch pine decline) in East Germany and European Russia during early 1970s
Oak decline in Germany and especially in France since early 1900s
Decline of *Pinus pinaster* on the Atlantic coast of France since early 1980s
Beech bark disease in northeastern United States and southeastern Canada
Littleleaf disease of shortleaf pine in the southeastern United States
Birch dieback in northeastern United States and southeastern Canada
Poleblight of western white pine in Rocky Mountains
Maple decline in northeastern United States and southeastern Canada
Oak decline in Pennsylvania, Virginia, and Texas
Ash dieback in northeastern United States and southeastern Canada
Sweetgum blight in southeastern United States

Note*:* A number of the declines attributed to biological or physical factors might involve toxic air pollutants, but studies are incomplete, and evidence is insufficient to make a strong connection.

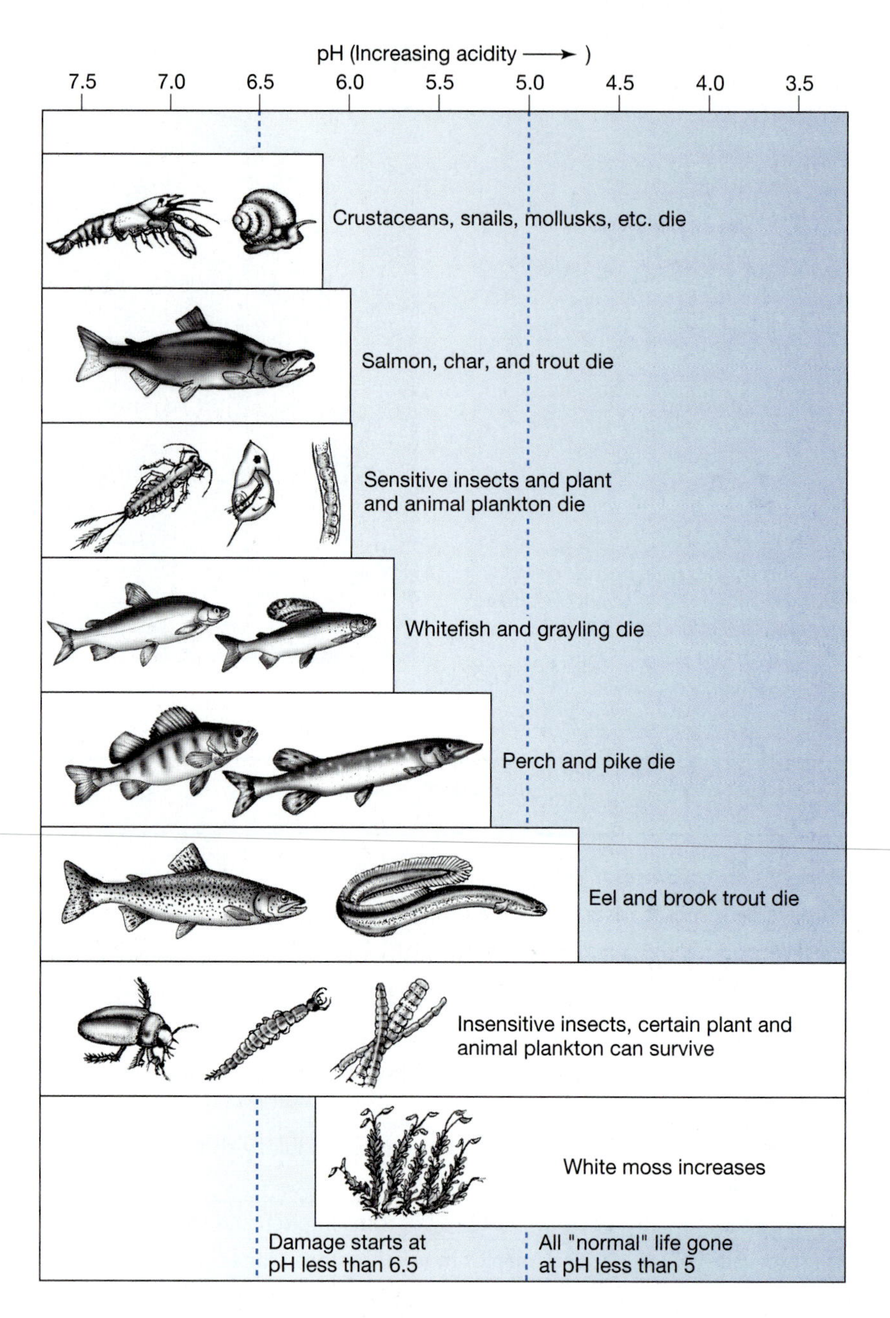

FIGURE 19.11 *Effect of water acidity on aquatic organisms.*

Clearly, acid deposition can have a profound impact on lakes, ponds, streams, and rivers. What is more, these impacts tend to work together, combining forces, to eliminate virtually all life from a lake. Some times of the year are worse than others, too. When snow containing acids begins to melt, for example, huge amounts of acid are released at the beginning of the snow melt, creating an **acid shock.** Concentrations of acids in the initial meltwater can be 5 to 10 times higher than during rainfall. Although adult fish are able to survive the sudden influx of acids, eggs and fry are more sensitive and often perish.

The impact of acid rain is not just biological. It is also economic. Acid deposition can literally destroy sport fishing in an area, eliminating millions of dollars in tourist revenues. Loss of revenues from the sales of fishing tackle and bait can also be substantial.

Effects On Human Health Fish are not the only organisms that are adversely affected by acid deposition. Humans are vulnerable as well. One of the greatest air pollution disasters the world has known occurred in London in 1952. The "killer fog" that shrouded the city for 5 days contributed to the death of more than 4,000 people from bronchitis, pneumonia, and heart disease. Although pH measurements of the fog were not recorded at the time, present-day scientists estimate that the fog was a highly diluted sulfuric acid mist with a pH ranging from 1.4 to 1.9! But that was almost four decades ago. Is acid deposition still a killer? According to the U.S. Office of Technology Assessment, more than 50,000 premature deaths are caused annually by sulfate-laden air. Acid precursors are the second largest cause of lung disease (after smoking) in the opinion of Dr. Phillip J. Landrigan of the Mount Sinai School of Medicine in New York City.

Toxic metals in drinking water can be harmful if ingested by humans. Today more than 40 million people in the United States, or about one in six, are drinking water with lead levels above 20 ppb—the level considered safe by the EPA. Where did the lead come from? It apparently was leached from galvanized water pipes and lead solder in copper pipes by the acidic drinking water. Such intake of lead by millions of Americans is cause for considerable concern because medical researchers have shown lead to be one of the factors responsible for high blood pressure and heart attacks in adults and for brain damage in children.

Some health experts believe that acid deposition may also be indirectly responsible for many cases of Alzheimer's disease. This is a disease of the elderly characterized by degeneration of the brain and severe loss of memory. Chemical analyses of the brains of people who died from the disease have revealed relatively high levels of aluminum. Some researchers believe that aluminum may have caused the disease. If so, where did it come from? How did it get into the body? The involvement of acid deposition is strongly suggested.

In Canada, more than 80 percent of the population lives in regions in which precipitation is high in acid. According to Environment Canada, thousands of people in this area have respiratory problems that may be caused by acid precursors. Children are especially vulnerable. In communities whose air is polluted by acids, sulfates, and nitrates children suffer from more chest colds, allergies, and coughs.

Controlling and Eliminating Acid Deposition

As in most issues, there are many ways to reduce the problem. This section looks at many of these solutions. It also examines ways to prevent the emission of acid precursors and eliminate the problem altogether. Many of these solutions will be further elaborated in Chapters 21 and 22, which discuss alternative energy strategies.

Switching from High-Sulfur to Low-Sulfur Coal The coal once burned in electric generating plants and industrial boilers varied greatly in sulfur content, from less than 1 percent to almost 6 percent. Switching to low sulfur coal cuts emissions substantially, and many utilities in the United States and Canada have chosen this strategy. Rather than install expensive pollution control devices, they were able to cut sulfur emissions as required by the Clean Air Act of 1990 (Chapter 18). Many did so at a huge cost savings, too.

The question arises, are our nation's supplies of low-sulfur coal sufficient for this purpose? Apparently, the answer is yes. A group of low-sulfur mining industries and environmentalists who have formed the Alliance for Clean Energy has determined that more than 14.1 billion metric tons of low-sulfur coal is available—adequate to meet the needs of our utilities and industries for many years.

As important as this strategy is, we must remember that switching to low-sulfur coal only eliminates sulfur dioxide. It does not affect nitrogen oxide emissions or carbon dioxide. At best, then, it is a temporary solution.

Using Smokestack Scrubbers Scrubbers were discussed in Chapter 18 as a means of pollution control. These devices remove particulates and sulfur dioxide. Leonard Kreisle, professor of engineering at the University of Texas—Austin, recently announced the development of a new smokestack device for removing sulfur dioxide emissions. Known as a **synergistic reactor,** it has several advantages over the smokestack scrubbers used to date: (1) it removes 100 percent of the sulfur dioxide; (2) it is much smaller; (3) it uses only one third as much energy; (4) it acts in seconds rather than minutes; and (5) gypsum, the only byproduct, has high commercial value for use in wallboard manufacture—that is, for making dry wall. Several companies are now producing drywall for building homes made from waste from conventional scrubbers (Figure 19.12).

It is believed that the scaled-up version of the synergistic reactor will remove 2,300 kilograms (5,000 pounds) of sulfur

FIGURE 19.12 *Scrubber waste (shown here) can be used to make drywall.*

dioxide per minute. As the gaseous emissions pass through the reactor, they enter a chamber filled with finely ground limestone and steam. It is here that the sulfur dioxide reacts with the limestone to form gypsum. The synergistic reactor has attracted considerable attention from the EPA, as well as environmental agencies in Canada, Europe, Russia, and Japan.

The installation of scrubbers, like switching to low-sulfur coal, is an important means of drastically reducing sulfur emissions. However, they're expensive to build and operate. Like its companion strategy, it does nothing to remove nitrogen dioxide, another component of acid deposition. Nor do scrubbers as currently designed remove carbon dioxide, a major contributor to the greenhouse effect.

Energy Conservation Writing for the highly respected Worldwatch Institute, Sandra Postel argues that energy conservation is both an effective and a relatively inexpensive strategy for controlling the emissions of sulfur dioxide and nitrogen oxide, the precursors of acid deposition, as well as carbon dioxide. It is, in fact, one of the most important strategies for creating a sustainable society. Consider a simple example.

In 1987, the U.S. Congress passed the **National Appliance Energy Conservation Act.** It required manufacturers to greatly reduce energy consumption by common appliances such as air conditioners, refrigerators, and water heaters. According to Postel, this law will reduce the nation's consumption of electricity by 70,000 megawatts per year, which is equal to the output of 140 large coal fired power plants. Postel estimates that if the United States could cut its electrical consumption by half, a goal attainable with current technology, the nation could reduce its coal combustion by 85 million metric tons of coal per year. The net result: a 4-million-ton (or 16 percent) annual reduction in sulfur dioxide emissions. Nitrogen dioxide emissions would also plummet, as would carbon dioxide releases. Moreover, this would cost only 1 percent of the $5 to $10 billion expenditure on the smokestack scrubbers, which would have the same effect. Additional energy conservation techniques are discussed in Chapter 22.

Renewable Energy Yet another preventive measure, and one that will become increasingly important in years to come as the human population and global economy expand, is renewable energy—solar energy, wind power, and the like. Although pollution is generated to make solar panels, wind machines and other devices to capture renewable energy, once operating these devices produce no carbon dioxide, nitrogen dioxide, or sulfur dioxide. Overall, they produce very little air pollution, and therefore are an excellent means of preventing pollution. Chapter 22 outlines many of the options and discusses their pros and cons.

Federal Legislation Environmental problems caused by acids in rain and snow have been understood since the 1880s. Lawsuits over acid rain caused by smelters were tried in the 1920s. It was not until the 1970s, however, that the true magnitude of the problem became well known in the 1970s. Still, it took nearly another 20 years for the Federal government to enact legislation, the Clean Air Act Amendments of 1990, to control acid deposition. As noted in Chapter 18, the **Clean Air Act Amendments of 1990,** called for reductions in sulfur dioxide emissions. It set limits on emissions from major sources such as power plants and developed a marketable permit system that allowed companies that were able to cut sulfur dioxide releases to sell allowances to others. Industry sources claimed that sulfur dioxide controls imposed by the Clean Air Act would cost as much as $3 to $11 billion.

Although this market-based system has been heralded as a success, most companies pursued other avenues to meet emission standards. Low-sulfur coal was the option of choice. Many companies were able to buy it cheaper or at the same price they were paying for higher sulfur coal. Because of the law, they were able to get out of contracts for high-sulfur coal that had built in escalation clauses—wording that ensured a steadily increasing cost for the coal the were buying. In short, many businesses benefitted from the law, as did our lakes, rivers, and people. The real cost was just slightly over $800 million a year.

Several states also passed acid rain legislation prior to the Clean Air Act Amendments of 1990. In 1991, after years of foot dragging, the U.S. government signed an agreement with Canada, known as the **Air Quality Accord,** that seeks to control the flow of acid pollutants across the border. In this agreement, total emissions of acid pollutants are capped at 13.3 million metric tons of sulfur dioxide for the United States and 3.2 million metric tons of sulfur dioxide for Canada. The agreement also calls for a reduction in nitrogen oxide emissions.

Canada, like the United States, limits the release of sulfur dioxide from industrial sources. Most companies have opted to use low sulfur coal to meet emissions requirements. Such efforts have resulted in some promising results. Clearwater Lake near Sudbury, Ontario was once heavily acidified with a pH of 4.1. Today, the water measures 4.7. Efforts in the U.S. are also starting to show some modest return.

19.3 Depletion Of Stratospheric Ozone

Ozone is a chemical component of the atmosphere that baffles many people. At ground level, it is a component of photochemical smog (Chapter 18). Here, it irritates the eyes and causes damage to lungs. You'll witness its effects on a polluted summer day in virtually any major metropolitan area. However, ozone is also found in the upper atmosphere.

As you may know from previous studies, the Earth's atmosphere is divided into several layers. The lowest region which is rich in oxygen is called the **troposphere** (tropo means to nourish). It extends from the Earth's surface up to about 6 miles or 10 kilometers (km) in altitude (Figure 19.13). Virtually all human activities take place in the troposphere, because it contains the oxygen we need to survive. Mt. Everest, the tallest mountain on the planet, is only about 9 km high. The next layer, the stratosphere, continues from 6 miles (10 km) to about 30 miles (about 50 km). Most commercial airline traffic occurs in the lower part of the stratosphere. Most atmospheric

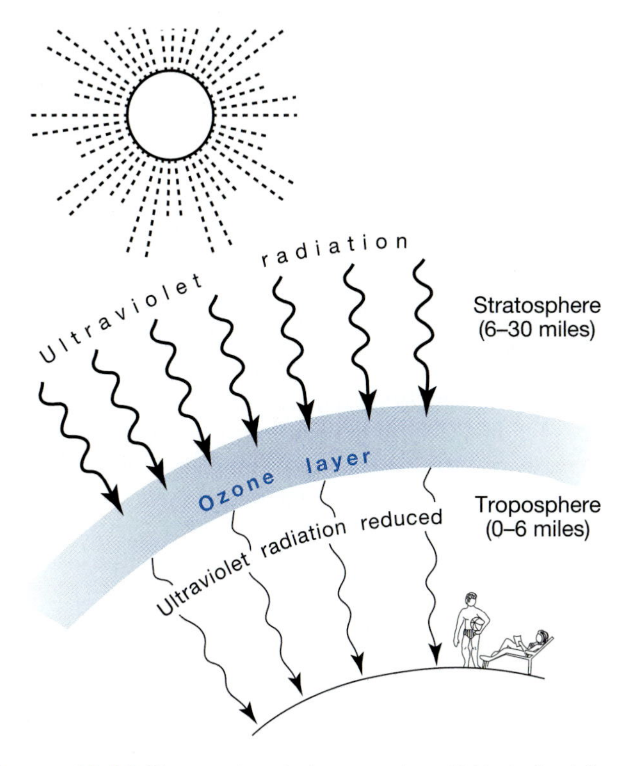

FIGURE 19.13 *The ozone layer in the stratosphere shields the Earth from potentially lethal ultraviolet radiation.*

ozone is concentrated in a layer in the stratosphere, about 9 to 18 miles (15–30 kilometers) above the Earth's surface. This is called the **ozone layer.**

Ozone is a molecule containing three oxygen atoms. Its chemical formula is O_3. Normal oxygen, the kind we breathe, has two oxygens (O_2). Ozone gas is blue and has a strong odor. Gaseous oxyen is colorless and odorless. Of the two, ozone is much less common. Out of each 10 million air molecules, about 2 million are normal oxygen, but only 3 are ozone.

All life on Earth is dependent on the presence of this ozone layer. Why? Ozone molecules filter out a portion of the sun's longer wave ultraviolet radiation, known as Ultraviolet B (UVB) radiation (Figure 19.13). Were this absorption screen of ozone not present, or were it to be thinned out, much more UVB radiation would reach the Earth's surface. The results, say the experts, would be catastrophic. This chapter deals with ozone in the stratosphere. Lower level ozone problems were discussed in Chapter 18.

Until a few years ago, the ozone layer was in a state of dynamic equilibrium. In other words, ozone-forming and ozone-depleting reactions proceeded at similar rates, and therefore the amount of stratospheric ozone remained fairly constant. It is true, however, that there might be variations in abundance as high as 10 percent from year to year. And even in a given year, the amount of ozone in the stratosphere might be two or three times greater in one place than another.

In 1974, F. S. Rowland, of the University of California, startled the scientific community with this speculation: **chlorofluorocarbon gases (CFCs)** used to propel deodorants, hair sprays, shaving creams, and insecticides from spray cans, and used in refrigerators and air conditioners could be depleting the ozone layer. Previously, it was thought that these chemicals were inert, hence their widespread use.

Widespread concern soon developed, because the total amount of CFCs released into the global atmosphere was considerable. In 1988, for example, CFC production reached an all-time high of 1.2 million tons (Figure 19.14).

Chlorofluorocarbons are rather simple but useful chemicals. They have a central carbon atom to which are attached chlorine and fluorine atoms. Two CFC compounds were once commonly used in the United States and abroad with Dupont Corporation being the major producer. One compound is CFC-11 ($CFCl_3$), a spray can propellant, and the other is Freon-12 (CF_2Cl_2), a refrigerant. Both of these compounds are extremely stable and inert under ordinary environmental conditions close to Earth. As a result, they will not react chemically either with the contents of the spray can or with the can itself. It was thought that they wouldn't react with chemicals in the environment, either, a characteristic that made these substances desirable as propellants.

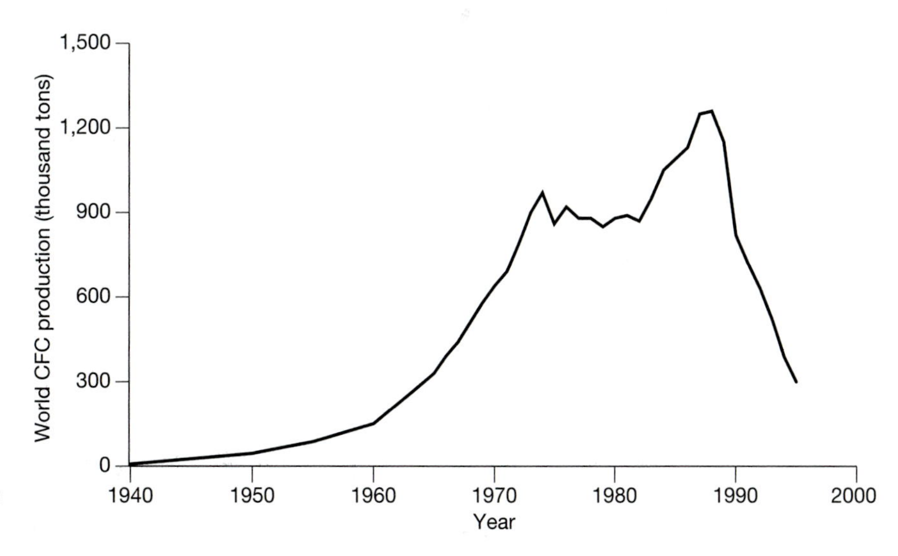

FIGURE 19.14 *World production of CFC's, from 1940 to 1995.*

But the propellant molecules used by men to spray foamy shaving cream on their beards escaped through the bathroom window, then floated into the sky. Within 5 years or less, these CFC molecules made their way into the stratosphere.

Leaky refrigerator cooling systems or automobile air conditioners also emitted CFCs that escaped into the atmosphere. Once in the stratosphere, though, these ordinarily stable compounds are exposed to intense radiation from UV light. This caused the molecules to decompose and release chlorine free radicals, highly reactive atoms. We shall follow this process by using $CFCl_3$ as an example:

$$CFCl_3 + \text{UV light} \rightarrow Cl\bullet \text{ (chlorine free radical)}$$

The Cl free radicals then react with ozone (O_3), causing it to form into molecular oxygen (O_2). One chlorine atom can cause the breakdown of 100,000 molecules of O_3. The reaction is as follows:

$$\underset{\text{(ozone)}}{O_3} + \underset{\text{(chlorine free radical)}}{Cl\bullet} \rightarrow \underset{\text{(chlorine oxide)}}{ClO} + \underset{\text{(oxygen)}}{O_2}$$

This reaction removes ozone from the ozone layer. Further reactions, too complex to include here, regenerate the chlorine free radical so that it can react with additional ozone molecules.

When the CFCs were first synthesized in the 1920s, they seemed too good to be true. They are stable, nonflammable, and nontoxic. Moreover, they vaporized at low temperatures. Furthermore, they can be produced very cheaply. As a result, they had many uses. They are used in refrigeration and air-conditioning systems, as solvents, and as sterilants, as well as in the production of plastic foams. For many years, the McDonald's fast-food chain used plastic foam containers. Most of the CFCs in the various products and applications have now found their way into the atmosphere. When foam-plastic burger holders crack open, for example, or when junked refrigerators rust and deteriorate, CFCs are released into the air and eventually contribute to ozone depletion.

CFC Accumulation and Thinning of the Ozone Layer

The atmospheric concentrations of the ozone-depleting gases are measured each day by scientists in Oregon, Ireland, Tasmania (off the coast of Australia), the Barbados (northeast of Venezuela), Samoa (the South Pacific), New Zealand, and other countries. In 1989, the atmospheric concentrations of CFC-11 and CFC-12 in the ozone layer were about 230 and 400 parts per trillion, respectively. The concentration of these gases was increasing at the rate of 5 percent annually. By 1998, the latest year for which data were available, CFC-11 and CFC-12 had reached 267 and 532 ppt, respectively (Figure 19.15).

Because the average life span of a $CFCl_3$ molecule is about 75 years and that of CF_2Cl_2 is 110 years, CFCs can disperse throughout the stratosphere all the way around the globe. As a result, they have become widely dispersed over both heavily industrialized and remote areas. Has the accumulation resulted in a decline in the ozone layer?

Yes. The most noticeable declines have been in Antarctica. In 1979, atmospheric scientists were both surprised and mystified by the appearance of a gigantic "hole" in the ozone shield during the fall and winter months over Antarctica (Figure 19.16). The level of ozone in this hole, which extended over the entire Antarctic continent, was 50 percent below normal. In 1988, the Ozone Trends Panel of the National Aeronautics and Space Administration (NASA) concluded: "The weight of evidence strongly indicates that man-made chlorine compounds are primarily responsible for the hole." Recent studies indicate that the ozone breakdown is accelerated in the presence of ice crystals

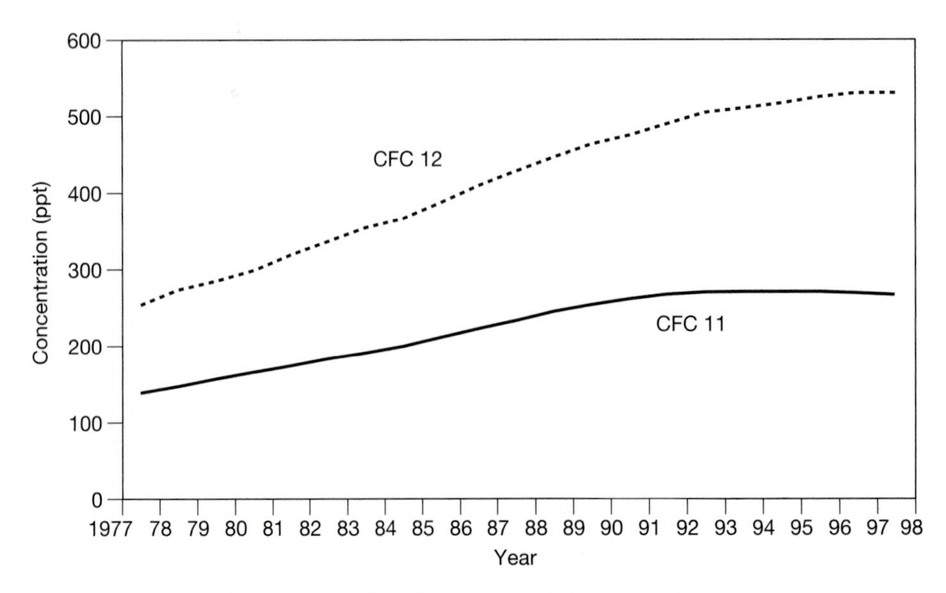

FIGURE 19.15 *CFC concentration in lower atmosphere.*

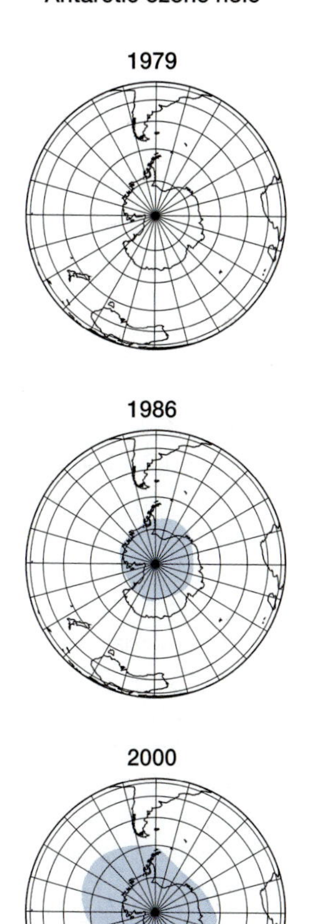

FIGURE 19.16 *Growth of the ozone hole (shaded area) over Antarctica.*

and sunlight. Although the greatest amount of ozone thinning has been caused by man-made CFCs, sometimes a natural event, such as a volcanic eruption, will contribute to the depletion of ozone. In 1992, Mt. Pinatubo in the Philippine Islands erupted and injected untold trillions of tiny sulfate droplets into the atmosphere. It is no coincidence that in the fall of 1993 the ozone hole over Antarctica was the largest ever recorded, according to David J. Hofmann of the National Oceanic and Atmospheric Administration. More than 70 percent of the ozone blanket was destroyed in a 23-million-square-kilometer area—almost the size of the entire North American continent. Scientists theorized that the trillions of tiny sulfate droplets injected by the eruption provided the surface on which the ozone-depleting reactions could occur. The implications are disturbing. As Rowland stated: "There's a distinct worry that what's happening in Antarctica could happen here."

In 1994, scientists reported a dramatic worsening of ozone depletion, not only in polar regions but over much of the Northern Hemisphere. Rumen Bujkov, of the U.N. International Ozone Commission, reported the ominous news at a 1994 meeting of the American Association for the Advancement of Science. According to Bujkov, levels of ozone have dropped 10 percent globally over the past 25 years. More serious for Americans, however, is that the ozone shield over the United States was thinned by 15 percent during the winters of 1992 and 1993. On some days the depletion was as high as 45 percent. Similarly high levels of ozone depletion have been recorded over Europe and Siberia.

Some scientists have maintained that even if the protective ozone blanket was partially destroyed, the increase of UVB radiation would be absorbed by clouds and atmospheric pollution. In some areas, this has been true. However, two Canadian scientists reported that the amount of UVB radiation in the air over Toronto, Ontario, increased 5 percent each year from 1989 to 1993, coincident with the annual thinning of the ozone shield. More and more studies are showing similar results.

Harmful Effects of UVB Radiation

Because CFCs take many years to migrate into the ozone layer and because they last for so long, ozone thinning is bound to get worse before it gets better. By one estimate, the ozone layer will deteriorate for another 100 years before it starts to mend. Others predict that the healing time will be much shorter, maybe 50 years. Only time will tell for sure. In the meantime, what can we expect from the loss of ozone?

Like many natural components of our environment, UVB light is beneficial in small amounts. Ultraviolet radiation tans light skin and stimulates vitamin D production in the skin. However, excess UVB exposure can cause serious problems. For example, it can cause serious skin burns, cataracts (clouding of the lens), and skin cancer in humans (Figure 19.17). Increased exposure to UV light may also suppress the human immune system, making us more susceptible to infectious diseases.

EPA researchers estimate that a 1 percent depletion of the ozone layer would lead to a 0.7 percent to 2 percent increase in UVB light striking the Earth. This, in turn, would lead to an

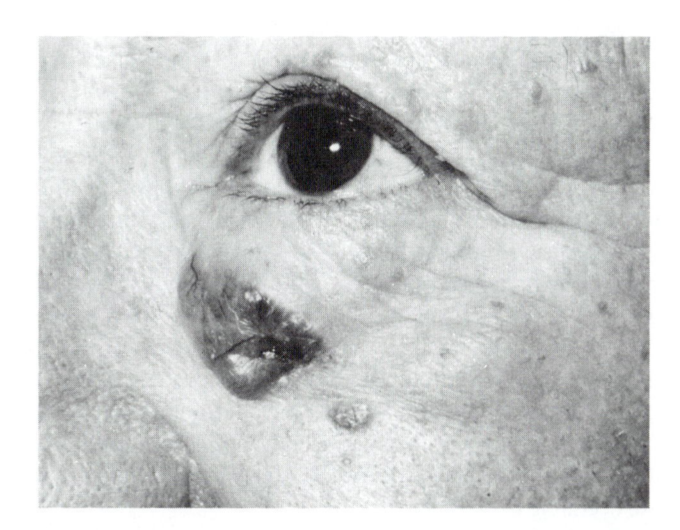

FIGURE 19.17 *Skin cancer caused by excessive exposure to ultraviolet radiation in sunlight. Thinning of the ozone shield will cause a sharp increase in the incidence of such cancers.*

increase in skin cancer rates of about 4 percent. The EPA estimates that ozone depletion will result in approximately 200,000 cases of skin cancer in the United States over the next five decades. Worldwide, the number is much higher. Especially hard hit will be countries such as New Zealand and Australia. In northern Australia, one city has already reported a dramatic increase in skin cancer.

In November 1991 a panel of scientists convened by the United Nations released a report on the predicted effects of ozone depletion throughout the atmosphere. They estimated that a 10 percent decrease in stratospheric ozone concentrations would cause 300,000 additional cases of skin cancer per year worldwide. They predicted that it will also cause an additional 1.6 million cases of cataracts (increase in opacity of the lens) each year.

Studies of skin cancer show that light-skinned people are much more sensitive to UVB radiation than more heavily pigmented individuals. In addition, some chemicals commonly found in drugs, soaps, cosmetics, and detergents may sensitize the skin to UVB radiation. Thus, exposure to sunlight may increase the incidence of skin cancers among light-skinned people and users of many commercial products.

Skin cancer has a low mortality, but because so many people will contract it, thousands of people will die from our use of CFCs. Many others will undergo surgery to remove tumors.

The study of the effects of UVB radiation on natural ecosystems such as forests, grasslands, lakes, streams, and estuaries has only recently begun. Studies of shallow-water ecosystems, however, indicate that UVB radiation can severely depress populations of phytoplankton, small crustaceans, and larval fish. Thus, say scientists, the depletion of the ozone layer will result in a diminished supply of animal protein for millions of hungry people in poor nations, where malnutrition is already a way of life.

Land- and water-dwelling plants could also suffer from increasing UVB radiation. Intense UVB radiation is usually lethal to plants; smaller, nonfatal doses damage leaves, inhibit photosynthesis, cause mutations, or stunt growth. Declining ozone and increasing UVB radiation could also cause dramatic declines in commercial crops such as corn, rice, and wheat, costing billions of dollars a year. It may also affect certain commercially valuable tree species like the loblolly pine.

In a hearing before the U.S. Senate in November 1991, Susan Weiler, head of the American Society of Limnology and Oceanography in Walla Walla, Washington, testified that studies in Antarctica by other scientists have shown that populations of phytoplankton (algae and other free-floating photosynthetic organisms) decrease about 6 percent to 12 percent when stratospheric ozone concentrations over the region drop by 40 percent. Since phytoplankton form the base of the aquatic food chain, damage to them could cause widespread ecological problems. One scientist thinks that ozone depletion and subsequent effects on the food chain may be the reason why two species of penguin are declining in the Antarctic.

Finally, UVB light is harmful to many products. Paints, plastics , and other materials deteriorate when exposed to UV light. Losses from further decreases in the ozone layer could cost society enormous amounts of money.

The Ban on Ozone-Depleting Chemicals

In the 1970s, fears caused by early projections of ozone depletion moved several nations, including the United States, Sweden, Finland, Norway, and Canada, to cut back on CFC-11. In 1978, for example, the United States banned the CFC used in spray cans. CFC-12, a refrigerant, coolant, and blowing agent, was not affected by the ban.

But the continuing accumulation of scientific evidence on the decline of the ozone layer made it evident a decade later that worldwide cooperation was needed. In 1987, the UN sponsored negotiations aimed at reducing global CFC production. In September of that year, 24 nations signed a treaty, called the **Montreal Protocol,** which would cut production of five CFCs in half by 1999 and freeze production of halons (used in fire prevention systems) at 1986 levels. (In halons, bromine atoms replace some or all of the chlorine atoms.) Although halons are used in much smaller quantities worldwide, they are far more effective in destroying ozone than CFCs.

This agreement paved the way for a gradual decline in CFC production in the industrial nations. But critics argued that it had too many loopholes. Like so many other pollution control strategies it would only slow the rate of destruction, not stop it. EPA computer projections showed that an 85 percent reduction in CFC emissions was needed to stabilize CFC levels in the atmosphere.

Before the Montreal Protocol went into effect, something unusual happened. In March of 1988, an international panel (described above) announced that ozone levels had fallen throughout the world. Two weeks later, DuPont, a leading major producer of CFCs, called for a total worldwide ban on CFC production. Two weeks earlier it had said that it would not support a ban.

Continuing bad news about ozone depletion brought negotiators to the table once again, this time in London, where in June 1990 they reached a new agreement. This treaty was signed by 93 nations and called for the complete elimination of CFCs and halons (another group of similar chemicals) by the year 2000, if substitutes were available by then. The signatories also agreed to phase out other ozone-depleting chemicals, among them carbon tetrachloride, methyl chloride, and even the hydrofluorocarbons (HCFCs), a class of chemicals (described below) once thought to be an excellent substitute for CFCs.

The news about the ozone layer continued to worsen. In 1992, a team of 40 scientists announced record-high concentrations of chlorine monoxide (ClO is produced from chlorine free radicals from the breakdown of CFCs) in the air above New England and Canada. Concentrations such as these had never been seen before, even in the Antarctic ozone hole. If levels of chlorine levels continue to climb, chances are good that a severe Arctic ozone hole will begin to appear with great regularity, exposing Canada and parts of the United States, Europe, and Asia to dangerous levels of UV radiation.

Aircraft measurements in 1992 also showed rather disturbing findings about global ozone outside the Arctic. In flights as far south as the Caribbean, scientists detected ClO concentrations of up to five times the amount they had anticipated.

In 1992, the nations of the world met in Copenhagen to sign another agreement calling for an acceleration of the phase-out of CFCs, carbon tetrachloride, and other ozone-depleting chemicals within 4 to 9 years. The projected effects of global efforts to phase out CFCs are shown in Figure 19.18, a graph of projected concentrations of ozone-depleting compounds (see Ethics in Resource Conservation Box 19.1).

Nearly a decade later, some encouraging results are beginning to appear. The production of CFCs and other ozone-depleting compounds has been greatly curtailed and CFC-11 concentrations in the lower atmosphere have begun to decline. CFC-12 concentrations appear to be leveling off. Globally, there's some indication that the decline in ozone levels may be over (Figure 19.19).

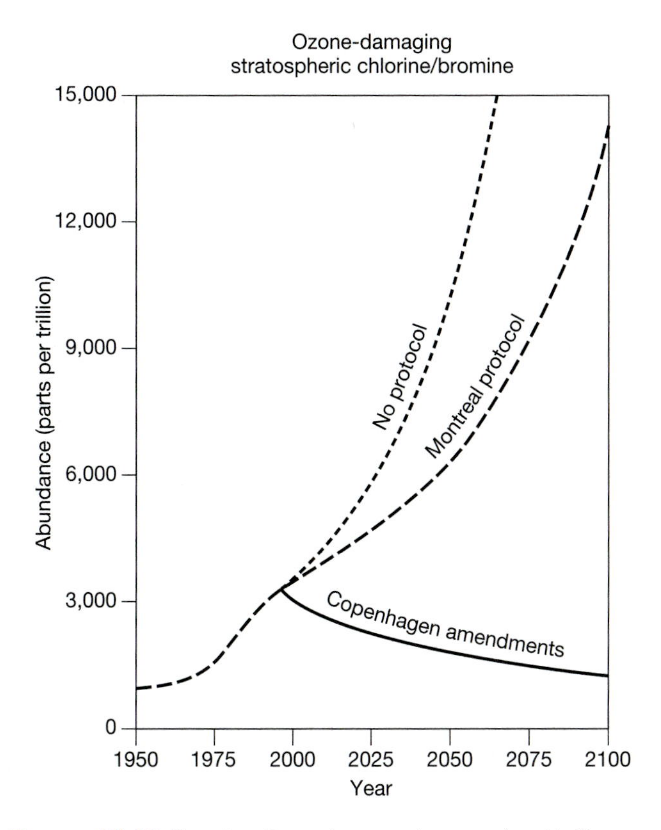

FIGURE 19.18 *The value of a good treaty and concerted world effort to stop an environmental nightmare.*

Substitutes for Ozone-Destroying CFCs

The development of replacement chemicals played a big role in the dramatic change in public policy regarding CFC production. It gave industry options and, in some cases, opportunities to profit from the shift to less harmful chemicals. Manufacturers have pursued two basic options: the use of less stable CFC compounds, which break down before they reach the stratosphere, and the production of non-CFC chemicals to be used as substitutes.

Consider the first option. By adding a hydrogen atom to the stable CFC molecule, researchers can make a group of CFCs–technically known as HCFCs or hydrochlorofluorocarbons) that break up in the lower atmosphere. In theory, the chlorine atoms released during this process are less likely to reach the stratosphere. In practice, some do, but they cause less damage. That is, they still deplete the ozone layer. Consequently, they are viewed as interim solutions.

Several less stable CFCs are already on the market. One of these, HCFC-22, is now used as a coolant in some home air conditioners. HCFC-22 is 20 times less destructive than the CFC-12 currently found in refrigerators and automobile air conditioners. Because of their effect on the ozone layer, though, the HCFCs are slated to be phased out by 2030.

The second most widely used CFC is CFC-11. Until recently, it was primarily used as a blowing agent for foam and, outside of the United States and a few other countries, as a spray-can propellant. HCFC-123 has been touted as a possible replacement. Manufacturers of foam insulation have eliminated it entirely.

Perhaps one of the most difficult challenges is finding a replacement for CFC-113. This compound is an all-purpose

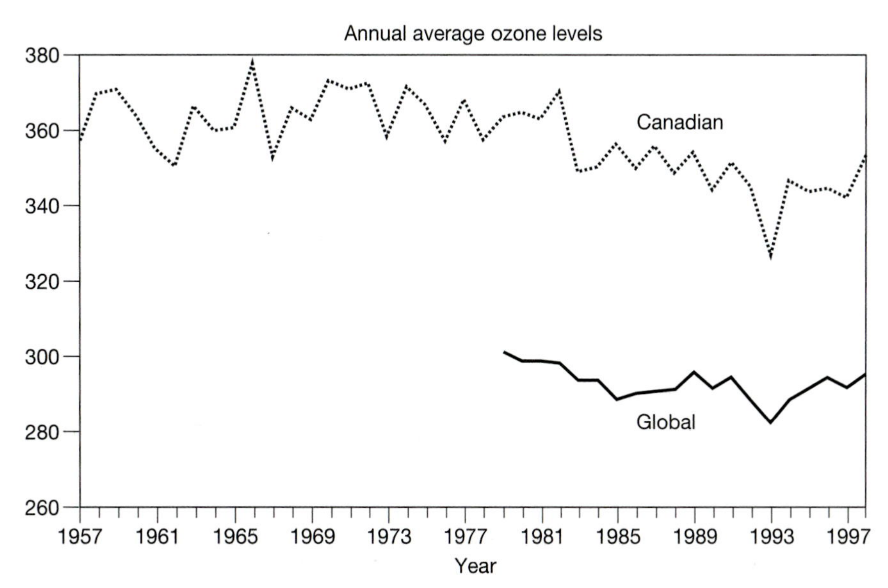

FIGURE 19.19 *Stratospheric ozone levels.*

ETHICS IN RESOURCE CONSERVATION 19.1

DEBATE OVER GLOBAL WARMING AND OZONE DEPLETION: DO WE HAVE AN OBLIGATION TO OTHER COUNTRIES?

In the negotiations over the treaties that are bringing an end to ozone-depleting chemicals, the less developed nations have repeatedly argued that the industrialized nations are discriminatory. The general line of reasoning goes as follows. Western nations proposing limits on CFCs or outright bans on ozone-depleting chemicals have reached a level of economic wealth and personal welfare only dreamed of in the less developed nations. We have developed various CFC-using technologies, among them refrigerators, air conditioners, and freezers, to promote our economic and personal welfare. But now that we have found out that these technologies (notably those using CFCs) are harmful, we are asking the less developed countries not to avail themselves of the same luxuries we enjoy. We are effectively restricting their development and prosperity, which we have no right to do.

In lengthy negotiations, the less developed nations have asked the developed nations to help them. China, for instance, was gearing up to manufacture and distribute tens of millions of CFC-cooled refrigerators to its citizenry. Thus, the government wanted financial assurances to offset the cost of alternative technologies as a condition of their signing the international treaty that would eventually eliminate all CFCs from the global marketplace.

More recently, negotiations over limits to carbon dioxide emissions have taken on a similar flavor. The less developed nations, which use only a fraction of the world's fossil fuels (about 24 percent), have 70 percent of the world's population. In order to limit carbon dioxide emissions and prevent a disastrous global warming, they are being asked to use alternative technologies.

Do the developed nations have an obligation to help the less developed nations pay the additional costs of developing alternative technologies? Why or why not? Is this an ethical debate?

Make a list of points in support of your viewpoint, then take the opposing viewpoint and summarize what you think it might be. Does this exercise change your viewpoint at all?

Can you think of any alternative means of promoting the betterment of people's lives in the less developed countries without financial assistance from the West?

cleaner for circuit boards produced for the computer industry. Because CFC-113 was not being considered for banning until a few years ago, the industry had not actively pursued replacements. At the signing of the Montreal Protocol, in fact, work to find a substitute had not even begun.

In January 1988, researchers announced the development of a compound called BIOCAT EC-7. It may be a partial replacement for CFC-113. This substance, isolated from orange peels, is very similar to kerosene and turpentine. EC-7 may replace a sizable share of the CFC-113 market, but it has its limitations. It is not versatile and is flammable. Industry representatives believe that no single compound will replace CFC-113 completely. Although options are available, intense research is under way to find ozone-friendly substitutes.

The Good News and Bad News About Ozone

The ozone story is one of humankind's greatest success stories. It not only illustrates how scientific knowledge can be used by society for the common good of humankind and all other species, it also shows how quickly changes can be made to bring about an end to actions responsible for the destruction of the global environment. Already, studies are showing that the rate of increase in the concentration of ozone-depleting chemicals is slowing. That's the good news. Unfortunately, however, many millions of tons of CFCs have already been released into the atmosphere. CFCs take 15 years or so to migrate into the stratosphere, and some CFCs last 110 years in the atmosphere. Because of this, scientists predict that the ozone layer will continue to thin at least to the end of the decade, before it begins to improve. But at least 100 years will be required to return the ozone layer to 1985 levels. Another 100 to 200 years may be needed for full recovery. A lot of people will contract skin cancer in the interim. Making matters worse, many ozone-depleting chemicals (and their replacements) are greenhouse gases and thus contribute to global warming and climate change.

Supporting the belief that we're not out of the woods yet, since 1979 ozone levels in the stratosphere have declined 4 to 6 percent over midlatitudes and 10 to 12 percent over higher southern latitudes. The ozone hole in 1997 and 1998 continues to present itself, ecompassing 27.3 million square kilometers. The hole in 1998 was the largest ever. Clearly, we've made a huge mess that will take a long time to correct itself.

Summary of Key Concepts

1. The greenhouse effect is caused by naturally occurring and human generated gases such as carbon dioxide and methane that trap heat (infrared radiation) radiating from the Earth's surface into outer space, warming the atmosphere. As a result of human activity, such as the consumption of fossil fuels Deforesation and the burning of forests that takes place when land is converted to human use, the carbon dioxide level has risen nearly 30 percent in the past century.
2. Fossil fuel consumption alone is responsible for the annual release of 6.4 billion metric tons of carbon (in the form of

carbon dioxide) into the air, or slightly less than 1 metric ton for each person on Earth.

3. The harmful effects of an increased global average temperature include (a) the inundation of coastal areas, (b) billions of dollars of property damage in the United States and other countries, (c) saltwater contamination of aquifers, (d) drought-triggered dust storms on the Great Plains of North America and in other countries, (e) inundation of major rice-producing regions, and (f) termination of economic growth throughout the world.
4. Many efforts are underway to reduce greenhouse gas emissions, although progress has been slow. One of the most significant efforts is the Kyoto Protocol, an international agreement to curb emissions by Industrial and former Eastern bloc nations.
5. Global climate change can be reduced by efforts to switch to more efficient combustion technologies and alternative non-fossil fuel technologies such as wind and solar energy.
6. Acid deposition is the deposition of acidic compounds in either wet or dry form. Wet deposition consists of acids deposited by rain, snow, fog, dew, or frost. Dry deposition of acid-forming materials takes place when dust particles containing nitrates and sulfates settle on the soil, water, or vegetation. Later those materials may react with water to form sulfuric or nitric acid. Acid deposition also includes the absorption of nitrogen and sulfur dioxide gases onto solid surfaces. These then combine with water to form acids.
7. The pH scale ranges from 0 to 14. A substance with a pH of 7 is neutral, neither acidic nor basic. Substances with pH values above 7 are alkaline or basic. Substances with pH values below 7 are acidic.
8. The pH of normal, nonpolluted rain is about 5.6. This slight acidity is the result of the small amount of carbon dioxide dissolved in the water to form carbonic acid.
9. Rainfall in much of the northeastern United States, southeastern Canada, and Scandinavia has a pH of 4.5 or lower because of the release of oxides of sulfur and nitrogen from upwind power plants, cities, and industrial sites.
10. Because emissions generated in one country may be transported by winds to others, control of the acid rain problem requires international cooperation.
11. Among the adverse effects of acid deposition on aquatic ecosystems are the following: (a) a reduction or termination of reproduction in many aquatic organisms, (b) disruption of the homing ability of salmon, (c) abnormal development of fish embryos, (d) a decline in fish food organisms, (e) the loss of desirable species of fish like bass and pike and an increase in less desirable species like carp and bullheads, (f) increased levels of aluminum, which interferes with normal gill function in fish, (g) an increase in mercury levels in fish to the point that they cannot be safely eaten, and (h) an inhibition of bacterial decay, which locks up nutrient elements in the dead remains of aquatic organisms.
12. Acid deposition has been addressed by national and international agreements to limit sulfur dioxide emissions at power plants, factories, and other stationary sources. Most sources have opted to switch to low-sulfur coal, although some have installed smokestack scrubbers.
13. Some reductions in acid deposition are occurring and some lakes are responding, but not until controls on nitrogen oxides are in place will we truly be successful in reducing this problem. Unfortunately, nitrogen oxide emissions are difficult to eliminate.
14. The ultraviolet (UV) light produced by the sun causes skin cancer in humans. However, under natural conditions, we are shielded from most of the sun's UV radiation by stratospheric ozone in the ozone layer.
15. In recent years, this ozone shield has begun to be depleted by gases such as chlorofluorocarbons, or CFCs, and nitrous oxides that result from human sources.
16. CFCs have been used as spray-can propellants, refrigerants, cleaning agents, and blowing agents.
17. The thinning of the ozone layer could have many adverse effects on people and the environment. It very likely will increase the incidence of skin cancer worldwide, increase the frequency of cataracts, and may reduce people's ability to fight off bacterial infections. It could also have a harmful impact on plant life. Corn, cotton, and wheat yields may drop. Aquatic ecosystems will be harmed as well.
18. In 1987, many nations agreed to reduce CFC emissions, but further research showed that the goals set in this agreement were not sufficient to prevent widespread damage. The agreement was modified in 1990 and tightened again in 1992, at which time the United States and 23 other nations signed an agreement to halt CFC production.
19. While significant progress has been made in reducing the production of ozone-depleting compounds, it may take 50 to 100 years for the ozone layer to recover due to the fact that CFCs have an extremely long lifespan, lasting up to 100 years.

Key Words and Phrases

Acid Deposition
Acid Precursors
Acid Rain
Acid Shock
Air Quality Accord
Alzheimer's Disease
Anthropogenic Source of Air Pollution
Buffer
Buffer Action
Carbon Dioxide
Chlorofluorocarbons (CFCs)
Clean Air Act Amendments (1990)
Deposition Program
Dry Deposition
Freons
Global Climate Change
Greenhouse Effect
Greenhouse Gas
Intergovernmental Panel on Climate Change
Kyoto Protocol
Liming Lakes
Methane
Montreal Protocol
National Appliance Conservation Act
National Atmospheric Deposition Program
Nitric Acid
Ozone
Ozone Layer
pH
pH Scale
Skin Cancer
Smokestack Scrubber
Stratosphere
Sulfur Dioxide
Sulfuric Acid
Synergistic Reactor
Troposphere
Ultraviolet Radiation
United Nations Framework Convention on Climate Change
Wet Deposition

Critical Thinking and Discussion Questions

1. Is there a possible cause-and-effect relationship between air pollution and the northward extension of the ranges of the armadillo, opossum, cardinal, mockingbird, and sea otter in the United States?
2. In what ways are human beings changing the normal flow of carbon through its elemental cycle?
3. Discuss both the benefits and the adverse effects of the carbon dioxide buildup in the global atmosphere.
4. What other pollutants contribute to global climate change?
5. Discuss the impacts of the continuing accumulation of greenhouse gases in the atmosphere.
6. Using your critical thinking skills, discuss the following statement: "The global climate has gone through warming periods before, so we shouldn't be concerned about any climate change occurring now."
7. List some of the evidence supporting the global climate change hypothesis.
8. What is acid deposition? What are the major sources? How do you contribute to this problem?
9. Discuss the harmful effects of acid deposition on aquatic ecosystems, soils, forests, materials, and human health.
10. Lakes A and B are located only 100 kilometers (60 miles) apart and receive the same amount of precipitation, the average annual pH of which is 4.5. Yet lake A is devoid of fish, whereas lake B abounds with them. Explain.
11. Discuss three strategies that can be used to bring the acid deposition problem under control. Which ones are the most sustainable? Why?
12. What can you do personally to reduce global climate change and acid deposition?
13. Using your critical thinking skills and knowledge you've gained in this course, analyze the statement, "Solutions to regional and global environmental pollution are too costly. We can't afford them."
14. Discuss the statement, "Ozone may be both beneficial and harmful to human health."
15. What is the ozone layer? How is it being altered?
16. What steps have been made to protect the ozone layer?
17. Why is international cooperation needed to control the problems of global warming, acid deposition, and ozone thinning in the stratosphere?
18. Many steps have been taken to reduce the production and release of ozone-depleting compounds, but ozone layer continues to thin. Why?

Suggested Readings

Baker, L. A., et.al. 1991. "Acidic Lakes and Streams in the United States: The Role of Acidic Deposition." *Science* 252(5007): 1151–1155.

Bird, E. 1993. *Submerging Coasts.* New York: Wiley. Examines the implications of rising sea level on coastlines and coastal communities.

Depledge, J. 1999. "Coming of Age at Buenos Aires: The Climate Change Regime after Kyoto." *Environment* 41(7): 15–20. Excellent discussion of climate change negotiations and treaties.

Flavin, C. 1996. "Facing Up to the Risks of Climate Change." In *State of the World 1996.* L. Starke (ed.). New York: W. W. Norton.

Flavin, C. 1994. "Storm Warnings: Climate Change Hits the Insurance Industry." *World-Watch* 7(6): 10–20. A very telling tale of how seriously the insurance industry is taking warnings of global climate change.

Flavin, C., and O. Tunali. 1995. "Getting Warmer: Looking for a Way Out of the Climate Impasse." *World-Watch* 8(2): 10–19. Excellent look at what nations aren't doing to address global climate change.

Kasemir, B., et. al. 2000. "Involving the Public in Climate and Energy Decisions." *Environment* 42(3): 32–42. Explores an aspect of change we don't discuss often in this book.

McDonal, A. 1999. "Combatting Acid Deposition and Climate Change: Priorities for Asia." *Environment* 41(3): 4–11, 34–41. Gives a great perspecitve on what other countries are doing and have to do to address these problems.

Moran, J. M., and M. D. Morgan. 1994. *Meteorology: The Atmosphere and the Science of Weather,* 4th ed. New York: Macmillan. Excellent nontechnical coverage of global pollution problems.

Moore, C. A. 1997. "Warming up to the Hot New Evidence." *International Wildlife* 27(1): 21–25. A shocking article that looks at global warming predictions and whether they are coming true.

Roberts, L. 1991. "Acid Rain Program: Mixed Review." *Science* 252(5004): 371. Penetrating analysis of the merits of the National Acid Precipitation Assessment program.

Skolnikoff, E. B. 1999. "The Role of Science in Policy: The Climate Change Debate in the United States." *Environment* 41(5): 16–20, 42–45. Interesting reading.

United Nations Environment Programme. 1994. *The Impact of Climate on Fisheries.* Nairobi, Kenya: UNEP. Examines the impacts of climate change on fisheries, based on several case studies.

Weiss, E. B., and Jacobsen, J. K. "Getting Countries to Comply with International Agreements." *Environment* 16–20, 37–45. An extremely important paper.

Web Explorations

Online resources for this chapter are on the World Wide Web at: **http://www.prenhall.com/chiras** *(click on the Table of Contents link and then select Chapter 19).*

Chapter 20

Minerals, Mining, and a Sustainable Society

The Earth's mineral wealth has been tapped for thousands of years. Today, the Earth's surface is scarred with mines and carelessly discarded mine wastes—telling signs of humankind's unrelenting search for minerals and its frequent carelessness (Figure 20.1). Where humans once mined the Earth's surface with primitive tools to extract valuable minerals, today, huge mining machines extract mineral resources to support a society so dependent on minerals that a shortage of any one of a few dozen would bring it to its knees. The sheer magnitude of mining and its impact on the environment, combined with our dependency on minerals,

FIGURE 20.1 *This open pit mine in the desert Southwest is a blatant reminder of the damage humans create in supplying their needs.*

make mining and mineral production an issue of extraordinary importance. Truly, the long-term future of modern civilization depends on the way we manage the Earth's mineral resources.

This chapter addresses two basic issues of concern: the supply of minerals and the impacts of mineral mining and processing. It answers four key questions crucial to modern society: (1) Are we running out of minerals? (2) Can we expand our mineral supplies? (3) What are the environmental impacts of our mineral-intensive lifestyle? (4) How can we create a more sustainable system of mineral extraction and production?

20.1 Supply and Demand

The automobile is, perhaps, the most visible sign of the industrialized world's dependency on minerals. In the United States, the automobile industry uses enormous amounts of metals, refined from minerals. Approximately 7 percent of all the copper, 10 percent of the aluminum, 13 percent of the nickel, 20 percent of the steel, 35 percent of the zinc, and 50 percent of the lead used by the United States each year goes into automobile manufacturing.

These minerals come from widely scattered parts of the world. Copper, for instance, comes from mines in Arizona, Chile, and Canada. Aluminum ore (bauxite) comes to the United States from Japan and Canada. Nickel is shipped from Australia, Norway, Botswana, and Canada. Iron ore comes primarily from U.S. mines, but also from mines in Canada, Liberia, and Brazil. Lead comes mostly from U.S. mines as well, with smaller contributions from other nations.

Some Features of Minerals

Unlike forests, wildlife, fisheries, and even soil, minerals are nonrenewable. That is, like oil, natural gas, and coal, minerals are a finite, or limited, resource. Each aluminum can carelessly tossed in a garbage can and later buried in a landfill depletes the world's supply of aluminum. But unlike oil and coal, minerals and metal can be recycled over and over, thus greatly extending their lifetimes. Unfortunately, many countries recycle only a small percentage of their scrap metals.

The great majority of minerals now used by society are extracted from the Earth's crust, the outer layer of the planet, which extends 24 kilometers (15 miles) below the Earth's surface. Some valuable minerals in the Earth's crust, such as gold, occur in a pure or elemental form. Most minerals, however, exist as chemical compounds consisting of two or more elements. For example, copper exists most commonly in the form of copper sulfide (CuS). Aluminum exists as aluminum oxide (Al_2O_3) and lead as lead carbonate ($PbCO_3$). Most of these compounds are found in rock that contains other minerals as well. A rock containing important minerals is called an **ore**—for example, iron ore and aluminum ore (bauxite). To extract metals from ore, the rock must be crushed and treated by heat or chemicals.

Although there are over 2,000 minerals in the Earth's crust, only a small number of them are abundant enough to be economically worth extracting. A deposit rich enough to be mined is called an **ore deposit.** Ores may be classified as high or low grade, depending on their mineral concentration. For instance, copper ore containing 3 percent copper is considered to be high-grade ore, whereas copper ore with only 0.3 percent copper is a low-grade ore.

U.S. Mineral Production and Consumption

The United States annually mines ores worth an estimated $39 billion. When refined into metals, however, these ores are worth $422 billion, or about 4.5 percent of our gross natural product (the total value of all goods and services produced by the nation, including income of U.S. companies operating in other countries). The leading mining states are Texas, Louisiana, California, and West Virginia. The leading mining nations are Canada, Australia, Russia, and the United States.

Although the United States has less than 5 percent of the world's population, it consumes about 20 percent of the world's nonfuel minerals each year. Enormous amounts of steel, copper, aluminum, and other metals are used to produce a variety of products that give Americans one of the highest standards of living in the world. To raise the rest of the world to our level of material wealth would require staggering amounts of minerals.

The United States' prodigious use of minerals has made it a great and prosperous nation but also one of the most vulnerable, because many of the materials it depends on for commerce and defense are imported from politically volatile regions of Africa (Figure 20.2). Minerals of extraordinary importance to an economy and a nation's national defense are called **strategic minerals.** So important are they, that they're stockpiled in case of a disruption in their supplies. The U.S. maintains a supply of numerous strategic metals, such as lead, copper, cobalt, and bauxite. In most cases, the goal is to have a two-year supply.

In 1999, the United States spent about $56 billion to import minerals and various metals. Many experts warn that widespread dependence on foreign sources—especially the politically volatile nations—could prove disastrous. Civil unrest or war between nations can temporarily halt the flow of valuable minerals.

Stockpiles (i.e., strategic reserves) may help in the short run but are not a long-term solution to vanishing mineral supplies. Nor is it an adequate solution to the development of mineral cartels—groups of mineral-exporting nations that band together to control supplies and prices of strategic minerals. China, for instance, controls most of the world's tungsten. What would happen if it and other tungsten-exporting nations joined forces to control tungsten exports to the United States or Japan or England? Or what would happen if Russia, which holds most of the world's palladium, decided to do the same?

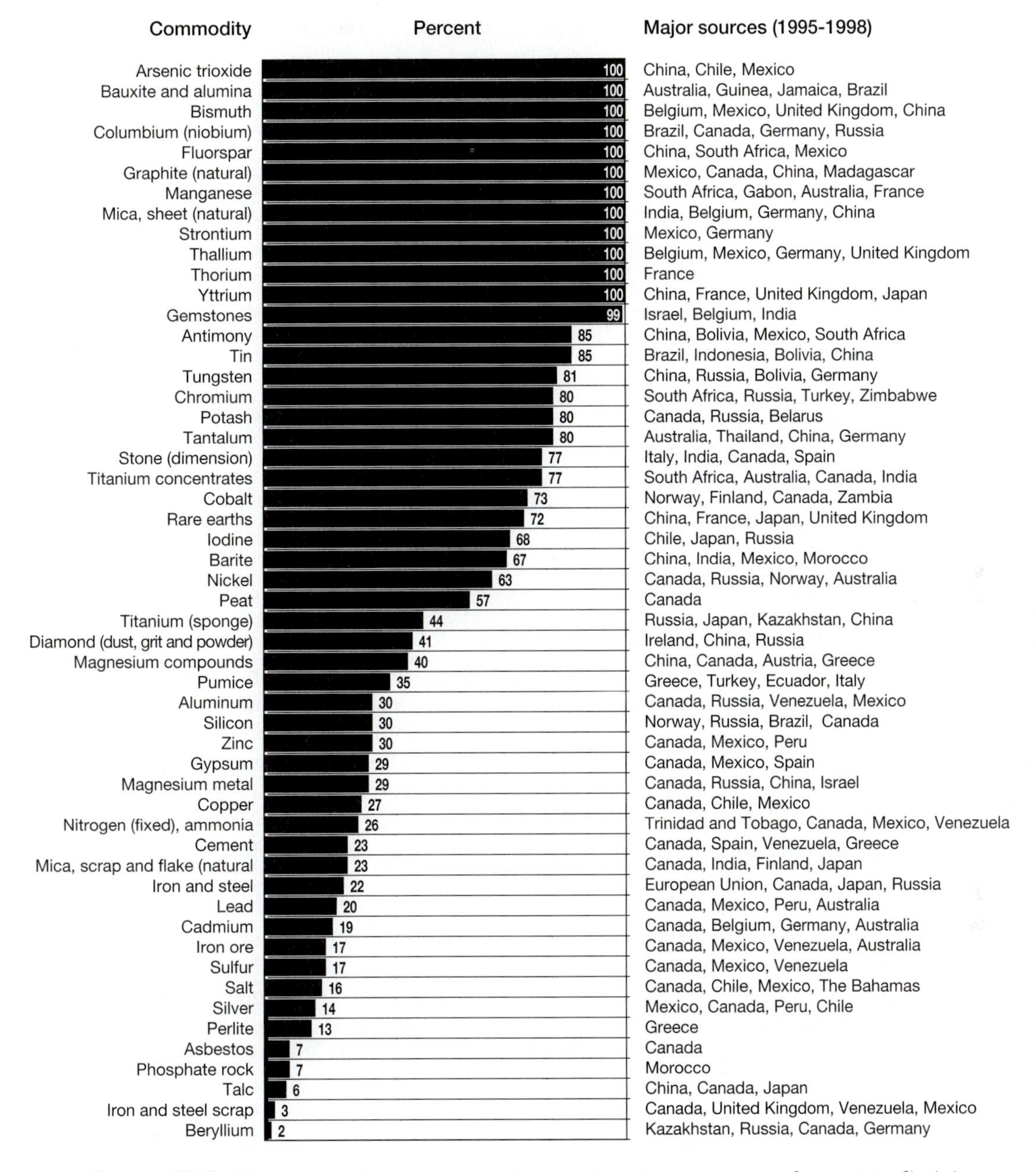

FIGURE 20.2 *U.S. net import reliance on selected minerals and metals as a percentage of consumption. Shaded area represents imports.*

Many experts argue that we need not fear mineral cartels because most producer countries depend mightily on steady exports of minerals for foreign exchange. Zambia's mineral exports, for instance, provide over half of that country's national income. Many industrialized nations import minerals from less developed countries, refine them, and sell the metals at ten times the price they paid for the minerals. Less developed nations often feel robbed of potential economic gains and have long urged industrial nations to import more refined metals from them to offset this imbalance. Such actions on the part of the industrialized nations could reduce tensions that might inspire mineral-exporting nations to band together in the first place, and indeed such actions are taking place. The United States, for instance, has increased its import of processed materials from $35 billion in 1994 to $62 billion in 1999.

Mineral Supplies: Are We Running Out?

With this background information in mind, let us look at an important question raised earlier in this chapter: Are we running out of minerals?

Determining the life span of existing mineral supplies is not easy. Geologists must first determine the rate of consumption and estimate potential growth in consumption, bearing in mind that even a modest increase in growth rate can result in a dramatic increase in the actual amount of mineral a nation consumes.

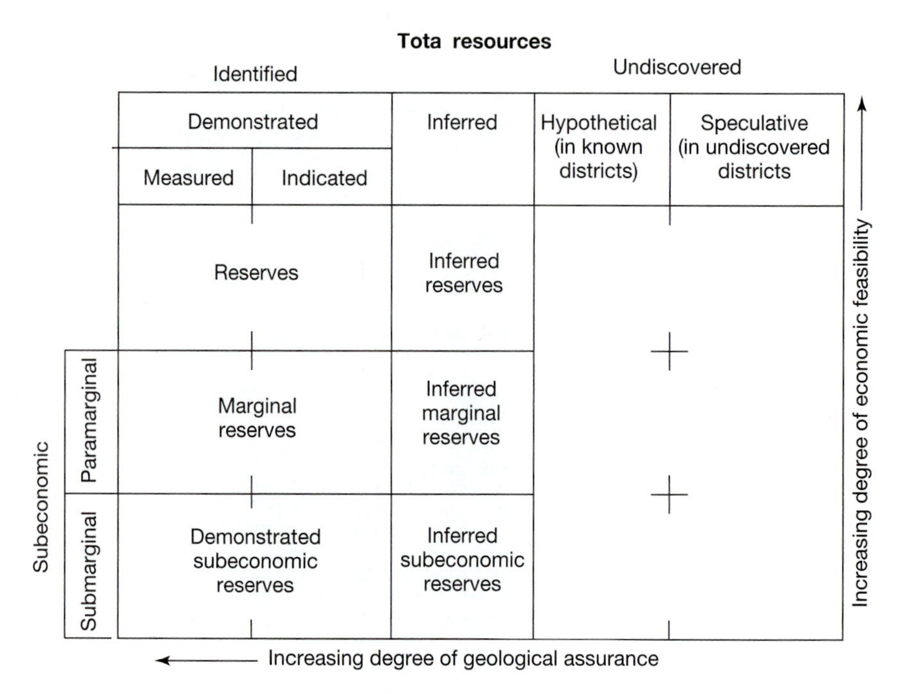

FIGURE 20.3 *Resources may be classified as reserves and total resources. Total resources include all the potentially minable, discovered and undiscovered deposits in the world. Resources may be divided into many categories. The demonstrated reserves are ones that geologists are fairly sure exist: They have been measured, or their presence is indicated by geological data, and they can be mined economically. (Based on the U.S. Dept. of Interior Classification Systems.)*

For instance, a resource with a life span of 1 billion years will last only 580 years at a 3 percent annual increase in the rate of consumption.

Next, geologists must determine the economically recoverable supply or reserve of each mineral in the Earth's crust. The reserve must not be mistaken for a similar figure, known as **total resources,** which is the total amount of mineral in the Earth's crust (Figure 20.3). The important difference is that the reserve includes only deposits that are feasible to mine, whereas total resources include all deposits and occurrences of a mineral, no matter how low the concentration may be.

The total resources of a mineral are often many times greater than it's reserve. To understand why, consider copper. A copper deposit 10 miles below the surface of the Earth is part of the total resources but not part of the reserve, because it would be too costly to mine. The world's total resources for copper are currently estimated at 1,600 million tons, whereas the reserve base is only 566 million tons.

Another factor to take into account is that the amount of a mineral reserve is not permanently fixed. That is, it can expand and contract, depending on a number of variables. For instance, new discoveries of economically recoverable minerals can expand the reserve base—what is economically recoverable. Economic incentives, such as from the government, lower the cost of production to a company (but not to society). Such subsidies can make it more profitable for a mining company to extract marginal or subeconomic ores. This artificial adjustment of the cost will shift some of the total resource into the reserve category.

The price of energy, heavily used in mining and processing minerals, also affects the reserve base. When energy is cheap, marginal ore deposits may become economical, thus expanding the reserve base. On the other hand, when energy prices rise, ores that were once economical to mine may become too costly. The reserve base therefore shrinks.

Environmental and worker protection laws can also affect the reserve base. For instance, environmental laws that require companies to reduce pollution from mines and reclaim mined land add to the cost of mining and may make marginally profitable reserves too costly to mine. Likewise, lax laws encountered in many less developed nations stimulate mining and expand the reserve base, often at a substantial cost to the environment, safety, and worker health. This distortion of the market may also occur in more developed countries, such as Canada, where government subsidies to the mining industry and lax laws lower the cost of production. In the United States, an antiquated mining law (the **Mining Act of 1872**) has a similar effect, argue environmentalists. This law, passed 120 years ago when the nation was young offers huge tracts of federal lands to mining companies for extremely low prices. But it is not just U.S. companies that benefit from this near-giveaway, companies from Canada, Japan, and other countries can take advantage of the U.S. government's generous gift. What is more, companies are not required to pay any kind of royalty on minerals extracted from public lands.

Labor costs can also profoundly shape the economic picture of mining and influence the reserve base. New technologies may cause the reserve base to expand. Highly efficient processes to extract and refine minerals, for instance, can reduce the price of mining marginal or subeconomic ores. This, like a variety of other factors, can increase the reserve base.

Based on existing estimates of world reserves and projections of consumption, it appears that three quarters of the 80 or so economically important minerals are abundant enough to meet our needs for many years. However, at least 18 economically essential minerals are bound to fall into short supply—some within a decade or two—even if nations greatly step up recovery and recycling. Gold, silver, mercury, lead, sulfur, tin, tungsten, and zinc belong to this group of endangered minerals. Don't be lulled by technological optimism; even if new discoveries and new technologies make it possible to mine five times the currently known reserves of these materials, this group will be 80 percent depleted by or before the year 2040, well within most of your lifetimes.

20.2 Can We Expand Our Mineral Supplies?

Something must be done, and quickly, to forestall the depletion of key mineral resources. But what can we do?

Unfortunately, there is no consensus on ways to satisfy future demand and prevent the economic turmoil that could result from widespread shortages of essential minerals. Some people, in fact, flatly dismiss the possibility of future shortages. They are called **technological optimists,** largely because they count on technological answers to this and a host of other environmental problems. The optimists are opposed by another group, often called the **pessimists,** who might be better labeled realists because they recognize the finite nature of the world's mineral resources and often propose innovative and cost-effective ways to avoid using up those resources.

This section looks at the views of the optimists and pessimists to answer the second question posed at the outset of this chapter: Can we expand our mineral supplies?

New Discoveries

A large portion of the Earth's crust has not been intensively explored for mineral deposits, say the optimists. By using current technologies, major finds are possible in Asia, Africa, South America, and Australia. The optimists point to substantial mineral discoveries in recent years as proof that current estimates of the world's mineral reserves fall far short of the world's real reserves. The pessimists, on the other hand, argue that the extremely rich deposits needed to substantially expand our reserve base simply do not exist. Even if they did, a fivefold expansion of the world reserves of critical minerals would only slightly offset the rapid depletion, in large part because of the rapid growth in human population and economic development. Development of mineral resources could also lead to massive environmental damage in remote, pristine wilderness areas like the Arctic tundra or tropical rainforests, displacing wildlife and people who have lived on the land sustainably for years.

Extracting Minerals From Seawater

William Page, a researcher from Great Britain, summarizes the viewpoint of the optimists: "Seawater is estimated to contain 1,000 million years' supply of sodium chloride; more than one million years of molybdenum (used to harden steel), uranium, tin, and cobalt; more than 1,000 years of nickel and copper. A cubic kilometer (0.25 cubic mile) of seawater contains approximately 11 metric tons each of aluminum, iron, and zinc." The oceans contain approximately 1,300 cubic kilometers (330 million cubic miles) of water, or about 14,000 million metric tons of each of these metals.

Not bad, say the pessimists, but there's a hitch. Even though the oceans contain vast quantities of dissolved minerals, except for bromine, magnesium, and table salt, most minerals are found in very low concentrations. The energy costs of extracting them would be prohibitive. In fact, just to extract 0.003 percent of the zinc the United States uses each year would require a plant that would process a water volume equal to the combined annual flows of the Hudson and Delaware rivers! Clearly, in this instance, limiting factors such as energy and cost must be taken into account.

Minerals From the Seafloor

Far-reaching optimists look to outer space to provide minerals for Earth, ignoring the exorbitant costs and energy requirements of these imaginative plans. Another group, looking closer to home, sees the ocean floor and the continental shelf as important sources of minerals for the future.

Minerals of the Continental Shelf Because the continental shelf is an extension of the continent that happens to be under water, it is not surprising that it could yield many of the minerals that are now extracted from mines on dry land. By one estimate, the continental shelf contains about 15 percent of the world's minerals. Tapping the economically feasible deposits, say the optimists, could expand our reserve base and postpone the day of depletion. Mining it on a large scale could create an environmental disaster.

Minerals of the Seafloor In 1990, an American marine expedition discovered curious nodules, called **manganese nodules,** on the floor of the Pacific Ocean (Figure 20.4). Rich in manganese, these nodules also contain nickel, iron, copper, cobalt, molybdenum, and aluminum. Manganese nodules cover about one fourth of the ocean's floor, mostly in international waters. Most are the size of potatoes, although they vary in size from tiny granules to the size of cantaloupes. The aggregate weight of these nodules on the Pacific seafloor alone is estimated at about 1,500 billion metric tons. Thus, while the land-based supply of copper could last only a few more decades,

FIGURE 20.4 *Manganese nodules taken from the floor of the ocean contain iron, nickel, copper, cobalt, and manganese.*

tapping the copper of manganese nodules could extend our reserve for thousands of years.

Problems with Mining the Seafloor Mining seafloor mineral deposits and manganese nodules, although attractive, is fraught with problems. To remove manganese nodules, ships would have to scour the ocean with huge devices, not unlike large vacuum cleaners, that suck the nodules up from the bottom. Recovering the solid nodules would be more costly than extracting oil and natural gas from offshore wells. In addition, the environmental impacts of such activities could be far-reaching. Dredging or scooping up the minerals from the seabed would increase the turbidity of the water, possibly affecting a wide array of sea creatures. Clouding of shallow waters could increase water temperature and make it unfit for many organisms adapted to cooler waters. Ocean-mining equipment would require enormous amounts of energy and cooling water and add further to global climate change. Heated water released into the ocean could further raise the temperature of the water. A final, and major, problem has more to do with politics than the environment.

Most nations claim ownership of waters 330 kilometers (200 miles) off their coasts. They presumably own the minerals in the continental shelf. But what of the manganese nodules on the ocean floor outside of territorial waters? Who owns the minerals in international waters?

Many people in the industrialized countries with the wealth and resources to mine manganese nodules believe they are legally entitled to these riches. However, the poorer nations wonder whether they shouldn't also share in the wealth, arguing that international waters are a common resource.

To settle this question of ownership and, more important, who will profit from the riches of international water, the United Nations began extensive negotiations in 1958. But progress has been slow. In 1982, 100 nations signed the **Law of the Seas Treaty.** The treaty places deep-sea mining outside territorial waters under international regulation. It also calls for a tax on seabed minerals, the proceeds of which would go toward helping less developed nations improve agriculture and develop economically. Believing that the seabed minerals belong to whomever can afford to mine them, many wealthy nations have refused to sign, among them the United States, Germany, and Great Britain. President Reagan refused to sign in 1981 because he felt the provisions of the treaty would jeopardize the ocean-mining interests of private U.S. firms. Unfortunately, the Reagan administration offered no alternatives. Today, the treaty has been signed by 109 countries and efforts are under way to gain Congressional approval in the United States. But at this writing (March, 2001), the U.S. has still not signed the treaty.

Currently, there is little mining of magnesium nodules or seafloor mineral deposits going on. Economics don't support extraction at this time. But seafloor minerals may someday help us meet the mineral needs of the world. Experts warn, however, that the cost may be high, both environmentally and economically. Seabed mining will be energy intensive and, in the light of declining fossil fuel supplies, the price of energy is bound to increase.

Improved Extraction Technologies

As new extraction technologies are developed, the optimists say, the mining industry will be able to extract more and more minerals from low-grade ores. The history of copper mining in the United States provides a good example. At the start of this century only the very high-grade ore, containing 30 kilograms (66 pounds) of copper per metric ton, was mined. As high grade ores were depleted, the efficiency of mining and processing improved so that progressively leaner ores could be used. Today, a very low-grade ore containing one-tenth as much copper (3 kilograms per metric ton) can be mined at a profit.

Pessimists admit that new technologies do permit the use of lower-grade ores. However, they point out that the lower the concentration of mineral in an ore, the greater is the amount of energy used for mining and processing. For example, twice as much energy is needed to produce a ton of copper from 10 percent bauxite as from 20 percent copper ore (Figure 20.5). Energy use increases environmental pollution and environmental damage from extraction and oil spills. This, in turn, increases economic costs (externalities) that are not included in the real cost of minerals.

Further reductions in mineral concentration will inevitably result in even larger increases in energy demand and greater environmental destruction. Making matters worse, the lower the concentration of the mineral, the greater is the amount of rock that must be mined and processed to produce a ton of mineral, and therefore the greater the amount of environmental damage, pollution from smelters, surface disruption at mines, and mine waste.

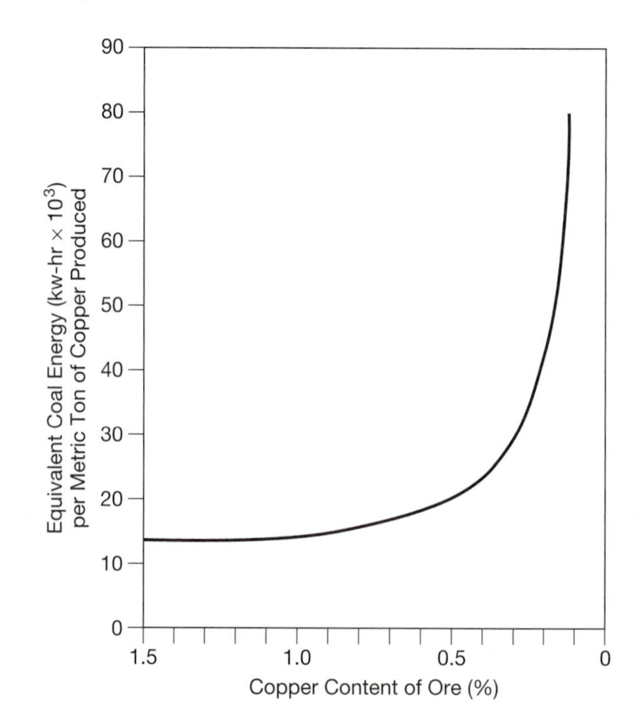

FIGURE 20.5 *Graphical representation of the amount of energy needed to extract a mineral as the concentration of the ore decreases. Note that once the concentration falls to a critical level, energy consumption increases drastically.*

Tapping Abundant Low-Grade Ores

As a general rule, optimists point out, for many minerals the total amount of ore increases as the grade decreases. Thus, the deeper miners dig into Earth, the more ore they find. This principle was forwarded by S. G. Lasky of the U.S. Geological Survey in 1950, and it appears to hold for a number of minerals, among them iron and aluminum. Optimists then note that as the technologies for deeper mining and for mineral extraction from lower-grade ores are developed, a literal bonanza of mineral wealth awaits the mining industry.

The pessimists respond that this principle does not apply to all ores, such as nickel, copper, and manganese. The amount of copper ore, for instance, increases up to a point as the grade decreases. Thus, there is more 1 percent ore than 2 percent ore, but below 1 percent the rule no longer holds. The 0.3 percent ore is one-fourth the amount of 1 percent ore.

Finding Substitutes

The substitution concept arose during World War II when, to save copper, government mints began turning out steel pennies. Within the past three decades numerous substitutes have been devised. Plastics have been a key alternative and have replaced wood, steel, and paper in a wide variety of products from fishing rods and speedboats to bath tubs and motor cars. Plastics are now even being used to replace metals, sometimes to the chagrin of those who end up with the products. Aluminum, another popular substitute, has nearly replaced steel in beverage cans and airplanes. Magnesium is being substituted for zinc in an increasing number of products, such as chemicals and pigments. Indeed, either naturally occurring or synthetic substitutes can be found for most metals used in today's society.

Pessimists argue that not all materials have adequate substitutes. For example, there is no other metal that has the unique chemical and physical properties of mercury, or the high melting point of tungsten, which makes this so valuable in high-speed tools and cutting edges. Furthermore, it may be impossible to find substitutes for the manganese used in desulfurized steel, the nickel and chromium used in stainless steel, and the tin used in solder. There's also no substitute for silver used in photographic film, but the inventors of the digital cameras may make film obsolete anyway!

Pessimists also note that some substitutes are inferior to the materials they replace—as anyone who has used a plastic snow shovel can attest. Aluminum, iron, magnesium, and titanium are among the most abundant elements in the Earth's crust and have great potential for becoming substitutes. However, they do not stand up to the task in all situations.

20.3 Mineral Conservation Strategies

A final problem is that some substitutes are scarce as well. This is certainly true of the molybdenum that is now being used in place of tungsten. The cadmium and lead that have replaced mercury in some types of batteries are also in short supply.

The optimist would have us believe that our mineral future is secure, and that worries over mineral supplies are needless. They assure us that new discoveries, seafloor deposits, lower-grade ores, new mining technologies, and substitutions will come to the rescue. The pessimist, however, knows that these avenues are limited at best, and argues that we would be well advised to look for additional ways to meet our needs. One of the easiest and most cost-effective strategies is to reduce demand.

Reducing Demand

Reducing our demand for minerals can be achieved by reducing population growth, cutting back per capita consumption, decreasing the size of products, and increasing product durability.

Population stabilization was discussed in Chapter 4. It is, as we have pointed out before, the cornerstone of all sustainable strategies. Beyond that, individuals can reduce per capita consumption by reducing unnecessary purchases and avoiding discardable items, a topic discussed in Chapter 17.

Manufacturers can contribute by reducing wasteful packaging and eliminating throwaways such as disposable pens, razors, and diapers. Yet another effective means of cutting back on mineral demand is to increase product durability—that is, to make products that last longer. Today, however, many manufacturers crank out flimsy tools, toys, appliances, and automobiles to a price-conscious public which, in its blindness to quality, gobbles them up, then throws them away when they break or are no longer in fashion.

That said, there is a movement, albeit a small one, for companies to lease products and then make their money on service. One major elevator builder, for instance, is now installing elevators in office buildings, but rather than installing it for a flat fee, they bear the construction costs themselves and then rent the elevator to the business, servicing it as needed. Obviously, in this instance, it makes sense to install the most durable system to minimize repair. One carpet manufacturer has a similar scheme. It leases carpet to businesses, but when a carpet becomes worn, the worn section is removed and replaced. There's no need to throw away 100 precent of a carpet when only 10 precent has gone bad!

In addition to increasing product durability, manufacturers could make products smaller, further stretching mineral supplies. Manufacturers have produced a variety of household appliances such as clocks, toasters, refrigerators, telephones, and stoves that are smaller and lighter than those on the market two decades ago. Similarly, the first calculators were heavy and cumbersome. They would fit into a briefcase but might take up a third of the space. Today, solar-powered calculators slip into a breast pocket with room to spare. Computers have also undergone a dramatic reduction in size and weight (Figure 20.6).

Recycling

Ruben L. Parson, a geography professor at St. Cloud State University, Minnesota, wrote that "[during] World War II we [Americans] reclaimed anything metallic, from abandoned

FIGURE 20.6 *Computer manufacturers are designing smaller computers and monitors that require less material to make, thus reducing production costs and environmental impacts.*

streetcar tracks and worn-out machinery to horseshoes and tin cans. Reclaimed metal gave us the machines that crushed Hitler's armored legions. We became scrap-conscious as never before. But we have too readily reverted to the reckless, wasteful discard of material that is typically American."

At least four factors are bound to spur greater interest in recycling in the United States and other countries: (1) a rise in oil prices (Chapter 21), (2) the depletion of strategic minerals, (3) the shortage of landfills (Chapter 17), and (4) environmental and economic concerns. Signs of this recycling revolution are already beginning to appear. In 1987, for instance, New Jersey, passed a statewide law that required all communities to recycle at least three commodities because of a shortage of landfill sites. Other states quickly followed. (For more on recycling programs, see Chapter 17.) In 1996, the most recent year for which data are available, there were 8,817 curbside recycling programs in the United States, serving 134.6 million people. Today, Americans recycle 27 percent of their trash, up dramatically from the past decade. Seven states—Maine, New York, New Jersey, South Carolina, Virginia, South Dakota, Minnesota—now recycle more than 40 percent of their municipal solid waste. Fifteen states have recycling rates from 30 to 40 percent.

Although recycling has taken off remarkably well in the past decade, Americans have only begun to tap its potential. Several obstacles seem to stand in the way of the full development of this strategy. In addition to the problems discussed in Chapter 17, notably a failure to develop adequate markets for recycled (secondary) materials, the federal government has proved to be a major impediment, for it still provides economic incentives to the mining industry that put the recycling industry at a grave disadvantage. What are these incentives? First, the federal government provides billions of dollars in **depletion allowances,** tax breaks that go to mining companies as they deplete their reserves. The tax breaks were designed to help mining companies invest in the exploration needed to unearth additional mineral supplies. The net effect of such incentives is to make virgin minerals artificially cheap and give them a competitive advantage over recycled materials in some instances. Second, federally mandated freight costs for shipping raw materials are, by law, lower than rates for metals bound for recycling plants. Ending these two unfair practices would benefit the recycling industry enormously.

Individual Efforts

Resource-conscious individuals can play a significant role in reducing mineral consumption by several simple measures. You can recycle all glass, aluminum, waste copper, and other metals (Figure 20.7). You can buy products made from recycled materials. You can also help reduce mineral consumption by cutting back on your consumption of unnecessary items. Most of us have closets and basements full of things we had to have but that now do little more than gather dust. Before you buy something, ask yourself if you really need it and if you will use it for more than a few weeks.

Individuals can contribute by purchasing durable clothes and goods. Avoid throwaways. Ditch your disposable pens (after you've used them up) and buy ballpoint pens, where all you have to throw away is the refill.

When you have a choice between a recyclable and nonrecyclable good, choose the recyclable product. When you have a choice between a product made from a renewable resource

FIGURE 20.7 *Stackable bins like these allow the homeowner to separate recyclables for easy recycling.*

(such as wood) and one made from nonrenewable materials (such as aluminum), choose the one made from a renewable resource.

The actions of individuals, when multiplied by many hundreds of thousands, can significantly reduce our reliance on minerals. If you are inclined toward action, you could organize a recycling program on your campus, in your community, or in your home. Armed with statistics on the benefits of recycling and with enthusiasm, you could become an instrument of social change in our transition to a sustainable society.

All of the techniques described in this section will help steer our society onto a sustainable course. But, as the next section points out, they are not enough to create a truly sustainable system of mineral supply.

20.4 Environmental Impacts of Mineral Production

A few miles from the scenic town of Aspen, Colorado, is an ugly scar called Climax. Climax is not a town. No one would want to live there, for it is the site of Amax's huge molybdenum mine and waste dump. Molybdenum (pronounced muh-LIB-de-num), or Molly-B for short, is a mineral that is used to harden steel in automobiles and other applications.

At Climax, miners have torn down nearly an entire mountain to extract this mineral (Figure 20.8). Found in a concentration of only about 0.2 percent, molybdenum is separated from the ore and trucked away for sale. The remaining material is dumped in a neighboring valley. Over the years, waste has transformed a once splendid mountain valley into a huge toxic moonscape, stark and ugly in a land otherwise breathtakingly beautiful.

This is but one example of the impact of our dependency on minerals. This section takes a broad look at the many impacts of mining and minerals processing and suggests ways to minimize them—that is, ways to create a more sustainable system of mineral supply.

FIGURE 20.8 *This mountain is being torn down by molybdenum miners in Climax, Colorado. The ore is then processed and waste is dumped in a nearby valley. Molybdenum is used to harden steel.*

Mining Impacts

Ninety percent of our nonfuel minerals is extracted from surface mines, excavations in the Earth's surface that allow miners to access underlying deposits. Surface mines are particularly destructive because the overlying soil and rock, called **overburden,** must first be removed and placed elsewhere. As a result, mining can quickly transform a scenic area into an ugly landscape. Reclamation is often difficult because many open-pit mines extend deep into the Earth's crust. (Surface coal mining is discussed in Chapter 21.)

Underground mines provide a smaller percentage of our minerals. As with surface mines, wastes must be removed from the mine and dumped elsewhere. Waste piles from both types of mine cause additional environmental problems if they are not vegetated and stabilized because heavy rains can wash the unstable soils into streams and lakes. Sediment in streams and lakes has many impacts. It increases water temperatures, which lowers the concentration of dissolved oxygen. This, in turn, can kill fish and other aquatic organisms, disrupting aquatic food chains. Sediment destroys spawning beds and detracts from scenic beauty and recreational uses. Sedimentation increases flooding because it reduces a stream's water-carrying capacity. When heavy rains come, water flows over the banks more easily, flooding nearby towns, farms, and pastures, and creating social, economic, and environmental damage.

Each year, approximately 1.7 billion metric tons of mine waste are produced from U.S. surface mines, with 5 percent coming from mineral surface mines and the rest from coal surface mines. Canadian mines produce 650 million metric tons of waste. Mine wastes contribute toxic metals to nearby waterways. For example, zinc, arsenic, lead, and other toxic metals may be leached from the spoils of iron ore mines by rainwater. In Canada and the United States, many mine wastes contain iron pyrite, a sulfur-containing compound. (Half of Canada's mine waste contains this chemical.) Rainwater combines with iron pyrite in gold and silver mine wastes in the West, creating sulfuric acid, which drains from the spoils as **acid mine drainage** into nearby streams, killing fish and other aquatic organisms. The economic damage is staggering. Unstable spoil piles can also form dangerous landslides when rainfall is heavy.

Certain minerals deposits that contain water soluble ores such as salt and potash may be removed by **solution mining.** In this technique, water is pumped into the deposit, where it dissolves much of the mineral, then pumped back to the surface. Although safe, this technique can pollute groundwater supplies. Some companies have used a more dangerous offshoot of the solution mining technique. They spray a deadly cyanide solution over mine wastes to extract gold from rock. Many environmentalists are justifiably alarmed by the prospects for contamination. In fact, in Colorado, one such mine is now costing the state and citizens about $40,000 a day to prevent pollution from draining into nearby waterways.

Besides creating eyesores, enhancing erosion, and polluting nearby streams, mining activities compete with other uses of wild lands. Mining in U.S. national forests and wilderness areas, for instance, disturbs wildlife and outdoor recreationists. Mining can destroy valuable timberland and grassland as well.

Mining can also use tremendous quantities of water. The lower the grade of ore, the greater the volume of water required. In the **hydraulic mining** of gold and silver, powerful blasts of water are used to wash soils from hillsides, which are then treated to extract the precious metals. In some areas, then, mining can divert water from ranches, farms, businesses, and municipalities.

Processing Minerals

Many ores are heated to high temperatures in specially built ovens that separate metals from the ore. A variety of toxic materials may be released into the atmosphere as a result of this process, called **smelting** (Figure 20.9). Smelters emit arsenic, mercury, zinc, and other toxic chemicals that are dangerous to bees and other animals. Especially harmful is fluoride gas, which is released from phosphate smelters and settles on vegetation around smelters. Fluoride can be ingested by cattle and other livestock. In high enough levels, fluoride causes a disease called fluorosis, characterized by pain in an animal's joints and softened bones and teeth, which leads to inability to stand or move around.

Perhaps smelters' best-known effect, though, results from the release of sulfur dioxide, a corrosive gas that combines with atmospheric moisture and oxygen to produce sulfuric acid, described in Chapter 19. Sulfuric acid and sulfur dioxide are lethal to plants and aquatic life and are suspected of having adverse effects on human health as well.

The huge copper and nickel smelter in Sudbury, Ontario—long a major producer of sulfur dioxide—has turned the neighboring area into a barren wasteland. Similar devastation has occurred in Montana and Tennessee. Recognizing this problem, the government pressed smelter owners to install tall stacks. This, they thought, would disperse pollutants, essentially diluting them in the atmosphere. The solution to pollution was dilution. However, the use of taller stacks only spread the pollutants out further, polluting distant lakes and killing fish and other aquatic organisms hundreds of kilometers from the source. Fortunately, many of the world's smelters like the one in Sudbury, Ontario, have installed pollution control devices and have cut sulfur dioxide emissions by 90 percent.

Creating a More Sustainable System of Mineral Production

Mining and mineral processing are two of the most environmentally damaging activities of humankind. They are also two of the world's most economically important industries. In many countries, mining and mineral processing are big business. Canada, for example, exports about $40 billion (Canadian dollars) worth of minerals each year. The mining industry pumps over $200 million a year into the nation's economy, accounting for about 4 percent of the nation's GDP. Because mining is such big business, the industry carries a lot of weight in political decisions. Change is not easy.

By reducing their demand for minerals, citizens the world over can reduce their need for virgin materials and reduce the associated impacts. But conservation is not sufficient. Minerals will continue to be extracted and fashioned into useful products even with the best recycling and conservation efforts. Thus, society must find ways to reduce the impacts of mining and mineral processing.

FIGURE 20.9 *Aerial view of a mineral smelter in New Mexico showing the tall stack that emits toxic gases into the atmosphere.*

One important way to reduce impact is through **reclamation,** the rehabilitation of land altered by mining (Figure 20.10). Surface mines, for instance, can be filled and the ground recontoured and planted to establish a vegetative cover that protects the soil. During mining, topsoil and wastes can be set aside and stabilized to minimize erosion, leaching, and landslides.

In 1977, U.S. Congress passed the Surface Mining Control and Reclamation Act, which requires coal mining companies to reclaim all surface-mined land. By law, companies must restore surface-mined land to its premining condition. Unfortunately, this law pertains only to coal mining; no specific federal legislation requiring reclamation on mineral lands exists, and state laws and regulations are often weak. According to the U.S.Bureau of Mines, between 1930 and 1980 only 8 percent of the land mined for metals and only 27 percent of the land mined for minerals was reclaimed. Showing how effective surface mining laws are, however, the Bureau noted that 75 percent of the land on which coal was mined has been reclaimed.

By one estimate, recontouring and planting the vast unreclaimed lands and leveling the spoil piles from all surface and underground mines would cost the United States approximately $30 billion. In Canada, the price tag is estimated to be $6 billion in Canadian dollars. Given budget constraints, it is unlikely that the United States will make much progress toward refurbishing the lands so recklessly mined by companies eager to make a profit and move on to other undisturbed land unless citizens put pressure on Congress.

In Canada, mineral mine reclamation falls under the purview of the provinces, each with its own set of rules. In the Northwest Territory and the Yukon, however, the federal government is in charge. Companies are required to develop and implement reclamation plans. As one source put it, "mines are no longer abandoned; they are closed following legislated or provincial procedures." Because of this, they often have few downstream impacts.

Mining companies in both countries are required to write **environmental impact statements** documents that list the potential impacts of mining and ways they're to be addressed. Impact statements often underestimate or ignore potential impacts. And despite this sometimes lengthy and costly procedure, companies ignore their plans to offset damage.

Monetary shortages have also weakened the inspection and enforcement of the Surface Mining Control and Reclamation Act. Without oversight, mining companies could return to earlier practices that left the land in a shambles.

The United States needs strong enforcement and needs money to reclaim abandoned lands. Concerned citizens can write their representatives to support continued funding of reclamation programs. To raise money, governments could increase taxes on minerals; these monies could be earmarked for land rehabilitation.

Smelting operations can be made more efficient. Cleaner fuels, such as natural gas, can be used. Pollution control devices can be installed to capture pollutants. Wherever possible, valuable minerals and other substances should be extracted from the sludge of pollution control devices and sold to willing buyers. In Canada, the Hudson Bay Mining has installed a hydrometallurgical process to extract zinc from ore, rather than using a smelter, which cut sulfur dioxide emissions by 98 percent. Canada federal government has started an other program, known as Accelerated Reduction/Elimination of Toxics (ARET) program to slash emissions at smelters. Nearly all of Canada's metal-producing companies have submitted plans that will cut emissions by 70 percent. These and other actions by the Canadian government and mining companies will help overcome a long history of environmental neglect by the industry.

In light of the limited supply of certain and minerals and the enormous impacts of mining on the environment, it is imperative that citizens the world over conserve what they have and find substitutes where they can. Recycling the minerals generously supplied by the Earth and reducing our demand can help us continue as we have. But eventually, many agree, unless adequate substitutes can be found, we will have to change our ways. These changes will involve reducing the global population, developing less resource-intensive lifestyles, and shifting our dependency toward renewable resources, which, if managed properly can ensure a sustainable society.

FIGURE 20.10 *Reclaimed land that has been surface mined to remove coal*

Summary of Key Concepts

1. The Earth's mineral wealth has been tapped for thousands of years. All around are the signs of our dependency on minerals and our often reckless exploitation of these resources. The sheer magnitude of human dependence on mineral and the impacts that mineral mining and processing have on the environment make minerals an important environmental issue and crucial to building a sustainable society.
2. Unlike forest and wildlife, minerals are nonrenewable, but they can be recycled. Achieving higher rates of recycling is essential to creating a sustainable society.
3. Minerals come from the Earth's crust. Most minerals exist as chemical compounds consisting of two or more elements. A rock containing minerals is called an ore.
4. The United States annually mines ores worth an estimated $39 billion. When refined, they become more than ten times more valuable.
5. Although the United States has only 4.5 percent of the world's population, it consumes about 20 percent of the world's nonfuel minerals. To raise the world to our levels of consumption would create an environmental disaster of epic proportions, and would be impossible because of a lack of reserves.
6. The United States' and other industrial countries' prodigious use of minerals has made them great and prosperous nations but also vulnerable ones, because many of the strategic minerals needed for commerce and defense come from unstable countries. To reduce this vulnerability, many governments have stockpiled strategic minerals.
7. Stockpiling may help in the short run but does nothing to protect nations from vanishing mineral supplies or the formation of cartels—groups of mineral-exporting nations that ban together to control the supplies and prices of minerals.
8. Many experts believe that cartels will not form around minerals because so many of the exporting nations need their mineral exports to produce foreign exchange.
9. Determining the life span of mineral reserves is not easy. First, scientists must determine the rate of consumption. Second, they must project future consumption levels, bearing in mind that increases in the consumption rate can greatly accelerate the depletion of a finite resource. Third, they must determine the economically recoverable supply, or reserve, of each mineral. The reserve must not be mistaken for the total resources, or the total amount of a mineral in the Earth's crust.
10. The reserve capacity of the world's mineral supply is not permanently fixed and can expand and contract, depending on a number of factors—for instance, new discoveries, economic incentives from governments, new technologies that allow miners to extract and process lower grades of ore more efficiently, the price of energy and labor, and the level of environmental controls.
11. Based on existing estimates, it appears that world reserves of three quarters of the 80 or so economically important minerals are abundant enough to meet our needs for many years, or, if they are not, adequate substitutes are available. However, at least 18 economically important minerals could fall into short supply, some within a decade or two.
12. Something must be done, and quickly, to provide much-needed minerals. Optimists believe that new discoveries, improved extraction technologies, the use of low-grade ores, and the use of substitutes will help to ensure sufficient supplies of important minerals.
13. Some observers believe that ocean floor minerals—deposits in the continental shelf and on the floor of the ocean—could help us stretch our mineral supplies. Manganese nodules, for instance, found on the ocean floor, could expand our reserve by thousands of years. Manganese nodules cover about one fourth of the ocean's floor and contain many important minerals, such as copper and iron.
14. Although attractive, mining seafloor mineral deposits and manganese nodules would be energy intensive and costly. It could increase the turbidity and temperature of ocean waters, and could upset the ecological balance of the oceans. Political problems also abound. The rich nations of the world want free access to the nodules in international waters and have refused to come to an agreement with the poorer nations, who want some of the proceeds of this international resource but have neither the money nor the resources to tap it.
15. Critics find serious fault in all of the strategies aimed at increasing supply. Pessimists believe that we must actively explore these strategies and that we must tap the generous potential of conservation.
16. One of the chief ways to conserve minerals is to reduce mineral demand. This can be achieved by reducing population growth, per capita consumption, and the size of various products. Increasing recycling and product durability can also assist. Economic incentives for nonrecycled materials could also be reduced or eliminated.
17. Considerable progress has been made in most of these areas in the past 25 years, but there is lots of room for improvement.
18. Individual efforts can also go a long way toward reducing mineral consumption.
19. The mining and the processing of minerals have numerous impacts on the environment. Ninety percent of our nonfuel minerals is extracted from surface mines, which are particularly destructive because the overlying soil and rock, or overburden, must be removed to get to the mineral deposits. As a result, mining can destroy scenic beauty.
20. Underground mines and surface mines produce enormous amounts of waste that are trucked away from the mine and deposited in huge piles. These piles are an eyesore and, if not vegetated, may be eroded by wind and rain. Sediment from waste heaps can fill streams, increase water temperature, disrupt food chains, destroy spawning beds, kill aquatic life, and increase flooding. Mine wastes also leach toxic chemicals, such as sulfuric acid, arsenic, mercury, and zinc.
21. Many ores are heated to high temperatures in specially built furnaces to separate metals from the ore. A number of toxic substances may be released into the atmosphere during this process, called smelting. Perhaps the most sig-

nificant is sulfur dioxide, which combines with oxygen and water in the atmosphere to produce sulfuric acid.

22. By reducing demand and increasing recycling, individuals can reduce their need for virgin materials and reduce the associated impacts, but conservation is not enough. Society must find ways to reduce the impact created by mining and processing.
23. One important way of reducing the impact of mining is by reclamation, the restoration of all mined land.
24. Strong enforcement of state laws and funds to reclaim abandoned lands are both vital to this task.

Key Words and Phrases

Acid Mine Drainage
Cartel
Continental Shelf
Depletion Allowance
Environmental Impact Statement
Finite Resource
Fluorosis
Improved Extraction technologies
Law of Supply and Demand
Law of the Seas Treaty
Manganese Nodules
Metal
Mineral
Mineral Conservation Strategy
Mineral Processing
Mining
Mining Act of 1882
Nonrenewable Resource
Ore
Ore Deposit
Overburden
Pessimist
Product Durability
Reclamation
Recycling
Reducing Demand
Reserve
Reserve Base
Smelter
Smelting
Solution Mining
Stockpiling
Strategic Mineral
Substitution
Surface Mine Control and Reclamation Act
Surface Mining
Technological Optimist
Total Resources
Waste Piles

Critical Thinking and Discussion Questions

1. Explain the paradoxical statement: "An American car is, in a sense, really a foreign car."
2. What is an ore?
3. In what form are most minerals found?
4. What percentage of the world's population lives in the United States? What percentage of the world's minerals do Americans consume?
5. Do you agree with the statement, "The United States' heavy reliance on imported minerals also makes it highly vulnerable"? Why or why not?
6. Define the terms reserve and total resources. How are they different? Why is it deceiving to calculate the life span of minerals based on total resources?
7. What factors cause the reserve base to shrink? What factors cause it to expand?
8. Debate the statement: "Our mineral supplies are adequate for many years to come, so we need not worry about shortages."
9. List and describe the major ways in which we can expand the reserve base. Describe the pros and cons of each technique.
10. Describe the mineral conservation strategy. In what ways can you be a part of this strategy?
11. Why is recycling often at a competitive disadvantage when compared to the use of raw materials?
12. Describe the major impacts of surface and underground mining and describe how they can be lessened.
13. Debate the statement, "The wealthy countries can afford to mine manganese nodules and should be allowed to do so without giving developing nations any of the profits."
14. List and describe the impacts of manganese nodule mining.
15. Using your critical thinking skills and your knowledge of ecology and related issues, discuss the following statement: "Recycling creates pollution. It's no better than producing materials from virgin resources."

Suggested Readings

Draper, D. 1998. *Our Environment: A Canadian Perspective.* ITP Nelson: Toronto. See the chapter on mining for an overview of Canadian issues and solutions.

Gardner, G., and P. Sampat. 1999. "Forging a Sustainable Materials Economy." In *State of the World 1999.* Starke, L (ed.). New York: W.W. Norton. Valuable reference.

Kane, H. 1996. "Shifting to Sustainable Industries." In *State of the World 1996.* Starke, L. (ed.). New York: W. W. Norton. Covers many important issues related to minerals and industry.

U.S. Bureau of Mines. 2000. *Mineral Commodity Summaries.* Washington, D.C.: U.S. Government Printing Office. Important source of data for all minerals. Published annually.

Young, J. E. 1992. "Free-Loding off Uncle Sam." *World-Watch* 5(1): 34–35. Excellent discussion of the unfair advantages given to the mining industry.

Young, J. E. 1992. "Mining the Earth." In *State of the World 1992.* Starke, L. (ed.). New York: W. W. Norton. Excellent source of information on the impacts of mining.

Young, J. E., and A. Sachs. 1995. "Creating a Sustainable Materials Economy" In *State of the World 1995.* Starke, L. (ed.). New York: W. W. Norton. Excellent source of information on recycling.

Web Explorations

Online resources for this chapter are on the World Wide Web at: **http://www.prenhall.com/chiras** *(click on the Table of Contents link and then select Chapter 20).*

Chapter 21

Nonrenewable Energy Resources: Issues and Options

Most college students alive today are only vaguely familiar with the twin oil crises of the 1970s. The first shock wave hit in 1973, when the **Oil Producing and Exporting Countries (OPEC)** imposed an embargo on oil, cutting back on exports and drastically raising the price. The brainchild of a Venezuelan multimillionaire, Juan Perez Alfonzo, who rode a bicycle to work and, reportedly, read by candlelight, the oil embargo was conceived in large part as a measure to reduce the waste of energy by the West and to slow down the rapid depletion of OPEC's vast but finite reserves. Iran's 1979 oil embargo, imposed largely for political reasons, dealt another blow to oil consumers.

In this perilous decade, crude oil prices shot up from $3 per barrel to $35. Inflation set in, reaching over 18 percent a year in the United States in 1978. High prices dampened consumer spending. One after another, manufacturers shut down operations, laying off workers by the thousands, until 70 percent of our industrial capacity lay idle.

The oil crises not only drove home the extent to which our economy and our lives were dependent on oil, they also taught us that virtually everything we bought or did required energy—lots of it. If it was priced too high, inflation, unemployment, and recession would emerge with devastating power.

In the year 2000, another oil crisis began to emerge, sending oil prices from $15 per barrel to $35 (as of September 2000). One of the causes of this crisis was overconsumption—demand exceeding supplies. High demand for automobile fuel, especially in the United States, for example, sent gasoline prices skyrocketing from $1 per gallon to over $2 per gallon in a year's time. Despite increases in production by certain OPEC countries, the price of oil remained high because of two essential shortages—a lack of refinery capacity and because of a lack of oil tankers to move oil from foreign sources to countries like the United States. Americans and others were simply using more oil and oil derivatives (like gasoline) than could be shipped and processed. At this writing, fears of inflation were beginning to appear, especially among those who had endured the 1970s. Concerns were mounting as to the effect heating oil shortages might have on Americans and others. Amongst all this, many utilities began to raise natural gas prices, setting up further worries among economists.

This chapter is about energy, specifically, nonrenewable energy sources such as coal, oil, natural gas, and nuclear energy. It describes each resource, projected supplies, and examines the pros and cons of each, especially the environmental pros and cons. It then looks at options for reducing our dependency on these fuel sources and steering onto a more sustainable energy path.

21.1 Global Energy Sources: An Overview

In the industrialized world, fossil fuels top the list of energy sources. So named because they were formed from plant and animal remains buried in the Earth millions of years ago, fossil fuels provide 85 to 90 percent of the energy demand of the industrialized world (Figure 21.1a) Even in the less developed nations, fossil fuels are a main source of energy (Figure 21.1b). China, for instance, gets 80 percent of its energy from coal. Three fossil fuels predominate—oil, gas, and coal. All of them are nonrenewable. A small amount of our energy also comes from nuclear fuels, also a nonrenewable source.

Of all the nonrenewable options, oil is the predominant energy source in the industrialized world, but oil supplies are on the decline. If current trends in population growth, economic growth, and energy use continue, most of us will live to see the end of oil. Some oil analysts experts claim that oil demand will exceed supply within the first decade of the new millennium (Figure 21.2).

After the oil crises of the 1970s, many countries turned to conservation and renewable energy sources, but in the 1990s many of them lost interest in these approaches. In this book, we use the term **conservation** in reference to a natural resource to mean two things: using only what you need (the frugality principle) and using all resources efficiently (the efficiency principle). Conservation strategies can save enormous amounts of energy, thus greatly extending the world's supplies of other fuels. John Herrington, former Secretary of Energy, calls energy conservation "our single largest resource." In easing demand by lowering speed limits or driving within the speed limit the most recent oil crisis could have freed up millions of gallons of gasoline, easing the crunch. But it was not an option visible to many politicians and users.

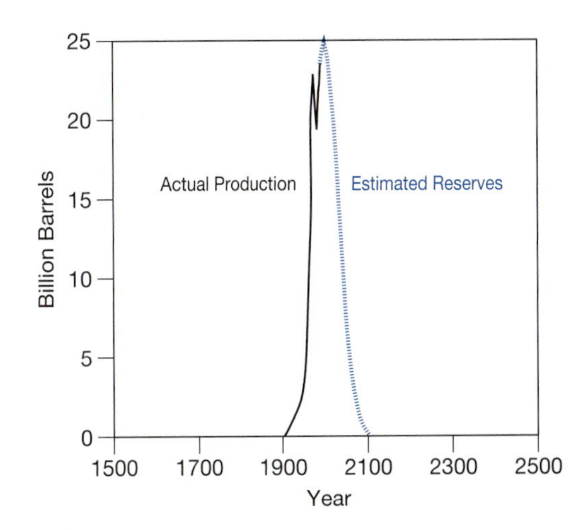

FIGURE 21.2 *Worldwide oil production. Note that oil demand is expected to exceed supply in the very near future. If reserves are larger than currently estimated, the peak in oil production could occur later.*

Of the renewable energy sources in use today, only two have made a significant contribution to the energy needs of the more developed countries. They are hydropower, or electricity from flowing rivers, and biomass, or crop residues, wood, and the like that can be burned or converted to gas or alcohol. In a few more developed nations such as Israel, rooftop solar panels provide a large percentage of the energy needed to heat water for domestic use. In others, wind energy has been tapped to produce electricity. In recent years, important strides have been made in generating electricity from sunlight by means of photovoltaics and a technology called solar thermal energy, both described in Chapter 22.

In sharp contrast, many of the less developed nations rely heavily on renewable fuels—wood, charcoal, cattle dung, and crop residues. In fact, about half the world's population—over 2.5 billion people—use wood as their primary source of energy, making it the major source of energy for human use on the planet. Although wood is renewable, excessive exploitation—

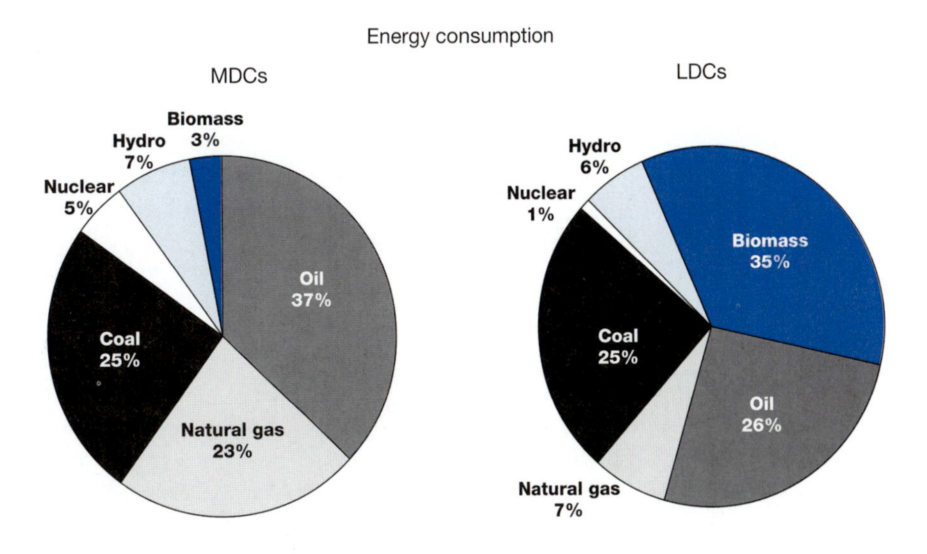

FIGURE 21.1 *Energy consumption by the more developed and less developed countries.*

caused largely by overpopulation—has resulted in widespread shortages and deforestation. In many rural villages in India, Bangladesh, and Nepal, for instance, villagers must travel great distances to find a few sticks of fuel to cook the evening meal. Trees around villages are stripped bare or missing altogether, having been torn down. Without wood, villagers have begun using dried cattle dung as a fuel, depriving farm fields of a once-rich source of fertilizer.

Struggling to industrialize, the less developed nations have begun to use increasing amounts of coal and oil. However, rising demand coupled with declining oil reserves could result in dramatic increases in prices that will cripple such efforts.

21.2 A Closer Look at Nonrenewable Energy Resources

This section examines our nonrenewable energy resources. It gives a brief history of their use, starting with coal; describes the benefits each provides; and looks at the impacts of their use.

Coal

Coal in use today originated as plant matter that grew in hot, muggy regions of the Earth 225 to 350 million years ago. These plants grew in lakes, streams, and coastal swamps and along their banks. Over the years, leaves and other plant matter fell into the water and accumulated on the bottom. Eventually, this rich organic material was covered by sediment that had eroded from the land. Over time, heat and pressure converted the organic material into peat, then coal, an organic material that, when burned, produces light and heat. The energy from coal, however, is ancient solar energy trapped by plants and stored in the bonds between the carbon atoms.

Today, coal is burned by electric utilities and in some factories and homes, where it releases the solar energy captured in the leaves many millions of years ago by photosynthesis. Today, coal supplies about 23 percent of the energy consumed each year in the United States. Similar percentages apply to other industrial nations. In Canada, however, coal contributes only about 13 percent of the total energy demand. In China, a country with rich supplies of coal and badly polluted air, coal meets 80 percent of the total energy demand.

The value of coal as a fuel was first discovered in the twelfth or thirteenth century by the inhabitants of the northeastern coast of England, who discovered that certain black rocks found along the shores, called "sea coales," burned. This monumental discovery led to coal mining for domestic heating and, much later, provided the impetus for the Industrial Revolution in the 1700s in England and the 1800s in the United States.

Coal mining started in the United States around 1860. Nevertheless, coal did not replace wood as a major fuel source until the early 1900s. Shortly after World War II, king coal began to be replaced by oil and natural gas because they were much easier to transport and were cheaper to extract (Figure 21.3).

Types of Coal Geologists recognize three major types of coal: lignite (brown coal), bituminous (soft coal), and anthracite (hard coal). These types vary in several respects, the most important being their carbon content, and their heat value, or the amount of heat they produce when burned per unit weight. Lignite has the lowest carbon content and heat value, anthracite the highest. As a general rule, the more heat and pressure that were applied to a bed of coal in the prehistoric past, the higher the grade. Most of America's coal is bituminous. Coal also varies in sulfur content. Today, many more developed countries such as Canada and the United States use low sulfur coal to reduce sulfur dioxide emissions and acid deposition.

Coal Reserves Because coal deposits lie just below the surface, scientists have been able to make fairly accurate estimates of worldwide coal reserves by drilling test holes in U.S. coal seams. A map of world coal deposits is shown in Figure 21.4. Surveys show that, of all the fossil fuels, coal is by far the most abundant. The proven coal reserves (those we're certain exist) worldwide are nearly 790 billion metric tons. At the current rate of consumption the proven reserves—those we know exist—would last approximately 200 years. Undiscovered reserves, those thought to exist, are believed to be massive as well. Some experts believe that the world's recoverable coal supply may last close to 1,700 years at the current rate of consumption.

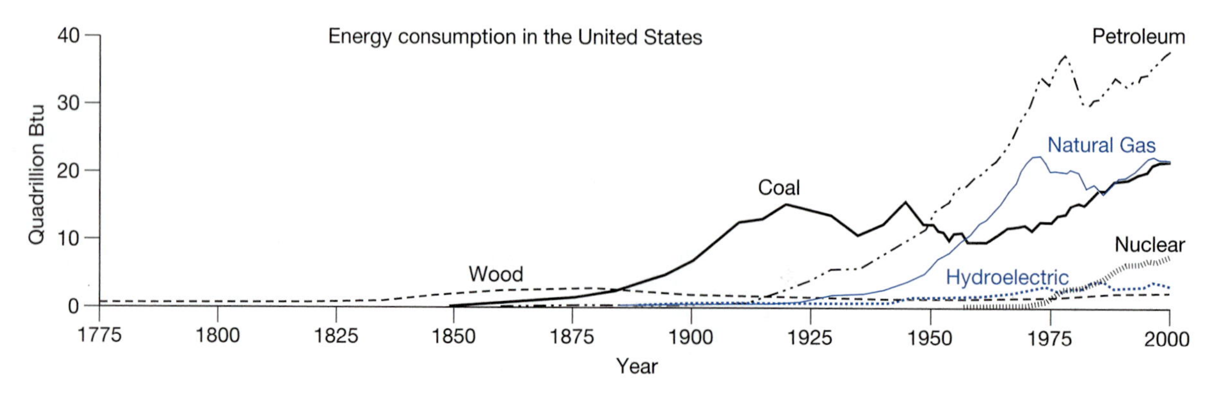

FIGURE 21.3 *The pattern of U.S. energy consumption from 1850, including a question about our future patterns of energy use. As supplies of petroleum become exhausted, what alternative sources will be exploited?*

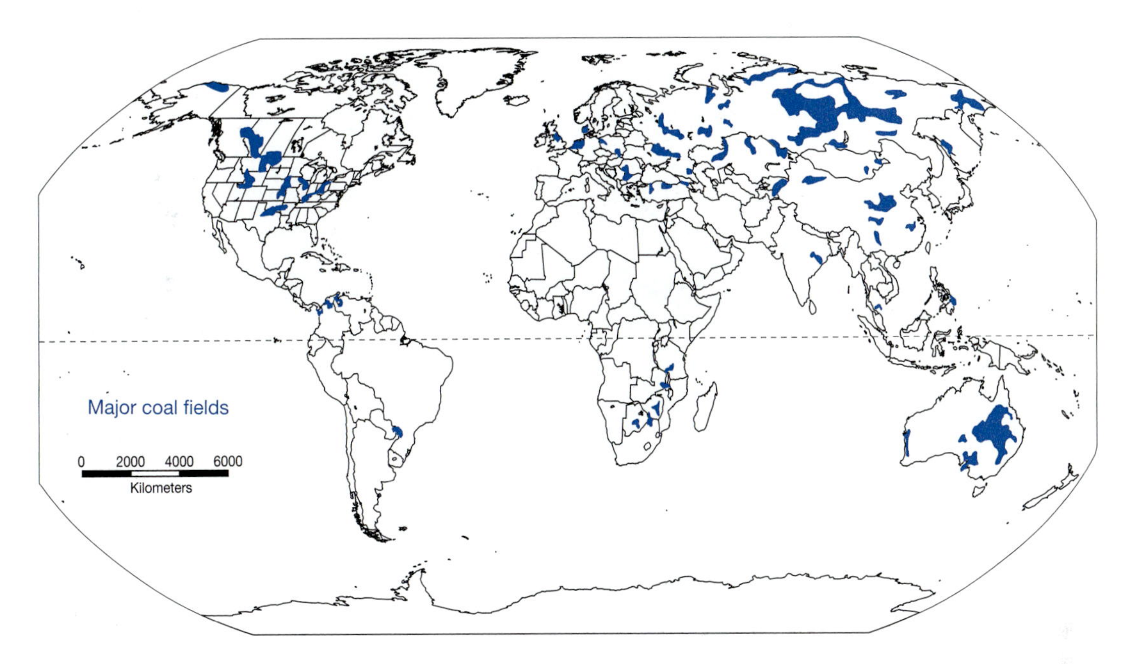

FIGURE 21.4 *Distribution of global coal deposits.*

The United States has about 30 percent of the world's coal and is sometimes called the "Saudi Arabia of coal." It should be pointed out, however, that not all fossil fuels are of equal value. In fact, coal is of limited use. It cannot easily or efficiently be used to power automobiles (except maybe electric cars) or jet planes, unlike liquid fuels derived from oil, the supplies of which are on the decline. Coal cannot easily or efficiently be used to heat homes or cook meals, unlike natural gas, whose supplies are better than oil's but are still limited. (Coal can be converted into synthetic gas and oil, as described later, but the processes are costly and dirty.)

Although coal is the world's most abundant fossil fuel, it is also the dirtiest one. Mining it causes enormous impacts. Burning coal to generate electricity produces incredible amounts of pollution and solid waste. It is one of the main sources of the greenhouse gas carbon dioxide. Unless we find much cleaner ways to burn it, much of our impressive coal deposits may have to remain forever buried in the Earth's crust.

Environmental Impacts of Surface Mining America consumes over 930 million metric tons of coal each year. About two thirds of that coal comes from surface mines, of which there are two basic types: contour mines and area strip mines. In the **contour mine,** which is found on hilly terrain in the eastern section of the country, coal seams are exposed by bulldozers or steam shovels that remove the overlying rock and dirt, called **overburden.** Before 1979, miners simply pushed the overburden over the side of the hill, destroying vegetation and creating serious erosion problems that often filled nearby streams with sediment (Figure 21.5A). Now, thanks to tougher mining regulations, miners must first remove the topsoil and set it aside to restore the site. Then they dig up the underlying layers and haul them away to a safe place, where they are held until they are trucked back to recontour the area when mining is completed.

The **area strip mine** is found on flat terrain in the Midwest and West in the United States and the central provinces of Canada (Figure 21.5B). As in contour mines, topsoil is first removed by bulldozers and set aside for later use. Then, huge shovels called **draglines** remove the overburden, which is placed in piles next to the excavation. The exposed coal seam is then dynamited, excavated, and loaded into trucks. After the coal has been removed from the first mined section, the process begins again, one strip at a time. The overburden from each new strip is placed in the previous excavation. Reclamation can begin on previously mined land after two or three strips have been cut. During reclamation, bulldozers first recontour the land, then spread the topsoil over the surface of the overburden, which is often replanted with native species or fast-growing cover crops that protect the soils from wind and rain.

Both forms of surface mining are fast and efficient ways of removing coal but they create eyesores. Mine waste (overburden) can erode away if proper precautions are not taken. Eroded sediment spills into streams and lakes and destroys fish habitat, recreational sites, and reservoirs that supply water for human use. By one estimate, there are 0.6 million hectares (1.5 million acres) of unreclaimed coal-mined land in the United States in need of reclamation, but only about a third of it must be reclaimed by law. The rest had been mined before the passage of the surface mining legislation and is exempt from reclamation.

Access roads can also be eroded away by rain and snowmelt, further adding to the sediment load of nearby streams and lakes. Surface mining creates dust and noise and destroys wildlife habitat, at least temporarily. Surface mines can also cause groundwater levels to fall considerably, drying up municipal and agricultural wells in neighboring areas.

FIGURE 21.5A *A contour mine in the eastern United States. Overburden dumped over the hill creates a highly erodible surface. Soil erosion at this site during rainstorms and snowmelt leads to serious siltation in nearby streams.*

Interestingly, underground mines actually disturb about as much land as surface mines per ton of coal removed because the materials removed to reach coal seams must be stored outside the mine. Underground mines can also collapse, killing workers and causing subsidence, a sinking of the surface above them. Since 1900, over 100,000 workers have been killed and over a million have been permanently disabled in underground mine accidents in the United States—a heavy price to pay for energy! Subsidence also damages buildings and roadways as mine shafts sink and cause huge cracks to develop in the surface. Imagine waking up one day to find your two-story house ripped in two because the ground underneath it had given way.

Cracks in the Earth's surface can divert streamwater into coal seams. Water seeping naturally into mines or as a result of subsidence cracks combines with naturally occurring iron pyrite (a sulfur-bearing mineral, also known as fool's gold) and oxygen to produce sulfuric acid. If not captured and neutralized, the acidic outflow, called **acid mine drainage,** leaks out of mines and pollutes nearby streams. How serious is this problem?

Abandoned underground mines in the United States produce 2.7 million metric tons of acid each year and pollute 11,000 kilometers (7,000 miles) of stream, most of which are in Appalachia. Stopping the flow from active mines is quite feasible, but acid mine drainage from abandoned mines has proved expensive and difficult, if not impossible, to control.

FIGURE 21.5B *Surface view of an area strip mine. Coal is extracted after a dragline removes overburden. Spoil piles will be recontoured and reseeded after mining.*

In 1977, the U.S. Congress passed the **Surface Mining Control and Reclamation Act,** which went into effect in 1979. This important law requires mining companies to reclaim their land—that is, restore the original contour of the land and replant it. It also calls on companies to control on-site erosion and acid contamination of nearby lakes and streams. The law also established state and federal watchdog agencies to monitor reclamation, and placed a federal tax on coal to finance the reclamation of land destroyed by previous strip-mining operations. Unfortunately, the estimated $4 billion the fund will yield before it expires is well below the $8 billion needed to complete the reclamation.

The Impacts of Coal Combustion A second major concern of many is the pollution produced by burning coal. Four pollutants are of particular interest: carbon dioxide, sulfur dioxide, nitrogen dioxide, and particulates.

As we pointed out in Chapter 19, the combustion of coal and other fossil fuels has resulted in the massive release of the greenhouse gas carbon dioxide. Much of it remains in the atmosphere. Although a little carbon dioxide is essential to maintaining the Earth's temperature, too much may cause a drastic increase in average global temperature that could disrupt our lives profoundly.

Sulfur dioxide and nitrogen dioxide, discussed in more detail in Chapter 19, are acid precursors—they form sulfuric and nitric acids, respectively, when they combine with oxygen and water in the atmosphere. These acids fall to the Earth in rain and snow and particulates, acidifying lakes and streams, killing trees and damaging crops, buildings, and statues. To control the emissions of sulfur dioxide, many utilities have switched to low-sulfur coal. Because it is a more expensive means of achieving the same end, fewer have installed smokestack scrubbers, pollution control devices that remove much of the gas from the smokestack. Such measures have helped the United States and other nations drastically reduce sulfur dioxide emissions.

Unfortunately, controlling nitrogen dioxide is no easy matter. The gas is not very soluble in water and therefore not removed by scrubbers. The only measure utilities use now is the control of combustion temperatures, but it has only a minor effect. Currently, U.S. utilities and factories produce about 23.6 million tons (21.25 metric tons) of nitrogen dioxide a year. Production is projected to climb even higher as the nation's fossil fuel dependency increases.

Coal combustion also produces enormous quantities of particulates. Like sulfur dioxide, they are relatively easy to control using various pollution control devices. Even with the best pollution control, however, a single power plant supplying a million people produces 1,500 to 30,000 metric tons of particulates a year.

For years, it was assumed that the best means of reducing pollution from coal-fired power plants was by pollution control devices. However, pollution control devices merely capture gaseous pollutants and convert them into a solid waste. This essentially shifts pollution from one medium (air) to another (ground), trading one nuisance for another. Creating a sustainable future will ultimately require efforts to reduce emissions through efficiency measures and renewable energy sources such as wind and solar energy.

Fluidized Bed Combustion Efficiency can occur at several levels. End-use efficiency refers to things that users like you and I can do to use energy more efficiently. Combustion efficiency is another option. Both can be combined to create tremendous savings in energy demand and pollution.

There are also ways to burn coal more efficiently. One of the most promising of the dozen or so alternative technologies is known as **fluidized bed combustion.** In this process, coal is crushed, then mixed with bits of limestone and propelled into a furnace in a strong current of air (Figure 21.6). The particles mix turbulently in the combustion chamber, ensuring very efficient combustion and therefore low carbon monoxide emissions. The furnace also operates at a much lower temperature than a conventional coal boiler, thus reducing nitrogen oxide emissions. The limestone reacts with sulfur oxide gases, producing calcium sulfite or calcium sulfate, reducing sulfur oxide emissions from the stack gases. These compounds, however, must be disposed of.

Several large demonstration projects are operating now in Colorado and Kentucky. Should they prove successful, they could pave the way for wider use of this technology. But don't expect an overnight transition: At least 30 to 50 years would be required for this technology to make a substantial contribution to our electrical generating capacity. And, as signs of global warming become more evident, many nations may find themselves switching entirely from coal to natural gas, a much cleaner fuel. Turbines similar to jet engines are now being used to produce electricity from natural gas. The process is much more efficient than electrical production from coal combustion and with far fewer pollutants. Carbon dioxide emissions alone are 50 to 70 percent lower. Over time, however, many nations may find themselves relying more on end-use efficiency measures, wind energy, solar energy, and clean-burning biofuels (for example, ethanol from corn) that do not contribute carbon dioxide to the atmosphere (Chapter 22).

Coal Gasification Another possible technology is **coal gasification,** a process in which combustible gas is produced from coal by one of several methods. This technology is viewed as a means of putting to use the world's enormous supplies of coal. Proponents argue that it could provide a supply of gas to replace declining reserves of natural gas.

Coal gasification facilities can be designed to be cleaner than conventional coal boilers now in place throughout the country. In this process, a coal—water mixture, called a slurry, is injected with oxygen into a heated chamber, producing three combustible gases: carbon monoxide, hydrogen, and methane. The heated gas is then cooled a bit and purified. The resultant gas burns as cleanly as natural gas. This particular technology is as efficient as natural gas combustion. Coal gasification, commonly used in the mid-1800s to produce gas from coal, could provide gas for home heating and other uses, a need that will become acute as domestic natural gas supplies become depleted. Like fluidized bed combustion, it produces less nitrogen oxide than conventional coal boilers and eliminates much of the sulfur dioxide as well. Both technologies, however, produce solid wastes that must

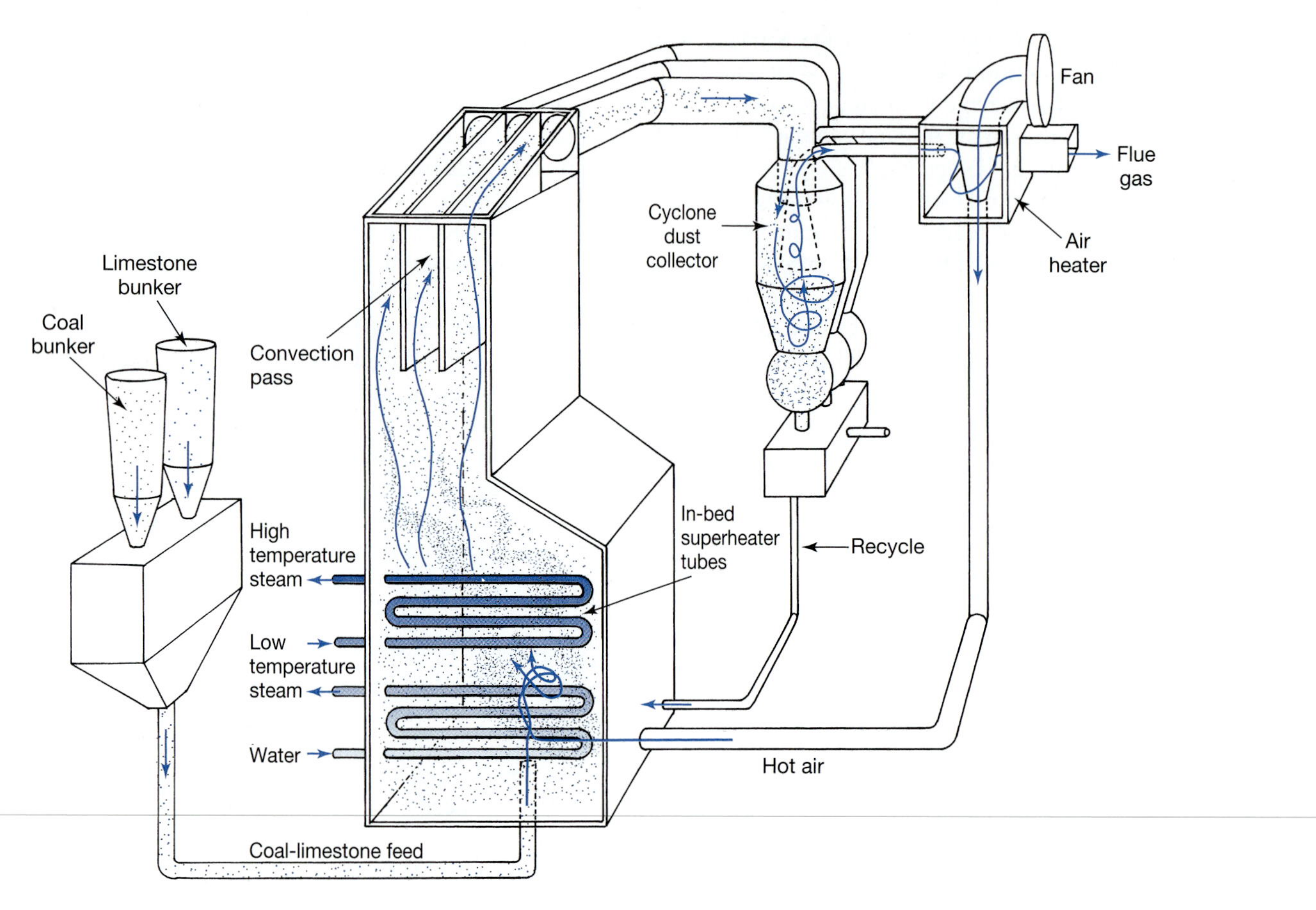

FIGURE 21.6 *Fluidized bed combustion. This process burns crushed coal blown into a furnace mixed with tiny limestone particles. The air turbulence in the furnace ensures thorough combustion, thus increasing efficiency. The limestone reacts with sulfur oxide gas, removing most of it from the smokestack. Steam pipes in the furnace help maximize heat efficiency.*

be carefully disposed of to prevent groundwater contamination. In addition, further combustion of fossil fuels adds more carbon dioxide to the atmosphere.

Researchers are exploring the possibility of converting coal to gas underground—or *in situ,* meaning in place. To do this, oxygen is pumped into coal beds that are subsequently ignited. Hydrogen, carbon monoxide, and methane gases are released from the smoldering fire and pumped back to the surface, where they can be burned to produce electricity. This process eliminates the need to surface mine the coal for above-ground gasification, but could result in widespread subsidence and uncontrollable underground fires that could smolder for decades.

Coal Liquefaction Coal can also be treated to produce a thick, oily substance in a process called **coal liquefaction.** At least four major processes now exist, each of which adds hydrogen to coal to produce oil. The oil can then be refined, just as crude oil is, to produce a variety of products, among them jet fuel, gasoline, kerosene, and many chemicals used to manufacture drugs, plastics, and a host of other products.

Coal liquefaction plants currently in operation show that this technology, while feasible, will be costly. It also generates numerous potentially harmful pollutants, such as phenol. And it would do nothing to reduce carbon dioxide levels.

Oil

The year was 1859 and the place was Titusville, Pennsylvania. When "Colonel" Edwin Drake's steel drill hit 20 meters (70 feet), a black, foul-smelling liquid came gushing from the well, signaling the dawn of a new energy era (Figure 21.7). After finds in several other countries, including Russia, less than a century later oil had become the world's most important source of energy.

Being liquid and relatively easy to transport long distances, either by ship or by pipeline, oil is often viewed as an ideal fuel, burning dirtier than natural gas but cleaner than coal. It provides about 40 percent of the United States' energy needs and a somewhat higher percentage of the demand in other industrial nations. Canada satisfies about 27 percent of its total energy demand from oil.

Oil Reserves Unlike coal, oil is in short supply. Some oil analysts believe that peak production, the time when production tops out and demand begins to exceed supply, could occur within this decade—between 2005 and 2010 (Figure 21.2). According to these analysts, the world's oil reserves could be gone by the year 2018.[1] Estimates of oil reserves are difficult and often

[1]Of course, this will never happen. As oil supplies fall, it will become prohibitively expensive. We'll never deplete the world's oil. We'll just run out of cheap oil and switch to other energy sources.

FIGURE 21.7 *America's first oil well, at Titusville, Pennsylvania. Photograph taken in 1864.*

FIGURE 21.8 *Oil well productivity. Note that oil production from U.S. wells have experienced a dramatic decline. Most of the good sites have been tapped out.*

inaccurate. Some oil geologists believe that the ultimate production, the amount of oil we will eventually be able to recover, is 2.5 times greater than once thought. Even so, rising demand (assuming a 5-percent per year increase) would wipe out these additional reserves, if indeed they exist, by the year 2038.

The outlook for oil in the United States is even grimmer. Oil production peaked in the early 1970s. Since then, domestic production has fallen steadily. Despite an increase in the number of wells drilled, daily output from domestic wells fell from 11 million barrels per day in 1973 to 6.2 million barrels in 1998. Figure 21.8 shows this trend but in barrels per well per day. Because of this, U.S. reliance on imported oil has increased dramatically to over 9.5 million barrels per day in 1999 (Figure 21.9).

Declining oil production in domestic wells means that the United States will become more and more dependent on imports. Unfortunately, much of the world's oil reserves are located in the politically unstable Middle East (Figure 21.10). Unless a clean, economical substitute is found, the United States could face frightening economic times, as could many other industrial nations like Japan and Germany.

The Impacts of Oil Production and Consumption Oil comes from wells on land and sea (Figure 21.11). The impacts of oil on the sea are discussed in Chapter 10. Land-based drilling can have many impacts as well. Roads and well sites, for instance, destroy wildlife habitat and wilderness and disturb sensitive species. They also increase soil erosion. Leaks from

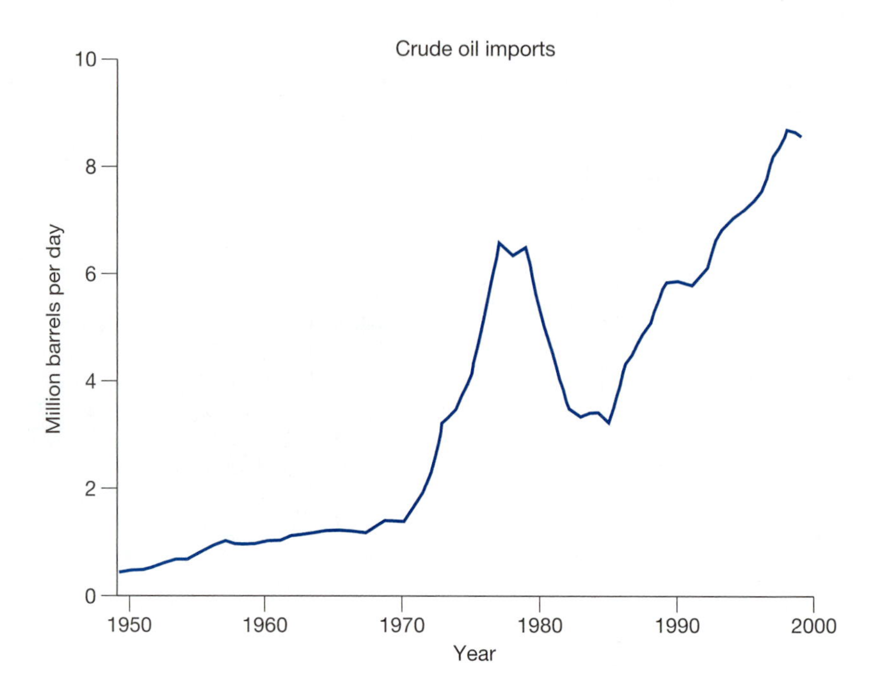

FIGURE 21.9 *Crude oil imports. Because of declining domestic production, the U.S. has grown steadily more dependent on foreign oil, much of it from unstable sources in the Middle East.*

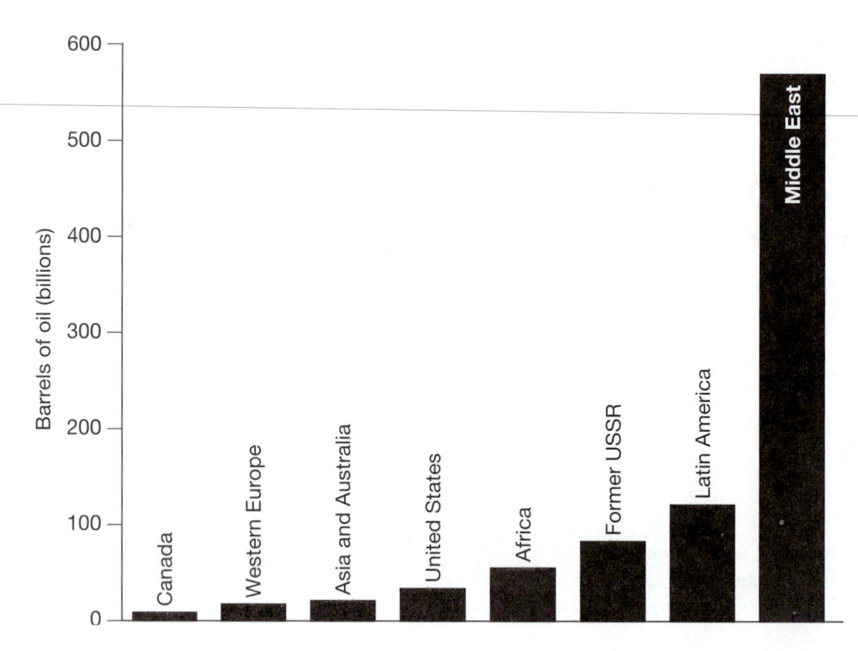

FIGURE 21.10 *World oil reserves. This graph shows where the world's oil reserves are located.*

wells or spills from pipelines can cover land with a thick, gooey residue that is difficult and costly to remove. Oil kills vegetation and seeps into the ground. Government plans to sell oil and gas leases in the 0.6-million-hectare (1.5-million-acre) coastal plain of the Arctic National Wildlife Refuge in Alaska's northeast corner that once created a storm of controversy. By one estimate, 20 to 40 percent of the caribou population would have been forced to move out of this 8-million-hectare (19.5-million-acre) refuge or perish as a result of habitat loss. Huge flocks of snow geese, estimated at 300,000 birds, that use the coastal plain as a staging ground on their annual migration would also have been forced to find new territory; many would likely have died. Musk oxen, only just reintroduced in 1969 and 1970 and now numbering about 500, would feel the ill effects of oil development.

No one knows how much oil, if any, lies underneath the sensitive coastal plain. The mean estimate is 3.2 million barrels with a 19 percent probability of occurrence. Full-scale development of the region would supply only about 4 percent of the projected oil demand of the United States by the year 2005 and less than 3 percent by 2010. Experience in nearby Prudhoe Bay, lying 160 kilometers (100 miles) to the west, suggests that the risks may be too great to support development. According to a report on Prudhoe Bay, between 400 and 600 oil spills are reported every year. In 1986, 240 million liters (64 million gallons) of toxic wastewater containing a variety of heavy metals, hydrocarbons, and chemical additives were released onto the tundra. Gas vapors and combustible liquids routinely burned in oil fields created plumes of black smoke that stretched 100 miles from the oil fields.

FIGURE 21.11 *Oil wells like these off the coast of Texas release oil into the ocean, which may soil nearby beaches and harm coastal wetlands.*

Nitrogen oxide and sulfur dioxide emissions from plants, biologists feared, would acidify the fragile tundra ecosystem. Making matters worse, the landscape around Prudhoe Bay is littered with broken-down vehicles, junked airplanes, used batteries, polyfoam insulation, tires, scrap metals, and so on. The damage, some environmental scientists fear, may be irreparable.

All of this illustrates the extent to which our desire for energy affects the environment and disrupts the lives of the many species that share this planet with us. Fortunately, this story ends on a happy note. In 1991, the U.S. Congress defeated a bill that would have opened up the refuge for drilling, despite then-President Bush's support. This victory for the environment is testimony to the power of environmental groups and citizens, whose letters and phone calls helped defeat what could have been a devastating project. But don't be lulled into complacency. It is unlikely that the oil companies and powerful representatives from oil-producing states will let this issue die. Many are lobbying president George W. Bush who supports oil development in the region to open up the Arctic Natural Wildlife Refuge as this book went to press.

Natural Gas

Natural gas is a combustible gas that consists primarily of methane (CH_4). Like coal and oil, it is a fossil fuel. Produced by decomposing plant and animal remains that were buried in the Earth for millions of years by sedimentary deposits, natural gas deposits often accompany coal and oil deposits.

Today, natural gas supplies about 23 percent of the energy needs of the United States and 31 percent of Canada's energy supply. Easily transported within a country by pipeline, it is used primarily for heating buildings, home cooking, industrial processes, and generating electricity.

Natural Gas Reserves

The picture for natural gas is a bit brighter than for oil globally, but not as good as that for coal. Estimates as to the amount of natural gas that will ultimately be recovered from the Earth before supplies run out—the ultimate production—vary considerably. By one fairly generous estimate, the global ultimate production is 10,000 trillion cubic feet. To date, global consumption has used up about 2,700 trillion cubic feet. Of the remaining 7,300 trillion cubic feet, only 4,400 are proven reserves (are known to exist). The rest are undiscovered reserves. The global proven reserves will last about 60 years at the current rate of consumption; the undiscovered reserves will last about 40 years. At best there is a 100-year supply, but increased consumption is likely, and with that so is a drastic reduction in global supplies.

The outlook for natural gas in the United States has recently brightened. In the 1980s, the total known reserves were about 187 trillion cubic feet. New studies, however, suggest that the total reserves could be seven times greater, or about 1,000 to 1,300 trillion cubic feet. At current rates of consumption, this would last about 60 years. Because it burns so cleanly, natural gas could become a transitional fuel—that is, a relatively clean fossil fuel that replaces oil and coal while the nation develops even cleaner, sustainable, renewable alternatives (Chapter 22). Again, however, despite the large U.S. supplies, even a modest increase in natural gas use could cut this supply in half. As with oil, something must be done about our declining natural gas reserves.

What could replace natural gas? Many options exist, but some are more sustainable than others. Among the least sustainable is coal gasification. More sustainable options are improved energy efficiency, hydrogen, and methane from organic wastes such as garbage, sewage, sludge, and livestock manure. These options are discussed in Chapter 22.

Oil Shale

Oil shale is a grayish brown sedimentary rock that was formed millions of years ago from the mud at the bottom of prehistoric lakes. Contained within the rock is a solid organic material known as **kerogen** (Figure 21.12). When heated to high temperatures, the rock gives off its oily residue, called **shale oil.** High-grade oil shale can produce up to 120 liters (30 gallons, or about three fourths of a barrel) of oil per ton of rock. Like petroleum, this oil can be refined to produce gasoline, jet fuel, kerosene, and a variety of feedstocks used by the chemical industry to produce fabrics, drugs, and plastics. The most valuable deposit in the United States is in the Green River Formation, a 43,000-square-kilometer (16,500-square-mile) region found in Colorado, Wyoming, and Utah (Figure 21.13). Eighty percent of this land is federally owned.

Scientists estimate that, with current technologies, 80 to 300 billion barrels of oil could be extracted from these deposits. To put things into perspective, that is enough oil to satisfy domestic demands (at the current rate of consumption) for 7 to 28 years. Doing so, however, would exact enormous economic and environmental costs.

Environmental Impacts of Oil Shale Production

Oil shale can be processed in **surface retorts,** large steel vessels in which crushed shale is heated to drive off the oil. Shale to supply surface retorts may come from either underground mines or surface mines, depending on the depth of the deposit. Surface mining operations, however, tear up large tracts of land. Much of the land overlying the Green River Formation is prime habitat for hundreds of thousands of mule deer. In addition, the retorts produce enormous amounts of waste, or spent shale (shale that has been burned). A modest operation producing 50,000 barrels per day would also produce 53,000 metric tons of spent shale, which would have to be disposed of properly to avoid contaminating groundwater and surface waters. Adding to the problem, crushing the shale before it can be retorted increases the volume of the shale by about 12 percent, so not all of it can go back into the excavation.

Retorts are notorious polluters. Sulfur oxides, nitrogen oxides, heavy metals, and various organic pollutants, all toxic to humans and wildlife, could foul the western skies if oil shale becomes a viable business. And surface retorts require tremendous volumes of cooling water—about 2.5 barrels per barrel of shale oil produced. Water, of course, is a hotly contested commodity in the rapidly growing arid West.

Some companies have experimented with another process called ***in situ* retorting,** to avoid the solid waste problems. In an *in situ* retort, the oil shale deposit is fractured with dynamite and set afire underground. In theory, the fire burns the shale (consuming some of the useful fuel). Heat from combustion, however, drives off the rest of the kerogen, which, in a vapor state, is pumped out of the deposit, condensed, and then refined. *In situ* retorting, tried in the late 1970s to avoid disturbing the land and generating solid wastes, has proved difficult to master. Fires go out easily because of incomplete fracturing. Complicating matters even more, groundwater often seeps into the oil shale bed, dousing the fire. Controlling pollution is also more difficult *in situ*. As a result, this technique has largely been abandoned.

Tar Sands

In some parts of the world, oil has migrated into neighboring layers of sandstone, creating **tar sands** or **oil sands.** Tar sands contain an organic material, called **bitumen,** that can be extracted from the sand, then refined to produce a variety of fuels and chemicals like shale oil. Significant deposits of bitumen are found in Alberta, Canada, and the United States. Although six states have economically attractive deposits, Utah is the leader, with over 90 percent of the commercially feasible tar sands.

Lying deep within the Earth's crust, U.S. deposits contain an estimated 27 billion barrels of bitumen, but are not readily surface mined. To recover the oil, hot steam is generally pumped down

FIGURE 21.12 *Block of oil shale and a beaker of the oil extracted from the shale.*

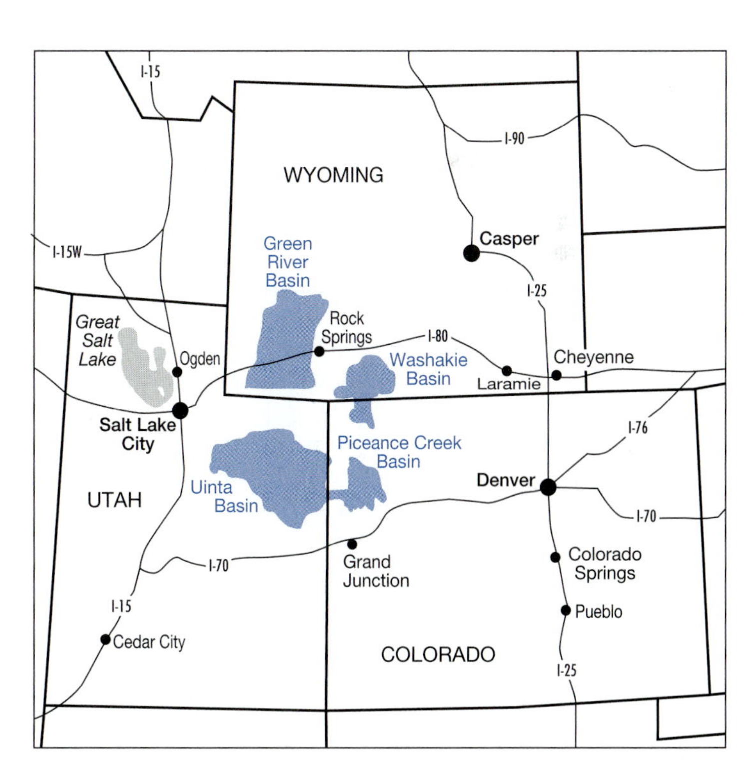

FIGURE 21.13 *Location of our nation's main oil shale deposits in Colorado, Utah, and Wyoming.*

into the deposits to free the bitumen, which is then pumped to the surface in recovery wells. When all is said and done, only about 1 billion barrels are likely to be recovered from our deposits, which is about one sixth of the United States' annual oil consumption.

Canadian deposits are much greater than those in the United States. In addition, the richer Canadian deposits are readily surface mined and are therefore processed in aboveground facilities. Containing an estimated 1.7 trillion barrels of oil. Some industry analysts believe that about 300 billion barrels of oil are recoverable, enough to supply Canada's oil demands for 200 years.

Unfortunately, tar sand production is very energy intensive. The energy needed to mine and process enough tar sand to produce a barrel of oil is equivalent to about 0.6 barrel of oil. The energy output of a fuel supply, or its **net energy efficiency,** takes into account what must be invested to get energy in the first place. Low net energy efficiency, say experts, will probably stifle tar sand production.

The Future of Fossil Fuels

The energy supply picture boils down to this: Oil, our most widely used fuel, is fast on the decline in the United States and abroad. Demand is expected to outstrip supply sometime in the early twenty-first century, causing a drastic increase in price and severe economic problems for countries that have not switched to alternative fuels.

The prospects for natural gas are better globally. U.S. domestic supplies seem adequate as well, but could be more quickly depleted as natural gas begins to be used as a transitional fuel. Eventually, natural gas supplies will run out and society will be forced to turn to other sources.

The prospects for coal are the best. The world has abundant supplies of coal, and the United States has about one third of the world supplies. But coal combustion, no matter how efficient, will continue to add to global carbon dioxide levels, contributing to global warming. Moreover, simply reducing the emissions of acid-generating pollutants from coal-fired power plants only shifts the problem: Instead of going up the smokestack, the pollutants are trapped and become a hazardous waste in need of careful disposal (Chapter 18).

It is time, say many experts, to begin to find a substitute for oil that can be used to heat homes, produce liquid fuels, and produce chemicals. Oil shale and tar sands are two alternatives, but their reserves are relatively small compared to global oil demand, and they are extremely costly to develop, both environmentally and economically. Biofuels, such as ethanol from crops, may be an essential component of a new, sustainable energy strategy. Ethanol, a renewable fuel made from corn and other crops, helps reduce net carbon dioxide production and could replace gasoline entirely.

Because shifting to alternative energy options takes time, we should begin now to develop substitutes. Of course, conservation (frugality and efficiency) will be an invaluable ally. Methane from trash or crop residues could help. Hydrogen from water could also help.

Hydrogen gas can be generated from one of the Earth's most abundant substances, water. Passing electricity through water causes the water molecules to split; in so doing, they produce oxygen and hydrogen. Hydrogen, in turn, can be burned to produce energy. When hydrogen gas burns it combines with oxygen, once again forming water. The only pollutant produced in the process is a small amount of nitrogen oxide. Unlike fossil fuels, hydrogen combustion produces no sulfur dioxide or carbon dioxide. This energy option is described more fully in Chapter 22.

Another option that is sometimes proposed as an answer to the declining supplies of polluting fossil fuels is nuclear energy. Is it a good alternative?

21.3 The Nuclear Energy Option: Is It Sustainable?

The world's first atomic bomb was dropped on Hiroshima, Japan, in 1945. This momentous and catastrophic explosion, which helped conclude the long and costly Second World War, also ushered in the Atomic Age—an age of nuclear weapons and nuclear power. This section is primarily about nuclear energy and radiation. It examines the impacts of nuclear power as well as its benefits, starting with some background information.

Understanding Atomic Energy and Radiation

To understand nuclear energy, you must understand a little bit about the atom.

Atomic Structure All matter, whether solid, liquid or gaseous, is composed of tiny particles called **atoms.** Put another way, atoms are the fundamental units of all living and nonliving matter. Each atom, in turn, is composed of a dense, centrally located **nucleus** containing positively charged particles called **protons** and electrically neutral particles called neutrons. The nucleus contains virtually all of the mass of an atom. Surrounding the nucleus is a region called the **electron cloud** where tiny, almost massless, negatively charged particles, known as **electrons,** are found. They orbit around the nucleus, ever attracted to its positive charge and always in motion at nearly the speed of light.

Atoms combine with other atoms to form molecules, such as water (H_2O), which contains two atoms of the element hydrogen and one atom of the element oxygen.

Isotopes and the Origin of Radiation Early scientists thought that the variety of matter they encountered resulted from the fact that pure substances, which they called **elements,** could combine to form a variety of different substances, a hypothesis that modern science has shown to be true. Today, scientists realize that elements are pure substances. Elements contain only one type of atom and scientists have identified over a hundred different types of elements, 92 of which occur naturally in the Earth's crust. Scientists have found that each element differs from the next in the number of protons it contains in its nucleus. Thus, the element carbon always has 12 protons and the element oxygen always has 16 protons. In fact, scientists classify the elements by the number of protons in their nucleus; this is called the **atomic number.** The atomic number of carbon is 12, and the atomic number of oxygen is 16.

In any given element, the number of electrons in the electron cloud of its atoms is always equal to the number of protons in the nucleus. Thus, atoms are electrically neutral. Although the number of electrons and the number of protons in the atoms of any given element are always the same, the number of neutrons may vary. For instance, all atoms of the element uranium have 92 protons in their nuclei. However, some uranium nuclei contain 143 neutrons and others have 146 neutrons.

Scientists add the number of protons and neutrons to obtain the **atomic mass.** Each proton and neutron has a mass of one unit. Therefore, uranium atoms with 92 protons and 143 neutrons have an atomic mass of 235. They're written as U-235. Uranium atoms with 146 neutrons have an atomic mass of 238. These different forms of the same element are called **isotopes.**

Because atoms of a given element differ with regard to the number of neutrons, small but measurable differences in mass exist within a group of similar atoms. How is mass determined?

For the most part, the mass of an atom is equal to the sum of the number of protons and neutrons. Although electrons are present, they are so light that they contribute little to the overall mass of an atom. To continue with the previous example, because some uranium atoms have more neutrons than others, their masses vary slightly.

Most elements are a mixture of isotopes. Scientists have found that many isotopes are stable, but others are not. Unstable isotopes give off radiation and are referred to as **radioisotopes.**

The Nature of Radiation In 1896, the scientist Henri Becquerel accidentally discovered that the element radium, when placed over a photographic plate in the dark, exposed the film. Pierre and Marie Curie, two French scientists, later discovered that radium was radioactive—that is, it emitted radiation, which was responsible for Becquerel's unexplained phenomenon. Three forms of radiation come from the nuclei of radium atoms: **alpha particles,** containing two protons and two neutrons; **beta particles,** negatively charged particles similar to electrons; and **gamma rays,** which are similar to x-rays.

Alpha particles are relatively heavy particles and can travel only 8 centimeters (3 inches) in air. They are easily stopped by material as thin as a sheet of paper or human skin (Figure 21.14). They are dangerous, however, if atoms that emit them are breathed into the lungs or ingested in contaminated food. In the lung or in the digestive tract they can irradiate cells, causing mutations and cancer.

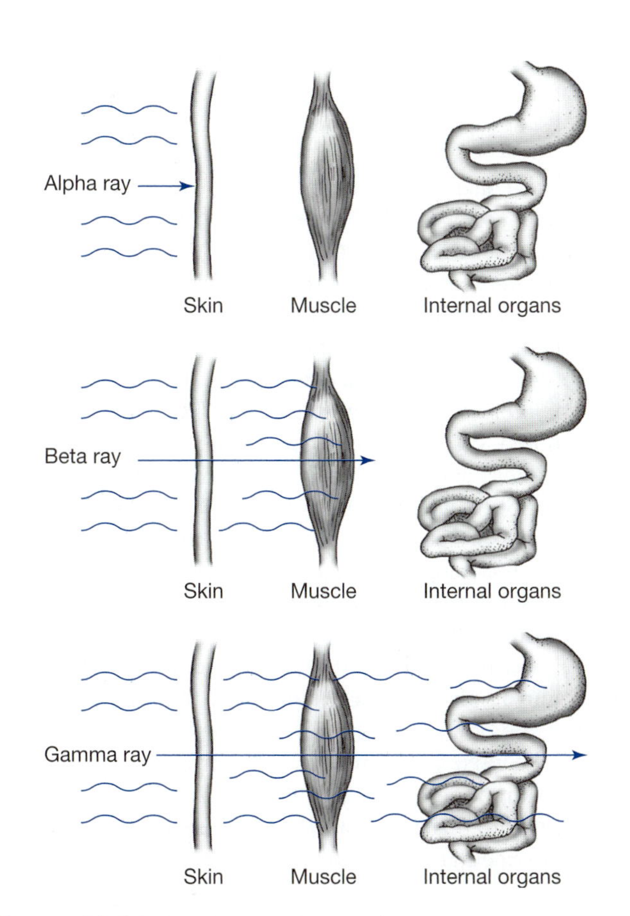

FIGURE 21.14 *Relative ability of alpha, beta, and gamma rays to penetrate the body.*

Beta particles, on the other hand, are much lighter particles and can travel much greater distances. They can even penetrate skin to irradiate underlying tissues. Wood or a thin layer of lead will stop them dead in their tracks.

Gamma rays are potentially the most dangerous form of radiation, for they can pass through walls and can easily penetrate the skin, reaching internal organs. A thick concrete or lead shield is needed to block them.

Radioactive Decay Radioisotopes emit radiation—alpha particles, beta particles, gamma rays, and others of lesser importance—from their nuclei to become more stable. Thus, over time, a mass of radioactive material actually decreases. The measure of its decrease in mass is called its half life. Technically, **half-life** is the time it takes a given mass of radioactive material to decay to one half its present mass. A kilogram of a radioactive compound with a half-life of 10 years, for instance, will weigh one half of a kilogram if left to sit from 2000 to 2010. By 2020, it will weigh one quarter of a kilogram.

The half-lives of isotopes vary from a fraction of a second to tens of thousands of years. For example, iodine-131 has a half-life of 8 days; cesium-137, 27 years; strontium-90, 28 years; carbon-14, 5,600 years; and plutonium-239, 24,000 years.

Nuclear Power

With this background information in mind, we now turn our attention to the nuclear energy option.

Atomic Fission and Chain Reactions Nuclear reactors and the first generation of nuclear bombs (the fission bombs) do not capture energy from radioactive decay, but instead draw their energy from **nuclear fission**—the splitting of certain atoms that occurs when they are struck by certain forms of radiation.

Nuclear power plants are fueled by **uranium.** Uranium-235 atoms are highly fissionable, or fissile. Uranium also gives off a form of radiation. When this radiation strikes a nearby uranium nucleus it can cause the latter to split into fragments, called **daughter nuclei** (Figure 21.15).

Enormous amounts of energy are given off during fission. So are additional neutrons. The neutrons create a cascade effect or **chain reaction** as they bombard additional uranium nuclei, causing still more fissions. In power plants, the rate of fission is carefully controlled by limiting the concentration of uranium in the fuel and by other means discussed later. Exceedingly complex in design, nuclear plants simply use the energy given off by this controlled chain reaction to boil water to make steam that turns electrical turbines. In atomic bombs, uranium is more highly concentrated and the chain reaction is uncontrolled or unmodulated, causing a huge explosion that releases enormous amounts of energy. Atomic explosions cannot occur in reactors because of the low uranium concentration. As we shall see in our discussion of the Chernobyl and Three Mile Island accidents, however, explosions at nuclear power plants can occur for other reasons and can rip apart a reactor, spewing radioactive material into the air.

Uranium contains lots of energy. In fact, 1 kilogram (2.2 pounds) of uranium produces the same explosive force as

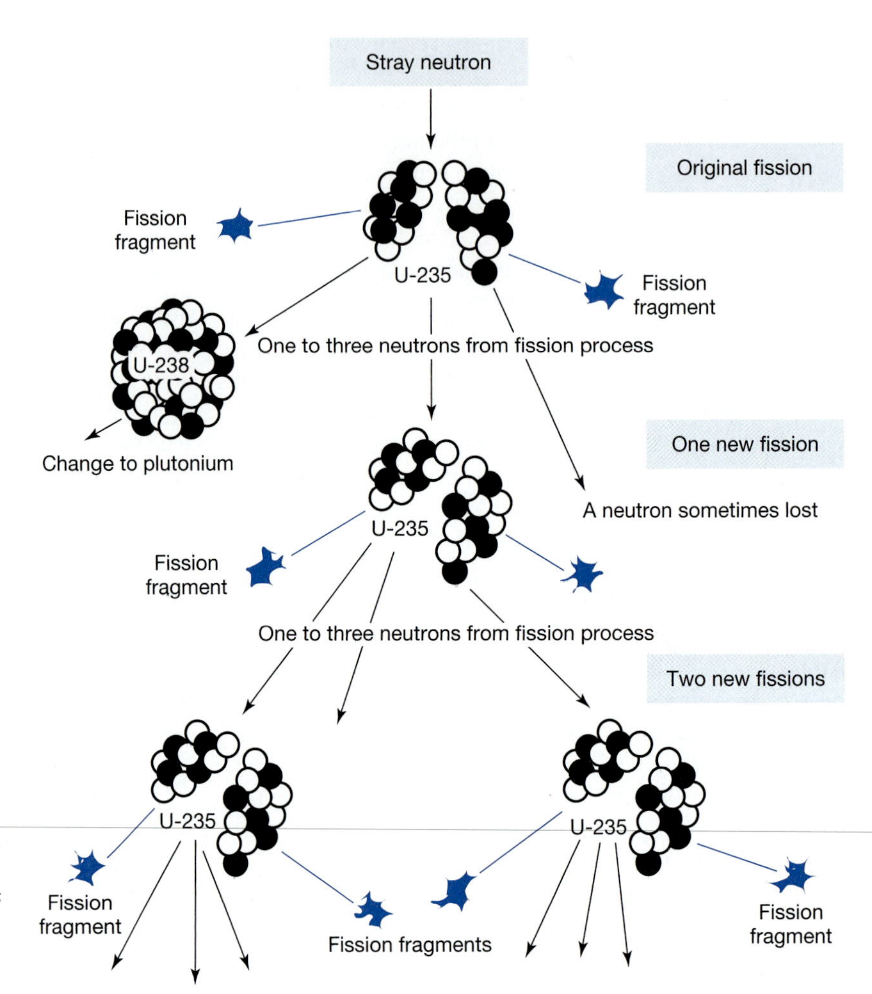

FIGURE 21.15 *Nuclear fission. The chain reaction starts when a uranium-235 atom is struck by a neutron given off by other uranium-235 nuclei. Uranium is split into fission fragments, or daughter nuclei, and heat. Additional neutrons are given off, thus splitting other atoms.*

9,000 metric tons of TNT. A kilogram of uranium releases as much fuel as nearly 3,000 metric tons of coal. It is no wonder that modern civilization once set its sights on this technology.

The Anatomy of a Nuclear Reactor

The uranium fuel used in all large U.S. nuclear power plants is an isotope called uranium-235. It comes from uranium ore. Uranium ore is a mixture of uranium-235, which is fissionable, and uranium-238, which is nonfissionable. In the United States, most uranium ore is found in the western states.

To create fuel for nuclear power plants like the one shown in Figure 21.16, the nuclear industry must first increase the concentration of uranium-235 to about 3 percent at a fuel enrichment plant. The enriched fuel, called **yellow cake,** is then formed into pellets. These are inserted into thin, stainless steel tubes, that when filled are known as **fuel rods.** About 4 meters (12 feet) long and about 1.2 centimeters (0.5 inch) in diameter, fuel rods are then bundled together, 30 to 300 per bundle, and lowered into the core of the reactor (Figure 21.17).

The **reactor core** provides a site where nuclear fission can be carefully controlled. Bathed in water to draw off heat, the reactor core is contained by a **reactor vessel** with a 15-centimeter (6-inch thick) steel casing (Figures 21.18–21.20).

In most countries, the reactor vessel is usually housed in a **containment building,** a domelike structure with 1.2-meter (4-foot) walls of concrete (Figure 21.21). Its function is to contain the radiation in case the reactor vessel or pipes rupture during an accident. Unfortunately, not all reactors have complete containment buildings. The nuclear reactor at Chernobyl, discussed shortly, had only a partial containment building, which proved inadequate in the fateful accident in 1986. Twenty of the former Soviet Union's 44 reactors now in operation and one U.S. reactor in the state of Washington (closed down shortly after the Chernobyl accident), used to make plutonium for nuclear weapons, have the same design.

The intensity of the chain reaction is controlled primarily by neutron-absorbing **control rods.** Made of cadmium or boron steel, the control rods fit in between the fuel rods and can be raised or lowered to control the rate of fission. Raising them starts the chain reaction and lowering them back into place can shut down the reaction almost entirely. Intermediate levels of fission occur between these two extremes.

The energy released from nuclear fission in the fuel rods heats the water bathing the reactor core. To avoid radioac-

FIGURE 21.16 *The largest operating nuclear power plant in New England. This is the Millstone Nuclear Power Station, in Waterford, Connecticut, located on the north shore of Long Island Sound.*

tive contamination, heat is then transferred to water circulating in huge pipes in the wall of the reactor vessel. The superheated water boils and produces steam (Figure 21.22). The steam is used to spin a turbine that turns an electrical generator.

Besides the reactor vessel and containment building, nuclear reactors are equipped with a host of safety devices, among them backup power units for the controls, redundant electrical wiring in case one circuit burns out, and redundant pipes in case one bursts. Perhaps the most important is the **emergency cooling system,** which provides cold water to the reactor core should it become overheated. The result is a technology whose complexity exceeds most others known to modern civilization.

Unfortunately, this complex technology can yield some unpleasant and costly surprises. The hydrogen bubble that built up in the Three Mile Island reactor, which exposed the reactor core and threatened to blow a hole in the reactor vessel, is one of those surprises. So was the catastrophic explosion at the Chernobyl reactor. Both were events that nuclear engineers had simply not foreseen. Interestingly, the explosion at Chernobyl was originally thought to have been caused by steam, but some recently released accounts indicate the explosion may have been a nuclear one, albeit a weak one. That was one of the more recent surprises. In addition, recent studies suggest too that more radiation was released into the environment than the official 3 percent figure given by the Russian government.

Pros and Cons of Nuclear Fission Few technologies have been so hotly debated as nuclear power. The following discussion summarizes the major views for and against nuclear power.

View in Favor of Nuclear Power

1. Nuclear power can provide an abundant supply of electrical energy. Although uranium-235 is a limited resource, proponents of nuclear power believe that a family of reactors, called **breeder reactors**, which convert abundant uranium-238 into fissionable plutonium-239, could provide electrical energy for hundreds of years. In fact, some experts predict that the spent fuel rods and wastes from uranium-processing plants in the United States alone, which contain abundant amounts of uranium-238, could last 1,000 years.

   The core of a breeder reactor contains small amounts of uranium-235 surrounded by a blanket of uranium-238. Neutrons given off by uranium-235 are absorbed by uranium-238, converting it to plutonium-239. Some of the plutonium-239 undergoes fission, producing energy.

   In theory, breeder reactors should create more plutonium than they consume. In fact, for every 100 atoms of plutonium-239 that fission, 130 atoms are produced.

FIGURE 21.17 *A nuclear fuel assembly storage rack containing loaded fuel rods at the Yankee Nuclear Power Station.*

FIGURE 21.18 *An 800-ton nuclear vessel, 72 feet long and 22 feet in diameter. Heat from the nuclear reactions contained in the vessel will produce steam sufficient to generate 800 megawatts of electrical power.*

This fissionable fuel could be extracted and used to generate power in other reactors.

2. Nuclear power helps reduce our dependency on foreign oil. Proponents of nuclear power believe that our dependency on oil could be further reduced by switching to electricity generated by nuclear power plants.
3. Nuclear power plants cause low radiation exposure. Nuclear power plants, when operating correctly, release very little radiation. In fact, one study showed that coal-fired power plants emit more radiation from their smokestacks. On the average, Americans receive 0.003 millirem (a measure of exposure) a year from nuclear power plants but as much as 100 to 250 millirems from x-rays, television sets, and naturally occurring radioactivity. The added risk of cancer to an individual is equivalent to smoking one cigarette per year.
4. It is environmentally clean. A properly functioning nuclear power plant is a much cleaner source of energy than a coal-fired power plant. In one year, a coal-fired plant producing 1,000 megawatts of electricity also produces 300,000 metric tons of ash, whereas a nuclear plant produces only about a ton of fission waste per year. Coal plants also release sulfur dioxide, nitrogen oxides, carbon dioxide, particulates, mercury, and other air pollutants. Solid wastes from coal plants contain selenium, mercury, benzopyrene, and some radioactive materials, such as uranium and thorium.
5. Nuclear power is a safe energy supply. The probability of an accident in the United States that would release large amounts of radioactivity, said supporters for many years, is not more than one in a million—about the probability of being struck by lightning. Other sources of electricity appear to pose far greater risks. For instance, each year 30 to 110 people lose their lives in coal mine accidents, mishaps during coal transportation, and from air pollution resulting from coal combustion.

View Opposed to Nuclear Power

1. Despite what proponents say, nuclear energy will not provide an abundant amount of energy. Supporters once hoped that breeder reactors would replace conventional fission reactors by the year 2000, allowing the nuclear industry to tap the generous supply of nuclear wastes from uranium enrichment plants and spent fuel from conventional fission reactors. The U.S. breeder program was canceled in 1983 by the U.S. Congress because of high cost and other problems. France, which now has breeder reactors in operation, has found that they are costly and not as effective as once hoped. The average breeder reactor requires about 30 years to reach the break-even point—that is, the point at which it produces as much plutonium fuel as it has consumed in the form of uranium-235. Because reactors have a 30-year life span at best, the breeder may never make good on its promise of generating more fuel than it consumes.

FIGURE 21.19 *Nuclear fuel is being lowered into its slot in the nuclear reactor. The reactor contains a grid guide structure (visible in the reactor vessel), control rods, and water.*

FIGURE 21.20 *Reactor vessel. This is the heart of the 540-megawatt nuclear-fueled electrical power generating plant near Monticello, Minnesota. A reactor vessel such as this may receive over 190,000 pounds of uranium oxide. Cooling water keeps the heat generated by the vessel at about 540°F.*

In addition to all of the problems of any fission reactors, discussed earlier, the breeder reactors now in use in France, and supposedly the prototype for future development, use liquid sodium as a coolant instead of water. Liquid sodium reacts violently with water and burns in air. Therefore, even a small leak in the cooling system could result in a disastrous accident, possibly leading to a meltdown of the reactor's core.

Despite the potential benefits of the breeder reactor, it seems to be a technology that may become obsolete before it has even gained a foothold. Without the breeder, the nuclear industry will be forced to rely on relatively scarce uranium-235, the supply of which may last only another 35 years or so.

2. Nuclear energy will not reduce the world's dependency on foreign oil. In fact, the argument that nuclear energy will reduce our dependency on oil is a false one because very little oil is currently used to produce electricity. U.S. dependency on foreign oil has increased dramatically in recent years because of falling domestic oil supplies and despite an expansion in the nation's nuclear capacity. The plain truth is that oil and nuclear energy are

FIGURE 21.21 *The Portland General Electric nuclear plant on the shores of the Columbia River at Rainier, Oregon. The dome-shaped containment building that shields the reactor is a characteristic feature of nuclear plants. Note the mammoth cooling tower to the left.*

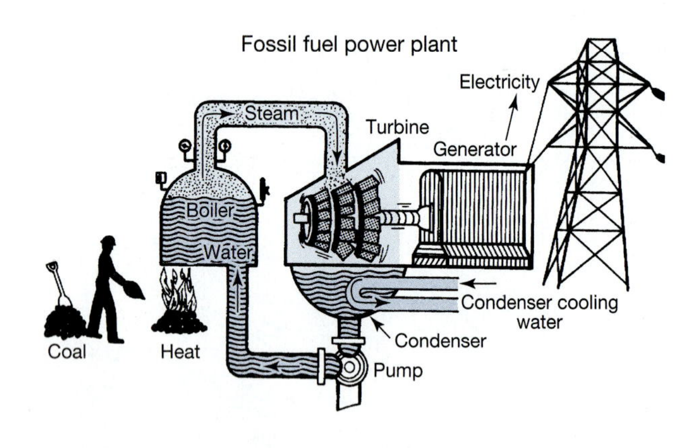

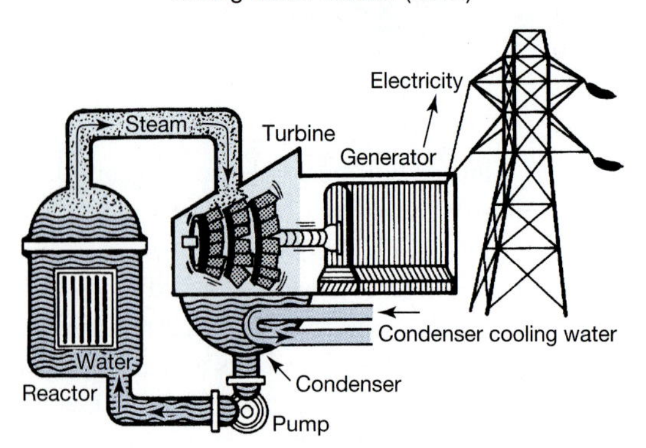

FIGURE 21.22 *Comparison of cooling systems for fossil-fuel power plant and nuclear power plant (boiling water reactor). Note water that is used to generate steam is not in direct contact with the coolant in the reactor core.*

two very different forms of energy that are used for very different purposes.

3. Nuclear power plants release harmful radiation. Although the radiation released from the routine operation of a nuclear plant is very small, some isotopes released from nuclear plants accumulate in animal tissues. For instance, cesium-137 concentrates in muscles and iodine-131 concentrates in the milk of cattle and in the human thyroid gland. Thus, low background levels can result in significant exposures to internal organs. Radiation is also released from uranium mines and mills, uranium mill tailings piles, processing plants, waste dumps, and transportation accidents, resulting in significant exposure.

   Serious accidents at nuclear power plants can release enormous amounts of radiation. For instance, the 1986 disaster at the Chernobyl reactor in the former Soviet Union released 7,000 kilograms (15,400 pounds) of radioactive material that contaminated many European countries. That accident injured 1,000 people, killed 31, and caused at least a $3 to $5 billion loss to the former Soviet Union. By various estimates, it may result in 5,000 to 1,000,000 cases of cancer in the former Soviet Union and Europe. If the Chernobyl accident had taken place at one of the many nuclear power plants closer to major urban centers, the damage would have been substantially higher. As a point of reference, a reactor accident at the Salem 2 nuclear reactor in Salem, New Jersey, could kill 100,000 people from radiation poisoning within a few months of the accident. Approximately 40,000 cases of cancer would result, and the economic damage would come to $150 billion (Table 21.1).

4. It is not environmentally clean. Nuclear power plants are generally cleaner than coal-fired power plants but produce 40 percent more thermal pollution per kilowatt of

electricity than a coal-fired power plant because fission releases more heat than coal, oil, and natural gas.

5. Safety. Several minor and two major nuclear reactor accidents have already occurred in the 50-year history of the nuclear industry. A near meltdown of the reactor resulted from a fire in 1975 at the Brown's Ferry plant in Alabama. A study by a prominent nuclear engineer and nuclear advocate, Norman Rasmussen, of the Massachusetts Institute of Technology, however, determined that the probability of an accident was not more than one in 10,000 reactor-years. Although that may sound reassuring, consider what it really means. In 2000, 452 nuclear reactors were operating worldwide, and with dozens more under construction or soon to be built. If the Rasmussen report estimate is correct, we can expect a major meltdown every 20 years or so. The Three Mile Island and Chernobyl accidents, discussed shortly, occurred only 7 years apart but, to be fair, there haven't been any major accidents since Chernobyl.

   The Rasmussen report is no longer considered valid. But don't be lulled into complacency. It was discredited largely because it greatly *underestimated* the probability of a nuclear accident. Moreover, it is turning out that many nuclear power plants were built in areas in which evacuation plans are ill-conceived or inadequate. Many contain faulty or overly optimistic assessments of how rapidly and efficiently a population could evacuate a region after an accident.

6. High cost. When nuclear power was first conceived, proponents argued that it would be so cheap to produce that it would not be economical to meter it. Forty years later, that dream is far from reality. Nuclear power is, in fact, one of the most expensive forms of electricity now commercially available, costing 10 to 15 cents per kilowatt, compared to 5 to 7 cents per kilowatt for coal (Table 21.2). One of the chief reasons why nuclear power plants are so costly is that they are expensive to build. A large plant costs $2 to $6 billion dollars. A comparable coal-fired power plant costs $0.5 to $1 billion. Investments in energy efficiency in these ranges would result in far more energy saved than could be produced by either type of plant! Moreover, repairs of nuclear power plants are extraordinarily costly and time-consuming as well, because of the danger of radiation exposure to workers. Thus, what would be a simple and inexpensive repair may take months and may cost millions of dollars, further adding to the cost of electricity. Finally, at the conclusion of a plant's useful life span of 20 to 30 years, it must be disassembled and disposed of. The estimated costs for "decommissioning" nuclear plants is about $0.5 million to $1 million per megawatt of power generation capacity. A 1,000-megawatt plant would cost an additional $0.5 billion to $1 billion to decommission. Unfortunately, the costs of decommissioning have not been added to fuel bills and will invariably be included as the older plants now reaching obsolescence are disassembled, making nuclear power even more expensive.

7. Lack of public and private support. In 1975, long before the Chernobyl accident, 64 percent of the American population polled supported nuclear power; in 1986, after this tragic and costly accident, public support nearly vanished, dropping to 19 percent. Similar trends have been witnessed in other countries as well.

   Long before Chernobyl, however, U.S. banks and insurance companies had withdrawn their support from the nuclear power industry. It was, they said, too costly and too risky an endeavor to support. Because of this, the

TABLE 21.1 *Potential Severity of Nuclear Reactor Accidents*

Plant*	Location	Early Deaths (Thousands)	Cancer	Financial Losses (Billion 1980 Dollars)
Salem 2*	Salem, N.J.	100	40	150
Peach Bottom 2	Peach Bottom, Penn.	72	37	119
Limerick 1	Montgomery, Penn.	74	34	213
Waterford 3	St. Charles, La.	96	9	131
Susquehanna 1	Berwick, Penn.	67	28	143
Shoreham	Wading River, N.Y.	40	35	157
Three Mile Island 1	Middletown, Penn.	42	26	102
Indian Point 3	Buchanan, N.Y.	50	14	314
Milestone 3	Waterford, Conn.	23	38	174
Dresden 3	Morris, Ill.	42	13	90

* Plants listed are those in densely populated areas.

Source: Sandia National Laboratory, "Estimates of the Financial Consequences of Nuclear Reactor Accidents." Prepared for the Nuclear Regulatory Commission, Washington, D.C., November 1982; Critical Mass Energy Project.

U.S. nuclear power industry has been on a downward spiral since 1979. Not one new nuclear plant has been ordered in the United States since then, and many more have been canceled, either in the planning stages or during construction. Despite this trend, some countries such as Canada, which currently derives about 12 percent of its total annual energy consumption from nuclear power plants, remain strong supporters of this otherwise waning technology.

In summary, despite its many benefits, nuclear power appears to be a technology in retreat. Even new reactor designs, intended to reduce costs and enhance safety, are not currently well received. In the next three sections, we examine in more detail three major issues regarding nuclear power: health effects, reactor safety, and the waste issue.

The Health Effects of Radiation

In this section we discuss two broad categories of effects: nongenetic and genetic effects. The effects of radiation, in general, vary with the age of the individual. Fetuses and newborn infants, for example, are much more sensitive to radiation than adults. Sensitivity also varies according to the type of tissue or organ. Fast-growing cells, such as those lining the intestines, those that produce hairs, and those in the bone marrow, are the most sensitive. Cells that do not divide, such as those in cartilage, muscle, and nervous tissue, are much more resistant.

The effects of radiation also vary with the intensity of the exposure. The higher the dose the more serious the effects, and the more quickly they are manifested. Radiation doses are frequently measured in **rads,** a measure of energy absorption by the tissue (rad stands for radiation absorbed dose). A rad is equal to about 100 ergs of energy per gram of tissue. An erg is an extremely small amount of energy, approximately the amount of energy imparted by a mosquito alighting on your arm.

Nongenetic Effects Radiation causes many effects, one of the most pronounced being death. As shown in Table 21.3, individuals exposed to 100 to 250 rads or less withstand such exposure levels, although they generally suffer a higher incidence of cancer later on in life. As the dose increases to 400 to 500 rads, however, only 50 percent of the people exposed will survive 3 weeks. At doses of 900 rads or greater, all people die within 3 weeks of exposure.

What are some of the symptoms displayed by people exposed to radiation? A person receiving less than 25 rads would not be aware of changes in his or her immediate health. (The likelihood of cancer, however, is increased considerably.) When the dose rises to 100 rads, the individual may suffer from weakness, fatigue, vomiting, and diarrhea. Eventually, however, these symptoms disappear and the individual begins to function normally. At a radiation exposure of 400 to 500 rads, individuals experience extreme nausea, vomiting, hair loss, and fatigue. Making matters worse, this level of radiation impairs red blood cell production, resulting in severe anemia. White blood cell production is also severely hampered. Because white blood cells help protect the body against infections, radiation victims are susceptible to infectious diseases caused by bacteria and viruses. This dose also destroys blood platelets, tiny elements in the blood that promote blood clotting. The result is massive hemorrhaging and excessive blood loss. Individuals exposed to 400 to 500 rads, therefore, usually die from one or a combination of the following: anemia, hemorrhage, and severe bacterial infections.

Much of the information on radiation's effects come from studies of the survivors of the nuclear bomb blasts in Hiroshima and Nagasaki at the end of World War II. Thousands of people were exposed to radiation doses ranging from 100 to 150 rads. The immediate effects included burns, fever, loss of hair, fatigue, intestinal bleeding, vomiting, and diarrhea. If the radiation damaged their bone marrow, many delayed effects were observed, including anemia, leukemia (cancer of the blood), and infections due to white blood cell suppression.

The children of pregnant mothers who were within 1,200 meters of the center of the explosion were born with severe birth defects, including mental retardation and deformations of the skull, heart, and skeleton. Many of the survivors and members of their families are shunned today by Japanese, fearful of the long-term genetic effects of radiation.

Genetic Effects: Cancer, Birth Defects, and Mutations Experiments on laboratory animals have shown that radiation causes cancer in a variety of organs such as the lymph glands, breasts, lungs, ovaries, and skin. Years of research confirm the connection between radiation and various forms of cancer in humans. Consider the following examples:

1. Skin cancer was prevalent among radiologists in the early days of the profession before adequate safeguards were instituted.
2. Bone cancer appeared in many women who painted radium dials on watches in the 1940s. Apparently, the women ingested radium when drawing paint brushes to a point with their lips.

TABLE 21.2 *Cost of Electricity from Different Sources*

	*(Cents per Kilowatt-Hour)**	
Source	***1983***	***2000***
Cogeneration	4–6	4–6
Coal	5–7	7–9
Small hydropower	8–10	10–12
Biomass	8–15	7–10
Nuclear	10–12	14–16
Wind Power	12–20	6–10
Photovoltaics	50–100	10–12

* Costs expressed in 1982 dollars.

3. Uranium miners have a high incidence of lung cancer, largely from breathing radioactive radon gas found in the air in and around mines.
4. Children irradiated when still in their mother's uterus (during routine maternal pelvic x-ray examinations) have a much higher cancer rate (50 percent higher) than children who were not x-rayed during gestation.
5. The rate of leukemia was substantially higher among Japanese citizens who had survived the bomb blast than in unexposed Japanese citizens from other cities. Leukemia often showed up 30 years after the bombings.

Most tumors result from genetic changes in the cells of the body. In general, these changes, known as **mutations,** may be caused by biological agents such as viruses, chemical agents such as toxic chemicals, or physical agents such as radiation. These mutagenic agents result in genetic changes that cause body cells to divide uncontrollably, forming tumors. The resultant tumors grow, debilitating individuals, and often spreading through the lymphatic vessels to other parts of the body where secondary tumors form. Exhausted and depleted by the rapidly growing tumor, the individual eventually dies—unless, of course, the tumor is excised or destroyed by drugs or radiation treatment.

Scientists have long known that radiation can also damage the chromosomes of germ cells—the sperm and ova. These mutations can be passed to an individual's offspring and manifest themselves as birth defects or as childhood cancer. Background radiation probably results in one mutation for every million sperm or ova produced. Additional radiation, such as from x-rays or from an accident at a nuclear power plant, can greatly increase the rate of mutation. James F. Crow, an eminent University of Wisconsin geneticist, suggests that any amount of radiation is potentially damaging to genetic material.

New and controversial research shows that exceedingly low levels of radiation have a more significant effect than previously thought. Studies also indicate that radiation is cumulative—that is, low levels add up over many years. As a result, a study by the National Academy of Sciences suggests that the current health standard—17 rads per year for the general public—may be too high. This level of exposure, the authors of the study assert, could increase cancer rates by 2 percent and increase the incidence of genetic defects by about 1 per 2,000 newborns. Against strong opposition, the Academy and other scientists have suggested reducing the allowable level of radiation to 0.017 rad per year.

Reactor Safety: Two Case Studies

To understand why nuclear energy has faltered one need only look at the two most horrifying experiences with it in recent history. Both accidents raise many questions and throw into question the wisdom of this approach.

Three Mile Island: The Beginning of the End? On March 28, 1979, disaster struck at the Three Mile Island nuclear power plant 10 miles from Harrisburg, Pennsylvania (Figure 21.23). The three-month-old reactor ran amok when a valve failed to close. This failure in turn triggered a series of events that resulted in the most costly and potentially dangerous accident in the history of the U.S. nuclear industry.

The details of the accident are difficult for all but trained nuclear engineers to understand, but the results were clear:

TABLE 21.3 *Noncancer Effects of Radiation on Humans*

Radiation Dosage in Rads	Percent That Will Die	Other Symptoms
10,000	100	Death in two days due to brain and heart damage
3,000	100	Permanent damage of brain, spinal cord, bone marrow and intestines; bleeding, infections
2,000	100	Permanent destruction of bone marrow and intestines; severe dehydration, bleeding, and infections
1,000	100	Permanent destruction of bone marrow and permanent damage to intestines
600	60	Permanent destruction of bone marrow
400–500	50	Fatigue, nausea, vomiting, loss of hair; reduced number of blood platelets and white blood cells; bleeding and infections
100–250*	0	Fatigue, nausea, vomiting, loss of hair, diarrhea, ulcers of gut; some may suffer permanent loss of fertility
25–100	0	No visible effects; lower number of white blood cells
0–25	0	No measurable effects

* Note: All individuals exposed to 100 rads or more will experience fatigue, nausea, vomiting, diarrhea, and hair loss.

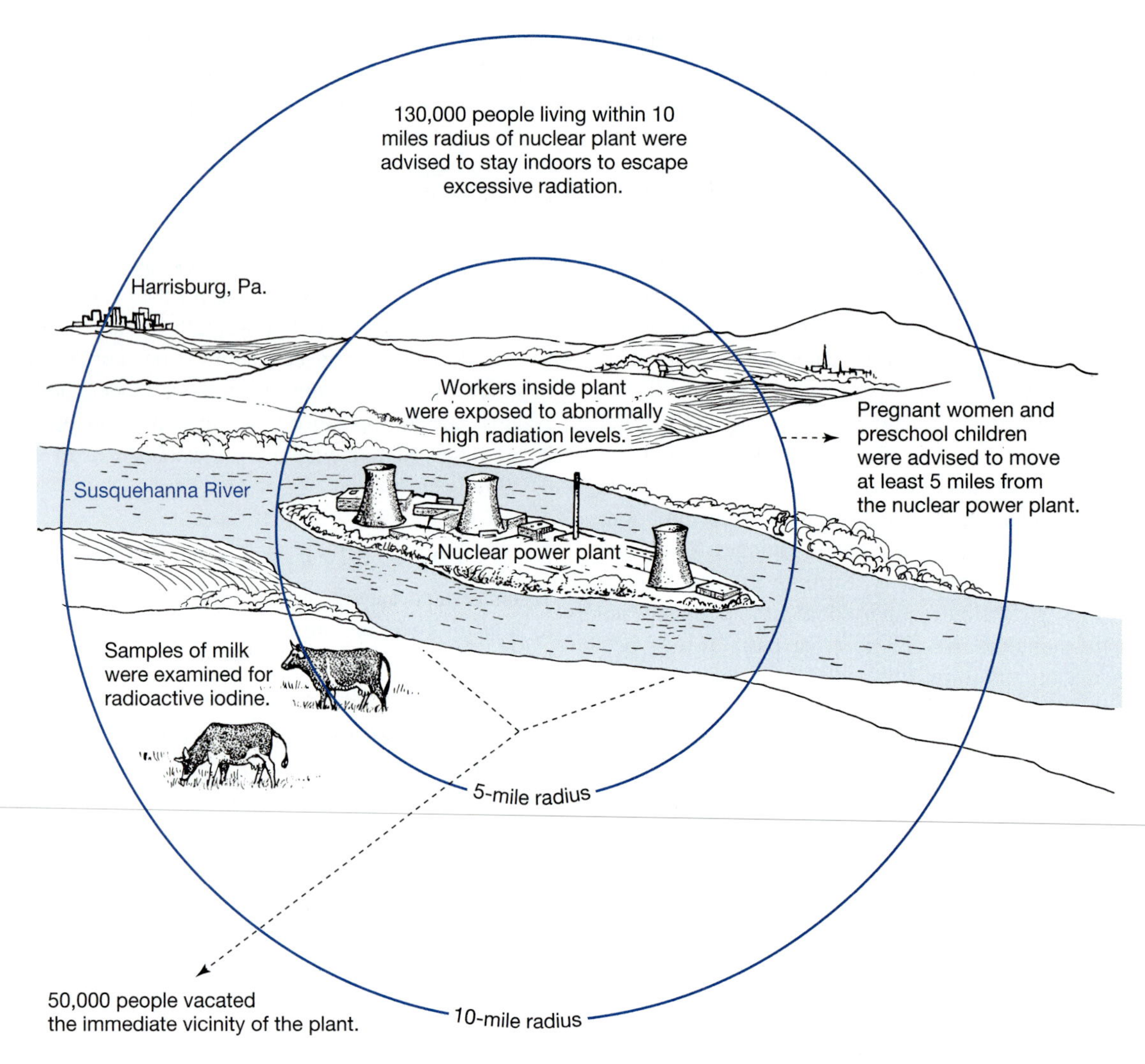

FIGURE 21.23 *Three Mile Island nuclear accident, March 28, 1979.*

Radioactive steam was dumped into the containment building of the reactor and into the air outside the plant. A cloud of radioactive steam drifted through the valley, sending residents scurrying for safety. Bursting pipes spilled radioactive cooling water into two buildings. In one of them, a million liters of radioactive water accumulated ankle-deep.

Something unusual then followed: Mysteriously, hydrogen gas began to accumulate in the reactor vessel, threatening to expose the core and cause a meltdown. Some experts feared that the bubble might result in an explosion of the reactor vessel and possibly the containment building that would release even larger amounts of radioactive material. Fortunately, the utility was able to draw off the hydrogen to prevent an explosion. A year later, photographs taken in the core by a remote camera showed that the nuclear fuel had melted down, rendering the reactor useless and adding to the cleanup costs.

The human toll—besides the agony and fear of those close to the reactor—was minimal, unless you were one of the victims. Eight workers inside the plant were exposed to high levels of radiation and are likely candidates for cancer. Fifty thousand people within the vicinity of the plant were evacuated, and 130,000 in outlying regions were warned to stay indoors to avoid being exposed to radiation. So far, no studies have shown any measurable increase in cancer as a result of this accident.

After several days of feverish efforts, nuclear engineers from the Nuclear Regulatory Commission and Metropolitan Edison, the company that owns the reactor, were able to shut down the plant, thus ending the release of radiation. Soon thereafter, the company began the long and expensive cleanup, which cost an estimated $1 billion.

Studies of the accident showed that human error, mechanical failures, and design flaws were the cause. Inexperienced operators working late at night, for instance, turned off the emergency core cooling system at the wrong time, closed valves on an emergency cooling system when they should have left them open, and disengaged water pumps that should have been left operable.

The Three Mile Island accident was a disaster of minor proportions, say utility officials and supporters of nuclear energy in general. No one died during the accident, and many health officials believe that the radiation released into the air and water will have little, if any, effect on local residents. Two noted (and controversial) radiation experts, however, believe that 300, possibly as many as 900, cases of cancer and leukemia will result from the incident. Many residents, they say, were exposed for 100 hours or longer to low levels of radiation, enough to noticeably increase the incidence of cancer in the region. Unfortunately, only time will tell if they are correct.

The Chernobyl Accident The accident at the Three Mile Island nuclear plant marked the beginning of the end of the nuclear power industry, already burdened by high construction costs and growing public skepticism. The accident at Chernobyl, however, may have sealed the fate of this troubled industry forever.

At 1:00 a.m. on April 26, 1986, the nuclear operators at the Chernobyl nuclear power plant in the former U.S.S.R. had just completed a full day of testing on the fourth and newest reactor. In the course of their test, the operators shut down safety systems and violated a number of operating procedures.

By 1:23 a.m. the reactor's power had fallen to only 6 percent of its operating level. With the emergency cooling system and other safety systems shut off and the control rods only part of the way out of the reactor core, power began to increase. However, when the operators pushed the button to drop the control rods back into place, the rods wouldn't budge. The core, experts now believe, had already begun to melt down, preventing the rods from falling into place. A few seconds later, shock waves about as strong as a mild earthquake tremor shook the plant. Two large explosions then shook the complex (Figure 21.24).

Experts believe that the explosions occurred when the fuel rods split open in the intensely hot reactor core. Enormous amounts of energy were released from the fuel rods, superheating the reactor cooling water and suddenly converting it into steam. The steam ripped open the partial reactor containment building. The 1,000-ton concrete slab above the reactor was thrown aside as steam, nuclear fuel, and graphite (one of the components of this nuclear reactor) were hurled skyward.

Helicopter pilots arriving soon after the explosion reported that they could see down into the glowing red core of the reactor through a gaping hole in the reactor building. To put out the fire and stop the spread of radiation, the pilots dropped 40 tons of boron carbide, 800 tons of limestone, 2,400 tons of lead, and thousands of tons of sand and clay on the damaged reactor over the next few days. The fire went out soon, but the reactor smoldered for several days, continuing to release millions of curies of radiation a day for nearly 2 weeks. On the eleventh day, liquid nitrogen was pumped under the reactor in an emergency cooling system installed after the accident. The molten fuel cooled enough to end the release of radiation.

News of the Chernobyl accident reached the West on April 29, several days after the explosion, and only after Sweden and other countries had reported high levels of radiation wafting in from the Soviet Union. The world press picked up the story on the 29th and hastily put together the few facts they could uncover while the world looked on in terror. The earliest reports from diplomats in the Soviet Union suggested that at least 2,000 people had been killed by the accident, the most serious accident in the history of nuclear energy.

When the dust settled and Western reporters were allowed access to the facts, it became clear that the initial reports of human fatalities were grossly exaggerated. Four months after the accident, the death toll had reached 31. Most of the victims were workers who had been exposed to high levels of radiation. All told, 237 people were hospitalized with acute radiation poisoning and burns. The high levels of radiation to which they were exposed will greatly increase their risk of cancer.

Some 135,000 people were evacuated from a 30-kilometer region, mostly north of the plant. Many of them will never be

FIGURE 21.24 *Aerial view of the damaged Soviet nuclear reactor building at Chernobyl after the fateful accident in 1986, which killed 31 people and may cause cancer in tens of thousands of others.*

able to return home but will live in a new city built by the government. A quarter of a million schoolchildren were sent away from Kiev, a city 80 kilometers south of the plant, on an early summer vacation a few weeks after the explosion. By various estimates 150 square kilometers (60 square miles) of prime farmland around the plant has been so badly contaminated with radiation that it will lie fallow for many decades. The reactor itself has been sealed in concrete to prevent further radiation leakage, and it may be several hundred years before workers can safely remove what is left of the reactor core.

The cost of this accident is difficult to determine. Soviet officials estimate their damages to be about $3 to $5 billion. But outside experts believe that the total costs could exceed $10 billion, when all the indirect effects are taken into account. Outside the former Soviet Union the costs are beginning to be known. In the United Kingdom, sheep farmers, whose livestock were contaminated by radiation and thus rendered unsuitable for food, estimate that they have lost $15 million. Swedish officials estimate their country's losses amount to $145 million, and the German government says it will pay farmers $240 million for lost crops and livestock.

In all, 20 countries were dusted with radiation, the long-term health effects of which no one really knows. Although it is difficult to estimate, scientists believe that the Chernobyl accident will result in 5,000 to 100,000 additional cases of cancer in the country. Outside of the former Soviet Union, there may be as few as 2,500 cancers to as many as 300,000, according to some experts. Radiation specialist John Gofman believes that the total number of cancer cases could range from 600,000 to 1,000,000. Half of these, he predicts, will be fatal. Russian scientists have already reported a thyroid cancer epidemic with rates 100 times greater than normal in areas near the accident.

The accident at Chernobyl raised questions about the wisdom of nuclear power. Although many experts believe that an accident of this magnitude could not happen in the United States, others are not so sure. Meanwhile, antinuclear sentiment has skyrocketed throughout the world. About half of the European population favor shutting down existing nuclear power plants. In the former West German state of Hesse, government officials have decided to push for a phaseout of nuclear power. In Poland, the nuclear program has been slowed. In Sweden, the government promised to phase out nuclear energy, while Swiss officials, long committed to nuclear power, have postponed further expansion of their nuclear power capability.

The Nuclear Waste Issue

After accidents, one of the biggest concerns of the environmental community, the public, and many government officials is radioactive waste disposal not only from power plants but also from weapons production facilities. The latter, not discussed in this chapter, could cost billions of dollars in the United States alone.

In nuclear power plants, fission byproducts begin to build up in fuel rods within a year or two of the time they are installed. These radioactive materials absorb free neutrons. As free neutrons build up, they slow the rate of fission, reducing a plant's operating efficiency. Therefore, each year about one third of the old fuel rods (40 to 60 fuel rods) are replaced by new ones. But what happens to the spent fuel?

In many countries, the spent rods are stored underwater and then transported to reprocessing plants that remove the highly radioactive byproducts (Figure 21.25). Reprocessing plants recover uranium-235 that has not been fissioned and also plutonium-239, which forms from uranium-238. At the reprocessing plant the contents of the spent rods are dissolved in acid. The usable uranium-235 and plutonium are then separated from the waste. During this process, however, about 460 liters (120 gallons) of highly radioactive liquid waste is produced for each metric ton of spent fuel. Reprocessing plants also emit approximately 100 times as much radiation as a properly operating nuclear power plant.

One of the greatest problems of the U.S. nuclear industry is that there are no reprocessing plants in operation in the country. As a result, many tens of thousands of metric tons of highly radioactive waste are now stored in water tanks at nuclear plants all over the country (Figure 21.26). Delays in establishing a permanent waste repository for high-level radioactive materials, according to California's Congressman George Brown, Jr., have turned our reactors into de facto long-term disposal sites. The mounting high-level waste potentially increases the severity of accidents. Making matters worse, many reactors are now reaching the end of their useful lifetime. "To continue the temporary storage of fuel rods at these sites would be asking for trouble," writes Congressman Brown. Since almost all of these soon-to-be-closed reactors are near rivers and lakes for easy access to cooling water, a catastrophic accident at any one could result in widespread contamination of local waterways.

In an effort to solve America's high-level waste problem, the U.S. Congress passed the **Nuclear Waste Policy Act** in 1982. It called on the Department of Energy to choose two sites for the disposal of high-level radioactive waste by 1987—one in the East and one in the West. Most experts believe that high-level wastes should be disposed of in deep caverns carved out of ancient salt beds, granite, volcanic tuff, or basalt. The experts hope that in these geologically stable sites, the waste will remain many thousands of years without leaking into groundwater and threatening human health.

Despite much initial optimism, site selection proved extraordinarily difficult. In 1986, the Department of Energy proposed three western sites: the Hanford nuclear reservation in southeastern Washington, Yucca Mountain in southwestern Nevada, and Deaf Smith County on the high plains of western Texas. Each of these sites, selected from a far larger field, was then to be studied in detail at a cost of $1 billion apiece to determine the best one. To save money, though, the U.S. Congress decided to focus efforts on the most likely candidate, the Yucca Mountain site in Nevada. It appears to be free from earthquakes and is dry. The repository would be constructed above the water table, which lies 700 meters (2,000 feet) below the surface. Unfortunately, Nevada's governor and three quarters of the state oppose the site. They are afraid that it is not as stable as government scientists believe, and they want to avoid the stigma of being the country's nuclear wasteland.

Nevertheless, research is now actively under way to determine if the site at Yucca Mountain will be suitable for long-term storage of wastes. This site would store nuclear waste from utilities and defense installations—mostly factories where atomic weapons were once made. After extensive research, officials are finding many problems that make even this site questionable.

Although many people believe that it was a mistake to build nuclear power plants in the first place, says California's Congressman George Brown, the truth is that by the year 2000, 42,000 metric tons of high-level waste had accumulated from commercial reactors. We must solve the waste problem that we have created. We have no other choice.

21.4 Fusion Reactors

The fission reactor is clearly a technology in decline. Although the nuclear power industry has proposed smaller, safer plants, high costs and skepticism on the part of the general public, insurance companies, and banks have not translated these new dreams into realities. For many years, the nuclear power industry has also banked some of its hopes on fusion reactors. What are they, and what are their prospects?

The sunshine that helps brighten your day has its origin in nuclear fusion reactions occurring in the sun. In a sense, **nuclear fusion** is the opposite of fission. Fission involves a splitting apart of atomic nuclei, whereas fusion unites two nuclei to form a new one. In nuclear fission, heavy nuclei like those of uranium are used because they tend to be unstable. In nuclear fusion, two lightweight nuclei are brought together by overcoming the mutual repulsion resulting from the positively charged protons in each nucleus. Light nuclei are used because they repel each other much less than larger nuclei. When two nuclei fuse, energy is released that eventually can be used by humans to generate electricity.

Hydrogen atoms are used in experimental fusion reactors because they are the lightest elements. Hydrogen atoms have one proton and no neutrons in their nuclei. The repulsive force is therefore quite small. Another advantage of hydrogen is that it is extremely abundant.

Hydrogen has two less common isotopes, deuterium and tritium. Deuterium has a proton and a neutron in its nucleus; tritium has a proton and two neutrons. A deuterium nucleus may be fused with another deuterium nucleus or with a tritium nucleus.

To overcome the repulsive force, the nuclei must be supplied with enormous amounts of energy. This can be done

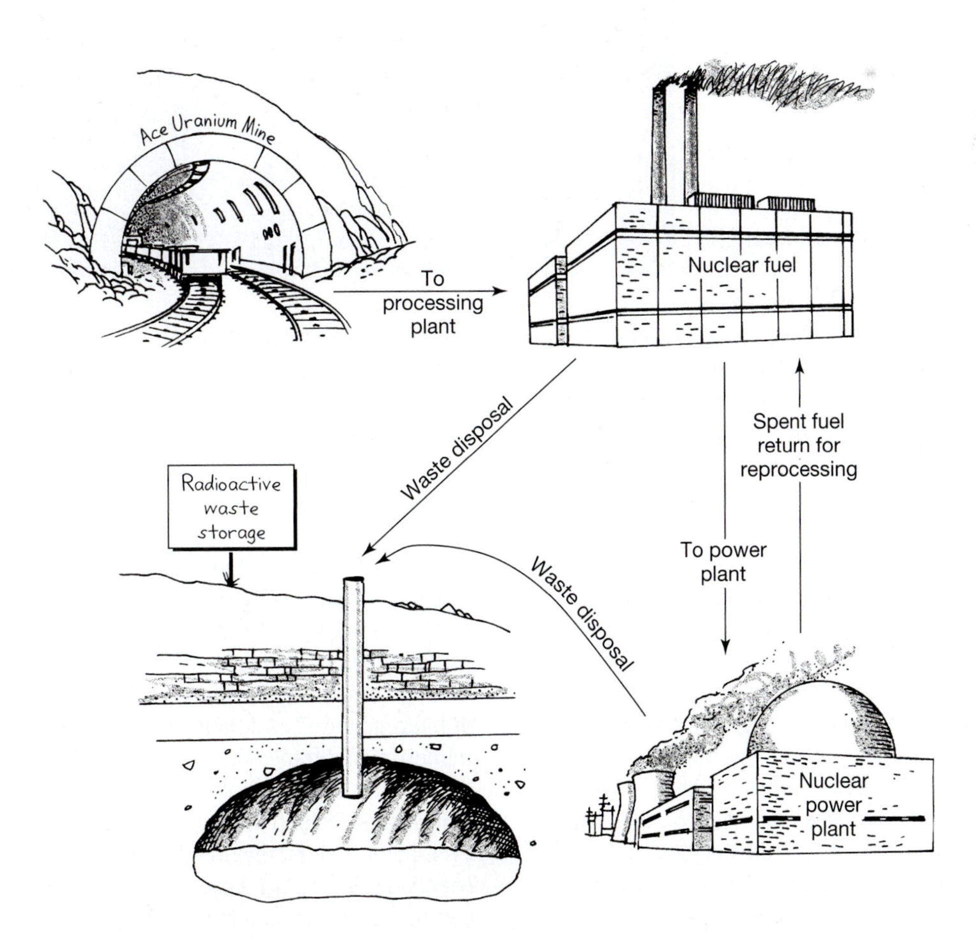

FIGURE 21.25 *Nuclear fuel-and-waste cycle (simplified view). Radioactive contamination of the human environment is most likely to occur during the transport of spent fuel for reprocessing and during the transport and disposal of nuclear waste.*

FIGURE 21.26 *Spent nuclear fuel "cooling off" under water. The spent fuel assemblies from the Savannah River Plant in Georgia illuminate a cooling basin. In the photograph, spent fuel of three different "ages" is shown under 20 feet of water. The brightest assemblies (top center) were just discharged from the reactor. In front of them are assemblies that were discharged a month earlier. Barely visible on the extreme lower left are assemblies that have "cooled" by radioactive decay for $3\frac{1}{2}$ months.*

by bombarding a fuel pellet containing deuterium and tritium with a laser beam. Thus energized, the fuel turns into a hot mixture of nuclei and electrons known as **plasma.** In the plasma, nuclei can speed toward each other, collide and fuse. The fusion of two nuclei releases enormous amounts of energy, potentially much more than is needed to initiate the reaction.

The 40 million °C heat required to start the reaction poses some design problems. No known metal can withstand temperatures anywhere near those required by fusion reactors. Therefore, scientists have proposed two major designs. The most popular is magnetic confinement. In this technique, a magnetic field suspends the plasma long enough for fusion to occur (Figure 21.27). The heat given off is picked up by liquid lithium, transferred to liquid potassium, and then transferred to water, which boils to generate steam and electricity.

Nuclear fusion offers several key advantages over nuclear fission. First, the fuel is abundant and inexpensive. The deuterium in only 1 cubic kilometer of seawater could provide us with as much energy as 1,500 billion barrels of oil, or 1.5 times as much oil as the world has consumed in the history of human civilization. Nuclear fusion is believed to be much safer as well. If a fusion reactor malfunctioned, the reaction would simply come to a stop. There is, say experts, no chance of an explosion. An accident would not release the massive quantities of radiation released by a fission reactor. Finally, because a fusion reaction is much more efficient than a fission reaction, less waste heat is released. As a result, a fusion reactor's potential for thermal pollution is considerably less. Less cooling water would also be needed.

Fusion has some drawbacks—some so significant that they could forever keep this form of energy out of our grasp. One of the most important hazards is tritium, the radioactive isotope of hydrogen used as a fuel. At high temperatures, tritium is extremely difficult to control and can pass right through metal. A second major problem is that, despite more than four decades of research, scientists have been unable to reach the break-even point, that is, they have been unable to produce as much energy as the experimental reactors consume. To be economical, fusion reactors must generate electricity at a price we can afford. Even though the costs cannot be predicted with great accuracy, it is quite possible that commercial fusion reactors could cost \$12 to \$20 billion, far in excess of breeder reactors and conventional fission reactors. That money invested in energy conservation and renewable sources such as wind could produce enormous amounts of clean energy.

Fusion reactors also produce highly energetic neutrons that will bombard the containment vessel, weakening the metal and requiring frequent replacement. The containment vessel would also become radioactive. Furthermore, if a vessel burst, it would release radioactive tritium and molten lithium, the coolant, which burns spontaneously and vigorously in air.

In 1980, Congress passed the **Magnetic Fusion Energy Emergency Act** to promote fusion energy. This law authorized

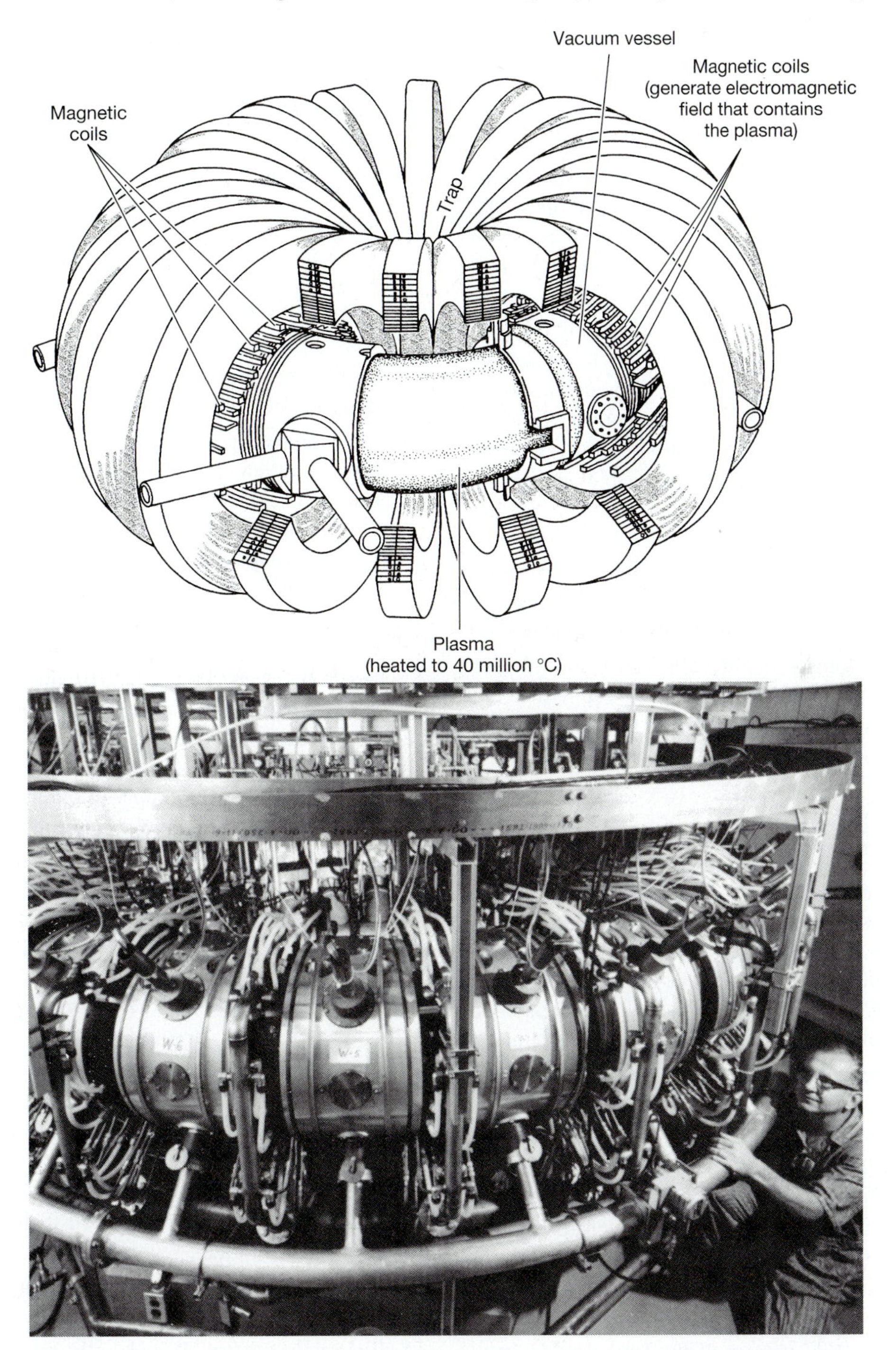

FIGURE 21.27 *Nuclear fusion reactor. (Top) Fusion research device. This diagram shows one research device that could bring the United States closer to controlling fusion, the force that powers the sun and the stars. Because it is easier to work with, this device will use hydrogen as its fusion fuel. Later-generation devices will employ deuterium and tritium as fuels. This device is known as the Princeton Large Torus (PLT). (Bottom) Nuclear fusion research. Magnetic coils used to confine and compress fusion plasma are visible in this fusion device. Microwave power levels are increased for the early-stage testing of a technique to heat plasma to the very high temperatures needed for fusion reactions. This device, known as the Elmo Bumpy Torus, is located at the Department of Energy's Holifield National Laboratory.*

the Department of Energy to spend roughly $1 billion per year from 1980 to 2000 on fusion research and development. The hard economic times of the 1980s, however, put an end to much of this money, further dimming the prospects for fusion energy. In 1996, the United States spent about $250 million on fusion research, a fraction of what is being spent in Europe and Japan. Plans to have a demonstration plant on line by 2025 have been abandoned because of the severe economic cutbacks.

21.5 America's Energy Future

The world is running out of oil, and natural gas is also on the decline. We have many options for replacing these fuels. Nuclear fission is one of them, but, given its high costs and poor safety record, it seems likely that fission will gradually die of its own accord. Fusion offers the promise of unlimited energy but because of costs may never become commercially feasible.

What about coal? Enormous deposits of this fossil fuel lie beneath the Earth's surface, but serious environmental obstacles lie in the way of much further development of this fuel, global warming and acid rain being the most significant. That leaves conservation and the host of renewable energy resources like solar energy, wind energy, geothermal energy hydropower, and biomass energy, all discussed in the next chapter. These sources of energy are currently underutilized and are likely to become increasingly more popular and less expensive as conventional fuel sources decline. Widespread government support of these technologies, akin to that now given to the fossil fuel and nuclear industries, could help realize the promise of these environmentally clean and renewable energy resources.

Summary of Key Concepts

1. The twin oil crises of the 1970s awakened the world to the extraordinary cost of our dependency on oil. It showed us that everything we bought or did required energy, and lots of it. High inflation created a near crippling of the world economy. In 2000, another energy crunch developed due primarily to levels of consumption that exceeded demand.
2. In more developed countries, fossil fuels such as coal, oil, and natural gas are the predominant sources of energy. Nuclear energy, conservation, and renewable energy sources have grown in importance since the 1970s, although in most countries nuclear power is being phased out.
3. In the less developed nations, renewable fuels, especially wood, form the mainstay of the energy diet, although industrialization in these countries is increasing their dependence on fossil fuels.
4. Coal supplies about 23 percent of the United States' and 13 percent of Canada's energy needs. It is burned today in electric utilities and in some factories and homes, where it releases the solar energy captured in plants many millions of years ago by photosynthesis.
5. Proven global coal reserves are estimated to be about 700 billion metric tons, enough to last about 200 years. Undiscovered global reserves could supply the world's needs for another 1,700 years. U.S. coal supplies could last 200 years at the current rate of consumption, making coal a likely energy source for many years to come.
6. Unfortunately, coal cannot easily or efficiently be substituted for oil and natural gas, and it is a dirty fuel. Mining it causes enormous impacts. Unless we find much cleaner ways to burn it, much of our impressive coal deposits may lie forever buried in the Earth's crust.
7. Fluidized bed combustion offers a way to burn coal cleanly and efficiently. Crushed coal is mixed with bits of limestone and propelled into a furnace in a strong current of air. The limestone reacts with the sulfur oxide gases, eliminating them from the smokestack gases. Despite the advantages of this technology, concerns over global warming may make it only a stopgap measure to a more sustainable system of energy.
8. Another technology is coal gasification, in which a combustible gas is produced from coal. New developments in this technology have greatly improved its efficiency.
9. Coal can be treated to produce a thick, oily substance in a process called coal liquefaction. The oily product can be refined in the same way that crude oil is refined to produce a variety of useful products, such as jet fuel, gasoline, kerosene, and the chemicals needed to make drugs and plastics.
10. Unlike coal, oil is in short supply. By various estimates production from proven global reserves will peak somewhere between 2005 and 2010, after which time demand will exceed supply, causing potentially crippling inflation. Some analysts believe that there's more oil than once thought, but even so, this additional supply, if it exists, will only extend the supply of oil another couple decades.
11. The picture for natural gas is brighter than for oil, but not as good as for coal. Although estimates vary, proven and undiscovered reserves will last 100 years at the current rate of consumption, but increased consumption is likely to reduce global supplies dramatically. Natural gas burns much cleaner than other fossil fuels and may become an important transitional fuel as we expand our reliance on renewable energy resources.
12. Oil shale is a sedimentary rock that was formed millions of years ago from the mud at the bottoms of lakes. It contains a solid organic material, known as kerogen, that, when heated, is released from the rock, producing shale oil, which can be refined much as petroleum is refined.
13. The most valuable deposit of oil shale in the United States occurs in Colorado, Wyoming, and Utah.
14. Unfortunately, oil shale development is economically and environmentally costly.
15. Tar sands contain a thick, oily residue, called bitumen, that can be extracted and refined much as shale oil is refined. U.S. tar sand deposits are not extensive. Canadian tar sand deposits are much larger. Unfortunately, tar sand production is very energy intensive.
16. An immediate substitute for oil is needed. It is also time to start developing alternatives to natural gas and coal, and es-

pecially to find some energy sources that do not add to global carbon dioxide levels. One possibility is nuclear energy. Nuclear energy is produced when atoms are split apart.

17. To understand nuclear energy, one must understand the structure of atoms. All matter is composed of atoms. Each atom is composed of a dense, centrally located nucleus containing protons, positively charged particles, and neutrons, particles without a charge. Surrounding the nucleus is the electron cloud where tiny, negatively charged particles, the electrons, are found.
18. Over 100 elements are known to science, each one differing from the next in the number of protons in its nucleus.
19. An element is a pure form of matter. In any given element, the number of protons in the atoms is always the same. Because the number of electrons in the electron cloud always equals the number of protons, atoms are electrically neutral.
20. Although all atoms of a given element have the same number of protons, atoms of the same element may vary in the number of neutrons. Thus, an element may have several alternative forms of atoms, called isotopes. If these forms are unstable, they are called radioisotopes.
21. Radiation consists of particles or electromagnetic waves given off by the nuclei of radioisotopes in an attempt to reach stability. The three common forms of radiation are alpha particles, beta particles, and gamma rays. Over time, radioactive elements decrease in mass as a result of the release of radiation. The measure of this decrease is called the half-life.
22. Nuclear reactors (and the first generation of nuclear bombs) capture energy from controlled nuclear fission, the splitting of atoms. In reactors, fission occurs when uranium atoms in fuel rods are bombarded by neutrons given off by other uranium atoms in the same rod. Enormous amounts of energy are released during this process.
23. Additional neutrons are also released during fission. These neutrons continue to bombard other nuclei, causing still other fissions in a chain reaction.
24. Nuclear power plants use the energy given off by controlled chain reactions to boil water to make the steam that turns electrical generators.
25. The uranium fuel used in all large nuclear power plants is a mixture of uranium-235 and uranium-238 that is packed in pellets and inserted in fuel rods. Fuel rods are inserted into the core of the nuclear reactor. The rate of fission is moderated by neutron-absorbing materials in the control rods. The entire assemblage of control and fuel rods is housed in a reactor vessel within a containment building.
26. Supporters of nuclear power argue that this technology offers many advantages that make it a desirable energy source: (a) Should the breeder reactor become feasible, the fuel supply would be abundant. (b) Nuclear power could help us reduce our dependency on foreign oil. (c) Properly operating plants release little radiation; in fact, a coal-fired power plant releases more radiation than a nuclear plant. (d) Nuclear plants release fewer solid wastes and air pollutants. (e) The probability of an accident, they once argued, was quite slim.
27. Opponents take exception. They note the following: (a) The breeder reactor is costly, fraught with safety problems, and plagued by a long payback period. (b) Nuclear power replaces very little imported oil. (c) Accidents at nuclear power plants can release enormous amounts of radiation with potentially devastating effects. (d) Nuclear power plants operating normally release more thermal pollution than coal-fired power plants. (e) Accidents at nuclear power plants occur with a rather high frequency. (f) Nuclear power is an expensive form of energy and is bound to increase in cost as we grapple with waste disposal and ways to decommission reactors. (g) The nuclear industry is fast losing its public and private support.
28. The effects of radiation vary with the age of the exposed individual. Fetuses and newborn infants are much more sensitive than adults. Fast-growing cells are much more sensitive than cells that do not divide.
29. Radiation exposure can be measured in rads (radiation absorbed dose). Individuals receiving 400 to 500 rads of radiation suffer extreme mortality; only half of them will be alive after 3 weeks. Individuals receiving doses less than 25 rads would not be aware of changes in their health. When the dose rises to 100 rads, however, they may experience weakness, fatigue, vomiting, and diarrhea. Eventually these symptoms disappear. These seemingly low levels of exposure nonetheless often result in an increase in cancer and birth defects in offspring.
30. Numerous studies show that radiation increases the incidence of a number of cancers, including leukemia, bone cancer, lung cancer, and skin cancer. Scientists believe that most tumors result from genetic changes in the cells of the body, causing them to divide uncontrollably. Radiation can also damage the chromosomes of germ cells; the resulting mutations can be passed to an individual's offspring and may show up as birth defects or cancer early in life.
31. One of the biggest concerns of many people is the release of radiation during normal, routine operations and, more important, during accidents.
32. Many people are concerned about radioactive waste disposal. In the United States, highly radioactive reactor and military waste has been building up for over four decades. The United States has neither reprocessing facilities to extract usable fuel from the waste nor any high-level radioactive waste disposal sites. To solve this dilemma, the U.S. Congress passed the Nuclear Policy Waste Act in 1982. It called on the Department of Energy to choose two sites for disposal, but site selection has been plagued with difficulties. The biggest problem is that few people want a high-level radioactive waste dump in their state. Although many people believe that it was a mistake to build nuclear power plants in the first place, the simple truth is that in 2000, 42,000 metric tons of high-level waste are stored at commercial reactors. We must solve the waste problem that we have created.
33. One alternative to nuclear fission is the fusion reactor. Fusion reactors are fueled by deuterium and tritium, isotopes of hydrogen that unite to form larger nuclei. To make them fuse, however, enormous amounts of energy must be supplied.

34. One of the main attractions of fusion power is that the fuel supply is exceptionally abundant. And fewer wastes would be produced than a conventional fission reactor. However, fusion reactors are likely to be prohibitively expensive and could be dangerous since the coolant, molten lithium, burns on contact with air. A tiny leak could destroy a reactor.

Key Words and Phrases

Fossil Fuels

Acid Mine Drainage
Acid Precursors
Anthracite
Arctic National Wildlife Refuge
Area Strip Mine
Bitumen
Bituminous
Coal
Coal Gasification
Coal Liquefaction
Coal Reserves
Conservation
Contour Mine
Dragline
Fluidized Bed Combustion
Gasification
Hydropower
In Situ Retort
Kerogen
Lignite
Natural Gas
Net Energy Efficiency
Oil
Oil Producing and Exporting Countries (OPEC)
Oil Reserves
Oil Sands
Oil Shale
Overburden
Proven Reserves
Retort
Shale Oil
Slurry
Smokestack Scrubbers
Spent Shale
Subsidence
Sulfuric Acid
Surface Mining Control and Reclamation Act
Surface Retort
Tar Sands
Ultimate Production

Nuclear Energy

Alpha Particle
Atom
Atomic Mass
Atomic Number
Beta Particle
Breeder Reactor
Chain Reaction
Containment Building
Control Rods
Daughter Nuclei
Element
Electron
Electron Cloud
Emergency Cooling System
Fuel Rods
Fusion Reactor
Gamma Ray
Half-life
Isotope
Magnetic Fusion Energy Emergency Act
Mutation
Neutron
Nuclear Fission
Nuclear Fusion
Nuclear Reactor
Nuclear Waste Policy Act
Nucleus
Plasma
Plutonium
Proton
Rad (Radiation Absorbed Dose)
Radioactive Decay
Radioisotope
Reactor Core
Reactor Vessel
Uranium
Yellow Cake

Critical Thinking and Discussion Questions

1. Describe the major sources of energy used by more developed countries. Which ones are in short supply? Which ones do you expect to be plentiful 50 years from now? Why? Are all of the constraints related to reserves or supply issues?
2. Using your critical thinking skills, analyze the statement, "The less developed countries rely primarily on renewable energy. Since it is renewable, they have nothing to worry about."
3. Define the terms: proven reserves, ultimate reserves, and undiscovered reserves.
4. How large is the global proven reserve of coal? How long will it last at the current rate of consumption?
5. Describe the environmental impacts of surface and underground mining of coal.
6. Describe fluidized bed combustion and draw a schematic diagram of the process. Why does it produce more energy and less airborne pollution than a conventional coal-fired power plant? Is it a sustainable solution to our nation's energy needs? Why or why not?
7. Debate the statement, "The world has abundant coal. Consequently, coal will become our major fuel source when oil runs out."
8. Define coal gasification and coal liquefaction.
9. How does exponential growth affect global oil supplies?
10. Discuss the pros and cons of drilling for oil in the Arctic National Wildlife Refuge. Use your critical thinking skills to analyze the various points.
11. What is the estimated ultimate global production of natural gas? How long will that last? What is the estimated proven reserve of natural gas for the United States?
12. Describe oil shale and tar sands. What are they? Where did they come from? How big are the deposits? What are the impacts of their production?
13. Describe the structure of an atom and define the terms isotope, radioisotope, radiation, and half-life.
14. Which cells of the body are most sensitive to radiation?
15. Describe how a nuclear power plant works, being sure to mention the following terms: fission, neutrons, daughter nuclei, uranium-235, fuel rods, control rods, reactor core, chain reaction, containment vessel, and emergency cooling system.
16. Do you agree with the following statement? "Nuclear fission is a clean, safe alternative to coal-fired power plants." Why or why not?
17. How does a breeder reactor differ from a conventional fission reactor? What are the pros and cons of the breeder reactor? What unique problems does it have?
18. How does a fusion reactor operate? What is its fuel? What are the pros and cons of nuclear fusion?
19. Suppose that a utility company proposed building a nuclear power plant near your campus. The construction of

the plant, the utility asserts, would revitalize the local economy. Critics agree that it would create a number of jobs, helping the economy, but argue that most of the workers would be brought in from outside. What would be your response to the proposed project?

Suggested Readings

Campbell, C. J. 1997. *The Coming Oil Crisis.* Petroconsultants Sa. Detailed and authoritative analysis of world oil supplies. Paints a rather dismal future.

Chiras, D. D. 1992. *Lessons From Nature: Learning to Live Sustainably on the Earth.* Washington, D.C.: Island Press. Describes transitional steps to a renewable energy economy.

DeCarolis, J. F., R. L. Goble, and C. Hohenemser. 2000. "Searching for Energy Efficiency on Campus: Clark University's 30-Year Quest." *Environment* 42(4): 8–20. Interesting story about efforts that could be carried out on many college campuses.

Flavin, C. 1999. "Rethinking the Energy System." In *State of the World 1999.* Starke, L. (ed.). New York: W. W. Norton. Part of an on-going series on ways to restructure global energy system that are good for people and the environment. Very important reference.

Hollander, J. M. 1992. *The Energy-Environment Connection.* Washington, D.C.: Island Press. Comprehensive survey of the problems created by energy use.

Kammen, D. M. 1999. "Bringing Power to the People: Promoting Appropriate Energy Technologies in the Developing World." *Environment* 41(5): 10–15, 34–41. Looks at some of the key issues involved in creating more sustainable energy systems in less developed countries.

Lenssen, M. 1992. "Confronting Nuclear Waste." In *State of the World 1992.* Starke, L. (ed.). New York: W. W. Norton. Detailed account of nuclear energy and its problems.

Lenssen, M. 1991. "Designing a Sustainable Energy System." In *State of the World 1992.* Starke, L. (ed.). (ed.) New York: W. W. Norton. Contains important information on the future of oil.

Roodman, D. M. 1997. "Reforming Subsidies." In *State of the World 1997.* Starke, L. (ed.). New York: W. W. Norton. Looks at the ways subsidies give fossil fuels and other environmentally unsound activities an advantage and how changes in policy can place other, more environmentally acceptable strategies affordable.

Segerstahl, B., A. Akleyev, and V. Novikov. 1997. "The Long Shadow of Soviet Plutonium Production." *Environment* 39(1): 12–20. Looks at the impacts of another aspect of radiation pollution, notably weapons production.

Youngquist, W. 1997. *Geodestinies. The Inevitable Control of Earth Resources Over Nations and Individuals.* Portland, OR: National Book Co. Fun to read and thought provoking look at renewable and nonrenewable resources.

Web Explorations

Online resources for this chapter are on the World Wide Web at: **http://www.prenhall.com/chiras** *(click on the Table of Contents link and then select Chapter 21).*

Chapter 22

Creating a Sustainable System of Energy: Efficiency and Renewable Energy

Imagine a world powered by clean energy from the sun or wind. Imagine, too, a prosperous economy and a world free of oil spills, toxic air pollution, acid rain, and greenhouse gases. Idyllic, you ask? Utopian and far from reality?

Not so, say proponents of energy efficiency and renewable energy sources. In fact, in a publication of the American Solar Energy Society, Dr. H. M. Hubbard asserted that by the year 2030, 50 percent of U.S. energy supplies could come from renewable sources. In subsequent decades, Americans could wean themselves nearly completely from fossil fuels, which lie at the root of many environmental problems, among them global warming, acid deposition, urban air pollution, habitat destruction, and water pollution. Several countries have already started down the same path.

Environmentalists and scientists interested in creating an Earth-friendly energy system envision a variety of renewable energy resources, many of which are already being used today. In addition to conservation, these alternatives include solar energy, wind energy, biomass, hydropower, geothermal energy, and possibly others. This chapter discusses the main options for a sustainable energy future, explaining each technology, its potential, and some of its main challenges. We begin with conservation measures—ways to reduce energy use and ways to use energy more efficiently.

22.1 Conservation

In most of the more developed nations, rising energy needs over the past 200 years have been met by simply producing more fossil fuel. To follow that strategy today means producing more coal, oil, and natural gas, and developing oil shale, tar sands, coal gasification, and coal liquefaction. Over the past 200 years, little attention has been given to finding ways to reduce energy consumption by using less (being more frugal) or through efficiency measures. In the Harvard Business School report entitled *Energy Future,* conservation is described as "no less an energy alternative than oil, gas, coal, or nuclear. Indeed in the near term," say the authors, "conservation could do more than any of the conventional sources to help the country deal with the energy problem."

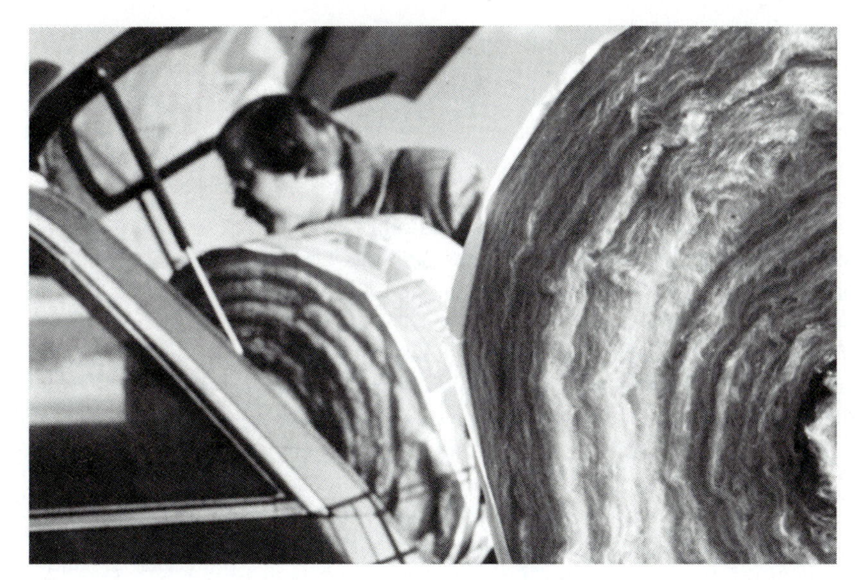

FIGURE 22.1 *Getting ready for winter. Adding insulation yourself will result in both energy and money savings. Many families in the northern states—New York, Pennsylvania, and Michigan—invest in insulation, caulking, weather stripping, and storm windows in the 1970s and early 1980s. Their homes will be snug despite the icy blasts of winter. Don't forget, however, that insulation also works in hot climates to keep a house cool.*

Conservation's Short-Lived Triumph

Energy conservation is not foreign to us. Following the oil crises of the 1970s, the United States took drastic measures to slash energy waste—at home, at work, in the air, and on the highways. Consumers added insulation to their attics, replaced leaky single-pane windows, caulked cracks around doors and windows, bought smaller cars, and turned down their thermostats, saving enormous amounts of energy (Figure 22.1). The average gas mileage of a new car increased from a paltry 14 miles per gallon in 1974 to 26 miles per gallon in 1986, thanks to congressional action. Over the same period factories cut their energy use, and consequently the amount of energy needed to produce a dollar of gross national product dropped by 25 percent. Between 1949 and 1999, efficiency measured by the amount of energy required to produce one dollar of GNP, has increased by 47 percent.

After a period of initial activity, however, the nation's fervor for energy conservation seemed to wane. Unsympathetic administrations under both presidents Reagan and Bush greatly slowed progress in energy conservation. Under Reagan, for example, federal energy conservation programs were cut 91 percent between 1981 and 1987. Both presidents resisted measures that would have improved energy efficiency in automobiles. By the time President Clinton took office, the average fleet mileage for new vehicles produced in the United States stood at only 27.5 miles per gallon.[1] During Clinton's term of office, attempts to increase automobile energy efficiency were thwarted year after year through congressional action—riders on bills that forbade the Department of Energy even from considering further increases in gas mileage! Cars got bigger, and new car gas mileage began to slip (Figure 22.2).

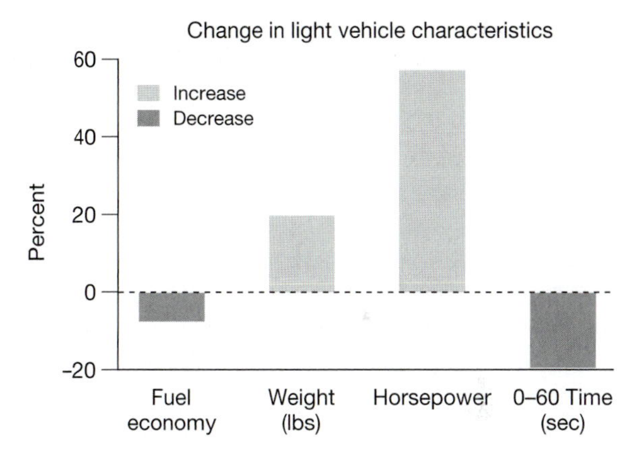

FIGURE 22.2 *American cars are getting heavier, more energy-intensive, and slower on acceleration.*

The loss of interest in conservation was all the more frustrating to those who recognized its many social, economic, and environmental benefits. One of the most pronounced, if least appreciated, of those benefits is the way conservation efforts stem the rise in oil prices and help the economies retain their vigor. In the early 1980s, conservation efforts in the United States and other countries, including Canada, lowered demand for gasoline resulting in a temporary glut of oil. This, in turn, stabilized oil prices, as one might predict from the supply-and-demand theory. Stable oil prices helped stabilize the cost of consumer goods, quelling the economic turmoil that had strangled the United States and much of the rest of the industrial world in the 1970s and early 1980s.

Another factor that contributed to the fall in oil prices was the fact that many non-Oil Producing and Exporting (OPEC) nations, such as Great Britain, increased exploration and oil production, causing OPEC's share of the world oil market to fall. In 1977, OPEC nations supplied two thirds of the world's oil. By 1985, their share had dropped to one third. To combat their loss in market share, in December 1985 the OPEC nations announced they would cut prices to regain their position in the market, and by April 1986 crude oil prices had been slashed in half. Today, OPEC's share of the world oil market is again on the rise, once again creating a potentially costly de-

[1]Nonetheless, energy consumption still increased dramatically, raising over 200 percent during that period while population only rose a little more then 80 percent.

pendency, for the OPEC nations currently house a majority of the world's oil reserves.

Through the latter half of the 1980s oil prices remained low, and while that was good for many sectors of the economy, it eroded much of our nation's resolve to conserve energy. As a sign of our nonchalance, in 1987 Congress passed a popular law that allows states to increase the speed limit on rural sections of interstate highways to 65 miles per hour, a measure that greatly increased fuel consumption. In 1996, the speed limit was raised again. In addition, in 1996 President Clinton lifted the extra tax on gas-guzzling cars. United States and foreign auto manufacturers began producing larger and larger vehicles with lower and lower gas mileage (Figure 22.3).

In summary, cheap energy prices and unsympathetic legislators and an apathetic public seemed to have turned the nation back to the wasteful pre-OPEC days. The American public and Congress appeared to have forgotten the chief lesson of the 1970s, that nations dependent on oil cannot afford to waste it.

It is a lesson we are beginning to realize again. At this writing (October, 2000), oil prices are sky high again and inflation is beginning to raise its ugly head. In the U.S., imports now exceed exports (Figure 22.4). High levels of consumption, say some experts, exceed the oil production system's ability to transport and refine oil, and may be a precursor of what is to come when oil demand exceeds supplies.

Oil is costly, in environmental and economic terms. Oil and oil byproducts are a major source of the greenhouse gas, carbon dioxide. We spend billions of dollars on oil, as well, money that could be invested in creating much more sustainable sources of fuel. The world's nations spent $61 billion to wage war in the Persian Gulf, a war fought largely to protect Kuwait's oil supplies from invading Iraquis. Although few people realize it, the United States today spends $50 billion a year for the military protection of oil tankers moving in and out of the Persian Gulf. If this subsidy were included in the price of Middle Eastern oil it would cost about $500 per barrel, much higher than the approximately $35 per-barrel price tag on the open market. If citizens of the United States paid the real cost of gasoline from Persian Gulf oil, it would cost over $25 per gallon.

FIGURE 22.3 *The Ford Excursion represents the epitome of fuel-consumption. This sport utility vehicle, which weighs nearly four tons, gets less than 10 miles per gallon. It was voted "Big Dumb Car of the Millenium" by PBS's Car Talk hosts, Tom and Ray Magliozzi aka Click and Clack, the Tappet Brothers, in 2000.*

Getting Back on Track

Contrary to popular belief, the energy crisis is still with us. Clearly, Americans must reverse the widespread nonchalance about energy and get back on the conservation track. We must become more efficient in energy use and we must develop and implement clean, nonpolluting alternatives.

We should point out that the benefits of conservation measures, the topic of this section, are more than simply a matter of conserving finite resources. Energy conservation drastically reduces many forms of pollution discussed in previous chapters such as acid deposition, greenhouse gas emissions, urban air pollution, and even water pollution. By helping us stretch world oil supplies, it gives researchers additional time to develop, and our society more time to install, affordable and environmentally sustainable alternatives to oil. Conservation helps families save money so they can have more to do the many things they need to do—such as pay for a college education. It helps businesses produce goods and services at a lower cost. And it can also help the world sustain lower oil prices and help the global economy retain its hard-won economic stability. Energy conservation also creates jobs. In short, energy conservation is to our economy what preventive medicine is to health care.

The Untapped Potential

Despite what some individuals claim, most nations have only begun to tap conservation. Enormous opportunities for conservation still exist in buildings, industry, motorized vehicles, and appliances.

Consider some examples. Portland, Oregon has adopted one of the nation's toughest building codes, requiring insulation in new as well as old homes, apartments, and office buildings. On another front, several auto manufacturers have test models that get 98 miles per gallon. The U.S. average new-car mileage, 23.8 miles per gallon, is the lowest in the developed world.[2] By increasing it to 40 or 50 miles per gallon, which some experts believe is easy to achieve, America could greatly stretch its oil supplies, creating more time in which to develop more efficient vehicles and alternative liquid fuels.

On still another front are enormous savings from lighting. The United States has approximately 4.7 billion square meters (50 billion square feet) of commercial space, requiring roughly 100 power plants just to supply it with electricity for lighting. In Seattle, officials predict they could cut their demand for electric lighting by 80 percent. How? By switching to 18-watt compact fluorescent light bulbs that produce as much light as 75-watt bulbs, by changing to energy-efficient light switches, by using brighter interior surfaces, and by using special sensors that shut off interior lights when natural light is sufficient.

[2]According to the EPA, the average mileage for domestic cars is currently 28.1 miles per gallon and 20.3 mpg for light duty trucks.

Nationwide, the economic savings and environmental benefits of such a plan boggle the mind.

New energy-efficient bulbs can be used in the home as well. Several manufacturers now offer compact fluorescent light bulbs that screw into incandescent sockets. These bulbs cost more but will outlast nine standard incandescent lamps and save enormous amounts of energy (Figure 22.5). The 15-watt bulb, for instance, replaces a 60-watt incandescent lamp and yields as much light. The $12 to $15 bulb will save $30 to $50 in electricity costs over its lifetime, depending on the cost of electricity in your locale. If you're paying 10 cents per kilowatt hour, for example, the bulb will save $45. And, if every home in America replaced just four of its frequently used light bulbs with these bulbs, the nation would save more electricity than is produced by six large nuclear power plants.

Finally, consider appliances. California cut the maximum allowable energy use in new refrigerators by half, a measure that went into effect in 1992. Another promising move, perhaps the only major gain in conservation nationwide in the less-than-promising 1980s, was the **National Appliance Energy Conservation Act** of 1987. The act calls on manufacturers to produce appliances that use 20 percent less energy than 1987 models, starting in 1992. In 20 years' time, this law alone will eliminate the need for 22,000 megawatts of electricity—the amount of energy produced by 22 large nuclear power plants. It will save Americans well over $28 billion in electric bills.

Energy conservation makes good economic sense for the consumer and producer. For example, improving the efficiency of machines and appliances—that is, reducing their energy demand—costs about 1 to 2 cents per kilowatt-hour of energy saved. In contrast, coal-fired power plants produce electricity for 5 to 7 cents per kilowatt-hour. Nuclear power plants produce it for 10 to 12 cents per kilowatt-hour. Energy expert Amory Lovins of the Rocky Mountain Institute estimates that oil conservation measures cost about $2 per barrel saved. So, for every $2 a company invests to save a barrel of oil, it saves $20 to $30, depending on the price of oil. Not a bad investment.

Resetting Our Priorities

Building a sustainable society will inevitably require a massive reduction in the use of fossil fuels—a worldwide cutback. This will require actions on the part of all citizens, businesses, and governments. To stir action, many observers argue that we must first raise awareness of the key issues, notably the potential seriousness of global warming, caused in large part by carbon dioxide emitted from the combustion of fossil fuels, and the limited supplies of oil. Through colleges and universities, through public schools, through government study and political leadership, we must raise the consciousness of the world's people regarding the dangers of our dependence on oil.

The federal government and the states can also take leadership roles in making energy conservation once again a national priority. Our long-term economic security depends on bringing energy conservation back into the American mainstream. Fortunately, there are some signs that the mood may be changing. For example, in 1991 the Environmental Protection Agency (EPA) instituted a nationwide energy conservation program aimed at voluntarily reducing electrical demand by the nation's largest corporations and by government buildings. Called the **Green Lights** program, this project has signed on hundreds of companies that, with the aid of the EPA, are working to cut electrical demand for lighting by installing compact fluorescent light bulbs, energy-efficient fluorescent tubes and ballasts, and other technologies.

More recently, the EPA started another voluntary project, the **Energy Star program.** This program calls on manufacturers of computers, printers, and monitors to reduce the energy demand

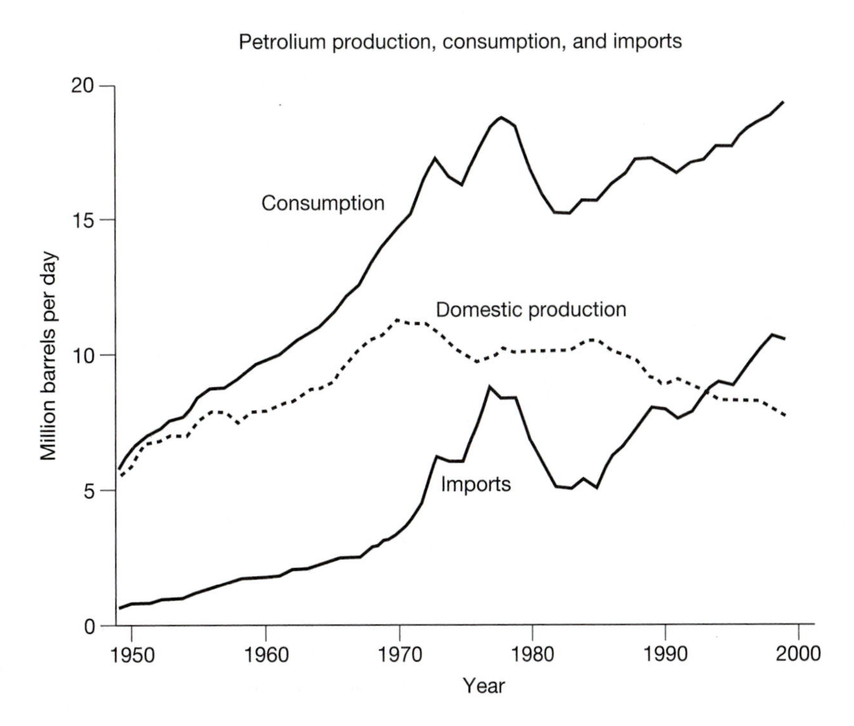

FIGURE 22.4 *U.S. oil imports are climbing rapidly again, making us more and more vulnerable to OPEC.*

FIGURE 22.5 *This compact fluorescent light bulb produces the same amount of light as a 100-watt bulb but with 75 percent less energy. Although it costs more, the bulb saves $40–$50 over its lifetime, so it actually saves you money.*

of their machines, primarily by providing power-down options—that is, automatic switches that lower energy demand when the machines are not in use, for example, when operators are on the phone or when they leave the room for a while. Because many operators (about 30 percent nationwide) leave their machines on all night, this feature could save enormous amounts of energy. Some companies have also found ways to reduce energy demand while computers are in use.

The Energy Star Program has recently been extended to other appliances like televisions and to homes. For homes, builders seek ways of building new homes that use at least 30 percent less energy than homes based on the national Model Energy Code (already a more efficient home than those built in earlier years). After a home's energy performance has been verified by an independent expert, it is given the **Energy Star label.**

Personal Actions

There is almost no end to the things an individual can do to reduce energy use. This abundance of options is both a blessing and a problem. It's a blessing because you have a wide assortment of areas to choose from; it's a curse because, faced with such choices, many people simply don't know which one or ones to choose, and so do nothing.

Accordingly, we have limited the following discussion to individual actions that do the most good—that is, actions that cost very little, yet make significant reductions in energy use. You may want to read through this material and then sit down and draw up a plan to reduce your own resource consumption.

Energy Conservation Around the House As individuals you can make significant contributions to resource conservation with little effort and monetary investment. You can begin by turning down the thermostat, which is one of the easiest and most cost-effective strategies available. A 6°F drop in room temperature can reduce fuel consumption by 15 to 20 percent. Keeping your furnace thermostat at 68°F during the winter can save enormous amounts of energy.

Try turning your thermostat down gradually. Drop it, on average, 1°F every few weeks until you reach the desired setting. This will give your body a chance to adjust. To help counterbalance the dropping temperature, you may want to dress more warmly. Thermal underwear, wool sweaters, and heavy socks can help you stay comfortable. Putting on a sweater is equivalent to raising the room temperature nearly 4°F. Wrapping up in a blanket while you are watching television or reading has the same effect.

You can drop the thermostat even further at night and stay perfectly warm without an electric blanket. Better yet, your body produces enormous amounts of heat (about as much as a 100-watt light bulb), and you can keep comfortable even in the coldest weather with blankets.

You can install automatic thermostats to adjust daytime and nighttime temperatures to save heat. Automatic thermostats, in fact, pay for themselves in reduced energy bills in as little as 2 or 3 months to as long as 3 years, depending on who installs them and other factors.

To some people, turning down the thermostat means they will end up being cold and uncomfortable. There's a good reason for this: Many houses are too drafty, and lowering the thermostat setting reveals the frigid internal winds created by leaks. In the 1970s, many people who tried to conserve energy found their homes frightfully uncomfortable and simply cranked up their heat to fight drafts, complaining that conservation was a bad idea. This strategy is short-sighted and wasteful. It's far cheaper and a far wiser use of resources to seal up the cracks through which air escapes with caulk and weather stripping. And it's amazing how many leaky spots you can find with a little effort.

Caulking and weather stripping have an astonishingly fast payback. In cold climates, caulking and weather stripping pay for themselves in reduced energy bills in 6 months to a year if you do the work yourself, a little longer if you hire someone to do the job for you. By reducing drafts, they will make your home more comfortable.

Turning down the setting on your water heater is another simple measure to reduce energy consumption. By turning your water heater to 140°F if you have a dishwasher (without a water heating function) and 120°F if you don't, you can save hundreds of kilowatt-hours of energy a year. A study by Oak Ridge National Laboratory, for instance, showed that lowering the setting from 160°F to 140°F saves 400 kilowatt-hours a year, as much electricity as many households use in a month. Lowering your water heater thermostat is fairly easy and won't affect personal hygiene at all.

The measures listed above achieve significant energy savings with virtually no expenditure. For homeowners willing to invest a little money to cut energy demands, four additional approaches with relatively quick paybacks are recommended. The first is ceiling insulation. Because most heat escapes through the ceiling, upgrading the insulation in your attic to R-30 to R-38 (R is a measure of heat resistance) can help cut energy consumption drastically. Depending on where you live and how warm you like your house, insulation can pay for itself in 3 to 7 years.

The second measure is storm windows. Self-installed storm windows may take 5 to 7 years to pay for themselves. If someone else does the work, count on a payback period that is doubled. However, the payback for insulation and storm windows in one respect, notably comfort, is immediate. Ceiling insulation, by far the easiest and cheapest to install in many homes, reduces heat loss. This slows down the movement of air in the house and, like caulking and weather stripping, creates a cozier domicile—a payback few people seem to consider when debating whether they can afford extra insulation. Storm windows have the same effect. They eliminate cold spots and drafts and greatly increase comfort levels. What is more, they help reduce energy consumption, with all its attendant environmental costs. You save money and help reduce environmental deterioration—not a bad investment!

The third inexpensive energy conservation measure, which also rapidly pays for itself, is insulation for water heaters and hot water pipes. Now available for $10 to $20, insulating blankets help hold in heat and reduce overall energy demand. Easy-to-install insulation for hot water pipes has a similar effect.

The fourth measure is energy-efficient light bulbs. Now available in most grocery stores or discount stores, the General Electric Miser series saves approximately 10 percent. As mentioned earlier, however, many manufacturers produce screw-in compact fluorescent bulbs that use 75 percent less energy than conventional incandescent bulbs. Many hardware and building supply stores and many discount stores carry these bulbs, as do some utility companies and specialty catalogue suppliers such as Real Goods and Jade Mountain. These bulbs are designed to produce a very appealing yellowish color of light, not the harsh blue of typical fluorescent bulbs.

Nationwide, automobiles consume approximately 350 billion liters (90 billion gallons) of gasoline each year. Much of that fuel is wasted by people driving erratically and at excessive speeds. By driving reasonably and sticking to the speed limit you can cut personal gasoline consumption by 10 percent, saving money and time spent at the pump, reducing pollution, and reducing the environmental impacts of oil production. Short trips that can be combined into one trip or replaced by walking or riding a bicycle could help further cut energy use.

Perhaps one of the biggest wastes of fuel is commuting—driving back and forth to work and school. Each year commuters travel billions of kilometers. Often only one or two passengers ride in each car. This common practice wastes fuel and creates unnecessary pollution and crowding on our highways. Individuals can help reduce these and other problems by joining car or van pools or taking buses to work or school. Van pooling and buses are, on average, nearly five times more energy efficient than the automobile. For example, it takes 400 kilojoules (a unit of energy) to move a passenger 1 kilometer (0.6 mile) by van pool, train, or bus. Passenger cars consume 1,800 kilojoules, and airplanes, incidentally, require 3,800 kilojoules. In other words, it is 4.5 times more efficient to take mass transit than your automobile. You're getting more or less the same service, only using 4.5 times less energy and creating 4.5 times less pollution. Air travel is nearly 10 times more wasteful. If you can join a van pool or ride the bus even occasionally, you should consider doing it. Your contribution, combined with that of other like-minded individuals, can add up quickly.

One of the wisest steps is purchasing an energy-efficient vehicle. Certain models on the road, like the Geo Metro, which gets 65 miles per gallon on the highway, provide a useful alternative for those who want to reduce their impact. The new hybrid cars are also a good investment. If you must have a larger vehicle, at least find the model that gets the best mileage.

22.2 Renewable Energy Strategies

The next step in building a sustainable energy system and a sustainable society is to install clean, renewable energy technologies. This section discusses the many exciting options.

Solar Energy

Sunlight is the ultimate source of energy that powers the global ecosystem, and it is destined to last several billion years. It comes to us free of charge and is a clean source of fuel. All we need to do is find ways to capture it and put it to use.

The thought of harnessing solar energy has piqued the interest of humans for ages. The Greeks used it 2,000 years ago to heat their homes—and considered those who didn't orient their homes to the south to tap the sun's generous heat to be barbarians. The Anasazi Indians of the desert Southwest used solar energy to heat their villages by building their homes in south-facing rock walls protected from the summer sun but open to the winter sun.

The amount of solar energy striking the Earth's surface on a cloudless day is over 100,000 times greater than the world's presently installed electrical capacity. The sunlight striking an area the size of Connecticut each year could provide all of the energy needed by the entire United States, yet Americans have tapped only a tiny portion of the sun's potential (Figure 22.6).

The sun's awesome power is amply illustrated by the world's largest solar furnace, in the French Pyrenees (Figure 22.7). This power plant has a 45-meter (150-foot) mirror that focuses sunlight on a boiler. Concentrated sunlight raises the temperature to 3,500° (6,300°)—sufficiently hot to melt a 1-foot-wide hole in a steel plate three-eighths of an inch thick in 60 seconds.

Solar Heating All buildings are solar heated to some extent. Without the sun, the Earth's average temperature would be a chilling -230°C (−450°F), so in many ways we already use the sun for heating, even if unintentionally. The intentional use of sunlight for heating takes two forms—active solar heating and passive solar heating.

Active solar systems, like the one shown in Figure 22.8, are generally mounted on rooftops and gather sunlight in collectors, closed boxes with black backgrounds to absorb light. Collectors are insulated and covered with glass to prevent heat loss. The heat is removed from the interior of the collector by air or a fluid flowing in pipes within the unit (Figure 22.9). The heat can then be transferred directly to the room or transported to a storage device for later use.

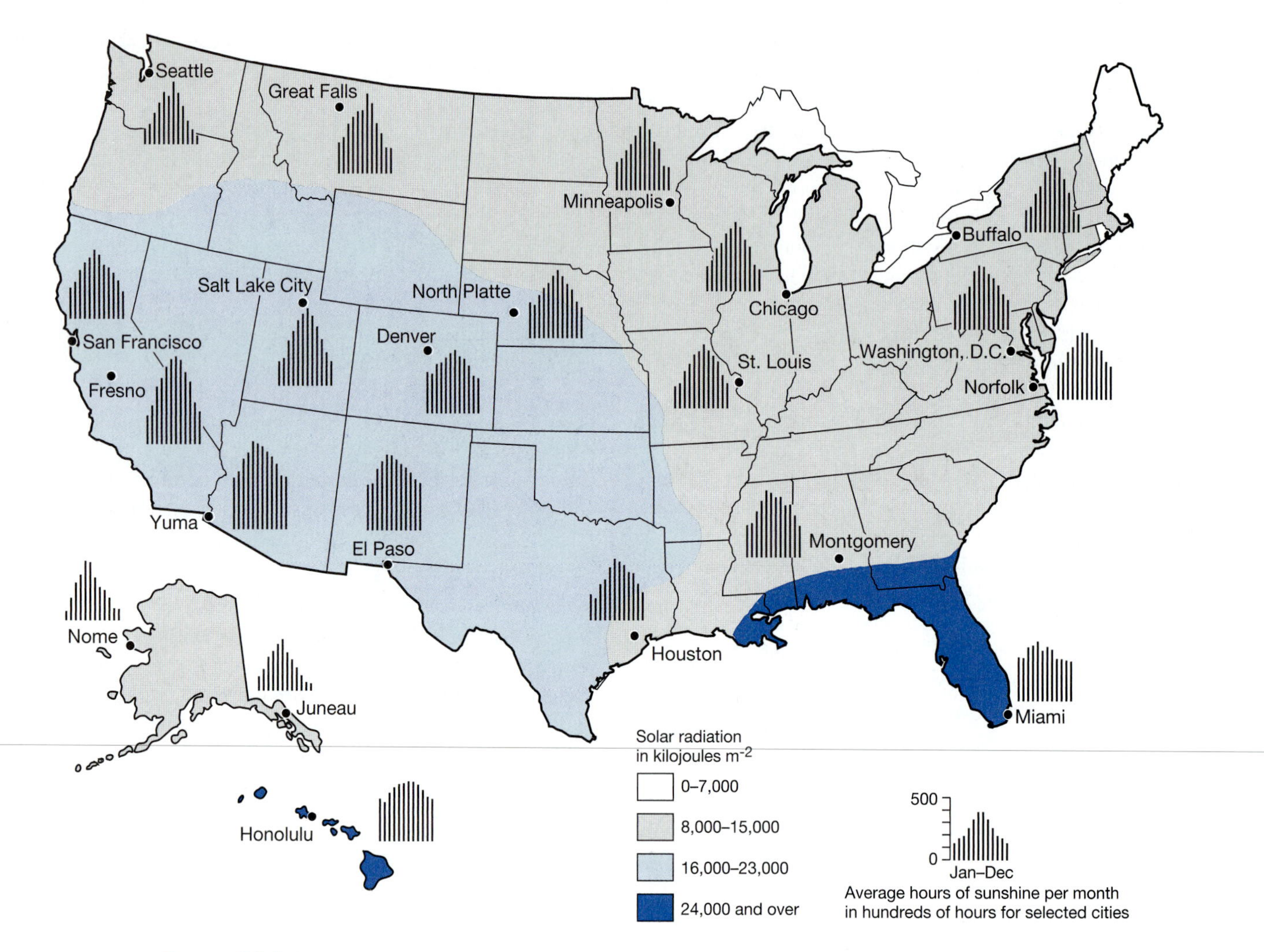

FIGURE 22.6 *Many parts of the United States and other countries have abundant sunlight for solar energy, both solar electricity and passive solar heating.*

Most active solar systems in use today provide hot water for domestic use—washing dishes, clothes, and people. Solar domestic hot water systems can provide 50 to 90 percent of a person's hot water needs, depending on one's location. Good, reliable systems are now available, too. But they tend to be expensive.

Passive solar systems are designed for space heating. In such systems, the building itself becomes a collector and heat-storing device. In the passive solar home shown in Figure 22.10, belonging to Dr. Chiras, south-facing windows and skylights let the winter sun penetrate the interior of the house. The sunlight strikes the walls and floors and is converted into heat. Special cement or block walls in the inside store the heat and radiate it out into the room at night. An overhang on south-facing windows prevents the summer sun from entering the house, keeping it comfortable all year round.

Passive solar homes must be well insulated; windows generally have double or triple panes and thick curtains to prevent heat loss during the night. Surprisingly, passive solar, well-insulated homes can function efficiently in fairly cold climates. One home built by the engineering department at the University of Saskatchewan had an annual heating bill of $40. By comparison, a conventional home of the same size would have a heating bill of $1,400. One hundred percent passive solar homes now operate in Maine, Vermont, and Wisconsin, among other places.

Passive solar energy can be used to heat commercial buildings as well. Ontario Hydro, a Canadian energy company, built a mammoth office building in Toronto, Canada, that relies entirely on solar energy and a system that captures waste heat from people, lights, and equipment and reuses it (Figure 22.11). Despite frigid winter weather, the building stays warm and comfortable inside. In Soldier's Grove, Wisconsin, the entire business community moved out of the floodplain to avoid the periodic flooding that wreaked havoc on their town. The town's citizens decided to convert the town into a solar town. They now have a solar post office, fire station, library, gas station, woodworking shop, American Legion hall, grocery store, and other community buildings. During the first 3 years of the new town's existence, the grocery store's heating bill was zero despite harsh winters—

FIGURE 22.7 *French solar furnace. This solar furnace, located near Odeillo in the Pyrenees Mountains of southern France, was built to test materials under extremely high temperatures. In the foreground, an array of 63 mirrors (heliostats), each measuring 6 meters by 7 meters, reflects sunlight onto the curved mirror surface of the office building in the background. This in turn focuses the sunlight on an aperture in the tower at the center, where temperatures of 7,000°F can be produced—enough heat to melt any known material.*

FIGURE 22.8 *Mt. Rushmore goes solar. Mt. Rushmore, South Dakota, site of the famed sculpted faces of former presidents, has a new solar energy system. Solar collectors on the roof of the Visitors Center transform the rays of the sun into energy for heating and air conditioning. The solar panels were developed by Honeywell. The system provides energy for 53 percent of the heating and 41 percent of the cooling for the 9,250-square-foot building.*

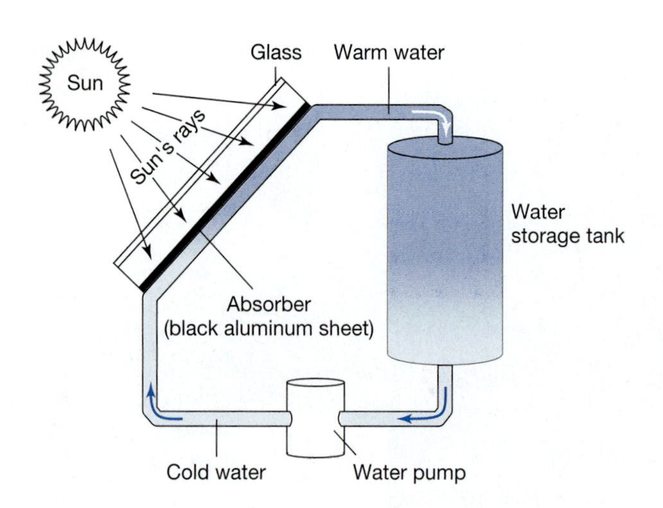

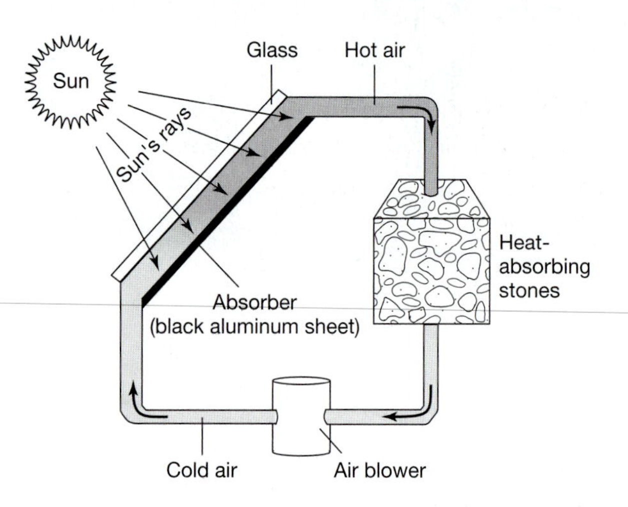

FIGURE 22.9 *Two types of active solar space-heating systems. Both rely on solar panels to absorb sunlight and produce heat.*

FIGURE 22.10 *Photograph of passive solar home at 8,000 feet in the foothills of the Rockies featuring south-facing windows and skylights that let the winter sun in. Superinsulation and high-efficiency gas backup heat keep winter heat bills low—about $100 a year.*

thanks to passive solar energy, superb insulation, and a system that captures waste heat from compressors and pumps it back into the 7,000-square-foot store. Remarkably, the entire system paid for itself in those 3 years.

Interestingly, many of the features that are used to create a passive solar heated building also contribute to natural cooling, or **passive cooling.** Thick insulation in walls and ceilings, for instance, which keeps houses warm in the winter, also keeps them cool in the summer. Thermal mass, including floors and interior walls made from bricks or cement, which absorb heat during the winter and radiate into the rooms at night, also soak up heat from people and other sources during the summer. In desert climates where nighttime temperatures drop dramatically, windows are opened and cool air flushes the accumulated heat out of the house, recharging it for the next day. **Earth sheltering** a home, that is pushing dirt up against back walls, sometimes covering walls and roof entirely with a layer of insulating earth, also helps solar homes operate more efficiently (Figure 22.12). Because the earth stays a constant 55°F below the frost line, the temperature of a house sheltered by earth remains fairly constant year round. During the winter, it is necessary to add only a small amount of heat to bring it into the comfort zone. Solar energy can do that admirably well. In the summer, the home stays naturally cool. The earth keeps it cool, warm air entering from open windows brings the temperature into the comfort zone.

Solar Electricity Solar energy can also be used to generate electricity. The French solar tower, described earlier and shown in Figure 22.7, heats water with sunlight and creates steam that runs an electric generator. This technology is therefore called **solar thermal electricity.** Somewhat similar systems are now operating in Sandia, New Mexico, and Barstow, California.

As a rule, these prototype power towers, completed in 1977, are small plants, producing only 1 megawatt of electricity—much smaller than conventional coal-fired power plants, which typically produce around 500 megawatts. The power tower stands about 15 stories high and is located in a large field of movable mirrors. Controlled by a computer, the mirrors track the sun across the sky and focus their beam of energy on the top of the tower. The intense heat produced there boils water, which is converted to steam to run an electric generator (Figure 22.13).

In recent years, engineers have created much simpler designs, which are proving quite successful (Figure 22.14). One company in California, for instance, developed a technology that produces electricity for 8.5 cents per kilowatt-hour, which is more expensive than coal (not counting coal's external costs) but cheaper than nuclear energy.

Electricity is more commonly produced by **photovoltaic cells,** thin wafers mounted in small arrays called modules on roof tops, on poles, or on the ground (Figure 22.15). Some are mounted on racks that track the sun across the sky from sunrise to sunset, which dramatically increases their electrical output. In recent years, several companies have begun producing roofing shingles and other roofing materials that contain PVs (Figure 21.16). That way, you get a roof and solar electricity.

FIGURE 22.11 *Ontario Hydro's energy-efficient, solar office building in Toronto.*

And one company is producing glass for windows that has PV material in it.

Photovoltaics convert sunlight energy directly into electricity. Each photovoltaic cell is a thin wafer of silicon or other material that, when struck by sunlight, emits electrons. Developed in 1954 by Bell Laboratories, photovoltaic cells were first used in 1958 to provide power for Vanguard I, America's second satellite. Since that time, photovoltaic cells have provided power for almost all other satellites and are now being used on Earth. One of the first applications was in remote locations where electricity is not available or is too expensive to install. River flow monitors in remote country, mountaintop radio relays, remote irrigation pumps, highway signs in unpopulated areas, lighthouses, and buoys all use them to provide energy. Photovoltaic cells also have some uses much closer to home, such as in calculators, watches, and other electronic devices. In 1980, photovoltaic cells powered the first flight of the Gossamer Penguin, a one-person solar airplane.

FIGURE 22.12 *Earth sheltering a house can reduce energy bills, works well with passive solar heating, and keeps it cool in the summer and warm in the winter.*

Despite these many uses, photovoltaics contribute very little to our overall energy consumption. As prices go down, however, experts predict that they will become more widely used. Dr. Chiras currently generates most of his electricity from a small array of solar panels and a small wind generator. Batteries, panels, and generator cost about $10,000.

Photovoltaic cells are an attractive energy source. The fuel is free and virtually inexhaustible. It is a clean technology as well, the only pollution being created during their manufacture. Widespread use would reduce acid deposition, greenhouse gas emissions, urban air pollution, strip mining, and all the other impacts created by the use of coal. They would also reduce the need for nuclear power, with all its impacts. Solar electricity could be used to recharge batteries in electric cars and could be used to generate hydrogen fuel that could be used in cars of the future as well. So, why haven't photovoltaic cells made it to the rooftop of every house in America?

The answer is cost. The silicon used to make solar cells comes from sand, one of the most abundant elements on Earth, but making solar cells is costly and tedious. Also, the solar

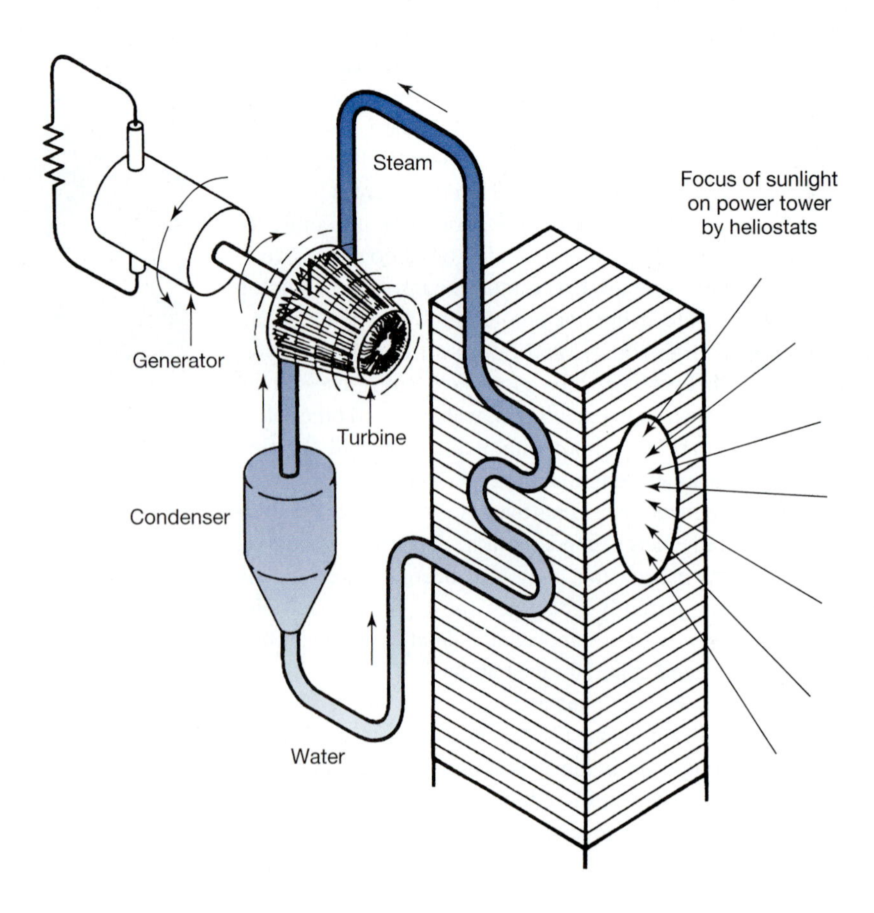

FIGURE 22.13 *(Top) View of solar power tower located at the Department of Energy's test facility at Sandia Laboratories in Albuquerque, New Mexico. (Bottom) The solar power tower. The concentrated rays of the sun convert water to steam, which in turn propels the turbine.*

FIGURE 22.14 *Solar thermal electric system. Sunlight reflected from these mirrors heats oil to 735°F. This heat is used to generate steam that drives a turbine to produce electricity.*

FIGURE 22.15 *Photovoltaic cells. These thin wafers of silicon absorb sunlight and generate an electrical current. They produce electricity with little impact on the environment and last 30 to 50 years.*

FIGURE 22.16 *Solar shingles. One of the newest developments, the solar shingle does double duty: it protects the home from rain and provides electricity.*

cells commercially available are fairly inefficient. A 1-square-meter panel is only enough to power a single 120-watt light bulb, and that panel costs about $300 to $500. Providing energy for a whole house with inefficient appliances and lighting might require a large array of solar cells costing $40,000 to $50,000. Making a house efficient in how it uses electricity, could cut the cost of a system to $10,000 to $20,000. Even as such, solar electricity costs three to five times more than commercially available power.

What is needed, say proponents, are more efficient and less expensive solar cells, both of which would help make this technology more competitive with other sources of electricity. At this writing, there are at least four on-line suppliers of renewable energy, offering generous discounts on panels. Researchers are also working on ways to improve efficiency. Tracking devices, mentioned earlier, are one of the results of this work. In 1986, researchers at Stanford University announced a new design that uses a parabolic mirror to concentrate sunlight on the solar cells. This could boost the efficiency from 10 to nearly 30 percent, a tremendous boon for the industry. Researchers have also produced multicrystalline wafers that are cheaper to manufacture than previous single-crystal cells. Multicrystalline wafers are now widely used. Another means of lowering prices would be mass production. Currently, photovoltaic panels are manufactured by small companies that cannot capitalize on economies of scale. Photovoltaic expert Jack Stone at the National Renewable Energy laboratory in Golden, Colorado, points out that photovoltaic technology is experiencing a catch-22. Small companies cannot justify building large plants without market demand, but market demand, in turn, is thwarted because prices are too high.

Photovoltaics could provide Americans with much of their electric needs. Theoretically, all of the electric power required by the United States could come from 12,000 square kilometers of solar cells—roughly the area occupied by all buildings in the lower 48 states. Brown University professor Joseph Loferski estimated that photovoltaics mounted on only 20 percent of Rhode Island's rooftops would provide all of the electric power needs of the state. The electric power would be generated only a few feet from where it would be used, thus removing the need for costly, inefficient, and unsightly transmission lines.

Photovoltaics do have some drawbacks. Sunlight is inconsistent, and some kind of storage system would be necessary. Batteries are the common means of storing electricity. While a PV array might last 30 to 50 years, batteries have to be replaced about every 7 to 10 years, depending on how well they're taken care of. Backup systems such as gasoline-powered generators are also needed to provide energy for long, sunless periods, although some people use wind generators to provide additional power in such instance. Some photovoltaic production practices also generate hazardous wastes.

Despite these problems, there are areas where solar electricity is thriving. In the less developed countries, for instance, solar electricity is proving to be a valuable source of electricity for thousands of villages located many miles away from power plants. What government officials in these countries have found is that it is far cheaper to import this technology that is too expensive for widespread use in the West than to string power lines to remote villages. Today, the lessdeveloped nations are the world's leading market for solar electricity. Solar electricity is also popular in remote areas of more developed countries, and if homes are being built more than half a mile away from a power line, it is cheaper to install a good solar electric system than to string wires.

Despite its difficulties, solar electricity is the world's second fastest growing source of energy, second to wind power. In the 1990s, sales of PVs increased on average about 16 percent per year. Although the total capacity of the world's photovoltaics is small. Another interesting fact about solar electricity that foreshadows its success is that Solarex, once one of the world's leading suppliers of PVs, is now owned by British Petroleum, one of the world's largest suppliers of oil. They're vigorously pursuing sales and have built 200 gas stations in Europe and North America that are powered by PVs. Dutch Royal Shell has committed to spending $500 million on renewable energy.

Our Solar Future

By supporting renewable energy sources and stepping up conservation, governments can become a positive force in the transition to a sustainable society. "The transition to the solar future," says Professor Kurt Hohenemser of Washington University, represents "a gigantic technological enterprise" that could require a century to complete. It would be much easier to accomplish if we started now, while we still have cheap fossil fuels. "The question," he continues, "should not be how to extend our growth-based industrial and economic system a little longer by still faster extraction of our finite resources, but rather how to spend some of our remaining resources wisely to accomplish a smooth transition to a solar-based society."

Geothermal Energy

The Earth stores enormous amounts of heat, or **geothermal energy,** which comes from the radioactive decay of naturally occurring radioactive substances in the Earth's crust and from molten rock in its interior (Figure 22.17). In some places, groundwater heated by the Earth's interior spews to the surface in remarkable displays called **geysers.** In other places the water merely bubbles up, filling pools (hot springs) or trickling into nearby streams. These regions are called **hydrothermal convection zones.** In other regions, heated groundwater may be trapped by impervious rock layers. These sites are called **geopressurized zones.** Heated by the molten rock underneath, the steam and superheated water can only be tapped by drilling holes into the pockets of steam. In still other areas, magma heats overlying rock, forming **hot rock zones.** Recent studies show that water can be pumped into these zones, then pumped out, where the energy is used for heating or electric power generation.

Most geothermal energy today comes from hydrothermal convection zones because they are the easiest and cheapest to tap (Figure 22.18). Steam or hot water from these zones can be used to heat all sorts of buildings. For example, almost all of the houses in Reykjavik, the capital city of Iceland, are heated by geothermal steam. Icelanders grow a variety of vegetables in steam-heated greenhouses as well. In the United States, at least 300 communities use geothermal heat for one reason or another. One rancher in South Dakota uses it to heat his home and several buildings. He also uses it to dry several thousand bushels of grain each year. All told, geothermal energy saves him over $5,000 a year in fuel bills.

Geothermal steam or hot water can be used, as mentioned earlier, to generate electricity. First used successfully in 1904 in Larderello, Italy, geothermal electricity is now produced in New Zealand, Japan, Mexico, the United States, the Philippines, Italy, Iceland, and the Russia. One of the world's largest projects is in northern California on the slope of an extinct volcano. Started in 1960, the project has a 900-megawatt capacity, supplying nearly a million people.

All in all, however, geothermal energy produces very little electricity worldwide. The total global capacity is now only about 6,000 megawatts, equivalent to six large nuclear power plants, but increases are predicted. According to one estimate, the United States alone could produce 27,000 megawatts of electricity from geothermal sources—enough electricity for 27 million people, or about one-tenth of the U.S. population.

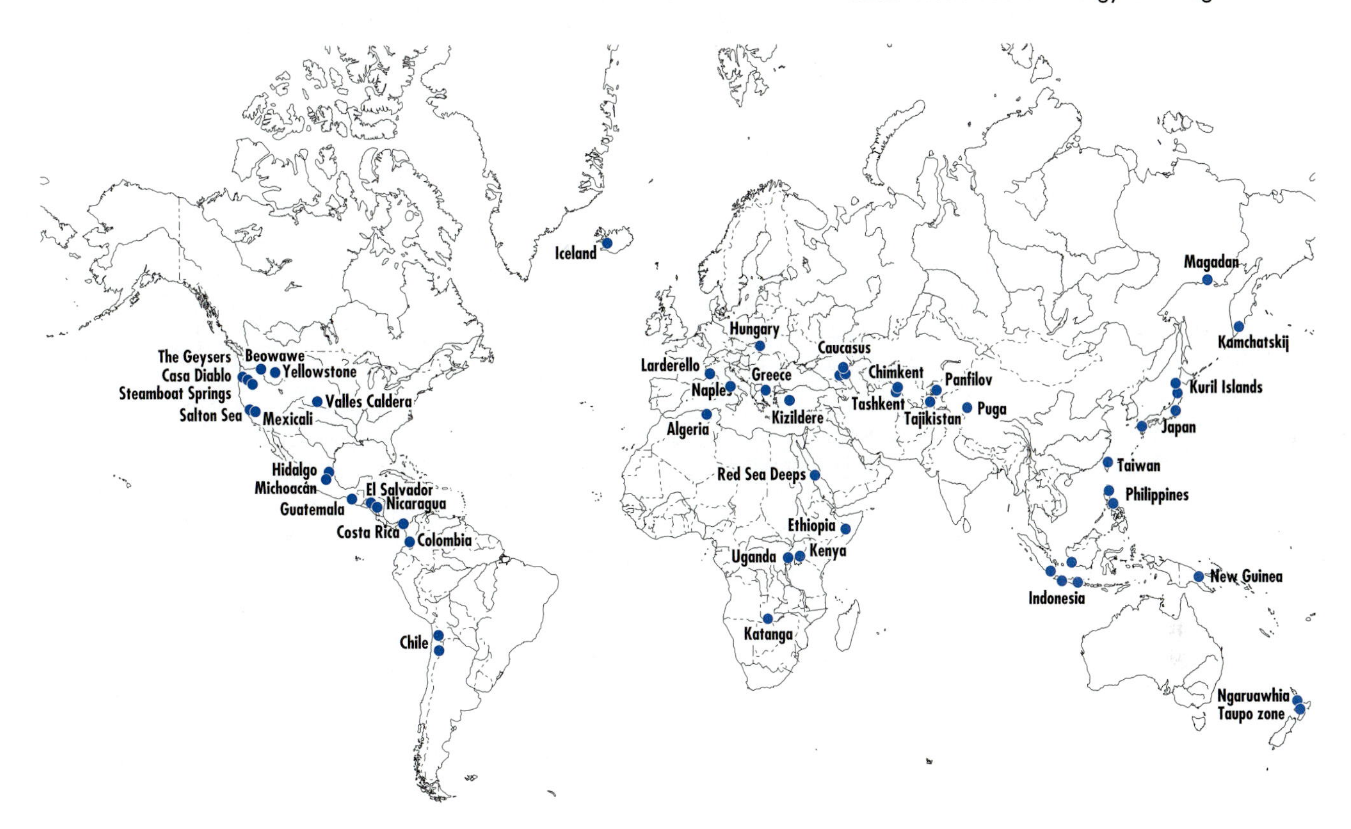

FIGURE 22.17 *Map of geothermal areas in the world. Most are located in regions of volcanic activity, past or present.*

Geothermal power is relatively inexpensive and much cleaner than coal-fired and nuclear power plants. Capital investment—the money needed to start a plant—is 40 percent less than for a coal-fired power plant and 300 percent cheaper than for a nuclear power plant. However, there are some drawbacks. First, minerals dissolved in the steam can corrode pipes and turbine blades. Second, geothermal plants can be very noisy. Third, geothermal plants release gaseous pollutants such as hydrogen sulfide, carbon dioxide, ammonia, and methane that often occur in the steam and require pollution control devices. Fourth, steam cannot be transported long distances without losing its heat. Finally, most of the hydrothermal convection zones in the United States are located along the Pacific Coast. The highly populated East Coast, where the demand for energy is high, would have to rely on hot rock zones, which are found throughout the United States but are more costly and difficult to tap.

Hydropower

Humans have tapped the power of flowing rivers, or **hydropower,** for thousands of years. In the United States, hydropower was used extensively in the 1800s to grind wheat and corn, saw logs, power textile mills, and water cattle. Today, hydropower is often used to generate electricity via turbines through which water flows, usually from the bottom of a dam. Hydroelectricity provides about 10 percent of America's total electricity, or about 4 percent of the nation's total energy consumption (Figure 22.19).

Hydropower offers several advantages: (1) it is relatively inexpensive, (2) it is pollution free, and (3) it is a potentially renewable source, provided reservoirs can be kept from filling with sediment. According to some estimates, enormous hydropower resources still lie untapped. Today, only about 17 percent of potential global hydropower is being used. You might assume, therefore, that when the supply of fossil fuels falls so low that it becomes too expensive, the world could turn to hydropower. Unfortunately, it is not that simple. Several problems cloud the prospects for hydropower.

First, the greatest hydropower potential lies in less developed countries in Africa, South America, and Asia. Because many of them lack financial and other resources, their hydropower potential is not likely to be developed. Second is the increasing cost of dam construction, with price tags ranging from $300 million to $2 billion, makes it increasingly less cost-effective. In addition, dams often have a short life span because of heavy sedimentation caused by poor land management—deforestation, overgrazing, and poor farming practices. This problem threatens many projects. For example, Egypt's Aswan Dam, the world's largest, probably will have a functional life of less than 200 years. In the United States, over 2,000 reservoirs have been filled with sediment, and some succumbed in fewer than 20 years. Pakistan's $1.3 billion Tarbella Dam, which took 9 years to build, could be filled by sediment in 20 years. Third, dams destroy the scenic beauty of wild canyons, used by anglers, kayakers, rafters, canoeists, and a variety of others. Fourth, reservoirs behind dams inundate forests, farmland, wildlife habitat, and habitat of indigenous human populations. Fifth, dams reduce

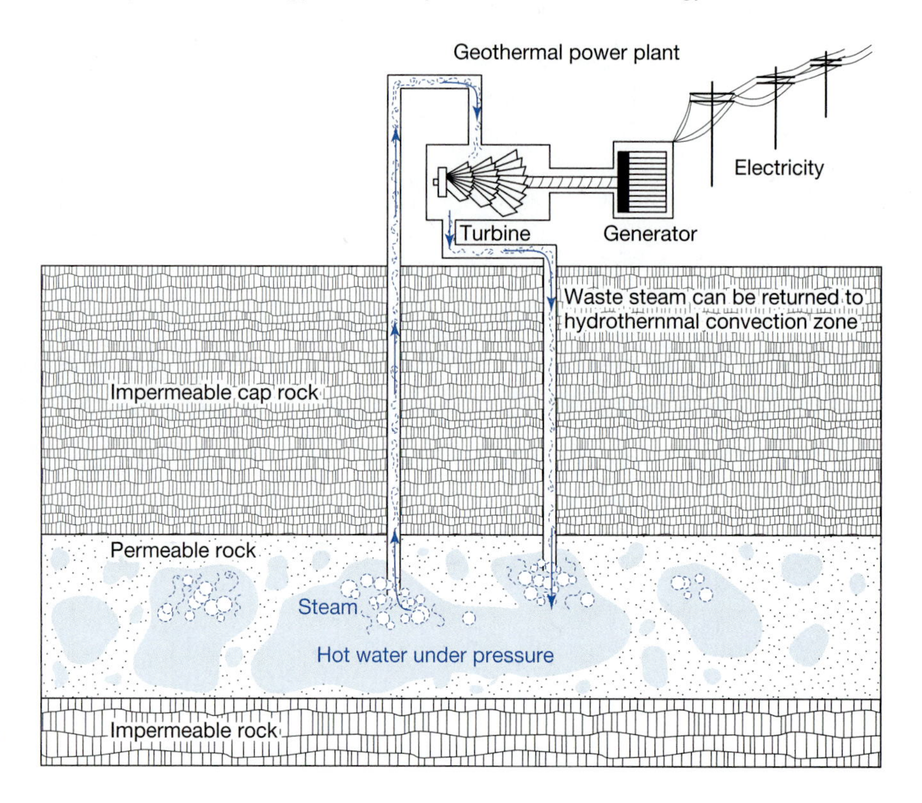

FIGURE 22.18 *Schematic view of a geothermal power plant operation.*

the natural flow of sediment bearing nutrients to estuaries and disrupt aquatic food chains in these ecologically and economically important zones. Sixth, dams can interfere with the migration of fish, such as salmon, and fluctuating levels can kill fish eggs laid in shallow water near shore. Seventh, in arid regions reservoirs accelerate evaporation. In some cases evaporation reduces the amount of the available irrigation water.

Despite the disadvantages, hydropower is here to stay, and its use is likely to increase, although only modestly, in the near future. But the generous estimates of untapped potential should be viewed cautiously. For instance, the United States currently generates 70,000 megawatts of energy from hydroelectric power and, according to estimates, has an additional 160,000 megawatts of untapped potential. However, this untapped energy lies in out-of-the-way places where dams may be economically and environmentally unfeasible. In fact, half of our untapped potential lies in Alaska, far from industrial and population centers. Canada generates 12 percent of its total energy consumption from hydropower, and has untapped potential, too. However, most of this resource lies in the northern reaches of this huge country far from Canada's population, which is concentrated along the southern portion of the country near the U.S.-Canada border. Much of the untapped potential in Africa, Asia, and South America is in remote locations as well.

Most hydropower projections concern themselves with large dams, with little consideration of smaller projects, those that could produce 1 to 10 megawatts of energy. It is these small projects, some proponents argue, that could provide enormous amounts of energy in countries throughout the world. China, for instance, has 90,000 small dams on streams and rivers providing electricity to remote rural villages—equivalent to one

FIGURE 22.19 *Hydroelectric power. Aerial view of Hoover Dam and Lake Mead on the Arizona–Nevada border. This world-famous dam spans the Colorado River. Built in 1935, the dam provides multiple benefits, such as flood protection, water storage for irrigation purposes, and hydroelectric power.*

third of the country's total electric production. France and the United States have numerous small dams as well.

Although these structures are inexpensive to build and operate and provide energy to consumers where it is needed without huge transmission losses, they do significantly alter streams, affecting fish and other aquatic organisms. Some environmentalists believe that a far better strategy would be to retrofit the 50,000 nonhydroelectric dams in the United States with small turbines to generate electricity. The environmental damage, they argue, has already been done, so why not make a dam built for recreation, water supply, or flood control, for instance, do double duty—that is, generate electricity as well?

Windpower

For many decades Great Plains farmers harnessed the winds sweeping across their rolling grasslands and wheat fields with a relatively simple device, the **windmill** (Figure 22.20). Used to pump water, grind grain, and generate electricity to light their barns, the windmill fell out of vogue in the 1930s as rural electric lines began to spring up, connecting remote farms with central power plants many miles away. Soon thereafter, the sight of windmills silhouetted against the blue prairie skies began to fade. In the 1970s, however, windmills began to pop up all across the nation, this time on special wind farms or in the backyards of suburban homes.

FIGURE 22.20 *The windmill of the past. Many-bladed windmills like these were once landmarks on American farms. They were used to pump water. Farmers stopped using them following the successful rural electrification program of the 1930s. Now farmers need alternative power sources to pump irrigation water because of high prices for gas and other fuels.*

Wind energy contributed little to global energy needs 20 years ago. In fact, in 1980 the global capacity was a paltry 600 megawatts. In 1986, however, 13,000 wind turbines in California provided 1,100 megawatts of power—enough electricity for a million people. Another 1,500 megawatts are soon to come on line. Most of these turbines are on specially built wind farms, located on mountain passes and connected to the existing electrical grid. Large wind farms are currently being planned or built in several other states, including Iowa, Minnesota, Montana, Colorado, New York, Maine, Texas, Oregon, Vermont, Washington, Wisconsin, and Wyoming.

Today, the U.S. wind energy industry produces enough energy to supply electricity to 2.5 million people. Global capacity, including that in the United States, is about 9,600 megawatts, or enough to supply nearly 10 million people, and up sharply in the past decade (Figure 22.21). In fact, wind is the fastest growing source of energy in the world, bar none!

Wind farms have also been built in Spain, Germany, Denmark, India, and the Netherlands. The Worldwatch Institute predicts that wind energy could provide 20 to 30 percent of the electricity needed by many countries. And some experts believe that wind pumps—similar to the windmills of the past—could be used in remote rural villages of India and Africa to provide drinking and irrigation water.

Many versions of the windmill are currently under study, ranging from huge units with 20-meter blades to tiny backyard wind machines that provide electricity for individual homes (Figure 22.22). One large wind generator in Boone, North Carolina, for example, sits atop a 40-meter (140-foot) tower. Its huge blades rotate 35 times a minute in a 50-kilometer (30-mile)-per-hour wind; the wind generator supplies electricity

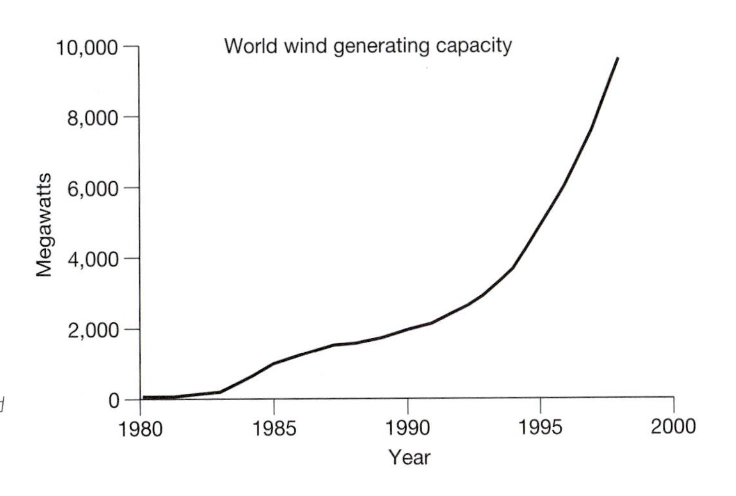

FIGURE 22.21 *Growth in wind energy. Wind energy is the world's fastest growing source of energy!*

FIGURE 22.22 *Experimental wind generator near Bushland, Texas, produces electricity from wind. This unit is 50 meters high and can produce about half the electricity consumed in the Bushland community when the winds are blowing.*

to 500 homes. Many companies are now manufacturing smaller wind turbines They're used for homes and farms.

Wind energy comes free of charge, is clean, and is renewable. It requires little land and does not preclude many other uses, such as grazing. Electricity from wind energy (wind farms in particular) currently costs 5 to 8 cents per kilowatt-hour and is therefore currently competitive with coal (5 to 7 cents) and cheaper than electricity from nuclear power (10 to 12 cents per kilowatt). Moreover, wind resources are fairly large in many parts of the world. In the state of Holstein, Germany, for example, wind satisfies 15 percent of the total electrical demand. In the United States, North Dakota, South Dakota, and Texas have sufficient wind capacity to provide electricity to the entire country.

But wind power is not a panacea. It has its own unique set of problems. Large generators are unsightly and can interfere with radio and television reception as well as with microwave transmission by telephone companies. Further, materials tend to wear out, necessitating replacement. Making matters worse, winds are not constant, making some means of storing electricity mandatory. However, electric storage is not well developed on a large scale. Nevertheless, given the expected rise in the cost of electricity from conventional power plants, wind could become a major source of electricity in the not too distant future.

Biomass

Wood, manure, crop wastes, and other forms of organic material are known as **biomass.** When burned to make energy they're sometimes called **biomass energy sources.** Together, biomass supplies about 14 to 19 percent of the world's energy. In many less developed nations, where fossil fuel consumption is still low, biomass may provide up to 90 percent of the total energy demand.

What makes biomass so attractive is that it is renewable, helps cut down on wastes, is often inexpensive, and does not add to the global carbon dioxide problem. Furthermore, biomass can be burned directly or converted into liquid and gaseous fuels, giving it a wide range of applications. Biomass fuels are also labor intensive—they require many workers to produce them. In the less developed nations, where employment is a major problem, labor-intensive fuels could provide work and income for people currently out of work.

Wood Wood is the most widely used form of biomass. In more developed countries, such as the United States, Norway, and Sweden, wood supplies about 10 percent of the home heating fuel. In the United States, 3 to 4 percent of the total energy consumed each year comes from wood. In Canada, it provides about 6 percent. Most of this wood is burned by the wood and wood products industry, hard hit by the high cost of fossil fuels in the 1970s.

Unfortunately, wood is a somewhat dirty fuel. Burned in wood stoves, it becomes a major source of particulates and other pollutants in many of America's urban areas. Now, many areas that do not meet National Ambient Air Quality standards restrict wood burning on high-pollution days. Some restrict the number of wood stoves and fireplaces, and some counties even require special EPA-rated woodstoves with catalytic converters or other design innovations that result in drastic decreases in pollution.

In the lessdeveloped nations, where as many as 2.5 billion people depend on wood for cooking and heating, shortages are becoming widespread. A report by the Food and Agriculture Organization of the United Nations, for instance, noted that 1.3 billion people are meeting their needs for wood by depleting existing supplies, cutting trees down faster than they can be replenished. Two thirds of these people live in Asia, especially near the Himalayas. Many of the rest live in arid parts of Central Africa and the Andean plateau of South America. In Africa, women and children may travel 50 kilometers (30 miles) a day in search of fuel wood. Depletion of wood creates enormous human suffering and also leads to widespread ecological damage like erosion, desertification, flooding, and habitat destruction.

What can be done? First and foremost, population growth must be curbed. Second, forests can be planted near villages and managed to produce a sustained yield. In the Philippines, for instance, the government launched a program to replant marginal rural land with trees to provide fuel wood for numerous small power plants aimed at providing outlying villages with electricity. Third, new, more efficient cooking stoves and heating stoves can be used by villagers to cut demand. Fourth, substitute energy sources such as wind energy, photovoltaic cells, and small hydroelectric power generators could be developed with the assistance of the United Nations.

Wood can also be burned more efficiently in more developed countries. Pellet stoves, which burn wood pellets made from sawdust from saw mills, burn hot and efficiently. Better yet are **masonry stoves** (Figure 22.23a). A masonry stove is a high-mass wood stove. Made from brick or stone, these devices burn wood in a highly insulated combustion chamber—so hot that it results in a near perfect burn. This eliminates unburned gases and prevents creosote from building up in the chimney. Creosote is a black, gooey material formed from volatile chemicals released when wood is burned. They deposit in flue pipes and in chimneys and can catch on fire. Because these gases are burned in a masonry stove, more heat can be wrung from the wood. Because creosote does not accumulate in the chimney, the chances of a chimney fire are eliminated. The other secret of a masonry stove is that it has high mass. Unlike conventional wood stoves made of steel, the thick masonry walls absorb heat given off by the fire and radiate slowly into the room. No baking in your easy chair while the woodstove is on! In addition, unlike conventional wood stoves where the heat escapes right up a straight pipe, the flue of a masonry stove is a labyrinth—that is, an elaborate, elongated pathway for hot gases to escape. As the hot gases escape, the heat is transferred to the masonry. Because of this, much of the heat is captured and radiated into the room (Figure 22.23b).

Other Biofuels Wood is only one of many forms of biomass that can be used to produce energy. Garbage can be incinerated to produce steam heat and electricity(Chapter 17). Manure, human wastes, and other organic refuse can also be used to produce methane gas, which is combustible. When mixed with water and heated in a closed container, organic material is degraded by anaerobic bacteria. Methane given off in the process is then burned to produce heat and electricity. In fact, many sewage treatment plants here and abroad now capture methane that was once vented to the atmosphere; they burn it and use the heat for offices or to generate electricity needed by the plant, and often sell the excess to local utilities. Plant material can also be used to generate ethanol, a liquid fuel that can be mixed with gasoline in a ratio of 1:9 and burned in automobiles. Called **gasohol**, it results in a more complete combustion and fewer pollutants. Alcohol can also be burned without dilution in specially designed cars and in factories. Most ethanol currently comes from corn, but virtually any crop material when properly fermented can produce alcohol. Brazil uses sugarcane to produce alcohol and cut its imports

(A)

(B)

FIGURE 22.23 *Masonry stoves are efficient and clean. You can heat a good sized room with a single burn. (A) Photo of masonry stove. (B) Drawing showing the interior. Note the flue gases must pass through an extensive labyrinth to escape. In the process, they lose much of their heat, which is absorbed by the masonry, then radiated slowly into the room.*

of foreign oil. It is now a major source of fuel for Brazilian trucks, cars, and busses.

Fuel farms could provide much of America's liquid fuel in the coming years. Excess wheat and other crops, for instance, could be used to produce substantial amounts of liquid fuel. As many developing nations become self-sufficient in food production, countries like Canada, Australia, and the United States, which currently export large amounts of food, could become major producers of ethanol fuel.

As promising as it may seem, ethanol production has a major obstacle that we must be overcome: The net energy yield of ethanol production is about zero. In other words, you get about as much energy out of it as you put into it. By using plants with a higher photosynthetic efficiency and by using solar and wind energy to operate the distillery, however, the energy output could be increased to make this process cost-effective.

Advances in ethanol production from waste wood also hold promise. At this writing, ethanol can be produced at a cost of $1.27 per gallon, cheaper than ethanol from corn. Further advances in this technology could drop the cost to 70 to 80 cents per gallon, making it competitive with oil at $25 per barrel. When one considers the external economic costs of oil, ethanol from wood wastes and other sources may be quite a bargain.

Hydrogen and Fuel Cells

Hydrogen gas is another potentially useful fuel that could someday be substituted for gasoline, jet fuel, and natural gas. Derived from water, hydrogen is one of the most abundant fuels available to humankind.

Hydrogen production is simple but rather costly. Hydrogen can be generated by passing an electric current through water in the presence of certain catalysts. Heat or sunlight can also break water down as long as appropriate catalysts are present.

Hydrogen gas burns readily in air, combining with oxygen to form water. In its favor, hydrogen combustion does not produce particulates, carbon dioxide, and sulfur oxides, those undesirable products of fossil fuel combustion. Instead, hydrogen combustion yields three products: energy, water, and a small amount of nitrogen dioxide, produced because the heat causes oxygen and nitrogen in the air to combine.

Unfortunately, in most hydrogen-producing processes, the net energy yield is negative, making production costly. Hydrogen is fairly explosive as well. And to be used in automobiles it would have to be liquefied. One way of doing so would be to cool it to very low temperatures, but this process consumes additional energy. Another is to pump hydrogen into a tank containing a metal hydride. This substance acts like a sponge, holding hydrogen and releasing it on demand. Unfortunately, the tanks don't hold much and they are heavy, which can be a disadvantage in a car.

Because of recent technological advances, hydrogen has become a cost-competitive source of energy in remote locations where electricity is expensive and storage is necessary. Further advances, say scientists, could make hydrogen a cost-effective substitute for environmentally damaging fossil fuels.

Hydrogen can also be pumped into a new device known as a fuel cell. Fuel cells produce electricity from hydrogen and oxygen. They look a lot like batteries, but they produce electricity, rather than store it. Electricity from fuel cells can be used to power generators, busses, and cars. The cities of Chicago, Illinois and Vancouver, British Columbia, Canada both have a limited number of fuel cell powered busses.

Automobile manufacturers are experimenting with fuel cell cars and several prototypes are currently being tested. In these models, however, hydrogen is generated from gasoline in a device known as a **reformer.** It strips hydrogen from the gasoline molecules and results in carbon dioxide emission. To be truly sustainable, advocates argue that we need to generate hydrogen from renewable sources, notably water, but the problems of storing hydrogen gas still seem to stand in the way of this approach.

Tidal Power

President Franklin Roosevelt often watched the rise and fall of the tides near his summer home on the Bay of Fundy. He was impressed with the potential power of the surging waters, for the tides of Fundy are the largest on Earth—up to 16 meters (50 feet). He wondered, no doubt, if the energy of those tides could somehow be captured, perhaps by finding a way to channel their energy through turbines similar to those used in power plants, where spinning blades produce electric energy (Figure 22.24).

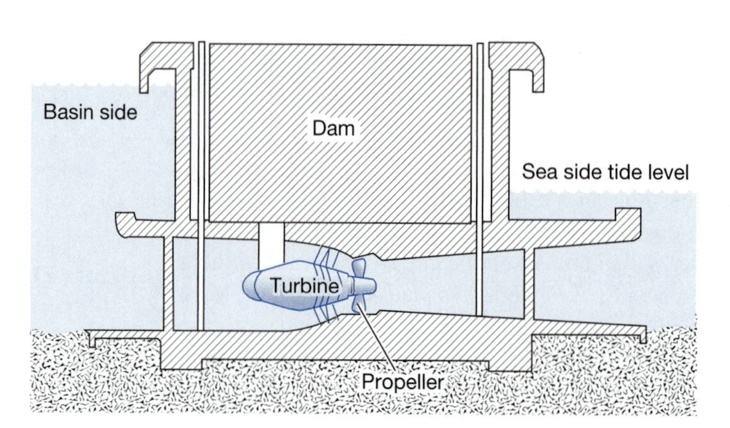

FIGURE 22.24 *In a tidal power plant, water flows through small openings in the dam, spinning a propeller that generates electricity.*

The world's first tidal-electric installation was built in 1966 in the La Rance estuary in France. It has a capacity of about 240 megawatts, equivalent to a small coal-fired power plant. A similar plant has been constructed in Russia. The power plant channels the tides through openings in a massive dam built at the mouth of the river. The water spins an underwater turbine, generating electricity.

Tidal power has many advantages. It is relatively inexpensive to tap, produces no toxic wastes or pollutants, and is renewable. On the downside, there are only about 24 good sites worldwide, and they would meet only about 5 percent of the United States' current electrical demand. The dams also impair ship travel and could upset aquatic life. Like hydropower, this technology may be of limited value in forging a sustainable future.

Ocean Thermal Energy Conversion

The ocean is a vast storehouse of heat, which could be tapped by specially designed **ocean thermal energy conversion (OTEC)** plants. These facilities exploit the temperature difference between warm surface waters and cold bottom waters in tropical areas. Each OTEC plant consists of a floating platform with enormous pipes that extend to a depth of 900 meters (3,000 feet). On the surface platform, ammonia, which boils at a relatively low temperature, is circulated through a series of tubes. The warm surface waters convert the ammonia into a gas that drives the blades of a turbine that drives a generator (Figure 22.25). The gas is then recondensed by cool water pumped up from the depths to start the cycle over again.

Electricity produced by OTEC plants can be transmitted to shore. Some plants could use it to desalinate salt water, thus providing drinking water for local populations.

OTEC plants could theoretically operate day and night in tropical waters, providing energy for nearby cities and towns. However, like many other technologically feasible ways to generate energy, this one has some problems. First is the question of efficiency. Because large amounts of water must be used, a considerable amount of energy must be spent to pump water to the surface. An experimental plant off the coast of Hawaii, for example, used 80 percent of the electricity it generated to pump water, giving it a low net energy yield. Second, suitable locations for OTEC plants are limited to a few places, such as the Gulf Coast and the waters off Hawaii, Guam, and Puerto Rico. These plants would also release considerable amounts of carbon dioxide into the atmosphere as the cold waters in which this gas was once dissolved are heated at the surface. Cooling surface waters could affect fisheries and local climate, increasing rainfall. Released on the surface, deeper waters, which are rich in nutrients, could also stimulate the growth of phytoplankton, upsetting the ecological balance.

No greater challenge exists for a democratic nation besieged with pressing problems than to sort out among the frantic rush of minor crises the long-term problems that could potentially cripple it—and to act on them. Energy is one of those problems. Supplies of several critical fossil fuels are limited, and fossil fuels are at the root of some of our most threatening environmental problems. In fact, it could be argued that our heavy dependency on fossil fuels is at the root of the current crisis of unsustainability.

At no time in our history have present actions been so important. Faced with impending oil shortages and a massive buildup of greenhouse gases, many observers believe that we must revamp our energy system to build a sustainable energy system. By combining energy efficiency and renewable energy with other important measures such as population control, recycling, and restoration of habitats, we can steer onto a sustainable path. The responsibility belongs to all of us, and to each of us. As Marshall MacLuhan once wrote, "there are no passengers on planet Earth, we are all crew." And lest we forget, good planets are mighty hard to find.

Summary of Key Concepts

1. Throughout history, satisfying the demand for future energy needs has resulted in methods of producing more fuels

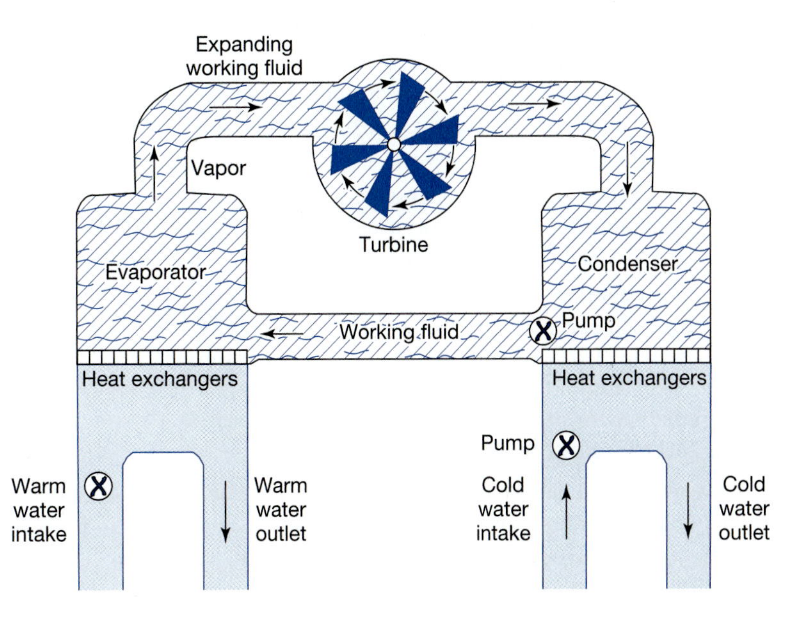

FIGURE 22.25 *Schematic diagram of a closed-cycle ocean thermal power plant.*

or developing alternatives. Very little attention has been given to finding ways to reduce energy consumption.

2. Nevertheless, conservation has created some enormous fuel savings. The average gas mileage of a new car sold in the United States, for instance, increased from 14 miles per gallon in 1974 to 26 miles per gallon in 1986 to around 28 mpg in 1995. Today it is around 28 miles per gallon. Over the same period, factories cut their energy use as well; the amount of energy needed to produce a dollar of gross national product dropped 25 percent. Between 1949 and 1999, efficiency measured by the amount of energy required to produce one dollar of GNP, has increased by 47 percent
3. Despite what many think, most of the world's people have only begun to tap the full potential of energy conservation. Enormous opportunities to improve efficiency exist in buildings, industry, motorized vehicles, appliances, and electronic devices.
4. Energy conservation makes good economic sense for consumers and producers. For example, improving the efficiency of machines and appliances costs about 1 to 2 cents per kilowatt-hour of energy saved, one-fifth of what it costs to produce electricity from coal and one-tenth of what it costs to produce electricity from nuclear energy.
5. Energy conservation will require actions by citizens, businesses, and governments.
6. Individuals can help by turning down their thermostats and wearing more clothes in the winter, caulking and weather stripping, turning down the setting on water heaters, insulating water heaters, driving at or below the speed limit, car pooling, taking mass transit, and buying energy-efficient vehicles.
7. Another important step required for the development of a sustainable energy system is the installation of clean, renewable alternatives to fossil fuels. The most likely renewable candidates include solar energy, wind energy, biomass, hydropower, and geothermal energy.
8. The sunlight that strikes an area the size of Connecticut each year could provide all of the energy needed by the United States, yet Americans have tapped only a tiny portion of the sun's potential.
9. Solar energy can be used to heat water for homes and industries, and to heat building interiors. Two major heating systems are in use today, active and passive.
10. Active solar systems consist of solar collectors that absorb sunlight and convert it to heat. The heat is then carried away and used to heat building interiors or water.
11. Passive solar systems are designed strictly for space heating. The building itself becomes a collector and heat-storage device. South-facing windows and skylights let the winter sun penetrate the interior of the house. Sunlight strikes the walls and floors and is converted into heat. Cement or block walls store the heat and radiate it out into the room at night.
12. Solar energy can also be used to generate electricity. Sunlight may be concentrated by mirrors to boil water to run an electrical turbine. This technology is called solar thermal electricity.
13. Electricity may also be generated by photovoltaic cells, thin wafers of silicon and other materials that, when struck by light, produce an electrical current. Solar energy trapped by photovoltaic cells could become a major source of energy as fossil fuels run out, but the cells are too highly priced now for routine use. Solar cells, however, make good economic sense in remote sites. More efficient solar cell and mass production would make them competitive with other forms of electrical generation.
14. The federal government currently provides approximately $50 billion in subsidies to the nuclear, coal, oil, and gas industries but almost no support to developing alternative energy sources. To build a sustainable future, we must level the playing field—either provide subsidies for renewable energy resources or eliminate subsidies for nonrenewables.
15. The Earth stores enormous amounts of heat, or geothermal energy, which comes from the radioactive decay of naturally occurring radioactive substances and from molten rock in the Earth's interior.
16. Most commercial geothermal energy today comes from hydrothermal convection zones, where heated underground water is ejected to the surface, emerging as geysers or hot springs. Steam or hot water from hydrothermal convection zones can be used to heat buildings and generate electricity. Despite the ease with which it can be used, global geothermal capacity is small.
17. Hydropower makes a significant contribution to global energy and is bound to increase in the future. Unfortunately, much of the untapped hydropower is in regions far from population centers and industry. Some environmentalists believe that retrofitting existing nonhydroelectric dams with small generators could provide additional electricity without the environmental impacts created by building new dams.
18. Wind energy is contributing more and more energy to global electricity needs. In fact, wind is now the fastest growing source of energy in the world with massive resources in many of the more developed countries. Wind energy may also provide electricity to rural villages in less developed nations where large electric generating plants are unfeasible.
19. Wood, manure, crop wastes, and other forms of biomass supply 14 to 19 percent of the world's energy demands. In less developed nations, where fossil fuel consumption is low, biomass may provide up to 90 percent of the total energy needed.
20. Wood is the most widely used form of biomass. In developed countries, like Norway and the United States, wood supplies about 10 percent of the home heating fuel. But most of the wood in these countries is burned by the forest products industry.
21. In developing nations, 2.5 billion people depend on wood as their primary fuel. Over half of them are depleting local wood reserves faster than they can be replenished.
22. Other forms of biomass can also be used to produce energy. Manure, human wastes, and other organic wastes, for instance, can be used to produce methane gas. Plant material can be used to generate ethanol. Fuel farms that pro-

duce crops to make ethanol could provide much of America's liquid fuel in coming years if the net energy yield of the process can be improved.

23. Hydrogen gas, produced from water, is a renewable fuel that burns very cleanly in cars and in stoves. Unfortunately, hydrogen is costly, explosive, and difficult to store.
24. The tides can be tapped by special dams that channel the water through openings equipped with turbines. Unfortunately, the global capacity of tidal energy is rather small.
25. The ocean is a vast storehouse of heat, which could be tapped by specially designed ocean thermal energy conversion plants. They use the warm surface waters to vaporize ammonia gas, which then runs an electric turbine. The gas is cooled by cold bottom waters that are pumped to the surface. Unfortunately, the energy yield of OTEC plants is poor, making the electricity uncompetitive.

Key Words and Phrases

Active Solar Systems
Biomass
Conservation
Fuel Farms
Gasohol
Geopressurized Zones
Geothermal Energy
Hot Rock Zones
Hydrogen
Hydropower
Hydrothermal Convection Zones
Microwaves
Ocean Thermal Energy Conversion
Passive Solar Systems
Satellite Solar Power Station
Solar Electricity
Solar Energy
Solar Photovoltaics
Tidal Power
Wind Farm
Windmill
Windpower

Critical Thinking and Discussion Questions

1. Make a list of the conventional energy sources and describe each one. Write a list of criteria to judge the sustainability of all potential energy resources. Now study your list of energy sources and rate each one for its sustainability.
2. What is an active solar energy system? What is a passive solar energy system? How are they similar? How are they different?
3. Describe two methods by which electricity can be generated from sunlight.
4. An engineering professor has discovered a new source of energy. Drawing on your critical thinking skills, what would you want to know about it before promoting it?
5. What is geothermal energy? What form of geothermal energy is the most readily accessible?
6. Using your critical thinking skills, debate the statement: "The United States has vast untapped hydropower potential that could produce enormous amounts of electricity."
7. Describe the potential of wind energy. Is it competitive with conventional fuel sources? What are the problems associated with wind energy, and how can they be overcome?
8. What is biomass energy? How important is it to our energy future?
9. Describe the pros and cons of tidal power, hydrogen as an energy source, and creating energy from ocean thermal energy conversion.
10. Do you agree with the following statement? "Conservation is no less an energy alternative than oil, gas, coal, and nuclear energy and is often much cheaper."
11. Make a list of ways you can cut your energy consumption by 25 percent.
12. You are appointed head of your city's energy department and asked to devise a long-term energy strategy that will carry your city well into the middle of the next century. Prepare your plan and defend it.
13. Make a list of criteria to judge the sustainability of all potential energy resources.
14. Debate the statement, "Electricity from coal is cheaper than solar thermal electricity."

Suggested Readings

Chiras, D. D. 1992. *Lessons From Nature: Learning to Live Sustainably on the Earth.* Washington, D.C.: Island Press. Describes steps to create sustainable energy and transportation systems.

Chiras, D. D. 2000. *The Natural House: A Complete Guide to Healthy, Energy-Efficient, Environmental Homes.* White River Junction, VT: Chelsea Green. Describes 14 natural building techniques and sustainable systems, including passive solar, photovoltaics, and wind energy, for home use.

Cole, N., and Skerret, 1995. P. J. *Renewables are Ready: People Creating Renewable Energy Solutions.* White River Jct.: Chelsea Green. Great collection of case studies.

Davidson, J. 1987. *The New Solar Electric Home.* Ann Arbor: aatec Publications. Great reading for anyone interested in photovoltaics.

Dunn, S., and C. Flavin, 2000. "Sizing up Micropower." In *State of the World 2000.* Starke, L. (ed.). New York: W. W. Norton. A critical examination of renewable energy.

Flavin, C., 1995 "Harnessing the Sun and the Wind." In *State of the World.* Starke, L. (ed.). New York: W. W. Norton. Important reading.

Flavin, C., and N. Lenssen, 1994. "Reshaping the Power Industry." In *State of the World 1994.* Starke, L. (ed.). New York: W. W. Norton. Important insights into changes under way in the power industry, especially those related to energy conservation.

Gipe, P. 1993. *Wind Power for Home and Business: Renewable Energy for the 1990s and Beyond.* White River Jct., VT: Chelsea Green. Somewhat technical but very useful introduction to small and medium-sized wind generators.

Johansson, T. B., et al. 1992. *Renewable Energy: Sources for Fuels and Electricity.* Washington, D.C.: Island Press. Detailed analysis of all renewable energy options.

Kachadorian, J. 1997. *The Passive Solar House.* White River Jct., VT: Chelsea Green. A very useful book on passive solar design. Outlines general principles and features an innovative design.

Potts, M. 1999. *The New Independent Home.* White River Jct., VT: Chelsea Green. Full of lots of interesting case studies about people who are living off grid on renewable energy.

Renner, M., 2000. "Creating Jobs, Preserving the Environment." In *State of the World 2000.* Starke, L., (ed.), New York: W. W. Norton. Shows the economic and employment benefits of a renewable energy strategy.

Roodman, D. M. 2000. "Reforming Subsidies." In *State of the World 2000.* Starke, L., (ed.). New York: W. W. Norton. Extremely important reading.

Schaeffer, J. 1999. *Solar Living Sourcebook* (10th ed.) Ukiah, CA. Superb reference! Showcases renewable energy products but also provides a great deal of background information on renewable energy.

Web Explorations

Online resources for this chapter are on the World Wide Web at: **http://www.prenhall.com/chiras** *(click on the Table of Contents link and then select Chapter 22).*

Afterword

Over the course of the past semester, you have learned a great deal about environmental and resource issues. If you've read the entire book, you have learned about ecology, economics, and ethics. You've studied human population growth, agricultural problems, pest control, water management, and water pollution. You've examined fisheries conservation, rangeland management, forest management, plant and animal extinction, and wildlife management. You've studied solid waste, hazardous waste, air pollution, mining, and energy.

Some issues you may have already been familiar with. Others you may have had no idea existed. No matter what your prior understandings were, we are confident that you have come to new realizations. We hope you've developed a deeper and broader understanding of the challenges that lie ahead, and ways to address the many pressing problems.

After reading this book, we hope that you recognize that environmental issues vary. Some are local. Others are regional or statewide. Some are global in scope. Some problems are minor; others may be more severe, even threatening to our future. Many problems have an impact on nonhuman species—in some cases, driving them to extinction.

In the course of your studies, you have learned about the many causes of environmental and resource issues, and the plethora of solutions available to individuals, businesses, and government. You have learned that solving issues often requires a cooperative approach with the participation of many stakeholders working together to find sustainable solutions. In many cases, solutions require international cooperation.

Sustainable solutions are often preventive in nature. Renewable energy such as solar power, for instance, prevents pollution from entering the atmosphere. Crop rotation reduces the need for insecticides, which in turn decreases the potential for pollution of ground and surface waters.

To be most effective, solutions must address the root causes of the problems—including our inefficient use of resources, linear thinking and linear systems design, our heavy dependence on nonrenewable resources, our failure to restore damaged ecosystems, our failure to manage systems for the long term, and our ever-expanding population.

Sustainable resource management seeks ways for us to thrive within the Earth's very real limits—that is, limits in its capacity to supply us with resources and assimilate our wastes. In this book, we've talked about learning to live within the carrying capacity of the ecosystems upon which we depend. Regulations on commercial fishing operations are a good example. They seek to limit harvest so that fisheries can remain productive well into the future. Timber management seeks to establish limits on commercial harvest of trees so we don't deplete our forests and ruin the soils upon which a healthy forest depends.

Sustainable solutions seek to achieve the long-term health and stability of natural systems as well as many human-dominated systems. Sustainable grazing, for instance, is carried out in ways that ensure the long-term health of grassland soils as well as the productivity of forage upon which our livestock depend. Agricultural practices seek to maintain the health of the topsoil, too, so that farmers can produce affordable, nutritious food *ad infinitum* without poisoning nearby waterways or beneficial species such as birds. Air pollution prevention seeks to ensure healthy air in our cities.

Another key in creating sustainable solutions is a shift toward management of entire ecosystems and watersheds. In the past, resource management has suffered from a kind of myopia, dealing with problems in limited ways. As a result, we didn't always devise solutions that could address the complex ecosystems from which many of our resources were derived. Ecosystem and watershed management broadens the scope of our actions and serves to protect entire systems with far better results.

Policies and practices that prevent problems, promote ways to live within limits, and ensure long-term health of ecosystems fully recognize our dependence on the Earth and its resources. They recognize that the Earth and its many and varied ecosystems are the biological infrastructure of our society. They're not frivolous concerns, as some people would like you to believe. Such polices also promote intergenerational equity, a fairness to future generations, and ecological justice, a fairness to the millions of species that share this planet with us.

Ultimately, to be sustainable, solutions must make sense from three perspectives: social, economic, and environmental. The challenge before us is to give equal priority to all three, not to draft policy that favors one over another. Hopefully, what you have found in this book is that there are many sustainable solutions to many problems that do make sense from these three perspectives. Unfortunately, a great many people have not realized the importance of natural systems to human welfare. They persist in old thinking, putting human interests, especially economic pursuit, above all else. You can help address the old thinking, and can become a positive influence on those around you. We need you to help spread the word that the rules of environmental protection and resource management have changed. We can live well and with little impact on the environment, if we seek sustainable solutions.

At the heart of sustainable solutions are five guiding tenets: conservation, recycling, renewable resource use, restoration, and population control. As the operating principles of nature, they helped ensure sustainability of undisturbed ecosystems for thousands of years; they can help us redesign the systems that provide us with the goods and services we require.

To create a sustainable future, we also need a new ethic, a sustainable ethic that recognizes that humans are a part of nature, that the Earth and its resources are limited, and that success comes from cooperation with natural forces, rather than domination and control.

We urge you to take an active part in building a sustainable future. Recycle. Buy recycled products. Use energy and other resources efficiently. Limit your consumption. Walk or ride a bus. When you drive, drive an energy-efficient vehicle. Contribute to organizations that seek to restore and protect the environment. Remember that planet care is the ultimate form of self care.

Glossary

Abortion. The premature expulsion of the fetus from the uterus.

Abstinence. Refraining from sexual intercourse either during key times of woman's menstrual cycle (around ovulation) or until after marriage.

Abyssal Zone. The bottom zone of the ocean, characterized by darkness, close to freezing temperatures, and high water pressures.

Accelerated Erosion. Wearing away of the land surface primarily as a result of human activities or, in some cases, animal activities. Accelerated erosion operates at a much faster rate than geological erosion.

Accelerated Eutrophication. Accelerated increase in the concentration of plant nutrients in water bodies caused by human actions, for example, excess fertilizer application on farmfields.

Acid Mine Drainage. Sulfuric acid produced by underground coal mines in areas with high levels of iron pyrite in the soil.

Acid Precipitation. Deposition of acids in rain, snow, mist, and fog.

Acid Precursors. A chemical such as sulfur dioxide that gives rise to acids (in this case, sulfuric acid) after reacting with other chemicals in the atmosphere.

Acid Rain. Rain that has a lower pH than "normal" rain; in other words, lower than pH 5.7. It is caused by the release of oxides of sulfur and nitrogen into the atmosphere.

Activated Sludge. The solid organic waste that has been intensively aerated and "seeded" with bacteria (in a secondary or tertiary sewage treatment process) to promote rapid bacterial decomposition.

Active Fishing Gear. Devices and methods for fishing that have mobility and are able to seek out and capture the fish or shellfish. Examples include trawl nets, purse seine nets, and large driftnets.

Active Solar System. System that gathers energy from the sun and stores it for heating water or rooms.

Adaptive Management. Scientific management of natural resources in which management strategies may be modified as the results of the assessment of management tools become known.

Additive Mortality. The concept that one kind of death of an animal population simply adds to the deaths caused by other factors.

Adsorption. The attraction of ions or compounds to the surface of a solid; for example, the attraction of ions to clay particles in soil.

Age-Structure Diagram (or Population Histogram). Graphical representation of a population according to age and sex. Aids in forecasting population trends.

Agent Orange. Defoliant used by American forces during the Vietnam War.

Aggregates. Soil particles grouped into a single mass or cluster.

Agroforestry. The practice of growing trees among crops or on pasture land.

Algal Bloom. Dramatic increase in algal growth in a lake or stream resulting from high levels of nutrient pollution.

Alpha Particle. A positively charged particle (proton) that is emitted from the nucleus of a radioactive atom.

Alternative Agriculture. Nonconventional approaches to agriculture that include, but are not limited to, organic, ecological, biodynamic, integrated, low-input, or no-till farming.

Altitudinal Migration. Seasonal movement of birds (e.g., grosbeaks and finches) and mammals (e.g., elk and bighorn sheep) up and down mountain slopes.

Ammonification. The process by which the bacteria of decay convert complex nitrogenous compounds occurring in animal carcasses and the excretions of animals, as well as the dead bodies of plants, into relatively simple ammonia (NH3) compounds.

Anadromous Fish. Fish, such as the Pacific salmon, that begin life in freshwater, travel to and mature in the sea, and return to their native stream to reproduce and die.

Animal Unit Month. The amount of forage needed to keep one head of cattle healthy for 1 month.

Annular Ring. A concentric ring, visible in the cross-section of a tree trunk, that is useful in determining the age of the tree.

Anthropogenic Pollutants. Pollution from human sources; for example, the combustion of fossil fuels. Compare with natural pollution.

Antimycin. A toxic substance that has been rather extensively used by fisheries biologists to eradicate carp.

Aquaculture. Cultivation of fish and other aquatic organisms in freshwater or saltwater for food, sport, and other goods and services.

Aquifer. A subterranean layer of porous water-bearing rock, gravel, or sand.

Aquifer Recharge Zone. A region in which surface water seeps into groundwater, replenishing it.

Artificial Insemination. The technique employed by cattle breeders in which sperm from a bull of one breed, such as a Hereford, might be refrigerated and used, over a period of time, to fertilize the eggs of the same breed or other breeds of cattle, possibly from widely separated localities.

Artificial Propagation. The captive breeding of fish, such as in hatcheries.

Artificial Reef. A reef constructed of housing debris, rubble, junked automobile bodies, tires, and so on, frequently placed in relatively shallow water near the coast. A method for increasing the number of breeding sites and providing more cover for marine fish.

Asbestos. A naturally occurring fibrous mineral once used in numerous ways, for example, for sound insulation and heat insulation in buildings. When inhaled, asbestos fibers can cause lung cancer and other diseases, which are most prevalent in smokers.

Aseptic Container. A sterile chamber.

Barrier Islands. Accumulations of coastal sediments parallel to and near the shore created by coastal wave, wind, and current action. These low-elevation islands are found worldwide.

Bathyal Zone. An ocean region of semidarkness in which photosynthesis cannot occur and green plants cannot survive. It is located between the euphotic and abyssal zones.

Beach Nourishment. An expensive human management practice of transporting sand from other sources and depositing it on a beach, temporarily replenishing eroded beach sand.

Bioaccumulation. Concentration of a chemical substance in an organism.

Biochemical Oxygen Demand (BOD). Same as biological oxygen demand.

Biogeochemical Cycle. See elemental cycle. Also known as a nutrient cycle.

Biological Control. Means of controlling pests using natural enemies or other potentially less environmentally harmful measures than chemical pesticides.

Biological Magnification (or Biomagnification). The increase in concentration of a chemical substance in a food web as it passes from lower levels to highest levels.

Biological Oxygen Demand (BOD). Measure of organic matter in water samples. Assesses oxygen used by decomposing bacteria.

Biome. Region of the Earth with characteristic climate and characteristic community of living organisms.

Biotic Potential (BP). The theoretical reproductive capacity of a species.

Birth Rate. Number of births per 1,000 people in a population.

Black Lung Disease. An occupational disease frequently contracted by coal miners.

Blackwater. Refers to waste water from houses containing feces and urine. Kitchen sink water, which carries large amounts of food debris, is also sometimes considered to be blackwater.

Blister Rust. A fungus-caused disease of the white pine that is characterized by the appearance of orange "blisters" on the bark.

Bog. A shallow depression filled with organic matter; for example, a glacial lake or pond basin filled with peat.

Botulism. A waterfowl disease caused by a bacterium and characterized by eventual respiratory paralysis and death.

Breeder Reactor. A nuclear reactor that uses a relatively small amount of uranium-235 as a "primer" to release energy from the much more abundant uranium-238. Produces plutonium-239 in the process.

Brown Lung Disease. An occupational disease frequently contracted by textile workers.

Brownfield. A term used to describe a site that has been previously contaminated

Browse Line. A line delimiting the browsed from unbrowsed portions of shrubs and trees in an area where the deer population exceeds the carrying capacity of the range.

Buffer. A chemical substance that protects soils and water from increases in pH.

Bycatch. The collective term used to describe captured marine organisms, including fish, shellfish, oceanic birds, and marine mammals which are not the target species of the fishery.

Carbon Absorption. A process employed by a tertiary sewage treatment plant in which dissolved organic compounds are removed from the effluent as they pass through a tower packed with small particles of carbon.

Carbon Dioxide. A chemical produced by the combustion of organic materials, including oil, coal, natural gas, and wood.

Carbon Dioxide Fixation. The incorporation of carbon dioxide into glucose molecules during the process of photosynthesis.

Carbon Monoxide. A clear, colorless, odorless gas produced by the incomplete combustion of organic substances such as coal or oil. In high concentrations, carbon monoxide exposure can be lethal.

Carcinogen. A cancer-causing chemical.

Carrying Capacity. The capacity of a given habitat to sustain a population of animals for an indefinite period of time.

Catadromous Fish. A type of fish, such as the American eel, that grows to sexual maturity in fresh water but migrates to the ocean for spawning purposes.

Catalytic Converter. Device attached to exhaust system of automobiles that oxidizes hydrocarbons to carbon dioxide and water and converts carbon monoxide to carbon dioxide.

Catch-and-Release-Only Restrictions. Fisheries management strategy for reducing angler pressure on a particular fish population by requiring the fish to be released (essentially unharmed) back into the water body from which it was caught.

Cellular Respiration. The complete breakdown of glucose in the cells of the body to produce energy needed by various cellular processes.

Central Arizona Project. A multi-million-dollar project to alleviate water shortage problems in Arizona by transporting water from the Colorado River.

Chain Reaction. The sequence of events that occurs when neutrons that have been emitted from a radioactive atom bombard another atom and cause it to emit neutrons that in turn bombard yet another atom, and so on.

Channelization. The process by which a natural stream is converted into a ditch for the ostensible purpose of flood control. Attendant environmental abuse is severe.

Chlorinated Hydrocarbon. A family of nondegradable pesticides such as DDT, dieldrin, and toxaphene. Hydrocarbons may have a harmful effect on nontarget organisms such as fish and birds. They persist for a long time in the environment and undergo biological magnification as they move through food chains.

Chlororganics. Potentially toxic organic compounds that form in water treated with chlorine. A good example is chloroform and carbon tetrachloride.

Clear-cutting. A method of harvesting timber in which all trees are removed from a given patch or block of forest. This is the method of choice when harvesting a stand composed of a single species in which all trees are of the same age.

Climax Community. The stable terminal stage of an ecological succession.

Closed Seasons. Fisheries management strategy that limits the take of a particular fish species at critical times (e.g. spawning, mating season) by not allowing fishing during that particular season.

Closed-Cycle Cooling System. A method of cooling power plants in which the cooling water is continuously recirculated instead of being discharged into a stream and causing thermal pollution.

Coal. An organic mineral produced from plant matter 250 to 300 million years ago. When burned, it produces energy that is harnessed to produce electricity.

Coal Gasification. Production of combustible gas from coal.

Coal Liquefaction. Production of oil from coal.

Co-composting. Mixing of compost with sewage sludge.

Coliform Bacteria. Bacteria that occur in the human gut. The coliform count is used as an index of the degree to which stream or lake water has been contaminated with human sewage.

Command Economy. An economy in which the production of goods is subject to central control, as in Cuba or the former Soviet Union.

Commons. A public area over which no single nation has sovereignty, such as the open ocean.

Community. All species living in a given area. Examples are the community of an oak woods, an abandoned field, or a cattail marsh.

Compensation Depth. The depth in a lake at which photosynthesis balances respiration. This level delimits the upper limnetic zone from the lower profundal zone.

Compensatory Mortality. The concept that one kind of mortality replaces another kind of mortality in animal populations.

Compost. Partially decomposed organic matter that can be used as a soil conditioner and fertilizer.

Condensation Nuclei. Particlates in the air that absorb moisture and can facilitate cloud formation.

Conservation District. The administrative and operative unit of the U.S. Department of Agriculture's Natural Resource Conservation Service. Conservation districts are organized and run by farmers and ranchers.

Conservation Tillage. The practice of restricting plowing of the soil to reduce erosion and leave enough of the previous crop residues so that at least 30 percent of the soil surface is covered when the next crop is planted.

Consumer. A term used for any animal link in a food chain.

Consumptive Use. The removal and alteration of natural resources.

Contour Farming. Plowing, seeding, cultivating, and harvesting at right angles to the direction of the slope, rather than down it.

Contour Mine. Mine used for coal and other minerals in hilly terrain. Cuts are made along the contour of the land.

Control Group. In scientific experiment, the control group is the untreated group. It is identical to the experimental group in every way except the lack of treatment.

Conventional Agriculture. An agricultural system that relies heavily on agrochemicals, new varieties of crops and labor-saving, energy-intensive farm machinery.

Cooling Tower. Device used to reduce thermal pollution before releasing cooling water from power plants and factories into lakes and streams.

Coral Reefs. Elaborate ocean structures, found primarily in tropical and subtropical waters, formed from the bodies of animals having calcareous skeletons (made from calcium materials) and certain species of algae that provide the sediment or "cement" that seals the coral framework.

Core Reserve An area set aside for wildlife protection in which no human activity is allowed. Surrounded by an buffer zone, a region permitting modest human activity.

Corridor, Terrestrial. A narrow strip of land that differs, usually in terms of dominant vegetation (such as forest or grassland), from the surrounding areas.

Creel (Catch) Limits. Fisheries management regulation that controls fishing pressure on a certain fish species by restricting the size or the number of fish an angler can take home.

Critical Population Size. Population size below which recovery is impossible.

Critical Thinking. A process by which one analyzes facts, assertions, and conclusions, attempting to discern their validity.

Crop Rotation. A planned sequence of various crops growing in a regularly recurring succession on one field.

Cross-media Contamination. The movement of a pollutant from one medium such as air to another such as water.

Crown Fire. A fire that spreads from treetop to treetop.

Cultivar. Refers to plants that are being cultivated to produce food or fiber.

Cyanosis. A disease characterized by a bluish discoloration of the skin. This is caused by the impaired effectiveness of hemoglobin to carry oxygen. An infant that drinks water carrying too high a level of nitrates may undergo chemical changes of its hemoglobin that in turn will result in cyanosis of the skin.

Cyclic Population. A population that peaks and troughs at regular intervals. Good examples are the 4-year cycle of the lemming and the 10-year cycle of the ruffed grouse.

Cyclone Filter. A type of air-pollution control device that removes particulate matter (dust) with the aid of gravity and a downward spiraling air stream.

Death Rate. Number of deaths per 1,000 people in a population.

Decreasers. A category of highly nutritious and extremely palatable range plants that generally decrease even under moderate grazing pressure.

Deepwater Swamp. Freshwater wetlands that support woody vegetation, primarily various species of cypress, gum, and tupelo trees, and remain flooded all or most of the year.

Deferred-Rotation Grazing. A grazing management system in which livestock are rotated between two or among more range areas to increase the long-term efficiency of range conversion into livestock production.

Demographic Transition. A change in a population that is characterized by decreasing birth and death rates. It usually occurs when a nation becomes industrialized.

Denitrification. The decomposition of ammonia compounds, nitrites, and nitrates by bacteria that results in the eventual release of nitrogen into the atmosphere.

Density-Dependent Factor. A population-regulating factor, such as predation or infectious disease, whose effect on a population is dependent on the population density.

Density-Independent Factor. A population-regulating factor, such as a storm, drought, flood, or volcanic eruption, whose effect is independent of population density.

Depletion Time. The time required until 80 percent of the available mineral supply is consumed.

Desalinization (also Desalination). The removal of salt from seawater in order to make it usable by humans, crops, and wildlife.

Desert Pavement. The stony surface of some deserts caused by excessive erosion of the thin topsoil resulting from water and wind action.

Desertification. The conversion of rangeland, rainfed cropland, or irrigated cropland to desert-like conditions caused by natural (namely climatic changes) and artificial factors (human-induced activities).

Detritus. General term referring to organic matter derived from dead bodies of animals, insects, plants, and so on.

Detritus (Decomposer) Food Chain. Sequence of organisms, each feeding on the one before it, starting with dead organic material (waste or animal and plant remains).

Deuterium. An isotope of hydrogen that may serve as fuel in nuclear fusion reactions.

Dioxin. An extremely toxic chemical occurring in the herbicide 2,4,5-T. In some areas it is suspected of causing birth defects and miscarriages.

Discards. The bycatch which are thrown back for various reasons including nontarget species, juvenile fish, endangered species, wrong size, inferior quality, or surplus to quotas.

Discount Rate. A number used by economists and business people to determine the present economic value of various business strategies.

Drainage Basin. See watershed.

Drawdown. Lowering of the water level in a reservoir for the purpose of flood control or hydropower.

Drawdown Phase. The occurrence of periodic low water levels common to many inland wetlands in regions that experience very little precipitation during the summer months or alternate wet and dry years.

Dredge Spoil. The sediment that has been scooped from harbor and river bottoms to deepen channels for navigation.

Dust Dome. A shroud of dust particles characteristically found over urban areas. It is caused by the unique atmospheric circulation pattern that results from the marked temperature differences between the urban area and outlying farmlands.

Dynamic Equilibrium. Describes any process or system in which change can occur but is corrected by natural mechanisms, so that the system remains more or less the same over long periods.

Ecological Island. Habitat cut off from surrounding area by natural features such as water or by farms, cities, roads, and so on. Highly vulnerable to species loss.

Ecological Justice. The concept that species other than humans have a right to exist.

Ecology. Study of the interrelationships that occur between organisms and their environment.

Economic Externality. An economic cost that is not factored into the determination of the cost of goods and services; for example, the economic cost of health effects of air pollution from factories.

Economic Incentives. Inducements to business to perform satisfactorily.

Economy. The sum total of commercial enterprises.

Ecosphere. The total area in which living organisms occur.

Ecosystem. A contraction for ecological system.

Ecosystem Simplification. The intentional or unintentional elimination of species from an ecosystem. Such changes may destabilize an ecosystem.

Ecotone (Edge). A transition zone within a landscape between two distinct ecosystems that usually has its own unique soil, vegetation, and hydrologic characteristics.

Electrolysis. Breakdown of water using electricity. Results in the formation of hydrogen, a useful fuel.

Electromagnetic Spectrum. Range of energy given off by the sun. At the lower end are low-energy radio waves; at the higher end are high-energy gamma rays. Visible light falls in the middle of the spectrum.

Electron. A negatively charged particle occurring in the orbit of an atom.

Electron Cloud. Region surrounding the atomic nucleus that contains electrons.

Electrostatic Precipitator. Device to remove particulates from smokestack gases.

Elemental Cycle. The cycle of movement of an element (e.g., nitrogen, carbon) from the nonliving environment (air, water, soil) into the bodies of living organisms and then back into the nonliving environment.

Emphysema. A potentially lethal disease characterized by a reduction in the number of alveoli in the lungs as well as a reduction in total respiratory membrane area.

End Point Separation. Process in which recyclables are removed from municipal trash at central stations.

Endangered Species. A species that is in immediate danger of extinction.

Energy. Defined by physicists as the ability to do work. Two basic forms exist: potential energy and kinetic energy.

Energy Laws. Also known as the laws of thermodynamics, they describe the origin and behavior of energy. See first and second laws of thermodynamics.

Energy Pyramid. Graphical representation of the energy in the various trophic layers in a food chain.

Entropy. Disorder or randomness in any system.

Environmental Limitations. Natural and human-induced occurrences that can cause mortality in fish populations. Natural environmental limitations include storms, soil mass movements, debris jams, winterkill, and predation by animals. Human-induced limitations include dams, pollution, riparian vegetation removal, and competition by exotic fish species.

Environmental Resistance (ER). Any factor in the environment of an organism that tends to limit its numbers.

Ephemeral Stream. Bodies of moving water which flow only during the wet season.

Epilimnion. The upper stratum of a lake, characterized by a temperature gradient of less than 1°C per meter of depth.

Essential Element. A chemical element required for the normal growth of plants.

Estuary. Semi-enclosed inlets (bays) that form transitional zones between coastal rivers and the sea.

Euphotic Zone. The open-water zone of the ocean, characterized by sufficient sunlight penetration to support photosynthesis. Located just above the bathyal zone.

Euryphagous. An organism having a highly varied diet, such as a pheasant or opossum, in contrast to an animal having a narrow or stenophagous diet, such as an ivory-billed woodpecker.

Eutrophication. The enrichment of an aquatic ecosystem with nutrients (nitrates, phosphates) that promote biological productivity (growth of algae and weeds).

Everglades. Large, unique freshwater marsh wetland located near the southern tip of Florida and dominated by a sedge called saw grass.

Exclusive Economic Zone (EEZ). The 330-kilometer (200-mile) belt of sea space extending from the coastline of each nation over which each respective nation has economic jurisdiction.

Exotic (nonnative) Fish Species. Fish that have been introduced into a water body in which they are not native, either intentionally or accidentally, from a different part of the country or from a foreign country.

Experimental Group. A group in a scientific experiment that receives the treatment or exposure under study.

Exponential Growth. Growth of any entity, such as population or resource demand, that occurs by a fixed annual percentage when the annual growth is added to the base amount.

Fabric Filter Baghouse. An air pollution control device that operates somewhat like a giant vacuum cleaner in removing solid particles from industrial smokestacks.

Fall Overturn. The thorough mixing of lake waters during autumn.

Faunal Collapse. Dramatic decrease in numbers and diversity of animal species.

Fibrous Root System. A complex root system, such as that of grass plants, in which there are several major roots and a great number of primary, secondary, and tertiary branches; useful in "binding" soil in place and preventing erosion.

Field Capacity. The water remaining in a soil two to three days after it has been saturated and after free drainage has practically stopped.

First Law of Energy (or Thermodynamics). Energy can be neither created nor destroyed but can be converted from one form to another.

Fish Ladders. A series of steplike rungs designed and built into hydroelectric dams for the purpose of providing safe passage to fish as they migrate upstream to their spawning grounds.

Floodplain. Flat areas of fine sediment deposition extending outward from the bank tops of lowland streams that receive floodwater and accompanying sediment when rivers and streams periodically overflow their banks.

Fluidized Bed Combustion. Process in which coal is crushed, mixed with bits of limestone, and air-blown into a furnace, where it is burned.

Fluorocarbons. A group of chemical compounds containing the elements carbon, chlorine, and fluorine. One group of these compounds, manufactured by Dupont under the trade name Freons, has been used in refrigerators, air conditioners, and aerosol spray bombs.

Fluorosis. A disease in animals caused by fluoride poisoning. Symptoms in livestock include thickened bones and stiff joints.

Flyway. One of the major migration pathways used by waterfowl; for example, Atlantic flyway and Pacific flyway.

Food Chain. The flow of nutrients and energy from one organism to another by means of a series of eating processes.

Food Web. An interconnected series of food chains.

Forage. Grasses, grasslike plants (sedges and rushes), forbs, and shrubs that supply food and energy for domesticated animals such as cattle, sheep, and horses, and for wild animals such as deer, giraffes, and wildebeests.

Fossil Fuel. Organic fuels derived from ancient plant or animal matter, including coal, oil, shale oil, and natural gas.

Freons. A group of fluorocarbon compounds manufactured by Dupont for use in refrigerators, air conditioners, and aerosol spray bombs. Unfortunately, Freons have contributed to the breakdown of the shield of ozone in the upper stratosphere that protects humans from ultraviolet radiation.

Freshwater Aquaculture. The controlled culturing, or "farming", of fish and other aquatic food that is practiced in inland, freshwater ponds.

Freshwater Marsh. An inland wetland characterized by nonwoody hydrophytes, shallow water depths, and shallow accumulations of peat material. The term describes a diverse group of inland wetlands including prairie potholes, vernal pools, the Everglades, the Great Lake marshes, and the Nebraska sandhill wetlands.

Fuel Cell. A device the produces electricity from oxygen and hydrogen.

Fuel Crops. Crops grown to generate ethanol or some other fuel that could replace traditional fossil fuels or other energy resources.

Fuel Rods. Rods packed with enriched uranium used for fueling nuclear power plants.

Fusion. See nuclear fusion.

Game Animals. Species that are harvested for recreational purposes.

Gamma Radiation. An intense type of radiation, similar to x-rays, that easily penetrates human tissues.

Gas-Liquid Chromatograph. Instrument used to determine the source of an oil spill.

Gene. A segment of the DNA that regulates some characteristic of an organism, including behavior.

Generalist. Species that can live anywhere and eat many different types of food.

Genetic Diversity. A term used to indicate a great variety of organisms (many different species) occupying a given area.

Genetic Engineering. A complex process that involves the isolation of genes, synthesis of genes, and their insertion into cells from the same or a different organism or species.

Genetic Erosion. A harmful effect on the gene pool of a given species when a large number of artificially propagated fish are used to enhance a population. Research has shown that the offspring of a hatchery fish bred with a native parent have a lesser chance of survival than pure native offspring.

Geographic Information Systems. A computer system consisting of hardware and software designed to assemble and store geographically reference information about the earth including vegetation, land disturbance, and human settlement patterns. This information can be displayed, manipulated, and analyzed.

Geological (Natural) Erosion. Wearing away of the Earth's surface by water, ice, wind, or other geological agents under natural environmental conditions.

Geopressurized Zones. Zones where heated groundwater is trapped by layers of impervious rock.

Geothermal Energy. Heat produced by the Earth from naturally occurring radioactive decay and from magma. Can be tapped to heat buildings or to produce electricity.

Glasphalt. A type of road-surfacing material that employs crushed glass in its manufacture rather than sand. It is more durable than ordinary asphalt.

Glassification. A method of disposing of radioactive waste by concentrating it and enclosing it in solid ceramic bricks.

Global Climate Change. Refers to the changes in the average weather conditions or climate of the planet brought about by pollutants such as carbon monoxide and methane produced by human activities.

Global Warming. A term that describes the warming of the Earth's atmosphere and oceans caused by a build up of greenhouse gases produced such as carbon dioxide and chlorofluorocarbons from human sources.

Grassland. Refers to land in which grasses are the predominant form of vegetation. Often flat or rolling terrain with few or widely distributed trees.

Graywater. Waste water from showers, washing machines, and sinks.

Grazer Food Chain. Sequence of organisms, each feeding on the one before it, starting with plants or algae.

Great Barrier Reef. Largest chain of coral reefs in the world off the coast of Queensland, Australia, reaching a length of over 2000 km and containing over 2,500 individual reefs.

Green Manuring. The practice of plowing under or surface-mulching plants (usually a grass or legume) while green, or soon after maturity, for improving the soil.

Green Revolution. The increased food production capability made possible in recent years because of selective breeding, the increased use of fertilizer, the development of seed banks, and the more intensive use of herbicides and insecticides.

Greenhouse Effect. The warming influence caused by the increased concentration of carbon dioxide and several other pollutants in the Earth's atmosphere.

Greenhouse Gas. Any one of several gases that trap heat escaping from the Earth's surface, making the Earth's atmosphere warm.

Groin. Piers of stone spaced about 30 meters apart that extend into the sea at right angles to the shoreline for the purpose of retaining beach sediments.

Gross Domestic Product (GDP). Sum total of all goods and services produced by an economy.

Gross National Product (GNP). The sum total of expenditures by governments and individuals for goods, services, and investments.

Gross Primary Production. The sum of all biomass production in an ecosystem, not taking into any account the losses due to cellular respiration.

Groundwater. Water that has infiltrated the ground, in contrast to runoff water, which flows over the ground surface.

Groundwater Overdraft. Refers to the removal of groundwater at rates that exceed their natural replenishment.

Gully Reclamation. The mending of a gully by either physical methods (check dams of boulders or cement) or vegetational means (planting of rapidly growing shrubs on the slopes).

Gyptol. The sex attractant produced by the female gypsy moth that serves to attract the male moth from considerable distances.

Habitat. The immediate environment in which an organism lives. A habitat includes such components as cover, food, shelter, water, and breeding sites.

Habitat Conservation Plans. Documents outlining steps to protect habitat to protect endangered or threatened species.

Habitat Fragmentation. The breaking up of contiguous areas of wildlife habitat often by home building or farming.

Habitat Restoration. The combination of various strategies, such as adding spawning gravels, revegetating streambanks and lake edges, establishing brush shelters, restoring the historic meanders of rivers, creating pools and riffles, and placing instream woody debris, for the purpose of improving habitat for fish and other aquatic life.

Half-life. The time required for one half of the radioactivity of a given radioactive isotope (e.g., uranium, strontium) to be dissipated.

Hardwood. A species of tree, such as oak, hickory, and maple, that has relatively hard wood, in contrast to the soft woods of the conifers such as spruce and pine. Synonymous with deciduous.

Harvestable Surplus. That portion of a population that can be taken by humans (hunters) without adversely impacting subsequent populations of that animal.

Hazardous Wastes. Substances produced by homes and factories that pollute our air, water, and soils and can have adverse effects on the environment and human health.

Headwater Stream. The beginning segment of a river system that is usually confined by a narrow valley, shallow bedrock, and coarse sediments and thus tends to have a narrow, straight channel pattern instead of wide and meandering.

Heartwood. The dark, central portion of a tree trunk characterized by the presence of dead xylem cells that have become filled with gums and resins.

Heat Island. The tendency for the atmosphere of a city to be warmer than the air in the surrounding farmlands. This is partly the result of the greater number of heat-generating sources (autos, factories, human bodies) in the city.

Home Range. The total area occupied by an animal during its life cycle-that is, the area required for feeding, breeding, loafing, and securing refuge from the weather and from predators.

Horizon. A layer of soil with specific properties and characteristics. The major horizons from the ground surface downward to bedrock are designated as horizons O (organic layer) A (topsoil), E (subsurface), B (subsoil), C (parent material), and R (bedrock).

Host. A term used to describe an organism that contains parasites.

Humus. The semistable, dark-colored material that represents the decomposition products of organic residues and materials synthesized by microorganisms.

Hybrid. The offspring that results from a cross of two different species or strains of animals or plants. For example, the Santa Gertrudis cattle resulted from a series of crosses involving two parental types, the Brahmin cattle and the shorthorn.

Hydraulic Mining. A mining technique used to extract gold and silver in which a powerful stream of water is directed against the face of the rock containing the minerals.

Hydrological Cycle. The circular movement of water from the ocean reservoir to the air (clouds), to the Earth in the form of rain and snow, and finally back to the ocean reservoir via streams and estuaries. Also known as the water cycle.

Hydrolysis. A type of chemical reaction in which a compound is broken down into simpler components by the action of water.

Hydroperiods. The various fluctuations in the duration and depth of flooding events throughout the year in an inland freshwater marsh.

Hydrophytes. Water-loving plants or those plants adapted to saturated or flooded soil conditions, including emergent, floating, and submergent species.

Hydroponics. The technique of growing crops in an aqueous nutrient solution without soil.

Hydropower. Power generated by the flow of water. Usually tapped by dams.

Hydroseeder. A machine employed to disperse grass seed, water, and fertilizer on steep banks.

Hypolimnion. The bottom layer of a lake.

Hypothesis. A tentative explanation of a particular phenomenon or observation.

Imprinting. The psychological or behavioral process by which migratory fish assimilate environmental clues to aid their return to their stream of origin as adults.

In Situ Retort. An underground site in which an oil shale deposit is broken up with dynamite and set afire to drive off kerogen.

Incidental Take. Refers to the inadvertent harvest of a species, for example, the harvest of turtles from shrimp fishing operations.

Increasers. A category of somewhat palatable but still highly nutritious range plants that tend to increase (at least temporarily) when a range is heavily grazed.

Individual Transferable Quotas (ITQs). A current marine fisheries management strategy which give fishers a guaranteed portion of the catch. In theory, ITQs essentially transform a common property resource into a private one; fishers have an economic incentive to protect the resource in order to maintain the same quota in subsequent years.

Indoor Air Pollution. Contamination of indoor air in homes, office buildings, and factories. Caused by combustion byproducts, paints, stains, finishes, and offgassing from building materials, carpets, and other materials.

Industrial Fixation. The "fixing" of nitrogen (in other words, combining it with hydrogen to form ammonia) by industrial means rather than by natural methods such as bacterial action.

Input Control. Any strategy that seeks to reduce the production of pollutants, such as modifying processes and reducing demand for goods.

Integrated Pest Management. A method of pest control that judiciously employs chemical, biological, and other methods (cultural), depending on the specific problem; the use of chemicals is minimized to avoid environmental damage.

Intercropping. A cultural method of pest control in which the farmer intermixes a number of different crops in a small area instead of devoting the entire area to a single crop.

Intergenerational Equity. Fairness to future generations.

Introduction. The bringing in of an exotic plant or animal to a new region-for example, the introduction of the European carp into the United States and of the coho salmon from the Pacific Coast to Lake Michigan.

Invaders. A term employed in rangeland management that refers to a pioneer species of plant, usually a noxious weed, that has become established in an overgrazed pasture.

Irruption. A sudden increase in the population of an organism, followed by a precipitous decline (crash). Frequently the carrying capacity of the habitat is reduced for many years thereafter. An example is the Kaibab deer irruption.

Isotope. A form of an element identical to the regular element except for a difference in atomic weight. Thus, deuterium is an isotope of hydrogen.

J-curve. An exponential growth curve which is shaped like the letter J.

Juvenile Dispersal. The dispersal of young animals (bald eagles, ruffed grouse, muskrats) from the general region of their hatching or birth site. The presumed function is to prevent overpopulation in the parental area.

Kepone. A highly toxic insecticide manufactured by a company in Hopewell, Virginia. As a result of carelessness the chemical was allowed to contaminate the James River and destroy much aquatic life.

Kerogen. The solid organic material that contains shale oil.

Krill. The crustaceans and other small marine organisms that are used as food by the baleen whales.

Kwashiorkor. Deficiency of protein intake.

Land Capability Classification System. A system that evaluates land according to its limitations for agricultural use. It classifies land into eight categories, with Class I having the least limitations and Class VIII having the greatest limitations.

Land Ethic. A view put forth by Aldo Leopold calling on humans to respect the land and all living things.

Landfill. A method of solid waste disposal in which the waste is dumped, compacted, and then covered with a layer of soil.

Large Woody Debris. Logs, branches, and root wads that have fallen into a stream (due to wind or erosional forces) that shape channel structure, provide fish habitat, and trap organic matter.

Latitudinal Migration. The north-south migration characteristic of caribou, gray whales, and many birds.

Law of Supply and Demand. Economic law describing the relationship of price, supply, and demand

Law of Tolerance. Says that organisms can live within a range of conditions; when conditions exceed that range, they are unable to survive.

Leach Field. An underground system of porous pipes that serves as a means of disposing of liquid draining from household septic tanks.

Legume. A pod-bearing member of the Leguminosae family, including peas, beans, and clover, that can fix atmospheric nitrogen.

"Let-It-Burn" Policy (or Prescribed Natural Fire Policy). A management policy followed by U.S. government agencies which permits wildfires to burn under careful surveillance.

Levee. A dike composed of earth, stone, or concrete that is erected along the margin of a river for purposes of flood control.

Life Cycle Costs. All costs from the point of origin of the material to the point of disposal.

Limiting Factors. Any factor in the environment of an organism, such as radiation, excessive heat, floods, drought, disease, or lack of micronutrients, that tends to reduce the population of that organism.

Limnetic Zone. The region of open water in a lake, beyond the littoral zone, down to the maximal depth at which there is sufficient sunlight for photosynthesis.

Littoral Zone. The shallow, marginal region of a lake, characterized by rooted vegetation.

Lowland Stream. Downstream segments of a river system that are usually wider and have more freedom to meander across the finer sediments of floodplain materials.

Malnutrition. Dietary deficiency resulting from inadequate intake of nutrients.

Malthusian Overpopulation. The type of overpopulation described by Robert Malthus; it results in an overtaxing of available food supplies and eventual massive starvation.

Manganese Nodules. Mineral-rich nodular accumulations found on much of the ocean floor. Contains manganese, nickel, iron, copper, cobalt, and other minerals.

Marasmus. Deficiency of proteins and calories.

Marine Aquaculture (Mariculture). The controlled culturing, or "farming", of fish and other aquatic food that is practiced in shallow bays or estuaries.

Market Economy. An economy that relies on monetary signals to control production and distribution of goods and services.

Mass Emigration. The mass movement of a given species from an area. An example is the mass movement of the snowy owl from the tundra to the United States during periods of lemming scarcity.

Mass Number. The mass number of an element is based on the total number of protons and neutrons present in the nucleus of the atom.

Maximization. The most efficient use of a resource that is possible with current technology. Waste is minimized.

Maximum Sustained Yield (MSY). The highest sustained fish or shellfish catch (number or weight) that can be allowed each season without impacting the reproductive capacity of that species. MSY is a mathematical calculation based on factors such as spawning mass, annual recruitment, annual growth in biomass, nonfishing mortality (natural death, predation), and fishing mortality.

Meandering Stream. The most common stream pattern that results when a channel naturally migrates laterally, eroding one bank and depositing material on the opposite bank. The stream channel is only straight for relatively short distances between curves, resembling a continuous string of S shapes.

Mesotrophic Lake. Any lake characterized by a moderate level of nutrients. Contrast with oligotrophic and eutrophic lakes.

Metabolic Reserve. The lower 50 percent of a grass shoot that is required by a grazed plant for survival; contains the minimum amount of photosynthetic equipment needed for food production purposes.

Microcephaly. A birth defect characterized by an abnormally small brain; associated with mental retardation; may be induced by radiation.

Microhabitat. The immediate, localized environment of an organism.

Micronutrient. Any one of dozens of nutrients needed in miniscule amounts, such as zinc and vitamins

Millirem. One thousandth of a rem, which is a unit for measuring the effect of radiation on the body of a living organism.

Mineral. Useful substance obtained by mining or digging in the Earth.

Mineral Reserves. Mineral deposits that are practical to mine.

Monocropping. The practice of growing a crop on the same field year after year.

Monoculture. Greatly altered ecosystem consisting of one or a few species. Generally more vulnerable to insects, disease, and adverse weather conditions than diverse ecosystems.

Monotype. An agricultural or forest planting composed of only one species.

Mulch. Dead plant material that accumulates on the ground surface; a reliable indicator of range condition.

Multiple-Use Management. A management principle that requires meeting a number of different needs on the same area of land. These needs or uses can include timbering, grazing, agriculture, mining, oil and gas leasing, hunting and fishing, recreation, soil conservation, wildlife protection, and watershed management.

Mutation. Any one of several changes in the genetic material (chromosomes) of an organism. Mutations may be caused by radiation, chemicals in the environment, and other agents.

Mycelia. The branching "root system" of a fungus.

Mycorrhiza. An intimate relationship between the root systems of trees and soil fungi.

Myxoma Virus. The virus that was used to control the rabbit outbreak in Australia; animals become infected when they consume contaminated forage.

National Park. An area of scenic beauty or historical importance that is to be maintained in unimpaired condition by a national government for the use of the people.

Native Fish Species. Fish that originated in the water body where they reside.

Natural Capital. A nation's ecological wealth.

Natural Resource. Any component of the natural environment, such as soil, water, rangeland, forest, wildlife, and minerals, that species depend on for their welfare.

Neritic Zone. The relatively warm, nutrient-rich, shallow water zone of the ocean that overlies the continental shelf; valuable in terms of fish production.

Net Energy Efficiency Yield. Amount of energy produced by a system when taking into account the total amount invested in the first place.

Net Primary Production. The total energy incorporated into the body of a plant as a result of photosynthesis minus the energy required for respiration.

Neutron. The electrically neutral particle in the nucleus of an atom.

Niche. Habitat and total functional role of an organism in an ecosystem—that is, the relationship to all biotic and abiotic factors.

NIMBY Syndrome. Not-in-my-backyard syndrome; general and widespread resistance to siting waste dumps or other potentially dangerous or unsightly facilities near someone's place of residence or community.

Nitrate Bacteria. Bacteria that have the ability to convert nitrites into nitrates; essential bacteria in the cycling of nitrogen.

Nitrogen Dioxide. Air pollutant produced during combustion of any organic material. Combines with water to produce nitric acid.

Nitrogen Fixation. Process of converting atmospheric nitrogen into an inorganic form that can be used by plants.

Nitrogen Intoxication. A condition which can cause fish mortality when the fish come in contact with the turbulent, nitrogen-enriched waters immediately below hydroelectric dams.

"No Net Loss" Wetland Policy. Federal land-use policy in the United States enacted in 1988 that protects wetlands from conversion to other land uses by requiring developers to create or restore wetland area and function in exchange for any wetland losses due to proposed development.

No-Till. A practice in which a crop is planted directly into a seedbed not tilled since harvest of the previous crop.

Nonbiodegradable Material. Material not susceptible to decomposition by bacteria. For example, DDT and other chlorinated hydrocarbon pesticides are relatively nonbiodegradable.

Nonconsumptive Use. The use, without removal or alteration, of natural resources.

Nongame Animals. Species that are not harvested for sport purposes.

Nonrenewable Resources. Resources such as coal, oil, and minerals that cannot be replenished within a reasonable period by natural processes. Nonrenewable resources occur in a fixed amount.

North American Water and Power Alliance (NAWAPA). Scheme for transferring water from the water-rich, low-population areas of northwestern Canada to the water-deficient areas of the United States and Mexico.

Nuclear Fusion. The generation of energy by causing the fusion of the nuclei of two atoms of a very light element such as hydrogen under temperatures of around 40 million mC.

Nutrient Flush. The sudden increase in water fertility that results when soil is newly submerged in the creation of a reservoir.

Ocean Thermal Energy Conversion. A method of using the temperature differential between different levels of the ocean to alternately gasify and condense a working fluid such as ammonia and in this way propel a turbine for the purpose of electrical power production.

Oil (Petroleum). Thick liquid containing numerous organic molecules and found in deep deposits in the Earth's crust. These molecules form jet fuel, gasoline, diesel, fuel oil, and many other useful products.

Oil Sands. Same as tar sands.

Oligotrophic Lake. A nutrient-poor lake occurring in the northern states and in high mountain areas, characterized by great depth, sandy or gravelly bottom, a sparse amount of rooted vegetation, and the low production of plankton and fish. Examples: Lake Superior and the Finger Lakes of New York.

Opportunity Cost. Cost of lost economic opportunities.

Optimum Yield (OY). A management principle that considers the biological, economic, social, and political values of a given fishery in order to maximize benefits to society. The chosen optimum level of use established for any given fishery is dependent on the goals and resource objectives of the associated society. Ideally, an optimum yield would be one that allows the greatest extrinsic yield (economic, sociocultural, nutritional benefits) to society with minimum loss of intrinsic yield (health) to the ecosystem.

Ore. Rocks that contain important minerals.

Organic Compounds. Any of many thousands of chemicals made principally from carbon, hydrogen, and oxygen.

Organic Farming. A farming system that combines traditional conservation-minded farming methods with modern farming technologies but excludes such conventional inputs as synthetic pesticides and fertilizers, emphasizing instead building up the soil with compost additions and animal and green manures, controlling pests naturally, rotating crops, and diversifying crops and livestock

Organic Fertilizer. A soil supplement made from organic sources such as manure or compost.

Organic Phosphorus Pesticides. A group of pesticides (malathion and parathion, for example) that are lethal to insects because they reduce the supply of cholinesterase at the junction (synapse) between two nerve cells in a nerve cell chain.

Organic Soil. Soil consisting of layers of plant material in various stages of decomposition. Organic soil contains at least 20 percent organic matter and is often referred to by the generic term, peat.

Output Control. Any strategy that seeks to reduce pollution after it has been produced, such as pollution control devices or disposal methods.

Overburden. The rock and dirt that overlies coal seams.

Overgrazing. Continued heavy grazing that exceeds the carrying capacity of the plant community and creates a deteriorated range.

Overnutrition. Refers to a condition in which one eats to excess. Leads to obesity and numerous medical problems such as heart attacks and diabetes.

Oxidation. The chemical union of oxygen with metals (iron, aluminum) or organic compounds (sugars). The former process is an important factor in soil formation; the latter process permits the release of energy from cellular fuels (sugars, fats).

Oxygen-demanding Wastes. A type of water pollutant that is broken down by naturally occurring microorganisms that consume dissolved oxygen, lowering the level of oxygen required by other species.

Oyster Watch. A well-coordinated program employed along our coasts that involves the use of oysters as biological monitors of marine pollutants.

Ozone. A gaseous component of the atmosphere; normally occurs at elevations of about 20 miles; important to humans because it shields us from the ultraviolet radiation of the sun; also represents one of the products resulting from the action of sunlight on the hydrocarbons emitted from the internal combustion engine.

Parasite. Refers to any one of numerous organisms that live on or in other organisms, living off nutrients in blood or tissue fluids.

Parent Material. The unconsolidated (soft and loose), weathered mineral or organic material from which soils form.

Particulate Matter. Minute solid and liquid particles in the atmosphere; soot.

Passive Fishing Gear. Devices and methods for fishing which are stationary and, therefore, depend on the fish or shellfish to come into contact with them. Examples include pots, traps, cages, hooks and lines, set nets, pound nets, and anchored gill nets.

Passive Solar System. Building designed to capture sunlight energy and produce heat for space heating.

Peat. Partially decomposed accumulations of organic matter under conditions of excessive moisture.

Peatlands. Land areas, such as bogs and fens, found in the colder northern, humid regions of the Northern Hemisphere characterized by extremely slow plant decomposition and thick accumulations of peat.

Perennial Stream. Stream that flows year-round and are sustained by groundwater during dry periods.

Permafrost. A permanently frozen soil or soil horizon.

Permanent Wilting Point. The water content of a soil after plants have removed all the water they possibly can and have begun to wilt.

Pesticide Treadmill. The use of ever-increasing amounts of pesticides because of the pests' ever-increasing resistance to the pesticides.

Pheromone. Chemical substance released by insects that affect other insects. Some of the best-known examples are sex attractant pheromones released by females to attract males.

Phloem. The elongate food-conducting cells of the trunk and branches of a tree. These cells convey food from the leaves downward to the root system.

Photochemical Smog. The type of smog that has plagued Los Angeles and other California towns, as well as other areas in the United States. Photochemical smog forms as a result of the action of sunlight on the hydrocarbon and nitrogen oxide emissions from motor cars and other sources. At nightfall the production of this type of smog ceases.

Photosynthesis. The process occurring in green plants by which solar energy is utilized in the conversion of carbon dioxide and water into sugar.

Photovoltaics. See solar cell.

Phytoplankton. Minute plants, such as algae, living in lakes, streams, and oceans, that are passively transported by water currents or wave action.

Pioneer Community. First community to become established on barren land.

Pittman-Robertson Act. Also known as the Federal Aid in Wildlife Restoration Act. This act levies an excise tax on the sales of sporting arms and ammunition; revenues from the federal tax are distributed to the states for habitat purchase, and wildlife management and research.

Plankton. Tiny plants (algae) and animals (protozoa, small crustaceans, fish embryos, insect larvae) that live in aquatic ecosystems and are moved about by water currents and wave action.

Point Bar. The sediment deposition areas which form on the inside bends of meandering streams.

Point Pollution. Pollution that is discharged from an extremely restricted area or "point", such as the discharge of sulfur dioxide from a smokestack or the discharge of carbon monoxide from the exhaust pipe of a motor car. This contrasts with nonpoint pollution, such as the runoff from a farm or urban area.

Pollution Control. Refers to measures like smokestack scrubbers that reduce pollution emitted by a pollution source. In many cases, it merely captures pollutants that must then be disposed of elsewhere.

Pollution Prevention. Refers to any strategy designed to prevent the production of pollution, for example, energy conservation or process changes in factories that eliminate pollutants.

Polychlorinated Biphenyls (PCBs). A class of chemicals used as electrical insulators. They persist in the environment and undergo biomagnification, causing problems in the food chain.

Pool. Deep areas on the streambed that have been scoured out by storm flows and are usually associated with the outside bend of a meandering stream.

Population. The individuals of a species occurring in a given area; for example, the population of deer in a cedar swamp or the population of black bass in a lake.

Population Growth Rate. Birth rate minus death rate plus net immigration multiplied by 100.

Pore Space. That portion of a soil occupied by both air and water.

Precautionary Approach. An international agreement in 1995 that will become part of the 1982 U.N. Convention on the Law of the Sea. The precautionary approach requires increased monitoring, inspecting, and reporting in an effort to protect fish stocks before they show signs of decline instead of managing by reacting to the decline itself.

Precision Farming. The integrated use of field observations, computers, geographical information systems (GIS), global positioning systems (GPS), remote sensing (aerial photography), and farm implements to practice site-specific management, so that farmers can improve efficiency, create cost savings, increase yields, and reduce environmental hazards.

Prescribed (Controlled) Burning. A type of surface burning (used by foresters, wildlife biologists, and ranchers) to improve the quality of forest, range, or wildlife habitat.

Primary Pollutant. Refers to a chemical pollutant produced by any one of many sources. It may undergo chemical reactions that result in the production of a new pollutant, called a secondary pollutant.

Primary Production. The total chemical energy produced by photosynthesis; on a global basis it amounts to about 270 billion tons annually.

Primary Sewage Treatment. A rudimentary sewage treatment that removes a substantial amount of the settleable solids and about 90 percent of the biological oxygen demand (BOD).

Primary Succession. An ecological succession that develops in an area not previously occupied by a community; for example, a succession that develops on a granite outcrop or on lava.

Process Redesign. Refers to a strategy taken by businesses to adjust manufacturing processed to reduce or eliminate hazardous waste and pollution.

Producer. A plant that can carry on photosynthesis and thus produce food for itself and indirectly for other organisms in the food chain of which it is a part.

Profundal Zone. The bottom zone of a lake, extending from the lake bottom upward to the limnetic zone and characterized by insufficient sunlight for photosynthesis.

Proton. A positively charged particle in the nucleus of an atom.

Purse Seine. A seine used by commercial fishermen that closes to entrap fish somewhat as a drawstring purse closes to "trap" money.

Put and Take Stocking. A management practice common in urban area water bodies where fishing pressure is heavy but there is no natural production. Fish, of catchable size, are planted at the beginning of fishing season so that most of them are caught in that same season.

Pyramid of Biomass. The graphic expression of the fact that there is a progressive reduction in total biomass (protoplasm) with each successive level in a food chain.

Pyramid of Energy. The graphic expression of the second law of thermodynamics as applied to the energy transfer in food chains. A certain amount of energy is lost in the form of heat as it moves through the links of a food chain, the greatest amount being present in the basal link (producer) and the least amount being present in the terminal link (carnivore).

Pyramid of Numbers. The graphic expression of the fact that the number of individuals in a given food chain is generally greatest at the producer level, less at the herbivore level, and least at the carnivore level.

Pyrolysis. The destructive distillation of solid waste.

Quadrillion. The number 1 followed by 15 zeros.

Quaking Bog. Naturally successional stage in the "filling in" of small, shallow lakes. Vegetation attaches to the lake edge and grows outward into the middle of the lake, resulting in the creation of a cushionlike, floating mat that bounces or "quakes" when walked upon.

Rabies. A disease caused by a virus that lives in the saliva of the host (carrier) animal.

Rad. A unit devised to measure the amount of radiation absorbed by living tissue; a rad is 100 ergs (an erg is a unit of energy) absorbed by one gram of tissue; about 1 roentgen, which is about the amount of radioactivity received from a single dental x-ray.

Radiation Inversion. Temperature inversion occurring because the ground cools faster than the air above it. Results in buildup of air pollution at ground level.

Radioisotope. Radioactive form of an atom.

Radon. Naturally occurring radioactive gas. Can leak into houses and other buildings, where it can cause lung cancer.

Range Management. Managing rangeland usually to maximize livestock or wild herbivore productivity without degrading rangeland quality.

Range of Tolerance. The tolerance range of a species for certain factors in its environment such as moisture, temperature, radiation, micronutrients, and oxygen.

Rangeland Condition. The current state of vegetation of a range site in relation to the potential natural plant community. In a sense, it is an estimate of how close a particular rangeland is to its productive potential.

Rangelands. Those areas of the world which are a source of forage (such as grasses and shrubs) for free-ranging native and domestic animals, as well as a source of wood products, water, energy, wildlife, minerals, recreational opportunities, natural beauty, open space, and wilderness.

Redd. The "nest" formed in the gravels by a spawning fish.

Rem. A unit of absorbed radiation dose taking into account the relative biological effect of various types of radiation; about 1 roentgen, which is roughly the amount of radiation received from a single dental x-ray.

Renewable Resources. Resources such as solar energy, trees, grass, and fish that replenish naturally through some biological or geophysical process.

Replacement-Level Fertility. Number of children a couple must have to replace themselves.

Reseeding. A practice where ranchers improve their rangelands by periodic seeding.

Reservoir. The artificial lake created behind a dam as river water is ponded and stored for use in hydropower generation and flood control.

Resident Fish. Fish that live their entire life in freshwater, sometimes confined to a lake, particular stream reach, or single tributary.

Resource Conservation and Recovery Act (RCRA). Under terms of this act, the EPA was given full authority to control pollution by solid waste, just as it had authority for controlling air and water pollution.

Respiration. The process by means of which cellular fuels are burned with the aid of oxygen to permit the release of the energy required to sustain life. During respiration oxygen is used up and carbon dioxide is given off.

Retort. Vessel used to heat crushed oil shale rock to produce shale oil.

Reverse Vending Machine. Automated recycling machine that pays operator for materials put into the machine.

Rhizobium. One of a genus of nitrogen-fixing bacteria that lives in the root nodules of legumes (alfalfa, clover).

Rhizome. An underground stem that occurs in grasses. It permits vegetative reproduction because the tip of the rhizome may develop a bud that can develop into a new plant.

Rhizosphere. The soil in the immediate vicinity of a plant root system.

Rhythm Method. A fairly unreliable form of contraception that relies on abstinence during a period just prior to and after ovulation.

Riffle. Mounds of sediment deposited on the streambed during storm flows that forms shallow water areas.

Riparian Wetlands. Wetlands that form along rivers and streams due to periodic flooding of these aquatic systems.

Riparian Zone. The entire linear strip of land on either side of a river or stream that forms a transitional area between the adjacent aquatic and upland ecosystems. A riparian zone has unique soil, vegetation, and hydrologic characteristics and includes riparian wetlands as well as land that may not meet the criteria for a jurisdictional wetland.

Root Nodule. Small bulbous attachments on the roots of certain plants, notably legumes (peas, beans, vetch, etc.) that contain bacteria that fix atmospheric nitrogen, that is, convert atmospheric nitrogen to a form plants can use.

Rotenone. A poisonous substance derived from the roots of an Asiatic legume. It has been extensively used to control rough fish populations.

Rough Fish. Undesirable trash fish such as garpike and carp.

Runoff Water. The water that flows over the land surface after rainfall or snowmelt and eventually forms streams, lakes, and marshes.

Salinization (also Salination). An adverse after-effect of irrigating land that has poor drainage properties. As a result, especially in the arid western states, evaporation of the salty water leaves a salt accumulation on the land, which renders the soil unsuitable for crop production.

Saltwater Intrusion. The contamination of freshwater aquifers with salt water as the result of excessive exploitation of those aquifers in coastal regions near the ocean.

Sanitary Landfill. A dump for municipal garbage. Refuse is covered daily with a layer of dirt to reduce flies, odors, and rodents.

Sapwood. The lighter, moist, more porous layer of xylem tissue immediately ensheathing the heartwood; composed of water and nutrient-transporting xylem cells.

Scrubber (also known as a smokestack scrubber). Device to remove particulates and sulfur dioxide from air pollution in smokestacks.

Seawall. An expensive structure, usually made of rock or concrete, placed along a beachfront to prevent erosion caused by the force of ocean waves.

Second Law of Energy (or Thermodynamics). During energy conversions, entropy increases.

Secondary Pollutant. A chemical pollutant derived from another pollutant after reacting with sunlight, moisture, or other pollutants.

Secondary Sewage Treatment. An advanced type of sewage treatment that involves both mechanical and biological (bacterial action) phases. Although it is superior to primary treatment, many of the phosphates and nitrates remain in the effluent.

Secondary Succession. An ecological succession that occurs in an area that had at one time already supported living organisms, for example, a succession developing in a burned-over forest or in an abandoned field.

Secured Landfill. Clay-lined landfill designed to hold hazardous wastes.

Seed-Tree Method. A silvicultural system in which trees are left during cutting to seed the next tree generation.

Selective Cutting. Procedure in which only certain trees are cut from a forest. Contrast to clear-cutting.

Septic Tank. A part of a rudimentary type of sewage treatment system used commonly by families who are located in rural areas. The sewage flows into the subterranean septic tank and is gradually decomposed by bacterial action.

Sewage. Human waste consisting of feces and urine.

Sewage Treatment Plant. A facility designed to remove most or all of the pollutants from sewage from homes.

Shade-Tolerant Plants. A species of plant, such as the sugar maple, that reproduces well under conditions of reduced light intensity.

Shale Oil. Oil produced by heating (retorting) oil shale. Can be refined and made into a variety of combustible products in much the same way that crude oil is refined.

Sheet Irrigation. A type of irrigation in which water flows slowly over the land in the form of a sheet.

Shelterbelt (Windbreak). Rows of trees and shrubs arranged at right angles to the prevalent wind for the purpose of diminishing the desiccating and eroding effects of the wind on crop and rangeland; commonly employed in the Great Plains.

Shelterwood Method. A silvicultural system that uses rotational selective cutting designed to leave a seed source and protective cover for forest regeneration.

Short-Duration Grazing. A grazing management system in which each pasture is grazed briefly and intensively, followed by a long period of nonuse.

Sigmoid Growth Curve. The S-shaped curve commonly followed by the population of an animal (deer, grouse, rabbit) when it has been newly introduced into a habitat with good carrying capacity.

Siltation. The filling up of a stream or reservoir with water-borne sediment.

Slash-and-Burn Agriculture. A type of agriculture maintained by natives of tropical rain-forest regions in which a patch of forest is cut and burned, crops are grown in the clearing for a few years until the fertility of the soil is exhausted, and then the area is deserted because the farmers move to another part of the forest to repeat the process.

Smelter. Device that melts ore to separate metals from the ore.

Smog. Term coined to refer to a combination of smoke and fog. Now used more broadly to include urban pollution.

Softwood. A species of tree such as spruce, pine, and fir that has softer wood than hardwoods such as oak and hickory; usually synonymous with conifer.

Soil. A natural system consisting of four components: mineral matter, organic matter, water, and air.

Soil Classification. The systematic arrangement of soils into groups on the basis of their characteristics.

Soil Erosion. A three-step process by which soil particles are detached from their original site, transported, and eventually deposited at a new location.

Soil Fertility. The ability of a soil to supply nutrients in amounts and forms required for maximum plant growth.

Soil Organic Matter. The living or dead plant and animal materials in the soil. It includes plant and animal residues at various stages of decomposition, small animals, microorganisms, and humus.

Soil Profile. A vertical section of the soil extending through all its horizons and extending into the parent material.

Soil Reaction. The degree of acidity, neutrality, or alkalinity of a soil. Soil reaction depends on the relative amounts of hydrogen ($H+$) and hydroxide ($OH-$) ions and is expressed as a pH value.

Soil Structure. The arrangement or grouping of soils particles into clusters or aggregates.

Soil Taxonomy. The soil classification system used in the United States (and some other countries).

Soil Texture. The relative proportions of sand, silt, and clay in a particular soil.

Soil-Forming Factors. The five factors responsible for kind, rate, and extent of soil development. They include (1) climate, (2) parent material, (3) organisms, (4) topography, and (5) time.

Solar Cell. A platelike device composed of two layers of silicon that converts solar energy directly into electricity.

Solar Energy. The radiant energy generated by the sun that "powers" all energy-consuming processes on earth, whether biological or nonbiological.

Solution Mining. Type of mining in which water is pumped into the deposit and dissolves the mineral; the mineral is then pumped to the surface.

Source Reduction. A term that refers to the reduction of waste at the source by reduced consumption, efficiency measures, process manipulation, recycling, or other measures.

Source Separation. Recyclables are separated at a source (home, factory, etc.) and are picked up by recyclers.

Spawning. Major life cycle stage of a fish that refers to egg deposition.

Specialist. Organism that feeds on only one or a few types of food and can live only in a certain habitat.

Splash Dams. Early logging practice of building dams to create a large volume of water in which to float logs down river.

Spoil Bank. The mounds of overburden that accumulate during a strip-mining operation. Unless properly limed and then vegetated, these spoil banks may become an important source of acid mine drainage and erosion.

Spring Overturn. The complete top-to-bottom mixing of water in a lake during the spring of the year when all the water is of about the same temperature and density.

Stamen. The club-shaped, pollen-producing part of a flower.

Stenophagous. Having a very specialized diet. Examples are the ivory-billed woodpecker, which consumes beetle larvae secured only from recently dead trees, and the Everglade kite, which feeds almost exclusively on the snail Pomacea caliginosa.

Sterilization. A method of human population control that involves cutting and tying the sperm ducts in the male and the oviducts in the female.

Stocking. The planting of artificially propagated fish into ponds, lakes, reservoirs, or streams.

Strategic Minerals. Minerals that are essential to the economic well-being and/or security of a nation.

Stream Order. A classification system, developed by Horton, based on the position of a stream within the entire network of tributaries.

Strip Cropping. The practice of growing crops that require different types of tillage, such as row and sod, in alternate strips usually along the contours or across the prevailing direction of the wind.

Strip Cutting. A method of harvesting timber in wide (about 80 meters) strips, allowing natural regrowth.

Strip Mining. The type of mining in which coal or iron, for example, is scooped from the earth by giant earth-moving machines.

Subsidence. Land collapse resulting from mining underground water reservoirs (aquifers).

Subsidence Inversion. Temperature inversion caused when a mass of high- pressure air settles over an area. Causes pollutants to build up at ground level.

Succession. The replacement of one community by another in an orderly and predictable manner. The succession begins with a pioneer community and terminates with a climax community.

Sulfur Oxides. One of two gases made from sulfur and oxygen and released during the combustion of sulfur-containing compounds such as those found in oil and gasoline.

Superfund Act (CERCLA). The act that provided a fund to the EPA to clean up extremely hazardous waste sites.

Surface Fire. A type of forest fire that moves along the surface of the forest floor, consuming litter, herbs, shrubs, and seedlings.

Surface Mining Control and Reclamation Act. Under terms of this act, which became effective in 1979, mining companies must restore a strip-mined area to its original condition, within the limits of available technology.

Surface Retort. Large steel vessel in which crushed shale is heated to drive off oil.

Surface Water. Any waterbody above the ground surface, including lakes, rivers, and ponds.

Suspended Sediment. Sediment carried in the water column.

Sustainable Agriculture. A system-level approach to understanding the complex interactions within agricultural ecologies. For a farm to be sustainable, it must produce adequate amounts of high-quality food, conserve resources, and be environmentally safe, profitable, and socially responsible.

Sustainable Society. Society based on conservation, recycling, the use of renewable resources, and population control. A sustainable society respects nature and seeks to find a sustainable relationship between humans and other species.

Sustained Yield. The concept that a forest or wildlife resource can be managed in such a way that a modest crop can be harvested year after year without depleting the resource as long as annual decrements are counterbalanced by annual growth increments.

Synergistic Effect. A condition in which the toxic effect of two or more pollutants (copper, zinc, heat) is much greater than the sum of the effects of the pollutants when operating individually.

Taiga. The northern coniferous forest biome, which is typically composed of spruce, fir, and pine.

Taproot. The type of root system characterized by one large main root; for example, that of a beet or carrot.

Technological Overpopulation. The type of overpopulation that results in massive pollution and resource exhaustion because of the high technological level maintained by the population. People die because of toxic contamination of the environment rather than because of food shortages.

Telemetry. The electronic technique involving a transmitter-receiver system in which the movements and behavior of animals (deer, elk, grizzly bear, salmon) are monitored from a distance.

Temperature invesion. See thermal inversion.

Terrace. Abandoned floodplains caused by the downcutting of a river that no longer receive overflow discharge. Terraces lie adjacent to but at a higher elevation than a current floodplain.

Terracing. A soil conservation technique in which steep slopes are converted into a series of broad-based "steps." The velocity of runoff water is thus retarded and soil erosion is arrested.

Territory. An area that is defended by a member of one species against other members of that same species.

Tertiary Sewage Treatment. The most advanced type of sewage treatment, which removes not only the BOD and the solids but also the phosphates and nitrates. The installation of such a plant at Lake Tahoe has arrested eutrophication of the lake.

Thermal Inversion. An abnormal temperature stratification of the lower atmosphere in which a layer of warm air overlies a layer of cooler air. Such an inversion frequently occurs at heights of 100 to 3,000 feet, resulting in stagnation of the air mass below the inversion. It contributes to air pollution problems, especially in industrial areas.

Thermal Plume. The warm water zone in a steam or lake caused by the release of heated water.

Thermal Pollution. An increase (or decrease) in water temperature that adversely affects aquatic organisms.

Thermocline. The middle layer of water in a lake in summer, characterized by a temperature gradient of more than 1°C per meter of depth.

Threatened Species. Species that is likely to become endangered in the near future.

Throughput. An economics term that relates to the amount of materials being produced and consumed in a given society.

Tidal Freshwater Marsh. A marsh located far enough inland to escape saltwater intrusion, but near enough to the coast to experience tidal fluctuations.

Tidal Salt Marsh. Marsh formed on the coast near river mouths, in bays, on coastal plains, around lagoons, or behind barrier islands. Tidal salt marshes are characterized by high salinity, tidal fluctuations, tidal creeks, salt-tolerant nonwoody vegetation, and extreme changes in daily and seasonal temperatures

Time Preference. An economic term that refers to one's preference for returns on investments or various actions such as soil conservation. A short time preference indicates a preference for receiving income in the near term.

Tolerable Soil Loss (or Soil Loss Tolerance). The maximum combined water and wind erosion that can take place on a given soil without degrading that soil's long-term productivity.

Topsoil. The uppermost and usually the most fertile part of the soil.

Total Fertility Rate. Projected number of children that females in a population will produce in their lifetimes, given current trends.

Toxic Substances Control Act (TOSCA). This act makes it mandatory for a company to notify the EPA 90 days in advance of its intention to manufacture a new chemical. If the EPA concludes that the chemical may be harmful to human health or may be environmentally destructive, it can deny the company permission to produce the chemical.

Tradable Permit. A sellable permit issued to a company that generates pollution. It allows the release of a certain amount of pollution. If a company can reduce pollution below the allotted amount, they can sell their emission allowance to another company. Also called markeable permit

Translocation. A practice used to restore a native fish population to its original water body. Large numbers of fertilized eggs are obtained directly from spawning adults (in their natural environment) of the target fish species which are well-established in an ecologically similar water body. The eggs are then immediately transferred to and placed in the original water body, thus creating a new population.

Transpiration. The evaporation of water from the breathing pores of a plant leaf.

Tree Farm. A private land area that is used to grow trees for profit under sound management principles.

Trickling Filter. Device used in secondary sewage treatment to reduce organic matter as well as nitrogen and phosphorus levels in human waste. Sewage is dripped over a bed of stones or bark coated with decomposer organisms.

Tritium. An isotope of hydrogen; a potential fuel for a nuclear fusion reactor.

Tropical Rainforest. A forest with 200 centimeters (79 inches) of annual rainfall spread evenly enough through the year to support broad-leaved evergreen trees, typically arrayed in several irregular canopy layers dense enough to capture more than 90 percent of the sunlight before it reaches the ground.

Tundra. The type of biome occurring in northern Canada and Eurasia north of the timberline. It is characterized by fewer than 10 inches of annual rainfall, subzero weather in winter, a low-lying vegetation composed of grasses, dwarf willows, and lichens, and a fauna consisting of lemmings, Arctic foxes, Arctic wolves, caribou, and snowy owls, among other species. The growing season lasts only six to seven weeks.

Turbidity. A measure of suspended sediment in water.

2,4-D. A herbicide that kills a weed because it mimics the plant's growth hormones, causing more rapid growth than can be sustained by its supply of oxygen and food materials.

2,4,5-T. A herbicide that operates on the same principle as 2,4-D. It kills a weed because it mimics the plant's growth hormones, causing more rapid growth than can be sustained by its supply of food, moisture, and oxygen.

Two-Story Reservoir. A reservoir that is sufficiently deep to become thermally stratified during the summer, complete with epilimnion, thermocline, and hypolimnion.

Ultimate Production. Amount of a natural resource that will probably have been removed after total supplies have been depleted.

Ultraviolet Radiation. One type of radiation produced by the sun. It is essential in small amounts, for it increases vitamin D production in the skin of humans. Excess levels can cause sunburn, cataracts, and skin cancer in people.

Undercut Bank. Bank shaped in the form of a "C" in cross section, maintained by rooted riparian vegetation, and provides excellent overhead cover and shade for fish.

Undernutrition. Dietary deficiency resulting from an inadequate intake of food.

Universal Soil Loss Equation (USLE). An equation used to predict the average annual soil loss per acre per year. Now revised, it is called the revised universal soil loss equation (RUSLE).

Upwelling. The movement of nutrient-rich cold water from the ocean bottom to higher levels by means of vertically moving currents.

User Fees. Levies placed on raw materials that are paid by the producers.

Vessel Element. The elongated xylem cell in a tree trunk or branch that has water transportation as a major function.

Vibert Box. A device designed to enhance trout populations by allowing eggs to develop under natural conditions while being protected from predation. This plastic box, filled with about 100 trout eggs, is planted in the gravel bed of a stream. Slots on all sides permit the free flow of stream water.

Volatile Organic Compounds. Any one of dozens of organic molecules produced during the incomplete combustion of fossil fuels and other organic materials such as wood.

Water Diversion. Refers to the removal of water from surface waters, especially streams, to areas of need often by pipes or canals.

Water Table. The upper level of water-saturated ground.

Waterlogging. Saturation of the root zone with water. Waterlogging chokes plants and kills them. It is usually caused by excess irrigation in poorly drained soils.

Watershed. The total area drained by a particular stream or river; may range from a few square miles in the case of a small stream to thousands of square miles in the case of the Mississippi River.

Weathering. The physical and chemical changes produced in rocks and soils by temperature and precipitation.

Wetland Delineation. The practice of defining the specific boundaries of a wetland based on field indicators of hydric soils, hydrophytic vegetation, and wetland hydrology.

Wetlands. Those areas that are inundated or saturated by surface or ground water at a frequency and duration sufficient to support, and that under normal circumstances do support, a prevalence of vegetation typically adapted for life in saturated soil conditions. Wetlands generally include swamps, marshes, bogs, and similar areas.

Whirling Disease. A condition currently infecting wild trout populations in the intermountain West. Whirling disease is caused by the fish parasite, *Myxobolus cerebralis,* believed to have been unintentionally introduced in 1955 from Europe. Visible symptoms of the disease are whirling behavior, black tail, or cranial and skeletal deformity.

Wilderness. A recognized, protected area where the earth and its community of life are untrammeled by humans, where humans themselves are visitors who do not remain.

Wildlife. All plants and animals on earth that are not domesticated. As it is generally used, the term is restricted to birds and mammals.

Wildlife Corridor. A protected area connecting two or more wildlife preserves that allows species to migrate to new habitat to find food, mates, or nesting.

Wildlife Management. The planned use, protection, and control of wildlife by the application of ecological principles.

Winterkill. A term referring to the fish kills that can occur during the long winters of the northern states when an icy barrier seals lakes off from atmospheric oxygen. Snow can form an opaque barrier that prevents sunlight penetration through the lake surface, resulting in a decrease in photosynthetic activity. Subsequently, the concentration of dissolved oxygen may drop to a level below which fish can survive.

Xerophytes. Specialized plants that are well adapted to survive in arid regions because of such water-conserving features as reduced leaves, recessed stomata, thick cuticles, accelerated life cycles, periodic dormancy, and the presence of water-storing (succulent) tissues.

Xylem. A type of tissue occurring in the trunks and branches of trees (and other plants) that serves to transport water and nutrients from the roots to the leaves and also provides support.

Zero Population Growth. Condition in which the growth rate of a country equals zero.

Zone of Deposition. Refers to the B horizon, or subsoil, that receives and accumulates the soluble salts and organic matter carried downward from the A horizon by percolating water.

Zone of Leaching. Refers to the A horizon, or topsoil, because many soluble salts are carried downward or leached from this horizon to the B horizon below it.

Zooplankton. Minute animals (protozoans, crustaceans, fish embryos, insect larvae) that live in a lake, stream, or ocean and are moved by water currents and wave action.

Illustration Acknowledgments

1.2A Peter Arnold, Inc. photo by Kelvin Aitken.
1.2B Peter Arnold, Inc. photo by Argus Fotoarchiv.
1.3 Photo Researchers, Inc. photo by Ulrike Welsch.
1.4 Photo Researchers, Inc. photo by David R. Frazier.
1.7, 5.9, 5.13, 7.2, 7.3, 7.25, 8.3, 8.9, 8.16, 8.17, 12.4, 13.23, 14.4, 14.17, 22.20, U.S. Department of Agriculture.
1.8, 17.10 Corbis photos by UPI.
1.9 Liason Agency, Inc. photo by Terry Ashe.
1.10 Corbis.
1.11 Liason Agency, Inc. photo by Chip Vanai.
2.1 AP/Wide World Photos photo by Mike Wintroath.
2.3 Pictor photo by T. Stephen Thompson.
2.6 Tri-Met (Tri-County Metropolitan Transportation District of Oregon).
2.7 Michael Collier.
2.8 Courtesy of the Aldo Leopold Foundation Archives.
3.15 Photo Researchers, Inc. photo by Hugh Spencer.
3.27 U.S. Fish and Wildlife Service photo by J. Malcolm Greary.
Case Study 3.1 Fig. 1, 20.10, 21.5b U.S. Geological Survey, Denver.
3.28, 6.19, 7.5, 7.6, 7.8, 7.16A, 7.16B USDA/NRCS/Natural Resources Conservation Service.
4.7 Corbis photo by Reuters/UPI.
4.8 World Food Programme.
4.9 AP/Wide World Photos photo by Stringer.
4.11 Photo Researchers, Inc. photo by Hank Morgan/Science Source.
5.2, 5.3 F.A.O. Food and Agriculture Organization of the United Nations photos by P. Pitter.
5.4 UN/DPI photo by John Isaac.
5.5 State of Virginia photo by Tim McCabe.
5.6, 5.14, 5.15 F.A.O. Food and Agriculture Organization of the United Nations.
5.8A Photo Researchers, Inc. photo by Richard R. Hansen.
5.8B Photo Researchers, Inc. photo by Jim Steinberg.
5.10 F.A.O. Food and Agriculture Organization of the United Nations. Photo by D. Mason.
5.11 USDA/ARS/Agriculture Research Service photo by Michael Licter Photography.
7.10, 7.10A, 7.24 John P. Reganold.
7.11 (c) Galen Rowell/Corbis.
7.13 Photo Researchers, Inc. photo by John Eastcott/Yva Momatiuk.
7.14 USDA/NRCS/NCGC/National Cartography and Geospatial Center photo by Quentin P. Bennett.
7.21 Photo Researchers, Inc. photo by Phillip Hayson.
7.23 Photo courtesy of John Reganold.
8.1 U.S. Department of Interior, Bureau of Reclamation. photo by E.E. Hertzog.
Case Study 8.1 Fig. 1 Corbis photo by Anthony Bannister; Gallo Images.
8.10 Photo Researchers, Inc. photo by Nigel Cattlin.
8.12, 13.24 Department of Interior, Bureau of Sport Fisheries and Wildlife.
8.13, 8.14, 15.10 Michigan Department of Natural Resources.
8.18, 8.20 USDA/APHIS/Animal and Plant Health Inspection Service.
9.3 Photo Researchers, Inc. photo by Jerry McCormick-Ray.
9.7, 12.29 New Jersey Division of Fish, Game and Wildlife.
9.19 Aerial photograph SW.H.69 71-2-14 used by permission of the Engineering Division Resource Mapping Section, Washington Department of Natural Resources.
9.21, 9.27 North Carolina Public Information Program photos by H.E. Riodervhizer III.
9.22 Visuals Unlimited photo by Bill Beatty.
9.24 Corbis photo by Marc Pesetsky/Reuters.
9.26 Courtesy of Dr. Erik Zobrist, Restorations Center, NOAA Fisheries Service.
10.8 AP/Wide World Photos photo by Dave Martin.
10.10 Photo Researchers, Inc. photo by Ron Curbow.
10.11 FEMA photo by Liz Roll/FEMA News Photo.
10.12, 10.15, 10.21, 10.24 Bureau of Reclamation photos by E. E. Hertzog.
10.13 USDA/NRCS/Natural Resources Conservation Service photo by Jim McConnell.
10.14 Bureau of Reclamation photo by Glade Wallker.
10.16 Visuals Unlimited photo by Jeff Greenberg.
10.17 Tsunami Wave.
11.1 Stock Boston photo by W. B. Finch.
11.4 Visuals Unlimited photo by Ted Whittenkraus.
11.5 Visuals Unlimited photo by W. Banaszewski.
11.6 Stock Boston photo by Bob Daemmrich.
11.7 Bureau of Reclamation photo by H.L. Personius.
11.9 USDA/NRCS/Natural Resources Conservation Service photo by Erwin W. Cole.
11.11 Franc Scarpace and Theodore Green, "Dynamic Surface Temperature Structure of Thermal Plumes," Water Resource Research 9(1):138–153, February 1973. (c)1973 by the American Geophysical Union.
11.12a, 11.12b, 11.12c Phototake/Carolina Biological Supply Company.

11.13 PP & L, Inc.
11.20, 12.8 Bureau of Reclamation.
11.23 John Todd Research and Design, Inc.
11.25 Sun Mar Corporation.
11.26 Milwaukee Metropolitan Sewage District.
Case Study 11.1 Fig. 1 Courtesy of GISGN Exotic Species Library/Ontario Ministry of Natural Resources.
11.27 The City of New York, Department of Sanitation
11.30, 11.31, 12.16, 13.11, 13.15, 13.20, 13.21, 13.22, 13.25, 14.5, 14.29, 19.17, 20.8, 22.10 Supplied by Author (Chiras).
11.34 Corbis/Sygma photo by Bill Nation.
11.35, 11.36 Corbis photos by Nick Didlick/Reuters.
11.37, 21.12, 21.18, 21.26, 22.1, 22.7, 22.8, 22.22 U.S. Department of Energy.
12.7, 15.1, 16.2 U.S. Fish and Wildlife Service.
12.12 Department of Natural Resources Bureau of Information and Education, State of Minnesota.
12.31 K. Fitzsimmons–University of Arizona.
Case Study 12.1 Figs. 1&2 University of Michigan Sea Grant Program.
Case Study 12.1 Fig. 5 Michigan Conservation Department.
Case Study 12.2 Fig. 1 Affordable Stock.
Case Study 12.2 Fig. 2 Courtesy of Portland District Army Corps of Engineers.
Case Study 12.2 Fig. 3 Photo Researchers, Inc. photo by Earl Roberge.
13.1, 13.2 Dr. Linda H. Hardesty.
13.10 Cartoon—"Tragedy of the Commons." Redrawn by John N. Smith, based on Bernard J. Nebel, Environmental Science, 2nd Edition, © 1987, p. 485. Reprinted by permission of Prentice.Hall, Inc. Englewood Cliffs, N.J.
13.14 EPA-Documerica photo by Gene Daniels.
Case Study 13.4 Fig. 1 USDA Forest Service.
14.7 Photo Researchers, Inc. photo by Ann Purcell.
14.9 Photo Researchers, Inc. photo by Calvin Larsen.
14.10 Photo Researchers, Inc. photo by Keith Gunnar.
14.14 Mobark, Inc.
14.16 U.S. Department of Agriculture photo by Miller Cowlin.
14.18 U.S. Department of Agriculture photo by Robert W. Neelands.
14.19 Photo Researchers, Inc. photo by Christian Grzimek/Okapia.
14.23 Yosemite National Park Service photo by Ralph Anderson.
14.25 Photo Researchers, Inc. photo by Renee Lynn.
14.26 Peter Arnold, Inc. photo by Klein/Hubert.
A Closer Look 14.1 Fig. 1 Tennessee Department of Environment and Conservation.
15.5a The Image Bank photo by Guido A. Rossi.
15.6, 15.7 State Historical Society of Wisconsin.
15.8 Photo Researchers, Inc. photo by M.H. Sharp.
15.9, Case Study 16.2 Fig. 1 Photo Researchers, Inc. photos by Tom and Pat Leeson.
15.11 George B. Schaller.
16.1 U.S. Department of Agriculture photo by Tom Beemers.
16.3 California Department of Fish and Game.
16.5 Minnesota Department of Natural Resources photo by Walter H. Wettschreck.
16.9 Michigan Department of Natural Resoures.
16.12 National Enforcement Investigations Center Laboratory, EPA.
16.15 Minnesota Department of Natural Resources.
16.17 Bureau of Reclamation photo by Lyle C. Acthelm.
16.19 U.S. Fish and Wildlife Service photo by Peter J. Van Huizen.
16.24 Visuals Unlimited photo by Tom Uhlman.
17.3a, 17.3b Courtesy of Owens-Illinois.
17.4 Eco.Cycle Recycling Center.
17.5 Woodland Park Zoo photo by Carol Beach.
17.6 Stock Boston photo by Michele Burgess.
17.8, Case Study 17.2 Fig. 2, 18.7a, 18.7b, 18.20 AP/Wide World photos.
17.9 3M Corporation.
Case Study 17.2 Fig. 1 Corbin/Sygma. Photo by Michael Phillippot.
18.1 U. S. Geological Survey/Geologic Inquires Group.
Case Study 18.2 Fig. 1 National Institute for Occupational Safety & Health
18.21 U.S. Department of Transportation.
18.22a Toyota Motor Sales, USA, inc.
18.22b American Honda Motor Co. Inc.
Case Study 18.3 Fig. 1 Corbis photo by Ralph Alswang.
Case Study 18.4 Fig. 1 Photo Researchers, Inc. photo by CNRI/Science Photo Library.
19.9 Photo Researchers, Inc. photo by De Sazo.Rapho.
19.10 Stock Boston photo by David Ulmer.
19.12 Courtesy of Knauf Fiber Glass.
20.1 U.S. Geological Survey, Denver photo by H.E. Malde.
20.4 National Institute for Occupational Safety & Health photo by David Barna.
20.6 Steelcase Inc.
20.7 Rocky Mountain News Library Co./NewsQuest photo by Cyrus McCrimmon.
20.9 Smithsonian Institution.
21.5a Photo Researchers, Inc. photo by Milton Rogouin.
21.7 U.S. Department of the Interior.
21.11 Corbis photo by W. Frerck/UPI.
21.16 Courtesy of C-E Power Systems.
21.17 Courtesy of Westinghouse Electric Corporation.
21.19 Nuclear Energy Institute.
21.20 Northern States Power Company.
21.21 Rollin R. Geppert.
21.24 SouFoto/EastFoto photo by TASS.
21.27 Courtesy of U.S. Department of Energy, Oakridge Operation.
22.3 AP/ WideWorld Photos
22.5 Photo Courtesy of GE Lighting
22.11 Ontario Hydro Archives.
22.12 Visuals Unlimited photo by Pat Armstrong.
22.13 Rainbow photo by Dan McCoy.
22.14 Pearson Education/PH College. photo by Susan Larson.
22.15 Photo Researchers, Inc. photo by John Keating.
22.16 Credit: DOE/NREL-PIX #06282.
22.19 Corbis photo by Marc Garanger.
22.23 Masonry Heater Association of North America photo by Gene Hedin.

Index

Page numbers with *f* indicate figures; page numbers with *t* indicate tables.

B

D

E

F

G

H

I

J

K

L

M

N

O

X

Y

Z

Common Conversions

METRIC TO ENGLISH

Metric Measure	Multiply by	English Equivalent
Length		
Centimeters (cm)	0.3937	Inches (in.)
Meters (m)	3.2808	Feet (ft)
Meters (m)	1.0936	Yards (yd)
Kilometers (km)	0.6214	Miles (mi)
Nautical mile	1.15	Statute mile
Area		
Square Centimeters (cm^2)	0.155	Square inches ($in.^2$)
Square meters (m^2)	10.7639	Square feet (ft^2)
Square meters (m^2)	1.1960	Square yards (yd^2)
Square kilometers (km^2)	0.3831	Square miles (mi^2)
Hectares (ha) (10,000 m^2)	2.4710	Acres (a)
Volume		
Cubic centimeters (cm^3)	0.06	Cubic inches ($in.^3$)
Cubic meters (m^3)	35.30	Cubic feet (ft^3)
Cubic meters (m^3)	1.3079	Cubic yards (yd^3)
Cubic kilometers (km^3)	0.24	Cubic miles (mi^3)
Liters (L)	1.0567	Quarts (qt), U.S.
Liters (L)	0.88	Quarts (qt), Imperial
Liters (L)	0.26	Gallons (gal), U.S.
Liters (L)	0.22	Gallons (gal), Imperial
Mass		
Grams (g)	0.03527	Ounces (oz)
Kilograms (kg)	2.2046	Pounds (lb)
Metric ton (tonne) (t)	1.10	Short ton (tn), U.S.
Velocity		
Meters/second (mps)	2.24	Miles/hour (mph)
Kilometers/hour (kmph)	0.62	Miles/hour (mph)
Knots (kn) (nautical mph)	1.15	Miles/hour (mph)
Temperature		
Degrees Celsius (°C)	1.80 (then add 32)	Degrees Fahrenheit (°F)
Celsius degree (C°)	1.80	Fahrenheit degree (F°)

ADDITIONAL ENERGY AND POWER MEASUREMENTS

1 watt (W)	=	1 joule/sec	1 W/m^2	=	2.064 $cal/cm^2/day$
1 joule	=	0.239 calorie	1 W/m^2	=	61.91 $cal/cm^2/month$
1 calorie	=	4.186 joules	1 W/m^2	=	753.4 $cal/cm^2/year$
1 W/m^2	=	0.001433 cal/min	100 W/m^2	=	75 $kcal/cm^2/year$
697.8 W/m^2	=	1 $cal/cm^2/minute$			

Solar constant:
1,372 W/m^2
2 $cal/cm^2/minute$